The Gene

Its Structure, Function, and Evolution

The Gene

Its Structure, Function, and Evolution

LAWRENCE S. DILLON

Texas A&M University
College Station, Texas

PLENUM PRESS • NEW YORK AND LONDON

Library of Congress Cataloging in Publication Data

Dillon, Lawrence S.
   The gene: its structure, function, and evolution.

   Bibliography: p.
   Includes index.
   1. Genetics. 2. Molecular genetics. I. Title.
QH430.D539   1986   575.1   86-25291
   ISBN 0-306-42319-7

© 1987 Plenum Press, New York
A Division of Plenum Publishing Corporation
233 Spring Street, New York, N.Y. 10013

Printed in the United States of America

Preface

For a long period in the early years of genetics, the gene was viewed as a hypothetical carrier, located in the cell, of a given hereditary trait. Then, with progress in cytogenetics, it became a specific locus on a certain chromosome, and still later, a sequence in the DNA molecule, after the chemical composition of chromosomes had been established. But that was back in a period of relative naïveté, before advanced technology permitted the rapid determination of the precise macromolecular structure of almost any gene. Yet, ironically, as understanding of chemical structure has become increasingly concrete, ideas of how the molecular unit induces given phenotypic traits—particularly those of multicellular plants and animals—have decreased in lucidity. Only in the most recent literature has any light been thrown on basic changes that can lead to mutations in the expression of, for example, eye color or wing length in *Drosophila*. All that is known is that a gene encoding a certain enzyme is modified, either directly or in the flanking regions, and that in specific cells that enzyme somehow leads to the production of white instead of the usual reddish eye color or to vestigial instead of normally functional wings.

The flanking regions just mentioned are assuming far greater functional importance in current molecular biological thought than was the case previously. Indeed, when the vast distances in the DNA between actual coding areas were first discovered, those "spacers" were often referred to as "junk DNA," being deemed to have no, or virtually no, active function in the cell. Gradually, however, the pendulum has swung toward the opposite extreme, so that much of the regions around a gene as far as 350, or even 1000, nucleotide base pairs from either end of the coding sector has often been implicated in the expression of some gene in one organism or another. "Boxes" of innumerable types have been described as active in triggering initiation or termination of transcription or other molecular activity. These signals, although supposedly universal, too often tend to be confused or absent when sought in other genes in the given organism. So many generalizations based on limited data have been made that future generations may well view the present period of exploration into the nature of the gene as a new age of naïveté which sought quick solutions to a problem that by then will have been most thoroughly demonstrated to be one of infinite complexity.

It is the author's aim to assist in the taking of the first steps across the threshold into understanding this most vital of all biological problems by thoroughly analyzing the genes and their surrounding sequences from a broad spectrum of organisms and the eukaryotic organelles. As many types as feasible are examined, particularly those found in both prokaryotes and eukaryotes, such as the transfer and ribosomal RNAs and cytochromes, along with others that are widespread and sufficiently characterized. No preconceptions are held as these gene structures are analyzed, in order to avoid bias, for the statement made by a geologist, "I will see it, if I believe it," unfortunately applies equally in all scientific investigations. Support of some particular dogma is not what is sought here, only an insight into how genes function insofar as the data permit. In short, the author attempts to follow the guidelines laid down by Thomas Henry Huxley: "Sit down before fact as a little child, be prepared to give up every preconceived notion, follow humbly wherever and to whatever abysses nature leads, or you shall learn nothing."

This being the first unified analysis of the gene on a broadly comparative basis, it is inevitable that new general characteristics and interrelationships among genes as entities in their own right should be disclosed. Genes, for example, have organizational properties unrelated directly to their encoded products that show them to fall into several major classes, each with its own distinctive properties. Indeed, on more than several occasions, the data indicate a need for hierarchical arrangements to be made, whether for the genes themselves, for their organization in the genome, or for the manners in which they at times overlap one another. At other points, the information does not permit firm conclusions to be drawn, but is sufficiently substantial to disclose new questions of extensive impact to be posed. Although evolutionary matters necessarily are interwoven with all aspects of the subject, the time is not yet here when it will be possible to demonstrate a mechanism for direct environmental influence upon the course of genetic changes, although faint hints at its existence do come to light.

Acknowledgment of the assistance of a number of persons is gladly made. In the first place, no study of this type could have even been begun without the innumerable researchers whose brilliant technological achievements patiently applied to solving particular problems in the laboratory have supplied the mass of detail needed for the present analysis. To all those persons, cited in the literature references or not, the author expresses his greatest admiration and deepest gratitude. Others have contributed in a more direct manner, particularly H. Faaren of Agrigenetics, who generously supplied information regarding certain seed proteins. Discussions with T. M. Hall and H. W. Sauers of Texas A&M University also have been of great value, and the author greatly appreciates their assistance, as he does that of many others who must remain unnamed here. Likewise, many persons have assisted with the preparation of the numerous tables, but Molly Allen and Esse Bakor have been especially helpful. Finally, my wife, as always, has been an indispensable collaborator throughout all the preparatory stages of the manuscript, literature search, preparation of tables, and interpretation of data.

LAWRENCE S. DILLON

Contents

The Gene

Its Structure, Function, and Evolution

1

Major Features of the Gene

Over the years, the gene has received numerous definitions, many of which became widely accepted. It has long been known as the unit of inheritance, or the factor that results in a given hereditary trait, definitions that continue in use today; it was also considered a point, or locus, on a chromosome that serves in the foregoing capacity. Later, as the molecular aspects of cell function began to unfold, a better understanding of the nature of the gene was gained, and the concept of "one gene, one protein" came into existence. Then, when proteins were perceived as being constructed of several subunits, each the product of a separate locus, that view became modified to "one gene, one polypeptide." All these ideas are both sound and unsound; even the most recent, which defines a gene as a sequence in the nucleic acid of the genome, contains an element of weakness, which will become apparent as the discussion of gene structure proceeds immediately below.

1.1. STRUCTURAL FEATURES OF THE GENE

The gene, like all biological units, has been found to be extremely diversified, even in its basic construction, but the numerous species can be perceived to fall into three major classes: simple, compound, and complex. The members of the last category, which according to present knowledge are confined to advanced metazoans, are extremely intricate, as is disclosed in a subsequent section. But even the first of these types is soon found to be simple only by comparison.

1.1.1. The Simple Gene

Far back in primordial times, the gene in the archetypal biological forms may have been merely a continuous sequence in a nucleic acid that specified a particular polypeptide or functional RNA, but that condition would have been confined to primitive organisms in which the genetic coding system was still undergoing completion. Perhaps that condition persists in some of the early descendants of those archaic living things, such as the simple RNA viruses (Chapter 10), but even in the better known bacteriophages of today, a gene is far more than just a continuous sequence of nucleotides encoding a particular macromolecule, as it certainly is also in cellular organisms.

The Coding Sequence. The primary functional portion of the gene obviously is the sequence of nucleotides that encodes a particular macromolecule which plays a role in the structure or metabolic processes of a cell. Thus the coding strand of this region of DNA is read by an enzyme that copies it in complementary fashion into RNA strands. These products then may be used, usually after further treatment called processing, either directly, as with the ribosomal and transfer RNAs, or, as in the case of messenger RNAs, after translation on ribosomes into proteins or their polypeptide subunits. Because the first products of the gene are in sequences of RNA that are complementary to the coding strands of DNA, it is customary in studies of the gene at the molecular level to present the complementary strand sequence of the duplex DNA, rather than the actual active one. Since that procedure simplifies comparison of the gene and its product, it is followed throughout the present study (Sharp, 1985).

Intergenic Spacers. It may be readily imagined that in the earliest protobionts that had genes coded in nucleic acid the latter consisted solely of coding sequences placed one after another. Perhaps these were arranged in uninterrupted series, for occasionally even in modern organisms two or more genes may abut against each other in this fashion. Indeed, as shown shortly, they sometimes even overlap. More typically, however, genomic regions that are not directly involved in the final product are found between the actual coding sectors. These portions, called intergenic spacers in a general sense, vary greatly in length, from a single base pair to many thousands (Federoff, 1979). Also in a nonspecific manner, the parts of these spacers adjacent to the mature gene are frequently referred to as flanking regions.

Since certain parts of these spacers exhibit specialization of function, separate terms are applied. That sector of the noncoding strand preceding the 5′ end of the gene proper (hereafter referred to as the mature gene) is called the leader (Baralle, 1983), while that following the 3′ end is the train (Figure 1.1A). If the habit of viewing the complementary strand is abandoned temporarily, it may be seen that the leader is actually located before the 3′ end of the functional region of the coding strand and that the train follows the 5′ end, because of the antiparallel arrangement of DNA strands (Figure 1.1A). This reversed situation should be borne in mind in order that the real nature of gene structure and function may be understood.

It also should be realized that one given strand of DNA does not necessarily bear the coding sequences of all the genes located in any single sector of the genome; sometimes some genes are located on one strand while others in the same series are on the opposite one. Since genes (or cistrons as they are frequently termed) are always read (transcribed) in the 3′ to 5′ direction on the coding strand, those on opposite strands also differ in polarity. Hence, it may be perceived that in instances where a number of genes form a cluster, as is the frequent case in both prokaryotes and eukaryotes, the train of one gene may be located under a mature gene of opposite polarity or even below its leader (Figure 1.2A). This topic receives full attention in Chapter 10.

Promoters. Before a gene can be read by the responsible enzyme, the correct coding region obviously must be located and its polarity determined. Since the nucleoids of prokaryotes and the chromatin of eukaryotes consist of many millions of nucleotides of only four major varieties, the task of locating the proper sector, which is comprised of but

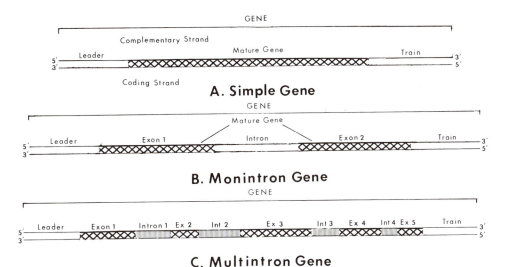

Figure 1.1. Varieties of simple genes. (A) Those genes that consist of the mature coding region plus essential parts of the flanks are considered members of the simple class. (B, C) Sometimes this basic pattern is modified by the presence of one or more introns.

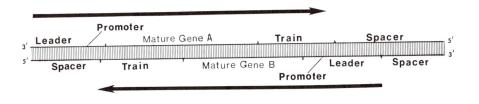

A. Polarity of Transcription

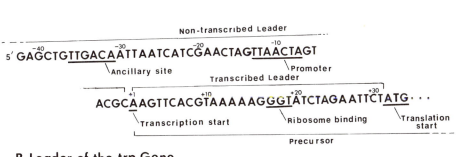

B. Leader of the trp Gene

Figure 1.2. Principal features of genes. (A) Not all genes are necessarily confined to a single strand of DNA, but may occur on either strand. Because transcription is always in the 3′ → 5′ direction, those on opposite strands have different polarities. (B) Structural relations at the 5′ end of a gene.

a few hundred nucleotides of the same four basic types, can be quite formidable. To assist in locating, reading, and processing the gene and its product, a number of signaling devices appear to exist. To delineate the nature of these signals, when they are actually present, is one main goal of this book. It will be found that, although the principle of having such devices present seems a simple necessity, in practice their identification by cellular processes is without equal insofar as complexity is concerned.

In the literature a number of different "boxes" have been proposed to serve in recognition purposes. The Pribnow or −10 box, of varying constitution, is often cited in studies of prokaryotic genes, while in those on eukaryotic cistrons the term Goldberg–Hogness box, of equally variable constitution, applies to a somewhat corresponding location. Various other names, including CAT-TA, ACT-TA, and TA-TA boxes, also have been in vogue. Since bacteriologists had named such functional sectors many years before sequencing DNA had become readily feasible, their term, promoters, is uniformly applied here to such signals to permit freer comparison from gene to gene and organism to organism. Promoters may be viewed as the specific site or sites of a given gene that serve in the latter's recognition and subsequent attachment by the transcribing enzyme (Figure 1.2B). Transcription usually begins a variable number of sites downstream (in the 3′ direction on the complementary strand), in *Escherichia coli* typically at a purine residue, but it varies widely from organism to organism.

Ancillary Sites. Another active site on the leader has frequently been reported to which several terms have been applied, as in the case of the promoter. Since this is frequently located between 30 and 40 sites upstream (in the 5′ direction on the complementary strand) from the transcriptional initiation site, it often is referred to as the −35 sequence, particularly with prokaryotic genes (Figure 1.2B). However, it is also called the CAP site in certain bacterial coding sequences, a topic investigated further in Chapter 2. For the sake of uniformity demanded in studies comparing prokaryotes, eukaryotes, and organellar structures, the name ancillary site is employed here. It is at such points that enzymes (ancillary proteins) that assist in the location and transcription of the gene may attach, as shown in the next chapter. This avoidance of the use of the bacteriological term CAP (referring to a cAMP-activated protein) has the additional advantage of reducing confusion, because many eukaryotic transcripts receive a cap in quite a different sense. Characteristically eukaryotic messengers receive a highly modified nucleotide on the 5′ terminus, usually methylated and linked to the main chain by 5′ to 5′ bonds (Adams and Cory, 1975; Perry and Scherrer, 1975). One of the commoner caps is $m^7GpppNm$-(Plotch *et al.*, 1981), often referred to as cap 1.

Enhancers. Still another site on the leader, quite recently discovered, is the enhancer, the prototype of which was found to be a 72-base-pair, tandemly repeated sequence (Benoist and Chambon, 1981; Gruss *et al.*, 1981; Weber *et al.*, 1984). This sequence was located 100 nucleotides upstream of the ancillary site in the DNA of simian virus (SV) 40. Mutational removal of this element reportedly reduced expression of early genes by a factor of at least 100, resulting in the loss of viability of the virus. A similar enhancer region has been described from other viral sources; a feature common to all of these is the series GGTGTGGAAAG, a sequence that occurs frequently in modified form (Khoury and Gruss, 1983). In actuality this series in SV40 is part of a 72-base-pair repeated sequence (Figure 1.3), but in other viruses it may be a portion of 50- to 100-base-pair repeated elements, only one of which is essential for enhancement.

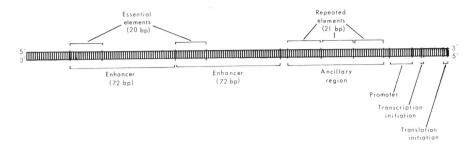

Figure 1.3. An enhancer region from SV40. (After Benoist and Chambon, 1981).

Enhancers have been detected associated with eukaryotic genes also, as shown frequently in subsequent chapters, studies on immunoglobulin genes proving to be especially productive (Ephrussi *et al.*, 1985; J. O. Mason *et al.*, 1985a; Mercola *et al.*, 1985).

Whether enhancers represent a single type or an entire class of related substances is a debatable point at this time, for they have been found to display a number of distinctive traits. Among those that indicate them to be a class of substances are the following characteristics: (1) They act most frequently when located on the coding strand, that is, in the *cis* configuration, but *trans*-acting ones also are known (Chapter 11). (2) They are effective when located either upstream or downstream from the promoter. (3) They are active whether arranged in the same or opposite polarity as the mature gene. (4) They are equally effective regardless of the organism from which the gene is derived when attached to foreign DNA. An example of the latter property is the finding that certain regions from a mouse immunoglobulin gene served as an enhancer when cloned to the SV40 early promoter (Mercola *et al.*, 1983). Other eukaryotic genes, however, have been shown to resist enhancement, including the α-globin gene of man.

Other Leader Signals. With the exception of the initiation site, all of the foregoing signals are frequently located in the nontranscribed region of the leader and thus are not part of the nascent RNA molecule (primary transcript) transcribed by the enzyme. Usually only one further signal follows the initiation site on the transcribed leader of messengers, those RNAs (mRNAs) that are subsequently translated into proteins (Figure 1.2B). Typically this feature, which clues the initiation point for translation, is the codon for methionine (ATG), but rarely certain codons for other amino acids can serve in this same capacity. As pointed out in greater detail later, in prokaryotes an additional signal is present here, because translation of mRNAs is required to occur concurrently with transcription. Accordingly, bacterial transcripts have a short sequence (the Shine–Dalgarno box) that provides a point of attachment for the ribosome, hereafter referred to merely as the ribosome binding site (Figure 1.2B).

The position and reading frame orientation of the ATG codon (AUG in the messenger) are apparently not the sole determinants for initiation of translation, as effectively demonstrated by a recent study (Johansen *et al.*, 1984). When the AUG was experimentally inserted into the leader of an mRNA, it had various effects on translation, depending upon the sequences that flanked it, the combination A(or G)CC before the AUG and a G after it being the most effective. It is pertinent here also to note that GUG or UUG may sometimes be substituted for the standard signal, as is seen in several instances in later chapters. However, these are uniformly translated as methionine, although the first alternate usually encodes valine and the second leucine. Rarely AUU (usually for isoleucine) may serve in this capacity, as in the carbamoyl phosphate synthetase of *Escherichia coli* (Piette *et al.,* 1984). Here, then, is the first example of a condition that recurs repeatedly in later chapters, the nonmechanistic behavior of mechanistic molecules, which has on earlier occasions provided the basis for a more realistic philosophy of living matter, referred to as the biomechanistic point of view (Dillon, 1978, p. 427; 1983, pp. 400–410).

Promoters of Prokaryotes. While the several chapters devoted to transcription provide an in-depth analysis of the nature of promoters from individual classes of genes, a preliminary examination of several examples at this point is essential to a more concrete understanding of the gene's structural features. Accordingly, Table 1.1 lists a number of leader sectors from the well-known bacterium *E. coli,* showing their promoters located about ten sites from the actual start of transcription indicated by +1. The predominant

Table 1.1
Promoters of Sets of Genes from *E. coli*[a]

		Ancillary Site		Promoter		
		−35	−30	−20	−10	+1
trp[b]	GAG–CTG	TTGACA	–ATTAATCA	TCGAACTAG–	T–TAACT	AGTACGCA
$lacUV5$[b]	CAGGC––	TTTACA	CTTTATGCT	TCCGGCTCG–	TATAAATG	TGT–GG–A
$pheU$[c]	––TTAGG	TTGACG	–AG–ATGTG	CAGATTACGG	TTTAATG	CG–CCC–G
$leuV$[c]	–––ACTA	TTGACG	AAA–A–GCT	GAAAACCAC–	TAGAATG	CGCCTCCG
$tyrT$[c]	––AACAC	TTTACA	–GCGGCGCG	CGTCATTTGA	TATGATG	CGCCCC–G
$tyrT$[d]	GTAACAC	TTTACA	–GCGG––CG	CGTCATTTGA	TATGATG	CGCCCC–G
$rrnA$[c]	–––CCTC	TTGTCA	GGC––CGGA	ATAACTCCC–	TATAATG	CGCCACCA
$rrnD$[c]	–––ATAC	TTGTGC	AAAAAATTG	GGA––TCCC–	TATAATG	CGCCTCCG
str[e]	ATATTTC	TTGACA	CCTTTTCGG	CATCG–CCC–	TAAAATT	CGGGC––G
$rpoB$[f]	ATATACT	GCGACA	GGACGTC––	CGTTCTGTG–	TAAATCG	CAATGA–A
$glnS$[g]	CTAACAG	TTGTCA	GCCTGTC–C	CGCTT–ATA–	AGATCAT	AC–CCC–G
$rpsT_1$[h]	GGAAAAG	CTGTAT	TCA–CACCC	GCAAGC–TGG	TAGAAT–	CCTGC––G
$rpsT_2$[h]	AAATCCA	TTGACA	AAAGAAGGC	TAAAA––GGG	CATA–TT	CCTCGG–C

[a]Promoters and ancillary sites are italicized; +1 marks the actual transcriptional starting site.
[b]De Boer *et al.* (1983).
[c]Schwartz *et al.* (1983).
[d]Sekiya *et al.* (1976); Berman and Landy (1979).
[e]Post *et al.* (1978).
[f]Post *et al.* (1979); An and Friesen (1980).
[g]Hawley and McClure (1983).
[h]Mackie and Parsons (1983).

Table 1.2
Promoters of Gene Sets of Various Prokaryotes[a]

	Ancillary Site				Promoter		
		-35	-30	-20		-10	+1
$rrnA_2^b$	TATTATGTA	*TTGACTT*	AGACAA	CTAAAGC-T	GT-	*TATTCT*	AATATAC-G
$rrnO_1^b$	TCATAACCC	*TTTACA-*	-GTCAT	AAAAATTAT	GG-	*TATAAT*	CATTT-C-G
$rrnA^b$	AAAAAGTTG	*TTGACA-*	GTAGCG	GCCGGTAAAT	GT-	*TATGAT*	AATAAA--G
$rrnO^b$	AAAAAAGTA	*TTGACCT*	AGTTAA	CTAAA-AAT	GT-	*TACTAT*	TAAGTA--G
$spoVG_L^c$	GGATTTCAG	AAAAAAT	CGT-*GG*	*AATTGATA-*	CA-	-*CTAAT*	GCTTT̲T--A
$spoVG_E^c$	TTAAAAACG	AGCAGGA	TTTCAG	-AAAAAATC	GT-	*GGAATT*	GATACA--C
$spoVC^c$	CATTTTTCG	AGGTTTA	AATCCT	TATCGTTAT	GGG	*TATTGT*	*TTGT*AAT-A
$penP^d$	AAAAAACGG	*TTGCAT*T	AAAATC	TTAC-ATAT	GT-	*AATACT*	TTCAAA--G
$malX^e$	AAAAAATAC	*TTGCAAC*	CGTTTT	CTAT-TTGT	GC-	*TATACT*	AAGCTC--A
$malM^e$	TTAAAACGC	*TTGCAAT*	TATGCG	TTGAAAAG-	GAG	*TATACT*	TATAAGT-A
$nifH^f$	ATACATAAA	CAGGCAC	GG*CTGG*	-TATGT-TC	CC-	*TGCACT*	TCTCTGC-T
$nifE^f$	AAAATCAAG	GCTCCGC	TT*CTGG*	-AGCGC-GA	AT-	*TGCATC*	TTCCCCC-T
$nifL^f$	CTGCACATC	ACGCCGA	TA*AGGG*	-CGCACCGG	TT-	*TGCATG*	GTTATCACC

[a]All experimentally established signals are italicized. +1, Actual transcriptional start site.
[b]*Bacillus subtilis;* Ogasawara *et al.* (1983).
[c]*Bacillus thuringiensis;* Wong *et al.* (1983).
[d]*Bacillus licheniformis;* McLaughlin *et al.* (1982).
[e]*Streptococcus pneumoniae;* Stassi *et al.* (1982).
[f]*Klebsiella pneumoniae;* Beynon *et al.* (1983).

base at each given site in these signals can be selected to give a "consensus" sequence, in the present case TATAATG. But such average compositions tend to give a distorted picture of how genes are recognized, for obviously the RNA polymerase does not react with a consensus through a wide spectrum of types, but with the particular combination that exists in each actual gene. For example, the cited consensus sequence is found in only two of the 13 that are listed, those for the two sets of rRNA genes, *rrnA* and *rrnD*. Such strong deviations from the norm as AGATCAT can be noted, in this case in the *ginS* gene, in which none of the bases corresponds to the consensus. While it is evident that transcription in *E. coli* usually is initiated with a purine nucleotide, a cytosine residue is employed in one of these examples.

Although the composition of the ancillary signal and promoter may thus vary over broad ranges, they are spaced at a markedly uniform distance from the initiation sites in *E. coli.* When samples of structure are compared from other bacteria, however, as in Table 1.2, variations in size and location as well as in composition are found to be rampant. This lack of uniformity is apparent even within a single genus, as among the three species of *Bacillus.* Whereas the signals of *B. subtilis* and *B. licheniformis* are identically located and similarly constructed, those of *B. thuringiensis* are widely disparate. In the first place, the promoters of this third species consist of 10 or 11 nucleotide residues, compared to six in the other two and to seven in *E. coli*, so that their 3′ ends are only one to three sites removed from the start of transcription (indicated by +1 in the table). Second, upstream signals are often lacking, being present only in the late form of *spoVG*, distinguished by the subscript L. During the early stages of sporulation this identical sector serves as the promoter, as shown in *spoVG*$_E$ in the table.

The two maltosaccharide utilization genes of *Streptococcus*, *malX* and *malM*, have signals quite comparable to *B. subtilis* and *B. licheniformis*, so that these three form a cluster of interrelated species not too remote in kinship from *E. coli*. On the other hand, *Klebsiella pneumoniae* is seen to be distinct from all the others given here in having both the promoter and the ancillary site shorter and of distinctive base composition. Moreover, the latter are situated downstream of those from bacilli and coliform bacteria. Whether these points of departure are suggestive of an advanced or more primitive condition cannot be made clear until the structures of protein-specifying genes of Archaebacteria and such primitive forms as *Clostridium*, *Beggiatoa*, and blue-green algae have been established in sufficient numbers.

Signals on the Trains. As might be expected, the chief signals on the 3′ trains of genes are those that are concerned with the cessation of transcription (Rosenberg and Court, 1979). Typically there exists a series of T–A base pairs, in prokaryotes usually six or more in number, while in metazoans four sometimes seem to suffice. Occasionally these series are accompanied on the upstream side by regions of dyad symmetry, so that a stem-and-loop structure could conceivably form in the transcript (Young, 1979; Holmes *et al.*, 1983). In some instances a combination of the stem-and-loop and the sequence of Ts seem to be requisite or at least most effective (Figure 1.4; Yanofski, 1981; Lau *et al.*, 1982)—the loop, of course, would form on the transcript, not in the DNA double strand (Platt, 1981), as illustrated by a *Bacillus subtilis* cluster of tRNA genes (Green and Vold, 1983). Such regions are rich in guanosine and cytidine residues.

But even in prokaryotes, termination is not always signaled in such a simple fashion, but often is dependent upon the presence of additional factors. While full discussion is more appropriate to the topic of polymerase activities, it can be mentioned now that two such factors are known, rho and NusA (Andrew and Richardson, 1985; Barik *et al.*, 1985; Bear *et al.*, 1985). Definite sequence requirements for attachment of these proteins have not been established, but sometimes are quite complex, being a series of tandem sectors

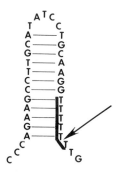

Figure 1.4. A possible terminator in a tRNA operon of *Bacillus subtilis*. (Green and Vold, 1983.)

that display little sequence homology (Holmes *et al.*, 1983). In some cases the active sites appear to be located 250–450 base pairs beyond the 3′ end of the mature gene (W. D. Morgan *et al.*, 1983). On occasion in prokaryotes, these terminators actually may occur on the transcribed portion of the leader preceding the translation signal, where they are known as attenuators and serve in regulation of the gene's activity.

In eukaryotes, signals for termination vary with the organism and with the class of RNA that results from transcription (Birnstiel *et al.*, 1985). Messenger RNAs in yeast have been shown in some instances to bear an eight-residue sequence TTTTTATA, which is sometimes repeated after a moderately long interval (Henikoff *et al.*, 1983). After activation by this control sequence, transcription terminates at the last T of the series CAATATTTG. In other examples termination has been found to occur upstream of the hexamer AATAAA; this same sequence serves also as a signal in its own right when on messengers, marking the point for cleavage of the train (Orkin *et al.*, 1985). It also marks the site for the poly(A) tail that is applied following transcription, but its presence is not requisite for those processes, whereas it is for cleavage (Montell *et al.*, 1983).

No single combination of nucleotides is sufficient universally and often several variations prove equally acceptable, as reported in a *Drosophila* tropomyosin gene (Boardman *et al.*, 1985) and in a *Xenopus* β-globin mRNA (P. J. Mason *et al.*, 1985). In a sea urchin, a conserved CAAGAAAGA sequence has been reported as essential for the formation of the 3′ terminus of histone H3 transcripts (Georgiev and Birnstiel, 1985). Furthermore, even the standard AATAAA signal needs specific sequences downstream from it for efficient operation—the combination YGTGTTYY has been described as possibly serving as an auxiliary in numerous mammalian genes (McLauchlan *et al.*, 1985).

Internal Control Signals. In certain classes of RNA, mostly small species such as tRNAs, 5 S rRNA, and viral products, promoter sequences have been described as being located within the mature gene sequence itself. Most of these internal control signals, as they have been called, are known from eukaryotic genes, but at least one has been described from *E. coli* (Irani *et al.*, 1983). Currently they provide the basis for such a controversial issue that their elucidation must be deferred to later sections covering the transcription initiation of specific types of macromolecules. Unfortunately, the topic also suffers from the same defect associated with the entire field concerned with the nature of gene regulation—too many dogmatic generalizations have been proposed. Such premature suppositions cannot fail to add confusion to the problem and result in further obfuscation of an already obscure subject, for they suggest quick, easy answers to some of the most profound questions in all of biology, pertaining to how genes are identified and transcribed.

1.1.2. Split Genes

One of the many surprises disclosed by the techniques permitting the relatively rapid sequencing of DNA was the discovery that many genes of viruses and eukaryotes were not simply continuous coding regions of the genome, but were interrupted by elements that were removed during processing. Such interrupting sectors, called introns, were first found in rRNA genes of *Drosophila* (Pellegrini *et al.*, 1977), but have since been described for genes for many classes of substances. In fact, the presence of introns is proving to be a characteristic, not an exceptional, condition of the genes of eukaryotes in general as well as some of those of organelles (Mahler, 1983). They have even been reported from some of the higher bacteria, such as *Sulfolobus* (Kaine *et al.*, 1983).

Monintron Genes. While the details of intron structure are best held for discussion until the part of the text concerned with the processing of transcripts (Chapter 12), a few observations are essential to the total picture of gene structure. These intervening sequences, as they are also called, are transcribed along with the actual coding sectors (referred to as exons) (Figure 1.1B) and hence are part of the precursor. The number present varies with the organism and class of substance encoded by the given gene. Some for large groups of macromolecular species are known not to contain more than a single intron, a condition illustrated by tRNA cistrons, which do not have more than one intron, located chiefly in the anticodon loop according to present knowledge. Among higher eukaryotes, although the major percentage of tRNA genes lack introns, a few have been reported to be thus equipped, including a tRNA^{Tyr} from *Xenopus* (Laski *et al.*, 1982), tRNA^{Leu} from *Drosophila* (Robinson and Davidson, 1981), and a tRNA^{Trp} from *Dictyostelium* (Peffley and Sogin, 1981). In contrast, about 20% of these genes in yeast contain introns. Among the reported instances are cistrons for four tRNAs^{Tyr}, whose inserts show no homology with the corresponding one from *Xenopus* (Goodman *et al.*, 1977; P. F. Johnson and Abelson, 1983), for tRNA$^{Leu}_3$ (Venegas *et al.*, 1979; Standring *et al.*, 1981; Klemenz *et al.*, 1982), for a minor species of tRNA^{Ser} but not the isoaccepting forms (Etcheverry *et al.*, 1979), and three for tRNA^{Phe} (Valenzuela *et al.*, 1978).

Occasionally it has been demonstrated that removal of the intron has no effect on the processing or functioning of the mature gene product (Gruss *et al.*, 1981; Treisman *et al.*, 1981; Langford and Gallwitz, 1983), but in other cases, including genes for SV40 (Gruss *et al.*, 1979; Lai and Khoury, 1979) and yeast tRNA^{Tyr} (P. F. Johnson and Abelson, 1983), removal results in lack of processing and loss of function. In the ciliate *Tetrahymena* the intron of the rRNA precursor is removed and the freed exons spliced autocatalytically, requiring no enzyme or other protein, at least *in vitro* (Kruger *et al.*, 1982).

Multintron Genes. The presence of more than one intron per gene is especially characteristic of those sequences whose transcript requires translation (Figure 1.1C), that is, mRNAs that encode proteins or polypeptides. For example, the rat muscle α-actin gene appears to contain six introns, while that rodent's cytoplasmic β-actin gene seems to have five (Nudel *et al.*, 1982). In these cases the interrupted genes encode entire proteins, but in others, such as the murine $A\beta^d$ gene, only a single polypeptide subunit of the protein is encoded (Malissen *et al.*, 1983). The cistron cited consists of six exons, which are, as is typical, of strongly contrasting lengths. The first is 93 nucleotide residues long, encoding 31 amino acids, the second contains 273 (for 91 amino acids), the third 282 (94), the fourth 121 (40+), the fifth 24 (8), and the last 12 (4). The introns, which constitute by far the greater part of the gene, are likewise of unequal lengths. The numbers of nucleotide residues found in each are, respectively, 1349, 1439, 307, 395, and 612. Total size of the introns in a given gene may reach unbelievable proportions. An especially remarkable instance of this disproportion is provided by the human thyroglobulin gene (van Omman *et al.*, 1983). Its five exons, which together consist of only 850 nucleotide residues, are dispersed over a 38,000-base-pair length of chromatin; in fact, the complete gene, including essential leader and train sectors, may exceed 100,000 base pairs.

The number of introns usually is a function of the species of macromolecule encoded by the gene, but can vary within family groups. Although some variation does occur from organism to organism, a surprising amount of conservation often is displayed. The range in number per gene is nearly unlimited, reaching a total of 52 introns in certain genes for collagen. But these aspects are better detailed in subsequent chapters as the structures of the several major types of macromolecules are examined.

1.2. TYPES OF COMPLEX GENES

In a sense the discussion of split genes also introduces a related phenomenon of gene structure that seems to be largely confined to metazoans, at least insofar as current knowledge indicates. This is the presence of three known types of complicated genes, the first of which is typified by the immunoglobulin coding sequences of vertebrates. In addition, a compound type also must be distinguished, which consists of two or more mature coding regions adjoining one another, without any spacer between them. This class, whose products are separated by processing immediately following translation, occurs frequently in viruses, prokaryotes, and cell organelles, as well as in the nuclear genomes of eukaryotes. In contrast, the two types of complex genes described shortly persist in the cell for a variable length of time following translation.

1.2.1. Immunoglobulins

Immunoglobulin Structure. To understand the structure of the most thoroughly documented of intricate genes, the assembled variety, it is first necessary to examine briefly the nature of the immunoglobulin molecules, a fuller discussion of which is provided elsewhere (Dillon, 1983, pp. 334–365) and in Chapter 8. Five great classes of immunoglobulins (Ig) are recognized, A, D, E, G, and M, some of which are represented by several subclasses. Basically all consist of four subunits, two heavy (H) and two light (L) chains. Since two types of the latter are known, referred to as kappa (κ) and lambda (λ), the general patterns of immunoglobulin structure thus might be expressed as either κ_2H_2 or λ_2H_2, mixtures of κ and λ in the same molecule being unknown.

The two types of light chains consist of 212–214 amino acid residues each, and both also are more or less equally divided into a pair of contrasting regions called domains. The C-terminal half is relatively free of variation and accordingly is known as the constant (C) domain, whereas the N-terminal portion, the variable (V) domain, is subject to great variation, and in large measure is involved in the specificity of antibodies. Since the five classes of these substances derive primarily from their differences, the heavy chains vary more extensively from one type to another than do the two light ones. Structurally they are similar to the latter in consisting of constant and variable domains, but have three or four of the C type, not just one as in the κ and λ varieties. The first two of these are separated by a short hinge region. In addition to the L and H chains, some mature immunoglobulin molecules (IgA and IgM) have a J chain that aids in joining some of the subunits, and an S chain is present in IgA when it is produced in glands.

Organization of Immunoglobulin Genes. Although the gene organization for L and H chains differs in a number of details, description of that of the latter class is sufficient to illustrate the nature of complex genes. Since, as already indicated, the five types of immunoglobulins differ only in their constant domains, there are separate coding sequences for each class, as well as for each of the three or four found in a given type. To distinguish the C domains, it is the usual practice to designate the several like components by Greek letter subscripts corresponding to the class. Thus the constant regions of IgM are designated as C_μ, those of IgA as C_α, those of IgD as C_δ, and so on. Because of the presence of multiple constant regions, the individual coding sequences are further distinguished by numerical subscripts, so that in the IgM set illustrated (Figure 1.5), the four sequences shown for constant regions are $C_{\mu1}$, $C_{\mu2}$, $C_{\mu3}$, and $C_{\mu4}$. In cases where a pseudogene intervenes, it is designated with a psi, as in $C_{\mu\psi}$. Since there are multiple copies of these domains for each of the five classes of immunoglobulins, their genomic organization can be seen to be of immense complexity.

Further complication is contributed by similar multiplicities of structures in the variable domain, the several parts of which form a separate complex. First there is a region (V_H) that encodes most of the variable portion of heavy chains, followed by a short D_H segment, and a number of J segments that eventually connect the variable domain to the constant region, distinguished by appropriate subscripts. Even the leader is a complex of parts, for only certain portions of its great length are present in the ultimate unit for transcription, a statement that is equally true both for the sectors that intervene between the two gene complexes and those separating the several components of each.

Preparatory Steps. Because the members of each complex encode neither an actual subunit of a definitive protein nor a functional RNA molecule, they do not fit any logical definition of the term gene. For this reason, as well as for clarity of discussion, two new terms have been applied to the coding sequences that exist prior to their assembly into the final transcriptional unit (Dillon, 1983, p. 359). First, each of the components of the

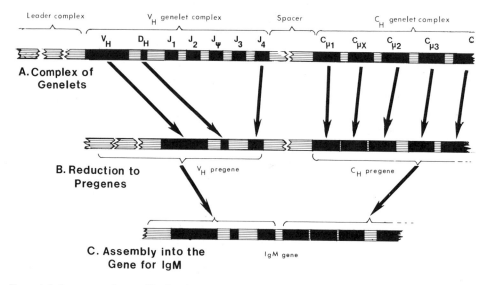

Figure 1.5. Structure and recombinational events in an assembled gene. The final immunoglobulin gene is assembled from a number of genelets in two stages, the intermediate product being a pregene.

variable and constant domains is referred to as a genelet (Figure 1.5). Since a given combination of V_H and D_H genelets is associated with only a single J genelet in the final gene, the genome in this region undergoes preparatory steps prior to transcription. Thus the V_H and D_H sectors may become joined to perhaps J_3 on the one hand, while in another cell at another period they may be combined with J_1, J_4, or some other member of that group, depending on environmental conditions. This assembled combination thus forms what is called a pregene encoding for the V_H domain.

Concurrently, depending upon cell age, place of formation, and other factors, the constant-region complex would likewise be undergoing certain poorly understood changes that appear primarily to result in reduction in size of the intervening segments that separate the several genelets. Furthermore, selection must be made by cellular activities as to the class of immunoglobulin that is to be prepared—IgM being the first to appear in many young cells. This early choice may be altered to a second class, often IgG, as the cell matures. At any rate, a separate pregene for the constant domain eventuates. Only after these two pregenes have been combined are the actual mature gene, leader, and spacer (that for IgM in Figure 1.5) formed ready for transcription. These essential and highly selective steps preparing the transcriptional unit are most likely carried out by the supramolecular genetic apparatus, under the influence of external and internal factors (Dillon, 1983, pp. 433–446).

1.2.2. Diplomorphic Genes

One of two types of complex genes, though far less complicated than those of the immunoglobulins, merits being set apart also from the others and is here given the name diplomorphic gene. As the term implies, the members of this group contain two coding regions, continuous with one another, but encoding two distinct polypeptides, which remain combined for a period after translation. Characteristically the product of the 5′ region is a transit (often "signal") protein, supposedly used for recognition purposes at its ultimate locus of function, as in membrane proteins of prokaryotes (but see Moreno *et al.*, 1980), or a point where addition of side chains commences, as in many eukaryotic enzymes, hormones, and other functional groups. Since the latter provide an ample overview of the nature of diplomorphic genes, their prokaryotic counterparts are reserved for appropriate sections later (Chapter 7).

To understand the structure of these dual genes, it is helpful to examine the translational product, rather than just the primary transcript or the coding region of the DNA itself, using the rat kallikrein gene as an example (Swift *et al.*, 1982). The primary mRNA is a sequence of 867 nucleotides, including 24 in a leader and 48 in the train; the remaining 795 residues encode a protein consisting of 265 amino acid residues (Figure 1.6). After translation, this preproenzyme, as the dual structure is called, is conducted by way of a vector system to the endoreticulum, where the first subunit of 17 amino acids interacts with a transit recognition particle. Hence, this short "pre" region functions as a transit protein, serving for recognition on the surface of this organelle by a specialized receptor (Walter and Blobel, 1983) and for transit through the membrane. The recognition particle then aids in translocating the entire protein into the lumen of the endoreticulum, where the transit peptide is removed enzymatically (Figure 1.6). After processing is completed the proenzyme is ready for secretion, but before it becomes activated the 11-

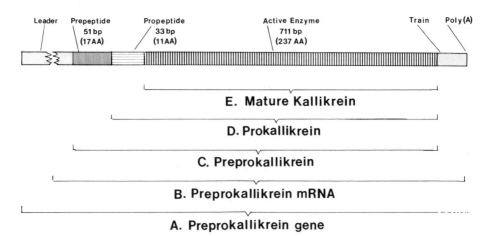

Figure 1.6. A complex diplomorphic gene. Diplomorphic genes have a presequence (encoding a transit or prepeptide), followed by a prosequence for an inhibitor or propeptide. Only when both of these products have been removed is the functional enzyme activated.

amino acid "pro" section is removed and catabolized. Thus the gene may be visualized to encode two functional units, the transit peptide and the enzyme, separated by an inactive pro region, which is without known function except in a passive way.

1.2.3. Cryptomorphic Genes

A number of genes are coming to light from eukaryotic sources that have a cryptic structure in that the ultimate active product or products are carried within the precursorial protein. These are released after enzymatic breakdown of the precursor and become functional following further processing. Perhaps the most lucid example is provided by the genes for a mating pheromone of baker's yeast known as α-factor (Kurjan and Herskowitz, 1982). The first of the genes to be sequenced, MFα1, was found to encode a messenger 495 nucleotide residues in length, plus a leader and train of undetermined lengths. When translated, the resulting protein proved to consist of 165 amino acid residues. Within this structure were included four tandemly arranged identical copies of the α-factor, each 13 amino acid residues in length. These were separated by short leader peptides, eight residues long, except for the first, which had only six (Figure 1.7A). Apparently this protein is stored in the yeast cells intact until unknown external and internal parameters induce a proteolytic breakdown of the precursor, releasing the α-factors and their leaders. Before being secreted, the latter, too, are cleaved enzymatically. It is thus evident that the gene encodes a large protein which is catabolized into the active pheromones contained in its carboxyl half. In a sense the amino half corresponds to the pro portion of diplomorphic genes, retaining the actual functional components in a quiescent stage and then being cleaved before the latter are utilized. Whether part of that upstream sector serves also as a transit protein has not been established.

In the higher eukaryotes, cryptomorphy can be carried to greater degrees of complexity than in the yeast pheromone, as exemplified by mammalian pancreatic glucagon genes

(Lopez *et al.*, 1983). In the present instance the first 20-amino acid sequence of the primary translational product, preproglucagon, is known to serve as a signal peptide, the precursorial protein proper (proglucagon) being 160 amino acid residues in length (Figure 1.7B). Within the latter are contained the structural sequences for glucagon and two nonidentical peptides said to be glucagonlike, all of which are immediately preceded and followed by basic dipeptides, such as lysylarginine or arginylarginine, potential proteolytic processing sites. But processing does not occur equally at all these similar points, for while the two glucagonlike peptides are liberated by the first processing activities, glucagon remains within a 69-residue sector known as glicentin. This hormone finally is freed for normal functioning only after a second round of enzymatic cleavage. It is interesting to note that in the yeast pheromone genes described previously, the spacer structures between the copies begin with basic dipeptides similar to those of the mammalian hormones and comparably are likely targets for proteolysis. However, the final removal of these appendages from the α-factors involves the tripeptide alanylglutamylalanine.

1.3. REDEFINITION OF TERMS

Because the foregoing analysis has disclosed gene structure to be far more complicated and variable than previously conceived, a summary of the major features appears desirable before organizational aspects receive attention. This procedure is particularly

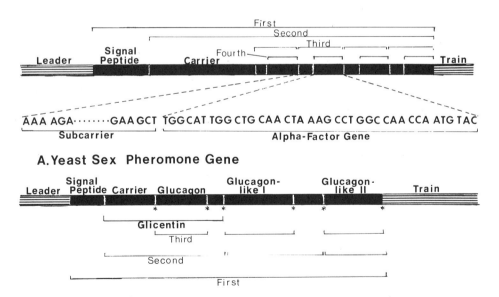

A. Yeast Sex Pheromone Gene

B. Mammalian Glucagon Gene

Figure 1.7. Two examples of cryptomorphic genes. These representatives of the class, at least five susses of which are known (Chapter 7), have extremely complex structures. The primary transcript is shown in black, but only parts of it actually encode the ultimate functional product.

essential in the case of the term gene itself, for none of the definitions given in the introductory paragraph can be considered to be completely satisfactory.

The Gene. A gene is a sequence in a nucleic acid (DNA, except in RNA viruses) that provides a code for a protein, polypeptide, or RNA of direct value to cellular metabolic processes, plus those parts, adjacent or internal, important to its being located, identified, transcribed, often translated, and processed. Internal transcribed regions, such as introns, are also a part of the gene proper when they occur.

Split Gene. Split genes are those that contain internal untranslated parts (introns) and are of two types, monintronic, with a single intron, and multintronic, with more than one.

Primary Transcript. The initial product of transcription is the gene proper, devoid of those portions of the leader and train involved entirely in recognition, initiation, and termination of transcription.

Mature Gene. The mature gene is that part of the coding region that corresponds to the primary structure of the polypeptide or RNA utilized directly in the cell, devoid of introns, leaders, and trains. Sometimes in polypeptides the amino acid used in initiating translation, typically methionine, is subsequently removed before it is of value in the cell. Consequently, in such cases the codon for that amino acid should not be considered part of the mature gene, but as a processing signal. It is to be expected that in genetic studies above the molecular level the mature gene will be referred to merely as the gene.

Compound Genes. Compound genes consist of two or more adjoining mature coding sequences for polypeptides, transcribed and translated as a single unit, but which become separated by processing almost immediately, without an intervening period in the cell.

Assembled Genes. In this type, represented by immunoglobulin cistrons, the final transcribed region of DNA is assembled from smaller subunits. In this case the smallest subunits, called genelets, are first united into larger combinations, the pregenes; two or more of the latter then are combined into the actual gene.

Diplomorphic Genes. A second type of complex gene is comprised of two or more coding regions, continuous with one another but each encoding a separate polypeptide. Characteristically the 5′ portion of the complementary strand encodes a transit (signal) protein, which is removed during passage through a membrane, releasing the actual functional molecule.

Cryptomorphic Genes. This final type is similar to the diplomorphic in being composed of continuous coding regions encoding two or more polypeptides. In this instance, however, the removal of the transit protein does not release the functional product; the actual active polypeptide is released only after extensive processing, following a usually extended period of existence in the cell or multicellular organism.

1.4. ORGANIZATION OF GENES

Some of the organizational characteristics of genes in general of necessity received passing mention in the foregoing discussion of immunoglobulins. The first of these is that the macromolecules of living things fall into families, IgA, IgD, IgE, IgG, and IgM being one such grouping. Many others are obvious families, the various histones forming one, the tRNAs another, the rRNAs a third, and so on. Some, however, are less evident, such as the human growth hormone gene family (Barsh *et al.*, 1983). Growth hormone (or somatotropin) is 85% related structurally to a second member, placental lactogen, but only 35% to a third component, prolactin. Closer kinship between the first two is also evidenced by their genes being clustered together at band q22–q24 on chromosome 17, whereas that for a third member is situated on chromosome 6. Clustering of related genes is a second common characteristic that was noted in the antibody analysis, to which such items as multiple copies of the same gene must be added. More than one copy of a gene is gained by processes referred to as gene duplication, when only a few of a kind are present, or as gene amplification, when dozens or hundreds of duplicates exist, as is often the case in higher eukaryotes.

1.4.1. Organization of Genes

Since the organizational aspects of genes in prokaryotes have been far more thoroughly explored than those from eukaryotic sources, it is logical to begin analysis of the present facet with various bacterial types, supplemented rarely with what is available from blue-green algae. Although viral and organellar genomes, particularly the simple mitochondrial ones, lack the complexity of arrangement found in prokaryotes, the organization of eukaryotic cytoplasmic genes is often very complicated and extremely variable.

Ribosomal RNA Operons. Probably the best known of all, the gene sets for the various rRNAs are favorable for introducing discussion because of the vast pool of information available and their relative simplicity. In bacteria in general, the genes for the three types of rRNAs are arranged into groups, called operons, that are transcribed as a unit. That is, each of the *rrn* operons, as they are distinguished, begins with a leader bearing recognition signals and the like, followed by a series of three coding regions, the first for the minor (16 S rRNA) species, the second for the major (23 S) species, and the third for the ancillary (5 S). Between the first two genes is a long intergenic spacer, on which is located either one or two cistrons for tRNAs. In *E. coli* seven such operons have been mapped and described (Figure 1.8), designated as *rrnA* to *rrnG*, and many of their sequences have been established; these, along with the eukaryotic sequences, receive detailed attention in Chapters 3 and 4. In some strains of the organism, *rrnF* is replaced by *rrnH*, which occupies a different sector of the genome. The major and ancillary rRNA genes are separated by a relatively short intergenic spacer, which is unoccupied by cistrons of other types. Following the 5 S rRNA gene, however, the train may bear various tRNA genes, as shown in Figure 1.8, there being anywhere from none to four present. In some instances, one or two of these sequences may be cotranscribed with the *rrn* operon, but in others, they may be transcribed either separately or, more typically, as pairs, as indicated by the thin brackets below each operon (Figure 1.8). In the illustration, two rRNA operons from eukaryotic sources are included to provide a preliminary insight into the organizational differences that have evolved.

Protein Genes of Eukaryotes. Not all operons are necessarily transcribed as a single unit, either in prokaryotes or eukaryotes; this condition is quite characteristic of protein-coding genes from all sources. One of the simpler examples is provided by the actin genes in the nematode *Caenorhabdites elegans*. Three of the four members of this family of contractile microfilament-forming proteins are arranged together as an operon,

Prokaryote
rrn

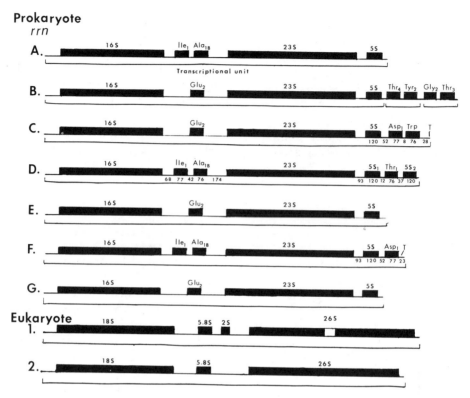

Figure 1.8. Ribosomal operons of prokaryotes and metazoans. In *E. coli* seven ribosomal RNA (*rrn*) operons exist (*A–G*), most of which have the same basic organization, but differ in the specific gene sets present. Among eukaryotes a 5.8 S rRNA gene is often present on the 18 S–26 S intergenic spacer, rather than one for a tRNA as in bacteria. In *Drosophila* (1) the gene for the major type is monintronic, whereas that for *Xenopus* (2) is undivided. (In part based on Federoff, 1979.)

while the fourth is located elsewhere in the genome (Files *et al.*, 1983). Each member of the set of three is transcribed independently, the first in the group (actin I) being of opposite polarity from the others, actins II and III.

The histones of eukaryotes are other representatives of this type of operon, but form units of slightly greater complexity, as seen more completely in Chapter 5. In *Drosophila* the order of the five different histone species is typically H1, H3, H4, H2A, H2B (Figure 1.9), but sometimes it is H4, H3, H1, H2B, H2A (Lifton *et al.*, 1977). Although the latter order is found in a few operons of *Xenopus laevis*, others are quite distinct, the major type having the arrangement H1, H2B, H2A, H4, H3 (Turner and Woodland, 1983), and in *X. tropicalis* still a different sequence prevails (Ruberti *et al.* 1982). This situation is made more complex by the presence of two different variants of H1 in *X. laevis* and by the multiplicity of copies present, as described later. Another arrangement, shown in Figure 1.9, from the sea urchin *Strongylocentrotus*, is H1, H4, H2B, H3, H2A; in this example and in *X. tropicalis*, the direction of transcription has been determined as being all of the same polarity (Cohn *et al.*, 1976; Kedes, 1979). But this is not a universal condition, because in the newt *Notophthalmus viridescens*, in which the genes are arranged H1, H3,

H2B, H2A, H4, the third one has a polarity opposite from the rest (Stephenson *et al.*, 1981), and similar disparities exist in the *Drosophila* operons just cited. Other variants in organization are known for this family. For instance, in the sea urchin *Lytechinus pictus*, the genes expressed during early development are in operons similar to the others listed here, whereas in those expressed in late development, H3 and H4 are arranged in operons separate from the other three species (Childs *et al.*, 1982). But these variants of histone gene organization are viewed in greater depth as transcriptive and processing activities are detailed in later chapters.

Nitrogen-Fixation Operons. A third type of operon is more typical of the original definition of that term, in that it is an operational unit consisting of a number of genes for products that interact to carry out one major activity of the cell. Typically each of this type includes one or more genes, called operators, that regulate the expression of the numerous others in the cluster. One of the most thoroughly documented instances of this sort is provided by the nitrogen fixation (*nif*) operons of many prokaryotes (Singh and Singh, 1981), that of *Klebsiella pneumoniae*, a facultative anaerobic bacterium, being especially well established (Figure 1.10). In this organism the operon consists of 17 genes for different polypeptides and proteins, arranged in eight transcriptional units. All of these appear to have the same polarity (Buchanan-Wollaston *et al.*, 1982), except *nifF* (Beynon *et al.*, 1983), but further sequencing of the various units may demonstrate that this condition is not as uniform as now seems to be the case.

However, that is not a major issue. The main features of this operon lie in the method of expressing the several transcriptional units. Under anaerobic conditions and the influence of lowered fixed-nitrogen levels in the cell, certain nitrogen-regulating genes become activated, that called *ntrC* being the most influential (Espin *et al.*, 1982). The product of this gene seems to serve as an ancillary protein, which aids in recognition of the *nifLA* transcriptional unit. In turn one of the ultimate products, the first operator (or *nifA* protein), acts in the same fashion on all the other units, enabling them to become tran-

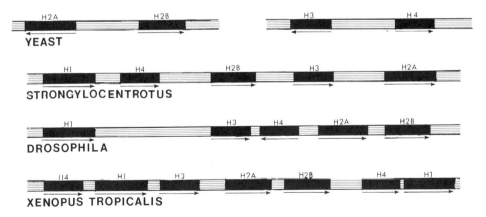

Figure 1.9. Representative operons for histones from various eukaryotes. In yeast two basic types of histone operons are present, each bearing two species of opposite polarity (Smith and Andrésson, 1983). Typically, in higher eukaryotes, a great multiplicity of histone operons is present, which vary in organization. Examples representative of common forms are shown; the arrows indicate the direction of transcription. (Kedes, 1979; Ruperti *et al.*, 1982; Turner and Woodland, 1983.)

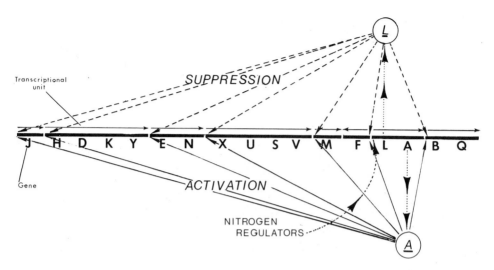

Figure 1.10. The nitrogen fixation (*nif*) operon of *Klebsiella pneumoniae*. The two controlling proteins A and L, products of *nifLA*, have opposing roles, the first activating, the second suppressing, transcription of all genes in the operon except their own. In turn, nitrogen-sensitive genes under certain conditions activate transcription of *nifLA*. The several transcription units that constitute the operon and their polarities are indicated by the small arrows.

scribed (Figure 1.10); the main consequence of these multiple activations is the production of the *nifHDKY* unit, the nitrogenase enzyme complex that reduces N_2 to NH_4^+ (McLean and Dixon, 1981). When the O_2 level becomes elevated, the second operator, the protein L encoded by the *nifL* gene, suppresses transcription of all the transcriptional units except its own, bringing about cessation of nitrogenase activity (Hill *et al.*, 1981; Merrick *et al.*, 1982).

Classes of Operons. It is self-evident from the foregoing discussion of gene organization that, in light of present information, three main types and two subtypes of operons may be considered to exist:

Class I. This class embraces all clusters of genes that are transcribed in unison. Thus the operons of this class are synonymous with transcriptional units, although the converse need not be true in all instances. Two subclasses are evident:

Subclass IA. Those transcriptional units that include genes for only one type of macromolecule fall here, the clusters of tRNAs that follow *rrn* operons being a familiar illustration (Figure 1.8B).

Subclass IB. Transcriptional units that contain genes for more than one type of macromolecule are classed here; examples are the *rrn* operons of *E. coli* that bear tRNAs on the intergenic spacers between the cistrons for the minor and major rRNA subunits, such as *rrnA*, *rrnB*, *rrnE*, and *rrnG* of *E. coli* (Figure 1.8). Further variations are those like *rrnC*, *rrnD*, and *rrnF*, whose transcriptional units include tRNA genes downstream from the rRNA coding sequences.

Class II. Clusters of structurally related genes that are not transcribed as a unit are considered to be class II operons. Among the better established examples are the clusters of histone genes, in which coding sequences for each species of these proteins are grouped into series; although typically duplicated tens or even hundreds of times in the genome, this condition is not essential to membership in this class. Each gene is transcribed independently of one another insofar as current knowledge indicates. However, it is not unlikely that some representatives may contain transcriptional units consisting of two or more genes. The distinctive feature of the class is the clustering of structurally related genes, largely transcribed separately.

Class III. In this class of operons are placed those clusters of genes that are functionally, but usually not structurally, interrelated. It corresponds most closely to the original concept of the operon in being an operational unit. The *nif* genes just discussed are obvious representatives of this class, in which the products of genes *A* and *L* operate the rest of the unit by activation or suppression of replication of the several transcriptional units. Their own transcriptional unit, however, is activated and suppressed by other factors in the cell or milieu. The *lys* operon, whose nine component genes are suppressed by lysine, is another clear example of this class (Cassan *et al.*, 1983). Comparable class III representatives are known for eukaryotes, but these have not been as fully characterized as those of bacteria. One member of this class is found in the mouse, with regard to the seven or more kallikrein operons active in the submaxillary gland, whose activities seem to center around the processing of such proteins as the epidermal growth factor and γ subunit of the nerve growth factor (Mason *et al.*, 1983).

At this relatively early stage in the knowledge of gene organization, the evolutionary steps in operon development are not yet visible. It is clear that the members of class I represent a more elementary type than the others in being simply a transcriptional unit of related genes, with or without others of contrasting nature. Whether the other two classes exemplify successive stages in phylogeny or separate lines of ascent, one among prokaryotes, the other in eukaryotes, can be clarified only by much additional information.

1.4.2. Amplified Genes

Another aspect of genomic organization has been alluded to on several occasions, namely, the multiplicity of copies of a given gene that frequently exists. Such amplification of genes, as the presence of multiple copies is called, falls into several categories. Not only do the genes themselves undergo amplification phylogenetically or ontogenetically, but many cases of reiterated sectors of intergenic spacers also have been reported. Since gene repetitiveness has been the more thoroughly documented aspect, the several categories of organization that can be detected receive attention first.

An important point in connection with the amplification of genes is that the mere presence of multiple copies of a given cistron may have little relationship to the number actually employed in cellular function. Particularly striking is the account of the genes for an important respiratory protein, glyceraldehyde-3-phosphate-dehydrogenase (Fort *et al.*, 1985). Although more than 200 copies of this and closely allied genes proved to be present in the rat genome, only one major mRNA species was expressed.

Category I of Gene Organization. The simplest condition of genomic structure is the absence of repetition, although organization into operons may exist. In addition to its lack of complexity, this category I organizational form is also evolutionarily elemental, for it is characteristic of the most primitive known genomes, those of the viruses. In no case in which the nucleotide sequence of an entire viral genome has been determined, including those of parvovirus, hepatitus B virus, tobacco mosaic virus, poliovirus, cauliflower mosaic virus, and many others, has the presence of a single repeated gene been revealed (Franck et al., 1980; Goelet et al., 1982; Nomoto et al., 1982; Astell et al., 1983; Ono et al., 1983). Single-copy genes are, however, by no means restricted to these simple organisms, but appear to characterize a basic condition in a broad spectrum of living things. Known examples among bacteria include several shown in connection with the rrn operons (Figure 1.8), such as the gene for tRNA^{Trp} on rrnC, which is the only copy known to occur in E. coli (E. A. Morgan et al., 1978). The deviant 5 S$_2$ gene of rrnD discussed in Chapter 3 is another instance of the unique condition. An additional extreme case of single genes in bacteria is that cited in the section on operons, the class III lys operon, whose nine or so members are scattered throughout the genome (Cassan et al., 1983).

Representatives from eukaryotic sources are probably abundant, but have not been thoroughly documented. One that has been demonstrated is the coding sequence of α_4-tubulin of Drosophila (Baum et al., 1983), while another is that for the α_2 type 1 collagen of the chick (Ohkubo et al., 1980; Yamada et al., 1983). In addition the latter reference shows that the gene for α_1 type III collagen is also represented by a single copy. Still other examples of uniqueness are represented by the genes for histones H3 and H4 in the fungus Neurospora crassa, whereas typically all of this type are reiterated in other eukaryotes (Woudt et al., 1983). Moreover, the three varieties of yolk proteins of Drosophila are encoded by separate single-copy genes (Hung and Wensink, 1981).

Category II Organization. Possibly the most basic type of gene duplication is that in which several to many copies of the same gene or operon are present, loosely scattered in the genome. For instance, with the α_4-tubulin of Drosophila just cited are three others (α_{1-3}), all of which differ from α_4 but are homologous to each other; they are located singly at multiple points in the genome (Baum et al., 1983). The seven or so sucrose-cleaving (invertase) genes, SUC1–SUC7, of yeast, at least some of which encode two different forms of the enzyme (Taussig and Carlson, 1983), present a similar situation. Probably the most thoroughly documented illustrations of this type of organization are two genes from D. melanogaster (Finnegan et al., 1977), known as 412 and copia genes. Although the functions of the products of neither gene have been established, the transcripts are exceedingly abundant in the cytoplasm of many different tissues—the mRNAs of copia constitute about 3% of all polyadenylated messengers. Copies of each are located at multiple sites dispersed throughout the genome. Still another illustration, a very simple one, is found in the genomic arrangement of the two types of histone operons of yeast, one consisting of H3 and H4, the other for H2A and H2B (Figure 1.9). Each type is represented by only two copies, but these are located at widely separated parts of the genome (Smith and Andrésson, 1983; Smith and Murray, 1983).

Category III Gene Amplification. The next more advanced type of multiple gene organization is that in which the same gene is reiterated one or more times on the same strand, one copy after the other. This tandem repetitiveness is exemplified by the 5 S rRNA gene of man and also those in Drosophila (Brown et al., 1971; Procunier and Tartof, 1976; Artavanis-Tsakonas et al., 1977). In the first instance the numerous copies are concentrated at a single site (L. D. Johnson et al., 1974), whereas in the second the arrangement of the approximately 160 5 S genes is not fully established, but they are known to be tandemly arranged and either in one or two groups (Grunstein and Hogness,

1975; Artavanis-Tsakonas *et al.*, 1977). One example of tandem repetitiveness is provided by the gene of the β subunit of human chorionic gonadotropin; at least eight closely linked copies are present, all of which are arranged in a group, four as tandem repeats, the others as inverted pairs (Boorstein *et al.*, 1982). It should not be supposed that identical degrees of amplification occur within a given family of genes, for this is certainly not the case. In the gene family for mouse α-amylase isozymes, *amy-1*, *amy-2*, and perhaps *amy-X*, only a single copy of the first is present, whereas at least two of the second and an undetermined number of the last have been reported (Schibler *et al.*, 1982). Still other examples are brought to light in later discussions. Arrangement of genes in multiple inverted pairs might be considered a special variant of this category.

Category IV Gene Organization. In this category are included the multiple copies of gene families or operons that are dispersed through the genome. Two extremely simple illustrations of this type of organization are provided by the histone genes of yeast, a primitive eukaryote that lacks histone H1 (Marian and Wintersberger, 1980). The cistrons of the others are arranged as short operons, H2A and H2B being grouped together, as are H3 and H4 (Figure 1.9) (Smith and Andrésson, 1983; Turner and Woodland, 1983). Each of these operons is repeated just one time (Hereford *et al.*, 1979). Recently the products encoded by the two H2B genes were found to differ at four sites in amino acid content, whereas the messengers diverged to an extent of 12.6%, 40 of the 390 sites having different nucleotide residues (Wallis *et al.*, 1980). Thus this case, as well as many others cited in subsequent chapters, indicates that reiterated genes are not necessarily identical to one another. Furthermore, the seven *rrn* operons of *E. coli* (Figure 1.8), which also fall into this category of organization, demonstrate that the contents of the repeated units are not consistently identical, differing in the tRNA genes that are included and length and content of the transcriptional units.

Category V Gene Reiteration. In this fifth and final class of gene organization, the operons or clusters are repeated in tandem fashion, a condition that present information suggests is more frequent than the diffuse type just described. The histone genes of the sea urchin *Lytechinus pictus* provide an especially informative case of this organization, because they are expressed in a developmental stage-related manner (Childs *et al.*, 1982). The genes functioning during the blastula stage, the α-subtype histones, are present in the form of several hundred nearly identical, tandemly repeated copies in the genome (Kedes, 1979), every one of which includes a single coding sequence for each histone species. Those that become expressed in the gastrula and later stages are present in far fewer repeats, moreover, each late cluster consists either of the cistrons for H1, H2A, and H2B, or those for H3 and H4. Both of these types of clusters are dispersed in the genome, typically separated by thousands of nucleotide residues (Childs *et al.*, 1982). Thus, even in the same organism, the genes for developmentally expressed products may differ in organization with the stage and perhaps with the tissue, too.

Another enlightening example is provided by the cistrons for the 20 or more different proteins that comprise the chorion of the *Drosophila melanogaster* egg. In the oocyte these are present in the category I mode of organization, but become amplified into the category II or IV condition in the follicle cells beginning with stage 8 and continuing into stage 12 of oogenesis (Spradling and Mahowald, 1980). One operon that contains several chorionic proteins formed in early oogenesis (stages 11 and 12) is located on the X chromosome and becomes amplified 14- to 16-fold in tandem style (Figure 1.11A). A

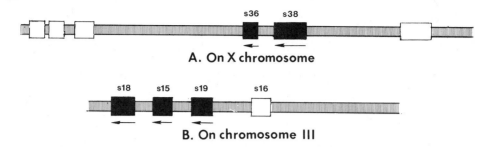

Figure 1.11. Chorion genes of *Drosophila* follicle cells. Sequences that have been fully characterized as such genes are shown as solid blocks, the others as open ones. (A) The operon on the X chromosome is reiterated 14- to 16-fold; (B) that on chromosome III is repeated about 60 times in tandem fashion. (Based on Osheim and Miller, 1983.)

second, on chromosome III, includes four genes for these proteins that are activated late (stages 13 and 14) in egg-shell formation (Figure 1.11B); it becomes tandemly repeated about 60 times (Osheim and Miller, 1983).

1.5. MOVABLE AND REPETITIVE ELEMENTS

Another of the many unexpected features of genomic organization that were revealed as sequencing studies became feasible is the confirmation of the presence of repeated sequences within the untranscribed and untranslated portions of the DNA. These re-petitive elements may be interspersed among single-copy genes (Davidson and Britten, 1973) and are often associated with heterogeneous nuclear RNA and the 3' trains of mRNA (Calabretta *et al.,* 1981; Davidson and Posakony, 1982). In a number of instances they have also been found in the introns of some coding regions as well as in the intergenic spacers. Among genes containing such intervening-segment-located repeated elements are those for the mouse dihydrofolate reductase (Feagin *et al.,* 1983), the silk fibroin of the silkworm (Pearson *et al.,* 1981), and conalbumin of the chicken (Cochet *et al.,* 1979). The majority, however, occur in the spacers and in satellite DNA and are most charac-teristic of larger viral and eukaryotic genomes. Great variation in number has been reported, ranging from relatively few, as in the 20-copy one of the mouse dihydrofolate reductase gene described in detail below (Feagin *et al.,* 1983) and the triple one in the Bermuda land crab (Bonnewell *et al.,* 1983), upward to around 1000, as in many dis-persed repeated elements in *Xenopus* (Davidson and Britten, 1973), and even to tens of thousands, as in others from the last source.

Many of these repeated elements in both prokaryotes and eukaryotes are relatively freely transposable from one region to another. While their true nature is just beginning to be exposed, these elements, under such varied names as transposons, inserted sequences, and transposable elements, are proving to exercise considerable and highly diversified influence on the expression of particular genes. They are frequently implicated in the production of gene mutations and chromosome aberrations, including duplications, inver-sions, and fusions (Kleckner, 1977; Farabaugh and Fink, 1980). Since they have been most thoroughly studied in *E. coli, D. melanogaster,* and the mouse, the examples for this introductory discussion are drawn primarily from those three sources; however, the topic is given more detailed attention in Chapter 3, Section 3.4.1, and in the whole of Chapter 9.

Repetitive Elements in the Mouse. The genome of the laboratory mouse has been explored for repetitive elements probably more extensively than that of any other eukaryote and thus is favorable as the starting point for discussion (for reviews, see Singer, 1982a,b). In many cases, the copy number of a given type may reach 10,000 or even 100,000; hence each such family can represent 1% or more of the total genome. One group, referred to as B1, consists of 130-base-pair units, while in that known as the R family the size is 475 base pairs (Krayev *et al.*, 1980; Gebhard *et al.*, 1982). Still another consists of elements of 5600 nucleotide residues (Meunier-Rotival *et al.*, 1982). In a more recently described family from this source, the elements, about 382 base pairs in length, were said to be nonidentical but highly homologous (Lueders and Paterson, 1982) and were represented in the genome by approximately 100,000 copies. When one of these, labeled pG3-28, was sequenced (Table 1.3), it was found to be highly homologous in structure with R1 of the R family of repeating elements. Of the 379 total sites, 312 were identical to those of R1 in the latter's region beyond site 123, for a rate of about 82%.

These repeated sequences were detected largely in association with immunoglobulin and β-globin gene clusters. The R family was originally described from six members associated with the former (Gebhard *et al.*, 1982), but was subsequently found adjacent to β-globin and a number of other genes (Lueders and Paterson, 1982). Moreover, related but divergent repeated elements have been demonstrated in several additional mammalian species, including the rat, Syrian hamster, and monkey. While the R family and the foregoing type of repeat of the mouse are closely related, kinship is not universal among repeated sequences of this sort. For one instance, the B1 family was shown not to be related to the R (Gebhard *et al.*, 1982), but relationship was not tested in regard to the other two or three families that have been reported in this mammal.

Homology among Repeated Spacer Elements. Very few sequence studies have been conducted on several members of the same family of repetitive spacer elements to determine the actual degree of relationship that exists. One of the exceptions analyzed three members of the same family of these structures from the Bermuda land crab (Bonnewell *et al.*, 1983). Two of these proved to be more closely related than the third, as can

Table 1.3
Nucleotide Sequence of a Repetitive Element in Mouse DNA

	10	20	30	40	50	60	70
GATGATGAT	TACAGTAAT	CTATGGATG	GAACACAGG	GCCCCAAGT	GGAGGAGCT	AGAGAACGT	ACCCAAGAG
80	**90**		**100**	**110**	**120**	**130**	**140**
CTGAAGGGG	CCTGCAACC	CTATAGGTG	GAACAACAA	AATAAACTA	ATCAGTAAC	CCCAGAGCT	CGTATCTCT
150	**160**	**170**	**180**		**190**	**200**	**210**
TGCTGCATA	TGTAGCAGA	AGATGGACT	AGTCAGCCA	TCATTGGAA	AGAGAGGCC	CCAAGGTCT	AGCAAACTT
220	**230**	**240**	**250**	**260**	**270**		**280**
TATATGTCT	CAGTACGGG	GACGCCGTT	GCCAGAAGT	GGGAGTGGG	TGGGTAAGG	GAGCAGGAG	ACAGGGAGG
290	**300**	**310**	**320**	**330**	**340**	**350**	**360**
AAATAGGGA	ACTTTCAGG	ATAGCATTT	GAAATGTAA	ATGAGGAAA	AATATAAAA	AAATTTTTT	TAAATTAAA
	370	**379**					
AAAAAACAG	ATACACTAA	A					

be noted in the comparison of their base sequences (Table 1.4). In addition to the 40 sites that differ, shown in italics, two major distinctions exist. The lesser of these is that the sequences referred to as RU and TRU contain a 24-nucleotide-residue sequence not found in the third, EXT, while a still greater point of departure is found in the sector enclosed in the box. In RU and TRU this segment is not repeated, but in EXT six copies occur, so that the element consists of a total of 1130 base pairs, in contrast to the 429 and 433 of RU and TRU, respectively. Although the six copies differ among themselves at four known sites plus a few still undetermined ones in the second repeat, they are virtually identical. Thus RU and TRU form a practically identical pair, differing at only 16 of their 433 sites (that is, 96% homologous), whereas both are less than 36% homologous with EXT, because of the latter's extensive repeated sequence. Hence, when adequate comparisons of their

Table 1.4
Three Members of a Repeat Family from a Crab[a]

	10	20	30	40	50	60	70
EXT	GACTCTGC*C*T	CACACCGCCG	ACTGCTACGC	*A*AGCCGCTAT	GCGGC-ATGG	CCTGTC*GA*AA	*A*AA*C*GAAA*A*C
RU	GACTCTGC*GT*	CACACCGCCG	ACTGCTACGC	*G*AGCCGCTAT	GCGGC-ATGG	CCTGTC*G*TAA	*GA*AGGAAT*AG*
TRU	GACTCTGC*GT*	CACACCGCCG	ACT*C*CTACGC	*G*AGCCGCTAT	GCGGC*T*ATGG	CCTGTC*A*TAA	*GA*AGGAAT*AG*

	80	90	100	110	120	130	140
EXT	ATGCGTACGT	TATTA*G*AATA	ACGAAATAAG	AAC*A*GG----	----------	----------	AACAAGAAGA
RU	ATGCGTACGT	TATTA*C*AATA	ACGAAATAAG	AAG*A*ACAACA	*GC*AAGAGCAG	GAACAAGAAC	AACAAGAAGA
TRU	ATGCGTACGT	TATTA*C*AATA	ACGAAATAAG	AAG*A*ACAACA	*A*GAAGAGCAG	GAACAAGAAC	AACAAGAAGA

	150	160	170	180	190	200	210
EXT	ACAAGA*C*GAA	GAAAGAAGAG	GAAT*A*A*C*ATC	AACAACACCA	AGAAG*A*GCA*C*	*G*GACGACTAC	AACAAGGAGA
RU	A------GAA	GAAAGAAGAG	GAA*C*AAC*G*TC	AACAACACCA	AGAAG*GA*CAT	*C*GACGACTAC	AACAAGGAGA
TRU	A----*GA*GAA	GAAAGAAGAG	GAA*C*A*G*CATC	AACAACACCA	AGAAG*GA*CAT	*C*GACGACTAC	AACAAGGAGA

	220	230	240	250	260	270	280
EXT	G*GG*GGAAG*GC*	AGGAAGGAAT	TGC*C*GCGTGG	CACATTCCGC	ATTTCA*C*CAA	GCT*C*CCCTCT	CCTCCTACCT
RU	G*GG*GGAAG*AC*	AGGAAGGAAT	TGC*T*GCGTGG	CACATTCCGC	ATTTCA*G*CAA	GCT*T*CCCTCT	CC---TACCT
TRU	G*C*GGGAAG*AC*	AGGAAGGAAT	TGC*T*GCGTGG	CACATTCCGC	ATTTCA*G*CAA	GCT*T*CCCTCT	CC---TACCT

	290	300	310	320	330	340	350
EXT	TGTCTAGGGC	ATGCGTGCCC	TCCTCGCG*AG*	TCGAGTTTAG	CGCGGAGAGT	TGCGGTGCC*G*	CGCGCC*GG*AAA
RU	TGTCTAGGGC	ATGCGTGCCC	TCCTCGCG*GG*	TCGAGTTTAG	CGCGGAGAGC	TGCGGTCG*C*A	CGCGCC*GG*AA
TRU	TGTCTAGGGC	ATGCGTGCCC	TCCTCGCG*AG*	TCGAGTTTAG	CGCGGAGAGC	TGCGGTGCC*A*	CGCGCC*GG*AA

	360	370	380	390	400	410	420
EXT	CAATCT*T*ACG	GCTGTCGAGG	GAGAAGGAG*A*	TGCAGTAC*AA*	*A*CTCTCTCGG	TGACTTGCGT	CGCTAACTCC
RU	CAATCT*G*ACG	GCTGTCGAGG	GAGAAGGAG*G*	TGCAGTGC*CG*	*C*CTCTCTCGG	TGACTTGCGT	CGCTAACTCC
TRU	CAATCT*G*ACG	GCTGTCGAGG	GAGAAGGAG*G*	TGCAGTGC*CG*	*C*CTCTCTCGG	TGACTTGCGT	CGCTAACTCC

EXT	AAGC*AAA*GGG	GATAG*CC*TCC	ATCGA*C*TC
RU	AAGC*CC*AGGG	GATAG*CG*TCC	ATCGA*C*TC
TRU	AAGC*CCAA*GG	GATA*CCG*TCC	ATCGA*NT*C

[a]Based on Bonnewell *et al.* (1983).

primary structures have been made, it is not unlikely that other families of such elements will also be found to be comprised of mixtures of nearly identical and only distantly related members.

This, like most repeated spacer elements, is G, C-rich as a unit, there being 223 Gs and Cs to 198 As and Ts in the complementary strand of EXT, not counting the five additional multiples of the boxed sector. This tally does not give a true picture, however, of the nucleotide constituency, for there are only 68 thymidine residues present, adenosine being by far the most frequent nucleotide, with 130 As to 112 Gs and 111 Cs. The first 70 sites of the internal repeat is especially rich in adenosine, over 50% consisting of that nucleotide. Moreover, the 24-base sector missing from EXT similarly is rich in A, to an extent near 65%. Thus, the total content of particular nucleotides or pairs gives only a partial view of these interesting, but still poorly understood, features of the genome.

1.5.1. Movable Elements in Bacteria.

Many of the movable elements in *E. coli* appear to be of complex structure, often including one or more genes within their boundaries that have functions aside from transposition. For instance, transposon *Tn21* carries genes for resistance against mercuric ions, sulphonamides, and streptomycin, *Tn501* also encodes resistance to mercuric ions, and *Tn1721* includes a cistron affording resistance to tetracycline (Bennett *et al.*, 1978; Schmitt *et al.*, 1979; de la Cruz and Grinsted, 1982). In addition, these and other members of the *Tn3* family of movable elements contain at least two genes required for transposition. The first of this pair, *tnpA*, encodes a protein involved in generating unions between donor and recipient regions of DNA that contain directly repeated copies of the movable element (Kleckner, 1981). The product encoded by the second member, *tnpR*, catalyzes recombination between specific sites in the two copies of the element. Recently, the primary sequences of *tnpR* from *Tn21* and *Tn501* have been established (Diver *et al.*, 1983). As may be seen in Table 1.5, the two representatives are not highly homologous, agreeing at only 418 of the 516 sites, a homology rate of about 74%. Thus two conclusions emerge: The genes involved in the recombination events differ significantly from one transposon to another, and, as a consequence, the coding sequences of this family of substances may react to different sites. Therefore generalizations in regard to their characteristics need to be proposed with great caution, and then only when sufficient sequence data are in hand.

Another translocatable element, *IS1*, is only 768 base pairs in length, the smallest such structure known in bacteria (Machida *et al.*, 1983). This unit, which consists of the two cotranscribed genes *insA* and *insB*, is able to mediate the simultaneous integration of two different plasmids into DNA. In one unit that carries it, the transposon *Tn9*, two copies of *IS1* are present, one on each side of a chloramphenicol-resistance cistron. That located upstream of this latter gene has been shown to be 20 times more efficient in its ability to produce cointegration of plasmids than its downstream counterpart, in spite of their identical primary structures. Recently this weakness has been demonstrated to result from readthrough when the chloramphenicol-resistance gene is transcribed (Machida *et al.*, 1983). *IS2* is somewhat longer, but still relatively small (1327 base pairs); several copies of this transposon are known to exist in the genome of *E. coli*, sometimes as mutation-causing inserts (Hinton and Musso, 1983).

Table 1.5
Gene Structure of tnpR of E. coli[a]

		10	20	30	40	50	60	70
Tn50l		GTGCAGGGGC	ACCGCATCGG	CTACGTCCGG	GTCAGCAGCT	TCGACCAGAA	CCCGGAACGC	CAGCTGGAAC
Tn21		ATGACTGGAC	AGCGCATTGG	GTATATCAGG	GTCAGCACCT	TCGACCAGAA	CCCGGAACGG	CAACTGGAAG
		80	90	100	110	120	130	140
Tn50l		AGACACAGGT	GAGCAAGGTG	TTCACCGACA	AGGCATCGGG	CAAGGACACC	CAGCGCCCCC	AGCTCGAAGC
Tn2l		GCGTCAAGGT	TGATCGCGCT	TTTAGCGACA	AGGCATCCGG	CAAGGATGTC	AAGCGTCCGC	AACTGGAAGC
		150	160	170	180	190	200	210
Tn50l		GCTGCTGAGC	TTCGTCCGCG	AAGGCGATAC	AGTGGTGGTG	CACAGCATGG	ACCGGCTGGC	CCGCAACCTC
Tn2l		GCTGATAAGC	TTCGCCCGCA	CCGGCGACAC	CGTGGTGGTG	CATAGCATGG	ATCGCCTGGC	GCGCAATCTC
		220	230	240	250	260	270	280
Tn50l		GATGACCTGC	GTCGCTTGGT	ACAGAAGCTG	ACTCAACGCG	GCGTGCGCAT	CGAGTTCCTG	AAGGAGGGCC
Tn2l		GATGATTTGC	GCCGGATCGT	GCAAACGCTG	ACACAACGCG	GCGTGCATAT	CGAATTCGTC	AAGGAACACC
		290	300	310	320	330	340	350
Tn50l		TGGTGTTCAC	TGGCGAGGAC	TCGCCGATGG	CCAACCTGAT	GCTGTCGGTG	ATGGGGGCCT	TCGCCTGAGTT
Tn2l		TCAGTTTTAC	TGGCCGAAGAC	TCTCCGATGG	CGAACCTGAT	GCTCTCGGTG	ATGGGCGCGT	TCGCCGAGTT
		369	370	380	390	400	410	420
Tn50l		CGAGCGCGCC	CTGATCCGCG	AGCGGCAGCG	TGAGGGCATC	ACCTTGGCCA	AGCAGCGTGG	CGCGTACCGG
Tn2l		CGAGCGCGCC	CTGATCCGCG	AGCGTCAGCG	CGAGGGTATT	GCGCTCGCCA	AGCAACGCGG	GGCTTACCGT
		430	440	450	460	470	480	490
Tn50l		GGCCGCAAGA	AAGCCCTGTC	CGATGAGCAG	GCTGCTACCC	TGCGGCAGCG	AGCGACGGCC	GGCGAGCCCA
Tn2l		GGCAGGAAGA	AATCCCTGTC	GTCTGAGCGT	ATTGCCGAAC	TGCGCCAACG	TGTCGAGGCT	GGCGAGCAAA
		500	510	520	530	540	550	560
Tn50l		AGGCGCAGCT	TGCCCGCGAG	TTCAACATCA	GCCGGGAAAC	CCTCTACCAG	TACCTCCGCA	CGGACGACTGA
Tn2l		AGACCAAGCT	TGCTCGTGAA	TTCGGAATCA	GTCGCGAAAC	CCTGTATCAA	TACTTGAGAA	CGGATCAGTAA

[a]Based on Diver *et al.* (1983).

1.5.2. Transposable Elements from Drosophila

Two transposable elements from *Drosophila* offer additional insights into their properties. One of them, the *P* element, is confined to the *D. melanogaster* P strain, in which the male contributes to dysgenesis when mated to females of the M (maternally contributing) strain, the reciprocal cross being normal. The resulting hybrids develop aberrantly, with reduced fertility, and show enhanced mutation and chromosomal aberration rates. Individuals of the P strain have multiple copies of the *P* element dispersed over all the major arms of the chromosomes, between 30 and 50 copies being present. However, according to a current study, many of these are mere fractions of the entire element, which is 2900 base pairs in length. This element appears to be involved in expression of the *white* locus, where it has been found inserted at two sites within a large open reading frame. Thus it was suggested that the white phenotype and its variations result from insertion of this or other transposable element within the *white* coding sequence (O'Hare and Rubin, 1983).

Another type of activity is exemplified by a recently described family of transposable elements of this fruitfly, named the *hobo* class (McGinnis *et al.*, 1983). The paper that first described it also reported its effects on an X-chromosome gene, *Sgs-4,* which is activated only during a short period of the developing insect's life. Its expression, confined to the approximately 200 cells of the larval salivary glands, results in the production of a glue used to secure the pupa case to a firm surface. Between 300 and 500 base pairs upstream of the mature gene is an essential control region which becomes hypersensitive to DNase I when the gene is activated. When the *hobo* element, 1300 base pairs long, was inserted upstream of the −35 signal, activity of the gene was greatly reduced, numerous new transcription starts were made, and the chromatin structure was altered. Nevertheless, the tissue specificity of expression of *Sgs-4* remained unchanged, as did its developmental timing. Hence, regulation of differentiation was unaffected by changes in the site of transcription initiation.

1.5.3. Transposons in Other Eukaryotes

An example of a transposable element that induces a chromosomal duplication is found in the yeast under the name of *Ty1*. This transposon, which is 5600 base pairs long and flanked by directly repeated sectors (δ regions) 250 base pairs in length, is represented by about 35 copies dispersed throughout the genome. The δ regions, totalling over 100 copies, occur both as flanking sectors of this element and as scattered independent units. When *Ty1* became inserted into the promoter of the gene *his4*, it induced the formation of a 5-base-pair duplication, consisting of the sequence -TAAGA, which arose during transposition (Farabaugh and Fink, 1980).

One transposable element from maize also induces gene mutations, but in addition, it serves as a site for chromosomal aberrations and in particular for chromosomal breakage. Its name, dissociation (*Ds*) element, is derived from the last of these properties. Several mutant and revertant strains involving the effects of this element when inserted at the *Shrunken* (*sh*) seed locus have been the subject of a recent investigation (Courage-Tebbe *et al.,* 1983). In one mutation, the *Ds* unit was located close to the *sh* gene and induced the production of a duplicated sequence of undetermined length. Although revertant strains lost the transposon sequence, the duplicated area remained, so that normal expression of *sh* was not fully regained.

1.6. STEPS IN GENE EXPRESSION—AN OVERVIEW

Now that the major features of the structure of the gene have received preliminary attention, the chief steps involved in its expression need brief summarization in order to make clear the organization followed in the remainder of the text. At each of these stages, the genes or their products of a number of diverse types, principally those that have been more thoroughly explored at the molecular level, receive detailed analysis. The principal representative classes include the tRNAs, 5 S and other rRNAs, histones, tubulins and microfilament-derived proteins, hemoglobins, immunoglobulins, and other proteins, especially those abundant in a large cross section of the living world.

1. *Transcription.* The very first step in the expression of a gene is the transcribing of the DNA into an RNA molecule. As shown in the preceding pages, this

involves the locating and identification of the DNA sector to be read, followed by attachment of the RNA polymerase and ancillary proteins, initiation of transcription, elongation of the RNA product chain, termination of transcription, and release of the transcript.

2. *RNA processing.* After the transcript is released, or even during its production, various extraneous parts are removed by a series of RNases, each of which performs one or more specific functions. Leaders and trains are removed, in whole or part, introns are excised and the several exons spliced together, and parts of the mature product that are not encoded in the DNA are added enzymatically as necessary. Other RNA processing events in particular cases include capping, addition of poly(A) tails, and the modification of nucleotides by enzyme activities.

3. *Translation.* Stated in simple terms, those RNAs that encode proteins receive ribosomes that translate the codons of the message into peptides or proteins. Here, as in transcription, the messengers must be recognized and attached to, and translation initiated at specific sites. The short amino acid chain that results must then be elongated, its translation terminated at the correct point, and finally the chain released.

4. *Protein processing.* The peptide or protein resulting from translation may then undergo processing through the addition of one or more amino acids enzymatically, or some of the amino acids may be modified, sometimes to the D-rotatory form and sometimes acetylated or methylated. Those molecules, typically diplomorph gene products and others that are to become glycoproteins, are transported in eukaryotes to the endoreticulum or dictyosomes for removal of the signal peptide and addition of sugar or other moieties; cryptomorphic gene products undergo proteolytic breakdown when the cell is ready to employ the concealed peptides.

5. *Degradation.* All the macromolecules of the cell, except perhaps the DNA chain itself, have well-marked half-lives, all being replaced by others exactly like themselves at a rate specific for each molecular type. This turnover of molecular constituents, carried out by various nucleases and proteases, is one of the most characteristic properties of living matter.

2

The Enzymes of Transcription and Transfer RNA Genes

The transcription of genes is undoubtedly the most complex and difficult undertaking of cells. Before transcription can begin, the enzyme must locate the specific, relatively brief sector of the genome that is to be transcribed. Out of the multimillion combinations of just the four different base pairs that comprise the DNA molecules, it is necessary to identify a sequence of as few as 100 or so pairs. Once this region has been recognized, the enzyme must attach to the DNA and separate the two strands. Transcription of the proper strand must then begin at a precise point and, after this initiation step, the RNA molecule thus begun must be elongated in accordance with the information in the genome. Then, finally, transcription must be terminated at the correct point to permit processing later, the new RNA then being released.

Rather than the four steps often recognized (Anthony *et al.*, 1966; Krakow *et al.*, 1976), transcription appears to involve six separate major events: (1) Recognition of the gene to be transcribed; (2) binding of the enzyme and other factors to the DNA; (3) local separation of the two DNA strands; (4) initiation of transcription at a precise point; (5) elongation of the RNA chain; and (6) termination of transcription and release of the RNA chain. Each of these processes is itself complex, most of them involving several factors which all too frequently remain unknown. However, the more important enzymes have received much attention, and it is with the chemical nature and properties of these substances that discussion begins.

2.1. DNA-DEPENDENT RNA POLYMERASES

The enzymes involved in transcribing genes into RNA form a complex group referred to as the DNA-dependent RNA polymerases to distinguish them from those that replicate RNA from templates of the same nature (Chamberlin, 1976; Losick and Chamberlin, 1976). As a group these enzymes have molecular weights of 500,000–700,000 in eukaryotes, in which 10–15 distinct polypeptide subunits are present, ranking among the most complex proteins of the cell. Typically, the subunits fall into two groups, large, with molecular weights of 100,000, two types of which are present, and small, in which the molecular weights range below 50,000. However, subunits of intermediate size occur in some classes.

CHAPTER 2

2.1.1. The Simpler DNA-Dependent RNA Polymerases

In prokaryotes, DNA viruses, and eukaryotic cell organelles, only a single class of DNA-dependent RNA polymerase usually is present, which transcribes the genes for all types of macromolecules. This absence of specialization in function, such as what is seen later to characterize the eukaryotic types, is one aspect of the enzymes from the present sources that indicates their relative simplicity. However, it is of considerable importance to note in this connection that two distinct RNA polymerases have been reported in rat mitochondria (Yaginuma *et al.*, 1982), so that the present seeming simplicity of eukaryotic organellar enzymes may prove to be an unsound generalization.

The Eubacterial RNA Polymerases. In the eubacteria a second condition suggestive of relative simplicity is the less complex subunit construction of these enzymes, for only four or five subunits are present per molecule of holoenzyme. The *E. coli* enzyme, which, like that of other eubacteria, is inactivated by rifampicin and streptolydigin, has a molecular weight of 440,000 and a protomeric structure of $\beta'\beta\sigma\alpha_2$. The subunit molecular weights are $\beta' = 160,000$, $\beta = 150,000$, $\sigma = 86,000$, and $\alpha = 40,000$ (Burgess, 1976; Chamberlin, 1976). The σ subunit readily separates from the holoenzyme, leaving the "core polymerase" $\beta'\beta\alpha_2$. By itself, the β' subunit is the only one capable of binding either to DNA or to σ, but a $\beta\alpha_2$ combination possesses limited abilities along those lines (Zillig *et al.*, 1976). The latter tripartite substructure in addition seems to possess the rifampicin-binding capacity of the holoenzyme (Huaifeng and Hartmann, 1983). The reported concentration of activity in the β' subunit is difficult to correlate with recently obtained direct evidence. By use of analogs of ATP and GTP as photoaffinity probes, two distinctly different active sites have been demonstrated in *E. coli* RNA polymerase, both located on the β subunit (Panka and Dennis, 1985). Seemingly, the two are employed alternately for the synthesis of phosphodiester bonds between nucleotides as they are added to the transcript.

Part of the ability of the σ subunit to dissociate from the holoenzyme is involved in its role in selection of the site to be transcribed. Thus it really is one of a number of site-specific proteins that stimulate transcription, to be discussed later (Losick and Pero, 1976). This role is accentuated in *Bacillus subtilis* by the existence of two σ subunits; the predominant form σ^{55} has a molecular weight approximating 55,000, while the scarcer type σ^{28} has a molecular weight around 28,000 (Gilman *et al.*, 1981). As a rule the former is supposed to interact with ancillary and promoter sequences having average constitutions of -TTGACA- and -TATAATA-, respectively, located at leader sites at -35 and -10 as mentioned in Chapter 1. In contrast, the smaller σ subunit reportedly reacts with -CTAAA- and -CCGATAT- located at corresponding sites. Still a third signal of unknown structure, located between 161 and 186 base pairs upstream, has been implicated in σ function (Johnson *et al.*, 1983).

RNA Polymerases of Advanced Bacteria. The DNA-dependent RNA polymerases of so-called Archaebacteria (Woese *et al.*, 1978; Fox *et al.*, 1980) have received extensive analysis in recent years, the enzyme from *Sulfolobus acidocaldarius* now being particularly well known (Zillig *et al.*, 1979, 1980; Prangishvilli *et al.*, 1982). These are far more complex in structure than those of the Eubacteria, approaching the enzymes of eukaryotes in number of subunits. In *Sulfolobus* two large and eight different small subunits are present, one of the latter being represented by four copies per molecule. The large pair has molecular weights of 122,000 and 101,000, respectively, while the small ones range from 44,000 to 10,800, yielding a total molecular weight of 423,500 for the holoenzyme. Actinomycin D and heparin interfere with the enzyme's ability to transcribe DNA, whereas rifampicin, streptolydigin, and α-amanitin have no effect; these resistant properties are suggestive of relationships to eukaryotic RNA polymerase I, as seen later. Although another advanced bacterium, *Halococcus morrhuae*, is similar in having ten subunits in its polymerase, only one of these exceeds 100,000 in molecular weight (Madon *et al.*, 1983). In a third advanced bacterium, *Thermoplasma acidophilum*, the polymerase has similar characteristics, except that only nine different types of subunits exist (Schnabel *et al.*, 1982). The E subunit, with a molecular weight of 22,000, seems to correspond functionally to σ of bacteria, dissociating readily from the holoenzyme and seemingly opening binding sites in native DNA.

The CAP Factor and Its Recognition Site. An affector known as CAP (cAMP-associated protein) is active in control of several operons of *E. coli*, including the *lac, ara,* and *gal.* It appears to be pleiotropic functionally, detecting the cyclic AMP level within the cell and, when the concentration becomes high, activating the several operons (Emmer *et al.*, 1970). CAP is a dimeric protein, with a molecular weight of 50,000, and binds to DNA only when complexed with cAMP. It induces a 50-fold increase in transcription of the *lac* operon, seemingly interacting with specific sequences in the vicinity of 60 sites prior to the actual coding region (Fried and Crothers, 1983). A region with complementary segments, reading in the noncoding DNA strand as GTGAG-TA-CTCAC, has been proposed as the CAP recognition site (Gilbert, 1976), but its mode of enhancing polymerase activity remains in doubt. One suggestion, which fits data derived from studies of mutations, is that CAP, σ, and the core polymerase form a linear group on the DNA, each component overlying its particular recognition site (Figure 2.1). When the several enzymes are drawn in proportion to size and visualized as ovals of 2:1 axial ratios, they appear to fit this concept well. However, in the activation of *ara* operon, the factor known as C protein must be present in addition to CAP and the holoenzymic polymerase (Green-

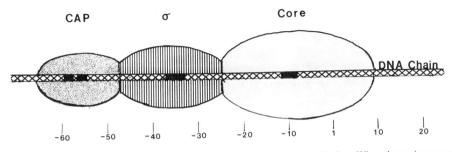

Figure 2.1. Possible relationships of factors in prokaryotic transcriptional initiation. When drawn in approximate size scale, the known active ingredients could lie together as shown along the DNA molecule. Site 1 marks the commencement site of transcription. The core polymerase reacts here with the -10 signal, the σ unit with the -35 box, and CAP with the -55. (Based on Gilbert, 1976.)

blatt and Schleif, 1971). *Bacillus subtilis* appears to have a site about 60 base pairs upstream that promotes transcription, but is not known to have the CAP factor (LeGrice and Sonenshein, 1982).

Other Ancillary Factors. Among the several factors from *E. coli* that enhance transcription, the first to be described was that termed the H factor (Jacquet *et al.*, 1971), which later proved to be a whole family of substances (Cukier-Kahn *et al.*, 1972). Only two members of the group, H_1 and H_2, appear to affect RNA polymerase activity. Both are of low molecular weight, H_1 being oligomeric and neutral, while H_2 appears to be monomeric and strongly basic. Their stimulatory effects, much stronger in combination than separately, seem to result from their changing the configuration of the DNA rather than from direct action upon the polymerases. Although H_1 stimulates the transcription of bacteriophage λ *lac* genes by *E. coli* RNA polymerase, it reduces that rate of synthesis of ribosomal RNA *in vitro* (Travers and Cukier-Kahn, 1974).

Like the two foregoing proteins, another factor, called HU, is heat-stable and of low molecular weight, 7000 daltons in the present case (Rouvière-Yaniv and Gros, 1975). Unlike H_1, which stimulates replication at some promoters to a greater extent than at others (Crépin *et al.*, 1975), HU acts unpreferentially. Electron microscopic examination disclosed that, upon binding, it induces thickening of the DNA chains and does not cause destabilization of the double helix; rather, with circular double-stranded DNA molecules like the genome of *E. coli*, it forms nucleosomelike structures (Rouvière-Yaniv *et al.*, 1976). Still another ancillary substance, referred to as the D factor, has been less fully characterized, but has been shown to increase the specificity of transcription of bacterio-phage λ DNA by *E. coli* RNA polymerase (S. Ghosh and Echols, 1972).

Actually, the HU protein also is proving to be a family of factors, which is present throughout the eubacteria. The *E. coli* representatives are known as NS1 and NS2 and have been reported also to be associated with the 30 S ribosomal subunits. A substance homologous to NS1 has now been described from *Pseudomonas aeruginosa*, but none related to the second protein has been found (Hawkins and Wootton, 1981). Whether HD, which consists of a tetramer of 9000-dalton subunits and resembles a histone in its reactions with DNA, is also a member of this family is not clear (Berthold and Geider, 1976). Still other, possibly related proteins resemble such histones as H2B, but do not promote transcription (Kishi *et al.*, 1982).

Requirement for Ribosome Attachment. In many bacteria, including *E. coli*, a peculiar requirement exists, not for initiation of transcription as in the preceding cases, but for continuation of that process, that is, elongation of the transcript. This requisite concurrent condition is active only in those cases in which the transcript is an mRNA, one that encodes a protein, and therefore needs to be translated into the final product (Siehnel and Morgan, 1983). To be more explicit, it is essential that ribosomes attach almost immediately to the growing messenger chain, otherwise it becomes abortive through premature termination of transcription (Stent, 1964; Imamoto and Schlessinger, 1975; Dennis, 1976; Jacobs *et al.*, 1978). Similar effects are noted under conditions of "polarity," that is, when termination arises through the presence of a stop codon, and in "coupling," termination induced by an antibiotic such as chloramphenicol. Certain sets of data permit the inference that polarity results from the arrest of the RNA polymerase activity (Korn and Yanofsky, 1976), but how the blockage of ribosomal action leads to this consequence has not received detailed attention.

Investigation of this problem has resulted in some insights into the nature of initiation of transcription by *E. coli* RNA polymerase. When transcription is initiated a ternary complex of DNA·enzyme·RNA is established, but the length of the chain of the last component varies greatly with time. At first, short, abortive chains of three to seven residues are formed and released from the complex (Hansen and McClure, 1980). When they have attained lengths of eight or nine residues, the σ subunit is released, and the ternary complex becomes fully stable and committed to transcriptive elongation until termination occurs. The transcript length at achievement of stability has been proposed to result from the number of residues made available to the polymerase by the duplex DNA molecule being separated into a single-stranded region of that length when the polymerase first attaches to the promoter. The enzyme then appears to recycle repetitively over this region, forming short, abortive RNA chains as it does so, until the DNA duplex is separated further, permitting the polymerase finally to break out of that cycle, fully committed to elongating the transcript until termination is signaled (Siebenlist *et al.*, 1980; Kinsilla *et al.*, 1982).

Additional details on the nature of the transcription complex have been provided (Hanna and Meares, 1983). In this study, radioisotopic RNA was produced by the polymerase of lengths varying between 4 and 116 nucleotide residues, the 5' end of each of which was then covalently attached to the adjacent polymerase subunit. When the transcript was 11 bases in length or less, it was found to be attached to the DNA, but when it had attained a length of 12 or more residues, it was always joined to the β and β' subunits. The σ subunit was not labeled by products exceeding three bases in length, and α was never labeled to any significant degree. Since the RNA chains are elongated by additions to the 3' end, it was concluded that the transcript at first lay in contact with the DNA within the polymerase, but began to protrude toward the enzyme's surface. This was reached when a 12-base length was attained. Thereafter, until 94 bases long, the transcript remained in contact with the β and β' subunits.

Termination Factors. Termination of transcript elongation may be effected in several ways. In many cases, the polymerase ceases activity when it encounters a terminator signal, as described in the first chapter. One well-established example of such termination is in the transcription of ribosomal RNA of *E. coli*, in which additional enzymic factors are unnecessary (Pettijohn *et al.*, 1970). However, in other instances the terminator may be ignored by the enzyme unless a second protein, the rho factor, is present (Roberts, 1976). In the absence of rho the polymerase reads through the terminator and transcribes genes downstream from that site, until a functional terminating sequence is encountered. As pointed out by Ward and Gottesman (1982), however, a termination is a far more complex process than previously considered.

Similar effects to the foregoing are brought about by another class of proteins, best referred to as antiterminators. One of the more thoroughly documented examples is the product of the *N* gene of bacteriophage λ (Luzzati, 1970). Since this protein N is functional both at rho-dependent and rho-independent sites, it does not act by inactivating that factor. In turn, the activity of N is controlled by still another set of factors, products of N-utilization genes. At least two such *nut* gene-encoded elements are present, *nutL*, which suppresses N activity in leftward-read genes, and *nutR*, which restricts its activity in the opposite direction (Friedman *et al.*, 1973; Adhya *et al.*, 1974).

A less well-known factor, called NusA, also has been found to be active in termination, but it resembles rho in having a positive action rather than the negative one of N and the other antiterminators (Greenblatt and Li, 1981). Moreover, it will most likely be found that there is a whole host of protein factors involved in termination, for reasons that become apparent in the discussions in the next section and following chapters.

2.1.2. The Eukaryotic DNA-Dependent RNA Polymerases

In eukaryotes the nuclear enzymes involved in transcription fall into three major classes, distinguished on the basis of chromatographic properties, subunit structure, and the type of gene transcribed (reviewed by Roeder, 1976; Dillon, 1978, pp. 181–183). Class I (or A) and class II (or B) DNA-dependent RNA polymerases are more closely related to one another than to class III (or C), in having a relatively simple subunit structure, but when tested with antibodies, all three show some degree of relationship (Engelke *et al.*, 1983). Functionally the classes display marked specificity, those of the first class transcribing genes for the three major RNA subunits of ribosomes, those of the second being active with hnRNA (and thus mRNAs), and those of the third acting upon 5 S, transfer, and other small RNA genes (Table 2.1).

Class I Polymerases. The subunit composition of class I nuclear DNA-dependent RNA polymerases is more variable than that of the others. While in yeast and

Table 2.1
DNA-Dependent RNA Polymerases

	Class I	Class II	Class III
α-Amantin	Resistant	Very sensitive	Mildly sensitive
Products	rRNA	Precursor mRNA (hnRNA)	5S rRNA: tRNA
Subclasses	2	2–3	1–2
Subunits			
Yeast (no.)	12	9	13
Yeast (mol. wt.)	185,000–12,300	170,000–12,500	160,000–10,700
Podospora (no.)	--	--	13
Podospora (mol. wt.)	--	--	174,000–10,000
Wheat germ (no.)	--	10	--
Wheat germ (mol. wt.)	--	220,000–16,000	--
Bombyx (no.)	--	--	9
Bombyx (mol. wt.)	--	--	155,000–18,000
Mouse (no.)	6	9	10
Mouse (mol. wt.)	195,000–19,000	240,000–26,500	155,000–19,000
Xenopus (no.)	--	--	14
Xenopus (mol. wt.)	--	--	155,000–17,000

Acanthamoeba, 11 different polypeptides are found (Thonart *et al.*, 1976; Valenzuela *et al.*, 1976), the corresponding holoenzyme in wheat germ reportedly is constructed of only seven and that of cauliflower of just six (Jendrisak, 1980; Guilfoyle, 1981). In spite of the abundance of this enzyme in the eukaryotic cell, it remains relatively poorly known, possibly as a result of its high degree of transcriptive specialization. In common with the other RNA polymerases, polymerase I transcription *in vitro* is dependent upon the presence of other cellular substances (Grummt, 1981; K. G. Miller and Sollner-Webb, 1981; Mishima *et al.*, 1981; Swanson and Holland, 1983). The last of these references shows that in yeast the product of transcription is a 35 S precursorial chain including transcripts of the genes for the 18 S (minor), 5.8 S (supplemental), and 25 S (major) rRNAs, plus a 7-kilobase leader. Interestingly, the latter includes an RNA direct copy (not complementary) of the gene for the 5 S (ancillary) rRNA, which is contained in the second strand of the DNA and has the opposite polarity. Thus in this case RNA polymerase III, which transcribes the 5 S gene, does so in the opposite direction from RNA polymerase I.

Although throughout the eukaryotes the rRNA genes are arranged in the same sequence (minor, supplemental, major), with the ancillary located elsewhere in the genome, the size of the nascent transcript varies widely. A 37 S molecule appears to be the original product in *Dictyostelium* (Batts-Young *et al.*, 1977); the low yield reported has prompted further researches, which demonstrate that processing occurs concurrently with transcription in this instance (Grainger and Maizels, 1980), as in the case in eubacteria (Abelson, 1979). Fuller details on the nature of the signals that are recognized by this class of RNA polymerases are provided in Chapter 4, where the rRNA genes receive attention.

Class II RNA Polymerases. The class II nuclear DNA-dependent RNA polymerases, all of which are inhibited by α-amanitin and are localized within the nucleoplasm, are concerned with the transcription of hnRNAs (mRNAs) (Roeder, 1976; Lewis and Burgess, 1982; Dahmus, 1983). As a whole they display a greater consistency of structure than do the class I proteins just described. In yeast nine subunits have been reported to be present (Buhler *et al.*, 1976; Hager *et al.*, 1976; Thonart *et al.*, 1976; Valenzuela *et al.*, 1976), the two large ones having molecular weights between 170,000 and 205,000 in one case and 145,000 in the other, and seven small ones ranging downward to 12,500. The *Acanthamoeba* enzyme consists of ten subunits, the two large ones having molecular weights of 178,000–193,000 and 152,000, while the smallest is under 10,000 (D'Alessio *et al.*, 1979). RNA polymerase II of wheat germ is constructed of 12 subunits (molecular weight ranging from 220,000 to 14,000), whereas that of cauliflower has 11 of molecular weight between 180,000 and 14,000 (Jendrisak *et al.*, 1976; Jendrisak, 1980; Guilfoyle, 1981; Paule, 1981).

However, multiple species of this class have been described from a number of organisms and diverse tissues. Kedinger and Chambon (1972) found two isozymes in calf thymus, IIA and IIB, similar to those that occur in *Xenopus* kidney and embryonic tissues (Roeder, 1974). In contrast, Sugden and Keller (1973) detected only a single variety in HeLa and KB cells, but this was probably an oversight. The isozymic pair from calf thymus appears to differ mainly in the largest subunit, which has a molecular weight of 214,000 in IIA and 180,000 in IIB (Kedinger *et al.*, 1974). In addition, this same study showed that rat liver contained three isozymes of this class, a condition found also in the mouse (Schwartz and Roeder, 1975). In the latter mammal, the distinctions between the two resulted from size and constitutional differences in the largest subunit (molecular weight 240,000 in II_0, 205,000 in II_A, and 170,000 in II_B). As in the case of polymerases of class I, the specific recognition and other properties are elucidated later when the synthesis of hnRNAs and mRNAs receives attention in Chapter 5 and subsequent chapters.

Class III Polymerases. The class III RNA polymerases have proven to be active in the synthesis of miscellaneous small RNA species, including 5 S ribosomal and transfer RNAs, as well as two small types known as VA from adenovirus-infected cells and other virus-related RNAs (Roeder, 1976; Weil *et al.,* 1979; Paule, 1981). Recently the structural properties have been established for enzymes extracted from various tissues of *Xenopus laevis* (Roeder, 1983). Under denaturing conditions, polyacrylamide gel electrophoretic analyses disclosed them to consist of at least ten distinct subunits ranging in molecular weight between 155,000 and 19,000, plus several that were not further characterized. Isozymic forms occurred in the ovaries and liver, whereas other tissues appeared to contain only one species. Because the major ovarian enzyme was observed to be functionally equivalent to that of somatic tissue, it was proposed that it was conserved and employed during development in somatic cells of the embryo.

Although the nucleus of the *Xenopus* oocyte contains an extraordinarily high level of RNA polymerases III, estimated to be between 3×10^9 and 5×10^9 per oocyte (Roeder, 1983), typically this type occurs only in low levels of concentration. Consequently, its characterization in most eukaryotes remains for the future (Paule, 1981). However, the structural properties have recently been described in the fungus *Podospora comata* (Barreau and Begueret, 1982). As in the amphibian species, great complexity was encountered, 13 subunits having molecular weights between 174,000 and 10,000 being identified. Each appeared to be represented once per holoenzyme, as was also the case with that of the clawed toad. Like other eukaryotic enzymes of the class that have been analyzed, there is no sharp division between large and small subunits, because one of molecular weight 87,000 provides an intermediate size range.

The simplest class III DNA-dependent RNA polymerase appears to be that from yeast, which consists of only ten subunits (Buhler *et al.,* 1976; Valenzuela *et al.,* 1976). The largest of these has a molecular weight of 160,000, followed by ones of 128,000 and 82,000 and seven others of decreasing size, the smallest having a molecular weight of 11,000. In *Acanthamoeba castellanii,* 14 subunits, ranging from molecular weight 169,000 to under 10,000, have been described, but in plants the picture of subunit complexity remains obscure. On the one hand, in cauliflower, only 11 subunits (molecular weight 150,000 to 17,500) were detected, whereas in wheat germ, 14 with a similar size range have been reported (Jendrisak, 1980; Guilfoyle, 1981).

2.1.3. Evolutionary Relationships of RNA Polymerases

Although too little is known of the structures of the DNA-dependent RNA polymerases to permit the erection of a meaningful phylogenetic tree, subunit composition data do provide a basis for preliminary deductions of relationships. Only after the amino acid or gene sequences of many of the subunits have been established from a sufficient number of prokaryotic, eukaryotic, and organellar sources will a more concrete basis be provided toward the construction of a definitive phylogeny.

Prokaryotic Beginnings. In spite of their being large molecules with high molecular weights comparable to those of eukaryotes, many prokaryotic RNA polymerases are obviously simpler in consisting of only four or five RNA subunits. Furthermore, their relatively primitive status is reflected in the lack of specialization in substrate, for only a single class is present per cell that transcribes genes for all types of RNAs. While the eubacterial molecule is currently the simplest known, it is quite probable that the enzyme of more primitive bacteria such as *Clostridium* and *Beggiatoa* and the blue-green algae will be still simpler, most likely in being smaller and consisting of less than five subunits. When the corresponding proteins of rickettsias, chlamydids, and viruses are eventually analyzed, they doubtless will shed further light on the origins of this group of enzymes. Because these polymerases from the so-called archaebacteria are constructed of 9–12 subunits, they clearly represent intermediate stages between those of the eubacteria and yeasts. This relationship is corroborated by the fact that their large components cross-react immunologically with those of the corresponding eukaryotic enzymes (Schnabel *et al.*, 1982). It is not improbable that as further researches are conducted into the problem of ancestry, some of the higher bacteria will be found to contain two, perhaps even three, classes of the DNA-dependent RNA polymerases, approaching the eukaryotic condition still more closely.

Relationships among Eukaryotic Polymerases. That the DNA-dependent RNA polymerases of eukaryotes are interrelated is clearly shown by the table presented by Paule (1981). There it is evident that three of the smaller subunits are shared by all three classes of the enzymes, although the respective molecular weights differ somewhat from taxon to taxon. In other words, the same three genes code for corresponding subunit polypeptides in each enzyme class in all eukaryotes, a condition most readily interpreted as signifying common descent. Furthermore, the same table demonstrates that several other small subunits are shared by polymerases I and III, but not by polymerase II. Consequently, the latter may be viewed as being the most primitive member of the trio and the two former as later derivatives. A possible sequence of events in the origins of the specialized classes seems to be as follows:

1. A single enzyme transcribes all classes of genes.
2. Later a derived enzyme (I) transcribes genes for all nontranslated RNAs, including ribosomal, transfer, and the various small RNAs, while the ancestral stock (II) becomes specialized for hnRNAs, most of which are translated into proteins.
3. A still later derivative (III), this time from class I, assumes transcriptive functions for the 5 S RNAs, tRNAs, and other small types, with class I enzymes retaining responsibility for the large, small, and 5.8 S subunits of rRNA.

It will be of interest to find how closely these preliminary deductions are actually corroborated when the higher bacteria polymerases ultimately receive investigation.

2.2. TRANSCRIPTION OF tRNA GENES

As pointed out in Chapter 1, genes no longer can be viewed merely as the series of nucleotide residues that encode a particular polypeptide or functional RNA. It is reemphasized that they need to be considered as an entire region of the genome, including a diversity of sequences adjacent to or interrupting the actual coding portions, for such parts are typically transcribed by the appropriate DNA-dependent RNA polymerase along with the mature gene. Parts of these appended regions, as will be recalled, are often involved in helping to identify a particular gene's location when there is a need for its product in the cell, for termination of transcription, and other functions discussed previously.

It is convenient to begin the analysis of transcription and the relevant signals with polymerase III activities of eukaryotes, along with corresponding functions in prokaryotes, viruses, and organelles, because its focal substances, chiefly tRNA and 5 S rRNA genes, have been more thoroughly explored than any of those for other major classes of mac-romolecules. In turn, beginning with initiation rather than termination is an obviously logical procedure, as is the use of the multitudinous sequences of tRNA genes instead of the less abundant ones of 5 S rRNAs. Although many prokaryotic tRNA cistrons have had their primary structures established, these are largely from a single organism (*E. coli*), and thus do not present the necessary comparative basis for introduction to what is shown to be an exceedingly complex subject. Termination then is considered in these coding sequences, while the genes for the 5 S rRNAs acted upon by the same RNA polymerase form the center for discussion in the following chapter. The frequent introns present in tRNA genes are not considered here, however, since they are more appropriately viewed with processing, rather than transcription, and thus are analyzed in Chapter 12.

2.2.1. Transcription Initiation in Drosophila tRNA Genes

The genes for eukaryotic tRNAs in this analysis of the initial transcriptive steps may be viewed, to paraphrase a familiar quotation, as the best of topics and the worst of topics. Best both because they provide an abundance of necessary data and they reveal most clearly the general situation that prevails; worst, as a result of the difficulties they present. Since many genes, for tRNAs as well as those of some other types of biochemicals, are represented by numerous copies per genome among advanced eukaryotes, they are more appropriate for present needs than those of yeasts or other simpler types, where gene amplification is absent or of lesser degree. Hence, representative gene structures from *Drosophila* are employed first, for this genus of insects has most of its tRNA genes greatly amplified, there being between 600 and 750 tRNA genes per haploid genome encoding at least 90 major and 33 minor species of tRNAs (Ritossa *et al.*, 1966; Tartof and Perry, 1970; Weber and Berger, 1976; Sharp *et al.*, 1981b; Leung *et al.*, 1984). As is universally the case among eukaryotes, the primary products of transcription vary widely; though they may at times embrace just a single cistron, they often are dimeric and not infrequently are polycistronic.

The Leaders of Drosophila Isoleucine tRNA Genes. The leader sequences have been established for a large number of different genes and for the same isoaccepting tRNA from a diversity of locations in the genome. For instance, Table 2.2 presents about 70 bases of each of six leaders from tRNA^{Ile} genes, five of which (marked A–E) are located in a single large sector of DNA (Robinson and Davidson, 1981), while the sixth (F) is from an area remote from the others (Hovemann *et al.*, 1980a). The sequences have been arranged so as to place more closely related varieties together, with individual sites spaced as necessary to bring out homologous regions to the maximum extent. Since the same polymerase is engaged in the transcription of each of the coding regions, it should be

Table 2.2
5' Leaders of Drosophila tRNA^Ile and tRNA^Glu Genes[a]

		Q		M		N		O				P
Isoleucine												
D	ACCAAT-TCT	ACGGCTGTT--	CGT-CG-	CATTTA	CTCTCA	AAA	A--GCG-------	TTT	-GG	CAA	AAAAGAACG	CAT----
C	TAAAAGCTCA	ATAT-TGT---	-GTTT--	CATTGG	C-GT--	AAA	-T-GCGTC----	TTT	TGA	CAG	AAATCAGCA	CAA----
B	A-CATGGAAG	GTAT-TGTTAG	-GTTCGC	TATTTA	--G-CG	AAA	ATCGCA------	TTT	TGT	TAA	AATTACAGA	CAT----
E	GTTAACAGAT	TTAT-C---AG	AATT-GC	TATTTG	--GT-A	AAA	CTGGCA------	TTT	TGG	CAG	CTGTGCATA	AAT----
A	GTTTT---GC	T-ATGATTTTC	AGTTT--	AGTACG	CCTTCC	AGT	CTGCTACGCC--	TCT	TG-	CAT	CCGTAAATA	CAT----
F	TGGAAAAATG	TTTTTATCTCT	AGCTTCA	CAAATT	--TTAC	TAA	-TA-T-TTT---	TTT	TCA	ATG	CATTCCA--	TGG----
Glutamic Acid												
C$_2$	--TTAAATA-	---C--GTTCAG	GTTG-GT	AATTAA	TTCAAT	AAG	AAA-ACC-GCG-	TTT	TCA	CAA	CAGGTTTGT	CGACG-
C$_1$	-TAGTTATAA	--CT-GTT-AG	-CAGAAA	CTTTGG	GACTGA	AAC	TAATATAGGAGA	TTG	ACT	AAA	TAGGTATGT	G------
C$_3$	AATCACTTAA	AGCTGATT-A-	-ATGTG-	CTTGCA	--CTT-	AAA	--ATACAGACGT	AAA	TCG	GAC	TTCGCTACA	GTA----
TR13$_2$	TCTTGATGaA	AATTTGTA-AT	TTTTT--	AATTAT	CAATT-	AT-	C-GTTAACC---	AAC	ACT	GAA	TGCAAGAAT	GTCATA
TR13$_3$	AGATCAGACA	AAG--GTAGCG	TGATTG-	TATTTA	TACAT-	AT-	C-GTTAACC---	AAC	ACT	GAA	TGCAAGAAT	GTCATA
TR113$_1$	ATAAATAAAT	ATTTAT-TTAG	-GATTT-	CTTGTA	AAACT-	ATA	CAATTATAG---	TTG	GCT	GAC	-ACGACACC	AAC----
TR113$_2$	TACGTTCTCT	TTTTAG--CCG	CAA-CGA	GTTGCT	TCGCTT	GTA	----TACAC-TT	TTT	CGC	CGC	GAC--GAAT	CGCAAG

[a] See text for references.

. activated by the same signal in all cases; thus an area in these leaders functioning in this capacity should have been evolutionarily highly preserved throughout this series. In short, one should expect identical, or nearly identical, short sequences to occur in a similar location in each and every leader of the tRNA genes.

One area that has been proposed as providing a signal for initiation of transcription is the so-called "ACT-TA box" located at region M of the table in the vicinity approximately 35 base pairs to the "left" (5') of the commencement of the actual gene (Indik and Tartof, 1982). In the first sequence given (tRNA^{Ile}D) a box similar to the model may be noted, consisting of the hexamer CATTTA that begins at the 39th occupied site from the gene's initial base. However, in the series of genes used here the gradual loss of this supposed signal is immediately evident in descending through the sequences, becoming in turn CATTGG, TATTTA, TATTTG, AGTACG, and CAAATT. Thus, instead of the expected universality of this proposed signal, the actual bases present are too highly variable from gene to gene to hold value as a promoter region. Moreover, hexamers of comparable quality often occur upstream from the foregoing, approximately in region Q.

The sectors marked N and O in the table are highly conserved and are more consistent than the ACT-TA box, O being deviant only in sequence A and then by only one transition of a T to a C. However, sequences similar to N and O, placed at comparable distances from one another, may be noted in several of the representatives given, especially upstream from Q. Still another sector, P, located just prior to the functional coding area, possesses a reasonable degree of consistency except in the TGG of sequence isoleucine F showing no resemblances to the CAT, CAA, and AAT of the rest. Other short, quasihomologous regions are readily observed in the isoleucine tRNA leaders as arranged here, but none show the requisite universality or the expected uniqueness.

The Leaders of Drosophila Glutamic Acid tRNA Genes. When the leaders of seven genes for tRNA^{Glu} of *Drosophila* are similarly aligned to bring out possible homologous regions, as in Table 2.2, great variation in the structure of the ACT-TA region (M in the table) again is observed. The last one in the column is especially irregular, only the third site corresponding to the sequence that is supposed to prevail. The terminal sector (P) is also obviously more irregular than it is among the isoleucine-isoaccepting species, ranging from six sites down to a single one. An especially clear indication that this variation is not an artifact of the sequence arrangement is provided by the first pair of species cited, both of which are from the same section of the genome (Indik and Tartof, 1982). In that referred to as C_2, six sites occur in region P, while in C_1, there is only one, yet the 15 bases preceding this region show the unusually high rate of homology for these leaders of 66%. The downstream end of the leaders of two adjacent genes in the cloned cluster TR13 likewise display extensive homology, whereas the first one of the trio (not cited) is quite disparate (Hosbach *et al.*, 1980). The two comprising a second cloned cluster (TR113$_1$ and TR113$_2$) similarly lack homology in this region.

In contrast to the high consistency found among the isoleucine-accepting types, region O in this series is extremely variable, TTT occurring only twice among the seven sequences and TTG an equal number of times. Two occurrences of AAC and one of AAA are the other extreme deviants. Region N is somewhat more consistent, at least one A occupying a site in every instance; nevertheless, the variation is still very marked. Consequently, any temptation that may have existed after examination of the six isoleucine species to suggest the AAAs of sector N together with the TTTs of region O as universal signals for transcription rapidly disappears.

Summary of Drosophila tRNA Gene Leaders. Even after leader representatives of a diversity of species are aligned to bring out regions of homology (Table 2.3), no firm solutions emerge. However, it becomes feasible to propose a possible explanation for both the regularity and inconstancy that exist. When the available leader sequences are surveyed, a number of species can be noted that have the trinucleotide TTT at a suitable distance from the actual coding areas of the genes to fall into column O, most of which also have AAA or at least AA- 6–12 sites upstream, enterable into column N. Eight, largely nonisoaccepting, species with this combination of traits are provided in the upper portion of Table 2.3, where they are labeled "T-activated" in accordance with their most consistent feature, the triple thymidine bases of region O. Below them are six other sequences which lack this characteristic, identified as "irregular" because of the inconsistency that occurs among them in the corresponding section. Even in certain of these varieties, however, region N shows a degree of correspondence to the T-activated ones in having triple A.

In all cases the ACT-TA box (M) shows the same high degree of variation found in the preceding ones. As can readily be observed, it is impossible from these examples to derive a consensus related to the ACT-TA, four Cs, six As, two Ts, and one G occurring in the first sites, four Cs, four As, three Ts, and two Gs in the second, five Ts, six As, no Cs, and two Gs in the third, and so on. Moreover, if this mixture of signals that is found in the region is assumed to trigger RNA polymerase activity, it is impossible to discern how that enzyme would fail to respond to the very similar sets of nucleotide residues that precede M in the 5' direction. For instance, one finds ACTATA and CATCAA upstream of that sector in the third example given, and AATGAT, AATGTT, and other similar hexanucleotides still farther in the 5' direction, which are not included in the table.

A Possible Solution. The uniformity of regions N and O in the one group of *Drosophila* leaders strongly suggests that those sites probably play a role in the activation of the RNA polymerase. To explain the absence of these signals from the irregular sequences, it is proposed that either several different species of RNA polymerase III transcribe the irregular leaders or distinct types of ancillary proteins assist in the recognition of the gene and activation of the polymerase, such as those shown to be involved in promoter selection with RNA polymerase II (Dynan and Tjian, 1983). That several different proteins are probably involved in the recognition of initiation sites is suggested by the combinations of trinucleotides that exist in, or are adjacent to, regions N and O. In two or three of the examples multiple A-occupied sites in or near the N zone are accompanied by similar triplet As in or around O. Other combinations are equally likely, but the issue cannot be resolved until additional leader sequences of tRNA genes of *Drosophila* have been established. Furthermore, much more needs to be learned about the kinds and activities of proteins that assist in the processes of transcription by DNA-dependent RNA polymerase(s) III. But it does seem definite that more than one polymerase–protein combination is involved in the initiation of transcription of these genes. Moreover, in the silkmoth, a large region including parts of both the leader and mature coding sectors, rather than specific short segments (Wilson *et al.*, 1985), have been noted to be essential to efficient transcription.

Table 2.3
5' Leaders of Various Drosophila tRNA Genes

				M		N		O				P
T-activated												
Ile D[a]	ACCAAT-TCT	ACGGCTGTT--	CGT-CG-	CATTTA	CTCTCA	AAA	A-------GCG-	TTT	-GG	CAA	AAAAGAACG	---CAT---
Glu C$_2$[b]	--TTAAATA-	-C--GTTCCAG	GTTG-GT	AATTAA	TTCAAT	AAG	AAA-ACC-GCG-	TTT	TCA	CAA	CAGGTTTGT	---CGACG-
Gly 56-6[b]	TACTATATA-	GTTTAGAAAAG	CAT-CAA	AATAAG	CTGG--	AAA	T--G--------	TTT	TGG	CAC	TTATCATTT	---CAACAA
Lys Reg 2[c]	TATATCGAAG	AAGAATGTTA-	-TTACAC	TGAAT-	CTCAGC	AAG	CTTAGGTC-CG-	TTT	TAA	TAT	-GGGT-----	---CATTCA
Asn Reg 4[c]	--TAT-GTCT	TTTCTTATGAC	AGATATT	CCAATG	AATATT	AAT	TAA------GCA-	TTT	CCA	TCC	-TAAGAAGG	AGCC-ATCT
Ile Reg 2[c]	--AAATTTGG	AAAAATGTT--	TTTATCT	CTAAGC	TTCAC-	AAA	TTTT-ACTAATA	TTT	TT-	TTT	CAATGCATT	---CCATGG
Arg 17D[d]	--ACATGCCA	AAGAATGTG--	GACGCTA	CCAAGC	AGTATC	AGA	--ATCA-TGGA-	TTT	TT-	GAA	ACGCCGCAA	CATCAAAG-
Leu B[e]	CCT-AATTTA	TTGGTTCAACG	CCCCATT	TTGTAT	ATACT-	TTT	ATGACACTAGAT	TTT	---	---	GGGGGCGGA	---ATCAGT
Irregular												
Leu A[e]	ACGAATAGAA	TTG-TTCT-AG	TACTGGC	ACATGC	GCTGGC	TCG	CGCGCTACTACG	GAA	-TC	GGA	CACGAGTTT	--GGCTAT-
Glu C$_3$[a]	TGTAAATCAC	TTAAAGCT--G	-ATTAAT	GTGCTT	GCACTT	AAA	---ATACAGACG	TAA	ATC	GGA	C-----TTC	GCTACAGTA
Lys Reg 4[c]	AAGAC--TAA	-TAAA----AT	TCCAAAA	AAAATA	TCCAAA	AAA	ATATTAAAAATG	TGG	TGT	GGC	AGTTTTTAC	CACCAAGAG
Glu C$_1$[a]	TAGTTATAAC	-TG----TTAG	CAG-AA-	ACTTTG	GGACTG	AAA	CTAATATAG---	GAG	ATT	GAC	TAAATAGGT	--TAT-GTG
Arg 35D[f]	----------	----------		---AGC	TTCAAA	GCA	AATACATAG-CA	CAA	AAC	---	TTAATCTTG	---ACAACC
Arg Reg 1[c]	TAACAATGTG	TACAA-ATA-A	TATTCG-	AGTTTG	ACTTT-	CTA	CTTG-TTTGAA-	-CT	GTT	ACA	CT-CGCACG	---TCAAGC
Arg YH48[g]	----------	----------	--------	-------	-------	---	--------------	-CT	GTT	ACA	CT-CGCACG	---TCAAGC

[a]From Table 2.2. [b]Hershey and Davidson (1980). [c]Hovemann et al. (1980a). [d]Dingermann et al. (1982). [e]Robinson and Davidson (1981). [f]Yen and Davidson (1980). [g]Silverman et al. (1979).

2.2.2. Initiation in Other Eukaryotic tRNA Genes

Perhaps examination of tRNA gene leader sequences from other eukaryotes will serve to clarify the situation to some extent. Unfortunately, only a few species of these organisms have had a sufficient number of full tRNA gene sequences established to provide an adequate picture, baker's yeast being the outstanding exception. However, *Xenopus* together with several mammals have had a sufficient number established to be at least suggestive of vertebrate conditions.

The Leaders of Saccharomyces tRNA Genes. Although no long series of leaders from the genes for any given type of isoaccepting tRNAs of the yeast has been sequenced, the more than 20 that are available should provide a good cross section. Some of these, however, are too brief to be meaningful, so only 15 are provided in Table 2.4. Here a condition comparable to that of *Drosophila* is revealed. In the first place the ACT-TA box mentioned earlier (column M in the table) is seen at once to be too variable to serve as a promoter, whereas there is a region O comparable to that of the fruitfly's leaders in which -AA is present in 12 of the tRNA species cited. The second region of consistency of *Drosophila*, N, is variable here, although TT- or -TT occur in eight instances and AAA and CCC in two sequences each. Thus two differences from the fruitfly condition are clear: First, the prime signal -AA in O differs from the TTT of the insect, and, second, only a single such signal is clearly present. This relative simplicity of structure certainly is in keeping with the overall primitive position of the yeast within the Eukaryota. Comparable signals, including CAA in O and CAT in N, are also present in more recently sequenced leaders, including two histidine tRNA genes (del Rey *et al.*, 1983).

Furthermore, although by far the greatest portion of cited tRNAs have the same signal sequence, more than one trinucleotide combination appears to be utilized here, as in the arthropod sequence. Although only one of such irregular types is given in the table, that for a tRNA^{Phe} YPT2 (perhaps plus leu$_3$ JB2K), at least several others are known, those for the leaders of four tRNA^{Tyr} genes, all of which lack -AA in the O region. However, this irregularity may be more apparent than real, for these leaders are at most 21 base pairs in length (Goodman *et al.*, 1977). Otherwise, they are quite in keeping structurally with others of those cited, three having the striking CAACAA hexamer found in a number of other representatives between the O and P regions shown in italics in the table. It is perhaps of some significance that the two AA- activated tRNA^{Phe} genes have identical leader sequences, whereas the irregular one exhibits little homology.

Careful comparisons of the sequences as they are aligned in Table 2.4 reveal a number of homologous segments, some of which are confined to two or three examples, while others are more widespread. Many of these occur in the section between columns O and P of the table. The first five leaders, from three different species of tRNA and from five contrasting regions of the yeast genome, show especially striking similarities here. Indeed, traces of the homologous elements can be detected also in the two tRNA^{Leu} sequences. The hexameric sector just preceding P in the two tRNA$_3^{Arg}$ genes offers another region of homology that is essentially repeated in the last five examples.

Comparable regions of identity and similarity are to be noted in the section lying between columns N and O. One hexameric element that is especially striking is GAGATA, repeated intact in four of the sequences and in modified form in four or five others. These and still others are in addition to those representatives given here that are located on the

Table 2.4
5' Leaders of Saccharomyces tRNA Genes[a]

AA-Activated	M	N	O	P
Glu₃ pY5[b]	CCAGTTTAA ACTTGA---	AATAGG AAA	-TC-TTTAAGGT- TAA ATC-AAACA	TGAGTT ATTA--
Glu₃ pY20[c]	GTATTGTCA TACTGA--C	GTATTC CAT	----TTTGAGATG CAA CAC-ATACG	TGTATT GTAA--
Arg₃ 18U[d]	AATTGG-TG AAATCAAAG	CATAG- TTT	-ACGTTTAAT-CC CAA TCT-*TCCAA*	*CAA*ACA GTA---
Arg₃ 19F[d]	AATTGAAAG CGTAATTAG	GTTTTA CTA	-TAATAAAGTAGT AAA ACT-TT*CAA*	*CAA*ATA GTA---
Ser₂ B[e]	TT-TCC CTGCCAGTG	GACAGT GTT	----GTCCTTACA CAA ACT---CAA	GAGATT AAA---
Ser₂ G[e]	-AGGAATCA	CAAAG- AAA	-A-CA*GAATAC*AC TAA ----T-CAA	-GT-TA GTAT--
Ser₂ J[e]		CCTAGG TTT	--GG*GATATA*TCT TAA GTCAATACA	TCATTA GTAA--
Met_f	TAAAATAG- TAGA--ATA	AAAACA TTC	ATATATCTACGTG CAA GCGTCAGAT	TGTAAT GTT---
Met₃[g]	CTCCATTGT TTCTTTTAT	TGAATA TTA	AAGCATTTAATGC TAG AATC-CTCC	ATAACA GATA--
Lys₂[h]	CATA	TCTCAT TTT	----*GAGATA*--- CAA CAC-GCTCC	GAAGAA CTCA--
Leu₃ JB2K[i]	ATCAGATG- --AGG-TAA	CATTAC TTG	-TACCGCCT---- CGC TTT--*CAAC*	*AAATAA* GT----
Leu₃ YLT2[j]	CATA	TCCCAT TTT	----*GAGATA*CAA CAA TTT--*CAAC*	*AAATAA* GT----
Phe YPT5[k]		CCC	-----AGA-GCG- CAA CTA--ATAT	ATGAAT ATAA--
Phe YPT15[k]		CCC	-----AGA-GCG- CAA CTA--ATAT	ATGAAT ATAA--
Irregular				
Phe YPT2[k]		-GGTTT GCA	-CGTTTTGA---- TGA TTTTCCTAT	AA-AAT ATAT--

[a]Important regions of homology are shown in italics.
[b]Eigel et al. (1981).
[c]Feldmann et al. (1981).
[d]Schmidt et al. (1980).
[e]Page (1981).
[f]Venegas et al. (1982).
[g]Olah and Feldmann (1980).
[h]Del Rey et al. (1982).
[i]Carrara et al. (1981).
[j]Standring et al. (1981).
[k]Valenzuela et al. (1978).

same sector of the genome, such as tRNAs[Phe] YPT5 and YPT15, which probably have been derived by direct gene duplication (but contrast Amstutz et al., 1985).

The Leaders of Vertebrate tRNA Genes. Known leaders from rat tRNA genes are comparable to those of the fruitfly and yeast in having a large number of species showing consistency of structure in the O region, plus a few irregular types (Table 2.5). In this case, the signaling sequence appears to be GT-, however, while the irregular forms have TG- or TT-. As may be noted in the table, the inconstancy, as elsewhere, is not associated with particular isoaccepting types, for two each of proline and lysine tRNAs

Table 2.5
5' Leaders of Rat tRNA Genes

					M	N	O			
GT-Activated										
Gly RT1dge[a]	GC	CGTGTGGTG	GCAGTC	GGC	GGCGAC	GTC	---GGG	GTG	TGGGTC	CG--GCTGTG---CG-
Asp RT1dge[a]	TTCCCT	AGCGTGG--	TAGACG	GGC	ACGGA-	GTC	---CGT	GTG	TTTGTG	---GCTGTCGTCG-
Glu RT1dge[a]	GTGT	-GTGTCT--	GTCTGT	TTC	CAGGAG	TTG	ATAGCC	GTG	AGCAT-	-GGGACCCA--CGC
Lys RT2-1C[b]		GAGCAGGAC	AATAAT	GGC	TCTGTA	ACT	CATTGG	TCA	TCACCT	TGC--TATG--CA-
Pro RT2-1C[b]			GGCC	NGC	AAGAAA	GTA	-TAA--	GTA	AAGTAG	CTG--TGT------
Pro RT2-1A[b]	GC-TGT	TTCAATTGC	CACCAT	AAA	GTGAAA	CTG	TGA---	GTA	---CTA	TTGG-TGT------
Lys RT2-1D[b]	TAC	CAACCTTGG	T--AAT	CAT	GATACT	CAG	CAA---	GTC	AATCGC	AGG-AC---------
Irregular										
Pro RT2-1D[b]	ACTCC	ACTTTA-TC	CCCC--	TTA	GAAG--	GAG	CAGA--	TGG	AACCAA	T--ACTCAA-------
Lys RT2-1A[b]	C	AGTTCAGTG	CCTTCA	GGT	CACG-T	GTG	TAAAC-	TTC	-GGCTC	AGAGGCAGTT----

[a]Sekiya et al. (1981); Shibuya et al. (1982).
[b]Sekiya et al. (1982).

have the GT- activator, the other in both cases being irregular. It is difficult to find an auxiliary signal corresponding to that of the N region in *Drosophila*. Undoubtedly additional sequences would serve to expose its nature, but for the present it must remain obscure.

Examination of the region lying 5′ to column M where ACT-TA or other boxes have suggested to exist in some organisms confirms their absence in these mammalian gene leaders. Nevertheless, short regions of homology can be detected there as well as in the sectors following column O. While such stretches of similar composition are suggestive of common origins in the evolution of some of these leaders, nowhere do they provide the constancy needed for a signaling device except in O, as already indicated. Even a region P, found in the *Drosophila* and yeast types, could not be located in these of the rat.

Too few tRNA genes from human sources have been sequenced to warrant full discussion. Visual analysis of the leaders from three such genes from a single DNA segment (Roy *et al.*, 1982) suggests that perhaps AGY serves as a signal, the Y representing either of the common pyrimidines C or T.

Leaders of Glutamic Acid tRNA Genes. Before leaving the structural aspects of leaders of eukaryotic tRNA genes, it might be well to compare analytically the sequences of those of a single isoaccepting type from a diversity of sources to determine whether they display any common feature. Since at least seven of such genes for species carrying glutamic acid have been established, that type can serve as the basis for discussion.

Although 5′ leader sequences are notoriously variable, surprisingly frequent homologies can be noted between the seven given in Table 2.6. No long series of identical sites are to be found, however, triple nucleotides being the maximum as a rule, and then largely between only two representatives. Despite the paucity of correspondences, the identical segments are so located and dispersed throughout the series that there can be no serious doubts about all having shared common ancestry. Nevertheless, nothing suggestive of a universal signal for initiation of transcription can be found, which in view of the species specificity of activity of RNA polymerases is what should be expected.

The interspecific correspondences are of special interest. In the region between columns N and O, it is noteworthy that the pY20 sequence of yeast has a pentameric segment identical to the corresponding portion of λ DmC_1 of the fruitfly, whereas the other two from that insect are quite distinct from both of the foregoing, but rather resemble one another. The members of the latter pair also display similarities to each other at points scattered throughout the sequences, but homologies to λ DmC_1 are scarce. The leader from the rat gene shows occasional short homologous series of sites with those of the fruitfly, especially to λ DmC_3—one region beginning 15 sites 5′ to M is particularly striking.

2.2.3. Transcription Initiation in Prokaryotic tRNA Genes

Among prokaryotes the signals for initiation of transcription of tRNA genes should not differ materially from those of other genes to be described later, because only a single type of DNA-dependent RNA polymerase occurs, as will be recalled. Hence, in theory, analysis of the leaders of this class of genes should provide for all the other major types that are examined in later sections.

Table 2.6

Comparison of 5' Leaders of tRNA^{Glu} from Various Eukaryotes

				M	N		O		P
Yeast pY5[a]	CTC-CAGT	TTAAAC	TTGAAATAG	GAAAT-	CTT	TAAGGT	TAA	ATCAAACAT	GAGTTATTA
Yeast pY20[b]	GGGTATTG	TCATAC	T-GACGTAT	TCCAT-	TTT	GAGATG	CAA	CACATACGT	GTATTGTAA
Fruitfly λ DmC$^{C}_{1}$	CAGAAACT	TTGG--	GACT-GAAA	CTAATA	TAG	GAGAT-	TGA	CT-AAATAG	GTA-TGTG-
Fruitfly λ DmC$^{C}_{2}$	TTG-GTAA	TTAA--	TTCA-ATAA	GAAAAC	-CG	CGTTTT	CAC	AACAGGTTT	GT-CGACG-
Fruitfly λ DmC$^{C}_{3}$	-TGATTAA	TGTGCT	TGCACTTAA	-AATAC	AGA	CGTAAA	TCG	GACTTCGCT	--ACAGTA-
Rat RT1dge[d]	GTGTGTG-	TCTGTC	TGTTTCC-A	GGAGT-	TGA	TAGCCG	TGA	GCATGGGAC	--CCA-CGC
Human[e]	AGCTGTC-	GCT-TC	TGACA----	GAAGA-	AGG	GAGAC-	AAA	GC-TCCCTG	-CTGTGTG-

[a]Eigel et al. (1981).
[b]Feldmann et a. (1981).
[c]Indik and Tartof (1982).
[d]Sekiya et al. (1981); Shibuya et al. (1982).
[e]Goddard et al. (1983).

In the literature a heptameric segment centered at the tenth site preceding the initiation site of transcription has been indicated to serve as a promoter (Pribnow, 1975; Scherer *et al.*, 1977; Schaller *et al.*, 1978). One especially informative investigation of initiation signals in *Escherichia coli* examined a series of mutant forms involving mutations in the promoter of a tRNA^{Tyr} gene (*tyrT*) (Berman and Landy, 1979). The wild-type promoter, TATGATG, was altered in two sites in certain mutants, one change resulting in a transversion of the first T to C, two others in altering the third T to C on the one hand and to A on the other. All these mutations resulted in reduced rates of transcription of the gene, but so did additional ones involving sectors at various distances from this heptamer. Hence, the pertinence of the −10 box is in question. Furthermore, additional current data demonstrate a diminishing view of that structure. In one study, promoter activity involving sequences up to 96 sites upstream from the initiation site for tRNA^{Tyr} were important in starting transcription and other controls were centered in the region no further than ten positions from the start site (Lamond and Travers, 1985). A second investigation, moreover, using a gene for tRNA^{Phe} of *E. coli*, reported the presence of two tandem promoters separated by about 60 residues (Caillet *et al.*, 1985), each of somewhat different structure.

Alignment of established leader sequences of tRNA genes from *E. coli* as in Table 2.7 discloses the same sort of difficulty in M found at the corresponding location in eukaryotic leaders, a striking absence of consistency. In fact, one notes the greater uniformity in the hexamer extending from site −9 to −4, more impressive as a possible signaling sequence because ATGCGC occurs in both the tRNA^{Tyr} and tRNA$_1^{Leu}$ leaders at that location. No other region displays any notable homologies that might serve as signals. The leader sequences from *Bacillus subtilis* and *Anacystis* likewise fail to add any firm substance to the region around −10 serving as a signal, for, similarly, in each of these prokaryotes no constancy of structure is to be found. The same lack of consistency is clearly shown in the list of known *E. coli* promoters that has been published (Hawley and McClure, 1983). However, those in the table do suggest a possible solution to the problem. In each of the three sequences from *B. subtilis* and one of the two from *Anacystis*, spaced sets of TTTs are found which could serve as triggering devices, those of the bacterial source having three such sets each, the blue-green alga only two. Moreover, it is readily perceptible that were all given sequences of sufficient length, one or more sets of triplet Ts might be found universally, since the short sequences show a number of other homologies with the longer ones. Thus the doubly echoed triplet could be part of a promoter in these organisms.

Referring back to the more advanced form, *E. coli*, one finds identical triplet Ts situated comparably to the foregoing sets in two of the sequences, neither of which is echoed. But when it is realized that the actual coding DNA strand has bases complementary to those of the noncoding strand given here, the presence of suitable triplet repeats of the same nature becomes obvious. In other words, it is possible that echoed triplet T–A base pairs function to activate the polymerase, regardless of which strand has a given component of such combinations, so long as the same nucleotide is repeated three times. Thus the AAA at the 5′ end of the segment of the tRNA^{Leu} shown in the table could work in conjunction with the TTT found downstream in the same strand. This proposal is, of course, still hypothetical, but is in accord with the data thus far available.

The situation is equally unclear in *Bacillus subtilis* tRNA leaders. Here, as in the *E. coli* cistrons, these DNA sectors are usually short, for most of this type of gene occur in

Table 2.7
5' Leaders for Bacterial tRNA Genes[a]

		M	N		O	P
Escherichia coli						
Tyr T[b]		TTACA	GCGGCGCG	TCA*TTT*GA	TATGATG	CGCCCC--
Ile[c]	AC*TTT*GCAGTGCTCACA	CAGATTGT	CTGATGAA	AATGAGCA	GT*AAAA*C	CTCTAC--
Leu pBR322[d]	*AAA*CCACTAGAATGCGC	CTCCGTGG	TAGCAATT	C*TTTTT*AA	GAATTGA	TGGTAT--
Leu$_1$[e]	TCGATAATTAAC	TATTGACG	*AAAA*GCTG	*AAAA*CCAC	TAGAATG	CGCC-TCC
Bacillus subtilis						
Ile[f]	GTTCCCTGTCTTG*TTT*A	G*TTTT*GAA	GGAAC*TTT*	GTTCCTTG	AATAAGT	TAAGAT
Gly 2[g]	ATTATTACCAAGG*TTT*C	-TCATAAG	GA---GAA	AGC-*TTTT*	*TTT*ATTG	CGATAT
Asn 1[g]	AGA*TTT*ATTAA	TTCTACCA	TG*TTT*GTT	TATGAAGC	*TTT*ATAT	C-TCAT
Arg 3[g]		T	TTACCTAA	CGGGATAT	TGT*AAA*T	GGAATT
Pro 4[g]			TATCTT	TTAATAGA	ATAGATA	GG*AAA*T
Lys[h]	TGAGAACTAGA*TTT*---	AAGTCGTT	TGC-TCTA	TAG*AAA*TT	CCGACAT	C*TTT*AT
Asp[h]	TCCTTA*TTT*GTCTGTGA	GAG-CTGA	CACGACAG	CTCTCCGG	GCAATTA	CTGTAA
Anacystis						
Ile[i]	T*AAA*CTAGTCTGGG*TTT*	TTCCTAGA	C*AAA*GATG	*TTT*GGGTC	AAGAGCA	GGATGT
Ala[i]		CTAG-	C*TTTTT*CA	TGAGAGTG	AAGAAGT	GATTGT

[a] Runs of three or more T or A are italicized.
[b] Sekiya *et al.* (1976a,b); Berman and Landy (1979).
[c] Sekiya and Nishimura (1979).
[d] Duester *et al.* (1981)
[e] Duester *et al.* (1982).
[f] Loughney *et al.* (1983).
[g] Wawrousek and Hansen (1983).
[h] Yamada *et al.* (1983).
[i] Williamson and Doolittle (1983).

clusters, probably just six in the present organism and those usually in association with rRNA genes. One such cluster, containing 21 tRNA genes representing 16 different species, has recently been sequenced from the genomic sector following the *rrnB* operon (Green and Vold, 1983). The longest intergenic spacer was 37 residues in length, located between the coding sequences for tRNA[Thr] and tRNA[Lys], and another 32 residues long preceding the former. But for the most part the spacers were between 3 and 15 nucleotide residues in length. One typical −10 signal was located on the longest spacer; the corresponding −35 sequence could not be found within the mature sequence that preceded this. Hence, it was suggested that the entire cluster, including the 16 S, 23 S, and 5 S rRNA genes, were transcribed as a unit.

2.2.4. Leader Segments of Organelle tRNAs

The organellar tRNA genes offer some unique opportunities for comparative studies. In the first place, an interesting concept has become widely accepted regarding the origins of chloroplasts and mitochondria, in which those structures are viewed as former "endo-symbionts," more correctly called endomutualists (Margulis, 1970; Raven, 1970; King, 1977). Second, the genes in organelles have not been so extensively amplified as in many organisms; in fact, in the mitochondrion an opposite trend can be noted, especially in that structure of higher eukaryotes, as appears later. Consequently, it is frequently possible to recognize identical genes and the associated DNA regions with a high degree of certainty, a condition that becomes especially clear in the discussions of chloroplast tRNA genes that follow immediately.

tRNA Leader Segments from Chloroplasts. Table 2.8 presents the noncoding leader segments of chloroplast genes for 16 species of tRNAs, drawn from tobacco, maize, and *Euglena,* together with one shown underscored from the blue-green alga *Anacystis.* To begin with, it is apparent that in no region of these leaders is there sufficient identity in any corresponding sectors of the sequences to be suggestive of a common signaling sequence. It is for this reason that the several regions are merely labeled from A to G, for only the 3'-terminal sector P can be considered truly homologous to those presented in the foregoing portions of this chapter. Even there, only three sets of sequences share identical series of four or more sites, a pair from tRNA^{Ile}, two others from tRNA^{Ala}, and a trio consisting of the methionine tRNAs of tobacco and maize and the leucine tRNA of maize. But nowhere is there correspondence to any of the M, N, or O sectors of earlier charts. Nor is it possible to bring out additional homologies to any extent by the usual device of inserting dashes appropriately, because to do so often increases homology at one point, only to decrease it at other locations up or downstream. Hence, at this time it appears necessary to fall back upon the proposal made in connection with the bacterial tRNA leader sequences, namely, that perhaps recognition or triggering depends upon echoed sets of triplet As or Ts, or, in some cases, even Gs or Cs.

The leaders for the tRNAs^{Ile} and tRNAs^{Ala} are especially informative, because all of those given in the table are derived from those operons that encode 23 S and 16 S rRNA discussed in Chapter 1 (Orozco *et al.,* 1980; Koch *et al.,* 1981; Orozco and Hallick, 1982; Takaiwa and Sugiura, 1982). Since these cistrons are transcribed together with those for the two species of rRNA, signals for their transcription are not essential, as are those for processing the given product. One might suggest that this observation explains the absence of signals noted above, were it not for the similar lack of appropriate sectors that exists in leaders from genes that are not associated with rRNAs in the genome, such as the first four sequences of Table 2.8. Examination of the isoleucine-accepting species from tobacco and maize chloroplasts shows that the two are largely homologous. The identical sites, italicized in the chart, amount to 50 of the 60 given, or 84%.

These observations take on even greater significance when a similar comparison of the two alanine-accepting tRNA leaders is made. Here the sites that show homology between the pair are italicized, as before; counts of identical sites in these two show 52 correspondences, for a homology rate of 87%. This extensive conservation of genic material is particularly remarkable, because such noncoding DNA segments as leaders,

Table 2.8
5' Leader Segments of Chloroplast tRNA Genes[a]

	A	B	C	D	E (−30)	F	G (−15)	P
Met (Tobacco)[b]	GATATTTAT	AATCCA	TCGACA-G-	ATGGGT	TTCATT	TGGTTCTCT	TTGGGATGA	TAAATG
Met (Zea)[c]	TCAATCTAT	AATCGA	TCGA-AGTA	ATGGG-	GCTTCT	T--TTGTTT	TGTGG-TGA	TAAATT
Tyr (Euglena)[d]	TTTTTGATC	TTAAAT	AACTTTTTC	TATGTT	AATTGT	TTTTGTTGT	TTAACATCT	AATTGC
Leu (Zea)[e]	AAAAACTAA	GAGATG	GATTAAATT	ATACAA	GGAATT	CATGCTTTC	AAAGAAAAG	TAAAAT
Leu₂ (Zec)[c]	GTCATATAT	TCCATA	TATCACATT	CGATAG	ATATCA	TATAATACG	ATTCACTTT	CAAGAT
Val (Tobacco)[b]	ATGGAATAT	TTATCT	TGACAAGAA	TTTATC	TACATG	ATAAAATAT	GTATCACAA	GCACTA
Ile (Anacystis)[f]	GGCTTCTAA	ACTAGT	CTGGGTTTT	TCCTAG	ACAAAG	ATGTTTGGG	TCAAGAGCA	GGATGT
Ile (Euglena)[g]	TAGATTGAA	AACAAT	GAAAAATAA	AAAAAA	TAAGTA	GGGAAACCT	CTTATTTTT	CCAAGA
Ile (Zea)[h]	AAGGGATGG	AGTTTT	TCTCGCTTT	TGGCGT	AGCGGC	CTCCCTTTG	GGAGGC-CG	CGCGAC
Ile (Tobacco)[i]	GAGGGATGG	GGTTTC	TCTCGCTTT	TGGCAT	AGCGGG	CCCCAGTG	GGAGGCTCG	CACGAC
Ala (Tobacco)[i]	GCCAGGGAA	AAGAAT	AGAAGAAGC	ATCTGA	CTACTT	CATGCATGC	TGGACTTGG	CTC-GG
Ala (Zea)[h]	CGCAGGGAA	AAGAAT	AGAAGAAGC	ATCTGA	CTCTTT	CATGCATAC	TCCACTTGG	CTCGGG
Ser₃ (Zea)[c]	CTGCAGGAA	-TACGA	AAACTCGCT	ATTCAC	TCAGTT	TATTTTCCA	TAATAAGAT	TATGTA
Phe (Zea)[c]	TTTTAGTCC	CTTTAA	TTGACATAG	ATGCAA	ATACTT	TACTAAGAT	GATGCACAA	GAAAGG
Thr₂ (Zea)[c]	ACATTTTAC	-AGATT	CCCTATATA	TATATT	TTTATT	TGTTACACT	ATTTCTGTT	ATGTAA
Thr (Euglena)[j]	TTGACAAAA	TAATTA	TAATGAGTA	AAATTA	AGTTCC	TATCATTAT	TATATTAAT	AATGAT
Gln (Euglena)[j]	TTGAAAACT	AAAAAT	TCGATTAGT	TTTCAA	CTAAAA	TTCTTTATA	CATAAATAC	AAAGTT

[a] Italicized and uncerscored sectors are explained in the text. [b] Deno et al. (1982). [c] Steinmetz et al. (1983). [d] Hollingsworth and Hallick (1982). [c] Steinmetz et al. (1982). [f] Williamson and Doolittle (1983). [g] Orozco et al. (1980). [h] Koch et al. (1981). [i] Takaiwa and Sugiura (1982). [j] Karabin and Hallick (1983).

introns, and trains have usually been considered to be subject to rapid evolutionary change. Yet although the monocotyledonous and dictyledonous branches of seed plants have been separated for many millions of years, these leaders from such distantly related forms nevertheless display a high level of conservation. But it should also be noted that identical genes from outside the ribosomal DNA region also can display large percentages of homology in their leaders. This is well illustrated by the two leaders of tRNA[Met] cistrons, one from tobacco, the other from maize chloroplasts (Deno *et al.*, 1982; Steinmetz *et al.*, 1983), in which 38 of the 60 corresponding sites are identically occupied, representing a 63% level of homology.

The more typical condition of leaders of the same type of gene from different organisms is revealed by a comparison of the threonine tRNA genes of maize and *Euglena*, in which only 18 of the corresponding 60 sites have identical occupants (Karabin and Hallick, 1983; Steinmetz *et al.*, 1983). This is a homology rate of only 30%, scarcely exceeding the 25% rate of random-chance combinations. The similar differences that exist between the tRNA[Ile] gene leaders of *Euglena* chloroplast and *Anacystis* are far more representative of their nature, only 13 of the 60 sites (or 21%) shown having corresponding occupants. Similarly, the disparities that occur between the tRNA[Ile] gene leaders from *Euglena* and maize or tobacco chloroplasts demonstrate the same situation, there being only seven identical sites.

tRNA Leader Sectors from Mitochondria. In mitochondria the genome is not as extensive as it is in chloroplasts, for the most part encoding only ribosomal proteins, rRNAs, tRNAs, and a number of proteins or polypeptides of the respiratory chain and membrane structures. As a whole the tRNA genes are relatively few in number, so that earlier it had been proposed that only a sufficient number is present to provide one type for each of the 20 common amino acids. Since DNA sequencing has become feasible, however, such has not proven to be strictly the case. In *Aspergillus* the 20 mitochondrial tRNA genes whose sequences have been established (Köchel *et al.*, 1981) include two each for glycine, serine, and leucine and three for methionine; in the isoaccepting species for the last of these, of course, the anticodon is CAU in each case, since this amino acid is provided with only one codon. Hence, the repetitions in species are not entirely for codon-recognition purposes.

Comparisons between the isoaccepting species of this leader given in Table 2.9 add emphasis to some of the conclusions drawn from the chloroplast species. Although the Annelida–Arthropoda lineage must have diverged from the Echinodermata–Chordata branch in Precambrian times, at least a billion years ago, the leader sequences for the valine tRNAs from representatives of the two lines, *Drosophila* and the mouse, show the remarkably high level of homology of close to 75%. As in the chloroplast, the homologous sites are not uniformly distributed, but the distribution here does not show the same pattern. In the present case, there are two chief regions of homology, A–D and F–P, separated by a narrow band (E) in which only two of the six corresponding pairs of sites have identical bases. Zones B and C are the centers of homology toward the 5′ end of the sequences, 14 of the 15 sites being identically occupied. D displays somewhat lesser conservation of bases, with a homology rate of 67%, whereas A has only a 34% level. On the 3′ side of E, zone F is the least variable, with eight of its nine sites identically occupied, P ranking next, if only those sites occupied in the fruitfly sequence are counted,

Table 2.9
5' Leaders of Mitochondrial tRNA Genes

	A	B	C	D	E	F	G	P
Val$_{UAC}$ Drosophila[a]	TATTAAGGT	AAGATA	AGTCGTAAC	ATA-GT	AGATGT	ACTGG-AAA	GTGTATCTA	GAATGA
Val$_{UAC}$ Mouse[b]	TTATGAGAG	GAGATA	AGTCGTAAC	A-AGGT	AAGCAT	ACTGG-AAA	GTGTGCTTG	GAATAAT---
Phe$_{GAA}$ Mouse[b]	TTCCGTGAA	-CCAAA	ACTC-TAAT	CATACT	CTATTA	CGCAATAAA	CATTAA---	CAA-------
Phe$_{GAA}$ Human[c]	CCCCTGCTA	ACC---	CCATACCCC	GAACCA	ACCAAA	CCCCAAAGA	CACCCCC--	CACA------
Lys$_{UUU}$ Yeast[d]				CCC	CAAAGG	AGTAATATA	TATTATGTA	TAAACAATA-
Lys$_{UUU}$ Aspergillus[e]	TTAAAATTT	ATCACA	TAGTATAAA	TTATAT	ATATTA	TATAGTTTT	AGATATATA	-AATATCTAA
Gly$_{UCC}$ Yeast[d]	CCCCATTTT	TAATTT	TATTAAGAA	GTTTAA	TTTACT	ATTTAATAA	TAAATGAAA	TAATA-ATA-
Gly$_{UCC}$ Aspergillus[e]								TATAG-ATA-
Gly$_{ACC}$ Aspergillus[e]	----AATA	ATATAA	TTATTTAAA	G--TAA	TATGAA	TATATATAC	TAAG--ATA	TATAATATAA
Ile Bacillus subtilis[f]	AAACAGAAC	GTTCCC	TGTCTTGTT	TAGTTT	TGAAGG	ATCTTTGTT	CCTTGAATA	AGTTAAGAT-
Ile$_{GAU}$ Aspergillus[e]	AAATAAGCA	AATAGG	AAGGGAATA	TATTTA	TTAAAA	TATTTTTTT	ATAAAAACT	GCGAATTGT-
Ile$_{GAU}$ Drosophila[g]	TTTTAAAAA	AAAAAT	TAGTAAATA	ATAAAA	AAAAAA	AAAAAAAAA	GATGAGTTT	TTTATTATT-
Ile Mouse[b]	TAGCATTAT	GTATGT	GACATATTT	CTTTAC	CAATTT	TTACACGGG	GAGTACCAC	CATACATAT-

[a]Clary et al. (1982). [b]Bibb et al. (1981). [c]Anderson et al. (1981). [d]Bonitz and Tzagoloff (1980). [e]Köchel et al. (1981).
[f]Loughney et al. (1983). [g]Clary et al. (1982).

while G is quite inconstant, with a homology rate of 34%. The main point here, nevertheless, is that indicated in an earlier discussion—the leaders of homologous genes from precisely corresponding sectors of the genome do not vary randomly, but can display a striking level of evolutionary conservation.

A second set of sequences (Table 2.9), those for the leaders of tRNA$_{GAA}^{Phe}$ from murine and human mitochondria, demonstrates the opposite condition, for the overall homology rate is a mere 26%, that is, it is at the strictly random-chance level of correspondence. Only in two of the zones, F and P, does the rate exceed randomness; in the first of these a rate of 56% is shown and in the second, a 50% level. Thus it is evident that either the closing statement of the preceding paragraph is totally erroneous or these two genes are not from corresponding regions of the genome. That the latter is more probably the case is suggested by comparison of two genes from murine mitochondrial sources, the present tRNA$_{GAA}^{Phe}$ and the tRNA$_{UAC}^{Val}$ examined earlier, a 35% homology being revealed. Similar comparison of the latter with the tRNA$_{GAA}^{Phe}$ from human organelle discloses these to be only 17% homologous.

The yeast and *Aspergillus* mitochondrial genes shown in the same table test this concept further, but first it must be recalled that the two organisms are not closely related as formerly considered. Current cellular and macromolecular data show that the yeast ranks among the most primitive eukaryotes, whereas fungi like *Aspergillus* are placed among the highly advanced, just below the metazoa (Dillon, 1962, 1978, 1981). Hence their respective lineages diverged probably 2 billion years ago. Unfortunately, two of the sequences have become reduced evolutionarily to mere remnants, one 36 residues in length, the other only nine; however, the reduction appears to have involved the 5' portion almost exclusively, regions F, G, and P being highly homologous.

The two lysine tRNA genes should certainly represent identical genes, for in each case they are located immediately toward the 3' side of the cistron for the small rRNA (Hovemann *et al.*, 1980a; Yen and Davidson, 1980), followed by tRNA$_{UCC}^{Gly}$ coding sequences. Although the two are obviously homologs, the amount of identity between site occupants is not very high, being just 36% based on the 36 sites of the yeast sequence. Similarly, the two tRNA^{Gly} genes with long leaders do not correspond strictly, for only a single glycine-accepting species is present in the yeast organelle. Thus the second one of *Aspergillus* is an advanced condition. Nonetheless, the leaders display extensive homology at 29 of 55 sites, for a 53% level. Once again the level of homology is not uniform through the sequences, the maximum rate (100% for the four sites) occurring in zone D. Zone P is next highest, with a 67% homology rate, G with 56% is next, followed by C and F, with 44% each, and finally by A, B, and E, the latter two having only a 22% rate.

The series of four tRNA^{Ile} leader sequences is provided in order to demonstrate the wide range of variation that can occur among leaders of seemingly homologous genes that are evidently not from identical regions of the genome. The highest total homology level, 39%, is between the sequences from *Aspergillus* and *Drosophila* mitochondria, followed by a 35% homology between the former and *B. subtilis*. The lowest, 20%, exists between the bacterium and the fruitfly organelle. The really valid conclusion that can be drawn from these four sequences is that they probably do not represent identically corresponding genes in most instances.

2.3. RECOGNITION IN THE MATURE tRNA GENE

In common with the other genes transcribed by polymerase III in eukaryotes, it has been proposed that "internal" signals (that is, within the sequence of the mature gene) exist that aid in recognition by the enzyme (DeFranco *et al.*, 1980; Schaack *et al.*, 1983). But experimental studies have yielded varying results, at times indicating an internal control region, including much of the entire portion of the mature gene (Sharp *et al.*, 1981a; Ciampi *et al.*, 1982; Ciliberto *et al.*, 1982), at others suggesting only the 5′ half of that sequence (Carrara *et al.*, 1981), and occasionally intimating both an internal and an external signal (DeFranco *et al.*, 1980; Sprague *et al.*, 1980; Dingermann *et al.*, 1982). Since clearly defined signals are not usually apparent in the 5′ leaders, as has just been seen, it may be profitable to explore the mature tRNA coding sequences themselves.

For present purposes a diversity of such genes from *Drosophila* are examined first, followed in turn by comparable series from baker's yeast, prokaryotes, and eukaryotic organelles. Insofar as is possible, the sequences are arranged in groups according to the initial base of the condon, those beginning with guanidine in group I, with cytidine in group II, adenine in group III, and uracil in group IV (Dillon, 1973, 1978). In all cases arm IV, the "extra" or "variable" arm, has been omitted, because its great variability in both site specificity and length makes it of improbable value for recognition purposes; nor has it ever been clearly implicated to serve in that capacity.

2.3.1. The Mature tRNA Genes of Eukaryotes

Throughout this presentation the regions of the tRNA molecules are named in accordance with the terminology presented previously (Figure 2.2; Dillon, 1978, pp. 140–142). Since the genes and their transcription form the topic here, not the mature products, discussion of tertiary structure is usually out of place, but where it arises, as it does briefly on several occasions, reference may be made to the fuller synopses provided elsewhere (Dillon, 1978, pp. 139–144, 235–348).

Comparisons of tRNA Genes of Drosophila. Comparisons of the 13 different tRNA gene sequences from *Drosophila* (Table 2.10) reveal the presence of relatively few invariable sites. Among the most constant of these are the T in the first site of bend 1, the G in the first position of the stem of arm II, and the A at each end and the paired Gs in the middle of the loop of the same arm. Then in arm V two adjacent G–C pairs always terminate the stem; following this at the 5′ end of the loop there almost always are two Ts. Since the first of these is replaced by an A in the two tRNA[Met] sequences (Sharp *et al.*, 1981b; Gauss and Sprinzl, 1983a, b), only the latter can appear among the invariant sites. In turn, that second T is uniformly followed by a cytidylic acid residue and two sites beyond by an adenylic acid one. Between these two a purine always occurs, usually a G (10 out of the 13 sequences), but occasionally an A. The precise base present there has no basis in chemistry, for although most of the tRNAs[Glu] have the A, one has a G. The remaining constant sites are by class only, pyrimidines or purines, and therefore are less attractive as possible signals.

Indeed, it is difficult to perceive any of these constant features in that light, at least with any degree of assurance, because most, if not all, are known to be involved in tertiary

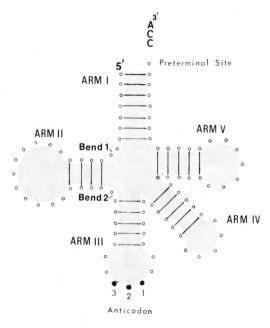

Figure 2.2. The structure of the transfer RNA molecule. The terminology is that of Dillon (1978).

structure. However, it is possible that in the gene, which is simply double-stranded, not looped, those same constancies may have a different functional significance than in the matured transcriptive product. But over and beyond that consideration, there is no sequence of sufficient length that is convincingly distinct enough to distinguish it from all the other combinations of the same sort that must occur thousands of times within the DNA. Hence, if all or part of these coding regions serve in enzyme recognition, it is through the total primary structure and not simply these few invariable sites.

Comparisons of tRNA Genes of Saccharomyces. In regard to the constant sites, the 16 tRNA genes from *Saccharomyces* shown in Table 2.11 differ but slightly from those just discussed. Yet these few differences are most striking. Bend 1 is more consistent than in the corresponding genes of *Drosophila,* uniformly consisting of T in the first site and a purine in the second, whereas the occupant of the latter is highly variable in the fruitfly. The constant sites of arm II are identical in both these eukaryotes.

Several distinctions between the sequences from the two sources can be noted in arm III, the anticodon arm. Whereas in the insect genes pyrimidines are universally present in the first site of the stem, no such consistency can be observed there in the yeast. However, it is only the serine and aspartic acid species that are exceptional to that rule in the latter organism, neither of which is represented among the *Drosophila* sequences. Thus this may prove not to be a valid distinction. At the penultimate position in this stem a G–C pair, with a rare G–T combination instead, is a persistent feature in all the sequences from the yeast, but the order of the combination varies frequently to a C–G or T–G. In the loop of this arm, pyrimidines always occupy the 5'-most location in the yeast, with a T uniformly present in the next site adjacent to the anticodon, whereas in the fruitfly the former is variable and the latter may be either common pyrimidine. To the 3' side of the anticodon, in contrast to the insect genes in which there is no consistent feature, in the yeast a purine occurs, usually an A. The last site in this loop is a function of the amino acid group and receives attention later.

Many of the constancies of arm V are familiar ones, particularly the G–C pair that terminates the stem in both organisms. However, in the fruitfly, but not here in the yeast, an identical combination precedes the final one. At the second site in the loop, a T is a universal trait, the catholicity of a like occupant in the first position being upset by the presence of an A in initiator methionine tRNA. The next three sites are also invariable in *Drosophila*, where CRA always occurs, but in yeast the first of this trio is not entirely constant, initiator methionine tRNA genes having a G there. Moreover, in the latter organism a pyrimidine always terminates the loop at the 3' side, a condition made inconsistent in the fruitfly by both known methionine tRNA cistrons.

Correlations to Amino Acid Group. When the sequences are arranged to accord with amino acid groups based on the 3'-terminal occupant of the anticodon, or conversely the initial site of the codon (Dillon, 1973, 1978), some sequence features can be noted to be consistent within, but not between, groups. Thus it is conceivable that certain of them may aid in recognition for transcriptive purposes. However, since the entire 3' half has been shown to be of no value in transcription, at least in certain tRNAs, attention is devoted solely to the opposite half, including the entire anticodon loop.

Some of the constant sequence features of the eukaryotic amino acid groups have already been pointed out (Dillon, 1978, pp. 319–337). Arm I features there were shown to involve site 7, which in group I amino acids was always a purine. The same condition holds in the yeast and *Drosophila* sequences provided here. In the other groups, pyrimidines as well as purines occur. Where genes for multiple types of isoaccepting species have been sequenced, as for the glutamic acid and lysine species in *Drosophila* (Table 2.10) and the serine in yeast (Table 2.11), the relative invariability of these arms, and the entire sequences for that matter, cannot fail to capture one's attention. In the fruitfly, five of the glutamate sequences are invariable here and the sixth differs only at site 4, where a T replaces a C. In the three lysine genes of the same organism, similar constancy prevails, the arm IA of lysine $tRNA_{2,4}$ having a C at site 7 and lysine $tRNA_{4,5}$ having an A. In the yeast, site 4 of the same arm is occupied by an A in three serine tRNAs and by a T in the other two. In contrast, the two methionine genes of the yeast differ markedly throughout this sector, the only identities being in sites 5 and 7, but such distinctions in structure are to be expected of these two isoaccepting species in keeping with the differences in their respective functions. Comparisons with the $tRNA_{1,2}^{Met}$ genes of *Drosophila* strongly suggest that both of the latter isoacceptors function solely in initiation of translation like the $tRNA_1^{Met}$ of the yeast.

Arm II also presents interesting correlations to both the amino acid groups and isoaccepting species, because here the intraspecies variation is even less than in arm I. Especially striking is the GT-T tetranucleotide of the 5' side of the stem in groups I and II, in contrast to the GC-C sequence of groups III and IV. The only exceptions among those given in Tables 2.10 and 2.11 are the GC-T of yeast alanine tRNA and the GT-C of the *Drosophila* glycine-accepting species. Also noteworthy is the unexpected variation in the complementary sequences in the 3' side of the stem. The loop of this arm displays similar

Table 2.10

Sequences of Genes for Drosophila tRNAs[a]

	Arm IA	Bend 1	Arm II	Bend 2	Arm III	Arm V	Arm IB
Group I							
Gly[b]	GCATCGG	TG	GTTC AGT–GG––TA GAAT	G	CTCGC CTGCCAC GCGGG	CCGGG TTCGATT CCCGG	CCGATGC A
Val$_4$[c]	GTTTCCG	TG	GTGT ACG–GG–TTA TCAC	A	TCTGC CTAACAC GCAGA	CCGG TTCGATC CCGGG	CGGAAAC A
Glu[d]	TCCCATA	TG	GTCT AGT–GG–CTA GGAT	A	TCTGG CTTTCAC CCAGA	CCGGG TTCGATT CCCGG	TATGGGA A
Glu$_{i,iiii-v}$[d]	TCCCATA	TT	GTCT AGT–GG–TTA GGAT	A	TCCGG CTCTCAC CCGGA	CCGGG TTCAATT CCCGG	TATGGGA A
Glu$_{ii}$[e,f]	TCCTATA	TT	GTCT AGT–GG–TTA GGAT	A	TCCGG CTCTCAC CCGGA	CCGGG TTCAATT CCCGG	TATGGGA A
Group II							
His[f]	(G)GCCGTGA	TC	GTCT AGT–GG––TTA GGAC	C	CCACG TTGTGGC CGTGG	CCAGG TTCGAAT CCTGG	TCACGGC A
Arg[g,h]	GGTCCTG	TG	GCGC AAT–GG–ATA ACGC	G	TCTGA CTACGGA TCAGA	CCAGG TTCGACT CCTGG	CAGGATC G

Group III

Asn[h]	GCCTCCG	*TG*	*GCGC AATTGG-TTA GGGC*	G	TTCGG CTG<u>TTAA</u> CCGAA	GG*TGG TTCGAGT CCCAC*	CCGGGGG	C
Lys[i] 5	GCCCGGA	*TA*	*GCTC AGTCGG--TA GAGC*	A	TTGGA CT<u>TTTAA</u> TCCAA	CAGG*G TTCAAGT CCCTG*	TTCGGGC	G
Lys[h,j] 2,4	GCCCGGC	*TA*	*GCTC AGTCGG--TA GAGC*	A	TGAGA CTC<u>TTAA</u> TCTCA	GTG*GG TTCGAGC CCCAC*	GTTGGGC	G
Met[k] 1,2	AGCAGAG	*TG*	*GCGC AGT-GG--AA GCGT*	G	CTGGG CC<u>CATAA</u> CCCAG	CGAG*G ATCGAAA CCTTG*	CTCTGCT	A
Ile[h,l]	GGCCCAT	*TA*	*GCTC AGTTGG-TTA GAGC*	G	TCGTG CT<u>AATAA</u> CGCGA	GCGG*G TTCGATC CCCTC*	ATGGGCC	A

Group IV

Leu[l] 3	GTCAGGA	*TG*	*GCCG AGC-GGTCTA AGGC*	G	CCAGA CTCAAGT TCTGG	GT*GG TTCGAAT CCCAC*	TTCTGAC	A

| Invariable Bases | T- | GY | ARY GG   A   RY | | Y   YY   R | GG -TCRA | CC | |

[a] Underscoring indicates the anticodons and italics represent internal promoters (boxes A and B), discussed in Chapter 3, Section 3.6.1.
[b] Hershey and Davidson (1980).
[c] Addison et al. (1982).
[d] Indik and Tartof (1982).
[e] Hosbach et al. (1980).
[f] Altwegg and Kubli (1980).
[g] Silverman et al. (1979); Yen and Davidson (1980); Dingermann et al. (1982).
[h] Hovemann et al. (1980a).
[i] DeFranco et al. (1982).
[j] DeFranco et al. (1980).
[k] Sharp et al. (1981b).
[l] Robinson and Davidson (1981).

Table 2.11

Genes for Mature tRNAs of Saccharomyces[a]

	Arm IA	Bend 1	Arm II	Bend 2	Arm III	Arm V	Arm IB
Group I							
Gly[b]	GCGCAAG	TG	GTTT AGT-GG--T-A AAAT	C	CAACG TTGCCAT CGTTG	CCCGG TTCGATT CCCGG	CTTGCGC A
Ala[c]	GGGCGTG	TG	GCGT AGTCGG--T-A GCGC	G	CTCCC TTAGCAT GGGAG	TCCGG TTCGATT CCCGA	CTCGTCC A
Asp[d]	TCCGTGA	TA	GTTT AAT-GG--TCA GAAT	G	GGCGC TTGTCGC GTGCC	CGGGG TTCAATT CCCCG	TCGCGGA G
Glu3 PY20[e]	TCCGATA	TA	GTGT AA-CGG-CT-A TCAC	A	TCACG CTTTCAC CGTGG	CGGGG TTCGACT CCCCG	TATCGGA G
Val[f]	GGTTTCG	TG	GTCT AGTCGG-TT-A TGGC	A	TCTGC TTAACAC GCAGA	CCCAG TTCGATC CTGGG	CGAAATC A
Group III							
Lys[g]	TCCTTGT	TA	GCTC AGTTGG---T-A GAGC	G	TTCGG CTTTTAA CCGAA	AGGGG TTCGAGC CCCCT	ATGAGGA G
Met1[h]	AGCCCCG	TG	GCGC AGT-GG--A-A GCGC	G	CAGGG CTCATAA CCCTG	CTCGG ATCGAAA CCGAG	CGGCGCT A
Met3[i]	GCTTCAG	TA	GCTC AGTAGG--A-A GAGC	G	TCAGT CTCATAA TCTGA	GAGAG TTCGAAC CTCTC	CTGGAGC A
Arg PJB18[d]	GCTCGCG	TG	GCGT AAT-GG---C-A ACGC	G	TCTGA CTTCTAA TCAGA	ATGGG TTCGACC CCCAT	CGTGAGT G

Group IV

Ser B,G,J[j]	GGCAACT	*TG*	*GCCG AGT-GG-TT-A* AGGC	G	AAAGA TTAGAAA TCTTT	GCA*GG TTCGAGT* CCTGC	AGTTGTC	G
Ser[k]_UCA	GGCTACA	*TG*	*GCCG AGT-GG-TT-A* AGGC	G	ACAGA CT<u>TGA</u>AA TCTGT	GCT*GG TTCAAAT* CCTGC	TGTGGTC	G
Ser[k]_UCG	GGCTACA	*TG*	*GCCG AGT-GG-TT-A* AGGC	C	AGAGA CT<u>CGA</u>AA TCTCT	GCT*GG TTCAATC* CCTGC	TGTGGTC	G
Trp[l]	GAAGCGG	*TG*	*GCTC AAT-GG--T-A* GACC	T	TTCGA CTC<u>CAA</u>A TCGAA	GCA*GG TTCAATT* CCTGT	CCGTTTC	A
Leu[m]	GGTTGTT	*TG*	*GCCG AGC-GGTCT-A* AGGC	G	CCTGA TT<u>CAA</u>GA TCAGG	AAGAG *TTCGAAT CTCTT*	AGCAACC	A
Phe[n]	GCCGATT	*TA*	*GCTC AGTTGG--G-A* GAGC	G	CCAGA CTGA<u>AGA</u> TCTGG	CTG*TG TTCGATC* CACAG	AATTCGC	A
Tyr[o]	CTCTCGG	*TA*	*GCCA AGTTGGTTT-A* AGGC	G	CAAGA <u>CTGAA</u> TCTTG	*GGGCG TTCGACT CGCCC*	CCGGGAG	A
Invariable sites	TR	GY	ARY GG A RY		YT R	G TCRA-Y C		

[a]Underscoring indicates the anticodons, and italics represent internal promoters (boxes A and B), discussed in Chapter 3, Section 3.6.1.
[b]Yoshida (1973).
[c]Penswick *et al.* (1975).
[d]Schmidt *et al.* (1980).
[e]Feldmann *et al.* (1981).
[f]Bonnet *et al.* (1974).
[g]Madison and Boguslawski, (1974).
[h]Venegas *et al.* (1982).
[i]Olah and Feldmann (1980).
[j]Page (1981).
[k]Ethcheverry *et al.* (1979).
[l]Ogden *et al.* (1979).
[m]Standring *et a'.* (1981).
[n]Valenzuela *et al.* (1978).
[o]Knapp *et al.* (1979); Colby *et al.* (1981).

correlations to groups in the tRNA genes of the fruitfly, but such correspondences are dilute or missing in those of the yeast.

In addition to a number of distinct trends in the respective groups, the stem of arm III shows several more marked features. In group I members, the first base on the 5' side of the stem is usually T and the second is C, but the order of these two bases may be reversed, as in the yeast alanine-accepting species. However, the yeast representatives are somewhat variable in that the first two residues may be CA, as in tRNA$^{Gly}_3$, or GG, as in tRNA^{Asp}. The single group II representative shows CC here; in group III TT predominates, but TC, TG, CA, and CT also occur; and in group IV, A is the more frequent occupant of the first site, with C the usual alternative, but T occurs occasionally. This stem always has either a G or C at site 5 on the 5' side, the latter being uniformly absent here in the remaining groups, although G occurs also in groups II and III. However, A is the most characteristic occupant of this site in groups III and IV, a trait that is completely consistent in the yeast sequences; T rarely is found here in these same two groups.

In the loop of this anticodon arm, several functional restraints may be operative whose discussion is more appropriate to another topic, translation. Thus CT occurs in all groups adjacent to the 5' side of the anticodons, which are italicized in the tables. The combination TT also may be observed here, particularly in yeast group I representatives, but elsewhere occasionally, too. In the fruitfly tRNA genes, CCC may be noted in tRNA$^{Met}_{1,2}$. Perhaps the most prominent of the group-correlated positions are found on the 3' side of the anticodon, where in the fruitfly AC occurs throughout group I, GC and GA in the two members of group II, AA constantly in group III, and GT in group IV's lone representative. In the yeast the group III occupants are identical to those of the insect, but the same doublet AA is also found in group IV types, along with GC or GA, and in group I the AC is sometimes replaced by AT or rarely by GC.

Comparisons of Multisource tRNAs of a Single Species. Thus it is clear that a seeming sufficiency of sequence order exists to provide recognition signals to the proteins of transcription. Some of the distinctive positional combinations could enable such a biochemical to identify the sequence as to group, while still others, particularly those of arm I, could identify the species. To pursue this possibility further, Table 2.12 provides the sequences of three series of isoaccepting types, each drawn from those currently available from a diversity of eukaryotes.

The 11 glutamic acid sequences provided in Table 2.12 show a remarkable degree of evolutionary conservation, the first three sites at the 5' end and the corresponding three before the 3' end being identical in all members of the series. Indeed, the entire arm I shows a high level of homology throughout its length. The initial TCC which marks each 5' terminus, however, also occurs in the yeast lysine tRNA gene, suggesting that that combination alone is not sufficient to identify these genes as to species, but certainly could in unison with group signals, as just seen.

Another highly conserved sector is found in the stem of arm II, which is invariable throughout the metazoan sequences and deviates at only one site each in the yeast and fission yeast primary structures on the 5' side and only in the former on the 3' side. In addition to the loop of this arm being highly conserved, its structure shows some correlation to differences in the anticodon. All of the tRNAs having the anticodon CTC have TTA following the nearly invariant double G, but this trait is present in a rat liver variety with TTC as the anticodon. In contrast, the remaining sequences with the latter triplet have CTA after the GG doublet. The *Drosophila* sequences with differing anticodons are further distinguished by a base difference to the 3' side of the guanosine pair, while the rat's differs on the 5' side.

In the stem of arm III, variation is too extensive to provide any possible recognition sites, the occupants of the last two sites being largely correlated to taxon, for almost all the metazoan sequences have GG here, in contrast to the CG of protistans. But it is of interest to note that the last rat sequence given here has reverted to the protistan combination. The loop of this arm, nevertheless, is nearly unvarying, AC being on the 3' side of the anticodon in all cases and CT on the 5' side, except in the rat liver example, where TT occurs.

With the exception of those of the yeast, the ten genes for tRNA^{Lys} show a parallel set of conserved sites at corresponding sectors; in fact, the first six sites—not just the first three—of arm I are highly conserved. Although the yeast sequence shows only three homologies with the rest in this region, in arm II the agreement among all ten genes is almost perfect, divergences being notable only at three points in the stem-and-loop regions combined. A parallel condition prevails as well among the nine tRNA^{Met} genes, with the occasional exception of the two elongator types, indicated in the table by subscript E. Even though this pair has a markedly different function from the initiator types, consistent differences in primary structure occur at only a few points, especially at the 5' end of arm I and in bend 1. Moreover, the loop of arm II is at once distinguished by the presence of an additional occupant (A) on the 5' side of the ubiquitous paired Gs. Thus one gains the distinct impression that this half of the tRNA gene could very well serve in recognition, not only of the type of gene, but also of the isoaccepting species and functional variety.

Genes for tRNAs Encoded by Two Code Classes. As is well known, three species of amino acids, arginine, leucine, and serine, are each carried by tRNAs of two contrasting codon classes. Although none of the tRNA isoaccepting types have been sequenced in the desired abundance, enough has been established to provide some insight into the nature of the problem. The purpose of the present comparisons is to elucidate the question of whether the carrier RNAs of each type differ in relation to the initial letter of the anticodon it bears, for it has been proposed (Dillon, 1978) that the presence of these amino acids in two code classes represents separate evolutionary origins in each case and provides the protobiont or cell with a means of distinguishing their respective biosynthetic pathways.

In the case of arginine isoacceptors (Table 2.13), the three genes from yeast immediately suggest that the proposal is sound, for the single representative bearing a group II anticodon (ACG) obviously differs from the two bearing a group III anticodon (TCT). The most outstanding feature is, of course, the initial TTC triplet at the 5' end of arm IA of the first of these and the GCT or GCG of the other two, with another prominent difference being the presence of the triplet TCA following the GG pair of the arm II loop in the first and only the doublet CA there in the others. Additional divergences are to be noted in the stems of loops II, III, and V. Thus it is apparent that sufficient differences occur between the group II and III species from the yeast to provide contrasts in signals for recognition by transcribing proteins. Although the two sequences from metazoan sources, both bearing group II anticodons, are amply different from the corresponding one of yeast, all three share the major distinctions found in the loop of arm II and the stem of arm V. It will be of

Table 2.12

Selected tRNA Genes from Multiple Eukaryotic Sources[a]

	Arm IA	Bend 1	Arm II	Bend 2	Arm III	Arm V	Arm IB
Glutamic acid							
Yeast[b]	TCCGATA	TA	GTCT AAC-GCCTA TCAC	A	TCACG CTTTCAC CGTGG	CGGGG TTCGACT CCCCG	TATCGGA G
Schizosaccharomyces pombe[c]	TCCGTTG	TG	GTCC AAC-GCCTA GGAT	T	CGTCC CTTTCAC CGACG	CGGGG TTCGACT CCCCG	CAACGGA G
Drosophila[d]	TCCCATA	TG	GTCT AGT-GCCTA GGAT	A	TCTGG CTTTCAC CCAGA	CCGGG TTCGATT CCCGG	TATGGGA A
Rat liver[e] $_{GAA}$	TCCCACA	TG	GTCT AGC-GGTTA GGAT	T	CCTGG TTTTCAC CCAGG	CCGGG TTCGACT CCCGG	TCTGGGA A
Rat liver[f] $_{GAG}$	TCCTTGG	TG	GTCT AGT-GGTTA GGAT	T	CGGCG CTCTCAC CGCCG	CCGGG TTCGATT CCCGG	TCAGGGA A
Drosophila[g] $_{i,iii,iv,v}$	TCCCATA	TT	GTCT AGT-GGTTA GGAT	A	TCCGG CTCTCAC CCGGA	CCGGG TTCCAATT CCCGG	TATGGGA A
Drosophila[g] $_{ii}$	TCCTATA	TT	GTCT AGT-GGTTA GGAT	A	TCCGG CTCTCAC CCGGA	CCGGG TTCCAATT CCCGG	TATGGGA A
Rat[h]	TCCTTGG	TG	GTCT AGT-GCTAA GGAT	T	CGGCG CTCTCAC CGCCG	CCGGG TTCGATT CCCGG	TCAGGGA A
Lysine							
Yeast[i]	TCCTTGT	TA	GCTC AGTCGGTA GAGC	G	TTCGG CTTTTAA CCGAA	AGGGG TTCGAGC CCCCT	ATGAGGA G
Drosophila[j] $_5$	GCCCGGA	TA	GCTC AGTCGGTA GAGC	A	TTGGA CTTTTAA TCCAA	CAGGG TTCAAGT CCCTG	TTCGGGC G
Mouse[k]	GCCTGGA	TA	GCTC AATTGGTA GAGC	A	TCAGA CTTTTAA TCTGA	CAGGG TTCAAGT CCCTG	TTCACGC G
Human[l]	GCCCGGA	TA	GCTC AGTCGGTA GACC	A	TCAGA CTTTTAA TCTGA	CAGGG TTCAAGT CCCTG	TTCGGGC G
Nematode[m]	GCCCGGT	TA	GCTC AGTCGGTA GAGC	A	CCAGA CTCTTAA TCTGG	GCGGG TTCGAGC CCCGC	ATTGGGC T
Drosophila[n] $_{2,4}$	GCCCGGC	TA	GCTC AGTCGGTA GAGC	A	TGAGA CTCTTAA TCTCA	GTGGG TTCGAGC CCCAC	GTTGGGC G
Rat[o] $_{1,2,3}$	GCCCGGC	TA	GCTC AGTCGGTA GAGC	A	TGAGA CTCTTAA TCTCA	GTGGG TTCGAGC CCCAC	GTTGGGC G

Methionine

Yeast[p]$_I$	AGCGCCG	*TG*	*GCGC AGT-GGAA*	GCCC	G	CAGGG CT<u>CAT</u>AA CCCTG	*CTCGG ATCGAAA CCGAG*	CGGCGCT A
Scenedesmus[q]$_I$	AGCTGAG	*TG*	*GCGC AGT-GGAA*	GCCT	G	ATGGG CT<u>CAT</u>AA CCCAT	*ACAGG ATCGAAA CCTGT*	CTCAGCT A
Tetrahymena[r]$_I$	AGCAGGG	*TG*	*GCGA AAT-GGAA*	TCGC	G	TTTGG CT<u>CAT</u>AA CTCAA	*AGAGG ATCGAAA CCTCT*	CTCTGCT A
Neurospora crassa[s]$_I$	AGCTGCA	*TG*	*GCGC AGC-GGAA*	GCCC	G	CNGGG CT<u>CAT</u>AA CCCGG	*ACTCG ATCGAAA CGAGT*	TGCAGCT A
Wheat germ[t]$_I$	ATCAGAC	*TG*	*GCGC AGC-GGAA*	GCGT	G	GTGGG CC<u>CAT</u>AA CCCAC	*CCAGG ATCGAAA CCTGG*	CTCTGAT A
Drosophila[u]$_I$	AGCAGAC	*TG*	*GCGC AGT-GGAA*	GCGT	G	CTGGG CC<u>CAT</u>AA CCCAG	*CGAGG ATCGAAA CCTTG*	CTCTGCT A
Human[v]$_I$	AGCAGAC	*TG*	*GCGC AGC-GGAA*	GCGT	G	CTGGG CC<u>CAT</u>AA CCCAG	*GATGG ATCTAAA CCATC*	CTCTGCT A
Yeast[w]$_E$	GCTTCAC	*TA*	*GCTC AGTAGGAA*	GAGC	G	TCAGT CT<u>CAT</u>AA TCTGA	*GAGAG TTCGAAC CTCTC*	CTGGAGC A
Mammalian[x]$_E$	GCCTCGT	*TA*	*GCGC AGTAGTA*	GCGC	G	TCAGT CT<u>CAT</u>AA TCTGA	*GTGAG TTCGATC CTCAC*	ACGGGGC A

[a]Underscoring indicates the anticodors, while italics represent internal promoters (boxes A and B), discussed in Chapter 3, Section 3.6.1.
[b]Kobayashi et al. (1974). [c]Wong et al. (1979). [d]Indik and Tartof (1982). [e]Chan et al. (1982). [f]Shibuya et al. (1982). [g]Hosbach et al. (1980). [h]Sekiya et al. (1981). [i]Madison and Boguslawski (1974). [j]DeFranco et al. (1982). [k]Han and Harding (1983). [l]Roy et al. (1982). [m]Tranquilla et al. (1982). [n]DeFranco et al. (1980); Yen and Davidson (1980). [o]Sekiya et al. (1982). [p]Simsek and RajBhandary (1972). [q]Olins and Jones (1980). [r]Kuchino et al. (1981). [s]Gillum et al. (1977). [t]H. P. Ghosh et al. (1982). [u]Silverman et al. (1979). [v]Santos and Zasloff (1981). [w]Gruhl and Feldmann (1976). [x]Piper (1975).

Table 2.13
Gene Sequences of Eukaryotic tRNAs with Dual Group Affiliations[a]

	Arm IA	Bend 1	Arm II	Bend 2	Arm III	Arm V	Arm IB	
Arginine_II								
Yeast[b]	TTCCTCG	TG	GCCC AATGG-TCA CGGC	G	TCTGG CTACGAA CCAGA	CCAGG TTCAAGT CCTGG	CGGGGAA	G
Drosophila[c]	GGTCCTG	TG	GCGC AATGG-ATA ACGC	G	TCTGA CTACGGA TCAGA	CCAGG TTCGACT CCTGG	CAGGATC	G
Mouse[d]	GGGCCAG	TG	GCGC AATGG-ATA ACGC	G	TCTGA CTTCGGA TCAGA	CTAGG TTCGACT CCTGG	CTGGCTC	G
Arginine_III								
Yeast[e]	GCTCGCG	TG	GCGT AATGG--CA ACGC	G	TCTGA CTTCTAA TCAGA	ATGGG TTCGACC CCCAT	CGTGACT	G
Yeast3[f]	GCGCTCG	TG	GCGT AATGG--CA ACGC	G	TCTGA CTTCTAA TCAGA	ATGGG TTCGACC CCCAT	CGAGTGC	G
Leucine_II								
Yeast[g]	GGGAGTT	TG	GCCG AGTGGTTTA AGGC	G	TCAGA TTTAGGC TCTGA	AAGGG TTCGAAT CCCTT	AGCTCTC	A
Human[h]	GGTAGCG	TG	GCCG AGCGGTCTA AGGC	G	CTGGA TTTAGGC TCCAG	GTGGG TTCGAAT CCCAC	CGCTGCC	A
Xenopus[i]	GTCAGGA	TG	GCCG AGCGGTCTA AGGC	G	CTGCG TTCAGGT CGCAG	GTGGG TTCGAAT CCCAC	TTCTGAC	A
Nematode[j]	GGAGAGA	TG	GCCG AGCGGTCTA A3GC	G	CTGGT TTAAGGC ACCAG	GTGGG TTCGAAT CCCAC	TCTCTTC	A
Bovine[k]	GGTAGCG	TG	GCCG AGCGGTCTA A3GC	G	CTGGA TTAAGGC TCCAG	GTGGG TTCGAAT CCCAC	CGCTGCC	A
Leucine_IV								
Yeast[l]	GGTTGTT	TG	GCCG AGCGGTCTA AGGC	G	CCTGA TTCAGC TCAGG	AAGAG TTCGAAT CTCTT	AGCAACC	A
Drosophila[m]	GTCAGGA	TG	GCCG AGCGGTCTA AGGC	G	CCAGA CTCAAGT TCTGG	GTGGG TTCGAAT CCCAC	TTCTGAC	A

Serine III							
Rat Liver[a]	GACGAGG	*TG*	*GCCG AGTCG*–TTA AGGC	G	ATGGA CTGCTAA TCCAT	GTG*GG TTCGAAT CC*CAT	CCTCGTC	G
Serine IV							
Yeast[o]	GGCAACT	*TG*	*GCCG AGTGG*–TTA AGGC	G	AAAGA TTAGAAA TCTTT	GCA*GG TTCGAGT CC*TGC	AGTTGTC	G
Yeast_CGA[p]	GGCACTA	*TG*	*GCCG AGTGG*–TTA AGGC	G	ACAGA ATCGAAA TCTGT	GCT*GG TTCAAAT CC*TGC	TGGTGTC	G
Rat hepatoma[q]	GTAGTCG	*TG*	*GCCG AGTGG*–TTA AGGC	G	ATGGA CTAGAAA TCCAT	GCA*GG TTCGAAT CC*TGC	CGACTAC	G
Schizosaccharomyces pombe[r]	GTCACTA	*TG*	*TCCG AGTGG*–TTA AGGA	G	TTAGA CTCGAAA TCTAA	GCA*GG TTCAAAT CC*TGC	TGGTGAC	G

[a] Underscoring indicates the anticodons and italics represent internal promoters (boxes A and B), discussed in Chapter 3, Section 3.6.1.
[b] Weissenbach et al. (1975).
[c] Silverman et al. (1979); Dingermann et al. (1982).
[d] Harada and Nishimura (1980).
[e] Schmidt et al. (1980).
[f] Kuntzel et al. (1972); Dillon (1978).
[g] Randerath et al. (1979).
[h] Roy et al. (1982).
[i] Galli et al. (1981).
[j] Tranquilla et al. (1982).
[k] Pirtle et al. (1980).
[l] Standring et al. (1981).
[m] Robinson and Davidson (1981).
[n] Rogg et al. (1975).
[o] Etcheverry et al. (1979); Page (1981).
[p] Piper (1978).
[q] Ginsberg et al. (1971).
[r] Mao et al. (1980).

great interest to see whether metazoan tRNA^{Arg} genes with group III anticodons will share in the distinctions that mark the two yeast genes of this group.

The five genes for tRNA^{Leu} of group II and the two of group IV show so few divergences from one another that no conclusions of any sort are permitted—even the customary distinctions between isoaccepting forms with different anticodons are lacking. Thus the several nonhomologous points that exist between the two genes from yeast could as readily provide signals for recognition of types bearing different anticodons as of those of different group memberships. However, the yeast and *Drosophila* genes of group IV do share several characteristics in the stem of arm III that may prove to be consistent traits when further sequences have been established.

In the five serine sequences that are currently available (Table 2.13), a situation parallel to that of leucine prevails, in that the representation of species is sparse and uneven, only one sequence of group III having been established, against four of group IV. Here, too, it is presently impossible to distinguish anticodon-correlated traits from those of the group. Yet the two rat genes, one each from the two groups, do appear to differ more markedly from one another than do the two from yeasts having different anticodons but belonging in the same group. Nevertheless, the evidence does not permit any firm conclusions to be drawn, except in the case of the arginine tRNA genes.

2.3.2. Prokaryotic and Viral Mature tRNA Genes

Because the analyses of eukaryotic tRNA genes have already provided an adequate insight into the possibilities for their recognition, the present study on the prokaryotic and viral genes is designed primarily to detect what basic differences and resemblances exist. Accordingly, attention is devoted to the known genes and mature tRNA primary structures of a single species of bacteria, *E. coli*, from which nearly 20 different sequences have been established, carrying a total of 14 distinct amino acids. After these have been analyzed, a similar examination of known viral tRNA genes completes the comparative aspects of the present class of biochemicals, except those of organelles.

Prokaryotic tRNA Genes. Table 2.14 presents the primary structures of mature tRNAs and their known genes of *E. coli*, all converted to gene characteristics. In this prokaryote, the sequence of the 5' side of arm I (arm IA) offers less promise for use in recognition than does the eukaryotic counterpart. In the first place, 16 of the 19 given there have G in the first site, two of the exceptions having A and one having C. Nor is the variety of occupants very great in sites 2 and 3—five sequences begin with GCG and three with GGG. However, the structure of the entire arm appears quite adequate for cognitive purposes.

In the three species of glycine-accepting tRNA genes, which differ from one another in the anticodons, it can be noted that the first five base pairs in arm I are identical, whereas the next two diverge sufficiently to provide clues for identification. In contrast, the two isoaccepting forms for valine, which have identical anticodons, differ in four of the seven base pairs in this arm. The same situation prevails between the two isoacceptors for methionine, the first of which (tRNA$_F^{Met}$) is involved only in initiation, the second entirely in elongation of the polypeptide chain. Thus, as in the eukaryotes, there could be a functional basis for this difference and the others that exist throughout their sequences. But the two tRNA^{Ile} forms, which have dissimilar anticodons, differ only in the first base pair, an identity of structure that is broken elsewhere in the molecule only in the last two base pairs of the stem of arm III. Hence, if transcription is selective for the genes of different isoaccepting tRNAs, these three points would have to suffice for recognition.

On the other hand, the three genes for tRNA^{Gly} mentioned above differ at a number of sites throughout their lengths, including bend 1, both the stem and loop of arm II, bend 2, the first two base pairs in the stem of arm III, and even in the first site of the loop of that arm, and in the second and fourth base pairs of the stem of arm V. The two tRNA^{Val} forms, although less variable than this trio, reveal several dissimilarities in the loop of arm II, and in the third base pair of the stems of both arms III and V.

Correlation of Prokaryotic tRNA Genes to Group. As was the case with the eukaryotic genes, correlations to the codon family can be noted in those of this prokaryote (Table 2.14). By way of example, in arm I the first base pair is always G–C in groups I and II, whereas in III and IV various combinations exist. Moreover, the fifth base pair in the same arm is variable in all groups except IV, in which a G–C pair is uniformly present. The stem of arm II shows several trends, as does the loop; for instance, GG combinations occur at the beginning of the latter only in group III sequences, but clear-cut correlations are lacking as a whole because of this arm's strong involvement in tertiary structure. Bend 2 likewise shows only trends, albeit these are more distinctly marked. Although A occasionally is the occupant of this site in all groups, it is the only nucleotide that occurs in group IV, while G is universal in group III with the sole exception of the elongator tRNA$_E^{Met}$, in which A is present.

The stem of arm III is unusually stable in the three representatives of group II in that a C–G pair always occupies the first site, T–A the second, and G–C the fourth, with this combination of occupants lacking in the three remaining groups. In group I, the fifth site of each stem is uniformly a C–G base pair, whereas in group IV either an A–T or T–A pair occupies that site. Here in the unmodified bases of genes, the loop displays no marked characteristics, except that the frequent TT on the 5′ side of the anticodon in group I components becomes nearly totally replaced by CT in the higher groups. The nearly uniform presence of A adjacent to the anticodon on the 3′ side in these genes among prokaryotes is lost after processing, for in group III components it becomes hypermodified to N-[9-(β-D-ribofuranosyl)-purine-6yl-carbamoyl]threonine, abbreviated A$_t$ (Dillon, 1978, p. 260), and in group IV 2-methylthio-N^6-($\triangle^6$-isopentenyl)-adenosine, or A$_s$, by processes that receive attention later. Here only the correlation to group is intended. The loop of arm V adds a bit more substance to this discussion, in that its 3′-terminal member is always T, except among group I members, in which C may be found instead. In this same group the preterminal site is consistently a pyrimidine, while purines prevail in the others, except for an occasional T in group IV.

The tRNA Genes of Viruses. The complete gene structures of tRNAs are known from only a few viruses aside from those in Table 2.15. In the case of bacteriophage T4 only eight genes for this class of biochemicals are present (Fukada and Abelson, 1980; Mazzara *et al.*, 1981), the primary sequences for all of which are given. In contrast to this degenerate condition, bacteriophage T5 has been shown to have a more intricate genome, which contains genes for tRNAs specific for all 20 amino acids plus several for isoaccepting types (Hunt *et al.*, 1980). However, relatively few of these have been sequenced, so it is possible to include only seven in the table. Also included are several from *E. coli* for comparative purposes, indicated by suitable subscripts in each case.

Table 2.14

Mature Transfer RNA Gene Structures from Escherichia coli[a]

	Arm IA	Bend 1	Arm II	Bend 2	Arm III	Arm V	Arm IB
Group I							
Gly1[b]	GCGGGCG	TA	GTTC AA--TGG--TA GAAC	G	AGAGC TTCCCAA GCTCT	GAGGG TTCGATT CCCTT	CGCCCGC T
Gly2[c]	GCGGGAG	TC	GTAT AA--TGG-CTA TTAC	C	TCAGC CTTCCAA GCTGA	GCGGG TTCGATT CCCGC	TGCCCGC T
Gly3[d]	GCGGGAA	TA	GCTC AG-TTGG--TA GAGC	A	CGACC TTGCCAA GGTCG	GCGAC TTCGAGT CTCGT	TTCCCGC T
Ala[e]	GGGGGCA	TA	GCTC AG-CTGG--GA GAGC	G	CCTGC TTGGCAC GCAGG	TGCGG TTCGATC CCGCG	CGCTCCC A
Glu[f]	GTCCCCT	TC	GTCT AG--AGGCCCA GGAC	A	CCCCC CTTTCAC GGGGG	AGGGG TTCGAAT CCCCT	GGGGGAC G
Val1[g]	GGGTGAT	TA	GCTC AG-CTGG--GA GAGC	A	CCTCC CTTACAA GGAGG	GCGGG TTCGATC CGTC	ATCACCC A
Val2[h]	GCGTCCG	TA	GCTC AG-TTGG-TTA GAGC	A	CCACC TTGACAT GGTGG	GGTGG TTCGAGC CCACT	CGGAGCC A
Group II							
Leu[i]	GCGAAGG	TG	GCGG AA-TTGGTAGA -CGC	G	CTAGC TTCAGGT GTTAG	GGGGG TTCAAGT CCCCC	CCCTCGC A*
His[j]	(G)GTGGCTA	TA	GCTC AG-TTGG--TA GAGC	C	CTGGA TTGTGAT TCCAG	GTGGG TTCGAAT CCCAT	TAGCCAC C
Arg[k]	GCATCCG	TA	GCTC AG-CTGG--TA GAGT	A	CTCGG CTACGAA CCGAG	GGAGG TTCGAAT CCTCC	CGGATGC A
Group III							
Ile1[l]	AGGCTTG	TA	GCTC AG-GTGG-TTA GAGC	G	CACCC CTGATAA GGGTG	GGTGG TTCAAGT CCACT	CAGGCCT A
Ile2[m]	GGGCTTG	TA	GCTC AG-GTGG-TTA GAGC	G	CACGA CTXATAA TCGTG	GGTGG TTCAAGT CCACT	CAGGCCC A
Ser3[n]	GGTGAGG	TG	GCCG AG--AGGCTGA AGGC	G	CTCCC CTGCTAA GGGAG	CGGGG TTCGAAT CCCCG	CCTCACC G
MetF[o]	CGGGGGG	TG	GAGC AGCCTGG--TA GCTC	G	TCGGG CTCATAA CCCGA	GTCGG TTCAAAT CCGGC	CCCCGCA A
MetM[p]	GGCTACG	TA	GCTC AG-TTGG-TTA GAGC	A	CATCA CTCATAA TGATG	ACAGG TTCGAAT CCCGT	CGTAGCC A

Group IV

Phe[q]	GCCCGGA	TA	GCTC-AG-TCGG--TA GAGC	A	GGGGA TTGAAAA TCCCC	CTTGG TTCGATT CCGAG	TCCGGGC A*
Ser[r1]	GGAAGTG	TC	GCCG AG--CGGTTGA AGGC	A	CCGGT CTTGAAA ACCGG	CAGAG TTCGAAT CTCTG	CGCTTCC G*
Tyr[s]	GGTGGGG	TT	CCCG AG-CGGCA-A AGGG	A	GCAGA CTCTAAA TCTGC	GAAGG TTCGAAT CCTTC	CCCCACC A
Trp[t]	AGGGGCG	TA	GTTC AA-TTGG--TA GAGC	A	CCGGT CTCCAAA ACCGG	GGGAG TTCGAGT CTCTC	CGCCCCT G

[a] Underscoring indicates the anticodons and italics represent internal promoters (boxes A and B), discussed in Chapter 3, Section 3.6.1. Asterisks mark those sequences in which the terminal-CCA is encoded by the gene. The G in parentheses in the histidine gene is added posttranscriptionally.

[b] Hill et al. (1973).
[c] Hill et al. (1970).
[d] Squires and Carbon (1971).
[e] Williams et al. (1974).
[f] Uziel and Weinberg (1975).
[g] Yaniv and Barrell (1969).
[h] Yaniv and Barrell (1971).
[i] Blank and Söll (1971).
[j] Singer and Smith (1972).
[k] Murao et al. (1972).
[l] Yarus and Barrell (1971).
[m] Harada and Nishimura (1974).
[n] Yamada and Ishikura (1973).
[o] Dube and Marcker (1969).
[p] Cory and Marcker (1970).
[q] Barrell and Sanger (1969).
[r] Ishikura et al. (1971).
[s] Küpper et al. (1978).
[t] Young (1979).

Table 2.15
Transfer RNA Gene Structures of Viruses[a]

	Arm IA	Bend 1	Arm II	Bend 2	Arm III	Arm V	Arm IB
Group I							
Gly[b]$_{E.C.}$	GCGGGCA	-TC	GTAT AAT–GGCT–A TTAC	C	TCAGC CTTCCAA GCTGA	GCGGG TTCGATT CCCGC	TGCCCGC T
Gly[c]$_{T4}$	GCGGATA	-TC	GTAT AAT–GG–T–A TTAC	C	TCAGA CTTCCAA TCTGA	GTGAG TTCGATT CTCAT	TATCCGC T*
Asp[d]$_{T5}$	GCGACCG	-GG	GCTG GCTT–GG–TAA TGGT	T	CTCCC CTGTCAC GGGAG	GTGGG TTCAAAT CCCAT	CGGTCGC G
Asp[e]$_{E.C.}$	GGAGCGG	-TA	GTTC AGTC–GG–TTA GAAT	A	CCTGC CTGTCAC GCAGG	GCGGG TTCGAGT CCCGT	CCGTTCC G*
Group II							
Pro[c]$_{T4}$	CTCCGTG	-TA	GCTC AGTTTGG–T–A GAGC	G	CCTGA TTTGGGA TCAGG	CAAGG TTCAAAT CCTTG	TATGGAG A
Pro[d]$_{T5}$	CTCCGAT	-TA	GCTC AATT–CGCT–A GAGT	A	CACCG TTTGGGG CGGTG	GAAGG TTCGAGT CCTTC	ATTGGAG A
Glu[c]$_{T4}$	TGGGAAT	-TA	GCCA AGTT–GG–T–A AGGC	A	TAGCA CTTTCAC TGCTA	AAAGG TTCGAGT CCTTT	ATTCCCA G
Glu[d]$_{T5}$	TGGGGAT	-TA	GCTT AGCTTGGCTCA AAGC	T	TCGGC CTTTTGAA GTCGA	ATTGG TTCAAAT CCATT	ATCCCCT G
His[d]$_{T5}$	(T)GTGGCTA	-TA	TCAT AATT–GGTT–A ATGG	T	CCTGA TTGTGAA TCAGG	GTGGA TTCGAAT TCTAC	TAGCCAC A
Leu[d]$_{T5}$	GGGGCTA	-TG	CTGG AACT–GG–TAG ACAA	T	ACGGC CTTAGAT TCCGT	GGGAG TTCGAGT CTCCC	TAGCCCC A
Leu[f]$_{E.C.}$	GCGAAGG	-TG	GCGG AATT–GGTAGA –CGC	G	CTAGC TTCAGGT GTTAG	GGGGG TTCAAGT CCCCC	CCCTCGC A
Group III							
Ile[c]$_{T4}$	GGCCCTG	-TA	GCTC AAT–GCTTAG CAGC	A	GTCCC CTCATAA GGGAA	ACCAG TTCAAAT CTGGT	CTGGGTC A
Thr[d]$_{T4}$	GCTGATT	-TA	GCTC AGTA–GG–T–A GAGC	A	CCTCA CTTCTAA TGAGG	GGCCG TTCGATT CCGTC	AATCAGC A*
Arg[g]$_{T4}$	GTCCCGC	-TG	GTGT AAT–GGAT–A GCAT	A	CGATC CTTCTAA GTTTG	CCTGG TTCGATC CCAGG	GCGGGAT A
Asn[d]$_{T5}$	GGTTCCT	-TA	GCTC TAAT–GGTT–A GAGC	C	GCACC TTGTTAA GTTGA	GCTGG TTCGAAT CCAGC	AGGGAACC G

Group IV

Ser$_{E.C.}^{h}$	GGAAGTG	-TG	GCCG -AGC--GGTTGA AGGC	A	CCGGT CTTGAAA ACCGG	CAGAG TTCGAAT CTCTG	CGCTTCC G
Ser 1$_{T5}^{i}$	GGAAGGT	-AG	GGCG TAGT--GG--TA CGCA	A	CTAGT CTTGAAA ACTAG	GATGG TTCGACT CCATT	ACCTTCC T*
Ser$_{T4}^{c}$	GGAGGCG	-TG	GCAG -AGT--GGTTTA ATGC	A	CCGGT CTTGAAA ACCGG	ATAGG TTCAAAT CCTAT	CGCCTCC G
Ser 2$_{T5}^{i}$	GGAAAAG	CAA	ATAG -ACT--GGCG-A CTAA	A	CCCGA TTGGAAA TCGGT	ATGGG TTCAACT CCCAT	CTTTTCC G*
Leu$_{T4}^{c}$	GCCGAGAA	-TG	GTCA -AATT-GG-TAA AGGC	A	CAGCA CTTAAAA TGCTG	GTGGG TTCGAGT CCCAC	TTCTCGC A*

[a] Underscoring indicates the anticodons and italics represent internal promoters (boxes A and B), discussed in Chapter 3, Section 3.6.1. An asterisk indicates that the gene encodes the -CCA terminal segment.
[b] Roberts and Carbon (1975).
[c] Fukada and Abelson (1980).
[d] Gauss and Sprinzl (1983a).
[e] Harada et al. (1972).
[f] Dube et al. (1970); Blank and Söll (1971).
[g] Mazzara et al. (1977).
[h] Ishikura et al. (1971); Yamada and Ishikura (1975).
[i] Kryukov et al. (1983).

Examination of the table shows at once that, as a whole, the tRNA genes of the two species of bacteriophage differ in no substantial way from those of the prokaryotes and higher organisms. At the same time it is clear from the differences between corresponding genes of the bacteriophages and the bacterium that there is no evidence of degeneracy in the sequences of the former* despite the evident loss of genome that characterizes phage T4. Although the two glycine sequences show differences at only 13 sites, for a homology rate of 81%, the pair of aspartic acid species are divergent at 30 positions, for a mere 56% homology. Similarly, the two leucine genes, one from T5, the other from *E. coli,* have differing occupants at 35 points (48% homology). Thus, further substance is added against the proposal that viruses represent escaped genes (Luria and Darnell, 1967), as well as against the old point of view that viruses are degenerate bacteria.

Where genes for isoaccepting species are available from both viruses, as is the case at present for just three sets, those from tRNA^{Pro}, tRNA^{Gln}, and tRNA^{Ser}, a number of identities are found to exist, accompanied by a considerable frequency of departure. The two proline tRNA sequences show a homology rate of 69%, with 21 differences, and the glutamine, one of 65%, with 24 contrasting sites. Thus, as shown by the differing morphology of the virions, bacteriophages T4 and T5 are indicated to be rather distantly related, as pointed out elsewhere (Dillon, 1978). This lack of close kinship is especially clearly demonstrated by the serine tRNA gene sequences. In the first place two isoaccepting genes from T5 have had their primary structures established recently (Kryukov *et al.*, 1983). With the interpretation of structure given here, using noncanonical bonding patterns at several points, a less radical departure from the standard is achieved with the two T5 genes than in the original article. Yet, in spite of that conservatism, each of this pair contains a number of unusual features, especially in bend 1 and arm II. Bend 1 in both cases has an unusual set of occupants, but is particularly unique in being comprised of three sites in tRNA^{Ser} 2, not four. The stem of arm II is also divergent in structure, consisting of ATAG on the 5' side and CTAA on the 3' side, combinations apparently found nowhere else in these macromolecules. Since similar, though less stringent, departures from the standard condition mark this stem in the histidine- and leucine-accepting species from this same organism, others from the same source may be expected to show comparable characteristics as progress is made in determining its tRNA genes. In the case of tRNA^{Ser} 1 from T5, it is the loop of this arm that is striking, for its first site is occupied by a T, not the universal A of all other known forms. Apparently this is the result of an addition to the loop, for the standard A and later nucleotides in the loop match those from other organisms. A count of identical sites indicates that 23 differences exist between T5's species 1 and that of *E. coli* (67% homology rate), and 33 between its species 2 and *E. coli* (53%), while that of T4 differs from the latter at 16 points, for a 77% homology rate. A comparison of the two bacteriophage sequences with the same anticodon reveals differences at 28 positions, or a 60% homology rate.

2.3.3. Sequences of Organellar tRNA Genes

The sequences of the tRNA genes from organelles are of high interest, not only for what they indicate, but also for what they do not. As has been the situation among the

---

*Degeneracies do occur in tRNA primary structures and in the genes encoding them, but known instances of this condition are confined to the genome of mitochondria, and then predominantly in higher Metazoa.

eukaryotic and prokaryotic species, strong possibilities for providing signals for transcription initiation are perceptible, without being conclusive, however, as in those of the other sources. Some of the lack of clarity stems from the relatively small number of sequences that are available from organelles, especially from chloroplasts; as a whole in a given class of isoaccepting species only two sequences from chloroplasts of different organisms are available at best, and in the majority of cases only one has been sequenced, if any. Nevertheless, enough is at hand to disclose the major features.

Mitochondrial tRNA Gene Sequences. Table 2.16 provides three sets of tRNA gene sequences from mitochondria and chloroplasts, along with bacterial and eukaryotic cytoplasmic types for comparison. As elsewhere in this series of studies, the variable arm (arm IV) portions are omitted to avoid meaningless distraction; insofar as is feasible, the sequences shown of like types are of species bearing the same anticodon. Subscripts on the sources are used to indicate the mitochondrial (M) and chloroplastic (C) genes, the cytoplasmic ones being unmarked. However, the absence of availability has necessitated inclusion among the glycine isoacceptors of a yeast cytoplasmic representative having different coding properties from the others and in the threonine series one from a chloroplast with a comparable divergent anticodon. Very few tRNA genes from plant mitochondria have been sequenced, but the situation has been improved by the recent establishment of the structures of tRNA^{Phe} and tRNA^{Trp}, both from the bean (Marechal *et al.*, 1985a,b).

The tRNA^{Gly} species of mitochondria are immediately revealing in failing to show an expected feature, close correspondence to the prokaryotic structure. For example, the arm I sequences from all sources differ radically from that of *B. subtilis,* whereas, unexpectedly, the yeast cytoplasmic gene exhibits strong homology to it. Indeed, the same relationships continue through arm II, especially in the loop, and reappear in arm V. Interestingly, the yeast mitochondrial gene is nearly identical with that of *B. subtilis* in arm V, as is the yeast cytoplasmic one, while the three vertebrate cistrons are noteworthy only for their heterogeneity in this region. The only important feature shared by this trio occurs in the loop in the form of one site fewer than elsewhere. This reduction in the bovine mitochondrial tRNA^{Gly} gene is the more remarkable, because in other bovine species that loop usually has an additional base present, as shown in the tRNA^{Thr} sequence.

In the threonine- and histidine-accepting types, again clear-cut relationships are wanting between the prokaryotic and mitochondrial representatives. Although a somewhat higher degree of homology exists between them than in the glycine isoacceptors, the level of identity is no greater than that which exists between the cytoplasmic and prokaryotic or between the former and mitochondrial. Thus no support is provided for the endosymbiont concept of organelle origins; rather, these data seem to substantiate the cellular evolution view of that topic (Dillon, 1962, 1963, 1981). By viewing any of the series provided in Table 2.16, it becomes quite evident that the mitochondrial tRNA genes have undergone extensive evolutionary change; such a degree of change would be unexpected were the mitochondrion and chloroplast actually species of endomutualistic prokaryotes within a series of eukaryotic hosts, as proposed by that theory.

The most prominent of unexpected alterations in these genes is the degeneracy that characterizes the loops of arms II and V in the mammalian examples. The loss of the paired guanosine residues from the central part of the first of these structures is especially remarkable, for they provide one of the universal traits of prokaryotic, viral, and eu-

Table 2.16
Representative tRNA Genes from Organelles[a]

	Arm IA	Bend 1	Arm II	Bend 2	Arm III	Arm V	Arm IB
Glycine							
Bacillus subtilis[b]	GCGGGTG	TA	GTTT AGT-GG-T-A AAAC	C	TCAGC CTTCCAA GCTGA	GTGAG TTCGATT- CTCAT	CACCCGC T
Yeast[c]$_M$	ATAGATA	TA	AGTT AATTGG-T-A AACT	G	GATGT CTTCCAA ACATT	GCGAG TTCGATT- CTCGC	TATCTAT A
Bovine[d]$_M$	ATTCTTT	TA	GTAT TAAC---T-A GTAC	A	GCTGA CTTCCAA TCAGC	TTCGG TCTAGT-- CCGAA	AAAGAT A
Rat[e]$_M$	ACTCCCT	TA	GTAT AAAC------A ATAC	A	ACTGA CTTCCAA TCAGT	TCTGA AAAAAC-- TCAGA	AGAGAGT A
Human[f]$_M$	ACTCTTT	TA	GTAT AAA-----T-A GTAC	C	GTTAA CTTCCAA TTAAC	TTTGA CAACAT-- TCAAA	AAAGACT A
Yeast[g]	GCGCAAG	TG	GTTT AGT-GG-T-A AAAT	C	CAACG TTGCCAT CGTTG	CCCGG TTCGATT- CCGGG	CTTGCGC A
Euglena[h]$_C$	GCGGGTA	TA	GCTC AGTTGG-T-A GAGC	G	TGGTC CTTCCAA GTCCA	GCGTG TTCGAAT- CACGT	TACCCGC T
Threonine							
Escherichia coli[i]	GCGGACT	TA	GCTC ACTAGG-T-A GAGC	A	ACTGA CTTGTAA TCAGT	ACCAG TTCGATT- CCGGT	AGTCGGC A
Yeast[j]$_M$	GTTATAT	TA	GCTT AATTGG-T-A GAGC	A	TTCGT TTTGTAA TCGAA	TGGGG TTCAAAT- CCCTA	ATATAAC A
Aspergillus[k]$_M$	GCCCGGT	TA	GCAT AAAA-G-T-A ATGT	A	TCCGT TTTGTAA TCGGA	ACAAG TGCGATA- CTTGT	ACTGGGC T
Mouse[l]$_M$	GTCTTGA	TA	GTAT AAA-----C-A TTAC	T	CTCGT CTTGTAA ACCTG	GAAGA TCT---TC- TCTTC	TCAAGAC A
Bovine[d]$_M$	GTCTTTG	TA	GTAC ATC----T-A ATAT	A	CTGGT CTTGTAA ACCAG	GGACA ACAACTAA CCTCC	CTAAGAC T
Spinach[m]$_C$	GCCCCTT	TA	ACTC AGT-GG-T-A GAGT	A	ACGCC ATGGTAA GGCGT	ATCGG TTCAAAT- CCGAT	AAGGGGC T
Yeast[n]	GCTTCTA	TG	GCCA AGTTGG-T-A AGGC	G	CCACA CTTGTAA TGTGG	ATCGG TTCAAAT- CCGAT	TGGAAGC A

Histidine

Escherichia coli[o]	GTGGCTA	*TA*	*GCTC AGTTGG-T-A GAGC*	C	CTGGA TTG̲T̲G̲A̲T TCCAG	*GTGGG TTCGAAT- CCCAT*	TAGCCAC C
Yeast[p] M	GTGAATA	*TA*	*TTTC AAT-GG-T-A GAAA*	A	TACCC TTG̲T̲G̲G̲T GCGTT	*CTGAG TTCGATT- CTCAG*	TATTCAC C
Aspergillus[q] M	GTGGGTG	*TA*	*GTTC AAA-GG-T-A GAAC*	A	GCTGT ATG̲T̲G̲G̲C ATAGT	*CCTAG TTCAATT- CTAGG*	TATCCAC C
Rat[r] M	GTAGATA	*TA*	*GTTT ACAA-----A AAAC*	A	TTAGA CTG̲T̲G̲A̲A TCTAA	*AGGAA ATCAAA-- TTCCT*	TATTTAC C
Mouse[z] M	GTGAATA	*TA*	*GTTT ACAA-----A AAAC*	A	TTAGA CTG̲T̲G̲A̲A TCTGA	*AGGAA ATAAAC-- CTCCT*	TATTCAC C
Human[d] M	GTAAATA	*TA*	*GTTT AACC-----A AAAC*	A	TCAGA TTG̲T̲G̲A̲A TCTGA	*AGAGG CTTACGA- CCCCT*	TATTTAC C
Drosophila[s]	GCCGTGA	*TC*	*GTCT AGT-GG-TTA GGAC*	C	CCACG TTG̲T̲G̲G̲C CGTGG	*CCAGG TTCGAAT- CCTGG*	TCACGGC A
Euglena[t] C	GTGGGTG	*TA*	*GCCA AGT-GG-T-A AGGC*	A	AAGGA CTG̲T̲C̲A̲C TCCTT	*GCGGG TTCGATC- CCCGT*	CATTCAC T
Maize[u] C	GCGGATG	*TA*	*GCCA AGT-GGATCA AGGC*	A	GTGGA TTG̲T̲G̲A̲A TCCAC	*GCGGG TTCAATT- CCCGT*	CGTTCGC C

[a]Subscripts M and C indicate mitochondrial and chloroplast genes, respectively. Underscoring indicates the anticodons, and italics represent internal promoters (boxes A and B), discussed in Chapter 3, Section 3.6.1. [b]Gauss and Sprinzl (1983b). [c]Bonitz and Tzagoloff (1980); D. L. Miller et al. (1980). [d]Anderson et al. (1982). [e]Cantatore et al. (1982). [f]Anderson et al. (1981). [g]Yoshida (1973). [h]Shibuya et al. (1982). [i]Berlani et al. (1980). [k]Köchel et al. (1981). [l]Bibb et al. (1981). [m]Kashdan and Dudock (1982). [n]Weissenbach et al. (1977). [o]Singer and Smith (1972). [p]Bos et al. (1979). [q]Netzker et al. (1982). [r]Grosskopf and Feldmann (1981). [s]Altwegg and Kubli (1980). [t]Hollingsworth and Hallick (1982). [u]Schwarz et al. (1981).

karyotic cytoplasmic species, as has already been seen. That the loss of this pair of nucleotides, which characterizes many of the mammalian and other metazoan representatives sequenced thus far, has been carried out evolutionarily gene by gene is indicated by many details of structure, a few examples of which suffice to establish the point. In the first place, not all metazoan mitochondrial tRNA genes have undergone this reduction in structure. The glutamine-, asparagine-, leucine-, and serine-carrying types of the organelle have not lost this doublet either in whole or in part; the asparagine types still carry at least one member of the pair, except in the mitochondrion of the mouse, in which neither is present (Bibb *et al.*, 1981; Grosskopf and Feldmann, 1981; Netzker *et al.*, 1982; Orozco and Hallick, 1982). In contrast, the glutamine-carrying species retain the GG combination in all mammals, including the mouse (Bibb *et al.*, 1981; Anderson *et al.*, 1982; Cantatore *et al.*, 1982); the corresponding gene from the fruitfly mitochondrion, however, is greatly reduced and has only one guanosine residue in the appropriate location (Clary *et al.*, 1982). Throughout the Metazoa, the leucine isoacceptors of this organelle show extensive variability, GG being present intact in the human tRNA$^{Leu}_{TAA}$ (Eperon *et al.*, 1980), but altered to an AG combination in the corresponding gene from the rat (Saccone *et al.*, 1981). Only a single member of the pair is found in the rat mitochondrial tRNA$^{Leu}_{TGA}$ gene (Grosskopf and Feldmann, 1981), as is the case in the mouse (Bibb *et al.*, 1981). In the serine tRNA sequences, the paired Gs are found in human, rat, and bovine mitochondria in those species bearing the anticodon TGA (Anderson *et al.*, 1981, 1982; Cantatore *et al.*, 1982), but is completely lost in those species from mouse and other mammalian mitochondria having the anticodon GCT (De Bruijn *et al.*, 1980; Bibb *et al.*, 1981; Anderson *et al.*, 1982).

A secondary indication of the individual basis for the loss of parts of mitochondrial genes lies in the extent of reduction. In some isoacceptors only a single site has been lost in the loop of arm II (as in the TGA-bearing serine species), whereas all but a few nucleotides are absent in others. For example, in tRNA$^{Ser}_{GCT}$ of mouse mitochondria the loop of arm II is reduced to the sequence TAG, although in another isoaccepting species from the same source and with identical coding properties, this reduction has occurred plus the complete loss of the stem. In other words, arm II here consists solely of the triplet TGT (Bibb *et al.*, 1981). Nearly as extensive reduction has been reported also in tRNA$^{Lys}_{TTT}$ from bovine mitochondria, with only two residues in the arm II loop (Anderson *et al.*, 1982), that of mouse, with but a single residue (Bibb *et al.*, 1981), and that of the rat, which has three residues in the loop, but lacks one pair in the stem (Cantatore *et al.*, 1982). Since separate mutagenic events by ordinary genetic means in each case seem most improbable, an internal control influence, possibly in the form of the supramolecular genetic mechanism (Dillon, 1981, 1983), seems to be involved.

The tRNA Genes of Chloroplasts. Relative to those of mitochondria, few tRNA gene sequences have been established from chloroplasts, and those are almost exclusively from *Euglena gracillis*, supplemented by an occasional one from maize or tobacco, plus one or so from spinach. But since at best less than half of the major types are represented, the total picture remains for future studies. However, those primary structures that have been established from chloroplasts show none of the striking departures from cytoplasmic sources that mark those of the mitochondrion. In the tRNA^{Gly} from *Euglena* (Table 2.16) arm I structure is identical to that of *B. subtilis* except in site 7. The remainder of the gene nevertheless shows a number of differences from the prokaryotic one, but nothing of particular note. It is especially interesting that in arm II, the yeast cytoplasmic gene is identical to the bacterial one except in one site, whereas the chloroplastic gene differs from the latter at three points.

A similar condition prevails between the spinach chloroplast tRNA^{Thr} gene and those of *E. coli* and yeast cytoplasm. Kinship to either of the latter sources is not in evidence on a firm basis, for the organellar sequence differs at 23 sites from that of the prokaryote and at 24 from the eukaryotic one. Since the yeast gene differs at only 27 points from the bacterial, no conclusion can be drawn. However, the two primary structures of tRNA^{His} from chloroplasts depart to a comparable degree from that of *E. coli,* but much more markedly from that of *Drosophila,* showing 37 or 38 contrasting sites to 23 between each of the chloroplast and prokaryote. Since the maize and chloroplast genes diverge from that of the *Euglena* organelle at 15 sites, the evidence again is inconclusive, for obviously the gene from the only eukaryotic cytoplasmic source now available, the highly advanced metazoan *Drosophila,* contributes greatly to the disparities that exist in the given comparisons. When an isoaccepting species from yeast has had its sequence determined, the affinities displayed by the other two types discussed here undoubtedly will be found to be repeated in this instance, too.

2.3.4. Transcription Initiation—A Clarification

Because of the depth of analysis permitted by the extensive information concerning the structure of tRNA genes, the main points derived therefrom may have become obscured by the details. Chief among the conclusions that may be drawn are the following:

1. Short signals and boxes in the leaders of tRNAs are ineffective by themselves, as are the internal controls of the mature coding sectors. Instead, extensive regions beginning with site −10 or −12 in the leader are involved, extending to near site +150 (Pratt *et al.,* 1985; Wilson *et al.,* 1985). In bacteria, regions still farther upstream may be active in promotion; these appear to be sites for ancillary proteins (Lamond and Travers, 1985). A similar situation may also prevail with chloroplast tRNA genes (Ohme *et al.,* 1985).
2. Part of this region in yeast is bound in a complex with a factor called τ (Camier *et al.,* 1985), but it is most probable that in prokaryotes and eukaryotes alike multiple factors are essential to transcription. In the latter group a greater number of species are probably involved.
3. The DNA-dependent RNA polymerase interacts with complexes formed by these ancillary proteins in combination with various regions of the DNA (Stillman *et al.,* 1985); perhaps it also reacts directly with particular points in the gene structure.
4. It is not unlikely that certain of the proteins recognize only genes for particular species of tRNA and others appear to be specific for the group to which each tRNA belongs.

2.4. TERMINATION OF TRANSCRIPTION OF tRNA GENES

One would suspect that termination of transcription should be a far simpler problem than the initiation of that process, for in this case the enzyme is already upon the proper gene, not seeking one particular coding sequence among the myriads that exist on the long, complex DNA molecule. All that would seem requisite is a sequence of sufficient length and character not likely to occur within a functional coding area. It would not be surprising from this preliminary theoretical view if such a terminating sequence would even be found to be nearly universal among genes of all types, regardless of the polymerase involved, because of the enzymes' probably having shared a common ancestry. Since the product of the present class of genes is not subjected to the attachment of the poly(A) tail that marks many others, some differences of detail may prove to exist in the point beyond the signal where termination is consummated, but the signal itself should in theory prove to be evolutionarily stable. These great expectations, like many other predictions of biological behavior, do not prove to be long-lived.

2.4.1. Termination of Eukaryotic tRNA Genes

Although quite a diversity of eukaryotic tRNA genes have had their sequences established through at least the proximal portion of the 3' trains, such data are available in greatest abundance from baker's yeast and *Drosophila*. Hence, as in the preceding sections, the 3' trains of genes from these two sources receive attention first.

The 3' Trains of Yeast tRNA Genes. In all the 3' trains of tRNAs that have been sequenced from yeast sources, a series of from six to eight T–A base pairs occurs (Table 2.17). Similar sequences of four or more such base pairs have been considered to serve as a termination signal, either alone or in combination with regions of dyad symmetry and the like (Rosenberg and Court, 1979; Young, 1979; Niles *et al.*, 1981; Kominami *et al.*, 1982; Henikoff *et al.*, 1983; Stroynowski *et al.*, 1983). Most frequently in this primitive eukaryote, these multiple identical base pairs begin close to the end of the gene proper, in 7 of the 12 shown in the table being separated from that terminus by only three to five sites, while in all three of the tRNA^{Tyr} sequences, the series begins immediately following the mature tRNA portion. However, in the tRNA$_2^{Lys}$ gene eight nucleotide residues precede the multiple Ts, in the first of the three tRNA^{Phe} cistrons, 32 residues separating the series from the actual gene. One of the three tRNA^{Tyr} genes has a second sequence of seven T–A base pairs nearly directly after the first, but there is no substantial reason to consider this a reserve termination signal.

Although diversity in primary structure is as rampant throughout the trains as it was in the 5' leaders, considerable homology among almost all the sequences is revealed when the sections beyond the signal are aligned as in the table. Especially notable is the CATT tetramer shown underscored. The only one listed that does not clearly align with the others is the tRNA^{Lys} gene, which also is mentioned above as being unique in placement of the signal. In contrast, the identity that exists between tRNAs^{Phe} YPT5 and YPT15 in this train for at least 28 sites is a forceful indicator of homology. Similarly, the obvious relationships shown by the three to five bases that intervene in the 5' end of the trains before the signals is suggestive of common descent or a common mechanism of origin. But the drawing of further conclusions is best held until the other eukaryotic and prokaryotic trains have been examined.

Trains of Drosophila tRNA Genes. In *Drosophila* much of the simplicity that marked the trains of yeast tRNA genes has been lost, for even a cursory look at Table 2.18 discloses that many of the sequences lack the series of six or more T–A base pairs found

Table 2.17
3' Trains of Yeast tRNA Genes[a]

Glu$_3$ Y20[b]	TAC*TTTTTT*GACACCATA----<u>CATA</u>CACTGTATGGTG
Glu$_3$ Y5[c]	TTA*TTTTTT*GTTTCTATA---C<u>CATT</u>ATTTTTCTTGA
Lys$_2$[d]	TTCTTTAC*TTTTTT*GCACGTTTTACAAATATGCA
Leu$_3$[e]	ATA*TTTTTT*AGT--TTTA----<u>CATT</u>TTTTTCGGGGA
Leu$_3$ JB2K[f]	ATAA*TTTTTTTTT*C
Phe YPT2[g]	TAATAAGCCATACGGGCTT---<u>CAAT</u>AAAGTTTT*ATTTTTTT*
Phe YPT5[g]	TTAA*TTTTTTTT*ACTTTT----<u>CATT</u>CGTTTTCCTCTTTT
Phe YPT15[g]	TTAA*TTTTTTTT*ACTTTT----<u>CATT</u>CGTTTT
Met YMT5[h]	ATAAA*TTTTTT*GAACTGT--C<u>CATT</u>AATAATA
Tyr YTA[i]	*TTTTTTT*GTTTTTTTATGTCTC<u>CATT</u>CACTTCCCAG
Tyr YTG[i]	*TTTTTTT*AAATCTTGCAG----<u>CTTT</u>CAATATTTAAATAT
Tyr YTC[i]	*TTTTTTT*AC-TTTTGATTA---<u>CCTT</u>-AATTTTTGAAGAT

[a]The farthest upstream runs of Ts are italicized, and important regions
of homology are underscored.
[b]Feldmann *et al.* (1981).
[c]Eigel *et al.* (1981).
[d]Del Rey *et al.* (1982).
[e]Venegas *et al.* (1979).
[f]Carrara *et al.* (1981).
[g]Valenzuela *et al.* (1978).
[h]Venegas *et al.* (1982).
[i]Goodman *et al.* (1977).

in the simpler eukaryote. The fruitfly's sequences are especially favorable for revealing this situation, because a number of multiple genes for the same isoaccepting tRNA have had their primary structures determined. Seven tRNA[Glu] trains can be noted in the table, eight tRNA[Lys], two tRNA[Asn] and tRNA[Leu], four tRNA[Arg], three tRNA[Met] (all of which are virtually identical), the same number of tRNA[Val], and six tRNA[Ile], plus a single one each for tRNA[His] and tRNA[Phe]. Nowhere else among eukaryotic genes is such broad coverage provided with such depth of information on the structure of the trains.

First of all it is evident that in few instances does the T–A series begin so close to the structural gene as was the frequent case in the yeast. Among the exceptions are the sequences of tRNA[Arg] R1 and T12, which begin in the third site beyond the termination of that cistron, and the tRNA[Leu] A, whose signal commences with the fourth site. In addition, the tRNA[Leu] series and that of tRNA[Ile] D starts with the fifth. The remainder are largely beyond the sixth to tenth base pairs. Indeed, all the typical signals in the tRNA[Glu] group initiate at sites 14, 15, or 16, as do those of several of the tRNAs[Lys]; that of the single tRNA[His] begins with site 34, that of tRNA[Ile] with site 41, and that of tRNA[Lys] 1, R2 with site 75. Consequently, the location of the signal is seen to be considerably more variable than in the yeast. In the case of tRNA$_2$[Arg], not shown in the table, the sequence of

Table 2.18
3' Trains of Drosophila tRNA Genes[a]

Glu 1[b]	TATGGTGAAGAAAT*CTTTTTTTTTTT*AGAAATTTTCACTTTAATTATAAG
Glu 2[b]	TATGGTAAAGAAATCTTCTTTTT--AGAAATTTTCACTTTAATTTATTT
Glu 3[b]	-ATGTGGTTGAATTG*TTTTTTTTTC*-GAAATTATTTTAAAAATTCAATG
Glu 4[b]	ACTACTGTGG--AGACA*TTTTTTTC*TCAACTCATCCATACTTGTATTTTT
Glu 5[b]	ACTTATGTGGATTTA*TTTTTTTT*AAGAGCTATAGTGGGATAGAAAATGCA
Glu_1[c]	AGCAACCAACTTTTTGTTGAGGTACTTTACATAATTTATTATTTGTTAC
Glu_2[c]	ATAATATTGTTTTGCTTAC*AAAAGAAACAAAGATCACTTTATATTCCCT*
His[d]	ATGTTGAAACAAACATTGTCACGGAGTTGGGTA*TTTTTTT*ACA
Lys_2 R3[e,f]	CATTATATATTTTATTTAATTTAATAATTTTTGATAGGGCTTGAG
Lys_4[g]	AATTTTC*TTTTT*TATATTTTAGAAAATAGTTTTATTAAAAATGTT
Lys_5 DT39[h]	GCAATCTTTTTGCC*TTTTTT*GGAAGTGTCGATTCGAAAGGCAATA
Lys_5 DT59[h]	TAACATCA*TTTTTTT*CCACTTTTTAAC*TTTTTT*CAGCGCAGACTA
Lys 1, R2[e,f]	AATTATTTTTGTTTGTAATAAAAATTTGGAAAAATGTTTTTATCT[l]*TTTTTTTT*
Lys 3, R4[e,f]	TGTTATATTTTTATATTAGTTAAATCATTTGTTTTGATAATGAAA
Lys 35D[f]	AAACTAATATTTTA*TTTTTT*ATGCG
Asn 1[e]	GAAATACG*TTTTTT*AAGTTTGA*TTTTTT*CCTTTAATTATATTTTC
Asn 2[e]	GCATACGAAATCAAAATTTTTGCCATGCGTTCTCCTTATTTAAAC
Arg R1, T12[e]	AA*TTTTTTT*GGCGTTATTTTAATTTTTATTATCCCCAATATGTGG
Arg 35D[f]	TGAAAAATAA*TTTTTTT*ATTTATTTACGTCGATGACAC
Arg 11F[f]	TATCCGAA*TTTTTTT*ATTTTAAAATACAAATGATACAA

the primary transcript indicates termination to occur at the 3' end of a series of six T–A base pairs that begins just two sites beyond the end of the mature gene (Hovemann *et al.*, 1980b).

Of more interest perhaps are the eight trains that entirely lack series of six or more T–A base pairs and hence have no recognizable termination signal. Since termination obviously must occur, there appear to be three alternatives as solutions to the problem. The first possibility is that only five T–A base pairs may suffice to trigger termination of transcription. In that case two of the three irregular tRNAs[Glu] and the second tRNA[Asn] trains may then be considered to have a suitable signaling device, but the remaining five of these trains have maxima of four such pairs in series. The second possibility, that a single interruption of the series of six or more T–A base pairs by one C–G or G–C combination has no effect, provides only tRNA$_2$[Glu] and tRNA[Lys] 1, R2 with a suitable device, but none of the remaining six. The third suggestion is that either six or more A–T

Table 2.18 (Continued)

Met$^i_{1,2,3}$	TGTGCTATATCA*TTTTTT*GGGAGATTTTTAAAAAATTGTGTATTG
Val$^i_{120R}$	GTTGGAATTTA*TTTTTT*GCTAAATATTTATTTATCATAATGTTCA
Val 14^j	TTGGAAATATTTATTTTAATGCATTTCCCAAAATTATTTTGCCTG
Val 55^j	GGTGATAAAC*TTTTTTTTTT*AGTTTTTATACAATTCGTATTTTAAG
Phe^j	ATAATAATTTTTGCACAAATTAGGCAGAACTCGCATAAAAAAAT
Ile^e	ACCGGTAATTATTTTAGAAAAATATAAGACGCTTTTTCACAC*TTTTTTT*
Ile A^k	GTCGAAC*TTTTTTT*ATAATGCTTTACG*TTTTTT*AGTAGATGCCTAA
Ile B^k	GGTGAA*TTTTTT*GTATCATCCCCCCCACAAGGTCTTTTAAAGATA
Ile C^k	AACATTC*TTTTTT*CTTGTTATATACATATTTACAGCTGAATAATA
Ile D^k	GATA*TTTTTT*ATTTCTGAACATCCTGATTACTTTGAGGTTCAAAG
Ile E^k	GAATAA*TTTTTTTTT*ATTACATTTCTAATGCACTTTGAAAATTAA
Leu A^k	AAA*TTTTTTT*AGTGGTTTTGTTTTTGAATGGTTGATAAA*TTTTT*
Leu B^k	AATA*TTTTTTT*AAGCCCAAAGAAATTAACAATGATTTGACAAACA

^{a}Runs of six or more Ts are italicized.
^{b}Hosbach et al. (1980).
^{c}Indik and Tartof (1982).
^{d}Cooley et al. (1982).
^{e}Hovemann et al. (1980a).
^{f}Yen and Davidson (1980).
^{g}DeFranco et al. (1980).
^{h}DeFranco et al. (1982).
^{i}Sharp et al. (1981b).
^{j}Addison et al. (1982).
^{k}Robinson and Davidson (1981).
^{l}Insert 29 base pairs.

or T–A bases, alone or in any combination of the two variations, can serve to induce termination. This would readily supply all the listed sequences with a functional device, but at the same time it would drastically change the concept of where the signals actually begin in each of these trains. In the tRNA^{His} sequence, for example, the triggering series would begin just two sites nearer the structural gene than now supposed, whereas in others, such as tRNA$^{Lys}_4$, tRNA^{Lys} 1, R2, or tRNA^{Phe}, the signal would commence immediately following the structural sector.

The possible difficulty with any of these proposals is that like series might exist within the functional coding region, especially in arm I. Actually the first suggestion that five T–A base pairs would be a signal could occur in the anticodon stem also, the stem of arm V being eliminated as a possiblity by virtue of the near universality of one or more G–C pairs there. Examination of compilations of known tRNA or tRNA gene sequences (Gauss and Sprinzl 1983a,b) shows that in fact a maximum of four T–A or A–T occurs in the loop of arm III, although several exist with one T–A and four A–T base pairs. While this combination is not in conflict with the first proposal of just five T–A pairings, it does negate the possibility of only four such pairings serving in that capacity. Within the established eukaryotic cytoplasmic tRNA genes and structural sequences almost no data suggest that any other of the proposals might not be valid. In the *Dictyostelium* gene for tRNA^{Trp}, arm II contains the hexamer TTTATT, thus invalidating the third proposal, for were that combination to serve as a signal, transcription would terminate far too soon in this case. The only other equally informative series is AAAAAA in arm V of the mouse gene for tRNA^{Gly}. This clearly indicates that the polymerase must be polarized specifically to recognize six T–A pairs, but not the same number of A–T pairs. Thus it is possible that five or more T–A pairs may serve in termination or series of six or more with a single interruption by a G–C or C–G pair, but mixtures of T–A and A–T pairs clearly do not function in that capacity in the fruitfly.

Trains of Other Metazoan tRNA Genes. The trains of tRNA genes from other metazoans do not differ significantly from those of the fruitfly. As is evident in Table 2.19, relatively few have sequences of six or more Ts in uninterrupted series, including those of fungi and seed plants. Since only five of the 16 provided there are thus equipped, series of five or more Ts are indicated by italics. Even so, one of the trains, from the human tRNA^{Lys} gene, has this combination situated at the unreasonably long distance of 60 base pairs from the actual end of the mature region. Absence of series of T–A base pairs of suitable length extends through the entire train of the tRNA^{Lys} of the mouse; no T–A base-pair series exceeding four is found in 213 sites. In the rat tRNA^{Gly} there is none in 428 and in rat tRNA^{Glu} none in 656. Regions of dyad symmetry also are lacking. Consequently, especially in combination with the similar situation just disclosed in insect genes, it appears doubtful that the actual functional signal for termination has yet been deciphered in the Metazoa in general insofar as the genes for this class of substance is concerned.

The soundness of this observation is substantiated, fortunately, from the 3′ train of a nematode tRNA^{Leu} gene. This, one of the few instances in which the sequences of both the tRNA gene and the primary transcript have been established together, demonstrates that termination is with the second T–A base pair in a series of four such pairs, shown underscored in the table. Thus series of five, six, or more are not essential in metazoan tRNA genes. However, the actual composition of the termination signal cannot be determined, because one or more of the C–G pairs that precedes the four T–A pairs may actually be a part of the signal, perhaps even along with one or both T–A pairs located 5′ to these C–G combinations. Knowledge of the primary structures of genes is highly essential, but that of the products is requisite in addition before the functioning of the genes and polymerases can be comprehended. This must be accompanied by data pertaining to the actual secondary structure of the coding regions of the DNA itself. For example, in the tRNA^{Lys} of mice (Han and Harding, 1983), the anticodon loop includes the sequence CTTTT. If the secondary structure of the DNA included such loops, that sequence would not serve for termination, since base pairs seem to be required, not single-stranded regions. On the other hand, if no such arms and loops exist in the intact DNA, as seems to be the actual condition, then the signal cannot be just this combination. It is because of the existence of such sequences within the mature coding areas that series of

Table 2.19
3' Trains of Higher Eukaryotic tRNA Genes[a]

Rat Gly[b]	GCGTGCCCAACTTTTGCCAAGGTCCTCACCCAGGACGCGCCC (0 in 428 bp)
Rat Asp[b]	ACATGATGTCC*TTTTT*GGACACCTAGGTGGCCCGCGGGCGCTA
Rat Glu[b]	GCCTCTCTTCTTAGTCCCTCCACTTCCCACCTCACCACCCTG (0 in 656 bp)
Human Glu[c]	TGTTTTACATGGCCGCCCTCCCGCAGGAATCTTCCTTCACTA (0 in 850 bp)
Human Lys[d]	GCATGTCTTTGCTTTTGGGTACCGCACTTCGCATAAAATGGT(18 bp)*TTTTT*A
Rat Lys 1[e]	TAGCTATTTTGCTATCGGAGATCTCATAAAGAAAGGAGAAACCT (0 in 64 bp)
Mouse Lys[f]	CTGATTCTCAAACTTTGAAAGTTAAACTTGATAGTAAGGTGG (0 in 213 bp)
Mouse His[g]	TTATCCTCTGGTCAC*TTTTTT*GCTCCACTCTCTCTCTGATGA
Human Gln[d]	TAAAGCTTTTCTTTTAATATCAGCATGTTGAATATTGTTAATTGA (0 in 488 bp)
Human Met 1[h]	AGAAGGGTGC*TTTTTTTTTTTTTTTT*CCCCC-CCCCCTTCTT-GAGGA
Human Met 2[h]	-----GGTCC*TTTTTTTTTTT*CCCCCCCCCGTCTATTTTCCT-GAGGA
Nematode Leu[i]	GTTCCTTTTGTCTTTG<u>TTTT</u>CGAAAAAAATTGAAAACGATTTACAGGAC
Human Leu[d]	GGCTGGAG*TTTTT*CTGGTGGATTACGCACCTGCTACAGTTCT(305 bp)*TTTTTTTTT*
Rat Pro 1[e]	TGGAGGGGGTCCAGTTTCCCAATACTTTTGCTTCCAACA*TTTTTTTT*
Petunia Asn[j]	*TTTTT*GAGTTATCGCTTTTCTGACCTAGCGCGACCCCTGTCCTTC
Dictyostelium Trp[k]	GTCCTTTTCGTCTTAAAAAAAAATAAGATCACAGGTTCGAAGTCAAGG (0 in 103 bp)

[a]Runs of five or more Ts are italicized. bp, base pairs. The underscored series of four Ts in the nematode sequence has been demonstrated to serve as a terminator (see text for discussion).
[b]Sekiya *et al.* (1981).
[c]Goddard *et al.* (1983).
[d]Roy *et al.* (1982).
[e]Sekiya *et al.* (1982).
[f]Han and Harding (1983).
[g]Han and Harding (1982).
[h]Santos and Zasloff (1981).
[i]Tranquilla *et al.* (1982).
[j]Bawnik *et al.* (1983).
[k]Peffley and Sogin (1981).

four T–A base pairs cannot serve alone for termination, as has been proposed (Garber and Gage, 1979; Tranquilla *et al.*, 1982).

2.4.2. Termination of Prokaryotic and Viral tRNA Genes

The devices that signal termination of transcription of tRNA genes in prokaryotes (Holmes *et al.*, 1983) and viruses are not as well established as could be desired, for two reasons. First, many of the cistrons whose primary structures have been determined are inserted between the genes for the major and minor subunits of rRNAs, with which they form a part of a continuous transcriptional unit (Nomura *et al.*, 1977; Young *et al.*, 1979; Loughney *et al.*, 1983). Second, many of the others that have been studied are arranged in clusters, often of six or more (e.g., Campen *et al.*, 1980; Wawrousek and Hansen, 1983; Yamada *et al.*, 1983), and are transcribed in groups, whose precise transcriptive delimitations have not been determined. Hence, the need for studies on the primary transcripts may be seen to be as pressing here as in the eukaryotes.

Termination in Virus tRNA Genes. As pointed out in a previous section, the eight tRNA genes of bacteriophage T4 are arranged in two groups, one containing seven species, the other including only tRNA^{Arg} (Fukada and Abelson, 1980). The latter is cotranscribed with two RNAs of unknown function, while the group of seven is transcribed as a single unit with a leader sequence perhaps 7000 base pairs in length (Black and Gold, 1971). All that is known of the train beyond the last gene of the cluster is the trimer TAT, so the termination signal, if present, remains undisclosed. An intergenic spacer of 500 base pairs follows this cluster before the start of the tRNA^{Arg} sequence, which is separated from a functionally unknown RNA (band D RNA) by the pentamer AATGA (Mazzara *et al.*, 1981).

Although a DNA sequence of the genes of two clustered tRNAs, both accepting serine, is quite complete and downstream includes two small RNA species of unknown function, it does not serve to clarify termination. The train, about 220 base pairs in length, contains two sets of six adjacent A–T base pairs, but three is the maximum length of series of T–A combinations (Kryukov *et al.*, 1983). Here again knowledge of the primary transcript is essential before progress can be made toward understanding the mechanism of termination of transcription.

Termination in Bacterial tRNA Genes. Several clusters of tRNA genes have been studied in *Bacillus subtilis,* the simpler of which is one reported from a sector of the genome devoid of rRNA cistrons (Yamada *et al.*, 1983). In this group are coding sequences for four different species, all oriented in the same direction, containing tRNA$_1^{Lys}$, tRNA^{Glu}, tRNA^{Asp}, and tRNA^{Phe}, in that order. The first two are separated by just nine base pairs, the next by 81, and the last by 30. Since the only recognizable termination signal (six T–A base pairs) is initiated 30 sites beyond the tRNA^{Phe} gene, it has been assumed, but not established, that the entire cluster is transcribed in unison.

A second group, containing six tRNA genes of identical orientation, was found associated with rRNA genes, being situated on a spacer between two ribosomal operons (Wawrousek and Hansen, 1983). Two of the six, those for tRNA^{Asn} and tRNA^{Thr}, were separated only by the sequence CATT; since the second of these was followed by a train 56 residues in length, including a set of six T–A base pairs beginning 40 sites downstream, they were considered to represent one transcriptional unit, and the remaining four, another. The latter included tRNAs carrying glycine, arginine, proline, and alanine, in that order. It is interesting that, whereas the tRNA^{Pro} and tRNA^{Ala} cistrons were separated by the sequence TTTTTTA, an apparent termination signal, the maximum number of adjacent T–A base pairs in the spacer between the last of these genes and that for the next 16 S rRNA was five. One such series began just five sites beyond the last tRNA gene and a second one 128 sites beyond. Thus, according to the actual signals, termination should occur before the last of the tRNAs, not following it as seems to be the actual case.

The situation in *E. coli* does not differ significantly from the foregoing. A tRNA$_1^{Asp}$ gene that follows the 5 S rRNA cistron in the ribosomal RNA operon, known as *rrnF,* contains a sequence of eight T–A base pairs commencing 32 sites beyond its 3′ terminus. Transcription termination for the entire operon was found to occur within this sector (Sekiya *et al.*, 1980) and proved to be dependent on the presence of the rho factor, at least *in vitro*. In another instance a cluster of three tandemly arranged tRNA$_1^{Leu}$ genes were studied (Duester *et al.*, 1981) that were separated by spacers 27 and 34 base pairs in length, respectively. Neither of these spacers nor the 234-base-pair train of the last gene in the series contained a sequence of T–A base pairs longer than five. Still another study on tRNA genes examined sections of the genome specifying the suppressor tRNA$_{suIII}^{Tyr}$ (Sekiya *et al.*, 1976a,b). In some of the plasmids that cistron occurred alone, in another it was associated with a functional tRNA$_1^{Tyr}$ cistron, and in a third it was followed by genes for tRNA^{Gly} and tRNA^{Thr}. Trains of various lengths, up to 23 base pairs, were located, but none included a recognizable terminator either in the form of multiple Ts or regions of dyad symmetry.

Consequently, it appears that six or more T–A base pairs in tandem can serve as a terminator in these bacteria, at least sometimes mediated by the rho factor; frequently, however, no recognizable signal is present, as in higher eukaryotes.

2.4.3. Termination of Organellar tRNA Genes

The tyrosine tRNA gene reported upon in the preceding section may not have had its train sequenced for a sufficient distance, as suggested by some of the studies on organellar DNA organization. Thus, analysis of these genes from mitochondria may throw additional light on the functioning of the cytoplasmic counterparts, yet at the same time they introduce many new problems innate to themselves.

Mitochondrial tRNA Gene Terminators. A few instances of the tRNA genes are quite straightforward as to their terminating signals. For example, a tRNA^{Tyr} sequence located about 200 base pairs beyond the end of a major rRNA coding sector in *Paramecium* mitochondria has a series of eight T–A base pairs beginning 41 sites downstream from the end of the gene proper that might serve in termination (Seilhamer and Cummings, 1981). Another example is a tRNA^{Gly} gene of *Drosophila yakuba* mitochondria that is located close to the coding sequence for cytochrome *c* oxidase subunit III. This one presents a problem, however. A series of six T–A base pairs is separated from the mature gene terminus by a single adenosine residue, and thus could provide for termination of transcription. The difficulty is that this sequence is part of a reading frame that encodes an unidentified product which begins immediately following its predecessor (Clary *et al.*, 1983). Thus the ATTTTTT heptamer would be involved in two distinct functions, termination of the tRNA gene and coding for amino acids of the unidentified sequence.

The same study reports a cistron of tRNA^{Ser}, located on the opposite DNA strand, that overlaps the 3′ terminus of another unidentified reading frame of opposite polarity; this, too, lacks any recognizable terminator. It is pertinent to note that the unidentified gene similarly lacks a transcription terminator, but includes *within the coding region* at least four series of six or more T–A base pairs in tandem.

Where tRNA genes are in clusters within the mitochondrial genome, distinct terminating signals are often lacking. In the *Aspergillus nidulans* organelle, two clusters have been reported, one on each side of the split gene for the major rRNA (Köchel *et al.*, 1981). In the upstream cluster of nine cistrons, only a single apparent terminator sequence can be noted and that is located 122 sites beyond the cistron of tRNA^{Trp} and just preceding that for tRNA^{Ile}. None follow the last gene of the cluster, one that codes for tRNA^{Pro}. Comparably, in the second cluster, which includes 11 of these genes, many of which are narrowly separated from one another—indeed those for tRNA$_1^{Leu}$ and tRNA^{Ala} are continuous—a standard signal for termination occurs only at one point. Six T–A base pairs in tandem occur 39 sites beyond the end of the tRNA^{Gln} cistron, just before that for tRNA$_2^{Met}$ located at the terminus of the cluster, but peculiarly, no such terminator is to be seen following the latter to end transcription of the cluster as might be expected.

In a cluster of seven tRNA genes from the yeast mitochondrion (Bonitz and Tzagoloff, 1980), the intergenic spacers are quite long. By and large the latter consist of highly repeated TA or TAA short sequences, with few longer series of Ts or As interspersed. However, at only several points are six or more T–A base pairs in evidence, the first being between the glycine and aspartic acid tRNAs about 500 sites removed from the end of the first of these. The train of the second gene contains three series of Ts of that length, the first located nearly 100 sites, the second roughly 450 sites, and the third more than 800 sites beyond the end of that gene. The serine and arginine tRNA sequences are subcontiguous, the train of the latter including the last evident terminator, at 230 sites beyond the gene's end. Thus, four of the seven genes in this cluster do not have terminators, although only one pair is obviously transcribed together as a single unit.

Chloroplast tRNA Gene Terminating Signals. As in prokaryotic and eukaryotic genomes, genes for tRNA^{Ile} and tRNA^{Ala} have been reported in the intergenic spacer between the minor and major rRNA cistrons of chloroplasts (Orozco *et al.,* 1980; Koch *et al.,* 1981; Takaiwa and Sugiura, 1982; Williamson and Doolittle, 1983). Since the entire segment is transcribed as a unit, no terminator sequences are to be expected. In contrast to the mitochondrial condition, the genes for this class of nucleic acids are dispersed through much of the genome, so that the clusters that occur consist of fewer coding sequences. Frequently, some of the genes in such clusters are followed by a terminator, although several in each group may be closely arranged and probably are cotranscribed. As an example, in the *Euglena* chloroplast one cluster consists of five tRNA genes, all arranged on the same DNA strand (Hollingsworth and Hallick, 1982). The first in the series, encoding tRNA^{Tyr}, has a train 63 base pairs in length, on which a series of six T–A base pairs is located, ending just three sites before the next gene, that for tRNA^{His}. Since the latter has a train just 14 sites in length separating it from the tRNA^{Met} cistron, which in turn is separated from the following one (for tRNA^{Trp}) by only four base pairs, it is likely that these three are transcribed as a unit. The train of the last of this trio has a series of eight T–A base pairs beginning five sites from its 3′ end, which could serve as a strong terminator.

The final two members of the cluster also appear to be transcribed as a unit, since only six residues separate the tRNA^{Glu} gene from that for tRNA^{Gly}. The apparent terminator for this pair, a sequence of six T–A base pairs, however, is located 106 sites downstream from the second gene. Another cluster in the same organism is comparable, but consists of only four components, three (encoding valine, asparagine, and arginine tRNAs) forming one transcriptional unit, whereas the fourth, for tRNA^{Leu}, is of opposite polarity (Orozco and Hallick, 1982). The triple unit has a set of six T–A base pairs beginning 21 sites beyond the tRNA^{Arg} gene, but the single one lacks any trace of a typical terminator.

A similar absence of terminators marks other genes for chloroplast tRNAs, as in the mitochondrion. For one instance, a region from the tobacco organellar genome that was sequenced was found to contain two tRNA genes, one for a methionine species and, on the opposite strand, one for a valine-acceptor species (Deno *et al.*, 1982). The 56-residue-long train of the first of these contained no tract of T–A base pairs, nor did the 141-site-long train of the second. Moreover, the latter gene was interrupted by an intron of 571 base pairs, including one sequence of six T–A base pairs in tandem and another of seven. Hence, if that primary structural feature is sufficient in itself to trigger termination of transcription, transcriptional activities would be expected to end within the intron and result in nonfunctional products. This identical situation prevails in a $tRNA_{UAA}^{Leu}$ gene of maize chloroplast that is interrupted by a 458-residue-long intron (Steinmetz *et al.*, 1982). Within this sequence is included one series of six T–A base pairs in tandem and another of eight. Finally, a spinach chloroplast gene, which, like the foregoing examples, is transcribed singly, but in this case is devoid of introns, likewise lacks a standard terminator (Kashden and Dudock, 1982). Although the train, which is 170 base pairs in length, contains several sequences of T–A base pairs, none of them exceeds five such combinations in length.

2.5. TRANSCRIPTION SIGNALS IN tRNA GENES—A SUMMARY

The results of this analysis of one class of RNA genes transcribed by RNA polymerase III in eukaryotes are best summarized before the second class (5 S rRNA) is described in the following chapter, in order that the main observations remain clear.

Initiation

1. The ACT-TA box does not serve as a promoter sequence in general for tRNA genes of *Drosophila;* in one group of tRNAs a triple A sequence followed shortly by a triple T possibly serves in that capacity. Other tRNA genes have no recognizable initiating sequence. It seems likely that several different ancillary proteins may be involved in initiation with RNA polymerase III in *Drosophila*.
2. A similar lack of universal, clear signaling sequences also marks the leaders from tRNA genes of other metazoans as well as those of prokaryotes and organelles. No trace of a Pribnow box is observable in tRNA genes of bacteria.
3. Short sequences, such as triple As or Ts or other trimeric combinations, sometimes echoed at different points, may serve as recognition signals in certain organisms.
4. In all cases it appears likely that several ancillary proteins are involved in the recognition of particular tRNA genes and that the polymerase interacts with these rather than with a signal.
5. The invariable sites within the mature tRNA coding region do not appear to provide sequences of sufficient length to serve as a unique signaling device, for an extensive region is requisite.
6. However, when the gene sequences are arranged according to their groups, consistent features in arms IA, II, and III are revealed that could serve in recogni-

tion on a group basis, with the remainder of the arm I sequence providing a species signal.

Termination

1. Termination for yeast tRNA genes may be signaled by sequences of six T–A base pairs, since all those studied to date bear such series at or beyond their 3' ends.
2. In other eukaryotes, prokaryotes, and mitochondria and chloroplasts, the process for transcription termination of tRNA genes remains obscure. Series of T–A base pairs are either absent or too short or are located unseemingly long distances beyond the end of mature genes.
3. Solution of the problem of termination depends on the establishment both of the mature gene and its primary transcript sequences.

3

The 5 S Ribosomal and Other Small RNAs

The second large category of substances transcribed in eukaryotes by DNA-dependent RNA polymerase III, in spite of the vast literature embracing it, is not nearly so satisfactory for discussions of gene structure as are the tRNA genes. Not that there is any paucity of sequencing studies; quite to the contrary, 5 S rRNA primary structures are probably more abundantly established than any other single species of macromolecule. More than 36 have been determined from eubacterial sources, 8 from archaebacterial ones, nearly 125 from eukaryotic cytoplasm, and 7 from eukaryotic organelles (Erdmann *et al.*, 1984). Their lack of favorableness stems from two factors. The first of these, the cotranscription in prokaryotes of 5 S rRNA genes with the minor and major rRNA species, restricts the effectiveness of the comparative approach to transcription promoters nearly entirely to eukaryotes. The second factor is that, despite the numerous sequence studies, the genes of 5 S rRNAs remain relatively poorly explored, for the great majority of research has focused on the structure of the transcription product, not upon the gene itself.

As a result of these limitations, it is possible to explore the leaders of this class of substances for presumptive promoters and ancillary signals less fully than was the case with the tRNA genes. To simplify comparisons of the genetic makeup of this class of macromolecules with the others, all sequences are presented in the tables in genelike terms, although some are actually derived from the transcriptive product, especially in the section concerned with the structure of the mature gene. In addition, other types of RNA genes known to be transcribed by RNA polymerase III are examined at the close of this chapter. The transcripts of all these class III genes, including those for tRNAs, appear to bear a simple 5' triphosphate-nucleotidal cap, in contrast to the complex one given in the introductory chapter as 2,2'-dimethyl,7-methylguanosine that characterizes class II genes transcribed by RNA polymerase II (Hellung-Larsen *et al.*, 1980; Zieve, 1981).

3.1. 5 S GENE LEADER SEQUENCES

Although the number of investigations into 5 S rRNA gene structure is undeniably small, fortunately those that have been conducted have largely concentrated on the chloroplast and two types of organisms, yeasts and *Xenopus*. Consequently, insight into the nature of possible promoters that may exist is provided both to an acceptable depth and breadth within a eukaryotic framework. Wherever appropriate, the leaders from prokaryotic 5 S genes are also provided in the tables to permit exposure of any homologous regions that may exist.

3.1.1. Yeast 5 S Gene Leader Sequences

In yeasts, as shown by the genomes of both *Torulopsis utilis* and *Saccharomyces cerevisiae,* the ribosomal genes are arranged in operons in the sequence 18 S, 5.8 S, 25 S, 5 S, with the last of these somewhat remotely placed from the others and of opposite polarity (Tabata, 1980). Because the 5 S gene is thus necessarily transcribed separately, its leader may be expected to contain possible promoters. Fortunately, the 5′ flanking regions from two species of these organisms have had their primary structures established and are included in Table 3.1 along with that of the 5 S gene of the *rrnD* operon of *E. coli.* The subdivisions M, N, O, and P of that chart correspond approximately to those of the tRNA leaders given in Chapter 2, while six additional sectors are assigned the designations A–F.

The 3′ half of region C coincides with the −35 position where the CAT-TA sequence has been suggested to serve as an ancillary signal, and, indeed, four of those six sites in that region in the *S. cerevisiae* leader correspond to that proposed signal. In contrast, only three of the sites in the *Torulopsis utilis* structure are as expected. On the whole, the two leaders show few homologous sites; while the points of agreement are mostly monomeric, several corresponding dimers can be noted at the beginning and middle of region C and at the very end of P. The only extensive sector of homology displayed by the two yeast leaders is that shown underscored in the table extending from the close of C through O and D, first pointed out by Tabata (1980). This sequence, TRT-ACCT, is both of sufficient length and in a location suitable for a promoter of transcription, but it has not actually been demonstrated to serve in that capacity.

It is of interest to compare the yeast sequences with that of *E. coli,* which does not need to enter into transcriptional initiation processes. Aside from occasional single sites, the *T. utilis* leader shows dimeric homologies with the bacterial one just preceding sector A and before the middle of C, while trimeric ones are situated at the end of B into N and again after the middle of C. Identical sites shared by *S. cerevisiae* and the bacterium are entirely otherwise situated, dimers occurring at the close of sectors B and P and in O; some triplets also can be found, such as that beginning in M and ending in B, and that located at the middle of C.

3.1.2. Advanced Eukaryotic 5 S Gene Leaders

Among advanced eukaryotes the leaders of 5 S rRNA genes of three species have been sequenced, two from *Xenopus borealis* and one each from *X. laevis* and *Drosophila melanogaster* (Table 3.1). Although an insufficient breadth of data is available to permit generalizations to be drawn, there is enough to provide for some insight into the possible existence of promoters.

Table 3.1
The Leaders of Eukaryotic 5 S rRNA Genes[a]

	A	M	B	N	C	O	D	E	F	P
Escherichia coli rrnB[b]	GCGGATGAG	AGAAGATTT	TCAGCC	TGATAC	AGA	TTAAATCAGAAC	GCA	GAA	GCG	GTCTGATAA	AACAGAATT
Torulopsis utilis[c]	CTTAGTAAG	CTGTTAGTA	AGCGGT	GTAAGC	AG-	TGTAAGCAG-TG	TAA	CCT	CCC	CCGGCGGCT	TCCCATTTC
Saccharomyces cerevisiae[d]	TCACTCCCA	CCTACTGAA	CATGTC	TGG-AC	C-C	TGCCCTCA--TA	TCA	CCT	CGC	--TTTCCGT	TAAACTATC
X. borealis 1[e]	TTTGCAAGG	TAAAGGTTT	TGCAC-	TTTTTT	CGT	CAAAGTCTTCAT	AGA	AGC	GTC	AAAAGTCTT	CACTCTGAT
X. borealis 3[e]	-TTTCCAGG	TCAAAATTT	TGCAGG	TTTTTT	CTT	CAAAGTCTTCAT	AGA	AGC	GTC	AAAAGTCAG	CAAACCTA-
X. laevis[f]	GTTTTCA-T	TTTCATTTT	TCCACA	GTGCCG	CTG	ACAAGTC---A-	AGA	AG-	CCG	AAAAGTCAG	CCTTGTGCC
Drosophila melanogaster[g]	-CAGTCTAT	TTCAGTCTA	TGGGCA	TAACTG	AAT	ATCAGAGTATAA	GGA	CA-	CTG	TTTAGCCCC	TCGACTTTC

[a]Underscoring marks regions of extensive homology. Additional sectors of homology among metazoans are italicized.
[b]Singh and Apirion (1982); Szeberényi and Apirion (1983).
[c]Tabata (1980).
[d]Maxam et al. (1977); Valenzuela et al. (1977a,b).
[e]Korn and Brown (1978).
[f]Fedoroff and Brown (1977, 1978); Miller et al (1978).
[g]Pirotta, cited in Korn and Brown (1978).

Xenopus 5 S rRNA Gene Leaders. The three leaders of 5 S rRNA genes from members of the genus *Xenopus* form the most solid basis for a comparative analytical search for homologous regions. In Table 3.1 only two of the three established sequences from *X. borealis* are provided, since those of genes 1 and 2 are nearly identical. As in all eukaryotic genes of this type that have been examined, those of the several species of *Xenopus* are arranged in clusters of tandem repeats (Brown *et al.*, 1971; Miller, 1983), three such clusters being present in *X. laevis* (Brownlee *et al.*, 1974; Fedoroff and Brown, 1978; Miller *et al.*, 1978). All the genes in a given cluster have been claimed to possess the same specific spacer sequence (Brownlee *et al.*, 1974; Brown *et al.*, 1977; Miller *et al.*, 1978; Peterson *et al.*, 1980), but determination of the actual primary structures, if the three members of a single cluster of *X. borealis* are representative (Korn and Brown, 1978), indicate the spacers to be similar among components of each cluster but not identical (Table 3.1). In addition, it has recently been demonstrated that many 5 S genes, sometimes referred to as "orphons" (Childs *et al.*, 1981), are scattered individually throughout the genome (Rosenthal and Doering, 1983).

Although the two leaders from a single cluster of genes from *X. borealis* differ at 15 of the 72 sites given in the table, there are several long stretches of homologous sites. One striking example of this condition is the AAARTTTTGCA of sectors A and M; another is the TTTTTTC-T of columns B and N, continuing thence through sectors C, O, D, and E and including the greater portion of F. As a consequence of this extensive homology, by themselves this pair of 5′ flanking regions provide few hints as to possible promoter regions. In combination with the corresponding sequence from *X. laevis,* however, which is little more than 50% homologous with that of *X. borealis* gene I, more concrete proposals can be made, some of which have been pointed out in whole or part by Korn and Brown (1978). Four larger areas of homology can be noted, shown in italics in the table. The first one, TTTT-CA, occurs in zones A and M; it may not be entirely fortuitous that the same hexanucleotide occupies approximately the same region in the *E. coli* leader. The second, AAGTC, if extended to read AARTC, is similarly found in identical location in the bacterial sequence, whereas the third, CA-AGAAG, is absent in the latter. Likewise, the final one, shown partly underscored, AAAAGTC, is unique to the eukaryotes. These four sectors are thus obvious candidates for service in an initiative signaling capacity, but none has yet been demonstrated actually to do so. Most appealing for such service is the third one, because its location from the end of C into D coincides closely with that suggested as a promoter in the yeasts.

Comparison with a Drosophila Leader. The reason for showing a portion of the fourth sector underscored is because a similar sequence is found in the leader of a 5 S rRNA from *Drosophila melanogaster* (Pirrotta, see Korn and Brown, 1978). This quadruplet, AGYC, is the only sector of any length homologous to both toad sequences. Perhaps the association with it of triplet Ts in place of the three As of *Xenopus* is of some significance in possible promoter function, but much more needs to be learned about activation of transcription by polymerase III to establish the validity of this suggestion. Although a number of singlet sites of identity also exist, only two doublets, shown in italics in sectors C and O, can be found. It is perhaps of some pertinence that they occur in regions that have been shown to be highly conserved evolutionarily in the three amphibian sequences.

3.2. POSSIBLE INTERNAL PROMOTERS

In contrast to the many differing interpretations of the secondary structure of the transcripts that have been advocated (Fox and Woese, 1975; Luoma and Marshall, 1978a,b; Hori and Osawa, 1979; Delihas and Andersen, 1982; Kjems et al., 1985), most investigators are in agreement as to extensive regions of the primary sequences being highly conserved evolutionarily. Indeed many series of nucleotides in these structures show high levels of homology throughout the living world, including prokaryotic sources as well as eukaryotic and organellar ones. Although many such regions doubtlessly represent areas of functional or structural significance, it is also possible that some could serve for recognition signals for use by the RNA polymerase or its ancillary proteins. Since the 5 S rRNAs from prokaryotic sources are the terminal part of a long, uninterrupted transcriptional unit, promoters are unnecessary for their synthesis. Therefore, any sequence consistencies that may be noted should be for functional or secondary or tertiary structural purposes, a conclusion that should be equally applicable to the structurally similar eukaryotic gene and its product.

3.2.1. Prokaryotic Mature 5 S rRNA Genes

The two tables of prokaryotic mature 5 S rRNA genes are not designed to suggest secondary structural characteristics as is frequently the case with such compilations (e.g., Erdmann et al., 1983). Here the sole purpose is to seek regions of complete or virtual constancy that might serve in structural or functional capacities. To aid in correlating the sequences with secondary structural features, Figure 3.1 shows the location of the various subdivisions used in Tables 3.2 and 3.3 on the E. coli sequence. Since the precise positions of loops and double-stranded stems varies slightly from one organism to another, care must be exercised in proposing structural functions to invariable sites, for the exact limits of such features are subject to change as mutations arise in one member of a given pair or the other.

The 5′ Portions. In Table 3.2 the 5′ portions of 18 prokaryote mature 5 S rRNA genes and one lower eukaryote (*Torulopsis*) are arranged in order of what appears to be increasing complexity (compare Vandenberghe et al., 1985). Examination of the chart, however, discloses few extensive regions of constancy that might serve in specific functions. The first three sites in column I are almost always GYG, which triplet is usually preceded by a G, although four exceptions can be noted. Column J affords some possibilities, ATAGCG being a very frequent combination there in spite of the variation that occurs. Although column K represents one side of a double-stranded area, which in theory should therefore be evolutionarily stable, it is highly variable. In contrast, region L consists of perhaps the most stable trimer in the entire molecule, even though it represents a single-stranded sector. Its CRC is 100% consistent throughout the prokaryotic sequences in the table and even in the yeast molecule. Moreover, the purine is nearly always A, only two types being exceptional. The next sector of high constancy is the CCC that begins column N, a trimer that extends from the blue-green alga *Anacystis* through most of the eubacteria that are represented, with *Prochloron* being slightly deviant. While the *Streptomyces* species is quite atypical, the same trimer or its kin, CTC, then continues into the

Table 3.2
Mature 5 S rRNA Genes of Prokaryotes—5' Portions[a]

	H	I	J	K	L	M	N	O	P	Q
Anacystis[b]	----TCCTG	GTGTCT	ATGGCG	GTATGGAAG-	CAC	TCTGAC	CCCATC	CCGAAC	TCAGTT-GTG	AAACATACCTGC
Prochloron[c]	----TTCCTG	GTGTCT	CTAGCG	CTATGGAAC-	CAC	TTCGAT	TCCATC	CCGAAC	TCGATT-GTG	AAACTTTGCTGC
Clostridium[d]	----TCC-A	GTGTCT	ATGACT		CAC	-TCCTT	CCCATT	CCGAAC	AGGCAG-GTT	AAGCTCTTAATGT
Spiroplasma BC3[e]	----TCT-G	GTGGCG	ATGGCA	TAGAGGTAA-	CAC	-CCGTT	CCCATC	CCGAAC	ACCGCC-GTT	AAGCACTATTAC
Bacillus brevis[f]	----TCT-G	GTGATG	ATGGCG	GAGGGGACA-	CAC	-CCGTT	CCCATA	CCGAAC	ACCGGC-GTT	AAGCCCTCCAGC
Bacillus megaterium[f]	----TCT-G	CTGGCG	ATAGCG	AAGAGGTCA-	CAC	-CCGTT	CCCATA	CCGAAC	ACCGAA-GTT	AAGCTCTTTAGC
Bacillus subtilis[g]	----TTT-G	GTGGCG	ATAGCG	AAGAGGTCA-	CAC	-CCGTT	CCCATA	CCGAAC	ACCGAA-GTT	AAGCTCTTCAGC
Paracoccus[c]	----GTCT-G	GTGGCC	AAAGCA	CCAGCAAAA-	CAC	-CCGAT	CCCATC	CCGAAC	TCGGCC-GTT	AAGTGCCGTAGC
Micrococcus[h]	---GTTAC-G	GCGGCT	ATAGCG	TCGGGGAAA-	CGC	-CCGGC	CGTATA	TCGAAC	CCGAA-GCT	AAGCCCCATAGC
Streptomyces[i]	---GTTTC-G	GTGGTC	ATAGCG	TGAGGGGAAA	CGC	-CCGGT	TACATT	CCGAAC	CCGGAA-GCT	AAGCCTTACAGC
Escherichia coli[j]	---TGCCT-G	GCGGCC	GTAGCG	CGGTGGTCC	CAC	-CTGAC	CCCATG	CCGAAC	TCAGAA-GTG	AAACGCCGTAGC
Proteus vulgaris[k]	---TGTCT-G	GCGGCC	ATAGCG	CAGTGTCC	CAC	-CTGAT	CCCATG	CCGAAC	TCAGAA-GTG	AAACATTGTAGC
Photobacterium[l]	---TGCTT-G	GCGACC	ATAGCG	TTATGGACC	CAC	-CTGAT	CCCTTG	CCGAAC	TCAGTA-GTG	AAACATAATAGC
Rhodospirillum[m]	---TGGCT-G	GTGGTC	ATTGCG	GGCTCGAAA-	CAC	-CCGAT	CCCATC	CCGAAC	TCGGCC-GTG	AAAGAGCCCTGC

Thermus thermophilus[n]	AATCCCC-C	GTGCCC	ATAGCG	GCCTGGAAC-	*CAC*	*-CCGTT*	*CCCAT*	CCGAAC	ACGGAA-*GTG*	AAGGACGCC**AGC**
Methanobrevibacter[o]	TAGGTT-G	GCGGTC	ATAGCG	ATCGGGTAT-	*CAC*	*-CTGGT*	*CTCGTT*	TCGATC	CCAGAA-*GTT*	*AAGTCT*TTTCGC
Thermoplasma[p]	--GGCAA--	-CGGTC	ATAGCA	GCAGGGAAA-	*CAC*	*-CAGAT*	*CCCATT*	CCGAAC	TCGACG-*GTT*	*AAGCCT*GCT-GC
Sulfolobus[q]	--GCCCA-C	CCGGTC	ACAGTG	GCGGGGTAT-	*CAC*	*-CCGGA*	*CTCATT*	TCGAAC	CCGGAA-*GTT*	*AAGCCC*TC-AC
Torulopsis utilis[r]	--GGTT---	GCGGCC	ATATCT	AGCAGAAAG-	*CAG*	*-CGTTT*	*CTCCGT*	CCGATC	AACTGT-*GTT*	*AAGCTT*CTAAGA

[a]The italicized sequences in L–N and in P–Q correspond to various A boxes discussed in Section 3.6.1. The underscoring in Q represents the beginning of box C, which is continued in Table 3.3.

[b]Corry et al. (1974).
[c]MacKay et al. (1982).
[d]Pribula et al. (1976).
[e]Walker et al. (1982).
[f]Woese et al. (1976).
[g]Marotta et al. (1976).
[h]Hori et al. (1980).
[i]Simoncsits (1980).
[j]Brownlee et al. (1968).
[k]Fischel and Ebel (1975).
[l]Woese et al. (1975).
[m]Newhouse et al. (1982).
[n]Kumagai et al. (1981).
[o]Fox et al. (1982).
[p]Luehrsen et al. (1981).
[q]Stahl et al. (1981).
[r]Nishikawa and Takemura (1974); Tabata (1980).

eukaryotic sequence. The recently established sequence from *Halobacterium volcanii* would be placed between *Thermus* and *Methanobrevibacter* in the structure of this segment (Daniels *et al.*, 1985).

Column O undoubtedly contains the longest sequence of uniformity in the entire molecule, a hexamer. Like the last trimer just discussed, the sequence CCGAAG is found in all the blue-green algae, the lower bacteria, the eubacteria, and even in a few archaebacteria, the TCGAAC of *Micrococcus* being the only variant. Moreover, those modifications that do exist among the higher bacteria and the lower eukaryotes are only slight ones. In columns P and Q are other trimeric combinations that because of their constancy may represent regions of functional significance, but on a more limited basis, the AAR in Q being especially appealing in this regard.

The 3' Portions. The distance from the transcription initiation site in the 3' portions of prokaryotic mature 5 S rRNA genes given in Table 3.3 makes it most unlikely that regions of constancy here might relate to possible promoters at corresponding sites in their eukaryotic counterparts. Consequently, any region of reasonable invariability should be suspected of playing a role either in structure or a function other than transcription. One combination that is especially striking in its constancy is provided by the last five nucleotide residues of column W plus the first four of region X, the sequence ARARTAGG. The only deviations from the nonamer among the lower representatives are found in the blue-green algae, in which a C replaces the final G, but the sequence breaks down entirely in most of the archaebacteria and the eukaryote. Another region of strong homology is the heptamer YGCYRRR found in the downstream region of column Y and the first three sites of Z, its final trimeric portion usually being AGG. The more advanced types deviate from this pattern, as might be expected, but the only exception elsewhere is the substitution in *Micrococcus* of a T for the second from last purine.

Evolutionary Features. The details of structure shown in both Tables 3.2 and 3.3 clearly support the phylogenetic conclusions drawn from comparisons of other macromolecular sequences from structural and functional aspects of the cell. To make the relationships among the various genes and species evident, stippling has been applied to the tables, with like densities indicating corresponding regions in the respective columns. To avoid confusion and to bring out major tracts of homology, not all series of identical sites are treated in this fashion, just those that appear to contribute to a clear picture of possible evolutionary kinship.

One major comparative feature of the macromolecule is its increasing length at the 5' end with evolutionary development. The lowest prokaryotes obviously lack four nucleotide residues found in the higher forms, as shown by that number of dashes at the beginnings of those primary structures. With increasing advancement, *Paracoccus* is seen to lack three sites there, and the remainder only two or none. Second, the sequences from the two genera of blue-green algae, *Anacystis* and *Prochloron*, may be noted immediately to rank as predecessors of such primitive bacteria as *Clostridium*, not as highly evolved forms derived from advanced bacteria, as so often is considered on an *a priori* basis (e.g., Hori and Osawa, 1979). The sharing of bases lacking in all the other sequences, such as the unique thymidines in the second to last site of column H and the first site of column M, is particularly convincing evidence, especially when taken along with the relatively short length of the genes and the high level of homology that exists between them and the primitive bacteria. Their kinship to *Clostridium* is clearly indicated by the presence of

Table 3.3

Mature 5 S rRNA Genes of Prokaryotes—3′ Portions[a]

	R	S	T	U	V	W	X	Y	Z
Anacystis	GGC--A	AC--GAT-A	GCTCCC-GG	GTAG	CCGGTCGC	TAAAAT	-AGCT	CGACGCC	AGGTC---
Prochloron	GGC--T	AA--GAT-A	CTTGGT-GG	GTTG	CTGGCTGG	GAAAAT	-AGCT	CGATGCC	AGGATT--
Clostridium	GCT--G	AT--GGT-A	CTGCAG-GG	GAAG	CCCTGTGG	AAGAGT	-AGGT	CGACGCT	GGGT----
Spiroplasma BC3	GCC--G	AC--GAT-A	G------CC	GCAA	GG---T--	--GAAT	-AGGG	CGATGCC	AGGT----
Bacillus brevis	GCC--A	AT--GGT-A	CTTGCT-CC	GCAG	GGAGCCGG	GAGAGT	-AGGA	CGTTGCC	AGGC----
Bacillus megaterium	GCC--G	AT--GGT-A	GTTGGG-AC	TTT-	GTCCCTGT	GAGAGT	-AGGA	CGTTGCC	AGGC----
Bacillus subtilis	GCC--G	AT--GGT-A	GTCGGG-GG	TTT-	CCCCCTGT	GAGAGT	-AGGA	CCCCGCC	AAGC----
Paracoccus	GCC--A	AT--GGT-A	CTGCG--TC	AAAA	GA-CGTGG	GAGAGT	-AGGT	CACCACC	AGACC---
Micrococcus	GCC--G	AT--GGTTA	CTGTAA-CC	GGGA	GGTTGTGG	GAGAGT	-AGGT	CGCGGCC	GTCGA---
Streptomyces	GCC--G	AT--GGT-A	CTGCAG-GG	GGGA	CCCTCTGG	GAGAGT	-AGGA	CGCGGCC	GAA-CT--
Escherichia coli	GCC--G	AT--GGT-A	GTGTGG-GG	TCT-	CCCCATGC	GAGAGT	-AGGG	AACTACT	AGGCAT--
Proteus vulgaris	GCC--G	AT--GAT-G	GTGTGG-GG	TCT-	CCCCATGT	GAGAGT	-AGGG	AACTACT	AGGCAT--
Photobacterium	GCC--G	AT--GGT-A	GTGTGG-GG	TCT-	CCCCATGT	GAGAGT	-AGGA	CATCATC	AGGCAT--
Rhodospirillum	GCC--A	AT--GGT-A	CTGCG--TC	TTAA	CC-GCTGG	GAGAGT	-AGGT	CGGTGCG	AGGCCT--
Thermus thermophilus	GCC--G	AT--GGT-A	CTGGC--GG	TCCG	GTCTATGG	GAATTT	-AGGT	AGCTGCG	GGGA----
Methanobrevibacter	GTTTTG	TTTGTGT-A	CTATGG-GT	GGGA	GGGTACGG	GAAGCG	CATTT	TGCTGTT	AGCTTTT-
Thermoplasma	GTATTG	CGT-TGT-A	CTGTATGCC	GGGA	GGATCCGC	-AGCCC	CAATA	AGCTGGG	ACCACT---
Sulfolobus	GTTAGT	GG--GGC-C	GTGGATACC	GTGA	GGATCCGC	GAAACT	CACTA	AGCTGGG	ATGGGTTTT
Torulopsis utilis	GGCT--G	ATCCAGT-A	GTGTAGTAG	GTGA	CCATACGC	GAAACT	CAGGT	-GCTGCA	AT--CT---

[a]References are given in Table 3.1; the underscored sequences in R–T represent the remainder of box C from Q of Table 3.2, as discussed in Section 3.6.1.

identical sectors such as those found in regions H, I, J, O, V, and Y (Tables 3.2 and 3.3). This conclusion, derived as it is from analysis of only a single molecular structure, would be of little moment by itself. However, it actually is still another piece of evidence confirming the same conclusions reached from extensive studies on comparative cell morphology, DNA structure, tricarboxylic acid cycle characteristics, and cell division traits detailed on other occasions (Dillon, 1962, 1981, 1983).

In the third place, the sequences make it obvious that a number of levels of phylogenetic development exist among the bacteria. While the increasing length at the 5' end already pointed out provides the clearest evidence, this is supported throughout by the stippled areas of homology shown in the two tables. At the lowest level above the blue-green algae, a branch includes *Clostridium, Spiroplasma* (and other myxoplasmids not shown), and *Bacillus*. Because later discussions necessarily must deal with these same relationships, convenience and brevity both are best served by provision of names for the several categories indicated by these tables. Accordingly this division is here named Protobacteria, the term Urbacteria being reserved for *Beggiatoa* and others already demonstrated to be even more primitive than the blue-green algae (Dillon, 1962, 1963, 1981). A branch to which the commonly used grouping Eubacteria is most logically confined is represented by the genera from *Micrococcus* through *Streptomyces;* since *Paracoccus* is intermediate between that level and the Protobacteria, it is the sole representative of a branch here called the Mesobacteria. Above the eubacterial division is a cluster of genera representing the most advanced bacteria, to which the inappropriate name Archaebacteria has been applied (Woese and Fox, 1977; Woese *et al.,* 1978). When other members of this taxon have had their 5 S rRNA genes sequenced and full data providing additional macromolecular primary structures have accumulated, it may be found advantageous to divide the group into two major components to reflect relationships more clearly.

When the possible origins of eukaryotes are sought by comparing the sequence of the primitive representative shown in the tables, *Torulopsis,* with those of the various bacteria and blue-green algae, examination of the evidence quickly indicates that numerous characteristics are shared by the yeast sequence and the four genera of Archaebacteria, the extra base (cytosine) in region X being especially significant. The most frequent homologies are quite clearly between the yeast and *Sulfolobus,* but many exist also with the other archaebacterial types, particularly *Methanobrevibacter smithi.* Further comparisons indicate in addition that very few important changes have been made to the 5 S rRNA gene among the eukaryotes, at least in this elementary type. The only prominent alteration is the insertion of a nucleotide residue (adenosine) in the seventh site of sector P.

This clear kinship between the lowest eukaryotes on the one hand and the highly advanced bacteria on the other runs counter to one of the basic assumptions on which the endosymbiontic concept is founded (Sagan, 1967; Margulis, 1970). The first tenet of that theory is that the eukaryotic cell arose in the primitive seas alongside the prokaryotic. Consequently, gene structures of macromolecules common to both types of cell organization should be expected to reflect kinships between primitive eukaryotes and the earliest of prokaryotes, not the most advanced of the latter. Similar affinities between these same two groupings have already been demonstrated in earlier discussions of the DNA-dependent RNA polymerase and appear repeatedly in sections that follow. It thus becomes evident that such chains of biochemical events as the electron-transport system, the complex tricarboxylic acid cycle, and others shared by higher prokaryotes and eukaryotes in general have been derived by means of ordinary hereditary processes and not by way of an invading endosymbiont.

3.2.2. Eukaryotic Mature 5 S rRNA Genes

Now that a few evolutionarily conserved areas have been located in the prokaryotic mature 5 S rRNA genes, it is of interest to see whether any of these are continued into the eukaryotic line of ascent, where they might logically be expected to serve as signals for transcriptive purposes. As in the case of prokaryotic sequences, the primary structures of the eukaryotic mature 5 S rRNA genes as given in the tables are not designed to reflect secondary structural features, but only to reveal areas of relative constancy that may be suspected of serving as internal promoters.

The 5' Portions. Table 3.4 gives the mature 5 S rRNA gene sequences from 19 representatives of eukaryotes plus one of the Archaebacteria for ease of comparison. As before, important regions of homologous sectors, usually confined to triplets or longer series, are stippled in various degrees of intensity. The 5'-terminal portion, region H, is observed to be quite variable in composition but rather uniform in length. The only deviations in the latter parameter are those of *Euglena,* which in some mutants of the gene have an extra guanosine, and the two green algae, which have a doublet (AT) appended. A second gene of *Tetrahymena* (Pederson *et al.,* 1984) shows one of its two deviations here, having a G in place of the A. One from *Trypanosoma* (not included) diverges strongly in region H, consisting of GGGT (Lenardo *et al.,* 1985), and three from *S. cerevisiae* are identical here to that of *Torulopsis* (Piper *et al.,* 1984).

The whole of columns I and J in combination shows the type of broad consistency one would expect of regions serving as internal promoters. With very few exceptions the sequence reads RCGRYCAYA-CY. Pyrimidines occur in the first and fourth sites in only two instances each and a T replaces the C of the second position on but a single occasion. The pyrimidine of the fifth site is C in 75% of the eukaryotes and that of the final one is the same base in about 60% of the representatives. The sequences in column L, where in prokaryotes the nearly invariable CAC is found, show the same relatively constant feature among the eukaryotes. However, in the present instance there are a number of deviations, especially among advanced protozoans and metazoans. Hence, while it may hold functional value in the prokaryotes and lower forms, that activity appears either lost or modified in higher eukaryotes.

The only other region of uniform structure is found in the later portion of column P and the beginning of Q, where the sequence GAAGTTAAGC occurs. The initial base is replaced by an A in one sequence and by a C in another, and the first A is altered to a T or C in two examples. While the second A is invariable, the C that follows it is only virtually so, for in the *Drosophila* molecule it becomes an A. The double Ts are likewise nearly invariant, and the second pair of As are totally uniform. The closing GC, however, is subject to somewhat greater variation.

The 3' Portions. Although, because of its distance from the initiation site of transcription, the 3' portion of the mature 5 S rRNA gene is unlikely to play a role in promotion, several important areas of constant structure exist within it. Some of these correspond to those found among the prokaryotes, such as the invariant guanosine in the first site of region R (Table 3.5). The remainder of that sector, however, is extremely

Table 3.4
Eukaryotic Mature 5S RNA Genes—5' Portions[a]

	H	I	J	K	L	M	N	O	P	Q
Methanobrevibacter[b]	TAGGTTT-G	GCGGTC	ATAGCG	ATGGGGTAT-	CAC	CTGGT	CTCGTT	-TCGATC	CCA-GAAGTT	AAGTCTTTTCGC
Torulopsis[b]	--GGTT---	GCGGCC	ATATCT	AGCAGAAAG-	CAC	CGTTT	CTCCGT	-CCGATC	AACTGTAGTT	AAGCTGCTAAGA
Euglena[c]	-GGCGT---	ACGGCC	ATA-CT	ACCGGGAATA	CAC	-CTGA	ACCCGT	-TCGATC	AAC-GAAGTT	AAGCCTCGTCAG
Crypthecodinium[d]	--GCTG---	ACGGCC	ATA-CC	GTGTCGAATG	CAC	-CGGA	TCTTTC	-TGACCT	CCG-GAAGTT	AAGCGGCACAGG
Acanthamoeba[e]	--GGAT---	ACGGCC	ATA-CT	GCGCAGAAAG	CAC	-CGCT	TCCCAT	-CCGAAC	AGC-GAAGTT	AAGCTGCGCCAG
Tetrahymena[f]	--GCTA---	TCGGCC	ATA-CT	AAGGTGAAAA	CAC	-CGGA	TCCCAT	-TCGAAC	TCC-GAAGTT	AAGCGCCTTAAG
Paramecium[g]	--GTTG---	GTGGCC	ATA-CT	AAGCCTAAAG	GAC	-CGGA	TCCCAT	-TCGAAC	TCC-GAAGTT	AAGCTGCGCCAG
Saprolegnia[h]	--GGAA---	TCGGCC	ATA-TT	AGCCTGACTA	CAC	-TGCA	TCCCGT	-CCGCTC	TGC-AAAGTT	AACCAGGCTCAA
Physarum[i]	--GGAT---	GCGGCC	ATA-CT	AAGGAGAAAG	CAC	-CTCA	TCCCGT	-CGGGATC	TGA-GAAGTT	AAGCTCCTTCAG
Porphyra[j]	--ACGT---	ACGGCC	ATATCC	GAGACACGCG	TAC	-CGGA	ACCCAT	TCCGAAT	TCC-GAAGTC	AAGGTCCGCGA
Caenorhabditis[k]	--GCTT---	ACGGCC	ATA-TC	ACGTTGAATG	CAC	-GCCA	TCCCGT	-CCGATC	TGG-CAAGTT	AAGCAACGTTGA
Drosophila[l]	--GCCA---	ACGACC	ATA-CC	AGGCTGAATA	CAT	-CGGT	TCTTCGT	-CCGATC	ACC-GAAATT	AAGAGGCGTCGG
Lytechinus[m]	--GCCT---	ACGACC	ATA-CC	ATGCTGAATA	TAC	-CGGT	TCTTCGT	-CCGATC	ACC-GAAGTC	AAGCAGCATAGG
Xenopus[n]	--GCCT---	ACGGCC	ACA-CC	ACCCTGAAAG	TGC	-CTGA	TCTTCGT	-CTGATC	TCA-GAAGTT	AAGCAGGGTCGG
Rattus[o]	--GTCT---	ACGGCC	ATA-CC	ACCCTGAACG	CGC		TCTTCGT	-CTGATC	TCG-CAAGCT	AAGCAGGGTCGG

Homo[p]	---CTCT---	ACGGCC	ATA-CC	ACCCTGAAGG	CGC	-CGGA	TCTCGT	-CTGATC	TCT-GAAGCT AAGCAGGGTCGG
Chlorella[q]	ATG-CT---	ACGTTG	ATA-CC	ACCACGAAAG	CAC	-CCGA	TCCCAT	-CAGAAC	TCG-GAAGTT AAAGTTGTTGG
Chlamydomonas[r]	ATGGAT---	-CGTTC	AAA-CC	TTCAAGGCCC	CTC	-CCCA	TCCCAT	-CAGCAC	TGG-GAAGAT AAGCCTGAATGG
Spinacia[s]	--GGGT---	CCGATG	ATA-CC	AGCACTAATG	CAC	-CGGA	TCCCAT	-CAGAAC	TCC-CCAGTT AAGCTTGCTTGG
Triticum[t]	--GGAT---	GGGATG	ATA-CC	AGGACTAAAG	CAC	-CGGA	TCCCAT	-CAGAAC	TCC-GAAGTT AAGCTTGTTGG

[a] The italicized sequences in L–N and in P–Q correspond to various A boxes. The underscoring in Q represents the beginning of box C, which is continued in Table 3.5. See Section 3.6.1 for discussion.

[b] See Table 3.2.

[c] Delihas et al. (1981c).

[d] Himmebusch et al. (1981).

[e] MacKay and Doolittle (1981).

[f] Luehrsen et al. (1980).

[g] Kumazaki et al. (1982).

[h] Walker and Doolittle (1982).

[i] Komiya and Takemura (1981).

[j] Takaiwa et al. (1982).

[k] Butler et al. (1981).

[l] Benhamou et al. (1977); Tschudi and Pirrotta (1980).

[m] Lu et al. (1980).

[n] Federoff and Brown (1977).

[o] Aoyama et al. (1982).

[p] Forget and Weissman (1969).

[q] Jordan et al. (1974); Luehrsen and Fox (1981).

[r] Darlix and Rochaix (1981).

[s] Delihas et al. (1981a,b).

[t] MacKay et al. (1980).

Table 3.5

Eukaryotic Mature 5S rRNA Genes—3' Portions[a]

	R	S	T	U	V	W	X	Y	Z
Methanobrevibacter	GTTTTG	TTTTGTGT-A	CTATGG-GT	TCCG	GTCTATGG	GAATTT	CA-TTT	TTAGCTGCG	AGCTTTTT
Torulopsis	GGCT-G	ATCGAGT-A	GTGTAGTAG	GTGA	CCATACGC	GAAACT	CA---G	GT-GCTGTA	AT--CT--
Euglena	GCCC-A	GTT-AGT-A	CTGAGGTGG	GCGA	CCACTTGG	GAACAC	TG---G	GT-GCTGTA	CG-CTT--
Crypthecodinium	GCCC-G	GAT-AGT-A	CTGGGGTGG	GGGA	CCGCCCGG	GAAGTC	CTTAGG	GT-GCTGTC	AG--CT--
Acanthamoeba	GCGG-T	GTT-AGT-A	CTGGGGTGG	GCGA	CCACCGG	GAATCC	AC---C	GT-GCCGTA	TC--CT--
Tetrahymena	GCTG-G	GTT-AGT-A	CTAAGGTGG	GGGA	CCGGTTGG	GAAGTC	CC---A	GT-GTCGAT	AG--CCT--
Paramecium	GCGA-T	GTT-AGT-A	CTAAGGTGG	GGGA	CCGGTTGG	GAAGTC	CT---C	GT-GTTGAC	AA--CCT--
Saprolegnia	GGGT-G	GGT-AGT-A	CTCAGGTGG	GTGA	CCACTGG	GAAGTC	CA---C	GT-GCTGAT	TC---T--
Physarum	GCGT-G	GTT-AGT-A	CTGGGGTGG	GGGA	CCACCTGG	GAATCC	CA---C	GT-GCTGCA	TT-CTT--
Porphyra	GTTG-G	GTT-AGT-A	ATCTCGTGA	AAGA	TCACAGGC	GAACCC	CC---A	AT-GCTCTA	CG--TC--
Caenorhabditis	GTTC-A	GTT-AGT-A	CTTGGATCG	GAGA	CGGCCTGG	GAATCC	TG---G	AT-GTTCTA	AG--CT--
Drosophila	GCGC-G	GTT-AGT-A	CTTAGATGG	GGGA	CCGCTTGG	GAACAC	CG---C	GT-GTTCTT	GG-CCT--
Lytechinus	GCTC-G	GTT-AGT-A	CTTGGATGG	GAGA	CCCCCTGG	GAATAC	CG---G	GT-GTTCTA	GG-CTT--
Xenopus	GCCT-G	GTT-AGT-A	CCTGGATGG	GAGA	CCGCCTGG	GAATAC	CA---G	GT-GTCGTA	GG-CTT--
Rattus	GCCT-G	GGT-AGT-A	CTTGGATGG	GAGA	CCCCCTGC	GAATAC	CG---G	GT-GCTGTA	GG-CTTT--
Homo	GCCT-G	GTT-AGT-A	CTTGGATGG	GAGA	CCCCGTGG	GAATAC	CG---G	GT-GCTGTA	GG-CTT--
Chlorella	GCTC-G	ACT-AGT-A	CTGGGTTGA	GGGA	TTAGCTGG	GAACCC	CG---A	GT-GAGTA	GT--GT--
Chlamydomonas	GCTG-A	ACT-AGT-A	GTACGGTGG	GGGA	CCACGTGC	GAATCC	TC---A	GT-GACGAC	CT-GGTT-
Spinacia	GCGA-G	AGT-AGT-A	CT-AGATGG	GTGA	CCTCTGG	GAAGTC	CT---C	GT-GTTGCA	CC-CCT--
Triticum	GCGA-G	AGT-AGT-A	GTAGGATCGG	GTGA	CCTCCTGG	GAAGTC	CT---C	GT-GTTGCA	TT-CCC--

[a] See Table 3.3 for references; the underscoring in R–T represents the remainder of box C from Q of Table 3.4, as discussed in Section 3.6.1.

variable. Another correspondence is in the later part of region S, where AGT-A is an invariant feature, but there the similarities cease. Some resemblance does occur in column W, GAA being a constant triplet there among eukaryotes, whereas GAG is a frequent sequence among prokaryotes. In other regions of low variability are the G-GA of region U, and the adjacent CCRCYYGG of region V, but only the GA combination in U is unvarying. It is of interest to note that in the eukaryotes a pyrimidine, almost always T, uniformly terminates the molecule, while in prokaryotes adenosine does. Still two other areas of fairly constant structure exist, the first of which is the early part of region Y. There RT-GYY prevails with striking regularity, the purine being guanosine in all except two instances; the first pyrimidine is slightly variable, being replaced by an adenosine in the two green algae. The second additional regular sector is the downstream two-thirds of area T, which consists of RGRTGG in a vast majority of the sequences given in the table.

Site-by-Site Comparisons. The changes that have occurred are most thoroughly brought out by site-by-site comparisons of the prokaryotic and eukaryotic two-dimensional structures shown in Figure 3.1A, B. Two levels of invariability are brought out by the use of contrasting type faces, boldface indicating sites that lack any variation throughout

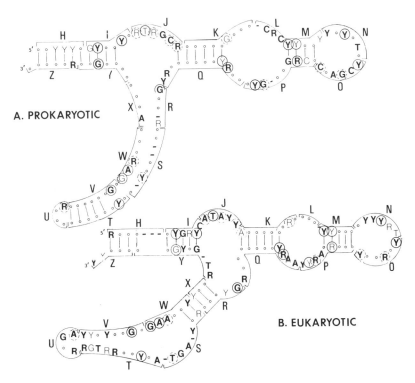

Figure 3.1. The secondary structure of 5 S rRNAs. The molecular structure of (A) prokaryotic and (B) eukaryotic species differs only in minor details. The large bold letters refer to the various regions of the molecules employed in the tables; smaller, bold capitals within the primary sequence represent sites that are invariable within that taxon, while lightfaced letters are sites that are nearly unvariable, having one or perhaps two known exceptions. Continuous circles enclose sites that are invariable in both prokaryotes and eukaryotes, and broken ones enclose those that are just nearly constant.

the sequences as given in Tables 3.2–3.5, and the smaller lightface indicating the existence of one, or at the most two, exceptional cases. In the first place the overall pattern of secondary structure can be noted to differ little between prokaryotes and eukaryotes. In the latter group, the stem formed by regions H, I, Y, and Z is slightly reduced, the gain of one additional pair of residues at the termini counterbalancing the loss of two sites in H; consequently, the stem consists of nine pairs instead of the ten found in prokaryotes. In the long stem, the chief major difference is noted to lie in the location of the paired region, in one case being in sectors T and V, in the other in R–S and W–X.

If evolutionary conservation is assumed to have a functional basis, one would expect extensive agreement between the two in the location of invariant sites, for in both cases the molecule forms part of the ribosomal RNA skeleton. On the illustrations universal conditions are indicated by a continuous circle if the prokaryotic and eukaryotic sites are identical and by a broken circle where a specific nucleotide is present in one, but a class representative (purine or pyrimidine) occurs in the other. Invariable pairs of specific bases are nonexistent, but two involving either a pyrimidine with a purine or with a guanosine exist, the former in M and P, the latter in I and Y. Moreover, the constant pyrimidine in T and the guanosine in V could very well form a pair in some sequences, as could the pyrimidine of S and the adenosine of W.

But lack of variability is far more characteristic of the loops and other single-chain regions. Sector J seems to be the most richly endowed portion of the structure, five of its six sites displaying a high level of homology between groups; consequently, this area must be suspected to be of special functional significance. The large terminal loop formed by N and O proves disappointingly sparse in constant sites, all four present involving classes rather than specific residues, at least in one source each. In fact, the smaller loop composed of P and Q contains a number of conserved sites equal to those of the larger, terminal one, which likewise are at the class level. Region W is the only other portion with a rich representation of invariant sites, three of its six falling into this category, with two of those being constant nucleotides (compare Hori *et al.*, 1985).

3.2.3. Mature 5 S rRNA Genes of Organelles

Why the 5 S rRNA sequences of bacteria and eukaryotes have attracted so much attention and those of the organelles so little is difficult to comprehend. Yet only five representatives of this class of substances have been sequenced from chloroplasts (Erdmann *et al.*, 1984). In mitochondria, the gene is usually missing, but strangely is present in the seed plant organelle, representatives having now been sequenced. Very much more needs to be accomplished in this area before a firm understanding of the nature of these organellar genes can begin to be achieved. But in order to present a full picture of the current status of transcription of genes for this type of substance, what little has been learned about the organellar representatives is analyzed in the following discussions.

Mature 5 S rRNA Genes of Chloroplasts. Because the 5 S rRNA gene from *Spirodela* chloroplast (Keus *et al.*, 1983) is identical to that from *Lemna*, and those of *Nicotiana* and *Spinacia* are identical, except that the latter source terminates in a nucleotide (G) absent in the former, only three of the five known sequences are provided here (Table 3.6). With those are included the primary structures of the substance from the blue-green alga *Anacystis* (Douglas and Doolittle, 1984) and the eubacterium *Photobac-*

Table 3.6
Mature 5 S rRNA Genes of Chloroplasts

	H	I	J	K	L	M	N	O	P	Q
Anacystis[a]	--TCCTG	GTGTCT	ATGGCG	GTATGGAAC	CAC	TCTGAC	CCCATC	CCGAAC	TCAGTT-GTG	AAACATACCTGC
Dryopteris[b]	TATTCTG	GTGTCC	CAGGCG	TAGAGGAAC	CAC	ACCGAT	-CCATC	TCGAAC	TTCGTG-GTG	AAACTCTGCCGC
Lemma[c]	TATTCTG	GTGTCC	TAGGCG	TAGAGGAAC	CAC	ACCAAT	-CCATC	CCGAAC	TTCGTG-GTT	AAACTCTACTGC
Spinacia[d]	TATTCTG	GTGTCC	TAGGCG	TAGAGGAAC	CAC	ACCAAT	-CCATC	CCGAAC	TTCGTG-GTT	AAACTCTACTGC
Photobacterium[a]	TGCTT-G	GGGACC	ATAGCG	TTATGGAAC	CAC	-CTGAT	CCCTTG	CCGAAC	TCAGTA-GTG	AAACGTAATAGC
Spinacia[e]$_{Cy}$	GGGT--	GCGATC	ATA-CC	AGCACTAAA	GCA	-CCGGA	TCCCAT	CAGAAC	TCC-GAAGTT	AAGCGTGCTTGG

	R	S	T	U	V	W	X	Y	Z
Anacystis	GGC--A	AC--GAT-A	GCTCCC-GG	GTAG	CCGGTCGC	TAAAAT	-A-GCT	CGACGCC	AGGTC-
Dryopteris	GGT--A	AC--CAA-T	ACTCGG-GG	GGGG	CCCTGCGG	AAAAAT	-A-GCT	CGATGCC	AGATA-
Lemma	GGT--G	AC--GAT-A	CTGTAG-GG	GAGG	TCCTGCGG	AAAAAT	-A-GCT	CGACGCC	AGAAT-
Spinacia	GGT--G	AC--GAT-A	CTGTAG-GG	GAGG	TCCTGCGG	AAAAAT	-A-GCT	CGACGCC	AGGATG
Photobacterium	GCC--G	AT--GGT-A	GTGTGG-GG	TCT-	CCCCATGT	GAGAGT	-A-GGA	CATCGCC	AGGCAT
Spinacia$_{Cy}$	GCGA-G	AGT-AGT-A	CTAGGATGG	GTGA	CCTCCTGG	GAAGTC	CT-CGT	-GTTGCG	CA-CCT

[a] From Tables 3.2 and 3.3.
[b] Takaiwa and Sugiura (1982).
[c] Dyer and Bowman (1979).
[d] Delihas et al. (1981a,b).
[e] From Tables 3.4 and 3.5.

terium, as well as that of the cytoplasmic type from spinach, to enable broad comparisons to be made readily. Close analysis discloses clear relationships to neither the blue-green algal, bacterial, nor eukaryote cytoplasmic sequences, although the last of these is distinctly more remote in kinship than the others. The very first sector, H, reveals the nature of the conflicting characteristics, for while its length and initial base are definitely bacterial and somewhat eukaryotic, the closing triplet CTG is obviously a blue-green trait (see Table 3.2). In sector J is a short series characteristic of the chloroplast types alone, for neither the CAG nor TAG found at its beginning occurs among prokaryotes of any sort. Although the first half of region K appears similarly unique, reference to Table 3.2 shows that the combination TAG occurs among the Protobacteria, suggesting an early origin for the chloroplast sequence. Area L tends to confirm prokaryotic relationships, as does the presence of six occupied sites in M, but these are promptly denied by the unique absence of the first base in region N. While region O is neutral, the sequences in P when viewed in a fuller evolutionary context (Table 3.4) show prokaryotic leanings, as do those in R and S, without offering anything conclusive. In section T only the eight occupied sites of prokaryotes are present rather than the nine of higher forms, but in W the first half, AAA, is confined to chloroplast representatives, the closing AAT resembling the blue-green types. Finally, the 3'-terminal section Z is strongly prokaryotic in its length as well as the composition of its first three nucleotides.

Thus, overall the inclinations are strongly prokaryotic, often toward the blue-green algae, but equally often not, a conclusion supported by counts of identical corresponding sites. With the *Anacystis* sequence, the *Dryopteris* chloroplastic 5 S gene shares 80 sites, whereas that from *Spinacia* has 82. The latter shows 70 sites homologous to that of *Photobacterium,* but only 45 with those of the spinach cytoplasmic gene.

Mature 5 S rRNA Genes from the Mitochondrion. Two of the few 5 S rRNA genes from mitochondria that have been sequenced to date, the nearly identical ones from the organelles of *Triticum* and *Zea* (Spencer *et al.,* 1981; Chao *et al.,* 1983), are shown in Table 3.7, together with sequences from *E. coli, Thermus thermophilus,* and *Triticum* cytoplasm for comparative purposes. Others not shown include those of soybean and *Oenothera* (Morgens *et al.,* 1984; Brennicke *et al.,* 1985). Here, as in the chloroplast, kinships are not positively establishable; however, they are still more obscure in the present that in the preceding case. In length of sector H and in its two initial bases, resemblances to the gene from *T. thermophilus* are quite evident, distinctive similarities that appear again in regions M and N. It is remarkable that region K, where the *Triticum* mitochondrial sequence closely parallels that of the archaebacterium, is also one of two sectors in which the *Zea mays* mitochondrial gene differs from the other. These are over and above those that are shared with the *E. coli* molecule. Counts of identical sites shared by the *Triticum* mitochondrial and *Thermus* sequences show the presence of 69 homologous positions, for a kinship ratio of 59%, whereas it holds only 60, or 51%, in common with the *E. coli* gene. In the present arrangement derived from the long phylogenetic series of Tables 3.4 and 3.5, only 41 identical sites are shared by the wheat mitochondrial and cytoplasmic 5 S sequences, but by rearranging the gaps and various sectors to maximize homologies this can be increased to 56, which number shows virtual agreement with that (55) reported by Gray and Spencer (1981). In the first case the relationships are at a 36% level, in the second, 48%.

Consequently it can be reasonably argued that the mitochondrial 5 S rRNA gene

Table 3.7

Mature 5 S rRNA Genes of Mitochondria[a]

	H	I	J	K	L	M	N	O	P	Insert	Q-
Escherichia[b]	--TGCCT-G	GCGGCC	GTAGCG	CGGTGGTCC	CAC	--CTGAC	CCCATG	CCGAAC	TCAGAA	------	GTGAAACGC
Thermus[b]	AATCCCC-C	GTGCCC	ATAGCG	GCCTGGAAC	CAC	--CCGTC	CCCATT	CCGAAC	ACGGAA	------	GTGAAACGC
Zea[c] M	AA-CCGG-G	CACTAC	GGTGAG	ACGTGTTTA	CAC	-CCGAT	CCCATT	CCGACC	TC-GAT	ATATAT	GTGGAATCG
Triticum[d] M	AAACCGG-G	CACTAC	GGTGAG	ACGTGAAAA	CAC	-CCGAT	CCCATT	CCGACC	TC-GAT	ATATAT	GTGGAATCG
Triticum[e] Cy	--GGTT---	GCCATC	ATACCA	GCACTAAAG	CAC	-CCGAT	CCCAT-	CAGAAC	TCCCA-	------	AGTTAAGCG

	-Q	R	S	T	U	V	W	X	Y	Z
Escherichia	CGT-AGC	GCC--A	AT--GGT-A	GTGTGG--GG	TCT-	CCCCATGC	GAGAGT	-AGGG	AACTGCC	AGGCAT---
Thermus	GCC-AGC	GCC--A	AT--GGT-A	CTGGC--GG	ACGA	CC-GCTGG	GAGAGT	-AGGT	CGGTGCG	GGGGA----
Zea M	TCT-TGC	GCC---	ATA-TGT-A	CTG-----AA	ATTG	---TTCGG	GAGACA	-TGGT	CAAAGCC	CGCCC----
Triticum M	TCT-TGC	GCC---	ATA-TGT-A	CTG-----AA	ATTG	---TTCGG	GAGACA	-TGGT	CAAAGCC	CGGAAA---
Triticum Cy	TGCTTGG	GGGA-G	AGT-AGT-A	CTAGGATGG	GTGA	CCTCCTGG	GAAGTC	CT---	CGT-GTT	GCATT-CCC

[a]Subscript M indicates mitochondrion; Cy, cytoplasm.
[b]From Tables 3.2 and 3.3.
[c]Chao et al. (1983).
[d]Spencer et al. (1981).
[e]From Tables 3.4 and 3.5.

shares kinship with the prokaryotic molecule, especially that of advanced types, as would be expected if the eukaryotes themselves have been evolved from similarly advanced members of that group. There is no evidence suggesting that the mitochondrion itself is a direct modified descendant of a prokaryotic symbiont, for there are too many unique features present. The inserted sequence ATATAT located between sectors P and Q is the most striking of these, made more thought-provoking by its being the antithesis of the trend noted among mitochondrial tRNA genes in the preceding chapter, representing an increase in length, rather than the decrease seen there. However, a few residues may be noted to have been lost in the 3' half of the molecule, but since these total only four in a region outstandingly variable, they can scarcely be convincingly considered a trend toward diminution.

3.3. TRANSCRIPTIONAL TERMINATION IN 5 S rRNA GENES

Data concerning the termination of transcription of 5 S rRNA genes, fortunately, are more abundant than those pertaining to initiation of the processes, but much more are needed before any firm conclusions may be drawn. Although limited in quantity, sequences of trains have been determined from bacterial, eukaryotic, and chloroplastic sources. Only those of mitochondria are not represented, an absence understandable in view of the apparent lack of 5 S genes from all but the seed-plant organelle.

3.3.1. Termination in Bacterial 5 S rRNA Genes

Except for one example from the blue-green alga *Anacystis nidulans,* the only terminator sequences of the 5 S rRNA genes that have been established among prokaryotes are those of *E. coli;* however, although restricted in breadth, the five available ones, being largely from different operons, do provide considerable depth (Table 3.8). As to the second one from a single operon, $rrnBt_L$, it is not clear how complete the sequence is nor why it differs so extensively from the first; perhaps the reason is that it is really an attenuator from the leader rather than a true terminator as it was called (Kingston and Chamberlin, 1981). All representatives share in one trait, the presence of regions of dyad symmetry, indicated by italicized letters in the table. Furthermore, each set of these regions, except that of the blue-green alga, is characterized by the existence of a noncanonical base pair, usually G–T, but possibly A–A in $rrnBt_L$. Additionally, in two cases one or two unpaired residues are found adjacent to the noncanonical combination, so that the hairpin loop resulting from the dyad symmetry would necessarily be bent to one side.

The differences in lengths of the regions of dyad symmetry are of particular interest, the range being from 8 residues to a maximum of 28. In *E. coli,* the two longest ones, *rrnB* and *rrnD,* lack unpaired bases on one side of the stem, whereas those sequences with only 8 or 10 on one side have 10 or 11 on the other, respectively. In most cases the loop consists of 4 residues, but in *rrnF* it has 6, and in *Anacystis,* 14. The amount of homology that exists between the different train sequences also is pertinent, especially between *rrnB* and *rrnD,* which differ at only 4 of the 43 sites, for a homology rate of over 90%. In contrast, the *rrnF* 3' sector shows homology to those of both *rrnB* and *rrnD* at a rate of 40%. That exhibited between the terminator and the attenuator sequences of *rrnB* is still

Table 3.8
Trains of Prokaryotic and Chloroplast 5 S rRNA Genes[a]

Anacystis[b]	CCTAATCAAAACATC*CTAACCCCCCTCTG*CCTCACTACAACAG*CGGAGGGGGG TTA*GTCTTTCGA
Escherichia rrnB[c]	CAAATAAAA*CGAAAGGCTCAGTC*-GAAAG*ACTGGGCCTTTCGT*
Escherichia rrnBt_L	---------TG*AACACGTAATT*C-*ATTACGAAGTT*T
Escherichia rrnD[e]	CAAAT*AAAACAAAAGGCTCAGTC*GGA-AGAC*TGGGCCTTTTGTTTT*ATCTGT
Escherichia rrnF[f]	CAAATTAAGCAGT*AAGCCGGT*CATAAA*ACCGGTGGTT*GTAAAAGAATTCGGT
Spirodela_C[g]	G*ATAAAA*AGCTTAACACCTCTTA*TT-TTAT*-AA------AATATGCAAAATC
Nicotiana_C[h]	G*ATAAAA*AGCTTAACACCTCTCA*TTCTTATC*AATTTTTCA*ATATG*-AAAA-C

[a]Italicized sequences represent regions of dyad symmetry; underscore represents noncanonical base pairs or single-stranded bases.
[b]Douglas and Doolittle (1984).
[c]Brosius et al. (1981); Singh and Apirion (1982); Roy et al. (1983).
[d]Kingston and Chamberlin (1981).
[e]Duestor and Holmes (1980); Szeberényi and Apirion (1983).
[f]Sekiya et al. (1980).
[g]Keus et al. (1983).
[h]Takaiwa and Sugiura (1980).

less, about 37%, and the blue-green sequence shows virtually no regions homologous with any of the others.

3.3.2. Termination in Eukaryotic 5 S Genes

Among eukaryotes, knowledge of the train structure of the 5 S rRNA genes embraces much of the entire group, but in very broken fashion. Three sequences have been established from yeasts and five from amphibians, but only one each from *Drosophila* and flax provide any diversity at higher levels, and none is available from any intermediate point in the phylogenetic tree. Those that have been determined, however, demonstrate high levels both of variability and evolutionary conservation.

Trains of Yeast 5 S rRNA Genes. Only the first of these seemingly conflicting qualities becomes apparent immediately in the trains of yeast 5 S rRNA genes (Table 3.9), for the three from *Saccharomyces cerevisiae* display only occasional homologous sites. Hence, it is evident that they represent the 3' flanking regions from different genes. On the other hand, the third one from this source is closely related to that from *Torulopsis utilis* in the first 11 sites and again in the terminal region. Since the intervening portions show scarcely any identical corresponding sites, these trains are perceived to be both evolutionarily stable and variable.

Two contrasting mechanisms for signaling termination seem to exist. In the second sequence from *S. cerevisiae* a stem-and-loop structure could form in the transcript as it is synthesized, involving the first seven sites and a second run of eight residues, both strands of this double-stranded region being indicated by italics in the table. Actual termination takes place downstream about ten sites between the pair of Ts shown underscored (Maxam

Table 3.9
Trains of Eukaryotic 5 S rRNA Genes[a]

S. cerevisiae[b]	GAT-A*GTTT*AACGG*AAAC*GCAGGTGATATGAGGGCAGGGTCCAGACATGTTCAGT
S. cerevisiae[c]	*ACCGAG*TAGTGTAGTGGGTGACCATACGCGAA*ACTCAGGT*GCTGCAATCT<u>TT</u>TAT
S. cerevisiae[d]	----TTATTTTTTTTTTTTTTTTTTTTTTTTTTTTCTAGTTTCTTGGCTTCCTATGC
T. utilis[d]	----TTTTTTTTA-CCAAGTTGGGTAACCGGCTTCCCAAAAGGGGCAGTCTACCC
Tetrahymena[e]	TTTTATTTTTTTTGTCAAGTTAAAGATTAAAAAT--AAAACTTAATTG
X. borealis 1[f]	TAAGACTTTTGCCAGGTC-AAGTTTTGCAGGGTTTTTCGTCAAAGTCTTCATAGA
X. borealis 3[f]	--------TTTCCAGGTCAAAATTTTGCAGGTTTTTTCTTCAAAGTCTTCATAGA
X. borealis oocyte[g]	GGTTATT*ACCTGGAT*GGGAGACCGCCTGGGA*AT*A<u>CC</u>AGG*T*GTCGTAGGCT<u>TTT</u>A
X. borealis oocyte[f,h]	GGTTAGT*ACCTGGAT*GGGAGACCGCCTGGGA*AT*A<u>CC</u>AGG*T*GTCGTAGGCT<u>TTT</u>C
X. borealis somatic[f]	GGTTAGT*ACTTGGAT*GGGAGACCGCCTGCGA*AT*A<u>CC</u>AGG*T*GTCGTAGGCT<u>TTT</u>G
Drosophila melanogaster[i]	GATGGGGGACCGTTGGGAACACCGCGT*TTA*GT<u>TTG</u>GCCTCGT*CCA*CAA*C*T<u>TTT</u>G
Flax[j]	TCCTTTTGCAATTTTTCTCCGGCGACGTTATGGGCACGCTTAGCGAGGTTACAT

[a]Italics represent sectors of dyad symmetry and underscoring indicates noncanonical or unpaired residues; underscored TT or TG combinations mark established points where termination occurs.
[b]Valenzuela *et al.* (1977a,b).
[c]Maxam *et al.* (1977).
[d]Tabata (1980).
[e]Pederson *et al.* (1984).
[f]Korn and Brown (1978).
[g]Ford and Brown (1976).
[h]Denis and Wegnez (1973).
[i]Jacq *et al.* (1977).
[j]Goldsbrough *et al.* (1982).

et al., 1977; Valenzuela *et al.*, 1977a,b); hence, this process occurs at some distance beyond the hairpin structure, not at its base as in the bacteria. Like the condition found in the eubacteria, an extra base is present on one strand that would introduce a bend in the tertiary configuration of the hairpin stem. The first sequence displays only very short regions of dyad symmetry; in neither this nor the other three species have the functional terminators been actually determined. In the last two cases termination possibly occurs within the long series of thymidine residues present in each (Tabata, 1980).

Terminators in Xenopus 5 S rRNA Genes. The three major subfamilies of the family of 5 S rRNAs mentioned earlier to occur in *X. laevis* include two expressed only in oocytes, referred to as the "major"- and "trace"-oocyte subfamilies, and one, the "somatic," expressed in all cell types. If a similar pattern of organization exists among the corresponding genes of *X. borealis,* as expected from an evolutionary standpoint, then two of the three groups are represented. Although the first two toad sequences given, those bearing the numerals 1 and 3, are members of a cluster of three from the oocyte, as in the case of the leader sequences given earlier, they do not support the concept that each subfamily has a specific spacer sequence peculiar to all its members. Here, as before, the gene 2 train sequence (not given in the table) differs but slightly from that of gene 1, but that of gene 3 is quite disparate at a number of sites. The greatest point of departure is in the absence of eight bases at the 5' end, a second marked one being the insertion of an additional adenosine residue at site 19 (Korn and Brown, 1978). In addition, four other positions show base substitutions, yielding a total homology rate of about 75%. Furthermore, it is obvious that both sequences labeled oocyte, one each from *X. borealis* and *X. laevis,* are members of the somatic subfamily as given in the table, for they differ from the adult somatic sequence at only two of the 54 sites shown there, just one more than they differ between themselves. Hence, it would appear that no representative of the second oocyte subfamily has yet been sequenced. Consequently, it is evident that the establishment of the primary structures of many more representative 5 S rRNA genes from the two species of *Xenopus* would be most profitable, especially if those of additional entire clusters could be determined. The data thus far in hand suggest both extensive homologies in some cases and great disparities in others.

The precise points of transcriptional termination were not determined with *X. borealis* genes 1–3, but it was suggested that one or more of the several series of four or more Ts could serve in that capacity (Korn and Brown, 1978). In the other three examples given in Table 3.9, termination occurred in the centers of groups of four Ts, that is, between the pairs printed underscored. In each case termination follows a stem-and-loop structure, the regions of dyad symmetry being indicated by italics. All three have one side of the paired region longer by one residue than the other, the unpaired member (A) appearing underscored. The underscored G and T of the somatic sequence point out a noncanonical base pair.

Termination in Other Eukaryotic 5 S rRNA Genes. The trains of only three additional 5 S rRNA genes have been determined from other eukaryotic sources, one each from *Drosophila melanogaster, Tetrahymena,* and flax (Jacq *et al.,* 1977; Goldsbrough *et al.,* 1982; Pederson *et al.,* 1984). Although no attempt has been made to maximize homologies, the 3' flanking region of the fruitfly may be noted to display a surprising number of sites identical to those of *Xenopus,* especially beginning with the 16th residue and again near the terminus of the region given. When the maximum amount of identical sites is induced in the usual fashion (not shown), the homology rate is increased to 48%, in contrast to the nearly 43% in the form listed in the table. Although the mechanism of termination of this organism is the same as in some of the other trains provided here, the regions of dyad symmetry shown in italics differ in location from those of *Xenopus,* and the unpaired base is located toward the 5' end, rather than the 3' as elsewhere. A third distinctive feature is that termination occurs between the last T of a series of five and its neighboring G, both residues being given underscored as before.

Little can be said about the sequences from *Tetrahymena* and flax, because the sites of transcriptional termination were not determined. No regions of dyad symmetry are present, but in both cases two series of four or more Ts are present near the 5' end that might serve as terminators. In addition, in the protozoan sequence, the series CAAGTT coincides closely with the same run in the *T. utilis* train.

3.3.3. Termination in Organellar 5 S rRNA Genes

Because of the absence already noted of 5 S rRNA genes from most eukaryotic mitochondria, it is not possible to discuss the organellar processes of transcription termination on a fully comparative basis. Moreover, the complete structures of only a few of this family of genes are available from chloroplasts, the trains of those that are known being included in Table 3.8 together with the *E. coli* and blue-green algal sequences. Since the chloroplast trains of *Nicotiana* and the green alga *Spirodela* show only one major region (six sites) of strong divergence from one another, evolutionary conservation is perceived to have been active at a high level. Forty-one of the 52 sites given in the table are identical, yielding a homology rate near 80%. Few corresponding sites are shared by either of these with any of the *E. coli* trains, the greatest degree of homology being with that structure from *rrnF*. But even in this case only 15 or the 52 sites are identical, the rate of homology, 29%, exceeding only slightly the 25% rate provided by random chance. Nor is any significant degree of homology evident between them and the *Anacystis* train.

Keus *et al.*, (1983) report the presence in these chloroplast genes of "perfect prokaryotic termination signals," but the regions of dyad symmetry, which they incorrectly refer to as palindromes (see Dillon, 1983), are in part misidentified in the case of *Spirodela* and partly overlooked in the *Nicotiana* sequence. One pair of complementary sectors is similar in the two representatives, as shown by the italicized portions, but in *Nicotiana* a second pair, likewise italicized, is also to be found. One extra member of the latter pair is missing from the *Spirodela* sequence, although the second member has been conserved. In no case, however, are the complementary sectors perfect copies of the prokaryotic terminators, all of which among these genes have one noncanonical base pair, a feature absent in those of the two chloroplasts. The first hairpin structure of the tobacco chloroplast contains an unpaired base (C), shown underscored. Thus it is more similarly constructed to many eukaryotic 5 S rRNA terminators than to prokaryotic ones. However, in no case has it been established where termination actually occurs, nor is it clear in even the complete train sequence of over 130 base pairs where that process could occur, for no series of four or more Ts exists beyond the sites given in the table. Thus the supposed existence of perfect prokaryotic terminators arose through prior beliefs, not observable facts.

3.4. TRANSCRIPTION OF OTHER RNA SPECIES

In eukaryotes, DNA-dependent RNA polymerase III transcribes a small number of types of RNA genes other than those of the tRNA and 5 S rRNAs, the majority of which are small nuclear species such as the *Alu* family and its relatives. Moreover, this same enzyme transcribes specific regions of the DNA of adenovirus and Epstein–Barr virus, whose products are usually referred to as viral-associated RNAs. Although many generalizations concerning the transcriptive processes of this miscellany of gene categories have appeared in print, the available data involving the factors remain relatively sparse. Since a greater number of studies have centered on the small heterogeneous RNAs, what is known of the processes in their genes is presented first, followed by the viral varieties.

3.4.1. Small Heterogeneous RNAs

The heterogeneous nuclear RNAs (hnRNA) of human cells include species up to at least 20,000 nucleotides in length (Robertson and Dickson, 1984); the small members that are only approximately 300 base pairs long are referred to as small nuclear RNAs (snRNAs). Actually not all such RNA species are confined to the nucleus, many being found in the cytoplasm as well. Hence, the recent acronyms for the miscellany of species, LINES for long and SINES for short interspersed nuclear elements, are equally inappropriate as catch-all terms (Singer, 1982). Accordingly, here the class is collectively referred to as the heterogeneous RNAs (hRNAs), with "long" (l), "small" (s), or "moderate" (m) added to designate the size categories. Attention at this time is confined to small hRNAs transcribed by polymerase III, since no long hRNA thus processed has yet come to light.

The Alu Family of Small hRNAs. It is thus evident that two former classes of substances have now been partially merged into one, a combination that has become increasingly in vogue during recent years (Rubin *et al.*, 1980; Weiner, 1980; Robertson and Dickson, 1984). Probably the most thoroughly investigated shRNAs are the members of the *Alu* gene family of primates, to which needs to be added several others of mammals in general (Li *et al.*, 1982). However, the latter lack several major traits of the former, particularly the sector sensitive to the restriction enzyme *Alu*I, from which the family name was derived, as well as the partial duplication of the 5' and 3' portions, as shown shortly.

The members of the *Alu* gene family proper are nearly 300 base pairs long and are represented in the haploid human genome by perhaps more than 400,000 copies, comprising about 5% of its total mass. They are so abundant that copies are separated from one another only by an average of about 7000 nucleotide residues. If they have a function in the cell's metabolism, it has not been firmly established to date, but several suggestions have been made, such as their serving as transposable elements or in transport of proteins. In the former proposed capacity, as in other transposons, they might be capable of influencing the mutation and expression of other genes. However, this possibility appears remote, for reasons made known later.

In what form such influence might become manifest has not been demonstrated, although the occurrence of *Alu* sequences in association with specific genes has been documented. While *Alu* transcripts, many of which bear polyadenylated trains, have been shown to be present in a large fraction of poly(A)-bearing mRNAs (Calabretta *et al.*, 1981), the best known examples of this arrangement are those associated with the β-globin family of genes in man. Included in the latter family are five cistrons in the order 5' ϵ, $^G\gamma$, $^A\gamma$, δ, β, the products of the first three occurring in fetal hemoglobin, and those of the latter two participating in the structure of adult A_2 and A hemoglobins, respectively (Dillon, 1983, p. 418). Only 20% of the sector occupied by this cluster consists of actual globin-coding sequences. Part of the remainder has been experimentally determined to be transcribable by an RNA polymerase III system, two transcription units having been detected and partial homology to these demonstrated in other regions included in the globin family sector (Duncan *et al.*, 1979). In that early study, one of the transcription units was shown to be located ~1500 base pairs upstream from the $^G\gamma$-globin gene, and one region of partial homology was found ~1000 base pairs proximal to the δ-globin gene, with another just downstream of that of the β-globin.

More recently, close association of eight *Alu* genes with that of β-globin has been revealed (Baralle *et al.*, 1980; Coggins *et al.*, 1980; Fritsch *et al.*, 1980; Jelinek *et al.*, 1980; Duncan *et al.*, 1981). Another study has also demonstrated that one of the two that flank the 5' side of the ϵ-globin coding sequence was of opposite polarity from the latter; transcripts appeared to be 350–400 nucleotide residues in length and nonpolyadenylated

(Allan and Paul, 1984). They were confined solely to nuclei of human K562 cells and were detectable only during expression of the ϵ-globin gene. In a separate but related investigation, the other *Alu* gene was cotranscribed with that of the ϵ-globin to form long, polyadenylated transcripts, but whether transcription was accomplished by RNA polymerase II or III separately or by teamwork processes involving both enzymes has not been determined (Allan *et al.*, 1983). An alternative possibility is that the products were generated by readthrough on the part of polymerase II, similar to that reported in the production of a heterogeneous group of poly(A)-bearing RNA species and a discrete type of 7 S RNA (Weiner, 1980; Elder *et al.*, 1981; Fuhrman *et al.*, 1981). Similar inverted pair arrangements seem to characterize *Alu* sequences associated with the δ-globin coding areas, and perhaps often with those preceding the β-globin (Maeda *et al.*, 1983; Poncz *et al.*, 1983), but it is too early to make generalizations. The whole picture of transcription of this class of movable elements needs extensive review in light of a recent investigation that made clear the fact that the primary transcript of *Alu* in man is a larger RNA known as 7SL (Ullu and Tschudi, 1984). Similar structures were also shown to exist in *Xenopus* and *Drosophila*. Since a much more extensive section of genomic DNA is thus involved, the original product of transcription would need to be subjected to considerable processing. At present it has not been ascertained whether such particles involve all *Alu* genes, including those that have been translocated, or just those that remain in their original locations in the genome.

Although sometimes it is claimed that these genes are flanked on both ends by short repeated sequences of about eight residues, this is a variable condition. In a few a repeat is found at both ends of a gene, in others two short repeats flank the 5' end of the mature coding sequence, in still others they flank the 3' end, and in many cases repeats are absent. In addition, one *Alu*-type sequence has now been demonstrated to be associated with an α-globin gene cluster in man (Shen and Maniatis, 1982), but it shows important structural differences pointed out in the following section. Two of typical composition have been sequenced following the *LDL* (low-density lipoprotein) receptor gene, separated by a shorter, unidentified component (Yamamoto *et al.*, 1984).

Primary Structure of Alu Genes. The primary structure of human *Alu* genes proper is quite complex. If attention is confined for the present to those from gene sequences that are associated with β-globins, the human *AluA–C* and *Alu53* in Table 3.10, each is found to consist of two partially identical segments that are separated by an inserted oligo(A) tract 15–17 residues in length (Schmid and Jelinek, 1982). As may be seen in the table, the differences in length of the inserts result from the varying numbers of adenosine residues present; otherwise the four that have been determined in this group are nearly identical.

The identities in sequence structure that exist even between corresponding 5' and 3' segments are everywhere limited in extent. One short sector that is almost universal among those given is the -TGGY located centrally in column II, another is the -CCAG of column IV, and a third is the -GG-G that closes column V. Still other constant sequences can be noted in columns VII and VIII, column IX being largely nonconforming insofar as similarity between 5' and 3' segments is concerned. In contrast, the 5' portions of *AluA–C* are entirely homologous here. Probably the greatest region of correspondences are those manifest in column III. It is of interest to note that, as far as present knowledge is concerned, the dual composition that characterizes the *Alu* family of genes is strictly a primate trait.

Two *Alu* genes of green monkey kidney cells have been sequenced recently (Krolewski *et al.*, 1984) and show a remarkable degree of homology with the four just discussed, but they are not included in the table, since they add little of pertinence here. One of the three, however, that of clone 4, is of particular interest because it possesses an extremely long insert 1 that embraces more than 45 nucleotide residues. But the various cistrons found preceding the human δ- and β-globin mentioned earlier also display important distinctions from those shown (Deininger *et al.*, 1981; Maeda *et al.*, 1983; Poncz *et al.*, 1983). All four of the sequences given by Schmid and Jelinek (1982) seem to resemble that *AluB* type as a whole, although various sectors display similarities to the *AluC* gene as well. But the greatest likeness to the *B* they reveal is in the presence of the terminal appendage that characterizes that type alone, an extra sequence that intervenes between the gene and the short flanking repeat. These appendages differ from the AGAAGAG of *AluB*, that before the δ-globin with reversed polarity being -A_7CA_3, and that of the second set AGAAGAGA$_4$GA$_5$; in contrast, the reversed one before the β-globin sequence is AGA$_{17}$GT, and its mate CGTCACA$_{11}$GA$_8$G$_7$.

One additional sequence that does merit detailed attention, however, is that referred to in Table 3.10 as human *Alu3'α1*, which was associated with α-family globin genes (Shen and Maniatis, 1982). While there can be no doubt that this cistron belongs in the *Alu* family, it has a number of distinctive characteristics. In the main, the striking features do not occur in the 5' portion. There are a few unusual base substitutions, such as a G for the typical C at site 1 of column IV, a T for the customary purine at site 4 of column VI, and an additional residue in column VIII. Moreover, the midportion shows a similar lack of significant deviation until after the sequence in column IX is reached. Unlike all other known *Alu* genes, in *3'α1* the second repeated series of residues is not followed by insert 2, but continues directly into a broken column X sequence. This in turn is followed by a recognizable column XI run, which then leads directly into the column XIII and XIV series of nucleotides and an insert 1. In short, this portion is constructed like the 5' portion, rather than the 3' of the other species. This second insert 1 of the gene is quite atypical in primary structure and in length, consisting of 15 nucleotide residues in addition to those shown in Table 3.10. Beyond that point the 3' gene structure corresponds to a partial column I and complete column II, followed by an insert of ten base residues. It then resumes in the usual fashion through columns III–VI and partially through VII, after which it becomes disparate again. Thus this *Alu* family member, consisting of the three more or less homologous segments, is an irregular trimer instead of a dimer like the others, each monomer of which is separated by an inserted region. Determining the structure of other *Alu* sequences associated with the α-globin genes should be viewed as a matter of utmost importance, in order to determine whether this unique pattern is followed by others from that source.

Comparisons with Related Gene Families. In addition to those just discussed, Table 3.10 provides several representative sequences from the mouse and chicken and a further type from human sources. Close comparison of the first three with the *Alu* types proper reveals an extent of homology that is clearly indicative of relationships among them, whereas in the murine B2 and the chicken type, kinship is scarcely detectable. Yet

Table 3.10
The Alu and Related Gene Families[a]

	I	II	III	IV	V	VI	VII	VIII
Human AluA (5')[b]	AGGCTGGGAG	TGGTGGCTCACG	CCTGTAATC	CCAGAA	TTTTGGGAG	GCCAAG	GC-AGGCAGA	-TCA-CCTGAGGTC
Human AluB (5')[b]	TGGCTGGATG	CGGTGGCTCAGG	CTTCGTAAAC	CCAGCA	CTTTCGGGAG	GCCAAG	GC-AGCCAGA	-TCA-CTTGAGGTC
Human AluC (5')[c]	-GGCTGGGCG	TGGTGGCTCACA	CCTCGTAATC	CCAGCA	CTTTCGGGAG	GCCGAG	GT-GGGTGGA	-TCA-CCTGAGGTC
Human Alu53 (5')[d]	-GGCTAGGCG	CGG--GTTCACG	CCTGTAATC	CCAGCA	TTTTGGGAG	GCTGAG	AC-GGGTGGA	-TCA---TGAGGTC
Human Alu3'α1 (5')[e]	-GCCCGGGCG	CGGTGGCTCACA	CCTGTAATT	GCAGCA	CTTTGCCAG	GCTTAG	GC-AGGTGGA	-TCAACCTGAAGTC
Human AluA (3')	-GA-CAGGCA	TGATGGCAAGTG	CCTGTAATC	CCAGCT	ACTTGGGAG	GCTGAG	GA-AGGAGAA	-TTG-CTTAAACCT
Human AluB (3')	--GCCCGGGCG	TG-TGGTCCATG	CCTGCAGTC	CCAGCT	ATTCAGGTG	GCTGAG	GC-AGGAGAC	-TTG-CTTGAACCC
Human AluC (3')	-GCCCGGGCG	TGGTGGCGCGCG	CCTGTAATC	CCAGCT	ACTCGGGAG	GCTGAG	GC-AGGAGAA	-TCG-CTTGAACCC
Human Alu53 (3')	-GCCAGCCGA	GTGTGGTGGCA	CCTGTAGTC	CCAGCT	ACTCAGGAG	GCTGAG	GC-AGGAGAA	-TGA-CTTGAACCT
Human Alu3'α1 (mid)	--TCCAGGTG	GATGACTCATG-	CCTGTAAAC	CTGGCA	CTTTGGGAG	GCGGAG	GTTGTAGTGA	GTCAAGATCT-GCC
Human Alu3'α1 (3')	A-GCA	GGTGGTTGGGCA*	CCTGTAATC	CCAGCA	CTTTGGGAA	GCCAAG	GT-G-GGCAG	ATCACAA---GGTC
Murine B1a[f,g]	--CCGGGCA	TGGTGGCCCATG	CCTTTAATC	CCAGC-	ACTCGGGAG	GCAGAG	GC-AGGGGGA	-TTT-CT-GAGTTC
Murine 7S[g]	----------	----------G	CCTGTAGTT	CCAGCT	ACTCGGGAG	GCTGAG	AC-ACGAGGA	-TCG-CTTGAGTCC
Human 7L1[h]	AGGGGCC-GGGCG	CGGTGGCGCGGTG	CCTGTAGTC	CCAGCT	ACTCGGGAG	GCTGAG	GC-TGGAGGA	-TCG-CTTGAGTCC
Murine B2[i]	GGGGC-	TGG-AGAG-ATG	GCTCAGTG	GTTAAG	AGCACCTGA	-CTGCT	CT-TCCGAAG	GTC-C--GAGTTC
Chicken CR1UI[j]	T-	TGGAGGAGGGAA	GATTGAGGT	TGGACA	-TCAGGG--	-------	-----GGAAG	--TT-CTTTACTAT
Human hY5[k]	AGT-	TGGTCCGACT--	-GTTGTCGG	TTATTG	TTAAGTTGA	TTTAAC	AT-TGTCTCC	CCCCACAACCGCGC
Human hY1[l]	GGC-	TGGTCCGAAG--	-GTAGTCGAG	TTATC-	TCAA-TTGA	TTGTTC	AC-AGTCAGT	TACAGATCGAACT-

	IX	Insert 2	X	XI	XII
Human *AluA* (5')	AAGAGTTCAAGA	-------------------	CCAACCTGG	CCAACATGG	----------
Human *AluB* (5')	AGGAGTTCAAGA	-------------------	CCAGCCTGA	CCAACATGG	----------
Human *AluC* (5')	AGGAGTTCAAGA	-------------------	CCAGCCTGG	CCAACATGG	----------
Human *Alu53* (5')	AGGAGATCGAGA	-------------------	CCATCCTGG	CTAACATGG	----------
Human *Alu3'α1* (5')	AGGGGTTCGAGA	-------------------	CCAGCCTAG	CCAACATAG	----------
Human *AluA* (3')	GGAAGGCAGGG-	TTGCAGTGACCGA-GATCATACCACTGCACT	CCAGCCTGG	-----GTGA	----------
Human *AluB* (3')	AGGAGGCAGAGG	TTGCGGTGAGCCTA-GATTGCACCATTGCACT	CTAGCTTGG	CTAGCTTGG	GCAATAGGGA
Human *AluC* (3')	AGGAGGTGGAGG	TTGCAGTGAGCCGA-GATCGCGCCCACTGCACT	CCAGCCTGG	CCAGCCTGG	GCAACAGAG-
Human *Alu53* (3')	-CGAGGTGGAGCC	TTGCAGTCGAGCCAACGATCGCGCCACTGTCAT	---------	-CATCATGG	GTGACAGAG-
Human *Alu3'α1* (mid)	ATCGCACTCCAG	-------------------	CTTG----G	GCAACAAGA	----------
Human *Alu3'α1* (3')	AGGAATTC				
Murine B1a	--GAGG------	-------------------	CCAGCCTGG	---------	----------
Murine 7S	AAGAGTTCTGGG	-------------------	CTGTAGTGC	---------	----------
Human 7L1	AGGAGTTCTGGG	-------------------	CTGTAGTGC	---------	----------
Murine B2	---AATTC----	-------------------	CCAGCA--A	CCA-CA---	----------
Chicken CR1U1	GAGAGTGGTGA-	-------------------	---------	---------	----------
Human *hY5*	TTGACTAGCTT-	-------------------	GCTGTTTT	---------	----------

(continued)

Table 3.10 (Continued)

	Insert 3	XIII	XIV	Insert 1	Appendage
Human $AluA$ (5')	------------------------	TGA-AATCCC	ATCTCTA-CA	AAAAATACAAAAAATTA	
Human $AluB$ (5')	------------------------	TGA-AACCCC	ATCTCTA-CT	AAAAATACA--AAATCA	
Human $AluC$ (5')	------------------------	TGA-AACCCC	GTCTCTA-CT	AAAAATACA-AAAATTA	
Human $Alu53$ (5')	------------------------	TGA-AACCCC	GTCTCTA-CT	AAAAATACA-AACAACA	
Human $Alu3'\alpha1$ (5')	------------------------	TGAAAACCCT	GTCTCTA-CT	AAAAAGAC-AAAAATTG	
Human $AluA$ (3')	------------------------	CAG-AACAA-	GACTCTGTCT		
Human $AluB$ (3')	------------------------	TGA-AACTCC	ATCTC-----	------------	AGAAGAG
Human $AluC$ (3')	------------------------	CGA-GACTCC	ATCTC		
Human $Alu53$ (3')	------------------------	AGA-GACTCC	GTCTC		
Human $Alu3'\alpha1$ (mid)	------------------------	GCG-A-----	AACTCTGTCT	CAAAAAAAATTTAATCTAATTTAA----	
Murine B1a	TCTTCAGAGTGAGTTCCAGGACACCAGGGCTACAC	AGA-AACCCT	GTCT		
Murine 7S	------------------------	------------	------------	------------	-------
Human 7L1	------------------------	------------	------------	------------	-------
Murine B2	------------------------	------------	------------	------------	-------
Chicken CRIU1	------------------------	------------	ATGCC-----	------------	-------

	XV	XVI	XVII	XVIII	XIX	XX	XXI	XXII
Murine 7S	CACCAGGTT	GCCTAAGGA	GGGGTGAAC	-GACCCAGG-CGGAA				
Human 7L1	CACCAGGTT	GCCTAACGA	GGGGTGAAC	CGGCCCAGGTCGGAA	ACGGAGCAG	GTCAAAACT	CCCGTGCTG	ATCAGTAGTG
Murine B2	TGGTGGCTC	ACAACCATC	CGTAATGA-	-GATCTGATGCCCTC	TTCTCGG-AG	TGTCTGAAG	ACAGCTACA	GTGG--AG--
Chicken CR1U1	CCGTCCATC	-CCT--GGA	GG--TGTTC	AAGGCCGGTTGGA-	-CGTGCCCT	GGGCAGCCT	GGGCTGGTA	CTGAAT-GTG

	XXIII	XXIV	XXV	XXVI	XXVII	XXVIII	XXIX
Human 7L1	GGATCGCGC	CTGTGAATA	GCCACTGCA	CTCCAGCCT	GGGCAACAT	AGCGAGACC	CCCGTCTCT

[a] The italicized sequences in columns II and III represent box A discussed in Section 3.5.1; those of column IX and onward correspond to box B. Asterisk indicates a 12-base-pair insert. r insert.

[b] Duncan et al. (1981).
[c] Deininger et al. (1981).
[d] Pan et al. (1981).
[e] Shen and Maniatis (1982).
[f] Krayev et al. (1980).
[g] Balmain et al. (1982).
[h] Ullu et al. (1982).
[i] Page et al. (1981); Krayev et al. (1982); Sakamoto et al. (1984).
[j] Stumph et al. (1981).
[k] Kato et al. (1982).
[l] Wolin and Steitz (1983).

all are transcribed by polymerase III. Doubtlessly the most striking distinctions of the first three are the differences in length in the 5′ direction, the murine 7 S RNA being strongly shortened and the human 7L1 considerably lengthened (Ullu *et al.*, 1982). Although the murine B1a sequence at this terminus is identical to the several *Alu* genes, it is somewhat elongated at the 3′ end (Krayev *et al.*, 1980). All five lack the inserted sectors and the structures in columns XI and XII, as well as the terminal appendage of *AluB*. The 7 S RNA of the laboratory rat, which is not included in the table, is largely homologous to that of the mouse (Li *et al.*, 1982), but its length is more comparable to that of the murine B1a substance (Balmain *et al.*, 1982). The human 7L1 has been shown to be involved in the secretion of proteins from cells and receives attention in that connection later (Walter and Blobel, 1982). In all these monomeric types homologization toward the 3′ terminus is not readily accomplished, for variation is more rampant there than in any other region. That of the murine B1a is especially distinguished by the existence of an inserted segment absent from the others. However, extensive homologies are at once evident between the human 7L1 and murine 7 S genes, but the former is far longer than the latter, having an additional 99-nucleotide extension.

Nevertheless, the extent of homology that is demonstrated to exist among many of these primary structures is obviously indicative of close kinship, and here it is proposed that a common name, probably Alu family, should be applied to the group as a whole. On the basis of the sequences currently known, the existence of four major groupings, apparently of subfamily status, are clearly evident. The first of these consists of those sequences that show extensive homology with the *Alu* genes proper but which are monomeric, such as the murine B1a and 7 S and the human 7L1. These represent most of the established sequences of a subfamily, here called the Monalu, whereas the various *Alu* genes from human and other primate sources seem to constitute a second group, dimeric in structure, which can be named the Diplalu subfamily accordingly. At the moment the third category has only a single established member, the trimeric *Alu3′α1*. This subfamily can be distinguished by the term Triplalu, while the fourth, which is probably composite, is not closely akin to the rest and could appropriately be referred to as the Analu. The murine B2 and the chicken CR1U1 may thus be placed here, together with a representative from Chinese hamster ovary cells (Haynes and Jelinek, 1981; Haynes *et al.*, 1981), until further knowledge is gained. In addition, the sequence found in an intron of the rat growth hormone gene (Page *et al.*, 1981) is closely allied to the murine B2 and, of course, is to be placed with it. Finally, the rabbit C family (Cheng *et al.*, 1984) shows occasional homology to *AluA* in regions II and III, but has no further detectable relationships; hence, it represents still an additional group.

The Cytalu Family. Still another shRNA type has recently been shown to be transcribed *in vitro* by RNA polymerase III, a variety called Ro RNA, present abundantly in the cytoplasm of mammals (Hendrick *et al.*, 1981). In human HeLa cells five varieties of the substance are found, while three are known in the rat and two in the mouse, the genes of man being referred to as *hY1–5*.

Surprisingly, the cistrons for these small RNAs seem to be present only as single copies (Wolin and Steitz, 1983). This reference also provided the primary structures of two genes, *hY1* and *hY3*, which occurred as neighbors, that of *hY5* having been established earlier (Kato, 1982). Two representatives appear at the end of the sequences given in Table 3.10, since *hY3*, being rather similar to *hY1*, is omitted. Although it is difficult to discern any regions homologous to any of the rest in the table, those few that are detected are confined to single bases or at most two in a row. However, there is one triplet sequence, the TGG beginning column II, that is identical to some of the others. Region IV is an especially good example of the lack of relationship, for the CCAGC- that marks the *Alu* genes is missing in each *hY* representative. The entire group, which is termed the Cytalu family here, thus is unrelated to any others of known composition. The name is in reference to their being found predominantly in the cytoplasm of cells.

A Look Toward the Future. The brief discussion of an organizational problem among *Alu* genes introduces another of far greater significance that is bound to concern all who explore gene structures. Although at the present time, the *Alu* family is the only evident group of genes whose interrelationships are in need of systematic arrangement, there undoubtedly will be many more of nearly equal complexity as knowledge of gene primary structure becomes increasingly available. But the members of that family abundantly suffice to bring the need for prompt attention to the forefront.

As alluded to in the preceding section, each of the five sequences of human *Alu* genes presented in Table 3.10 clearly represents a different species; the first four, being dimeric, were placed in a different subfamily from the fifth, *3'α1*, which is trimeric. The two *Alu* genes from green monkey kidney, briefly discussed earlier, likewise show marked kinship with the human *AluA–C* and *Alu53* (Krolewski *et al.*, 1984), whereas the four others located near the human δ- and β-globin genes are obviously kin to *AluB*. Since there are thus two kinship groups readily apparent among these few Diplalu genes, as visualized in Figure 3.2, how many will there prove to be when adequate sequencing has been accomplished among the 400,000 *Alu* genes present in the human genome alone? Moreover, how does one express the structural relationships between the various types of *Alu* genes proper with the human 7L1, the murine B1, and the 7 S cistrons of the mouse and rat, let alone the *hY1* and *hY3?* Most of these, too, are present in very high numbers in the genomes of all mammals and possibly in those of the majority of vertebrates. Although here they have been grouped loosely in families and subfamilies, this treatment doubtlessly will prove inadequate later, for their diversity can be expected only to become much greater as more representatives are isolated and sequenced.

Since classification is not a matter of great appeal, the impending problem disclosed here is not likely to receive attention until chaos resulting from increased knowledge compels those concerned to take appropriate action. A format for its solution already is at hand, created by biologists more than 200 years ago to solve a similar problem, when the complexity involving the countless species of living organisms compelled a search for such a system. With systematic biology's hierarchy of categories already familiar to all workers at the molecular level, a comparable scheme should present few problems for its erection. But whether in its application to genes, numerical or nominal designations are to be selected remains for the future. It is to be hoped that, whatever the system that eventuates, it be begun before the confusion resulting from the absence of a format becomes too deeply embedded in the literature. One need only recall that interspersed elements of many kinds are already known from all the types of eukaryotes whose genomes have been explored, and that there are around 10 million extant species of higher living things. Thus, with a mean of only 1000 such sequences present in each species, there are 10 billion interspersed genes represented in the genomes of the organisms of today. Those sequences that have already been identified incorrectly as members of the

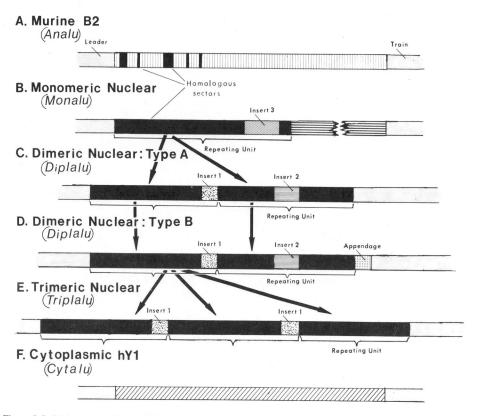

Figure 3.2. Major types of class III repeated elements. Five major types plus one variety are recognizable among these elements whose sequences have been established, the most thoroughly established being (C–E) the *Alu* type. These consist of imperfect repeating units, either (C,D) dimeric as in *AluA* and *AluC*, sometimes (D) with an appended segment as in *AluB*, or (E) trimeric as in *Alu53*. (B) Others, such as murine B1, show extensive homology to the *Alu* repeating unit but are monomeric, and (A) still others, including the murine B2, show only sporadic homology to those monomers. (F) Finally, those that occur chiefly in the cytoplasm display no homology.

Alu family and have led to misinterpretation of its possible functions (Schon *et al.*, 1981; Kato *et al.*, 1982; Watanabe *et al.*, 1982; Schimenti and Duncan, 1984) are mild examples of the confusion that may result from the absence of a systematic treatment if it persists too long.

Functional Considerations. The possible functions of such a diversified family of genes as the *Alu* can scarcely escape speculation. Among the first and most repeated suppositions is that the members are active in the transpositioning of structural genes about the genome (Jelinek and Schmid, 1982; Di Giovanni *et al.*, 1983; Krolewski *et al.*, 1984). In other words, as pointed out before, they are said to serve as transposons and thus may be associated with the origins of duplications, inversions, and similar chromosomal aberrations. Others have suggested that they may be involved in the processing, transport, or splicing of messengers (Lerner and Steitz, 1981; Robertson and Dickson, 1984), but no experimental evidence supports these views. Furthermore, a role in transposition on the part of these genes has been negated recently on the basis of structural considerations (Di

Giovanni *et al.*, 1983). As a whole, elements that have been directly implicated in the dispersal of genes, such as the *copia* and foldback groups (Spradling and Rubin, 1981; Potter, 1982; Scherer *et al.*, 1982), have symmetric ends, which is not uniformly the case here.

A more current suggestion is that *Alu* repeats have an evolutionary function in stabilizing gene structure, thereby maintaining diversity by inhibiting gene conversion (Schimenti and Duncan, 1984), based largely on observations of globin gene structure, particularly the differences that exist between the α- and β-globin gene families. Two main difficulties exist, the first of which pertains to the entire concept of gene conversion. The question raised is: How can genes undergo diversification after their duplication has occurred if gene conversion is an active force that maintains homology between related cistrons? One would expect that if a conserving force is meaningful, it would act immediately upon newly duplicated sequences, not just later after diversification has occurred. The second is that the evidence offered in support of this supposed role of *Alu* elements is derived from one of the cases of misidentification listed above. Neither of the introns ascribed to the Alu family by the investigators shows any significant homology to established members (Schimenti and Duncan, 1984). This aspect of the problem receives closer attention in a chapter on introns and their processing (Chapter 12), to which the discussion more correctly appertains.

A different functional basis for *Alu* inserts appears to emerge as their immense numbers and diversification are viewed. Perhaps the most likely role for their frequency and distinctions is that they serve in recognition of particular gene sets. The distinctiveness between the trimeric gene from the α-globin family from those dimeric ones associated with β-globin sequences is particularly convincing of the possibility of this proposal, and the presence of an appendage on all that flank the δ-globin gene is added evidence. Still further solid support is provided by the isolation of a factor that reacts specifically with a heatshock gene. This protein, which is essential for RNA polymerase II transcription of the cistron, binds to the regulatory site following activation by elevated temperatures (Parker and Topol, 1984). But perhaps even more supportive of this proposal is the occurrence of the repetitive element sigma, consisting of about 340 base-pairs each and flanked by 8-base pair inverted repeats (del Rey *et al.*, 1982). This peculiar structure is known to occur only in association with tRNA genes.

Obviously the complex metabolic processes of the cell require the presence of adequate control mechanisms. While many such controls are exercised at the other levels of anabolism, such as translation and processing, regulation of transcription also forms a major part of the whole picture. And how can individual genes or even operons be singled out for transcription when a need arises if there is no way of recognizing them for specific polymerase activity?

3.4.2. Viral-Associated Genes

Several viral genes specifying small ribonucleic acids have been found to be transcribed by cellular RNA polymerase III, the four known ones of which have now had their primary structures established (Akusjärvi *et al.*, 1980; Rosa *et al.*, 1981). Two of the cistrons, those for VAI and II, are from the genome of adenovirus 2, and the other pair, known as EBER1 and 2, are from that of the Epstein–Barr virus. Since the genes are fairly small (156–168 residues) their structures can be readily compared, as in Table 3.11.

Table 3.11
Virus-Associated Genes[a]

	I	II	III	IV	V	VI
VAI Leader[b]	GCTCTGCCGCGTG-AG	GCG-TGCCGA	GTCGTTGACGCT	CTAG-AC	CCTGCAAAA	GGAGAGCCTGTAAGC
VAII Leader[b]	GGGCTAGCTTTTTTGG	CCACTGGCCG	CGCGGGCCGTAA	GCGGTTA	GGCTGGAAA	GCGAAAGCCATTAAGT
EBER1 Leader[c]	CCCCGCCCCGTCA-CG	GTGACGTAGT	CTGTCTTGAGGA	GATG-TA	GACTTGTAG	ACACTGCAAAACCTC
EBER2 Leader[c]	CCTACAACCGTGA-CG	TAGCTGTTTA	CCAGCATGTATA	GA-GTTA	---CGGTTC	GCTACATCAAAC---

	VII	VIII	IX	X	XI	XII	XIII	XIV
VAI Gene	GGGCACTCTTCC	GTGGTCTGG	TCGATAAAT	TCGCAAGGG	TATCATGGC	GGACGACCG	GGGTTCGAA	CCCCGGATC
VAII Gene	GGCTCGCTCCCT	GTAGCCGGA	GGGTTATTT	TC-CAAGGG	TTGAGTCGC	AGGACCCCC	-GGTTCGAG	TCTCGGGCC
EBER1 Gene	AGGACCTACGCT	GCCCTAGAG	GTTTTGCTA	GGGAGGAGA	CGTGTGTGG	CTGTAGCCA	CCCGTCCCG	GGTACAAGT
EBER2 Gene	AGGACAGCCGTT	GCCCTAGTG	GTTTCGGAC	ACACCGCCA	ACGCTCAGT	GCGGTGCTA	CCGACCCGA	GGT-CAAGT

	XV	XVI	XVII	XVIII	XIX	XX	XXI	XXII
VAI Gene	CGGCCG	TCCGCCGTG	ATCCATGCGGTT	ACGGCCCGC	GTGTCGAACCCA	GGTGTCGCGA	CGTCAGAGACA	ACGGGGGAG
VAII Gene	-GGCCG	GACTGCGGC	GAACGGGGGTTT	GCCTCCCCG	TCATGCAAGACC	CCGGCTTGCA	AATTCCTCC	GGAAACAGG
EBER1 Gene	CCCGGG	TGGTGAGGA	CGGTGTCTGTGG	TTGTCTTCC	CAGACTCTGCTT	TCTGCCGTC	TTCCGTCAA	GTACCAGCT
EBER2 Gene	CCCGGG	GGAGGAGAA	GAGAGGCTTCCC	GCCTAGAGC	ATTTGCAAGTCA	GGATTCTCT	AATCCCTCT	GGGAGAAGG

	XXIII	XXIV	XXV	XXVI	XXVII	XXVIII
VAI Train	CGCTCC---	----------	-TTTTGGCTT	CCTTCCAGG	CGCGGC	GGCTGCT
VAII Train	GACGAGCCC	C----------	TTTTTTGCTT	-TTCCCAGA	TGCATC	CGGTGCT
EBER1 Train	GGTGGTCCG	CATG------	-TTTTGATCC	AAACTTTTG	TTTTAG	GATTTATG
EBER2 Train	GTATTCGGC	TTGTCCGCTA	-TTTTTTTGT	GGCTAGTTT	TGCACC	CACAACAT

[a] Italicized sequences in columns VII–IX approximate box A, and the underscoring in XII–XIV or XV correspond to box B.

[b] Akusjärvi et al. (1980).

[c] Rosa et al. (1981).

The Leader Sequences. In each of the viral genomes the genes occur in pairs, the train of the first merging into the leader of the second. An intergenic spacer 160 base pairs in length separates the two EBER sequences, but only a 100-residue region lies between those of the adenovirus. Upon examination of the table, occasional regions of homology can be noted at corresponding sites, all of which, however, involve only two of the four sequences. Nevertheless, one minor exception does exist—after the middle of column I the trinucleotide CCG occurs in VAI and the two EBER leaders. This is really a portion of a longer homology between the latter pair, which may be seen to be CCGT-ACG, undoubtedly the longest such sequence in this flanking region. In addition, the tetranucleotide AAAG is located at the end of column V and beginning of VI in the VA cistrons and a set of the pentanucleotide TAAGY brings their leaders to a close. EBER2 shows occasional agreement with VAII as in column IV, and only a little more frequently with EBER1, but VAI displays no sequences of three or more nucleotides that match corresponding sites of either of the Epstein–Barr viral leaders.

The Mature Coding Regions. A similar lack of meaningful homology characterizes the mature coding regions, and those that are present are largely confined to the pair from a particular virus. Thus the entire sector of column X is identical in the two adenoviral genes, whereas in those of the other virus no agreement can be noted, either between the two EBER sequences or with the others. Among the outstanding harmonious sectors of the VA genes are a heptanucleotide in column XIII, a triplet in XIV, a pentanucleotide in XV, and a final one of like size in XVIII.

The EBER sequences have their most extensive area of homology in columns VIII and IX, where the dodecanucleotide GCCCTAG-GGTTT is an obvious feature. An octonucleotide, GGT-CAAGT, in column IX, continued by a six-base region in column XV, may hold significance as a promoter, as discussed shortly. Further important identities are to be observed only in column XVI. In addition, as in the leaders, occasional segments of three or more residues show homologies between the coding regions of VAII and EBER2, whereas such is not the case between the two number one genes. The tetranucleotide GCCT in column XVIII is the first example of these identities, followed by TGCAAG in column XIX, AATYCCTCYGG in columns XXI and XXII, and AGG at the close of that region.

Possible Promoter Sequences. As with the *Alu* and other class III genes, a number of promoter regions have been suggested for these viral-encoded units. One of these, proposed by Galli *et al.* (1981), is shown in italics in Table 3.11 in columns VII–IX; this box A was derived, along with a box B, from comparisons of the VAI sequence with homologous regions in tRNA, 5 S rRNA, and others, not by experimentation. In the present case the former box is too interrupted and variable from one sequence to another to be convincing as a site for transcriptional recognition. A second proposal, underscored in Table 3.11 in columns XII–XV, is somewhat more uniform in location (Rosa *et al.*, 1981); it embraces the box B of Galli and co-workers (1981), but is more extensive. However, as with the first of these supposed promoters, its variability in the several representatives raises questions as to the validity of its service in that capacity. In contrast, by use of site-directed mutagenesis, two additional teams of researchers found two regions active in transcriptional control (Fowlkes and Shenk, 1980; Guilfoyle and Weinmann, 1981). One such sector, which was unusually long in that it began at site 9 or 10 and extended to 72–76 of the mature VAI coding sequence, affected transcription in an all-or-none fashion. A second control signal, not further identified, was determined as being in the leader in that same investigation. Later, in the summary closing this chapter (Section 3.5.1), the entire problem of recognition of these and the related genes is more fully treated.

Termination in Viral-Associated Genes. Termination apparently is signaled by a series of thymidine residues in each of the viral sequences, but not in a precise fashion. In the pair from adenovirus, transcription is closed following any one of the Ts lying immediately adjacent to the mature gene, as shown in column XXV of Table 3.11. Thus the 3' ends of the primary transcript are heterogeneous (Akusjärvi *et al.*, 1980). The sequences from the Epstein–Barr virus are slightly less variable, for transcription terminates after either the third or fourth T of the EBER1 train and after either the third, fourth, or fifth of the seven that lie immediately adjacent to the terminus of the mature coding sequence of EBER2 (Rosa *et al.*, 1981).

Function of Viral-Associated Genes. Little is known of the functions of these viral gene products, the most thoroughly documented species being VAI. In the late period of viral infection this substance is made in much greater quantities than VAII, in a ratio approaching 40:1, but both types seem to become combined into ribonucleoprotein particles in association with one, or perhaps several, cellular antigens (Lerner *et al.*, 1981). A mutant adenovirus that failed to synthesize VAII grew normally, but a second genetic variant that could not provide VAI RNA grew at a reduced rate (Thimmappaya *et al.*, 1982). The conclusion was drawn that this substance is essential for efficient translation of viral mRNAs synthesized during late stages of infection, by unblocking the ribosomal complex to permit selective translation of late adenovirus mRNAs (Thimmappaya *et al.*, 1982; Katze *et al.*, 1984).

3.5. SUMMARY OF TRANSCRIPTION OF CLASS III GENES

Although, as stated earlier, in a strict sense prokaryotic genes cannot be divided into classes on the basis of the DNA-dependent RNA polymerase that transcribes them as in eukaryotes, it is essential to any broad understanding of the nature of gene signals to treat them along with their eukaryotic counterparts in this manner. All the RNA polymerases are related, as clearly shown by their structure, and in every organism, they need parallel, if not identical, mechanisms for the control of their transcription and subsequent processing. Since the class III genes of eukaryotes, those transcribed by RNA polymerase III, have now been seen to represent quite a diversity of types, a summary of the analyses of their transcriptive processes seems basic to clarity.

3.5.1. Summary of Recognition and Initiation

A number of studies have been published that are concerned with the problem of recognition of class III genes in eukaryotes alone, primarily based on those from human or amphibian sources, along with a few from viruses. Among the common elements frequently proposed in those analyses is the existence of one or two "boxes" within the mature coding regions (Kressmann *et al.*, 1979; Fowlkes and Shenk, 1980; Galli *et al.*, 1981; Hoffstetter *et al.*, 1981; Ciliberto *et al.*, 1982a,b, 1983; Traboni *et al.*, 1982;

Murphy and Baralle, 1983). Such "internal promoters" will be recalled from the preliminary attention accorded them in connection with the tRNAs (Chapter 2, Sections 2.3.1 and 2.5); only now after all known class III gene types have received attention is it feasible to scrutinize them closely on the broad comparative basis the concept requires.

In the search for a common element that might be expected to signal the initiation of transcription by RNA polymerase III in eukaryotes, one should suspect that, in addition to evolutionary conservation of sequence composition, it should be placed at similar distances from the actual initiation point. Moreover, the existence of one or more ancillary signals to aid in distinguishing the several types from one another seems an appropriate secondary requirement, possibly to be waived in the case of viral genes because of their relative simplicity. Indeed, researches by some laboratories have already uncovered such factors active with this class of genes (Burke *et al.*, 1983; Lassar *et al.*, 1983). Here the various boxes that have been proposed are examined, followed by presentation of the problems associated with the structures, discussion of the known ancillary proteins, and suggestions for further experimental studies.

The Major Boxes A and B. One of two sequences that have been suggested to be common to all genes transcribed by RNA polymerase III has been called box A. This proposed internal promoter has the structure (T)RRYNNARY-GG, in which, as usual, R signifies purine, Y indicates either common pyrimidine, N denotes a nucleotide of any type, and a hyphen indicates either any nucleotide as in N or an occasional absence of an occupant. As suggested by the parentheses, the T is often included in the proposed box (Galli *et al.*, 1981; Hoffstetter *et al.*, 1981; Folk *et al.*, 1982), but is sometimes excluded by others (Ciliberto *et al.*, 1983). The second major common internal promoter is known as box B, located in arm V of the mature transcriptional product of tRNA genes, that is, approximately between sites 50 and 62 in typical small-type molecules. The suggested sequence in tRNA$_{CUG}^{Leu}$ of X. laevis was GGTTCGAATCC (Galli *et al.*, 1981), generalized to GGTTCGANNCC by the same research team.

Although attempts have been made to suggest that these two boxes are universal among the class III genes, primarily they were derived from studies on those of tRNAs. In applying them to 5 S cistrons, two different boxes corresponding to A have been proposed, the earlier being suggested to embrace the portions in italics in columns L–N of Tables 3.2 and 3.4 (Ford and Brown, 1976). While this proposal has subsequently been largely abandoned, it has been followed in at least one important recent analysis (Galli *et al.*, 1981). The majority of reports, however, suggest a sequence that more closely corresponds to that given earlier for this box, shown in italics in columns P and Q of those tables. In Table 3.12 the standard tRNA sequence is compared to this 5 S region of all the metazoans given in Table 3.4, along with a summarizing metazoan sequence derived therefrom and a similar generalized one for the protistan genes. It is notable that the homologies to the tRNA box A are not precise. For example, whereas the tRNA promoter begins with TRR, every cited metazoan and protistan 5 S gene has AAG, suggesting that this combination rather than the other is essential for promotion in the present cistrons. The sixth to ninth constituents appear equally significant, for the AAGC of those sites is quite as invariant as the beginning trio. Among metazoan types this is followed by an additional A, but this and the remainder of the box must be of little importance, to judge from the variability that is observed. Hence, the requirements for transcriptional recogni-

Table 3.12
"Boxes" of 5 S Genes

	Box A[a]	Box C[b]
Proposed[c]	TRRYNNARY-GG	GTCGGGCCTG GTT-AGTA CTTGGA[d]
Caenorhabditis	AAGTTAAGCAAC	GTTGAGTTCA GTT-AGTA CTTGGA
Drosophila	AAATTAAGCAGC	GTCGGGCGCG GTT-AGTA CTTAGA
Lytechinus	AAGTCAAGCAGC	ATAGGGCTCG GTT-AGTA CTTGGA
Xenopus	AAGTTAAGCAGG	GTCGGGCCTG GTT-AGTA CCTGGA
Rattus	AAGCTAAGCAGG	GTCGGGCCTG GTT-AGTA CTTGGA
Homo	AAGCTAAGCAGG	GTCGGGCCTG GTT-AGTA CTTGGA
Metazoan	AAGYYAAGCARN	RTNGRCYNYR GTT-AGTA CYTRGA
Protistan	AAGTTAAGCNNN	NNNRRGNNNN RNYGAGTG NTNNGA

[a]From Table 3.4
[b]From Tables 3.4 and 3.5.
[c]Ciliberto *et al.* (1983).
[d]Sakonju and Brown (1982) indicate 14 additional sites as being involved in this
box.

tion of 5 S rRNAs are to be viewed as similar, but decidedly not identical, to those for tRNAs, a point that needs cognizance in future researches on this problem.

A 5 S Promoter—Box C. As progress was made, it became evident that the 5 S genes lack a promoter corresponding to the box B of tRNAs; instead a region immediately following box A came to be suspected of playing an important role. This box C, as it is called, has been suggested to be GTCGGGCCTGGTTAGTACTTGGA-, the 3' terminus not having been firmly established as yet [see Sakonju and Brown (1982) for details]; the corresponding sector is shown underscored in Tables 3.2–3.5. Consequently, the cited study thus showed the two italicized portions of Tables 3.2 and 3.4 to be inactive in promotion. On the other hand, Table 3.12 suggests somewhat contrary results based on evolutionary conservation, for while only the T of the first three sites in the suggested sequence is actually uniformly conserved, a four-residue tract in column S near the middle of the sequence, TAGT, is invariable among metazoans; in other eukaryotes it differs only by the substitution of a G or a pyrimidine for the initial T and by a change in the final purine. Indeed, among protistans much of this entire box is extremely variable, except for that postcentral tetranucleotide. Thus it may prove to be of particular importance in recognition.

Boxes in Alu-Related Genes. The attempt has been made also to correlate boxes A and B of tRNAs with certain sectors of *Alu* genes and their relatives (Galli *et al.*, 1981). In Table 3.10 these regions are shown in italics, box A in column II and the very beginning of III and box B in column IX and the first two residues of X or the intervening insert. To bring out the resemblances and differences between the two boxes of tRNAs and 5 S rRNA and those of the present class of substances, the sequences have been

Table 3.13
Internal Boxes of Alu-Related and VA Genes[a]

	Box A	Box B
tRNA	TRRYNNARY-GG	GGTTCGANNCC
Metazoan 5S	AAGTTAAGCAAC	-----------
Human *AluA* (5')	TGGCTCACG-CC	AGTTCAAGACC
Human *AluB* (5')	TGGCTCAGG-CT	AGTTCAAGACC
Human *AluC* (5')	TGGCTCACA-CC	AGTTCAAGACC
Human *Alu53* (5')	--GTTCACG-CC	AGATCGAGACC
Human *AluA* (3')	TGGCAAGTG-CC	AGGCAG-GGTT
Human *AluB* (3')	TGGTGCATG-CC	AGGCAGAGGTT
Human *AluC* (3')	TGGCGCGCG-CC	AGGTGGAGGTT
Human *Alu53* (3')	TGGTGGGCA-CC	AGGTGGAGCTT
Murine B1a	TGGCGCATG-CC	-----GAGGCC
Murine 7S	--------G-CC	AGTTCTGGGCT
Human 7L1	TGGCGCGTG-CC	AGTTCTGGGCT
Generalized *Alu*	TGGYNNRNR-CY	AGNYNNRGNYY
VAI	TCTTCCGTGGTC	ACCGGGGTTCGAACCC
EBER1	TG-CCCTAGAGG	TCCCGGGTACAAGTCC

[a]The tRNA and 5 S sequences are from Table 3.11, while the *Alu* boxes are from Table 3.10; VAI signals are from Bhat *et al.* (1983) and those of EBER1 are deduced from Rosa *et al.* (1981).

arranged together in Table 3.13. The uniformity that exists in the box A sectors of the various *Alu*-related genes is particularly striking, especially the initiating TGGY combination and the paired Cs that close the proposed signal, for the first of these series is universal and the second has only one exception to constancy in the presence of a T in the 5' portion of *AluB*. However, when correlation with the corresponding region of tRNAs is sought, the results are disappointing, because only the initial quartet of nucleotides is convincingly homologous, the consistent A in the seventh site of the tRNA type being replaced by a G in a number of the transpositional RNA cistrons. Most disconcerting is the closing pair of Cs in the latter located where dual Gs are a catholic character of the former. Further discordances are displayed by the boxes of two *Alu* sequences reported by Perez-Stable *et al.* (1984).

Still fewer homologs can be detected in the box B regions. An AG combination exists at the beginning of the *Alu*-related signals that is quite as evolutionarily conserved as the GG of the tRNAs; in addition, the site that follows it is of no consequence in the former type, offering marked contrast to the ever-present T of the latter. Almost no other similarities can be noted between the two sets of promoters except that double Cs frequently terminate the present sequences as they do in the others; however, the dual Ts or CT combinations that often replace them tend to dilute the similarity. In summary, it does not appear likely that the enzymes that react to these boxes in the *Alu* genes and congeners are identical to those that respond to the signals of tRNAs, a statement equally applicable to the 5 S rRNA genes.

A particularly troublesome point of departure among Monalu types is the nearly complete absence of box A in the mouse 7 S gene and a remarkably similar absence of box B in the B1 gene from the same animal. One cannot help but wonder, since both murine cistrons are transcribed by the RNA polymerase III that acts on the tRNA genes, how much these conserved regions are actually involved in that activity. Because G-(- -)CC is apparently all that is requisite for recognition and transcription of these two murine genes, why are the remaining sites of any consequence in those processes elsewhere? These observations are strongly reenforced by the murine B2 and chicken Analu sequences that have only scattered points of homology with any of the others shown in Table 3.10. As pointed out in a preceding section, in B2 a few sites in columns I–III have occupants corresponding to those of other genes, homologies that are then absent until column VIII. There the closing hexanucleotide GAGTTC is identical to that of murine B1a and largely so to others adjacent to it in the table. Beyond that point, however, a complete absence of homology exists, including the italicized sequences in columns IX and X that have been suggested to be promoters. Yet this gene has been demonstrated to be transcribed by RNA polymerase III (Sakamoto *et al.*, 1984), as have the chicken and Cytalu genes, which lack even these few sections of similarities.

Signals in Virus-Associated Genes. Since the genes for the virus-associated RNAs VAI and II and EBER1 and 2 also have been demonstrated to be transcribed by RNA polymerase III, their sequences, too, should be expected to include regions similar to one or more of the suggested promoters. As may be noted following the *Alu*-related sequences in Table 3.13, the box A sectors of the VA and EBER genes differ both from one another and from all the other types transcribed by this polymerase, an initial T being the sole feature common to all. The TCT parenthetically separated from the remainder of the sequence is not part of the experimentally determined promoter (Bhat *et al.*, 1983), but is included in the table to show that realigning the entire box in the 3′ direction would not overcome the discrepancies that exist in the first three sites. No purines are present here to correspond to those universally present elsewhere. As a whole, the VA gene structure more closely resembles that of certain *Alu* cistrons, whereas the EBER (Rosa *et al.*, 1981) approaches the tRNA abstraction, but the similarities in each case are few in number.

In contrast, the box B series of nucleotides displays a much better fit to the abstractions of the three other types of genes, the VA promoter corresponding to the tRNAs in a most striking fashion. On the other hand, the EBER signal shows a greater likeness to several in the *Alu* 5′ portion, but not to the general formula of that class. In both viruses, the proposed promoter is longer than typical of current practice, but each more closely complies to the present trend of enzyme-protection experiments (Sakonju and Brown, 1982; Bhat *et al.*, 1983). Although the viral sequences show precise homologies at only 33% of the sites, there are enough additional ones dispersed throughout their lengths of a class type (purine or pyrimidine) that their possible use as a site for attachment of the same (but unknown) enzyme can be perceived. To the contrary, the two boxes A provided in the table show so few meaningful correspondences that the feeling cannot be avoided that perhaps this region is not strictly essential to cognizance on the part of the enzymes.

3.5.2. Factors Involved in Initiation of Transcription

As might be anticipated from the complexity of the signals seemingly involved in the initiation of transcription by RNA polymerase III, more than a few factors aside from that enzyme have been found essential to those processes. Although not all results of experimental studies are completely in harmony, there is frequent agreement on major points, if not on the finer details. Because a factor that is active with the transcription of 5 S genes has received the greater share of attention, it is convenient to start discussion with it before providing some insight into the nature of the others, including at least one that has an inhibiting effect.

A Transcription Factor for 5 S Genes. During the early stages of its development in *Xenopus laevis,* the oocyte is enriched in RNA, ~75% of which consists of tRNA and 5 S rRNA, with only a small amount of other rRNA species intermingled. The two abundant classes of this nucleic acid are not free, but occur bound as ribonucleoprotein (RNP) particles, one type of which sediments with a coefficient of 7 S, the other at 42 S (Ford, 1971; Denis and Mairy, 1972; Picard and Wegnez, 1979). The smaller consists of a molecule each of 5 S rRNA and a protein with a molecular weight of 39,000, whereas the larger consists of undetermined quantities of tRNAs and 5 S rRNA, combined with two proteins, with molecular weights of 48,000 and 40,000 (Picard *et al.,* 1980). Although the larger protein binds to the tRNA and smaller to the 5 S, the latter is not identical to the 39,000-dalton species, which has become known as transcription factor IIIA (TFIIIA). Both particles have been clearly demonstrated to be confined to the cytoplasm of *Xenopus* oocytes, although by itself 5 S rRNA enters the nucleus to become concentrated in the nucleolus where ribosomes are formed (Mattaj *et al.,* 1983).

One species of TFIIIA is reactive only with the oocyte type of 5 S rDNA, present in about 20,000 copies per cell and expressed only in oocytes and early embryos, whereas a distinct but immunologically related factor acts upon the somatic gene, which is represented by around 400 copies per haploid genome. To keep terminology precise but simple, it is proposed that the two species of transcription factors A should be distinguished by superscripts, A[o] designating the oocytic and A[s] the somatic varieties. *In vitro* TFIIIA[s] acts upon both types of genes, but may or may not do so *in vivo* (Pelham *et al.,* 1981). According to an early account, either species binds specifically to the intragenic region of 5 S genes, and to that class only, not to cistrons of tRNAs or others (Hanas *et al.,* 1984b). Binding takes place approximately from sites 45 to 96, that is, from the beginning of column P of Table 3.4 through column V of Table 3.5 (Engelke *et al.,* 1980; Sakonju and Brown, 1982). However, a more detailed account of this protein's reaction with the 5 S gene or transcript has been made possible by subdividing it proteolytically into three functional domains (Smith *et al.,* 1984). At one end of the protein a domain comprising about one-fourth of the molecule was found to be requisite for transcription, but not for binding to the DNA. Adjoining this was a second domain of similar size that bound to the 5' portion of the internal control region, while the remaining half of the whole factor reacted only with the 3' portion of that signal. The last of these proved inactive in promoting transcription, but the first two were essential to that process. Since the reaction is between an ancillary protein and the box, the latter is seen not to be an "internal promoter" as usually designated, but an ancillary site. Although not so clearly demonstrated as in the present case with 5 S rRNA genes, a similar condition undoubtedly will prove to prevail also in others of the class III category. The actual promoters thus remain unidentified.

The foregoing binding is neither dependent on nor affected by the presence of RNA polymerase III. Since TFIIIA° thus reacts with both the more widely accepted box A and the adjacent box C of 5 S genes only, any resemblance the former of these may have to the box A of tRNA cistrons and others is of no real significance. The 5' end of the given region, moreover, does not interact specifically even with large amounts of the protein in the absence of the 3' region (Sakonju, 1981), suggesting that the ancillary site probably is restricted to the box C sequence and perhaps a few adjacent sites in the 5' direction. Additionally, the two regions need to be correctly oriented for efficient transcription initiation (Bogenhagen, 1985) and in the silkworm a sequence on the leader must be present (Morton and Sprague, 1984). A need for additional factors was secondarily demonstrated by a quantative assay of the 5 S transcriptive processes in *Xenopus*, for the results indicated a fourfold difference in competition strength between oocytic and somatic genes, the former being the weaker (Wormington *et al.*, 1981). Although the observed distinctions were alleged to stem from differences in promoter structure between oocyte and somatic cell sources, this has not been fully established, since all sequences determined from *Xenopus* are oocytic in origin (Erdmann *et al.*, 1984).

Before leaving the TFIIIAs for an examination of other factors, a few additional observations of their properties need to be made. Each definitely binds to the 5 S rRNA genes at the internal control region, for they fail to react with them if that sector has been removed (Sakonju *et al.*, 1981; Hanas *et al.*, 1983). One effect of these factors' attachment to the promoter is the partial relaxation of the DNA molecule (Reynolds and Gottesfeld, 1983). The observed extent of uncoiling, 0.2–0.4 helical turn per binding site per total 5 S rDNA molecule, may be too slight to be of great significance to the acts of recognition and initiation. That action suggests that two to four nucleotide pairs in the reactive site are denatured, but as Hanas *et al.* (1984a) have pointed out, unwinding of the DNA is not an important aspect of these enzymes' binding actions.

Other Transcription Factors III. The presence of at least two additional factors involved in the activation of 5 S rDNA has been revealed in a search for control of these genes (Reynolds *et al.*, 1983; Setzer and Brown, 1985). A chromatin fraction isolated from *Xenopus* blood or kidney cells contained ~50% of the oocyte 5 S rRNA genes, but only around 2% of the total chromatin cDNA, all the genes remaining in a transcriptionally inactive state. However, they could be activated either by salt extraction of some of the chromosomal proteins or by incubation with an extract prepared from oocytes of the amphibian. The latter preparation was not specific for the 5 S cistron, but stimulated transcription of tRNA and other small RNA genes.

In an extensive investigation into transcription by RNA polymerase III of 5 S, tRNA, and VA genes, but not the *Alu* family, Lassar *et al.* (1983) found a need for three factors in addition to the polymerizing enzyme. TFIIIA, as already noted, proved to be active only with 5 S cistrons, but transcription also required the presence of TFIIIB and TFIIIC. These latter two factors were suggested to be sufficient for recognition by the polymerase of both tRNA and VA genes, but in view of the differences of structure of the promoter shown in Table 3.13 and in extent noted by other workers (Fowlkes and Shenk, 1980; Guilfoyle and Weinmann, 1981), this may be an overgeneralization. No search for additional factors active on *Alu* family genes appears to have been made, but some success has been attained with those of 5 S rRNA. In one analysis, a factor with a molecular weight approximating 22,000 was isolated (Shi *et al.*, 1983), and additional, but indirect, evidence also was gained that a third factor was present in the transcriptional complex, much larger than the rest, with a molecular weight approaching 100,000. Still more recently an additional factor with a molecular weight of 64,000–68,000 has been detected by use of autoimmune sera (Gottesfeld *et al.*, 1984), which inhibits RNA polymerase III transcription. This factor, called SS-B, was demonstrated to be an RNA-binding protein that reacted with 5 S rRNA transcripts *in vitro*.

Finally, in one of the few researches on the nature of polymerase III transcription complexes in eukaryotes other than *Xenopus*, it has been shown that in *Drosophila* 5 S and tRNA genes shared at least one associated factor (Burke *et al.*, 1983). One additional investigation into the present problem using nonvertebrate sources examined the nature of the processes in sea urchins (Morris and Marzluff, 1983). The factor, extracted from the eggs of these animals, has a negative influence specifically on homologous polymerase III activity, inhibiting transcription by that enzyme when injected into the nuclei of embryos at the blastula or gastrula stage. Additionally, in yeast a protein fraction has been found that specifically binds to blocks A and B and an intermediate area of tRNA genes (Ruet *et al.*, 1984; Stillman and Geiduschek, 1983). Even chloroplast tRNA genes seem to require an extensive leader section for transcription (Gruissem *et al.*, 1983).

3.6. ANALYTICAL SYNOPSIS

That the present state of knowledge concerning the processes of transcription by RNA polymerase III is less than completely satisfactory is indeed self-evident from the preceding discussions. Particularly unclear is the nature of the ancillary proteins that interact with the various types of genes, a facet of the problem made even more obscure by premature generalizations and the frequent failure to identify accurately the particular factor under investigation. To expose the current situation more fully so that future researches in this area may be expedited through the maze of complexities that characterize the problem, some of the more pressing difficulties associated with polymerase III transcription are presented, followed by suggested solutions.

3.6.1. Present Difficulties with Polymerase III Transcription

Since the *Alu* and viral-associated genes remain relatively poorly explored as to their processes of transcription, this exposition of difficulties associated with recognition and initiation of transcription by RNA polymerase III necessarily centers primarily on the tRNA and 5 S rRNA cistrons. However, the results of this analysis prove to be equally applicable to all the types of cistronic DNA transcribed by the present enzyme.

Problems with tRNA Gene Transcription. Undoubtedly the greatest problem in understanding the initiation of transcription of tRNA genes derives from oversimplification of the location, structure, and activities of various signaling sequences. Many studies persist in assigning all initiating activities to two internal "promoter" regions, actually ancillary sites, the first usually ranging from positions 8 or 13 to 20 and the second occupying sites 51–64 in typical tRNAs such as those that carry methionine, leucine, or tyrosine (Kressmann *et al.*, 1979; Galli *et al.*, 1981; Koski *et al.*, 1982). The earlier papers envisioned the polymerase as equal in radius to the entire tRNA gene and therefore capable of interacting with the 5' and 3' internal signals simultaneously. This view overlooks one well-established feature of tRNA structure—the many species often differ greatly in length, an observation resulting from two variable features in primary structure. The first inconstancy is the presence of arm IV between the two signaling regions; this variable arm, as it is usually called, ranges in length from a minimum of four residues, as in most tRNA^{Gly} genes and tRNA^{Glu} genes, to as many as 16 or 18, as exemplified by tRNA^{Leu} genes and tRNA^{Ser} genes (Sprinzl and Gauss, 1984). The second modification that influences the distance between the two ancillary sequences in genes is the frequent presence of introns, characteristically located in the vicinity of the anticodon. Such intervening segments have been shown to range in length from 13 residues in a tRNA^{Tyr} from *Xenopus* and 14 in a tRNA^{Trp} from Dictyostelium to 691 in a tRNA^{Gly} from tobacco chloroplast (Peffley and Sogin, 1981; Laski *et al.*, 1982; Deno and Sugiura, 1984). These deviations from a model length make it extremely unlikely that the 3' signal (box B) can possibly serve in any direct role in initiating transcription, at least in every tRNA gene. The inference is supported by a paper typically not referred to in studies on this problem, which showed that the 3' half of the yeast tRNA$_3^{Leu}$ cistron was unessential to initiating the processes (Carrara *et al.*, 1981), for an inserted sector of suitable length was found to serve efficiently in its stead. Thus box B conceivably could be involved in an interaction with a secondary factor associated with the processes, but close involvement with the polymerase itself seems totally excluded as a possibility, as shown in 5 S rRNA genes.

In addition, recent investigations have revealed the need for part of the 5' leader for efficient transcription of tRNA genes, as reported in less detail in Chapter 2. Here as elsewhere in the entire picture of the problem of recognition and transcription, results of experimental investigations are often contradictory. Among the first to show the need for part of the 5' flanking section was one on a silkworm tRNA$_2^{Ala}$ gene transcribed in a *Xenopus* germinal vesicle extract (Garber and Gage, 1979), the results of which suggested that a six-base-pair portion of the leader immediately preceding the mature gene was sufficient to support proper initiation. A paper published by a second group of researchers merely implicated the 5' leader without going into further detail (Kressmann *et al.*, 1979). To the contrary, a study by DeFranco *et al.* (1980) indicated that a 5'-external control repressed transcription in a *Drosophila* tRNA^{Lys} cistron, rather than simulating it. Later a second investigation into the silkworm tRNA^{Ala} gene reported that a 5' flanking region beginning 11 base pairs upstream from the initiation point was essential to transcription (Sprague *et al.*, 1980). Similarly a recent study demonstrated the need for a 5' flanking sequence with the *Drosophila* tRNA^{Arg} gene, but in a heterologous system, either 5' or 3' "half" molecules promoted transcription (Sharp *et al.*, 1983). The inability of the internal control signals to promote transcription by themselves alone has vividly been shown by an investigation of tRNA genes in *Schizosaccharomyces pombe* (Pearson, see Gamulin *et al.*, 1983). Three serine species that were sequenced proved to be followed in each case by a cistron for an initiator tRNA^{Met}. Although each of the latter had typical internal promoters, they could not be transcribed independently by RNA polymerase III, only when cotranscribed with the preceding tRNA^{Ser} gene.

The more penetrating investigation of *Drosophila* tRNA^{Arg} conducted by Dingermann *et al.* (1982) served to clarify the situation to some extent. Three of four identical repeated cistrons in a single cluster were found to be transcribed in a homologous extract, while the fourth, *p17DArg*, was not; but in all cases transcription was modulated by the 5′ flanking sequence. In a heterologous system from HeLa cells, all four were efficiently transcribed, but even small amounts of *Drosophila* extract had an inhibiting effect when introduced into the HeLa system. The latter reaction obviously resulted from incompatibility of substances within the respective extracts. This observation in turn indicates that studies under heterologous conditions may not truly be indicative of the transcriptive requirements of *in vivo* or homologous systems. While the coding regions of the four were identical, the 5′ sequences were not, and deletion mutations of this region in *pYH48* drastically reduced transcriptional efficiency, whereas in *p17DArg*—the one not transcribed in *Drosophila*—mutations in that leader did not enable it to be copied unless the entire flanking region was replaced by that from one of the other three. Thus recognition requirements by the RNA polymerase may approach those of bacteria (Lamond and Travers, 1983). This same laboratory has now demonstrated that 33 base pairs in the leader preceding the mature tRNA gene are essential for their efficient transcription in this insect (Schaack *et al.*, 1984), a finding echoed in a study of silkworm tRNA transcription (Fournier *et al.*, 1984). These results have been confirmed to some extent by researches on a yeast tRNA$_3^{Leu}$ gene, in which replacement of the section between 2 and 12 sites preceding the structural sequence greatly reduced transcriptional efficiency (Johnson and Raymond, 1984). Thus the functional promoter of tRNA genes appears to be located on the leader, relatively close to the mature coding sequence.

The Length Problem and 5 S Gene Transcription. Since many recent investigations into the nature of 5 S gene transcriptions have recognized the absence of box B from this class of cistrons, not all the foregoing considerations are directly applicable to this aspect of the problem. But there is an indirect inference that may be drawn that pertains not only to the 5 S rRNA type, but to the tRNA variety and all the others. Since factors of various sizes, ranging from 20,000 to perhaps 100,000 daltons, have been demonstrated to react with boxes A and C of 5 S rRNA, their presence on those genes would seem to preclude the immense molecule of RNA polymerase III from overlying the same region at the onset of transcription. Their masses suggest sizes equivalent to between 70 and 330 nucleotide residues. Hence, it appears likely that the polymerase occupies the leader during initiation, as in the eukaryotic polymerase I and II and the prokaryotic enzyme.

However, with the polymerase preceding the mature coding region, a length problem of somewhat different nature can then be perceived. Box A of tRNA genes is located consistently at eight sites from their 5′ ends, whereas in 5 S rRNA cistrons, the corresponding box is situated beginning around site 48 (Table 3.4). Assuming the interaction to be similar between identical components, one would expect initiation to occur about 40 sites further upstream with tRNA genes than with the 5 S. In other words, the nascent tRNA transcripts should uniformly bear a leader segment that is longer by 40 residues than those of unprocessed 5 S rRNA transcripts. Actually, primary transcripts of eukaryotic tRNA genes have leaders only 2–11 residues long, whereas those of 5 S do not have any (Dingermann *et al.*, 1982). The problem is, therefore, why do not the tRNAs have much longer leaders?

The most obvious answer is that the interacting substances are not identical in both cases. Because the polymerase is the same, the difference must lie in the second reactant, the one that is active with the first ancillary signal. Therefore the ancillary proteins involved with the respective genes must differ. That they are similar to a large extent is reflected in the homologies that exist between these signals in the several classes of genes; that they are different is correspondingly mirrored in the distinctions between those same structures. Hence, there is a separate protein factor involved in the recognition of the ancillary signals of tRNA, 5 S rRNA, *Alu*-related, and viral-associated genes, provided that that particular conserved sector of DNA proves to be actually involved in the recognition steps. In the case of the 5 S rRNA mature coding sequences (Table 3.12) boxes A and C may act as a single unit with a complex formed of several distinct factors, or A may not actually be concerned with this activity, the latter seemingly being the case, as noted earlier.

A Duality Problem in Alu Genes. One well-known fundamental characteristic of the *Alu* subfamily of genes has escaped attention in the various proposals of internal promoters for its transcription—its dual or triploid structure. As Table 3.10 clearly demonstrates, the 3' portion of true Diplalu cistrons is largely homologous to the entire 5' portion, icluding the italicized sequences identified as corresponding to boxes A and B. If recognition depends either on direct interaction of the polymerase with these promoters or on indirect activity involving transcription factors and the polymerase, there is nothing to prevent the frequent transcription of 3' portions devoid of the 5'. In other words 3'-"half" transcripts should be produced possibly at a rate comparable to complete ones, but no 5'-"halves" should be expected. Following the same line of reasoning, with Triplalu genes entire, two-third, and final one-third transcripts should be produced.

Members of other related gene subfamilies, such as the Monalu, Analu, and Cytalu genes, which lack this repeated structuring, would not be exposed to such partial transcript production. The murine members of the first of these subfamilies, however, have defective promoter sequences, as pointed out earlier. In the 7 S cistron box A is reduced to only the last three of the nucleotide residues, whereas in the B1a representative, box B consists of but five of the usual 11. Since a different box is defective in each case, it is not possible to explain the problem away by supposing that the genes of the mouse are regulated by factors distinct from those of man; furthermore, no evidence supports that point of view. Again referral to Table 3.10 reveals that the Analu and Cytalu types of genes completely or virtually lack homology with the signal regions of the others. Perhaps in all cases the real active ancillary site has been overlooked in attempts at correlating regions of supposed activity in these genes with those of tRNA cistrons. Studies involving areas of the coding regions protected against RNase activity by specific replication factors appear the most productive method of attacking this problem.

Viral-Associated Gene Difficulties. The first difficulty associated with understanding the transcription of viral genes lies in the structure of the two internal signals. Not only do the two from each viral source differ strongly from one another, but they contrast sharply, both individually and collectively, from those of other class III genes (Table 3.12). If one summarizes the entire collection of boxes A given in that table, the resulting consensus is found to be NNNYNNNNNRNN, a totally nonsensical combination. The result of summarizing just the two VA sequence signals, TNTYCCNNGRNN, shows only a little improvement, because they differ at too many points to provide the needed extensive homology. Although the boxes B display a far greater number of corresponding sites, both between themselves and that of tRNAs, the frequent differences make the pair less than completely convincing. Hence, the conclusion seems inevitable that the VA genes are transcribed by way of a separate set of associated factors, differing from those of any others transcribed by RNA polymerase III.

The second problem with the transcription of these viral genes is why a eukaryotic cell should carry genes for distinctive transcription factors used only to produce the coding regions for foreign substances. Although solid evidence is lacking, the final answer to this question most likely will be demonstrated to be that it does not do so, that those ancillary proteins are encoded in the respective viral genomes. This possibility explains the variation between the adenovirus and Epstein–Barr sequence signals in addition to providing a plausible answer as to why the necessary enzymes are produced.

3.6.2. The Transcriptive Processes

The processes of transcription by DNA-dependent RNA polymerase III are complex, in keeping with the several highly divergent types of genes on which it is active. Hence, it may be no accident that it is both the largest and most complicated of the three types of eukaryotic RNA polymerases (Chapter 2, Section 2.1.2) in consisting of a greater number of subunits than either of the others. As a consequence, the number of distinct ancillary proteins suggested by the foregoing paragraphs should have been expected, if any control is to be exercised by the cell over the production of the various types of macromolecules embraced here.

As in any first attempt at providing a generalized picture of a highly involved activity, the following proposals must of necessity be hypothetical to a large degree and need confirmation by future experiments. Although conjectural, the suggestions fit the available data and thereby are to be considered valid to that extent.

A General View of the Processes. Because DNA-dependent RNA polymerase III is unable to recognize any gene without assistance from other factors, as has been demonstrated by several studies (Gruissem *et al.*, 1981; Lassar *et al.*, 1983), it seems self-evident that the ancillary protein or proteins must then be responsible for the recognition processes of transcription. In the case of 5 S rRNA genes, transcription factor IIIA[s] or IIIA[o] would thus locate the internal promoter region of a somatic cell or oocyte and attach to it (Hanas *et al.*, 1983). Whether this factor is actually a single substance or a complex has not been determined, but the latter is a possibility because of the length of the promoter sequence to which it interacts. Two additional factors, IIIB and IIIC, bind to the 5 S gene before the RNA polymerase can attach and initiate transcription (Lassar *et al.*, 1983; Shi *et al.*, 1983). As just pointed out, the transcribing enzyme seemingly attaches to an unidentified promoter in the 5' end of the noncoding sequence—actually the 3' end of the coding DNA—after it has reacted with the ancillary proteins bound to the gene. Only then can it proceed downstream along the DNA in its synthesis of the RNA transcript. What fate the ancillary proteins experience as transcription proceeds has not been investigated.

But it must be noted that TFIIIA binds to the entire promoter region of 5 S genes, extensive though that area may be (Hanas *et al.*, 1983; Stillman *et al.*, 1984). Hence, there is no presently recognized point on the mature coding region suitable for attachment of factors IIIB and IIIC, if those identical proteins are likewise active with tRNA, *Alu*, and VA genes, as has been reported (Lassar *et al.*, 1983). It will be recalled that besides IIIA a protein of a molecular weight approximating 22,000 and another of perhaps over 100,000 are needed before the polymerase can act on 5 S genes (Shi *et al.*, 1983), echoing the three factors that had been proposed in a prior investigation (Segall *et al.*, 1980). One feature that proves particularly troublesome with the several reports is the rather wide range of molecular weights that have been ascribed to factor IIIA, varying from 37,000 (Engelke *et al.*, 1980) to 45,000 (Shi *et al.*, 1983). Although there may be factors common to the transcriptive needs of 5 S rRNA, tRNA genes, and the others, as described on occasion (Burke *et al.*, 1983; Lassar *et al.*, 1983), at present the absence of common signals to provide the requisite binding sites makes this proposal improbable.

Related but Distinct Factors—A Prediction. On the basis of the numerous conflicting results of investigations with regard to properties and molecular weights, and the lack of like signals in the several gene classes, a prediction of numerous factors being associated with polymerase III transcription seems a logical step. It is not improbable that the several ancillary proteins will often prove to be interrelated as behooves an evolving system, yet they would be expected to be so distinct that under *in vivo* conditions each would interact with one class of genes and no other. Thus factor IIIA may be found to be active solely with 5 S rRNA cistrons, assisted by a particular variety each of factors IIIB and IIIC that do not associate with tRNA or *Alu* genes. In turn, tRNA gene transcription would require the presence of varieties of factors IIIB and IIIC that do not aid in recognition of 5 S rRNA or *Alu* coding regions. And similarly with *Alu* genes, still further species of transcriptive factors may be expected.

But even this degree of complexity may prove to be simplistic. In view of the very numerous types of tRNAs that exist and the obvious need for a cellular control mechanism over their synthesis, a number of species of factors IIIB and IIIC doubtlessly are involved in tRNA transcriptional control, and this over and above variation associated with ontogenetic development and tissue differences. Some of these may be active at the group level of tRNA structure, as proposed in Chapter 2, Section 2.3.1. If the *Alu* genes eventually are shown to fall into numerous familial types, as appears probable from the little that is currently known, then a multiplicity of factor species also seems implicit to their transcriptional processes, to provide the necessary control mechanisms. Much more rapid progress will be made in the future concerning the intricacies of transcription by polymerase III when the possible existence of multiple, similar classes of ancillary factors is borne in mind and searched for carefully in experimental investigations.

4

The Remaining Ribosomal RNA Genes

As was pointed out in the opening chapter of this section, the ribosomal genes of eukaryotes other than the 5 S species are transcribed by DNA-dependent RNA polymerase I. Despite the large size of the mature genes, considerable progress has already been made in establishing their primary structures, including those of the flanking regions, intergenic spacers, and introns. Consequently, although the data available are understandably not overly abundant, they are amply sufficient to disclose both several features peculiar to the transcription of this class of biochemicals and their evolutionary relationships.

4.1. INITIATION OF TRANSCRIPTION OF PROKARYOTIC rRNAs

Because of the complexity of the rRNA operons and their included genes, the discussion follows a somewhat enlarged format of text organization relative to other chapters. The greatest divergence is that the prokaryotic and eukaryotic aspects of the several main topics are occasionally treated under separate major headings, rather than in a more unified manner. Not only is this treatment more feasible for presentation of current knowledge, it also facilitates comparisons between the two great classes of cellular types.

4.1.1. Nature of Prokaryotic rRNA Operons

Throughout the blue-green algae and bacteria, the rRNA genes are arranged in operons consisting of those for the minor (19 S), major (23 S), and ancillary (5 S) species, in that order. As shown in Chapters 1 and 2, genes for tRNAs typically are located on the minor–major spacer and one or more may follow that of the ancillary cistron (Figure 4.1A). Differences in organization do appear in the higher ranks of the Prokaryota, however. For example, in a species of Archaebacteria (*Methanococcus vanniellii*), one 5 S gene has been reported to occupy an operon containing genes for tRNAs but none for other species of rRNA (Jarsch *et al.*, 1983); in addition, there are four others with the typical pattern of organization of the three usual species. In another representative of that

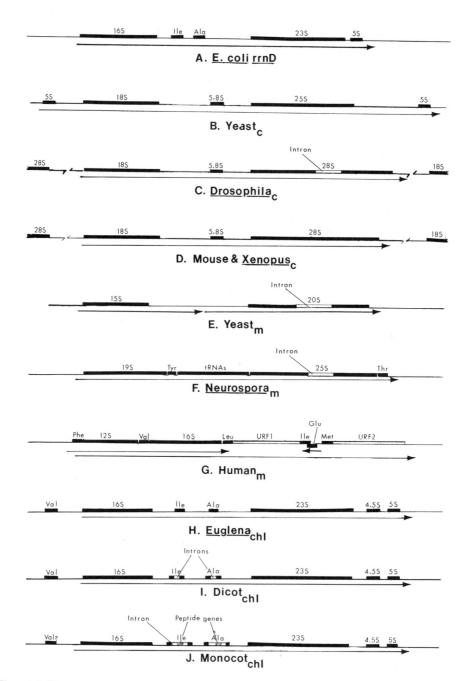

Figure 4.1. The genomic organization of rRNA genes. Since Figure 1.8 illustrates the genomic arrangement of the several prokaryotic rRNA operons, only a single one is presented here for comparison with these of the eukaryotes. Subscript c indicates cytoplasmic species, m denotes mitochondrial species, and chl denotes chloroplastic forms.

group (*Halobacterium halobium*), only the typical arrangement has been reported to exist (Mankin *et al.,* 1984).

The number of these structural units per cell varies extensively from one species to another. In *E. coli* seven are present, some of which have polarities opposite to those of others; *rrnD,* for a case in point, is oriented differently than *rrnB* or *rrnE* (Hill and Harnish, 1981; Morgan, 1982). *Bacillus subtilis,* with ten sets, is somewhat more richly endowed (Kobayashi and Osawa, 1982; Loughney *et al.,* 1983), but that number is often reduced in mutant strains (Gottlieb *et al.,* 1985). *Caulobacter crescentus,* and such members of the mycoplasmas (Mollicites) as *Mycoplasma mucoides capri, Acholeplasma laidlawii,* and *Ureaplasma,* have only two each (Ohta and Newton, 1981; Amikan *et al.,* 1982), although a small number of that group have but a single copy of this gene set (Amikam *et al.,* 1984). The number of these operons present in the blue-green algae appears to have been investigated only in *Anacystis nidulans,* in which two operons have been found (Douglas and Doolittle, 1984). Because all known representative prokaryotic sets contain one or more tRNA cistrons and are transcribed as a unit at least through the ancillary rRNA species, these operons are clearly of class IB, diversified by varying degrees of amplification.

None of the archaebacteria that have been studied appear to have many multiple copies of these *rrn* operons. Perhaps the most abundant sets are found in *Thermus thermophilus,* which has at least two and perhaps more (Ulbrich *et al.,* 1984). *Methanobacterium thermoautotrophicum* possesses two full operons per genome, whereas *Sulfolobus acidocaldarius* and *Thermococcus celer* have one plus an extra 5 S rRNA gene (Neumann *et al.,* 1983). Others, including *Halobacterium halobium, Thermoproteus tenax, Thermophilum pendens, Desulfurococcus mobilis,* and *D. mucosus,* have only a single copy of these operons. In the genera *Methanobacterium, Halobacterium,* and *Sulfolobus,* the genes in each set are closely linked, but in the others, the intergenic spacers appear to be somewhat lengthened.

4.1.2. Promotion of Prokaryotic rRNA Synthesis

Promoter Location. Another peculiarity of prokaryotic rRNA transcription initiation is in the location and number of promoters. Wherever investigated among these organisms, two promoters have consistently been found (Glaser and Cashel, 1979; Young and Steitz, 1979; Henckes *et al.,* 1982), situated about 100 or so base pairs from one another upstream of the minor rRNA gene. The distance between them and the 5' end of that mature gene is also remarkable, for the most remote one, referred to as promoter 1 (P1), is situated between 297 and 305 sites upstream in the several operons of *E. coli* and between 242 and 278 in *B. subtilis* (Table 4.1). P2 is more uniformly placed in the first of these organisms, at sites 186–189, whereas in the second they are more variable in location, that of *rrn315* being at −146 and that of *rrnB* at −187. In addition, operon *rrnB* of *E. coli* has been demonstrated to contain two further promoters, P3 and P4, located more than 1000 base pairs upstream from P1. At least *in vitro* these were shown to serve as functional initiation signals (Boros *et al.,* 1983); however, their actual participation in *in vivo* initiation of transcription has still to be determined. Nevertheless, the location of these possible promoters about 1400 base pairs before the mature gene should be borne in mind for later comparisons. Recently a search was made for the promoter in an archaebacterial leader (Mankin *et al.*, 1984); although the 5′ flanking region sequenced was nearly 800 base pairs in length, no actual initiating signal nor start site was determined.

Promoter Structure. Table 4.1 lists the sequences of promoter regions that have been established, six from *E. coli* and two from *B. subtilis*, aligned at corresponding regions of activity insofar as is feasible in order to permit ready comparison. Also included are the two additional ones from *rrnB* of *E. coli*, identified as P3 and P4; because of their structural characteristics, both are included with the P1 sequences following the remainder. Even superficial examination at once discloses the uniformity of structure that prevails in each type of promoter, offering marked contrast to the heterogeneity found among the tRNAs. In every operon of *E. coli*, P1 consists of the sequence TATAATG, both preceded and followed by a cytidine residue, and in the two of *B. subtilis* it has the sequence TATATT; AT consistently is adjacent downstream from that run, but in this case the upstream residue varies. Actual initiation of the transcriptive processes wherever specifically established has been with a purine triphosphate; experimentally determined sites in the table are indicated by underscoring, while presumptive sites are in italics.

The second promoter is less uniform in that two different structures occur in *E. coli* and *B. subtilis* alike. In the former, TATTATG and TAATATA can be noted, all preceded by a guanosine residue and followed by a cytidine, and in the latter, TATATT (as in its P1) and TATAGT, with thymidine and adenosine being the adjacent residues. Transcription starts with CTP in five cases from *E. coli* and in the other probably with GTP, the latter trinucleotide being characteristic of the two *B. subtilis* operons. P3 and P4 differ structurally from all others, respectively having the sequences TAAAATG and TAAAATA, with variable residues upstream and with G on the downstream side. Hence, there are only three invariable sites in these promoters of *E. coli*, TA---T-, and one must suspect that it is to these that the polymerase reacts insofar as this generally accepted region is concerned. Nevertheless, it remains unclear as to the actual effects of the differences in signal sequences, for experiments have demonstrated that *E. coli* RNA polymerase can utilize *B. subtilis* promoters *in vivo* (Wong and Doi, 1984).

But more than this sector is involved in the reactions with the transcribing enzyme, as effectively shown by a number of experimental studies (e.g., Siebenlist *et al.*, 1980). In these researches approximately six residues lying on the 5′ side of the promoter were shown to interact with the RNA polymerase. Thus the evolutionarily conserved TCC-C at the 3′ end of sector B of Table 4.1 together with P1 might well be a part of the recognition signal, perhaps along with the equally conserved CGCC at the beginning of sector C. With P2, the constant GC- and the CRC-A of regions B and C, respectively, may play a similar role. However, in *rrnG* the presence of CRAC in sector C weakens the likelihood that downstream sites participate in recognition activities. Comparable consistencies and variabilities can be noted also with the coliform P3 and P4 and with the two promoters of the bacillus.

The Ancillary Signal. As a whole, ancillary signals have been poorly investigated, those that have been suggested in the literature being italicized in Table 4.1. However, such remarkable sequence uniformity exists in the region indicated with both P1 and P2 that its involvement with initiation appears self-evident. In the first promoter series, two variations in both species of bacteria appear to exist, the established CTTGT being present in all the *E. coli* representatives except *rrnE*. In the latter TTG-CRR is a strong possibility, for it corresponds to one of those of *B. subtilis,* the other in all likelihood being TTGACC. In the case of the P2, only a single signal has been determined, the TTGACAA found in *rrnB* of the bacillus. But again in this region such striking consistency prevails that proposals of TTGACT as a second ancillary site for *B. subtilis* and TTGACTC for *E. coli* seem self-evident. It may not be without significance that near identity exists within each of the two series of sequences in the entire B region of P2, so that this sector, too, may eventually be found to play some role in the recognition processes.

Functional Relationships. Although the number of experiments on promoter function has been relatively small, the results stemming from them are of much interest. In one study it was shown that the RNA polymerase reacted similarly with both P1 and P2, unwinding the DNA molecule to about the same extent (12 base pairs) in each case (Siebenlist *et al.,* 1980). The relative activity of the two promoters was the subject of a recent thorough investigation (Glaser *et al.,* 1983). When small amounts of supercoiled DNA (0.1 μg/ml) from *rrnA* were used as template *in vitro* at 20°C, full-length transcripts from P2 appeared first, followed after 1 min by those from P1. After 2 min, transcripts from the latter were produced at a much greater rate than from P2, whose production efficiency decreased with time—after 4 min, transcription from P1 was four times that from P2. If such higher concentrations of *rrnA* DNA as 10 μg/ml were used, however, transcription production after 4 min from the second promoter was twice as great as from the first.

The presence of ppGpp (3'-diphosphate guanosyl-5'-diphosphate) is known to have an inhibitory effect on rRNA transcription in *E. coli* (Gallant, 1979), for it is involved in the so-called stringent response (cessation of rRNA synthesis when the bacteria are starved for a required amino acid) (Cashel and Gallant, 1969). In experiments with *rrnA* DNA, the effects varied with the promoter, transcription from P1 being retarded 80–90%, whereas that from P2 was inhibited only 50% (Glaser *et al.,* 1983). Rifampicin at concentrations of 0.1 μg/ml showed effects comparable to those of ppGpp. Moreover, the effects of ppGpp varied from one operon to another (Oostra *et al.,* 1977, 1980); for example, *rrnE* proved to be far more sensitive to the substance than *rrnX* (Hamming *et al.,* 1980). This resulted from a greater rate of complex formation with the promoter in the former than in the latter.

Only a single report on the action of prokaryotic ancillary enzymes appears to have been published (Muto, 1981). In an *in vitro* system consisting of *E. coli* RNA polymerase and a plasmid carrying an *rrnC* operon, rRNA synthesis was stimulated by a crude protein fraction prepared from *E. coli* cells. Subsequently, the activity was recovered by fractionation on a Sephacryl S200 column and found to result from a protein with a molecular weight of around 50,000. The factor specifically increased the initiation of transcription rate and was inhibited by the addition of ppGpp. In view of the varying effects found with different promoters, it is highly essential that more extensive investigations be conducted, both with the ppGpp and ancillary proteins using each of the several *rrn* operons and their respective separate promoters. The effects of rRNA levels of the cell also require further investigation, for it is these, not those of tRNA as once believed, that control transcription of rRNA operons as well as the gene sets of tRNAs (Gourse and Nomura, 1984).

Table 4.1
Promoter Regions of Prokaryotic rrn Operons[a]

	A	Ancillary	B	P1	C	Initiation	D
E. coli A (-305)[b]	TAAAITTTCC	TCTTGTCAG	GCCGGAAT-AACTGC-	CTATAATGC	GCCAC	CACT	GACACGGAACAACGGCAA
E. coli B (-301)[c]	TAAAITTTCC	TCTTGTCAG	GCCGGAAT-AACTCC-	CTATAATGC	GCCAC	CACT	GACACGGAACAACGGCAA
E. coli D (-299)[d]	CAAAAAAT	ACTTGTGCA	AAAAAATTGGATCC-	CTATAATGC	GCCTC	CGTT	GAGACG--ACAACGTG--
E. coli E (-297)[b]	AATTTTTCT	A-TTG-CGG	CCTCGGAGAACTCC-	CTATAATGC	GCCTC	CATC	GACACGGCGG-ATGTGAA
E. coli G (-297)[e]	TATTTTTC-	GCTTGTCAG	GCCGGAAT-AACTCC-	CTATAATGC	GCCAC	CACT	GACACGGAACAACGGCAA
E. coli X (-300)[d]	CATTTTTCC	GCTTGTCTT	CCTGAGCCG-ACTCC-	CTATAATGC	GCCTC	CATC	GACACGGCGG-ATGTG--
B. subtilis B (-278)[f]	TAAAAAACT	A-TTG-CAA	TAAATAAATACAGGTG	TTATATTTAT	TAAA-	CGTC	-----GCTG-ATG-CAC
B. subtilis 315 (-242)[g]	TAAAITGGG	TGTTGACCT	TTTGATAATATCCGTG	ATATATTAT	TATT-	CGTC	-----GCTG-ATAAACA
E. coli B P3[c]	TAAGGATTA	CTCATCTTA	TCCTTATCAAACCGT-	-TAAAATGG	GCGG-	TGT-	GAG---CTTG--TGGGTG
E. coli B P4[c]	GTATCCGGT	CACCTCTCA	CCTGACAGTTCGTGG-	-TAAAATAG	CCAA-	CCTG	-----TTCG-ACAAAGA

	A	Ancillary	B	P2	C	Initiation	D
E. coli A (-186)	AATAAATGC	TTGACTCTG	TAGCGGGAAGGC-	*GTATTATGC*	AC-A-	CCCC	GCGCC-GCTG-AGAAAAA
E. coli B (-187)	AATAAATGC	TTGACTCTG	TAGCGGGAAGGC-	*GTATTATGC*	AC-A-	CCCC	GCGCC--CTG-AGAAAAA
E. coli D (-189)	AATTCAGGG	TTGACTCTG	AAAGAGGAAAGC-	*GTAATATAC*	GCCA-	CCTG	GCGACAG-TG-AGCTGAA
E. coli E (-187)	AATTCAGGG	TTGACTCTG	AAAGAGGAAAGC-	*GTAATATAC*	GCCA-	CCTC	GCGACAG-TG-CGCTAAA
E. coli G (-188)	AAGAAATGC	TTGACTCTG	TAGCGGGAAGGC-	*GTATTATGC*	ACAC-	CGCC	GCGCC-GCTG-AGAAAAA
E. coli X (-189)	AATTCAGGG	*TTGACTCTG*	AAAGAGGAAAGC-	*GTAATTATC*	GCCA-	CCTC	GCGACAG-TG-AGCTGAA
B. subtilis B (-187)	AAAAAGTTG	*TTGACAAAA*	AAGAAGCTGAATG	*TTATATTAG*	TAAAG	CTGC	TTCCAT--TG-AAGAAGT
B. subtilis 315 (-146)	AAAAAGTTG	TTGACTTAA	AAGAAGCTAAATG	*TTATAGTAA*	TAAAG	CTGC	TTCGTT--TA-AGCGGCA

[a]Experimentally established start sites are underscored. Presumptive regions of importance are italicized.
[b]De Boer et al. (1979); Gilbert et al. (1979).
[c]Boros et al. (1983).
[d]Young and Steitz (1979).
[e]Shen et al. (1982).
[f]Stewart and Bott (1983).
[g]Loughney et al. (1983).

4.2. INITIATION OF TRANSCRIPTION OF EUKARYOTIC rRNAs

The multiplicity of copies of rRNA genes in eukaryotic genomes described shortly has a bearing on the comparative aspects of the problem that has not yet been recognized in the literature. As in the genes for tRNAs and 5 S rRNAs, the wide disparity that exists among initiation regions of the operons for the present biochemicals has been attributed merely to lack of evolutionary conservation (Niles *et al.*, 1981; Harrington and Chikaraishi, 1983; Michot *et al.*, 1983), whereas in reality it more than likely is related to the chance sequencing of genes from noncorresponding sectors of the genome. A later discussion presents most convincing support for this suggestion.

4.2.1. Organization of rRNA Genes

Except in two particulars, the organization of rRNA genes in eukaryotes parallels that of the prokaryotes. Here, as there, the cistrons for several types are arranged as operons, but that for the ancillary is widely separated from, and transcribed independently of, the others. In addition, a 5.8 S (here called the supplemental type) gene is present in the spacer between the minor and major species, which is devoid of any tRNA cistrons. Hence, the operons belong in class IA, not in IB as in prokaryotes, a distinction that may be evidence that the class IB operon is evolutionarily more primitive than the IA. Throughout the eukaryotic world, the sequence of genes in the operons on the noncoding strand of DNA is 5'-minor–supplemental–major-3'.

The rRNA Genes of Protistans. Also as in many prokaryotes, the operons of protistans are typically present in multiple copies, but in the present organisms amplification is of the category V type, not IV, the sets being tandemly arranged, not dispersed. Among the true yeasts, the degree of amplification is ~140-fold per haploid cell (Bell *et al.*, 1977). The major portion (70%) of these are in chromosome XII, not in I as once believed, in addition to which there are two small populations in satellite DNA (Bollon, 1982). As the result of transcription, a precursor is produced having the sedimentation coefficient of 35 S.

Although these genes have currently been studied in only a few other protistans (Berger and Schweiger, 1982; Blackburn, 1982), those of *Tetrahymena* have received much attention (Cech *et al.*, 1982; Yao, 1982; Challoner *et al.*, 1985). In this ciliate the operons are located in multiple copies of DNA molecules, separate from the chromatin of the chromosomes, as shown by their being packaged in a number of extrachromosomal nucleoli (Engberg *et al.*, 1976; Karrer and Gall, 1976; Vavra *et al.*, 1982). Each molecule has a molecular weight of 13×10^6, containing two operons arranged in inverted fashion, with both initiation regions located toward the center. Because of this arrangement, these structures have consistently been erroneously referred to as palindromes. Misapplication of this term with its very precise meaning to segments that do not read identically in each direction is a technical error equivalent to the misidentification of lipoproteins as glycoproteins or to referring to purine bases as steroids. A sector of DNA that consists, for example, of a sequence like GAGTTACATTGAG or CTAGCCTATCCGATC is a palindrome, because either combination is the same whether read from the 5' or 3' end. In contrast, a section in the noncoding strand consisting of GGACATGCATGTCC is not a palindrome, even though its complement in the coding strand reads identically in the seemingly opposite direction. Actually both together may read the same, but *each alone* does not do so. Hence, the latter example and its complement together form a region of dyad symmetry, not a palindrome. A major functional distinction is that in the case of a palindrome, its complementary strand or sector of a stem-and-loop structure, being also a palindrome but of complementary bases, complements the other in either a parallel or antiparallel fashion, whereas the nonpalindromic pair can pair only in the latter mode. Because of the limited numbers of different bases in nucleotides, short actual palindromes abound in both DNA and RNA, but how does one distinguish real ones if the term is misused, as occurs all too frequently in the literature? In the case of the rRNA operons of *Tetrahymena*, the two sets of genes should be referred to as being arranged as inverted pairs, with their 5' ends placed medially but separated by a long spacer.

The rRNA Genes of Metazoans. Varied patterns of amplification have been described in several types of metazoans. For example, in the nematode *Caenorhabditis elegans*, 55 copies of the operon exist, each consisting of about 7000 base pairs, except one of 4100 that has a defective minor rRNA species gene (Files and Hirsh, 1981). To date, the number of copies of the *rrn* operons in *Drosophila* does not appear to have been established, but the multiple copies sometimes include 28 S rRNA genes that contain introns, while others have those genes uninterrupted (Beckingham, 1982). Since two classes of introns based on relative lengths exist, as well as varying lengths of the so-called nontranscribed spacer, the operons have been found to be highly heterogeneous in length (Wellauer and Dawid, 1978; Wellauer *et al.*, 1978; Pavlakis *et al.*, 1979; Long and Dawid, 1979). The sea urchin *Paracentrotus lividus* shows a similar variation in size, likewise stemming from the inconstancy of the spacer (Passananti *et al.*, 1983). But this variability in length was not reported in a related species (*Lytechinus variegatus*) when its spacer was sequenced (Hindenach and Stafford, 1984).

During the polytenization that occurs in certain tissues of *Drosophila*, the ribosomal operons are underreplicated relative to others (Henning and Meer, 1971; Spear and Gall, 1973; Kunz *et al.*, 1982). In these processes the gene sets containing cistrons for major species that are devoid of introns are favored relative to those that bear such intervening segments (Franz *et al.*, 1981), but the latter are increased as the number of intronless genes decreases from one mutant genome to another. Because only a small percentage of intron-containing major species transcripts could be found in the nuclei, it was supposed that that type of gene was not transcribed and, hence, was functionless (Long and Dawid, 1979; Long *et al.*, 1981) The recent finding in *Tetrahymena* that the introns acted spontaneously to remove themselves and unite the two resultant fragments (Kruger *et al.*, 1982; Zaug *et al.*, 1983) affords an alternative explanation—splicing (probably by enzymes in this insect) is accomplished so rapidly that few examples bearing the introns remain. The so-called nontranscribed spacers mentioned above have been shown actually to be transcribed in this genus of organisms, as well as in *Xenopus* and the mouse (Rungger *et al.*, 1979; Kohorn and Rae, 1982a,b; Reichel *et al.*, 1982; J. R. Miller *et al.*, 1983).

The rRNA Genes of Fungi and Seed Plants. Among the fungi the ribosomal operons are arranged as in the ciliates, being in multiple copies of large extrachromosomal molecules, each bearing two copies with reversed polarities (Ferris and Vogt, 1982). In *Dictyostelium* 180 copies are present (Hoshikawa *et al.*, 1983), but in *Physarum* only ~100 are found, located on linear DNA segments about 60,000 base pairs in length

(Seebeck and Braun, 1982), and in *Coprinus* as few as 60–90 have been reported (Cassidy *et al.*, 1984). These are arranged as inverted pairs that are transcribed separately into large precursors containing one copy of each of the three usual genes (Weiner and Emery, 1982).

Unlike amphibians and certain other metazoans, the seed-plant rDNA is confined to chromosomal chromatin, the number of copies per genome varying between 1000 and 30,000, depending on the species (Ingle *et al.*, 1975; Leweke and Hemleben, 1982). The usual genes, minor (18 S), supplemental (5.8 S), and major (25 S), are in tandem arrays, the repeat unit lengths ranging from 8000 to 18,500 base pairs (Gerlach and Bedbrook, 1979; Yakura *et al.*, 1983), and heterogeneity resulting from different lengths of spacers (Yakura *et al.*, 1984). Transcription by RNA polymerase I in the organizer region of the nucleolus produces long precursors containing one copy of each of the genes, which are then processed by multiple steps to form the mature products.

4.2.2. Promotion of Eukaryotic rRNA Operons

The state of knowledge concerning promotion of eukaryotic rRNA operons is diametrically opposite that of the prokaryotes, for while considerable breadth of coverage exists, there is little depth. In only two cases has more than a single promoter region been sequenced from the same organism; as may be noted in Table 4.2, in both yeast (*S. cerevisiae*) and *Tetrahymena*, two or three different gene sets have had their primary structures established, that of a third example, not included, from the ciliate differing strongly from the other two (Spangle and Blackburn, 1985). What is most unfortunate is that the majority of investigators have overlooked the possibility that dual promoters may exist, a condition that has been firmly documented for the prokaryotic counterparts. One operon alone has been sufficiently explored to expose the existence of duality of transcription initiation, that shown in the table as *Tetrahymena* 2 (Niles *et al.*, 1981). Moreover, promotion involves the presence of a complex of proteins and DNA sequences, along with polymerase I, not just the latter enzyme alone (Iida *et al.*, 1985); some modification of the polymerase may also prove essential for transcription of the ribosomal RNA genes (Paule *et al.*, 1984).

A little depth to promoter knowledge has been gained recently by the results of two studies on structural aspects of yeasts on a comparative basis. In one investigation the entire region around the transcriptional initiation site was sequenced in a second species of *Saccharomyces* (*S. carlbergensis*) (Klootwijk *et al.*, 1984). The results showed complete agreement at each point with the sequence given in the table as yeast 2 (Valenzuela *et al.*, 1977; Bayev *et al.*, 1980; Klemenz and Geiduschek, 1980). In the second set of researches, leaders of four species of the superfamily Saccharomycetoidea were compared (Verbeet *et al.*, 1984). All of these, *Kluyveromyces lactis, Hansenula wingei, S. carlbergensis,* and *S. rosei,* displayed a high level of homology in much of the region near initiation site 1 of the yeast 2 sequence, as well as in the regions upstream and downstream from that point. However, no correspondences from other sources to yeast 1 have been reported. Another difficulty with defining promoters is often in their remote location— one from *S. cerevisiae* has proven to lie 2000 base pairs upstream (Elion and Warner, 1984).

Absence of Homology. The lack of appreciation just mentioned of the applicability of the prokaryotic structural condition to the promoter regions of these more advanced organisms accounts in part for the slow progress that has been made in establishing the details of transcription of these genes by polymerase I. But it is only one of several contributing factors. The obvious lack of homology that exists among their leader sequences (Table 4.2) reflects a condition pointed out in previous chapters. Whenever operons or genes are highly reiterated, as here, extensively homologous leaders can be expected only when corresponding genes are compared, that is, when the genes are taken from comparable operons. The two sequences from *Tetrahymena* provide concrete evidence of the reduced number of identical sites in leaders that have not been taken from corresponding genomic sectors, for they agree at only 42 of the 111 sites, for a 37% homology rate. In contrast, the representatives from the mouse and rat (Urano *et al.*, 1980; Mishima *et al.*, 1981; Rothblum *et al.*, 1982; Harrington and Chikaraishi, 1983) evidently are from closely corresponding genes, for they agree at 61 of the first 75 sites, for a homology rate of 81%, as indicated by the italicized regions. The human gene also appears to be somewhat related, for it shows occasional short sectors (italicized) that agree with the evolutionarily conserved regions of the other two mammals. It is not unlikely, when adequate sampling and sequencing have been accomplished, that a leader sequence will be found present in human cells showing a level of homology with these others from mammalian sources nearly equivalent to that between the rat–mouse pair.

Possible Promoter Regions. Because of the inverted, outwardly oriented arrangement of the dimeric operons that constitute the extrachromosomal DNA segments, transcription proceeds from near the center in both directions; that is, each of two sets of enzymes transcribes a different DNA strand (Gall *et al.*, 1977; Niles, 1978). In *Physarum polycephalum*, transcription is initiated 18,000 base pairs upstream of each mature 19 S gene and results in precursors 13,400 base pairs in length (Ferris and Vogt, 1982). A similar length (equated to around 35 S, but this appears incorrect) seems to characterize the *Tetrahymena* primary product (Michot *et al.*, 1983), but in vertebrates, variable lengths occur. In *Xenopus laevis*, the precursor is about 8000 nucleotide residues long, with a sedimentary coefficient of 40 S, the transcription start lying about 900 base pairs before the mature gene (Wellauer and Dawid, 1974; Sollner-Webb and Reeder, 1979). The transcript is slightly longer in mammals, since it sediments at 45 S (Financsek *et al.*, 1982; Harrington and Chikaraishi, 1983). Consequently, the organization of the rRNA genes of eukaryotes is similar to that of prokaryotes in having the initiation sites widely removed from the first mature gene, but the distance involved is considerably greater here, in keeping with the much larger size of the genome.

In *Xenopus* and the mouse, the precise starting point of transcription from P1 has been established as being with an adenosine immediately following the AGGT of column Q (K. G. Miller and Sollner-Webb, 1982; Sollner-Webb *et al.*, 1982). However, no conclusion has been reached as to the location and nature of the promoter. One report located it in *Xenopus* within the region lying between 320 sites upstream and 113 downstream of the initiation start (Bakken *et al.*, 1982), but this extensive area has been reduced on a preliminary basis to that between 12 sites upstream and 16 downstream from the initiation point (K. G. Miller and Sollner-Webb, 1982). However, ancillary signals lying close to either 40 or 150 sites upstream also seem to be essential for transcription. Further, in the mouse only a limited number of nucleotides around the initiation site have proven essential; a change from G to A at site -7 completely eliminated promoter activity, whereas other mutations at nearby locations had no effect (Kishimoto *et al.*, 1985).

Table 4.2
Regions Surrounding the Initiation Sites (IS) of Eukaryotic rRNA Genes[a]

P1 Sequences

				M	N	O	P1	Q
Escherichia rrnA[b]								CTATAATCG	GCCACC
Saccharomyces 1[c]						--	-------	-------	-----
Saccharomyces 2[c]					GT	GAGGA-AAAGTA	GTTGGGAGG	TACTTC
Tetrahymena 1[d]	GACTTTTGAGAC	TTAGAGAAA	ATTTTTCTGGCA	AAAAAAAAA	AAAAAA	GTATCAGGGGGG	TAAAAATGC	ATATTT
Tetrahymena 2[e]	GAAACAGAAAT	ACAAGGAAA	CCAGAAATTTTC	TCTAAACAA	AGACTT	TTTTTGTGGCAA	AAAAAAAA	AAATAT
Dictyostelium[f]	CAAAAAAATACT	GACTACCAT	ACAAACTAGTGT	GACCATTAG	CCATTG	AGCTCCGTGACT	ACCCCAAAT	ACATAT
Drosophila[g]	AGAATAGCCCGT	ATGTTGGGT	GCTAATGGAATT	GAAAATACC	CGCTTT	GAGGACAGCGGC	TTCAAAAAC	TACTAT
Xenopus[h]	CGGCCGCCCGGC	CTCTCGGGC	CCCCCGCACGAC	GCCTCCATG	CTACGC	TTTTTTGGCATG	TGCGGGCAG	GAAGGT
Mouse[i]	CCTGTCACTTTC	CTCCCTGTG	TCTTTTATGCTT	GTGATCTTT	TCTATC	TGTTCC-TATTG	GACCTGGAG	ATAGGT
Rat[j]	CCTGTCCATGCTT	ATCCCTGTG	TCTTTTACACTT	TTCATCTTT	GCTATC	TG-TCCTTATTG	TACCTGGAG	ATATAT
Human[k]	GGTCGGTGACGC	GACCTCCCG	GCCCGGGGGAGG	TATATCTTT	CGCTCC	CAGTCGGCATTT	GGGCCGCCG	GGTTAT

P2 Sequences

	Q	IS 1	P2	R	IS 2	S
Saccharomyces 1	------	GTG	TGAGGAAAA	GTAGTTGGG	AGG	TACTTC
Saccharomyces 2	TACTTC	ATG	CGAAAGCAG	TTGAAGACA	AGT	TCGAAAAG
Tetrahymena 1	ATATTT	AAG	AAGGGGAAA	CATCTCCGG	ATA	AAAAATAAAATA
Tetrahymena 2	AAATAT	AGT	AGACCGTCC	GGACTTTTG	AGA	CTTAGAGAAAAT
Dictyostelium	ACATAT	ACA	AGAAGAGTG	AGCAAGCAG	ATG	CAAGTGAAAGAA
Drosophila	TACTAT	AGG	TAGGCAGTG	GTTGCCGAC	CTG	GCATTGTTCGAA
Xenopus	GAAGGT	AGG	GGAAGACCG	GCCCTCGGC	GCG	ACGGGCGCCCGA
Mouse	ATAGGT	ACT	GACACGCTG	TCCTTTCCC	TAT	TAACACTAAAGC
Rat	ATATAT	GCT	GACACGCTG	TCCTTTTGA	CTT	CTTTTTGTCATT
Human	GGTTAT	GCT	GACACGCTG	TCCTCTGGC	GAC	CTGTCGCTGGAG
Escherichia coli rrnA	GCCACC	ACT	GACACGGAA	CAACGGCAA		

[a] Experimentally established promoters and start sites are underscored. Important regions of homology are italicized.
[b] Table 4.1.
[c] Valenzuela et al. (1977); Bayev et al. (1980); Klemenz and Geiduschek (1980).
[d] Saiga et al. (1982).
[e] Niles et al. (1981).
[f] Hoshikawa et al. (1983).
[g] Long et al. (1981); Coen and Dover (1982); Kohorn and Rae (1982b, 1983).
[h] Sollner-Webb and Reeder (1979).
[i] Urano et al. (1980); Mishima et al. (1981).
[j] Rothblum et al. (1982); Harrington and Chikaraishi (1983).
[k] Financsek et al. (1982); Miesfeld and Arnheim (1982).

What has been established in *Drosophila* differs to a considerable extent from the vertebrate processes. Initiation, for instance, has been reported to begin at the first adenosine of AGGT, not at that residue following this sequence (Kohorn and Rae, 1983). As shown in the table, the start site follows CTAT of column Q. Further, this same article presented evidence that that tetranucleotide was a major component of the initiation processes. In addition, another report by this same team showed that a second component of the promoter was located in the region between sites 23 and 47 upstream of the starting point of transcription (Kohorn and Rae, 1982b).

Spacer Promoters. One of the complicating factors in investigations of promotion in vertebrates is the presence of sequences on the intergenic spacers of *rrn* operons that nearly perfectly duplicate the leader sequence that has been demonstrated to be active in initiation, at least in *Xenopus* (Boseley *et al.*, 1979). In that amphibian each such spacer contains two or more copies of the region extending from ten sites downstream of the transcription start site to 150 upstream of that point. Recently five such spacer sectors have been sequenced and tested for their promotional activity by injection into oocytes (Morgan *et al.*, 1984). The results indicated that the differences in activity of these intergenic regions observed *in vitro* did not arise from primary structural distinctions between them, for all tested were equally effective, duplicating the performance of the leader region. These results contrasted strongly with *in vivo* observations, for in life the spacers are inactive, except in rare individuals. The primary structure varied at most only at one or two sites between intergenic spacer promoter regions and were nearly 90% homologous with the corresponding normal leader sequence, differing at but 18 positions. However, one structural peculiarity was not noted that may interfere with transcriptional functions *in vivo* by these spacers. Instead of the AGGT that precedes the start site in both *Xenopus* and the mouse, as pointed out above, the tetranucleotide was TTGT in all the spacers tested. Until the actual promoter has been determined, the suppressor of activity in the normal cell must remain conjectural.

Evolutionary Relationships. Close kinship to the prokaryotic genes can be demonstrated by a comparison with a known promoter from *E. coli rrnA*. If that initiation signal is placed near that of *Tetrahymena* 1 as in Table 4.2, where both are shown underscored, a striking degree of homology is revealed at once, the bacterial TATAATG comparing to the ciliate's TAAAAATG. In each case these sequences are followed by a C. No claim is being made that this run of nucleotides in the eukaryote is actually the P1 promoter; it is merely being suggested that a strong possibility exists of its serving in that capacity, along with the others listed in the P1 column. The proposal is considerably confirmed when all the sequences are continued toward the initial mature gene through the first initiation site and the P2 region, and then into and beyond the second start site (Table 4.2). Comparisons of the *E. coli rrnA* sequence with those of the three mammals show the presence of an eight-base series, CTGACACG, that is repeated precisely and at the same location immediately following initiation site 1 in all of the latter species. In the table, these three as well as the corresponding sector of *E. coli* are underscored. Since this degree of homology seems much more likely to represent evolutionary conservation of an important functional region rather than pure happenstance, at least part of it may prove to be involved in promoter activities.

This relationship of eukaryotic sequences with those of such moderately advanced prokaryotes as *E. coli* corroborates once more the evident derivation of the former type of cell from bacterial stock, thus again casting doubt upon the validity of the endosymbiontic theory of its origins, as pointed out earlier.

4.3. INITIATION OF TRANSCRIPTION IN ORGANELLAR rRNA GENES

What has been accomplished with the problem of transcription of rRNA genes in organelles has proven to be unexpectedly diversified—and incomplete. Among the surprising features is the great reduction in length of the leader sequence compared to that of prokaryotes, not only in mitochondria, where loss of parts is commonplace, but also in chloroplasts, in which it is unusual. Many of the initiation and promoter sites have not been positively identified; thus those that have been proposed, being presumptive only, may be subject to change. Furthermore, no ancillary sites have been detected in either organelle, but one would assume that this is the result of oversight, at least in the chloroplast. As in the *rrn* operons of prokaryotes and eukaryotic cytoplasmic types, the transcriptive products are compound and require processing to produce the ultimate RNAs and proteins, when the latter are encoded in intergenic spacers or introns.

4.3.1. Chloroplast rRNA Genes

Organization of Chloroplast rRNA Genes. The chloroplast rRNA genes are arranged in two or more operons in the several organisms in which they have been investigated; although all are of class IB organization, there are a number of differences between those of this organelle from lower and higher eukaryotes. In *Euglena*, there are three operons arranged in tandemly repeated fashion, each consisting of cistrons for the minor rRNA, two tRNAs, the major rRNA, and 5 S rRNA, in that order in the 5′ to 3′ orientation (Graf *et al.*, 1980; Keller *et al.*, 1980; Orozco *et al.*, 1980). The intergenic tRNAs are the same as those of *rrnD* and certain other *E. coli* operons, those for isoleucine and alanine (Figure 4.1). In addition to the three operons, there is a single copy of the 16 S species (Jenni and Stutz, 1979), which apparently has the identical orientation and possibly the same sequence as the other minor rRNAs of the complete operons (Koller and Delius, 1982). At one time it was believed that each of the three operons was followed by genes for tRNAs, but actual sequencing of the spacer between *rrnB* and *rrnC* revealed a sector of DNA nearly 1000 residues long that was too rich in A–T base pairs to encode any such nucleic acid (El-Gewely *et al.*, 1984).

In the chloroplast of all the green plants, including protistan and seed-plant types, much the same structural arrangement prevails, but only full operons are present, and those are oriented as inverted repeats (Bedbrook *et al.*, 1977; Whitfeld *et al.*, 1978a; Jurgenson and Bourque, 1980; Kusuda *et al.*, 1980). The gene sets in these organisms, however, are longer than those of *Euglena*, containing 22,000 base pairs compared to 5600. As a whole the sequence of genes is like that of the simpler protistan, being 5′-16 S, tRNA^Ile, tRNA^Ala, 23 S, 4.5 S, 5 S-3′, but there are several marked distinctions. The most important is the presence of a 4.5 S rRNA, a type of nucleic acid that receives further attention later in this chapter, which is not a separate entity in the euglenoid. A second, less fundamental, but more striking difference is the occurrence of long introns in the two intergenic tRNAs, a condition now known also to be prevalent in the chloroplasts of tobacco (Takaiwa and Sugiura, 1982a) and maize (Koch *et al.*, 1981).

A third contrasting condition is the frequent presence of a gene for tRNA^{Val} in the 5′ flanking sequence of the chloroplast rDNA in such seed plants as maize, tobacco, and spinach (Schwarz *et al.*, 1981; Todoh *et al.*, 1981; Briat *et al.*, 1982), as well as in the green alga *Spirodela* (Keus *et al.*, 1983). None was detected in the leader of *Chlamydomonas*, but since the tRNA gene typically ends about 250 sites before the mature rRNA cistron, the ~220 sites sequenced were not of sufficient length to determine its presence (Dron *et al.*, 1982). In *Euglena gracilis* there may be some variation from one operon to the other, for reports differ. For example, Keller *et al.* (1980) described the presence of either a tRNA^{Trp} or tRNA^{Glu} gene preceding all four of the minor rRNA coding sequences based on hybridization studies, whereas sequencing disclosed a fragment of a tRNA^{Trp} (mostly the 3′ half) and a complete but atypical tRNA^{Ile} farther upstream (Roux *et al.*, 1983). As just reported, no tRNA genes were found in the spacer between *rrnB* and *rrnC* when it was sequenced (El-Gewely *et al.*, 1984).

Initiation in Chloroplast rDNA. Study of the initiation of transcription of the rDNA in chloroplasts is still in a state of flux, for few thorough investigations of the subject have been completed. Consequently, promoter and ancillary sites have not been firmly established, nor have many actual starting points been experimentally determined. However, a small number of preliminary proposals have been advanced that seem to provide a solid basis for the more exacting researches that certainly will follow. One such has already appeared, in which the promoter for an rRNA operon from *Zea mays* chloroplasts was firmly identified as located 117 sites upstream from the mature coding region and downstream from that of tRNA^{Val} (Strittmatter *et al.*, 1985). Hence, the latter is not included in the primary transcript of this *rrn* operon.

The five chloroplast rRNA operon leader sequences listed in Table 4.3 represent the present state of the art rather fully and include the corresponding regions of the *E. coli rrnA* operon for ready comparison. In most cases two possible promoters (P1 and P2) can be detected, the structures of each of which are aligned with those of the bacterium. Each of the four P1 sites from chloroplasts closely resembles that of the prokaryote, except that the fifth position in each case is a pyrimidine rather than a purine, thymidine being found in the algae and dicotyledonous representatives and a cytidine in the monocotyledonous maize. The P1 actual start point was determined in the tobacco chloroplast and was said to be initiated consistently with ATP (Briat *et al.*, 1982), as indicated by underscoring. Others shown in the same manner are those proposed either in the literature or here. The *Euglena* leader shows only a single promoter, since the region farther upstream is occupied by partial genes for tRNA^{Trp} and tRNA^{Ile} as reported earlier.

The several P2 sequences also agree in most respects with that of *E. coli*, but they are more variable than the P1 sites. The occupant of the first site is a guanidine residue in the organelles from the two dicotyledonous plants, instead of the thymidine of the others, and the third is an adenosine in all chloroplasts, rather than the thymidine of the prokaryote. The maize structure differs from the others in the fifth site in having a cytidine, while the euglenoid has a thymidine there in addition to another of the same type in the seventh. Although still hypothetical, the start of transcription from this promoter seems to be with CTP as in the bacterium, at least in *Spirodela* (Keus *et al.*, 1983). Neither upstream nor downstream sites for the attachment of ancillary factors have been determined, but a

Table 4.3

Leaders from Organellar Minor rDNA[a]

	B	P1	C	Initiation	D	P2	C'	Initiation	D'	E
Escherichia rRNA[b]	GGAATAACT--C	TATAAT	GCC	ACCACTGAC-	ACG-(+97)-G	TATTATG	CAC	ACCCCG	CGCCGCTGAGAA	AAAGCG (+63)
Euglena_C[c]					G	TAAATGT(+15)ACCAAG			AGGGTGAAAGGA	TTTGAC (+15)
Spirodela_C[d]	GGTAGGGATGGC	TATATT	GCT	GGGAGCCGAA	CCTC-CAG-GC	TAATATG	AAG	CGCATG	GATACAAGTTAT	GCCTTG (+66)
Spinach[e,f]_C	GGTAGGGATGGC	TATATT	TCT	GGGAGC-GAA	CTC--CAGGC-	GAATATG	AAG	CCCATG	GATACAAGTTAT	GCCTTG (+27)
Tobacco[e]_C		TATATT	-CT	GGCAGC-GAA	CTC--CGGGG-	GAATATG	AAG	CCCATG	GATACAAGTTAT	GCCTTG (+28)
Maize_C		TATACT	GCT	GGTGGC-GAA	CTCAGGCTAA-	TAATCTG	AAG	CGCAT-	GATACAAGTTAT	-CCTTG (+28)
Saccharomyces[e,g]_M	TATTTATTATTA	TATAAG	TA-	(+0)		T ATATAAG	TA	(+0)		
Kluyveromyces[g]_M	TTATATATAATA	TATAAG	TA-	(+0)		T ATATAAG	TA	(+0)		
Human[h]_M	CCCCGAACCAAC	CAAACC	-CC	AAAGACACCC	CCC (+61)	CACCCCCA	TAA	ACAA(A)		

[a]Columns and *E. coli* sequence from Table 4.1. Experimentally established start sites are underscored. Presumptive promoters and important regions of homology are italicized.

[b]Insert 97 residues.

[c]Roux *et al.* (1983).

[d]Keus *et al.* (1983).

[e]Briat *et al.* (1983).

[f]Briat *et al.* (1982).

[g]Osinga *et al.* (1982).

[h]Montoya *et al.* (1983); Chang and Clayton (1984).

rather long sequence is shown in italics that has been highly conserved evolutionarily, following the probable start points in the green plants, but not in *Euglena*.

That much still remains to be learned about transcription of the rRNA genes of chloroplasts is especially clearly brought out by the investigation of the processes in the spinach chloroplast (Briat *et al.*, 1983). In this *in vitro* study four transcript classes were obtained, differing in length; with either *E. coli* or chloroplast RNA polymerase, chains of about 850, 550, 350, and 260 residues were obtained. Although the 550-site chain appeared to be a gene for a protein arranged in the opposite polarity, the other two long ones clearly must have been initiated upstream of the tRNA^{Val} cistron. Because the DNA studied included only leader and a sector 140 base pairs in length of the mature minor rRNA gene, it is possible that even greater variability of transcript length could result when entire operons are transcribed.

4.3.2. Mitochondrial rRNA Genes

Arrangement of rRNA Genes in Protistan Mitochondria. The rRNA genes of mitochondria do not lend themselves to a concise description of their arrangements as did their counterparts in chloroplasts, for great differences in organization characterize the various types of organisms. The sole common feature is that only a single operon exists. Nor does their constitution add any support to the concept of endosymbiontic origins. In those primitive eukaryotes, the yeasts, where the clearest reflections of such origins should be manifest, scarcely any semblance can be noted to the bacterial *rrn* operons, except that the minor rRNA cistron lies 5' to the one for the major nucleic acid. First on the list of differences is the discrepancy in size, the minor species sedimenting with a coefficient of 15 S rather than 16 S, and the major with 21 S in contrast to the 23 S of bacteria. But a more notable contrast occurs in a strain known as ω^+. At least in this variety the cistron for the major species is split by an intron consisting of 1143 base pairs (Borst *et al.*, 1977; Borst and Grivell, 1978; Dujon, 1980; Clark-Walker *et al.*, 1983), a feature absent in the alleles. The most remarkable departure, however, is that tRNA genes are lacking from the regions adjacent to both rRNA cistrons and also from the intergenic spacer. Instead of a cistron for this type of substance, the minor species is flanked on the 5' side by the coding sequence for the third subunit of cytochrome oxidase (Clark-Walker and Sriprakash, 1983).

Although higher protistans like *Trypanosoma* share in some of the yeast characteristics, such as the absence of tRNA cistrons in the vicinity of the rDNA, there are traits that lead into those of the metazoans. Among the more striking of these is the reduction in size of the rRNAs in these protozoans, for the minor species sediments at 9 S and the major at 12 S (Eperon *et al.*, 1983; de la Cruz *et al.*, 1985a,b). The intergenic spacer is greatly diminished, too, consisting of only seven base pairs. Even in the less advanced group, the Euciliata as represented by *Paramecium*, some reduction in size of the mitochondrial rRNAs can be noted, for the minor species sediments with a coefficient of 13 S and the major with 20 S (Seilhamer *et al.*, 1984). However, some of the reduction is a consequence of processing, an insert of 83 sites and adjacent regions being removed after transcription to form a small, separate segment. The entire mitochondrial genome of these organisms is unique in being linear rather than circular as elsewhere.

Mitochondrial rRNA Genes in Metazoa. Among the various phyla of metazoans the arrangement is similarly diversified. What is known, for example, of the echinoderm arrangement recalls that of the yeast to some extent in that a subunit of cytochrome oxidase lies adjacent to the rRNA genes (Goddard *et al.*, 1982; Roberts *et al.*, 1983). Only in the present instance, the subunit encoded is the first, not the third, and the neighboring rRNA gene is that for the major species, not the minor. That this organization is not prevalent throughout the Metazoa in general is indicated by that of the fruitfly *Drosophila yakuba*, which shares many features with the vertebrate ribosomal operon (Clary and Wolstenholme, 1985). In the mammalian mitochondrion the arrangement of the cistrons is 5′-tRNA^{Phe}, minor rRNA (12 S), tRNA^{Val}, major rRNA (16 S), tRNA^{Leu}-3′, followed by an unidentified reading frame and genes for three tRNAs. The latter encode species accepting isoleucine, glutamine, and methionine (initiating), in that same order (Van Etten *et al.*, 1980; Bibb *et al.*, 1981; Dubin *et al.*, 1982; Anderson *et al.*, 1981, 1982a,b). The insect arrangement differs in having these last three genes upstream of an A, T-rich sector that lies before the tRNA^{Phe} at the head of each operon (Clary *et al.*, 1982). Also, in mammalian mitochondria the arrangement is quite compact, so that the tRNA^{Phe} gene is separated from the 12 S cistron by a single nucleotide residue, if any. The latter coding sequence abuts against that of the tRNA^{Val}, which in turn is contiguous to the 5′ end of the major species. Finally, this latter cistron terminates against the tRNA^{Leu} without an intervening sector (Van Etten *et al.*, 1980). Hence, there are virtually no intergenic spacers in the entire operon.

Mitochondrial rRNA Genes in Green Plants. Because the genome of green plant mitochondria is 20–30 times as large as that of the metazoan organelle and five times or more that of the yeast (Iams and Sinclair, 1982), it lacks the former's reductional features. The organization of the rRNA genes is not fully established, but it obviously differs from all others that have been described. In the first place, the cistrons for the major and minor species do not form an operon as is typical elsewhere, but are widely separated, having an intergenic spacer of about 16,000 base pairs between them in maize (Iams and Sinclair, 1982; Chao *et al.*, 1983, 1984). Although only a single copy of the major (26 S) species is present, the minor species is duplicated, either in whole or part, in various portions of the genome; the number of copies varies with the individual, for it seems often to be involved in recombination events resulting in duplication (Stern and Palmer, 1984). In wheat, rye, and broad bean mitochondria, a similar unusual arrangement appears to prevail (M. W. Gray *et al.*, 1982). At least in wheat, the 5 S gene seemingly is associated with the minor species, rather than with the major as is usual, following the former by perhaps 500 nucleotide residues. It will be recalled that only the genomes of mitochondria of these plants are known to include a cistron for the 5 S species. Transfer RNAs occur nearby on the genome, but no more specific information is available.

Mitochondrial rRNA Genes of Fungi. Unique features are not absent among fungal mitochondrial organization of the present class of genes, although as a whole they are not unlike those of most eukaryotes. In size the minor (17 S) and major (24 S) approach the corresponding ones of plant mitochondria just described, the latter being interrupted by a long intron in the organelle of *Neurospora crassa* (Heckman and RajBhandary, 1979; Yin *et al.*, 1982). This intron, 2295 nucleotide residues in length, has been demonstrated to encode a reading frame for an unknown protein (Burke and RajBhandary, 1982). However, the most outstanding trait is found in the presence of ten or so genes for tRNAs on the intergenic spacer. Among those that have been identified here are the coding sequences for tRNA^{Tyr}, tRNA^{Val}, and tRNA^{Trp}. In addition, approximately the same number follow the cistron for the major species, one for tRNA^{Thr} being known to neighbor that gene.

Transcription Initiation in Mitochondrial rRNA Genes. Although the view has prevailed that each round of transcription in the mitochondrion involves the entire genome, the single transcript subsequently being processed into the various units (Ojala *et al.*, 1980, 1981; Beilharz *et al.*, 1982), more recent experimental investigations do not entirely support that concept. The results of these researches have demonstrated that the transcriptive processes of mitochondrial rRNA genes are as highly varied from one major taxon to another as are their organizational aspects. In the first example of this, the two species of yeast (*Saccharomyces cerevisiae* and *Kluyveromyces lactis*), it was found that the genes for the major and minor rRNAs are transcribed separately (Levens *et al.*, 1981a,b; Osinga *et al.*, 1982)—in fact, a total of at least 17 transcriptional initiation sites have been detected in the yeast mitochondrial genome (Christianson *et al.*, 1982; Edwards *et al.*, 1982). The transcript of the minor species has a leader of about 40 residues in length and requires processing at the 5' end, but that of the major type begins with site 1 of the mature coding sequence. Although the actual promoter site has not been fully established, the identical nonanucleotide sequence lies directly before the 5' end of the mature gene in the two species of rRNA in both kinds of yeast (Osinga *et al.*, 1982). This has been proposed to serve in all cases as part of the promoter, but in view of the length of the transcribed leader of the minor rRNA, it can scarcely be involved in that capacity directly. However, it could conceivably hold value as an ancillary site. The nonamer of each yeast is shown twice in Table 4.3 to bring out its homologies with presumptive promoters 1 and 2 of other genes. All capping of the transcripts was with an adenylic residue, the cap being guanosine triphosphate (Osinga *et al.*, 1982).

Even less abundant precise information appears available concerning transcription in the mitochondria of other organisms. With the *Neurospora* organelle serving as the example of the fungal processes, it has been determined that the polymerase activity begins with the minor species and proceeds uninterruptedly through the numerous tRNA cistrons that fill the intergenic spacer, continuing at least through the gene for the major type (Green *et al.*, 1981). Little else has been established thoroughly, aside from the nature of the primary transcript of the large species, which proved to be a 35 S precursor, containing the long intron in addition to the mature gene (Garriga *et al.*, 1984). Processing probably involving three steps then produced the mature nucleic acid by activities that receive detailed attention in Chapter 12.

Initiation in Vertebrate Mitochondria. To date, information regarding the precise points of initiation and the signals associated with transcription of these genes in the mammalian mitochondria have also been largely lacking. However, several current reports on the human organelle serve to illuminate the events to a considerable extent. One important feature involved in the processes is the location of the tRNA^{Phe} gene mentioned earlier as being just upstream of the 5' end of the DNA coding for the minor species (12 S) of rRNA. This is not actually continuous with the latter in man as once reported (Crews and Attardi, 1980), but is separated from it by a single nucleotidal residue, an adenosine (Anderson *et al.*, 1981; Chang and Clayton, 1984). A similar condition occurs in the murine organelle, in which a C intervenes (Bibb *et al.*, 1981), but not in the bovine, which has the two cistrons contiguous (Anderson *et al.*, 1982a). By means of a capping enzyme (guanylyl transferase) it was shown that there were two points for initiation of transcription of the heavy (H) strand of the human mitochondrial DNA, one upstream of the phenylalanine tRNA cistron and a second within it, just prior to the 12 S rRNA gene (Montoya *et al.*, 1982). Transcription beginning at the latter site continued along almost the entire H strand, but when initiated before the tRNA^{Phe} gene, it terminated just after the sequence coding for the major rRNA (Montoya *et al.*, 1983). Obviously this arrangement has the advantage of permitting the mitochondrion to produce an abundance of the ribosomal nucleic acids without synthesizing an excess of other macromolecules.

More recently the sequences of the presumptive promoter site preceding the tRNA^{Phe} gene on the human mitochondrial H strand and that nearby for the light (L) strand were established (Chang and Clayton, 1984). That of the H strand is reported in Table 4.3 as P1, initiation occurring within two to four sites following its termination. The CATACCGCCAAA of the L-strand promoter is somewhat homologous to the P2, actual initiation occurring at any of the three terminating adenosine residues. However, no promoter within the tRNA^{Phe} coding region was sought in the report cited. If the primary structure of that gene is examined, a similar sequence may be detected, which is shown in Table 4.3 as P2. When these three presumptive promoters are aligned to bring out homologies and differences, the relationships become readily apparent:

Heavy strand (P1)   CAAACC-CC<u>AA</u>A
Light strand   CATACCGCC<u>AA</u>A
tRNA^{Phe} (P2)   CAC-CCCATAAACAA(<u>A</u>)

In the last case transcription begins with the initial adenosine of the mature minor rRNA gene shown in parentheses, whereas in the others it commences with the bases presented underscored. Here correspondences are both readily notable and fairly numerous, but this is not the case with the presumed promoters of mitochondria from other sources and those of chloroplasts and *E. coli* (Table 4.3). Promoter 1 of human mitochondria is seen to have two adenosine residues at sites corresponding to those of the others, but nothing else, while its P2 has at most a single adenosine placed in agreement with the rest. Somewhat comparable heterogeneity in length of primary transcripts has been reported for bovine mitochondria (Hauswirth *et al.*, 1984).

Possible Multiple RNA Polymerases. One possible basis for the often conflicting results and overlooked promoter regions is found in a report on transcription in the rat mitochondrion. In an investigation of the so-called D-loop, a region closely preceding the tRNA^{Phe} gene just discussed, that also includes the DNA replication origin (Anderson *et al.*, 1981, 1982a,b; Walberg and Clayton, 1981), two peaks of DNA-dependent RNA polymerase were detected, indicating the presence of two such enzymes (Yaginuma *et al.*, 1982). The peak I fraction was equally active with heat-denatured DNA and the native molecule, whereas the peak II material transcribed the latter preferentially. Such multiplicity of transcribing enzymes could very well account for a portion of the inharmonious data that exist, a condition that should be recognized in future investigations.

Ancillary Factors. To date, transcription of the mitochondrial genome is too poorly understood to allow definite knowledge of the nature—or even the presence—of ancillary factors. However, at least one study has suggested a possible need for their presence in the yeast organelle (Edwards *et al.*, 1982). If such supplementary agents are eventually demonstrated to exist, a search for their binding sites then must be conducted. In the yeast mitochondrion, in which initiation immediately follows the promoter, such signals have to be located either within the mature coding sequence, as with class III genes, or preceding the promoter. This requirement would likewise hold for P2 of the human mitochondrion, while that of P1 could either precede or follow it, perhaps even within the tRNA[Phe] cistron.

4.4. THE MATURE MINOR rRNA CODING REGION

The tremendous task of establishing the sequences of a number of the minor species of rRNA has been completed for several representatives each of the four great source categories, prokaryote, eukaryote, chloroplast, and mitochondrion. The majority of those currently available are shown together in Table 4.4, an arrangement that is both conve- nient for the essential comparisons and economical of page space. Fortunately, the cytoplasmic type from all major groups of organisms, including the green plants and fungi, can now be represented in that table, since the necessary sequencing has recently been completed. Consequently, the essential features of the gene for this important type of macromolecule are certainly disclosed on a sufficiently broad scale to provide a deep insight into the similarities and contrasts that exist among at least the larger groupings of organisms.

4.4.1. The Prokaryotic Minor-rRNA Genes

The three prokaryotic minor-rRNA genes of Table 4.4 are those of representatives of different principal subdivisions of the Prokaryota, a blue-green alga (*Anacystis nidulans*), a eubacterium (*E. coli rrnB*), and an archaebacterium (*Halobacterium volcanii*) (Brosius *et al.*, 1981; Gupta *et al.*, 1983; Tomioka and Sugiura, 1983). Additional cistrons, one from the archaebacterium *Halococcus morrhua* and another from *Proteus vulgaris,* have also been sequenced (Carbon *et al.*, 1981; Leffers and Garrett, 1984), but since they agree in all major points with those of *Halobacterium* and *E. coli*, respectively, they are not included in the tabulation. Although homologies of three or more adjacent nucleotide residues among all three organisms are occasionally evident, the greater majority by far of such identical sequences are shared by the blue-green alga and *E. coli,* as is made clear by the blocked sectors. Among such common regions involving all three forms are those located in row 1, columns B and G–I; row 2, column N; row 3, columns R–T; row 4, columns E–G; row 5, columns I and K–N; row 6, columns O and P; row 7, columns A, B, F, and G; row 8, columns H–J and L–O; row 9, columns P, R, and S; row 10, columns C, D, F, and G; and row 11, columns J–O. The sequences in columns M–O of this last series are among the longest of the conserved regions and must be suspected of playing a role of unusual importance, either in secondary structure or in the functioning of the molecule. A similar but somewhat briefer evolutionarily conserved sector follows shortly thereafter, the TAAACGATG of row 12, column Q. Only triplets or the like are then to be found in immediately ensuing sectors, examples being provided by row 12, column W, and row 13, columns A–G. In row 14, columns H–J, an additional extensive homologous area is to be noted; this reads YGGTTTAATT-GAY-CAACG and must similarly be suspected of being a region of prime significance. The same row subsequently is virtually devoid of triplet or longer combinations, the ACC in column L being the sole exception.

Beginning in row 15, column R, is an additional region of frequent, short homologies—column S being completely homologous—that continues through columns T–W and into row 16, column A, which is highly conserved in the three prokaryotes. Identical triplet or longer combinations again become sporadic, columns C, D, and H each containing one or more, as does row 17, column I. The adjacent column J consists of a totally conserved monomer, whereas the subsequent sectors, K–N, include only scattered identities. From the end of column O and through P and brokenly through Q is a homologous sequence of considerable length. Then row 18 returns to scattered short, identical combinations, but row 20 begins with TTGYACACACCGCCCGTCA in columns L–O, the next to longest region of constancy common to all three prokaryotes in the entire molecule. The most extensive conserved sequence, three residues longer than this latter one, can be noted in row 21, columns D–G; this consists of AAGTCGTAACAAGGTARCCGTA. In the remainder of the row is one of the outstanding sectors where *E. coli* shows kinship with *Halobacterium* rather than to *Anacystis* as usual, to be noted in columns G and H. Only slight differences in length are apparent, the *Halobacterium* sequence being scarcely shorter than that of *Anacystis*, and the *E. coli* example being clearly the longest.

4.4.2. Eukaryotic Cytoplasmic Minor-rRNA Genes

Relatively few genes for the eukaryotic cytoplasmic minor rRNA have become available, only a small number having been sequenced. Of these, just four, those for the yeast, rat, rice, and *Dictyostelium* (Rubtsov *et al.*, 1980; McCarroll *et al.*, 1983; Torczynski *et al.*, 1983; Chan *et al.*, 1984; Ozaki *et al.*, 1984; Takaiwa *et al.*, 1984) are presented (Table 4.4), because those from the clawed toad, mouse, and rabbit (Salim and Maden, 1981; Connaughton *et al.*, 1984a,b; Raynal *et al.*, 1984) are largely homologous to that of the included rodent. In addition, one of the crustacean *Artemia* is available (Nelles *et al.*, 1984). Even the several given show the presence of a high level of evolutionary conservation (blocked areas), in spite of the great phylogenetic distances that exist between the source organisms. Since the identities they share are thus very numerous, comparisons between the eukaryotic and prokaryotic species prove sufficient.

The 5'-Terminal Region. At the extreme 5' end of the eukaryotic molecules (row 1, column A), suggestions of relationships with *Halobacterium* are visible, but that is all, and columns B and G are too universal to hold other than functional significance. However, in columns E and F triplets occur that correspond to those of the archaebacterium, but not to the other two prokaryotes. A similar statement is true for column H, in which seven of the nine sites of rat cytoplasmic species match those of *Halobacterium*. Comparable kinship implications can be found only in row 2, column M, and row 7, column A, and thus are limited in number. In this same region, however, resemblances closer to *Anacystis* and *E. coli* than to *Halobacterium* do not occur at all.

The occasional absence of nucleotides in a given sector is also of interest, a trait that characterizes all known sequences at certain points, even those of eukaryotes, which have the largest molecules. Reduction at the extreme 5' end is the first of a series in the eukaryotes, a feature reflected also in the archaebacterium, as well as in the three from

Table 4.4
Sequences of the Minor rRNA Species[a]

Row 1	A	B	C	D	E	F	G	H	I
Euglena[b]	TGGAAATGAAGAGT	TTGATC	CTT	GCTCAGGGTGAA	CGCTGG	CGGTA-TG-CT	TAA	CACATGCAA	GTTGAACGAAAT
Chlamydomonas$_C$[c]	-ATCCATGGAGAGT	TTGATC	CTG	GCTCAGGACGAA	CGCTGG	CGGCA-TG-CT	TAA	CACATGCAA	GTCGAACGAGCA
Tobacco$_C$[d]	-TCTGATGGAGAGT	TCGATC	CTG	GCTCAGGATGAA	CGCTGG	CGGCA-TG-CT	TAA	CACATGCAA	GTCGGACGGGAA
Anacystis[e]	--AAAATGGAGAGT	TTGATC	CTG	GCTCAGGATGAA	CGCTGG	CGGCG-TG-CT	TAA	CACATGCAA	GTCGAACGGTAA
Escherichia[f]	--AAATTGAAGAGT	TTGATC	ATG	GCTCAGATTGAA	CGCTGG	CGGCA-GG-CC	TAA	CACATGCAA	GTCGAACGGTAA
Halobacterium	-----ATTCC---GG	TTGATC	CTG	CCGGAGGTC-AT	TGCTAT	TGGGGTCCGAT	TTA	GCCATGCTA	GTTGCACGACTT
Yeast$_M$[h]	-ATTTATAAGAAT	ATGATG	TTG	TTTCAGATTAAG	CGCTA-	AATAAGGA-CA	TGA	CGCATACGA	GTCATACGTTTA
Bovine$_M$[i]	-ACTAGGT-------	TTGGTC	GCA	GCCTTCCTGTTA	ACTCTT	AATAA-ACT--	TA-	CACATGCAA	GCATCTACACCC
Murine$_M$[j]	--AAAGGT-------	TTGGTC	CTG	GCCTTATAATTA	ATTAGA	GGTAA-AA---	TTA	CACATGCAA	ACCTCCATAGAC
Yeast$_{Cy}$[k]	-----TATC--TGG	TTGATC	CTG	CCAGTAGTCATA	TGCTTG	TCTCA-AAGAT	TAA	GCCATGCAT	GTCTAAGTATAA
Fungus$_{Cy}$[l]	-----TAAC--TGG	TTGATC	CTG	CCAGTAGTCATA	TGCTTG	TCTCA-AAGAT	TAA	GCCATGCAT	GTCTAAGTATAA
Rat$_{Cy}$[m]	-----TACC--TGG	TTGATC	CTG	CCAGTAG-CATA	TGCTTG	TCTCA-AAGAT	TAA	GCCATGCAT	GTCTAAGTACGC
Rice$_{Cy}$[n]	-----TACC--TGG	TTGATC	CTG	CCAGTAGTCATA	TGCTTG	TCTCA-AAGAT	TAA	GCCATGCAT	GTGCAAGTATGA

Row 2	J	K	L	M	N	O	P
Euglena_C	TACTAGCAA	TAGTAAT--	TTAG-TGGC-GG	AGGCGGACG	GGTGAGTAA	T-ATGTAAGAA-	TCTCGGCTTGGCGGA
Chlamydomonas_C	---AAGCAA	---------	TTTG-TG-T-AG	TGGCGAACG	GGTGGCTAA	C-GCGTAAGAA-	CCTACCTATCGGAGG
Tobacco_C	GTGG-----	------T-G	TTTCC-----AG	TGGCGGACG	GGTGAGTAA	C-GCGTAAGAA-	CCTGCCCTTGGGAGG
Anacystis	---------	----GCT-C	TTCGGAGCT-AG	TGGCGGACG	GGTGAGTAA	C-GCGTGAGAA-	TCTGCCTACAGGACG
Escherichia	CAGGAAGAA	GCTTGCTCC	TTTGCTGACCAG	TGGCGGACG	GGTGAGTAA	T-GTCTGGAA-	ACTGCCTGATGGAGG
Halobacterium	CA-------	---------	------TACTCG	TGGCGAAAA	GCTCAGTAA	CACGTGGCCAA-	ACTACCCTACAGAGA
Yeast_M	TTATTGATA	AGATAATA-	AATATGTGGTGT	AAACCTGAG	TAATTTTAT	TAGGAA-TTAAT	GAACTATAGAATAAG
Bovine_M	CAGTGAGAA	TGCC-CTCT	AGGTTATTA---	AAACTAAGA	GGAGCTGGC	ATCAAGCCACACA	CCCTGTAGCTCACGA
Murine_M	CGGTGTAAA	ATCC-CT-T	AAACATTTACTT	AAAATTTAA	GGAGAGGGT	ATCAAGCACATT	AAAA-TAGCTTAAGA
Yeast_Cy	----GCAA-	--TT----T	A-TACAGTGAAA	CTCGGAATG	GCTCATTAA	ATC-AGT-TAT-	CGTTTATTTGA----
Rat_Cy	----ACGG-	--CC-----	GGTACAGTGAAA	CTCGGAATG	GCTCATTAA	ATC-AGT-TAT-	GGTTCCTTTGGTCGC
Fungus_Cy	----ATT--	--CT----T	-GTACGATGAAA	CTCGAGACG	GCTCATTAC	AAC-AGT-GATA	AA---CT-AATAGA
Rice_Cy	----ACTAA	--TT-----C	GA-ACTGTGAAA	CTCGGAATG	GCTCATTAA	ATC-AGT-TAT-	AGTTTGTTTGA----

(continued)

Table 4.4 (Continued)

Row 3	Q	R	S	T	U	V	W
Euglena$_C$	GGAATAACA	GATGGAAA-	CGT-TTGCTA	ATGCCTCATAAT	^{p}AGTAGGTA--GT	TA--------AG	AA-TCTCGCCTA
Chlamydomonas$_C$	GGGATACAT	TG-GGAAA-	CTG-TTGCTA	ATACCCCATACA	GCTGAGGA--GT	GA-AAG---GTG	AAAAACCGCC-G
Tobacco$_C$	GGAACAACA	GCTGGAAA-	CGG--CTGCTA	ATACCCCGTAGG	CTCAGGAG--CA	AA-AGG----AG	GAA-TCCGCCCG
Anacystis	GGGACAACA	GTTGGAAA-	CGA-CTGCTA	ATACCCGATGTG	CCCAGAGG--TG	AA-ACA---TT	TA---TGGCCTG
Escherichia	GGGATAACT	ACTGGAAA-	CGG-TAGCTA	ATACCGCATAAC	GTCGGCAAGACCA	AAGAGG----GG	GACCTTCGGGCC
Halobacterium	ACGATAACC	TCGGGAAA-	CTG-AGGCTA	ATAGTTCATACG	GGAGTCATGCTG	GAATGCCGACTC	CCCGAAACGCTC
Yeast$_M$	CTAAATACT	TAATATAT-	TAT-TATATA	AAAATA-AT-TT	ATA-TAATA--A	AAAGGA----TA	TATATATAATAT
Bovine$_M$	---------	-CGCCTTG-	-CT-TAACC-	ACAC--CCCACG	GGAAACA-G-CA	--GTGA---CAA	AAA-TTAAGCCA
Murine$_M$	---CAC---	----CTTG-	-CC-TAGCC-	ACACC-CCCACG	GGACTCA-G-CA	--GTGA---TAA	ATA-TTAAGCAA
Yeast$_{Cy}$	TAGTT-CCT	TTACTACAT	GGTATAACCG	-TGCTAATTCTA	-GAGCTAATACA	--TGCTTAAAAT	CTC---GACCC-
Rat$_{Cy}$	TCGCT-CCT	CTCCTACTT	GG-ATAACTG	-TGGTAATTCTA	-GAGCTAATACA	--TGCCGACGGG	CGC--TGACCCC
Fungus$_{Cy}$	---CT-TT-	C-GGGTTTT	AC-C---TTT	-TGG-ATAACCG	-CAG-TAAATCG	GGGCTAATA---	CATACAAGCGA-
Rice$_{Cy}$	TGGTA-CG-	T-GCTACTC	GG-ATAACCG	-TAGTAATTCTA	-GAGCTAATACG	--TGCAACAAAC	CCC---GACTT-

Row 4	A	B	C	D	E	F	G
Euglena_C	-GGCATGAGCTTG	-CA-TCT--	------GA	TTAGCTT---GT	TGGTGAGGTAAA	GGCTTA-CCA	AGGGCACGATCA
Chlamydomonas_C	ATAGAGGGGCTTG	-CG-TCT--	------GA	TTAGCTA---GT	TGGTGGGGTAA	CGGGCCTCCCA	AGGGCCACGAGCA
Tobacco_C	-AGGAGGGGCTCG	-CG-TCT--	------GA	TTAGCTA---GT	TGGTGAGGCAAT	AGCTTA-CCA	AGGGCATGATCA
Anacystis	-TAGATGAGCTCG	-CG-TCT--	------GA	TTAGCTA---GT	TGGTGGGGTAAG	GGCCTA-CCA	AGGGCACGATCA
Escherichia	-TCTTGCCATCGG	ATG-TGCCC	AGATGT-GA	TTAGCTA---GT	AGGTGGGGTAAC	GGCTCA-CCT	AGGGCACGATCC
Halobacterium	-AGGGCCTGTAGG	ATG-T-GGC	TGCGGCCCA	TTAGCTA---GA	CGGTGGGGTAAC	GGCCCA-CCG	TGCCGATAATCG
Yeast_M	ATATTTATCTATA	GTCAA-GCC	AATAATGT	TTAGGTA---GT	AGGTTTATTAAG	AGTTAAACCT	AGGCAACGATCC
Bovine_M	TAAACGAAAGTTT	GAC-T----	AAG--T-TA	TATTAATTA-G-	-GGTTGG-TAAA	---TC--TCG	TGCCAGCCACCG
Murine_M	TAAACGAAAGTTT	GAC-T----	AAG--T-TA	TACCTCTTA-G-	-GGTTGG-TAAA	---TT--TCG	TGCCAGCCACCG
Yeast_Cy	---TT---T---	---GGA---	-AGA-GATG	TATTTATTA-GA	TAA-----AAAA	-TCAA-T--	-GTC------T
Rat_Cy	-CCTTCCCGTGGG	---GGG---	AACG-CGTG	CATTTATCA-GA	TCA------AAA	-CACAA-CCC	GGTCAGCCCCCT
Fungus_Cy	--T--------	---GGG---	-------TG			------	----------
Rice_Cy	-CC--------	---GGG---	-AGG-GGCG	CATTTATTA-GA	TAA-----AAGG	--CTGA-CG-	----------

(continued)

Table 4.4 (Continued)

Row 5	H	I	J	K	L	M	N
Euglena$_C$	GTA-GCTGA	TTTGAGAGGATG	ATCAGCCAC	ACTCGGATTGAG	A-ACGCAACAGA	CTTCTACGG	AAGGCAGCAGTG
Chlamydomonas$_C$	GTA-GCTGG	TCTGAGAGGATG	ATCAGCCAC	ACTCGGACTGAG	ACACGGCCCAGA	CTCCTACGG	GAGGCAGCAGTG
Tobacco$_C$	GTA-GCTGG	TCCGAGAGGATG	ATCAGCCAC	ACTCGGACTGAG	ACACGGCCCAGA	CTCCTACGG	GAGGCAGCAGTG
Anacystis	GTA-GCTGG	TCTGAGAGGATG	ATCAGCCAC	ACTCGGACTGAG	ACACGGCCCAGA	CTCCTACGG	GAGGCAGCAGTG
Escherichia	CTA-GCTGG	TCTGAGAGGATG	ACCAGCCAC	ACTCGAACTGAG	ACACGGTCCAGA	CTCCTACGG	GAGGCAGCAGTG
Halobacterium	GTA-CGGGT	TGTTGAGAGCAAG	AGCCCGGAG	ACGGAATTCTGAG	ACAAGATTCCGG	GCCCTACGG	GCGCAGCAGGC
Yeast$_M$	ATAAATCGAT	AATGAAAGTTAG	AACGATCAC	GTTGACTCTGAA	ATATAGTCAATA	-TC-TATAA	GATACAGCAGTG
Bovine$_M$	CGG-TCA--	TACGATTAACCC	AAGCTAACA	GGAGTACGGCGT	AAAACGTGTTAA	AGC-A-CAT	ACCAAATAGGGT
Murine$_M$	CGG-TCA--	TACGATTAACCC	AAACTAATT	ATCTT-CGGCGT	AAAACGTGTCAA	CTATAAATA	AATAAATAGAAT
Yeast$_{Cy}$	---------	T--------	--------	--CGGAC-TCTT	---TGATGATTC	ATAATA-AC	TTTTC-G--AAT
Rat$_{Cy}$	CCC---GGC	TCC---GGCCGG	GGGTCGGGC	GCCCGGCGGCTT	---TGGTGACTC	TAGATA-AC	CTCCG-GCCGAT
Fungus$_{Cy}$	ACT--GGC	AAC--GGAAGC	TCAGCGATT	ATTAGCATTCTA	CCAATGCCTTCG	GGTTTT-GG	GTGATACCGAAT
Rice$_{Cy}$	CG----GGC	TCC---GCCCGC	TGATCC----	-----GATGATT	-----------C	ATGATA-A-	CTCG-ACGGAT-

Row 6	O	P	Q	R	S	T	U
Euglena_C	--AGGAATTTTCCG	CAATGGGCGCAA	GCCTGACGG	AGGAATACCGCG	TGAAGAAGACG	GCCTTTGGG	TTGAAAACC
Chlamydomonae_C	--AGGAATTTTTCG	CAATGGGCGCAA	GC--GACGG	AGGAATGCCGCG	TGCAGGAAGAAG	GCCTGTGGG	TCGTAAACT
Tobacco_C	--GGGAATTTTCCG	CAATGGGCGAAA	GC--GACGG	AGCAATGCGCGC	TGGAGGTAGAAG	GCCCACGGG	TCGTGAACT
Anacystis	--GGGAATTTTCCG	CAATGGGCGCAA	GC--GACGG	AGCAACGCGCGG	TGGGGGAGGAAG	GTTTTTGGA	CTGTAAACC
Escherichia	--GGGAATATTGCA	CAATGGGCGCAA	GCCTGATGC	AGCCATGCCGCG	TGTATGAAGAAG	GCCTTCGGG	TTGTAAAGT
Halobacterium	--GCGAAACCTTTA	CACTGCACCCAA	GTCCGATAA	GGGGACCCCAAG	TGCCGAGGGCATA	TA------	---GTCCTC
Yeast_M	--AGGAATATTGGA	CAATGATCCGAAA	GATTGATCC	AGTTAC-TTATT	AGGATG-ATATA	TAAAAATAT	TTTATTT--
Bovine_M	-TAAAATTCTAACTA	AGCTGTAAAAAG	CC---ATGA	TTAAAA---TAA	AAATAA-ATGAC	GAAAGTGAC	CCTACATA
Murine_M	-TAAAATCCAACTT	ATATGTGAAAAT	TC---ATTG	TTAGGA-CCTAA	A--CTCAATAAC	GAAAGTAAT	TCTAGTCAT
Yeast_Cy	-CGCATGGCCTT-G	TGCTGGCGAT-G	GT-TCATTC	AAA---TTT---	---CTGCCCTAT	CAACTTTCG	ATGGTAGGA
Rat_Cy	-CGCACGTCCCC-G	TGGCCGCGAC-G	AC-CCATTC	GAA----CGT--	---CTGCCCTAT	CAACTTTCG	ATGGTAGTC
Fungus_Cy	AATATTGCAGATCG	AGGATTT-ATCT	TCCACAAGT	CTACT-GTGTCA	---CTGCCCTAT	CAACTTTCG	ATGGTACGG
Rice_Cy	-CGCACGGCCCTCG	TGCCGCGAC-G	-CATCATTC	AAA---TTT---	---CTGCCCTAT	CAACTTTCG	ATGGTAGGA

(continued)

Table 4.4 (Continued)

Row 7	A	B	C	D	E	F	G
Euglena$_C$	TCTTTTCTCAAA	GAAGAAG-----	AAA---------	-----------TGA	CGGTATTTG	AGGAATAAG	CATC-GGCT
Chlamydomonas$_C$	GCTTTTCTCAGA	GAAGAAG-----	------------	------TTCTGA	CGGTATCTG	AGGAATAAG	CACC-GGCT
Tobacco$_C$	TCTTTTCCCGGA	GAAGAAG-----	------------	-----CAATGA	CGGTATCTG	GGGAATAAG	CATC-GGCT
Anacystis	CCTTTTCTCAGG	GAAGAAG-----	AAAG--------	-----------TGA	CGGTACCTG	AGGAATAAG	CCTC-GGCT
Escherichia	ACTTTCAGCGGG	GAGGAAGCGAGT	AAAGTTAATACC	TTTGCTCATTGA	CGTTACCCG	CAGAAGAAG	CACC-GGCT
Halobacterium	GCTTTTCTCGAC	CGT-AAGCCG--	------------	------------	----GTCG	AGGAATAAG	AGCTGGGCA
Yeast$_M$	TATTTAGTT--C	CGGGGCCCGGCC	AC----GGAGCC	GAACCCGAAAGG	AGAAATATT	AAATATTTA	TAATAAT^q--
Bovine$_M$	GCC-GACGCACT	ATAGCTAAGACC	CAAACTGGGATT	AGATACCCC--A	C-TATGCTT	AGCCCTAAA	CACAGATAA
Murine$_M$	TTATAATACACG	ACAGCTAAGACC	CAAACTGGGATT	AGATACCCC--A	C-TATGCTT	AGCCATAAA	CCTAAATAA
Yeast$_{Cy}$	TAGTGCCT-AC	CATGGTTTCAAC	GGGTAACGCGGA	ATAAGGGTTCGA	TTCCGGA-G	AGGGAGCCT	GAGAAACGG
Rat$_{Cy}$	GCCGTGCCT-AC	CATGGTGACCAC	GGGTGACGGGGA	ATCAGGGTTCGA	TTCCGGA-G	AGGGAGCCT	GAGAAACGG
Fungus$_{Cy}$	TATTGGCCT-AC	CATGGTTGTAAC	GGGTAACGGGGA	ATTAGGGTTCGA	TTCCGGA-G	AGGGAGCCT	GAGAAATTGG
Rice$_{Cy}$	TAGGGCCCT-AC	CATGGTGGTGAC	GGGTGACGAGA	ATTAGGGTTCGA	TTCCGGA-G	AGGGAGCCT	GAGAAACGG

Row 8	H	I	J	K	L	M	N	O
Euglena_C	-----AATTCCGTG	-CCAGCAGCC	GCGGTAATA	CGGGAGATG	CGAGCGTTA	TCCGGAATT	ATTGGGCGT	AAAGAGTTTGTA-
Chlamydomonas_C	-----AACTCTGTG	-CCAGCAGCC	GCGGTAATA	CAGAGGGTG	CAAGCCGTTG	TCCGGAATG	ATTGGGCGT	AAAGCGTCTGTA-
Tobacco_C	-----AACTCTGTG	-CCAGCAGCC	GCGGTAATA	CAGAGGATG	CAAGCCGTTA	TCCGGAATG	ATTGGGCGT	AAAGCGTCTGTA-
Anacystis	-----AATTCCGTG	-CCAGCAGCC	GCGGTAATA	CGGGAGAGG	CAAGCCGTTA	TCCGGAATT	ATTGGGCGT	AAAGCGCCTGCA-
Escherichia	-----AACTCCGTG	-CCAGCAGCC	GCGGTAATA	CGGGAGGTG	CAAGCCGTTA	ATCGGAATT	ACTGGGCGT	AAAGCGCACGCA-
Halobacterium	-----AGACCGGTG	-CCAGCCGCC	GCGGTAATA	CCGGCAGCT	CAAGTGATG	ACCGATATT	ATTGGGCGT	AAAGCGTCCGTA-
Yeast_M	-----AGTCCTGAC	-TAATATTTG	ATATTTGTG	CCAGCAGTC	CAAGCGTA	ACACAAAGA	---GGGGA	---GCGTTAATC-
Bovine_M	ATTACATAAACAA	-AATTATTCG	CCAGAGTAC	TACTAGCAA	CAGCTTAAA	ACTCAAAGG	ACTTGGCGG	TGCTTTATATCC-
Murine_M	-TTAAATTTAACA	AAACTATTTG	CCAGAGAAC	TACTAGCCA	TAGCTTAAA	ACTCAAAGG	ACTTGGCGG	TACTTTATATCC-
Yeast_Cy	----CTACCACAT	-CCAAGGAAG	GCAGCAGGC	GCGCAAATT	ACCCAATCC	TAATTCAGG	GAGGTAGTG	ACAATAAATAAC-
Rat_Cy	----CTACCACAT	-CCAAGGAAG	GCAGCAGGC	GCGCAAATT	ACCCACTCC	CGACCCGG	GAGGTAGTG	ACGAAAAATAAC-
Fungus_Cy	----CTACCACTT	-CTACGGAAG	GCAGCAGGC	GCGCAAATT	ACTCAATCC	CAATACGGG	GAAGTAGTG	ACAATAAATATCA
Rice_Cy	----CTACCACAT	-CCAAGGAAG	GCAGCAGGC	GCGCAAATT	ACCCAATCC	TGACACGGG	GAGGTAGTG	ACAATAAATAAC-

(continued)

Table 4.4 (Continued)

Row 9	P	Q	R	S	T	U	V
Euglena$_C$	GCCGGTCAA	GTGTGTTTAATG	TTAAAAGTCA	AAGCTTAACTTT	-GGAAGG--G	CATTAAA-AACTG	CTAGACTTGAGT
Chlamydomonas$_C$	GGTGGCTCG	TAAAGTCTAATG	TGAAATACCA	GGGCTCAAACCTT	-GGACCG--G	CATTGGAGTACTG	ACGAGCTTGAGT
Tobacco$_C$	GGTGGCTTT	TTAAGTCCGCCG	TGAAATCCCA	GGGCTCAACCCT	-GGACAG--G	CGGTGA-AACTA	CCAAGCTGGAGT
Anacystis	GCCGGTTAA	TCAAGTCTGTTG	TGAAAGCGTG	GGGCTCAA-CCT	CATACAG--G	CAATGGA-AACTG	ATTGACTAGAGT
Escherichia	GCCGGTTTG	TTAAGTCAGATG	TGAAATCCCC	GGGCTCAA-CCT	-GGAAAC-TG	CATCTGA-TACTG	GCAAGCTTGAGT
Halobacterium	GCCGGCCAC	GAAGGTTCATCG	GGAAATCCGC	CAGCTCAA-CTG	-GCCGGGGTC	CGGTGAA-AACCA	CGTGGCTTGGA
Yeast$_M$	--ᵖ-GGTTTA	AAGGATCCGTAG	AATGAATTA-	TATATTATAATT	-TAGAGT-TA	ATAAAATTAATTA	AAGAATTATAAT
Bovine$_M$	TTCTAGAGG	AGCCTGTTCTAT	AATCGATAA-	ACCCCGATAAAC	-CTCACCAAT	TCTTGCT-AATAC	AGTCTATATACC
Murine$_M$	ATCTAGAGG	AGCCTGTTCTAT	AATCGATAA-	ACCCCGCTCTAC	-CTCACCATC	TCTTGCT-AATTC	AGCCTATATACC
Yeast$_{Cy}$	GATACAGGG	CCCATTCG-GGT	CTTGTA-ATT	GGAATGAGTACA	-ATGTAAATA	CCTTAAC-GAGGA	ACAATTGGAGG-
Rat$_{Cy}$	AATACAGGA	CTCTTTCGAGGC	CCTGTA-ATT	GGAATGAGTCCA	-CTTTAAATC	CTTTAAC-GAGGA	TCCATTGGAGGG
Fungus$_{Cy}$	ATACCTATC	CTTTTTGGAGG-	---GCA-ATT	GAAATGAACACA	-AATTAAAAC	TCTTAAT--TAAC	ACAATTGGAGGG
Rice$_{Cy}$	AATACCGGG	CGCTTTAG-TGT	-CTGGTAATT	GGAATGAGTACA	-ATCTAAATC	CCTTAAC-GAGGA	TCCATTGGAGGG

Row 10	A	B	C	D	E	F	G
Euglena$_C$	ATGGTA-CGGGT	GAAGGGAATTTC	CAGTGTAGC	GGTGAA-ATG	CGTAGAGATTGG	AAAGAACACCAA	TGGCGAAGG
Chlamydomonas$_C$	ACGGTA-CGGGC	AGAGGGAATTCC	ATGTGGAGC	GGTGAA-ATG	CGTAGAGATATG	GAGGAACACCAG	TGGCGAAGG
Tobacco$_C$	ACGGTA-CGGGC	AGAGGGAATTTC	CGGTGGAGC	GGTGAA-ATG	CGTAGAGATCGG	AAAGAACACCAA	CGGCGAAAG
Anacystis	ATGGTA-CGGGT	ACCGGGAATTCC	AGTGTCTAGC	GGTGAA-ATG	CGTAGATATCTG	GAAGAACACCAG	CGGCGAAAG
Escherichia	CTOGTA-CAGGG	GGGTACAATTCC	AGTGTCTAGC	GGTGAA-ATG	CGTAGAGATCTG	GAGCAATACCCG	TGGCGAAGG
Halobacterium	CCCGAA-GGCTC	GAGGGTACGTC	CGGGGTAGG	AGTGAA-ATC	CCGTAATCCTGG	ACGGACCACCGA	TGGCGAAAG
Yeast$_M$	AGTAAA-GATGA	AATA---ATTAT	AAGACTA--	-GTGAA-AA-	TATTAA-TTAAA	TA-TTA-AC---	TGAC-ATTG
Bovine$_M$	GCCATCTTCAGC	AAACCCTAAAAA	GGAAAAAAA	GTAAGC-GTA	ATTATGATACAT	AAAAACGTTAGG	TCAAGGTGT
Mouse$_M$	GCCATCTTCAGC	AAACCCTAAAAA	GGTATTAAA	GTAAGC-AAA	AGAATCAAACAT	AAAAACGTTAGG	TCAAGGTGT
Yeast$_{Cy}$	CAAGTCTGGTGC	CAGCAGCAGCCG	TAATTCCAG	CTCCAA-TAG	CGTATATTAAAG	TTCTTGCAGTTA	AAAAGCTCG
Rat$_{Cy}$	CAAGTCTGGTGC	CAGCAGCCCCGG	TAATTCCAG	CTCCAA-TAG	CGTATATTAAAG	TTCCTGCAGTTA	AAAAGCTCG
Fungus$_{Cy}$	CAAGTCTGGTGC	CAGCAGCCCCGG	TAATTCCAG	CTCCAATTAG	CATATACTAAAG	TTCTTGCAGTTA	AAAAGCTCG
Rice$_{Cy}$	CAAGTCTGGTGC	CAGCAGCCCCGG	TAATTCCAG	CTCCAA-TAG	CGTATATTTAAG	TTCTTGCAGTTA	AAAAGCTCG

(continued)

Table 4.4 (Continued)

Row 11	H	I	J	K	L	M	N	O
*Euglena*_C	CACTTTTCT	AGGCCAATA	CTGACGCTGAGAAAC	GAAAGC	TGAGGGAGCAAACAG	-GATTAGATA	CCCTG-TAG	TCT
*Chlamydomonas*_C	CGCTCTGCT	GGGCCGAAA	CTGACACTCAGAGACAC	GAAAGC	TGGGGAGCGAATAG	-GATTAGATA	CCCTAGTAG	TCC
*Tobacco*_C	CACTCTGCT	GGGCCGACA	CTGACACTGAGAGAC	GAAAGC	TAGGGGAGCGAATGG	-GATTAGATA	CCCCAGTAG	TCC
Anacystis	CCCGCTACT	GGGCCATAA	CTGACGCTCATGGAC	GAAAGC	TAGGGGAGCGAAAGG	-GATTAGATA	CCCCTGTAG	TCC
Escherichia	CGGCCCCCT	GGACGAAGA	CTGACGCTCAGGTGC	GAAAGC	GTGGGGAGCCAAACAG	-GATTAGATA	CCCTGGTAG	TCC
Halobacterium	CACCTCGAG	AAGACGGAT	CCGAGGGTGAGGGAC	GAAAGC	TAGGGTCTCGAACCG	-GATTAGATA	CCCGGGTAG	TCC
*Yeast*_M	---------	--A--GGGA	TTAAAACTAGAGTGC	GAAA-C	--G---------	-GATTCGAATA	TTCCTGTAG	TTT
*Bovine*_M	---------	--A--ACCT	ATGAAA---TGGGAA	GAAA-T	--GGGCTACA----	----------	--------	TTC
*Murine*_M	---------	--A--GCCA	ATGAAA---TGGGAA	GAAA-T	--GGGCTACA---	----------	--------	TTT
*Yeast*_Cy	TAGTTGAAC	TTTGGGCCC	-------GGTTGGCCG	GTCCGA	TTTTTTCGTGTAC-T	-GGATTTCCA	ACGGGGCCT	TTC
*Rat*_Cy	TAGTTGGAT	CTTGGGAGC	GGGCGGGGGTCC-G	CCCGGA	GGCGGCTCAGCGCCC	-TGTCCCAGC	CCCTG-CCT	CTC
*Fungus*_Cy	TAGTTGAAG	TTTAAGGTT	TACCGGGTTTATGTC	ATTTAC	CACTTCGTTGGTTA--	-AATCGACAC	CGGTATCTC	TTT
*Rice*_Cy	TAGTTGGAC	CTTGGGCGC	GGGCGGGGCGGTCCG	CCTCAC	GGCAGGGCACCGA-CC	----TGCTCG	ACCCT-TCT	GCC

Row 12	P	Q	R	S	T	U	V	W
Euglena_C	-TGCCCG	TAAACTATG	GATACTAAG	TGG-TGCTG	AAA----GTCCAC	T------GC	TGTAGTTAA	CACGTTAAGTAT
Chlamydomonas_C	-CAGCCCG	TAAACTATG	GAGACTAAG	TGC-TGCCG	C-AA----GCA-	-----GTGC	TGTAGCTAA	CGCGTTAAGTCT
Tobacco_C	-TAGCCCG	TAAACGATG	GATACTAGG	CGC-TGTGC	G-ATCGACCCGTG	CA----GTGC	TGTAGCTAA	CGCGTTAAGTAT
Anacystis	-TAGCCCG	TAAACGATG	AACACTAGG	TGT-TGCGT	GAATCGACCCGCG	CA----GTGC	CGTAGCCAA	CGCGTTAAGTGT
Escherichia	-ACGCCCG	TAAACGATG	TCGACTTGG	AGGTTCTGC	C-CTTGAGGCGTG	G-----CTTC	CGGAGCTAA	CGCGTTAAGTCG
Halobacterium	-TAGCTG	TAAACGATG	CTCGCTAGG	TGTGACACA	G-GCTACGAGCCT	GTGTTGTGC	CGTAGGGAA	GCCGAGAAGCGA
Yeast_M	CTAGTAG	TAAACTATG	AATACAATT	ATTTATAAT	A-TATATTATATA	TAAATAATA	AATGAAAAT	GAAAGTATTCCA
Bovine_M	-TC-TAC	ACCAAGAGA	ATC^u ACGAA	AGTTATTAT	G-AAACCAATAAC	CAAAGGAGG	ATTTAGCAG	TAAACTAAG---
Murine_M	-TC-TTA	TAAAAGAAC	ATTACTATA	CCCTTTA-T	G-AAACTAAAGGA	CTAAGGAGG	ATTTAGTAG	TAAATTAAG---
Yeast_Cy	CTT-CTG	GCTAACCTT	GAGTCCTTG	TGGCTCTTG	G-CGAACCAGGAC	--TTTTACT	TTGAAAAAA	TTAGAGTGTTCA
Rat_Cy	-GGGGCC	CCCTCGATG	CTCTTAGCT	GAGTGTCCC	G-CGGGGCCCGAA	GCGTTTACT	TTGAAAAAA	TTAGAGTGTTCA
Fungus_Cy	CTTAATA	GTTCAGGTT	GTATTATCT	T-TGATAGT	G-CTTGTTTGGA-	CATTTCACT	GTGAGAAAA	TTGTGGTGTTTA
Rice_Cy	GGGCGATG	CGCTC-CTG	GCCTTAACT	GGCCGGGTT	CGTGCCTCC-GGC	GCCGTTACT	TTGAAGAAA	TTAGAGTGCTCA^v

(continued)

Table 4.4 (Continued)

Row 13	A	B	C	D	E	F	G
Euglena$_C$	CCCGCC	TGGGGAGTACGC	TTGCACAAG	TGAAACTCA	AAGGAATTGACG	GGGGCCCGC-AC	AAGCGGTGGAGCATG
Chlamydomonas$_C$	CCCGCC	TGGGGAGTATGC	TCGCCAAGAG	TGAAACTCA	AAGGAATTGACG	GGACCCG----AC	AAGCGGTCGATTATG
Tobacco$_C$	CCCGCC	TGGGGAGTACGT	TCGCCAAGAA	TGAAACTCA	AAGGAATTGACG	GGGGCCCGC-AC	AAGCGGTGGAGCATG
Anacystis	TCCGCC	TGGGGAGTACGC	ACGGCAAGTT	GGAAACTCA	AAGGAATTGACG	GGGGCCCGC-AC	AAGCGGTGGAGTATG
Escherichia	ACCGCC	TGGGGAGTACGG	CCGCAAGGT	TAAAACTCA	AATGAATTGACG	GGGGCCCGC-AC	AAGCGGTGGAGCATG
Halobacterium	GCCGCC	TGGGAAGTACGT	CCGCAAGGA	TGAAACTTA	AAGGAATTGGGCG	GGGGAGCACTAC	AACCGGAGGAGCCTG
Yeast$_M$	----CC	TGAAGAGTACGT	TAGCAATAA	TGAAACTCA	AAACAATAGACG	GTTACAGAC-TT	AAGCAGTGGAACATG
Bovine$_M$	AATAGA	GTGCTTAGTTGA	ATTAGGCCA	TGAAGCACG	CACACACCGCCC	GTCACCCTCCTC	AAATAGA-TTCAGTG
Murine$_M$	AATAGA	GAGCTTAATTGA	ATTCAGCAA	TGAAGTACG	CACACACCGCCC	GTCACCCTCCTC	AAATTAAATTAAACT
Yeast$_{Cy}$	AAGCAG	GC--GTATTGCT	CGAATATAT	TAGCATGGA	ATAATAGAATAG	GACGTTTGTTC	TATTTTGTTGGTTTC
Rat$_{Cy}$	AAGCAG	GCCCGAGCCCGCC	TGGATACCG	CAGCTAGGA	ATAATGAATAG	GAC-CGCGGTTC	TATTTTGTTGGTTTT
Fungus$_{Cy}$	AAGCAG	GGGTCT-CGCCT	GATCTTTTG	CAGCATGGT	ATGATGAAACAT	GACATTTTACGC	TATTGG-TTTG----
Rice$_{Cy}$	AAGCCA	TCG-CT-C----	TGGATACAT	TAGCATGGG	ATAACATCATAG	GAT-TCCGGTCC	TATTGTGTTGGCCTT

Row 14	H	I	J	K	L	M	N	O	P
Euglena C	TGGTTT	AATTCG	ATGCAACAC	GAAGAA	CCT-TACCAG	GATTTGACA	GGATCTAGGAA	GTTTGGAAG	AACGCA---GTA
Chlamydomonas C	TGGATT	AATTCG	ATACAACGC	GAAGAA	CCT-TACCAG	GGTTTGACA	-TGTCAA-GAA	-CCTCTCAG	AAATGGGAGGGTG
Tobacco C	TGGTTT	AATTCG	ATGCAAAGC	GAAGAA	CCT-TACCAT	GGCTTGACA	-TCCCGC-GAA	TCCTCT-TG	AAAGAGAGGGGTG
Anacystis	TGGTTT	AATTCG	ATGCAACGC	GAAGAA	CCT-TACCAG	GGTTTGACA	-TCCCCC-GAA	TCTCTTGGA	AACGAGAGA-GTG
Escherichia	TGGTTT	AATTCG	ATGCAACGC	GAAGAA	CCT-TACCTG	GTCTTGACA	-TCC-ACGGAA	GTTTTCAGA	GATGAGAAT-GTG
Halobacterium	CGGTTT	AATTCG	ACTCAACGC	CGGACA	TCT-CACCAG	CTCC-GA--	-CTA-CAGTGA	TGACGATCA	GGTTGATGA-CCT
Yeast M	TTATTT	AATTCG	ATAATCCAC	GACTAA	CTT-TACCAT	ATTTTGAAT	-ATT-ATAATA	ATTATTATA	-ATTATTAT-ATT
Bovine M	CATCTA	ACCCTA	TTTAAACGC	ACTAGC	TACATGAGAG	GAGACAAGT	--CGTAACAAG	GTAAGCATA	C-(to row 21)
Murine M	TAACAT	AATTAA	TTTCTAGAC	ATCCGT	TTA-TGAGAG	GAGATAAGT	--CGTAACAAG	GTAAGCATA	C-(to row 21)
Yeast Cy	TAGGAC	CATCGT	AATGATTAA	TAGGGA	CGG-TCGGGG	GCATCGGTA	-TTCAATTG--	TCGAGGTGA	AATTCTTGG-ATT
Rat Cy	CGGAAC	TGAGGC	CATGATTAA	GAGGGA	CGG-CCGGGG	GCATTCGTA	-TTGCGCCGC-	TAGAGGTGA	AATTCTTGG-ACC
Fungus Cy	CGTTTA	AAGTGT	AATGATTAA	TAGGGA	TGG-ATGGGG	GTGTTCATA	-TTGGTGGGC-	GAGAGGTGA	AATTCGTTG-ACC
Rice Cy	CGGGAT	CGGAGT	AATGATTAA	TAGGGA	CAG-TCGGGG	GCATTCGTA	-TTTCATAGT-	CAGAGGTGA	AATTCTTGG-ATT

(continued)

Table 4.4 (Continued)

Row 15	Q	R	S	T	U	V	W
Euglena_C	CCTTCGGGTATCTAG	ACACAGGTGGTG	CATGGC	TGTCGTCAG	CTCGTGTCG	TGAGATGTTTGGG	TTAAGTCCCGCA
Chlamydomonas_C	CCCTAACGGACTTGA	ACACAGGTGGTG	CATGGC	TGTCGTCAG	CTCGTGTCTG	TGAAGTGTATAG	TTAAGTCTCATA
Tobacco_C	CCTTCGGGAACGCGG	ACACAGGTGGTG	CATGGC	TGTCGTCAG	CTCGTGCCG	TAAGGTGTTTGGG	TTAAGTCCCGCA
Anacystis	CCTTCGGGACCGGGG	AGACAGGTGGTG	CATGGC	TGTCGTCAG	CTCGTGTCG	TGAGATGTTTGGG	TTAAGTCCCGCA
Escherichia	CCTTCGGGAACCGTG	AGACAGGTGGTG	CATGGC	TGTCGTCAG	CTCGTGTTG	TGAAATGTTTGGG	TTAAGTCCCGCA
Halobacterium	TATCACGACCCTGT-	AGAGAGGAGGTG	CATGGC	CGCCTCAG	CTCGTACCG	TGAGGCGTCCTG	TTAAGTCAGGCA
Yeast_M	------------	--ACAGCGTTA	CATTGT	TGTCTTTAG	TTCGTGCTG	CAAAGTTTTAGA	TTAAATGTG-CA
Yeast_Cy	TATTGAAGACTAACT	ACTCGGAAAGCG	TTTGCC	AAGGACGTT	TTCGTTAAT	CAAGAACGAAAG	TTGAGGG------
Rat_Cy	GGCGCAAGACGAACC	AGACCGAAAGCA	TTTGCC	AAGAATGTT	TTCATTAAT	CAAGAACGAAAG	TCGGAGGTTCGA
Fungus_Cy	CTATCAAGATGAACT	TCTCCGAAAGCA	TTCACC	AAATACTTC	CCCATTAAT	CAAGAACGAAAG	TTTGGGGATCGA
Rice_Cy	TATGAAAGACGAACA	ACTCCGAAAGCA	TTTGCC	AAGGATGTT	TTCATTAAT	CAAGAACGAAAG	TTGGGGGCTCGA

Row 16	A	B	C	D	E	F	G	H
Euglena C	--ACGAGCGCAACC	CTTTTTT-T	TAATTA	ACGCTTGTC	ATT-------	TAGAAATAC	TG------C	TG---GTT-
Chlamydomonas C	--ACGAGCGCAACC	CTCCTCT-T	TAGTTG	CCATT----	----------	-TGGTTCTC	TAAAGAGAC	TGCCAGTGT
Tobacco C	--ACGAGCGCAACC	CTCGTGT-T	TAGTTG	CCATCGTT-	CAGT-----	TTCGAACCC	TGAACAGAC	TGCCGGTGT
Anacystis C	--ACGAGCGCAACC	CACGTTT-T	TAGTTG	CCATCATT-	CAGT------	TGGGCACTC	TAGAGAAAC	TGCCGGTGA
Escherichia	--ACGAGCGCAACC	CTTATCC-T	TTGTTG	CCAGCGGTC	CGGC------	CGGGAACTC	AAAGGAGAC	TGCCAGTGA
Halobacterium	--ACGAGCGAGACC	CCCACTT-C	TAATTG	CCAGCAGCA	GTTTCGACTGGC	TGGGTACAT	TAGAAGGAC	TGCCGCTGC
Yeast M	TAAACGAGCAAAACT	CCATATATA	TAATTT	TATATTATT	TATT-AAT---	--A-TA---	AAAGAA---	----------
Yeast Cy	-------ATCTGAT	ACCGTCGTA	GTCTTA	ACCATAAAC	TATGCCGACTAG	--ATCGGGT	GGTGTTTTT	TTAATGACC
Rat Cy	---AGACGATCAGAT	ACCGTCGTA	GTTCCG	ACCATAAAC	GATGCCGACTGG	CGATGCGGC	GGCGTTATT	CCCATGACC
Fungus Cy	---AGACGATCAGAT	ACCGTCGTA	GTCCAA	ACTATAAAC	TATGTCGACCAG	GGATCGGTT	AAAATT-TT	TTCAAAATT
Rice Cy	---AGACGATCAGAT	ACCGTCGTA	GTCTCA	ACCATAAAC	GATGCCGACCAG	GGATCGGCG	GATGTTGCT	TATAGGACT

(continued)

Table 4.4 (Continued)

Row 17	I	J	K	L	M	N	O	P	Q
Euglena C	ATTACC	GGAGGAAGG	TCAGGAACGA-CGT	CAAGTC-ATC	ATG	CCCCTT	ATATCCTGG	GCTACACAC	GTGCTACAATGG
Chlamydomonas C	-AAGCT	GGAGGAAGG	TCAGGAATGA-CGT	CAAGTC-AGC	ATG	CCCCTT	ACATCCTGG	GCTTCACAC	GTAATACAATGG
Tobacco C	TAAGCC	GGAGGAAGG	TGAGGAATGA-CGT	CAAGTC-ATC	ATG	CCCCTT	ATGCCCTTG	GCGACACAC	GTGCTACAATGG
Anacystis C	CAAACC	GGAGGAAGG	TGTTGGACGA-CGT	CAAGTC-ATC	ATG	CCCCTT	ACATCCTGG	GCTACACAC	GTAGTACAATGC
Escherichia	TAAACT	GGAGGAAGG	TGGGGATTGA-CGT	CAAGTC-ATC	ATG	GCCCTT	ACGACCAGG	GCTACACAC	GTGCTACAATGG
Halobacterium	TAAAGC	GGAGGAAGG	AACGGCCAA-CGG	TAGGTC-AGT	ATG	CCCCGA	ATGAGCTGG	GCTACACGC	GGGCTACAATGG
Yeast M	------	------	-AGCAATTA-AGA	CAAATC-ATA	ATG	ATCCTT	ATAATATCG	GTAATAGAC	GTGCTATAATAA
Yeast Cy	CACTCG	GTACCTTAC	GAGAAATCA-AAG	TCTTTG-GGT	TCT	GGGGGG	AGTATGGTC	GCAAAGGCT	GAAACTTAAAGG
Rat Cy	CGCCGG	GCAGCTTCC	GGGAAACCA-AAG	TCTTTG-GGT	TCC	GGGGGG	AGTATGGTT	GCAAAG-CT	GAAACTTAAAGG
Fungus Cy	TAATCG	GCACCTTGT	GAGAAATCATGAG	TGTTTA-GAT	TCC	GGGGGG	AGTATGGTC	GC-AAGTCT	GAAACTTAAAGG
Rice Cy	CCGCCG	GCACCTTAT	GAGAAATCA-AAG	TCTTTGGGGT	TCC	GGGGGG	AGTATGGTC	GC-AAGGCT	GAAACTTAAAGG

Row 18	R	S	T	U	V	W	X	Y	Z
Euglena C	TTAAG-	ACA	ATAAGTTGCAAT TTT	GTGAAAATGAGG	TAAT-- CTTAAA	CTTAGCCTA	AGT	TCGGATTGTAGG	CTG
Chlamydomonas C	TTGGG-	ACA	ATCAGAAGCGA- CTC	GTGAG----AGG	TAGGGG[2] CTCAAA	CCCAACCTC	AGT	TCGGATTGTAGG	CTG
Tobacco C	CCGGG-	ACA	AAGGGTCGCGAT CCC	GCCAGGGTGAGG	TAACCC CAAAAA	CCCGTCCTC	AGT	TCGGATTGCAGG	CTG
Anacystis	TCCGG-	ACA	GCGAGACGCGAA GCC	GCGAGGTGAAGG	AAATCT CCCAAA	CCGGGCTC	AGT	TCAGATTGCAGG	CTG
Escherichia	-CGCAT	ACA	AAGAGAAGCGAC CTC	GCCAGAGCCAACG	GGACCT CATAAA	GTGCGTCGT	AGT	CCCGATTGGACT	CTG
Halobacterium	TCGAG-	ACA	ATGGGTTGCTAT CTC	GAAAGAGAACGC	TAATCT CCTAAA	CTCGATCGT	AGT	TCCGATTGAGGG	CTG
Yeast M	AAT-G-	ATA	ATAAAATTATAT[b] CAT	TTTAATTTTTAA	TATATT TTTTTA	TTATATATT	AAT	ATCAATTATAAT	
Yeast Cy	AATTGA	CGG	AAGGGCACCACT AGG	AGTGGACCCTGC	GGCT-A ATTTGA	CTCAACACG	GGG	AAACTACCAGG	TCC
Rat Cy	AATTGA	CGG	AAGGGCACCACC AGG	AGTGGACCCTGC	GGCTTA ATTTGA	CTCAACACG	GGA	AACCTCACCCGG	CCC
Fungus Cy	AATTGA	CGG	AAGGGCACACAA TGG	AGTGGACCCTGC	GGCTTA ATTTGA	CTCAACTCG	GGA	AAACTTACCAAG	CTA
Rice Cy	AATTGA	CGG	AAGGGCACCACC AGG	CGTGGCGCCCTGC	GGCTTA ATTTGA	CTCAACACG	GGG	AAACTTACCAGG	TCC

(continued)

Table 4.4 (Continued)

Row 19	A	B	C	D	E	F	G	H	I	J	K
Euglena C	AAA	CTC	GCCTAC	ATGAAG	CCCGAATCG	CTAGTAATC	GCCGG-TCAGCT	ATACGG	CGGTGAATA	CGTTCT	CGGGCC
Chlamydomonas C	CAA	CTC	GCCTAC	ATGAAG	CCCGAATCG	CTAGTAATC	GCCAG-TCAGCT	ATATGG	CGGTGAATA	CGTTCC	CGGGTC
Tobacco C	CAA	CTC	GCCTGC	ATGAAG	CCCGAATCG	CTAGTAATC	GCCGG-TCAGCC	ATACGG	CGGTGAATT	CGTTCC	CGGGCC
Anacystis C	CAA	CTC	GCCTGC	ATGAAG	GCCGAATCG	CTAGTAATC	GCAGG-TCAGC-	ATACTG	CGGTGAATA	CGTTCC	CGGGCC
Escherichia	CAA	CTC	GACTCC	ATGAAG	TCCGAATCG	CTAGTAATC	GT-GGATCAGA-	ATGCCA	CGGTGAATA	CGTTCC	CGGGCC
Halobacterium	AAA	CTC	GCCCTC	ATGAAG	CTGGATTCG	GTAGTAATC	GC-ATTTCAAT-	AGAGTG	CGGTGAATA	CGTCGC	TGCTCC
Yeast M	AAA	TTC	GATTAT	ATGAAA	AAAGAATTG	CTAGTAATA	CGTAAATTAGT-	ATGTTA	CGGTGAATA	TTCTAA	CT-GTT
Yeast Cy	AGA	CAC	AATAAG	GATTGA	CAGATTGAG	AGCTCTTTC	TTGATTTTGTGG	GTGGTG	GTCCATGGC	CGTTTC	TCAGTT
Rat Cy	GGA	CAC	GGACAG	GATTGA	CAGGTTGAT	AGCTCTTTC	TCGATTCCGTGG	GTGGTG	GTCCATGGC	CGT-TC	TTAGTT
Fungus Cy	AGA	TAT	AGTAAG	GATTGA	CAGACTAAA	AGATCTTTC	ATGATTCTATAA	GTGGTG	GTCCATGGT	CGT-TC	TTAGTT
Rice Cy	AGA	CAT	AGCAAG	GATTGA	CAGACTGAG	AGCTCTTTG	TTGATTCTATGG	GTGGTG	GTCCATGGC	CGT-TC	TTAGTT

Row 20	L	M	N	O	P	Q	R	S
Euglena C	TTG	TACACACCG	CCCGTC	ACACCATGG	AAGTCGGCTCGTCGCCC	--GAAGTTATTATCTT	GCCTGAA --	AAGAGGGAA
Chlamydomonas C	TTG	TACACACCG	CCCGTC	ACACCATGG	AAGCTCGTTCTGCTC	--CAAGTCGTTACCCT	AACCTTC --	GGGAGGGGG
Tobacco C	TTG	TACACACCG	CCCGTC	ACACTATGG	GAGCTGCCCATGCCC	--GAAGTCGTTACCTT	AACCCCA --	AGG-GGGGG
Anacystis	TTG	TACACACCG	CCCGTC	ACACCATGG	AAGTTGGCCATGCCC	--GAAGTCGTTACCCT	AACCGTT CG	CGGAGGGGG
Escherichia	TTG	TACACACCG	CCCGTC	ACACCATGG	GAG-TGGGTTGCAAA	-AGAAGTAGGTAGCTT	AACCTTC --	GGGAGGGCG
Halobacterium	TTG	CACACACCG	CCCGTC	AAAGCACCC	GAG-TGAGGTCCGGA	-TGA-GGCCACC---	---------	---ACACG
Yeast M	TCG	CACTAATCA	CTCATC	AGGCGTTGA	AACATATTATTATCT	-TATTATTTATAT---	AATATTT TT	TAATAAATA
Bovine M	(ACG	CACACACCG	CCCGTC	ACCCTCCTC	AAATAGATTCAG)			
Yeast Cy	GGT	GGAGTGATT	TGTCTG	CTTAATTGC	GATAACGAACGAGAC	-CT---TAACCTA-CT	AAATAGT GG	TGCTAGCAT
Rat Cy	GGT	GGAGCCATT	TGTCTG	GTTAATTCC	GATAACGAACGAGAC	TCTCGGCATGCTAACT	AGTTACG CG	ACCCCCGAG
Fungus Cy	GGT	GGAGCCATT	TGTCTG	GTCAATTCC	GATAACGGACGAGAC	-CTCGACCTGCTAACT	AGT-AGT$^{3'}$TC	TGACTCGAT
Rice Cy	GGT	GGAGCCATT	TGTCTG	GTTAATTCC	GTTAACGAACGAGAC	-CTCAG-CTG----CT	AACTAGC TA	TGCG--GAG

(continued)

Table 4.4 (Continued)

Row 21	A	B	C	D	E	F	G	H	I
Euglena$_C$	A-TACCTAAGGC	CTGGC T GGT	GACTGGGGT	GAAGTCGTA	ACAAGG	TAGCCG	TACTGG	AAGGTGTGG	CTGG AACAA
Chlamydomonas$_C$	G-CGCCTAAAGC	AGGGC T AGT	GACTAGGGT	GAAGTCGTA	ACAAGG	TAGGGC	TACTGG	AAGGTGGCC	CTGG CTCAC
Tobacco$_C$	A-TGCCGAAGCG	G-GGC T AGT	GACTGGAGT	GAAGTCGTA	ACAAGG	TAGCCG	TACTGG	AAGGTGCGG	CTGG ATCAC
Anacystis$_C$	G-CGCCGAAGT	AGGGC T GAT	GACTGGGGT	GAAGTCGTA	ACAAGG	TAGCCG	TACGGG	AAGGTGTGG	CTGG ATCAC
Escherichia	CTTACCACTTTG	T-GAT T CAT	GACTGGGGT	GAAGTCGTA	ACAAGG	TAACCG	TAGGGG	AACCTGCGG	TTGG ATCAC
Halobacterium	GTGGTCGAATCT	G-GCT T CGC	AAGGGGCT	TAAGTCGTA	ACAAGG	TAGCCG	TAGGGG	AATCTGCGG	CTGG ATCAC
Yeast$_M$	TAAATAATTATT	A-ATT^{2}A -TG	AATTAATGC	GAAGTTG-A	ATACAG	TTACCG	TAGGGG	AACCTGCGG	TGG^{122} TAAAT
Bovine$_M$	-------	-------	-------	-------	-------	-------	-------	-------	TGGA AAGTG
Murine$_M$	-------	-------	-------	-------	-------	-------	-------	-------	TGGA AATTG
Yeast$_{Cy}$	TTGCTGGTTATC	C----- A CTT	CTTAGAGGG	ACTATCCGT	-TTCAA	GCCGAT	GGAAGT	TTGAGGCAA	TAAC AGGTC
Rat$_{Cy}$	CGG-TCGGCGTC	CCCCA A CTT	CTTAGAGGG	ACAACTGGC	GTTCA-	GCCACC	CGAGAT	T-GAG-CAA	TAAC AGGTC
Fungus$_{Cy}$	AGGTACGAA-TT	A--AA A CTT	CTTAGAGGG	ACTACCTGC	-CTCAA	GCAGGC	GGAAGT	CCGAGGCAA	TAAC AGGTC
Rice$_{Cy}$	CCAT--CC-CTC	CCGCA G CTT	CTTAGAGGG	ACTATGGCC	GTTTAG	GCC-AC	GGAAGT	TTGAGGCAA	TAAC AGGTC

	J	K	L	M	N	O
Chlamydomonas $_C$	CTCC					
Tobacco $_C$	CTCCTTC					
Anacystis	CTCCTTT					
Escherichia	CTCCTTT					
Halobacterium	CTCCTTA					
	CTCCT--					
Yeast $_M$	ATTCTTACATAGGTA	TTAATCTAAATA	TTGAATATGAGG	TGTTNGTAGTT----	-----------	-------CCCG
Bovine $_M$	TG-CTTGGA-TAAAT					
Murine $_M$	TG-CTTGGAATAAT-					
Yeast $_{Cy}$	TGTGATGCCCTTAGA	ACGTTCTGGGC	GCACGCGCGCTA	CACTGA-CGGAGCCA	GCCAGT--C	TAACCTTGGCCG-
Rat $_{Cy}$	TGTGATGCCCTTAGA	-TGTCCGGGCT	GCACGCGCGCTA	CACTGAACTGGTTCA	GCCGTGTCCC	TACCCTACGCCG-
Fungus $_{Cy}$	TGTGATGCCCTTAGA	-TACCTTGGGC	GCACGCGCGCTA	CAATGT-AGGAAACA	A--AAAGGC	--TCCTGGTCCG-
Rice $_{Cy}$	TGTGATGCCCTTAGA	-TGTTCTGGGC	GCACGCGCGCTA	CACTGA-TGTATCCA	ACGAGTATA	TAGCCTGGTCCG-

(continued)

Table 4.4 (Continued)

Row 23

	Q	R	S	T	U	V	W
Yeast Cy	AGAGGTCTT	GGTAATCTT	GTGAAACTCCGT	CGTGCTGGGAT	AGAGCATTGTAA	TTATTGCTCTTC	AACGAGGAA
Fungus Cy	GAAGGATTG	GGTAATCAT	TTGAATTTCCTA	CGTAACTGGGCT	TGATCTTTGTAA	TTATTGATCATA	AACGAGGAA
Rat Cy	GCAGGCGCG	GGTAACCCG	TTGAACCCCATT	CGTGATGGGGAT	CGGGGATTCCAA	TTATTCCCCATG	AACCAGGAA
Rice Cy	ACAGGCCCG	GGTAATCTT	GGCAAATTTCAT	CGTGATGGGGAT	AGATCATTGCAA	TTGTTGGTCTTC	AACGAGGAA

Row 24

	A	B	C	D	E	F
Yeast Cy	TTCCTAGTAAGC	GCAAGTCATCAG	CTTGCGTTGATT	AGCTCCCTGCCC	TTTGTACACACCCGCC	CGTCGCTAGTAC
Fungus Cy	TTCCTTGTAAGC	GTAAGTCATTAC	CTTATGCTGAAT	ATTCTCCCTGCCC	TTTGTACACACCCC	CGTCGCTCCTAC
Rat Cy	TTCCCAGTAAGT	GCGGGTCATAAG	CTTGCGTTGATT	AAGTCCCTGCCC	TTTGTACACACCCGCC	CGTCGCTACTAC
Rice Cy	TGCCTAGTAAGC	GCGAGTCATCAG	CTCCGGTTGACT	AGCTCCCTGCCC	TTTGTACACACCCGCC	CGTCGCTCCTAC

Row 25

	G	H	I	J	K	L	M
Yeast Cy	CGATTG	AATGGCTTAGTG	AGGCCTCAGGAT	CTGCTTAGAGAA	GGG-GGCA-AC-	TCCATCTC--AG	AGCGGGAGAA
Fungus Cy	CGATCG	AATGATACGGTA	AAGTTAAACGGTA	CGTTTTATCT--	-GT-GGCA-AC-	--ACTGA--TAT	AAATTAAAA
Rat Cy	CGATTG	GATGGTTTAGTG	AGGCCCTCGGAT	CGGCCCCGCCGG	GGTCGGGCCCAGC	GCCTTGCGGAG	GCCTGAGAA
Rice Cy	CGATTG	AATGGTCCCGTG	AAGTGTTTCGGAT	CGCGGCGACGGG	GGCGGTTCGCCCG	-CCCCGAC--G	TCGCCGAGAA

Row 26	N	O	P	Q	R	S	T	U	V
Yeast Cy	TTTCGACAA	ACT	TGGTCATTTGGA	GGAACT	AAAAGTCGTAAC	AAGGTTTCCGTA	GGTGAACCTGCG	GAAGGATCA	TTA
Fungus Cy	GTTATTTAA	ATC	TCATTGTTTAGA	GGAAGG	AGAAGTCGTAAC	AAGGTATCCGTA	GGTGAACCTGCG	GATGGATCA	TTTT
Rat Cy	GACCGTCGA	ACT	TGACTATCTAGA	GGAAGT	AAAAGTCGTAAC	AAGGTTTCCGTA	GGTGAACCTGCG	GAAGGATCA	TTA
Rice Cy	GTCCATTGA	ACC	TTATCATTTAGA	GGAAGG	AGAAGTCGTAAC	AAGGTTTCCGTA	GGTGAACCTGCG	GAACGATCA	TTG

[a]Subscript C indicates chloroplast, M indicates mitochondrion, Cy indicates cytoplasm. [b]Graf et al. (1982). [c]Dron et al. (1982). [d]Tohdoh and Sugiura (1982). [e]Tomioka and Sugiura (1983). [f]Brosius et al. (1981). [g]Gupta et al. (1983). [h]Sor and Fukuhara (1982). [i]Anderson et al. (1982a). [j]Bibb et al. (1981). [k]Rubtsov et al. (1980). [l]Ozaki et al. (1984). [m]Torczynski et al. (1983). [n]Takaiwa et al. (1984). [o]Preceded by a 493-nucleotide-residue sequence. [p]Insert 15 residues. [q]Insert 144 residues. [r]Insert ATAAT. [s]Insert ATAATAATA. [t]Insert ATATAT. [u]Insert AAG. [v]Insert TAAGC. [w]Insert 37 residues. [x]Insert CTCTGTT. [y]Insert 79 residues. [z]Insert 25 residues. [aa]Insert CTAT. [bb]Insert 13 residues.

mitochondria, in which such loss is of frequent occurrence throughout their genomes. That such deletions are not artifacts of the alignment procedure is at once confirmed by the two adjacent sectors of columns B and C, which are of nearly uniform constitution in all the sequences given. At the beginning of row 2, loss of residues is widespread, occurring even in the blue-green alga; then it is not until row 6, columns R and S, that a loss of occupied sites in the eukaryotic gene can be found, the last of this feature in the 5′-terminal region. However, there are several areas in which the yeast gene has undergone reduction independently, as in row 2, column P; row 3, column W; and row 4, columns A, F, and G. In row 5, columns H–K, particularly strong losses have occurred, where 26 nucleotide residues are absent that are present in the rat cytoplasmic species and the others.

The Middle Regions. In dealing with such long sequences as the present, it aids discussion to consider a midregion separately, here taken to be represented by rows 9–16. At its commencement there are few homologies except between members of the same taxon or type of organelle and in many columns even such limited homologies are virtually lost. This condition prevails through much of the entire region and may represent incorrect alignment of the sequences as an outgrowth of the increment in length of the eukaryotic macromolecule mentioned earlier.

But addition of new sites has been accompanied also by loss of others, particularly in the yeast gene. Such independent deletions are scarcer here than in the 5′ end, taking place on a large scale only in row 11, column J; row 15, column W; and row 16, column A; and at scattered positions on a small scale. In general this area seems to be a region of lesser importance, where deletion of parts has little impact; in fact, the mitochondrial genes of vertebrates appear to have lost much of this midsection beginning with the end of row 14. Many more sequences from all types of sources must be determined before it will be possible to establish homologies in these genes more firmly.

The 3′-Terminal Sections. The remainder of the molecules, beginning with row 17, may be viewed as the 3′-terminal sections. Here, as in the midregion, very little of note is seen, since the sectors of homology in the eukaryotic cytoplasmic species are confined to themselves. Exceptions do occur, but these nearly entirely represent single sites, although dimeric combinations shared with *Halobacterium* exist in row 17, column L; row 18, column W; and row 19, columns E and G. In addition, similar pairs of identical nucleotides shared with all the prokaryote or the two bacterial sequences are located in row 17, columns P and Q; row 18, columns T and U; and row 19, column F. Moreover, a few are common to the eukaryote cytoplasmic and *E. coli* or *Anacystis* genes alone, such as in row 18, column V; row 19, column H; and row 21, column C. Homologous triplets are restricted to row 19, column J, a CGT being shared with all the prokaryotes.

Homology among the Minor rRNAs. The obvious lack of strong homology undoubtedly stems from the numerous insertions that have occurred in eukaryotic sequences, since that source taxon parted from their advanced bacterial forebears (M. W. Gray *et al.*, 1984; Van Knippenberg *et al.*, 1984). To judge from the persistent homology that prevails between the yeast and rat sequences, the insertions must have been made early in the history of these organisms. The two genes are predominantly identical from the very beginning of row 1 until near the extreme 3′ terminus at row 27, column V, the final 32 sites being totally inharmonious. One final point of interest can be noted—the yeast sequence ends with multiple Ts as in the case with *Anacystis*, whereas the others have TTR.

However, in view of the identity of function of this type of rRNA that prevails throughout the cellular living world, homologous sites between major taxa of organisms are surprisingly few. The prokaryotic types, including even that of *Halobacterium*, share a sufficient number of identities that the sequences can be aligned fairly easily. Similarly, the several cytoplasmic representatives of eukaryotes agree among themselves to a large extent. But comparisons between these two groups are made on a less than firm basis, a view shared by others who have attempted them (Gupta *et al.*, 1983; Leffers and Garrett, 1984). Undoubtedly the sequences of this gene from additional members of the higher bacteria could assist in establishing homologous sites and locations of insertions and deletions on a much firmer basis. Nevertheless, as both of those references also point out, the kinship between the Archaebacteria and the Eukaryota is distinctly closer than between the Eubacteria and the latter group. So once more the data point toward the derivation of the eukaryotic cell from an advanced prokaryote, thereby again negating the basic premises of the endosymbiontic concept.

4.4.3. The Organellar Mature Minor-rRNA Genes

Although in both the prokaryotic and eukaryotic minor-rRNA genes a high level of evolutionary conservation exists, one class of organellar cistrons, those of mitochondria, affords quite a contrast to that in general. On the other hand, the chloroplast representatives do not diverge to any extent from the general condition; indeed, they are shown to possess strong homology to the prokaryotic counterparts. Because of this conservation, it appears advisable to discuss them before the more heretical mitochondrial cistrons are examined.

Chloroplastic Mature Minor-rRNA Genes. Only three of four available mature gene sequences for the minor rRNA of chloroplasts are included in Table 4.4, since that from the maize organelle (Schwarz and Kössel, 1980) is virtually identical to that of the tobacco (Tohdoh and Sugiura, 1982). Those included are homologous to a remarkable degree, especially in view of one (that of *Euglena*) being only distantly related to the other source organisms. Since this isolate is from a member of the more primitive branches of the eukaryotic line of ascent, it should be expected to reflect the ancestral stock more fully than the two from the highly advanced green plant lineage. At its 5'-terminal region, however, it is more divergent than the latter from those prokaryotic sequences given in the tabulation in that it is longer by two sites, while the green algal and tobacco chloroplastic sequences are only one site longer than the blue-green and *E. coli* genes. Phylogenetic resemblances are rarely clear-cut in these nucleotide chains. Column B of row 1, for example, is absolutely identical in all the chloroplastic and prokaryotic representatives of Table 4.4, but they are also identical in the two eukaryotic cytoplasmic species, while the three mitochondrial ones diverge. In the adjacent column C agreement in part is shown with both prokaryote and eukaryote sequences, but the *Euglena* chloroplast type is irregular, as is that of *E. coli*. As progress through the sequences is made away from this relatively stable sector, the three from the chloroplast are found to display an increasing homology with the prokaryotic, and to a lessened degree with the eukaryote cytoplasmic. But as stated before, the relationships are never clear-cut on a long-term basis.

The occasional loss of particular sectors is of special interest. In row 2, columns J and K, both the *Chlamydomonas* and tobacco sequences have undergone deletions, losses that have evidently occurred independently in the two organelles, for they differ in each case. Column L reflects a similar loss. Comparable processes have been active in row 3, columns U and V, and row 7, columns C and D. In the latter two places and in the B column that precedes, closer kinships to *Anacystis* than to *E. coli* are distinctly manifest. Additional deletions are relatively rare and are sometimes confined to a single representative, as in row 12, columns T and U; row 16, columns A, G, and H; row 18, column U; and row 20, column R. In the last instance, mixed kinship relations are displayed. Thus, as a whole, these chloroplast genes have not experienced any great reduction in length.

Mitochondrial Minor-rRNA Genes. As was observed with the tRNAs and 5 S rRNAs in preceding chapters, reduction in length is a prime characteristic of mitochondrial genes, and that for the minor rRNA is no exception. But that statement holds only in general, for the yeast mitochondrial coding sequence follows an opposite trend through the acquisition of eight inserted regions, placed at irregular intervals throughout its length. These locations are indicated by appropriate signals, while the actual inserted sequences are provided at the end of the table.

The first such insert precedes the ATTT- shown in row 1, column A (Sor and Fukuhara, 1980, 1982); it is largely through its great length (493 nucleotide residues) that the total yeast gene of 2179 base pairs exceeds the 1544 of that of *E. coli*. Much of the remainder is contributed by the second insert, whose 144-base-pair sequence is placed at the end of row 7, column G. This, like all the added sectors of this organellar gene, consists almost entirely of A–T base pairs, being interrupted by only four Gs and a like number of Cs. In the central region of the gene, the inserts are short, the third (row 9, column P) being merely ATAAT, the fourth (row 10, column B) ATAATAATA, and the fifth (row 10, column C) ATATAT. After a region of considerable extent in which a series of deletions occurs, notably in rows 11, 13, and 15–17, the last three inserts are to be found. The longest of these, the sixth, in row 18, column T, consists of a run of 37 residues, while the seventh and eighth in row 21, columns B and I, respectively, just precede the 3' terminus. The wheat mitochondrial minor-rRNA gene with 1968 base pairs also exceeds in length that of *E. coli* (Chao *et al.*, 1984), but its longest insert, of 336 residues, is located toward the 3' end.

In broad terms, in contrast to those of the yeast and wheat just discussed, the murine and bovine mitochondrial mature coding sequences more closely adhere to the general pattern of reduction that has characterized the other genes of preceding chapters, as does also that of the human organelle which is omitted from the table because of its similarity to the bovine (Anderson *et al.*, 1982b). All of the mammalian sequences are alike in having strong deletions in row 11 and then one of great length, extending from the end of row 14 to row 21, column I.

Homologies of Mitochondrial Minor-rRNA Genes. Although those two minor-rRNA genes of mammalian mitochondria are extensively homologous to one another, their relationships to that of the yeast organelle, like those of the others included in the table, are extremely difficult to establish. In fact, beyond the first three or four rows of Table 4.4, it is virtually impossible to locate regions of comparable structure between the mammalian mitochondrial gene and those either of the prokaryotes or eukaryotic cytoplasm, not to mention that of the yeast mitochondria.

Perhaps the greatest contribution to this lack of identifiable harmonious sectors is made by the replacement of G and C residues by Ts and As. This trend is made especially clear by the yeast mitochondrial gene, beginning with row 3. There in the 75 sites that are available only one pair each of Gs and Cs is present, the remainder being As, Ts, and vacant sites. In contrast, the *E. coli* sequence contains 21 Gs, 17 Cs, and six vacancies, along with 22 As and only ten Ts. This preponderance of thymidine and adenosine residues appears throughout the entire yeast molecule in longer or shorter sectors. Outstanding examples of the trend are apparent in row 6, columns S–U; row 7, columns E–G; row 8, columns I and J; row 9, columns R, S, U, and V; row 16, columns C–G; row 18, columns T–Y; and row 20, columns P–S. That these areas of divergence are not entirely artifacts of alignment, which in the present case largely agrees with that given by Sor and Fukuhara (1982), is made implicit by sectors of strong homology that immediately follow the T, A runs. Although many instances occur, especially convincing ones are found in row 8, columns K and L; row 10, column D; row 11, the end of column J into K, and again in M; row 15, columns S–U and W; and row 17, columns J and L–Q.

After row 4, the two mitochondrial sequences from mammalian sources proved to be so divergent that all serious attempts at finding sectors homologous to any of the others had to be abandoned, except at the very close of the mature genes. Only one conclusion can be drawn at this time—the mammalian mitochondrial minor-rRNAs are related as a whole to neither the prokaryotic nor eukaryotic cytoplasmic species.

4.5. THE SPECIAL rRNA GENES

Two rRNA genes are special in that they are confined to the genome of only a single large category of sources. The better known of these is that here named the supplemental (5.8 S), located on the intergenic spacer of eukaryotic *rrn* operons (Figure 4.1B–D); a second type, termed the appendicular (4.5 S) intervenes between the major and ancillary rRNA genes in chloroplast DNA (Figure 4.1H–J). Among the insects there is an additional component, called the 2 S rRNA, that is associated with the 5.8 S, of which it seems to be a fragment (Pavlakis *et al.*, 1979). In neither special type have functions been established, except that the supplemental, when tightly attached to the major species in ribosomes, appears to bear six of the ribosomal proteins (Lee *et al.*, 1983; Liu *et al.*, 1983), those known as L14, L19, L21, L29, L33, and L39.

4.5.1. The Genes of Supplemental rRNA

When a study of numerous RNA species is made on a broadly comparative basis as here, so many varieties are encountered named solely on the basis of their sedimentation coefficients that their identities often become confused. Moreover, the same species from different organisms may vary in that property, as in the case of the minor-rRNA type just discussed. Furthermore, the coefficient designation may change over the years, as witness the 5.8 S rRNA, which formerly was referred to as the 7 S rRNA. It is for these reasons that the latter type of rRNA, as indicated earlier, is here named the supplemental species and the 4.5 S is called the appendicular.

Current knowledge of the supplemental rRNA genes suffers from the same characteristics that afflicted understanding of the ancillary, for although the macromolecule from a wide sample of the eukaryotic world has been subjected to sequence studies, all except a few such investigations have focused solely on the mature product. To date seven vertebrate and a like number of other metazoan, one poriferan, four green plant, and 12 protistan 5 S rRNA sequences have been established (Erdmann et al., 1984; Dorfman et al., 1985), but of these 30 structures only seven include the flanking regions and the gene proper. Six of the latter are supplied in Table 4.5, that of Drosophila (Pavlakis et al., 1979) being omitted, since it is nearly identical to that of Sciara (Jordan et al., 1980). Thus those available are unusual in being rich in fungal (three) and insect species (two), rather than vertebrate, which is not represented. The sole additional one is that of Saccharomyces (Bell et al., 1977), but this is incomplete in that it lacks the 3′ train. In the table, sequences of the mature genes of man (Khan and Maden, 1977) and part of that of the major species of E. coli are added to provide comparative material. As indicated by a report on the gene from rat, some heterogeneity in structure will probably be found as further investigations are completed (Smith et al., 1984).

When the major-rRNA species from eukaryotic sources is treated with heat or urea to disrupt hydrogen bonds, the supplemental type is released (Pene et al., 1968). Bonding between the two macromolecules has been proposed to approach the pattern shown diagrammatically in Figure 4.2A (Walker and Pace, 1983). When free, as new transcripts might be, the supplemental rRNA molecule acquires a looped secondary structure, for which a number of somewhat differing configurations have been proposed (Van et al., 1977; Pavlakis et al., 1979; Darlix and Rochaix, 1981; Fujiwara et al., 1982; Nazar, 1982; Olsen and Sogin, 1982; Liu et al., 1983; Vaughn et al., 1984). One of the more currently popular structures is shown in Figure 4.2B.

Structural Properties of the Leader. Examination of the five leader sequences in Table 4.5 fails to disclose any notable degree of homology between them, despite the fact that three are from a single taxon, the fungi. In column G, just preceding the 5′ end of the principal mature genes, the third to last site is consistently occupied by a T, even in the E. coli sequence, but the lupine gene is exceptional. In the cistrons from Saccharomyces cerevisiae and Schizosaccharomyces pombe, this region in part often forms a portion of the mature product after processing. The latter organism is known to synthesize two mature products, a more abundant one commencing with the AAA of column I and a less common one that initiates with ATTATTTA from column G (Schaak et al., 1982). In S. cerevisiae there are three forms. The principal one begins as in the fission yeast with AAA of column I, a second less abundant type has TATTAA of column G appended to it, and the third has ATATTAA (Rubin, 1974). Since the leader is transcribed along with the mature gene, these varieties reflect differing processing procedures, not varieties of genes. Also, in the original description of the lupine gene (Rafalski et al., 1983), the transcript is given as beginning with the tetranucleotide CTAA, but the sequence aligns more harmoniously with the other primary structures when those four residues are considered as part of the leader as given in Table 4.5. Thus it is not unlikely that heterogeneity at the 5′ end of the green-plant mature rRNA may exist, as in the others in which the condition has been confirmed. Further examination of the leaders discloses no evidence of relationship between any of them and the E. coli major-rRNA gene. Moreover, the lupine gene leader's lack of even the few homologies noticeable among the others suggests it

Table 4.5
Representative Supplemental rRNA Genes[a]

Leaders	A	B	C	D	E	F	G
Escherichia 23S[b]	AACACTGAACAACGAAAG	TTGTTCCTGAGT	CTCTCAAAT	TTTCGCAAC	TCTGAAGTGA	AACATCTTC	GGGTTGTTGA
Saccharomyces[c]						AAATTTA	AAATATTAA
Schizosaccharomyces[d]	GCAAAATTAAATTATAAA	CCTTGAAATTTG	TTTTTGAAG	TCTGAATTA	ATTATA-TCT	AATATATAA	AATTATTTA
Neurospora[e]	G	TGCCGAAACTAA	ACTCTTGAT	ATTTTATGT	CTCTCTGAGT	AAACTTTTA	AATAAGTCA
Physarum[f]	GACTGCTTTCCCC	TTACCCGTTTCC	CCCGCTTAG	CAAACTTTA	ACCGTT-AAA	TTAAACAAC	GGAACGTAC
Sciara[g]	TGTCGTATTTTACGTCCT	TAGCTGGGCGTA	AATGCCCTT	TATGTTAAC	AATAAA-TTT	GTAAAAAAA	ATTTAATTC
Lupinus[h]	TGAAATCGTTTAGTTCGC	CCCCGCCGGCCC	GGAGACGGT	GCTCGTGCG	GGCGGC-GTT	GCGACACGC	TTATCCTAA

Mature Genes	I	II	III	IV	V	VI	VII	VIII
Escherichia 23S(+12)[i]	AAGGCT	ACACGG-TG	GATGCCCTG	GCAGTCAGA	GGGCATGAAGAAC	GTGCTA--ATCT	GCCATAAGCGT	CGGTAAGGT
Saccharomyces[j]	AAACTT	TCAACAACG	GATCTCTTG	GTTCTCGCA	TCG-ATGAAGAAC	GCAGCGAAATGC	GAT--AGGTAA	TGTGAATTG
Schizosaccharomyces	AAACTT	TCAGCAACG	GATCTCTTG	GCTCTCGCA	TCG-ATGAAGAAC	GCAGCGAAATGC	GAT--ACGTAA	TGTGAATTG
Neurospora	AAACTT	TCAACAACG	GATCTCTTG	GTTCTCGGCA	TCG-ATGAAGAAC	GCAGCGAAATGC	GAT--AGGTAA	TGTGAATTG
Physarum	ACCGTT	GGG-CGATG	GATTGCTTG	GTGCCTGCT	TCG-ACGAAGAGC	GCAGTGAAACGC	GAT--AACTTT	TGTGACTCG
Sciara	AACCCT	AAG-CGGGG	GATCACTTG	GTTTGTGGG	TCG-ATGAAGAAC	GCAGCAAACTGC	GTG--TTGACA	TGTGAACTG
Human[k]	CGACTC	TTACGGGTG	GATCACTCG	GCTCGTGCG	TCG-ATGAAGAAC	GCAGCGCCTAGCCT	GCGAGAATTAA	TGTGAATTG
Lupinus	AGACTC	TCGGCAACG	GATATCTCG	GCTCTTGCA	TCG-ATGAAGAAC	GTAGCGAAATGC	GAT--ACTTGG	TGTGAATTG

(continued)

Table 4.5 (Continued)

Mature Genes	IX	X	XI	XII	XIII	XIV	XV
Escherichia	GATATGAACCGTT	ATAACCGCGGATT	TCCGAA	TGGGGA---AAC	CCAGTGTCT	TTCGACACA	CTATCATTAACTGAA
Saccharomyces	CAG-AATTCCGTG	AATCATCGAAT-C	TTTTGAA	CGCACATTGCGC	CCCTTGGTA	TTCC--AGG	-------------
Schizosaccharomyces	CAG-AATTCCGTG	AATCATCGAAT-C	TTTTGAA	CGCACATTGCGC	CTTTGGGTT	CTACCA-AAG	-------------
Neurospora	CAG-AATTCAGTG	AATCATCGAAT-C	TTTTGAA	CGCACATTGCGC	TCGCCAGTA	TTCTGG-CGA	-------------
Physarum	CAC-TCTCTG-TG	A-TCAAGGTCT-C	CTTGAA	CATTAGTCGGGC	CTTGCCTTC[k]	GGGCAC-TGG	CCCCCT-------
Sciara	CAG-GACACA-TG	A-ACATTGACA-T	TTTTGAA	CGCATATTGCGG	TCCATACTG	*TGTAT----	-------------
Human	CAG-GACACA-TT	GATCATCGACA-C	TTCGAA	CGCACT-TGCGG	CCCCGGGTT	CCTCCCG---	-------------
Lupinus	CAG-AATCCCGTG	AACCATCGAGT-C	TTTTGAA	CGCAAGTTGCGC	CCGAAGCCA	TTAGGCCGA-	-------------

Mature Genes	XVI	XVII
Escherichia	TC-CATAGG-	TTAATG
Saccharomyces	GGGCATGCC-	TGTTTGAGGGTCATTT
Schizosaccharomyces	--GCATGCC-	TGTTTGAGTGTCATT
Neurospora	--GCATGCC-	TGTTCGAGGGTCATTT
Physarum	TGGGATGCC-	T-------TGTCCTT
Sciara	GGACCACACA	TGGTTGAGGGTCGTT
Human Hela	GGGCTACGCC	TGTCTGAGCGTCGCT
Lupinus	GGGCACGCC-	TGCCTGGGTGTTGCAC

Trains	M	N	O	P	Q
Schizosaccharomyces	ACAATCTTCTCA	CAAAAATGTTT	TTTTTAAATATT	TTTGATGAGGTG	TTGAACGAAAAT
Neurospora	-CAACCATCAAG	CTCTGCTTGCGT	TGGG		
Physarum	TTACAGTGCTTA	CGACCGACTCGG	GGGGATTAACCT	TTTCTCGCGTTA	TACCACCTTTAA
Sciara	AGA--TTACTTA	ATAAAAATTGCA	TTATTACGA---	TTGGATGCTTTT	TTGG
Lupinus	ATCGTTGCCCCC	GTGCCCTTGGCCA	CGTGCAGGCACG	AAACGGGGC	

[a]Boxed sections denote regions of evolutionary conservancy.
[b]Young *et al.* (1979)
[c]Bell *et al.* (1977); Bollon (1982).
[d]Schaak *et al.* (1982).
[e]Selker and Yanofsky (1979).
[f]Otsuka *et al.* (1982)
[g]Jordon *et al.* (1980).
[h]Rafalski *et al.* (1983).
[i]Jacq. (1981).
[j]Rubin (1975).
[k]Nazar *et al.* (1975).
[l]Insert ATTTGTTTCTTTTATTAGAAGC.

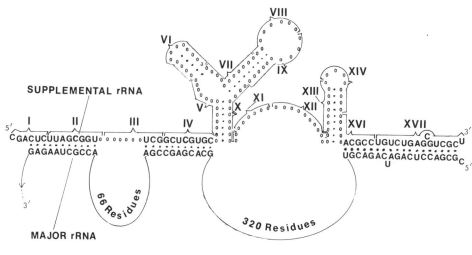

A. Interaction Sites

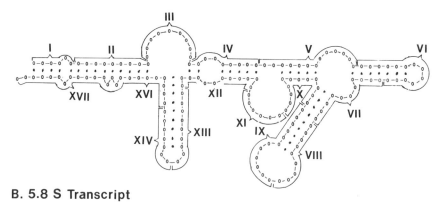

B. 5.8 S Transcript

Figure 4.2. Structural features of the supplemental (5.8 S) rRNA. (A) The molecule of supplemental rRNA possibly is hydrogen-bonded to a particular region of the major species through extensive base-pairing. (Walker and Pace, 1983.) (B) One of several models that have been proposed for the secondary structure of supplemental rRNA. (Vaughn *et al.*, 1984.)

belongs to an entirely different gene set, one especially striking distinction being the paucity of A–T base pairs in E and F, a region abounding with them in all the rest.

Structural Properties of the Mature Gene. When the mature coding sequences of this type of rRNA are scanned (Table 4.5), considerable variation is at once apparent, along with the several regions of complete invariability indicated by the boxes, the first such constant region being the opening triplet (GAT) of column III. Almost the entire column V shows this type of stability, the only variants being the substitution of a C for a T in the sixth site in *Physarum* and a G for an A in the penultimate position in that same organism. The sixth column is nearly as stable, although disparities may be noted in the *Sciara, Physarum,* human, and lupine examples. After a rather broad, relatively inconstant zone, the triplet GAA is found to comprise the latter half of column XI. This is the last indication of evolutionary stability until near the 3' terminus, where GTC is found in all except the lupine sequence. Finally, two or three Ts bring the mature coding region to a close, the human and flowering plant again being exceptional.

The asterisk in Table 4.5 at the beginning of column 14 marks the location of an inserted region that lies between the 5.8 S sequence of *Sciara* and the so-called 2 S rRNA, whose coding region commences at that point. The numerous sites of homology between the latter in this region and correspondingly in the others appear to confirm that this minute type is merely a fragment of the supplemental species. In the mature product the supplemental and this fragment do not become covalently bonded, so that the inserted portion does not strictly correspond to an intron. However, the insert is removed by processing in the cytoplasm after the two larger parts have become joined by hydrogen bonding. One pattern of attachment has been proposed to lie in the area extending from the final GG of column XII to the asterisk on the one hand and that part following the asterisk and continuing into the GGACC of column XVI (Jordan *et al.,* 1980), as follows:

$$\text{5.8 S rRNA} \quad \text{5'-GGTCCATACTG-3'}$$
$$\text{2 S rRNA} \quad \text{3'-CCAGGTATGT-5'}$$

The remainder of each portion extends thence to the left of the diagram. As far as is firmly established, this peculiar subdivisioning of the rRNA is strictly a dipteran characteristic, having been reported from *Drosophila* as well as *Sciara* (Pavlakis *et al.,* 1979). In contrast, in neither lepidopteran whose supplemental rRNA genes have been sequenced (*Bombyx* and *Philosamia*) is a subdivision of any sort found (Feng *et al.,* 1982; Fujiwara *et al.,* 1982). However, in preliminary studies the supplemental rRNA of many insects have been demonstrated to undergo cleavage, but at various points each different than that of the flies, a condition that seemingly occurs in dragonflies, grasshoppers, earwigs, and termites (Beckingham, 1982).

Characteristics of the Trains. The extreme variability that is immediately apparent in the train sequences is probably correlated to their lack of functional significance, for all these genes are cotranscribed with that for the major-rRNA type, which follows closely. Yet in spite of the inconstancy that exists, some traces of homology still can be found, particularly among the several from fungi. But such kinships in structure are to be expected of multiple representatives of a single major taxon; what is unexpected are the striking homologies that exist between the *S. pombe* sequence and that of *Sciara,* an insect. Similarities of that type are evident between the two entire trains, but the most compelling are those in the area from the third site in column N through the sixth position of column O. Here again the lupine sequence is so divergent as to show almost no homologies and to suggest, as in the case of the leaders, that it is from a different part of the genome than the others.

Possible Origins of the Gene. A fair abundance of articles speculating on the origin of this supplemental rRNA gene has appeared over the last decade or more. Before a large number of sequences of the various types of rRNA had been established, the earlier conjectures proposed that the 5.8 S species had been derived from the 5 S by the familiar processes of duplication of the latter and subsequent evolution of the duplicate in the various taxa of eukaryotes (Doolittle and Pace, 1971; Erdmann, 1975). However, this notion was demonstrated to be unsupported by comparisons of the primary structures of the two classes of rRNAs (Cedergren and Sankoff, 1976), for they showed only the random-chance percentage of homologies. Furthermore, the genes proved to be located on opposite sides of the major-rRNA cistron. More recently, a proposal was advanced somewhat parallel to the concept of origins of the 2 S rRNA outlined in a previous paragraph, to the effect that the supplemental rRNA gene is a fragment of that for the major-rRNA species of prokaryotes (Nazar, 1980; Jacq, 1981). Later this idea was expanded and further developed by Clark and Gerbi (1982). In the several studies the extent of homology depended upon the mode of alignment of the contrasting types, ranging from 47–50% by Jacq, to 53% by Nazar, to 55% by Clark and Gerbi. In Table 4.5, the alignment follows that of the first author to the extent that the first 12 residues of the *E. coli* sequence are omitted.

Although it has been shown in preceding chapters that the relationships between the prokaryotic and eukaryotic cell are at a level above that of *E. coli* in the phylogenetic tree, the limited number of primary structures of the major-rRNA genes that have been established compels the use of that organism's cistron as the basis of comparison. Rather than making counts of corresponding sites, which may or may not hold significance, the regions that have been demonstrated to be evolutionarily conserved seem to afford greater relevance. Moreover, it is more realistic that such comparisons be made of the bacterial with the yeast sequences, rather than with that of the highly evolved vertebrates, as has been the custom in other studies.

Insofar as universally conserved triplet or longer sequences are concerned, the prokaryotic gene shows agreement at the first such point, shown heavily stippled in Table 4.5, column III, and further demonstrates partial agreement with the second and third ones, located in column V. In this same region homology is also noted in the near-constant sectors demarcated by light stippling; however, in the adjacent column VI, no such agreement with the lightly stippled runs is found. In columns VIII and IX there exists little convincing identity between the coliform bacterial sequence and the near-constant sectors of the rRNAs, but in column XI, where the last area of perfect constancy occurs, the same GAA is present in the bacterial sequence as elsewhere. As pointed out earlier, the remainder of the molecular structure is too variable to provide for regions of constancy of three or more adjacent residues. Thus the bacterial and eukaryote genes show some identity at these regions of seeming particular importance, but the amount of homology is insufficient to permit any firm conviction. Perhaps when the major rRNA from an archaebacterium has been sequenced, comparisons of this sort may prove more productive.

In the meantime, one further check on the proposed origin of this gene can be made. As Clark and Gerbi (1982) have pointed out, if the supplemental rRNA gene has been derived as postulated from that for the major rRNA of bacteria, then the corresponding gene for the latter species in eukaryotes should have that much of the 5′ end missing. In other words, if part of the bacterial coding region has gone into making the 5.8 S portion, then the large species should be truncated to the same extent, and accordingly should lack about 190–200 nucleotides. Since this involves analyses of the major rRNA sequences, the comparisons, and any conclusions derived therefrom, need to be held for Section 4.6, where those sequences are discussed.

4.5.2. The Appendicular Species of rRNA

The overall picture of a small species (4.5 S) of rRNA from chloroplasts, here called the appendicular as noted previously, is much the same as the preceding type, for its functions and evolutionary origins are still subject to question. In the plastids of most flowering plants that have been investigated since its discovery (Dyer and Bowman, 1976; Whitfeld *et al.*, 1978b), as well as in those of mosses and liverworts (Troitskii *et al.*, 1984), the mature product consists of ~105 nucleotide residues, but in that of the broad bean it includes only ~80 (Bowman and Dyer, 1979). Hence, the sedimentation factor, 4.5 S, is not a suitable term for the class as a whole. In addition to its size, a particularly outstanding property of this species is the presence of an unphosphorylated 5' end.

Sequence Comparisons. Another parallel to the supplemental type is in the relative absence of structural information concerning the entire gene, but in the present case, even the number of available sequences of mature coding regions and their product is low. Only one species additional to the four of Table 4.6 has had its sequence established, that from wheat (Wildeman and Nazar, 1980); it is omitted from that tabulation because of its virtual identity to the maize sequence. From the little that is known of the present type, it appears to be a characteristic of the higher green plants, including ferns (Takaiwa *et al.*, 1982); perhaps mosses or their allies, too, and maybe even the stoneworts, will prove to possess it when the necessary analyses have been made or the sequences have been determined. But it is absent in the organelles of their close algal forebears, such as *Chlamydomonas* (Rochaix and Malnoe, 1978), as well as in those of more primitive eukaryotes like *Euglena* (P. W. Gray and Hallick, 1979). This latter information is particularly relevant to a later discussion.

Because of this confinement to a single major class of organisms and their close relatives, a high degree of homology is to be expected among the sequences, nor is one disappointed when the four of Table 4.6 are examined. Even the two leaders show sequence agreement to an unexpected extent—one run of 28 residues conserved in all four leaders can be seen in columns A–C and another of 21 is found downstream from it in columns D–F. However, immediately preceding the 5' end of the mature gene (columns G and H), where those for many types of macromolecules like tRNAs, 5 S rRNAs, and others are often homologous, much variation is evident. Since the appendicular gene is cotranscribed with that for the major species, no region of promoter activity needs to be searched for, either in the two sequences given or in the entire intergenic spacers of which these are a part. That spacer of tobacco chloroplasts is 101 residues in length, and that of maize, 79 (Edwards and Kössel, 1981; Takaiwa and Sugiura, 1982b).

In the mature coding sequences such long regions of homology are few in number, the two longest continuous areas being 14 residues in length. One of these is located from the third site in column III into column V, and the other from the eighth position of column VIII extending almost through the entire column X. An additional continuous region of constancy is present in the last nine places of column XI, another of just six sites terminates the genes, and one of five sites closely follows the initial nucleotides at the 5' terminus. Several others that are completely conserved can be noted (heavy stippling),

Table 4.6

Sequences of Appendicular rRNA Genes of Chloroplasts

Leaders	A	B	C	D	E'	F	G	H
Maize[a]	AGCACAGCC	GAGACAGCGACG	GGTTCTCCA	CCCACACGG	GGATGGAGCGAC	AGAAGTAGT	ATCGGAAA--	-TAGGA
Tobacco[b]	AGCACAGCC	GAGACAGCGACG	GGTTCTCTG	CCCCTCCGG	GGATGGAGCGAC	AGAAGTTTT	TTTGAGAAT	TCAAGA

Mature Genes	I	II	III	IV	V	VI	VII	VIII	IX	X
E. coli[a] (3' end)	---GGAGGA-	ACGTT	GAGAGGAC	GACCT	TG-GATAGGCCG	GGTGTG--	----TAAG	CCCAGCCATCGGT	TG-A	GCTAAC
Fern[c]	---TTAGGTC	AGGGC	-AAGAGGA-	GCCGT	TTATCACCACGA	TACGTGCT	AAGTGGACG	TGCAGTAATGTA-	TGCA	GCTGAG
Maize	---TTAAGGTA	GCGGC	GA-GAGGA-	GCCGT	TTAAATAGGTGT	CAAGTG--	----GAAG	TCCAGTGATGTA-	TGGA	GCTGAG
Tobacco	---GAAGGTC	AGGGC	G-AGAGGA-	GCCGT	TTATCATTACGA	TACGGTTC	AAGTGGAAG	TGCAGTGATGTA-	TGCA	GCTGAG
Spinach[d]	AGAGAGGTC	AGGGC	G-AGACGA-	GCCGT	TTATCATTACGA	TACGTGTC	AAGTGGAAG	TCCAGTGATGTA-	TGCA	GCTGAG

Mature Genes	XI	XII	XIII	Trains	M	N	O	P
E. coli	CGGTACTAA	TGAAACCGTGA	-GGCTT-AAC		CTT			
Fern	GCATCCTAA	TAGACCGAGA	-GGTTTCAAC					
Maize	GCATCCTAA	CCAACGAA--	CCAATTCGAAC		CTTGTTCCT	ACAGGACCTG		
Tobacco	GCATCCTAA	CAGACCCGTA	-GACTTGAAC		CTTGTTCCT	ACATGACCTGAT	CAATTCGAT	CAGGCACTCGCC
Spinach	GCATCCTAA	CAGACCCACA	-GACTTCAAC		CTTGTTCCT			

[a] Edwards and Kössel (1981); Edwards et al. (1981).
[b] Takaiwa and Sugiura (1982a).
[c] Dryopteris acuminata: Takaiwa et al. (1982a).
[d] Kumagai et al. (1982).

along with some of near universal identity (light stippling), but these are mostly confined to triplet sets.

So in spite of the close kinship that exists among the source organisms, variation is still relatively extensive. Many of the deviant sites are contained in the maize sequence, the fern coding region unexpectedly often being more harmonious with the two from dicotyledonous plants. In fact, the fern–maize pairs of sequences share only 76 homologous sites, whereas the fern and tobacco genes have 88 sites occupied identically. Thus on the basis of percent homology of the genes, the dicotyledonous plants are more closely allied to ferns than are their monocotyledonous counterparts. The phylogenetic implication, then, is that the latter type of seed plant diverged from the primitive ancestral stock prior to the advent of ferns and that the pteridophytes gave rise later to the dicotyledonous plants. This set of conclusions is drawn out merely to show the fallacy of making evolutionary deductions from a single set of data, as too frequently mars the literature today.

Origins of the Appendicular Type of rRNA. As in the case of the supplemental species, direct origins of the appendicular type have been proposed to be from the major species of rRNA of such prokaryotes as *E. coli* (Edwards and Kössel, 1981; Edwards *et al.*, 1981). Since the gene for this type follows that of the major species in the chloroplast genome, the 3' end of the latter class would be expected to be involved, not the 5'-terminal region as in the speculated origins of the supplemental type. When the necessary comparisons are made as between the sequences of *E. coli* and chloroplasts, a 67% homology can be calculated between the *E. coli*–maize pair and 71% between the *E. coli*–tobacco pair (Edwards *et al.*, 1981). Alignment of the relevant part of the prokaryotic variety with the four sequences of plant chloroplasts is provided in Table 4.6, with the very 3' end of the bacterial sequence extending three residues into the train of the seed-plant chloroplast.

When aligned sequences are examined, with particular attention devoted to the heavily stippled, fully conserved sectors in that table, rather extensive homology is to be noted in columns I–IV, a condition repeated to a lesser extent in column VI and from the end of VII into VIII and again in VIII–X. Further sites with identical occupants are found in columns XI–XIII, but in *E. coli* these runs are often interrupted by nonidentical residues where full constancy exists between the green plant chloroplast genes. Thus there appears to be a basis for considering the appendicular types as a possible derivative of the 3' end of the major species of *E. coli, insofar as these comparative data are concerned.* But other pertinent sets of facts show that, although the data are firm, the interpretation is invalid. As reported in the introduction to the present section, the chloroplast genomes of both *Chlamydomonas* and the relatively primitive *Euglena* lack genes for this class of rRNA—that is, the cistrons for the major rRNA type are intact into the green algae. Hence, the resemblances between the *E. coli* major rRNA and the present species that have been discussed derive from the relative conservatism of this rRNA since the time of the eukaryotic origin from an advanced archaebacterium. Thus the homologies are a reflection of prolonged evolutionary preservation of a gene for an important, mutation-sensitive type of ribonucleic acid, not of direct ancestor–descendant relationships. The splitting off of this sector from the conserved gene apparently did not occur until the advent of ferns or their immediate ancestors.

4.6. THE GENES FOR THE MAJOR rRNA

Nothing bespeaks the high level of sophistication of the present state of the art and level of skill in today's biochemistry more clearly than the establishment of the primary structures of the huge gene for the major species of rRNA. This accomplishment has been achieved not only once, but in more than a dozen organisms and organelles. The 4718 residues of the longest of those thus sequenced, that of the rat (Chan *et al.*, 1983), forms the backbone of the series presented in Table 4.7, the several other eukaryotic, prokaryotic, and organellar species then being aligned to it. Reflecting the vital importance to living cells of this macromolecule, the primary structures show extensive homology, with many areas remaining evolutionarily constant through the entire range of organismic and organellar types. Such regions of constancy are blocked off in the table, wherever combinations of three or more residues are invariable in both the prokaryotic and eukaryotic worlds. When appropriate the organellar sequences are included in the blocked-off sectors, a condition that occurs often in the case of the chloroplast, but very rarely with the mitochondrion, as is shown shortly.

4.6.1. The Prokaryotic Major-rRNA Genes

Only two primary structures from prokaryotic sources are included in Table 4.7, since few are currently available. Since one of these is from the blue-green alga *Anacystis nidulans* (Kumano *et al.*, 1983) and the other from the standard bacterium *E. coli* (Brosius *et al.*, 1980), two major subdivisions of this group of organisms are represented. Hopefully the sequence of this gene from one—or several—of the Archaebacteria will be completed so that a fuller picture of the evolutionary relationships of this coding sequence can be presented. Then when those of *Clostridium, Beggiatoa,* and a rickettsia have been determined, understanding of the primitive levels of cellular life will be greatly enriched. The primary structures of the genes from other operons of *E. coli* would also be highly instructive in shedding light on the amount of variation that occurs within this organism and the possible effects of the differences on heterogeneity of the ribosomes.

Considering the phylogenetic distance that exists between the two forms presented in the tabulation, an amazing amount of homology is displayed. In preparing the table, in both cases the sequences had to be truncated at the 5′ end to allow for the formation of the supplemental species in the eukaryotes, thus confirming that the latter type had been derived from the primitive major form. At the very onset of the genes thus abbreviated, shown in row 1, column A, is evidence of the frequent sectors of identity, a condition reflected again in the conserved blocks in column I of that same row. To list all such sectors, however, would be an exercise in tedium, for they are made self-evident by the demarcation of related areas throughout the entire length of the tabulation.

Certain peculiarities and distinctive features common to both organisms nevertheless can be enumerated. It is noteworthy that, by and large, the similarities of structure among prokaryotic sequences also embrace the regions present in eukaryotic genes that are vacant in the present representatives. For example, both the *Anacystis* and *E. coli* types have vacant regions in row 2, columns Q–S; however, that of the blue-green alga closes with row 3, column A, whereas that of the bacterium continues into column C of that same row. Few such traces of lengthier deletions are to be found in the bacterial sequence;

rather, for the greater part its filled sequences are slightly longer than those of the alga. Evidence of this sort is to be found in row 22, column W; row 23, column F; row 24, column J; row 27, column T; row 29, columns A, B, and E; row 34, column R; row 37, columns S–U; row 45, columns D and E; and row 62, columns Q and R. Moreover, inequalities of sectorial length also are provided by occasional "insertions" apparently gained independently after the respective branches had separated. These are marked by superscripts wherever they occur and refer to footnotes at the end of the table. It should be noted that the sequences of both genes for this rRNA in the genome of *Anacystis* have now been established, the second differing from that given at only eight sites (Douglas and Doolittle, 1984).

4.6.2. The Eukaryotic Major-rRNA Gene

Six sequences of the coding regions for the major rRNA from eukaryotes have been established, all of which, except that of the mouse and *Dictyostelium*, are contained in Table 4.7. The murine gene (Hassouna *et al.*, 1984) is so similar to that of the rat that its inclusion would contribute only to the length of an already very long tabulation; the *Dictyostelium* sequence is omitted for similar reasons (Ozaki *et al.*, 1984). Here as in the prokaryotes homologies are both frequent and extensive—they are, in fact, so much the rule that it is needless to enumerate them. Almost every row in the table shows constancy among a majority of the sites included within its limits. As is to be expected, the two vertebrate genes (Chan *et al.*, 1983; Ware *et al.*, 1983) are much more similar to one another than to either that of the yeast or the fungus (Georgiev *et al.*, 1981; Otsuka *et al.*, 1983).

The Vertebrate Sequences. The evolutionary trend in the development of the cistron for this type of rRNA very obviously involves an increase in length, for that of the mammal embraces over 4700 nucleotide residues and that of the amphibian, *Xenopus*, slightly exceeds 4100. In contrast, the coding regions of the *Physarum* species is nearly 3800 residues in length and that of the yeast is just under 3400—in *Dictyostelium*, it is still shorter, containing just 3240 sites (Ozaki *et al.*, 1984). By way of comparison, the *E. coli* gene includes 2904 and the blue-green alga has 2869.

For much of the alignment of the amphibian, *Physarum*, and *E. coli* primary structures, the author is indebted to the report on the *Xenopus* gene (Ware *et al.*, 1983). In that article it was pointed out that the amphibian differs from the *E. coli* cistron principally through the insertion of nine major sectors in the former. In the present chart, ten such sectors were found, located in rows 2 and 3, rows 7–17 (in part), rows 18, 19, 23, 24, 29, 31–33, 37–45, 51, 52, and 60–62, with that in row 27 a possible addition to the list. Although the gene of this primitive quadripedal vertebrate has thus gained in size through relatively long insertions, the cistron of the mammal has increased through even lengthier accessions. The longest such insertions include one of over 200 residues in rows 12–15, another of 102 in rows 27 and 28, and a third of ~225 in rows 39–41. In addition, there are three important lesser regions of gain, two of 36 residues, respectively, in rows 16 and 50, and another of 19 sites in row 22. But the insertions have not been all on the part of the rat sequence, for that of the amphibian has gained several independently, for example, in rows 18 (12 residues) and 24 (five sites). Still other unilateral increases are indicated by superscripts as in the prokaryotic primary structures.

Table 4.7
Mature Coding Sequences of the Major rRNA Species[a]

Row 1	A	B	C	D	E	F	G	H	I	J
Tobacco[b,c]	TGAATCCATGGG	CAGGCAAG	AGA-C-A	ACC	TGGCGA	ACTGAAA	CAT	CTTAGTGAGCCA	GAGGAAAAGAAA	GCA
Anacystis[b,d]	TGAATCCATAGG	GTGGCGCG	AG--CGA	ACC	CGGCGA	ATTGAAA	CAT	CTTAGT-AGCCG	GAGGAAGAGAAA	ACA
Escherichia[b,e]	TGAATCCATAGG	TTA--ATG	AGG-CGA	ACC	GGGGGA	ACTGAAA	CAT	CTAAGT-ACCCC	GAGGAAAAGAAA	TCA
Saccharomyces[f]	TTTGCACCTCAAA	TCA--GGT	AGG-AGT	ACC	CGCTGA	ACTTAAG	CAT	ATCAAT-AAGCG	GAGGAAAAGAAA	CCA
Physarum[g]	CGGGATGGCAG-	ACC--AGG	GCG-TTC	ACC	CGCTCA	ATTTAAG	CAT	ATTAGT-CAGCG	GAGGAAAAGAAA	TCA
Xenopus[h]	TCAGACCTCAGA	TCA--GAC	CCG-GCG	ACC	CGCTGA	ATTTAAG	CAT	ATTACT-AAGCG	GAGGAAAAGAAA	TCA
Rat[i]	CGGCACCTCAGA	TCA--GAC	GTG-GCG	ACC	CGCTGA	ATTTAAG	CAT	ATTAGT-CAGCG	GAGGAAAAGAAA	CTA
Human$_M$[j]	ACCAAGCATAAT	ATA--GCA	AGGACTA	ACC	CCTATA	CCTTCTG	CAT	--AATG-AA-TT	AACTAGAAATAA	CTT
Yeast$_M$[k]	TAAACCTTTATA	TTA--ATA	ATGTTAT	TTT	TTATTA	TATTTA-	TAT	A-AGAATAA-TT	ATTAATAA-TAA	TAA
Paramecium$_M$[l]	ACCAAATTTTG-	ATA--GAA	AATAAGT	ACC	GTGAGG	GAAAGGG	TGA	AAAGAAAT--TT	TGTAGAATGCT-	TAA
Mouse$_M$[m]	AACAAGCAAAG-	ATT--AAA	CCT-TGT	ACC	TTTTGC	ATAATCA	ACT	AACTAGAAAACT	TCTAACTAAAA-	GAA

Row 2	K	L	M	N	O	P	Q	R	S
Tobacco_C	AAAGCG	ATTCCCGT	AGTA	GCGGCGAG	CGAAATCGGA	GCAGCC	TAA---------	----------	----------
Anacystis	AAAGTG	ATTCCCTC	AGTA	GCGGCGAG	CGAACGGGGA	CCAGCC	TAA--------	----------	----------
Escherichia	ACCGAG	ATTCCCCC	AGTA	GCGGCGAG	CGAACGGGGA	GCAGCC	CAG--------	----------	----------
Saccharomyces	ACCGGA	-TTGCCTT	AGTA	ACGGGCGAG	TGAAGCGGCA	AAAGCT	CAAATTTGAAAT	CTGGTACCT	TCGGTGCCCGAG
Physarum	ATCGAG	ATTCCCGT	AGTA	GCTGCGAG	CGAACAGGGA	AAAGCC	CGAGACCGAAT-	CCCTTCCCC	CCTAGCGGGGGT
Xenopus	ACCACG	ATTCCCCC	AGTA	ACGGGCGAG	TGAAGAGGGA	AGAGCC	CAGCGCCGAAT-	CCCG-CGCC	CGCCG-GGCGCG
Rat	ACCACG	ATTCCCTC	AGTA	ACGGGCGAG	TGAACAGGGA	ACAGCC	CAGCGCCGAATC	CCCGCGCCG	CGCCGCGCGCGCG
Human_M	TGCACG	GAGAGCCA	A--A	GCTAAGAC	CCCG-GA---	AAA--C	CAGAC--GAGCT	ACCTAAGAA	CAGCTAAAAGAG
Yeast_M	ACTAAG	TGAACTGA	AA-C	ATCTAAGT	--AACTTAAG	GATAAG	AAATC--AACAG	AGATATTAT	GAG-TATTGGTG
Paramecium_M	AAGA-T	CTGAAATC	--TA	-GTCC-AG	TGAAACAGTT	AAAGCG	TGTTGTTTTAAC	GTACCTTTT	GTA-TAATGGGC
Mouse_M	TTACAG	CTAGAAAC	CCCG	AAACCAAA	CGAGCTACCT	AAAAAC	AATTTTATGAAT	CAACTCGTC	-TA-T-GTGGCA

(continued)

Table 4.7 (Continued)

Row 3	A	B	C	D	E	F	G	H
Tobacco_C	[n]AAAACGG--G	TTGTG---GG	----AGAGGA-A	TAC--AAGCGT	CGTG-CTGC	TAGG---CGA	AGACGCCCG	AATGCTGCACC-
Anacystis	[o]TCCACGGAG	TTTCG---GG	TCGTGGGA-CAGC	AAT--GTGGAC	TCTGAATGT	TAGA---CGA	AGCAGCTG-	AAAACTGCACC--
Escherichia	--------	--------	---------AGC	CTG--AATCAG	TGTGTGT	TAGT---GGA	AGCGTCTGG	AAAGCGCCCCG-
Saccharomyces	---TTGTAA	TTTGGAGAG	GGCAACTTTGGG	GCC---GTTCCT	TGTCTATGT	TCCTTGGAA	CAGGACGTC	ATAGAGGGTGAG
Physarum	CGCGGACCT	GTAGTCTAA	GGGTTGGGACCG	CGA--GTCAAA	GACTTACAG	CACACAGC	TAAGATCCT	G---GAG-CACAG
Zenopus	---GGACGT	GTGGCGTAC	GGGAGACCGGAC	CCCCCCGGCGC	GGCTCG---	GGGGCCCAA	GTCCTTCTG	ATCGAGGCCCAG
Rat	---GGAAAT	GTGGCGTAC	GGAAGACCCACT	CCC--CGGCGC	CCCTCCTGG	GGGGCCCAA	GTCCTTCTG	ATCGAGGCCCAG
Human_M	--------	--------	CACACCCGTCTA	TGT-AGCAAAA	TAGTGG---	------GA	AGATTTATA	-----GGTAGA

Row 4	I	J	K	L	M	N	O	P
Tobacco_C	--------	CTAGATGGC	GAAAGTCCAG	TAGCCGAAAGC	AT-CACTAGCTTA	-TGCTCTGA[P]GAGT	AGCATG	GGGCACGTG
Anacystis	--------	AGAGAAGGT	GAAAGTCCTG	TAGTCGAAAAT	TGAAAC-AGCCTA	--GCTGAAT[P̄]GAGT	AGCACG	GAGCACGTG
Escherichia	--------	ATACAGGGT	GACAGCCCC3	TACACAAAA-	T-GCACATGCTGT	GAGCTCGAT GAGT	AGGGCG	GGACACGTG
Saccharomyces	AATCCCG-T	GTGGCGAGG	--AGT--GCG	GTT-CTTTGT-	-AAAGTGCCTTCG	------A-AG -AGT	CGAGTT	GTTTGGGAA
Physarum	GGGGTT-AC	ACGCGACCT	--GGTAGGGC	GGGCACCAGA-	-GGGACTGACCCG	CGAGTA-GG CCAG	CTT---	------GAA
Xenopus	CCCCCGGAC	GGTGTTTAGG	CCGGTGGGCC	GCCCCCGGG--	-CGGCGGGACCCG	GTCTCCTCG GAGT	CGGGTT	GTTTGGGAA
Rat	CCCGTGGAC	GGTGTGAGG	CCGGTA-GCG	GCC-CCGGCG-	-CGCCGGGCCGGG	TCTTCC-CG GAGT	CGGGTT	GCTTGGGAA
Human_M	---GGCGAC	AAACCTACC	GAGCCT---G	GTGATAGCTG-	-GTTGTCCAAGAT	AGAATC-TT A---	---GTT	CAACTTTAA

Row 5

	Q	R	S	T	U	V	W
Tobacco_C	---GAATCCCGTCTGA	-ATCGACAAGACC	A-CCTTGCAAG	GCTAAATA	CTC--CTGGGT	GACCCGATAG	CGAAGTAGTTACCGTG
Anacystis	---AAATTCCGTGTG-	AATCCCGCGAGGAC	-ACCTCGTAAG	GCTAAATA	CTC--CTGTGT	GACCCGATAG	TGAACCAGTACCGCG
Escherichia	---GTATCCTGTCTGA	ATATGGGGGACC	ATCCTC-CAAG	GCTAAATA	CTC--CTGACT	GACCCGATAG	TGAACCAGTACCGTG
Saccharomyces	---TGCAGCTCT-AAG	-TGGGTGGTAAAT	TCCATC-TAAA	GCTAAATA	TTG--GCGAGA	GACCCGATAG	CGAACAAGTACAGTG
Physarum	-AGTGCTGGCTG-AAG	ATAGGTGCTGCTG	GCCATC-TAAG	GCTAAATA	TGAAACCAGGC	AACCCGATAG	CAAACAAGTACCGTA
Xenopus	---TGCAGCCCA-AAG	-CCGGTGGTAAAC	TCCATC-TAAG	GCTAAATA	CCG--GCACCA	GACCCGATAG	CGGACAAGTACCGTA
Rat	---TGCAGCCCA-AAG	-CGGGTGGTAAAC	TCCATC-TAAG	GCTAAATA	CCG--GCACGA	GACCCGATAG	CCAACAAGTACCGTA
Human_M	ATTTGC----CCA-CAG	--AACCCTCTAAA	TCCCCT-TCTA	AATTTAAC	TGT--TAGTCC	AAAGAGGAA	CAGCTCTTTGGACAC

Row 6

	A	B	C	D	E	F	G	H
Tobacco_C	AGGGAAGGG	TGAAAAGAA	CCCCATC--	GGGGACTGAA-	ATAGAACATGAA	A-CCGTAAGC	TCCCAAGCA	GTGGGAGGC
Anacystis	AGGGAAAGG	TGAAAAGAA	CCCCGG-AA	GGGGAGTGAA-	ATAGAACATGAA	A-CCGTGAGC	TTACAAGCA	GTCGGAGCC
Escherichia	AGGGAAAGG	CGAAAAGAA	CCCCGGCGA	GGGGAGTGAA-	AAAGAACCTGAA	A-CCGTGTAC	GTACAAGCA	GTGGGAGCA
Saccharomyces	AATGGAAAGA	TGAAAAGAA	CTTTGAAAA	GA-GAGTGAA-	AAAGTACGCTGAA	ATTCTTGAAA	GGGAAGGGC	ATTTGA-TC
Physarum	AGGGAAAGC	TGAAAAGGA	CCTCGTTGA	G--GAGTTAAA	AGAG--CATGAA	ATCCCCCAA	TGAGAA-CG	GTAAATCTA
Xenopus	AG-CAAAGT	TGAAAAGAA	CTTTGAAGA	GA-GAGTTCA-	AGAGGGCGTGAA	A-CCGTTAAG	AGGTAAACG	GGTGGGGCC
Rat	AGGCAAAGT	TGAAAAGAA	CTTTGAAGA	GA-GAGTTCA-	AGAGGGCGTGAA	A-CCGTTAAG	AGGTAAACG	GGTGGGGTC
Human_M	------T	AGGAAAAAA	CCTTGTAGA	GA-GAGTAAA-	AAATTTAAC---	A-CCCAT---	-AGT--AGG	CCTAAAAGC

(continued)

Table 4.7 (Continued)

Row 7	I	J	K	L	M	N	O	P
Tobacco$_{C}$	CA-----GG	GCTCTG	AC^{q}G	CCTGTTGA-	AGAATGAGCCG	GCGA	to row 17	
Anacystis	CGA-TTCAA	CGGGTG	AC^{r}G	CCTGTTGA-	AGAATGAGCCG	GCGA	to row 17	
Escherichia	CGC-TTAGG	CGTGTG	AC^{s}A	CCTTTTGT-	ATAATG-GTCA	GCGA	to row 17	
Saccharomyces	-AGACATGG	TGTTTT	GT^{t}T	CTCGCATTT	CACTGG-GCCA	GCATCAGTTTTGG	TGGCAGGAT-AA	ATCCATAGG-AATGT
Physarum	-TACGGCTC	GTCGAA	AG^{u}A	CGGCGCTGA	GTACGG-CACA	GCTGCGGCTGAGG	CGACGCTGGGAA	AGGTGACCT-AACGG
Xenopus	-GTGCGGTC	CGCCCG	GA G	GATTCAACC	CGGCGG-GTCA	GCGCCGCCGGGAC	CGGGCCACTCGG	CGGACCCCC-CCGCC
Rat	CGCGCAGTC	CGCCCG	GA G	GATTCAACC	CGGCGG-CGC-	GCGCCGGCCGG-C	CGGTGGTCCCGG	CGGATCTTTCCCGCT
Human$_{M}$	AGCCACCAA	TTA---	AG A	AAGCGTTCA	AGC-----TCA	ACA	to row 17	

Row 8	Q	R	S	T	U	V
Saccharomyces	AGCTTGCCTCGGTAAG	TATTATAGCCTG	TGGGAATACTGC	CAGCT-GGGACT	GAGGACTGCGAC	GTAAGTCAAGGAT
Physarum	CCAATCTCGCTTACGG	-GCGAGGACGGT	CAAGGACACCGG	AACCGCGGGACC	TCTCCGTCTAGG	CTTACACTCACG-
Xenopus	GGCCCCCTCCCCTCGC	CGGGAGGGCGGT	CGCGCGGGGGGG	ACGCGGCCCGGG	CGSCCCGGCGCC	SPSPGCGSTTTC-
Rat	CCCCGT-TCCTCCCGA	CCCCTCCACCCG	CGCGTCTCTCTC	CCCCCTCCCCGC	GTCCCGCCGTCG	CCGTCCCCGCTC-

Row 9

	A	B	C	D	E	F	G	H
Saccharomyces	GCTGGCATAATG	GTTATATG-	to row 17					
Physarum	CGTGAATCGTT	CGCCCCATA	ACCGTGAGACGT	ACAGCGTGT	CAGAGC	TTAGTTT	to row 17	
Xenopus	CTCCGCGGCGT	GCGCCGCC	GGCTCCGGGCCG	CGTCGGAAG	GCCCAG	GGGTTCAGG	GGAAGGTGG	CCGGCCGCC
Rat	CTCCCTCCGGGG	GGGTGTCGG	CGGGGCCTCCGG	CGGCCGGCG	CGGGGT	GTGGTGGGG	GCGGGCCGGG	CGGGGCCGG

Row 10

	I	J	K	L	M	N
Xenopus	CCCGCGCGGC	TACAGCCCCCC	CCAPCGCACCAGCACT	CGCCGTCCGCCG	GGGCCGAGGGAG	ACGCCGGCCTCC
Rat	GGGTGGCGTCGG	CGGGGGACCCCC	CCCCGTCCGGCGACCG	GCCCCGCCGCGGG	CCGCACTTCCACC	GTGGCGGTGCGC

Row 11

	O	P	Q	R	S	T
Xenopus	GCGGTCCTCCTC	CCCGGAGCGGCT	CCCGCCGCTCCC	CCCCGGGGGGGGC	GCCCGGGGCGGG	GAAGGGGAAGG
Rat	CGCGACCGGCTC	CGGGACGGCTGG	GAAGGCCCGCG	GGGAAGGTGGCTCGG	GGGGGGCCGGCT	CACCCGTGGGCG

(continued)

Table 4.7 (Continued)

Row 12

	U	V	W			
Xenopus	GGCCCCCCGCTC	CCGGCGCGGCTG	TCA to row 15			
Rat	CCGGACCACCCC	GCCCCGAGTGTT	ACAGCCCCCCGG	CAGCAGGGCTCGCCG	AATCCCGGGGCC	GAGGGAGCCGGA

Row 13

Rat	TACCCGTCGCCG	CGCTCTCCCCCC	GGCCTCTCCCCTCCC	GCCCCTCCCCGT	GGGGTGGCGGAA	AGGGGGGCGGTC

Row 14

Rat	GCGGGGGGCGGG	CCGCCCCTCCCA	CGGGCCGGACCGGCTCT	CCCACCCCCCGT	CGCCTCCGTCGT	CCCTCTCGGGGG

Row 15

	A	B	C	D	E	
Xenopus	-----------	---ACCGGGGCG	GACTGCCCCCAGTGC	GCCCCG-----TC	CGC-CGAGGGCGG---	
Rat	TCCGGGGGCCCG	GGGGGCGGGGCG	GACTGTCCCCAGTGC	GCCCCGGGGGCTC	GTCGGCGCCGTCG	GGCCCGGGGGGGCCG

Row 16

		F	G	H	I	J
Xenopus	---------	---------	-----GAGGG	---CCGCCGGGAG	CCGCCRAGGG	TCCGCGGCG	ATGTCGGTG
Rat	TCGTCACGCGCT	CTCCCTCCCTT	CTCGGGGTGGGGGGG	AGCGAAGCC-GAG	-CGCACGGGG	TCGCCGGCG	ATGTCGGCT

Row 17

	A	B	C	D	E	F	G
Tobacco$_C$			C	TCATAGGCAGTG	GCTTGGTTAA	GGG--------	AACCCACCG
Anacystis	(from row 7)		C	TTATAGGCACTG	GCA-GGTTAA	GGCC-------	GAAATGCCG
Escherichia			C	TTATATTCTGTA	GCAAGGTTAA	CC---------	GAATAGG-G
Saccharomyces	CCG--CC	CGTCTTGAAACA	CGGACCAAGGAG	TCTAACGTCTAT	GCGAGTGTTT	GGGT-----GT	AAAACCC-A
Physarum	ACCACC	CGTCTTGAAACA	CGGACCAAGGAG	TCTAACGTACAT	GCAAGTCGAA	CAGCTTTCACGC	TGAGACC-G
Xenopus	TCCCACCCGACC	CGTCTTGAAACA	CGGACCAAGGAG	TCTAACGCCGC	GCGAGTCGGA	GGGACTCGCGGC	GAAACCC-T
Rat$_M$	ACCCACCCGACC	CGTCTTGAAACA	CGGACCAAGGAG	TCTAACGGCTGC	GCGAGTCAG-	GGGCTCGTCCGA	-AGCCGCC
Human$_M$	(from row 7)		C	CCACTACCTAAA	AAATCCCAAA	CA---------	---------

(continued)

Table 4.7 (Continued)

Row 18	H	I	J	K	L	M	N
Tobacco_C	GAGCCGTAGC	GAAAGCGA-GTC	TTCATAGGGC--	------------	------------	------------	------------
Anacystis	AAGCCAAAGC	GAAAGCGA-GTC	TGAATAGGGC--	------------	------------	------------	------------
Escherichia	GAGCCGAAGG	GAAACCGA-GTC	TTAACTGGGC--	------------	------------	------------	------------
Saccharomyces	TACGCGTAAT	GAAAGTGA-ACG	TAGGTT----GG	-GGC--CTC	GCAAGAGGTGCA	CAATCGACCGAT	CCTGATCGTCTTC
Physarum	CGAAGCTAAT	GCAAAAAG-CAC	CCAGGG----CGA	-GCC--TAT	GGC--GGACCCCG	CAATCCTCTCCT	AACCGGGCTCTC
Xenopus	GTGGCGCAAT	GAAGGTGA-GGG	CCCGGGCGCCCC	-GGC--TGA	GGTGGGATCCCG	CCGCCCCTCCCT	CCCCCCCCCCG
Rat	GTGGCGCAAT	GAAGGTGAAGGG	CCCCGTTCCCGG	GGGCCCCGA	GGTGGGATCCCG	AGG--------	---CCTCTCCAG
Human_M	------TAT	--AACTGAACTC	CTCACACCCA--	------------	------------	------------	------------

Row 19	O	P	Q	R	S	T	U
Tobacco_C	------------	------------	------------	------------	------------	AATT-GTCA	CT-GCTTATCGAC
Anacystis	------------	------------	------------	------------	------------	GATA-GTCA	GT-GTTTATAGAC
Escherichia	------------	------------	------------	------------	------------	GTTAAGTTG	CAGGG-TATAGAC
Saccharomyces	------------	------------	------------	----------G	GATGGATTTGAG	TAAGAGC-A	TAGCTCTTGGGAC
Physarum				----------GGG		TTTGAGCAT	GTCATGTTAGGTC
Xenopus	_v CGCCGCGGGC	GCACCACCG	GCCCGTCTC	GCCCGCCCCGTC	GGGGRGGTGGNG	CGTGAGCGC	GCGCGATTAGGAC
Rat	T CCGCCCGAGGGC	GCACCACCG	GCCCGTCTC	GCCCGCCGCGCC	GGGGAGGTGGAG	CACGAGCGT	ACGCG-TTAGGAC
Human_M	------------	------------	------------	------------	------------	ATTGGA to row 47	

Row 20	A	B	C	D	E	F	G
Tobacco_C	CCGAAACCTG	GGTGATCTATCC	ATGACCAGGATG	AAGCTTGGGTGAA	ACTAACTGG	AGGTCCGAA-CCG	ACTGATGT-
Anacystis	CCGAACCCG	GGTGATCTAACC	ATGGCCAGGATG	AAGCTTGGGTAAC	ACCAACTGG	AGGTCCGAA-CCG	ACCGATGT-
Escherichia	CCGAAACCC	GGTGATCTAGCC	ATCGGCAGGTTG	AAGGTTGGGTAAC	ACTAACTGG	AGGACCGAA-CCG	ACTAATGT-
Saccharomyces	CCGAAAGAT	GGTGAACTATGC	CTGAATAGGGTG	AAGCCAGAGAAA	CTCTGGTGG	AGGCTCGTA-GCG	GTTCTGACG
Physarum	CCGAAAAAC	GACGAGCTATGC	TTGAGTAGGCCG	AAGCCAGGAGAAA	TCTTGGTGG	AAGGTCGTACGCA	GTACTGACG
Xenopus	CCGAAAGAT	GGTGAACTATGC	CTCGGCAGGGCG	AAGCCAGGAGAAA	CTCTGGTCG	AGGTCCGTA-GCG	GTCCTGACG
Rat	CCGAAAGAT	GGTGAACTATGC	TTGGGCAGGGCG	AAGC-AGAGGAAA	CTCTGGTCG	AGGTCCGTA-GCG	GTCCTGACG

Row 21	H	I	J	K	L	M	N	O
Tobacco_C	TGAAGAATC	AGCCGGATGA	GTTGTGGTT	AGGGGTGAAATG	CCACTC	GAACCCAGA	GCTAGCTGGTTC	TCCCCGAAAT
Anacystis	TGAAAAATC	GGCGGATGA	GCTGTCTGGTT	AGGGGTGAAATG	CCAATC	GAACCCCGA	GCTAGCTGGTTC	TCCCCGAAAT
Escherichia	TGAAAAACT	AGCCGGATGA	CTTGTCGCCT	CGGGGTGAAAGG	CCAATC	AAACCGGGA	GATAGCTGGTTC	TCCCCGAAA-
Saccharomyces	TGCAAATCG	ATCGTCGAA	TTTGGGTAT	AGGGGCGAAAGA	CTAATC	GAACCATCT	AGTAGCTGGTTC	CTGCCGAAG-
Physarum	TGCAAATCG	TTTGTCAAA	CTTGCAGTA	AGGGGCGAAACA	CCAATC	GAGTCGTTT	AGTAGCTGGTTT	CCACCGAAG-
Xenopus	TGCAAATCG	GTCGTCCGA	CCTCGGTAT	AGGGGCGAAAGA	CTAATC	GAACCATCT	AGTAGCTGGTTC	CCTCCGAAG-
Rat	TGCAAATCG	GTCGTCCGA	CTTGGGTAT	AGGGGCCAAAGA	CTAATC	GAACCATCT	AGTAGCTGGTTC	CCTCCGAAG-

(continued)

Table 4.7 (Continued)

Row 22	P	Q	R	S	T	U	V
Tobacco_C	GC-GTTGAG	GCGAGCAGTTGA	CTGGA----	-----C-------	----ATCT----	AGG--GGTAA	AGCAC-TGT
Anacystis	AC-GTTGAG	GCGTAGCGGTA--	-TGGAT----	----T-------	-----ATAGCGG	TGG-GGTAG	AGCAC-TGA
Escherichia	GC1ATTT-A	GG--TAGCGCCTC-	GTGAAT----	----T-------	-----CATCTCC-	GGG--GGTAG	AGCAC-TGT
Saccharomyces	TTTCCCTCA	GGATAGCAGAAG-	CTCGTA----	----T-------	-----CAGTTTT-	ATGAGGTAA	AGCGAATGA
Physarum	TTTCCCTCA	GGATAGCAAAGG--	AAAAAG----	--TGT-------	---CG-AGTATG-	GGGCGGTAA	AGACAATGA
Xenopus	TTTCCCTCA	GGATAGCTGGCG-	CTSGTC----	--CGT-------	---CCGAGTTTT-	ATCCGGTAA	AGCGAATGA
Rat	TTTCCCTCA	GGATAGCTGGCG-	CTC-TCGCAA	CGCGGTTCGCTCGACA	ACCCGCAGTTTT-	ATCCGGTAA	AGCGAATGA

Row 23	A	B	C	D	E	F	G	H
Tobacco_C	TTCGGT---	GCGGG--CCG	CGAGAGCGGTAC	CAAATCGA-	GGCAAACTC	----TG---	AATA-----	---------
Anacystis	TTCGGT---	GCGGG-C?G	CGAGAGCGGTAC	CAAATCGA-	GTCAAACTC	----CG---	AATA-----	---------
Escherichia	TTCGGCA-A	GGGGGTCAT	CCCGACT--TAC	CAACCCGAT	G-CAAAC--	---TGCG--	AATA-----	---------
Saccharomyces	TTAGAGGTT	CCGGGGTCG	AAAT-GA--CCT	TGACCTATT	CTCAAACTT	TAAATATGT	AAGAGTCC	TTGTTACTTAATTGA
Physarum	TTAGGAG-C	ACCGGGCCG	TTTGACC--GTT	CGGCTCATT	CTCAAACTT	TTAATGCTC	TA--AAGCCA	TTGACCACGCGTCGT
Xenopus	TTAGAGGTC	TTGGGGCCG	AAATCGA--TCT	CAACCTATT	CTCAAACTT	TAAATGGGT	AAGAGCCC	GGCTCGTGGCTTGG
Rat	TTAGAGGTC	TTGGGGCCG	AAA-CGA--TCT	CAACCTATT	CTCAAACTT	TAAATGGGT	AAGAGCCC	GGCTCGTGGCGTGG

Row 24

	I	J	K	L	M	N	O	P
Tobacco_C	------ CTAGAT	---ATG^x CAG^y-TC	GGCTAGTGA	GA-CGATGGGGG	ATAAGC	TTCATCGTC	GAGAGGGAA	ACAGCCCGG
Anacystis	------ CGCCGT	---GTA CAC -CA	TGCCAGTCA	GA-CTGTCGGGGG	ATAAGC	TCCATCGTC	AAGAGGGAA	ACAGCCCAG
Escherichia	------ C-CGGA	GAATGT TAT -CA	CGGGAGACA	CA-CGGCGGGTG	CTAACG	TCCGTCGTG	AAGAGGGAA	ACAACCCAG
Saccharomyces	ACGTGG ACATTT	GAATGA AGA GCT	TT---TAGT	GGGCCATTTTTG	GTAAGC	AGAACTGCC	GATGCGGGA	TGAACCGAA
Physarum	AGCAGC^w TCTGGG	CAGTAT CGG GTC	TCCTTTAGT	GGGCCACGTT-C	GTAAGC	AAGGATGCC	AAAAAGGGT	TCAAACTTG
Xenopus	AGCCGG GGCCTG	GAATGC GNV GCA	CGCCATAGT	GGGCCACTTTTG	GTAAGC	AGAACTGCC	GCTGCGGGA	TGAACCGAA
Rat	AGCCGG -GCC--	---TGG ATG CGA	GTGCCTAGT	GGGCCACTTTTG	GTAAGC	AGAACTGGC	GCTGCGGGA	TCAACCCAA

Row 25

	A	B	C	D	E	F	G	H
Tobacco_C	ATCACCAGCT	AAGGC-CCC	TAAATG-ATCGC	TC--AGTGA---	TAAAGGAGG	TAGGGGTGC	AGAGACAGCC	AGGAGG
Anacystis	ACCACCAGCT	AAGGT-CCT	CAAATC-AGAAC	TT--AGTGA---	TAAAGGAGG	TGGGAGTGC	ATAGACAACC	AGGAGG
Escherichia	ACCGCCAGCT	AAGGT-CCC	AAAGTC-ATGGT	TA--AGTGG---	GAAACGATG	TGGGAAGGC	CCAGACAGCC	AGGATG
Saccharomyces	CGT-AGAGTT	AAGGTCGCCG	GAATAC-ACGCT	CATCAGACACCA	GAAAAGGTG	TTAGTTCAT	CTAGAC-AGC	CCGACG
Physarum	TGC-CCGGTT	A-CGTTCTC	GAAGAC-ATTGG	CA-ATGACCACT	GAAAAGGTA	TTGCCTCAT	GATGAC-AGC	TTGACA
Xenopus	CGC-CGGTT	AAGCGCCC	GATGCCGACGCT	CATCAGACCCCA	GAAAAGGTG	TTGGTTGAT	ATAGAC-AGC	AGGACG
Rat	CGC-CGGTT	AAGCGCCC	GATGCCGACGCT	CATCAGACCCCA	GAAAAGGTG	TTGGTTGAT	ATAGAC-AGC	AGGACG

(continued)

Table 4.7 (Continued)

Row 26

	I	J	K	L	M	N	O	P
Tobacco_C	TTTGCC--TA	GAAGCAGCC	ACCCTTGAAAGAG	TGCGTAATA	GCTCAC	TGATCGAG-	CGCTCTTGC	GCCGAAGATG-AA
Anacystis	TTTGCC--TA	GAAGCAGCC	ATCCTTAAAAGAG	TGCGTAATA	GCTCAC	TGGTCAAG-	CGCTCCTGC	GCCGAAAATG-AA
Escherichia	TTGGCT-TA	GAAGCAGCC	ATCATTTAAAGAA	AGCGTAATA	GCTCAC	TGGTCGAGT	CGGCCTGC-	GCGGAAGAGATGTAA
Saccharomyces	GTGGCCATG	GAAGTCGGA	ATCCGCTAAGGAG	TGTGTAACA	ACTCAC	CCGCCGAAT	GAACTAGCC	-CTGAAAATGGAT
Physarum	GTGGCCATG	GAAGCCGCC	ATCTGCTAAGGAG	TGCGTAACA	GCTCAC	CAGCCGAAT	GGGTAGTC	-CTGAAAATGGAC
Xenopus	GTGGCCATG	GAAGTCGGA	ATCCGCTAAGGAG	TGTGTAACA	ACTCAC	CTGCCGAAT	CAACTAGCC	-CTGAAAATGGAT
Rat	GTGGCCATG	GAAGTCGGA	ATCCGCTA-GGAG	TGTGTAACA	ACTCAC	CTGCCGAAT	CAACTAGCC	-CTGAAAATGGAT

Row 27

	Q	R	S	T			
Tobacco_C	CGGGGCTAAGCG	ATCTCCCGAAGC	TGTGGGATG---	------ --TA--			
Anacystis	CGGGGCTAAGTT	CTGTACCGAAGC	TGTGGAATT---	------ --G--			
Escherichia	CGGGGCTAAACC	ATGCACCGAAGC	TCCGGCAGC---	------ --GACG			
Saccharomyces	GGCGCTCAAGGG	TGTTACCTATAC	TCTACCGTC---	------A GGGTTG			
Physarum	GAGGTTCAACCA	GTGCAACCATAC	C-GGCCAAGGCT	CTTATA[2]TTCGGC			
Xenopus	GGCGCTGGAGCG	TCGGGCCCATAC	CCGGCCGTCGCC	GGCGGCT GGGTCA			
Rat	GGCGCTGGAGCG	TCGGGCCCATAC	CCGGCCGTCGCC	GGCAGT CGGAAC	GGGACGGGA	GCGCCGCCG	GGCGCGCGA

to row 29

Row 28

	U	V	W	X	Y	Z
Rat	CCCCGGGGCCG	GGGGGCTCGGC	TTCGCCGGCCG	CCGCCCGTCCAC	CCCCGGGGCTCC	CCCCGCCGGTCGGG

Row 29

	A	B	C	D	E	F	G
Tobacco_C	AAAATAC-----	----------	-ATCGGTAGGGGA	GCGGTTCC--	----------	GCCTTA-GA-	GAGAAGCCTCC-
Anacystis	CTG-TG---	----------	AATTGGTAGGGGA	GCGTTCCGT	---CGT----AG	GG-TGAAGCG	GTAGCGGAAGC-
Escherichia	CTTATGCG----	--------TT	GTTGGGTAGGGGA	GCGTTCTGT	AAGCCTGC-GAA	GG-TGT-GCT	GTGAGGCATGCT
Saccharomyces	ATAT--------	GATGCCCTG	AC-GAGTAGGCAG	GCCTGGAGG	TCAGTGACGAAG	CC-TAG-ACC	GTAAGGTCGGGT
Physarum	TCGGCAA3CTATT	CCATGAGTA	AACGAGTAGGGCG	GATACCCCA	GGTCGGGTTGAA	GC-GTG-CTG	TTGACAGCACGT
Xenopus	GTCCCC3GGGGC	TAGGCCCGG	ACTGAGTAGGAGG	GCCCCGGCG	GTGCCCCGGAA	GC-GCG-CGC	GAGGGCCCGGGT
Rat	CCCCGC3GAGCC	TACGCCGCG	AC-GAGTAGGAGG	GCCGCTGCG	GTGAGCCTTGAA	GCCTAG-GGC	GCGGGCCCGGGT

Row 30

	H	I	J	K	L	M	N
Tobacco_C	aa GTCGACCAAGC	GGAAGCGAG	AATGTCGGC	TTGAGTAACGCA	AACATTGGTGAG	A-ATCCAATGCCC	CGAAAACCT-AA
Anacystis	aa GTCGACCAAAC	GGAAGTGAG	AATGTCCGC	TTGAGTAGCGAA	AACATGGGTGAG	A-ATCCATGCCC	CGAAATCCC-AA
Escherichia	GGA-3GT-ATC	AGAAGTGCG	AATGCTGAC	ATAAGTAACGAT	AAAGCGGGTGAA	A-AGCCCGCTCGC	CGGAAGACC-AA
Saccharomyces	CGA-ACG-GCC	TCTAGTGCA	GATCTTGGT	GGTAGTAGCAAA	TATTCAAATGAG	A-ACTTTGAAGAC	TGAAGTGGGGAA
Physarum	GGA-3CG-GCC	GGGGCAGCA	GATCTTGGT	AGTAGTAGCAAG	TATTCATATGCA	ACACTCTGAAGGC	CGAAGTGGAGGA
Xenopus	GGA-3CC-GCC	GCGGGTGCA	GATCTTGGT	GGTAGTAGCAAA	TATTCAAACGAG	A-ACTTTGAAGGC	CGAAGTGGAGAA
Rat	GGA-3CC-GCC	GCAGGTGCA	GATCTTGGT	GGTAGTAGCAAA	TATTCAAACGAG	A-ACTTTGAAGAA	CCAAGTGGAGAA

(continued)

Table 4.7 (Continued)

Row 31	O	P	Q	R	S	T	U	V
Tobacco$_C$	GGGTTCCTC	CGCAAGGTT	CGTCCACGG	AGGGTGAGTC-AG	GGCCTA	AGATCAGGCC	GAAAGGC-TAG-	----------
Anacystis	GGGTTCCTC	CGGAAGGCT	CGTCCCGCGG	AGGGTTAGTC-AG	GTCCTA	AGGCCGAGGCA	GAAGTGCGTAG-	----------
Escherichia	GGGTTCCTG	TCCAACGTT	AATCGGGGC	AGGGTGAGTC-GA	CCCCTA	AGCCGAGGCC	GAAAGGCGTAG-	----------
Saccharomyces	AGGTTCCAC	GTCAACAGC	AGTTGGACG	TGGGTTAGTC-GA	TCCTAA	GAGATGGGG-	AAGCTCCGTTTC	AAA----GG
Physarum	GGGTTCCTC	AACACACAGC	AGTTGATTG	GGGGTTAGTCGGA	ATCTAA	GCGTCAAGG-	GAAACCTAGGTT	AACCTCGGG
Xenopus	GGGTTCCAT	GTGAACAGC	AGTTGAACA	TGGGTCAGTC-GG	TCCTAA	GAGATGGGC-	GAGCGCCGTTCG	GAGGGACG
Rat	GGGTTCCAT	GTGAACAGC	AGTTGAACA	TGGGTCAGTC-GG	TCCTGA	GAGATCGGC-	GACTGCCGTTCC	GAGGGACG

Row 32	A	B	C	D	E	F	G	H
Tobacco$_C$	---------	---------	---------	-TCGA-TCGACA	ACAGGT	GAATATTCC	TGTACT^{aa}CCT	TGTTGGTCC----
Anacystis	---------	---------	---------	-TCGA-TCGACA	ACAGGT	TAATATTCC	TGTACC -GA	TTTTGGATT----
Escherichia	---------	---------	---------	-TCGA-TCGGAA	ACAGGT	TAATATTCC	TGTACT TGG	TGTTACTGC----
Saccharomyces	CCTGATTTT	ATGCAGGCC	ACC------	ATCGAAAGGGAA	TCCGGT	-AAGATTCC	GGAACT TGG	ATATGGATT----
Physarum	CTCTCGTCC bb	CCGTGGCTT	GGGCTCGGA	CCCCACAGAGAA	GCGGGT	TAATATTCC	TGCACT CAA	CTCGATCTG----
Xenopus	GGCGATGGC	CTCCGTCGC	CCTCGGCCG	ATCGAAAGGGAG	TCGGGT	TCAGATCCC	CGAACC CGG	AGTGGCCGA----
Rat	GGCGATGGC	CTCCGTTGC	CCTCAGCCG	ATCGAAAGGGAG	TCGGGT	TCAGATCCC	CGAATC CGG	AGTGGCCGAGAT

Row 33	A	B	C	D	E	F	G
Tobacco$_C$	-------------	-------------	-------------	--------CGAG	GGACGGAGGAG	GCTAGGT	TAGCCGAAAG
Anacystis	-------------	-------------	-------------	-----GTGCGAG	GGACGGAGAAG	GCTAGGC	CAGCAGGATG
Escherichia	-------------	-------------	-------------	------GAAGGGG	GGACGGAGAAG	GCTATGT	TGGCCGGGCG
Saccharomyces	-------------	--CTTCAC-GGT	AACGTAACTGAA	TGTGGAGACGTC	GG-CGCGAGCC	CTGGGAG	GAGT--TATC
Physarum	-GA-------CAG	TGGGCGAC-CTT	AAAGCGATATCC	GAAACCAGTGAC	GTAACTGGGAA	CCGTGAG	AGGGA-TTTT
Xenopus	^{dd}GGG--------GCC	TCCAGTGC-GGC	GACGCGACCGAT	CCCGGAGAAGCC	GSGSGGGAGSC	CCGGGAG	AAGAGTTCTC
Rat	GGCGGCGCCGAGGCG	TCCAGTGCCGGT	AACGCGACCGAT	CCCGGAGAAGCC	G-GCGGGAG-C	CCGGG-G	A-GAGTTCTC

Row 34	P	Q	R	S	T	U	V
Tobacco$_C$	ATGGTTATC	--GGTTCAA^{ee}	ACG--TAAGG^{ff}CTGC	TTTGTCAGG--	GTAA^{gg}GAG--AAA	ATGGCT^{ii}GCCAATG	TTCGAATACCAG
Anacystis	TTGGTTA-C	CTG-TCCAA	--------G	TGTCCGAGGC	GTT-^{hh}CGGCGAAA	ACCGTC^{ii}GCTGAGG	CGTGAGTGCGA-
Escherichia	ACGGTTGTC	CCGGTTTAA	GCGTGTA-GG	TTTTCCAGG-	CAAA	ATCAAG GCTCAGG	CGTGATGCACGAG
Saccharomyces	TTTTCTTCT	T-AACAGCT	-TATCAC-CC	ATTGGTTTA-	TCCG	GTCTTA TG--GCT	GGAAGAGGCCAG
Physarum	CCCTCCTTG	TTACGGGAG	TTAACTG-CT	TAAGCTTGC-	ATTG	CCCAG^{jj}CTACTCT	CGAGAGCCATGTC
Xenopus	TTTTCTTTG	TGAAGGGCA	GGGCCCC-CC	ATGGGTTCG-	CCCC	GGGCC GCGCCCGTT	GGAAAGCGTCGC
Rat	TTTTCTTTG	TGAAGGGCA	GGG-CGC-CT	ATGGGTTCG-	CCCC	GGGCC GTGCC-TT	GGAAAGCCGTCGC

(continued)

Table 4.7 (Continued)

Row 35	A	B	C	D	E	F	G
Tobacco$_C$	GCGGCTACCGCC-	GGAA-GTAACCCAT	GCCATACTCCCA	GGAAAAGC-	-TCGAACGACT-	--TTGA	GCA--AGAG[ll]
Anacystis	CCGGCTACCGCGG	GGAA-GTGGTTGAT	GTCAAGCTTCCA	AGAAAAGC-	-TCTAAACACG-	--TTAA	TCC-AAAAT[mm]
Escherichia	GCACTACCGTGC	-TGAAGCAACAAAT	GCCCTGCTTCCA	GGAAAAGCC	-TCTAAGCATCA	--GGTA	ACATCAAAT
Saccharomyces	CACCTTTCTGG	-CTCCGGTGCGCTT	GT-GACGGCCCG	TGAAAATCC	ACAGGAAGGAA-	-TA-	ATCCTAGGT
Physarum	GTTCTTCCATGT	-GTTCCGCGGTCTCC	TTCCGTTGCCCT	AGAAAAGCT	GGCAGA-TGGGT	GAAA[kk] GTCCTT	CGGTTGAAC
Xenopus	GGTTCCGGCGGC	-GTCCGGTGAGCTT	CTCGCTGGCCCT	TGAAAATCC	GGGGGAGAGGGT	GTAA ATCTCT	GCCCGGGGC
Rat	AGTTCCGGCGGC	-GTCCGGTGAGCT-	CTCGCTGGCCCT	TGAAAATCC	GGGGGAGAGGGT	GTAA ATCTC-	GCCCGGGGC

Row 36	H	I	J	K	L	M	N	O
Tobacco$_C$	TGTACCCGA	AACCGA	CACAGGTGG	GTAGGTAGA	GAATACCTAG	GG-GCGCGAGAC	AACTCTC-TCTAA	GGAACTCGGC
Anacystis	TGTACCCTA	AACCGA	CACAGGTGG	GACGGTAGA	GTATACCAAG	GG-GCGCGAGGT	AACTCTC-TCTAA	GGAACTCGGC
Escherichia	CGTACCCGA	AACCGA	CACAGGTGG	TCAGGTAGA	GAATACCAAG	GC-GCTTGAGAG	AACTCCG-GTGAA	GGAACTAGGC
Saccharomyces	CGTACTGAT	AACCGC	AGCAGGTCT	CCAAGGTGA	ACAGCCTCTA	GT-TGATAGAAT	AATGTAG-ATAAG	GGAAGTCGGC
Physarum	CGTACCTA-	ATCCGC	AGCAGGTCT	CCAAGATGA	GCAGTCTCTG	GC-GCATAGAAC	AAAGTAGCGTAAG	GGAATTCGGC
Xenopus	CGTACCCAT	ATCCGC	AGCAGGTCT	CCAAGGTGA	ACAGCCTCTG	GCATCGTTAGAAC	AATGTAG-GTAAG	GGAAGTCGGC
Rat	CGTACCCAT	ATCCGC	ACCAGGTCT	CCAAGGTGA	AC-GCCTCTG	GCATCTTGGAAC	AATGTAG-GTAAG	GGAAG-CGGC

Row 37

	P	Q	R	S	T	U	V	W
Tobacco	AAAATAGCC	CCGTAACTTCGG	GAGAAGG-	--------	--------	-GGTG---CC	TCCT-CAC-	to row 45
Anacystis	AAAATGACT	CCGTAACTTCGG	GAGAAGG-	--------	--------	-AGTG---CC	CAC--CTA-	to row 45
Escherichia	AAAATGGTG	CCGTAACTTCGG	GAGAAGGCA	C--CCT---	--GATATCT	AGGTG-AGT	CCCT-CGC-	
Saccharomyces	AAAATAGAT	CCGTAACTTCGG	GATAAGGAT	TGGCTCTAA	GGGTCGGGT	AGTGA-GGGC	CTTGGTCAG	ACG-CAGCG
Physarum	AAGCCGGAT	TCGTAACTTCGG	GATAAGGAT	TGGCTCTAT	AGGCTGGGT	GTCGCTGGGC	TGGGTAAGG	CTC-GCGGG
Xenopus	AAGTCAGAT	CCGTAACTTCGG	GATAAGGAT	TGGCTCTAA	GGGCTGGGT	CCGTC-GGGC	TGGGGCCG	AAG-CGGGG
Rat	AAGCCGGAT	CCGTAACTTCGG	GATAAGGAT	TGGCTCTAA	GGGCTGGGT	CCGTC-GGGC	TGGGGCCG	AAGGCGGGG

Row 38

	A	B	C	D	E	F
Saccharomyces	GGGCGTG-CT	to row 45				
Physarum	ATGTGC---	to row 45				
Xenopus	CTGGGC-CGCGC	CGCGGGCTGGACGAAGGCGC	CCCCGTGGC---	G---------	-----------C	TC-TTTCCCTCC
Rat	CTGGGGCGCGC	CCCGGGCTGGACG-AGGCGC	CGCCCCCCCCT	CCCACGTCC	GGGGAGACCCCC	TCCTTTCCGCCC

Row 39

Rat	GGGCCCGCC	CTCCCCTCTCCC	CGCGGGGCCCCG	CCGTCCCCCCCG	TCGTCGCCGTGGTCC	CCTCCTCCCTCCTT

(continued)

Table 4.7 (Continued)

Row 40

Rat   CTTCCCCGTCCGCGG GGGGGACGGGGC GGGTGCGGGGGG GCGCGCGCGCGC GCGGCCCAGGGG CGGCGGTCCAA

Row 41

Xenopus   from row 38   -TC

Rat   CCCCGCGCGGGC CGGAGCGGGGGG AACCCGCGGGCCCCC GGTGGGGGGGG CCCCGACACCCG GGGGGACCCGC

Row 42

	G	H	I	J	K	L
Xenopus	CCGCCCCCTCT	CTCTCCGGCCCC	CCCTCCGGGGG	GCCCCGGGGGCGGGG	GGGCGCCCGGGG	GCCGGGAGCCCC
Rat	GGCGGCGGCGAC	TCTGGACGCGAG	CCGGGCCCTTCC	CGTGGATCGCCCCAG	CTGCGGCGGGCG	TCCGCGGCCGGTC

Row 43

	M	N	O	P	Q	R	S
Saccharomyces			from row 38				
Physarum	--TTTAGT---	---AAT------	---GGCCGAG--	-TGTGGA--	---------	--------	
Xenopus	CGGCGGCGGCGA	CTCTGGACGCGCC	GCCGGG---CCC	TTCCTGTGGATC	G-CCCC----	AGCTGCGG-	CGCCCGCCT
Rat	CCGGGGAGCCCG	GCGGGTCGCCCG	GCGGGGTTTTCC	TCCGGCCTCGTC	CTCCCCCTT	CCCCCTCCG	CGGGGTCGG

Row 44

	T	U	V	W	X	Y	Z
Saccharomyces	CTGCTTGGT	GGGGCTTGC	TCTGCTAGG	CGGACTACT--T	GCCTGCCTTGTT	GTAGA-CGGCCT	TGGTAGCTCTC-
Physarum	GCGCTTTCG	GGTGCCGAG	AGCGTCAAC	GGG-CCCCT--C	GGGGCTGAAAAC	-TAGATCATTGC	GGAGGTTCGCCC
Xenopus	CTCCCCGC	GCCGTCCCC	CTCCTCGC	CTCCCCCG--T	CAGGGGACGG-	GCGC-GWGCSGC	GGGCGCGCCGGG
Rat	GGGTTCCCG	GGGTTCGGG	GTTCTCCTC	CGCGGCGCGGTT	CCCCCGCCGGGT	GCGCCCCCCGGG	CGCGGTTTCCCG

Row 45

	A	B	C	D	E	F	G
Tobacco$_C$			-AAAG	GGG----------	--GTCGCAGTGA	CCAGGCCCGGGC	GACTGTTTACC-
Anacystis	from row 37		AGACG	TGG----------	--GTCGCAGTGA	AGAGGCCCAGGC	GACTGTTTACC-
Escherichia			GGATG	GAGCT-GA--AAT	CAGTCGAAGATA	CCAGCTCGCTGC	AACTGTTTATTA
Saccharomyces	TTGTAGACC	GTCGCTTGC	TACAATTAA	CAGATCAACTTAG	AACTGGTACGGA	CAAGGGAATCT	GACTGTCTAATT
Physarum	GAGTGGTGA	AAACCGGCA	CACCGGCTAA	CAGCT-AACGTAG	AACTTACAAAGG	CTAGGGAATCC	AACTGTATAATT
Xenopus	GCGGCCCG	GCCTCGCCC	GGGCCCTAG	CAGCT-GACTTAG	AACTGGTGCGGA	CTAGGGAATCC	GACTGTTTAATT
Rat	CGGCCGCCC	GCCTCGGCC	GGGCCCTAG	CAGCC-GACTTAG	AACTGGTGCGGA	CTAGGGAATCC	GACTGTTTAATT

(continued)

Table 4.7 (Continued)

Row 46	H	I	J	K	L	M	N
Tobacco_C	AAAAACACAG	GTCTCCGCAAAG	TCGTAAGACCAT	GT-ATGGGG-CT	GACGCCTGCCC	AGTGCCG[nn]CCT	GAT-GACAG
Anacystis	AAAAACACAG	GTCTCCGCTAAG	TCGTAAGACGAT	GT-ATGGGGGCT	GACGCCTGCCC	AGTGCCG[oo]GGT	CAGCGCCAAG
Escherichia	AAAAACACAG	CACTGTGCAAAC	ACGAAAGTGGAC	GT-ATACGTGT	GACGCGTGCCC	GGTGCCG[pp]GGT	TAGCGCCAAG
Saccaromyces	AAA-ACATAG	CATTGCGATGGT	CAGAAAGTGATG	TTGACGCAATGT	GATTTCTGCCC	AGTGCTC TGA	ATGTCAAAG
Physarum	AAA-ACATAG	CGATTTGTTGGT	GCCAAA-GCCTG	TAAACGAATCGT	GATTTCTGCCC	AGTGCTC TGG	ATGTTAAAA
Xenopus	AAA-ACAAAG	CATCGCGAAGGC	CCGAGCGGGTG	TTGACGCCATGT	GATTTCTGCCC	AGTGCTC TGA	ATGTCAAAG
Rat	AAA-ACAAAG	CATCGCGAAGGC	CCGCGCGGGTG	TTGACGCCATGT	GATTTCTGCCC	AGTGCTC TGA	ATGTCAAAG

Row 47	O	P	Q	R	S	T	U
Tobacco_C	GGGAGCCGGCGACCG	AAGCCCCGGTGA	ACGGGCGGCC	GTAACTATA	ACGGTCCTAAGG	TAGCGAAAT	TCCTTGTCG-
Anacystis	TGAAGCTCGGCGACCG	AAGCCCCGGTGA	ACGGGCGGCC	GTAACTATA	ACGGTCCTAAGG	TAGCGAAAT	TCCTTGTCG-
Escherichia	CGAAGCTCTTGATCG	AAGCCCCGGTAA	ACGGGCGGCC	GTAACTATA	ACGGTCCTAAGG	TAGCGAAAT	TCCTTGTCG-
Saccaromyces	TGAAGAAATTCAACC	AAGGCGCGGAGTAA	ACGGGCGGGA	GTAACTATG	ACTCTCTTAAGG	TAGCCAAAT	GCCTCGTCA-
Physarum	TGGCGAAATCCAACC	AAGCTCGGGGTAA	ACGGGCGGGA	GTAACTATG	ACTCTCTTAAGG	TAGCCAAAT	GCCTCGTCA-
Xenopus	TGAAGAAATTCAATG	AAGGCGCGGGTAA	ACGGGCGGGA	GTAACTATG	ACTCTCTTAAGG	TAGCCAAAT	GCCTCGTCA-
Rat	TGAAGAAATTCAATG	AAGGCGCGGGTAA	ACGGGCGGGA	GTAACTATG	ACTCTCTTAAGG	TAGCCAAAT	GCCTCGTCA-
Human_M	CCAATCTATCACCCT	ATAGAAGAACTA	ATGTTAGTA	-TAAGTA-A	CATGAAAAACATT	CTCCTCCGC	ATAAGCCTGC

Row 48

	A	B	C	D	E	F	G	H
Tobacco_C	GGTAAGTTC	CGACCCGCA	CGAAAGGCG	TAACGATCTGG	GCA--CTGTCTC	GGAGAGAGG	CTCCGTGAA	ATAGACATGT
Anacystis	GGTAAGTTC	CGACCCGCA	CGAAAGGCG	TAACGATCTGG	GCG--CTGTCTC	AGAGAGAGG	CTCGGCGAA	ATAGGAGTGT
Escherichia	GGTAAGTTC	CGACCTGCA	CGAATGGCG	TAATGA--TGG	CCAGGCTGTCTC	CACCCGAGA	CTCACGTGAA	ATTGAACTCG
Saccharomyces	TCTAATTAG	TGACGCGCA	TGAATGGAT	TAACGA--GAT	TCCCACTGTCCC	TATCTACTA	TCTAGCGAA	ACCACA-GCC
Physarum	TTTAATTTG	TGACGCGCA	TGAATGGAT	TAATGA--GAT	TCCCACTGTCCC	TACCTACTA	TCTAGCGAA	ACCACA-GCC
Xenopus	TCTAATTAG	TGACGCGCA	TGAATGGAT	GAACGA--GAT	TCCCACTGTCCC	TACCTACTA	TCTAGCGAA	ACCACA-GCC
Rat	TCTAATTAG	TGACGCGCA	TGAATGGAT	GAACGA--GAT	TCCCACTGTCCC	TACCTACTA	TCCAGCGAA	ACCACA-GCC
Human_M	GTCAGATTA	AAACACTGA	ACTGACAAT	TAA-------	CAGCCCAATATC	TAC--AAT-	CAACCAACA	AGTCAT-TAT

Row 49

	I	J	K	L	M	N	O	P
Tobacco_C	CTGTCAAGATGC	GGACTACCT	GCACCTAGACAG	AAAGACCCT	ATGAAGCTT	CACTGTTCC	CTGGGA	TTGG-CTTT
Anacystis	CTGTCAAGATAC	GGACTACCT	GCACCCGGACAG	AAAGACCCT	ATGAAGCTT	TACTGTAGC	TTGGTA	TTGG-CTTC
Escherichia	CTGTCAAGATGC	AGTGTACCC	GCGGCAAGACGG	AAAGACCCC	GTGAACCTT	TACTATAGC	TTTGACA	CTGAACATT
Saccharomyces	AAGGCAACGGGC	TTGGCAGAA	TCAGCGGGGAAA	GAAGACCCT	GTTGAGCTT	GACTCTAGT	TTGACA	TTGTGAAGA
Physarum	AAGGCAACGGGC	TTGGCACAA	TTAGCGGGGAAA	GAAGACCCT	GTTGAGCTT	GACTCTAGG	CATAGA	C-GGAGGT
Xenopus	AAGGCAACGGGC	TTGGCGGAA	TCAGCGGGGAAA	GAAGACCCT	GTTGAGCTT	GACTCTAGT	CTGCAA	CTGTGAAGA
Rat	AAGGCAACGGGC	TTGGCGGAA	TCAGCGGGGAAA	GAAGACCCT	GTTGAGCTT	GACTCTAGT	CTGGCA	CGGTGAAGA
Human_M	TACCCTCACTGT	CAACCCAAC	ACAGGCATGCTC	ATAAGGAAA	GGTTAAAAA	AAGTAAAAG	GAACTC	GGCAAATCT

(continued)

Table 4.7 (Continued)

Row 50	Q	R	S	T	U	V	W
Tobacco_C	GGGCCTTTCC	TGCCCAGCT	TAGGTGGAA	GGGCAAGAAGGC	-----CT------	----CCTTCCG--	----------
Anacystis	GGGCTTTGAC	TGCCCAGGA	TAGGTGGGA	GGCTATGAGACT	-----TT------	-----CCTTGTG--	----------
Escherichia	GAGCCTTGA-	TGTGTAGGA	TAGGTGGGA	GGCTTTGAAGTG	-----TG------	-----GACGCC---	----------
Saccharomyces	GACATAGAG-	GGTGTAGAA	TAAGTGGGA	G-CTTC-------	-----------	-----------	----------
Physarum	GATTCTAAA-	GGTGTAGCA	TAGGTGGGA	GGGCCCATG----	-----------	-----------	----------
Xenopus	GACATGAGA-	GGTGTAGGA	TAAGTGGGA	GGCCCCG-CGC	-----------	-------TCG	----------
Rat	GACATGAGA-	GGTGTAGAA	TAAGTGGGA	GGCCCCGGCGC	CCCCCCGTTCCC	CGCGAGGGGTCG	GGGCGGGGTCCG
Human_M	TACCCCGCCT	GTTTACCAA	AAACATCAC	CTCTAG---CAT	-----CA------	----CCAGTAT--	----------

Row 51	A	B	C	D	E	F	G
Tobacco_C	GGGGGGCCCGAG	-CCATCAG	TGAAATACCAC	TCTCGAAGGGC	TAGAATTCT	AACCTT-GTGTCA	GGAC--------
Anacystis	GGGGAAG-TGGA	GCCAACGG	TGAAATACCAC	TCTGTCAAAGC	TAGAAGTCT	AAC-TTTGAGCCG	TT----------
Escherichia	AGTCTGCATGGA	GCCGACCT	TGAAATACCAC	CCTTTAATCTT	TGATGTTCT	AACGTTTGA-CCG	TA----------
Saccharomyces	------------	GGGCCCAG	TGAAATACCAC	TACCTTTATAG	TTTCTTTAC	TTATTCAAT-GAA	------GCGGAGC
Physarum	------------	--CCGTCAA	TGAAATACCAC	CACTTTCGACA	TCGCTTTGC	TAATGCTGT-AAC	GAACGAACGGAA
Xenopus	TCGCAAAGGGGC	GCCGCCGG	TGAAATACCAC	TACTCTTATCG	TTTTTTCAC	TTACCCGGT-GAG	GCGGGGGGGGGA
Rat	CCCGCCTCGCGG	GCCGCCGG	TGAAATACCAC	TACTCTCATCG	TTTTTTCAC	TGACCCGGT-GAG	GCGGGGGGCCGA
Human_M	TAGAGGCACGGC	CTGCCCAG	TGACACATCTT	TAACGGCCGCG	--GTACCCT	AAC-CGTG---CAA	AG----------

Row 52

	H	I	J	K	L	M	N
Tobacco$_C$	-------	--	-------	-------	-------C	TACGGGCCA	AGGGACAGTC
Anacystis	-------	--	-------	-------	-------A	TCCGGCGA-	AGG--CAGTA
Escherichia	-------	--	-------	-------	-------A	TCCGGGTTG	CGGA-CAGTG
Saccharomyces	TGGAATTCATTT	AG	CATTCAAGGTCC-	C-------ATTC	GGGGC-TGA	TCCGGGTTG	AAGA-CATTG
Physarum	CCCCGTCCCTCT	99TC	GGC-CTCTGTGG-	GCTTCATGCCG-	GTACGTAAT	TGCGTGTAG	CAGA-CTATC
Xenopus	GCCCCGAGGGGC	GG	ACC--CAAGCGCS-	CGGGCCCCGCGC	CGGGCGCGA	CCCGCTCCG	AGGA-CAGTG
Rat	GCCCCGAGGGGC	GG	CGC--CGAACGCGT	CCCGCGCGCGGG	CGGGCGCGA	CCCGCTCCG	GGGA-CAGTG
Human$_M$	-------	---	-------	-------	-------G	TAGCATAAT	CACT-TGTT-

Row 53

	O	P	Q	R	S	T	U	V
Tobacco$_C$	TCAGCTAGA	CAGTTTCTA	TGG-GGCGTA	-GGCCTCCCAAA	AGGTAACGG	AGGCGTGCA	AAGGTTTCC	TCGGGGCGG-A
Anacystis	TCAGGTGGG	CAGTTTCAC	TGGCGGGC--	---CCTCCTAAA	AGGTAACGG	AGGCGCGCCA	AAGGTTCCC	TCAGGCTGG-T
Escherichia	TCTGGTGGG	TAGTTTCAC	TGG-GGCCGT	-CTCCTCCTAAA	GAGTAACGG	AGGAGCACG	AAGGTTGGC	TAATC-CTG-G
Saccharomyces	TCAGGTGGG	CAGTTTCGC	TGG-GGCGGC	ACATCTGTTAAA	CGATAACGC	AGATGTCCT	AAGGGGGGC	TCATG-GAG-A
Physarum	TATG-TGGG	CAGTTTCGC	TGG-GGCGGA	AAA-CTGCTACA	CGGCAACGG	CAGTCTCCT	AAGGTCCAC	TCAGA-GAC-G
Xenopus	GCAGGTGGG	CAGTTTCAC	TGG-GGCGGT	ACACCTGTCAAA	CCGTAACGC	AGGTGTCT	AAGGCGAGC	TCAGGGGAGCG
Rat	CCAGGTGGG	CAGTTTCAC	TGG-GGCGGT	ACACCTGTCAAA	CGTAACGC	AGTGTCT	AAGGCGAGC	TCAGG-GAGG-
Human$_M$	-CC----TTA	AATAGGGAC	CTG-TATGAA	-TGGCTCCACGA	GGGTTCAGC	T-GTC-TCT	-TACTTT--	TA---ACC-A-

(continued)

Table 4.7 (Continued)

Row 54	A	B	C	D	E	F	G	H
Tobacco$_C$	-CGGAGATT	GGCCCTCGA	GTGCAAAGG	CAGAAGGGA	GCTTGACTGCAA	GACCCACCC	GTCGAGCAG	-GGACGAAA
Anacystis	-TGGAAATC	AGCCGACGA	GTGCAAAGG	CATAAGGGA	GCTTGACTGCAA	GACCTACAA	GTCGAGCAG	-GGACGAAA
Escherichia	TCCGGACATC	AGGAGGTTA	GTGCAATGG	CATAAGCCA	GCTTGACTGCGA	GCCGTGACGG	CGCGAGCAG	-GTCGCAAA
Saccharomyces	ACAGAAATC	TCCAGTAGA	ACAAAAGGG	T-AAAGCCC	-CTTAGTTT-GA	TTT-CAGTG	TGAAATACAA	ACCATGAAA
Physarum	ACAGAAACG	TCTCGTAGA	GA-TAAAGG	CAAAGTGG	GCTTAACTCGCA	TTTTTCAGTA	GTAATGTGA	AGCAAGAAA
Xenopus	ACAGAAACC	TCCCGTCGA	GCAGAAGGG	CAAAAGCTC	GCTTGATCTTGA	TTTTCAGTA	TGAATACAA	ACCGTGAAA
Rat	ACAGAAACC	TCCCGTGGA	GCAGAAGGG	CAAAAGCTC	GCTTGATCTTGA	TTTTCAGTA	CGAATACAG	ACCGTGAAA
Human$_M$	-GTGAAATT	GACCTCCCC	GTG------	---AAGA--	---GGCGGGCAT	AACACAGCA	AGACGAGAA	GACCCTATG

Row 55	I	J	K	L	M	N	O	P
Tobacco$_C$	-GTCGG--CCTT	AGTGATCCGACG	GTGCCGACT	GGAAGGGCCG	TCGCTCAAC	GGATAAAAG	TTACTCTAG	GGATAACAG
Anacystis	-GTCGG--CCTT	AGTGATCCGACG	GTTCTGACT	GGAAGGGCCG	TCGCTCAAC	GGATAAAAG	TTACTCTAG	GGATAACAG
Escherichia	-GCAGG-TCAT	AGTGATCCGGTG	GTTCTGAAT	GGAAGGGCCA	TCGCTCAAC	GGATAAAAG	GTACTCCGG	GGATAACAG
Saccharomyces	-GTGTG-GCCT	ATCGATCCTTTA	GTCCCTCGG	AATTTG-AGG	CTAGAGGTG	CCAGAAAAG	TTACCACAG	GGATAACTG
Physarum	-TTGCG-GCTT	AACGATCCTTAC	CGGCGGGTG	CCAGCCCACG	CTTGAGGTG	AGAGAAAAG	TTACCACAG	GGATAACTG
Xenopus	CGCGGGWGCCT	CACGATCCTTCT	GACTTTTTG	GGTTTT-AAG	CAGGAGGTG	TCAGAAAAG	TTACCACAG	GGATAACTG
Rat	-GCGGG-GCCT	CACGATCCTTCT	GACCTTTTG	GGTTTT-AAG	CAGGAGGTG	TCAGAAAAG	TTACCACAG	GGATAACTG
Human$_M$	-GAGCT-TTAA	TTTATTAATGCA	AACAGTACC	TAACAA-ACC	CACAGGTCC	TAAACTACC	AAACC-TGC	ATTAAAAAT

Row 56

	Q	R	S	T	U	V	W
Tobacco_C	-GCTGATCTT-CCC	-CAAGAGTCACA	TCGACGGGAAGG	TTTGGCACC	TCGATGTCGGCTC	TTCGCCAC	CTGGGGCTG
Anacystis	-GCTGATCTC-CTC	-CAAGAGTTCACA	TCGACGAGGAGG	TTTGGCACC	TCGATGTCGGCTC	ATCGCAAC	CTGGGGCTG
Escherichia	-GCTGATACC-GCC	-CAAGAGTTCATA	TCGACGGCGGTG	TTTGGCACC	TCGATGTCGGCTC	ATCACATC	CTGGGGCTG
Saccharomyces	-GCTTGTCGC-AGT	-CAAGCGTTCATA	GCGACATTGCTT	TTTGATTCT	TCGATGTCGGCTC	TTCCTATC	ATACCGAAG
Physarum	-GCTTGTCGC-CGC	-CAAGCGTTCATA	GCGACGTGGCTT	TTTGATCCT	TCGATGTCGGCTC	TTCCTATC	ATACTAAAG
Xenopus	-GCTTGTCGCCGGC	-CAAGCGTTCATA	GCGACGTCGCTT	TTTGATCCT	TCGATGTCGGCTC	TTCCTATC	ATTGTGAAG
Rat	-GCTTGTCGC-GGC	-CAAGCGTTCATA	GCGACGTCGCTT	TTTGATCCT	TCGATGTCGGCTC	TTCCTATC	ATTGTCAAG
Human_M	TTCGGTTGGGGCGA	CCTCGGAGCAGAA	CCCAACC--TCC	AGCACTACA	TGCTAAGACTTCA	CCAGTCAA	AGCGAACTA

Row 57

	A	B	C	D	E	F	G	H
Tobacco_C	TAGTATGT-	CCAAGGGTT	GGGCTGTTC	GCCCATTAA	AGCGGTACG	TGAGCTGGG	TTCAGAAACG	TCGTGAGACAGT
Anacystis	AAGTCGGTC	CCAAGGGTT	GGGCTGTTC	GCCCATTAA	AGCGGTACG	TGAGCTGGG	TTCAGAAACG	TCGTGAGACAGT
Escherichia	AAGTAGGTC	CCAAGGGTA	TGGCTGTTC	GCCATTTAA	AGTGGTACG	CGAGCTGGG	TTTAGAACG	TCGTGAGACAGT
Saccharomyces	CAGAATTCG	GTAAGCGTT	GGATTGTTC	ACCCACTAA	TAGGGAACA	TGAGCTGGG	TTTAGACCG	TCGTGAGACAGG
Physarum	C-GAATTCA	GTAAGTGTT	GGATTGTTC	ACCCTCTTA	-AGGGAACG	TGAGCTGGG	TTTAGAC-G	TCGTGAGACAGG
Xenopus	CAGAATTCA	CCAAGCGTT	GGATTGTTC	ACCCACTAA	TAGGGAACG	TGAGCTGGG	TTTAGACCG	TCGTGAGACAGG
Rat	CAGAATTCA	CCAAGCGTT	GGATTGTTC	ACCCACTAA	TAGGGAACG	TGAGCTGGG	TTTAGACCG	TCGTGAGACAGG
Human_M	CTATACTCA	ATTGATCC-	AATAACTTG	ACCAAC---	GGAACAAGT	TACCCTAGG	GATAAACAGC	GCAATCCTATTC

(continued)

Table 4.7 (Continued)

Row 58

	I	J		K	L	M	N	O
Tobacco_C	-TCGGTCCATATC	CGGTGTGGG	CGTT	AGAGCATTG	AGAGGACCTTTC	CCTAGTACG	AGAGGACCG	GGAAGGACCAC
Anacystis	-TCGGTCCATATC	CGGTGCAGG	CGTA	AGAGTATTG	AGAGGATTTCTC	CCTAGTACG	AGAGGACCG	GGAAGGACCAC
Escherichia	-TCGGTCCCTATC	TCCCGTGGG	CGCT	GGAGAACTG	AGGGGGGCTGCT	CCTAGTACG	AGAGGACCG	GAGTGGACGCAT
Saccharomyces	-TTAGTTTTACCC	TACTGATGA	-ATG	TTACCAGCA	ATAGTAATTGAA	CTTAGTACG	AGAGGAACA	GTTCATTCGGAT
Physarum	-TTAGTTTTACCC	TCCTACTTG[???]	-CGG	TGTGCCCTG	ATAGTAGATCAC	TTCAGTACG	AGAGGAACC	AGTGATTCAGAC
Xenopus	-TTAGTTTTACCC	TACTGATGA	-TGT	GTTGTTGCA	ATAGTAATCCTG	CTCAGTACG	AGAGGAACC	GCAGGTTCAGAC
Rat	-TTAGTTTTACCC	TACTGATGA	-TGT	GTTGTTGCC	ATGGTAATCCTG	CTCAGTACG	AGAGGAACC	GCAGGTTCAGAC
Human_M	TAGACTCCATATC	AACAATAGG	-GTT	TACGACCTC	GATGTTGGATCA	GGACATCCC	GATGGTGCA	GCCCTATTAAA

Row 59

	P	Q	R	S	T	U	V	W
Tobacco_C	-CTCTGGTTGACC	AGTTATCGTGCC	CACGGTAAAC	GCTGGGTAGCCAA	GTGCCGG-AGCG	GAT	AACTGCTGA	AAGCATCTA
Anacystis	-CGCTGGTGTACC	AGTTATCGTGCC	-AACCTAAAC	GCTGGGTAGCTAC	GTGTGG-AGTG	GAT	AACCGCTGA	AAGCATCTA
Escherichia	-CACTGGTGTTCG	GGTTGTCATGCC	-AATGGCACT	GCCCGGTAGCTA-	AATGCGGAAGA	GAT	AAGTGCTGA	AAGCATCTA
Saccharomyces	-AATTGGTTTTTG	CCGGCTCTCTGAT	-CAGGCCATTG	CCGGCA-AGCTAC	CAT-CCG-CTG	GAT	TATGGCTGA	ACGCCCTCTA
Physarum	-CAATGGTGTAAG	CCCTGCTCGAAC	-GGG-CAGAG	CGCTAG-AGCTAC	-GT-CTG-TTG	GAT	GAAGGCTGG	AAGCATCTA
Xenopus	-ATTTGGTGTATG	TGCTTGGCTCAG	-GAGCCAATG	GGGGCA-AGCTAC	CAT-CTG-TGG	GAT	TATGACTGA	ACGCCTCTA
Rat	-ATTTGGTGTATG	TGCTTGGCTCAG	-GAGCCAATG	GGGGCA-AGCTAC	CAT-CTG-TGG	GAT	TATGACTGA	ACGCCTCTA
Human_M	GGTTCGTTTGTTC	AACGGATTAAAGT	-CCTACGTGA	TCTGAGTTCAGAC	CGG-AGTAAT-	CCA	GGTCGGTTT	CTATCTACC

Row 60	A	B	C	D	E	F	G
Tobacco_C	AGTAGTAAG	CCCACCCCAA-	GATGAGTGC	TCTCCTATTC			
Anacystis	AGTGGGAAG	CCCACTC-AA-	GATGAGTAC	TCTCATGGCA	to row 62		
Escherichia	AGCACGAAA	CTTGCCCCGA-	GATGAGTTC	TCCCTGACCC	TT		
Saccharomyces	AGTCAGAAT	CCATGCTAGAA	CGCGGTGAT	TTCTTTGCTC	CACACAATATA	GAT-GGATACGA	ATAA-GGGGTCCT---
Physarum	AGCCCGAAG	CCATGCTAGA-	TAGGGGACG	TCGCAATCAA	AGATACGCTTG	CTTTTAAGTCTG	GCAACGCGGTCTTTTA
Xenopus	AGTCAGAAT	CCCCCCTAAA-	CGTGACGAT	ACCGCAGCGC	CGCGGAGCCTC	GGTCGCGCCTCGG	ATTA-GCCGGCGCCCC
Rat	AGTCAGAAT	-CCGCCCAAG-	CGGAACGAT	ACGGCAGCGC	CGAAGGAGCCT	CGGTTGGCCCCG	GATA-GCCGGCTCCCC
Human_M	-TTCA-AAT	TCC-TCCCTG-	TACGAAAGG	ACAAGAGAAA	TAAGGC	to row 62	

Row 61	H	I	J	K	L	M	N
Saccharomyces	--------TGTG	GCGTCGCTG	AACC-ATAG--C	AG--GCTAG	C-AACGGTGC AC	TTGGCGGAA	AGGCCTTGGGT-
Physarum	CCATCAACAGCC	AGTCAACTG	AGCAGGACG--C	GA-TCCAAA	-GAG--GTCG[3.3]GC	C----CGTAA	AAAGGTCAGTT-
Xenopus	CCCGGCGGCG	CCGGGCGGC	AGAG--CCGCTC	GCC---TCG	GGA-CCGGAG CG	CGGACG--AA	AGGGGGCCGCC-
Rat	GTCCGTCCCGT	CCGGCCGGGT	CCCCGCCTCGTC	GCCCCCCCG	GGAAACGGGG TG	CGCCCGGAA	AGGGGCCCGCC

(continued)

Table 4.7 (Continued)

Row 62	O	P	Q	R	S	T	U	V	W
Tobacco$_C$	----------	----------	----------	^{tt}GAA	----GG---- |	-TC	ACGGCGAGAC	GA-GCC----	GTT^{vv} GATAGG
Anacystis	----------	----------	---TAAG---	----CCG TAA	----GG---- |	-TC	ACGGGTAGAA	CA-CCC----	GTT AATAGG
Escherichia	----------	----------	---TAAGGG	--TCCT GAA	----GG---- |	-AA	C----GTTGAA	GACCACGAC	GTT GATAGG
Saccharomyces	GCTTGC----	TGG--CGAA	TTGCAATG-	CATTTT G--	CGTGGGGA T	AAA	TCA-TTTCTA	TACCACTTA	GAT GTACAA
Physarum	CCTCGTTAA	GCCGCCCCC	ACGGGGGCA	TCAACG GCG	CTCCGCCT^{uu} T	AAA	CCACACAGCA	TACGA-CTG	TGC GTAGCG
Xenopus	TCTC--TCC	CGGAGCGCA	CCGCACGTT	CGTGGG GAA	CCTGGTGC T	AAA	TCA-TTCCTA	TACGA-CTG	ATT CTGGGT
Rat	TCTCGCCCG	TCACGCTTA	ACGCACGTT	CGTGTG GAA	CTTGGCGC T	AAA	CCA-TTCCTA	GACGACCTG	CTT CTGGGT
Human$_M$	----------	----------	----------	-------- CTA	---CT---- |	-TC	ACA------AA	G---CGC---	CTT CCC----

Row 63	A	B	C	D	E	F	G	H	I
Tobacco_C	TGTCATGTG	GAAGTGCAG	TGATGTATG	CAGCTGAGGCAT	CCT-AACAGA	CCGGTAGAC	TTGAA-C		
Anacystis	CCCTATGTG	GAAGTTCAG	CAATGGATG	AAGCTGAGGCGT	ACT-AATAGA	CCGAG-GGC	TTGAC-CTC		
Escherichia	CCGGGTGTG	TAAGCGCAG	CGATGCGTT	GAGCTAACCGGT	ACT-AATGAA	CCGTGAGGC	TTAAC-CTT		
Saccharomyces	CGGGTATT	GTAAGCGGT	AGAGTAGCC	TTGTTGTTACGA	TCTGCTGAGA	TTAAGC--C	TT--TGT-T	GT-CTGATT	TGT
Physarum	GAAATGGTG	TTAGTGATC	AAGCTGGCT	GGCCGTCTAGCA	TTGCGAGACG	CTGAGCCAG	TGTTT----	GGGTTTTTT	TGTCGGC
Xenopus	CAGGGTTTC	GTCGGTAGC	AGAGCAGCT	ACCTCGCTCGGA	TCT-ATTGAA	AGTCATCCC	TTGGC-CAA	GC	
Rat	CGGGGTTTC	GTACGTAGC	AGAGCAGCT	CCCTCGCTCGGA	TCT-ATTGAA	AGTCAGCCC	TCGACACAA	GGGTTTGT	
Human_M	CCG----TA	AATGATATC	ATCTCAACT	TAGT-ATT-ATA	C---C-CAC-	ACC---CAC	CCAAGAACA	GGGTTT	

[a] Subscript C indicates chloroplast, M indicates mitochondrion.
[b] The tobacco chloroplast sequence begins at site 142, the *Anacystis* at site 143, and the *E. coli* at 158, in order to compensate for the appendicular rRNA gene.
[c] Takaiwa and Sugiura (1982). [d] Kumano *et al.* (1983); Douglas and Doolittle (1984). [e] Brosius *et al.* (1980, 1981). [f] Georgiev *et al.* (1981). [g] Otsuka *et al.* (1983). [h] Ware *et al.* (1983). [i] Chan *et al.* (1983). [j] Anderson *et al.* (1981). [k] Sor and Fukuhara (1983). [l] Seilhamer and Cummings (1981). [m] Van Etten *et al.* (1980); Bibb *et al.* (1981). [n] Insert ACCGTG. [o] Insert ACCAAAC. [p] Insert CCC. [q] Insert CGCGT. [r] Insert GGCGT. [s] Insert TGCGT. [t] Insert 25 residues. [u] Insert 28 residues. [v] Insert GGGCGGGGGGGGGG. [w] Insert GCTGG. [x] Insert 19 residues. [y] Insert 12 residues. [z] Insert 17 residues. [aa] Insert 11 residues. [bb] Insert 31 residues. [cc] Insert GCC. [dd] Insert 48 residues. [ee] Insert GA. [ff] Insert TGTCC. [gg] Insert GAAGGGGTA. [hh] Insert GGAGCGGAG. [ii] Insert CGA. [jj] Insert GCAAC. [kk] Insert CGTGTT. [ll] Insert GGTACC. [mm] Insert TGCC. [nn] Insert 22 residues. [oo] Insert 16 residues. [pp] Insert 15 residues. [qq] Insert 147 residues. [rr] Insert 25 residues. [ss] Insert AAATTA. [tt] Insert 101 residues. [uu] Insert 20 residues. [vv] Insert TATCATTAC.

The Protistan Genes. The yeast and fungus gene structures contribute greatly to an understanding of how the eukaryotic condition was acquired from their prokaryotic ancestors. However, the *Physarum* cistron is at once distinct in being interrupted by two long introns located in its 3' half (Nomiyama *et al.,* 1981a,b; Otsuga *et al.,* 1983), a trait that occurs in certain other eukaryotes. Among these are several insects, including *Drosophila* and *Bombyx* (Jamrich and Miller, 1984, Lecanidou *et al.,* 1984), and the euciliate *Tetrahymena* (Wild and Gall, 1979; Wild and Sommer, 1980; Kan and Gall, 1982). Although for by far the greater part the *Saccharomyces* sequence agrees more fully with those of the vertebrates than with the bacterial, rare sectors reflect relations to the latter. One such instance is situated in row 6, column E, and others can be noted in row 18, column I; row 27, columns Q–S; row 33, column M; row 34, column V; row 37, column P; row 46, column J; row 49, column O, row 50, column T; row 52, column M; and row 60, column D. In addition, kinship is expressed, also to a limited extent, by the correlations between vacant regions in the two genes, as in row 17, column F; row 19, columns O–R; row 27, columns S and T; row 28, column A; and row 33, column I. But the phylogenetic distance between these two organisms is too great to reveal much convincing data. Undoubtedly when the sequence of this gene from an advanced prokaryote such as *Halobacterium* or *Thermus* has been determined, the steps in development of this macromolecule will become evident.

4.6.3. The Organellar Major-rRNA Gene

Here in the sequences for the major-rRNA gene from the two types of eukaryotic organelles, great contrasts both in structure and phylogenetic relationships are obvious. That from the chloroplast is compellingly prokaryotic throughout its length, with occasional deviations indicative of its separate evolution. On the other hand, those from mitochondrial sources have diverged so strongly from their prokaryotic forebears that almost no correlation remains between the respective sequences. Perhaps a fuller discussion will make the contrasting conditions clear.

The Chloroplast Gene. Two major-rRNA genes from different sources of chloroplasts have had their sequences established, but that of the tobacco (Takaiwa and Sugiura, 1982b) given in Table 4.7 brings out the essential features so thoroughly that inclusion of the gene from the maize plastid (Edwards and Kössel, 1981) would serve no useful purpose. Here, as in the case of the blue-green alga and *E. coli* genes, the 5'-terminal region is excluded to compensate for the loss of that part to the supplemental gene, as discussed earlier. As a whole the homology is so extensive that alignment with corresponding regions from the two prokaryotic sources is readily accomplished, a condition testified to by the frequent inclusion of all three of these genes in the blocked sectors indicative of universal identity.

Comparisons extending throughout the entire lengths of the three chloroplast genes offer no clear proof of specific kinships. On the one hand, sectors occur that are identical to the corresponding region of the *E. coli* gene, others that indicate common ancestry with the blue-green alga, and still others that are distinct from both. However, those shared with *Anacystis* are undoubtedly predominant, as one should expect between photosynthetic organelles and a possible predecessor. Why direct descent of these structures from a nonphotosynthetic species like the coliform bacterium should ever be suspected is difficult to comprehend, yet comparisons between those two types are frequent in the literature. The overall picture of similarities and differences that are revealed in the table is more harmonious with a concept that the chloroplast genome is probably a fragment (plasmid) of the genome of an advanced photosynthesizing prokaryote passed on to the ancestral stock of the Eukaryota along with the remainder of the genome.

The Mitochondrial Gene. To show the impossibility of drawing equally firm conclusions regarding the phylogenetic origins of the mitochondrion on the basis of primary structure of this gene, the entire sequence from the human organelle is included in Table 4.7, together with ~150 nucleotides from each of the corresponding regions of mitochondria from yeast, *Paramecium,* and mouse (Anderson *et al.,* 1981; Bibb *et al.,* 1981; Seilhamer and Cummings, 1981; Sor and Fukuhara, 1983). As in the chloroplast, all begin after due allowance for deletion of the 5' end to form a supplemental gene in eukaryotes. In this opening region, homologies between these organellar sequences and that of *E. coli* are occasionally detectable, as in row 1, column B, where relationships between the prokaryote and the yeast mitochondrion can be noted, and in column D, a constant region, where kinships between the bacterium and the *Paramecium* and murine mitochondrial sequences are evident. That of the ciliate organelle displays three correspondences with conserved areas in row 2, where the others show none. The full human structure similarly has rare regions that are homologous to the invariant sites, as in rows 6, 47, 51, and 53. But as a whole, the primary structures are seen largely to lack meaningful correspondences, between both themselves and the others that have been established, an observation made also by Machatt *et al.* (1981).

This lack of recognizable homologous sectors stems from two major processes that have been active in mitochondrial gene evolution. First are the changes in size that have occurred here as in so many other areas of the organellar genome. For example, the mouse mitochondrial gene of 1583 residues is obviously nearly 50% shorter than the 2904 of *E. coli,* and the 2204 residues of the *Paramecium* organelle represent nearly a 25% reduction from that of the prokaryote. In the yeast, in contrast, the gene has increased in size so that it embraces 3273 nucleotide residues, representing an approximately 10% growth (Sor and Fukuhara, 1982). In addition, the mitochondrial genes of many strains of the latter organism contain a large intron of 1132 base pairs, and that of *Neurospora crassa* has one of 2284 (Burke and RajBhandary, 1982).

But a still greater contribution to the large degree of uniqueness on the part of these organellar genes is their reversal from a predominantly $G + C$ content that characterizes the prokaryotic and vertebrate sequences—ranging between 80 and 85% in the latter (Hassouna *et al.,* 1984)—to nearly 80% $T + A$ in the yeast organelle gene (Sor and Fukuhara, 1983). Thus on the basis of this complete reversal of nucleotide proportions, there can exist at most a 20–30% homology between the contrasting source types, that is, approximately the rate expected from random-chance combination.

4.7. TRANSCRIPTIONAL TERMINATION IN CLASS I GENES

Despite the sophistication and skills of the biochemist, the processes of termination in those genes transcribed by RNA polymerase I in eukaryotes and their relatives in prokaryotes and organelles remain virtually unknown. In the prokaryotes transcription terminates beyond the gene for the ancillary rRNA and thus already has been discussed in Chapter 3. In mitochondria the transcriptive terminals have in many cases not been fully determined and in others they end beyond a tRNA gene, as was shown in Chapter 2. Hence, tabulations of the 3' trains can meaningfully include only those from eukaryotic and chloroplastic sources.

4.7.1. Termination in Prokaryotic rRNA Operons

Even these restricted resources provide a few trains that might include termination signals; those that are available are listed in Table 4.8 together with corresponding sectors from *rrnF* and *rrnD* of *E. coli* (Sekiya *et al.*, 1980; Morgan, 1982). Transcription termination has been demonstrated in the first of these bacterial operons to be dependent, at least *in vitro*, on the presence of the rho factor and to occur in the region of the eight adjacent thymidines (underscored) following a hairpin formation, italicized in the table. This region follows a tRNA[Asp] gene that lies slightly downstream from the ancillary rRNA cistron of the operon. The putative terminator of a second operon from this organism, that of *rrnD*, differs somewhat in being placed downstream at some slight distance from the 3' end of the second ancillary rRNA gene which characterizes that structure (Duester and Holmes, 1980). Nevertheless, the two proposed terminators show frequent homologous sites and general agreement in the placement of the stem-and-loop dyad symmetry and run of eight Ts. However, in the present instance the latter are interrupted in the center by a guanosine residue.

4.7.2. Termination in Eukaryotes

In the *Physarum* train sequence provided in Table 4.8, there is neither an evident stem-and-loop structure nor a run of thymidine residues to indicate a possible point for termination. However, in the region 13–7 base pairs upstream from the end of the mature coding region, there is a series of seven Ts, as shown in Table 4.7, row 63. That this sector could be involved in termination is suggested by the report on the processes of transcription in *Xenopus* (Sollner-Webb and Reeder, 1979). In that investigation the primary transcript, sedimenting at 40 S, was demonstrated to terminate at the same point as the mature product; hence, according to these data termination must depend on a signal close to the last site of the mature gene. But what signals the end of transcription is not clear. Perhaps it is the row of four Ts that immediately follows the coding region, for the cistron lacks a series of Ts in excess of three adjacent ones in at least the 260 sites at its 3'-terminal region.

In other copies of the *Xenopus* rRNA gene, termination was normal when clusters of only three Ts were present in the train, but reduction to two or less interfered with that process (Bakken *et al.*, 1982; Sollner-Webb *et al.*, 1982). Thus there must be some processing at this end of the transcript, despite the earlier reports. The transcripts of mouse rRNA operons likewise have a train; in the single case that has been investigated it was 30 sites in length. As may be noted in Table 4.8, the T marking the stop point abuts directly against a stem-and-loop structure, as in many other terminators, but in this organism only a single T apparently serves in the latter capacity (Kominami *et al.*, 1982). In *Tetrahymena thermophila* termination of transcription occurs in a more orthodox

Table 4.8
Trains of the Major rRNAs or Their 4.5 S Adjuncts[a]

Escherichia coli rrnE[b]	CTTATTAAGAA	CGGTCGAGTT--AAC	*GCTCGAGGT*	TTTTTTTCG	TCTGTATAT	CTATTATTG
Escherichia coli rrnD[c]	ATAAACAAAAG	*G-CTC-AGTCGGAAG*	*ACTGGGCCT*	TTTGTTTTA	TCTGTTGTT	TGTCGGTGA
Physarum polycephalum[d]	--GCCTGGGGA	GGGCCTAATA--ATG	CCCCGTCCC	TTGTGTTAA	CGGGGCGTT	ACAAGTCGT
Xenopus[e]	TTTTGTCGGAA	GGAGCAGGCGCGGAAG	GGGCGCCCC	GCCGCCGGC	CGGCGCGAC	GTCCCGTCC
Mouse[f]	----------TC	T-CTGCGGGCTTTCC	CGTGGCACG	*CCCGCTCGC*	TCGCACGCG	ACCGTGTCG
Tetrahymena thermophila[g]	----------	----------	--CTCATCT	CCCTTT--A	TTTTTTT	
Tobacco chloroplast[h]	CTTGTTCCTAC	ATGACCTGATCAACT	*TGATCAGGC*	ACTCGCCAT	CTATTTTCA	TTGTT
Hamster mitochondrion[i]	----------	---ACCCTAG-ACA-	*-AGGGTTTA*	TTAGGGTGG		

[a]Regions of dyad symmetry are italicized; underscored sequences are terminators.
[b]Sekiya *et al.* (1980).
[c]Duester and Holmes (1980).
[d]Otsuka *et al.* (1983).
[e]Sollner-Webb and Reeder (1979).
[f]Kominami *et al.* (1982).
[g]Din *et al.* (1982).
[h]Takaiwa and Sugiura (1980, 1982a).
[i]Kotin and Dubin (1984).

fashion, taking place at the adenosine preceding a run of six adjacent Ts and following a cluster of three of those residues (Din *et al.*, 1982). Thus while cessation occurs at a determined site, it has not been established which series of thymidines triggers that process.

4.7.3. Termination in Organelles

The train of the tobacco chloroplast gene similarly fails to throw much light on terminating transcription, for the longest run of Ts has only four members (Takaiwa and Sugiura, 1980). This sequence, underscored in the table, lies about 13 sites downstream from the end of a hairpin structure (indicated in italics). Thus the chloroplast termination processes may prove to parallel those of the prokaryote and eventually may even be demonstrated to be dependent upon the presence of some factor similar to rho.

In the mitochondria of hamster, as perhaps also those of other mammals, transcription termination has been proposed to be triggered by a signal within the 3'-terminal region of the mature major-rRNA cistron (Kotin and Dubin, 1984). Beginning at the 17th site preceding its very end, there is a potential stem-and-loop structure ending in a run of three Ts, the last of which marks the site where the majority of the transcripts terminate. Other sites of termination are similarly shown underscored. Thus the proposal made earlier for the *Physarum* gene receives indirect support. However, it is quite evident that the control of the processes of transcription in class I genes and their allies has been poorly explored and that much more needs prompt attention.

5

Nucleic Acid-Associated Protein Genes

This chapter begins a rather lengthy section devoted to the genes transcribed in eukaryotes by DNA-dependent RNA polymerase II and their prokaryotic and organellar counterparts that continues through Chapter 9. By far the majority of these class II genes encode proteins of numerous types, and hence the transcripts are messengers. In addition, the group contains those for small nuclear RNAs best represented by the U family in the human genome, along with a heterogeneous mixture of cistrons for inserted elements, but these are not viewed until the steps of processing are discussed in Chapter 12, for they are concerned in that activity. The first protein genes to be explored here are those that are associated *in vivo* with DNA, namely the histones and relatives, or with RNA, such as the ribosomal proteins. The treatment of the latter types at this point close to the examination of the rRNAs appears a logical step; moreover, the genes for histones seem equally essential prior to a discussion of any other proteinaceous substance, since their products have often been claimed to be active in the control of transcription. Since such regulation has also been considered to be manifested by way of those histone-containing bodies called nucleosomes that adhere to DNA strands (Olins and Olins, 1974), it is expedient to discuss those structures first and then proceed to the genes themselves.

5.1. HISTONES AND THE NUCLEOSOME

There are five virtually universal types of histones present in eukaryotes, H1, H2A, H2B, H3, and H4, plus a sixth one, H5, which is confined to blood cells of vertebrates. Since the first and last of these do not enter directly into the formation of the chromatin-associated nucleosome, they need receive no further mention for the present. When the respective members of the family are analyzed later, their structural characteristics are fully detailed, but in order to understand nucleosomal properties, it needs to be pointed out now that two of the histone family, H2A and H2B, supposedly resemble one another in being rich in lysine, and that two others, H3 and H4, are viewed as having an abundance of arginine, both of which amino acids are basic in reaction. However, structural analyses later cast doubt upon the validity of these categorizations.

5.1.1. The Nucleosome and Chromatin Structure

As is the universal practice in preceding discussions of transcription of eukaryotic genes, the DNA is viewed simply as double linear sequences of deoxyribonucleotide residues. However, a truer view is that large portions of the duplex molecules are wound around a cluster of histones to form the nucleosomes just mentioned. When such chromatin is subjected to extensive digestion with nuclease, "core" nucleosomes are released, consisting of an octamer comprised of two molecules each of histones H2A and H2B, and of H3 and H4. Around this complex is wound 1 3/4 turns of duplex DNA, requiring a total of 140–186 base pairs of this latter substance (Barnes *et al.*, 1982). The resulting body is in the form of a low cylinder, 110 Å in diameter and 55 Å high (Finch *et al.*. 1977; Pardon *et al.*, 1977; Klug *et al.*, 1980; Derenzini *et al.*, 1983a). In intact particles, the ends of the DNA strands are secured by either histone H1 or one of several nonhistone proteins, typically protamines or a member of the "high-mobility group" (HMG). Because of the off-center emergence of the DNA duplex molecule, the nucleosomes appear, not like beads on the genomic structure, but more like the charms of certain bracelets. It is of special interest to note that nucleosomelike bodies are found in the complex DNA viruses, including adenovirus. These particles, however, do not contain octamers of histones, but are constructed solely of a viral protein referred to as polypeptide VII (Vayda *et al.*, 1983).

The typical eukaryotic organization of chromatin just described is subject to great changes, depending on the activity of a given sector of the genome. Actually the nucleosomal bodies have been believed by many workers to be characteristic only of transcriptionally inactive strands, whereas regions being transcribed have been considered to be in an open configuration, free of nucleosomes, at least in ribosomal RNA genes (Franke *et al.*, 1978, 1979; Mathis *et al.*, 1981). More recently, however, the extended regions of nucleosome-free chromatin have been demonstrated to be a permanent condition, unrelated to transcriptional activity (Derenzini *et al.*, 1983a,b). Nevertheless, it remains well-established that active genes are more readily digested by DNase I and micrococcal nuclease (Mathis *et al.*, 1981; Spinelli *et al.*, 1982; Weisbrod, 1982; Bryan *et al.*, 1983), so that some change in tertiary structure or supercoiling of the chromatin fibers appears to accompany transcription. A role in this alteration of configuration has been ascribed to histone H1, possibly by interaction between adjacent molecules of this protein. The feasibility of cross-linking of these molecules has lately been explored experimentally, resulting in a demonstration that such connections could occur between the COOH-terminal regions of neighboring H1 molecules (Ring and Cole, 1983). Moreover, the NH$_2$ end also was found to form bridges with the base of the COOH portion, so that an arrangement perhaps similar to Figure 5.1 might result. Other workers in this area, it should be noted, have demonstrated that in addition to histone H1, two others, H2A and H2B, together or separately play an important part in determining the length of the spacer DNA (Kornberg, 1977; Oudet *et al.*, 1977; Zalenskaya *et al.*, 1981). However, a more recent investigation employing a high-resolution scanning transmission electron micro-

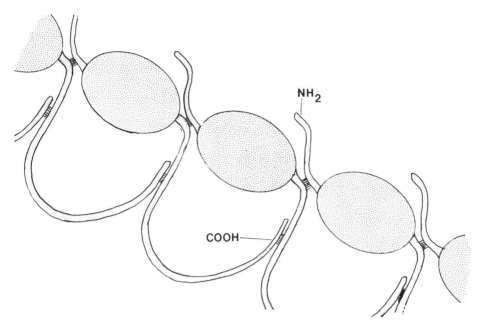

Figure 5.1. Possible arrangement of H1 histone molecules on DNA. Only the former are shown, their 5' ends bonded to the 3' and the trains also bound to one another. (Based on Ring and Cole, 1983.)

scope (Stoeckert *et al.*, 1984) has shown that H2A along with H3, rather than H2B, are associated with H1 (or H5) at the points of departure of the DNA from the core nucleosome (Figure 5.2).

The range in base-pair count given in a preceding paragraph is an artifact of differing techniques rather than a condition innate to the nucleosome (Simpson, 1978; Allan *et al.*, 1980; Zalenskaya *et al.*, 1981; McGhee *et al.*, 1983a; Ramsay *et al.*, 1984; Bavykin *et al.*, 1985). Variation in length also is associated with the configuration of the DNA, being shorter in the B form than in the Z form (Miller *et al.*, 1985). Moreover, it is sometimes proposed that the DNA molecule makes two complete turns around each core nucleosome, rather than the 1 3/4 (Burlingame *et al.*, 1985). Regardless of which set of details may prove to be correct, native chromatin appears to have a well-defined structure in the form of shallow supercoils known as "solenoids" (Finch and Klug, 1976). The DNA filaments have a pitch of ~11 nm, a diameter of ~30 nm, and include an average ~6 nucleosomes per solenoid turn. But this number, too, is highly variable, because the DNA spacer between these bodies ranges from perhaps as little as one or two to nearly 80 or 100 base pairs, depending on the source material (Stein and Bina, 1984); in vertebrate tissues the range in length provides ~1/4 to ~1/2 turn of the duplex DNA between nucleosomes. Apparently the resulting 30-nm supercoil is identical to the thick fiber that is characteristic of eukaryotic chromatin *in vivo* (Olins and Olins, 1979; McGhee *et al.*, 1983a; Butler, 1984).

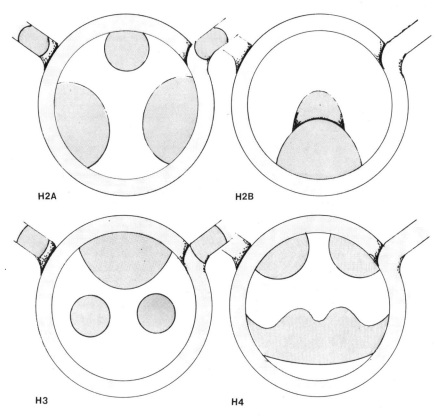

Figure 5.2. Distribution of histones within the nucleosome and on the adjacent DNA duplex molecules, as demonstrated by electron microscopy. (Modified from Stoeckert *et al.*, 1984.)

5.1.2. Histone Synthesis and Nucleosome Assembly

Since the suspicion exists that histones play important roles in control of transcription, examining the synthesis of these proteins and their subsequent incorporation into nucleosomes and chromatin offers some advantages. But as in the preceding section, knowledge of these subjects is in a state of flux, with many contradictory experimental results, depending on the model and techniques employed.

The Synthesis of Histones. In broad terms, the synthesis of histones is coupled to the cell cycle, taking place in the S phase when the DNA is being replicated (Gallwitz and Mueller, 1969; Gurley *et al.*, 1972; Schofield and Walker, 1982). This condition has been found to prevail in a number of organisms, including, by way of example, *Physarum polycephalum,* Chinese hamster, and the human (Robbins and Borun, 1966; Gallwitz and Mueller, 1969; Wu and Bonner, 1981; Kelly *et al.*, 1983). In mammalian cells at least, two histone synthetic events occur, depending on the class of these substances being produced. One group, referred to as replication variants, are synthesized in conjunction with DNA replication in rapidly dividing cells, and thus are confined to the S phase

(Kinkade and Cole, 1966; Franklin and Zweidler, 1977). However, the second category, the replacement variants, are present in enhanced levels in differentiated tissues, whose cells are not undergoing mitotic division, and are synthesized principally in the G_1 phase (Sittman *et al.*, 1983a). This latter set of processes is essential to the continuous turnover of histones that takes place in such cells, along with all the macromolecules of every type (Djondjurov *et al.*, 1983). Regulation of the synthetic steps appears to be at the levels of transcription and mRNA degradation (Sittman *et al.*, 1983a; Alterman *et al.*, 1984). More recent developments, moreover, have demonstrated that the replacement varieties, now better known as the "basally expressed" or "constitutive" types, differ in sequence structure from the cell-cycle-dependent ones (see Sections 5.3.1 and 5.3.4), have introns, and their mRNAs are polyadenylated (Wells and Kedes, 1985). In yeasts, too, synthesis of histones, although coupled to DNA synthesis in vegetative cells, proceeds continuously during sporulation, that is, in meiotic specimens (Marian and Wintersberger, 1980). There is some evidence now that the time of synthesis may also be related to the specific type of histone. One example of this sort is provided by H4 of *Physarum polycephalum,* which is transcribed in late G_2 as well as in S phase, although translation appears to be confined to the latter period (Wilhelm *et al.*, 1984). Moreover, coupling of histone and DNA synthesis has been reported to be dependent upon the production of another, unknown protein (Helms *et al.*, 1984).

The Assembly of Nucleosomes. The assembly of histones into the form of core nucleosomes involves the distortion of the DNA double helix to produce a condensed superhelical structure, the forces involved sometimes being suggested to be largely electrostatic. The negatively charged phosphate groups of the DNA duplex molecule tend to interact with the positively charged amino acids lysine and arginine that occur abundantly in the histones and other nucleic acid-binding proteins. According to one account (Jordano *et al.*, 1984a,b), the two histones H3 and H4, that are usually considered especially rich in arginine residues, are electrostatically more strongly bound to DNA than are the other two (H2A and H2B), that are particularly rich in lysine and are also essential to the formation of nucleosomes. Thus, in this hypothesis, the first step in the assembly of components for these particles is the formation of a tetramer of H3 and H4, followed by the distortion of the DNA molecule into the form of a prenucleosomal body. However, the same studies showed that the next step, the assembly of two copies of each lysine-rich histone into a tetramer and the latter's incorporation into the prenucleosome, was also essential for the formation of the particles as well as to their subsequent stabilization. In other investigations, the core histones are described as being arranged singly along the DNA molecule at regular intervals (Schick *et al.*, 1980). But these are only the basic reactions.

One especially pertinent aspect of the formative processes of these nucleosomes is the relationship of newly developed particles on freshly synthesized DNA. In researches into this subject, the genome of simian virus 40 (SV40) has proven to be especially profitable, for its relatively small DNA macromolecule is supercoiled and complexed with 21 octameric histone units to form what is known as a minichromosome. During replication, one strand of the DNA is synthesized continuously, whereas the other is produced discontinuously as Okazaki fragments (Närkhammar-Meuth *et al.*, 1981). Examination of the minichromosomal replicatory processes with various reagents determined that the continuous strand of DNA retained the parental nucleosomes and that all newly created particles became associated with what was synthesized by way of Okazaki intermediates (Roufa and Marchionni, 1982). Recently, however, some doubts have been cast on the validity of these conclusions, using the same experimental model. The more recent results indicate that both parental and nascent nucleosomes are distributed to the two strands of DNA during replication (Cusick *et al.*, 1984). In nascent and parental DNA strands alike, the histone-containing bodies are distributed in a nonrandom fashion, for the end of the minichrosome encompassing the origin and replication sites and the transcriptional control along with its enhancer element are devoid of nucleosomes (Cereghini and Yaniv, 1984). In the ribosomal genes of *Tetrahymena*, which will be recalled as being arranged as inverted pairs on linear DNA plasmids, it has proven feasible to show that the nucleosomes were likewise nonrandomly arranged (Palen and Cech, 1984); a similar condition exists in yeast (Thoma *et al.*, 1984). Because the transcriptional initiation sites on each strand are downstream from the centrally located origin of replication, it was possible to demonstrate that the latter area contained nucleosomes, whereas the former regions did not. Consequently, it was made evident that more is embraced in the assembly of these bodies than purely mechanistic interactions involving electrostatic forces. Later this statement is found to be reinforced by consideration of the ribosomal proteins, which are similarly rich in lysine and arginine, but *in vivo* bind only to RNA and not to DNA.

Other Facets of Assembly. A number of other facets of nucleosome formation require consideration. Since histone synthesis takes place on ribosomes in the cytoplasm, while chromatin formation proceeds within the nucleus, transportation of those proteins through the nuclear membrane is requisite. But such transport is not necessarily immediate, for histone storage pools have been detected in both embryonic and somatic cells (Groppi and Coffino, 1980; Shih *et al.*, 1980). Moreover, after transport into the nucleus, the various species of core histones do not interact with the DNA on the same time scale. In contrast to the sequence of events predicated on the basis of electrostatic charges, the two so-called lysine-rich types enter into chromatin formation immediately, but those considered arginine-rich experience a prior lag period (Seale, 1981). Nor are all the histones found to be associated with the DNA in the form of nucleosomes. About 30% of H2B, half of H4, and all of H3 were shown to be readily dissociated, whereas those fractions in core nucleosomes were not.

In connection with the nonmechanical association of histones and DNA, it is pertinent to note that proteins other than histones are involved in nucleosome formation, as shown by a small number of investigations. In the oocytes of *Xenopus*, a substance called nucleoplasmin, having a molecular weight of 29,000, was demonstrated to interact with histones to induce the assembly into those bodies (Earnshaw *et al.*, 1980). This substance has also been found in mammalian cells (Krohne and Franke, 1980). More recently a protein of 59,000 daltons extracted from HeLa cells was found to be active in nucleosome formation when combined with octamers of the four core histones and DNA (Ishimi *et al.*, 1983). Moreover, it has recently been clearly demonstrated that the assembly of nucleosomes and the supercoiling of the DNA duplex molecule are separate events. By use of an extract from *Xenopus* oocytes, supplemented with ATP and Mg^{2+}, the latter process was shown to be an active complex reaction, driven by ATP and requiring at least two enzymes (Glikin *et al.*, 1984; Ryoji and Worcel, 1984). Hence, it is probable that nucleosome formation, like nearly every other known cell activity, is enzyme-mediated, not strictly mechanistic (Dillon, 1981, 1983).

5.1.3. Nucleosomes and Transcriptional Control

As already stated, histones and the nucleosomes they form have been proposed to be active in the control of transcription, at least to the extent of loosening the supercoils of the DNA of prokaryotes, as well as eukaryotic chromatin. One factor that has been suggested as being involved in the latter substance is that the chemical composition of the nucleosome itself may play a role, for these structures are highly heterogeneous in composition as a result of the differing accessory proteins associated with the cores. Another concept is that the nucleosomes unfold in areas undergoing transcription (Moyne *et al.*, 1982). Apparently histone H1 is the most prevalent in regions not being transcribed (Albright *et al.*, 1980; Egan and Levy-Wilson, 1981), whereas in areas undergoing transcription the HMG nonhistone proteins 14 and 17 abound (Weisbrod and Weintraub, 1981). These are never present with H1, but other proteins also frequently enter into nucleosome formation. Furthermore, there are two major types of all the histones (aside from the replication and replacement variants referred to earlier), one associated with euchromatin, the other with heterochromatin (Levy *et al.*, 1982).

Unfortunately these differing classes are rarely considered in many experimental studies. Among those that overlooked the heterogeneity of structure was an investigation on the localization of 5-methylcytosine in DNA. The results showed that 80% of that modified nucleoside is found in nucleosomes that contain histone H1, while those that have HMG proteins instead are undermethylated about twofold (Ball *et al.*, 1983). Researches that have escaped this failing deal with the phosphorylation of H1, in particular stressing the distinctions in this property between two subtypes, histones H1A and H1B (Ajiro *et al.*, 1981a). During G_1 of the cell cycle, H1A molecules contained at most one phosphate group, while H1B had from 0 to 3.1. With the approach of S phase both varieties gained about one phosphate per mole. Then during mitosis further increases of three to four groups per mole were noted in both types, but the level of phosphorylation was always lower in H1A than in H1B, reflecting a possible distinction in function. Another study had demonstrated that the several subtypes of H1 are associated with chromatin having different degrees of condensation, H1C being related to a more strongly compacted chromatin than were H1A or H1B (Huang and Cole, 1984). The possibility exists that the various levels of condensation resulted from the subtypes, rather than being an effect of the condition of the DNA complex.

Chemical Modification. Another chemical reaction with the core proteins, in this case with their amino ends, has long been correlated with control of gene activity (Allfrey, 1980). The acetylation that occurs posttranslationally at specific lysine residues has been considered to loosen the DNA supercoiling, thus exposing the genes for transcription. However, a current study has demonstrated that acetylated chromatin was only slightly less compacted than the control substance, so that hyperacetylation of these proteins did not open the solenoids to any extent (McGhee *et al.*, 1983b). Further, transcribing chromatin has been demonstrated not to be preferentially enriched with acetylated histones (Yukioka *et al.*, 1984), and the level of modification has been reported to decrease during embryonic development of the sea urchin (Chambers and Shaw, 1984). In fact, the increased sensitivity to DNase activity that results from loosening of the solenoid coils is now being viewed as only one of several conditions needed for transcription to occur (Mathis *et al.*, 1981; Kuo *et al.*, 1982). One recently demonstrated requirement is the presence of nuclear actin, at least in the transcription of lampbrush chromosomes (Scheer et al., 1984).

A second type of chemical modification that occurs *in vivo* is the methylation of lysine residues near the 5' end of arginine-rich histones, particularly H3 (Branno et al., 1983), a reaction carried out by a methylase in the presence of S-adenosyl-[methyl-^{3}H]-methionine. The enzymic activity was associated with specific embryonic stages of the sea urchin, particularly the mesenchymal blastula and early gastrula. Whether the changes observed were correlated to transcriptional activity could not be established.

Still another type of modification has been shown to occur to histones, one that is strongly correlated to the mitotic processes (Mueller et al., 1985). In *Physarum*, H2A and H2B react with ubiquitin to form the conjugated proteins designated as uH2A and uH2B, which interact with one another and enter into nucleosome formation in normal fashion. Through use of specific antibodies, it was found that the ubiquinated species disappeared at metaphase but were reformed at anaphase.

5.2. CONTROL OF HISTONE GENES

Since there are so many types of histones, clarity of discussion may be gained by examination of their common structural features, followed by discussion of the differences that exist between the several types. Such shared characteristics are in the main confined to the control signals. Consequently, the promoters and associated signals of the 5' leaders receive prime attention; then, after the mature gene sequences have been examined, the terminators of the 3' trains become the center of focus. But all these topics need to be prefaced by a discussion of the arrangement of the genes within the genome.

5.2.1. The Organization of Histone Genes

Histones have been a subject for investigation over several decades, so that the organization and structure of their genes from many organisms have become well established (Hentschel and Birnstiel, 1981; Maxson et al., 1982). Indeed, because of this extensive knowledge, their arrangements have already received limited discussion in the introductory chapter (Chapter 1, Section 1.2.1; Figure 1.9).

Once thought to be strictly a eukaryotic biochemical, histonelike proteins have now been reported from several prokaryotes. First among these was that called HU from *E. coli* (Rouvière-Yaniv and Gros, 1975; Berthold and Geider, 1976); subsequently, several others have been described from the same source (Lathe et al., 1980). Still more recently, others with greater histonelike qualities have been demonstrated in several advanced bacteria, including *Thermoplasma* and *Sulfolobus* (Lathe et al., 1980; Green et al., 1983). As yet, however, nothing resembling a nucleosome has been reported from any prokaryotic source, but the newer histonelike protein NS1 and NS2 structures act on DNA, producing a scaffolding effect (Lammi et al., 1984).

The Histone Genes of Yeast. In the first place, the organization of histone genes in *Saccharomyces* and close allies is distinct from that of all other known eukaryotes in lacking H1, as reported in Chapter 1, Section 1.2.2 (Marian and Wintersberger, 1980; Certa et al., 1984). This deficiency is one of the many traits that establishes those organisms as being at the most primitive level of the Eukaryota. Moreover, it undoubtedly contributes to a second of its primordial characters, the absence of chromosomes. Although the genome of yeasts is divided into several distinct linkage groups, the chromatin has never been demonstrated to condense into the typical chromosomal structures of more advanced forms (Dillon, 1962, 1981, p. 561). Much information concerning the role of this species of histone in formation of chromosomes and nucleosides could be gained if these organisms were used as the models for appropriate experimentation.

Aside from this absence, little has been established as to the genomic organization of the histone genes. However, those for H2A and $H2B_1$ form one operon 6000 base pairs in length and those for H2A and $H2B_2$ another of 13,000 base pairs, neither of which is repeated (Hentschel and Birnstiel, 1981). H3 and H4 are united in a separate operon, present in two copies (Turner and Woodland, 1983). In each gene set the cistrons are of opposite polarities.

Histones from a few other lower eukaryotes have been investigated, but the species of these proteins have not always been identified. In one of the exceptional studies, four types were identified, including H1, H3, and H4, along with a novel form (H01) that took the place of H2A and H2B in the nucleosome core (Rizzo et al., 1985). In a second, treating these proteins in *Euglena,* the five usual ones of vertebrates were isolated, although H2A had a higher lysine-to-arginine ratio than typical and H1 appeared to be deficient in aspartic acid (Jardine and Leaver, 1978). In the dinoflagellates, whose cellular phylogenetic level is close to that of the euglenoids (Dillon, 1981, p. 561), the chromosomes are permanently condensed, but lack histones of all types (Rizzo, 1981; Rizzo and Morris, 1984). Instead they have one (or more?) histonelike proteins that may serve in a similar capacity as the present type; the species known as HCc has had its amino acid sequence established.

Histone Gene Organization in Metazoans. Among the nonvertebrate metazoans, two types of organisms have been explored quite thoroughly in regard to the organization of their histone genes, *Drosophila* and the echinoderms. In the fruitfly, the cistrons are arranged in operons, each containing a single gene for all five of the common species (Figure 5.3), and are present in 100–110 copies (Lifton et al., 1977). These fall into two size categories, one of 4750 base pairs, the other of 5000, the latter usually being the more abundant class (Strasbaugh and Weinberg, 1982). In each case the gene arrangement is the same, the size difference being the result of an addition of 250 base pairs to the long intergenic spacer (Kedes, 1979). Since the operons are dispersed, but contain related genes, they fall into class II, category IV of gene organization. Two of the cistrons, those for H4 and H2B, are of opposite polarity from the others (Figure 5.3).

The histone coding organization in echinoderms has been thoroughly documented in two sea urchins, but unfortunately some premature generalizations were made in the earlier literature that have not withstood the test of time. This is true in particular concerning length and sequence heterogeneities, which were considered lacking at one period (Kedes, 1976; Overton and Weinberg, 1978). In *Lytechinus pictus,* three different operons have been demonstrated among the early genes alone (Cohn and Kedes, 1979; Holt and Childs, 1984). All of these have basically the same order of coding regions, but two (AD and BC) are fragments of the third (E), as suggested in Figure 5.3. Hence, the fundamental gene order is (H2A, H1, H4, H2B, $H3_2$), all of which have the same polarity (Holt and Childs, 1984). Among the major distinctions of the sea urchin early genes from

A.-C. Lytechinus

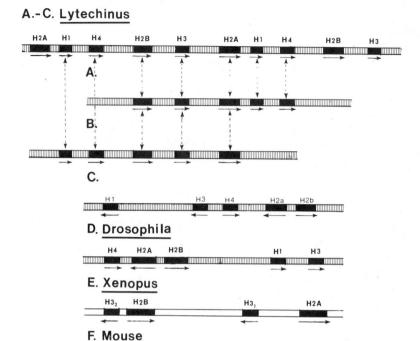

Figure 5.3. The organization of histone genes in the genomes of various metazoans. (In part based on Kedes, 1979; Old *et al.*, 1982; Alterman *et al.*, 1984; and Holt and Childs, 1984.)

those of *Drosophila* is that they are of category V arrangement, being tandemly arranged; each species is represented by 300–600 copies (Old and Woodland, 1984). The late genes of *Strongylocentrotus purpuratus*, however, have been shown to be like those of the insect in belonging to category IV. Not only are the operons dispersed, but they also are distinct from the early ones in being represented by only 5–12 copies per genome (Maxson *et al.*, 1983). Few of these studies have taken into account any possible variation in gene sequences within a given histone species, such as has been so clearly demonstrated to exist in *Tetrahymena* (Allis *et al.*, 1979, 1980), as well as other organisms. One of the exceptions is the report by Spinelli *et al.* (1979), which presents evidence for the existence of two different early sets of these cistrons in *Paracentrotus lividus*.

Associated with these few developmental aspects of the histone genes is one that may reflect phylogenetic origins. In the unfertilized eggs of *S. purpuratus*, H1 mRNA was demonstrated to be strongly underrepresented relative to those of the other types (Mauron *et al.*, 1982). Transcripts of that species were present at a concentration only 10% as high as those of H2B and H3. Similar low titers of H1 have been reported during early embryogenesis of *Xenopus* as a result of a reduced rate of synthesis (Woodland, 1980). In that paper it was suggested that the chromatin of the embryo may be deficient in this histone. If the proposal proves to be correct, the eggs and early embryos of these and perhaps other metazoans should afford unusual opportunities for exploration of the functions of this type of protein. The parallel to the yeast for a similar use advocated earlier as well as the latter's lack of H1 appear to be too self-evident to require further comment.

Histone Gene Organization in Vertebrates. As is the frequent case, the major portion of the researches on histone gene organization has been conducted on two species of *Xenopus*, and to a smaller extent on the newt *Notophthalmus viridescens*. In the latter animal, the operons, which contain sequences for all five histones on 9000-base-pair segments, are represented by 600–800 copies per haploid genome (Stephenson *et al.*, 1981). Another salamander, the axolotl, has been reported to have 1600 copies of these gene sets (Old and Woodland, 1984). Among the unusual features is the separation of the individual repeats by extensive, but variable, amounts of satellite DNA. The two species of clawed toad, *X. laevis* and *X. borealis*, while basically alike in histone gene organization, show extensive variation from one individual to another (Old *et al.*, 1982). About 70% of the operons had the arrangement H4, H2A, H2B, H1, H3, with the second member having a polarity opposite to that of the others (Figure 5.3). During the early development of *X. laevis*, four variants of H2B and two of H4 were detected by electrophoresis (Van Dongen *et al.*, 1983). Although previously a reiteration frequency between 20 and 50 had been reported (Van Dongen *et al.*, 1981), 90 copies were noted later (Turner and Woodland, 1983); furthermore, three major varieties of H1 have been uncovered (Destrée *et al.*, 1984). Some evidence in the latter report suggests that the variant may be correlated to the operon type.

The organization of these genes in other vertebrate genomes has not been thoroughly established; however, it is clear that they are arranged into operons, which are not tandemly arranged, but are dispersed. Approximately 15 copies are present in the chicken and up to four times that number in mouse and man (Sugarman *et al.*, 1983; Old and Woodland, 1984). Both in the chicken and human genomes—and probably those of all mammals—the genes in each operon are of mixed polarities, so that both strands of the DNA are transcribed (Bruschi and Wells, 1981). In man at least three types of arrangements of the several histone cistrons have been detected, including one that bears two copies each of H3 and H4 (Sierra *et al.*, 1982); the latter apparently is located on the long arm of chromosome I (Green *et al.*, 1984).

5.2.2. The Leaders of Histone Genes

In Table 5.1 are aligned the leader sectors of more than 30 histone genes, largely from various vertebrates, but with an admixture of those of yeast, fungi, and sea urchins, and even one from a seed plant, to provide as broad a comparative base as possible. All are centered not only at the established or presumptive promoter sequence, but also at ancillary and cap sites insofar as feasible. The starting points for transcription that have been experimentally determined are shown underscored, while the other regions of importance in transcription initiation are in italics.

The Promoters. The promoters provided in Table 5.1 have the same features that have characterized those of preceding chapters—much agreement with the general formula TATA- and great divergence from one specific example to another. All these representative types have a pyrimidine at the first site (the second site in many of the actual signals), which is most frequently a T. At the beginning of the chart in H1 and following types, the second of those sites promises to be universally occupied by an A except for the presence of a T in the yeast II and *Neurospora* H3 sequences, but later in the leaders of H4 and H5 species, several Ts occur, along with a G in the *Physarum* represen-

Table 5.1
Leaders of Eukaryotic Histone Genes[a]

	Ancillary	Promoter	Cap site
H1			
Strongylocentrotus[b]	CTG TCTCCTC CCACGTACGCA ACAATGCC	*TTATATTGA* GCGGTTGCCGA	GCCGA-TGG --------- TTATTCGT +32
Xenopus XLHW8, first[c]	CTG GCCAATC AGGGACCCAGA AAAGATGA	*ATATAAGGA* GGTGCATATA	AACTGAAAG TTTAGATTT TAGTTCGC +21
Xenopus XLHW8, second[c]	TCA ACCAATG AAGTTT------ AA-------	*CTATAAAAG*	
Chicken[d]	GCG CCGTGCG GCGGGGGGGC TCTGCAGC	*GCACCAATC* ACCGGCGGGC	TCCGCTCTA TAA+25GCC CAGTGGTT +31
H2A			
Strongylocentrotus[e]	--- ------- --------- --CACCAG	*GTATAAATA* GCCACCAAAA	CGGTGCTGG GCA-----T CCATTCAA +64
Xenopus h22, first[f]	TGT CACAATG CC+19GATCCC GACGTTTG	*GTATAAATA* GCCACCAAAA	AAGATAGGT GGTCA---A CCATTCAA +64
Xenopus h22, second[f]	--- CACCAAC A--------GAT GGCGGTAT	*TCATGAAAT* C	
Human h5G[f]	--- TATGAAT TCCT +8 ACC TCCAGTCA	*GTATAAATA* CTTCTCTGCC	TTGCCTTCT AATGTAGTT TCATTACA +40
Chicken[g]	ACC CACCAAA TA+57ACGCCT TTCCTCCC	*TTATAACTG* CTTTTCATTG	GTTCAAATT CGATTC-+6 TCATTGGC +128
H2B			
Strongylocentrotus[h]	--- ------- --------- ----CCGT	*GTATAAAAA* GGAAAGGTTC	TCGC--TGG --------- CCATTCAC +72
Mouse[h]	CTC TGACAA- ------GGACAG CCACCGCT	*TTATTTAAA* GAGCAGG--AA	AGGAACGGA ACAGTT+30 TCATTCTC +10
Human h4c[f]	TTG CATAAG- --------- -CGATTCT	*ATATAAA-A* GGGCCTTGTC	ATACCCTGC --------- TACGCTT +26
Chicken[i]	TTA TCCAATC AG+20TCGATT -TTGCCCC	*TTAAATAG* GCGAGAGTGC	TCGCAGCCG --------- GCACTCCG +30

H3

Yeast I[j]	TTC	CTCAACC	TT-+5TTCTTT	CTTTCTAG	*TTAATAAGA*	AAAACATCTA	ACATAAATA	----------	----------	-TATAAAC	+7
Yeast II[j]	CCT	TCCAACT	-----GTTCTT	CCCCTTTT	*ACTAAAGGA*	TCCAA-----	----------	----------	----------	GCAAACAC	+7
Neurospora crassa[h]	-TC	GACCACC	AG------CGC	AGGTCATC	*ACTTAATCA*	TCGCTCA---	----------	---------T	CCATCAAA	+31
Strongylocentrotus first[b]	CAC	CAGGATC	C-------CGC	AGC---AC	*ATATAAATA*	GCTGAAAATT	GCCAG-TGG	TT------C	TCATTCAT	+49
Strongylocentrotus second[b]	TTG	ACCAATC	AA------GAG	AGC---TT	*TTACAAACG*	G				
Lytechinus 21[l]	TGA	ACCAATG	GG+29GCACGA	TACCATAG	*CTATAAGA*	TCAGCCGGGA	A--TTTCGA	AA------A	TCAGTTTT	+26
Lytechinus 13[l]					*TATAAATA*	GCGGTCGCGG	TTACTTTCG	AAAATCTTT	TCAGTTAT	+28
Lytechinus A[l]			GATCTCT	CGGAGCTC	*TTATAAATA*	GCGGTCATAT	TTTCAGCGG	CA------G	TCACTCAT	+51
Xenopus[m]	TTA	CATCATT	TA-+5TGTTTA	ACTCCATT	*GTAAATAAA*	CTTTCCACTA	CCAAAAAAA	AAA+44--G	CCAACCAG	+46
Mouse 3.1[h]	TGG	GAAGGAG	GGGTA+30GCC	ACCCAAGC	*GTACTTAAA*	GGCCAAAGTG	CGCTACTTA	-------G	GTATCTCA	+22
Mouse 3.2[h]	TCA	ACCAATC	AGGA-----GCA	TGTTCCTT	*CTATAAAGG*	AACCAGAACC	TAACCTC--	-------T	GCATTTCC	+15
Human h5B[f]	GGG	TCCAATA	GTT----GGTG	GTCTGACT	*CTATAAAAG*	AAGAGTAGCT	CTTTCCTTT	CCTC-----C	ACAGACGT	+31

H4

Yeast I[j]	AAG	ACCAATT	TGATG+25AAT	TCTTTTTC	*CTATAAATA*	CCAGATATT	TTTTCTATA	TGA-------	TGGTTTCC	+69
Yeast II[j]	TTG	TTCACTC	GCGCC+37TGC	GTATGTAG	*TTATATCAT*	ATATAAGTAT	ATTAGGA--	----------	TGAGGGGG	+104
Physarum[n]	TCA	CGAGCCC-	AG---------	CGGCAAGA	*GCGTTAAG*	ATGGCCCGAC	GTGCGATGT	CCGGTGC--	TCAACATC	+18
Neurospora crassa[h]	TGC	CGCAACC-	GC----------	--CGTGGG	*GTACTTAAA*	GTCCGCTCTC	TTCCCCGCC	TTCT----T	CCCACCTT	+64
Strongylocentrotus[o]	TCA	CCCAAGT-	CC----------	GCAATGGT	*GTAACAATA*	CTCGCTGCAA	TCCGGTTGA	GGC-----A	TCATTCGC	+62
Lytechinus 21[l]	TTG	ATGAACAG	A+39 CGTCCG	GAATAGGG	*GTATATATA*	CCGGTCGAAA	TCCTATCAC	G-------T	CAATTGA	+26
Lytechinus 19[l]					*TATATATA*	CCGCACGAAC	AGCAGAATT	GAGTA---T	CAGTTTGA	+26

(continued)

Table 5.1 (Continued)

	Ancillary	Promoter	Cap site
Human lambda 41, first[b]	GGT TTCAATCT -----GGTCCG ATATCTCT	*GTATATTAC* GGGGAAGACG GTGACGCTC CGATCGAXC	XXCTATCG +13
Human lambda 41, second[b]	GGG GACAATTG +11 CGCCGG CGCCCTCG	*GTTTTCAAT* C	
Human u4A[f]	TCC GCCAACTG TC-------- --------	*GTATAAAGG* CGCTGCCTCA GGTCAGAGG ------CC	A<u>C</u>AAAAGC +27
Chicken[d]	CCC TCCAGG-- +35 GGGCCC- CGCCCCTG	*GTTTCAATC* AGGTCCGACC ATACGCCAT AACACG+21	TCACT<u>G</u>GT +25
Wheat[q]	GCA TCCACGG +17 CAACCT- CTCGACCC	*CTTTAAGAC* GCCCTTCGCC CCACCCAGC AAATC-----	ACA<u>G</u>CACC +61
H5			
Chicken[r]	ACA GGCAGTCC T +6 GGTCCG TGCCGCAC	(*uCTTAAATu*) CCTGCTCGTG GCGACCGCGC GGCC------	GCAGA<u>C</u>GC +99

[a] See text for explanation of italics and underscore.
[b] Sures et al. (1978).
[c] Hentschel and Birnstiel (1981).
[d] Sugarman et al. (1983).
[e] Grosschedl and Birnstiel (1980).
[f] Zhong et al. (1983).
[g] D'Andrea et al. (1981).
[h] Sittman et al. (1983a,b).
[i] Grandy et al. (1982).
[j] Smith and Andrésson (1983).
[k] Woudt et al. (1983).
[l] Roberts et al. (1984).
[m] Ruberts et al. (1982).
[n] Wilhelm and Wilhelm (1984).
[o] Grunstein et al. (1981).
[p] Sierra et al. (1983).
[q] Tabata et al. (1983).
[r] Ruiz-Carrillo et al. (1983).

tative. The third site appears not to be critical in recognition by RNA polymerase II, for it is occupied by A or either pyrimidine, G alone being unacceptable here. Even more freedom of structure is permissible at the fourth site, for all of the common nucleotide residues are found here in one or more of the examples given. At the fifth position, A predominates strongly, but T is present there in 10 of the 38 sequences, while a lone C occurs in the second promoter of the human λ41 leader. Finally, in the sixth site all four common bases are again seen to be occupants. Thus the general formula for this promoter, even with the omission of rare exceptional occupants, becomes:

$$\begin{matrix} & A\,T & \\ T & T\,C & A \\ C & & - \\ & C\,A & T \end{matrix}\, -$$

If recognition depends solely or predominantly on the present signal, the transcribing enzyme must consequently be able to react equally with any one of 36 possible different combinations. Even in those cases where a second promoter sequence can be found farther upstream from the first, there usually are several differences in structure. For example, the second one for the *Xenopus h22* H2A gene has (T)CATGAA(AT), in comparison to the (G)TATAAA(TA) of the first. Similarly the human λ41 H4 leader has a combination (C)TATATT(AC) for the first promoter and (G)TTTTCA(AT) for the upstream one.

The Ancillary Protein Signal. In the representatives given, the signal that should interact with an ancillary protein is far more variable than the promoter, to such an extent at times that it cannot be recognized. Indeed, the very first example in Table 5.1 illustrates this condition, for no sequence present in the vicinity of its typical location even vaguely resembles the -CAAT combination that typifies it. As a matter of fact, few of the remaining species from *Strongylocentrotus* or other sea urchins correspond, the two exceptions being from histone H3 genes. Further examination of the leader sections discloses that no consistency prevails in these ancillary sites, either with regard to any given species of histone or the source organism.

The location of the proposed signal, too, is subject to wide variation. The area intervening between the upstream element and the promoter may be as short as five sites, as in the human *u4A* gene for histone H4, or as long as 73, illustrated by the H2A sequence from the chicken. These variations similarly lack correlation with either the protein species or the source organism. At present no explanation is available for all these variables, except that perhaps there are a number of different ancillary proteins present in each organism. But until greater effort is devoted to the search for these multiple factors and their respective sequence requirements, little progress can be made toward understanding the processes of transcription of the genes for this most important class of biochemicals.

The Cap Site. The location of the actual starting point of transcription where a cap is placed has been established for a number of these genes from a variety of sources. In the majority of cases where this cap site has been experimentally determined (shown underscored in the table), the principal or sole reactant is ATP. As may be noted, present data do indicate this to prevail uniformly in histones H1, H2A, H2B, and H3. However, among the genes for H4 considerable latitude in the requirements for cap formation is obvious, for the start sites of the yeast leaders at once suggest that any of the four common nucleotides can serve in this capacity. Although adenosine is employed in the majority of the remaining sequences, the cap site in the chicken H4 leader is a guanosine and in the H5 from the same source a cytidine.

The location of the cap sites with respect to the 5' ends of the mature coding sequences also is variable, with a minimum distance of seven sites in the two yeast H3 leaders to a maximum of 128 in the chicken H2A. For the greater part, nevertheless, its placement is in the range of 20–50 sites upstream from that point. This contrasts strongly both with the much lengthier leaders that were seen with the rRNA genes transcribed by polymerase I and the very short ones in those acted upon by polymerase III.

5.3. THE MATURE HISTONE CODING SEQUENCES

The current state of knowledge concerning the mature gene structures for the several species of histones is uneven, some types having examples from numerous sources established and others having only a few. This contrasts strongly with the primary structures of the polypeptides, for the amino acid sequences of most of these proteins have been abundantly determined. But even in those genes that have been most thoroughly explored structurally, the data do not cover all the principal aspects of the topic, because it is becoming increasingly evident that each species, at least in the advanced eukaryotes, actually embraces a number of subtypes that differ strongly from one another, as the discussions of vertebrate histones H1 and H3 demonstrate.

5.3.1. The Histone H1 Mature Gene

Subtypes of Histone H1. This variability within a given species is becoming an especially striking feature of the H1 histones. As shall shortly be seen, many subtypes should actually be considered separate species, rather than mere variants, because of the obvious lack of homology. For instance, a form known as H1°, found in nondividing cells, and thus a constitutive type (Section 5.1.2), has a region of 37 amino acid residues that reportedly resembles the corresponding sector of H5 rather than that of the typical H1 histones. Consequently, it was suggested that H1° and H5 should be placed in a category separate from the rest (Pehrson and Cole, 1981). However, it has also been demonstrated that H1° is represented by two subtypes in Chinese hamster ovary cells, H1°a and H1°b, that differ strongly in composition of at least one 20- to 30-amino acid sector (D'Anna *et al.*, 1981). H1° not only differs in sequence structure, but even in the rate of turnover (Pehrson and Cole, 1982) and in chromatin has a different distribution (Jin and Cole, 1985).

In the H1 fraction proper, at least five subtypes are generally recognized, the genes of which are expressed in correlation with the mitotic activities of the cell. H1A and H1B are the predominant types in early embryogenesis and in rapidly dividing cells. In the mouse, H1C, H1D, and H1E are produced by both dividing and nondividing cells, the second of which decreases in relative abundance with cell maturity (Lennox and Cohen, 1983); the first three are most effective in condensing DNA (Jin and Cole, 1985). H1A and H1B from HeLa cells have been reported to differ strongly in structure, the former being smaller and more hydrophobic than the latter; therefore the pair could be expected to differ in the reactions with DNA or other proteins, or both (Ajiro *et al.*, 1981b). Although specific details are lacking, the operons containing H1A were found to diverge in gene arrangement from those including H1B (Zernik *et al.*, 1980). Similar relations of the five variants have been described in *Xenopus* (Risley and Eckhardt, 1981) and trout somatic tissues (Seyedin and Cole, 1981). However, it has not been established how these various forms correlate to the replication and replacement types given in a previous section (see also Plumb *et al.*, 1984). Mammalian testicular tissue, moreover, has been reported to produce a specific variant, that from the boar being distinguished as H1T (Cole *et al.*, 1984).

As intimated in part in the opening section of the discussion of histones, H1 is displaced in chromatin to some extent and in a specific fashion by other proteins, mostly unknown. In sperm, however, the replacement is of a more specific and less varying nature, especially in lower vertebrates and nonchordate metazoans. For a case in point, the spermatozoan chromatin of the marine worm *Chaetopterus variopedatus* contains only two basic proteins, histone H1 and a species of protamine (DePetrocellis *et al.*, 1983).

The Mature Gene Structures. The coding sequences provided in Table 5.2 reflect still an additional way of categorizing the H1 histones. The first two of these, from the sea urchins *Psammechinus miliaris* and *Strongylocentrotus purpuratus*, represent a form, otherwise unidentified, associated with euchromatin (Levy *et al.*, 1982). On the other hand, the remaining pair, from *Xenopus laevis* and the chicken, are from heterochromatin (Sugarman *et al.*, 1983; Turner *et al.*, 1983). Both types are characterized by being rich in codons for lysine (underscored), as in all other histones, and for alanine (in italics), which is replaced in relative abundance by arginine in the remaining species. The relatively few codons for this last amino acid are boxed in the tabulation. A high level of alanine in this species (or these species) does not appear to have been pointed out previously. Only one codon for tyrosine is present per chain.

It is at once apparent that the coding sequences of *Strongylocentrotus* and *Psammechinus*, which is incompletely established, are distinctly homologous to a large degree. Nevertheless, it is also evident that the two genes differ to a surprising extent, considering that they represent the DNA coding for an important protein from two closely related species of organism. Although the majority of codons for the two characteristic amino acids are located at comparable sites, there are a number of exceptions. In row 1, for instance, the *Strongylocentrotus* sequence has a lysine codon unmatched in the *Psammechinus*, and in row 2, only one of the three for this amino acid that is present in each is correspondingly located in the other. Similar discrepancies exist throughout the pair insofar as both are represented. In fact, the part of the *Psammechinus* coding region given includes codons for only 15 lysines, while that of the other sea urchin has 21 in the same region. Similarly, the codons for alanine are more abundant in the latter than in the former, in a ratio of 15:12. On the other hand, the first sequence of the pair is slightly richer in arginine codons, having five against the three of the other.

The *Xenopus* and chicken mature coding regions are so similar that there can be no doubt that they represent the same species of histone H1, one that differs strongly from that of the echinoderms. Among the distinctions, aside from the eight scattered codons deleted from the avian type, is the absence of the initiating ATG in the mature product from that same source, the methionine residue being removed during processing, as indicated by the lowercase letters. But the greatest distinction is a feature disclosed for the

Table 5.2

Mature Gene Sequences for Histone H1[a]

Row 1

Psammechinus miliaris[b]
ATG ACT GAC ACT GCC --- --- AAG AAA GTT ACC CAG AAG AAG CCG GCG GCT CAC CCA CCT GCT GCC GAA ATG GTG

Strongylocentrotus purpuratus[b]
ATG GCT GAG AAT AGC TCT AAG AAG GTG ACT ACT AAG AAG CCG GCC CAC CCA CCG GCT GCC GAG ATG GTT

Xenopus[c]
ATG ACA GAA ACT GCT GCA ACT GAG ACA ACT CCC GCC GCT CCC GCG GCA GAA CCC AAA CAG AAG AAG CAG CAG

Chicken[d]
atg TCG GAG ACC CCC GTT GCC GCG GCG --- CCC GCG GTG TCT GCG CCC GGC GCC AAG GCC GCC GCC AAG ---

Row 2

Psammechinus
ACT ACA GCA ATC AAA GAG CTC AAG GAA CGC AAG GGG TCT TCT CGT CAA GCA ATC GCG AAC TAC ATC AAG GCC CAT

Strongylocentrotus
GCT ACA GCA ATC ACC GAG TTG AAG GAC CGC AAT GGC TCC TCG CAA GCA ATA AAG AAG TAT ATC GCT ACC AAT

Xenopus
CCT AAG AAG GCA GCG GGA GGC GCT AAG GCC AAA CCC TCC GGA CCG GCA TCT GAG CTG ATC GTG AAA TCC

Chicken
CCG AAG AAG GCG GCG GGC GGC GCC AAG CCC AAG GCG GCG CCC GGC GTC ACC AGC CTG ATC ACC AAG AAG GCC

Row 3

Psammechinus
TTC GAT GTA GAG ATA GAT CAA CAG CTG GTA TTC ATC AAG AAG GCC CTG AGA TCT GGG GTC GCG AAG GGC ACG TTG

Strongylocentrotus
TTC GAT GTG CAG ATG GAC CGA CAG CTG CTA TTC ATC AAG CGG CTA AAG TCT GGC GTG GAG AAA GGC AAA CTA

Xenopus
GTG TCC GCC TCT AAG GAG CGT GGT GGG GTG TCC CTG GCC GCT CTC AAG AAG GCC TTG GCT GCC GGA GGT TAC AAT

Chicken
GTG TCC GCC TCC AAG GAG CGC AAG GGG CTC TCC CTC GCC GCG GCG CTC AAG AAG GCC GCC GGC GGC TAC GAC

Row 4

Psammechinus
GTC CAG ACG AAA GGC ACG GGG GCA TCG GGA TCC ATC AAG CTA --- ACA AGA TTG GAC CGT ACA CTG --- --- ---

Strongylocentrotus
GTG CAG ACG AAG GGG AAA AAG GGA TCG GCG GGT TCT TTC AAG GTG AAT GTG CAG GCA CAG GCG TCG GAG

Xenopus
GTG AGG AGG AAC AAC AGT CGC CTC AAG GCT CTC AAG AAA GGG ACT CTC ACC CAA GTC AAA

Chicken
GTG GAG AAG AAC AAC AGC CGC ATC AAG CTG GGG CTC AAG AAG AGC CTC AGC GGC ACC CTG CTG CAG ACC AAG

Row 5

Psammechinus
--- TCC AAG AAA GTC ACC --- CAG CCG AAG (incomplete)

Strongylocentrotus
AAG GCC AAG AAG GAG AAG GCA AAA CTG CTA GCA CAG CGT GAG AAG GCC AAG GAA AAA GGC TGC AGC GAA

Xenopus
GGC AGC GGA GCC TCT GGA TCC TTC AAG CTG AAC AAG CAG CTG GAG ACC AAG GCG GTG AAG GCC AAG AAG

Chicken
GGC ACC GGC GCC TCG GGC TCT TTC AAG CTG AAT AAA AAG CCG GGT GAG ACA AAA GCG AAA GCG ACT AAG AAG AAG

Row 6

Strongylocentrotus
GAA GGA GAA ACT GCA GAA GGC AGC CGC CCA AAG AAA GTC AAG GCA GCC CCC AAG AAA GCG AAG AAG CCA GTA AAG

Xenopus
AAG CTC GTG GCG CCC AAA GCC AAA AAA CCC GTC GCG GCA AAG AAA CCC AAA TCC CCT AAA AAG CCC AAG AAG

Chicken
--- CCC GCG GCC --- AAG CCC AAG AAG CCG --- GCG GCC CCT GCT GCT GCC AAG AAG CCC AAG AAG AAG

(continued)

Table 5.2 (*Continued*)

Row 7

Strongylocentrotus	AAA	ACG	ACX	GAG	AAA	GAG	AAG	AAG	ACT	CCA	AAG	AAG	GCA	CCC	AAG	AAG	CCA	GCA	GCC	AAG	AAG	AAA	TCA	ACA
Xenopus	GTC	TCG	*GCA*	*GCA*	*GCA*	AAG	AGC	CCC	AAG	AAG	*GCG*	AAG	AAA	CCG	GTA	AAG	*GCC*	CCC	AAA	AGC	CCC	AAG	AAG	CCC
Chicken	---	*GCA*	*GCG*	*GCG*	CTG	AAG	AGC	CCC	AAG	AAG	AAA	*GCC*	AAG	AAG	CCG	*GCA*	*GCC*	ACC	AAG	AAG	AAG	*GCG*	*GCC*	AAG

Row 8

Strongylocentrotus	CCA	AAG	ACG	CCC	AAG	AAG	*GCA*	*GCC*	*GCA*	AAG	AAA	CCC	AAG	AAG	ACT	*GCC*	---	AAG	CCC	AAG	AAG	CCT	*GCG*	GXX	AAG
Xenopus	AAA	*GCT*	GTT	AAA	CCC	AAG	AAG	GTG	ACC	AAG	AGT	CCA	*GCT*	AAA	AAG	*GCC*	ACT	AAG	CCC	AAA	*GCT*	*GCC*	AAG	*GCC*	AAA
Chicken	*d*AAG	*GCT*	GGC	CGC	CCC	AAG	AAG	ACT	*GCC*	AAG	AGC	CCG	*GCC*	---	AAG	*GCA*	---	AAG	*GCG*	GTG	AAG	AAG	CCC	AAA	*GCT*	---

Row 9

Strongylocentrotus	AAG	*GCT*	*GCA*	AAG	TCC	AAG	TGA																	
Xenopus	ATA	*GCC*	AAA	CCC	AAA	ATA	---	*GCC*	AAA	*GCG*	AAG	*GCG*	*GCT*	AAG	GGG	AAG	AAG	*GCT*	*GCG*	*GCT*	AAA	AAG	TAA	
Chicken	---	*GCC*	AAG	---	TCA	AAG	*GCG*	*GCC*	AAA	CCC	AAG	*GCG*	*GCC*	AAG	*GCA*	*GCG*	*GCG*	ACC	AAA	AAG	AAG	TAA		

---

[a]Lysine codons are underscored, those for alanine are italicized, and those for arginine are boxed.

[b]Levy *et al.* (1982).

[c]Turner *et al.* (1983).

[d]Sugarman *et al.* (1983).

[e]Insert AGC CCC AAG AAG *GCT* ACC.

first time by this analysis, the presence of an 18-base-pair insert in the chicken gene, placed just before the 3' terminal sector. The amphibian gene encodes for one more lysine residue in the ratio of 64:63, the missing member being replaced by an arginine codon in the chicken. On the other hand, the latter is superior in number of codons for alanine, with 52 vs. 45 of the toad. Thus the two vertebrate genes are of the same species, but represent different subtypes. Only one codon for tyrosine is in evidence, in both cases located near the 3' end of row 3; the single one of the two sea urchin sequences is placed near the corresponding end of row 2.

It is of interest to compare the euchromatinic variety of the sea urchin to the two heterochromatinic ones of the vertebrates. In the 5' portion delimited by the end of the *Psammechinus* sequence in row 5, the *Strongylocentrotus* sequence encodes for 22 lysines and 16 alanines, whereas that from *Xenopus* has 18 and 18, respectively, and the chicken 18 and 20. The total content of lysine codons for the region occupied by the entire sea urchin sequence is 59 for the latter and 58 for the amphibian, compared to 53 for the chicken. Thus the echinoderm gene product is more strongly basic in the 5' portion than in the others, but slightly more weakly so in the 3' part. Hence, its reaction with DNA would differ markedly from those of the others.

5.3.2. The Mature Histone H2A Gene

Unlike the foregoing histone, the H2A sequences in Table 5.3 conspicuously represent a single species, even including the highly aberrant variety from the chicken embryo. That complete uniformity is lacking in the others is made equally obvious by the numerous dashed inserts needed to bring out the homologies that exist, but for the greater part, the insertions are confined to the 5'-terminal sector. The consistency in structure shown here is illusionary, for in mammalian cells this histone type has been reported to consist of eight distinct subtypes (West and Bonner, 1980; Hatch *et al.*, 1983). Wheat embryos also have been demonstrated to contain several heteromorphic variants of H2A (Rodrigues *et al.*, 1979; Spiker, 1982), and starfish gonads and *Tetrahymena* have at least two variants each (Fusauchi and Iwai, 1983; Martinage *et al.*, 1983), and in the ciliate one minor type is confined to the macronucleus (Allis *et al.*, 1980). Little of the polymorphism can be attributed merely to the presence of multiple genes, for the two from yeast (Table 5.3) differ at only 23 of the 379 sites, a mere five of which induce an alteration in amino acid content (Choe *et al.*, 1982).

The Two Yeast H2A Genes. The two genes of H2A from yeast just discussed afford several observations of interest. The first of these is that none of the changes in nucleotides affects either of the important amino acids lysine and arginine. One consistent pattern in the codons for the first of these is the altering of its final nucleotide G of the yeast 1 gene to A in yeast 2. This condition can be noted near the beginning of row 2, twice at the center of row 4, and close to the 3' terminus, while none is found in the opposite direction. For a basic protein, there is a remarkable paucity of arginine codons present, not only in these two, but in all members of this class. In the yeast this amino acid is encoded entirely by AGA, except in the first of these codons in row 4, where an A → G mutation has occurred in the final position. The sole perceptible inserted region of any length that is absent in mature tissues of the metazoans is that located near the close of row 1, where the nonanucleotide GCTTCTCAA occurs.

Table 5.3
Mature Gene Sequences for Histone H2A[a]

Row 1

```
Yeast 1[b]        atg TCC GGT --- GGT GGT AAA GCT GGT TCA GCT GCT AAA GCT --- TCT CAA TCT AGA TCT
Yeast 2[b]        atg TCC GGT --- GGT GGT AAA GCT GGT TCA GCT GCT AAA GCT --- TCT CAA TCT AGA TCT
Psammechinus[c]   atg TCT GGC AGA GGA AAG --- AGT GGA AAG GCC ACC AAG GCA AAG --- --- --- --- ACG CGC TCA
Xenopus[d]        atg TCT GGA AGA GGC AAA CAA GGC GGC AAG ACT CGC GCT AAG GCA AAG --- --- --- --- ACT CGC TCA
Human[e]          ATG TCT GGT CGC GGC AAA CAA GGC GGC AAG GCT CGC GCC AAG GCA AAG --- --- --- --- ACT CGG TCT
Chicken[f]        atg TCG GGG CGC GGA AAG CAG GGC GGC AAG GCG GCC AAG --- --- --- --- --- --- TCG CGC TCG
Chick embryo[g]   atg GCA GGT --- GGG AAG --- GCT GGG GAC AGC GGG AAG GCG AAG GCG GTG TCT CGC TCG
```

Row 2

```
Yeast 1           GCT AAG GCT GGT TTG ACA TTC CCA GTC GGT AGA GTG CAC AGA TTG CTA AGA GGT AAC TAC GCC ---
Yeast 2           GCT AAA GCT GGT TTA ACA TTC CCA GTT GGT AGA GTG CAC AGA TTG CTA AGA GGT AAC TAC GCC ---
Psammechinus      TCC CGT GCA GGG CTC CAG TTT CCA GTC GGA CCT GTT CAT CGG TTT CTC CGA AAG GGC AAC TAT GCA ---
Xenopus           TCT CGG GCG GGG CTG CAG TTC CCA GTC GGC GGC GTT CAC CGG CTC TTG AGG AAG GGC AAT TAT GCC ---
Human             TCT CGT GCA GGT TTG CAG TTT CCT GTG GGC CGA GTC CAC CGC CTG GGC AGG G-- --C TAC TCC ---
Chicken           TCG CGG GCC GGG CTG CAG TTC CCC GTG GGC CCC GTG CGC CTG GGC CAG CTG CGC AAG GGC AAC TAC GCG ---
Chicken embryo    CAG AGA GCC GGA TTG CAG TTC CCC GTG GGC CGC ATC CAT CAT CGG CAC CTG --- AAG ACG CGC ACC ACG AGC
```

Row 3

Yeast 1	---	CAA	AGA	ATT	GGT	TCT	GCT	CCA	GTC	TAC	TTG	ACT	GCT	GTC	TTG	GAA	TAT	TTG	GCC	GCT	GAA	ATT
Yeast 2	---	CAG	AGA	ATT	GGT	TCT	GCT	CCA	GTC	TAT	CTG	ACT	GCT	GTC	TTA	GAA	TAT	TTG	GCT	GCT	GAA	ATT
Psammechinus	---	AAG	AGG	GTC	GGC	GGT	GGA	GCT	CCT	GTC	TAC	ATG	GCT	GCC	GTC	CTA	GAG	TAC	CTC	ACT	GCC	GAA	ATC
Xenopus	---	GAG	CGG	GTG	GGA	GCC	GGA	GCT	CCG	GTC	TAT	CTG	GCC	GCA	GTG	CTC	GAG	TAT	CTG	ACC	GCT	GAG	ATC
Human	---	GAG	CGC	GTC	GGC	GCT	GGC	GCC	CCG	GTG	TAT	CTC	GCG	GTG	GCG	CTT	GAG	TAC	CTG	ACC	GCC	GAG	ATC
Chicken	---	GAG	CGG	GTG	GGC	GGC	GCC	CCG	CTG	GTG	TAC	CTA	GCG	GCC	GTG	CTG	GAG	TAC	CTG	ACG	GCC	GAG	ATC
Chicken embryo	CAT	GGG	GTC	GGG	GCC	ACC	GCC	GCC	GCG	GTG	TAC	AGC	GCT	GCC	ATC	CTC	GAG	TAT	CTC	ACT	GCT	GAG	GTC

Row 4

Yeast 1	TTA	GAA	TTA	GCT	GGT	AAT	GCT	GCT	AGG	GAT	AAC	AAG	AAG	ACC	AGA	ATT	ATT	CCA	AGA	CAT	TTG	CAA	TTG
Yeast 2	TTA	GAA	TTG	GCT	GGT	AAT	GCT	GCT	AGA	GAT	GAT	AAC	AAA	AAA	ACC	AGA	ATT	ATT	CCA	AGA	CAT	TTA	CAA	TTG
Psammechinus	TTG	GAA	CTC	GCA	GGC	AAC	GCT	GCC	CGC	GAC	AAC	AAG	AAA	TCT	AGG	ATC	ATC	CCA	CGC	CAC	CTT	CAA	CTC
Xenopus	TTG	GAG	TTG	GCC	GGC	AAC	GCT	GCT	GCG	GAC	AAC	AAG	AAG	ACC	CGC	ATC	ATC	CCC	AGG	CAC	CTG	CAG	CTC
Human	CTA	GAG	CTG	GCG	GGC	AAT	GCG	GCC	GCC	GAC	AAC	AAG	AAG	ACC	CGC	ATC	ATC	CCG	CAC	CTG	CAA	TTG
Chicken	CTG	GAG	CTA	GCG	GGC	AAC	GCG	GCC	GCC	GAC	AAC	AAG	AAG	ACG	CGC	ATC	ATC	CCC	CAC	CTG	CAG	CTG
Chicken embryo	CTG	GAG	TTG	GCA	GGC	AAC	GCC	TCC	AAG	GTG	AAG	CTG	AAG	ATC	ACT	CCC	CGC	CAT	TTG	CAG	CTG

(continued)

Table 5.3 (Continued)

Row 5

```
Yeast 1         GCT ATC AGA AAT GAT GAC GAA TTG AAC AAG CTA TTG GGT AAC GTT ACC ATT GCC CAA GGT GGT GTT TTG
Yeast 2         GCC ATC AGA AAT GAT GAT GAA TTG AAC AAG CTA TTG GGT AAT GTT ACC ATC GCC CAA GGT GGT GTT TTG
Psammechinus    GCT GTG CGT AAT GAA GAA CTC AAC AAG CTT TTG GGT GGT GTG ACG ATC GCT CAA GGT GGT GTT CTG
Xenopus         GCT GTG CGC AAC GAT GAG GAG CTC AAC AAA CTG GTC GGA GTC ACT ATC GCT CAG GGC GGG GTT CTG
Human           GCC ATC CGC AAT GAC GAG GAG CTT AAT AAA CTT TTG GGG CGT GTG ACC ATC GCG CAG GGC GGT GTT TTG
Chicken         GCC ATC CGC AAC GAC GAG CTC AAC AAG CTG CTG GGC GTG ACC ATC GCG CAG GGC GGG GTG CTG
Chicken embryo  GCG ATC CGC GGC GAC GAA GAG --- TTG GAT TCC CTC ATC AAA GCC ACC ATA GCG GGG GGA GGC GTC ATC
```

Row 6

```
Yeast 1         CCA AAC ATC CAT CAA AAC TTG TTG CCA AAG AAG TCT GCC AAG GCT ACC AAG GCT TCT CAA GAA TTA TAA
Yeast 2         CCA AAC ATT CAC CAA AAC TTG TTG CCA AAG AAG TCT GCC AAG GCT TCT CAA GAA CTG TAA
Psammechinus    CCC AAC ATC CAA GCC GTG CTG CTT CCC AAG AAG AAA ACT --- --- --- GCT AAA TCA AGC TAG
Xenopus         CCC AAC ATT CAG TTC GTG CTG CTG CCC AAG AAG AAA TCC --- --- AAG TCG GCC AGC AAG TGA
Human           CCT AAT ATT CAG GCG CTG CTG CCT AAG AAA ACT GAG AGC CAT CAT AAG GCC AAG GGA AAG TGA
Chicken         CCC AAC ATC CAG GCC CTG CTG CCC AAG AAG AAG GCC ACC GAC CAC --- AAG GCC AAG TGA
Chicken embryo  CCC CAC ATC CAC AAG TCT CTG ATC GGG AAG AAG --- --- GGC CAG CAG --- --- AAA ACC GCG TAG
```

[a] Codons for lysine are underscored, those for arginine are italicized, and those for tyrosine are stippled.
[b] Choe et al. (1982).
[c] Busslinger et al. (1980).
[d] Moorman et al. (1982).
[e] Zhong et al. (1983).
[f] D'Andrea et al. (1981).
[g] Harvey et al. (1983).

The Chicken Embryonic H2A Gene. The variant H2A gene of the chick embryo that is listed, designated H2A.F, is the only one sequenced thus far that merits rank as a distinct subtype. As a whole it corresponds more closely to those from yeast rather than the other metazoans, at least with regard to deletions and insertions. However, in place of the nonanucleotide just mentioned, there is the dodecanucleotide GCGAAGGCGGTG. In addition, there is a hexanucleotide insert AGCCAT at the close of row 2 and beginning of row 3. Although thus resembling the protistan sequence, its affinities with regard to the nature of the arginine codons accord with the structures from vertebrates and the sea urchin, for the CGN family of codons is employed, rather than AGR of the yeasts, an important point that receives further attention in Table 5.5. A single exception to this trend may be observed near the 5' end of row 2, at the site where CGG or CGT occur in metazoan genes and lysine codons in the yeast.

The Chemical Properties. The present histone is most distinctive from the others more in what it lacks than in any feature it possesses. Whereas H1 has a plethora of alanine codons, as just seen, H2B has three distinctively located tyrosine codons, and H4 has an unusual abundance of those for glycine, nothing of prominence is visible in the present case. As observed earlier, arginine triplets are not as numerous as might be expected, nor for that matter are those for lysine. The two sequences from yeast have 11 of the latter and 10 for the former, while that from human sources has equal numbers (13) of each. This and the *Xenopus* gene with 13 and 12, respectively, encode a slightly more basic produce than the yeast's, as does that for the chicken, with 14 and 12. At one extreme is the chicken embryo coding region, with 13 lysine and 9 arginine triplets, while those of the two sea urchins *Psammechinus* (shown) and *Strongylocentrotus* (Sures *et al.*, 1978), with 12 and 12, respectively, signal arginine-rich products as equally as lysine-rich ones. As a consequence of the frequent near balance in content of codons for this pair of amino acids, the designation of H2A as a lysine-rich species is unrealistic, as is the resultant generalization of such properties as the sole basis for the nature of their interactions with DNA and other histones. This point receives reemphasis in the discussion of histone H4.

5.3.3. The Mature Histone H2B Gene

As attested by the relatively small number of sequences presented in Table 5.4, the gene for histone H2B has not been subjected to very many analytical studies, although a number of amino acid sequences of its product have been established. This paucity of gene structure is all the more surprising in view of the role of this species of histone, not only in nucleosome-core formation, but also in affecting the length of the interspace DNA (Zalenskaya *et al.*, 1981; Jordano *et al.*, 1984b). Moreover, in conjunction with H2A, this species has been found to function as the homeostatic thymus hormone (Reichhart *et al.*, 1985a,b).

Like the histone H1 described previously, the H2B species is represented by a number of variants, the frequency of which within a given organism is related directly, but not consistently, to the number of operon duplications. In most sea urchins, including *Parechinus angulosis*, as many as 600 gene sets may be found, but only eight recognized subtypes of H2B exist (Brandt *et al.*, 1979). It is of interest to note that in the common

Table 5.4
Mature Gene Sequence for Histone H2B[a]

Row 1

Yeast 1[b]	ATG TCT GCT AAA GCC GAA AAG AAG CCA --- GCC TCC AAA GCC CCA GCT GAA AAG AAA CCA GCC GCT AAA AAG
Yeast 2[b]	ATG TCC TCT GCC GCC GAA AAG AAA CCA --- GCT TCC AAA GCT CCA GCT GAA AAG AAG CCA GCT GCC AAG AAA
Strongylocentrotus[c]	atg GCT CCA ACA GCT CAA GTT GCT AAG AAA GGC TCC AAG GCA GTC AAA GGC ACC AAG ACG GCC XGC GGT
Mouse[d]	ATG CCT GAG CCC GCC AAG TCC GCT CCT --- GCC CCG AAG AAG GGC AAG AAG GCC AAG GCC ACC AAG GCC CAG
Chicken[e]	atg CCT GAG CCG GCC AAG TCC GCA CCC --- GCC CCC AAG AAG GGC AAG AAG GCG AAG GTC ACC AAG ACC CAG

Row 2

Yeast 1	ACT TCC ACT GAT GGT AAG AAG AGA AGC AAG GCT --- AGA GAA ACA TCT TCT TAC ATT
Yeast 2	ACA TCA ACC TCC GTC GAT GGT AAG AAG AGA TCT AAG GTT --- AGA GAG ACC TAT TCC TCT TAT ATC
Strongylocentrotus	--- --- GGC AAG AAG --- --- --- AAG --- AGG AAG GAG AGT TAT GGA ATC TAC ATC
Mouse	--- --- --- AAG AAG GAC GGC AAG AAG CGC AGC CGC AAG GAG AGC TAC TCG GTG TAC GTG
Chicken	--- --- --- AAG AAG GGC GAC GAC AAG AAG CGC AAG AAG CGC AAG GAG AGC TAC TCG ATC TAC GTG

Row 3

Yeast 1	TAC AAA GTT TTG AAG CAA ACT CAC CCT GAC ACT GGT ATT TCC CAA AAG TCC ATG TCT ATC TTG AAC TCT
Yeast 2	TAC AAA GTT TTG AAG CAA ACT CAC CCA GAC ACT GGT ATT TCC CAG AAG TCT ATG TCT ATT TTG AAC TCT
Strongylocentrotus	TAC AAA GTC CTC AAG CAG GTT CAT CCA GAT ACC GGC ATC TCC AGT CGG GCC ATG GTC ATC ATG AAC AGC
Mouse	TAC AAG GTG CTG AAG CAA GTG CAC CCC CAC ACC GGC ATC TCC TCC AAG GCC ATG GGC ATC ATG AAC TCG
Chicken	TAC AAG GTG CTG AAG CAG GTG CAC CCC CAC ACG GGC ATC TCG TCC AAG GCC ATG GGC ATC ATG AAC TCG

```
Row 4

Yeast 1            TTC GTT AAC GAT ATC TTT GAA AGA ATC GCT ACT GAA GCT TCT AAA TTG GCT GCG TAT AAC AAG AAG TCT
Yeast 2            TTC GTT AAC GAT ATC TTT GAA AGA ATT GCT ACT GAA GCT TCT AAA TTG GCC GCT TAT AAC AAG AAA TCC
Strongylocentrotus TXX GTY AAC GAC ATC TTC GAG CGA ATT GCC GGC GAA TCT TCC CTC GCT CAG TAC AAC AAA AAG TCG
Mouse              TTC GTG AAC GAC ATC TTC GAG CGC ATC GCG GGA GAG GCG TCC CTA GCG CAT TAC AAC AAG CGC TCG
Chicken            TTC GTC AAC GAC ATC TTC GAG CGC ATC GCC GCG GCG TCG GAG GCG CTG GCG CAC TAC AAC AAG CGC TCG

Row 5

Yeast 1            ACT ATC TCT GCT AGA GAA ATT CAA ACC GCT GTT AGA TTG ATC TTA CCA GGT GAA TTG GCT AAG CAT GCT
Yeast 2            ACT ATT TCT GCT AGA GAA ATC CAA ACA GCC GTT AGA TTG ATC TTA CCT GGT GAA TTG GCT AAA CAT GCC
Strongylocentrotus ACX XTC AGC AGT CGC GAG ATT CAG ACC GCC GTG CGC CTC ATC CTC GGA GAG CTG GCA AAG CAC GCT
Mouse              ACC ATC ACG TCC CGG GAG ATC CAG ACG GCC GTG CGC CTG CTG CTG CCC GGG GAG CTG GCC AAG CAC GCC
Chicken            ACC ATC ACG TCG CGG GAG ATC CAG ACA GCC GTG CGG CTG CTG CTG CCC GGC GAG CTG GCC AAG CAC GCG

Row 6

Yeast 1            GTC TCT GAA GGT ACT AGA GCT GTT ACC AAG TAC TCT TCC TCT ACT CAA GCA TAA
Yeast 2            GTC TCC GAA GGT ACT AGG GCT GTT ACC AAA TAC TCC TCC TCT ACT CAA GCC TAA
Strongylocentrotus GTG AGC GAG GGT ACC AAG GCA GTG ACG AAA TAC ACT ACC TCC AAG TAG
Mouse              GTG TCG GAG GGC ACC AAG GCT GTC ACC AAG TAC ACC AGC TCC AAG TGA
Chicken            GTC TCC GAG GGT ACC AAG GCG GTC ACC AAG TAC ACC AGC TCC AAG TAG
```

aCodons for lysine are underscored, those for arginine are italicized, and those for tyrosine are stippled.
bWallis et al. (1980).   cSures et al. (1978).   dSittman et al. (1983a,b).   eGrandy et al. (1982).

yeast *S. cerevisiae,* which possesses only two operons containing this gene, the two products differ in four amino acids of the 130 that are present, as shown immediately below.

The Histone H2B Genes of Yeast. Although the products of the two yeast H2B proteins thus vary but slightly, the sequences have 49 of their 390 sites occupied differently (Wallis *et al.,* 1980); hence, this pair of genes is less stable than those of H2A, whose members diverge at only 23 of their 379 sites. Among the more outstanding distinctions between the two genes of this type is the loss in yeast 2 of the first lysine codon from the yeast 1 sequence (row 1, fourth codon), which has one for alanine in its stead. Both possess the dodecanucleotide AC-TC-AC-TCC located at the beginning of row 2, which is absent from the two vertebrate cistrons and largely from the sea urchin. To judge from the amino acid sequence of H2B$_3$ from *Parechinus angulosis,* some variants from echinoderms have an even greater insert here (Strickland *et al.,* 1978). In these invertebrates and the yeast, the entire 5'-terminal sectors are revealed as being subject to extensive variation in structure, whereas the 3' half is quite uniform. As a whole the former cistron appears to be transitional between the protistan and vertebrate types.

The Vertebrate Histone H2B Gene. The two genes for histone H2B from vertebrates are highly homologous on a site-to-site basis, despite the source species being from near-endpoints of two divergent lines from a remote reptilian ancestor. No major differences can be noted, although mutations resulting in amino acid changes occur at five sites and the translational termination signal is TGA in the mouse gene and TAG in that of the chicken (Grandy *et al.,* 1982; Sittman *et al.,* 1983b). Comparisons of these with the two from yeast bring out specific examples of the variability of the 5' end alluded to earlier, with row 1 the chief center of attention, as it was also in H2A. Since the lysine residues play such an important role in the product's interaction with DNA, as shown in the discussion opening this chapter, one must suspect that their number and disposition in the molecule would be subject to strong evolutionary constraints and therefore be unvarying.

But examination of row 1 in Table 5.4 reveals considerable fluctuation both in their number and location. The yeast 1 gene has eight codons for that amino acid, the yeast 2 seven, and the sea urchin and vertebrates just six; in none of the sequences is there a coding signal for an arginine in this sector. In the product of the variant sea urchin gene four each of lysine and arginine residues are found in the comparable section. Row 2 offers quite a contrast. Here the mouse sequence has seven lysine codons, the chicken eight, and the echinoderm and the two from yeast only five. Furthermore, the latter couple have triplets for only two arginines, while the vertebrates have three each, as does the sea urchin. Although beyond this sector all the genes are fairly uniform in structure, some important distinctions may be noted in row 4. There the yeasts show codons for three lysines and one arginine, while the vertebrate cistrons have only one for the first of these amino acids and three for the second, the sea urchin being intermediate with two for each.

Distinctive Features of H2B genes. Because of this variability in the 5'-terminal portion of the gene and inconstancy in location of the lysine and arginine codons, another trait of a less variable nature is of aid in establishing homology. Such an unusual trait can be noted in row 2 and beginning of 3, where three codons for tyrosine (TAC or TAT) are lightly stippled. While the function of this amino acid in the mature product is unknown, it must play an important role, for the presence, arrangement, and relative location of the

three codons or their products are invariable in all the nucleotide and amino acid sequences of H2B that have been examined, as are those of two other sets of triplets for
tyrosine.

Unlike H2A, the length of the gene is subject to considerable variation, largely as a
result of the differences in the 5' sector. Those of the two vertebrates are the shortest,
consisting of 378 base pairs, the yeast being slightly longer with 390; in contrast, that of
the variant sea urchin cistron is 429 base pairs in length. The overall totals of lysine and
arginine condons show considerable consistency, those for the first of these ranging from
19 in yeast 2, to 20 each in yeast 1 and the mouse gene, to 21 in that of the chicken. In
addition, there are five codons for the second amino acid in the two sequences from yeast,
six in that of the chicken, and seven in the mouse cistron. On a combined basis, the
vertebrates are seen to have gained a total of two additional codons for these important
molecules and are thus slightly more basic in reaction than those of the primitive eukaryote. Consequently, the present genes encode a truly lysine-rich product, whereas that
of H2A does not.

5.3.4. The Mature Histone H3 Gene

Histone H3 will be recalled as the first of the pair of these proteins that are considered
arginine-rich. In view of this supposed sharp contrast in fundamental amino acid content,
it could be suspected that many other distinctions in structure should also occur, at least to
the extent permitted within the framework of the histone class of proteins. Whether this
expectation is supported can readily be discovered, for the present species has attracted a
modest number of analytical studies at the level of the gene, permitting more meaningful
comparisons than was the case with the H2B type. Moreover, although polymorphism
abounds here, the differences that occur are more restrained, enabling homologies to be
readily established.

Although the variability is doubtlessly correlated to the multiplicity of repeated
copies that prevails, more factors are involved in their differentiation than mere mutagenesis within a given genus or major category of organisms, a statement made more
explicit by the several sequences from the mouse, as shown shortly.

The 5'-Terminal Portion. Because of the uniformity that is exhibited by the seven
mature genes of H3 given in Table 5.5, a more unified analysis can be made than was
possible with the several preceding species. In the 5' portion represented by rows 1–3 is
found the preponderance of the lysine codons present in this type of histone, 11 of the total
of 14–16 being located here—the second row with five being especially richly endowed.
Moreover, half of the total of 14 arginine codons is situated in this sector, so that it is both
lysine- and arginine-rich. In the present set of coding sequences, it is the fourth row that
holds a slight plurality over the rest, three encoding arginine, in contrast to the two apiece
of the others. Unlike H2B, this histone cistron has no concentration of tyrosine codons,
but has the three that it contains rather scattered, rows 2, 3, and 5 each having a single
copy. However, the two yeast sequences lack the second one. No codons for acid amino
acids (GAN) occur in the first two rows, but two are located in the third.

The 3'-Terminal Portion. Since only half of the arginine codons are arrayed in the
3'-terminal portion, along with three to five of those for lysine, this part can scarcely be
said to be outstandingly rich in any of the important amino acids. Their distribution within

Table 5.5
Mature Gene Sequences for Histone H3[a]

Row 1

Yeast 1[b]   ATG GCC *AGA* ACA CAA CAA AAG AAG TCC ACT GGT GGT AAG GCC CCA *AGA* AAG CAA TTA GCT TCT

Yeast 2[b]   ATG GCC *AGA* ACT AAA CAA ACT *AGA* AAG TCC ACT GGT GGT AAA GCC CCA *AGA* AAA CAA TTA GCC TCC

Neurospora[c]   ATG GCC *CGC* ACT AAG CAG ACC *CGC* AAG TCC ACC GGT GGC AAG GCC CCC *CGT* AAG CAG CTC GCT TCC

Lytechinus[d]   ATG GCC *CGT* ACC AAG CAG ACC GCA *CGT* AAA TCT ACT GGA GGA AAA GCT CCA *CGC* AAG CAG CTC GCC ACC

Mouse 1[e]   ATG GCT *CGT* ACT AAG CAG ACT GCT *CGC* AAG TCT ACC GGC GGC AAG GCC CCG *CGC* AAG CAG CTG GCC ACC

Mouse MEL[f]   ATG GCT CAT ACA AAG CAG ACT GCC CCC *CGC* AAA TCC ACC TGT GGT AAA GCA CCT *AGG* AAA CAA CTA GCT ACA

Wheat[g]   atg GCC *CGC* ACC AAG CAG ACG GCG *AGG* AAG TCG ACC GGC GGC AAG GCG CCG *AGG* AAG CAG CTG GCG ACC

Row 2

Yeast 1   AAG GCT GCC *AGG* AAA TCC GCC CCA TCT ACC GGT GGT GTT AAG AAG CCT CAC *AGA* TAT AAG CCA GGT ACC

Yeast 2   AAG GCT GCC *AGA* AAA TCC GCC CCA TCT ACC GGT GGT GTT AAG AAG CCT CAC *AGA* TAT AAG CCA GGT ACT

Neurospora   AAG GCT GCC *CGC* AAG TCC GCC TCG GCC TCC ACC GGT GTC AAG AAG CCC CAC *CGT* TAC AAG CCC GGT ACC

Lytechinus   AAG GCT GCC *CGC* AAA TCC GCC CCA GCC GGA GTC AAG AAG CCC CAT *CGT* TAC *CGT* CCC GGA ACC

Mouse 1   AAG GCC GCC *CGC* AAG AGC GCC CCG GCC ACC GGC GGC GTG AAG AAG CCT CAC *CGC* TAC *CGT* CCC GGC ACT

Mouse MEL   AAA GCT GCT XGC AAG AGT GCG CCC TCT ACT GGA GGG GTG AAG AAA TCT CAT *CGT* TAC *AGG* CCA GAT ACT

Wheat   AAG GCC GCT *CGC* AAG AAG TCC CCG GCC ACC GGC GGC AAG AAG CCG CAC *CGC* TTC *CGC* CCC GGC ACC

Row 3

Yeast 1	GTT GCT TTG *AGA* GAA ATC *AGA* *AGA* TTC CAA AAA TCT ACT GAA CTG TTG ATC *AGA* AAG TTG CCT TTC CAA
Yeast 2	GTT GCC TTG *AGA* GAA ATT *AGA* *AGA* TTC CAA AAA TCT ACT GAA CTG TTG ATC *AGA* AAG TTA CCT TTC CAA
Neurospora	GTC GCT CTC *CGT* GAG ATT *CGT* *CGC* CAG AAG TCC ACT GAG CTT CTG ATC *CGC* AAG CTC CCC TTC CAG
Lytechinus	GTC GCT CTC *CGT* GAG ATC *CGT* *CGC* CAG AAG AGC ACC GAG CTT CTC ATC *CGC* AAG CTC CCC TTC CAG
Mouse 1	GTG GCA CTG *CGC* GAG ATC *CGG* *CGC* **TAC** AAG TCG ACC GAG CTG CTG ATC *CGC* AAG CTG CCG TTC CAG
Mouse MEL	GTG GCA CTC CTT GAA ATC *AGA* *CGC* **TAT** --- TCC ACT GAA CTT CTG ATT CAT *CGC* AAG CTC CCC TTT CAG
Wheat	GTG GCG CTC *CGC* GAG ATC *CGC* AAG **TAC** CAG AAG AGC ACG GAG CTG CTC ATC *CGC* AAG CTC CCC TTC CAG

Row 4

Yeast 1	*AGA* TTG GTC *AGA* GAA ATC GCT CAA GAT TTC AAG ACC GAC TTG *AGA* TTT CAA TCT TCT GCC ATC GGT GCC
Yeast 2	*AGA* TTG GTC *AGA* GAA ATC GCT CAA GAT TTC AAG ACC GAC TTG *AGA* TTT CAA TCT TCT GCT ATC GGT GCT
Neurospora	*CGT* *CTC*[h] GTC *CGT* GAG ATT *CGC* *CGC* AAG TCC GAC CTC *CGC* AAG ACC GAC CTC *CGC* TTC CAG AGC TCT GCC ATC GGC CTC
Lytechinus	*CGT* CTG GTC *CGT* GAG ATT GCC CAG GAC TTC AAG ACC GAG CTC *CGC* AAG TCT GCC AGC TCC GCC ATC GGC CTC
Mouse 1	*CGC* TTG GTG *CGC* GAG ATC GCG CAG GAC TTC AAG ACC GAC CTG *CGC* TTC CAG AGC TCC GCC TCG GCC GTC ATG GCT
Mouse MEL	*CGT* CTG GTG CAA GAA ATT GCT CAG GAC TTC AAA ACA GAT CTG *CGC* TTC CAG AGT GCA GCT ATT GGT GCT
Wheat	*CGC* CTA GTG *CGC* GAG ATC GCC CAG GAC TTC AAG ACC GAC CTC *CGC* TTC CAG AGC ACC TCC GCC GTC TCC GCC

(continued)

Table 5.5 (Continued)

Row 5

Yeast 1	TTG CAA GAA TCT GTC GAA GCC TAC TTA GTC TCT TTA TTT GAA GAT ACC AAC TTG GCT GCC ATT CAC GCC
Yeast 2	TTG CAA GAA TCC GTC GAA GCA TAC TTA GTC TCT TTG TTT GAA GAC ACT CTG GCT GCT ATT CAC GCT
Neurospora	CTC GAG TCC GTC GAG TCT TAC CTC GTC TCT CTC TTC GAG GAC ACC AAC CTC TGC GCT ATC CAC GCT
Lytechinus	CTC CAG GAG GCC AGC GAA GCT TAC TTG GTC GGT CTT TTC GAG GAC ACC AAC CTG TGT GCC ATC CAC GCT
Mouse 1	CTG CAG GAG GCC TGT GAG GCC TAC CTC GTG GGT CTG TTT GAG GAC ACC AAC CTG TGC GCC ATC CAC GCC
Mouse MEL	TTA CAG GCA AGT GAG GCC AAT *CGG* GTT GGC CTT TTT GAA GAT ACC AAT CTG TGT GCT ATC CAT GCC
Wheat	CTG CAG GAG GCC GCC GCC TAC CTC GTG GGC CTC TTC GAG GAC ACC AAC CTC TGC GCC ATC CAC GCC

Row 6

Yeast 1	AAG *CGT* GTC ACT ATC CAA AAG AAG GAT ATC AAG TTG GCT *AGA AGA* TTA *AGA* GGT GAA *AGA* TCA TAG
Yeast 2	AAG *CGT* GTT ACT ATC CAA AAG AAG GAT ATC AAA TTG GCC *AGA AGA* CTA *AGA* GGT GAA *AGA* TCA TGA
Neurospora	AAG *CGT* GTC ACC ATC CAG AGC AAG GAC ATC CAG CTC GCC *CGC* CTC *CGC* GGT *CGC* GAG *CGC* AAC TAA
Lytechinus	AAG *CGT* GTC ACC ATC ATG CCA AAG GAC ATC CAG CTT GCC *CGT* ATC *CGT* GGC GAG *CGT* GCC TAA
Mouse 1	AAG *CGT* GTC ACC ATC ATG CCC AAG GAC ATC CAG CTG GCC *CGT CCC* ATC *CCC* GGG GAG *AGG* GCT´ TAA
Mouse MEL	AAA TGT GTA ACA GTT ATT CCA AAA GAT ATC CAG TTA GCA CAC AGC ATA CTC GGA GAA *CGT* GCT TAA
Wheat	AAG *CGC* GTC ACC ATC ATG CCC AAG GAC ATC CAG CTC GCC *CCC* ATC *CCT* GGC GAG *AGG* GCC TAG

a Codons for lysine are underscored, those for arginine are italicized, and those for tyrosine are stippled.
b Smith and Andrésson (1983).
c Woudt et al. (1983).
d Roberts et al. (1984).
e Sittman et al. (1983a,b).
f Sittman et al. (1981).
g Tabata et al. (1984).
h Insert the intron of 67 nucleotidyl residues.

this sector, however, is illuminating. Especially striking is the complete absence of codons for both lysine and arginine from the row 5 fraction, where four for acid amino acids are located. While row 4 is poor in triplets specifying the first of these biochemicals, that deficiency is compensated to a degree by the presence of five signaling an arginine. But it is the latter segment of row 6 that can really be stated to be arginine-rich, for it is this preterminal region that encodes nearly a third of the total in the whole macromolecule. Hence, this tail must be considered to perform an especially pertinent role in the functioning of the present histone, in basicity resembling the gene of H2A, in which a preponderance of lysine codon exists.

The Mouse H3 Sequences. The two sequences from the mouse given in Table 5.5 afford, at least in a preliminary way, an insight into the possible evolutionary origins of the polymorphism that characterizes each species of these proteins. The one referred to as mouse 1 is a representative of a near-identical pair from a single operon known as MM221 (Sittman *et al.*, 1983b); because its mate (mouse 2) differs at a mere 13 sites with only one change in amino acid specifications (a cysteine → serine change), it is omitted from the table. The second one given, here named mouse MEL from its source, murine erythroleukemia tissue, obviously diverges from the other at innumerable points (Alterman *et al.*, 1984). The same statement, of course, applies to the identical gene called MH3-6 (Sittman *et al.*, 1981). Far from enumerating these distinctions, however, present purposes are better served by examination of their nature.

An aspect of codon relations alluded to in the discussions of the H2A genes can now be given fuller consideration. The arginine codons are especially significant, because that amino acid is encoded by two different sets, one containing two beginning with A (AGA and AGG) and the other having four with the initial nucleotide C (CGA, CGC, CGG, and CGT). If the two yeast sequences (Smith and Andrésson, 1983) are scrutinized for these codons, it can be noted that either AGA or AGG are consistently employed, usually the former and never any of the C series, quite as in histone H2A. Comparisons with the others disclose that neither the *Neurospora* nor *Lytechinus* sequences (Woudt *et al.*, 1983; Roberts *et al.*, 1984) use any except the latter series for arginine and that these are restricted to CGC and CGT. A similar condition prevails in the mouse 1 gene, except that in the very last arginine codon of row 6 is AGG and on two adjacent occasions CGG occurs, as may be noted near the beginning of row 3. In contrast, the mouse MEL sequence has combinations like those of the yeast, an AGG near the end of row 1 and an AGA close to the onset of row 3. Without elaboration on the details that may be readily noted in the table, comparable retention of the primitive yeast usage can be perceived in the case of the six each of leucine and serine codons, the former being CTN and TTR and the latter TCN and AGY.

Hence, the appropriate explanation appears to be that a small number of operons, beginning with the two of yeast, has been present in the genomes of eukaryotes over the millenia, with diversification occurring gradually among the several copies. Perhaps as few as eight or ten basic sequences in a given histone type thus came into being, maybe a greater number, but more likely fewer. Some of these retained a primitive codon here, others at a different site, but all developed along semiindependent lines after the basic diversifications had occurred. Consequently, the polymorphisms represent various combinations of the basic set, suitably mutated among the major groups, plants, metazoans,

Table 5.6
Mature Gene Sequences for Histone H4[a]

Row 1

Yeast 1[b]
ATG TCC GGT *A GA* GGT AAA GGT GGT AAA GGT CTA GGT AAA GGT GGT GCC AAG *CGT* CAC --- *AGA* AAG ATT

Tetrahymena[c]
atg GCC GGT - -- GGT AAA GGT GGT AAA GGT ATG GGT AAA GTC GGA GCC AAG *AGA* CAC TCC *AGA* AAG TCT

Physarum[d]
atg TCT GGA *C GT* GGT AAA GGA GGC AAG GGA CTC GGC AAC GGA GGC GCC AAG *AGG* CAC --- *AGG* AAG GTG

Neurospora[e]
ATG ACT GGA *C[k]GC* GGC AAG GGC GGC AAG GGC CTC GGA AAG GGC GGT GCC AAG *CGC* CAT --- *CGC* AAG ATT

Strongylocentrotus[f]
atg TCA GGT *C GA* GGA AAA GGA GGA AAG GGA CTC GGA AAG GGT GGT GCC AAA *CGT* CAT --- *CGC* AAG CTT

Xenopus[g]
ATG TCT GGA *A GA* GGC AAG GGC GGA AAG GGT CTG GGC AAA GGA GGA GCT AAG *CGC* CAC --- *AGG* AAG GTG

Human[h]
ATG TCC GGC T GT GCA AAG GGC GGA AAG GGC TTA GGC AAA GGT GGC GCT AAG *CGC* CAC --- *CGC* AAG GTC

Chicken[i]
atg TCT GGC *A GA* GGC AAG GGC GGG AAG GGG CTC GGC AAA GGG GGT GCC AAG *CGC* CAC --- *CGC* AAG GTG

Wheat[j]
atg TCC GGG *C GC* GGC AAG GGA GGC AAG GGC CTA GGC AAG GGC GGC GCC AAG *CGC* CAC --- *CGG* AAG GTC

Row 2

Yeast 1
CTA *AGA* GAT AAC ATC CAA GGT ATT ACT AAG CCA GCT ATC *AGA AGA* --- TTA GCT *AGA AGA* GGT GGT GTC

Tetrahymena
AAC AAG GCT TCC ATT GAA GGT ATT ACT AAG CCC GCT ATC *AGA AGA* --- TTA GCT *AGA AGA* GGT GGT GTT

Physarum
CTC *CGT* GAT AAC ATC CAG GGT ATT ACC AAG CCT GCT ATC *CGC AGA* --- TTG GCT *CGC CGT* GGT GGT GTG

Neurospora
CTT *CGT* GAC AAC ATC CAG GGT ATC ACC AAG CCC GCT ATC *CGC CGT* --- CTC GCT *CGT CGT* GGT GGT GTC

Strongylocentrotus
CTA *CGA* GAT AAC ATC CAA GGC ATC ACC AAG CCT GCA ATC *CGT CGA* --- CTX GCT *AGA AGG* GGA GGT C͞T͞C

Xenopus
CTG *CGG* GAT AAC ATC CAG GGC ATC ACT AAG CCC GCC ATC *CGC CGC* CTG GCA *CGC AGA* GGG GGA --- GTC

Human
TTG *AGA* GAC AAC ATT CAG GGC ATC ACC AAG CCT GCC ATT *CGG CGT* --- XTA GCT *CGG CGT* GGC GGC GTT

Chicken
CTG *CGC* GAC AAT ATC CAG GGC ATC ACC AAG CCG GCC ATT *CGC* --- *CGC* CTG GCG *CGG CGC* GGC GGC GTC

Wheat
CTC *CGC* GAT AAC ATC CAG GGC ATC ACC AAG CCG GCG ATC *CGG* --- *CGG* CTG GCG *CGG CGG* GGC GGC GTC

Row 3

Yeast 1
AAG *CGT* ATT TCT GGT T TG ATC TAC GAA GAA GTC *AGA* GCT GTC TTG AAA TCC TTC TTG GAA TCC GTC ATC

Tetrahymena
AAG *AGA* ATT TCC TCT T TC ATT TAC GAC GAC TCC *AGA* CAA GTC TTG AAG TCT TTC TTA GAA AAC GTT GTT

Physarum
AAG *CGT* ATC[k] AGC AAC A CC ATC TAC GAG GAG ACC *CGT* GGA GTC CTG AAG ACC TTC TTG GAG AAC GTC ATC

Neurospora
AAG *CGT* ATC TCT GCC A[k] TG ATC TAC GAG GAG ACC *CGT* GGT GTC CTC AAG ACC TTC CTC GAG GGT GTC ATC

Strongylocentrotus
AAG *AGG* ATC TCT GGT C TC ATC TAC GAA GAG ACA *CGC* GGT GTA CTG AAG GTC TTC CTG GAG AAT GTC ATC

Xenopus
AAG *CGC* ATC TCC GGC C TC ATC TAC GAG GAG ACT *CGC* GGG GTG CTG AAA GTG TTC CTG GAG AAC GTT ATC

Human
AAG *CGG* ATC TCT GGC C TC ATT TAC GAG GAG ACC *CGC* GGT GTG CTG AAA GTG TTC TTG GAG AAT GTG ATT

Chicken
AAG *CGC* ATC TCG GGG C TC ATC TAC GAG GAG ACG *CGC* GGC GTG CTC AAG GTC TTC CTT GAG AAC GTC ATC

Wheat
AAG *CGC* ATC TCG GGG C TC ATC TAC GAG GAG ACC *CGC* GGC GTG CTC AAG ATC TTC CTC GAG AAC GTC ATC

(continued)

Table 5.6 (Continued)

Row 4

Yeast 1	*AGA* CAC TCT GTT ACT TAC ACC GAA CAC GCC <u>AAG</u> *AGA* <u>AAG</u> ACT CTT ACT TCT TTG GAT GTT GTT --- ---
Tetrahymena	*AGA* GAC GCT GTC ACT TAC ACT GAA CAC GCT *AGA AGA* <u>AAA</u> ACC GTC ACT GCT ATG GAC GTT GTC --- ---
Physarum	*CGT* GAC GCT GTG ACC TAC ACT GAG CAT GCC *CGC CGC* <u>AAG</u> ACA GTG ACT GCC ATG GAC GTT GTC --- ---
Neurospora	*CGT* GAT GCC GTC ACC TAC ACC GAG CAC GCC <u>AAG</u> *CGC* <u>AAG</u> ACC GTC ACC TCC CTC GAG GTT GTC --- ---
Strongylocentrotus	*CGT* GAT GCA GTC ACC TAC TGC GAG CAC GCT --- --- AAG *AGG* <u>AAG</u> ACT GTC ACA GCC ATG GAC GTG GTG
Xenopus	*CGC* GAC GCG GTC ACC TAC ACC GAG CAC GCC --- --- AAG *AGG* <u>AAG</u> ACC GTC ACC GCT ATG GAT GTG GTG
Human	*CGC* GAC GCA GTC ACC TAC ACC GAG CAC GCC --- --- <u>AAG</u> *CGC* <u>AAG</u> ACC GTC ACA GCC ATG GAT GTG GTG
Chicken	*CGC* GAC GCC GTC ACC TAC ACC GAG CAC GCC --- --- <u>AAG</u> *AGG* <u>AAG</u> ACG GTC ACG GCC ATG GAC GTG GTC
Wheat	*CGC* GAT GCC GTC ACC TAC ACC GAG CAC GCC *CGC CGC* <u>AAG</u> ACC GTC --- --- ACC GCC ATG GAC GTC GTC

Row 5

Yeast 1	TAT GCT TTG <u>AAG</u> *AGA* CAA GGT *AGA* ACC TTA TAC GGT TTC GGT GGT TAA
Tetrahymena	TAC GCC CTC <u>AAG</u> *AGA* CAA GGC *AGA* ACT CTC TAT GGT TTC GGT GGT TGA
Physarum	TAT GCC CTC <u>AAA</u> *CGC* CAG GGA *CGC* ACT CTG TAC GGA TTC GGC GGC TAA
Neurospora	TAC GCC CTC <u>AAG</u> *CGC* CAG GGC *CGT* ACC CTC TAC GGT TTC GGT GGT TAA
Strongylocentrotus	TAT GCA CTA <u>AAG</u> *AGG* CAG GGT *CGT* ACA TTG TAC GGC TTC GGC GGC TAA
Xenopus	TAT GCT CTG <u>AAG</u> *CGC* CAA GGA *CGC* ACT CTG TAC GGC TTC GGA GGT TAA
Human	TAC GCG CTC <u>AAG</u> *CGX* CAG GGG *AGX* ACC CTC TAC GGC TTC GGA GGC TAG
Chicken	TAC GCG CTC <u>AAG</u> *CGC* CAG GGA *CGC* ACC CTC TA' GGC TTC GGC GGT TAA
Wheat	TAC GCG CTC <u>AAG</u> *CGC* CAG GGC *CGC* ACC CTC TAC GGC TTC *CGC* GGC TAA

[a]Codons for lysine are underscored, those for arginine are italicized, those for tyrosine are lightly stippled, and those for glycine are darkly stippled.
[b]Smith and Andrésson (1983).
[c]Bannon *et al.* (1984).
[d]Wilhelm and Wilhelm (1984).
[e]Woudt *et al.* (1983).
[f]Grunstein *et al.* (1981).
[g]Turner and Woodland (1982).
[h]Sierra *et al.* (1983).
[i]Sugarman *et al.* (1983).
[j]Tabata *et al.* (1983).
[k]Interrupted by an intron at this point.

fungi, and the like, and again in their respective phyla, classes, and orders. Thus, even in genomes like those of sea urchins that contain several hundred histone operons, the heterogeneity of the genes they contain will be found to embrace relatively few basic variants. This statement is borne out by the mouse H3 genes that have been sequenced, with mouse 2 basically like mouse 1 and with mouse MEL and MH3-6 being identical.

5.3.5. The Mature Histone H4 Gene

The last of the core histones and the second of the so-called arginine-rich types, H4, in being amply supplied, compensates to a degree for the paucity of sequences established for some of the preceding types. As may be seen in Table 5.6, the coding sequences for this species are available from a unicellular organism in addition to the usual yeast, plus a pair from fungi and one from a seed plant, over and above the standard sea urchin, amphibian, mammalian, and avian forms. In the table it can be noted, too, that the underscored codons for lysine and those italicized for arginine fall nicely into regular arrays with only minor exceptions, recalling the uniformity found in H2A. Further, it is also quite obviously the smallest of these proteins, its gene subtending only five rows instead of the customary six.

Distribution of Codons for Basic Amino Acids. Because of the constancy of structure, it is sufficient just to enumerate the overall characteristics, as in the case of histone H3. The immediate 5′-terminal portion represented in row 1 of Table 5.6 is especially impressive with its array of codons for the active amino acids. No fewer than five of its total of ten triplets specifying lysine are found here, along with three for arginine. This combination, together with the codons for four arginines and one lysine in row 2, coupled with only a single triplet for an acid amino acid, make this end of the encoded product strongly basic in reaction. In the remainder of the cistrons, these triplets are more evenly distributed, with two each for lysine and arginine in row 3, one for lysine and three for arginine in row 4, and one and two, respectively, in row 5. Hence, unlike H2A and H3, there is no highly active tail region in the present species.

The tyrosine codons are unevenly distributed between lower and higher eukaryotes, only one site thus occupied near the beginning of row 3 being shared by all. At the end of row 4, and medially in row 5, the protistan sequences have two more, both of which are absent from the metazoan and seed-plant representatives. However, the latter cistrons then gain one in a brief insert that is missing from the protistan genes, a hexanucleotide addition located just before the extreme 3′ end of the mature gene. Among the other outstanding features of the sequences encoding this histone is the presence in row 1 of a large concentration of codons for the amino acid glycine, no fewer than eight (stippled) being located here. This contrasts with the occurrence of two such codons in the corresponding sector of H3, four in H2A, and none in H2B.

Because none of the source organisms are represented in the table by two variants of this gene, the observation made in the preceding section cannot be carried through here in comparable fashion. Nevertheless, that the idea may prove equally applicable to the present substance is intimated by a few series of arginine codons that similarly show the employment of triplets with an initial A in the yeast (and often in *Tetrahymena*) and those beginning with C among the higher forms. This is exemplified by the very first series in row 1, in the four of row 2, and most of the remainder. The few leucine and serine codons present vary too greatly from one representative to another to offer either contradictory or supportive evidence.

5.3.6. The Mature Gene for Histone H5

The gene for histone H5 has a number of unusual characteristics that set it apart from the common species. In the first place it is expressed solely in the erythroid cells of vertebrates, including fish, but not mammals; however, it is known well only from the chicken and other domestic fowl. Unlike the other types, all of which are represented in the genome by multiple copies, the H5 gene is unique and unlinked; like the remaining types, except for some from fungi, no introns are present (Krieg *et al.*, 1983). Following transcription, the mRNA for this type, in contrast to most others, acquires a poly(A) sequence (Krieg *et al.*, 1982), in spite of the absence of the typical signal AATAAA. The presence of such a tail is suggestive of this gene having arisen relatively recently in some highly advanced metazoan, probably a vertebrate. In the chromatin of erythroid cells, the H5 product replaces a part of the H1, so that their H5 + H1 content is in equal molar proportions to that of the latter alone in somatic tissues in general. Consequently, the chromatin of this tissue differs from that of all others, but the specific effects of this species on nucleosome spacing and structure of chromatin supercoiling remain unexplored.

Characteristics of the Mature Gene. One of the few mature coding sequences available for histone H5 is that from chicken erythrocytes, from which it has been established thrice (Krieg *et al.*, 1982, 1983; Ruiz-Carrillo *et al.*, 1983). This, plus the recently sequenced cistron from duck (Doenecke and Tönjes, 1984), is the basis for Table 5.7, but the histone H1 cistron from the chicken is also included for comparison. As may be readily observed, some support exists for considering the two species to be related, as was proposed earlier on the basis of the amino acid sequence of the polypeptide (Yaguchi *et al.*, 1979). However, the number of distinctive features appears to outweigh the similarities.

To begin with, one of the most outstanding distinctions of histone H1 is missing, the alanine-codon-rich 5'-terminal sector. In row 1 of the present species, only two (italicized) are present in the chicken and six in the duck, compared to the nine in H1, and in row 2 there are four in the duck and five in the other, a series of three being located medially, whereas H1 has three scattered and two adjoining triplets. Additionally, the distribution of lysine and arginine codons is quite distinct from that in the other sequence. In the chicken one of the three in each of rows 1 and 4 is correlated with the one of the comparative material, and in the duck two of the four; none in row 2 does so; two of three in row 3 are identically located; and in row 5 three of its seven match those of H1, three of six in row 6, four of nine in row 7, and three of eight in row 8. Finally, while row 9 of the sequence closes with a series of three lysine codons, quite as in H1, as a whole the majority of these triplets throughout the two genes cannot be clearly homologized. The correspondences in location of the arginine codons are even fewer, for H1 is not as abundantly supplied with them as H5. Row 5 seems to be the sector of greatest similarity, there being four identical corresponding sites and six related ones. Perhaps the chief resemblance between the two is in the length, for each is about 40% longer than any of the core types, but H1 is distinctive even in this aspect.

Table 5.7
Mature Gene Sequences for Histone H5[a]

Row 1

Chicken H5[b]   atg ACG GAG AGC CTG GTC CTA TCC CCA *GCC* CCA GCC --- AAG CCC AAG AAG CGG GTG --- AAG *GCA* TCG CGG CGG
Duck H5[c]   atg ACG GAC AGC CCC ATC CCG *GCC* CCG *GCC* GCC --- AAG CCC AAG AAG CGG *GCG* --- AAG *GCT* CCT CGG AAG
Chicken H1[d]   atg TCG GAG ACC CCC GTT GCC --- *GCG* CCC *GCG* --- GTG TCT *GCG* CCC GGC *GCC* AAG *GCC* *GCC* *GCC* AAG

Row 2

Chicken H5   TCG *GCA* TCG CAC CCC TAC CCC GAG ATG *GCG* *GCG* *GCC* ATC CGT *GCG* GAA AAG AGC GGC GGC
Duck H5   CCG *GCG* TCC CAC CCC AGC TAC TCG GAA ATG CTG *GCG* *GCC* ATC CGG *GCC* GAG AAG AGC GGT GGC
Chicken H1   AAG CCG AAG AAG *GCG* *GCG* GGC GGC *GCC* CCC *GCG* AAG CCC AGC GTC ACC GAG CTG ATC

Row 3

Chicken H5   TCC TCG CGG CAG TCC ATC CAG AAG TAC ATC AAG AAG GTG GGC CAC AAC *GCC* GAT CTG CAG
Duck H5   TCC TCC CGG CAG TCC ATC CAG AAG TAC GTG AAG AAG GTG GGC CAC CAG CAG *GCC* GAC CTC CAG
Chicken H1   ACC AAG GTG TCC TCC AAG GAG GAG *GCC* CGC AAG GGG CTC TCC *GCC* *GCC* CTC CTC AAG AAG *GCG* CTT *GCC*

Row 4

Chicken H5   ATC AAG CTC TCC ATC CGA CGT CTC CTG *GCT* GCC GGC GTC CTC AAG CAG ACC AAG AAA GGG GTC GGG *GCC* TCC
Duck H5   ATC AAG CTC TCC ATC CGG CGC CTG CTC *GCC* GGC GGC GTC CTC AAG CAG ACC AAG AAA GGG GTC GGG *GCC* TCC
Chicken H1   *GCC* GGC TAC GGC GTG GAG AAG AAC AAC AGC CGC ATC AAG CTG GGG CTC ATC AAG AGC GTC AGC AAG

Row 5

Chicken H5 GGC TCC TTC CGC TTG *GCC* AAG AGC GAC AAG *GCC* AAG AAG AGG TCC CCC --- GGG --- AAG AAG AAG ---

Duck H5   GGC TCC TAC CGC CTG *GCC* AAG GGG GAC *GCC* AAG AAG AAG TCC CCA GCT GGG AGG AGG AAG AAG AAG AAG

Chicken H1 GGC ACC CTG GTG CAG ACC AAG GGC ACC GGC TCG GGC TCT TTC --- AAG --- CTG AAT AAA AAG ---

Row 6

Chicken H5 GCC CTC AGG AGG TCC ACG TCT CCC AAG AAG *GCG* *GCG* AGG CCC AGG AAG *GCC* AGG TCA CCG *GCC* AAG AAG

Duck H5   GCT CCC CGA AGG TCC ACG TCA CCC AGG AAG *GCG* *GCG* AGG CCC AGG AAA *GCC* AGG TCG CCG *GCC* AAG AAG

Chicken H1 CCG GGT GAG ACA AAA *GCG* AAA *GCG* ACT AAG AAG AAG CCC AAG *GCG* *GCG* AGG CCC AGG *GCG* *GCC* *GCC* AAG

Row 7

Chicken H5 CCC AAA *GCC* ACC *GCC* AGG AAG AAG *GCC* AAG AAG TCG *GCA* AGC CCC AAG AAG *GCC* AAG AAG CCA AAG

Duck H5   CCC AAA *GCC* *GCC* *GCC* AGG AAA *GCC* AGG AAG AAG TCG *GCC* AGC CCC AAA AAA *GCC* AAA AAA CCA AAG

Chicken H1 AAG *GCT* GCG *GCT* AAG AAG CCC AAG AAG *GCA* *GCG* *GCG* GTG AAG AAG AGC CCC AAG AAG *GCG* AAG AAG

Row 8

Chicken H5 ACT GTT AAG GCC AAG TCG CGG AAG *GCC* TCC AAG *GCC* AAG AAG GTG --- AAG CGG TCG AAA CCC AGA *GCC*

Duck H5   ACT GTT AAG GCC AAG TCG CTG AAG ACA TCC AAG GTG AAG AAG *GCA* --- AAG CGG TCA AAA CCC AGA *GCC*

Chicken H1 GCG *GCA* *GCT* GCT GCC ACC AAG AAG AAG *GCG* *GCG* *GCC* AAG *GCT* GGC *GCGC* CCC AAG AAG ACT *GCC* AAG AGC CCG

(continued)

Table 5.7 (Continued)

Row 9

Chicken H5	AAG	TCT	GGC	*GCC*	CGG	AAA	TCG	CCC	---	---	---	---	---	---	---	---	---	---	---	---			
Duck H5	AAA	TCC	GGC	*GCC*	CGA	AAA	TCG	CCC															
Chicken H1	*GCC*	AAG	*GCA*	AAG	*GCG*	GTG	AAG	CCC	AAA	*GCT*	*GCC*	AAG	TCA	AAG	*GCG*	*GCC*	AAA	CCC	AAG	*GCG*	*GCC*	AAG	*GCA*

Row 10

Chicken H5	---	---	---	---	AAG	AAG	AAG	TGA
Duck H5	---	---	---	AAA	AAG	AAG	AAG	TGA
Chicken H1	AAG	AAG	*GCA*	ACT	AAA	AAG	AAG	TAA

[a]Codons for lysine are underscored, those for alanine are italicized, and those for arginine are boxed.
[b]Krieg *et al.* (1982).
[c]Doenecke and Tönjes (1984).
[d]Sugarman *et al.* (1983).
[e]Insert 18 nucleotide residues.

5.4. TERMINATION IN HISTONE GENES

As a separate process, termination of transcription of histone genes has rarely been studied, but fortunately an abundance of their trains has been determined. Although in only a very few instances have the actual termination sites (underscored) been established, Table 5.8 presents about 30 of the pertinent sectors of the 3'-flanking regions. Wherever they occur, runs of six or more Ts are italicized, and regions of dyad symmetry are underlined with arrows that point toward the loop of the folded structure. Also indicated is whether or not a polyadenylation signal (AATAAA) is present, the sequence of which is occasionally modified.

Termination Signals. The standard termination signal consisting of six or more adjacent Ts rarely occurs in those histone genes that have been sequenced. Indeed, just three in Table 5.8 show a structure of that sort, two H1 trains (*Xenopus* 1ch4 and the murine) and one H4 (*Neurospora*). Since the first two of these are provided also with the region of dyad symmetry to be discussed shortly, it is evident that this classical symbol is of only occasional value in the cessation of transcription among the histone genes, except perhaps those of the fungi.

The trains from yeast genes as a whole seem to be devoid of any recognizable terminator, unless a series of four Ts can serve in that capacity. This number can be found in the H2A and H4 genes from that organism in column D and in its H3 gene in column B. In the *Neurospora* H3 train, a like number similarly located can be found, but its H4 3' flank possesses two series of six Ts each, separated from one another by a single C. In the lone representative from *Tetrahymena*, the train of an H4 coding region, at most three Ts are adjacent, situated in column A. Consequently, no apparent standard combination for triggering transcription cutoff can be noted among the protistans.

A Possible Metazoan Terminator. Nearly all of the histone gene trains from metazoan sources shown in Table 5.8 have a region of dyad symmetry, located in columns E and F—the only exception is the chicken F H2A gene (Harvey *et al.*, 1983). Although this stem-and-loop structure has been pointed out previously in occasional gene sequences, its frequency of occurrence and near constancy have never before been suspected; however, Busslinger *et al.* (1979) demonstrated its presence in all the sea urchin histone genes. The double base-pairing sectors consist of (G)GCYCT(T) and (A)AGRGC(C), typically with TT- as the loop structure; the 5' portion of the stem is identical in 18 of the 22 trains thus equipped, whereas the 3' part is constant in 17. The four in the 5' stem sequence that are disparate is really one exception copied four times, for it occurs in three variants of the H5 gene trains from the chicken and the one from the duck. The same statement applies equally to four of the five divergences in the 3' part. In this set, the dyad reads (C)TCTT(C)CA (G)YA TG(G)AAGA(G), so that it is not convincingly homologous to the others. The remaining exceptional structures on the 3' half are the GGGCC of the human λ41 cistron for H4 and the GGAGCC of the chicken H2B, if the noncanonical base pairing (underscored in the table) is found to take place.

It is interesting to note the presence of a comparable, largely homologous stem-and-loop structure in the train of the wheat H4 gene. Even if the T–G noncanonical base pairing does occur, this sector of dyad symmetry is quite short, consisting at most of four base pairs. Despite this green plant's being only remotely related to the metazoans, its similarity to the hairpin loop of the latter is too striking to be ignored, for the sites·

Table 5.8
Trains of Histone Genes

	A	B	C	D	E	F	G	Poly(A) Signal[a]
H1								
Strongylocentrotus[b]	TTTGCACGCCAACTT	TCCCATCCTACC	AAAAC----	----------	GGCTCTT	TTCAGAGCC	ACCACA	No
Xenopus Xlch4[c]	AGGCTCGCTTCGGTT	TTTTCTGTTTTT	TGCCCCCA	CAATCAAGATA	AGCCCTT	TTAAGGGCC	ACAAAA	No
Xenopus Hw19[d]	GCAGCTGCTCCCTC	GCTCGCTCACTA	GTGGCCCAT	TCAACCAAA--	GGCTCTT	TTAAGAGCC	ACCACA	No
Chicken[e]	GATGACAGAAGAAAC	TCGAGTCTGCTC	ATTTAAAAA	CCCCA-----AA	GGCTCTT	TTAAGAGCC	ACCCAT	No
H2a								
Yeast 2[f]	AAAGAAACAAAGCCG	GTCAAACATTCT	TGTCATTTG	GGGTTTTAAAG	TAGGTCA	TATGAGGAA	GACTGG	Yes
Murine[g]	GCCAGTGAGCTAAGT	TTTTTTTTTTTTTT	T+17Ts AA	ACAAAACCCAA	GGCTCTT	TTCAGAGCC	ACCACT	No
Human b5G[h]	AGAGTTAACGCTTCA	TGCACTGCTGTT	T+4GTCAGC	AGACA+16 AA	GGCTCTT	TTCAGAGCC	ACCTAC	No
Chicken xho[i]	GCACCGCGGCAGGCAG	CGCTCTCTCGAGA	GAACAGTCC	--------AA	AGTCCTT	TTCAGAGCC	ACCCAC	No
Chicken F[j]	AGGACGCGGGGGTCCC	ACCCGCGCGCCCG	TCCGGGGCC	CAACCGGCCCC	AACGCGG	CCGGGCTGC	GCGGCG	No
H2B								
Murine[g]	GTGCTCAAGACTCAG	CTCTTAACCCA-	----------	----------AA	GGCTCTT	TTCAGAGCC	ACTCAA	No
Human h4A[h]	ACAGTGAGTTGGTTG	CAAACTCTCAAC	CCTAAC---	----------	GGCTCTT	TTAAGAGCC	ACCCAT	No
Chicken[k]	AGGGGTGCGGATTAC	TGATTTTAACCC	AAA-------	----------	GGCTCTT	TTCGGAGCC	ACCATT	Yes
H3								
Yeast 1[l]	TTTGTTGATTGTCAT	CAGTTTTAGTAA	AAA+4 ACA	AAAACACAATA	AAATATA	AATCAATAT	ATTTAG	No
Neurospora[m]	GCGACTCTTCGATAT	GGAGTAGTTTGC	TTTGGGTTT	TCGGGGTAGTC	TAGTCAG	ATTCTGGGG	TTATAT	No

Lytechinus [n]	GTCGTTTAACCTTGT	TTGAAGCCAAAA	---------	----------CC	GGCTCTT	TTCAGAGCC	ACTAAA	Yes
Murine 1 [g]	GGGTTTCTTGTTAATC	CACACAACCACT	TTA------	----------AA	GGCTCTT	CTTAGAGCC	ACCCAT	No
Murine 2 [g]	TAGGCAGGCTTTCTA	CACTGGCACGTA	AACCAAA--	----------AC	GGCTCTT	TTAAGAGCC	ACCTCC	No
Human h5B [h]	ATGTAAAGTTACTTT	TTGATCAGTCTT	AAAACCCA-	----------AA	GGCTCTT	TTCAGAGCC	ACCCAC	No
H4							
Yeast 2 [l]	ACAA-TCGGTGTTA	AACAATCGGTGT	TTGAAATTA	-TTTTCATGCCT	TTCAAAA	AATAAAATA	ACA	Mod
Tetrahymena [o]	ACAAAATATTTATCT	TAAAAAATTAAA	AAGTAAAAA	-GCTGCATGCTT	ACTCAAA	GGTAATAGT	GTAATT	No
Neurospora [m]	ATGTCTCGCCGTCAT	TAGCAGCAGCGT	GTCTTTTTT	-CTTTTTTCTTC	ACCATAA	CGACGACGA	ATAATC	No
Lytechinus [n]	GAAGTTCATCTCTTC	AAACCCATCAAC	AAA------	----------CC	GGCTCTT	TTCAGAGCC	ACCAAA	No
Strongylocentrotus [p]	TAGAATAACAA---	----------	----------	----------AC	GGCTCTT	TTCAGAGCC	ACCAAA	No
Human 4A [q]	GGGTCCTTCTCTACC	AATAA------	----------	----------GA	GGCCCTT	TTCAGGGCC	CCTACT	No
Human B41 [r]	GCCCCGCCTCCAGCT	TTTGCACGTTTCG	ATCCCA---	----------AA	GGCCCTT	TTTGGGCCG	ACCACT	No
Chicken [e]	ACTCGCTCCGATTC	CGGCCACCCGAA	CTCGTTTTT	-AGCAACCCAAA	GGCTCTT	TTCAGAGCC	GCCCAC	Mod
Wheat [s]	GGGCCGGCCGGCCGA	CGGGAGTCACTC	TTTGTCGCC	-GCCTGC+24CC	GACTTGT	TT-AGTTCG	CTATTT	No

(continued)

Table 5.8 (Continued)

	A	B	C	D	E	F	F	Poly(A) Signal[a]
H5								
Chicken 1[t]	GCAGCCCGGGGGCTT	-GCCCAGGCTCT	CCCCATTGG	TTTTCT+116 CT	TCTTCCA	TATGGAAGA	GTTCCC	Mod
Chicken 2[u]	GCAGCCCGGGGCTTT	-GCCCAGGCTCT	CCCCATTG-	GTTTCT+118CT	TCTTCCA	TATGGAAGA	GTTCCC	Mod
Chicken 3[v]	GCAGCCTGGGGGCTT	-TGTCCAGGCTC	TCCCCATTG	GTTTCT----CT	TCTTCCA	TATGGAAGA	GTTCCC	Mod
Duck[w]	GAAGTCTGGAGGCCT	TTGTCCAGACTC	CCCCCGTTG	GTTTCT+135CA	TCTTCAG	CATG-AAGA	GTTCCC	

[a]Modified as AATAA or AACAAA. Arrows indicate sectors of dyad symmetry, whereas underscoring points out possible noncanonical base pairs and experimentally established termini.

[b]Levy et al. (1982).
[c]Zernik et al. (1980).
[d]Turner et al. (1983).
[e]Sugarman et al. (1983).
[f]Choe et al. (1982).
[g]Sittman et al. (1983a,b).
[h]Zhong et al. (1983).
[i]D'Andrea et al. (1981).
[j]Harvey et al. (1983).
[k]Grandy et al. (1982).
[l]Smith and Andrésson (1983).
[m]Woudt et al. (1983).
[n]Roberts et al. (1984).
[o]Bannon et al. (1984).
[p]Grunstein et al. (1981).
[q]Heintz et al. (1981).
[r]Sierra et al. (1983).
[s]Tabata et al. (1983).
[t]Ruiz-Carrillo et al. (1983).
[u]Krieg et al. (1983).
[v]Krieg et al. (1982).
[w]Doenecke and Tönjes (1984).

homologous between the two types are not confined to that structure alone, but extend into the adjacent regions.

Problems with Origins. A very unusual problem with origins is presented by the stem-and-loop structure, one that must be resolved before any full understanding of the processes of evolution at the cellular level can be gained. To summarize, the facts related to this enigma are these:

1. There are five different genes of histones, plus a sixth in certain specialized tissues of vertebrates.
2. With the exception of the gene for histone H1, which arose after the true yeasts had branched off, these genes are homologous to a high degree among all eukaryotes that have been studied. That for histone H1 shows a similar degree of homology wherever it is present.
3. Hence, in the Metazoa the individual histone genes have been inherited from the protistan ancestral stock.
4. Yet the trains of all five types have a homologous structure, even though the remainder of those 3'-flanking sectors show little evidence of common descent.
5. Thus the stem-and-loop must have been applied to these genes separately in each case in some early metazoan type or at least in the echinoderm–chordate line of ascent.
6. Furthermore, a similar structure arose in the seed plants or their forebears, the degree of homology between which and the metazoan's being suggestive of origins by a common mechanism.
7. In the case of the H5 genes, which have a comparable, but not homologous, structure, the differences arise as a result of these cistrons having a comparatively recent origin, probably in an early vertebrate.

There is only one mechanism proposed thus far in the literature capable of programming nucleotidyl additions to the genome or specifically modifying the sequence of selected regions of the DNA. This is the supramolecular genetic mechanism, the existence of which was proposed on two previous occasions (Dillon, 1981, 1983) because of the obvious need for programming during the assembly of such organelles as the centrioles, microtubules, and microfilaments and in the genomic changes that occur during immunogenesis and development. Since it may have induced these dyad symmetric regions, this apparatus also may be effective in evolutionary events at the molecular level as predicted (Dillon, 1983). It will be of great interest to see if other comparable sets of genes for proteins that are analyzed in subsequent chapters also reveal the presence of such a mechanism.

5.5. RIBOSOMAL PROTEIN GENES

Since the genes remain unestablished for a second class of proteins that bind nucleic acids, the high-motility group, its members cannot be discussed at this point as they logically should. However, those for a third type, the ribosomal proteins that bind ribonucleic acid, have been determined in fair abundance. These are a heterogeneous lot, as is shown in connection later with their mature gene sequences, whose individual functions are difficult to establish. Some beginnings have been made toward the determination of specific functions of a certain few representatives. That known as S1 in *E. coli* is essential to the binding of the messenger to the small ribosomal subunit (Suryanarayana and Subramanian, 1984) and L16 binds the tRNA, seemingly through a reaction with the -CACCA 3′-terminal sequence (Maimets *et al.*, 1984). In addition, L23 has been established as being located in the peptidyl-transfer area of the ribosome, where it probably plays a specific role in that process (Vester and Garrett, 1984). Nevertheless, as a whole, knowledge of this critical aspect of ribosome function is extremely limited. Generally speaking, the chief obstacle to comprehending their separate roles is that the 50–80 or more types that exist work as a unified team to translate mRNAs into polypeptides, the details of which processes are beyond the scope of this book. Here are viewed such topics as the genomic organization and structure of their cistrons, initiation and termination of their transcription, and evolutionary considerations that may arise.

5.5.1. The Organization of Ribosomal Protein Genes

Because of the large number of individual species of these ribosomal protein genes, discussion of their organization in the genomes of principal experimental organisms can scarcely be as succinct as was the case with the less varied types examined earlier. Nor have the arrangements in the DNA been so thoroughly explored. Consequently, meaningful discussion of this aspect is necessarily limited largely to the genes of the two models that have been studied most, the familiar *E. coli* among the Prokaryota and the (baker's) yeast (*S. cerevisiae*) of the Eukaryota. However, wherever available, brief details of other forms are also presented. In all cases the species of these proteins are named first on the basis of the ribosomal subunit of which they form a part, large (L) or small (S), combined with a number in order of their mobility during electrophoresis in decreasing order. Thus L1 and S1 are the two largest and slowest moving in each source organism.

Gene Organization in E. coli. In *E. coli* the genes for its 52 species of ribosomal proteins are organized into a minimum of 16 operons. Among the better known of these are *spc, str, s10,* and α (Figure 5.4). The first of these has recently had its entire nucleotide sequence established and was shown to contain the genes for ten species (L14, L24, L5, S14, S8, L6, L18, S5, L30, and L15, in that order) plus several reading frames for a protein-export and other unidentified proteins (Cerretti *et al.*, 1983). That operon referred to as *s10* includes coding regions for S10, L3, L4, L23, L2, S19, L22, S3, L16, L29, and S17 (Figure 5.4; Dean and Nomura, 1982). Members of the *str* gene set encode S12, S7, and a number of various other proteins, including certain ones also involved in translation. Finally, the α operon contains cistrons for S4, S11, S13, and L17, along with one for the α subunit of RNA polymerase (Bedwell *et al.*, 1985). In some cases the operons are quite short; l11, for example, includes just L11 and L1, as far as is currently established. None of these genes appears to be repeated in the genome. Thus, in broad terms these bacterial cistrons fall into class IB, category I.

Gene Organization in Yeast. As a whole the genes for the approximately 80 ribosomal proteins of yeast are not clustered as they are in the bacteria, and, moreover, frequently, but not uniformly, they are represented by multiple copies (Fried *et al.*, 1981;

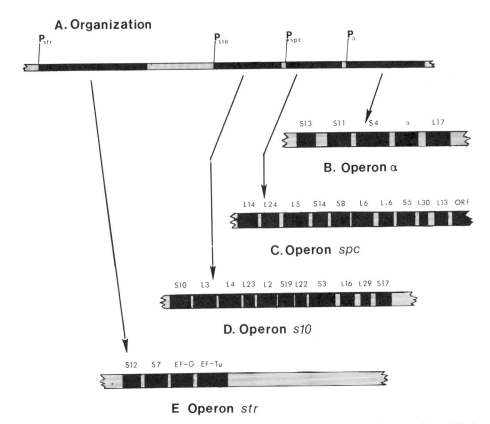

Figure 5.4. Four of the 16 or more operons for ribosomal proteins in *E. coli*. These four are clustered in the genome, but are transcribed separately.

Monk *et al.*, 1981; Woolford and Rosbash, 1981). For instance, the *CRY1* gene, which encodes ribosomal protein 59, is present only as a single copy, located apart from any other for these substances (Larkin and Woolford, 1983), as is that for L25 (Leer *et al.*, 1985b), whereas the single genes for S24 and L46 are adjacent and divergently transcribed (Leer *et al.*, 1985a). However, the majority are duplicated. For a case in point, S16a, which forms a short operon with an unidentified ribosomal protein called rp28, is present twice in the haploid genome (Molenaar *et al.*, 1984). That for S10 similarly is represented by two functional copies (Leer *et al.*, 1985b).

    Gene Organization among Metazoans. Very little has been established concerning the organization of ribosomal protein genes in metazoans, not even in *Drosophila*. Available information indicates that certain (unidentified) representatives for this class of substance occur as a single copy (Vaslet *et al.*, 1980); however, at least one has been found to be repeated an indefinite number of times (Fabijanski and Pellegrini, 1982). In the latter case, the multiple copies were associated with ribosomal genes, which were similarly repeated.

    Among the vertebrates there appears to be an increase in reiteration of these cistrons with evolutionary advancement. In *Xenopus* and other amphibians, most of these genes seem to be present as single copies, scattered through the genome (Pierandrei-Amaldi *et al.*, 1982), but L1, L14, and S19 are represented by two copies each, and S1, S8, and L32 by perhaps as many as four or five (Bozzoni *et al.*, 1981). Mammals alone have been demonstrated to have a fair number of each species present. In the mouse, for instance, 7–20 duplications per haploid genome have been detected, and in the human even greater repetition has been suggested (Monk *et al.*, 1981; Pierandrei-Amaldi *et al.*, 1982).

Gene Organization in Organelles. So little knowledge has been gained concerning the organization of the ribosomal protein genes from the two types of DNA-containing organelles that clarity is best served by treating them in a unified fashion, as here. In the yeast mitochondrion one gene, *var1*, has been shown to encode a ribosomal protein of the small subunit that has not been further identified (Zassenhaus *et al.*, 1984). This, like most other genes of the organelle, is present as a single copy, being situated in a poly-cistronic precursor which also contains the cistron for ATPase subunit 9 and $tRNA^{Ser}_{UCN}$. In the bovine, murine, and human mitochondrial genomes, whose entire sequences have been established as reported earlier (Anderson *et al.*, 1981, 1982; Bibb *et al.*, 1981), the cistron for the corresponding $tRNA^{Ser}$ is located between those for cytochrome oxidase subunits I and II, along with several others likewise encoding various tRNAs. Since no coding region of ATPase subunit 9 exists in these organelles of mammals, only that for subunit 6, apparently the yeast ribosomal protein gene *var1* has similarly been deleted during the reductional processes to which their DNA has been subjected. Moreover, because only nine unidentified reading frames exist within the genomes, it is obvious that most, if not all, of the codons for the 85 species of proteins of the mammalian organelle must be located in the nucleus (Matthews *et al.*, 1982). Consequently, their presence must be borne in mind as the nuclear genome is probed for the ribosomal protein genes.

This suggested nuclear location of these mitochondrial cistrons is supported by analogy to the situation for the corresponding coding sequences of the chloroplast. In that organelle of *Chlamydomonas*, in which genomic reduction has not occurred, five to six of the large-subunit and ~14 of the ribosomal small-subunit proteins are encoded by chloroplast genes, the remaining 26 or 27 of the large and seven of the small being products of nuclear cistrons (Schmidt *et al.*, 1983). Nothing, however, has been established concerning their organization in the genome.

5.5.2. Transcriptive Initiation of Ribosomal Protein Genes

In the initiative processes of transcription of ribosomal protein genes, a modest beginning has been made, especially with those from *E. coli* and, to a smaller extent, from yeast. A fair representation of the current state of knowledge is contained in Table 5.9, where 16 sequences are made available. That existing information is strongly limited, however, is made obvious by the inclusion of only a single representative each from a mitochondrion, indicated by a subscript M, and a chloroplast, marked with a subscript C. In the majority of cases, furthermore, the promoter sequences are presumptive only, the yeast mitochondrial *var1* promoter alone having been experimentally determined among eukaryotes. Even the actual start site remains speculative in a few of the examples cited, those that have been established firmly being shown underscored. When only a region denoting start sites has been determined, italics are employed.

Table 5.9

Possible Promoters of Ribosomal Protein Genes[a]

	A	B	C	Promoter	D	E	F	G	Initiation	
E. coli str[b]	ATATTTCTT	GACACCTTT	TCGGCATCG	CCCTAAAATTCG	G--------	------	------	------	---CGTC	+90
E. coli rpsA P1[c]	CAAGAACGT	GCGCATCGC	CGCATGCTA	CAGTTGCAGGTG	AAGGGCTTT	AGTGTT	-AACTT	T------	GAGCGCC	+875
E. coli rpsA P2[c]		GTGAAG	GGCTTTAGT	G-TTAACTTTGA	GCG--CCTTT				---TGG	+868
E. coli rpsA P3[c]	CTTTAGTGT	TGGATTCCA	CCACCTTAA	GCATTGAGCAAG	TGATTGAAA	AAGCGC	TACAAT	------	ACGCGCG	+132
E. coli rpsA P4[c]		CAAGTGA	TTGAAAAAG	CGCTACAATACG	CGCGCCAGA	AATTGG	CTCTCG	CAT+8	--GAATTT	+98
E. coli spc L14[d]	ATAATGCCG	CGCCCTCGA	TATGGGAT	TTTTAACGACCT	GATTTTCGG	GTCTCA	GTAGTA	GTT+6	--CATTAG	+11
E. coli a[e]	CTTTTCTTG	CAAAGTT-G	GGTTGAGCT	GGCTAGATTAGC	--------	------	------	------	--CAGCC	+93
Yeast[M] var1[f]				ATATAAGTA	+630 and +550					
Yeast rp29 P1[g]	GATATATTC	AACGGATGG	TGTGTATTT	TAGTACGTTTAC	GCAGGTTTG	CGCTGT	TGCTA-	------	-TTGCCA	+500
Yeast rp29 P2[g]				AGTTAAGCGAA	--------	------	------	------	--GACACCA	+482
Yeast S33[h]	TGAAATTTC	AGTCTAATA	GATGATTTA	TTGTAAATTACA	GTTGTGTTC	GTTTTC	GATTCT	TCT+9	AAAACCA	+11
Yeast L25[i]	TACAATCTT	TTAGATATT	ATCTTTTAA	AATTATTTTAAA	ACAATTTTA	AATGTA	TCTCAT	+25GCT	AAACAAA	+19
Yeast L17a[i]	TGTTAGTAG	CTAAAAATC	TCTTACTTT	CTTTATTTTGAT	TCTTGGGTC	TTACAA	GCAATA	---CAA	AACCAAC	+19
Tobacco[C] S19[j]	CGCCGTAGT	AAATAGGAG	AGAAAATCG	AATTAAATTCTT	CGTTTTTAC	AAAAAA	AAA---	------	AAAAATA	+11
Drosophila rp49[k]	TGCATTAGT	G +17 TTA	GCTTGATAA	GTGATATTTCCA	GTGGGTCAG	TGCACT	AATGGC	+14CCT	ACCAGCT	+4
Mouse L32[l]	TCATCCCTC	CAGTATCCC	AAGTAACAT	CATTATAATACA	CATTACAAA	TTAGCT	GCTG--	------	---CTTC	+48

[a]Subscript M indicates mitochondrion, C indicates chloroplast. Presumptive promoters and ancillary sites are italicized. Underscoring marks establ shed start sites.
[b]Post et al. (1978); Post and Nomura (1980).   [c]Schnier and Isono (1982); Pedersen et al. (1984).   [d]Cerretti et al. (1983).   [e]Post et al. (1980).   [f]Zassenhaus et al. (1984).   [g]Mitra and Warner (1984).   [h]Leer et al. (1983).   [i]Leer et al. (1984a).   [j]Sugita and Sugiura (1983).   [k]O'Connell and Rosbash (1984).   [l]Dudov and Perry (1984).

Initiation in E. coli. A number of experimental studies have been conducted on the ribosomal proteins of *E. coli*, so that the *in vitro* promoters of several operons have been uncovered. Moreover, in the *str* and *spc* gene sets those structures have been demonstrated to initiate transcription also *in vivo* (Ikemura *et al.*, 1979; Post *et al.*, 1980). As the seven examples of Table 5.9 make clear, the conjectural initiating signals (italicized) show the same characteristics as those of the genes examined in preceding sections, a similarity to the standard TAATA- and great variation around that sequence. The other real consistency in structure is in their being T, A-rich, but even that trait is subject to contradiction—for example, both *rpsA* P1 and P3 have as many G, C residues as T, A. In a number of instances given by Post *et al.* (1980), the promoter is located at or near the center of a region of dyad symmetry, typically invovling G, C nucleotides to a large degree. But no such stem-and-loop structures are in evidence in those shown in the table. Moreover, such conformations would not come into existence until after the transcribing enzyme had produced an RNA strand of suitable length, so it remains impossible to perceive how their presence could possibly be of service in locating the given promoter in the first place. As a whole the actual start site of transcription in this prokaryote has been found to be a purine, G being more frequent than A; however, in the β operon initiation was at a C (Post *et al.*, 1979). No ancillary signal of any sort appears to have been established as yet.

The operon *rpsA* is the only apparent gene set from which multiple promoters have been reported (Pedersen *et al.*, 1984). On the leader of this genomic sector four such signals have been described, given as P1–P4 in Table 5.9, the first one being the farthest upstream from the mature coding region. P2 and P4 lie between P1 and P3 and their respective start sites, so that they appear within column F of the latter pair as well as individually. The lack of uniformity in base composition just described is displayed by these four as elsewhere, as is the absence of a clear ancillary signal. It is not unlikely that such multiple promoters will be found to prevail among many of these operons besides *rpsA*, in parallel to the situation that characterized genes transcribed in eukaryotes by RNA polymerase I. In each of the present quartet, the adjacent nucleotide, on both the upstream and downstream sides, is variable, any of the common nucleotides being present. On a combined basis the sequence to which the RNA polymerase would apparently respond is as follows:

$$T \begin{matrix} A \\ T \end{matrix} \begin{matrix} A \\ G \\ C \end{matrix} \begin{matrix} A \\ C \\ G \end{matrix} T N N$$

Initiation in Eukaryotes and Organelles. The presumed and established promoters of the mitochondrion and chloroplast so closely resemble those of the cytoplasm of eukaryotes that treatment together is economical of space without sacrificing clarity. As comparisons of the promoters listed in Table 5.9 quickly disclose, there is little to distinguish the italicized portions for the eukaryotic signals from corresponding parts of the prokaryotes. Viewed as a unit, the hypothesized structure to which RNA polymerase II would react reads

$$\begin{matrix} T A \\ A T \end{matrix} \begin{matrix} T C A T \\ C A T G N \\ A T G C \end{matrix}$$

Thus the enzyme of these organisms requires less specificity than that of *E. coli*, for the fourth site may be occupied by any one of three, rather than just two, of the common nucleotides. In contrast, its requirements for flanking nucleotides are more exacting, because only a G or a T occurs on the 5' side and G is excluded from the 3' end.

In most cases the established sequences flanking the mature gene on the 5' end are too short to expose any multiple promoters that might exist, that known as *rp29* from yeast being one of the exceptions (Mitra and Warner, 1984). In this case the transcriptive start sites were found to lie at about 506, 490, 485, and 480 sites upstream from the functional gene, the leader being shortened posttranscriptionally by removal of an intron of 458 nucleotide residues. Although a number of alternatives exist, the two most promising sequences for possible promoters lying close to start sites are those given in the table. The P2 signal begins five positions downstream from the starting point at position −506. Comparable homologies were detected at similar distances upstream from the coding section in six of eight yeast sequences tested, but were not identified as being actual promoters (Teem *et al.*, 1984).

In the mitochondrial gene two promoters also have been detected, but in this case they proved to be identical in composition (Zassenhaus *et al.*, 1984). This pair was found to include nine sites on the active sector, not just the six or seven of the prokaryotes. Both were located at unusually long distances before the mature gene for mitochondrial cistrons, one 630, the other 550, sites upstream of that point, but it must be recalled that the yeast mitochondrial genome is longer than the prokaryotic, not shorter as in the metazoan organelle.

5.6. THE MATURE GENES FOR RIBOSOMAL PROTEINS

Because of the large number of species for ribosomal proteins involved, in each organism running between 50 and 100 or more, the impression of abundance given by Tables 5.10 and 5.12 that bear the known sequences is misleading. Although the primary structures of all the ribosomal proteins of *E. coli* have been determined (Wittmann-Liebold *et al.*, 1984), actually less than one-fourth the total of their genes have been sequenced, and this bacterium is by far the most thoroughly explored organism. In addition, a good beginning has been made with the yeast cistrons and a smattering from the chloroplasts, but none from mitochondria have been established. Nevertheless, those that are provided in the tables are amply sufficient for present purposes, although many presently unanswerable problems are apparent. Numerous homologies between certain sequences may be perceived in one region that are lost in another, accompanied by the appearance of new relationships between still other sets of nucleotides. So while it may be too early to suggest evolutionary relationships on any firm basis, one's curiosity cannot fail to be stimulated. This is especially true since extensive homology between genes from two source organisms is disclosed by the tables for the first time.

5.6.1. The Ribosomal Protein Genes of the Small Subunit

In keeping with what must be highly varied roles in protein synthesis, the ribosomal proteins of both the large and small subunits are a heterogeneous lot, a trait readily perceptible in these genes. The adjective small as applied to these coding units and their

Table 5.10

Gene Sequences of Representative Small-Subunit Ribosomal Proteins[a]

Row 1

E. coli S1[b]	ATG ACT GAA TCT TTT GCT CA A CTC TTT GAA GAG TCC TTA AAA GAA ATC GAA ACC CGC GCG CGT TCT ATC
E. coli S5[c]	ATG GCT CAC ATC GAA AAA CA A GCT GGC GAA CTG CAG GAA AAG CTG ATC GCG GTA AAC CGC GTA TCT AAA
E. coli S7[d]	atg CCA CGT CGT CGC GTC AT T GGT CAG CGT AAA ATT CTC CCG GAT CCG AAG TTC GGA TCA GAA CTG CTG
Euglena_C S7[e]	ATG TCT CGA AGA AGA GC A AAA AGA AGA ATA ATA TCA CAA GAT CCT ATC TAT AAC AGT ACT TTA GCA
E. coli S8[c]	ATG AGC ATG CAA GAT CCG AT C GCG GAT ATG CTG ACC CGT ATC CGT AAC GGT CAG GCC GCG AAC AAA GCT
E. coli S10[f]	ATG CAG AAC AAA AGA ATC CG T ATC CGC CTG AAA GCG TTT GAT CAT CGT CTG ATC GAT GAA GCA ACC GCG
Yeast S10[g]	ATG AAG* TTG AAC ATT TCT TA C CCA GTC AAC GGG TCT CAA AGG ACC TTC GAA ATT GAT GAT CAC CGT
E. coli S12[d]	atg GCA ACA GTT AAC CAG CT G GTA CGC AAA CCA CGT CGT GCT CGC AAA GTT GCG AAA AGC AAC GTG CCT GCG
Euglena_C S12[e]	ATG CCT ACA TTA GAA CAT TT A ACA CGA TCA CCG AGA AAA ATA CGA AAA ACT AAA TCA CCA GCA
E. coli S14[c]	ATG GCT AAG CAA TCA ATG AA A GCA CGC GAA GTA AAG CGC GTA GCT TTA GCT AAA TCA GCG AAA
Yeast S16A[h]	ATG CCA GGT GTT TCC GTT AG*A GAC GTT GCT GCT CAA GAT TTC ATT AAT GCT TAC GCT TCT TTC TTG CAA
Tobacco_C S19[i]	GTG ACA CGT TCA CTA AAA AA A AAT CCC TTT GTA GCC AAT CAT TTA TTA AAA ATT GAT AAG CTT AAC
E. coli S20[j]	TTG GCT AAT ATC AAA TCA GC T AAG AAG CGC GCC ATT CAG TCT GAA GCT CGT AAG CAC AAC CCA AGC
Yeast S33[k]	atg GAT AAC AAA ACC CCA GT C ACT TTA GCC ATC AAG GTC ATC AAA GTT TTA GGA AGA ACC GGT TCT CGT GGT

Row 2

```
E. coli S1      GTT CGT GGC GTT GTT GCT ATC GAC AAA GAC GTA GTA CTG GTT GAC GCT GGT CTG AAA TCT GAG TCC
E. coli S5      ACC GTT AAA GGT GGT CGT ATT TTC TCC ACA GCT CTG ACT GTA GTT GGC GAT GGT AAC GGT CGC GTT
E. coli S7      GCT AAA TTT GTA AAT ATC CTG ATG GTA GTA TCT ACT GCT GAA TCT ATC GTA TAC AGC GCG
Euglena_C S7    AGT AAA GTA ATT AAT AAA ATA TTA TTG AAT GGA AAA ACT CTA GCT CAG TAT ATT TTT GAA ACA
E. coli S8      GCG GTC ACC ATG CCT TCC TCC AAG CTG AAA GTG GCA ATC GCC AAC GTG CTG AAG GAA GGT TTT ATT
E. coli S10     GAA ATC GTC GAG ACT GCC AAG CGC ACT GGT CGT GGT CGC ATC CCG GTT CCG ACA CGC AAA
Yeast S10       ATT CGT GTT TTC TTC GAC AAG AAG AGA ATC GGT CAA GAA GTC GAT GGT GAT GAA TTC AAG
E. coli S12     CTG GAA GCA TGC CCG CAA AAA CGT GGC GTA TGT GTA CGT GTA ACT ACC ACT CCT AAA AAA CCG AAC
Euglena_C S12   TTA AAA GGA TGC CCG CAA AAA CGC GCA ATA TGC ATG CGG GTT TAC ACA ACA CCA AAA AAA CCA AAT
E. coli S14     CGC GCT GAA CTG AAA GCG ATC ATC TCT GAT GTG AAC GCT TCC GAC GAT CGT TGG AAC GCT GTT CTC
Yeast S16A      AGA CAA GGT AAG CTA GAA GTT CCA GGT GTT GAC ATT GTC AAG ACC TCT TCT GGT AAC GAA ATG CCA
Tobacco_C S19   ACA AAA GCA GAA AAA GAA ATA ATA GTA ACT TGG TCC CGG GCA TCT ACC ATT ATA CCC ACA ATG ATC GGT
E. coli S20     CGT CGC TCT ATG ATG CGT ACT TTC ATC AAG AAA GTA TAC GCA GCT ATC GAA GCT GGC GAC AAA GCT GCT
Yeast S33       GGT GTC ACC CAA GTC CGT GTC GAA TTC TTG GAA GAC ACT TCC AGA ACT ATT GTC AGA AAC ACT GTC AAG AGC GGC
```

(continued)

Table 5.10 (Continued)

Row 3

E. coli S1	GCC ATC CCG GCT GAG CAG TTC AAA AAC GCC CAG GGC GAG CTG GAA ATC CAG GTA GGT GAC GAA GTT GAC
E. coli S5	GGT TTT GGT TAC GGT AAA GCG CGT GAA GTT CCA GCA GCG ATC CAG AAA GCG ATG GAA AAA GCC CGT CGC
E. coli S7	CTG GAG ACC CTG GCT CAG CGC TCT GGT AAA TCT GAA CTG GAA GCA TTC GAA GTA GCT CTC GAA AAC GTG
Euglena_C S7	ATG AAA AAT ATA CAG GAA ATT TAT AAA AAA GAC CCC TTA GAC ATT CTA AGA AAA AAA GCG ATA AAA AAC GCA
E. coli S8	GAA GAT TTT AAA GTT GAA GGC GAC ACC AAG CCT GAA CTG GAA CTT ACT CTG AAG TAT TTC CAG GGC AAA
E. coli S10	GAG CGC TTC ACT GTT CTG ATC TCC CCG CAC GTC AAC AAA GAC GCG CGC GAT CAG GAA ATC CGT ACT
Yeast S10	GGC TAC GTC TTC AAG ATC TCT GGT GGT AAC GAC AAA CAA GGT TTC CCA ATG AAG CAA GGT GTT TTG TTG
E. coli S12	TCC GCG CTG CGT AAA GTA TGC CGT GTT CGT CTG ACT AAC GGT TTC GAA GTG ACT TCC TAC ATC GGT GGT
Euglena_C S12	TCC GCC TTG CGC AAA GTA ACA AGG GTA ACA CTT TCT TCA GGG TTA GAA GTT ACA GCT TAT ATA CCA GGA
E. coli S14	AAG CTG CAG ACT CTC CCC CGT GAT TCC AGC CCG TCT CAG CGT AAC CGT TGC CGT CAA ACA GGT CGT
Yeast S16A	CCA CAA GAT GCC GAA GGT TGG TTC TAC AAG GCT GCC TCT GTT GCC AGA CAC ATT TAC ATG AGA AAA
Tobacco_C S19	CAT ACG ATT GCT ATC CAT AAT GGA AAA GAG CAT TTG CCT ATT TAT ATA ACG GAT AGT ATG GTA GGC CAC
E. coli S20	GCA CAG AAA GCA TTT AAC GAA ATG GCA CCG ATC GTG GAC GCT GCT CAG GCT GCT AAA GGT CTG ATC CAC AAA
Yeast S33	CCA GTT AGA GAA AAC GAC ATT TTG GTT CTA ATG GAA TCT GAA CGT CGT CGT TTG CGT TAG

Row 4

```
E. coli  S1     GTT GCT CTG GAC GCA GTA GAA GAC GGC TTC GGT GAA ACT CTG CTG TCC CGT GAG AAA AAA CGT CAC
E. coli  S5     AAT ATG ATT AAC AAC GTC GCG CTG AAT AAC GGC ACT CTG CAA CAC CCT GTT AAA GGT GTT CAC ACG GGT TCT
E. coli  S7     CGC CCG ACT GTA GAA GTT AAG TCT CGC CGC GTT GGT (incomplete)
EuglenaC S7     TCT CCA CAA ATG GAA ACA AGA AAG CGT CGT ATC GGA GGA ACA ATA TAT CAA GTT CCT GTA GAA GTA AAA
E. coli  S8     GCT GTT GTT GAA AGC ATT CAG CGT GTC AGC CGC CCA GGC TTG CGC ATC TAT AAA CGT AAA GAT CAG CTG
E. coli  S10    CAC TTG CGT CTG GTT GAC ATC GTT GAG ACC AAG AAC AAC GTT GAT GCT CTG CTG CTG GAT CTG
Yeast    S10    CCA ACT AGA ATC AAG AAG TTG TTG ACC AAG AAC AAC GTT TCT TGT TAC AGA CCA AGA CGT GAT GGT GAA AGA
E. coli  S12    GAA GGT CAC AAC CTG CAG GAG CAC TCC GTG ATC CTG ATC CGT GGC GGT CGT GTT AAA GAC CTC CCG GGT
EuglenaC S12    ATA GGG CAT AAT TTA CAA GAA CAT TCA GTC CTT ATC ATC AGA GGT GGA AGA GTT AAA GAT TTA CCA GGA
E. coli  S14    CCG CAT GGT TTC CTG CGA AAG TTT GGG TTG AGC CGT ATT AAG GTC CGT GAA GCC GCT ATG CGC GGT GAA
Yeast    S16A   CAA GTT GGT GTT GGT GGT AAA TTG AAC AAA TTA TAC GGT GGT GCC AAG AGC AGA GTT AGA CCA TAC AAG
TobaccoC S19    AAA TTG GGA GAA TTT GCA CCT ACA TTA AAT TTT AGA GGA CAT GCA AAA AGC GAT AAT AGA TCT CGT CGT
E. coli  S20    AAC AAA GCT GCA CGT CAT CAT AAG GCT AAC CTG ACT GCA CAG ATC AAC AAC AAA CTG GCT TAA
```

(continued)

Table 5.10 (Continued)

Row 5

E. coli S1	GAA GCC TGG ATC ACG CTG GAA AAA GCT TAC GAA GAT GCT GAA ACT GTT ACC GGT GTT ATC AAC GGC AAA
E. coli S5	CGC GTA TTC ATG CAG CCG GGT TCC GAA GGT ACC ATC GCC GGT GGT GCA ATG CGC GCC GTT CTG
Euglena_C S7	GAA GAT CGT GGA ACT AGC TTA GCA TTA AAA TTC ATA ATA GAA AAA GCT AGA GAA AAA GGA AGA GGA
E. coli S8	CCC AAA GTT ATG GCG GGT CTG GGT ATC GCA GTT GTT TCT ACC TCT AAA GGT GTT ATG ACT GAT CGT GCA
E. coli S10	GCT GCC GGT GTA GAC GTG CAG ATC AGC CTG GGT TAA
Yeast S10	AAG AGA AAG TCC GTC AGA GGT GCC ATT GTT GGT CCA GAT TTG GCT GTC TTG GCT TTC GTC ATT GTC AAG
E. coli S12	GTT CGT TAC CAC ACC GTA CGT GGT GCG CTT GAC TGC TCC GGC GTT AAA GAC CGT AAG CAG GCT CGT TCC
Euglena_C S12	GTA AAG TAC CAC GTA ATA CGT GGA TGT TTA GAC GCA GCA AGT GTA AAA AAT CCC AAA AAT GCA AGA TCT
E. coli S14	ATC CCG GGT CTG AAA AAA GGC TAG
Yeast S16A	CAC ATT GAC GCT TCC GGT TCT ATC AAC AGA AAG GTT TTG CAA GCT TTG GAA AAG ATT GGT ATC GTC GAA
Tobacco_C S19	TAA

Row 6

E. coli S1	GTT AAG GGC GGC TTC ACT GTT GAG CTG AAC GGT ATT CGT GCG TTC CTG CCA GGT TCT CTG GTA GAC GTT
E. coli S5	GAA GTC GCT GGG GTT CAT AAC GTT CTC GCT AAA GCC TAT GGT TCC ACC AAC CCG ATC AAC GTG GTT CGT
Euglena_C S7	ATA TCC ACA AAA CTA AAA AAT GAA ATA ATT GAT GCA TCG AAT AAC ACA GCT GAA GCT GTA AAA AAA AAA
E. coli S8	GCG CGC CAA GCT CTT GGT CTT GGT GGC GAA ATT ATC TGC TAC TAC GTA GCC TAA
Yeast S10	AAG GGT GAA CAA TTG GAA GGT CTA ACT GAC ACT GTT CCA AAG AGA TTG GGT CCA AAG AGA GCT
E. coli S12	AAG TAT GGT GTG AAG CGT CCT AAG GCT TAA
Euglena_C S12	AAA TAC GGT CTA AAA AAA CCA AAA CCC AAA TAA
Yeast S16A	ATC TCT CCA AAG GGT GGT AGA ATC TCT GAA AAC GGT CAA AGA GAT TTG GAT CGT ATT GCC GCT CAA

Row 7

E. coli S1 CGT CCG GTG CGT GAC ACT CTG CAC CTG GAA GGC AAA GAG CTT GAA TTT AAA GTA ATC AAG CTG GAT CAG

E. coli S5 GCA ACT ATT GAT GGC CTG GAA AAT ATG AAT TCT CCA GAA ATG GTC GCT GCC AAG CGT GGT AAA TCC GTT

Euglena S7 GAG CAG ATA CAT AAA ACA GCT GAA GCA AAT AAA GCA TTC TCA AAT ATG AAA TTT TAA

Yeast S10 AAC AAC ATC AGA AAG TTC TTC GGT TTG TCC AAG GAG GAT GAC GTT CGT GAT TTC GTC ATC AGA AGA GAA

Yeast 16A ACT TTG GAA GAA GAC GAA TAA

Row 8

E. coli S1 AAG CGC AAC AAC GTT GTT TCT CGT CGT GCC GTT ATC GAA TCC GAA AAC AGC GCA GAC GAT CAG CTG

E. coli S5 GAA GAA ATT CTG GGG AAA TAA

Yeast S10 GTC ACC AAG GGT GAA AAG ACT TAC ACC AAG AAG GCT CCA AAG ATC CAA AGA TTG GTT ACT TCT CAA AGA TTG

Row 9

E. coli S1 CTG GAA AAC CTG CAG GAA GGC ATG GAA GTT AAA GGT ATC AAA AAG ATC GTT AAG AAC CTC ACT GAC TAC GGT GCA TTC

Yeast S10 CAA AGA AAG AGA CAC CAA AGA GCT TTG AAG GTC AGA AAC GCT CAA AGA GCT GCT GCC GAA

Row 10

E. coli S1 GTT GAT CTG GGC GGC GTT GAC GGC CTG CAC ATC ACT GAC ATG GCC TGG AAA CGC GTT AAG CAT CCG

Yeast S10 TAC GCT CAA TTG TTG GCT AAG AGA TTG TCT GAA AGA AAG GCT GAA AAG GCC GAA ATC AGA AGA AGA AGA

Row 11

E. coli S1 AGC GAA ATC GTC AAC GTG GGC GAC GAA ATC ACT GTT AAA GTG CTG AAG TTC GAC CGC GAA CGC GAA CGT ACC CGT

Yeast S10 GCT TCT TCT TTG AAG GCT TAA

(continued)

Table 5.10 (Continued)

Row 12

E. coli S1   GTA TCC CTG GGC CTG AAA CAG CTG GGC GAA GAT CCG GTA GCT ATC GCT AAA CGT TAT CCG GAA GGT

Row 13

E. coli S1   ACC AAA CTG ACT GGT CGC GTG ACC AAC CTC ACC CAC TAC CCC TCC TTC CTT CAA ATC CAA CAA CCC CTT

Row 14

E. coli S1   GAA GGC CTG GTA CAC GTT TCC GAA ATG GAC TGG ACC AAC AAA AAC ATC CAC CCG TCC AAA GTT GTT AAC

Row 15

E. coli S1   GTT GGC GAT GTA GTG GAA GTT ATG GTT CTG GAT ATC GAC GAA GAA CGT CGT CGT ATC TCC CTG GGT CTG

Row 16

E. coli S1   AAA CAG TGC AAA GCT AAC CCG TGG CAG CAG TTC GCG AAC ACC CAC AAC AAG GGC GAC CGT GTT GAA GGT

Row 17

E. coli S1   AAA ATC AAG TCT ATC ACT GAC TTC GGT ATC TTC ATC GGC TTG GAC GGC GGC ATC GAC GGC CTG GTT CAC

Row 18

E. coli S1   CTG TCT GAC ATC TCC TGG AAC GTT GCA GGC GAA GAA GCA GTT CGT GAA TAC AAA AAA AAA GGC GAC GAA ATC

Row 19

E. coli S1   GCT GCA GTT GTT CTG CAG GTT GAC GCA GAA GAA CGT GAA CGT ATC TCC CTG GGC GTT AAA CAG CTC GCA GAA

Row 20

E. coli S1   GAT CCG TTC AAC AAC TGG GTT GCT CTG AAC AAG AAA GGC GCT ATC GTA ACC GGT AAA GTA ACT GCA GTT

Row 21

E. coli S1   GAC GCT AAA GGC GCA ACC GTA GAA CTG CCT GAC GGC GTT GAA GGT *TAC* CTG *CGT* GCT TCT GAA GCA TCC

Row 22

E. coli S1   *CGT* GAC *CGC* GTT GAA GAC GCT ACC CTG GTT CTG AGC GTT GGC GAC GAA GTT GAA GCT AAA TTC ACC GGC

Row 23

E. coli S1   GTT GAT *CGT* AAA AAC *CGC* GCA ATC AGC CTG TCT GTT *CGT* GCG AAA GAC GAA GCT GAC GAG AAA GAT GCA

Row 24

E. coli S1   ATC GCA ACT GTT AAC AAA CAG GAA GAT GCA AAC TTC TCC AAC AAC GCA ATG GCT GAA GCT TTC AAA GCA

Row 25

E. coli S1   GCT AAA GGC GAG TAA

[a] Asterisks mark the locations of introns in two yeast genes, one each in S10 and S16A. Codons for lysine are underscored, those for arginine are italicized, and those for tyrosine are stippled. Subscript C indicates chloroplast.
[b] Schnier et al. (1982).
[c] Cerretti et al. (1983).
[d] Post and Nomura (1980).
[e] Montandon and Stutz (1984).
[f] Olins and Nomura (1981b).
[g] Leer et al. (1982).
[h] Molenaar et al. (1984).
[i] Sugita and Sugiura (1983).
[j] Mackie (1981).
[k] Leer et al. (1983).

products applies strictly to the relative size of the ribosomal subunit of which they are a part. For instance, in *E. coli,* the largest of these small-subunit protein genes for S1 is 1671 base pairs in length, whereas the longest of the large subunit, for L1, is merely 705 sites long. Each set, then, needs to be viewed as a separate class, whose common features, whatever they may prove to be, derive from their general functions and their mutual binding to the rRNA. It is for these reasons that the present proteins and their genes are treated in separate sections.

Chemical Properties. Since the component proteins of the small subunit bind rRNA in a reaction probably not unlike that of the histones with DNA, it is to be expected that lysine and arginine codons would be plentiful in these mature coding regions. Nor is disappointment encountered. However, the degree of arginine- and lysine-richness is widely disparate from one species to another, and, on the limited data available, also from one source organism to another. As a general trend, the larger the molecule, the less basic it is. In the gene for S1 in *E. coli,* there are 38 codons for lysine and 24 for arginine, while in the smallest thus far sequenced from that same source, that for S20 with 264 base pairs, there are 14 and 6, respectively. Still smaller is that for S33 from yeast, with only 204 base pairs; this proves to be a decidedly arginine-rich species, with only 4 lysine codons but 10 arginine.

Perhaps a more meaningful procedure is to reduce the counts to percentages, based on the total number of triplets for each given type. The total counts and percent conversions of a selected representation are given in Table 5.11. There it becomes instantly evident that the product of the S1 gene is by far the weakest in alkalinity of all those cited, with a total of only 11% codons for basic amino acids, and that the most strongly alkaline of the *E. coli* types is S14, with 26%. In that quality this nearly equals histone H2B, which has 28%.

The three representative genes from chloroplasts are of particular interest, all of which encode a product that has a high content of basic amino acids. The two of the *Euglena* organelle have totals of 21 and 25%, while the tobacco chloroplast S19 is at 37%. Although the homologous *E. coli* and *Euglena* chloroplast genes for S12 share a fairly high level of identical sites, they nevertheless differ extensively in their fundamental chemical properties. That of the prokaryote is slightly more arginine-rich, with 15 codons for that amino acid, compared to 13 for lysine. On the other hand, the corresponding chloroplast cistron is decidedly lysine-rich, with 19 codons for that substance against 12 for arginine. It would be of great interest to have the sequence of a blue-green alga ribosomal protein for comparison to note whether a similar level of basicity exists; perhaps that condition will be found to characterize these proteins from all photosynthesizing organelles and organisms.

Basic Dominions. Since it was noted in the discussions of the several histones that the pertinent codons were not evenly distributed throughout the respective molecules, it is instructive to examine the genes for the present class of proteins also to locate possible domains of activity. In S1 five regions are especially richly endowed with codons for basic amino acids, rows 5 and 10 each encoding three lysines and two arginines and rows 11 and 14 having two for the former and three for the latter amino acid. Row 21, however, has still a greater total number of such codons, having three for each of the pair.

In the *Euglena* chloroplast gene for S7, which is a homolog of that same species from *E. coli,* the codons for this basic constituent are very unevenly distributed, none of the ten

Table 5.11
Basicity of Selected Ribosomal Proteins[a]

Protein	Number of Codons			Percent		
	Total	Lys	Arg	Lys	Arg	Basic
E. coli S1	557	38	24	7	4	11
E. coli S5	168	12	13	7	8	15
Euglena$_C$ S7	157	24	10	15	6	21
E. coli S12	125	13	15	10	12	22
Euglena$_C$ S12	126	19	12	15	10	25
E. coli S14	100	11	15	11	15	26
E. coli S20	88	14	6	16	7	23
Tobacco$_C$ S19	43	10	6	23	14	37
Yeast S33	68	4	10	6	15	21
Histone H2B	129	27	8	21	7	28

Small-Subunit Proteins

Protein	Number of Codons			Percent		
	Total	Lys	Arg	Lys	Arg	Basic
E. coli L1	235	23	10	10	4	14
Yeast L3	388	41	31	10	8	18
E. coli L5	180	15	14	8	8	16
E. coli L10	166	12	12	7	7	14
E. coli L11	143	15	3	10	2	12
E. coli L29	105	16	6	16	6	22
Yeast rp29	156	24	15	15	10	25
Yeast L25	138	21	7	15	5	20
Mouse L32	135	18	11	13	8	21
Histone H4	104	11	13	11	13	24

Large-Subunit Proteins

[a]Subscript C indicates chloroplast.

for arginine being in the closing third of the molecule and rows 1 and 5 together having nearly one-half of the total of both kinds. In contrast, the first two and the final rows of the *E. coli* S12 cistron hold 20 of the total of 28 codons for lysine and arginine, the third and fourth having only a few each. Similarly with each of the others of Table 5.10, the two terminal portions, either alone or combined, are more heavily equipped with these codons than are the central rows.

5.6.2. The Genes for Large-Subunit Ribosomal Proteins

The 18 genes for the large-subunit ribosomal proteins included in Table 5.12 are from a somewhat greater diversity of sources than are those for the small-subunit assemblage of Table 5.10. Here are included one from a mammal and another from an insect, in addition to the abundance from *E. coli* and baker's yeast. However, none from chloroplasts is available here, in contrast to the three of the preceding section from such sources. In spite of that deficiency, those listed here will prove to be quite sufficient to supply a preliminary insight into their nature and even their evolutionary history. Regrettably, these—as will many, many more of the future—fail to throw light on their individual functions. In determinations of such activities great caution must be exercised to ensure correct identification of the component being investigated. For example, in a current study on the RNA binding activity of L3 of yeast, the amino acid sequence given under that designation is totally different on a codon-to-codon basis from the nucleotide sequence of the gene for the supposedly identical protein (Schultz and Friesen, 1983; Yaguchi *et al.*, 1984). Thus, obviously a misidentification has occurred that could actually retard, rather than enhance, progress in this difficult field.

Chemical Properties. Quite as among the ribosomal proteins of the small subunit, the chemical properties of the large vary widely; in the present case, however, the range of variation is not so great, nor is the correlation between size and total alkalinity so clear (Table 5.11). By way of illustration, L1, the largest of this set from *E. coli*, is equally as basic as L10, and more strongly so than the even smaller L11. Moreover, L3 of yeast, with a length of over 1000 nucleotide residues and a total codon content for basic amino acids of 18%, nearly equals the alkaline content of its L25 (20%), although the latter consists of only ~40 nucleotides. Hence, that generalization does not apply to this family of genes. Unlike the small subunit's members, no arginine-rich species has yet had its gene sequence established, although one from *E. coli*, L10, has equal numbers of codons for lysine and arginine (Post *et al.*, 1979), and L5 has nearly the same quality, with 15 codons for the former and 14 for the latter (Cerretti *et al.*, 1983). The most abundantly equipped with codons for these alkaline biochemicals is *rp29* of yeast, which has 24 for lysine and 15 for arginine, so that 25% of its total of 156 codons are for this pair (Table 5.11).

Examination of Table 5.12, which presents the mature genes for this family of proteins, reveals wide variation in the distribution of these codons, quite like that found in the preceding group. One pattern that recurs with some degree of frequency is a concentration of codons for the basic amino acids in the 5' half or so of the molecule, with L1 of *E. coli* and L32 of the mouse being exceptionally clear examples. In some instances of this sort, including L24 of the bacterium, there is a strong secondary resurgence of concentration just before the 3' terminus. The last species mentioned has ten lysine and three arginine codons in the first 50 of its total; row 3 diminishes to three, all for the

Table 5.12

Gene Sequences of Representative Large-Subunit Ribosomal Proteins[a]

Row 1

E. coli L1 [b]	atg GCT AAA CTG A CC AAG CGC ATG CTG GTT ATC CGC GAG AAA GTT GAT GCA ACC AAA CAG AAG TAC GAC ATC
Yeast L3 [c]	ATG TCT CAC AGA A AG TAC GAA GCA CCA CGT CAC GGT TTC TTG CCA AGA AAG AGA GCT GCC
E. coli L5 [d]	ATG GCG AAA CTG C AT GAT TAC GAA GCA GTA GTT AAA AAA CTC ATG ACT GAG TTT AAC TAC AAT
E. coli L6 [d]	ATG TCT CGT GTT G CT AAA GCA CCG GTC GTT CCT GAC GTT AAA GGT GTT GAC GAC GGT CAG GTT
E. coli L10 [b]	atg GCT TTA AAT C TT CAA GAC AAA GCA ATT GTT GCT GAA GTC AGC GAA GTA GCC AAA GGC GCG CTG
E. coli L11 [b]	atg GCT AAG AAA G TA CAA GCC TAT GTC AAG CTG CAG GTT GCA GCT ATG GCT AAC CCG AGT CCG CCA
E. coli L7/12 [b]	atg TCT ATC ACT A AA GAT CAA ATC ATT GAA GCA GCA GTT GCA GCT ATG TCT GTA ATG GAC GTT GTA GAA CTG
E. coli L14 [d]	ATG ATC CAA GAA C AG ACT ATG CTG AAC CTG GCC GAC AAC GTC TCC GGT GCA CGT CGT GTA ATG TGT ATC AAG
E. coli L15 [d]	ATG CGT TTA AAT A CT CTG TCT CCG GCC GAA GGC TCC AAA AAG GCG GGT AAA CGT CTG GGT CGT GGT ATC
Yeast L16 [e]	ATG TCT GCC AAA G CT CAA AAC CCT ATG CGT GAT TTG AAG ATC CGT GGT TTG GTC TTA AAC ATT TCT
E. coli L18 [d]	ATG GAT AAG AAA T CT GCT CGT ATC CGT CGT GCG ACC CGC GCA CGC CGC AAG CTC CAG GAG CTG GGC GCA
Yeast L17a [f]	ATG TCC GGT AAC G GT GCT CAA GGT ACT AAG AAG TTT AGA ATC TCA TTA GGT CTA CCA GTC GGT GCC ATC ATG
E. coli L24 [d]	ATG GCA GCG AAA A TC CGT CGT GAT GAC GAA GTT ATC GTG TTA ACC GGT AAA GAT GCT CCA AAG GTT AAA GGT GGT
Yeast rp28 [g]	ATG GGT ATC GAT C AC ACT TCC AAG CAA CAA CAC AAG AGA TCT GGT CAC AGA ACT GCT CCA AAG TCT GAC AAT
Yeast rp29 [h]	ATG AAG GTT GAA A TC GAT TC GAT TCT TTT TCA GGT GCC AAA ATC AGA GGT ACC TTG TTT TTC GTC CGT
Yeast L25 [f]	ATG GCT CCA TGT G CT AAG GCT ACT AAG GCT GCT AAA GCT GTC GTT AAG AGT GGT AAG AAG AAG GCT
Mouse L32 [i]	ATG GCC CTC C GG CCT CTG GTG AAG CCC AAG ATC GTC AAA AAG AGG ACC AAG AAG TTC ATC ATC AGG GAC
Drosophila rp49 [j]	ATG --- ACC ATC C GC CCA GGA TAC AGG CCC AAG AAG ATC GTG AAG AAG CGC ACC CAC CGC CAC
E. coli L30 [d]	ATG GCA AAG ACT A TT AAA ATT ACT CAA ACC CGC AGT GCA ATC GGT CGT CCG CTG CCG AAA CAC CAC ACG

(continued)

Table 5.12 (Continued)

Row 2

```
E. coli L1      AAC GAA GCT ATC GCA CTG CTG AAA GAG CTG GCG ACT GCT AAA T TC GTA GAA AGC GTG GAC GCT GTT
Yeast L3        TCC ATC AGA GCT AGA GTT CCT TTT CCA AAG GAT GAC AGA T CC AAG CCA GTT GCT CTA ACT TCC TTC
E. coli L5      TCT GTC ATG CAA GTC CCT CGG GTC GAG AAG ATC ACC CTG AAC A TG GGT GTT GGT GAA GCG ATC GCT GAC
E. coli L6      ATT ACG ATC AAA GGT AAA AAC GGC GAG CTG ACT CGT ACT CTC A AC GAT GCT GTT GAG GTT AAA CAT GCA
E. coli L10     TCT GCA GTA GTT GCG GAT TCC CGT GGC GTA ACT GTA GAT AAA A TG ACT GAA CTG AAA CGT AAA GCA GGT CGC
E. coli L11     GTA GGT CCG GCT CTG GGT CAG CAG GGC GTA AAC ATC ATG GAA T TC TGC AAA GCG TTC AAC GCA AAA ACT
E. coli L7/12   ATC TCT GCA ATG GAA GAA AAA TTC GGT GTT TCC GCT GCT G CT GTA GCT GTA GCT AAA GGC CCG GTT
E. coli L14     GTT CTA GGG GGC TCG CAC CGA CGA TAC GCA GGC GTA GGC GAC A TC ATC AAG ATC ACC ATC AAA GAA GCA
E. coli L15     GGT TCT GGC CTC GGT AAA ACC GGT GGT CGT GGT CAC AAA GGT C AG AAG TCT CGT GGC GGT GGC GTA
Yeast L16       GTT GGT GAA TCT GGT GAC AGA TTG ACC AGA GCG TCC AAG GTT T TA GAC CAA TTA TCT GGT CAA ACT CCA
E. coli L18     ACT CGC CTG GTG GTA CAT CGT ACC CCG CGT CAC AAT TAC GCA C AG GTA ATT GCA CCG GAA CGT TCT GAA
Yeast L17a      AAC TGT GCT GAC AAC AGT GGT GCC ACA AAC TTG TAC ATT ATC G CC GTG AAA GGC TCT GGT TCC AGA TTG
E. coli L24     AAA GTT AAG AAT GTC CTG TCT TCC GGC AAG GTC ATT CTT GAA G GT GT ATC AAC AAC TTG GTT AAG AAA CAT CAG
Yeast rp28      GTC TAC TTG AAA TTG TTA GTC AAA TTA ACT TTC TTA GCT CGT C GT CGT ACT GAT GCT GCT CCA TTC AAC AAG
Yeast rp29      GGT GAC TCC AAA ATC TTC AGA TTC CAA AAC TCC AAA TCT GCC T CT CGT TTG TTC AAG CAA AAG AAC CCA
Yeast L25       TTG AAG GTC AGA ACT TCT GCT ACC TTC AGA CTA CCA AAG TTG G CT AGA TTG GCT AAA AGA GCT GCT CAG
Mouse L32       CAG TCA GAC CGA TAT GTG AAA ATT AAG CGA AAC TGG CGG AAA C CC AGA GGC ATT GAC AAC AGG GTG CGG
Drosophila rp49 CAG TCG GAT CGA TAT GCT AAG CTG TCG CAC AAA TGG CGC AAG C CC AAG ATC GAC AAC AGA GTC GGT
E. coli L30     CTG CTT GGC GTG CTG CGT CGT CGT CAC ACC GTA CAG GC GAG GAT ACT CCT GCT ATT CGC GGT
```

Row 3

```
E. coli L1        AAC CTC GGC ATC GAC GCT CGT AAA TCT GAC CAG AAC GTA CGT GGT GCA ACT GTA CTG CCG CAC GGT ACT
Yeast L3          TTG GGT TAC AAG GCT GGT ATG GCT AAG ACC ATT GTC AGA GAT TTG GAC AGA CCA GGT TCT AAG TTC CAC AAG
E. coli L5        AAA AAA CTG GAT AAC GCA GCA GAC GAC CTG GCA ATC TCC GGT GCA GAC GGT GCT GGT GCT ACC CCG TTC
E. coli L6        GAT AAT ACC CTG ACC TTT GGT CCG GAT GGT GTT GTT AAC ACC CTG CTG CTG CTG GAA GGT ACT CCG TTC
E. coli L10       GAA GCT GGC GTA TAC ATG CGT GTT GTT CCG ATT CCG GTT GTT TAC GCT GCT GGT GTT TCT TTC ACT TTC
E. coli L11       GAT TCC ATC GAA GAA AAA GGT CTG CCG ATT CCG GTA GTA ATC ACC GTT TCT TTC ACT TTC
E. coli L7/12     GAA GCT GCT GAA GAA AAA ACT GAA TTC GAC GTA ATT CTG AAA GCT GCT GGC GCT AAC AAA GTT GCT GTT
E. coli L14       ATT CCG CGT GGT AAG GTC AAA AAG GGT GAT GTG CTG AAG GCG GTA GTG CGC ACC AAG AAG GGT GTT
E. coli L15       CGT CGC GGT TTC CAG GGT GGT CAG ATG GTA CCT CTG TAC CGT CGT CTG CTG CCG AAA TTC GGC TTC ACT TCT CGT
Yeast L16         GTT CAA --- TCC AAG GCC AGA TAC ACT GTC AGA AGA AGA ACT GGT ATC AGA AAC GAA AAA ATT GCT GTT
E. coli L18       GTT CTG GTA GCT TCT GCT TCT ACT GTA GAA AAA GCT ATC GCT GAA CAA CTG AAG TAC ACC GGT AAT AAA GAC
Yeast L17a        AAC AGA TTG CCA CCC GCG TCT CTA GGT GAT ATG GTT ATG GCC ACC GTT AAG AAG GGT AAG CCA GAA TTG
E. coli L24       AAG CCG GTT CCG GCC TTG AAC CAA CCG GGT ATC GGC GGT AAA GAA GCC GCT ATT CAG GTT TCC AAC
Yeast rp28        GTT GTC TTG AAG GCT TTG TTC TTG AAG ATC AAC AGA CCA CCT GTT TCT GTC TCT AGA ATT GCT AGA
Yeast rp29        AGA AGA AGC ATC GCT GGT TCG ACT GTC TTA TTC TTC AGA AAG CAT CAC AAG AAG GGT ATC ACC GAA GAA GTT GCT AAG
Yeast L25         GCT TCC AAG GCT GTT CCA CAC CAC TAC AAC AGA TTG GAC TCA AAG GTC ATT GAG CAA CCA ATC ACT TCC
Mouse L32         AGA AGG TTC AAG GGC GGA CAG TAT CTG ATG GCC AAC ATC GTT GAC CCC AAC ATC GGT TAT GGG AGC AAC ATC GGT TCC AAC
Drosophila rp49   CGC CGC TTC AAG GGA CAG TAT CTG ATG GCC AAC ATC GTT GAC CCC AAC ATC GGT TCG AAC AAG ACC CGC ACC CAC ATG
E. coli L30       ATG ATC AAC GCG GTT TCC TTC ATG GTT GTT AAA GTT GAG GAG TAA
```

(continued)

Table 5.12 (Continued)

Row 4

E. coli L1	GGC *CGT* TCC GTT *CGC* GTA GCC GTA TTT ACC CAA GGT GCA AAC GCT GAA GCT GCT <u>AAA</u> GCT GCA GGC GCA **TAC**
Yeast L3	*CGT* GAA GTT GTC GAA GCT GTC ACC GTT GAC ACT CCA CCA GTT GTC GTT GGT GTT GTC GGT GTT ACT CTG *CGT*
E. coli L5	GCA *CGC* <u>AAA</u> TCT GCA GGC TTC <u>AAA</u> ATC *CGT* CAG GGC **TAT** CCG ATC CGC GGC TGT <u>AAA</u> GTA ACT CTG GTA
E. coli L6	GCC CTG CTG AAC TCA ATG GTT ATC GGT GTT ACC GAG GAC TTC ACC <u>AAG</u> AAC CTG CAG CTG GTT GGT GTT GTA
E. coli L10	GAG TGC CTG <u>AAA</u> GAC GCG CCG CCG GCA GTT TTT GGT GTT CCG ACC ATT GCA **TAC** TCT ATG GAA CAC CCG GGC GCT GCT
E. coli L11	GTT ACC <u>AAG</u> ACC CCG GCA GCA GTT CTG CTG GGT <u>AAA</u> <u>AAA</u> GCG GCT <u>AAA</u> GAC GCT ATC TCT GGT TCC GGT <u>AAG</u>
E. coli L7/12	ATC <u>AAA</u> GCA GTA *CGT* GGC GCA ACT GGC CTG GGT CTG <u>AAA</u> GAA GCT <u>AAA</u> GAC GCT TTG GTA GAA TCT GCA CCG
E. coli L14	*CGT* *CGC* CCG GAC GGT TCT GTC ATT *CGC* TTC GAT GGT AAT GTT CTT CTG AAC AGC GGC GTA GTA AGC AGC GAG
E. coli L15	<u>AAA</u> GCT GCG ATT ACA GCC GAA ATT *CGT* TTG TCT GCT <u>AAA</u> GTA GAA GTA GGC GGT GTA GTA GAC CTG
Yeast L16	CAC GTT ACC GTC *AGA* GGT GTC CCA <u>AAG</u> GCT --- GAA GAA ATT TTG GAA *AGA* GGT TTG <u>AAG</u> GTC <u>AAG</u> GAA **TAC**
E. coli L18	GCG GCT GCA GCT GTG GGT <u>AAA</u> GCT GTC GCT GTC GAA *CGC* GCT CTG GAA <u>AAG</u> GCC ATC <u>AAA</u> GAT GTA TCC TTT
Yeast L17a	*AGA* <u>AAG</u> <u>AAG</u> GTT ATG CCA GCT ATT GTT GTC *CGT* CAA GCT TCT TGG *AGA* *AGA* *AGA* GAC GGT GTC TTT
E. coli L24	GTA GCA ATC TTC AAT GCG GCA ACC GGC <u>AAG</u> GCT CAC *CGT* GTA GGC TTT GCT *AGA* TTC GAA GAC GGT <u>AAA</u> <u>AAA</u>
Yeast rp28	GCT TTG <u>AAG</u> CAA GAA GGT GCT GCT AAC <u>AAG</u> ACT GTT GTC GTT GGT ACT GTT GTT TTG GAC TTG ATC <u>AAG</u> GCC *AGA*
Yeast rp29	<u>AAG</u> *AGA* TCT *AGA* <u>AAA</u> <u>AAA</u> ACC GTT <u>AAG</u> GCC CAA *AGA* CCA ATT ACC GGT GCT TCT TTG GAC TTG ATC <u>AAG</u> GAA
Yeast L25	GAA ACC GCT ATG <u>AAG</u> <u>AAG</u> GTT GAA GAT GGT AAC ATT TTG GTT TTC CAA GTT TCC ATG <u>AAA</u> GCT AAC <u>AAA</u>
Mouse L32	CTG CCC AGC GGC TTC *CGC* <u>AAG</u> TTC CTG GTC CAC AAT GTC <u>AAG</u> GAG CTG GTG CTG CTC ATG TGC AAC
Drosophila rp49	CTG CCC ACC GGA TTC <u>AAG</u> <u>AAG</u> TTC CTG GTG CAC AAC GTG *CGC* GAG GTC CTG CTC ATG CAG AAC

Row 5

E. coli L1	GAA	CTG	GTA	GGT	ATG	GAA	GAT	CTG	GCT	GAC	CAG	ATC	CAG	AAG	AAA	GGC	GAA	ATG	AAC	TTT	GAC	GTT	GTT	ATT
Yeast L3	GTC	GAA	ACC	CCA	AGA	GGT	TTG	ACC	GTC	TGG	GCT	GAA	CAT	TTG	TCT	GAC	GAA	GTC	AAG
E. coli L5	GGC	GAA	CGC	ATG	TGG	GAG	TTC	TTT	GAG	GGC	CTG	CTC	ATC	ACT	ATT	GCT	CCT	GTA	CCT	CGT	ATC	CCT	GAC	TTC	CGT
E. coli L6	GGT	TAC	CGT	GCA	GCG	GTT	AAA	GGG	AAT	GTG	ATT	AAC	CTG	TCT	TCT	CAT	CCT	GTT	GAC	CAT
E. coli L10	GCT	CGT	CTG	TTC	AAA	GAG	TTC	GCG	AAT	GCG	AAT	GCA	AAA	TTT	GAG	GTC	AAA	GCC	GCT	CCG	TTT	GAA	GGT
E. coli L11	CCG	AAC	AAA	GAC	AAA	GTG	GGT	GGT	AAA	AAA	ATT	TCC	CGC	GCT	GCG	CAG	ATC	GCG	CAG	ACC	GCG	AAA	GCT	GCC
E. coli L7/12	GCT	GCT	CTG	AAA	GAA	GGC	GTG	AGC	AAA	GAC	GAC	GCA	CTG	AAA	AAA	CTG	GAA	GAA	GCT	GGC
E. coli L14	CAG	CCT	ATC	GGT	ACG	CGT	ATT	TTT	GGG	CCG	GTA	ACT	CGT	GAG	CTT	CGT	AGT	GAG	TTC	ATG	AAA	ATT
E. coli L15	AAT	ACG	CTG	AAA	GCG	GCT	AAC	ATT	ATC	ATC	GGT	ATC	CAG	ATC	CAG	GAG	TTC	GCG	AAA	GTG	ATC	CTG	GCT	GGC	GAA
Yeast L16	CAA	TTG	AGA	GAC	AGA	AAC	TTC	TCT	GCT	ACC	GGT	AAC	TTC	GGT	ATT	GAG	GAA	CAC	ATT	GAC	TTG
E. coli L18	GAC	CGT	TCC	GGG	TTC	CAA	TAT	CAT	GGT	CGT	GTC	CAG	GCA	CTG	GCA	GAT	GCT	GCC	CGT	GAA	GCT	GGC	CTT
Yeast L17a	TTG	TAC	TTC	GAA	GAC	AAT	GCT	GGT	GTC	ATC	GCT	AAT	CCT	AAG	GGT	GAA	ATG	AAG	TCC	GCC	ATC	ACT
E. coli L24	GTC	CGT	TTC	TTC	AAG	TCT	AAC	AGC	GAA	ACT	ATC	AAG	TAA
Yeast rp28	ATC	TTT	GAA	TTC	CCA	AAG	ACC	ACT	GTT	GCT	GGT	CTT	GCT	GGT	GCC	AGA	GCC	AAG	ATT	GTT
Yeast rp29	AGA	AGA	TCT	TTG	AAG	CCA	GAA	GTT	AGA	AAG	GCT	AAC	AGA	GAA	GAA	AAA	TTG	AAG	GCC	AAC	AAA	GAA	AAG
Yeast L25	TAC	CAA	ATC	AAG	AAG	GCC	GTC	AAG	GAA	TTA	TAC	GAA	GTT	AAC	ATT	TTG	GTC	AGA	CCA	CCA	AAC	GGT	ACC	AAG
Mouse L32	---	AAA	TCT	TAC	TGT	GCT	GAG	ATT	GCT	CAC	AAT	GTG	TCC	TCT	AAG	AAC	CGA	AAA	GCC	ATT	GTA	GAA	AGA
Drosophila rp49	CCG	CGT	--T	TAC	TGC	GC-	GAG	ATG	-CC	CAC	GGC	GTC	TCC	TCC	AAG	AAG	C--	AAG	GAG	ATT	ATC	GAG	CGC

(continued)

Table 5.12 (Continued)

Row 6

E. coli L1	GCT TCT CCG GAT GCA ATG *CGC* GTT GTT GGC CAG CTG GGC CAG GTT CTG GGT CCG *CGC* GGC CTG ATG CCA
Yeast L3	*AGA AGA* TTC **TAC** AAG AAC TGG **TAC** AAG TCT AAG AAG AAG GCT TTC ACC AAA **TAC** TCT GCC AAG **TAC** GCT
E. coli L5	GGC CTG TCC GCT AAG TCT TTC GAC GGT *CGT* AAC ATG GGT GTC *CGT* GAG CAG CAG ATC ATC TTC
E. coli L6	CAG CTG CCT GCG GTC ATC ACT GCT GAA TGT CCG ACT CAG GAA ATC GTG CTG AAA GGC GCT GAT AAG
E. coli L10	GAG CTG ATC CCG GGG TCT CAG ATC GAC *CGC* CTG GCA ACT CTG CCG ACC **TAC** GAA GAA GCA ATT GCA *CGC*
E. coli L11	GAC ACT GGT GCC GAC ATT GAA GCG ATG ACT *CGC* TCC ATC GAA GGT GCA *CGT* TCC ATG GGC CTG
E. coli L7/12	GCT GAA GTT GAA GTT AAA TAA
E. coli L14	ATC TCT CTG GCA CCA GAA GTA CTC TAA
E. coli L15	GTA ACG ACT CCG GTA ACT GTT *CGT* GGC CTG *CGT* GTT ACT AAA GGC GCT GCT ATC GAA GCT GCT
Yeast L16	GGT ATC AAG **TAT** GAC CCA TCC ATC GGT ATT TTC GGT ATG ATG GAT TTC **TAT** GTC GTC ATG AAC *AGA* CCA GGT
Yeast L17a	GGT CCA GTC GGT AAG GAA TGT GCC GAT TTA TGG CCA *AGA* GTT CCA TCT AAC TCC GGT GTT GTT GTG TAA
E. coli L18	CAG TTC TAA
Yeast rp28	AAG GCT GGT GAA TGT ATC ACT TTG GAT CAA TTA GCT GTC *AGA* GCT GTC CCA AAG GGT CAA AAC ACT TTG
Yeast rp29	AAG GCT AAA GCT GCT GCT *AGA* AAG GCT GAA AAG GCT AAG TCT GCT GGT ACT AAG GCT CAA AGT TCT AAG TTC
Yeast L25	AAG GCT **TAC** GTT *AGA* TTG ACT GCT GAC **TAC** GAT GCT TTG GAC ATT GCT AAC *AGA* ATC GGT **TAC** ATT TAA
Mouse L32	GCA GCA CAG CAG CTG GCC ATC *AGA* GTC ACC AAT CCC AAC GCC *AGG* CTA *CGC* AGC --- GAA GAA GAA TAG
Drosophila rp49	GCC **AAG** CAG CTG TCG GTC *CGC* -TC ACC AAC CGC AAC GGT CGC CTC CGT CTC A-A GAA GAA *CGA* GG- TAA

Row 7

E. coli L1   AAC CCG AAA GTG GGT ACT GTA ACA CCG AAC GTT GCT GAA GCG GTT AAA AAC GCT AAA GCT GGC CAG GTT
Yeast L3   CAA GAT GGT GCT GGT ATT GAA AGA GAA TTG GCT AGA ATC AAG AAG GCT TCC GTC GTC AGA GTT TTG
E. coli L5   CCA GAA ATC GAC TAC GAT AAA GTC GAC GTT CGT GGT TTG GAC ATT ACC ACT ACT GCG AAA
E. coli L6   CAG GTG ATC GGC CAG GTT GCA GCG GAT CTG CGC GCC TAC CGT CCT GAG CCT TAT AAA GGC AAG GGT
E. coli L10   CTG ATG GCA ACC ATG AAA GAA GCT TCG GCT GGC AAA CTG GTT GCT CTG GCT GTA CGC GAT GCG
E. coli L11   GTA GTG GAG GAC TAA
E. coli L15   GGC GGT AAA ATC GAG GAA TAA
Yeast L16   GCT --- AGA GTC ACT ACA AGA AAG AGA TGT AAG GGT ACC GTT GGT AAC TCC CAG AAG ACA ACT AAG GAA
Yeast rp28   ATC TTG AGA GGT CCA AAC TCC AGA GAA GCT GTC AGA CAC TTC GGT ATG GGT CCA CAC AAG GGT AAG
Yeast rp29   TTC AAG CAA GCT AAG GGT GCT GCT TTC CAA AAG GTT GCT GCT ACT TCT CGT TAA

Row 8

E. coli L1   CGT TAC CGT AAC GAC AAA AAC GGC ATC ATC CAC ACC ATC GGT AAA GTG GAC TTT GAC GCT GAC AAA
Yeast L3   GTC CAC ACT CAA ATC AGA AAG ACT CCA GCT GCT CAA AAG GCT CAT TTG GCT GAA ATC CAA TTG AAC
E. coli L5   TCT GAC GAA GAA GGC CGC GCT CTG CTG GCT GCC TTT GAC TTC CCG TTC CGC AAG TAA
E. coli L6   GTT CGT TAC CGT GCC GAC GAA GTC GTC CGC ACC AAA GAG GCT AAG AAG AAG TAA
E. coli L10   AAA GAA GCT GCT TAA
Yeast L16   GAC ACC GTC TCT TGG TTC AAG AAG CAA AAG TAC GAT GCT GAT GTT TTG GAC AAA TAA
Yeast rp28   GCT CCA AGA ATC TTG TCC ACC GGT AAG AAG TTC GAA AAG GCT AGA GGT AGA AGA TCT AAG GGT TTC

(continued)

Table 5.12 (Continued)

Row 9

E. coli L1   CTG AAA GAA AAC CTG GAA GCT CTG CTG GTT GCG CTG AAA AAA AAA GCA AAA CCG ACT CAG GCG AAA GGC GTG

Yeast L3   GGT GGT TCC ATC TCT GAA AAG GTT GAC TGG GCT GCT CGT GAA CAT TTC GAA AAG ACT GTT GCT GTC GAC AGC

Yeast rp28   AAG GTG TAA

Row 10

E. coli L1   TAC ATC AAG AAA GTT AGC ATC TCC ACC ACC ATG GGT GCA GGT GTT GAC GTT GAC CAG GCT GGC CTG AGC

Yeast L3   GTT TTT GAA CAA AAC GAA ATG ATT GCT ATT GAC GTC ACC AAG GGT CAC GGT TTC GAA GGT GTT ACC

Row 11

E. coli L1   GCT TCT GTA AAC TAA

Yeast L3   CAC AGA TGG GGT ACT AAG AAA AAA TTG CCA AGA AAG ACT CAC ACA GGT CTA AGA AAG GTT GCT TGT ATT GGT

Row 12

Yeast L3   GCT TGC CAT CCA GCC CAC GTT ATG TGG AGT GTT GCC AGA GCT GGT CÁA AGA GGT TAC CAT TCC AGA ACC

Row 13

Yeast L3   TCC ATT AAC CAC AAG ATT TAC AGA GTC GGT AAG GGT GAT GAT GAA GCT AAC GGT GCT ACC AGC TTC GAC AGA

Row 14

Yeast L3   ACC AAG AAG ACT ATT ACC CCA ATG GGT GGT TTC GTC CAC TAC GGT GAA ATT AAG AAC GAC TTC ATC ATG

Row 15

Yeast L3   GTT AAA GGT TGT ATC CCA GGT AAC AGA AAG AGA ATT GTT ACT ACT TTG AGA AAG TCT TTG TAC ACC AAC ACT

Row 16

Yeast L3   TCT *AGA* AAG GCT TTG GAA GAA GTC AGC TTG AAG TGG ATT GAC ACT GCT TCT AAG TTC GGT AAG GGT *AGA*

Row 17

Yeast L3   TTC CAA ACC CCA GCT GAA AAG CAT GCT TTC ATG GGT ACT TTG AAG AAG GAC TTG TAA

"Codons for lysine are underscored, those for arginine are italicized. The arched lines for *Drosophila* in rows 5 and 6 are explained in the text. Asterisks designate locations of introns.

former, but the last two rows show a strong increase to five for the first of these and three for the latter (Cerretti *et al.*, 1983). Rarely is the distal end more strongly enriched for basic codons than the proximal, but that arrangement does occur in yeast *rp29*, where nine lysine and three arginine codons are found in the penultimate sector, against two for each at the 5' portion.

A Case of Identity. Because of the large number of ribosomal proteins and the paucity of genes that have been sequenced from sources other than bacterial ones, relatively little chance exists at the moment for finding a homolog of any of these in the table from *E. coli*. Frequently, the same triplet occurs at the identical site in a number of the species—for instance, TCT follows the opening ATG in yeast L3 and L16 and in *E. coli* L6 and L7/12, and its mate, TCC, is found in yeast L17a. But further comparisons fail to expose any other significant correspondences, although almost every vertical column affords multiple copies of the same or related codons. As an additional illustration of these happenstance resemblances, near the center of row 3 are six duplications of the triplet GAT or GAC, in L1, L5, L6, L7/12, L14, and L17a. While not without significance, these similitudes consistently fail to be supported by additional homologies among these several genes. Computer programs using the amino acid sequences likewise failed to expose any new homologies among all of *E. coli's* products, except the familiar L7 = L12 and S20 = L26 (Wittmann-Liebold *et al.*, 1984).

However, there is one exceptional pair, which is the more unexpected since it involves the only representative sequence from each of two organisms, the mouse and *Drosophila*. At the extreme 5' end of these two cistrons, placed adjacent in Table 5.12, few homologies are noted, but beginning just before the first codon for lysine (the AAG near the center of row 1), there is a series of identical or related codons that continues with only minor interruptions through the remaining lengths of the genes. Thus here for the first time the genes for the murine L32 and the fruitfly *rp49* species are shown to encode the homologous ribosomal protein.

It is highly informative to compare this pair from distantly related metazoans more closely, especially with regard to the codons for their characteristic active amino acids, lysine and arginine. This chance sequencing of corresponding genes is especially fortunate in that both source organisms are highly specialized, for the insect is quite as advanced in its phylum Arthropoda as the mammal is in the Chordata. When the codons for lysine are examined first, it can be noted that they are largely conserved between the two genes, although exceptions do occur. In row 2, three such disparities appear, followed by two in each of rows 3–5 and one each in rows 1 and 6; in some instances, a substitution of an arginine for a lysine has taken place. Furthermore, at several points sections have been deleted from the insect gene (or, contrarily, insertions have been acquired by the murine example), in five cases involving loss or gain of an entire triplet. On the other hand, several such mutations embrace only one or two of the nucleotides of a given codon, so that frameshift alterations are observed in one sequence or the other. Here, it is convenient to consider that such changes represent deletions in the fly cistron; accordingly, the corresponding nucleotide residues are arranged beneath those of the mouse, the actual codons in the *Drosophila* sequence being indicated by arched lines. One example of this mutational type is found at the beginning of row 5, and the second before the 3' terminus, where virtually the whole final row is involved. One further feature is of importance to understanding the evolutionary history of the genes—an intron is located at corresponding points in the two homologs, indicated by an asterisk at the center of row 2.

In contrast, a second insert in the mammalian representative is absent from that of the insect. Hence, it may prove that the first intron is a characteristic of an ancestral metazoan stock, whereas the second may characterize just the chordate or vertebrate lineage.

A New Principle for Molecular Evolution. The structures presently available for this group of proteins are certainly far too few in number and too restricted as to source organisms to permit the suggesting of a phylogenetic history and will doubtlessly remain so for some years. This pessimism is made all the more acute when Table 5.12 is surveyed for underscored lysine and the italicized arginine residues, for no columns of similar styles appear. True, there is a short file here and another there, but each disappears after extending through three or at most four sequences. Hence, the impression gained is that these genes do not represent a related family, but merely a group that encodes substances whose only common trait is that of being located on the same organelle. But is this a valid assessment of these cistrons?

Some information that suggests a conflicting point of view is the fact that they are arranged in several operons in *E. coli,* sets of genes that not only coexist, but are transcribed in unison. It is much more readily perceptible that the individual components of such operons arose by duplication and subsequent evolutionary adaptation of an ancestral gene than that a number of phylogenetically unrelated cistrons were brought together to form the unified chains. Consequently, there would seem to be a strong likelihood that all the ribosomal proteins arose from a few—or even a single—progenitory type, from which they subsequently deviated as individual functions became more marked between the increasing diversity of types. This increase in number is an obvious distinction between the ribosomes of *E. coli* and those of yeast, as well as between those of lower protistans and advanced eukaryotes. When these substances and their genes are investigated in the blue-green algae, rickettsias, and chlamydids, more than likely further reductions in ribosomal protein species complexity will be revealed, probably in the given sequence of organisms.

Since the location of the codons for the more active ingredients is correlated more closely to functional requirements than to ancestral influences, a question arises as to what then can be depended upon for clues as to their phylogeny. In the first place, there can be little doubt that such interrelationships will be clearer among species derived from more primitive organisms like *Beggiatoa* and the rickettsias, but even in those gene sequences, it will be the triplets encoding rarer and noncritical amino acids that will provide the principal clues to the kinships. Sites for active, essential constituents are the most reliable indicators of homology between corresponding genes for the same species of substance from different organisms, whereas the chance evolutionary relics of space-filling amino acids are the providers of data for multispecies descendants of a given ancestral type. Therefore the column-by-column identities that exist in Tables 5.10 and 5.12 are not entirely vagaries of random mutations, but in many cases represent true homologies of structure reflecting evolutionary development. But at the present time an insufficient representation of complete sequences is available to permit a reading of such a history.

5.7. TERMINATION IN RIBOSOMAL PROTEIN GENES

Because of the intermixing of large- and small-subunit ribosomal genes in those operons, where the members have been determined, and because of their probable unified

Table 5.13
Trains of Ribosomal Protein Genes[a]

E. coli rpsA[b]	TTCTCTGACTCT	TCGGGATTTTA	TTCCGAAGTTTG--	TTG	AGTTTACTTGAC	AGATTCCAGGTT	TCGTCCCTCTAA
E. coli spc[c]	TTTTTTCGGCATA	------TTTTTC	TTGCAAAGTTGG--	TTG	AGCTGGCTAGAT	TAGCCAGCCAAT	CTTTTGTATGTC
E. coli S20[d]	TCGCCAATTTGC	TGAAGCTTTGTG	AAAAAGCCCGCG--	CAA	GCGGGTTTTTTT	ATGCCTGCTGCT	TTTCTGATGCGT
$Euglena_C$ S7[e]	TATATATAAATT	TTTTAAGCGTTA	AAAAAAAATAAAA-	AAT	AAAATTTGACCA	AAATCTTTCTAA	TAAAATGAGTT
$Tobacco_C$ S19[f]	TATTAATAAAAA	AAATCTAGATGC	TTATGATTCAGTA-	GTA	GGAGGCAAACCT	TATGCTAAAGAA	GAAAAAAACAG
Yeast S10[g]	GTATGTCTTGA	CCTGCATAATT	TTATACTTTTTCC-	TTT	TAGATATATGTA	CTTTTGGCTTAA	TTTAATATAAT
Yeast S16A[h]	ATATAAAATCTA	TAATTTATATAT	ATATACTACTATA-	CAA	TTATCATCATAC	AAGTAATAAA	TTTTTTTTAA
Yeast rp28[h]	GTTAACTGAAAT	GAAAATTTCATA	TTTACTTTTTTA-	TTG	TTACTCATTTGT	AATTCATAAAC	
Yeast L3[i]	GAAGTTTTGTTA	GAAAATAAATCA	TTTTTTAATTGAGC	ATT	CTTATTCCTATT	TTATTTAAATAG	TTTTATGTATT
Yeast L16[j]	TTTAATTAGTTT	GTGAAGAAATAT	AATACATTTATAT-	ACT	CATATCTATGTT	TTTTTTGTAACC	AATTAATAAAAA

Yeast L25[k]	TCTAATTGGTTT	AATTAATAATTT	AATATTA*TTTTA*–	AA*T*	*TTTT*CTTTAA*AT*	*ATA*CAATAAATC	TT+9GTTAA<u>AT</u>
Yeast rp29[l]	GATTTATGCTCG	AACTTATTATGT	ACAATGAATA*TTT*–	*TT*C	TTTTAAATCA*TT*	*TTT*AAATATTTC	AA+60CATTA<u>T</u>
Mouse L32[m]	---ATGGCTTGT	GTGCATGTTTTA	TGTTTA*AATAAAA*–	TCA	CAAAACCTGCA$_{27}$ A	CAAATTAGCT	GC*TTTTTTTTTC*
Drosophila rp49[n]	GCTTAAGAATTCT	TGAGAGTTCTTG	TAACGTGGTCGGA–	ATA	CACATTTGTAAA	ACGT*TAATATAC*	CGGACTTTTAC

[a]Runs of five or more T are italicized, as are putative polyadenylation signals; underscored letters indicate actual termination sites and noncanonical base pairs; arrows underline regions of dyad symmetry. Subscript C indicates chloroplast. Asterisks designate locations of introns
[b]Schmier et al. (1982).   [c]Cerretti et al. (1983).   [d]Mackie (1981).   [e]Montandon and Stutz (1984).   [f]Sugita and Sugiura (1983).   [g]Leer et al. (1982).   [h]Molenaar et al. (1984).   [i]Schultz and Friesen (1983).   [j]Leer et al. (1984b).   [k]Leer et al. (1984a).   [l]Mitra and Warner (1984).   [m]Dudov and Perry (1984).   [n]O'Connell and Rosbash (1984).

evolutionary history, it is beneficial to examine the trains of both classes together. As a result of their frequently being transcribed as entire gene sets, only a small proportion of the individual cistrons are equipped with trains bearing transcriptional termination and related signals. In Table 5.13, where the available 3′ flanks are listed, rows of five or more adjacent Ts are italicized, as are poly(A) signals.

Termination Signals. The trains of the three ribosomal protein genes of *E. coli* set the stage for those of eukaryotes and chloroplasts, for each represents one of the trio of conditions that might exist in those others. That of the operon called *spc* has the standard terminator of prokaryotes in that six adjacent Ts are found, in the present instance directly after the stop codon for translation. Since that location close to the gene's end is quite unusual, the actual signal may prove to be the five adjoining Ts that follow six sites later. The likelihood that this second set could function as suggested is enhanced somewhat by the location at a comparable point in the bacterial *rpsA* train of a run of Ts of identical length, for no other recognizable signal is present in the latter. In the third member of the trio the run of Ts is a terminal feature on a sector of dyad symmetry (underlined by arrows) that contains in typical fashion one noncanonical pair (underscored) as in many of the rho-dependent signals seen in prior chapters. The obvious homologies that exist between the first two bacterial trains are of particular interest in their suggestion of common descent for the two operons they represent.

The two sequences from chloroplasts have little of note to offer. That from the euglenoid organelle, S7, is transcribed along with its linked mate S12 into a dicistronic precursor devoid of the gene for EF-Tu that follows it (Montandon and Stutz, 1984), so that a terminator should be present in its train. This probably is represented by the six italicized Ts located ten sites downstream from the transcriptional stop codons, but that cluster has not been established as serving in that capacity. In the short train that has been determined for the tobacco organelle cistron, no terminator of any sort can be detected. Although runs of As abound, the longest series of adjoining Ts consists of three (Sugita and Sugiura, 1983).

Like those of the prokaryote, the trains of these genes from baker's yeast are a varied lot, but here the variation is less extensive. No region of dyad symmetry is present, neither here nor in the others from eukaryotes. Several, including the genes for rp28, L3, and L16, have six or more neighboring Ts located in standard fashion and therefore present no problem, at least on a preliminary basis. But the cistrons for S10, L25, and rp29 are distinctive in lacking any sequences of six or more Ts in their trains, instead having two sets each of five consecutive Ts. In all three cases, these are separated by relatively few residues, 11 being the maximum. The gene of yeast for S16A and that for L32 of the mouse share an unusual feature. In each instance there is a run of Ts of unquestionable length that could be a standard promoter; however, these sequences *follow* the poly(A) signal discussed below, rather than preceding it in classical fashion. Actual 3′ termini of the primary transcript have been determined in only two of these genes, those for yeast L25 and rp29, both of which terminate in the adenosine residue shown underscored.

Polyadenylation Signals. The standard poly(A) signal, AATAAA (italicized in the table) is present in a number of the eukaryotic sequences, whether or not the individual example is known to bear that type of tail. In fact, only the *Drosophila* rp49 has been stated to have such an appendage. However, the combination AATATA given in the original article as serving in that capacity for this insect gene is somewhat aberrant, as is one of the two thus designated from other sources. That for the yeast rp28, with ATAAA, is one instance of this sort and another is that for L25, which has AATATA as in the *Drosophila* gene. In one case, the cistron for yeast S16A, there is a standard poly(A) signal followed by a terminator 50 base pairs downstream as mentioned earlier; this latter is preceded directly by a second poly(A) indicator.

6

Genes for Energy-Related Proteins

Proteins and protein systems involved directly with the production or employment of such energy-storing molecules as ATP and the other triphosphorylated nucleosides are of numerous types. Some of these form large families, such as the familiar cytochromes, while others comprise small groups of related varieties, and still others work as clusters of diversified composition. But even the large family just cited, the cytochromes, are not unified as to their specific function. Although all conduct electrons, the best known members of that group comprise the greater portion of an electron-transport chain whose end product is ATP or kin substance, but others, notably the cytochromes P-450, are concerned with syntheses that break down that type of nucleoside to liberate required energy. Still another category of energy-using compounds contains the varied components of a group involved in using such energy to convert free nitrogen into ammonia and thereby "fix" that element. Thus this chapter centers both on substances that create stores of energy and others that draw upon that supply. No relationship between the several classes involved here is to be expected; their sole common denominator is a direct interaction with high-energy nucleosides. But the possibility that this varied lot may show some features in common nevertheless needs to be explored—the two unrelated classes examined in the preceding chapter certainly were found to share some very significant traits.

The genes for the cytochromes are both the better known and most widely distributed among organisms of almost all sorts, *Beggiatoa* and its congenors, along with rickettsias and chlamydids, being the only taxa known to be devoid of this class of biochemical. Consequently, this large group needs to be analyzed to establish its characteristics before the comparatively few of parallel function are examined.

6.1. THE CYTOCHROMES

The cytochromes are a complex class of proteins containing well over 20 different species, whose chief common structural feature is the presence of an iron porphyrin (heme) moiety. Most are reddish in color, except the a/a_3 type, which tends to be greenish from the presence of an atom of bound copper. Because most are constituents of membranes, cytochrome c alone being soluble, the several species have remained difficult to isolate and purify. As first stated, the major common functional trait is the transport of electrons, a property dependent entirely upon the heme group, together with copper in a few varieties. Reduced cytochromes are often referred to as ferrocytochromes, while the oxidized molecules are called ferricytochromes.

As already noted, the most familiar species are those active in the electron-transport chain, an exceedingly complicated association of multiple types of proteins, some of which still remain unidentified. In simple terms, this chain is arranged into five complexes, whose properties and composition have been detailed elsewhere (Dillon, 1981). The second outstandingly important association of cytochromes already mentioned, the cytochromes P-450, is comprised of more than eight varieties. Characteristically these are membrane-bound oxidases functioning at the termini of NADPH-dependent electron transport pathways (Dillon, 1981, pp. 349–353). The most thoroughly documented members of this category are derived from metazoans, those of the liver, gonads, and adrenal cortex of vertebrates being the most frequently investigated. Many from these three sources metabolize steroids, such as the conversion of 17-hydroxy-progesterone to 11-deoxycortisol and the synthesis of cortisol from cholesterol, the latter reaction requiring no fewer than four different representatives of this group for its completion.

As usual in this set of analytical studies, initiation and transcription of the cytochrome genes of all types together is examined first, followed by separate discussions of the mature coding regions of the more important species groups, beginning with those of the electron-transport chain. Finally, what is known of transcription termination is summarized, before analyses of other energy-related types are begun.

6.1.1. Initiation of Transcription of Cytochrome Genes

Two factors contribute to the current scarcity of data regarding initiation of transcription of the cytochrome genes, the location of several types in close proximity to one another in the genomes of chloroplasts or mitochondria and the frequent presence of long, A,T-rich segments in many other leaders. As will be recalled from previous discussions involving organellar cistrons, often those for cytochromes may similarly be cotranscribed with neighbors and thus may lack promoters altogether. Additionally, a third element may also be responsible for the existing lack of information, an insufficient length of leaders that have been sequenced, for in two cases discussed shortly, the transcriptional start sites have been demonstrated to lie 2500 or more nucleotide residues upstream from the mature genes.

The net result of these several aspects of structure is that the table of promoters (Table 6.1) contains only five sequences of leader sectors for comparative purposes, two from cytochromes c, one from a cytochrome c_1, and the remainder from cytochromes P-450. In the last pair alone have the start sites (underscored) been fully determined (Mizukami *et al.*, 1983; Sogawa *et al.*, 1984), and all the promoters and other signals, both of which types are italicized, are presumptive only.

The Putative Promoters. All five of the leaders whose sequences are provided here are from nuclear genes (Table 6.1). In the *S. pombe* sector, the cited signal was selected

Table 6.1
Sections of Leaders of Cytochrome Genes

	Putative ancillary	Putative promoter						Start site
Cytochrome c								
S. pombe[a]	ACCAATCC	TGTTTTTAC	-------T	TTAACGAAGG	TTATCGCACGCA	AAAAA-GGTAAACA	TTGG	+216
Chicken CC9[b]	TGCAATCT	TCAAACCCA	AGA----G	CTATTAACTCG	TGGCTGGAAGGC	ATTCAGGAGAGTTA	CCAA	+122
Cytochrome c₁								
Yeast[c]	TTCAATTGA	TTAGTTTGA	ACT+10-A	ATAATATTTT	ACAATTTGCATT	TTCATTACACTATA	TCAT	+42
Cytochrome P-450								
Rat P-450[d]	GCCAACAGA	CA--------	-------G	TCATATAAAG	GTCCTGGTCCCT	TCACCCTA----AC	CATC	+2562
Rat P-450[e]	CATAACTGA	GTGTAGGGG	CAGATTCA	GCATAAAAGA	TCCTGTGGAGA	GCA--------TG	CACT	+3400

[a]Russell and Hall (1982).
[b]Limbach and Wu (1983).
[c]Sadler et al. (1984).
[d]Sogawa et al. (1984).
[e]Mizukami et al. (1983).

from among a number of other possible combinations in the nearly 500-site-long section given in the original contribution (Russell and Hall, 1982), largely because of its proximity to a promising candidate for an ancillary signal. A similar statement applies to the cytochrome c_1 gene of the yeast (Sadler *et al.*, 1984), while that of the chicken CC9 was suggested by Limbach and Wu (1983). On the other hand, the two from rat, P-450$_c$ and P-450$_e$, were proposed on the basis of their location relative to the determined start sites (Mizukami *et al.*, 1983; Sogawa *et al.*, 1984). As may be noted, both of these are situated 2000 or more nucleotide residues before the mature gene, whereas those of the other three are just between 40 and 250 sites upstream from that point. Whether further investigation will reveal comparably long leaders for both classes of cytochromes or whether this will prove to be a species characteristic remains for the future.

The suggested promoters are in keeping structurally with those of preceding pages in showing resemblance to the canonical TA-TA- combination. However, as is so frequently the case, the similarities primarily stem from the presence there of at least three As and one or more Ts. Also as elsewhere, the other residues may be Cs or Gs, the actual sequence being highly inconstant.

Possible Ancillary Signals. The possibility that the ancillary signals given in Table 6.1 will be demonstrated to be valid is even more tenuous than the likelihood that the promoters will, for their selection was necessarily based on the latter. Their only claim for attention here lies in their similarity of structure to those that were more firmly established and discussed in previous chapters. Moreover, they lie at distances from the suggested promotion signals comparable to actually established cases. Of special interest is the presence of two nearly adjacent similar structures of the chicken CC9 leader. The proposed ancillary sites lie 8–21 nucleotide residues farther upstream; almost all open with the sequence CCA-, the single exception having the C replaced by a T. In all these cases, the downstream end of the signal is either a C or a T.

6.1.2. The Genes for Cytochromes b

Although the current literature typically refers to the subject of this section merely as cytochrome *b*, that term actually embraces an entire class of biochemicals, ten of the known varying types of which have been characterized elsewhere (Dillon, 1981, pp. 325–327, 341–349). Hence discussion should refer to these varied substances as the cytochromes *b*, a point that permits a better comprehension of the heterogeneity that is found later when their mature gene sequences are surveyed.

Characteristics of Cytochromes b. The respiratory cytochromes *b* of mitochondria, moreover, involve two representatives, both of which are components of subunit III of the respiratory chain. One of these, the classical type, is reduced by succinic acid even in the absence of ATP and is distinguished as cytochrome b_K. On the other hand, cytochrome b_T is reduced by that reagent only in the presence of ATP (Davis *et al.*, 1973; Dillon, 1981, p. 342). Since only a single gene for this complex (cytochrome b_K) exists in the fully sequenced mitochondrial genome of mammals and *Drosophila*, it is self-evident that the other must be encoded by a nuclear gene, as yet undetected.

Also included among the cytochromes *b* are at least two types involved in electron transport in photosynthetic processes. Both the lesser one of these, often referred to as cytochrome b_{559}, and the major species, cytochrome b_6, are represented in Table 6.2 by

Table 3.2

Mature Gene Sequences for Cytochromes b[a]

Row 1

Yeast CBP1[b]	ATG TTT TTA --- CCT CGT CTC GTT CGG TAC AGG ACC GAG AGG TTT ATA AAA ATG CCT ACC AGG ACC
Yeast D273[c]	ATG GCA TTT --- AGA AAA TCA --- --- --- AAT GTG TAT --- TTA AGT TTA GTG AAT AGT TAT
Trypanosoma[d]	TAA ATT TTA --- TAT AAA AGC GGA GAA AAA AAG AAG GGT CTT TTA ATG TCA GGT TGT TTA TAT AGA ATA
Aspergillus[e]	ATG AGA ATT --- --- TTA --- --- AAA AGT --- CAT CCT TTA CTA AAA --- ATA GTA AAT TCG TAT
Drosophila[f]	ATG CAT AAA CCT TTA CGA AAT TCC CAC CCT TTA TTT AAA ATT GCT AAT AAT GCT TTA GTT GAT TTA CCA
Cyprinus[g]	ATG GCA AGC --- CTA CGA AAA ACA CAC CCT CTC ATT AAA ATC GCT AAC GAC GCA CTA GTT GAC CTA CCA
Human[h]	ATG ACC --- CCA ATA CGC AAA ATT AAC CCC CTA ATA AAA AAA TTA ATT AAC CAC TCA TTC ATC GAC CTC CCC
Maize[i]	ATG ACT ATA --- AGC AAC CAA CGA TTC TCT CTT CTT AAA CAA CCT ATA TAC TCC ACA CTT AAC CAG CAT
Spinach b_6[j]	ATG ATT GGT --- TCG AAG AAC GTC TCG AGA TTC AGG CGA TTG CGG ATG ATA ATA ACT AGT AAA TAT GTT
Spinach b_559[k]	ATG TCT GGA --- AGC ACA --- GGA GAA CGT TCT TTT GCT GAT ATT ATT ACC AGT ATT CGA TAC TGG GTT

Row 2

Yeast CBP1	TTG CGA CGA ATC AAC CAC AGC AGC AGG GAT CCA ATT CAA AAA CAG GTC TTG GCC CTT ATC AAA GCA AAT
Yeast D273	ATT ATT GAT TCA CCA CAA CCA TCA TCA ATT AAT TAT TGA TGA AAT ATG GGT TCA TTA TTA GGT TTA TGT
Trypanosoma	TAT GGG GTA GGT TTT AGT TTA GGA TTT TTT ATA GCA TTG CAA ATA ATT TTC GGA TCA TTA TTA GCT TTA TGT
Aspergillus	ATA ATA GAT TCA CCT CAA CCA GCT AAT TTA AGT TAT GGA TCA CTT GGA TTA TGT TTA TTA GCT TTA TGT
Drosophila	ATT AAT ATT --- --- TCA AGA TGA TGA GCA TGA TTT GGA AAC TTT GGA TCA CTA TGC TTA ATT ATT ACC CAA ATT
Cyprinus	ACA CCA TCC AAC ATC TCA GCA TGA TGA AAC TTT GGC TCA CTC CTA GGA CTA TGC TTA ATT ACC CAA ATT
Human	ACC CCA TCC AAC ATC TCC GCA TGA TGA AAC TTC GGC TCA CTC CTT GGC GCC TGC CTG ATC CTC CAA ATC
Maize	TTA ATA GAT TAT CCA ACC CCG AGC AAT CTT AGT TAT GGG TGG GGG TTC GGT TGC TTA TTA GCT ATT TGT
Spinach b_6	CCT CCT CAT GTC AAC ATA TTT TAT TGT CTA GGC GGA ATT ACG CTT ACT TGT TTT TTA GTA CAA GTA GCT
Spinach b_559	ATT CAT AGC ATT ACT ATA CCT TCC CTA TTC ATT GCG GGT TGG TTA TTC GTC AGC ACA GGT TTA GCT TAC

Table 6.2 (Continued)

Row 3

Yeast CBP1	GCG AAT TTA AAT GAC AAT GAC AAG TTG AAA ATA *CGG* AAA TAT TGG TCT GAC ATG GCG GAC TAC AAA AGT
Yeast D273	TTA GTT ATT CAA ATT GTA ACA GGT ATT TTT ATG GCT TAT TCA TCT AAT ATT GAA TTA GCT TTT
Trypanosoma	TTA TTT TTT AGT TGT TTT ATT TGT TCA AAT TGA TAT TTT GTA TTA TTT TGA GAT TTT GAT TTG GGT
Aspergillus	TTA GGT ATA CAA ATA GTA ACA GGT GTT ACA TTA GCT ACA` CCT AGT GTA TCA GAA GCA TTT
Drosophila	TTA ACT GGA TTA TTT TTA GCT ATA CAC TAC ACA GCA GAT GTT AAC TTA GCT TTT TAT AGT GTT AAT CAT
Cyprinus	TTA ACC GGC CTA TTC CTA GCC ATA CAC TAC ACC TCA GAC ATC TCA ACC GCA TTC TCA TCT GTT ACC CAC
Human	ACC ACA GGA CTA TTC CTA GCC ATG CAC TAC TCA CCA GAC GCC TTT TCA TCA ATC GCC CAC
Maize	TTA GTC ATT CAG ATA GTG ACT GGC GTT TTT TTA GCT ATG CAT TAC ACA CCT CAT GTG GAT CTA GCT TTC
Spinach b_6	ACA GGG TTT GCG ATG ACT TTT **TAC TAT** *CGT* CCA ACC GTT ACT GAT GCT TTT GCT TCT GTT CAA **TAT** ATA
Spinach b_{559}	GAT GTG TTT GGA AGC CCT *CGT* CCA AAC GAA **TAT** TTC ACA GAG AGC *CGA* CAA GGA ATT CCA TTA ATA ACT

Row 4

Yeast CBP1	CTT *CGG* AAA CAA GAA AAT AGC TTA CTG GAA AGC TCT ATA TTA CAC GAG GTC AAG ATC GAA GAT TTC ATC
Yeast D273	TCA TCT GTT GAA CAT ATT ATA *AGA* GAT GTG CAT AAT GGT **TAT** ATT TTA *AGA* TTA CAT GCA AAT GGT
Trypanosoma	TTT GTG ATA *AGA* AGT GTA CAT ATA TGT TTT ACA TCT TTA TTA TTA **TAT** ATC CAT ATA TTT
Aspergillus	AAT TCT GTA GAG CAT ATT ATG *AGA* GAT GTA AAT AAT GGA TGA TTA GTA GTA **TAC** TTA CAC TCT AAT ACA
Drosophila	ATT TGC *CGA* GAT GTA AAT **TAT** GGT TGA TTA TTA *CGA* ACT TTA CAC GAT AAC GGT GCA TCA TTT TTT TTT
Cyprinus	ATC TGC *CGA* GAC GTA AAT **TAC** GGC TGA CTA ATC *CGT* AAT G (incomplete)
Human	ATC ACT *CGA* GAC GTA AAT **TAT** GGC TGA ATC ATC *CGC* **TAC** CTT CAC GCC AAT GGC GCC TCA ATA TTC TTT
Maize	AAC AGC GTA GAA CAC ATT ATG *AGA* GAT GTT GAA GGG GGC TGG TTG CTC *CGT* **TAT** ATG CAT GCT AAT GGG
Spinach b_6	ATG ACT --- GAA GTC AAC TTT GGT TGG TTA ATA *CGA* TCC GTT CAT *CGG* TGG TCG GCA AGT ATG ATG GTT
Spinach b_{559}	*GGC* *CGT* TTT GAC TCT TTG GAA CAA CTT GAT GAA TTT AGT *AGA* TCC TTT TAG

Row 5

Yeast CBP1	ACT	TTC	ATC	AAT	CGC	ACA	AAA	ACC	TCA	TCT	ATG	ACT	ACA	ACA	GGA	ATT	TAT	AGA	AGA	GAA	TGT	TTG	TAC
Yeast D273	GCA	TCA	TTC	TTT	TTT	ATG	GTA	ATG	TTT	ATG	CAT	ATG	GCT	TTA	TAT	TAT	GGT	TCA	TAT	AGA	TCA		
Trypanosoma	AAC	TCA	ATA	ACG	TTA	ATA	ATA	TTG	TTT	GAC	ACA	CAT	ATA	TTA	GTA	TGA	TTT	ATA	TTG	TTT			
Aspergillus	GCT	TCA	GCT	TTC	TTC	TTT	TTA	GTA	TAC	TTA	CAC	ATA	GGA	ACA	GGT	TTA	TAT	TAT	GGA	TCT	TAC	AAA	ACA
Drosophila	ATC	TGT	ATT	TAC	TTA	CAT	ATT	GGT	CGA	GGA	ATT	TAT	TAC	GGA	TCA	TAT	TTA	TTT	ACA	CCA	ACT	TGA	TTA
Human	ATC	TGC	CTC	TTC	CTA	CAC	ATC	GGG	CGA	--	GGC	CTA	TAT	TAC	GGA	TCA	TTT	CTC	TAC	TCA	GAA	ACC	TGA
Maize	GCA	AGT	ATG	TTT	CTC	ATT	GTG	GTT	CAC	CTT	CAT	ATT	TTT	CGT	GGT	CTA	TAT	CAT	CGG	AGT	TAT	AGC	AGT
Spinach b_6	CTA	ATG	ATG	ATC	CTG	CAT	GTA	TTT	GTG	TAT	CTT	ACC	GGT	GGG	TTT	AAA	AAA	CCT	CGA	GAA	TTG	ACT	

Row 6

Yeast CBP1	CAA	TGC	AAG	AAA	AAC	TTG	GAT	CTA	AAT	CAA	GTG	TCC	CAA	GTT	TCA	TCC	GTA	AGA	CAT	CAA	AAG		
Yeast D273	CCA	AGA	GTA	CTA	TTA	TGA	AAT	GTA	GGT	GTT	ATT	ATT	TTC	ATT	ATT	TTA	ACT	ATT	GCT	ACA	GCT	TTT	TTA	GGT
Trypanosoma	GTA	TTT	ATA	ATA	ATA	GCT	TTT	ATA	GGA	TAT	GTA	CTG	CCT	TGT	ACA	ATG	ATG	TCA	TAC	TGA	GGT	TTA	
Aspergillus	CCT	AGA	ACT	TTA	ACA	TGA	GCT	ATT	GGA	ACA	GTA	ATA	CTA	ATA	GTT	ATG	ATG	GCC	ACA	GCC	TTC	TTA	GGT
Drosophila	GTA	GGA	GTA	ATT	ATT	TTA	TTT	TTA	GTA	ATA	GGA	ACA	GCT	TTT	ATA	GGT	TAT	GTT	TTA	CCT	TGA	--	GGA
Human	AAC	ATC	GGC	ATT	ATC	CTC	CTG	CTT	GCA	ACT	ATA	GCA	ACA	GCC	TTC	CTA	TTA	ATG	GTC	CTC	CCG	TGA	GGC
Maize	CCT	AGG	GAA	TTT	GTT	TGG	TGT	CTC	GGA	GTT	GTC	ATA	TTC	CTA	TTA	ATG	ATT	GTG	ACA	GCT	TTT	ATA	GGA
Spinach b_6	TGG	GTT	ACA	GGC	GTG	GTT	CTG	GGT	GTA	TTG	ACC	GCG	TCT	TTT	GGA	GTA	ACT	GGT	TAT	TCC	TTA	CCT	TGG

(continued)

Table 6.2 (Continued)

Row 7

Yeast CBP1   CCC TTG ACT ACG CAA TTG GAT ACT ATG *CGC* TGG TGT GTT GAT GAT GCC ATC GGC ACA GGA GAC ATA GTT

Yeast D273   TAT TGT TGT GTT TAT GGA CAG ATG TCA CAT TGA GGT GCA CTA GTT ATT ACT AAT TTA TTC TCA GCA ATT

Trypanosoma   ACG GTG TTT AAT ATT ATA GCA ACA GTA CCA ATT TTA GGT ATA TGA TTA TGT TAT TGA ATT TGG GGA

Aspergillus   TAT GTT TTA CCT TAT GGT CAA ATG GCT ACT GTA ATT ACT ACT AAC CTA ATG AGT GCT ATA

Drosophila   CAA ATA TCA TTT TGA GGA GCC ACA GTA ATT ACT ACA AAC TTA TTG TCA GCT ATC CCT TAT TTA GGT ATA GAC

Human   CAA ATA TCA TTC TGA GGG GCC ACA GTA ATT ACA AAC TTA CTA TCC GCC ATC CCA TAC ATT GGG ACA GAC

Maize   TAC GTA CCA CCT TGG GGT CAG ATG AGC TTT TGG GGA GCA ACA GTA ATT ACA AGC TTA GCT AGC GCC ATA

Spinach b_6 --- GAC CAA ATT GGC TAT TGG GCA GTA AAA ATT GTA ACA GGC GTA CCG GAT GCT ATT CCT GTA ATA GGA

Row 8

Yeast CBP1   ATG GCT GCC GAC CTT TTC CTG CTG TAC TAC *AGA* TTA TTT ACA GAT GAT AAA AAG CTA GAC GAA CAA TAT

Yeast D273   CCA TTT GTA GGT AAC GAT AT⁻ GTA TCT TGA TTA TGA GGT GGG TTC TCA GTA TCT AAC CCT CTA ATC CAG

Trypanosoma   AGT GAA TTT ATA AAC GAT TTT ACA TTA TTA AAG TTA CAT GTA TTA TTA CCA TTT ATA TTA

Aspergillus   CCT TGA ATA GGT CAA GAT ATT GTT GAG TTT ATT TGA GGA GGT TTC TCT GTA AAT AAT GCA ACT TTA AAC

Drosophila   TTA GTA CAA TGA TTA TGA GGA GGA TTT GCT GTA GAT AAT GCT ACT *CGA* TTT TTC ACA TTT CAT

Human   CTA GTT CAA TGA ATC TGA GGA GGC TAC TCA GTA GAC AGT CCC ACC CTC ACA *CGA* TTC TTT ACC TTT CAC

Maize   CCA GTA GTA GGA GAT ACC ATA GTG ACT TGG GGT TTC TCC GTG GAC AAT GCC ACC TTA AAT

Spinach b_6 TCG CCT TTG GTA GAG TTA TTA *CGT* GGA AGT GCT GCT GTG GGA CAA TCT ACT ACT TTG *CGT* TTT TAT AGT

Row 9

```
Yeast CBP1    GCT AAG AAA ATA ATA TCA GTA TTA GCG TAC CCA AAC CCA CTG CAT GAT CAT GTT CAT CTA CTC AAA TAT
Yeast D273    AGA TTC TTT GCG TTA CAT TAT TTA GTA CCT TTT ATC ATT GCT GCA ATG GTT ATT ATG CAT TTA ATG GCA
Trypanosoma   CTA ATA ATA TTA ATT TTA CAT TTT TGT CTA CAT TAT TTT ATG AGT TCT GAT GCA TTT TGT GAT AGG
Aspergillus   AGA TTC TTT GCA TTA CAT TTC TTA TTA CCT TTT GTA TTA GCT GCT TTA GCA TTA ATG CAT TTA ATA GCT
Drosophila    TTT ATT TTA CCT TTT ATT GTT CTT GCT ATA ACT ATA ATT CAT CTA TTT TTA CAT CAA ACA GGA TCT
Human         TTC ATC TTG CCC TTC ATT ATT GCA GCC CTA GCA ACA CTC CAC CTC CTA TTC TTG CAC GAA ACG GGA TCA
Maize         CGT TTT TTT AGT CTC CAT CAT TTA CTC CCC CTT ATT TTA GCA GGC GCC AGT CTT CTT CAT CTG GCT GCA
Spinach b6    TTA CAC --- ACT TTT GTA TTG CCT CTT --- CTT ACT GCC GTA TTT ATG TTA ATG CAC TTT CTA ATG ATA
```

Row 10

```
Yeast CBP1    TTA CAA --- CTG AAC TCT CTG TTC GAA AGT ATA ACC GGA GGC GGA ATA AAG TTA ACG AGG TTT CAA TTA GAA
Yeast D273    TTA CAT --- ATT CAT GGT TCA TCT AAT CCA TTA TCT ATT ACA GGT AAT TTA GAT AGA ATT CCA ATG CAT TCA
Trypanosoma   TTT GCA TTT TAT TGT GAA AGA TTA AGT TTT TGT ATG TGG TTT TAT TTG AGA GAT ATG ATG TTT TTA GCA TTT TCA
Aspergillus   ATG CAT GAT GAT ACA GTA GGA TCA GGT AAT CCT TTA GGT ATT TCT GCT AAT TAC GAT AGA TTA CCT TTT GCT CCT
Drosophila    AAT AAC --- CCT ATT GGT TTA AAT TCT AAT ATT GAT AAA ATT CCT TTT CAC CCA TAC TTC ACA TTT AAG GAT
Human         AAC AAC --- CCC CTA GGA ATC ACC TCC CAT TCC GAT AAA ATC ACC TTC CAC CCT TAC TAC ACA ATC AAA GAC
Maize         TTG CAT --- CAA TAT GGA TCA AAT AAT CCA TTG GGT GTA CAT TCT GAG ATG ATG TCT TTA GCA ATT GCT TCT TAC CCT
Spinach b6    CGT AAA --- CAA GGT ATT TCT GGT --- CCC TTA TAG
```

(continued)

Table 6.2 (Continued)

Row 11

```
Yeast CBP1    ACT CTT TCT AAT AAG GCC CTC GGC TTA AGT AAT GAA GCC CCG CAA TTA TGC AAG GCT ATA CTG AAC AAA
Yeast D273    TAC TTT ATT TTT AAA GAT TTA GTA ACT GTT TTC TTA TTT ATG TTA ATT TTA GCA TTA TTT GTA TTC TAT
Trypanosoma   ATA TTA TTA TGT ATG ATG TAT GTT ATA TTT ATA AAT TGG TAT GTA TTT CAT GAG GAA TCT TGA GTT
Aspergillus   TAT TTT ATA TTT AAA GAT TTA ATA ACT ATA TTT ATA TTC TTT ATT GTA TTA TCA ATA TTT GTT TTC TTT
Drosophila    ATT GTA GGA TTT ATT GTA ATA ATT TTT ATT CTA ATT TCA TTA GTT TTA ATT AGA CCA AAT TTA TTG GGA
Human         GCC CTC GGC TTA CTT CTC CTC TCC TTA ATG ACA TTA ACA CTA TTC TCA CCA CTC CTA GGC
Maize         TAT TTT TAT GTA AAG GAT CTT GTA GGT CGG GTA GCT TCT GCT ATC TTT TTT TCC ATT TGG ATT TTT TTT
```

Row 12

```
Yeast CBP1    CTG ATG --- AAT ATA AAC TAT TCT TTG ACT AAC GAT TTG AAG CTT CGG GAT GAT CAA GTG CTG CTT GCG TAC
Yeast D273    TCA CCT --- AAT ACT TTA GGT CAA AAT ATG GCC TTA TTA TTA ATT ACA GTA ATT AAT ATT TTA TGT GCT
Trypanosoma   ATA GTA --- GAT ACA CTA AAA ACA TCA GAT AAA ATA TTA CCA GAA TGA TTT TTT TTG TAT TTA TTC GGT TTT
Aspergillus   ATG CCT --- AAT GCT TTA GGT --- GAT AGT GAA AAT TAT GTT GCT ATG CAA GCT CAC ATT CAA ACT CCA --- CCT
Drosophila    GAC CCA GAT AAC TTT ATT CCT GCT AAT CCT TTA GTA ACA CCA CCT CAA CCA GAA TGA TAT TTT TTA
Human         GAC CCA GAC AAT TAT ACC CTA GCC AAC CCC TTA AAC ACC CCT CCC CAC ATC AAG CCC GAA TGA TAT TTC CTA
Maize         GCT CCA --- AAT GTT TTG GGG --- CAT CCC GAC AAT TAT ATA CCT GCT AAT CCG ATG CCC ACC CCG CCT CAT
```

Row 13

Yeast CBP1   AAC TCC ATT GAT GAA AAT TAT AGA AGA GGA AAT GTT GCA AGT GTG TAT TCT ATT TGG AAC AAA ATC AAA

Yeast D273   GTA TGC TGG AAA TCT TTA TTT ATT AAA TAT CAA TGA AAA ATT TAT AAT AAA ACT CTA TAT TAT TTT ATT

Trypanosoma   TTA AAG GCA ATC CCA GAT AAG TTT ATG GGT TTG TTT TTA ATG GTT ATT TTA TTA TTC TCA TTA TTT TTA

Aspergillus   GCT ATA GTT CCA GAA TGA TAT CTT TTA CCT TTC GCT ATT TTA AGA TCT ATA CCT AAT AAA TTA TTA

Drosophila   TTT GCT TAC GCA ATT CTT CGT TCA ATT CCT AAT AAA TTA GGA GGA GTT ATT GCA TTA GTT TTA TCA ATT

Human   TTC GCC TAC ACA ATT CTC CGA TCC GTC CCT AAC AAA CTA GGA GGC GTC CTT GCC CTA TTA CTA TCC ATC

Maize   ATT --- GTG CCG GAA TGG TAT TTC CTA CCG ATC CAT GCC ATT CTT CGC AGT ATA CCT GAC AAA GCG GGG

Row 14

Yeast CBP1   GAG CAC TAT GTT TCC ATT TCT GCA CAT GAT TCC AGA ATC ATT TAT AAA GTC TTC AAG ATT TGT ACC CAT

Yeast D273   ATT CAA AAT ATT TTA AAT ACA AAA CAA TTA AAT TTC GTA TTA AAA TTT AAT TGA ACA AAG CAA TAT

Trypanosoma   TTT ATA TTG AAT TGT ATA TTA TGA TTT GTG TAT TGT AGA AGT TCA TTA TTA TGA TTA ACA TAT TCG TTA

Aspergillus   GGT GTT ATA GCT ATG TTT GCT GCT ATA TTA GCA TTA ATG CCT ATA ACT GAT TTA TCT AAA TTA

Drosophila   GCA ATT TTA ATA ATT TTA CCT TTT TAT AAT TTA AGA AAA TTC CGA GGA ATC CAA TTT TAT CCA ATT AAC

Human   TTC ATC CTA GCA ATA ATC CCC ATC CTC CAT ATA TCC AAA CAA CAA AGC ATA ATA TTT CGC CCA CTA AGC

Maize   GCT GTA GCC GCA ATA GCA CCA CCA GTT TTT ATA TCT CTC TTG GCT TTA CCT TTT TTT AAA GAA ATG TAT GTG

(continued)

Table 6.2 (Continued)

Row 15

Yeast CBP1	AAT AGA GCC TAT AGA TCT ATA TGT AGC GAA ATG TTT TGG CAA TTA ACT CCA GAG TAC TAT TGT AAT AAC
Yeast D273	AAT AAA ATA AAT ATT GTA AGT GAT TTA TTT AAT CCC AAT AGA GTA AAA TAT TAT TAT AAA GAA GAT AAT
Trypanosoma	ATA TTA TTT TAT AGT ATA TGA ATG AGT GGT TTT TTA GCA TTA TAT GTA GTA TTA GCA TAT CCA ATA TGA
Aspergillus	AGA GGA GTA CAA TTT AGA CCT TTA AGT AAA GTA GTA TTC TAT ATT TTT GTA GCT AAC TTC TTA ATA TTA
Drosophila	CAA ATT TTA TTT TGA TCT ATA TTA GTT ACA GTA ATT TTA TTA ACA TGA ATT GGA GCT CGA CCA GTT GAA
Human	CAA TCA CTT TAT TGA CTC CTA GCC GCA GAC CTC CTC ATT CTA ACC TGA ATC GGA GGA CAA CCA GTA AGC
Maize	CGT AGT TCA AGT TTT CGA CCG ATT CAC CAA GGA ATA TTT TCG CTT TTG GCG GAT TGC TTA CTA CTA

Row 16

Yeast CBP1	CCT TTG ATA TTA CCG GCA ATT ATT GAC TTC ATT ACA AAG CAA GAC TCT TTA ACA ATG GCC AAG GAA CTC
Yeast D273	CAG CAG GTA ACC AAT ATA AAT TCT TCT AAT ACT CAC TTA ACG AGT AAT AAA AAG AAT TTA TTA GTA GAT
Trypanosoma	ATG GAA TTA CAA TAC TGA GTA TTA TTA TTT TTC TTG ATA GTG TGT AGG TTA GAT TAG
Aspergillus	ATG CAA ATA GGT GCA AAA CAC GTT GAA ACT CCA TTT ATT GAA TTT GGA CAA ATT TCT ACT ATT ATT TAT
Drosophila	GAA CCT TAT GTA TTA ATT GGA CAA ATT TTA ACT ATT ATT TAT TTT TTA TAT TTA ATT AAC --- ---
Human	TAC CCT TTT ACC ATC ATT GGA CAA GTA GCA TCC GTA CTA TAC TTC ACA ACA ATC CTA --- --- ---
Maize	GGT TGG ATC GGA TGT CAA CCT GTG GAG GCA CCA TTT GTT ACT ATT GGA CAA ATT TCT --- TCT TTC TTT

Row 17

Yeast CBP1 ATG CAG AAC ATT AAC AGA TAC ACT TTA CCC GAA AAC CAT CAT ATT GTC TGG CTT AAC AAG AGA TGT CTT

Yeast D273 ACT TCA GAG ACT ACA CGC ACA CTA GAA AAT AAA TTT AAT TAT TTA TTA AAT ATT TTT AAT ATA AAA AAA

Aspergillus TTT GCA TAT TTC TTT GTA ATA GTT CCT GTT AGT TTA ATT GAA AAT ACT TTA GTA GAA TTA GGA ACT

Drosophila --- --- --- CCA CTA GTT --- --- ACA AAT TGA TGA --- --- --- GAT AAT TTA TTA AAT

Human --- --- --- ATC CTA ATA CCA ACT --- --- ATC TCC CTA ATT GAA --- --- --- AAC AAA ATA --- CTC

Maize TTC TTC TTG TTT GCC ATA ACG CCC ATT CCG GGA GTT GGA AGA GGA ATT CCA AAA TAT TAC ACG

Row 18

Yeast CBP1 TCT TCA TTG CTA AGA ATG CAT TTG AAA TTT AAC GAT TCT AAC GGT GTA GAT AGG GTT TTG AAG CAA ATA

Yeast D273 ATA AAT CAA ATT ATT CTT AAA AGA CAT TAT AGT ATT TAT AAA GAT AGT AGT ATT AGA TTT AAC CAA TGA

Aspergillus AAA AAA AAC TTT TAA

Drosophila TAA

Human AAA TGG GCC TAA

Maize GAA TAG

Row 19

Yeast CBP1 ACA ACA AAT TTC AGG GCG CTT TCG CAA GAA AAT TAT CAA GCA ATA ATT ATT CAC CTT TTC AAA ACA CAA

Yeast D273 TTG GCC GGT TTA ATT GAC GGA GAT GGT TAT TTT TGT ATT ACT AAA AAT AAA TAT GCA TCT TGT GAA ATT

(continued)

Table 6.2 (Continued)

```
Row 20
  Yeast CBP1   AAC CTC GAT CAT ATC GCT AAG GCA GTC AAA TTA CTC GAT ACT CCC CCC GGA CAA GCA ATG TTA GCC
  Yeast D273   CTT GTA GAA TTA AAA GAT GAA ATG TTA AGA CAA ATC CAA GAT AAA TTT GGT TCT GTA AAA TTA

Row 21
  Yeast CBP1   TAT GGG TCA ATA ATT AAC GAA GTA GTT GAT TGG AAA TTG GCT TCA AAG GTC AAG TTC ACC GAT AAT TTG
  Yeast D273   AGA TCA GGT GTT -- AAG ACT ATT AGA TAT AGA TTA CAA AAT AAA GAA GGT ATA ATT AAA TTA ATT AAT

Row 22
  Yeast CBP1   ATG GCA CTT GTA AAC GAT TTG TTG ACG AAG GCA CAT GAT TTT GAT CCT GAC CAC AGA AAC TCT CTT TGG
  Yeast D273   GCC GTT AAT GGT AAT ATT CGT AAT AGT AAA AGA TTA GTA CAA TTT AAT AAA GTA CGT ATT TTA TTA AAT

Row 23
  Yeast CBP1   AAT GTG GTT TCC GCT TTA TAC ATT AAA AAA CTT TGT CAT TAT AAA AAG CGA GAT GGT AAA TTT GTT GCC
  Yeast D273   ATC GAT TTT AAA GAA CCT ATT AAA ACT AAA GAT AAT GCT TGA TTT ATA GGG TTC TTT GAT GCT GAT

Row 24
  Yeast CBP1   AAT GCC AAG AAG GAT ATC GAT TTG GCA AAA CTA CTT TAT ATA AAT GCT GCA AAG AGA AGT AAA ACA
  Yeast D273   GGT ACT ATT AAT TAT TAT TAT TCC GGT AAA TTA AAA ATT AGA CCT CAA TTA ACT ATT AGC GTT ACA

Row 25
  Yeast CBP1   TAC TGG ACA AAA TCG AAC TGT AAC CCA TTC ATT GCA TCC TCC CCA TGT GAT GTC AAA TTA AAA
  Yeast D273   AAT AAA TAT TTA CAT GAT GTT GAA TAC TAT AGA GAA GTA TTT GGT GGT AAT ATT TAT TTT GAT
```

Row 26

Yeast CBP1   GTG AAT AAT CAA AAC AGG TTT ACT ATT TTA AGG AAT ATT GCA TTA AGC GCA CTG CAG ATA GGA

Yeast D273   AAA GCT AAA AAT GGT TAT TTT AAA TGA TCT ATT AAT AAT AAA GAA TTA CAT AAT ATT TTT TAT

Row 27

Yeast CBP1   AGA ACA GAC ATT TTT CTT TGG GCG TGC GCA GAA CTA TAC CAG AAC GGT ATG ACG ATT GAG GAA

Yeast D273   CTT TAT AAT AAA AGT TGT CCT TCT AAA TCT AAT AAA GGT AAA CGT TTA TTT TTA ATT GAT AAA

Row 28

Yeast CBP1   TTG AAG TTA GAC TGG AAT TTC ATC TTA AAA CAT CAA ATT AGA AAT TCA GAG TTC AAA ACA AAC

Yeast D273   TTT TAT TAT TTA TAT GAT TTA TTA GCT TTT AAA GCA CCT CAT AAT ACT GCT TTA TAT AAA GCT

Row 29

Yeast CBP1   AAG GAG ATC ATA CAA GAT ATT AAA AAG CAT GGT GTG TCG GCT GTC AAA CGT TAC TTA AGA TGA

Yeast D273   TGA TTA AAA TTT AAT GAA AAA TGA AAT AAT AAT TAA

[a]The yeast CBP1 gene is nuclear and the pea and spinach are chloroplast; all the rest are mitochondrial. Codons for arginine are italicized, those for lysine are underscored, and those for tryptophan are stippled. The frequent canonical stop-codon (TGA) encodes amino acids in these mitochondrial genes.

[b]Dieckmann et al. (1984).

[c]Nobrega and Tzagoloff (1980).

[d]Benne et al. (1983).

[e]Waring et al. (1981).

[f]Clary e' al. (1984).

[g]Araya et al. (1984).

[h]Anderson et al. (1981).

[i]Dawson et al. (1984).

[j]Heinemeyer et al. (1984).

[k]Herrmann et al. (1984).

gene sequences from spinach chloroplasts (Heinemeyer *et al.*, 1984; Herrmann *et al.*, 1984). Regrettably, no coding sequence for any member of the cytochromes, including the present group, has been established from bacterial or blue-green algal sources.

One additional cistron is included in the table, one that does not encode a cytochrome of any sort. This is the primary structure of the gene *CBP1* from the nucleus of the common yeast, the product of which is responsible for conferring a stable 5′ end on the messenger of cytochrome *b* (Dieckmann *et al.*, 1984). Although nearly twice the size of the longest actual cytochrome *b*, its gene sequence shows frequent homologies to those of that protein class, as becomes evident shortly.

Homologies among Mitochondrial Cytochromes b. Because the mitochondrial genes for the cytochromes *b* comprise the great bulk of the tabulation of their mature coding regions (Table 6.2), they may serve as a weathervane to suggest the direction the others may be taking. If the codons for lysine (underscored), those for arginine (italicized), and those for tyrosine (stippled) are compared first, the immediate impression is that little evolutionary conservation has been active here. In row 1, however, the sixth and seventh sites are perceived to be similar in the yeast D273, cyprinid, and human sequences (Nobrega and Tzagoloff, 1980; Anderson *et al.*, 1981; Araya *et al.*, 1984), as they are three positions later in the trypanosome and aspergillid representatives (Waring *et al.*, 1981; Benne *et al.*, 1983). Then centrally four lysine codons form a column. Just prior to the end of that same row, two sets of tyrosine codons are aligned, those from the trypanosome and the spinach chloroplast first, followed by those of yeast D273 and the aspergillid.

Because in earlier chapters such variation in structure of the 5′ terminus of the genes was a commonplace occurrence, no conclusions can be drawn, not even on a preliminary basis. In contrast, row 2 provides a convincingly firm basis for suggesting relationships. Beginning with the first site of that row in the D273 and aspergillid representative, there is a series of identical and similar codons that runs interruptedly through the remainder of that sector. In this series 11 identical and five related triplets can be counted, broken by only four nonhomologous ones. Comparison of the yeast and maize (Dawson *et al.*, 1984) cistrons here shows six identities, five similarities, and nine noncorrespondences, yielding a slightly greater than 50% relationship. The recently sequenced wheat mitochondrial gene adds confirmation to the generality of this structure among the green plants, as that of *Schizosaccharomyces* contributes to the firming of fungal relationships, for it closely resembles that of *Aspergillus* (Boer *et al.*, 1985; Lang *et al.*, 1985). In contrast, the human, *Xenopus* (not given), and cyprinid sequences (Dunon-Bluteau *et al.*, 1985), which are closely homologous as expected, differ strongly from the yeast structure in showing no identical and only four similar codons. Additionally, in this same region further examination shows that the *Drosophila* triplets (Clary *et al.*, 1984) compare closely to those of the two vertebrate species. Thus, row 2 suggests the existence of at least two kinship groups, the yeast–fungal–seed-plant genes on the one hand and the metazoan on the other. These two sets of relationships continue throughout the table and require no further discussion.

Other homologies are less pronounced, those involving the trypanosome gene being especially obscure. Because of its phylogenetic position in the hierarchy of living things, this mitochondrial gene should indicate an intermediate position between the yeast on the one hand and the fungus or metazoan structures on the other, but usually these expected kinships are mostly lacking. Nevertheless, it remains a possibility that the gene from *Trypanosoma* may represent the subtype cytochrome b_T, rather than the cytochrome b_K of the other six, which could explain the low level of homology that it displays.

Chloroplast Cytochromes b. As discussed earlier, only a single example of the two species of cytochromes b (b_6 and b_{559}) from chloroplasts has had the nucleotide sequence of its gene established, both from the spinach organelle (Heinemeyer *et al.*, 1984; Herrmann *et al.*, 1984). Cytochrome b_6 is part of the chloroplast b/f complex, but little is known of the functions in photosynthesis of cytochrome b_{559}. Nevertheless, many proposals have been made. Among them are that it carries out a redox activity in the splitting of water, serves as an electron or proton carrier, either in the main chain of events or in a cycle around photosystem II, and serves as a redox carrier on the side path involving the plastoquinone pool.

Since the physiological activity of both species is thus quite different from the mitochondrial, extensive homology between the several types is scarcely to be expected, nor is much to be perceived. However, rows 2 and 4 show occasional correspondences of the b_{559} to the maize mitochondrial sequence, the significance of which is difficult to comprehend. That species is considerably shorter than any of the remainder in the table, consisting of only 84 triplets, compared to ~390 of the maize mitochondrial form, while cytochrome b_6 is 220 codons in length. Perhaps the most outstanding feature of the first gene is the existence of an extensive region of condons for hydrophobic sectors. This begins with the tyrosine codon (TAC) near the close of row 1 and extends throughout row 2. Such regions are strongly suggestive of their being embedded within a membrane, in this case probably a thylakoid. In the chloroplast genome this cistron lies adjacent to an unidentified coding frame known as URF-39, with which it is cotranscribed (Herrmann *et al.*, 1984).

Possible Influences of Nuclear Species. That influences upon the structure of several organellar genes may come from nuclear sources is strongly suggested by the codon composition of the yeast *CBP1* gene. This encodes a protein, quite unrelated to the cytochromes, which processes cytochromes b to some extent, as already stated. Even in row 1, which showed only low homology levels among obviously related mitochondrial genes, there are numerous correspondences between the codons for *CBP1* and the others. The TTT near the 5′ terminus, for instance, shows kinship to the TAT, TTA, ATT, and TTG that occur below it, and its neighbor ATA is similarly matched by ATG, ATT, and ATA. The next codon, AAA for lysine, is duplicated in the *Aspergillus* representative, and the AAT of *Drosophila* is related.

Comparable examples of kinship (or influence) abound throughout the chart, even though no attempt at homologization has been made. But to avoid belaboring the point unduly, only one other region of such relationship is cited, the latter portion of row 5. Beginning with AGA, which encodes arginine, which is matched by ATA in *Aspergillus* and AAA in the yeast mitochondrion, progress can be made through GGA (matched in four sequences), ATT, TAT, AGA, and on through TTG. Only the final codon TAC fails to show homology with any of the others. Thus, either common ancestry or exposure to the same genetic influences seems to provide the only evident explanation for the shared structural features here demonstrated.

Table 6.3
Mature Coding Sequences for the Cytochromes c[a]

Row 1

Yeast CYC7[b]	ATG GCT AAA GAA AGT GGA TTC ACG GGA TTC AAA CCA GGC TCT GCA AAA AAG GGT GCT ACG TTG TTT AAA ACG ACG
Yeast CYC1[c]	atg --- --- --- ACT GAA TTC AAG GCC GGT TCT GCT AAG AAA GGT GCT ACA CTT TTC AAG ACT AGA
S. pombe[d]	ATG --- --- --- CCT --- TAC GCC CCT GGA GAC GAA AAG AAG GGT GCT TCC CTG TTC AAG ACT CGT
Rat[e]	atg --- --- --- GGT GAT GTT GAA AAA GGC --- --- --- ATT TTT GTT CAA --- AAG
Chicken[f]	ATG --- --- --- GGA GAT ATT GAG AAG GGC --- --- --- ATT TTT GTC CAG --- AAA
Yeast[g]	ATG --- --- ACC GCA GCT GAA CAC GGA TTG CAC --- --- GCC CCA GCA TAT GCT TGG TCC CAC AAT GGG

Row 2

Yeast CYC7	TGT CAG CAG TGT CAT ACA ATA GAA GAG GGT CCT AAC AAA GTT GGA CCT AAT TTA CAT GGT ATT TTT
Yeast CYC1	TGT CTA CAA TGC CAC ACC GTG GAA AAG GGT GGC CCA CAT AAA AAG GTT GGT CCA AAC TTG CAT GGT ATC TTT
S. pombe	TGC GCT CAA TGT CAT ACC GTT GAA AAA GGC GGC GCC AAC AAG AAG GTC GGT GGT CCC AAT TTG CAC GGT GTA TTT
Rat	TGT GCC CAG TGC CAC ACT GTG GAA AAA GGC AAG CAT AAG ACT GGA CCA AAC CTC CAT GGT CTG TTT
Chicken	TGT TCC CAG TGC CAT ACG GTT GAA AAA GGA GGC AAG AAG CAC ACT GGA CCC AAC CTT CAT GGC CTG TTT
Yeast c_1	CCT TTT GAA ACA TTT GAT CAT GCA TCC ATT AGA AGA GGT TAC CAG GTT TAC CGT GAA GTT TGT GCC GCC

Row 3

```
Yeast CYC7  GGT AGA CAT TCA GGT CAG GTA AAG GGT TAT TCT TAC ACA GAT GCA AAC ATC AAC AAG A  AC GTC AAA TGG
Yeast CYC1  GCC AGA CAC TCT GGT CAA GCT GAA GGG TAT TCG TAC ACA GAT GCC AAT ATC AAG AAA A  AC GTG TTG TGG
S. pombe    GGC CGT AAG ACC GGT CAA GCT GAG GGT TTC TCT TAC ACC GAA GCC AAT CGC GAT AAG G  GT ATT ACT TGG
Rat         GGG CGG AAG ACA GGC CAG GCT GCT GGA TTC TCT TAC ACA GAT GCC AAC AAG AAA G[h]GT ATC ACC TGG
Chicken     GGA CGC AAA ACA GGA CAA GCT GAG GGC TTC TCT TAC ACA GAT GCC AAT AAG AAG AAA G[h]GT ATC ACT TGG
Yeast c₁    TGC CAT TCT CTT GAC AGA GTT GCT TGG AGA ACT TTG GTT GTT TCT CAT ACC AAC G  AA GAG GTT CGT
```

Row 4

```
Yeast CYC7  GAT GAG GAT AGT ATG TCC GAG TAC TTG ACG AAC CCA AAG AAA TAT ATT CCT GGT ACC AAG ATG GCG TTT
Yeast CYC1  GAC GAA AAT AAC ATG TCA GAG TAC TTG ACT AAC CCA AAG AAA TAT ATT CCT GGT ACC AAG ATG GCC TTT
S. pombe    GAT GAA GAA ACT CTT TTC GCC TAC CTC GAA AAC CCC AAG AAG TAT ATC CCC GGT ACT AAG ATG GCC TTT
Rat         GGA GAG GAT ACC CTG ATG GAG TAT TTG GAA AAT CCC AAA AAG TAC ATC CCT GGA ACA ACA AAG ATG ATC TTC
Chicken     GGT GAG GAT ACT CTG ATG GAG TAT TTG GAA AAT CCA AAG AAG TAC ATC CCA GGA ACA AAG ATG ATT TTT
Yeast c₁    AAT ATG GCC GAA GAA TTT GAG TAC GAT GAC GAA CCT GAT CAA GGT AAC CCT AAA AAG AAG AGA CCA GGT
```

Row 5

```
Yeast CYC7  GCC GGG TTG AAG AAG GAA AAG GAC AGA AAC GAT TTA ATT ACT TAT ATG ACA AAG GCT --- GCC AAA TAG
Yeast CYC1  GGT GGG TTG AAG AAG GAA AAA GAC AGA AAC GAC TTA ATT ACC TAC TTG AAA AAA GCC --- TGT GAG TAA
S. pombe    GCT GGC TTC AAG AAA CCG GCT GAT CGT AAC AAC GTC ATT ACG TAT TTG AAG AAG GCC ACC TCT GAG TAA
Rat         GCT GGA ATT AAG AAG AAG GGA GAA GAA AGG GCA GAC CTG ATA GCT CTT AAA AAG GCT ACT AAT GAA TAA
Chicken     GCG GGT ATC AAG AAG AAG TCT GAG AGA GTA GCA TTA ATA GCA TAT CTC AAA GAT GCC ACT TCA AAG TAA
Yeast c₁    AAG TTG TCC GAT TAC ATC CCT GGC CCA TAC CCA AAC GAA CAG GCT GCA AGA GCT GCT AAT CAA GGT GCC
```

(continued)

Table 6.3 (Continued)

Row 6

Yeast c_1 TTG CCA CCT GAT CTA TCT TTG ATC GTG AAA GCT AGA CAC GGT GGT TGT GAC TAC ATT TTC TCT TTG TTG

Row 7

Yeast c_1 ACC GGT TAT CCT GAT GAA CCT CCT GCT GTG GCT TTA CCA CCA GGT TCT AAT TAT AAC CCT TAC TTC

Row 8

Yeast c_1 CCA GGT GGT TCC ATT GCA ATG GCA AGA GTC TTG TTT GAT GAC ATG GTT GAG TAC GAA GAT GGT ACC CCC

Row 9

Yeast c_1 GCA ACG ACA TCT CAA ATG GCA AAG GAC GTT ACC ACC TTT TTA AAC TGG TGT GCC GAA CCT GAA CAT GAC GAA

Row 10

Yeast c_1 AGA AAG AGA TTG GGT TTG AAA ACG GTG ATA ATC TTA TCA TCT TTG TAT TTG CTA TCT ATC TGG GTG AAG

Row 11

Yeast c_1 AAG TTC AAA TGG GCC GGT ATC AAA ACC AGA AAA TTC GTT TTC AAT CCA CCA AAA CCA AGA AAG TAG

[a]Codons for lysine are underscored, those for arginine are italicized, and those for tyrosine are stippled.
[b]Montgomery et al. (1980).
[c]Smith et al. (1979).
[d]Russell and Hall (1982).
[e]Scarpulla et al. (1981).
[f]Limbach and Wu (1983).
[g]Sadler et al. (1984).
[h]Sequence is interrupted by an intron at this point.

6.1.3. The Mature Coding Regions for the Cytochromes c

Because of its relatively small size and ease of extraction, the ubiquitous cytochrome c component of the electron-transport chain has had its primary structure established from more than 100 species of organisms (Fitch, 1976; Russell and Hall, 1982). In strange contrast, its gene sequences have been determined from only four, the two variants from baker's yeast bringing the total available to five. Two representatives from the chicken have similarly been sequenced, but since the coding regions are identical, only one is shown in Table 6.3 (Limbach and Wu, 1983). None from either bacterial or blue-green algal sources are available, so that comparative data are restricted. However, one cistron for the distinctive, much longer species, cytochrome c_1, has been sequenced from yeast (Sadler *et al.*, 1984), and is included in Table 6.3 with the others.

The Cytochrome c Gene. Although in eukaryotes the product of the cytochrome c gene functions in the mitochondrion, the cistron itself is located in the nucleus; translation of the transcript and addition of the heme prosthetic group are completed in the endoreticulum before the protein is transported to the organelle (Clark-Walker and Linnane, 1967). Among protistans no introns interrupt the coding sequence, but in vertebrates— and possibly other metazoans—an insert of 103 (rat) to 174 (chicken) residues is found, an asterisk in row 3 indicating its location in both cases (Scarpulla *et al.*, 1981; Limbach and Wu, 1983). As a whole the genes are as constant in construction as their remarkably uniform primary structures would lead one to expect, but the degree of homology varies from region to region. One sector that has been shown to be particularly highly conserved evolutionarily is the one that encodes amino acids 70–80 in the yeast *CYC1* cistron, that is, in row 4 from the GAG triplet preceding the tyrosine codon TAC to the ATG beyond the third lysine codon (Montgomery *et al.*, 1980). In addition, a shorter invariant sector is found near the beginning of row 2.

Specific Homologies. The two varieties from baker's yeast, one from each of the two genes it possesses for this protein (Smith *et al.*, 1979; Montgomery *et al.*, 1980), differ strongly at only a single point, the insertion in the *CYC7* gene of a dodecanucleotide just after the initial ATG. Although the third nucleotide residue in each triplet rarely agrees with the corresponding one in the other, the first two deviate but rarely. By count, only 19 distinctions can be noted in the two entire gene sequences. The *Schizosaccharomyces pombe* representative (Russell and Hall, 1982) shows only one addition to the cistron (just before the 3' terminus) and a single deletion (near the 5' terminus). Examination of the coding region shows the existence of two sets of homologies where all are not identical. On the one hand, the structure resembles that of baker's yeast, differing strongly from those of the two vertebrates. This relationship is observed in the tenth triplet in row 1 and again in the fourth and fifth triplets from the 3' end of that same row; another pair of instances is provided by the AAC and GTC on the sides of the lone lysine codon of row 2. In row 4, the opening GAT, along with a CTT four codons later, the TAC for the first tyrosine, and a GCC just before the end are further examples, as are the TTC, ATT, and ACG of row 5.

On the other hand, kinships may be displayed with the vertebrates, the first indications of which occur in row 3. Just after the 5' end is a nonanucleotide CGTAAGACC that is in harmony with the same region in the rat and chicken sequences, but not with either from the yeast. Later in the same row is a TTC and before the end a hexanucleotide

GGTATT, and in row 4 are three more examples, CTT, GAA, and ATC. Consequently, the gene sequence demonstrates what has been proposed elsewhere (Dillon, 1962, 1981), that *Schizosaccharomyces* is a true fungus and thereby occupies an intermediate position between the metazoans and those primitive eukaryotes, the true yeasts.

The two vertebrate mature coding sequences are largely homologous to one another, including regions of deletions. Three short examples of the latter condition (besides the long one just after the initiating codon) are obvious features of row 1, but these and the intron mentioned earlier are the only notable characteristics that distinguish them from the three more primitive types.

The Cytochrome c_1 Gene. The cytochrome c_1 gene, encoded in the nuclear genome, is unusually complex and leads into the topic that is analyzed in the following chapter. Instead of the usual simple leader, the mature coding section is prefaced in addition by a long presequence; thus the product of transcription is a complex protein of the type here called diplomorphic (Chapter 1, Section 1.2). In the present case, the intact early transcript is conducted across the mitochondrion into the inner membrane, where two processing events occur, addition of a heme unit first, followed by proteolytic removal of the presequence (Wakabayashi *et al.,* 1982). Much of the bulk of the mature product projects into the intermembrane space of the organelle, where it accepts electrons from the Rieske nonheme iron clusters and passes them to cytochrome c (Ludwig *et al.,* 1983).

Only a single cytochrome c_1 gene has been fully sequenced, that of yeast, whose mature coding region is given in Table 6.3 (the presequence is shown in Table 7.6). Occasionally sites homologous to corresponding ones in the cytochromes c may be observed, but these are scattered and transient, rarely exceeding a single triplet in length. Moreover, the present species is twice as long as the preceding one, even without the presequence.

In the study of the gene structure encoding this protein, it is claimed that the molecule is anchored in the intermembrane space by a hydrophobic region of 19 amino acid residues in the presequence (Sadler *et al.,* 1984). But similar regions are present in most such transit or signal peptides, as they are also called, as shown in the succeeding chapter. They appear to serve during transit through membranes, however, rather than as anchors, for they must be removed before the mature protein becomes functional. Because of the frequent presence in the mature sequence of codons for charged amino acids, either acidic or basic, there appears to be only one hydrophobic region of sufficient length to penetrate a membrane and serve as a holdfast. This sector occupies most of row 7, beginning with the first CCT and continuing for a total of 25 triplets to the second GCA of row 8; this, then, rather than the presequence, possibly represents the true transmembrane portion of the gene.

Evolutionary Relationships. Nothing further can be deduced concerning the evolutionary relationships of cytochrome c until numerous additional gene sequences have been established, especially those from protistans of all types, seed plants, and, most particularly, from bacteria and blue-green algae. However, the single representative of cytochrome c_1 is sufficient to support two observations that have not been previously made. The first regards the presequence that characterizes this nuclear-encoded species. Frequently the suggestion is made in support of the endosymbiontic theory of mitochondrial origins that genes for essential respiratory and other proteins now lacking from the organelle's genome had once existed there but had since been transferred to the cell's nucleus. The need for a complex prefatory sequence encoding a peptide in order to transfer the product into the mitochondrion after translation in the cytoplasm makes it most evident that much more is involved in the transpositioning of genes between two cell parts than appears on the surface. Not only do those cistrons need to be moved from one genome to the other on a broad basis involving whole populations of organisms, but they must then undergo elaboration immediately by addition of a prefatory sector to permit the product to be transported into and positioned correctly in the supposed original inhabitant—again on a populational basis. In plainer terms, the cytochrome c_1 gene would have had to gain the elaborate presequence either prior to or immediately upon removal to the nuclear DNA, for without it the gene in its new location would have produced a useless product, resulting in lethality through the loss of function. Hence, unless preadaptation had occurred, there would have been no time to gain this vital appendage through the usual slow processes of evolution.

The second point applies equally to the coding sequence and its product. Cytochrome c_1 has recently been shown to be present in the membranes of *Paracoccus denitrificans* (Ludwig *et al.*, 1983). However, it is not present in *E. coli* nor in most other eubacteria that have been explored. Consequently, the members of the present genus may have to be considered more highly advanced prokaryotes than was evidenced by the 5 S rRNA genes of Tables 3.2 and 3.3. Reexamination of those sequences suggests that the *Paracoccus* species could, in spite of its slight shortening of the 5′ end, fit quite comfortably with *Rhodospirillum* or *Photobacterium*, the -CCGAT of column M of Table 3.2 being especially supportive of this proposal. That *Paracoccus* is transitional, rather than a possible invader of the ancestral eukaryotic cell, is further suggested by the presence in it of nonheme iron clusters, while the other components of complex III of the mitochondrion are absent (Ludwig *et al.*, 1983).

6.2. GENES FOR THE CYTOCHROME c OXIDASE COMPLEX

Cytochrome c oxidase, often called cytochrome a/a_3, is a complex protein of the inner mitochondrial membrane, consisting of two hemes, two copper units, and nine peptide subunits, three of which are encoded by mitochondrial genes, the rest by nuclear ones (Coruzzi and Tzagoloff, 1979; Maarse *et al.*, 1984). Although all three of the organellar cistrons have been sequenced from several sources each, few of the nuclear components have been determined, among which are those of the yeast subunits IV and VI and partial sequence of the former from bovine sources (Lomax *et al.*, 1984; Maarse *et al.*, 1984; Wright *et al.*, 1984). As a result of this one-sidedness, no initiation or other signals have been established, for these mitochondrial sequences suffer from the same limitations as previous types, the existence of overly short or A,T-rich leaders or the lack of recognizable promoters. Hence, it is possible only to report the mature coding sections of this family of genes. Since the major features are brought out adequately by the genes for subunits II and III, that for I requires no separate attention.

6.2.1. The Mature Coding Sequence of Subunit II

Subunit II (CoII) of cytochrome c oxidase is of particular interest because it has been shown to bind one of the two a hemes and one of the copper atoms and to lie in close

Table 6.4

Mature Gene Sequences for Subunit II of Cytochrome c Oxidase[a]

Row 1

Yeast oxi1[b]	ATG	TTA	---	GAT	TTA	---	TTA	AGA	TTA	CAA	TTA	ACA	ACA	TTC	ATT	ATG	AAT	GAT	GTA	CCA	ACA	CCT	TAT	GCA
Trypanosoma[c]	ATG	AGT	TTT	ATA	TTA	ACT	TCA	ATG	ATA	TTT	TTA	ATG	GAT	TCA	ATA	ATT	---	GTA	TTA	ATA	TCT	TTT	TCA
Drosophila[d]	ATG	TCT	ACA	TGA	GCT	AAT	TTA	GGT	TTA	CAA	GAT	AGA	GCT	TCT	CCT	TTA	ATG	GAA	CAA	TTA	ATT	TTT	CAT
Human[e]	ATG	GCA	CAT	GCG	CAA	GTA	GGT	CTA	GGT	CTA	CAA	GAC	GCT	ACT	TCC	CCT	ATC	ATA	GAA	GAG	CTT	ATC	ACC	TTT	CAT
Murine[f]	ATG	GCC	TAC	CCA	TTC	CAA	CTT	GGT	CTA	CAA	GAC	GCC	ACA	TCC	CCT	ATT	ATA	GAA	GAG	CTA	ATA	AAT	TTC	CAT
Bovine[g]	ATG	GCA	TAT	CCC	ATA	CAA	CTA	GGA	TTC	CAA	GAT	GCA	GAT	GCA	ATC	CCA	ATC	ATA	GAA	GAA	CTA	CTT	CAT	TTT	CAT
Wheat[h]	ATG	ATT	CTT	CGT	TCA	---	TTA	TCA	TGT	TGT	CTC	ACA	ATC	GCT	CTT	TGT	GAT	GCT	GCG	GAA	CCA	CAA	TGG	CAA
Rice[i]	ATG	ATT	CTT	CGT	TCA	---	TTA	GAA	TGT	TGT	CTC	ACA	ATC	GCT	CTT	TGT	GAT	GCT	GCG	GAA	CCA	CAA	TGG	CAA

Row 2

Yeast oxi1	TGT	TAT	TTT	CAG	GAT	TCA	GCA	ACA	CCA	AAT	CAA	GAA	GGT	ATT	TTA	GAA	TTA	CAT	GAT	AAT	ATT	ATG	TTT
Trypanosoma	ATA	---	TTT	CTA	---	TCT	GTA	TGA	---	ATA	---	---	TGT	GCA	TTG	ATT	ATA	GCA	ACA	GTA	TTA	ACT	GTA
Drosophila	GAT	CAT	GCA	TTA	---	---	---	---	---	---	TTA	ATT	TTA	GTA	ATA	ATT	ACA	GTA	TTA	GTA	GGA
Human	GAT	CAC	GCC	CTC	---	---	---	---	---	---	ATA	ATC	ATT	TTC	CTT	ATC	TGC	TTC	CTA	GTC	CTC
Murine	GAT	CAC	ACA	CTA	---	---	---	---	---	---	ATA	ATT	GTT	CTA	ATT	AGC	TCC	TTA	GTC	CTC
Bovine	GAC	CAC	ACG	CTA	---	---	---	---	---	---	ATA	ATT	GTC	TTA	ATT	AGC	TCA	TTA	GTA	CTT
Wheat	TTA	GGA	TCT	CAA	GAC	GCA	GCA	ACA	CCT	ATG	ATG	CAA	GGA	ATC	ATT	GAC	TTA	CAT	CAC	GAT	ATC	TTT	TTC
Rice	TTA	GGA	TCT	CAA	GAC	GCT	GCA	ACA	CCT	ATG	ATG	CAA	GGA	ATC	ATT	GAC	TTA	CAT	CAC	GAT	ATC	TTT	TTC

Row 3

Yeast *oxi1*	TAT	TTA	TTA	GTT	ATT	TTA	GGT	TTA	GTA	TCT	TGA	ATG	TTA	TAT	ACA	ATT	GTT	ATA	ACA	TAT	TCA	AAA	AAT
Trypanosoma	ATA	AAA	ATA	AAT	AAT	ATA	TAT	TGT	ACA	---	TGA	GAT	TTT	ATA	TCA	TCA	AAA	TTT	ATA	GAT	ACA	TAT	TCG
Drosophila	TAT	TTA	ATG	TTT	---	ATA	TTA	TTT	TTT	AAT	AAT	TAT	GTA	AAT	CGA	TTT	CTT	TTA	CAT	---	GGA	CAA	CTT
Human	TAT	GCC	CTT	TTC	---	CTA	ACA	CTC	ACA	ACA	AAA	CTA	ACT	AAT	ACT	AAC	ATC	TCA	GAC	---	GCT	CAG	GAA
Murine	TAT	ATC	ATC	TCG	---	CTA	ATA	TTA	ACA	ACA	AAA	CTA	ACA	CAT	ACA	ACA	ATA	GAT	---	GCA	CAA	GAA
Bovine	TAC	ATT	ATT	TCA	---	CTA	ATA	CTA	ACG	ACA	AAG	CTG	ACC	CAT	ACA	ACG	ATA	GAT	---	GCA	CAA	GAA
Wheat	TTC	CTC	ATT	CTT	ATT	TTG	GTT	TTC	GTA	TCA	CGG	ATG	TTG	GTT	CGC	GCT	TTA	TGG	CAT	TTC	AAC	GAG	CAA
Rice	TTC	CTC	ATT	CTG	ATT	TTG	GTT	TTC	GTA	TCA	CGG	ATG	TTG	GTT	CGC	GCT	TTA	TGG	CAT	TTC	AAC	GAG	CAA

Row 4

Yeast *oxi1*	CCT	ATT	GCA	TAT	---	AAA	TAT	ATT	AAA	CAT	GGA	CAA	ACT	ATT	GAA	GTT	ATT	TGA	ACA	ATT	TTT	CCA	GCT
Trypanosoma	TTT	GTA	CTT	GGA	---	ATG	ATG	TTT	ATA	TTG	TCT	TTA	TTG	TTA	AGG	TTG	TGT	TTG	TTG	TAT	TTT	ACT	
Drosophila	ATT	GAA	ATA	ATT	---	TGA	ACT	ATT	CTC	CCA	GCT	ATT	ATT	TTA	TTA	TTT	ATT	GCT	CTT	CCT	TCA	TTA	CGA
Human	ATA	GAA	ACC	GTC	---	TGA	ACT	ATC	CTG	CCC	GCC	ATC	ATC	CTA	GTC	CTC	ATC	GCC	CTC	CCA	TCC	CTA	CGC
Murine	GTT	GAA	ACC	ATT	---	TGA	ACT	ATT	CTA	CCA	GCT	CTA	ATC	CTT	ATC	ATT	GCT	CTC	CCC	TCT	CTA	CGC	
Bovine	GTA	GAG	ACA	ATC	---	TGA	ACC	ATT	CTG	CCC	GCC	ATC	ATC	TTA	ATT	GCT	CTT	CCT	TCT	TTA	CGA	
Wheat	ACT	AAT	CCA	ATC	CCA	CAA	AGG	ATT	GTT	CAT	GGA	ACT	ACT	ATT	ATT	GAA	ATT	ATT	CGG	ACC	ATA	TTT	CCA	AGT
Rice	ACT	AAT	CCA	ATC	CCG	CAA	AGG	ATT	GTT	CAT	GGA	ACT	ACT	ATT	ATT	GAA	ATT	ATT	CGG	ACC	ATA	TTT	CCT	AGT

(continued)

Table 6.4 (Continued)

Row 5

Yeast *oxi1*	GTA	ATT	TTA	TTA	ATT	AIT	GCT	TTC	CCT	TCA	TTT	ATT	TTA	TTA	TAT	TTA	TGT	GAT	GAA	GTT	ATT	TCA	---
Trypanosoma	---	TGT	---	---	---	---	---	---	---	---	---	---	---	ATA	AAT	TTT	GTG	AGT	---	---	TTT	GAT	---
Drosophila	---	TTA	---	---	---	---	---	---	---	---	---	CTT	TAT	TTA	GAT	GAA	ATT	AAT	GAA	---
Human	---	ATC	---	---	---	---	---	---	---	---	CTT	TAC	ATA	ACA	GAC	GAG	GTC	AAC	GAT	---
Murine	---	ATT	---	---	---	---	---	---	---	---	CTA	TAT	ATA	GAC	GAA	ATC	AAC	AAC	---
Bovine	---	ATT	---	---	---	---	---	---	---	---	CTA	TAC	ATA	GAT	GAA	ATC	AAT	AAC	---
Wheat	GTC	ATT	CTT	TTG	ATT	GCT	ATA	CCA	TCG	TTT	GCT	CTG	TTA	TAC	TCA	ATG	GAC	GGG	GTA	TTA	GTA	GAT
Rice	GTC	ATT	CCT	TTG	ATT	GCT	ATA	CCA	TCG	TTT	GCT	CTG	TTA	TAC	TCA	ATG	GAC	GGG	GTA	TTA	GTA	GAT

Row 6

Yeast *oxi1*	CCA	GCT	ATA	ACT	ATT	AAA	GCT	ATT	GGA	TAT	CAA	AAA	TAT	GAA	TAT	TCA	GAT	TTT	ATT	AAT
Trypanosoma	---	---	TTG	---	TGT	AAA	GTA	ATA	GGT	TTT	CAG	GTA	TAT	TTG	GAG	AAA	CCA	CGA
Drosophila	CCA	TCA	ACT	TTA	AAA	AGT	ATT	GGT	CAT	CAA	TGA	TAC	TGA	TAT	TCA	GAT	TTT	AAT	AAT
Human	CCC	TCC	CTT	ACC	ATC	AAA	TCA	ATT	GGC	CAC	CAA	TGA	TAC	GAG	TAC	ACC	GAC	GGC	GGA
Murine	CCC	GTA	TTA	ACC	GTT	AAA	ACC	ATA	GGG	CAC	CAA	TGG	TAC	GAG	TAT	ACT	GAA	TAC	GGA
Bovine	CCA	TCT	ACA	GTA	AAA	ATA	ACC	ATA	GGA	CAT	CAG	TGA	TAC	TGA	TAT	ACA	GAT	TAT	GAG	GAC
Wheat	CCA	GCC	ATT	ACT	ATC	AAA	GCT	ATT	GGA	CAT	CAA	TGG	TAT	*CGG*	AGT*	TAT	TCG	GAC	TAT	AAC	AGT
Rice	CCA	GCC	ATT	ACT	ATC	AAA	GCT	ATT	GGA	CAT	CAA	TGG	TAT	*CGG*	AGT*	TAT	TCG	GAC	TAT	AAC	AGT

Row 7

Yeast *oxi1*	---	GAT	AGT	GGT	GAA	ACT	GTT	GAA	TTT	GAA	TCA	TAT	GTT	ATT	CCT	GAT	GAA	TTA	TTA	GAA	GAA	GGA	CAA
Trypanosoma	---	TAT	---	---	---	TAG	---	TAA	TTT	AA?	-?-	TAT	TAG	AA?	AGT	GAT	TAT	TTA	ATA	GGA	GAT	---	---
Drosophila	---	---	---	---	ATT	---	GAA	TTT	GAT	TCA	TAT	ATA	ATT	CCT	ACA	AAT	GAA	TTA	GCA	ATT	GAT	GGA
Human	---	---	---	---	CTA	---	ATC	TTC	AAC	TCC	TAC	ATA	CTT	CCC	CCA	TTA	CTA	GAA	CCA	GGC	GAC
Murine	---	---	---	---	CTA	---	TGC	TTT	GAT	TCA	TAT	ATA	ATC	CCA	ACA	GAC	CTA	AAA	CCT	GGT	GAA
Bovine	---	---	---	---	TTA	---	AGC	TTC	GAC	TCC	TAC	ATA	ATT	CCA	ACA	TCA	GAA	TTA	AAG	CCA	GGG	GAG
Wheat	TCC	GAT	GAA	CAG	TCA	CTC	---	ACT	TTT	GAC	AGT	TAT	ACG	ATT	CCA	GAA	GAT	GAT	CCA	GAA	TTG	GGT	CAA
Rice	TCC	GAT	GAA	CAG	TCA	CTC	---	ACT	TTT	GAC	AGT	TAT	ACG	ATT	CCA	GAA	GAT	GAT	CCA	GAA	TTC	GGT	CAA

Row 8

Yeast *oxi1*	TTA	*AGA*	TTA	TTA	GAT	ACT	GAT	ACT	TCT	ATA	GTT	CCT	GTA	GAT	ACA	CAT	ATT	*AGA*	TTC	GTT	GTA	ACA
Trypanosoma	TTA	*AGA*	ATA	TTA	CAG	TGT	AAC	CAT	GTA	TTG	ACA	TTG	TTA	AGT	TTG	GTT	ATT	TAT	AAA	TTA	TGA	GTA	TCT
Drosophila	TTT	*CGA*	TTA	TTA	GAC	GTT	GAT	AAT	*CGA*	GTA	ATT	TTA	CCA	ATA	AAT	TCA	CAA	ATT	*CGA*	ATT	TTA	GTA	ACA
Human	CTG	*CGA*	CTC	CTT	GAC	GTT	GAC	AAT	*CGA*	GTA	GTA	CTC	CCG	ATT	GAA	GCC	CCC	ATT	*CGT*	ATA	ATA	ATT	ACA
Murine	CTA	*CGA*	CTG	CTA	GAA	GTT	GAT	AAC	*CGA*	GTC	GTT	CTG	CCA	ATA	GAA	CTT	CCA	ATC	*CGT*	ATA	TTA	ATT	TCA
Bovine	CTA	*CGA*	CTA	TTA	GAA	GTC	GAT	AAT	*CGA*	GTT	CTA	CCA	ATA	CCA	ATA	ACA	ATC	CTA	*CGA*	ATG	CTA	TTA	GTC	TCC
Wheat	CCA	*CGT*	TTA	TTA	GAA	GTT	GAC	AAT	*AGA*	GTG	GTT	GTA	CCA	GCC	AAA	ACT	CAT	CTA	*CGT*	ATG	ATT	GTA	ACA
Rice	CCA	*CGT*	TTA	TTA	GAA	GTT	GAC	AAT	*AGA*	GTG	GTT	GTA	CCA	GCC	AAA	ACT	CAT	CTA	*CGT*	ATG	ATT	GTA	ACA

(continued)

Table 6.4 (Continued)

Row 9

Yeast *oxi1*	GCT	GCT	GAT	GTT	CAT	GAT	TTT	GCT	ATC	CCA	AGT	TTA	GGT	ATT	AAA	GTT	GAT	GCT	ACT	CCT	GGT	*AGA*
Trypanosoma	GCA	GTA	GAT	ATA	CAC	TCA	TTT	ACA	ATA	TCA	AGT	TTA	GGT	ATA	AAA	GTA	GAG	AAC	CTG	GTA	GGT	GTA
Drosophila	GCC	GCA	GAT	GTA	ATT	CAT	TCT	TGA	ACA	GTC	CCA	GCT	TTA	GGA	GTA	AAG	GTT	GAC	GGA	ACT	CCT	GGA	*CGA*
Human	TCA	CAA	GAC	GTC	TTG	CAC	TCA	TGA	GCT	GTC	CCC	ACA	TTA	GGC	TTA	AAA	ACA	GAT	GCA	ATT	CCC	GGA	*CGT*
Murine	TCT	GAA	GAC	GTC	CTC	CAC	TCA	TGA	GCA	GTC	CCC	TCC	CTA	GGA	CTT	AAA	ACT	GAT	GCC	ATC	CCA	GGC	*CGA*
Bovine	TCT	GAA	GAC	GTA	TTA	CAC	TCA	TGA	GCT	GTG	CCC	TCT	CTA	GGA	CTA	AAA	ACA	GAC	GCA	ATC	CCA	GGC	*CGT*
Wheat	CCC	GCT	GAT	GTA	CCT	CAT	AGT	TGG	GCT	GTA	CCT	TCA	CCT	TCA	GGT	GTC	AAA	TGT	GAT	GCT	GTA	CCT	GGT	*CGT*
Rice	CCC	GCT	GAT	GTA	CTT	CAT	AGT	TGG	GCT	GTA	CCT	TCA	CCT	TCA	GGT	GTC	AAA	TGT	GAT	GCT	GTA	CCT	GGT	*CGT*

Row 10

Yeast *oxi1*	TTA	AAT	CAA	GTT	TCT	GCT	TTA	ATT	CAA	*AGA*	GAA	GGT	GTC	TTC	TAT	GGG	GCA	TGT	TCT	GAG	TTG	TGT	GGG
Trypanosoma	ATG	AAA	TAA	TGT	TTG	CTA	CAA	ATA	ACG	CAA	CTC	CTC	TTN	TAC	GGA	CAA	TGT	ACT	GAA	TTG	TGT	GGT
Drosophila	TTA	AAT	CAA	ACT	AAT	TTT	ATT	AAC	*CGA*	CCA	GGG	TTA	TTT	TAT	GGT	CAA	TGT	TCA	GAA	ATT	TGC	GGG
Human	CTA	AAC	CAA	ACC	ACT	TTC	ACC	GCT	ACA	CCG	ACA	*CGA*	GTA	TAC	TAC	GGT	CAA	TGC	TCT	GAA	ATC	TGT	GGA
Murine	CTA	AAT	CAA	GCA	ACA	GTA	ACA	TCA	AAC	*CGA*	CCA	GGG	TTA	TAT	GGC	CAA	TGC	TCT	GAA	ATT	TGT	GGA
Bovine	CTA	AAC	CAA	ACA	ACC	CTT	ATA	TCG	TCC	*CGT*	CCA	GGC	TTA	TAT	TAC	GGT	CAA	TGC	TCA	GAA	ATT	TGC	GGG
Wheat	TCA	AAT	CTT	ACC	TCC	ATC	TCG	GTA	CAA	*CGA*	GGA	GTT	TAC	TAT	GGT	CAG	TAC	GAG	ATT	*CGT*	GGA
Rice	TCA	AAT	CTT	ACC	TCC	ATC	TCG	GTA	CAA	*CCA*	GAA	GTT	TAC	TAT	GGT	CAG	TAC	TAT	GAG	ATT	TCT	GGA

Row 11

Yeast *oxi1*	ACA	GGT	CAT	GCA	AAT	ATG	CCA	ATT	AAG	ATC	GAA	GCA	GTA	TCA	TTA	CCT	AAA	TTT	TTG	GAA	TGA	TTA	---
Trypanosoma	GTA	TTA	CAC	GGT	ATG	CCT	ATT	GTA	ATA	AAT	TTT	ATA	TAG	---	AAA	GGT	ATA	TAA				
Drosophila	GCT	AAT	CAT	AGT	ATG	CCA	ATT	GTA	ATT	GAA	AGT	GTT	CCT	GTA	AAT	AAT	TTT	ATT	AAA	TGA	ATT	TCT
Human	GCA	AAC	CAC	AGT	TTC	ATG	CCC	ATC	GTC	CTA	GAA	ATC	CCC	CTA	AAA	ATC	TTT	GAA	ATA	GGG	CCC	GTA
Murine	TCT	AAC	CAT	AGC	TTT	ATG	CCC	ATT	GTC	CTA	GAA	ATG	GTT	CCA	CTA	AAA	TAT	TTC	GAA	AAC	TGA	TCT	GCT
Bovine	TCA	AAC	CAC	AGT	TTC	ATA	CCC	ATT	GTC	CTT	GAG	TTA	GTC	CCA	CTA	AAG	TTT	GAA	AAA	TGA	TCT	GCG
Wheat	ACT	AAT	CAT	GCC	TTT	ACG	CCT	ATC	GTC	GTA	GAA	GCA	GTG	ACT	TTG	AAA	GAT	TAT	GCG	GAT	TGG	GTA	TCC
Rice	ACT	AAT	CAT	GCC	TTT	ACG	CCT	ATC	GTC	GTA	GAA	GCG	GTG	ACT	TTG	AAA	GAT	TAT	GCG	GAT	TGG	GTA	TCC

Row 12

Yeast *oxi1*	AAT	GAA	CAA	TAA					
Drosophila	AGA	AAT	AAT	TCT	T--				
Human	---	TTT	ACC	CTA	TAG				
Murine	---	TCA	ATA	ATT	TAA				
Bovine	---	TCA	ATA	TTA	TAA				
Wheat	AAT	CAA	TTA	ATC	CTC	CAA	ACC	AAC	TAA
Rice	AAT	CAA	TTA	ATC	CTC	CAA	ACC	AAC	TAA

[a]Codons for lysine and arginine are underscored and italicized, respectively; those for tyrosine (TAC, TAT) are lightly stippled, those for phenylalanine (TTC, TTT) are moderately stippled, and those for tryptophan (TGA, TGG) and cysteine (TGC, TGT) are heavily stippled. Asterisks denote locations of introns.
[b]Coruzzi and Tzagoloff (1979); Fox (1979).   [c]Hensgens et al. (1984).   [d]Clary and Wolstenholme (1983).   [e]Anderson et al. (1981).
[f]Bibb et al. (1981).   [g]Anderson et al. (1982).   [h]Bonen et al. (1984).   [i]Kao et al. (1984).

conjunction with cytochrome *c*. The gene for this peptide from metazoans, yeast, and fungi does not contain introns, whereas those of three monocots (maize, wheat, and rice) do (Fox and Leaver, 1981; Bonen *et al.*, 1984; Kao *et al.*, 1984). In contrast, the two representatives sequenced from dicotyledonous plants, *Oenothera* and pea, are uninterrupted (Hiesel and Brennicke, 1983; Moon *et al.*, 1985). Even though the trains, like the leaders, provide little information concerning control mechanisms, they are not totally devoid of interest. In the four metazoan sequences, the mature coding sector is followed closely by a gene for tRNA^{Lys}, being separated from the former by 25 sites in the human representative and by three each in the other two from mammals; that from *Drosophila* lies so close to the other that its translation-termination signal is reduced to a single T (Clary and Wolstenholme, 1983).

In Table 6.4, where the eight known gene sequences of subunit II are provided, the same stratagems are employed as in the preceding tables, lysine codons being presented underscored, those for arginine in italics, and those for tyrosine in light stippling. In addition, codons for phenylalanine (TTT, TTC) are demarcated by moderate stippling and those for tryptophan (TTG, TGA) and cysteine (TGC, TGT) are brought out by dark stippling, for they display a peculiar feature of many mitochondrial genes.

Unusual Codons for Tryptophan. The unusual trait just mentioned is in the use of TGA as a codon for tryptophan, whereas in both prokaryotes and eukaryotic cytoplasmic and chloroplast cistrons that triplet serves as one of the three translational termination signals. In the "universal" code catalog, TGG is considered the only codon for that amino acid. Even in itself the usage given would be strange enough, but the unusualness is heightened by the selectivity with which it occurs. For some time it appeared that this distinctive function of TGA was a catholic trait of mitochondrial genes, but the present sequences demonstrate quite effectively that that conclusion is not true.

In each example given a total of four to six codons for tryptophan are present, but they do not align with those of the others in every instance. The *Drosophila* representative shows a TGA in the fourth site quite independently, and the two plant sequences are provided with one each just before the end of row 1, a feature repeated in row 3. Then in row 4, the four metazoan cistrons alone have a triplet for this amino acid, whereas that of the yeast has one near the end of that same row. In fact, only in row 6 is an entire column of those codons notable; there are, however, two additional ones that are nearly uniform. The first of these is in row 9, where the yeast is exceptional, the second being located in row 11, where the human sequence is irregular. Additionally, in row 6, a single column is broken by the presence of an arginine codon in the two plant types. Thus it would appear that the precise placement of tryptophan is not critical. If these codons are examined, however, it is noted that in all cases the combination TGA is employed, except in the wheat and rice, where the standard TGG is uniformly in use. But there is one additional exception in row 6 on the part of the human coding sequence, which is the sole unbroken column of these codons, where TGG is employed in that example. Consequently, this amino acid is encoded by two triplets in mitochondria, TGA being the most frequent, except in the seed plants, where only TGG seems to be employed.

The mechanical basis for the altered usage is no mystery, for obviously no change in codon–anticodon interaction is involved. All that has happened is that the anticodon of the single mitochondrial tRNA^{Trp} (UCA) is made to interact with both UGU and UGC in the messenger RNA, so one of the enzymes of the ribosome may have become modified. In like fashion the same sort of variation underlies the usage of the CG- family of codons in metazoan mitochondria in place of the AGA and AGG; in those organelles no tRNAs exist that are capable of interacting with the latter pair, so that termination of translation results. Hence, they have become termination codons in the mammalian mitochondrial codon catalog (Anderson *et al.*, 1981). In the latter case the change has arisen as a result of the shortening of the mitochondrial genome among metazoans, but there is an evolutionary element involved, too, that can be brought out more appropriately in a later section.

Other Distinctive Features. As a whole the series of sequences of Table 6.4 shows the usual trends among mitochondrial genes, lengthy cistrons in the yeast and seed plants, and reduction in those of metazoans. It is interesting to notice that the four of the latter category begin with an increase in size, for all have an inserted codon at site 6 of row 1 that is lacking in the rest. Thereafter, however, the tendency is in the usual opposite direction. Decrease in size is accomplished largely through the loss of three major sectors, though deletions of single codons may be observed in rows 3, 5, 7, and, in vertebrates, 9. The longer lost parts are found in row 2, where a deletion of eight triplets occurs, in row 5, with a loss of 11 codons, and in row 7, with five. A minor opposite trend is found among the seed plants, in which the yeast gene is exceeded by six codons, chiefly through additions at the extreme 3' terminus.

For the greater part homologies are readily establishable among the sequences, regardless of the source. Although that of yeast often shows closer kinship to the two seed-plant cistrons, it frequently displays relationship to metazoan types rather than with those monocotyledonous genes. Highly conserved sites, nevertheless, are relatively few in number. In fact, in the entire first four rows, there are only three such constancies, all ATY for isoleucine, one in row 2 and two in row 4. At the end of row 5 begins the sector that must be the major arena of activity, for that region with two columns of identical codons leads into row 6, where 11 are located. Shortly thereafter, a second important region occurs, for row 7 has three constant columns and row 8 five, rows 9 and 10 have nine each, and row 11 just five. Six of these invariant positions encode tyrosine, only two lysine, and four arginine, while triplets for glutamic acid and proline, with six each, and glycine, with five, make up the remaining major portion. Two columns of codons largely constant for cysteines (TGC, TGT) mark an ideal location for attachment of the heme *a* prosthetic group, situated in the midst of the evolutionarily conserved region in row 10, but actual sites and means of such attachment remain unestablished, a statement equally true for the copper atom.

6.2.2. The Mature Genes for Cytochrome c Oxidase Subunit III

The structures of genes for subunit III of cytochrome *c* oxidase have not been quite as broadly established as have those for subunit II, in that none are available from seed-plant sources. However, one from the fungal mitochondrion that has been determined (Netzker *et al.*, 1982) compensates to some extent for that deficiency. In addition to the six gene sequences currently at hand, the first row of subunit II from the human mitochondrion (Anderson *et al.*, 1981) is added to Table 6.5 for comparative purposes. The same conventions are employed as in the preceding section; codons for lysine are underscored and those for arginine are italicized; those for tyrosine are lightly stippled, those for phenylalanine are moderately stippled, and those for tryptophan and cysteine are heavily stippled.

Examination of the tables reveals again that the relationships between the several

Table 6.5

Genes for Subunit III of Cytochrome c Oxidase[a]

Row 1

```
Yeast oxi2[b]     ATG --- ACA CAT TTA GAA AGA AGT AGA CAT CAA CAA CAT CCA CAT ATG GTT ATG CCT TCA CCA TGA
Aspergillus_M[c]  GTG --- ATA TAT CAA TCT AAA AGA AAT TTT CAA AAC CAT CCG CAT TTA GTA TCA CCA TCA CCT TGA
Drosophila_M[d]   ATG TCT ACA CAC TCA AAT CAC CCT CAT TTA GTT GAT TAT AGC CCA TGA CCT TTA ACA GGT GCT ATT
Human_M[e]        ATG --- ACC CAC CAA TCA CAT GCC TAT CAT ATA GTA AAA CCC AGC CCA CCT TTA ACA GGG GCC CTC
Murine_M[f]       ATG --- ACC CAC CAA ACT CAT GCA TAT CAC ATA GTT AAT CCA AGT CCA TGA CCA TTA ACT GGA GCC TTT
Bovine_M[g]       ATG --- ACA CAC CAA ACT GCT CAT CAT CAT ATA GTA AAC CCA AAC AGC CCT TGA CCT CTT ACA GGA GCT TTG
Human II[e]       ATG --- GCA CAT GCA GCG CAA GTA GGT CTA CAA GAC GCT ACT TCC CCT ATC ATA GAA GAG CTT ATC ACC
```

Row 2

```
Yeast oxi2        CCA ATT GTA GTA TCA TTT GCA TTA TTA GCA TTA TCA CTA GGA TTA ACA TTA ACA ATG CAT GGT TAT ATT
Aspergillus_M     CCA TTA TTT ACT AGT ATT TCA TTA ATA CTT ACA ACA GTA TTA TTT ATG CAT GGT TTC GAA
Drosophila_M      GGA GCT ATA ACA ACT GTA TCA GGT ATA GTA TGA CAT CAA TAT CAT CAA TTC TTA TTA
Human_M           TCA GCC CTC CTC CTA ATG ACC TCC GGC CTA GCC ATG TGA TTT TTT CAC TTC CAC TTT CAT CAA TAT TCA TTA ACG CTC CTC ATA CTA
Murine_M          TCA GCC CTC CTT CTA ACA TCA GGT CTA GTA ATA TGA TTT TTT CAC TTC ATG AAT AAT TCA ATT ACA CTA TTA ACC CTT
Bovine_M          TCT GCC CTC CTT ATA ACA TCC GGC CTA ACC ATG TGA TTT TTT CAC TTT CAC TCA ATG AAC TCA ATG ACC CTG CTA ATA ATT
```

Row 3

```
Yeast oxi2      GGT AAT ATG AAT ATG GTA TAT TTA GCA TTA TTT GTA TTA TTA ACA AGT TCT ATT --- --- --- TTA TGA
Aspergillus_M   GGA TTC CAA --- --- TAT TTA GTA GTA CCA GTA CTA TGA TGA CCA GAT GTT TCA --- --- GTA ATG GGT TTA TGA
Drosophila_M    GGT AAT ATT ATT ACT ATT TTA ACA GTT TAT CAA TGA TGA CGA GAT GTT TCA --- --- --- CGA GAA
Human_M         GGC CTA CTA ACC AAC ACA CTA ACC ATA TAC CAA TGA TGG CGC GAT GTA ACA --- --- --- CCA GAA
Murine_M        GGC CTA CTC ACC AAT ATC CTC ACA ATA TAT CAA TGA TGA CGA GAC GTA ATT --- --- --- CGT GAA
Bovine_M        GGC CTA ACA ACA AAT ATA CTA ACA ATA TAC CAA TGA TGA CCA GAT GTT ATC --- --- --- CGA GAA
```

Row 4

```
Yeast oxi2      TTT AGA GAT ATT GTA GCT GAA GCT ACA TAT TTA GGT GAT CAT ACT ATA GCA GTA AAA AAA GGT ATT AAT
Aspergillus_M   TTT AGA GAT GTA ATA TCA GAA GGA ACA TAT TTA GGA AAT CAT CAT ACT AAT GCT GTA CAA AAA GGA TTA AAC
Drosophila_M    GGA ACT TAC CAA GGA TTA CAT --- --- ACT TAC GCA CTA --- --- ACT ATT GGT TTA CGA GGA ATA ATT
Human_M         AGC ACA TAC CAA GGC CAC CAC --- --- ACA CCA GTC --- --- CAA AAA GGC CTT TGA TAC GGG ATA ATC
Murine_M        GGA ACC TAC CAA GGC CAC CAC --- --- ACT CCT ATT GTA --- --- CAA AAA GGA CTA CGA TAT GGT ATA ATT
Bovine_M        AGC ACC TTC CAA GGG CAC CAT --- --- ACC CCA GCT GTC --- --- CAA AAA GGC CTC CGT TAT GGA ATA ATT
```

Row 5

```
Yeast oxi2      TTA GGT TTC TTA ATG TTT GTA TTA TCT GAA GTA ATC TTT GCT GGT TTA TTC TGA GCT TAT TTC CAT
Aspergillus_M   TTA GGA GTA GGT TTA TTT ATT ATC TCT GAA GTA TTC TTT TTA GCA ATA TTC TGA GCA TTC TTC CAT
Drosophila_M    TTA TTT ATT TTA TCA GAA GTT TTT TTA GTT TTT GTT GCA TTT TTT CAT CAC AGT TTA TCT
Human_M         CTA TTT ATT ACC TCA GAA GTT TTT TTC TTC GCA GGA TTT TTT GCC TTT TAC CAC AGC CTA GCC
Murine_M        CTA TTC ATC GTC TCG GAA GTA TTT TTC TTC GCA GGA TTC TTC GCG TTC TAT CAT AGC CTC GTA
Bovine_M        CTT TTT ATT ATC TCC GAA GTA CTA TTT TTC TTT ACC GGA TTT TTC GCT TTC TAC CAC AGC CTC GCC
```

(continued)

Table 6.5 (Continued)

Row 6

Yeast *oxi2*	TCA	GCT	ATG	AGT	CCT	GAT	CTA	TTA	GGT	GCA	TGT	TGA	CCA	CCC	GTA	GGT	ATT	GAA	GCT	GTA	CAA	CCT
Aspergillus$_M$	AGT	GCT	ATC	TCA	CCT	AGT	GTA	GAA	TTA	GGT	GCA	CAA	TGA	CCA	CCA	TTA	GGT	ATA	CAA	GGA	ATA	AAT	CCT
Drosophila$_M$	CCA	GCA	ATT	---	---	---	GAA	TTA	GGA	GCT	TCA	TGA	CCT	CCT	ATG	GGA	ATT	ATT	TCA	TTT	AAT	CCA
Human$_M$	CCT	ACC	CCC	---	---	---	CAA	TTA	GGA	GGG	CAC	TGG	CCC	*CGA*	ACA	GGC	ATC	ACC	CCG	CTA	AAT	CCC
Murine$_M$	CCA	ACA	CAT	---	---	---	GAT	CTA	GGA	GGC	TGC	TGA	CCT	CCA	ACA	GGA	ATT	TCA	CCA	CTT	AAC	CCT
Bovine$_M$	CCC	ACC	CCT	---	---	---	GAA	CTA	GGC	GGC	TGC	TGA	CCC	CCA	ACA	GGC	ATT	CAC	CCA	CTA	AAC	CCC

Row 7

Yeast *oxi2*	ACC	GAA	TTA	CCT	TTA	TTA	AAT	ACT	ATT	TTA	TCT	TCT	GGT	GCT	ACT	GTA	ACT	TAT	AGT	CAT	CAT
Aspergillus$_M$	TTT	GAA	TTA	CCT	TTA	CCT	TTA	AAC	ACA	ATA	ATT	TTA	TCA	TCA	GTT	GCT	ATT	ACA	TAT	GCA	CAT	CAT
Drosophila$_M$	TTT	CAA	ATT	CCT	TTA	TTA	AAT	ACA	GCT	ATT	CTT	TTA	GCT	TCA	GGA	GTT	ACA	GTA	ACT	TGA	GCT	CAT	CAT
Human$_M$	CTA	GAA	GTC	CCA	CTC	CTA	AAC	TCC	GTA	TTA	CTC	GCA	GTA	TCA	ATC	ACC	TGA	GCT	CAC	CAT
Murine$_M$	CTA	GAA	GTC	CCA	CTA	CTT	AAT	ACT	TCA	GTT	CTA	CTT	TCA	GTT	TTA	ATT	ACA	TGA	GCT	CAT	CAT
Bovine$_M$	CTA	GAA	GTC	CCA	CTG	CTC	AAC	ACC	TCT	GTC	CTA	TTG	GCT	TCC	GGA	GTT	TCT	ATT	ACC	TGA	GCC	CAT	CAT

Row 8

Yeast *oxi2*	GCC	TTA	ATC	GCA	GGT	AAT	*AGA*	AAT	AAA	GCT	TTA	TCA	GGT	TTA	TTA	ATT	ACA	TTC	TGA	TTA	ATT	GTT	ATT
Aspergillus$_M$	TCA	TTA	ATA	CAA	GGA	AAT	*AGA*	AAA	GGT	GCA	TTA	TAT	GGT	ACA	GTT	GTA	ACA	ATA	CTA	TTA	GCA	ATA	GTA
Drosophila$_M$	*AGA*	TTA	ATA	GAA	*AGA*	AAT	CAT	TCA	CAA	ACT	ACT	CAA	GGA	TTA	TTT	TTT	ACA	GTT	TTA	CTT	GGG	ATT	TAT
Human$_M$	AGT	CTA	ATA	GAA	AAC	AAC	*CGA*	AAC	CAA	ATA	ATT	CAA	GCA	CTG	CTA	ATT	ACA	ATT	ACC	ATT	ACA	CTA	GGT	CTC	TAT
Murine$_M$	AGC	CTT	ATA	GAA	GGT	AAA	*CGA*	AAC	CAC	ATA	AAT	CAA	CAA	GCC	CTA	CTA	ATT	ACC	ATT	ACA	CTA	GGA	CTT	TAC
Bovine$_M$	AGT	TTA	ATA	GAA	GGG	GAC	*CGA*	AAG	CAT	ATA	TTA	CAA	GCC	CTA	ATC	ACC	ATC	ACA	TTA	GGA	GTI	TAC

Row 9

Yeast oxi2	TTT	GTT	ACT	TGT	CAA	TAT	ATT	GAA	TAT	ACT	AAT	GCT	GCA	TTC	ACT	ATC	TCT	GAT	GGT	GTT	TAT GGT TCA
Aspergillus$_M$	TTC	ACA	TTC	TTC	CAA	GGA	GTA	GAA	TAT	ACA	GTA	TCA	TCA	TTC	ACA	ATA	TCT	GAT	AGT	GTT	TAT GGA TCA
Drosophila$_M$	TTC	ACA	ATT	TTA	CAA	GCT	TAT	GAA	TAT	ATT	GAA	GCT	CCA	TTT	ACT	ATT	GCT	GAT	TCA	GTT	TAT GGT TCA
Human$_M$	TTT	ACC	CTC	CTA	CAA	GCC	TCA	GAG	TCT	CCC	TTC	GAG	TCT	CCC	TTC	ACC	ATT	TCC	GAC	GGC	ATC TAC GGC TCA
Murine$_M$	TTC	ACC	ATC	CTC	CAA	GCT	TCA	GAA	TAC	TTT	GAA	ACA	TCA	TTC	TCC	ATT	TCA	GAT	GGT	ATC	TAT GGT TCT
Bovine$_M$	TTC	ACA	CTA	CTA	CAA	GCC	TCA	GAA	TAC	TAT	GAA	GCA	CCT	TTT	ACT	ATC	TCC	GAC	GGA	GTT	TAC GGC TCA

Row 10

Yeast oxi2	CTA	TTC	TAT	GCT	GGT	ACA	GGA	TTA	CAT	TTC	TTA	CAT	ATG	GTA	ATG	TTA	GCA	GCT	ATG	TTA	GGT GTT AAT
Aspergillus$_M$	TGT	TTC	TAT	TTT	GGA	ACA	GGT	TTC	CAC	GGT	TTA	CAC	GTT	ATA	ATT	GGT	ACA	GCA	ACT	TTT	TTA GCA GTA GGT
Drosophila$_M$	ACT	TTT	TAT	ATG	GCC	ACT	GGA	TTC	CAT	GGA	GTT	CAT	GTT	CTA	ATT	GGA	ACA	ACT	TTC	TTA	TTA GTA TGC
Human$_M$	ACA	TTT	TTT	GTA	GCC	ACA	GGC	TTC	CAC	GGA	CTT	CAC	GTC	ATT	ATT	GGC	TCA	ACT	TTC	CTC	ACT ATC ATC
Murine$_M$	ACA	TTC	TTC	ATG	GCT	ACT	GGA	TTC	CAT	GGA	CTC	CAT	GTA	CTA	ATT	GGA	TCA	ACA	TTC	CTT	ATT GTT TGC
Bovine$_M$	ACT	TTT	TTT	GTA	GCC	ACA	GGC	TTC	CAC	GGC	CTC	CAC	GTC	ATT	ATT	GGG	TCC	ACC	TTC	CTC	ATT ATT GTC

Row 11

Yeast oxi2	TAT	TGA	AGA	ATG	AGA	AAT	TAT	CAT	TTA	ACA	GCT	GGA	CAT	CAT	GTT	GGA	TAT	GAA	ACA	ACT	ATT ATT TAT
Aspergillus$_M$	TTA	TGA	CGT	TTA	GCT	GCT	TAC	CAC	TTA	ACT	GAT	CAT	CAT	CAT	GTT	GGA	TAC	GGA	TCA	GGA	ATC TTA TAC
Drosophila$_M$	TTA	TTA	CGT	CAT	TTA	AAT	AAT	CAT	TCA	AAA	AAT	CAT	CAT	TTT	GGA	GTA	GCA	GCT	GCA	TTT	GGA TGA TAC
Human$_M$	TTC	ATC	CGC	CAA	CTA	CTA	ATA	TTT	CAC	TTT	ACA	TCC	AAA	TTT	CAC	TTC	GGA	TAT	TTT	GAA	ACA GCC GCC GCC GCA TGA TAC
Murine$_M$	CTA	CTA	CGA	CAA	CTA	CAT	AAA	TTT	CAC	TTT	ACA	TCA	AAA	TTT	CAC	TTC	GGA	TTT	GAA	GCC	GCC GCA GCA TGA TAC
Bovine$_M$	TTC	TTC	CGC	CAA	TTA	AAA	TTT	CAT	TTT	ACT	TCT	AAC	CAC	CAC	TTC	GGA	TTT	GAA	GCC	GGT	GCC TGA TAC

(continued)

Table 6.5 (Continued)

Row 12

Yeast oxi2	CTA	CAT	GTT	TTA	GAT	GTT	ATC	TGA	TTA	TTA	TAC	GTA	CTA	TTT	TAT	TGA	TGA	GGT	GTT	TAA
Aspergillus_M	TGA	CAT	TTT	GTT	GAT	GTT	GTA	TGA	TTA	TTA	TAT	ATA	TCA	GTA	TAT	TGA	TAC	TGA	GGT	TAT	TAA
Drosophila_M	TGA	CAT	TTT	GTT	GAT	GTA	GTT	TGA	TTA	TTA	TAT	ATC	ACA	ATT	TAC	TGA	TGA	GGA	GGG	TAA
Human_M	TGG	CAT	TTT	GTA	GAT	GTG	GTT	TGA	CTA	TTT	CTG	TAT	GTC	TCC	ATC	TAT	TGA	TGA	GGG	TCT	T--
Murine_M	TGA	CAT	TTT	GTA	GAC	GTA	ATC	TGA	CTT	TTC	CTA	TAC	GTC	TCC	ATT	TAT	TGA	TGA	GGA	TCT	T--
Bovine_M	TGA	CAT	TTC	GTA	GAC	GTA	GTC	TGA	CTT	TTC	CTC	TAT	GTT	TCT	ATC	TAT	TGA	TGA	GGC	TCC	T--

[a] Codons for lysine are underscored and those for arginine are italicized; those for tyrosine (TAC, TAT) are lightly stippled, those for phenylalanine (TTC, TTT) are moderately stippled, and those for cysteine (TGC, TGT) and tryptophan (TGA, TGG) are heavily stippled. M, mitochondrion.

[b] Thalenfeld and Tzagoloff (1980).

[c] Netzker et al. (1982).

[d] Clary and Wolstenholme (1983).

[e] Anderson et al. (1981).

[f] Bibb et al. (1981).

[g] Anderson et al. (1982).

representatives are not close, except among the three from mammalian mitochondria. Also as elsewhere, the 5'-terminal regions are perceived to be variable, although here not to the extent that marked the initial section of subunit II and many others. Underscored triplets are noticeably lacking in numbers, as are italicized ones. There is no complete column encoding lysine anywhere in these genes, and only a single one of triplets for arginine, located near the beginning of row 11. While these codons for the basic amino acids thus are scarce, those for phenylalanine are relatively abundant, and highly conserved sectors (entire vertical columns of TTT and TTC) are distinctive features of rows 5, 9, 10, and 12. In some cases, additional columns are comprised of a combination of triplets for tyrosine and phenylalanine, but near the 3' terminus is a series entirely occupied by codons for tyrosine. In other instances the TGA of tryptophan may form complete vertical rows, often alone, but sometimes in combination with the TAC or TAT of tyrosine.

As aligned in the table, all the sequences terminate at exactly corresponding points, if the abbreviated translational termination signals of the three genes of mammalian mitochondria are compensated for by dashes. All of the latter are followed immediately by tRNA-encoding sequences that overlap with this final codon. Although the spacing is identical, the tRNA species is not, for those of the mouse and cattle are glycine-carrying and that of man is glutamic acid-carrying. In *Drosophila* the gene for tRNA^{Gly} lies downstream about 20 sites from the termination signal of subunit III, whereas in *Aspergillus*, a tRNA^{Lys} cistron follows the present one after 64 nucleotide residues. In the yeast representative no gene is in evidence for more than 2000 sites downstream from termination, most of the residues in this train being As or Ts. The shortening of the intergenic spacers is readily followed, for it is part of the well-known diminution of the mitochondrial genome that characterizes the metazoan organelle. However, the mechanism of change in the tRNA species remains obscure.

One other feature of the evolutionary development of the mammalian mitochondrion is particularly well brought out by the present series of sequences (Table 6.5). At one point of the *Aspergillus* cistron the phylogenetic differences are especially striking, a location centering on its second phenylalanine codon (TTT) in row 2. Here the next triplet downstream is ATA, which here as elsewhere in the universal code specifies isoleucine. However, beneath the TTT is also ATA, but in this case, as in all metazoan mitochondrial systems, methionine is encoded instead. Thus in this organelle of animals methionine is signaled by both ATA and ATG; only the latter serves that purpose in the nonmetazoan world. As pointed out in the preceding discussion, the change is in the properties of the tRNAs and the enzymes that charge and discharge them with amino acids, while the chemical nature of nucleotidyl residues of codons and anticodons remains constant.

6.3. CYTOCHROME f GENES

Although in a strict sense it is not involved in cell respiration, cytochrome f nonetheless is a member of an electron-transport chain of a related nature fundamental to photosynthesis. Properly this type is a member of the cytochrome c class, similarly having a heme a linked by thioester ligands from cysteines located near the 3'-terminal region. Thus it corresponds structurally to other members of the group, but in the photosynthetic

Table 6.6
Mature Genes for the Cytochrome f of Chloroplasts[a]

Row 1

Wheat[b]: TAT CCC ATT TTT GCG CAG CAG GGT TAT GAA AAC CCA ATT GGA CGA GAA GCA ACT GGA ATT GTA TGT GCC AAT TGT

Pea[c]: TAT CCC ATT TTT GCC CAA CAA GGT TAT GAA AAT CCT CGA GAA GCT ACC GGC CGG ATT GTA TGT GCT AAT TGC

Pea protein[d]: ATG GGT CAT AAT AAT TAT TAT GGC GAA CCC GCA TGG CCT AAC GAT CTT TTA TAT ATT TTT CCA GTA GTG ATT CTG

Row 2

Wheat: CAT TTA GCT AGC AAG CCC CTC GAT ATT GAA GTT CCC CAA GCT GTG CTT CCC GAT ACT GTA TTT GAA GCA

Pea: CAT TTA GCT AAT AAG CCC CTA GAT ATT GAG GTT CCA CAA GCG GTA CTT CCC GAT GTA TTT GAA GCA

Pea protein: GGG ACT ATT GCC TGT AAC GTG GGC TTA GCG GTT CTC GAA CCA TCA ATG ATT GGG GAA CCC GCG GAT CCG

Human b (row 12)[e]: GAC AAT TAT ACC CTA GCC AAC

Row 3

Wheat: GTT CTT CGA ATT CCT TAT GAT ATG CAA TTG AAA CAA GTT CTT GCT AAT GGA AAA AAG GGA GGG TTG AAT

Pea: GTT GTT CGA ATT CCT TAT GAT ATG CAA GTG AAA CAA GTT CTT GCT AAT GGG AAA AAG GGG GCT TTG AAT

Pea protein: --- TTT GCA ACT CCT TTG GAA ATA TTA CCC GAA TGG TAT CCC GTA TTT CAA ATA CTT CGT ACC

Human b (row 13): ACA ATT CTC CGA TCC

Row 4

Wheat: GTA GGT GCT GTT CTT ATT TTG CCC GAG --- GGA TTC GAA TTA GCG CCG CCC GAT CGT ATT TCC CCT GAG TTA

Pea: GTA GGA GCT GTT CTT ATT TTA CCA GAG --- GGT TTT GAA TTG GCC CCT CCT CAT CTT TCG CCC CAG ATT

Pea protein: GTA CCC AAT AAA TTA TTG GGC GTT CTT TTA ATG GTT TCA GTA CCC GGA TTA TTA ACA GTA CCC TTT TTG

Human b (rows 13 and 14): GTC CCT AAC AAA CTA GGA GGC GTC CTT --- GCC CTA TTA CTA TCC ATC CTC ATC CTA GCA ATA

Row 5

```
Wheat          AAA GAA AAG ATA GGA AAT CTT GCT TTT CAG AGT TAT CGT CCC GAT AAA AAA AAC ATT CTT GTG ATA GGC
Pea            AAA GAA AAG ATA GGT AAT TTG TCT TTT CAA AGC TAT CGT CCC ACA AAA AAA AAT ATT CTT GTC ATA GGC
Pea protein    --- GAG AAT --- GTT AAT AAG --- TTC CAA AAC CCA TTT CGG CGT CCA GTA GCA ACA ACT GTC TTT TTG
Human b        ATA ATA TTT --- CGC CCA CTA AGC TGA CTC CTA
(rows 14 and 15)
```

Row 6

```
Wheat          CCT GTT CCC GGT AAG AAA TAT AGT GAA ATT GTC TTT CCC ATT CTT TCC CCT GAT CCT GCT ACG AAG AAA
Pea            CCC GTT CCT GGG AAA AAA TAT AGT GAA ATT ACC TTT CCT ATT CTT TCT CCG GAT CCT GCT ACT AAG AGA
Pea protein    ATT GGT ACC GTC GTG GCT CTT TTG GGT ATT GGA GCA ACA TTA CCT ATT ATT GAA AAA TCC CTA ACG TTA
```

Row 7

```
Wheat          GAT GCT CAT TTC TTA AAA TAT CCC ATA TAT GTA GGG GGA AAC CGA GGA AGA GGA CAG ATC TAT CCT GAT
Pea            GAT GTT TAC TTC TTA AAA TAT CCT CTA TAC GTA GGC GCG AAC ACG GGA ACG GGT CAG ATT TAT CCC GAC
Pea protein    GGT CTT TTT TAA
```

Row 8

```
Wheat          GGT AGC AAG AAG AGT AAC AAT ACA GTC TAT AAT GCT ACG TCA ACA GGT ATA ATA AAA ATA CTA CGT AAA
Pea            GGA AGC AAG AAG AGT AAT AAT AAT GTT TCT AAT GCT ACA GCA ACA GGT GTA GTA AAA CAA ATA ATA CGA AAA
```

Row 9

```
Wheat          GAA AAA GGG GGG TAT GAA ATA TCC ATA GTT GAT GCA TCA GAT GCA CCC CAA GTG ATT GAT ATT ATA CCT
Pea            GAA AAA GGT GGA TAT GAA ATA ACG ATA GAT GCA TCA GAT GCA AGT GCA GTG ATT GAT ATT ATA CCG
```

(continued)

Table 6.6 (Continued)

Row 10

Wheat: CCC GGA CCA GAA CTT CTT GTT TCA GAG GGG GAA TCC ATC AAG CTT GAT CAA CCA TTA ACA AGC AAT CCT

Pea: CCA GGA CCA GAA CTT CTT GTT TCA GAG GGT GAA TCT ATC AAA CTT GAT CAA CCA TTA ACG AGT AAT CCT

Row 11

Wheat: AAT GTG GGA GGT TTT GGT CAG GGG GAC GCA GAA ATC GTG CTT CAG GAT CCA TTA *CGT* GTC CAA GGC CTT

Pea: AAT GTG GGT GGA TTT GGT CAG GGG GAT GCA GAA ATA GTG CTT CAA GAT CCA TTA *CGT* GTC CAA GGT CTC

Row 12

Wheat: TTG TTC TTC GCA TCC GTT ATT TTG GCA CAA GTT TTT TTG GTT CTC AAA AAG AAA CAG TTT GAA AAG

Pea: TTG CTC TTC GCA TCT ATT ATT TTG GCA CAA ATC CTT TTG GTT CTT AAA AAG AAA CAA TTT GAG AAG

Row 13

Wheat: GTT CAA TTG TAC GAA ATG AAT TTC TAG

Pea: GTT CAA TTG TCC GAA ATG AAT TTT TAG

[a] Codons for cysteine (TGY) and lysine (AAR) are underscored, those for arginine are italicized, those for tyrosine are stippled. The underlined sector near the 3' end demarks a transmembrane section.
[b] Willey *et al.* (1984b).
[c] Willey *et al.* (1984a).
[d] Phillips and Gray (1984).
[e] Table 6.2.

electron-conducting chain it seems functionally to displace both cytochromes c and c_1. In chloroplasts the present species is part of a complex bound in the thylakoids, which operates as an oxidoreductase that provides an electron-transport chain between the plastoquinone and plastocyanin of photosystem II. Typically the complex is comprised of cytochromes f and b_{563}, the Rieske iron–sulfur center (Bonner and Prince, 1984), and one or more additional proteins; the sequence of one gene of the latter is included in Table 6.6, with those of two of cytochrome f (Phillips and Gray, 1984). In the location stated, cytochrome f receives electrons from the Rieske iron–sulfur center and passes them to the copper-containing protein plastocyanin. Translation of the genes for the present species of cytochrome is on the ribosomes of the thylakoids (Minami and Watanabe, 1984).

6.3.1. The Mature Gene for Cytochrome f

The mature gene for cytochrome f has had its structure determined from just three sources, the chloroplasts of pea, wheat, and spinach (Alt and Herrmann, 1984; Willey *et al.*, 1984a,b), the sequences of the first two of which are included in Table 6.6. All three are prefaced by a coding region for a transit peptide, but discussion of that feature is reserved for the next chapter along with others of this type. The presence of the prefatory region in this chloroplast thylakoid-bound gene contrasts to the condition just observed in mitochondrial inner membrane representatives, which lack any prefacing sequence unless encoded by a nuclear gene. Although one example given is from a monocot and the other a dicot, the two sequences are seen to be largely homologous throughout their lengths, to the extent of 85% (Willey *et al.*, 1984b).

The codons for the two cysteines (underscored) that are involved in binding the heme prosthetic group are located near the end of row 1. As before, those for lysine are also shown underscored, those for arginine are italicized, and stippling covers those for tyrosine. In addition, a region is underlined in rows 11 and 12 consisting of triplets for hydrophobic amino acids that is believed to lie within the thylakoid membrane (Willey *et al.*, 1984a). Tyrosine triplets are unusually abundant, there being ten in the pea and 11 in the wheat sequence; these are concentrated in the 5' half, only one or so occurring downstream of row 7. Since 20 codons for lysine and six to eight for arginine can be noted, these cistrons encode products that are as basic as many of the nucleic acid-binding species examined in preceding pages. In a number of instances those triplets are arranged in pairs, and in row 12 a set of three occurs. In contrast to a frequent condition in histones and ribosomal proteins, the lysine triplets are rarely bordered by codons for arginine, but sometimes have those for tyrosine as neighbors.

6.3.2. A Protein of the Cytochrome b–f Complex

The protein just mentioned that is associated with the cytochrome b–f complex of chloroplasts is of particular pertinence in the present discussion for two reasons. Perhaps the more important of these is the absence of a structural feature that characterizes the preceding component. Although this 15,200-dalton protein is located in the thylakoids in close conjunction with the foregoing type and its gene is similarly located in the organelle's genome, it lacks the transit peptide supposedly needed for passage into the membrane (Phillips and Gray, 1984). Consequently, the mature substance may eventually prove to be located within the interthylakoid spaces on the surface of the membranes.

The second feature pertains to the evolutionary origins of this protein. In the article detailing its structure, it was indicated that the midsection of the polypeptide showed a 50-amino acid-residue homology with the carboxyl end of mitochondrial cytochrome *b*, without offering further details (Phillips and Gray, 1984). Since, as has already been shown, no gene structures for the cytochrome have been established from that organelle, it is not possible to confirm those correspondences. However, portions of rows 12–14 of the human cistron for that protein from Table 6.2 are aligned with the present sequence in rows 3–5. Although the relations are often tenuous, there are several most convincing sectors. Among these is the 15-residue section of row 3 beginning with the tenth triplet (CCC); another commences just before the end of that same row and continues for nearly half the length of row 4. In row 5 is a short series of related triplets that for all practical purposes completes the homologies between the two proteins.

In addition, a number of homologous sectors exist between this chloroplast protein and the cytochrome *f* that have not received prior recognition. The correspondences are not concentrated in any single area, nor are any long sequences homologous, but there are frequent identical matching sites in rows 2 and 3 and again in row 5, as well as related codons that have two of their three components alike. No serious attempt has been made to bring out additional similarities by insertion of hyphens, because the real problem here is how these relationships arose, not the degree of kinship.

What is the significance of the pronounced resemblance to cytochrome *b* and the lesser one to cytochrome *f?* The scant data available certainly do not permit any firm answers to be made, but several principal possibilities can be perceived. Of these the more obvious is that a fragment of a cytochrome *b* gene provided the ancestral framework for a protein that later was elongated by addition of parts from a cytochrome *f* cistron. The resulting gene then evolved by ordinary processes to form the modern version.

6.4. GENES FOR CYTOCHROMES P-450

In the secretory tissues of metazoans a class of cytochromes of great complexity abounds that is known to occur in few other organisms. However, since one representative of these cytochromes P-450 has been well documented from the bacterium *Pseudomonas putida* when grown on camphor (Eble and Dawson, 1984), the seeming sparse distribution reflects gaps in current knowledge rather than a real absence of these proteins. When fungal, seed-plant, protozoan, and algal sources receive adequate attention, this class of substances will probably be found to be of universal occurrence. Typically, its members are key enzymes in mixed-function oxidase systems, in eukaryotes characteristically located in the endoreticulum and Golgi apparatus. Insofar as gene structure is concerned, present knowledge is confined to tissues of insect and mammalian origin (Kulkarni and Hodgson, 1980). The oxidase systems exhibit extremely broad substrate specificity, acting on steroids, bile acids, camphor, cholesterol, coumarin, and even such exogenous substances as drugs, pesticides, and carcinogens (Thomas *et al.*, 1981; Kaipainen *et al.*, 1984). At least certain of the metazoan members of the complex act in conjunction with cytochrome b_5, NADPH, and NADPH-reductase (Ingelman-Sundberg and Johansson, 1980). The broad substrate capabilities of the systems stem from two factors, the multiplicity of molecular species of cytochrome P-450 and the wide latitudes in specificity of the individual types of the latter.

In many tissues the synthesis of the various species is preferentially induced by the presence of certain chemical compounds. Two that have been widely employed in researches on rat and mouse liver genes for these cytochromes are methylcholanthrene and phenobarbitol (Sogawa et al., 1984). In the rat the former induces the production of cytochromes P-450$_{c,d}$ and the latter induces that of cytochromes P-450$_{b,e}$; but these species are also referred to as cytochromes MC$_{1,2}$ and PB$_{1,2}$, the identities between the several groups remaining unestablished (Kuwahara et al., 1984). Furthermore, the reference cited demonstrated that actually four species were induced by phenobarbitol. But the synthesis of these heme-bearing proteins may be selectively induced by a diversity of chemicals, certain types, moreover, being controlled by the physiological state of the organism. One example of the latter condition is provided by a species (P-450$_{PG\omega}$) purified from lungs of pregnant rabbits that is active in the ω-hydroxylation of prostaglandins, for during late gestation its production rate becomes enhanced over 100-fold (Williams et al., 1984). Recently this group of cytochromes has been shown to be even more complex by the description of several new families. One of these is represented by a pregnenolone 16α-carbonitrile-induced species from rat liver (Gonzalez et al., 1985) and another by a *Pseudomonas* form that demethylates guaiacol (Dardas et al., 1985). Moreover, the existence of a greater diversity than generally suspected is evidenced by the gene sequences that show little homology to others bearing the same designation. Included among these is a phenobarbitol-inducible P-450$_b$ from rat liver (Suwa et al., 1985), a dioxin-inducible P$_1$-450 from human tissues (Jaiswal et al., 1985), and two 2,3,7,8-tetrachlorodibenzo-*p*-dioxin-inducible forms P$_4$-450 and P$_6$-450 from rabbit (Okino et al., 1985).

6.4.1. Methylcholanthrene-Induced Cytochromes P-450 Genes

Table 6.7 gives a number of the mature coding regions of various cytochrome P-450 genes, including four that are induced by methylcholanthrene and one by phenobarbitol. The former group is represented by P-450$_{c,d}$ from rat liver (Kawajiri et al., 1984; Yabusaki et al., 1984a,b) and P$_1$-450 and P$_3$-450 from the mouse (Gonzalez et al., 1984); the second group is exemplified only by one sequence from the rat, P-450$_e$ (Mizukami et al., 1983). For the present, focus is primarily on the first of the two sets. Examination of the first four sequences discloses that the upper pair, P$_3$-450 and P-450$_d$ (Gonzalez et al., 1984; Kawajiri et al., 1984; Kimura et al., 1984; Sogawa et al., 1984; Yabusaki et al., 1984a,b), are identical throughout row 1, but beginning with the seventh site in row 2, they become nearly completely different for a short space. After the first several positions in row 3, however, they resume the earlier, obviously homologous condition throughout the remainder of their lengths, including terminations at corresponding positions. In like fashion the short segment of P$_1$-450 that has been sequenced may be observed to show similar relationships to the rat cytochrome P-450$_c$.

Cytochrome P-450$_c$. The second gene from rat liver P-450$_c$ (Yabusaki et al., 1984a,b; Hines et al., 1985), although largely homologous to the first pair, differs at so many points that it must indeed be considered a separate and distinct species. In the first place it is markedly longer than the preceding, not only by the 18 additional nucleotide residues at the 3' end, but also through two additions at various internal locations. The first gain is by means of the nine-base-pair insert in row 1, and the second, through one of identical length in row 11. Another gain of three residues in row 13 is nullified later by a deletion of the same size in row 20. Moreover, in some instances the gene for the present species is purportedly interrupted by six introns, although only five are shown in the actual sequence (Sogawa *et al.*, 1984), but this feature may vary (Yabusaki *et al.*, 1984a). Despite these differences and the frequent deviations between corresponding sites, the similarities between the P-450$_d$ and P-450$_e$ types demonstrate that they are close relatives nevertheless. The identical locations of many of such important codons as the italicized ones for arginine and the underscored triplets for lysine are clear indicators of such kinship.

Chemical Properties. Since the three are related, it may be expected that they should share similar chemical properties. Surprisingly, despite the murine sequence's being the identical species as the rat P-450$_d$, it displays a number of dissimilarities in structure, especially with regard to the codons for lysine and arginine, features that are invariant in P-450$_{c,d}$. For example, in row 2, P$_3$-450 shows the presence of one triplet for each of those amino acids, whereas the rat pair has two for each. In the mouse gene, row 4 begins with a codon for arginine instead of one for lysine as in the other two, and near the end of row 15 a coding signal for either of these charged monomers is lacking.

For the greater part, however, P-450$_c$ is more frequently deviant with regard to these basic amino acids, especially in the central portion of the molecule. Occasionally it lacks a codon for one or the other of these substances at points where they are present in the first pair listed, concrete instances of the sort being found near the middle of both rows 6 and 9, and near the termini of rows 10 and 11. Moreover, row 12 has a triplet for arginine that does not coincide with those of the other genes, besides showing two pairs for lysine that are likewise unmatched, and row 13 has a number of disparities in this regard. The latter section is less irregular, but row 19 is deficient in two codons for lysine, while row 23 lacks one codon each for arginine and lysine. Thus, although the two species are similar in structure, the distribution of triplets for the basic monomers implies that distinctions in chemical properties of the inner domains exist.

6.4.2. Phenobarbitol-Induced Cytochromes P-450 Genes

Although the existence of multiple forms of phenobarbitol-induced P-450 has been ascertained (Kuwahara *et al.*, 1984), the gene for only a single representative has been fully sequenced, that for rat liver P-450$_e$ (Table 6.7; Fujii-Kuriyama *et al.*, 1982). Accordingly, comparisons can be made only to the set just discussed.

The sequence is deficient in 15 nucleotide residues at the 5' end, a lack that is fortunate because it has assisted in aligning this gene with the others. Thus arranged, it becomes evident that the present type probably shared ancestry with the 3-methylcholanthrene-induced species, an observation made especially convincing by the correspondences found in the first half of row 1. Thereafter homologous sites are scarce, but where present they often occur as groups of six or more positions, like the ACACCTG near the end of row 3, the AC-ATC of the beginning of row 5, and the -AGGA-GCC of row 7. Moreover, near the middle of rows 8 and 9 are found GCYAAC and TTGG-G-TGT, respectively, and from the end of row 11 into the start of row 12 is the rather long, broken series AAR--G-G-ACYAC---ACA. Another sequence of fair length is ATYCA-GAG at the opening of row 16 and still another is provided by the AAG-RCATTGR from the terminus of row 20 and beginning of row 21. As a whole it is evident that the P-450$_e$

cistron is more closely allied to the P-450$_c$ than to the others, the main differences to be noted being the extension at the 5' end and the shortened 3' terminus.

6.4.3. Genes for Other Types of Cytochromes P-450

In addition to the two principal sets from mammalian liver endoreticulum described in the preceding sections, there are a number of other types of cytochromes P-450, several located in the nucleus, some in the mitochondrion, and still more in the endoreticulum of various organs. Such insects as the housefly (*Musca domestica*) are known similarly to possess a multiplicity of this protein type (Kulkarni and Hodgson, 1980), but no information has been garnered as to the properties of their genes, a statement equally true of the mammalian nuclear species. However, one additional representative from the endoreticulum remains for discussion, in this case from cells of the adrenal gland, along with those from mitochondria whose gene sequences have been established.

Adrenal Cytochrome P-450$_{c21}$. The cytochrome P-450 of the adrenal gland that is of particular concern is one of a group of five involved in the metabolism of steroids. A different type is essential at each of the five steps in the chemical conversion of cholesterol, which are as follows (White *et al.*, 1984):

1. In the mitochondrion the C-22,27 side chain of cholesterol is first cleaved from that molecule to produce pregnenolone.
2. Following conduction to the endoreticulum, the 3β-hydroxyl group of the pregnenolone undergoes dehydrogenation to result in progesterone.
3. Still in the latter organelle, the progesterone then undergoes two hydroxylations, the first at the 17α position.
4. The second hydroxylation occurs at carbon 21.
5. After transport back to the mitochondrion, the steroid receives a final hydroxylation at the 11β position to yield corticol.

Thus far only a single representative of the cholesterol-modifying cytochromes P-450 has had its gene sequence established, that known as cytochrome P-450$_{c21}$, which is active in step 4. Even that one is incomplete, for, lacking much of both the 5' and 3' sectors, it begins in row 7 of Table 6.7 and ends in row 14. Comparison with the other cistrons establishes beyond reasonable doubt that the present protein is most closely related to the phenobarbitol-induced species given in that table, P-450$_e$. Although the kinship level is low, there is a sufficient number of identical and related sites present to make the aligning of the two sequences a fairly straightforward process.

Thus incomplete, the midportion that has been established nevertheless displays some unique features, especially in having an unusually high number of codons for cysteine (heavy stippling). In the short section available five such triplets can be noted, equal to the totals of each of the entire liver sequences—in fact, the murine P$_3$-450 gene has only four. When compared to the rat P-450$_e$, the present coding series is seen to have a number of short deletions, three each in rows 10 and 13, two in row 11, and one in row 14. Moreover, it appears to lack any introns, such as characterize the phenobarbitol-induced species; however, it does have a 21-site-long inserted coding section that is absent from any other of these genes, the location of which in row 9 is appropriately indicated.

Table 6.7
Mature Coding Regions of Various Cytochromes P-450 Genes[a]

Row 1

```
Mouse P₃-450ᵇ            ATG GGG TTC TCC CAG TAC ATC --- --- --- TCC --- --- TTA GCC CCA GAG CTG CTA CTG GCC ACT GCC ATC
Rat P-450_d ᶜ            ATG GGG TTC TCC CAG TAT ATC --- --- --- TCC --- --- TTA GCC CCA GAG CTG CTA CTG GCC ACT GCC ATC
Rat P-450_c ᵈ            ATG CCT --- TCT GTG TAT GGA TTC CCA GCC ACA TCA GCC ACA GAG CTG CTC CTG GCC GTC ACC ACA
Mouse P₁-450ᵉ            ATG --- --- TAT --- --- TAT GGA CTT CCA GCC TTC GTG TCA GCC ACA GAG CTG CTC CTG GCT GTC ACC GTA
Rat P-450_e ᶠ            TTG CTC CTC GCT --- CTC CTC GTG GGC TTC TTG TTA CTC TTA GTC AGG GGA CAC CCA AAG TCC CGT
Bovine_M P-450+ᵍ_scc     ATG CTA GCA AGG GGG CTT CCC CTC CGC --- TCA GCC CTG TGC CCA ATC CTG AGC ACA
Bovine_M P-450-ᵍ_scc     --- --- ATC TCC ACA AAG ACC CCT CGC CCC TAC AGT GAG ATC CCC TCC CCT GGT GAC AAT GGC TGG CTT
```

Row 2

```
Mouse P₃-450            TTC TGT TTA GTG TTC TGG ATG GTC CAG AGC CTC AAG GAC CCA AGC CTC GGT TCC CAA AGG CCT GAA GAA TCC ACC
Rat P-450_d             TTC TGT TTA GTG TTC TGG GTG TTG AGA GGC ACA AGG ACC CAG GTT CCC AAA GGT CTG AAG AGT CCT CCC
Rat P-450_c             TTC TGC CTT GGA TTC TGG GTG GTT AGA GTC ACA AGA ACC TGG GTT CCC AAA GGT CTG AAG AGT CCA CCC
Mouse P₁-450            TTC TGC CTT GGA TTC TGG GTG GTC AGA GTC GTC AGC GCC ACA (Incomplete)
Rat P-450_e             GGC AAC TTC CCA CCA GGA CCT CGT CCC CTT CCC CTC TTG GGG AAC CTC CTG CAG TTG GAC AGA GGG GGC
Bovine_M P-450_scc+    GTG GGG GAG GGC TGG GGC CAC CAC AGG GTG CAC ACT GGA GAG GGA GCT GGC ATC TCC ACA AAG ACC CCT
Bovine_M P-450_scc-    AAC CTC TAC CAT TTC TGG TGG AGG TCA CAG GGC TCA CAG AGA ATC CAC CAC TTT CGC CAC ATC GAG AAC TTC CAG
```

Row 3

```
Mouse P₃-450            CGG ACC CTG GGG CTT CCC TTC ATT GGG CAC ATG CTG ACT GTG GGG AAG AAC CCA CAC CTG TCA CTG ACA
Rat P-450_d             GGA CCC TGG GGC TTG CCC TTC ATT GGG CAC ATG CTG ACC CTG GGG AAG AAC CCA CAC CTA TCT CTG ACA
```

```
Rat P-450d           GGA CCC TGG GGC TTG CCC TTC ATT GGG CAC ATG CTG ACC CTG GGG AAG AAC CCA CAC CTA TCT CTC ACA
Rat P-450c           GGA CCC TGG GGC TTG CCC TTC ATA GGG CAC GTG CTG ACC CTG GGG AAG AAC CCA CTG TCA CTG ACA
Rat P-450e           CTC CTC AAT TCC ATG CAG CTT CGA GAA AAA TAT GGA GAT GCG TTC ACA GTA CAC CTG GGA CCA AGG
BovineN P-450scc+    CGC CCC TAC AGT GAG ATC CCC TCC CCT GGT GAC AAT GGC TGG CTT AAC CTC TAC CAT TTC TGG AGG GAG
BovineN P-450scc-    AAG --- TAT GGC CCC ATT TAC AGG GAG AAG CTT GGC AAT TTG GAG TCA GTT TAT ATC ATT CAC CCT GAA

Row 4

Mouse P3-450         CGG CTG AGT CAG CAG TAT GGG GAC GTG CTG CAG ATC CGC ATC GGC TCC ACT CTG GTG GTG CTG AGC
Rat P-450d           AAG CTG AGT CAG CAG TAT GGG GAC GTG CTG CAG ATC CGC ATT GGC TCC ACA CCC GTG GTG CTG AGC
Rat P-450c           AAG CTG AGT CAG CAG TAT GGG GAC GTG CTG CAG ATC CGT ATT GGC TCC ACA CCC GTG GTG CTG AGC
Rat P-450e           CCT GTG GTC ATG CTA TGT GGG ACA GAC ACC ATA AAG GAG GCT CTG GGC CAA GCT GAG GAT TTC TCT
BovineM P-450scc+    AAG GGC TCA CAG ATC CAC TTT CGC CAC ATC ATC GAG AAC TTC CAG AAG TAT GGC CCC ATT TAC AGG GAG
BovineM P-450scc-    GAC GTG GCC CAT CTC TTC AAG TTC GAG GGA TCC TAC CCA --- GAG AGA TAT GAC ATC CCG CCC TGG CTG

Row 5

Mouse P3-450         GGC CTG AAC ACC ATC AAG CAG GCC CTG GTG AGG CAG GGA GAT GAC TTC AAG GGC CGA CCA GAC CTC TAC
Rat P-450d           GGC CTG AAC ACC ATC AAG CAG GCC CTA GTG AAG CAG GGG GAT GAC TTC AAA GGC CGG CCA GAC CTC TAC
Rat P-450c           GGC CTG AAC ACC ATC AAG CAG GCC CTG GTG AAA CAG GGG GAT GAC TTC AAA GGC CGG CCA GAC CTC TAC
Rat P-450e           GGT CGG GGA ACA ATC GCT GTG ATT GAG CCA CCA ATC TTC AAG GAA TAT GGT GTG ATC TTT GCC AAT GGG GAA
BovineM P-450scc+    AAG CTT GGC AAT TTG GAG TCA GTT TAT ATC ATT CAC CCT GAA GTC GGA GTC TTC AAG AAG TTC GAG
BovineM P-450scc-    GCC TAT CAC CGA CCA TAT TAT CAG AAA CCC ATT GGA GTC CTG TTT AAG AAG ACC GGA ACC TCA AAG AAA GAC
```

(continued)

Table 6.7 (Continued)

Row 6

Mouse P₃-450   AGC TTC ACA CTT ATC ACT AAC GGC AAG AGC ACT TTC AAC CCA GAC TCT GGA CCC GTG TGG GCT GCC

Rat P-450_d   AGC TTC ACA CTT ATC ACT AAT GGC AAG AGC ATG ACT TTC AAC CCA GAC TCT GGA CCG GTC TGG GCT GCC

Rat P-450_c   AGC TTC ACA CTT ATC GCT AAT GGC CAG AGC ATG ACT TTC AAC CCA GAC TCT GGA CCG CTG TGG GCT GCC

Rat P-450_e   CGC TGC AAG GCC CTT CGG CGA TCT CTG GCT ACC ATG AGA GAC TTT GGG ATG GGA AAG AGG AGT GTG

Bovine_M P-450_scc + GGA TCC TAC CCA GAG AGA TAT GAC ATC CCG TGG CTG GCC TAT CAC CGA TAT TAT CAG AAA CCC ATT

Bovine_M P-450_scc - CGG GTG GTC CTG AAC ACG GAG GTG ATG GCT CCA GAG GCA ATA AAG AAC TTC ATC CCA CTG AAT CCA

Row 7

Mouse P₃-450   CGC CGG CGC CTG GCC CAG GAT GCC CTG AAG AGC TTC TCC ATA GCC TCG GAC CCG ACG TCA GCA TCC TCG

Rat P-450_d   CGC CGG CGC CTG GCC CAG GAT GCC CTG AAG AGT TTC TCC ATA GCC TCA GAC CCC ACA TCA TCC TCT

Rat P-450_c   CGC CGG CGC CTG GCC CAG AAT GCC CTG AAG AGT TTC TCC ATA GCC TCA GAC CCA ACA CTG GCA TCC TCT

Rat P-450_e   GAA GAA CGG ATT CAG GAG GAA GCC CAA CTG TGT TTG GTG GAA CTG AAA TCC CAG GCC GCC CCA CTG

Bovine P-450_c21 h   (Incomplete)   CAG GCC GGT GCC CCC GTG

Bovine_M P-450_scc + GGA GTC CTG TTT AAG AAG TCA GGA ACC TGG AAG AAA GAC CCG GTG GTC CTG AAC ACG GAG GTG ATG GCT

Bovine_M P-450_scc - GTG TCT CAG GAC TTC GTC AGC CTC CTG CAC AAG CCC ATC AAG CAG CAG GGC TCC GGA AAG TTT GTA GGG

Row 8

Mouse P₃-450   TGC TAT TTG GAG GAG CAC GTG AGC AAG GAG GCT AAC CAT CTC GTC AGC AAG CTT CAG AAG GCG ATG GCA

Rat P-450_d   TGC TAC TTG GAG GAG CAC GTG AGC AGC GCT AAC GAG GCT AAC CAT ATC AGC AAG TTC CAG AAG CTG ATG GCA

Rat P-450_c   TGC TAC TTG GAA GAG CAC GTG AGC GAC GCC GAA TAC TTA ATC GCA GAG GCC AAG AAG TTC CAG AAG CTG ATG GCA

```
Rat P-450_e              GAT CCC ACC TTC CTC TTC CAG ATC ACA GCC AAC ATC ATC TGC TCC ATT GTG TTT GGA GAG CGC TTT
Bovine P-450_c21         ACC ATC CAT ACG GAA TTC TCT CTG CTT ACC TGC AGC ATC ATC TGT TAC CTC ACT TTT GGA AAC AAG GAG
Bovine_M P-450_scc +     CCA GAG GCA ATA AAG AAC TTC ATC CCA CTG AAT CCA GTG TCT CCA GAC TTC GTC AGC CTC CTG CAC
Bovine_M P-450_scc −     GAC ATC AAG GAA GAC CTG TTT CAC TTT GCC TTT GAG TCC ATC ACC AAT GTC ATG TTT GGG GAG CGC CTG

Row 9

Mouse P_3-450            GAG GTT GGC CAC TTC GAA CCA CAG GTC GTG GAA TCG GCT AAC GTC ATT GGT GCC ATG TGC
Rat P-450_d              GAG GTT GGC CAC TTC GAA CCA GTC AAC CAG GTG GTG GAA TCG GCT AAT GTC ATC GGA GCC ATG TGT
Rat P-450_c              GAG GTT GGC CAC TTC GAC CCT CCT TTC AAG TAT CAG GTG GTG TCA GTG GCC AAT GTC ATC ATC TGT GCC ATA TGC
Rat P-450_e             GAC TAC ACA GAC CGC CAG TTC CTG CTG GAG CTG TTC TAC CGG ACC TTT TCC CTC CTA AGT TCA
Bovine P-450_c21         GAC ACC TTA GTA CAT GCC TTT CAC GAC TCT GTT CAG GAC TTG CAT GAA GAC CTG GGT TCC
Bovine_M P-450_scc +     AAG CGC ATC AAG AAG CAG CAG GGC TCC GGA AAG TTT GTA GGG GAC ATC AAG AAG TTT ATT GAT GCC CTG TAC AAG ATG TTC GCC
Bovine_M P-450_scc −     GGG ATG CTG GAG GAG GAG ACA ACA GTG AAC CCC GAG AAC CCC CAG AAG AAG GCC GTC TAC AAG ATG CAC

Row 10

Mouse P_3-450            TTT GGG AAG AAC TTC CCC CGG AAG AGC GAG GAG ATG CTG AAC ATC GTG AAT AAC AGC GAC AAG TTT GTG
Rat P-450_d              TTT GGG AAG AAC TTC CCC AGG AAG AGC GAG GAG ATG CTG AAC CTC AAC CTC GTG AAG AGC AGC GAC GAC TTT GTG
Rat P-450_c              TTT GGG AGA CGT TAT GAC CAC CAC GAT GAC CAA GAG CTG CTC AGC ATA AGC AAT CTA AGC AAT GAG TTT GGG
Rat P-450_e              TTC TCC AGC CAG GTG TTT GAG TTC TTC TCT GGG TTC TCT CTG AAA TAC TTT CCT GGT GCC CAC AGA CAA ATC
Bovine P-450_c21         TTT TCT --- CAG GTT CTT CCC CAA CCC TGG GCT --- CTG GAG GCT GAA GCA GGC --- CAT AGA GAA CAG
Bovine_M P-450_scc +     TTT GAG TCC ATC ACC AAT GTC ATG TTT GGG GAG CGC CTG GGG ATG CTG GGG ACA GTG AAC CCC GAG
Bovine_M P-450_scc −     ACC AGT GTC CCT CTG AAC GTC CTC CCA GAA CTG TAC CTA TTC AGA ACC AAG AAG ACT TGG AGG GAC
```

(continued)

Table 6.7 (Continued)

Row 11

Mouse P₃-450	GAG --- --- --- GTC ACC TCA GGG AAT GCA GTG GAC TTC TTC CCG GTC CTG CGC TAC CTG CCC AAC
Rat P-450_d	GAG --- AAT --- --- GTC ACC TCA GGG AAT GCT GTG GAC TTC TTT CCG GTC CTG CGC TAC CTG CCC AAC
Rat P-450_c	GAG GTT ACT GGT TCT GGA TAC CCA GCT GAC TTC ATT CCT ATC CTC CGT TAC CTC CCT AAC TCT TCC CTG
Rat P-450_e	TCC AAA AAC CTC CAG GAA ATC CTC GAT TAC ATT GGC CAT ATT GTG GAG AAG CAC AGG GCC ACC TTA GAC
Bovine P-450_c21	GGA CCA CAT GGT GGA GAA GCA GCT GAC --- GCC CCA CAA --- GGA GAG CAT GGT GGC CGG CCA GTG GAG
Bovine_M P-450_scc+	GCC CAG AAG AAG TTC ATT GAT GCC GTC TAC AAG ATG TTC CAC ACC AGT GTC CCT CTG CTC AAC GTC CCT CCA
Bovine_M P-450_scc-	CAT GTA GCC GCA TGG GAC ACA ATT TTC AAT AAA GCT GAA AAA TAC ACT GAG ATC TTC TAC CAG GAC CTG

Row 12

Mouse P₃-450	CCC GCC CTC AAG AGG TTT AAG ACC TTC AAT GAT AAC TTC GTG CTG TTT CTG CAG AAA ACT GTC CAG GAG
Rat P-450_d	CCA GCC CTC AAG AGG TTT AAG AAC TTC AAT GAT AAC TTT GTG CTG TCT CTG CAG AAA ACA GTC CAG GAA
Rat P-450_c	GAT GCC TTC AAG GAC TTG AAT AAG AAG TTC TAC AGT TTC ATG AAG AAG CTA ATC ATC AAA GAG CAC GAA
Rat P-450_e	CCA AGC GCT CCA CGA GAC ACT ACA TTC ATC GAC ACT TAC CTT CTG CGC ATG GAC AGA GTA GGG AAG AAG
Bovine P-450_c21	GGA TAT GAC CGG ACT ACA TGC TCC AAG GGG GTA GGG AGG CAA AGA GTA GAA GAG GGC CGG GGA CAG CTC
Bovine_M P-450_scc+	GAA CTG TAC CGT CTA TTC AGA ACC AAG ACT TGG AGG GAC CAT GCC GCA TGG GAC ACA ATT TTC AAT
Bovine_M P-450_scc-	AGA CGG AAA ACA GAA TTT AGG AAT TAC CCA ATC CTC TAC TGC CTC CTC AAA GAG AAG ATG CTC

Row 13

Mouse P₃-450	CAC TAC CAA GAC TTC AAC AAG AAG AAC AGT ATC CAA GAC ATC ACA AGT GCC CTG TTC AAG CAC --- AGC ---
Rat P-450_d	CAC TAT CAA GAC TTC AAC AAG AAG AAC AGT AAC AGT ATC CAG GAC ATC ACA AGC GCC CTG TTC AAG CAC --- AGT ---
Rat P-450_c	ACA TTT GAG AAG GGC CAC ATC CGG GAC ATC ACA ACA GAC AGC CTC ATT GAG CAT TGT CAG GAC GAC AGG AGG ---

```
Rat P-450_c21        ACA GAG TTC CAT CAT GAG AAC CTC ATG ATC TCC CTG TCT CTC TTC TTT GCT GGC ACT GAG ACC ---
Bovine P-450_c21     CTG GAA GGA CAC GTG --- --- CAC ATG --- TCT GTG GTG GAC CTT TTC ATC GGG GGC ACT GAA ACC ACG
Bovine_M F-450_scc+  AAA GCT GAA AAA TAC ACT GAG ATC TTC TAC CAG GAC CTG AGA CGG AAA ACA GAA TTT AGG AAT TAC ---
Bovine_M F-450_scc-  TTG GAG GAT GTC AAG GCC AAT ATT ACG GAG ATG CTG GCA GGG GGT GTG AAC ACG ACA TCC ATG ACA ---

Row 14

Mouse P_3-450        GAG AAC TAC AAA GAC     CTC ATC CCC GAG GAG AAG ATT GTC AAC ATT GTC AAT GAC ATC TTT
Rat P-450_d          GAG AAC TAC AAA GAC GAC AAC GGT GGT CTC ATC CCT CAG GAG AAG ATT GTC AAC ATT GAC ATC TTT
Rat P-450_c          CTC GAC GAG AAT GCC AAT --- GTC CAG CTC TCA GAT GAT AAG GTC ATT ACG ATT GTT TTT GAC CTC TTT
Rat P-450_e          ATG AGC ACC ACA CTC CGC TAT GGT TTC CTG CTC ATG CTC AAG TAC CCC CAT GTC GCA GAG AAA GTC CAA
Bovine P-450_c21     GCG AGC ACC CGT CTC --- CTG GGC TGT GGC GTT CCT ACT TCA CCA CCC CGA GAT TCA G---(Incomplete)
Bovine_M 2-450_scc+  CCA GGC ATC CTC --- --- TAC TGC CTC CTG AAA AGT GAG AAG ATG CTC TTG GAG GAT GTC AAG GCC AAT
Bovine_M 2-450_scc-  TTG CAA TGG CAC --- --- TTG TAC GAG ATG GCA CGC AGC CTG AAT GTG CAG ATG CTG CGG GAG GAG

Row 15

Mouse P_3-450         GGA --- GCT G GC TTT GAC ACA GTC ACA ACC ACA GCC ATC ACC TGG AGC ATT TTG CTA CTT GTG ACA TGG CCT
Rat P-450_d           GGA --- GCT G GA TTT GAA ACA GTC ACA ACA GCC ATC TTC TGG AGC ATT TTG CTA CTT GTG ACA GAG CCC
Rat P-450_c           GGA --- GCT G GG TTT GAC ACA ATC ACA ACT GCT ATC TCT TGG AGC CTC ATG TAC CTG GTA ACC AAC CCT
Rat P-450_e           AAG --- GAG A TT GAT CAG GTG ATC GGC TCA CAC CGG CTA CCA ACC CTT GAT GAC CGC AGT AAA ATG CCA
Bovine_M P-450_scc+   ATT ACG GAG A TG CTG GCA GGG GGT GTG AAC ACG ACA TCC ATG ACA TTG CAA TGG CAC TTG TAC GAG ATG
Bovine_M P-450_scc-   GTT --- CTG A AT GCC CGA CGC CAG GCA GCA GAG GGA GAC ATA --- AGC AAG ATG CTG CAA ATG GTC CCA CTT
```

(continued)

Table 6.7 (Continued)

Row 15

Mouse P₃-450	GGA --- GCT G GC TTT GAC ACA GTC ACC ACA GCC ATC ACC TTC TGG AGC ATT TTG CTA CTT GTG ACA TGG CCT
Rat P-450_d	GGA --- GCT G GA TTT GAA ACA GTC ACA ACA GCC ATC TTC TGG AGC ATT TTG CTA CTT GTG ACA GAG CCC
Rat P-450_c	GGA --- GCT G GG* TGT GAC ACA ATC ACA ACT GCT ATC TCT TGG AGC CTC ATG TAC CTG GTA ACC AAC CCT
Rat P-450_e	AAG --- GAG A TT GAT CAG GTG ATC GGC TCA CAC CTT GAT GAC CGG AGT CGC AGT AAA ATG CCA
Bovine_M P-450_scc⁺	ATT ACG GAG A TG CTG GCA GGG GGT GTG AAC ACG ACA TCC ATG ACA TTG ACA TTG CAA TGG CAC TTG TAC GAG ATG
Bovine_M P-450_scc⁻	GTT --- CTG A AT GCC CGA CGC CAG GCA GGA GAC ATA --- AGC AAG ATG CTG CAA ATG GTC CCA CTT

Row 16

Mouse P₃-450	AAC GTG CAG AGG AAG ATC CAT GAG GAG CTG G AC ACG GTG GTT GGC ACG GAT CGG CAA CCA CGG CTT TCT
Rat P-450_d	AAG GTG CAG AGG AAG ATT CAT GAG GAG CTG G AC ACG GTG ATT GGC AGA GAT CGG CAG CCA CGG CTT TCT
Rat P-450_c	AGG ATA CAG AGA AAG ATC CAG GAG GAG TTA G*AC ACA GTG ATT GGC AGA GAT CGG CAG CCC CGG CTT TCT
Rat P-450_e	TAC ACT GAT GCA GTT ATC CAC GAG ATT CAG A GG TTT TCA GAT CTT GTC CCT ATT GGA GTA CCA CAC AGA
Bovine_M P-450_scc⁺	GCA CGC AGC CTG AAT GTG CAG ATG CTG C GG AGA GAG GTT CTG AAT GCC CGA CGC CAG GCA GAG GGA
Bovine_M P-450_scc⁻	CTC AAA GCT AGC ATC AAG GAG ACG CTG AGA C TC CAC CCC ATC TCC GTG ACC ATC CTG CAG AGA TAC CCT GAA

Rat P-450c CCA TTC ACC ATC CCC CAC AG*C ACC ATA AGA GAT ACA AGT CTG AAT GGC TTC TAT ATC CCC AAG GGA CAC

Rat P-450e AGT TCA GCT CTC CAT GAC CC A CAG TAC TTT GAC CAC CAC AGC TTC AAT CCT GAA CAC TTC CTG GAT

BovineMP-450scc+ CAC CCC ATC TCC GTG ACC CT G CAG AGA TAC CCT GAA AGT GAC TTG GTT CTT CAA GAT TAC CTG ATT CCT

BovineMP-450scc- GGC CGA GAC CCT GCC TTC TT C TCC AGT CCG GAC AAG TTT GAC CCA ACC ACC AGG CTG AGT AAA GAC AAA

Row 19

Mouse P3-450 TGT ATC TAC ATA AAC CAG TGG CAG GTC AAC CAT GAT GA G AAG CAG AAA GAC CCC TTT GTG TTC CGC

Rat P-450d TGC ATC TTC ATA AAC CAG TGG CAG GTC AAC CAT GAT GA G AAG CAG AAA GAC CCC TTT GTG TTC CGC

Rat P-450e TGT GTC TTT GTG AAC CAG TGG CAG GTT AAC CAT GAC CA G GAA CTA TGG GGT GAT CCA AAC GAG TTC CGG

Rat P-450e GGC AAT GGG GCA CTG AAG AGT GAA GCT TTC ATG CC C TTC TCC ACA GGA AAG CCC ATT CTT GGC

BovineMP-450scc+ GCC AAG ACA CTG GTG CAA GTG GCC ATC TAT GCC ATG GG C CGA GAC CCT GCC TTC TTC TCC AGT CCG GAC

BovineMP-450scc- GAC CTC ATC CAC CGG AAC CTG GGC TTT GGC GG A GTG CGG CAG TGC GTG GGC --- --- CGG CGG

Row 20

Mouse P3-450 CCA GAG CGG TTT CTT ACC AAT AAC TCG GCC ATC GAC AAG ACC CAG AGC GAG AAG --- GTG ATG CTC

Rat P-450d CCA GAG CGG TTT CTT ACC AAT GAC AAC ACG GCC ACT --- CTG GAC AAA CAC CTG AGT GAG AAG --- GTG ATG CTC

Rat P-450e CCT GAA AGG TTT CTT ACC TCC AGT GGC ACT --- CTG GAC AAA CAC CTC AGT GAG AAG --- GTC ATT CTC

Rat P-450e GAA GGC ATT GCC CGA AAT GAA TTG TTC CTC --- TTC TTC ACC ACC ATC CTC CAC AAC --- TTC TCT GTG

BovineMP-450scc+ AAG TTT GAC CCA ACC AGG TGG CTG AGT AAA --- GAC AAA GAC CTC ATC CAC TTC CGG AAC CTG GGG TTT

BovineMP-450scc- ATC GCC GAG CTG GAG ATG ACC CTC CTC --- ATC CAC ATT CTG GAG AAC TTC AAG --- GTT GAA ATG

(continued)

Table 6.7 (Continued)

Row 21

Mouse P$_3$-450	TTC GGC TTG GGA AAG *CGG* TGC ATT GGG GAG ATC CCG GCC AAG GAA GTC TTC CTC TTC TTA GCC
Rat P-450$_d$	TTC GGC TTG GGA AAG *CGG* TGC ATT GGG GAG ATC CCG GCC AAG GAA GTC TTC CTC TTC TTA GCC
Rat P-450$_c$	TTT GGT TTG GGC AAG *CGA* TGC ATT GGG GAG ACC ATT GGC *CGA* CTG GAG GTC TTT CTC TTC CTG GCC
Rat P-450$_e$	TCA AGC CAT TTG GCT CCC AAG GAC ATT GAC CTC ACG CCC AAG GAG AGT GGC ATT GGA AAA ATA CCT CCA
Bovine$_M$P-450$_{scc}$ +	GGC TGG GGA GTG --- *CGG* CAG TGC GTG GGC *CGG* ATC CCC GAG CTG GAG ATG ACC CTC TTC CTC ATC
Bovine$_M$P-450$_{scc}$ −	CAG CAT ATC GGT GAC GTG GAC ACC ATA TTC AAC CTC ATC CTG ACG CCG GAC AAG CCC ATC TTC CTT GTC

Row 22

Mouse P$_3$-450	ATC CTG CTG CAG CAT CTG GAG TTT AGT GTG CCA CCG GGT GTG AAG GAC CTG ACA CCC AAC TAT GGG
Rat P-450$_d$	ATC CTC CTG CAT CAG CTG GAG TTC ACT GTG CCA CCG GGC GTG AAG GAC CTG ACA CCC AGC TAT GGG
Rat P-450$_c$	ATC TTG CTG CAG CAA ATG GAA TTT AAT GTC TCA CCA GGC GAG AAG GTG GAT ATG ACT CCT GCC TAT GGG
Rat P-450$_e$	ACG TAC CAG ATC TGC TTC TCA GCT *CGG* TGA
Bovine$_M$P-450$_{scc}$ +	CAC ATT CTG GAG AAC TTC AAG GTT GAA ATG CAG CAT ATC GGT GAC GTG GAC ACC ATA TTC AAC CTC ATC
Bovine$_M$P-450$_{scc}$ −	TTC *CGC* CCC TTC AAC CAG GAC CCG CCC CAG GCG TGA

Row 23

Mouse P₃-450   TTG ACC ATG AAG CCC GGG ACC TGT GAA CAC GTC CAG GCA TGG CCA CGC TTT TCC AAG TGA

Rat P-450_d   CTG ACC ATG AAG CCC AGA ACC TGT GAA CAC GTC CAG GCC TGG CCA CGC TTC TCC AAG TGA

Rat P-450_e   CTG ACT TTA AAA CAT GCC CGC TGT GAG CAC TTC CAA GTG CAG ATG CGG TCT TCT GGT CCT CAG CAT CTC

Bovine_M P-450_scc+ CTG ACG CCG GAC AAG CCC ATC TTC CTT GTC TTC CGC CCC TTC CGC AAC CAG GAC CCG CCC CAG GCG TGA

Row 24

Rat P-450_e   CAG GCT TAG

a Codons for lysine are underscored and those for arginine are italicized; those for tyrosine (TAC, TAT) are lightly stippled, those for phenylalanine (TTT, TTC) are stippled moderately, and those for tryptophan (TGG) and cysteine (TGC, TGT) are stippled heavily. M, mitochondrion. (+), Indicates that coding includes a transit peptide; (−) indicates that the coding lacks a presequence.
b Kimura et al. (1984).
c Kawajiri et al. (1984); Yabusaki et al. (1984b).
d Sogawa et al. (1984); Yabusaki et al. (1984a).
e Gonzalez et al. (1984).
f Fujii-Kuriyama et al. (1982); Mizukami et al. (1983).
g Morohashi et al. (1984).
h White et al. (1984).
i Insert a 21-residue sector.

Relative to the corresponding region of P-450$_e$, it is deficient in codons for the two basic amino acids, having only two for lysine and five for arginine, compared to five and nine, respectively, of that rat sequence. More properly, the section given encodes an acidic product, for a total of 21 triplets for aspartic and glutamic acids is found here, against 16 in its kin. It shows no other propensity for enrichment with codons for any other amino acid, having just two for tyrosine and one for tryptophan.

Mitochondrial Cytochromes P-450. The mitochondrial cytochromes P-450 differ from their endoreticular counterparts in a number of details. In the first place, they, being nuclear genes like the others of that source that function within the mitochondrion, require the presence of a region encoding a transit peptide at the 5′ end so that their product may be conducted into the organelle and be inserted on a membrane. As elucidated further in the following chapter, such sectors are removed by enzyme action after transport. Second, the electron-transport chain of which they are a part differs somewhat in constituency, so that these types receive electrons from adrenodoxin, an iron–sulfur protein, which serves as the NADPH-adrenodoxin reductase, rather than from NADPH-cytochrome P-450 reductase, which is a flavoprotein (Morohashi *et al.*, 1984).

The total number of species of these cytochromes confined to mitochondria has not been finally established. As pointed out earlier, two types, P-450$_{scc}$ and P-450$_{11}\beta$, are active in steroid modification, but others seem to be essential in bile acid synthesis and vitamin D$_3$ metabolism (Niranjan *et al.*, 1984). One or more of the latter appear also to be involved in the activation of such carcinogens as aflatoxin B and benzo(α)pyrene to an electrophilic reactive form.

The sole member of the group whose coding sequence has been established is that for P-450$_{scc}$, only a single gene for which appears to be present in bovine tissues (John *et al.*, 1984; Morohashi *et al.*, 1984). In Table 6.7, this appears twice, the first sequence having the transit peptide-coding portion in place and the second lacking that prefatory part. This double appearance is designed to shed light on the possible origins of that region, it being one of the few opportunities known where such comparisons can be made. Were it derived by modification of the ancestral gene for a cytochrome, then the entire coding area should display homology to the others in the table, whereas if it represents an independent addition, then the mature coding sequence alone should show the relationships. The results, however, are too obscure to permit the drawing of any firm conclusions, because the extent of homology is insufficient with both alignments, although the evidence somewhat inclines in favor of the second alternative. Further sequences from lower organisms possibly will enable this problem of origins to be resolved.

Relationships of the mature coding sequence appear to be equally divided between such 3-methylcholanthrene-induced species as rat P-450$_d$ and the phenobarbitol-induced ones like P-450$_e$. Quite unexpectedly, comparisons of the primary structures of the encoded products show a similar kinship, rather than to bacterial cytochromes P-450 (Morohashi *et al.*, 1984). Thus, the evidence favors the view that the mitochondrial types have always been encoded by nuclear genes.

6.5. ATPase GENES

The ATPases, enzymes that use ATP as a substrate, are a complicated group of largely unrelated proteins whose sole common feature appears to be their involvement in the metabolism of the triphosphorylated nucleoside that stores and supplies much of the energy requirements of the cell. Among the miscellany of types sharing in the function are ATP-transport molecules, including *Trk* and *Kdp* membrane systems of *E. coli*, which derive energy from ATP for the transport of K^+ ions. Some of these appear to be related, at least distantly, to cation-transporting enzymes of eukaryotes, among which is the Ca^{2+}-carrying ATPase of the endoreticulum (Hesse *et al.*, 1984). But the most thoroughly investigated types are several components of an ATPase complex found in prokaryotes and in chloroplasts and mitochondria of eukaryotes.

6.5.1. Genes of the ATPase Complex

ATPase, more properly called ATP-synthase, is well-recognized as being a complex, rather than a single multisubunit protein, for many of its components have specialized individual functions, which become apparent as the whole association is described. In simple terms, the complex is in the form of a mushroom-shaped body (Figure 6.1), whose base is embedded in the membrane of which it is a part (Borst and Grivell, 1978; Walker *et al.*, 1984). It is this enzyme that frequently is involved with the cytochrome chain, some of whose members have just been elucidated; it induces the phosphorylation of ADP to produce ATP in a reaction driven by the membrane proton-potential gradient generated by the electron-transport system.

Composition. The membrane-embedded portion, often referred to as the F_0 segment, contains a proton-conducting channel that leads the protons to the outer globular segment, called the F_1 part, where the phosphorylation of ADP is catalyzed (Senior, 1973; Tybulewicz *et al.*, 1984). The stalklike portion is part of the F_1 in bacteria and is shortened, as in the illustration; this, along with a number of other structural differences,

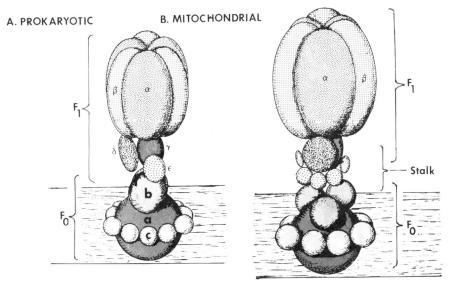

Figure 6.1. (A) Prokaryotic and (B) mitochondrial ATPase complexes. In the organellar structure, the number and varieties of particles in each section have not been fully established, so the diagram is suggestive only. (Part A is based on Walker *et al.*, 1984; part B is adapted from data in Senior, 1973.)

distinguishes the organellar variety from the prokaryotic. In the former, F_0 is constructed of four subunits, in the latter of only three, named simply as a, b, and c. The actual base in bacteria consists of one a and 10–12 copies of c, the two b subunits providing part of the stalk (Fillingame, 1981; Walker *et al.*, 1984). The stoichiometry of F_1 in prokaryotes has similarly been determined, its five subunits being combined as three each of α and β and one each of γ, δ, and ϵ (Figure 6.1A). Among eukaryotic organelles F_1 likewise consists of five subunits, but the actual composition of the region remains unknown, as does the constitution of the stalk.

Gene Organization. The organization of the genes for this substance has been determined in three species of bacteria, *E. coli, Rhodospirillum rubrum,* and *Rhodopseudomonas blastica* (Tybulewicz *et al.*, 1984; Walker *et al.*, 1984; Falk *et al.*, 1985). In the first of these, the establishment of the complete nucleotide sequence demonstrates all eight genes for the F_0 and F_1 subunits to be combined into a single set, called the *unc* operon, arranged as in Figure 6.2A; these are preceded by a reading frame for an unknown protein referred to as unc1. All are transcribed as a single unit, but there is an additional promoter following the γ cistron that serves for the β and ϵ components. In the third bacterium, the F_0 and F_1 members form separate operons, those of the former unit remaining unexplored. The latter group is arranged as in *E. coli,* including the second promoter for β and ϵ. However, in the present species there is an open reading frame of opposite polarity that encodes an unidentified protein (Figure 6.2B; Tybulewicz *et al.*, 1984).

Although the genomic arrangement of most of these genes in the eukaryotes remains unknown, the greater part are encoded in nuclear DNA. Even the situation in the mitochondrion remains obscure, because, as in other instances mentioned before, premature generalizations have been made on insufficient data. Now the complete establishment of the mitochondrial DNA sequences of three mammals has demonstrated that only a single ATPase component (subunit 6) is encoded there (Anderson *et al.*, 1981, 1982; Bibb *et al.*, 1981), not two as previously believed. In the fungi two components of F_0 have been recognized (Borst and Grivell, 1978), but in yeast the situation is particularly confused.

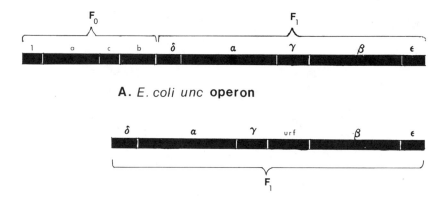

A. *E. coli* unc **operon**

B. *R. blastica atp* **operon**

Figure 6.2. Structure of ATPase operons in prokaryotes. (A) The ATPase operon (*unc*) contains the genes for all subunits. (B) In *Rhodopseudomonas blastica* an unidentified reading frame (*urf*) interrupts the ATPase gene sequence for the F_1 subunit. (Both based on Tybulewicz *et al.*, 1984.)

The gene *oli1* codes for subunit 9, a proteolipid (Macino and Tzagoloff, 1979), and *oli2* encodes subunit 6, both of which are components of F_0; in addition, *aap1* has been identified as the gene for subunit 8 (Macreadie *et al.*, 1983). Subunit 5, formerly believed to be represented in the organellar genome, has proven to be a contaminant there (Orian and Marzuki, 1981; Orian *et al.*, 1981). In addition, the sequence for a cistron proposed as being for either subunit 3 or 6 (Macino and Tzagoloff, 1980) is here identified as actually encoding the former, as shown later. To date the chloroplast ATP-synthase genes have been only slightly investigated, but all that have been investigated appear to be parts of the F_1 region, rather than the F_0 of the other organelle. The one that encodes subunit β seems to be firmly established as present here (Stern and Palmer, 1984).

6.5.2. The Structure of ATPase Genes

Except for three or so from bacterial sources, all the genes for ATPases have been from mitochondrial genomes. Among the eight of the latter origins given in Table 6.8, three are from the yeast organelle, one each from that cell part of *Neurospora* and *Drosophila*, and the remainder from mammals. Five of those have been identified as cistrons for subunit 6, while the three of *Saccharomyces* represent subunits 3, 8, and 9. Whereas all of the foregoing are components of the enzyme involved in the respiratory electron-transport chain, that of *E. coli*, the *kdpC* gene, encodes a subunit of an ATP-driven K^+-transport protein. Subunit 9 from maize mitochondria has now been sequenced (Dewey *et al.*, 1985).

Genes for Electron-Chain Subunits. Because as a whole lysine and arginine residues are not particularly abundant in the products of electron-chain ATP-synthase genes, the codons for both of these basic amino acids are shown underscored in Table 6.8. However, leucine is of outstanding importance in most of the subunits, and accordingly its codons are accentuated by being italicized. Those for phenylalanine, which amino acid is also of exceptional importance, are lightly stippled, and the triplets for tyrosine and tryptophan are heavily stippled. When the five sequences for mitochondrial subunit 6 (indicated by the subscript M6) are compared, much of the flavor of all the known cistrons that have been established is recognized, in that homologous regions are not especially prominent except in the three from the mammalian organelle. In length, the four from metazoan sources are identical (Anderson *et al.*, 1981, 1982; Bibb *et al.*, 1981; Clary and Wolstenholme, 1983), each consisting of 681 nucleotide residues, providing 227 codons. Moreover, they are alike in sharing a peculiar condition at their 3' ends, in that the final A of the translational termination signal (TAA) serves also as the first site of the ATG of the adjacent gene, which encodes subunit III of cytochrome *c* oxidase. The missing 3'-terminal A has been stated to be added by polyadenylation during processing, but the problems this simplistic explanation raises must remain for discussion in Chapter 12. Here the only point is that the members of this quartet are obviously homologs. Even the sequence from the *Neurospora* mitochondrion shows clear indications of kinship, especially in the not infrequent correspondences in location of such key codons as those for leucine, tyrosine, and the two basic amino acids. The low level of homology, along with the cistron's greater length (759 residues or 253 codons), indicate the relationship to be distant and that the enzyme's structure is not highly conserved evolutionarily.

The Three Genes from Yeast Mitochondria. Because the three ATPase subunits of yeast represent components of the active enzyme different from all the remainder, little

Table 6.8
Genes for Subunits of ATPases[a]

Row 1

Yeast$_{M_9}$ [c]	ATG	CAA	TTA	G	TA	TTA	GCA	GCT	AAA	TAT	ATT	GGA	GCA	GGT	ATC	TCA	ACA	ATT	GGT	TTA	TTA	GGA	GCA	GGT
Neurospora$_{M_6}$ [d]	ATG	TTT	AAT	A	TC	CTT	AGT	CCA	TTA	AAT	CAA	TTA	GAA	ATA	AGA	GAT	TTA	TTA	GAT	ATA	GAT	ACT	TTA	GGA
Drosophila$_{M_6}$	ATG	ATA	ACA	A	AT	TTA	TCT	GTA	TTT	GAC	CCT	TCA	GCA	ATT	TTT	AAT	TTA	TCA	TTA	AAT	TGA	TTA		AGA
Murine$_{M_6}$ [e]	ATG	AAC	GAA	A	AT	CTA	TTT	GCC	TCA	TTC	ATT	ACC	CCA	ACA	ATA	GGA	TTC	CCA	ATC	GTT	GTA	GCC	ATC	
Human$_{M_6}$ [f]	ATG	AAC	GAA	A	AT	CTG	TTC	GCT	TCA	TTC	ATT	GCC	CCC	ACA	ATC	CTA	GGC	CTA	CCC	GCA	GTA	CTG	ATC	
Bovine$_{M_6}$ [g]	ATG	AAC	GAA	A	AT	TTA	TTT	ACC	TCT	TTT	ATT	ACC	CCT	GTA	ATT	TTA	GGT	CTC	CCT	GTA	ACC	CTT	ATC	
E. coli kdpC [h]	ATG	AGT	GGA	T	TA	CGT	CCG	GCA	TTA	TCA	ACA	TTT	ATC	TTT	CTG	---	TTA	TTG	ATT	ACT	GGC	GGC	GTT	TAC
Yeast$_{M_3}$ [i]	ATG	TTT	AAT	T	TA	TTA	AAT	ACA	AAT	ACA	TCA	CCA	TTA	GAT	CAA	TTT	GAG	ATT	AGA	CTA	TTA	TTT	GGT	
Yeast$_{M_8}$ [j]	ATG	CCA	CAA	T	TA	GTT	CCA	TTA	TAT	TTT	ATG	AAT	CAA	TTA	ACA	TTA	TTT	GGT	TTC	TTA	TTA	ATG	ATT	CTA	TTA

Row 2

Yeast$_{M_9}$	ATT	GGT	ATT	GCT	ATC	GTA	TTC	GCA	---	GCT	TTA	ATT	AAT	GGT	GTA	TCA	AGA	AAC	CCA	TCA	ATT	AAA	GAC	CTA
Neurospora$_{M_6}$	AAT	TTA	CAC	ATT	TCT	ATA	ACT	AAT	---	ATT	GGT	TTT	TAT	TTA	ACA	ATA	GGA	GCT	TTC	TTT	TTC	TTG	GTT	ATA
Drosophila$_{M_6}$	ACA	TTT	TTA	GGA	CTT	TTA	ATA	ATT	---	CCT	TCA	ATT	TAT	TGA	TAT	TAT	ATA	CCT	TCT	CGT	TAT	AAT	ATT	TTT	TGA
Murine$_{M_6}$	ATT	ATA	TTT	CCT	TCA	ATC	CTA	TTC	TTC	---	CCA	TCC	TCA	AAA	CTA	TAT	CTC	CGT	CAT	TCT	TTC	CAA
Human$_{M_6}$	ATT	CTA	TTT	CCC	CCT	CTA	TTG	ATC	---	CCC	ACC	TCC	AAA	CTA	TAT	CTC	ATC	ACC	ACC	ACC	CAA
Bovine$_{M_6}$	GTA	CTA	TTC	CCA	AGC	CTA	CTA	TTC	---	CCA	ACA	TCA	AAC	CTA	GTA	AGC	AAT	CGC	TTT	GTA	ACC	CTC	CAA
E. coli kdpC	CCG	CTG	CTG	ACC	ACC	GTA	CTG	GGG	---	CAA	TGG	TGG	TTT	CCC	CAG	GCC	AAT	GGT	TCG	TTG	ATT	CGT	GAA
Yeast$_{M_3}$	---	TTA	CAA	TCA	TCA	TTT	ATT	GAT	TTA	---	AGT	TGT	TTA	---	AGT	ACA	ACA	TTA	TCA	TTA	TAT	ACT	ATT	ATT
Yeast$_{M_8}$	TTA	ATT	TTA	TTC	TCA	CAA	TTC	TTT	TTA	CCT	ATG	ATC	AGA	AGA	TTA	GTA	TCT	TAT	AGA	TTT	ATT	TCT	AAA

Row 3

Yeast_M9: GTA TTC CCT ATG GCT ATT TTT GGT TTC GCC TTA TCA GAA GCT ACA GGT TTA TTC TGT TTA ATG GTT TCA TAT

Neurospora_M6: AAT CTT TTA AGT ATA AAT TAT AAT AGA TTA GTT AAA GAA TTT AAA AGT CAA GAA TCT TTA TAT

Drosophila_M6: AAT TCA ATT TTA ACA CTT ATT ATC AAA CAA GAA TTT AAA CCT TCA GGT CAT AAT GGA TCT

Murine_M6: CAC TGA GTT AAA CTT ATT ATC AAA CAA ATA ATG CTA ATC CAC ACA CCA AAA GGA CGA ACA TGA ACC

Human_M6: CAA TGA ATC AAA CTA ACC TCA AAA CAA ATA ATG AGT ATA ATG CAC AAC CAC AAT TCT AAA GGA CGA ACC TGA TCT

Bovine_M6: CAA TGA ATA CTT CAA CTT GTA TCA AAA CAA ATA ATG AGT ATC CAC AAC AAT TCT AAA GGA CAA ACA TGA ACA

E. coli kdpC: GGT GAT ACG GTG CGC GGT TCG GCA TTA ATC GGG CAG AAT TTT ACC GGC AAC GGC GGC TAT TTT CAT GGT CGC

Yeast_M3: GTA TTA TTA GTT ATT ACA AGT TTA TAT CTA TTA ACT AAT AAT AAT ATT ATT GGT TCA AGA TGA

Yeast_M8: =TTA TAA

Row 4

Yeast_M9: TTC TTA TTA TTC GGT GTA TAA

Neurospora_M6: GCT ACT ATT TAT AGT ATA GTA ACA AGT CAA ATA AAT GGT CAA ATA TAC TTT CCA TTT ATT

Drosophila_M6: ACT TTT ATT TTT ATT TCT TTA TTT TCA TTA ATT TTT AAT AAT TTA ATA GGT TTA TTT CCT TAT ATT

Murine_M6: CTA ATA ATT GTT TCC CTA ATC ATA TTT ATT GGA TCA ACA AAT CTC CTA GGC CTT TTA CCA CAT ACA TTT

Human_M6: CTA ATA CTA GTA TCC TTA ATC ATT TTT ATT GCC ACA ACT AAC CTC CTC GGA CTC CTG CCT CAC TCA TTT

Bovine_M6: TTA ATA TTA ATA TCT CTG ATC CTA TTT ATT GGA TCA ACA AAC CTA CTA GGC CTA TTA CCC CAT TCA TTC

E. coli kdpC: CCG TCG GCA ACG GCA GAA ATG CCC TAT AAT CCA CAG GCT TCT GGC GGG AGC AAT CTG GCG GTC AGT AAC

Yeast_M3: TTA ATT TCA CAA GAA GCT ATT TAT GAT ACT ATT ATA AAT ATG CTT AAA GGA CAA ATT GGA GGT AAA AAT

(continued)

Table 6.8 (Continued)

Row 5

$Neurospora_{M_6}$	TAT	ACT	TTA	TTT	ATT	ATT	TTA	ATA	AAC	AAT	CTT	ATA	GGA	ATG	GT*	T	CCT	TAT	AGT	TTC	GCA	AGT	ACA
$Drosophila_{M_6}$	TTT	ACA	AGA	ACA	AGT	CAT	---	TTA	ACT	TTA	TCT	TTA	GCT	CT	T	CCT	TTA	TGA	TTA	TGT	TTT	ATA
$Murine_{M_6}$	ACA	CCT	ACT	ACC	CAA	CTA	TCC	ATA	AAT	CTA	AGT	ATA	GCC	ATT	CCA	CT	A	GCT	GGA	GCC	GTA	ATT	ACA
$Human_{M_6}$	ACA	CCA	ACC	ACC	CAA	CTA	TCT	ATA	AAC	CTA	GCC	ATG	GCC	ATC	CCC	TT	A	TGA	GCG	GGC	ACA	GTG	ATT	ATA
$Bovine_{M_6}$	ACA	CCA	CCA	ACA	CAA	CTA	TCA	ATA	AAC	CTA	GGC	ATA	GCC	ATC	CCC	CT	G	TGA	GCA	GGA	GCC	GTA	ATT	ACA
E. coli kdpC	CCT	GAG	CTG	GAT	AAA	CTA	ATA	GCC	GCA	CGC	GTT	GCT	GCA	TTA	CGG	GC	C	GCT	AAC	CCG	GAT	GCC	AGC	GCG
$Yeast_{M_3}$	TGA	GGT	TTA	TAT	TTC	CCT	ATG	ATC	TTT	ACA	TTA	ATG	TTT	ATT	TT	T	ATT	GCT	AAT	TTA	ATT	AGT	ATG

Row 6

$Neurospora_{M_6}$	AGC	CAT	TTT	GTA	GTG	ACA	TTT	GCT	CTT	AGT	TTC	ACT	ATA	GTT	TTA	GGA	GCA	ACT	ATT	TTA	GGT	TTC	CAA
$Drosophila_{M_6}$	TTA	TTC	GGT	TGA	ATT	AAT	CAT	ACA	CAA	CAT	ATA	TTT	GCT	CAC	TTA	CCT	CAA	GTA	CCT	ACA	CCT	GCA	ATT
$Murine_{M_6}$	GGC	TTC	CGA	CAC	AAA	CTA	AAA	AGC	TCA	CTT	GCC	CAC	TTC	CCA	CAA	GGA	ACT	CCA	ATT	TCA	CTA	ATT
$Human_{M_6}$	GGC	TTT	CGC	TCT	AAG	ATT	AAA	AAT	GCC	CTA	GCC	CAC	TTC	CCA	CAA	GGC	ACA	CCC	ACA	CCC	ATC
$Bovine_{M_6}$	GGA	TTC	CGC	AAT	AAA	ACT	AAA	GCA	TCA	CTT	GCC	CAT	TTC	CCA	CAA	GGA	ACA	CCC	ACT	CCA	ATC
E. coli kdpC	AGC	GTT	CCG	GTT	GAA	CTG	GTG	ACG	GCA	TCG	GCA	GGG	CTG	GTA	TTT	ATT	ATC	TCT	TTA	AGT	ATT	ATT	TGA	GCG	GCG
$Yeast_{M_3}$	ATT	CCA	TAT	TCA	TTT	GCA	TTT	TCA	GCT	CAT	TTA	GTA	TTT	ATT	ATC	TCT	TTA	AGT	ATT	AGT	TTA	TTA

Row 7

$Neurospora_{M_6}$	AAA	CAT	GGA	TTA	GAA	TTT	TTC	TCT	CTA	TTA	GTT	CCA	GCA	GGT	TGT	CCT	TTA	GCC	CTT	CTT	CCT	CTG	TTA
$Drosophila_{M_6}$	TTA	ATA	CCT	TTT	ATA	GTA	TGT	ATT	GAA	AAT	ATT	AGA	TTT	ATT	CAA	CCA	ATG	GCA	GTC	GCA	TTA	CCG	GGA	ACT	TTA	GCT	GTT	GTT	CGA
$Murine_{M_6}$	CCA	ATA	CTT	ATT	ATT	ATT	GAA	ACA	AGC	CTA	TTT	ATT	CAA	CCA	ATG	GCA	GTC	GCA	TTA	CGC	CTT	ACA

Human_M5   CCC ATA *CTA* GTT ATT ATC GAA ACC ATC AGC ATC *CTA* *CTC* ATT CAA CCA ATA GCC GCC GTA *CGC* *CTA* ACC

Bovine_M6   CCA ATA *CTA* GTA ATT ATT GAA ACT ATC AGC *CTT* TTT ATT CAA CCT ATA GCC *CTC* GCC GTG *CGG* *TTA* ACA

E. coli kdpC GCC TGG CAA ATC CCA *CGC* GTG GCG AAA GCG *CGT* AAT *CTC* AGC GTT GAA CAG *CTC* ACG CAA *CTG* ATC GCA

Yeast_M3   GGT AAT ACT ATT *TTA* GGT *TTA* **TAT** **AAA** CAT GGT **TGA** GTA TTC TTC *TTA* TCA *TTA* *TTA* GTA CCT GCT GGT ACA

Row 8

Neurospora_M6   GTT *TTA* ATT GAA **TTC** ATT TCT **TAT** *TTA* GCA **AGA** AAT ATA *TTA* GGA *TTA* **AGA** *TTA* GCA GCT AAC ATC

Drosophila_M6   *TTA* ACA GCT AAT ATA ATT GCT GGA CAT *CTT* *CTA* *TTA* ACC *TTA* *TTG* GGA AAT ACA GGA CCT TCT ATA TCT

Murine_M6   GCT AAC ATT ACT GCA GGA CAC *TTA* *TTA* ATA CAC *CTA* ACT *CTA* GTA *TTA* ATA AAT ATT

Human_M6   GCT AAC ATT ACT GCA GGC CAC *CTC* ATG CAC *CTA* ATT GGA AGC GCC ACC *CTA* GCA ATA TCA ACC ATT

Bovine_M6   GCT AAC ATC ACT GCA GGA CAC *CTA* *TTA* AAT CAC *CTA* ATC GGA GGA GCT ACA *CTT* GCA *CTA* ATA AGC ATT

E. coli kdpC **AAA** **TAC** AGC CAA CAA CCG *CTG* GTG AAA **TAT** ATC GGC CAG CCG GTT GTC AAC ATT GTT GAA *CTC* AAT *CTG*

Yeast_M3   CCA *TTA* CCA *TTA* GTA CCT *TTA* *TTA* GTT ATT ATG GAA ACT *TTA* TCT **TAT** ATT GCT **AGA** GCT ATT TCA *TTA*

Row 9

Neurospora_M6   *TTA* TCA GGT CAT ATG *TTG* *TTA* CAT ATT *TTA* GCA GGA TTT ACT **TAC** AAT ATA ATG ACA AGC GGT ATT ATC

Drosophila_M6   **TTC** *TTA* *CTA* GTA ACA TTT *TTA* *TTA* GTA GCC CAA ATT GCT *TTA* *TTA* --- GTT *TTA* GAA TCA GCT GTA ACT

Murine_M6   AGC CCA CCA ACA GCT ACC ATT ACA **TTT** ATT ATT *CTC* ACA ATT *CTA* GAA ATC GCA ATT GCA GTA GCA

Human_M6   AAC *CTT* CCC TCT ACA *CTT* ATC ATC TTC ACA ATT *CTA* *CTG* *CTA* ATT *CTA* GAA ATC GCT GTC GCC

Bovine_M6   AGC ACT ACA ACA GCT *CTA* ATT ACA TTC ACC ATT *CTA* ACA ATT *CTA* ACA ATT *CTA* GAG **TTT** GCA GTA GCT

E. coli kdpC GCG *CTG* GAT **AAA** *CTT* GAT GAA TAA

Yeast_M3   GGT *TTA* **AGA** *TTA* GGT --- TCT AAT ATC *TTA* GCT GGT CAT *TTA* *TTA* ATG GTT ATT *TTA* GCT GGT *TTA* *CTA*

(continued)

Table 6.8 (Continued)

Row 10

Neurospora M_6	TTC TTC TTT TTA GGT TTA ATA CCT TTA GCT TTT ATT ATA GCT TTC TCA GGA TTA GAG TTA GGA ATT GCC
Drosophila M_6	ATA ATT CAA TCC TAT GTA TTT GCT GTT TTA AGA ACT TTA TCT AGA GAA GTA AAT ACA TAA
Murine M_6	TTA ATT CAA GCC TAC GTA TTC ACC CTC CTA AGC CTA TAT CTA CAT GAT AAT ACA TAA
Human M_6	TTA ATC CAA GCC TAC GTT TTC ACA CTT CTA GTA AGC CTC TAC CTG CAC GAC AAC ACA TAA
Bovine M_6	ATA ATC CAA GCC TAT GTA TTC ACT CTC CTA GTC AGC CTA TAT CTG CAT GAC AAC ACA TAA
Yeast M_3	TTT AAT TTT ATG TTA ATT AAT TTA TTT ACT TTA GTA TTC GGT TTT GTA CCT TTA GCT ATG ATT TTA GCT

Row 11

Neurospora M_6	TTC ATC CAA GCT CAA GTT TTT GTA GTT TTA ACT AGC GGA TAC ATT AAA GAC GCA TTG GAT CTA CAT TAG
Yeast M_3	ATT ATG ATG ATT TTA GAA TTG GCT ATT GGT ATT ATC CAA TCT TTT GTT TGA CTT ATC TTA ACA GCA TCA TAC

Row 12

Yeast M_3	TTA AAA GAT ACA TTA TAC TTA CAT TAA

---

[a] Codons for both arginine and lysine are underscored, and those for leucine are italicized; those for tyrosine (TAT, TAC) and tryptophan (TGG, TGA) are lightly stippled and those for phenylalanine are heavily stippled. M, mitochondrial subunit. Asterisks indicate locations of introns, where established.

[b] Macino and Tzagoloff (1979); Edwards et al. (1983).

[c] Morelli and Macino (1984).

[d] Clary and Wolstenholme (1983).

[e] Bibb et al. (1981).

[f] Anderson et al. (1981).

[g] Anderson et al. (1982).

[h] Hesse et al. (1984).

[i] Macino and Tzagoloff (1980).

[j] Macreadie et al. (1983).

indication of kinship can be expected, nor is much to be found. Nevertheless, close scrutiny of the subunit 3 cistron (subscript M3) along with that of *Neurospora* for subunit 6 reveals some surprising similarities and identities that are not readily explained away. Moreover, its total length exceeds that of the fungal gene by only 27 nucleotide residues; hence, it is not beyond the pale of possibility that subunit 3 of the yeast will eventually be recognized as a homolog of subunit 6 of higher organisms.

An amazing number of identities in occupants of corresponding sites is likewise notable between subunits 3 and 8, although no attempt at homologization has been made. These resemblances particularly seem to involve leucine or tyrosine triplets, so that the gene for the smaller component may prove to have been derived from a fragment of that for the larger one. Furthermore, the coding sequence for subunit 9 displays some apparent degree of kinship with that for 8, but because of the latter's sequence here being also involved with subunit 3, in the table the two necessarily remain out of phase by one triplet. When properly aligned by insertions and deletions, it becomes apparent that this gene, too, may have arisen by fragmentation of one for a larger component, possibly that of subunit 3.

Bacterial ATPase Genes. Since the gene from *E. coli* is part of an enzyme of entirely different function from the mitochondrial species, comparisons with the other on a site-to-site basis are futile. The sequence given, *kdpC*, is provided in the table more to bring out contrasts than to suggest homologies. Actually it is the smallest of the three subunits of the active enzyme that have been sequenced (Hesse *et al.*, 1984)—B is 2049, A is 1674, and C is 573 nucleotide residues in length.

In addition, the five genes of the *atp* (*unc*) operons of *Rhodopseudomonas blastica* have had their nucleotide sequences established (Tybulewicz *et al.*, 1984), all of which are members of the F_1 complex. Since they thus afford no opportunity for comparison with the mitochondrial subunits, which form a part of F_0, none are provided in the table.

Chemical Properties of ATPases. As already suggested by the remarks in explanation of the table, the various components of the ATPases show some unique chemical properties that differ with the function of the enzyme. In terms of percentage composition on the basis of total numbers of codons of those in the table, the most heavily enriched with triplets for arginine and lysine are the mouse$_{M6}$ and *E. coli kdpC* sequences, each of which contains 11, for a 5% rate. The human$_{M6}$ and bovine$_{M6}$ are slightly lower in content, with totals of ten and eight for the same amino acids, respectively. Of those from yeast, the genes for subunits 8 and 9 have only one lysine and one arginine codon each, while that for subunit 3 contains only eight. In contrast, the F_1 subunits of *R. blastica*, not included in the table, have a higher content of both arginine and lysine, the subunit-α gene encoding a total of 24 of those amino acids, for a rate of 13%. Since the γ subunit also seems equally equipped, the F_1 components as a whole may prove to be more basic than the membrane-embedded F_0 constituents.

Of particular interest in illustrating the low level of evolutionary conservation displayed by these genes is the relative frequency of the codons for phenylalanine, TTT and TTC. Since on a random-chance basis, the rate of occurrence for the two triplets combined should approximate 3%, the products of all of those sequences in the table would have to be considered phenylalanine-rich, except for that of *E. coli*, which has a rate close to 2.5%. The five mitochondrial sequences for subunit 6 are especially informative, for their ratings range from a low of 4% in the human gene, through 6% in both the murine and bovine cistrons and 7% in that of the fruitfly, to 10% in the fungal gene. The most richly endowed is that for the small subunit 8, which shows ~13% codons for this substance. As a whole neither triplet is discriminated against, usage of each being fairly equal; the *Drosophila* gene is exceptional, however, for 17 TTTs occur and no TTCs.

Codon usage, as well as percent composition, is also of much interest with the six triplets that encode leucine. While on the basis of random-chance combination a mean of about 10% of a gene sequence should encode leucine, these cistrons for ATPases are strongly enriched in that respect. Almost all have a content of 16% or more, the *kdpC* gene of *E. coli* being the sole exception, with only 12%. Indeed, most run well above that level, the yeast$_{M3}$ and human, bovine, and *Drosophila* sequences for subunit 6 showing a composition of 20%, with the yeast$_{M8}$ being particularly rich, with a 25% ratio. In codon usage, there are strong biases in all sources studied. This is especially true of the three genes of yeast, in which 69 of their total of 76 are represented by the codon TTA, six by CTA, and one by CTT, the remaining ones, TTT, CTC, and CTG, not being employed at all. The *Neurospora*$_{M6}$ cistron shows a similar preference for TTA, 30 of its 42 leucine signals being that combination. CTT is its strong second preference, with a total of six, while TTG, CTA, and CTG account for the remainder. In the *Drosophila* sequence, the preferences are nearly identical, 35 of its 42 leucine triplets being TTA, with CTC and CTG receiving no usage. On the other hand, the mammalian trio shows trends completely opposite to these others, all displaying a marked preference for CTA, which comprises nearly 50% of the total in each case. TTA is next most abundant in the bovine (8 of 44), CTC in the human (7 of 44), and CTT in the murine (9 of 39); TTG finds use only once in the three sources combined. CTG varies greatly from one of these mammalian types to the other, being employed twice in the mouse, three times in the bovine, and at six points in the human sequence. These distinctions in codon employment levels would not be so difficult to comprehend if the genes were of nuclear origin, for they reflect overall trends that have been well established. But how can they be explained if the mitochondrion is considered a modified prokaryotic symbiont, as it often is thought to be?

6.6. RIBULOSE-1,5-BISPHOSPHATE CARBOXYLASE GENES

The present and following sections center on enzymes that are involved with the fixation of carbon dioxide on the one hand and nitrogen on the other. The net final result of the fixing of the first substance is the synthesis of sugar, and that of the second, the production of amino acids. Accordingly, the two enzyme systems together are responsible for the creation of all foodstuffs, not only for mankind, but for the entire living world. Neither of the enzyme classes carries out this heavy responsibility without assistance, for both sets draw upon ATP as the source of requisite energy. And the synthesis of that biochemical involves enzymes of many kinds.

The enzyme of concern here, ribulose-1,5-bisphosphate carboxylase/oxygenase, often abbreviated to "rubisco," is reputed to be the most abundant protein in existence (Miziorko and Lorimer, 1983). Its anabolic function of adding CO_2 to ribulose-1,5-bisphosphate, using energy from ATP to yield two molecules of 3-phosphoglyceric acid, is the reaction that leads to the conversion of the inorganic gas into a net gain in organic compounds. But it is a double-edged sword, for it also serves as catalyst in the oxygenation of rubilose-1,5-bisphosphate to produce one molecule each of 3-phosphoglyceric acid and 2-phosphoglycolic acid, the latter of which enters into the photorespiratory cycle ultimately to be oxidized to CO_2 and water.

6.6.1. Properties of the Genes

The constitution of the active ribulose-1,5-bisphosphate carboxylase protein varies extensively with the source, but in general it has the subunit structure $\alpha_8\beta_8$, the large subunit having a molecular weight around 56,000 and the small having molecular weight between 12,000 and 14,000 (Shinozaki et al., 1983). The enzyme of *Rhodospirillum rubrum* (Tabita and McFadden, 1974; Nargang et al., 1984), however, is a dimer of identical subunits, having a molecular weight resembling that of the large polypeptide of the others. Although this difference in constitution of the protein is quite striking, the distinction functionally is not, for only the large subunit contains a functional site and no activity is known to exist in the small subunit (Reeck and Teller, 1983).

The location of the genes for the two subunits has not been established in the blue-green algae; they appear to be parts of a single operon, although they are separated by a spacer of 545 base pairs in *Anabaena* (Curtis and Haselkorn, 1983; Shinozaki and Sugiura, 1983; Shinozaki et al., 1983; Nierzwicki-Bauer et al., 1984). In seed plants, *Euglena*, and green algae, the large subunit is encoded in the chloroplast genome, where it is present as a single copy in the DNA, but since the genome of the organelle is represented by several hundreds or thousands of copies, the total gene dosage per cell is extremely high (Chelm et al., 1977; Koller et al., 1984). The cistron is uninterrupted in the chloroplasts of seed plants and green algae (McIntosh et al., 1980; Zurawski et al., 1981; Dron et al., 1982), but it is broken by nine introns in *Euglena* (Koller et al., 1984). That for the small subunit is located in the nucleus of eukaryotes and is transcribed in the cytoplasm. Then, after transport into the chloroplast the transit peptide is removed during processing (Coruzzi et al., 1983); discussion of the latter feature is reserved for the next chapter.

6.6.2. Comparisons of the Gene Structures

In addition to the cistron sequences for four representatives of the large subunit and three of the small in Table 6.9, gene structures of several others of the major component are known, including those of *Chlamydomonas* and spinach (Zurawski et al., 1981; Dron et al., 1982), and a minor component from maize (Mazur and Chui, 1985), all of which closely resemble those that are cited. In addition, that of the unicellular blue-green alga *Synechococcus* PCC6301 has been established (Reichelt and Delaney, 1983); it is closely homologous to that of *Anacystis*. Since the structures of the two types of genes show virtually nothing in common, comparisons must be made separately.

The Genes for the Large Subunit. The sequences of the enzyme's large subunit from two species of blue-green algae, a bacterium, and a seed-plant chloroplast are remarkably similar in total length, differing by a maximum of ten codons. If allowance is made for deletions, the *Rhodospirillum* gene is found to be the shortest and that of *Anabaena* the longest, exceeding that of the maize by two codons. Except at the extreme 5' end, where variation frequently occurs in genes for many different substances, the two

Table 6.9

Mature Genes for Ribulose-1,5-Bisphosphate Carboxylase Large (L) and Small (S) Subunits[a]

Row 1

Anacystis L[b]	ATG	CCC	---	---	---	AAG	ACG	CAA	TCT	GCC	GCA	GGC	TAT	AAG	GCC	GGG	GTG	AAG	GAC	TAC	AAA CTC
Anabaena L[c]	ATG	TCT	TAC	GCT	CAA	ACG	AAG	ACT	CAG	ACA	AAA	TCT	GGG	TAT	AAA	GCC	GGG	GTT	CAA	GAT	TAC *AGA* CTA
Rhodospirillum[d]	ATG	GAC	---	---	---	CAG	TCA	TCT	*CGT*	TAC	GTC	AAT	CTG	GCG	CTC	AAG	AAG	GAG	GAT	ATC	GCC GGC GGC
*Maize*_C L[e]	ATG	TCA	---	CCA	GAA	ACA	ACT	AAA	GCA	AGT	GTT	GGA	TTT	AAA	GCT	GGT	GTT	AAG	GAT	TAT	AAA TTG
Anacystis S[f]	ATG	AGC	ATG	AAA	ACT	CTG	CCC	AAA	GAG	*CGT*	TTC	TCG	GAG	ACT	TTC	TCG	CCT	CCC	CTC	AGC	GAT
Anabaena S[g]	ATG	---	---	CAA	ACC	TTA	CCA	AAA	GAG	*CGT*	AAG	ACC	CTT	TCT	TCT	TCC	TTA	CCC	CTC	ACC	GAC
*Pea*_N S[h]	ATG*	CAG	GTG	TGG	CCT	CCA	ATT	GGA	AAG	AAG	TTT	GAG	ACT	CTT	TCC	TAT	TTG	CCA	CCA	TTG	ACC *AGA*

Row 2

Anacystis L	ACC	TAT	TAC	ACC	CCC	GAT	TAC	ACC	CCC	AAA	GAC	ACT	GAC	CTG	CTG	GCG	GCT	TTC	*CGC*	TTC	AGC	CCT CAG
Anabaena L	ACT	TAT	TAC	ACA	CCT	GAT	TAC	ACA	CCT	AAA	CCT	AAA	GAT	ACA	GAT	ATT	CTG	GCG	GCA	TTC	*CGT* GTT	ACA CCC CAG
Rhodospirillum L	GAG	CAT	GTG	CTT	TGT	GCC	TAT	ATC	ATG	AAG	CCC	AAG	GGA	TAT	GGC	TAT	GTG	GCG	ACC	GCG	GCG	CAT
*Maize*_C L	ACT	TAC	TAC	ACC	CCG	GAG	GAG	TAC	GAA	ACC	AAG	GAT	ACT	GAT	ATC	TTG	GCG	GCA	TTC	*CGA*	GTA	ACT CCA CAG
Anacystis S	*CGC*	CAA	ATC	GCT	GCA	ATC	GAG	TAC	ATG	ATC	GAG	CAA	GGC	TTC	CAC	CCC	TTG	ATC	GAG	TTC	AAC	GAG
Anabaena S	GTT	CAA	ATC	GAA	AAG	CAA	GTC	CAG	TAC	ATT	CTG	AGC	CAA	GGC	TAC	ATT	CCA	GCC	GTT	GAG	TTC	AAC GAA
*Pea*_N S	GAT	CAC	TTG	TTG	AAA	GAA	GTT	GAA	GTT	CTC	CTC	*AGA*	GGA	AAG	GGA	TGG	GTT	CCT	TGC	TTG	GAA	TTT GAG TTG

Row 3

Anacystis L	CCG	GGT	GTC	CCT	GCT	GAG	GAA	GCT	GGT	GCG	GCG	ATC	CCG	GCT	GAA	TCT	---	TCG	ACC	GGT	ACC	TCG ACC ACC
Anabaena L	CCC	GGA	GTT	CCC	TTT	GAG	GAA	GCG	GCT	GCG	GCT	GCA	GCG	GCT	GAG	TCT	---	TCT	ACT	GGT	ACT	TGG ACG ACC
Rhodospirillum L	TTC	GCC	GCC	GAG	AGT	TCG	ACG	GGC	ACG	AAC	GTC	GAG	GTC	TGC	ACC	ACC	---	GAC	GAT	TTC	ACC	*CCG* GGC GTC

Maize$_C$ L   CTC GGG GTT CCG CCT GAG GAA GCA GGG GCT GCA GTG GCT GCG GAA TCT TCT GCT GCT GGT ACA TGG ACA ACT

Anacystis S   CAC TCG AAT CCG --- --- --- --- --- --- --- --- --- --- --- --- --- GAA GAG TTC TAC TGG ACG ATG

Anabaena S   GTT TCT GAA CCT --- --- --- --- --- --- --- --- --- --- --- --- ACC GAA CTT TAT TGG ACA CTG

Pea$_N$ S   GAG* AAA GGA TTT GTG TAC GCT GAG CAC AAC AAG TCA CCA GGA TAC TAT --- GAT GGA AGA TAC TGG ACA ATG

Row 4

Anacystis L   GTG TGG ACC GAC TTG CTG ACC GAC ATG GAT CGG TAC AAA TAC GGC AAG TAC CAC ATC GAG CCG GTG CAA

Anabaena L   GTA TGG ACA GAC CTG TTA ACC GAT CTA GAT CGT TAC AAA GGT TGC GAT ATC GAA CCA GGT CCC

Rhodospirillum L   GAC GCC CTG GTC TAT GAG GTG GAC GCC CGC GAG CTG ACC AAG ATC GCC TAT CCG GTG GCT TTG TTC

Maize$_C$ L   GTT TGG ACT GAT GGA CTT ACC AGT CTT GAT CGT TAC AAA GGA CGA TGC TAT CAC ATC GAG CCC GTT CCT

Anacystis S   TGG AAG CTC CCC CTG TTT TTT GGT GCT AAG AGC CCT CAG CAA GTC CTC GAT GAA GTG CGT GAG TGC CGC AGC

Anabaena S   TGG AAG CTA CCT TTG TTT GCT AAA ACA TCC CGT GAA GTA TTG GCA GAA GTT CAA TCT TGC CGT TCT

Pea$_N$ S   TGG AAG CTT CCT ATG TTT TTT GGT ACC ACT GAT GCT TCT AAG AAG TTG AAG CTT GAT GAA GTT GTT GCC

Row 5

Anacystis L   GGC GAA GAG AAC TCC TAC TTT GCG TTC ATC GCT TAC CCG CTC GAC CTG TTT GAA GGG TCG GTC ACC

Anabaena L   GGC GAA GAC AAC CAA TCC ATT GCC TAC ATC GCT TAT CCT TTG GAT CTG TTT GAA GAA GGC TCC ATC ACC

Rhodospirillum L   GAC CGC AAC ATC ACC GAC GGC AAG ATG GCC ATC GCC AAG GCG CTC ACG TTC CCA TTA GAC CTA TTT GAA GAG ATG GGA AAC AAC CAG

Maize$_C$ L   GGG GAC CCA GAT CAA TAT ATC TGT GTA GCT TAT CGT GTT GTA GGA TTT GAC AAC ATC TTT GAA GAG GGT TCT GTT ACT

Anacystis S   GAA TAC GGT GAT TGC TAC ATC CGT GTC GCT GGC TTC GAC AAC AAG TGC CAA ACC GTG AGC TTC

Anabaena S   CAA TAT CCT CAC CAC TAC ATC CGT GTT GTA GGA TTT GAC AAC TGC CAA ATC CTG AGC TTC

Pea$_N$ S   GCT TAC CCC CAA GCT TTC GTT GTT CGT ATC ATC GGT GTT GAC AAC GTT CGT CAA TGC ATC AGT TTC

(continued)

Table 6.9 (Continued)

Row 6

Anacystis L	AAC ATC CTG ACC TCG ATC GTC GGT AAC GTG TTT GGC TTC AAA GCT ATC CGT TCG CTG CGT CTG GAA GAC
Anabaena L	AAC GTT TTG ACC ATT GTA GGT AAC GTA TTT GTT TTT AAA GCA TTA CGC GCA TTG CGT TTG GAA GAC
Rhodospirillum L	GGT ATG GGC GAC GTG GAA TAC GCC AAG GTG CCC GAG GCT TAT CGC GCC CTG TTT
Maize_C L	AAC ATG TTT ACT TCC ATT GTG GGT AAC GTA TTT GGT TTC AAA GCC CTA CGC GCT CTA CGT TTG GAG GAT
Anacystis S	ATC GTT CAT CGT CCC GGC CGC TAC TAA
Anabaena S	ATC GTT CAC AAA CCC AGC AGA TAC TAA
Pea_N S	ATT GCG CAC ACA CCA GAA TCC TAC TGA

Row 7

Anacystis L	ATC CGC TTC CCC GCC TTG GTC AAA ACC TTC CAA GGT CCT CCC CAC GGT ATC CAA GTC GAG CGC GAC
Anabaena L	ATT CGC TTT CCT GTT GCT TAC ATC AAG ACC TTC CAA GGC CCT CAC CAC GGT ATC CAA GTT GAG CGT GAC
Rhodospirillum L	GAT GGC CCG AGC GTC AAT ATC TCG GCC CTG TGG AAA GTG CTG GGG CGG CCC GAG GTC GAC --- --- ---
Maize_C L	CTA CGA ATT CCC CCT GCT TAT TCA AAA ACT TTC CAA GGT CCA CCA CGT ATG CAA GTT GAA AGG GAT

Row 8

Anacystis L	CTG CTG AAC AAG TAC GGC CGT CCG ATG CTG GGT TGC ACG ATC AAA CCA CTC GGT CTG TCG GCG AAA
Anabaena L	AAA TAA AAC AAA TAT GGC CGT CCT CTG TTG GGT TGT ACC ATC AAA CCA TTA GGT CTG TCT GCT AAG
Rhodospirillum L	--- --- --- --- --- --- GGC GGT CTG CTC GTC GGC ACG ATC ATC ATC AAG CCG AAG CTC GGC CTG CGT CCC AAG
Maize_C L	AAG TTG AAC AAG TAC GGT CGT CCT TTA TTG GGA TGT ACT ATT AAA CCA AAA TTG GGA TTA TCC GCA AAA

Row 9

```
Anacystis L        AAC TAC GGT CGT GCC GTC GAA TGT CTG CGC GGC GGT CTG GAC TTC ACC AAA GAC GAC GAA AAC ATC
Anabaena L         AAC TAC GGA CCC GCT GTA TAC GAG TGT TTG CGC GGT GGT TTG GAC TTC ACC AAA GAC GAC GAA AAC ATT
Rhodospirillum L   CCC TTC GCC GAG GCC TGC CAC GCC TTC TGG CTG GGC GGC -- -- GAC TTC ATC AAG AAC GAC GAG CCC CAG
Maize_C L          AAT TAC GGT AGA GCG TGT TAT TAT GAG TGT CTA GGT GGA CTT GAT TTT ATC ACC AAA GAT GAT GAA AAC GTA
```

Row 10

```
Anacystis L        AAC TCG CAG CCG TTC CAA CCC TGG CGC GAT CGC TTC CTG TTT GTG GCT GAT GCA ATC CAC AAA TCG CAA
Anabaena L         AAC TCC GCA CCA TTC CAA AGA TGG CGC GAT CGC TTC TTG TTT GTA GCT GAT GCC ATC ACC AAA GCA CAA
Rhodospirillum L   GGC AAT CAG CCC TTC GCC CCC TTG CGC GAC ACC ATC GCC CTG GTC TTT TGT GCC GAC ATG AGG CGG GCC CAG
Maize_C L          AAC TCA CAA TTT ATG CCG CGC TGG AGA GAC CGT TTC GTC TTT GCC GAA GCA ATT TAT AAA TCA CAA
```

Row 11

```
Anacystis L        GCA GAA ACC GGT GAA ATC AAA GGT CAC TAC CTG AAC GTG ACC GCG CCG ACC TGC GAA GAA ATG ATG AAA
Anabaena L         GCA GAA ACA GGC GAA ATC AAA GGT CAC TAC CTA AAC GTG ACC GCT CCT ACC TGT GAA GAA ATG CTA AAA
Rhodospirillum L   GAC GAG ACC GGC GAG GCC AAG CTG TTC TCG GCC AAT ATC ACC GCC GAC GAT CCC TTC GAG ATC ATC GCC
Maize_C L          GCC GAA ACT GGT GAA ATC AAG GGG CAT TAC TTG AAT GCG ACT GCA GGT ACA TGC GAT GAA ATG ATT AAG
```

Row 12

```
Anacystis L        CGG GCT GAG TTC GCT AAA GAA CTC GGC ATG CCG ATC ATC CAT GAC TTC TTG ACG GCT GGT TTC ACC
Anabaena L         CGG GCT GAG TAC GCT AAA GAA CTC AAA CAG CCC ATC ATC ATG CAC GAC TAC CTG ACC GCA GGT TTC ACA
Rhodospirillum L   CGT GGC GAG TAT GTG CTG GAG ACC TCG GCC AAC GCC TCG ATG CAT GAC TAC CTG GCC GTC CTC GGC GGC TAT
Maize_C L          GGA GCT GTA TTT GCA AGG CAA TTA GGG GTT CCT ATT GTA ATG CAT GAC TAT CTA GGT ACA GGT TTC ACC
```

(continued)

Table 6.9 (Continued)

Row 13

Anacystis L	GCC AAC ACC ACC TTG GCA AAA TGG TGC CGC GAC AAC GGC GTC CTG CTG CAC CTG CAC ATC CAC CGT GCA ATG CAC
Anabaena L	GCT AAC AAC ACC TTG GCT CGT TGG TGT CGT GAC AAC GGT CTT CTA CTG CAC ATC CAC CGC GCG ATG CAC
Rhodospirillum L	GTC CCC GGC GCC GCG ATC ACC ACG GCG CGC CGC CGC TTC CCC GAT AAC TTC TTG CAT TAT --- CAC
Maize$_C$ L	GCA AAT ACT ACT TTG TCT CAT TAT TCC CGC GAC AAC GGC CTA CTT --- CAC ATT CAC CGA GCA ATG CAT

Row 14

Anacystis L	GCG GTG ATC GAC CGT CAG CGT AAC CAC GGG ATT CAC TTC CGT GTC GTC TTG GCC AAG TGT TTG CGT CTG TCC
Anabaena L	GCA GTA ATC GAC CGT CAA AAG AAC CAC CAC TTC CGT GTA TTG GCT AAA GCC CTA CGT CTA TCT
Rhodospirillum L	CGG GCT GGC CAC GGC GTC ACC TCG CCC --- CAG TCC AAG CGC GGC TAC GCC TTC GTC CAT TGC
Maize$_C$ L	GCA GTT ATT GAT AGA CAG AAA AAA CAT GGT ATG CAT TTC CGT GTA TTA GCT AAA GCA TTG CGT ATG TCG

Row 15

Anacystis L	GGT GGT GAC CAC CTC CAC TCC GGC ACC GTC GTC GGC AAA CTG GAA GGC AAA GCT TCG ACC TTG GGC ---
Anabaena L	GGT GGT GAC GAC ATC CAC ATC CAC GGT ACC GTA GTA GGT AAA AAA TTG GAA GGT GAA CGC GGT ATC ACA ATG GGC ---
Rhodospirillum L	AAG ATG GCC CGC CTG CAG GGC GCC AGC GGC ATC CAC ACC GGC TTT GGC AAG ATG GAA GGC GAG
Maize$_C$ L	GGG GGA GAT CAT ATC CAC TCC GGT ACA GTA GTA GGT AAG TTA GAA GGG GAA CGC GAA ATA ACT TTA GGT ---

Row 16

Anacystis L	TTT GTT GAC TTG ATG CGC GAA GAC CAC ATC GAA GCT GAC CGC GGG GTC TTC TTC ACC CAA GAT
Anabaena L	TTC GTT GAC CTA CTA CGT GAA AAC TAC GTT GAG CAA GAC TCT CGC GGT ATT TAC TTT ACC CAA GAC

Rhodospirillum L TCC AGC GAC CGC GCC ATC GCC TAT ATG CTG ACC CAG GAC GAG GCC CAG GGG CCG TTC TAC CGT CAA TCC
Maize_C L   TTT GTT GAT TTA TTG CGC GAT GAT TTT ATT GAA AAA GAT CGT TCT CGC GGT ATC TTT TTC ACT CAG GAC

Row 17

Anacystis L   TGG GCG TCG ATG CCG GGC GTG CTG CCG GTT GCT TCC GGT ATC CAC GTG TGG CAC ATG CCC GCA CTG
Anabaena L   TGG GCT TCT CTA CCT GGT GTA ATG GCA GTT GCT GCT TCC GGT GGT ATC CAC GTA TGG CAT ATG CCA GCG TTG
Rhodospirillum L TGG GGC GGC ATG AAG GCT TGT ACG CCG ATC ATC AGC GGC ATG AAC GCC CTG CGC ATG CCC GGC TTC
Maize_C L   TGG GTA TCC ATG CCA GGT GTT ATA CCG GTG GCT TCT GGG GCT ATT CAT GTT TGG CAT ATG CCA GCT CTG

Row 18

Anacystis L   GTG GAA ATC TTC GGT GAT GAC TCC GTT CTC CAG TTC GGT GGC GGC --- ACC TTG GGT CAC CCC TGG GGT
Anabaena L   GTA GAA ATC TTC GGT GAT GAC TTC GTA CTA CAA TTC GGT GGT GGT --- ACA CTC GGA CAC CCT TGG GGT
Rhodospirillum L TTC GAG AAC CTG GGT AAT GCC AAT GTC ATC TTG ACC GCC GGC GGC CCC TTC GGC CAT ATC GAC GGC
Maize_C L   ACC GAA ATC CTT GGA GAT GAT TCA GTA GTA TTA CAA TTT GGT GGA GGA --- ACT TTA GGA CAT CCT TGG GGA

Row 19

Anacystis L   AAT GCT CCT GGT GCA ACC GCG AAC CGT GTT GCC TTG GAA GCT TGC GTC CAA GCT CGG AAC GAA GGT CGC
Anabaena L   AAC GCT CGT GGT GCA ACC GCT AAC CGT GTA GCT TTG GAA GCT TTG GTC CAA GCA GTC AAC GAA GGT CGT
Rhodospirillum L CCG GTG GCC GGG GGG CGG TCG TTG CGT CAA GCC TGG --- --- --- CGG GAC GGG GTT CCG
Maize_C L   AAT GCA CAT GGT GCA GCA GCT AAT CGT GTA CAA GCC TGT GTA CAA GCT CGT AAC GAA GGG CGC

(continued)

CHAPTER 6

Table 6.9 (Continued)

Row 20

Anacystis L	GAC CTC TAC *CGT* GAA GGC GGC GAC ATC CTT *CGT* GAA GCT GGC AAG TGG TCG CCT GAA CTG GCT GCT GCC
Anabaena L	AAC TTG GCT *CGT* GAA GGT AAC GAC GTT ATC *CGT* GAA GCT GCT AAG TGG TCT CCT GAA TTG GCT GTC GCT
Rhodospirillum L	GTT CTG GAC TAT GCC *CGC* GAG CAC AAG GAA CTG GCC *CGC* GCC TTC GAG TCC TTC CCC GGC GAC GCC GAC
Maize$_C$ L	GAT CTA GCT *CGC* GAG GTA CAA ATT ATC AAA GCA GCT TGC AAA TGG AGT GCC --- GAA CTA GCC GCA GCT

Row 21

Anacystis L	CTC GAC CTC TGG AAA GAG ATC AAG TTC GAA TTC GAA ACG ATG GAC AAG CTC TAA
Anabaena L	TGC GAA CTG TGG AAA GAA ATC AAG TTC GAG TTT GAG GCA ATG GAT ACC GTC TGA TAA
Rhodospirillum L	CAG ATC TAT CCG GGC TGG *CGC* AAG GCC CTG GGC GTC GAG --- GAC ACC *CGC* AGC GCC CTT CCG GCG TAA
Maize$_C$ L	TGC GAA ATA TGG AAG GAG ATC AAA TTC GAT GGT TTC AAA GCG ATG GAT ACC ATA TAA

[a] Codons for arginine are italicized, those for lysine are underscored, and those for tyrosine are stippled. Asterisk indicates location of intron. C, Chloroplast, N, nucleus.
[b] Shinozaki *et al.* (1983).
[c] Curtis and Haselkorn (1983).
[d] Reeck and Teller (1983).
[e] McIntosh *et al.* (1980).
[f] Shinozaki and Sugiura (1983).
[g] Nierzwicki-Bauer *et al.* (1984).
[h] Coruzzi *et al.* (1983).

from blue-green algae are largely homologous and are readily aligned. The region from row 5 to the 3' terminus is especially well conserved evolutionarily, except that an arginine codon in some genes replaces one for lysine in others at frequent points. As a whole, the cistron from the maize chloroplast is obviously related to the two of the cyanophyceans, but the kinship is far more distant than it is between the latter. In general a distinct trend in the former toward greater homology with the *Anabaena* sequence is manifest, but this is a tendency only, not a fixed condition. Interestingly, the chloroplast gene employs AGG or AGA for arginine, whereas in the two of blue-green algal sources the CG- series is uniformly used. The gene of the tobacco chloroplast that is not given in the table has these same traits, since it is 90% homologous to the others from this organelle (Shinozaki and Sugiura, 1982).

Since the bacterial enzyme has such a different quaternary structure from the others, its nucleotide sequences, too, might be suspected to diverge strongly, and indeed examination of the table proves this to be the case. Although similarities at corresponding points are frequent, out and out homologous sites are sparse. However, such codons as those for the two basic amino acids are consistently similarly located and conserved, as are those for tyrosine. Hence, the bacterial gene may be considered a subtype of the gene, although not a homolog.

The Genes for the Small Subunit. Because of the less essential physiological role of their encoded product, the genes of the small subunit have been reputed to be more widely divergent than those for the large component (Zurawski *et al.*, 1981). Scrutiny of the three given in Table 6.9, however, reveals that the diversification is not as great as suspected. Although all three display frequent differences, they also show many points of identity or kinship, so that aligning the several examples is fairly straightforward. The greatest point of departure from the two blue-green examples (Shinozaki and Sugiura, 1983; Nierzwicki-Bauer *et al.*, 1984) is the presence in the chloroplast representative of a 36-nucleotide-residue insert at about the middle of the gene (row 3), which codes for a very basic and tyrosine-rich segment.

6.7. GENES FOR NITROGEN-FIXATION ENZYMES

The fixation of atmospheric nitrogen to produce ammonia is a complex process (Shanmugam *et al.*, 1978) requiring the products of 17 genes in *Klebsiella pneumoniae* (Ow and Ausubel, 1983), the only organism for which a detailed genetic map of the cistrons has been constructed. However, by far the greater part of this total is involved in electron-transport activation of the enzymes, regulation of the genes, and formation of the prosthetic group (Roberts and Brill, 1981). The substances concerned directly with the processes are two enzymes, dinitrogenase and dinitrogenase reductase. Consequently, attention here centers nearly exclusively on the three genes that encode these two molecules, except in the discussions of gene organization and initiation of transcription. The dinitrogenase product, also known as "component 1" or the "MoFe protein," is an $\alpha_2\beta_2$ tetramer, the α subunit having a molecular weight of 56,000 and the β having a molecular weight of 59,000. The former is a product of the gene *nifD* and the latter is a product of *nifK*. The dinitrogenase reductase, referred to also as "component 2" or the "Fe" protein, is a dimer of two identical subunits of ~35,000 daltons, which is encoded by *nifH*

(Lammers and Haselkorn, 1983; Kaluza and Hennecke, 1984). Its function is to supply electrons to the dinitrogenase, which in turn employs them to reduce N_2 to NH_3.

The prosthetic groups are complex, especially in the dinitrogenase. There are at least three different metallic ion centers, including 30–32 iron atoms. Of these 12–16 are contained in two copies of a molybdenum–iron complex ($MoFe_{6-8}S_{8/9}$); most of the remainder are in Fe_4S_4 clusters, and two others possibly in a poorly understood Fe_2S_2 unit. In addition, the reductase contains one Fe_4S_4 cluster; all of these are attached to the proteins by way of cysteine ligands.

6.7.1. The Organization of Nitrogen-Fixation Genes

Only in *Klebsiella pneumoniae* have the details of organization been fully elaborated, so this organism must be employed as a model for comparisons of others. In this pathogenic bacterium many facets of control of the operon have also been determined (Ausubel, 1984), but since the main points have already been outlined (Chapter 1, Section 1.4.1), discussion is confined to the genomic arrangement of these genes. The 17 genes in the operon are arranged into seven or eight transcriptional units, all of which, except *nifF*, are of the same polarity (Figure 6.3A; Beynon *et al.*, 1983). The reason for doubt about the number of transcriptional units is because that known as *nifM* may possibly be transcribed separately from *nifUSV*, at least under certain conditions. It is to be noted that the genes that encode the products that are directly active in nitrogen fixation form one of the transcriptional units *nifKDHY*.

This arrangement together of the nitrogenase-reductase components seems to be of widespread occurrence among the prokaryotes, but the remainder of the operon appears to be scattered. In *Rhizobium meliloti*, the symbiont of alfalfa root nodules, for example, *nifHDK* forms a separate operon; although *nifA* is known to be present and to serve a similar function as in *K. pneumoniae*, it is located elsewhere in the genome (Ow and Ausubel, 1983; Szeto *et al.*, 1984). In the present species of *Rhizobium*, the *nifHDK* operon is located on a plasmid, but among other species or strains, that arrangement does not exist (Haugland and Verma, 1981). Indeed, in *R. japonicum nifH* forms an operon separate and distinct from *nifDK* (Figure 6.3B) (Fuhrmann and Hennecke, 1984; Kaluza and Hennecke, 1984).

Among the blue-green algae, the gene arrangement varies with the genus. In many filamentous types, including *Anabaena*, nitrogen fixation under aerobic conditions proceeds in specialized cells called heterocysts (Mevarech *et al.*, 1980). Here, the genes, though seemingly individually transcribed, are somewhat clustered but loosely so, *nifH* being followed 115 nucleotides downstream by *nifD*. Then *nifK* lies about 11,000 base pairs farther downstream of the latter (Lammers and Haselkorn, 1983). The heterocyst-producing *Nostoc muscorum* appears also to have this arrangement (Singh and Singh, 1981), as do the members of another filamentous genus, *Calothrix*, even though they usually lack heterocysts and accordingly can fix nitrogen only under anaerobic conditions (Wyatt *et al.*, 1973). In contrast, many unicellular genera, including *Gloeothece*, do not produce heterocysts, but nevertheless carry out nitrogen fixation in the presence of air (Kallas *et al.*, 1983). In these forms, the genes have been shown to be clustered, with *nifK* heading the group, so that the sequence reads *nifK, nifD, nifH* (Figure 6.3C). Some evidence indicates the latter two to be cotranscribed. Unfortunately, nothing is known of

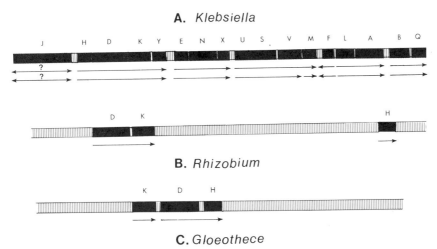

A. *Klebsiella*

B. *Rhizobium*

C. *Gloeothece*

Figure 6.3. Operons of nitrogen-fixation genes. (A) The most compact arrangement characterizes the *nif* operon of *Klebsiella;* direction of transcription has not been fully determined for the *J* and *M* genes. (Based on Beynon *et al.,* 1983.) (B) Two operons involved are located on a single plasmid in *Rhizobium japonicum.* (Based on Weinman *et al.,* 1984.) (C) In certain blue-green algae, the genes are clustered as here in *Gloeothece.* (Based on Kallas *et al.,* 1983.)

the nitrogen-fixation gene structures of the important bacterial genus *Clostridium,* although the amino acid sequences of their products have been established (Hase *et al.,* 1984).

6.7.2. Initiation of Transcription of Nitrogen-Fixation Genes

The transcriptional initiation processes have received a fair amount of attention in the literature, but as a whole the data are unevenly distributed, because the studies pertain almost exclusively to two bacterial genera, *Klebsiella* and *Rhizobium.* The processes in *Clostridium* have been completely neglected, as they have in the blue-green algae. Although leader sequences of some length have been established from *nif* genes of *Anabaena,* no possible signals comparable to the rest can be noted in the two representatives given in Table 6.10, for not even a start site has been experimentally determined.

Quite in contrast is the situation with the leaders of the two genera of bacteria just mentioned. In the table all seven from this pair of sources can be noted to have the starting sites of transcription (underscored) established. As is immediately apparent, no nucleotide is especially favored in this capacity, since all four common residues are employed to approximately equal extent. Upstream from these at nearly equal distances lie the dual elements (italicized) of the promoter. Like other sequences that function in promotion, the precise structure varies rather widely from one gene to another. Of the six sites in the 5′ portion, four are invariant, while the first two are almost always YT, the only exception being the AG of *Klebsiella nifLA* (Drummond *et al.,* 1983). But since those two nonconforming nucleotides are nevertheless part of the established sequence, this element must be perceived to consist of NNGGYR.

Table 6.10
Leaders of Nitrogen-Fixing (nif) Genes[a]

Anabaena D[b]	CAC	TCTCCC	TTCCCG	ACTCCT	CACTCT	CCCAAA	TATACT	TCTATTCCC	CCATTCGTA	AGAGTCACT	GAGGCAGAT
Rhizobium japonicum DK[c]	ACC	CTGGCA	TGCCGG	*TTGCAA*	AGTCTT	GGATCA	AGAAGC	CGCCCTCCC	AACAGCTAA	CCTTTTAAA	GGACACCAG
R. japonicum H[d]	ACC	TTGGCA	CGGCTG	*TTGCTG*	ATAAGC	GGCAGC	AACAC	(+143)			
R. parasponiae H[e]	GTG	TTGGCA	TGCCGA	*TTGCTG*	TTGAGT	TGCAGC	AACAC	(+142)			
R. meliloti HDK[f]	CGG	CTGGCA	CCACT-	*TTGCAC*	GATCAG	CCCTG-	GCGCG	(+62)			
Klebsiella pneumoniae HDKY[g]	CGG	CTGGTA	TGTTCC	*CTGCAC*	TTCTCT	GCTGGC	AAACA-	---------	-----CTCA	ACAACAGGA	GAAGTCACC
K. pneumoniae F[h]	AAC	CTGGCA	CAGCCT	*TCGCAA*	TACCCC	TGCGAG	AACGC	(+35)			
K. pneumoniae LA[i]	ATA	AGGGCG	CACGGT	*TTGCAT*	GGTTAT	CACCGT	TCGA	(+65)			
Anabaena K[j]	GAG	GGGGAG	TGAACC	TCCCAG	GCTATC	CTCACT	CATCAC	TTACAAACC	AACCAGCAA	GCGTAGAGA	GATACAACA

[a]Underscored nucleotides are transcription start sites where experimentally established; italicized regions are conserved in bacterial sequences.
[b]Lammers and Haselkorn (1983).
[c]Kaluza and Hennecke (1984).
[d]Fuhrmann and Hennecke (1984).
[e]Scott et al. (1983).
[f]Török and Kondorosi (1981).
[g]Scott et al. (1981); Bitoun et al. (1983).
[h]Beynon et al. (1983).
[i]Drummond et al. (1983).
[j]Mazur and Chu (1982).

The downstream portion of five nucleotides is obviously more uniform structurally than the foregoing, with the central GC combination being invariant insofar as these seven genes are concerned. When expressed to cover all the deviations shown in the table, the present element reads YYGC^A_T, the pyrimidines in both instances being nearly uniformly T. The location of this possible promoter is relatively stable, ranging from ~40 to ~150 base pairs before the translational initiation site. If the *Anabaena nifK* sequence is examined, some slight agreement can be noted in those sites that correspond to the 5' element, particularly in the two central Gs. Moreover, in the region corresponding to the downstream portion some degree of homology also is noticeable. Whether these resemblances hold any real significance or are merely of happenstance origin remains for the results of investigations into promotion in the blue-green algal nitrogen-fixation genes. Before leaving the subject, it is pertinent to observe that in the 300-residue leader of a *nifB*-like gene of *Rhizobium* (Rossen *et al.*, 1984), nothing comparable to this bipartite promoter can be found.

6.7.3. The Mature Nitrogen-Fixation Genes

It is fortunate that at least one representative gene for each of the major nitrogen-fixation peptides has had its nucleotide sequence established and that two of them are represented by three or four such analyses (Table 6.11). In the case of *Anabaena*, structures of *nifD*, *nifK*, and *nifH* have now been determined, so the trio can be compared free of influence from source differences. Recently established gene sequences that are not included in the table are those of the dehydrogenase from *Desulfovibrio vulgaris* (Voordouw and Brenner, 1985) and *nifK* from *Rhizobium japonicum* (Thöny *et al.*, 1985).

The Gene Sequences of nifD and nifK. Except at the very 5' end, where inconstancies often exist between the same gene from different source species, the mature coding sequences of *nifD* from the blue-green alga and the two bacteria are readily arranged to align homologous sites. Even a cursory examination of the structures reveals the three to be close allies. Surprisingly, kinship between the two bacterial types is not clearly greater than the relationship displayed by either with the cyanophycean.

Disparities often exist in the placement of the codons for the strongly basic amino acids lysine and arginine, especially in the cistron from *Klebsiella*, where they are particularly frequent in the beginning rows. In the first row, the *Anabaena* sequence has an additional codon for lysine, while the other two have one for arginine (but at different points) that is lacking in the blue-green alga. In the second row of the *Klebsiella* representative three triplets for lysine found in the alga are absent, while *Rhizobium* lacks one. In its place the latter has a codon for arginine and *Klebsiella* two, but again their locations are different in each case. Comparable differences in the *Klebsiella* gene are also obvious features in rows 3, 7, and 8, beyond which point its incompleteness prevents further comparisons. However, in regard to the codons for these two amino acids, *Rhizobium* is outstandingly more closely related to *Anabaena* than it is to the other bacterium.

Because of the involvement of cysteine with secondary structure and attachment of the prosthetic group in the encoded product, codons (heavy stippling) specifying that amino acid are usually regarded as being highly conserved evolutionarily, nor are the present genes exceptional in this respect. As a whole the three sequences agree in their placement throughout the available lengths, except at one point in row 3. There the triplet

Table 6.11
Mature Coding Sequences for Nitrogen-Fixing Proteins[a]

Row 1

Anabaena D[b]
ATG ACA CCT --- --- CCT GAA AAC AAG AAT TTT GTA GAT GAA AAT --- --- --- AAG GAA CTT ATT CAA

Rhizobium japonicum D[c]
ATG AGT CTC --- --- GCC ACG ACC AAC AGC GTC GCA GAA ATC AGG GCT CGC AAC AAA GAG CTG ATC GAG

Klebsiella pneumoniae D[d]
ATG --- --- --- --- --- ATG ACC AAC GCA ACG GGC GAA CGT --- --- --- GCG CTG ATC CAG

Anabaena K[e]
ATG CCT CAG --- --- AAT CCA GAA ACT GTA GAC CAC GTT --- GAT CTA TTC --- AAA --- --- --- CAG

Anabaena H[f]
ATG ACT GAC GAA AAC ATT AGA CAG ATA GCT TTC TAC GGT AAA GGC GGT ATC GGT AAA TCT ACC ACC TCC

Rhizobium meliloti H1[g]
ATG GCA GCT CTG --- --- CGT CAG ATC GCG TTC TAC GGT AAG GGG GGT ATC GGC AAG TCC ACG ACC TCC

Rhizobium japonicum H2[h]
ATG GCT TCA CTA --- --- AGA CAA ATC GCC TTC TAC GGG AAG GGC GGA ATC GGC AAG TCC ACC ACT TCG

Klebsiella pneumoniae H[h,i]
ATG ACC ATG --- --- --- CGT CAA TGC GCT ATT TAC GGT AAA GGC GGT ATC GGT AAA TCC ACC ACC ACG

Row 2

Anabaena D
GAA GTT CTG AAA GCT TAT CCC GAA AAA TCT CGC --- AAA AAG CGC GAA AAA CAC --- --- CTC AAC GTC

Rhizobium D
GAG GTG CTG AAG GTC TAT CCG GAG GAA ACC GCG --- AAA AGG CGT GCC AAG CAC --- --- CTC AAC GTG

Klebsiella D
GAA GTC CTG GAG GTG TTC CCG GAA --- ACC GCG CGA AAA GAG CGC AGA AAG CAC ATG CAC ATG GTC AGC GAT ---

Anabaena K
CCA GAA TAC ACC GAG CTA TTT GAA AAC AAG AGA --- AAG AAC TTT GAA GGC GCT --- --- CAT CCT --- ---

Anabaena H
CAA AAC ACC CTT GCA GCT ATG GCA GAA ATG GGT CAA CGC ATC ATG ATT GTA GGT --- --- TGC GAC --- ---

Rhizobium H1
CAA AAT ACA CTC GCC GCG CTT GTC GAC CTG GGG CAA AAG ATC CTT ATT GTA GGC --- --- TGC GAT --- ---

Rhizobium H2
CAG AAC ACG CTA GCG GCG CTG GCA GAG ATG GGT CAG AAG ATC ATC CTG ATT GTA GGG --- --- TGC GAT --- ---

Klebsiella H
CAG AAC CTC GTC GCC GCC CTG GCG CTG GGG GAG ATG GGT AAG AAA ATC ATG ATC GTC GGC --- --- TGC GAT ---

Row 3

```
Anabaena D      CAT GAA GAA AAC AAG TCT GAT TGC GGC GTT AAG TCT AAC ATC AAA TCC GTT CCT GGT GTA ATG ACC GCC
Rhizobium D     CAT CAA GCA GGT AAG TCG GAC TGC GGG GTG AAG TCC AAC ATC AAA TCC ATA CCC GGC GTG ATG ACG ATA
Klebsiella D    CCT AAA ATG AGC GTC GGC AAG TGC ATT ATC TCT AAC CGC AAA TCA CAA CCC GTA ATG ACC GTA
Anabaena K      CCT GAA GAA GTT GAA AGA GTG TCT GAA TGG --- --- ACA --- AAA TCT TGG GAC TAC CGG GAA AAG AAC
Anabaena H      CCT AAA GCT GAC TCC ACC CGT CTG ATG CTT --- --- --- CAC TCC AAA GCT CAA ACC GTA CTA CAC TTA
Rhizobium H1    CCG AAA GCG GAC TCC ACG CGC CTC ATC CTG --- --- --- AAC GCA AAG GCA CAG CAG ACC GTA CTG CAT CTT
Rhizobium H2    CCG AAA GCG GAC TCG ACT CGC CTT ATT CTG --- --- --- CAC GCC AAG GCT CAA GAC ACG ATT TTG AGT CTT
Klebsiella H    CCG AAG GCG GAC TCC ACC CGT CTG ATT CTG --- --- --- CAC GCC AAA GCA GAC AAC ACC ATT ATG GAC ATG
```

Row 4

```
Anabaena D      CCT GGT TGT GCT TAT GCA GGT TCT AAG GGT GTT TGG GGT CCT ATT AAG GAC ATG ATC CAC ATC AGC
Rhizobium D     AGA GGG TGC GCC TAT GCA GGG TCG AAG GGG GTG GTC TGG GGA CCA ATC AAG GAC ATG GTT CAT ATT AGC
Klebsiella D    CGC GGC TGC GCC TAC GCC GGT TCC GGT AAA GGG GTG GTA TTT GGG CCG ATT AAG GAC ATG GCC CAT ATT TCG
Anabaena K      TTC GCT CGT GAA --- --- --- GCT TTA ACC GTT AAC CCT GCT AAA GGT TGC CAA CCT GTA GGC GCG ATG TTC
Anabaena H      GCT GCT GAA CGC --- --- --- GGT GCA GTA GAA GAC CTC CAC GAA CTC ATG TTG ACC GGT TTC CGT
Rhizobium H1    GCG GCA ACC GAA --- --- --- GGT TCC CTC GAA GAT CTC GAG GAC CTC CTC AAA GTG GGT CTC TAC AGA
Rhizobium H2    GCG GCG AGC GCC --- --- --- GGC AGC GTG GAG GTG GAG CTC GAG CTC GAT GTA ATG AAG GTT GGC TAC CAG
Klebsiella H    GCC GCG GAA GTC --- --- --- GGC TCG GTC GAG GTC GAA CTC GAG CTC CTG CAA ATT GGC TAC GGC
```

(continued)

Table 6.11 (Continued)

Row 5

```
Anabaena D    CAC GGG CCT GTA GGT TGC GGT TAC TGG TCT TGG TCT GGT CGT AAC TAC TAC GTT GGT GTA ACT GGT
Rhizobium D   CAT GGC CCG GTT GGC TGC GGC CAA TAT TCA TGG GGC TCG CGG CGC AAC TAT TAC GTT GGC ACC ACG GGC
Klebsiella D  CAC GGA CCG GCT GGC TGC GGC CAG TAT TCC CGC GCC GAA CGA CGC AAC TAC TAC ACC GGA GTC AGC GGC
Anabaena K    GCT GCT TTG GGT TTT GAA GGT ACT CCT TTC GTA CCT CTA CAA GGT TCT GTT GCT TTC CGT
Anabaena H    GGC GTT AAG TGC GTA GAA TCT GGT GGT CCA GAA CCC GGT GTA GGT TGC GCC GGT CGT GGT ATC ATC ACC
Rhizobium H1  GGC ATC AAG TGC GTG GAG TCC GGT GGC GGC CCG GGC GTC TGC GCC GGA CGC GGC GTT ATC ACC
Rhizobium H2  GAC ATT CGC TGC GTT GAG TCC GGT GGC CCT GAG CCA GGT GTC TGC GCC GGC CGC GGT GTC ATC ACC
Klebsiella H  GAT GTG CGC TGC GCG GAA TCC GGC GGC GGG CCA GGC GTC TGC GCG GGA CGC GGC GTG ATC ACG
```

Row 6

```
Anabaena D    ATC AAC TCT TTC GGT ACC ATG CAC TTC ACC TCA GAC TTC CAA GAA CGT GAC ATC GTG TTC GGT GGT GAC
Rhizobium D   ATC GAT AGC TTC GTG ACT CTG CAG TTC ACC TCC GAC TCT GAT TTT CAG GAG GAA AAG GAT ATC GTA TTT GGC GGC GAC
Klebsiella D  GTC GAT AGC TTC GGC ACG CTG AAC TTC ACC TCT GAT TTT CAG GAG CGC GAC ATC GTC TTC GGC GGC GAT
Anabaena K    AGA CAC CTC AGC CGT CAC TAC AAA GAG CCT TGC --- TCC GCA GTA TCT TCT TCC ATG ACA GAA GAT
Anabaena H    GCC ATT AAC TTC TTA GAA GAA AAC GGC GCT TAC --- --- CAA GAC CTA GAC TTC GTA TCC TAC GAC GTA
Rhizobium H1  TCG ATC AAC TTC CTG GAA GAG AAC GGC GCT TAC --- --- AAC GAT GTC GAT ATT GAC TAC GTC TCA TAC GAC GTG
Rhizobium H2  TCG ATC AAT TTT CTT GAA GAG GCC GGA GCC TAC GAG --- AAC --- --- ATT GAC TAT GTT TCT TAC GAT GTG
Klebsiella H  GCG ATC AAC TTT CTT GAA GAA GGA GGC GCC TAC CAG --- --- GAC GAT CTC GAT CTC GTG TTC TTC TAT GAC GTG
```

Row 7

Anabaena D	AAA	AAA	CTC	ACT	AAA	CTC	ATT	GAA	GAA	---	CTC	GAT	GTT	CTT	TTC	CCT	CTC	AAC	CGT	GGT	GTT	TCC ATT
Rhizobium D	AAG	AAA	CTG	GAC	AAA	ATC	CTT	GAT	GAA	---	ATC	CAA	GAG	CTG	TTT	CCA	CTC	AAC	AAC	GGC	ATT	ACG ATA
Klebsiella D	AAA	AAG	CTC	AGC	AAG	CTG	ATT	GAA	GAG	---	ATG	GAG	TTG	CTG	TTC	CCG	CTC	ACC	AAA	GGG	ATC	ACC ATT
Anabaena K	---	---	GCA	GCA	GTA	TTC	GGT	GGT	TTG	AAC	AAC	ATG	ATC	---	GAA	GGT	ATG	---	CAG	GTT	TCA	TAC CAA
Anabaena H	TTG	GGT	GAC	GTT	GTA	TGT	GGT	GGT	TTC	GCT	ATG	CCT	ATC	ATC	CGT	GAA	GGT	AAA	GCA	CAA	GAA	ATC TAC ATC
Rhizobium H₁	CTA	GGG	GAC	GTA	GTA	TGC	GGC	GGC	TTT	GGC	ATG	CCT	ATT	CGC	GAA	AAC	AAG	GCT	CAG	GAA	ATC	TAC ATC
Rhizobium E₂	CTT	GGC	GAC	GTT	GTT	TGC	GGT	GGC	TTT	GCG	ATG	CCA	ATC	CGC	GAA	AAC	AAG	GCG	CAG	GAG	ATC	TAC ATC
Klebsiella H	CTC	GGC	GAC	GTG	GTC	TGC	GGC	GGC	TTC	GCC	ATG	CCG	ATC	CCG	CGC	GAA	AAC	AAA	GCC	CAG	GAG	ATC TAC ATC

Row 8

Anabaena D	CAA	TCT	GAA	TGT	CCC	ATT	GGA	TCT	ATT	GGG	GAT	GAC	ATC	GAA	GCT	GTT	GCT	AAG	AAA	ACT	TCT	AAG CAA
Rhizobium D	CAA	TCA	GAG	TGC	CCG	GTA	GGC	TTG	ATC	GGT	GAC	GAT	ATC	GAG	GCG	GTG	TCA	AGG	GCG	AAA	TCC	AAA GAA
Klebsiella D	CAG	TCG	GAA	TGC	CCG	GGG	CTG	ATC	GGT	GAT	GAT	ATC	AGC	GCG	GTG	GCC	AAC	GCC	AGC	AGC	AAG	---
Anabaena K	CTG	TAC	AAG	CCT	AAG	ATG	ATT	GCT	GTT	TGC	---	ACC	ACC	TGT	ATG	GCG	GAA	GTT	ATC	GGA	GAT	GAC TTG
Anabaena H	GTT	ACC	TCT	GGT	GAA	ATG	GCG	ATG	TAT	GCT	GCT	AAC	AAC	ATC	GCT	CGC	GGT	ATT	TTG	AAA	TAT	GCT
Rhizobium E₁	GTC	ATG	TCC	GGT	GAG	ATG	ATG	CTC	TAT	GCC	GCC	AAC	AAC	ATC	GCG	AAG	GGT	ATC	CTG	AAG	TAC	GCC
Rhizobium E₂	GTG	ATG	TCT	GGT	GAA	ATG	GCA	ATG	TAT	GCC	GCA	AAC	AAT	ATT	TCC	GGG	ATC	CTG	AAA	TAC	GCG
Klebsiella H	GTC	TGC	TCC	GGC	GAA	ATG	ATG	GCG	ATG	TAC	GCG	GCC	AAC	AAT	ATC	TCC	AAA	GGG	ATC	GTT	AAA	TAC GCC

(continued)

Table 6.11 (Continued)

Row 9

Anabaena D	ATT --- GGT <u>AAG</u> CCT GTT GTA CCC TTA *CGT* **TGC** GAA GGT TTC *CGT* GGT GTG TCT CAG TCC TTA GGA CAC
Rhizobium D	**TAT** GGA GGC <u>AAG</u> ACC ATC GTC CCG GTC *CGT* **TGT** GAG GGC TTT *CGG* GGT GTG TCG CAG TCA CTA GGC CAT
Klebsiella D	GCG CTG GAT <u>AAA</u> CCG GTG ATC CCG GTA *CGC* **TGC** GAA GGC TTT *CGC* GGC GTG TCG CAG TCT CTG GGG CAC
Anabaena K	GGC GCG TTC ATC ACC AAC TCC <u>AAG</u> AAC GCT GGT TCT ATT CCT CAA GAT TTC CCC GTA CCC TTT GCT CAC
Anabaena H	CAC TCC GGT GTA *CGT* TTA GGT GGT TTG ATC **TGT** AAC AGC *CGT* <u>AAG</u> GTT GAC *CGT* GAA GAC GAG TTA
*Rhizobium H*₁	CAT GCG GGC GGC GTG *CGG* CTG GGG GGG TTG ATT **TGC** AAC GAG *CGC* CAC ACC GAT *CGG* GAG CTC GAC CTC
*Rhizobium H*₂	AAC TCA GGT GGG GTG *CGG* TTG GGC GGC CTG ATC **TGC** AAC GAG *CGG* CAG ACC <u>AAG</u> GAC GAC *CGG* AAG GAA TTG GAA CTG
Klebsiella H	<u>AAA</u> TCC GGC <u>AAG</u> GTG *CGC* CTC GGC GGC CTG ATC **TGT** AAC TCA *CGT* CAG ACC GAC ACC GAC *CGT* GAA GAC GAA CTG

Row 10

Anabaena D	CAC ATC GCT AAC GAC GCT ATC *CGT* GAC TGG ATT TTC CCA GAA **TAC** GAC AAG CTC <u>AAG</u> <u>AAG</u> AAA GAA ACC *AGA*
Rhizobium D	CAC ATT GCA AAC GAT GCG GTA *CGC* GAT TGG ATT TTC GGG CAT ATC GAG GCC GCC GAG GGC <u>AAA</u> --- CCA <u>AAG</u>
Klebsiella D	CAT ATC GCC AAC GAC GTG GTG *CGC* GAC TGG GAC TGG ATC C-- (Incomplete)
Anabaena K	ACA CCT --- AGC TTT GTT GGT TCC CAC ATC ACT GGC **TAC** GAC AAC ATG ATG <u>AAG</u> GGT ATT CTG TCT AAC
Anabaena H	ATC ATG --- AAC TTG GCT GAA *CGT* TTG AAC ACC CAA ATG CAC TTC GTA CCT *CGT* GAC AAC ATC GTT
*Rhizobium H*₁	GCC GAG --- GCA CTT CCC GCC *CGC* CTC AAT TCC <u>AAG</u> CTC CAC CTC GTC CCG CCG CCG *CGC* GAC AAT ATC GTT
*Rhizobium H*₂	GCG GAA --- GCG TTG GCC <u>AAG</u> CTT GGC ACT CAA CTG ATC **TAC** TTC GTG CCG CCG *CGT* GAC AAT GTG GTG
Klebsiella H	ATT ATT --- GCC CTG GCG GAA <u>AAG</u> CTC GGC ACT CAA CTG ATC **TAC** TTC GTG CCG CCC *CGT* GAC AAC ATC GTG

Row 11

Anabaena D	CTT	GAC	TTC	GAG	CCA	AGC	CCC	TAT	GAT	GTA	GCT	CTA	ATC	GGT	GAC	TAC	AAC	ATC	GGT	GGT	GAC	GCT	TGG
Rhizobium D	---	---	TTC	GAG	CCG	ACA	CCA	TAC	GAT	GTT	GCG	ATC	ATC	GGA	GAC	TAC	AAT	ATC	GGC	GGC	GAT	GCT	TGG
Anabaena K	TTG	ACA	GAA	GGT	AAG	AAG	AAA	GCT	ACC	AGC	---	---	AAC	GGC	AAA	ATT	AAC	TTC	ATT	CCT	---	GGT	TTT
Anabaena H	CAA	CAC	GCA	GAA	TTG	CGT	CGT	ATG	ACC	GTT	---	---	AAC	GAG	TAC	GCA	CCA	GAC	AGC	AAC	CAA	GGT	
Rhizobium H₁	CAG	CAC	GCA	GAG	CTC	AGA	AAG	ATG	ACA	GTG	---	---	ATC	CAA	TAT	GCG	CCG	AAC	TCT	AAG	CAA	GCC	
Rhizobium H₂	CAG	CAT	GCA	GAG	CTG	CGT	CGC	ATG	ACG	GTG	---	---	CTT	GAA	TAT	GCA	CCC	GAT	TCC	AAG	CAG	GCT	
Klebsiella H	CAG	CGC	GCG	GAG	ATC	CGC	CGC	ATG	ACG	GTT	---	---	ATC	GAG	TAC	GAC	CCC	GCC	TGT	AAA	CAG	GCC	

Row 12

Anabaena D	GCC	AGC	CGG	ATG	CTG	TTG	GAA	GAA	ATG	GGC	TTA	CGT	GTA	GTA	GCT	CAG	TGG	TCT	---	GGT	GAT	GGT	ACA	CTC
Rhizobium D	TCA	TCG	CGA	ATT	CTG	CTT	GAA	GAG	ATG	GGA	CTA	CGG	GTA	ATC	GCG	CAG	CAG	TGG	TCC	---	GGC	GAC	GGT	TCA	CTG
Anabaena K	GAT	ACC	TAT	GTA	GGT	AAC	AAC	GCC	GAA	TTG	---	AAG	CGC	ATG	ATG	GGT	GTA	---	ATG	GGT	GTT	GAC	TAC	ACC
Anabaena H	CAA	GAG	TAC	CGC	GCA	TTA	GCT	AAG	AAG	ATC	---	AAC	AAC	---	GAC	AAG	CTC	ACC	ATT	CCT	ACA	CCA	ATG	GAA
Rhizobium H₁	GGG	GAA	TAT	CGC	GCC	CTG	GCT	GAA	AAG	ATC	CAT	GCA	AAT	TCC	GGC	GGC	AGC	ACC	GTC	CCT	ACA	CCG	ATC	ACT
Rhizobium H₂	GAT	CAC	TAT	CGG	AAA	CTA	GCG	GCC	AAG	GTT	CAC	AAT	AAT	GGC	GGC	AAG	GGC	ATC	ATT	CCG	ACC	CCG	ATC	TCA
Klebsiella H	AAC	GAA	TAC	CGC	ACC	CTG	GCG	GCG	CAG	AAG	GTC	AAC	AAC	ATG	GTG	GTG	---	CCG	ACG	CCC	TGC	ACC		

(continued)

Table 6.11 (Continued)

Row 13

Anabaena D   AAC GAG TTG ATC CAA GGC CCT GCT AAG TTA GTC CTC ATC CAC TGC TAC CGT TCT ATG AAC TAC ATC

Rhizobium D   GCC GAG CTC GAA GCA ACG CCG AAG GCA CTC AAC ATT CTG CAT TGC TAC CGT TCC ATG AAC TAT ATC

Anabaena K   ATC CTG TCT GAC AGC AGC GAC TAC TTT GAT TCA CCT AAC ATG GGT GAA TAC GAA ATG CCA AGT GGT

Anabaena H   ATG GAT GAA CTA GAA GCT CTG AAG ATC GAA TAC GGT CTA TTA --- GAC GAC ACC AAG CAC TCT GAA

Rhizobium H₁   ATG GAG GAA CTG GAG GAC ATG CTC GAC TTT GGA ATC ATG AAG AAG AGC GAC GAG CAG ATG CTT GCC GAA

Rhizobium H₂   ATG GAT GAG CTC GAG GAC ATG CTG ATG GAG CAT GGC ATT ATA AAG GCC CTG GAT GAA TCA ATC ATC GGC

Klebsiella H   ATG GAT GAG CTG GAA TCG CTG ATG GAG TTC GGC ATC ATG ATG GAA GAG GAA GAC ACC AGC ATC ATT GGC

Row 14

Anabaena D   TGC CGT TTG GAA GAA CAA TAC GGT ATG CCT TGG ATG GAG TTC AAC TTC TTC GGC CCC ACC AAG ATT

Rhizobium D   TCA CGC CAC ATG GAA GAG AAG TTC GGC ATC CCT TGG TGC GAG TAC AAC TTC TTC GGA CCT TCA AAG ATC

Anabaena K   ACA AAG CTG GAA GAT GCG GCT TCT ATC AAC GCT AAA GCA ACT GTT GCT CTC CAA GCT TAC ACC ACA

Anabaena H   ATC ATC GGT AAG CCC GGA GAA GCT ACC AAT AGG TCA TGC CGT --- AAT TAG

Rhizobium H₁   CTC CAC GCC AAG GAA GCC AAG GTA ATA GCC CCC CAC TGA

Rhizobium H₂   AAA ACC GCC GCC GAA CTC GCA GCC TCG TAA

Klebsiella H   AAA ACC GCC GCC GAA GAA AAC GCG GCC TGA

Row 15

Anabaena D	GCT GCT TCT TTA CGT GAA ATC GCA GCT AAG TTT GAT TCT AAG CAA GAA AAC GCT GAG AAG GTA ATT
Rhizobium D	GCG GAC TCA CTG CGC AGG ATT GCG GGT TAT TTT GAC GAC AAG ATC AAG GAA GGC GCC GAG CGA GTG ATC
Anabaena K	CCT AAG ACC --- CTC GAA TAC ATC AAA ACC CAG TGG AAG CAA GAA ACA CAA GTA TTG CGC CCC TTC GGT

Row 16

Anabaena D	GCT AAG TAC ACA CCA GTA ATG AAT GCT GTA CTA GAT AAA TAC CGC CCT CGC TTG GAA GGT AAC ACC GTA
Rhizobium D	GAG AAG TAT CAG CCG CTG GTG GAC GCC GTG ATT GCA AAA TAT CGC CCG CGC CTC GAG GGC AAG ACG GTG
Anabaena K	GTT AAG GGT ACT GAC GAG TTC TTG ACT GCT GTT TCT GAA TTG ACC GGT AAA GCT ATT CCT GAA GAA TTG

Row 17

Anabaena D	ATG TTG TAC GTA GGT GGT CTA CGT CCT CCT CTT CCC GCT TTT GAA GAC CTG GGT ATC AAA GTA
Rhizobium D	ATG CTG TAC GTC GGC GGC CTT CGT CCG CGG CAT GTG ATT GGC GCG TAC GAG GAC CTC GGG ATG GAC GTC
Anabaena K	GAA ATC GAA CGC GGT CGT TTA GTT GAT GCT ATC ACT GAC TCC TAC GCT TGG ATT CAT GGT AAG AAG TTC

Row 18

Anabaena D	GTT GGT ACT GGC TAT GAA TTC GCT CAC AAT GAC GAT TAC AAA CGT ACC ACC CAC TAC ATC GAT AAC GCC
Rhizobium D	ATT GGC ACT GGC TAC GAG GAG TTC GGT CAC AAC GAC GAC TAT CAG CGC ACA GCT CAG CAC GTG AAG GAC
Anabaena K	GCT ATC TAC GGC GAT GAT TTG ATC ATC ACC AGC TTC TTG TTA GAA ATG GGT GCT GAA CCA

Row 19

Anabaena D	--- ACC ATC ATT TAC GAT GAC GTT ACC GCC TAC GAA TTT GAA GAG TTC GTA AAA GCT AAA AAG CCT GAT
Rhizobium D	AGC ACC CTC ATC TAT GAT GAC GTC AAT GGC TAT GAG TTC GTC CAG CGC TTC GTC AGA CTC CAG CCT GAT
Anabaena K	GTA CAC ATC CTC TGC AAC AAC GGT GAT GAC ACC --- TTC AAG AAA GAA ATG GAA GCT ATC CTC GCT GCT

(continued)

Table 6.11 (Continued)

Row 20

Anabaena D	TTA ATT GCT TCT GGT ATT AAA GAG AAG AAG TAC GTC TTC --- CAA AAG ATG GGT CTT CCC TTC *CGT* CAA ATG CAC
Rhizobium D	CTT GTC GGC TCA GGC ATC AAG GAA AAG TAC GTT TTC --- CAA AAG ATG AGT GTG CCG TTC *CGG* CAG ATG CAT
Anabaena K	AGC CCA --- TTT GGT --- AAA GAA GCT AAA GTC TGG ATT CAA AAA GAC TTG TGG CAC TTC *CGT* TCC TTG TTG

Row 21

Anabaena D	TCT TGG GAT TAC TCC GAA CTT GGC GAC GGG GTG CAG --- --- --- ATG --- --- TCA GAT --- --- GAG
Rhizobium D	TCG TGG GAC TAT TCG GGT CCA TAT CAC GGT TAT GAC GGC TTT GCG ATC TTC GCG *CCC* GAC ATG GAC ATG
Anabaena K	TTC ACC GAG CCT GTA GAC TTC TTC ATC GGT AAC TCC TAC GGT AAG TAC CTG TGG *CCC* GAT ACC AGC ATC

Row 22

Anabaena D	GTA *AGG* TTT TTT TGT GAG GGG *AGA* AAA --- AAG AGT CTA TTT TTA --- GCC TAA
Rhizobium D	GCC GTC AAC TCG CCA ATT TGG AAA *AGA* ACG ACG AAA GCT CCC TGG AAG GAC GCC GAG *CGC* CAA GAC TCC *AGG*
Anabaena K	CCA ATG *CGG* ATT GGT TAT CCT CTC TTC GAT *CGC* CAC CAC TTA CAC *CGC* TAT TCT ACC CTC GGC TAC

Row 23

Rhizobium D   CTG CAG AAT AAC GCA ACT *CGT* CTC GCT CTC *CGG* GAA AGC CCG GGG ATT CCA ATC TGA

Anabaena K   CAA GGT GGT CTA AAT ATC CTC AAC TGG GTT GTT AAC ACC CTG TTG GAT GAA ATG GAT *CGC* AGC ACC AAC

Row 24

Anabaena K   ATC ACT GGT <u>AAG</u> ACC GAT ATC TCC TTT GAC TTG ATC *CGC* TAG

aCodons for arginine are italicized, those for lysine are underscored, those for tyrosine are lightly stippled, and those for cysteine are heavily stippled.
bLammers and Haselkorn (1983).
cKaluza and Hennecke (1984).
dScott et al. (1981).
eMazur and Chui (1982).
fMevarech et al. (1980).
gTörök and Kondorosi (1981).
hFuhrmann and Hennecke (1984).
iSundaresen and Ausubel (1981).

encoding cysteine in *Klebsiella* is displaced downstream one site relative to the others. Similarly with the occurrence of codons for tyrosine (light stippling), agreement among the three representatives of *nifD* is nearly universal, but again there is one exceptional point. In this case the first triplet for the amino acid in row 5 of the *Anabaena* sequence is discordant, being situated upstream one site from the others.

The relationships of *nifK* to either *nifD* or *nifH* are not readily discernible. In rows 3 and 5 an occasional tendency toward kinship with the *Anabaena nifD* gene is displayed, but beginning with row 7, the leaning is toward the *nifH* sequences, if anything. All in all, however, this cistron would have to be declared to represent a lineage separate from either of the other types. It is far longer than any of the *nifH* group and even slightly exceeds the *nifD* cistron of *Rhizobium* in length.

The nifH Genes. The two *nifH* genes from *Rhizobium* and the single ones from *Anabaena* and *Klebsiella* are more uniform in structure than are the foregoing, particularly in regard to the codons for charged and other important amino acids. Not infrequently, corresponding sites of all four sequences are identical. Nevertheless, there are two marked inclinations, the first and most pronounced being for the *Anabaena* representative to differ from the others. The second and lesser tendency is for the *Rhizobium nifH$_1$* cistron to correspond more closely to that of the blue-green alga and for the *nifH$_2$* to resemble the *Klebsiella* gene, so that *Rhizobium* may be considered to occupy a position intermediate to the others. While through row 3 the codons for charged amino acids typically are aligned with those of *nifD* and *nifK,* beyond that location, not a single instance of such correspondence is to be noted. This observation holds equally for the triplets for tyrosine, there being one exception near the middle of row 11, and not a single instance of homology among cysteine codons appears to exist. In view of the importance of this amino acid in the attachment of the prosthetic group, this lack of homology is telling evidence that the genes for the three types of peptides did not have recent common origins.

6.7.4. Termination among Nitrogen-Fixing Protein Genes

The processes of terminating transcription of the nitrogen-fixing protein genes have received no attention in any of the source organisms, so that experimental evidence is completely lacking. Furthermore, as already observed, in many cases the genes are part of a transcriptional unit whose components are transcribed together, and in other instances the train sequences that have been established are too short to expose the possible presence of recognizable terminators. Nevertheless, the small number of trains of seeming suitable length that exist in the literature are presented in Table 6.12.

Scanning of that list reveals only one fairly typical terminator, the series of five Ts (italicized) in the train of the *nifD* gene of *Anabaena*. The 3'-flanking regions of the *nifK* and *nifH* cistrons from the same source show remarkably little homology with the first one, in addition to lacking any suggestion of a recognizable terminator. The only other structure that shows promise of possible employment in termination is a region of dyad symmetry in the train of *Rhizobium nifH.* As is immediately apparent, the stem-and-loop is not accompanied downstream by a series of Ts as in typical bacterial terminators, nor does it possess such other of those features as the noncanonical base pairs and an excess of sites on one side of the stem regions that frequently characterize rho-dependent structures. Consequently, the significance to termination of this sector of dyad symmetry is not evident at this time.

Table 6.12

Trains of Nitrogen-Fixing Protein Genes[a]

Anabaena D[b]	ATACAGGTT	GTAGGGTTATC	TGG	GAACAATACA	*TTTTT*GCGGCTT	AAGACAATATAA	CTGTTATTGGGT
Anabaena K[c]	AAATTAATG	CAGCGTGCCAT	TGA	AAGGTAGAAC	TTAGGACTGGG	GATTGGGTATTG	GGTACTAGGAAT
Anabaena H[d]	GAGACACGG	AGACAGGAGAT	GAG	GAGCAATTCC	TCTTCCCACTCT	CCCTTCCCGACT	CCTCACTCTCCC
R. meliloti H[e]	CGCCGCCGA	GAGGGTGGCGC	AGC	TGGACGCGGC	GTGCCATTCCAC	AAACGCGGCCAT	TGATGAGGTGTC
R. japonicum H[f]	AGGCCGCGG	GTCGCCGCCTT	GCG	AAGGCGGCGGA	CGATGCCGGTCT	CCCTCACCCCCC	TTCCCGGGGACC
Klebsiella pneumoniae H[g]	GCACAGGAC	AATTAT					

[a]Arrows indicate regions of dyad symmetry.
[b]Lammers and Haselkorn (1983).
[c]Mazur and Chui (1982).
[d]Mevarech et al. (1980).
[e]Török and Kondorosi (1981).
[f]Fuhrmann and Hennecke (1984).
[g]Scott et al. (1981); Sundaresan and Ausubel (1981).

6.8. AN ENIGMA AND A SUGGESTION

Innate to the foregoing discussion of the genes for nitrogen-fixation proteins is the existence in modern seed plants of an unexplainable condition, the evolutionary value of which is difficult to perceive. However, there is now a methodology in molecular biology sufficient to correct the situation insofar as plants of particular economic value are concerned.

The Evolutionary Enigma. It is evident that many species through a broad spectrum of the Prokaryota have the capacity to fix free nitrogen under various conditions. Even the familiar *E. coli* has the ability to reduce enteric nitrogen to ammonia, and *Klebsiella pneumoniae* carries out the same reaction with the atmospheric element in the host lungs. Hence, this chemical capability is confined to neither photosynthetic nor nonphotosynthetic types, but exists in organisms of a wide diversity of habitats, including freshwater blue-green algae, soil inhabitants, parasites, and free-living forms. Indeed, the greatest portion of the nitrogen atoms in the biosphere of today ultimately has been derived through these fixative processes, for the nitrogen cycle of organic matter involves loss to the atmosphere, through the action of denitrifying bacteria, as well as gain (Ausubel and Cannon, 1981).

Since by one means or another the eukaryotes are the phylogenetic derivatives of prokaryotes, the ability to carry out this fundamental process evidently was not inherited from the ancestral stock along with the remainder of the genome. And therein lies the evolutionary enigma. Why were the genes for nitrogen-fixing substances lost, when the capability to use the free element holds such obvious survival value under low nitrate or nitrite conditions in the environment?

Other questions arise. If the chloroplast is a direct descendant of an invading blue-green alga, as has often been proposed, why were not the genes for nitrogen fixation retained along with the rest of the genome? As pointed out on preceding occasions, this absence appears to corroborate the belief that the organelle did not have its origin by way of an endosymbiont. It is much more readily perceived that the genes for the photosynthetic apparatus were once part of a plasmid that ultimately was incorporated into internal membranes of the cell to produce the chloroplast. Most plasmids whose molecular structure has been explored consist of groups of genes for functionally related products; consequently, as the primitive thylakoid system became membrane-enclosed in the evolving cell, the plasmid(s) encoding genes for photosynthesis became incorporated into the primordial chloroplast, while those carrying unrelated genes, such as those for nitrogen fixation, were not. Although the deficiency in the organelle may thus be explained, that of the eukaryotic cell cannot and may always remain an unsolved problem of evolutionary biology.

A Suggestion. From a human standpoint, the absence in cultivated plants of the capacity to produce their own ammonia and nitrates from atmospheric sources holds great economic significance. In all nations, but especially in less developed countries, the production of agricultural products would be vastly improved if the dependence upon fertilizers could be diminished. Obviously labor and costs would be greatly decreased, factors that in depressed lands certainly must limit the needed level of production.

But biochemists and molecular biologists now have the technology to overcome nature's shortcomings. Probes for the nitrogen-fixation genes have been developed, so that the cloning and transplantation of a *nif* operon from an organism like *Klebsiella* into cells of economically important plants should lie well within the realm of possibility. Perhaps first attention should be concentrated on maize, rice, and other grain crops, but potatoes, yams, tomatoes, and fiber plants, along with endless numbers of others, could also be improved by the use of genetic engineering along these lines.

7

Complex Genes

Thus far in this analysis, only those genes have been examined that simply encode a single product, whether an RNA or a protein. Although often the coding sequences have been found to be interrupted by untranslated sectors, the presence of such introns had no effect on the immediate product of translation, because of their removal before that process occurred. But now the point has been reached where the dictum "one gene, one peptide" loses its validity, for in the several major classes of genes that receive attention in this chapter two or more distinct proteins are encoded in every case. Each of these products must undergo processing before the main component (or components) is able to function. The simpler of the diverse complex genes is that class, earlier named diplomorphic (Chapter 1, Section 1.1.3), that codes for a double translational product. As a rule, but not always, the bulk of the transcript becomes translated into an active enzymic or structural protein; in addition, this bears a prefatory peptide that appears to be requisite for the protein to pass through a membrane. The latter may be either the cytoplasmic covering or the sheath that encloses an organelle, such as a mitochondrion, chloroplast, Golgi body, or endoreticulum.

As a result of the analyses of Chapters 2–6, full gene sequences that encode the entire mature product have been surveyed in depth sufficient to provide an adequate understanding of their nature, unique features, and interrelatedness. Consequently, attention here focuses solely on the characteristics of the adjuncts, except as a prominent, unique trait demands otherwise, beginning with those of the simplest nature and confronting the remainder in turn as their complexity increases.

7.1. DIPLOMORPHIC GENES

Diplomorphic genes encode many contrasting families of macromolecules, but in broad terms the class may be considered to embrace two major groups. One includes those whose products are secreted through the cytoplasmic membrane, either into the environment or an enclosing capsule such as exists around many bacteria and algae. The second category viewed here has already received some mention in the preceding chapter, for it embraces those nuclear genes that encode a product that functions in the mitochondrion or chloroplast, like those of cytochromes f and c_1. Since more attention has been devoted to researches on bacterial products than those of other prokaryotes and protistans, these provide a major segment of those examined here, along with hormones, blood proteins, and a diversity of mammalian molecules. To those two large groups are added the few available from insects and fungi, and a larger, but still minor, fraction from seed plants.

7.1.1. Simple Diplomorphic Genes from Mammalian Sources

The primary transcripts and coding regions of many mammalian genes for secretory products have been established, so that they provide a firm foundation for some of the complexities that are to follow. Although the gene structures and the prefatory portions have been determined in great abundance, almost no special attention has been devoted to transcriptional problems, such as initiation and termination. However, near the close of this section is summarized what can be gleaned from the occasional promoter and other signals that have been incidentally garnered along with the main studies.

Although the term "signal peptide" is frequently applied to the temporary portion (presequence) preceding the protein proper, another name in general use, "transit peptide," appears superior in that it describes the main function of the appendage, service in passage through a membrane (Figure 7.1). Throughout much of this chapter the conventions used in the tables are uniform, underscoring indicating codons for polar charged (hydrophilic) amino acids, and italics for the apolar (hydrophobic) ones, groups of the latter being especially characteristic of membrane-enclosed structures. To bring out the critical features in spite of the variations in length that exist, the presequences are aligned both at their 5′ termini and again at the 3′ point where cleavage takes place during processing. The first two codons of the mature coding region are also included with the latter to disclose any clues possibly employed by the cleaving enzymes.

The Glycoprotein Family of Hormones. In vertebrates a unique family of hormones is found, whose members are closely related in quaternary structure. All consist of αβ dimers, the α subunit of which appears to be shared by all types, while the β subunit is distinctive and specifies the biological activity of the particular mature protein. The two types of subunits share one common feature, that of being glycosylated (Ramabhadran *et*

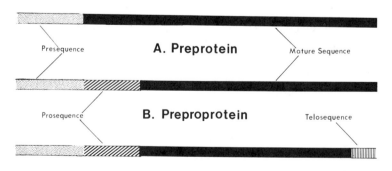

C. Modified Preprotein

Figure 7.1. (A) Simple and (B,C) complex diplomorphic genes. Complex diplomorphs bear both a pre- and a prosequence, and occasionally (C) a telosequence as well.

Table 7.1
Transit Peptide Genes of the Glycoprotein Hormones[a]

Row 1

Subunit α

Mouse TSH[b]	ATG GAT TAC TAC AGA AAA TAT GCA GCT GTC ATT CTG GTC ATG
Rat GP[c]	ATG GAT TGC TAC AGA AGA TAT GCG GCT GTC ATT CTG GTC ATG
Human CG[d]	ATG GAT TAC TAC AGA AAA TAT GCA GCT ATC TTT CTG GTC ACA
Bovine GP[e]	ATG GAT TAC TAC AGA AAA TAT GCA GCT GTC ATT CTG ACC ATT

Subunit β

Mouse TSH[f]	ATG AGT --- --- --- --- GCT GCC GTC CTC CTC TCC GTG CTT
Bovine TSH[g]	ATG ACT --- --- --- --- GCT ACC TTC CTG ATG TCC ATG ATT
Human LH[h]	ATG GAG --- --- --- --- ATG TTC CAA* GGG CTG CTG CTG TTG
Rat LH[i]	ATG GAG --- --- --- --- AGG CTC CAG* GGG CTG CTG CTG TGG

Row 2

Subunit α

Mouse TSH	CTG TCC ATG TTC CTG --- CAT ATT CTT CAT TCT	CTT CCT
Rat GP	CTG TCC ATG GTC CTG --- CAT ATT CTT CAT TCT	CTT CCT
Human CG	TTG TCG GTG TTT CTG --- CAT GTT CTC CAT TCC	GCT CCT
Bovine GP	TTG TCT CTG TTT CTG --- CAA ATT CTC CAT TCC	TTT CCT

Subunit β

Mouse TSH	TTT GCT CTT GCT TGT --- GGG CAA GCA GCA TCC	TTT TGT
Bovine TSH	TTT GGC CTT GCA TGT --- GGA CAA GCA ATG TCT	TTT TGT
Human LH	CTG CTG CTG AGC ATG GGC GGG ACA TGG GCA TCC	AAG GAG
Rat LH	CTG CTG CTG AGC CCA AGT GTG GTG TGG GCC TCC	AGG GGC

[a]Codons for apolar (hydrophobic) amino acids are italicized and those for charged ones (hydrophilic) are underscored. Cleavage sites are indicated by the vertical arrow. Asterisk denotes the location of an intron. TSH, thyrotropin; GP, glycoprotein; CG, chorionic gonadotropin; LH, luteinizing hormone.
[b]Chin et al. (1981).
[c]Godine et al. (1982).
[d]Fiddes and Goodman (1979).
[e]Erwin et al. (1983).
[f]Gurr et al. (1983).
[g]Maurer et al. (1984).
[h]Boorstein et al. (1982); Policastro et al. (1984).
[i]Chin et al. (1983); Jameson et al. (1984).

al., 1984). Three species of the mature proteins are produced in the pituitary, including luteinizing (LH or lutropin), follicle-stimulating (FSH or follitropin), and thyroid-stimulating hormones (TSH or thyrotropin) (Van Heuverswyn *et al.,* 1984), while the fourth, chorionic gonadotropin (CG), is secreted by the placenta of mammals.

Table 7.1 includes four representative presequences of the genes encoding each subunit, although examples only from two hormone species can be listed for the β peptide. When the structures of the α presequences are examined, their virtual identity is at once apparent, despite the fact that their sources are from four diverse species of mammals and from different hormones. Consequently, there can be no doubt that the same α subunit serves in each dimeric protein, ragardless of the latter's function. Very little variation from one sequence to another can be noted; the only difference of any consequence is in the bovine glycoprotein (GP) presequence just before the end of row 1, where a neutral amino acid is encoded instead of the apolar one of the others. A similar distinction occurs in the human CG peptide at the termination of that same row. In contrast, the first codon for the mature coding sequence is seen to vary widely, whereas the second is constant.

There is some confusion in the literature regarding the correct identity of the two β subunits given here as luteinizing hormones, as well as the precise location of the cleavage site. Two articles pertaining to the human gene originally described it as encoding chorionic gonadotropin (Boorstein *et al.,* 1982; Policastro *et al.,* 1983), but in a footnote at the end of the first of these references, the encoded product was reidentified as luteinizing hormone. Since the rat LH cistron given is perceived to correspond closely to that of the human (Chin *et al.,* 1983; Jameson *et al.,* 1984), it appears that the structure of chorionic gonadotropin remains unestablished. The location of the cleavage site was arbitrarily decided as being at the position suggested in the table, but perhaps it will prove to be before the TTC and TCC, as in the reference on the human gene (Policastro *et al.,* 1983).

The presequences for the four β subunits are also highly conserved evolutionarily, but not to the extent found in the preceding ones. The two representative genes of TSH transit peptides are obvious homologs, as are those encoding that peptide for LH, but identities between the pairs are rare. The most universal feature is the length, which is almost identical in all four sequences. Near the end of row 1 are two columns of similar codons, one involving CTC, ATG, and CTG, the other GTG, ATG, and CTG, and in row 2 there is one more, containing CTT and CTG. The only other resemblance in codon makeup is at the very 3' end, where TCT, TTC, and TCC are found. Because the TSH structures represent cDNA corresponding to the mRNA, not the gene proper, any consistency in the placement of the intron with those of the two LH genes that may exist is not disclosed at present.

Although the mature sequences of these genes are of no concern here, it is pertinent to note that that of human chorionic gonadotropin β subunit stands out from the remainder of the family in having an extension of 87 nucleotides at the 3' end (Lentz *et al.,* 1984). This addition encodes 29 amino acids, including four serines to which oligosaccharide moieties become attached after translation has been completed. The function of this peculiarity has not been determined.

The Prolactin Family of Hormones. The prolactin family of hormones includes a small number of polypeptides that have rather extensive sequence homology as well as a degree of overlapping in biological activity (Martial *et al.*, 1979). Embraced in this category in addition to the type form are growth hormone (GH), chorionic somatomammotropin (CS; placental lactogen), and proliferin (PLF), all except the placental being secreted by the pituitary gland (Table 7.2). The importance of this group of hormones is reflected in the fact that in the bovine anterior pituitary the messenger for prolactin constitutes 60% of the total mRNA on polysomes (Sasavage *et al.*, 1981). At least in cattle, there appear to be multiple loci for these genes, but their location and arrangement in the genome have yet to be determined.

The degree of homology between genes depends on both the species being compared and the source organisms (Linzer and Talamantes, 1985). Among the prolactin presequences, the level of agreement between corresponding sites is particularly outstanding near the end of row 1 and again just after the onset of row 2. Comparable consistency can be observed in the transit peptide cistrons of the growth hormones at similar points, both among themselves and with those of prolactins. Here, as in the preceding family, the greatest points of fluctuation are those ending the presequence and the pair beginning the mature coding region, sites where least variation might be expected because of their involvement in cleavage during processing. As a whole in the entire presequence the proportion of codons for apolar amino acids is surprisingly low for a peptide of this function, running close to 55%, except in the bovine growth hormone sequence, in which a level of 73% is found.

The unexpected high degree of identity between the human growth hormone and chorionic somatomammotropin presequences led to a comparison of the two mature coding regions (Table 7.3). After viewing numbers of highly diverse macromolecules under identical names distinguished only by subscripts, as in the cytochromes P-450, finding two very similar ones bearing such distinctive designations comes as a jolt. As may be noted in Tables 7.2 and 7.3, not only do the presequences and mature genes display close resemblances, but so do the 3′ trains, including the rare, true palindromic sequences of some length shown boxed. Moreover, the level of kinship extends even into the 5′ leader (Table 7.10), typically a region of low evolutionary conservation. Hence, it would appear to be far more realistic to refer to these closely related hormones as growth hormones (or somatotropins) A and B, or perhaps pituitary and chorionic growth hormones. With the suggested change in names the similarities between the pair certainly could be better appreciated than under their current designations. Previously the somatomammotropin gene had been compared only with a prolactin sequence, from which it is obviously distinct (Cooke *et al.*, 1981).

Other Simple Diplomorphic Genes from Vertebrates. Quite a few genes of vertebrates encode products that bear transit peptides, and many more are becoming apparent as additional sequences are established. Among those also known to fall here is α_1-antitrypsin (Long *et al.*, 1984a), thymosin β_4 (Goodall *et al.*, 1985), several varieties of lipoproteins, such as E, A-IV, C-I, and C-II (Boguski *et al.*, 1984; Fojo *et al.*, 1984; Knott *et al.*, 1984a; Zannis *et al.*, 1984), the β subunit of muscle acetylcholine receptor (Tanabe *et al.*, 1984), and many other receptors. It should be noted, however, that not all lipoprotein genes are to be classified here, because some, including A-I and A-II, are more complexly structured and are accordingly examined in a later section. Still further vertebrate genes that current knowledge suggests belong here are those of the various caseins (L. Hall *et al.*, 1984b; Stewart *et al.*, 1984), but in reality they may not. Because

Table 7.2
Presequence Genes of the Prolactin Family of Hormones[a]

Row 1

Prolactins

Rat[b]
ATG AAC AGC --- CAG *GTG* TCA *GCC* CGG <u>AAA</u> *G GG* --- ACA --- --- *CTC CTC CTG CTG CTG ATG*

Human pituitary[c]
ATG AAC *ATC* <u>AAA</u> <u>AAA</u> *GGA* TCG *CCA* TGG <u>AAA</u> *G*GG* --- TCC --- --- *CTC CTG CTG CTG CTG CTG*

Human decidua[d]
(Incomplete) TCC --- --- *CTC CTG CTG CTG CTG CTG*

Bovine[e]
ATG <u>GAC</u> AGC <u>AAA</u> --- *GGT* TCG TCG CAG <u>AAA</u> *G GG* --- TCC <u>CGC</u> *CTG CTC CTC CTG CTG GTG*

Growth hormones

Bovine[f]
ATG ATG GCT *GCA* --- *GGC* --- *CCC CCG* --- ACC --- TCC --- --- *CTG CTC CTG GCT TTC*

Human[g]
ATG --- *GCT* ACA --- *GGC* TCC --- <u>CGG</u> --- ACG --- TCC --- --- *CTG CTC CTG GCT TTT*

Proliferins

Murine[h]
ATG CTC CCT --- --- --- TCT *TTG ATT* CAA *CCA* TGC TCC --- --- TGG *ATA CTG CTC CTA*

Chorionic somatomammotropin

Human[i]
ATG --- *GCT CCA* --- *GGC* TCC --- <u>CGG</u> --- ACG --- TCC --- --- *CTG CTC CTG GCT TTT*

Row 2

Prolactins																	↓		
Rat	*ATG*	TCA	AAC	*CTT*	*CTG*	*TTC*	TGC	---	---	---	---	CAA	AAT	*GTG*	CAG	ACC		*CTG*	CCA
Human pituitary	*GTG*	TCA	AAC	*CTG*	*CTG*	*CTG*	TGC	---	---	---	---	CAG	AGC	*GTG*	GCC	CCC		*TTG*	CCC
Human decidua	*GTG*	TCA	AAC	*CTG*	*CTC*	*CTG*	TGC	---	---	---	---	CAG	AGC	*GTG*	GCC	CCC		*TTG*	CCC
Bovine	*GTG*	TCA	AAT	*CTA*	*CTC*	*TTG*	TGC	---	---	---	---	CAG	*GGT*	*GTG*	*GTC*	TCC		ACC	CCC
Growth hormones																		
Bovine	*GCC*	---	---	*CTG*	*CTC*	TGC	*CTG*	*CCC*	TGG	---	ACT	CAG	*GTG*	*GTG*	*GGC*			*GCC*	*TTC*
Human	*GGC*	---	---	*CTG*	*CTG*	TGC	*CTG*	*CCC*	TGG	*CTT*	CAA	GAG	*GGC*	AGT	*GCC*			*TTC*	CCA
Proliferin																		
Mouse	*CTA*	*CTG*	---	*GTG*	AAC	AGC	TCG	*TTA*	*TTG*	TGG	---	AAG	AAT	*GTT*	*GCC*	TCA		*TTT*	CCC
Chorionic somatomammotropin																		
Human	*GCC*	---	---	*CTG*	*CTC*	TGC	*CTG*	*CCC*	TGG	*CTT*	CAA	<u>GAG</u>	*GCT*	*GGT*	*GCC*			*GTC*	CAA

[a] Codons for hydrophobic amino acids are italicized and those for hydrophilic ones underscored. Vertical arrow points to cleavage sites. Asterisks indicate location of introns.
[b] Gubbins et al. (1979); Cooke et al. (1980).
[c] Takahasi et al. (1984).
[d] Cooke et al. (1981); Truong et al. (1984).
[e] Sasavage et al. (1981).
[f] Miller et al. (1980).
[g] Martial et al. (1979); Roskam and Rougeon (1979).
[h] Linzer and Nathans (1984).
[i] Selby et al. (1984).

Table 7.3
Relationships between Growth Hormone (GH) and Somatomammotropin (SMT) Genes[a]

Row 1

Human GH[b]: TTC CCA ACC ATT CCC TTA TCC AGG CCT TTT GAC AAC GCT ATG CTC CGC GCC CAT CCT CTG CAC CAG CTG

Human SMT[c]: GTC CAA ACC GTT CCC TTA TCC AGG CTT TTT GAC CAC GCT ATG CTC CAA GCC CAT CGC GCG CAC CAG CTG

Row 2

Human GH: GCC TTT GAC ACC TAC CAG GAG TTT GAA GAA GCC TAT ATC CCA AAG GAA CAG TAT TCA TTC CTG CAG

Human SMT: GCC ATT GAC ACC TAC CAG GAG TTT*GAA GAA ACC TAT ATC CCA AAG GAC CAG AAG TAT TCA TTC CTG CAT

Row 3

Human GH: AAC CCC CAG ACC TCC CTC TGT TTC TCA GAG TCT ATT CCG ACA CCC TCC AAC AGG GAG GAA ACA CAA CAG

Human SMT: GAC TCC CAG ACC TCC TTC TGC TTC TCA GAC TCT ATT CCG ACA CCC TCC AAC ATG GAG GAA ACG CAA CAG

Row 4

Human GH: AAA TCC AAC CTA GAG CTG CTC CGC ATC TCC CTG CTG ATC CAG TCG TGG CTG GAG CCC GTG CAG TTC

Human SMT: AAA TCC*AAT CTA CAG CTG CTC CGC ATC TCC CTG CTC ATC GAG TCG TGG CTG GAG CCC GTG CGG TTC

Row 5

Human GH: CTC AGG AGT GTC TTC GCC AAC AGC CTG GTG TAC GGC GCC TCT GAC AGC AAC GTC TAT GAC CTC CTA AAG

Human SMT: CTC AGG AGT ATG TTC GCC AAC AAC CTG GTG TAT GAC ACC TCG GAC AGC GAT TAT CAC CTC CTA AAG

Row 6

Human GH GAC CTA GAG GAA GGC ATC CAA ACG CTG ATG GGG AGG CTG GAA GAT GGC AGC CCC CGG ACT GGG CAG ATC

Human SMT GAC CTA GAG GAA GGC ATC CAA ACG CTG ATG GGG*AGG CTG GAA GAC CGC CGG AGC CGG ACT GGG CAG ATC

Row 7

Human GH TTC AAG CAG ACC TAC AGC AAG TTC GAC ACA AAC TCA CAC AAC GAT GAC CTA CTC AAG AAC TAC GGG

Human SMT CTC AAG CAG ACC TAC AGC AAG TTT GAC ACA •AAC TCG CAC AAC CAT GAC GCA CTG CTC AAG AAC TAC GGG

Row 8

Human GH CTG CTC TAC TGC TTC AGG AAG GAC ATG GAC AAG GTC GAG ACA TTC CGC ATC GTG CAG TGC CGC TCT

Human SMT CTG CTC TAC TGC TTC AGG AAG GAC ATG GAC AAG GTC GAG ACA TTC CTG CGC ATG GTG CAG TGC CGC TCT

Row 9

Human GH GTG GAG GGC AGC TGT GGC TTC TAG CTG CCC GGG TGG CAT CCC TGT GAC CCC TCC CCA GTG CCT CTC

Human SMT GTG GAG GGC AGC TGT GGC TTC TAG GTG CCC GAG TAG CAT CCT -GT GAC CCC TCC CCA GTG CCT CTC

Row 10

Human GH CTG GCC CTG GAA GTT GCC ACT CCA GTG CCC ACC AGC CTT GTC CTA ATA AAA TTA AGT TGC ATC AAA AAA

Human SMT TTG GCC CTG -AA GGT GCC ACT CCA GTG CCC ACC AGC CTT GTC CTA ATA AAA TTA AGT TGT ATC ATT TCA

[a]Asterisk indicates the location of an intron. Vertical arrow indicates cleavage site.
[b]Roskam and Rougeon (1979).
[c]Selby et al. (1984).

of the complex cleavages that can be produced by plasmin, at least β-casein may eventually prove to be encoded by a member of the cryptomorphic genes (Section 7.6.1). A similar condition is prevalent in the haptoglobin genes of man (Bensi *et al.*, 1985), and possibly also in that for human pancreatic polypeptide (Boel *et al.*, 1984).

7.1.2. Simple Diplomorphic Genes from Seed Plants

In the seed plants, too, there are a number of newly translated proteins that bear transit sequences. These include not only those encoded by nuclear genes for use in the mitochondrion or chloroplast, as mentioned in the preceding chapter, but numerous types transcribed or processed within the endoreticulum or dictyosomes also have been shown to possess that feature. In addition, it is not unlikely that some types that are excreted through the cell membranes also have appended presequences, but no gene structures for any in that category have been determined, except perhaps the lectin cistron discussed below. Undoubtedly the most thoroughly documented are the seed-storage proteins, which receive attention first, while some of the energy-related types then supplement earlier discussions.

Seed-Storage Proteins. The seed-storage proteins comprise several large, complex families, since each major group of plants appears to have one or more distinctive types. Some of these are of a more complex organization and are analyzed in Section 7.6.2, but many appear to belong in the present category. In maize, zein, accounting for 50% of the total endosperm protein, is the primary representative, being synthesized in the endosperm from 14 to 55 days after fertilization (Pedersen *et al.*, 1982; Spena *et al.*, 1983). Within the haploid genome of this monocot are found about 120 copies of the gene, located on at least three of the ten chromosomes, with much deviation from one copy to another. The two presequences given in Table 7.4 are from a pair of adjacent cistrons. It is improbable that the translation termination signal TAA actually exists at the 3' end of this sequence in the E19 species as given in the original reference (Spena *et al.*, 1983), for that would end the translational processes. In all likelihood this represents either a typographical error or a misreading of the chromatographic blocks and probably should be TAC as in the E25 structure.

The next most abundant protein in the endosperm of maize, glutelin-2, accounts for 15% of the total. Its presequence is quite unlike that of the zeins in being much shorter, in this respect more closely resembling the corresponding parts of dicotyledonous seed-storage proteins (Prat *et al.*, 1985), with which it is placed in the table. Homology with those sequences is perceived to be only occasional, so that no kinship is in evidence. A second monocot, barley, has the seed endosperm enriched in an alcohol-soluble class of proteins known as hordeins; these comprise a complex mixture of polypeptides that fall into four principal subdivisions, the B and C types providing 95% of the total present (Kreis *et al.*, 1983; Forde *et al.*, 1985). However, the gene structure has yet to be fully determined. In a third grain, hexaploid wheat, gliadins translated on the endoreticulum of endosperm cells provide the major storage component. These, too, are a complex group, being separable into 35–50 components that can be arranged into three subfamilies (Kasarda *et al.*, 1984; Okita *et al.*, 1985). Only two of the trio have had a gene of a representative sequenced, one of the α subfamily and another of the α/β type (Rafalski *et al.*, 1984; Sumner-Smith *et al.*, 1985), both of which are included in Table 7.4. The

Table 7.4
Transit-Peptide Genes of Seed Storage Proteins[a]

Row 1

Zein E19[b]	ATG GCA GCC	AAA ATA TTT	TGC CTC CTT	ATC CTC CTT	GGT CTT TCT	GCA AGT GCT	GCT ACG GCG				
Zein E25[b]	CTG GCA GTC	AAA ATA TTT	TGC CTC ATG	CTC CTT GCT	CTT TCT GCA	AGT GCT GCT	AAC GCG				
Gliacin α/β[c]	ATG --- ---	AAG ACC TTT	CTC ATC CTT	GTC CTC CTT	GCT ATT GTG	GCG --- ---	--- ---				
Gliacin α[d]	ATG --- ---	AAG ACC TTT	CTC ATC CTT	GCC CTC CTT	GCT ATC GTG	GCA --- ---	--- ---				
Phaseolin[e]	ATG ATG AGA	GCA AGG GTT	CCA CTC CTG	TTG CTG GGA	--- ATT CTT	--- --- ---	--- ---				
Lectin[f]	ATG CAT GAT	CAT GGC TTC	CTC CAA GTT	ACT CTC CCT	AGC CCT CTT	CCT --- ---	--- ---				
Glutelin-2[g]	ATG --- AGG	--- GTG TTG	CTC --- GTT	GCC CTC ---	GCT CTC TTG	GCT --- ---	--- ---				

Row 2 (Arrow indicates cleavage site →)

Zein E19	ACC ATT TTC	CCG CAA TGC	TCA CAA GCT	CCT ATA GCT	TCC CTT CTT	CCC CCG TAA	CTC TCA
Zein E25	ACC AAT TTT	CTG CAA TGC	TCA CAA GAT	CCA ATT GCT	TCC CTT CTT	CCC TCA TAC	CTC TCA
Gliadin α/β	ACC ACC ---	--- --- ---	--- GCC ACA	ACT GCA ---	--- --- ---	--- --- ---	GTT AGA
Gliadin α	ACC ACC ---	--- --- ---	--- GCC ACA	ACT GCA ---	--- --- ---	--- --- ---	GTA AGA
Phaseolin	--- --- TTC	CTG GCA ---	TCA --- CTT	TCT GCC ---	TCA TTT GCC	--- ACT TCA	CTC CGG
Lectin	--- --- ---	--- --- TGC	GCT TCT CAG	CCA --- CGC	--- AAA CTC	--- --- ---	AGC CAC
Glutelin-2	--- --- ---	CTC --- GCT	--- GCG ---	AGC GCC ---	--- --- ---	--- TCC	ACG CAT

[a] Codons for hydrophobic amino acids are italicized and those for hydrophilic ones are underscored. Arrow indicates the cleavage site.
[b] Spena et al. (1983).
[c] Rafalski et al. (1984).
[d] Kasarda et al. (1984).
[e] H. Pearen, personal communication (1985).
[f] Hoffman et al. (1982).
[g] Prat et al. (1985).

genomic numbers often vary with the cultivar and have not been firmly documented; the location of one gene, that of α-1Y, has been demonstrated to be on chromosome 6A of wheat (Anderson *et al.*, 1984).

The dicotyledonous seed-storage proteins have not been so thoroughly explored at the level of the gene as have their counterparts; however, in addition several gene structures that have been established are of the more complex type mentioned earlier. Hence, only a single representative of the simple group from this source is currently available. That one, from the French bean (*Phaseolus vulgaris*), is one of a group of polypeptides termed phaseolin that comprises about half of the storage proteins of the seed (Talbot *et al.*, 1984). These are deposited rapidly, beginning when the cotyledons are about 7 mm in length and continuing until they have attained 17–19 mm (Slightom *et al.*, 1983). Now that the presequence as well as the full coding region of the gene has been determined (H. Paaren, personal communication), the former can be noted to be quite distinct from the four from monocotyledonous sources. Among the especially outstanding distinctions is the presence of two methionine codons at the 5' end, followed by two for arginine separated by one triplet, but none for lysine. Despite the sequence being aligned as fully as possible with the others, very few corresponding sites can be noted. Still one other dicotyledonous member is known in part, conglycinin of the soybean, a trimeric protein of $\alpha\alpha'\beta$ constitution; while much of the gene structure has been determined, the presequence is missing (Schuler *et al.*, 1982). Since this protein is trimeric rather than monomeric like the rest, it obviously represents a family separate from those given here. Finally, among the albumins that are also abundant in beans is a substance called lectin, a protein widely spread throughout the living world. This really serves as an immune-related substance rather than as a source of food for the growing embryo (Hoffman *et al.*, 1982), but nevertheless, shares many features with the others of Table 7.4.

Considering the diversity of the sources, an unexpectedly high level of homology among the transit-peptide genes of the three species of seed-storage proteins is revealed by examination of the tabulation, but it is particularly those from grains that display close kinship (Reeck and Hedgcoth, 1985). The initiating codon, which is CTG for leucine in zein E25, is usually followed shortly by a codon for a charged amino acid, even in the lectin. Thereafter in row 1 a number of further correspondences can be noted among the triplets, nearly all of which specify hydrophobic monomers. In the second row the level of evolutionary conservation is greatly reduced, as is the proportion of codons for apolar amino acids. One feature that particularly merits notice is the relative frequency of the CT- family of codons for leucine, while the others for that amino acid, TTA and TTG, are not employed at all.

The Ribulose-Bisphosphate Carboxylase Family. In the preceding chapter one of the important energy-related families of proteins was seen to be that of ribulose-1,5-bisphosphate carboxylase, one of the principal genes involved in carbon dioxide fixation. This enzyme will be recalled as being comprised of eight copies each of large and small subunits, the former being encoded in the chloroplast genome, the latter in the nucleus. Hence, it is the minor component whose gene is provided with a presequence, and thus requires attention here. Only four full gene structures for the transit peptide have been determined, plus an additional partial one. Since three of the total are from wheat, only three different source organisms are thus represented in Table 7.5.

Here, as in the seed proteins of monocotyledons, sequence homology is at a high

Table 7.5
Gene Sequences for the Transit Peptides of Ribulose-1,5-Bisphosphate Carboxylase[a]

Row 1

Pea *rbcS*[b]	*ATG* *GCT* TCT *ATG* *ATA* TCC TCT TCC *GCT* *GTG* ACA ACA *GTC* AGC CG̲T̲ *GCC* TCT AG̲G̲ *GGC* CAA TCC
Lemna G-3[c]	*ATG* *GTT* TCC ACC *GCC* *GCC* *GTG* *GCC* CG̲C̲ *GTC* *GCC* CCT *GCC* CAG ACC AAC *ATG* *GTG* *GGC* --- ---
Wheat *W9*[d]	*ATG* *GCC* --- --- --- --- --- --- --- --- CCA *GCC* *GTG* *ATG* *GCT* TCT TCC
Wheat *WS4.3*[d]	*ATG* *GCC* --- --- --- --- --- --- --- --- CCC *GCC* *GTG* *ATG* *GCT* TCG TCG

Row 2

Pea *rbcS*	*GCC* --- *GCA* *CTG* *GGC* *CCA* *TTC* *GGC* *GGC* *CTC* AA̲A̲ TCC *ATG* ACT *GGA* *TTC* *CCA* *GTG* --- AA̲G̲ AA̲G̲
Lemna G-3	*GCC* --- --- --- *TTC* AAC *GGG* TGC CG̲C̲ TCC TCC *GTC* *GCC* *TTC* *CCC* *GCC* ACC CG̲C̲ AA̲G̲
Wheat *234*[e]	(Incomplete) *GTC* *GCT* *CCT* *TTC* *CAG* *GGG* *CTC* AA̲G̲ TCC ACC *GCC* *GGC* *CTC* *CCC* *GTC* AGC TGC CG̲C̲
Wheat *W9*	*GCC* ACC ACC *GTC* *GCG* *CCC* *TTC* *CAG* *GGG* *CTC* AA̲G̲ TCG ACC *GCC* *CTC* *CCC* *ATC* AGC TGC CG̲C̲
Wheat *WS4.3*	*GCT* ACC ACC *GTC* *GCA* *CCC* *TTC* *CAG* *GGT* *CTC* AA̲G̲ TCC ACA *GCC* *GGC* *CTG* *CCC* *GTC* AGC CG̲C̲ CG̲C̲

Row 3

Pea *rbcS*	*GTC* AAC ACT GA̲C̲ --- *ATT* ACT TCC *ATT* ACA AGC AAT *GGT* *GGA* AG̲A̲ *GTA* AA̲G̲ TGC → *ATG* CAG*
Lemna G-3	*GCC* AAC AAC GA̲T̲ *TTG* TCG ACT --- *CTC* *CCC* AGC TCC *GGC* AG̲G̲ *GTT* AGC TGC *ATG* CAG
Wheat *234*	--- TCC AAC *GGC* *GCT* AGC *CTC* *GCC* AGC *GTC* AGC AAC *GGT* *GGA* AG̲G̲ *ATC* AG̲G̲ TGC *ATG* CAG
Wheat *W9*	--- TCC *GGC* AGC ACC *GGC* *CTC* AGC AGC *GTC* AGC AAT *GGC* *GGA* AG̲G̲ *ATC* AG̲A̲ TGC *ATG* CAG
Wheat *WS4.3*	--- TCC --- AG̲G̲ *GGC* AGC *CTC* *GGC* AGC *GTC* AGC AAC *GGC* *GGA* AG̲G̲ *ATC* AG̲G̲ TGC *ATG* CAG

[a] Codons for hydrophilic amino acids are underscored and those for hydrophobic ones are italicized. Asterisk indicates the location of an intron. The arrow indicates cleavage site.
[b] Coruzzi et al. (1983, 1984). [c] Stiekema et al. (1983). [d] Broglie et al. (1983). [e] Smith et al. (1983).

Table 7.6
Presequences of Cytochromes and Related Proteins[a]

Row 1

Yeast$_N$ cyt c_1 [b]	ATG	TTT	TCA	AAT	CTA	TCT	AAA	CGT	TGG	GCT	CAA	AGG	ACC	CTC	TCG	AAA	AGT	TTC	TAC	TCT	ACC	GCA	ACA
Pea cyt f [c,e]	ATG	GAT	AGG	GAA	GAA	CTG	---	---	AGT	AAC	CTA	CCT	AAT	CTT	ATT	GTA	GAA	ATT	TTC	AGG	ATC	AAG	GAT
Spinach$_c$ cyt f [d]	GTG	GAT	AGG	GAA	GAA	CTT	---	TAC	---	TAG	CAA	CCT	ACC	CAA	TTT	ATT	GTA	TAA	ATT	TTC	GGA	ATC	AAT	GGT
Wheat$_c$ cyt f [e]	GTG	TAT	AGG	GAA	CTA	GAT	GTT	AGC	TAC	CTA	TCT	AAT	XTT	ATT	GTA	GAA	AXX	TTC	TGG	ATC	TGC	GAT
Yeast cyt c peroxidase [f]	ATG	ACT	ACT	GCT	GTT	AGG	CTT	TTA	CCT	TCA	CTG	GGC	AGA	ACC	GCC	CAT	AAG	AGG	TCT	CTC	TAC	CTG	TTC
Bovine cyt P-450$_{scc}$ [g]	ATG	CTA	GCA	AGG	GGG	CTT	CCC	CTC	CGC	TCA	GCC	CTG	GTC	AAA	GCC	TGC	CCA	CCC	ATC	CTG	AGC	ACA	GTG
Yeast cyt c oxidase IV [h]	ATG	CTT	TCA	CTA	CGT	CAA	TCT	ATA	---	---	---	AGA	TTT	TTC	---	AAG	CCA	GCC	---	ACA	---	AGA
Bovine cyt c oxidase IV [i]	ATG	TTG	GCA	ACC	AGA	GTA	TTT	ACC	---	---	CTG	ATT	GGT	---	AGG	CGT	GCA	---	ATC	---
Yeast cyt c oxidase VI [j]	ATG	TTA	TCA	---	AGG	GCC	---	ATA	---	---	TTC	AGA	AAT	---	---	CCA	GTT	---	ATA	AAT	AGA

Row 2

Yeast$_N$ cyt c_1	GGT	GCT	GCT	AGT	AAA	TCT	GGC	AAG	CTT	ACT	CAA	AAG	CTA	GTT	ACA	GTT	GCT	GCC	GCC	GGT	ATC
Pea cyt f	TGT	ACC	ATG	CAA	ACT	AGA	AAT	GCT	TTT	TCT	TGG	ATA	AAG	AAA	GAG	ATT	ACT	CGA	TCT	ATT	TCC	GTA	TTG
Spinach cyt f	TGG	ACT	ATG	CAA	ACT	ATA	AAT	ACC	TTT	TCT	TGG	ATA	AAA	GAA	CAG	ATT	ACT	CGA	TCC	ATT	TCC	ATA	TCA
Wheat cyt f	TGG	ACT	ATG	GAA	AAT	AGA	AAT	ACT	TTT	TCT	TGG	GTA	AAG	GAA	GAA	CAG	CAG	ACT	CGA	TCG	ATT	TCT	GTA	TCG
Yeast cyt c peroxidase	TCC	GCT	GCT	GCT	GCT	GCT	GCT	GCA	ACT	TTT	GCT	TAC	TCG	CAA	TCC	CAC	AAG	AGA	TCA	TCG
Bovine cyt P-450$_{scc}$	GGG	GAG	GGC	TGG	GGC	CAC	CAC	AGG	GTG	GGC	---	---	---	---	---	---	---	---	---	---	---
Yeast cyt c oxidase IV	---	ACT	TTG	TGT	---	AGC	---	---	---	---	---	---	---	---	---	---	---	---	---	---
Bovine cyt c oxidase IV	---	---	---	---	---	TCC	---	---	---	ACC	---	---	TCG	GTG	---	---	---
Yeast cyt c oxidase	---	ACT	TTA	TTG	---	AGA	GCC	AGA	CCT	GCT	GCT	TAT	CAT	GCA	ACT	AGA	TTG	ACT	---	AAA	---	---	---

Row 3

```
Yeast cyt c₁           ACC GCA TCG ACT TTA CTC TAT GCA GAC TCA TTA ACT --- --- --- --- --- --- --- GCC --- GAA GCT | ATG ACC
Pea cyt f              CTC ATG ATC TAT ATA ATA ACT CGA GCA CCC ATT TCA --- --- --- --- --- --- --- --- --- AAT GCA | TAT CCC
Spinach cyt f          CTT ATA TAT ATA ATA ACT CGG TCA TCC ATT GCG --- --- --- --- --- --- --- --- --- --- AAT GCC | TAT CCC
Wheat cyt f            ATC ATC ATA TAC GTA ATA ACT CGG ACA TCT ATT TCA --- --- --- --- --- --- --- --- --- AAT GCA | TAT CCC
Yeast cyt c peroxidase TCT TCT CCT GGG GGT AGT AAC CAC GGA TGG AAC AAC TGG GGG AAG CCA GCT GCT TTG GCT TCC | ACT ACA
Bovine cyt P-450scc    --- --- --- --- ACT GGA GAG GGA --- --- --- --- --- --- --- --- --- GCT GGC | ATC TCC
Yeast cyt c oxidase IV --- --- --- --- TCT AGA TAT --- --- --- --- --- --- --- --- --- CTG CTT | CAG CAA
Bovine cyt c oxidase IV --- --- --- --- --- --- --- --- --- --- --- --- --- --- --- --- TGT GTT CGG | GCC CAT
Yeast cyt c oxidase VI --- --- --- --- AAT ACG TTT ATT CAA AGT --- --- --- --- --- --- --- AGG AAG TAT | TGT GAC
```

[a] Codons for apolar (hydrophobic) amino acids are italicized and those for hydrophilic ones are underscored. N, nuclear; C, chloroplast; cyt, cytochrome. Vertical arrow shows the cleavage site.

[b] Sadler et al. (1984).
[c] Willey et al. (1984a).
[d] Alt and Herrmann (1984).
[e] Willey et al. (1984b).
[f] Kaput et al. (1982).
[g] Morohashi et al. (1984).
[h] Maarse et al. (1984).
[i] Lomax et al. (1984).
[j] Wright et al. (1984).

level, especially at the 3' end, where a number of sites are invariant or virtually so. In addition, several unusual features for a peptide gene of this sort are to be observed, chief among them being the consistent placement of codons for arginine, two double columns of which occur, one set at the end of row 2, and another, which is interrupted, similarly placed in row 3. The greatest distinctions between sequences from the dicotyledonous and monocotyledonous sources are in the 5' end, where the pea and duckweed representatives are greatly elongated relative to the others. All these presequences are obviously longer than any that have been viewed earlier. As a whole, codons for hydrophobic amino acids are more prolific in the third row, but the CT- family for leucine is not especially abundant proportionately with the rest.

7.1.3. Presequences of Cytochrome-Related Genes

As pointed out in the preceding chapter, many cytochromes and related proteins that function in the mitochondrion or chloroplast are encoded in the nuclear genome, such as the ribulose-1,5-bisphosphate carboxylase small subunit in the foregoing section, and similarly are translated in the cytoplasm of the cell. Hence, many of the genes whose structures were examined in Chapter 6 receive attention here in connection with their presequences, whose products function in the penetration through the organellar membrane. As may be seen in Table 7.6, quite a diversity of types bear this appendage, the nine structures shown representing six different species.

General Characteristics. The coding regions for the transit peptides vary extensively in length, the yeast cytochrome *c* peroxidase, with 67 codons, being the longest (Kaput *et al.*, 1982), and the bovine cytochrome *c* oxidase subunit IV, with 22, the shortest (Lomax *et al.*, 1984). Even the three representatives of cytochrome *f* from chloroplasts display unexpected slight differences in length, having a range from 57 in the pea to 60 in the wheat (Willey *et al.*, 1984a,b). The spinach example as given in the table differs from that indicated in the paper describing the gene sequence (Alt and Herrmann, 1984), since the investigation apparently believed the first ATG in row 2 to be the initiating codon, whereas in all likelihood it is the GTG of row 1, as in the wheat example (Willey *et al.*, 1984b).

Homologies of Structure. Homologies among the entire assemblage of nine are obviously lacking. Despite the differences in length, which permit much freedom in aligning the shorter components with the larger, it is not possible to bring sites together to produce a single full column of either identical or related codons. Not even the underscored triplets for lysine and arginine form vertical series in excess of four representatives. Furthermore, as a whole the cytochrome and related gene presequences are not outstandingly rich in codons of apolar amino acids. Indeed, one of those cited, the cistron for yeast cytochrome *c* oxidase subunit VI, consists of only 40% such triplets, and almost all have 50% or less codons for those hydrophobic amino acids. The sole exception is the gene for bovine P-450$_{scc}$, which encodes at a 70% level of apolarity (Morohashi *et al.*, 1984). The highest concentration of these units is medially, where the yeast cytochrome *c* peroxidase, for an extreme example, has an unbroken series of ten. As a general rule, the codon in the ultimate position at the cleavage site specifies a small apolar amino acid, but that of bovine cytochrome *c* oxidase subunit IV is for arginine and that of yeast subunit VI signals tyrosine. Similarly, the character of those in the penultimate site is variable, four being for hydrophobic, four for polar uncharged, and one for polar charged amino acids.

The three representatives for cytochrome *f* do not display a high level of evolutionary conservation, but vary rather freely. Perhaps the strangest condition is the rather frequent homologies between the pea sequence and the wheat, the spinach example often differing from the two others. While variation exists, the spinach sequence resembles that of wheat at eight sites, and that of the pea the same number of times; in contrast, the pea agrees with the wheat at 14 sites. As on other occasions, single characteristics here are disclosed as being totally unreliable as indicators of phylogenetic kinships.

Between the two representatives of cytochrome *c* oxidase subunit IV the level of homology is astonishingly low, complete agreement between corresponding sites occurring only at two or three triplets. Perhaps the two are actually two different subunits, which happen to show a like chromatographic pattern in the complete enzyme and consequently bear the same numerical designation.

7.1.4. Simple Diplomorphic Genes from Miscellaneous Sources

Comparatively few gene sequences have been established among eukaryotes other than yeasts, vertebrates, and seed plants, and of those that have been established, most either cannot be fitted into the "simple" category or lack transit-peptide portions, or the latter are too incomplete to be meaningful. Thus the four of Table 7.7 containing a pair each from fungal and insect sources must be considered representative on a preliminary basis. Only the two from the insect (*Bombyx mori*) are related, but even there the level of homology is not high. Nevertheless, those available contribute to the fabric of the total picture of presequence structure.

Two Unrelated Fungal Genes. Cutinase, one of the two simple diplomorphic genes whose sequences have been established from fungi, is an enzyme secreted by cells of a pathogenic species involved in the penetration of the host seed plant. The sequence of the transit peptide of this glycoprotein, which hydrolyzes the cutin coat of leaf cuticle and is essential for successful invasion, is from *Fusarium solani,* a pest of potatoes (Soliday *et al.,* 1984). The second is from a basidiomycete, *Schizophyllum commune,* and encodes an unnamed glycoprotein involved in the development of fruiting bodies (Dons *et al.,* 1984).

One unusual feature of the *Fusarium* sequence is the location of an intron between the transit peptide and mature coding portions. Of course, this would have no effect upon the proteolytic removal of the presequence, because the intron would have been removed prior to translation. Both genes are remarkable for the high concentration of codons for hydrophobic amino acids in their 5′ portions (row 1), where only one triplet for a polar charged type is present and two or four for uncharged polar amino acids. To the contrary, on the 3′ portion half or less of the codons encode apolar species. Here, too, only four triplets encoding a charged monomer are to be noted in the *Fusarium* gene and three in that of *Schizophyllum.*

Two Related Insect Genes. As stated before, the pair of genes from insects encode chorion proteins of the silkmoth, but the two presequences do not display the expected high level of homology (Iatrou *et al.,* 1984). Because of the extreme richness in cysteine (30%), the substances encoded by the mature genes are believed to contribute to the formation of the outer shell of the egg. Although both share this compositional feature, the mature coding sequences differ extensively, A being rich also in codons for glycine, while B is not so heavily marked with triplets for that amino acid. Thus they really form a small family of genes, rather than multiple copies of a single unit. As a result, it is less

Table 7.7
Presequence Genes from Miscellaneous Eukaryotes[a]

Row 1

Fungal genes

Schizophyllum 1G2[b]
ATG CCC TTC TCG CTC GCC ATC CTT GCT CTC CCC GTC CTC GCG GCT GCG ACT GCG GTT CCC CGC GGC

Fusarium cutinase[c]
ATG AAA TTC TTC --- GCT --- --- --- --- CTC ACC ACA CTT CTC GCC GCC ACG GCT TCG

Insect chorion genes

Bombyx A[d]
ATG TTT ACC TTC --- --- --- --- --- --- --- --- --- --- GCT CTT CTC

Bombyx B[d]
ATG --- --- --- GCC --- --- --- --- --- --- --- GCT --- ---

Row 2

Fungal genes

Schizophyllum 1G2
GGC GCT TCC AAG TGC AAC AGC GGT CCC GTC CAG TGC AAC ACC CTG GTC GAC* → ACT AAG

Fusarium cutinase
GCT CTG CCT ACT TCT AAC --- --- CCT GCC CAG GAG CTT GAG GCG CGC CAG CTT   GGT AGA

Insect chorion genes

Bombyx A
CTT CTC TGC GTT CAG GGC TGC CTG ATC CAA* AAT GTG TAC --- --- --- GGT   CAG TGC

Bombyx B
AAA CTC ATT GTC TTC GTC TGC GCC ATC GCC CTC GTG GCT CAG*TCC GTT TTG GGC   ACT GGT

[a]Codons for hydrophobic amino acids are italicized and those for hydrophilic ones are underscored. Asterisk indicates the location of an intron. Arrow indicates cleavage site.
[b]Dons et al. (1984).
[c]Soliday et al. (1984).
[d]Iatrou et al. (1984).

difficult to understand the low grade of homology that exists between the sequences for the transit peptides shown here. However, the correspondences are sufficient, especially in the similar—but not identical—location of an intron, to suggest that the two have had common ancestry. Codons for hydrophobic amino acids are numerous and distributed fairly uniformly through both presequences, but only one triplet for a charged monomer can be observed in the B gene and none in the other.

7.2. SIMPLE DIPLOMORPHIC GENES FROM PROTISTS AND PROKARYOTES

Although many interesting variations in presequence structure were found in the foregoing pages, no common theme was in evidence among the higher forms described there. However, it is possible that at the lower levels of phylogeny, the simpler organization that might exist in transit-peptide structures may better reveal the fundamentals. Accordingly, this section centers on presequences from the true yeasts and bacteria as the basis for comparison, using genes that encode products secreted from the cells into the environment. Since not all such secretions are provided with the simpler type of gene being considered at this time, only those cistrons whose products do not undergo further processing after removal of the transit protein receive attention now. Eventually many lower eukaryotic genes of this category should be available, for most, like the green alga *Chlamydomonas,* release an abundance of substances into the milieu (Voigt, 1985). Moreover, many antigens, either secreted or membrane-embedded, doubtlessly will prove to bear prepeptides, as one from *Plasmodium falciparum* has already proven to do (Hope *et al.*, 1985).

7.2.1. Diplomorphic Genes of Yeast and Bacteria

Yeast Secreted-Protein Genes. The first yeast genes for proteins secreted through the cell membrane consist of two cistrons for acid phosphatase, *PHO3* and *PHO5*, a tandemly repeated duplication (Arima *et al.*, 1983; Bajwa *et al.*, 1984). In addition, a third copy may be present, but its existence has not been firmly determined. No organization into operons exists, the *PHO* genes being located on linkage group II, while those for their controlling elements are dispersed throughout the genome. After translation (on endoreticulum?) the acid phosphatase is glycosylated and then transported through the cell membrane into the periplasmic space that lies beneath the cell wall. As may be observed in Table 7.8, the sequences of the transit peptides display approximately the same level of homology (82%) that has been reported for the mature coding portion (Bajwa *et al.*, 1984). Despite the similarity of structure, which extends deeply into the flanking regions, the controls differ widely, for *PHO3* is independent of exogenous influences, whereas *PHO5* is governed by a series of positively and negatively acting products of regulatory genes. The noteworthy features of the presequences include a codon for a basic amino acid near the 5' end and a level of 75% or less presence of triplets for hydrophobic monomeric units. Of the latter class, those for valine (GTN) and alanine (GCN) are outstandingly abundant.

The presequence of the third gene (*SUC2*) in Table 7.8 has much the same structural

Table 7.8
Transit Peptide Sequences of Yeast Secreted-Protein
Genes[a]

Row 1

Yeast PHO3[b]	ATG TTT --- AAG --- TCT GTT GTT TAT TCG GTT
Yeast PHO5[b,c]	ATG TTT --- AAA --- TCT GTT GTT TAT TCA ATT
Yeast SUC2[d]	ATG CTT TTG CAA GCT TTC CTT TTC CTT TTG GCT

Row 2

Yeast PHO3	CTA GCC GCT GCT TTA GTT AAT GCA GGT ACA
Yeast PHO5	TTA GCC GCT TCT TTG GCC AAT GCA GGT ACC
Yeast SUC2	GGT TTT GCA GCC AAA ATA TCT GCA TCA ATG

[a]Codons for hydrophilic amino acids are underscored and those for hydrophobic ones are italicized.
[b]Bajwa et al. (1984).
[c]Arima et al. (1983).
[d]Taussig and Carlson (1983).

properties as those just described, including the relatively short total length and single codon for a basic amino acid (Taussig and Carlson, 1983). The latter, however, is located close to the 3' end, rather than near the 5' end. Additionally, the level of the hydrophobicity is higher in the present case, all but three of the 19 codons being for apolar monomers. Triplets for leucine (CTN, TTA, and TTG) constitute more than 25% of the entire structure and those for alanine (GCN) about 20%, but none for valine (GTN) is to be seen. SUC2 encodes invertase, which is glycosylated when secreted, but not when retained intracellularly.

Bacterial Secreted-Protein Genes. Any illusions about primitive presequences being consistently short as implied by the yeast examples are quickly dispelled when the gene structures for bacterial secreted proteins are viewed. While the first three listed in Table 7.9 are of comparable length (18–20 codons), that of the *E. coli* gene for F pilin (*traA;* Frost *et al.,* 1984) has 50 and that of the *Bacteroides nodosus* pilin cistron only seven (Elleman and Hoyne, 1984). Nevertheless, some similarity in structure can be noted, particularly in the presence of at least one codon for a basic amino acid in proximity to the 5' end. But again there are exceptions, for three of those given (*Pseudomonas aeruginosa* exotoxin A and *E. coli ompT* and *traA*) lack that feature (Gordon *et al.,* 1984; Gray *et al.,* 1984). The first of these lacks such triplets completely, but *traA* has a pair medially and *ompT* has two spaced ones before and at the 3' terminus. Thus the presequences of bacteria are not consistently divided into charged and hydrophobic sectors as sometimes described (Michaelis and Beckwith, 1982; Emr and Silhavy, 1983).

Among those included in the table are several enterotoxins secreted through the cell walls into the environment, such as that of *Vibrio cholerae* (*V.c.*) and two from *E. coli* (*E.c.*). The cholera enterotoxin is comprised of a single subunit A and five of the much smaller B, the transit peptide portion of each of which has been established (Gennaro and

Table 7.9

Presequences of Bacterial Genes for Secreted Products[a]

Row 1

Toxins

V.c. enterotoxin AI[b]	ATG	---	---	GTA	AAG	---	---	---	---	---	---	---	---	---	---	---	---	---	ATA	ATA	---	TTT
V.c. enterotoxin B[c]	ATG	---	---	ATT	AAA	TTA	AAA	---	---	---	---	---	---	---	---	---	---	---	---	---	---	TTT
E.c. enterotoxin LTA[d]	ATG	---	---	---	AAA	---	AAT	---	---	---	---	---	---	---	---	---	---	ATA	ACT	---	TTC
E.c. enterotoxin LTB[d]	ATG	---	---	AAT	AAA	GTA	AAA	---	---	---	---	---	---	---	---	TGT	TAT	---	GTT
P.a. exotoxin A[e]	ATG	---	CAC	CTG	---	---	ATA	CCC	CAT	---	TGG	---	ATC	CCC	CTG	GTC	GCC	AGC	CTC
C.d. toxin 228[f]	GTG	---	AGC	---	AGA	AAA	---	---	---	CTG	TTT	---	GCG	TCA	ATC	TTA	ATA	GGG	GCG

Enzymes

B.p. xylanase A[g]	ATG	AAT	TTG	---	AGA	AAA	TTA	AGA	CTG	TTG	TTT	GTG	ATG	---	TGT	ATT	GGA	CTG	---	---
E.c. lamB[h]	ATG	ATG	ATT	ACT	CTG	CGC	---	AAA	---	CTT	CCT	CTG	GCG	GTT	GCC	GTC	GCA	---	---
B.s. α-amylase[i]	ATG	---	TTT	GCA	AAA	CGA	TTC	AAA	ACC	TCT	TTA	CTG	CCG	TTA	TTC	GCT	GGA	TTT	TTA
B.e. α-amylase[j]	ATG	---	AAA	CAA	CAC	AAA	AAA	CGC	CTT	TAT	GCC	CGA	TTG	CTG	CCG	CTG	TTA	TTT

Outer membrane proteins

E.a. ompA[k]	ATG	---	---	AAA	AAG	ACA	---	---	GCT	---	ATC	---	GCG	ATT	GCA	---	GTG
S.t. ompA[l]	ATG	---	---	AAA	AAG	ACA	---	---	GCT	---	ATC	---	GCG	ATT	GCA	---	GTG
E.c. ompF[m]	ATG	ATG	---	---	AAG	CGC	---	AAT	---	ATT	CTG	GCA	---	GTG	ATC	GTC	---	CCT
E.c. ompC[n]	ATG	---	---	AAA	---	GTT	AAA	GTA	CTG	TCC	---	---	CTC	CTG	---	GTC	---	CCA
E.c. ompT[o]	ATG	TGT	CTC	AGT	---	---	TTT	---	GTC	CCT	CTT	---	---	---	---	---	---	TTT

(continued)

Table 7.9 (Continued)

```
E.c. traA^p         ATG AAT GCT GTT TTA AGT GTT CAG GGT GCT TCT GCG CCC GTC AAA AAG AAG TCG TTT^u
B.n. pilin^q        ATG --- --- AAA AGT --- --- --- --- --- --- --- --- --- --- --- --- ---
E.c. fimA^r         ATG --- --- AAA --- ATT AAA ACT CTG GCA --- ATC GTT GTT --- --- CTG TCG
E.c. fimpA^s        ATG --- --- AAA AAA GCA --- --- TTC TTA --- --- TTA GCA        GTT TTT TTT
S.a. protein A^t    TTG --- --- AAA AAG --- AAA AAC ATT TAT TCA ATT CGT AAA CTA GGT GTA GGT^v

Row 2                                                                       →

Toxins

V.c. enterotoxin A1   GTG --- TTT TTT ATT --- TTC TTA TCA TCA TTT --- TCA TAT GCG     AAT GAT
V.c. enterotoxin B    GGT GTT TTT ACA GTT TTA CTA TCT TCA GCA TAT GCA CAT GGA         ACA CCT
E.c. enterotoxin LTA  ATT --- TTT TTT ATT --- TTA TTA GCA TCG CCA --- TTA TAT GCA     AAT GGC
E.c. enterotoxin LTB  TTA --- TTT ACG GCG --- TTA CTA TCC TCT CTA TGT GCA TAC GGA     GCT CCC
P.a. exotoxin A       GGC --- CTG CTC GCC --- GGC GGC TCG TCC --- --- GCG TCC GCC     GCC CAG
C.d. toxin 228        CTA CTG GGG ATA GGG --- GCC CCA CCT TCA --- --- GCC CAT GCA     GGC GCT

Enzymes

B.p. xylanase A       ACG CTT --- ATA CTG --- --- ACG GCT GTA CCA --- GCC CAT GCG     AGA ACC
E.c. lamB             --- --- --- GCG GGC --- GTA ATG TCT GCT CAG --- GCA ATG GCT     GTT GAT
B.s. α-amylase        TTG CTG TTT TAT TTG GTT --- CTG GCA GGA CCG --- GCG GCT GCG     AGT GCT
```

B.ℓ. α-amylase	GCG CTC --- ATC TTC TTG CTG CCT CAC TCT GCA --- GCT GCG GCG	→	GCA AAT
Outer membrane proteins			
E.c. ompA	GCA CTG GCT GGC TTC --- GCT --- ACC GTA --- --- GCG CAG GCC		GCT CCG
S.t. ompA	GCA CTG GCT GGT TTC --- GCT --- ACC GTA --- --- GCG CAG GCC		GCT CCG
E.c. ompF	GCT CTG TTA GTA GCA --- GGT --- ACT --- --- GCA AAC GCT		GCA GAA
E.c. ompC	GCT CTG CTG GTA GCA --- GGC --- --- GCA --- GCA AAC GCT		GCT GAA
E.c. ompT	--- --- TGT ACT --- --- --- AAA AAC ATA --- GTA TTG AGG		ATA ACC
E.c. traA	CCG GCT GCT GTT CTG ATG ATG TTC TTC CCG CAG CTG GCG ATG GCC		GCC GGC
B.ℓ. pilin	--- --- --- --- --- --- TTA --- CAA --- AAA GGT		TTC ACC
E.c. fimA	GCT CTG TCC --- CTC --- --- AGT TCT --- ACA GCG GCT CTG GCC		GCT GCC
E.c. fimpA	CTC --- --- ACT GGG --- GGC GGG --- GTT --- --- TCT CAC GCT		GCG GTT
S.a. protein A	GGT ACA TTA CTT ATA TCT GGT GGC GTA ACA CCT GCT GCA AAT GCT		GCG CAA

aCodons for hydrophilic amino acids are underscored and those for hydrophobic ones are italicized. Abbreviations: V.c., Vibrio cholerae; E.c., Escherichia coli; P.a., Pseudomonas aeruginosa; C.d., Corynebacterium diphtheriae; B.p., Bacillus pumilus; B.s., Bacillus subtilis; B.l., Bacillus licheniformis; E.a., Enterobacter aerogenes; S.t., Salmonella typhimurium; B.n. Bacteroides nodosus; S.a., Staphylococcus aureus. bLackman et al. (1984). cGennaro and Greenaway (1983). dYamamoto et al. (1984a,b). eGray et al. (1984). fKaczorek et al. (1983). gFukusaki et al. (1984). hEmr and Silhavy (1983). iOhmura et al. (1984). jSibakov and Palva (1984). kBraun and Cole (1983). lFreudl and Cole (1983). mInokuchi et al. (1982). nMizuno et al. (1983) oGordon et al. (1984). pFinlay et al. (1984); Frost et al. (1984). qElleman and Hoyne (1984). rKlemm (1984). sMooi et al. (1984). tUhlén et al. (1984). uInsert TTT TCC AAA TTC ACT CGT CTG AAT ATG CTT CGC CTG GCT CGC GCA GTG ATC. vInsert ATT GCA TCT GTA ACT TTA.

Greenaway, 1983; Lockman *et al.*, 1984). Actually the first of these gene products undergoes further processing into smaller components and thus is not truly a simple gene, but since this action is conducted only after the enterotoxin has entered a host cell, it appears valid to consider it simple as here, at least on a tentative basis. The two genes from *E. coli, LTA* and *LTB,* encode subunits corresponding to those from the cholera organism, which unite in the identical AB_5 ratio to produce the holoenzyme (Yamamoto *et al.,* 1984a,b). Since the end product has a comparable effect on the host, the two proteins are obviously members of the same family, a condition reflected in the presequences (Table 7.9). Although all four are distinctly related, the structures within each corresponding pair display a higher level of kinship to one another than between contrasting types. One of the outstanding chemical properties of this quartet is the inclusion of numerous codons for phenylalanine (TTT, TTC), especially in the A subunit of *V. cholerae.*

Two additional toxins are represented in the table by sequences of their transit peptide genes, one from *Pseudomonas aeruginosa* (*P.a.*; Gray *et al.,* 1984), the other from *Corynebacterium diphtheriae* (*C.d.*; Kaczorek *et al.,* 1983). Despite the similarities in subunit structure and activities that appear between the components of this pair and the preceding, very few homologies in nucleotide sequences exist. The diphtherial gene is somewhat similar to the others in having codons for a basic amino acid near the 5' terminus, but here two are found side by side. Among the numerous additional divergences evidenced is the nearly complete lack of codons for phenylalanine, which are displaced by those for serine (TCN, AGC, AGT), alanine (GCN), and leucine (CTN). Glycine triplets are also moderately abundant in the *Pseudomonas* cistron.

Bacterial Simple Enzyme Genes. Among the enzymes secreted into the environment or deposited in the outer membrane are several from diverse bacteria that are encoded by simple genes including presequences (Table 7.9). The first of these, from *Bacillus pumilus* (*B.p.*), encodes a xylanase that breaks down xylans extracellularly (Fukusaki *et al.,* 1984); two for α-amylase from *Bacillus subtilis* and *B. lichenformis* that break down starches are similarly deposited in the medium (Ohmura *et al.,* 1984; Sibakov and Palva, 1984). On the other hand, a fourth representative, *lamB* from *E. coli,* is a component of the outer membrane that facilitates the passage of maltose from the surroundings into the cell (Emr and Silhavy, 1983). All are seen to encode largely hydrophobic amino acids, and the trio secreted into the milieu are alike in having three codons for basic components near the 5' end. That which remains in the cell wall is distinct in having only two, but these are correspondingly located. It is of great importance to note that, when cloned in *E. coli,* the α-amylase gene of *B. subtilis* was produced and transported to the exterior in normal amounts, but the xylanase gene of *B. pumilus,* although transcribed at usual levels, accumulated in the cytoplasm instead of being secreted (Fukusaki *et al.,* 1984; Ohmura *et al.,* 1984).

Outer Membrane Proteins. Another large group of proteins that bear presequences are those of the cell coat, which must pass through the cytoplasmic membrane to attain their ultimate location. Out of the fairly large number whose gene structures have been determined, ten have been selected for Table 7.9 to show both the similarities and differences that characterize them. The two extremes of length of the presequences that have already been mentioned are found here and afford a first view of the contrasts that exist. Seven of the ten are of typical structure in having two or three codons for basic amino acids close to the 5' end; that of *B. nodosus* for pilin has only one, but an additional one occupies the penultimate site. And, as already noted, the *E. coli* cistrons for outer membrane protein T (*ompT*) and for pilin F (*traA*) are likewise irregular in the placement of such triplets.

The two sequences for outer membrane protein A (*ompA*) from *Enterobacter aerogenes* and *Salmonella typhimurium* that head the list exemplify the high degree of evolutionary conservation that usually exists between corresponding transit peptides, for they differ at only two sites (Braun and Cole, 1983; Freudl and Cole, 1983). In addition, they illustrate the conceptual model of a presequence, for they have, in order, the initiation codon ATG, a pair of triplets for lysine, followed by a single one for a neutral amino acid, an uninterrupted series of 12 or 13 for hydrophobic amino acids, and an absence of any for additional charged monomeric units. Similar continuous runs of 12 or 13 codons for hydrophobic amino acids occur in the *E. coli ompT* and *ompC* sequences (Inokuchi *et al.*, 1982; Mizuno *et al.*, 1983) and *traA* for pilin (Finlay *et al.*, 1984). The remaining four presequences do not fit this supposed standard pattern so well. Although largely endowed with codons for hydrophobic units, there are no such long, uninterrupted series, the longest stretch being six in *E. coli's fimA*, five in its *fimpA*, and six also in the *S. aureus* cistron for protein A (Klemm, 1984; Mooi *et al.*, 1984; Uhlén *et al.*, 1984). In contrast, the *B. nodosus* pilin gene has only a single such triplet in its interior—three of its total of seven codons are for hydrophobic, two for basic, and two for neutral amino acids. Consequently, this presequence is less than 45% hydrophobic.

Although not included in the table, presequences of related bacterial genes are not without interest. That of a strain of *E. coli* pathogenic in the urinary tract of man, encoding the $F7_2$ fimbrial subunit, consists of 21 codons, all but four of which are for hydrophobic amino acids (van Die and Bergmans, 1984). In contrast, that for a vitamin B_{12} receptor from the same organism has only 11 of its 20 codons for that type of monomer and two for basic amino acids near its 5' end instead of only one (Heller and Kadner, 1985).

7.2.2. Transcriptional Control Signals of Diplomorphs

While there is no firm reason to suspect that initiation and termination of transcription in diplomorphic genes would differ from those of the simple type, the possibility exists nevertheless, and hence requires investigation. Regrettably, however, the same situation prevails here as in too many preceding aspects, in that too little experimental work has been conducted to present a clear picture, even preliminarily. Since the reports of the vast majority of eukaryotic diplomorphic genes provide neither promoter nor terminator sequences, even of a presumptive nature, attention here is necessarily confined to those from bacterial sources. Almost all of those provided in Table 7.10 are proposed signals and are given in italics, while the very few that have been experimentally determined are underscored. To the contrary, in the case of the trains, underscoring indicates nucleotide residues that are unpaired, while the arrows display regions of dyad symmetry.

Possible Promoters of Diplomorphic Genes. In Table 7.10 the promoters in the first 11 leader sections from simple diplomorphic genes of bacteria are presumptive only, being based on a resemblance to the TA-TA- box mentioned in many preceding discussions. Similarly, the ancillary signals are based on these suggested promoters and may or

Table 7.10
Flanking Sequences of Bacterial Simple Diplomorphic Genes[a]

	Ancillary site		Promoter	Start site
Leaders				
E.a. ompA P1[b]	GAGTTCACA CTTGTA--	---------- AGTTTCTAAC TAAGTT	GTAGAC TTT ACATCG	
E.a. ompA P2[b]	CACTTGTAA GTTTCT--	---------- AACTAAGTTG TAGACT	TTACAT CGC CAGGGG	
S.t. ompA P1[c]	GAGTTCACA CTTGTA--	---------- AGTTTCCAAC TACGTT	GTAGAC TTT ACATCG	
S.t. ompA P2[c]	CACTTGTAA GTTTCC--	---------- AACTACGTTG TAGACT	TTACAT CGC CAGGGG	
E.c. ompT[d]	GACTTAGAA GTTCCTAG	---------- AACGACATT- TTAAGT	CAACAA CTT ACCGCG	
E.c. fimA[e]	TGTTTGATA TGTAAATT	---------- ATTTCTATTG TAAATT	AATTTC AC- ATCACC	
E.c. fimpA[f]	GAGTTGTGT ATTC----	---------- GCTGGCACCT TATTAT	GCAGAT CCG GGCAAG	
B.l. α-amylase[g]	TATTTGTT- AA------	---------- AAATTCAAAA TATTTA	TACAAT AGC ATGTGT	
S.a. protein A P1[h]	ATTTTAGTA TTGCAATA	---------- CATAATTCGT TATATT	ATGATG ACT TTACAA	
S.a. protein A P2[h]	GTATTGCAA TACATAA-	---------- TTCGTTATAT TATGAT	GACTTT ACA AATACA	
P.a. exotoxin A[i]	ACATTCACC ACTCTG--	---------- CAATCCAGTT CATAAA	TCCCAT AAA GCCCTC	
E.c. malK[j]	GGATTTAAG CCATCT--	---------- CCTGATGACG CATAGT	CAGCCC A(+48)	
E.c. malE[j]	AGGAGGATG GAAAGAGG	TT-------- GCCGTATAAA GAAACT	AGAGTC CGT (+45)	
Trains				
B.n. pilin[k]	TAGCTAGCTCT TAAAATGC GAAAGCCTCTC TCT TGAGAGGCTTTT TTATGGTTTATTGTT			
E.a. ompA[b]	GTTTCCTACGA TAA----- -AAAACCCGCT CGA TGCGGGTTTTTT TTGGCCTGATTCTTG			
S.t. ompA[c]	GTTTCCGTCTG ATA----- AAAAACCCCGC GTC -GCGGGTTTTTT GCTCTGGTCTGGATG			
E.c. fimA[e]	CCTACCCAGGT TCAGGA[l] +8 CGGGCAGGG ATG CCCACCCTTGTG CGATAAAAAATAACGA			
S.a. protein A[h]	CCTTAGGTGCA CGCT[m]+128CTAAATGCACG AGC AACATCTTTTGT TGCTCAGTGCATTTT			
P.a. exotoxin A[i]	CTGCCGCGACC GGCCGGCT CCCTTCGCAGG AGC CGGCCTTCTCGG GGCCTGGCCATACAT			

[a]Abbreviations: *E.a.*, *Enterobacter aerogenes*; *S.t.*, *Salmmella typhimurium*; *E.c.*, *E. coli*; *B.l.*, *Bacillus licheniformis*; *S.a.*, *Staphylococcus aureus*; *P.a.*, *Pseudomonas aeruginosa*; *B.n.*, *Bacteroides iodosus*. P1, promoter 1; P2, promoter 2. Italics indicate presumptive, and underscores established, signals.
[b]Braun and Cole (1983).
[c]Freudl and Cole (1983).
[d]Gordon *et al.* (1984).
[e]Klemm (1984).
[f]Mooi *et al.* (1984).
[g]Sibakov and Palva (1984).
[h]Uhlen *et al.* (1984).
[i]Gray *et al.* (1984).
[j]Bedouelle *et al.* (1982).
[k]Elleman and Hoyne (1984).
[l]Insert seven residues.
[m]Insert 128 residues.

may not have any relation to the actual functional sequence. Hence, they should be viewed only as interesting possibilities and, as such, do not merit detailed attention at this time. Examination of the two gene leaders of the *E. coli* maltose system, *malE* and *malK*, is more meaningful, since the promoters and ancillary signals are deduced from actual experimentally determined start sites of transcription (Bedouelle *et al.*, 1982). Consequently, although they are not indisputably established, at least they have some basis in fact.

The location of the two start sites ~50 positions upstream of the initiator codon of translation is one reason for not taking the 11 presumptive ones too seriously, for most of those lie closer to that point. In neither of the two *mal* promoters is much resemblance displayed to the canonical TA-TA-, nor can standard features be found in the abutting nucleotides. The upstream ancillary signal shows still less constancy between the two underscored, and no similarity of any sort to the CAAT- box that is supposed to mark this site. As the result of all the uncertainties that exist, it is not possible to deduce whether these simple diplomorphs are transcribed by special mechanisms or whether the usual processes are active.

Possible Terminators of Transcription. The terminators of transcription are similarly hypothetical, since no experimental studies of any sort have been conducted on termination in this type of gene. Nevertheless, the ones that have been proposed have structural features identifiable with those that have been determined *in vitro* or *in vivo*. In all six examples of trains from diplomorphic genes given in Table 7.10, a stem-and-loop region is present, which almost always ends downstream in a series of Ts. Additionally, the first three representatives show a surprisingly high level of homology within this structure, but not elsewhere in the train. The implication of this degree of localized conservation is that these regions of dyad symmetry have a particularly important function, probably in termination as proposed. Only in three of them are unpaired bases (underscored) present, the *fimA* train of *E. coli* being unusual in having such unmated sites on each side of the stem.

7.3. ANALYSIS OF PRESEQUENCES OF SIMPLE GENES

In order to bring out any general trends that may exist in the transit-peptide portion of simple diplomorphic genes, the chief characteristics of those that have been reviewed are analyzed in Table 7.11. For the sake of clarity, apolar (hydrophobic) amino acids (Rose *et al.*, 1985) are indicated by the abbreviation A, charged polar (hydrophilic) ones by the letter C, and polar uncharged (neutral) ones by N. At the cleavage point, the nature is indicated of the monomers encoded by two codons on each side, that is, the last two of the presequence and the first couple of the mature coding sector of the gene. In the table, the presequences are divided into approximate halves to bring out differences in distribution that may exist, with the percent codons for apolar amino acids calculated for each "half" separately and again for the total structure.

Mean Properties of Presequences. When the properties of the presequences are viewed as described, it is noted that, in the vertebrate hormones of Table 7.11, the 3' half encodes a higher proportion of hydrophobic amino acids than does the 5' half, except for the β subunits of the two LH genes. This condition, however, is exceptional, for the mean

Table 7.11
Summary of Presequence Structure of Simple Diplomorphic Genes[a]

Table	Gene[b]	Number of codons	5'-Half				3'-Half				Total % Apolar	Transit peptide		Mature sequence	
			A	N	C	%A	A	N	C	%A		Pen	Ult	Site 1	Site 2
7.1	Murine TSH α	24	7	3	3	54	7	4	0	64	58	N	N	A	N
	Rat GP α	24	7	3	3	54	7	4	0	64	58	N	N	A	N
	Human CG α	24	7	3	3	54	7	4	0	64	58	N	N	A	N
	Bovine GP α	24	7	3	3	54	7	4	0	64	58	N	N	A	N
	Murine TSH β	20	6	3	0	67	8	3	0	73	70	A	N	A	A
	Bovine TSH β	20	5	4	0	56	8	3	0	73	65	A	N	A	A
	Human LH β	20	7	1	1	77	8	3	0	73	75	N	A	A	A
	Rat LH β	20	7	1	1	77	7	4	0	64	70	N	A	N	C
7.2	Rat prolactin	28	9	5	2	56	5	7	0	44	50	N	N	A	A
	Human pituitary prolactin	29	5	5	3	53	7	5	0	56	55	A	A	A	A
	Bovine prolactin	30	5	5	4	50	7	5	0	56	53	A	N	N	A
	Bovine GH	26	11	2	1	79	8	4	0	63	73	A	A	A	A
	Human GH	26	8	4	1	62	7	5	1	52	58	A	A	A	A
	Murine proliferin	29	9	6	0	60	7	6	1	50	55	N	A	A	A
	Human somatomammotropin	26	9	3	1	70	9	3	1	68	69	A	A	A	N

7.4	Zein E19	39	15	5	1	70	11	7	0	61	67	A	N	A	N
	Zein E25	39	15	5	1	70	8	10	0	44	58	N	N	A	N
	Gliadin α/β	20	12	1	1	86	2	4	0	33	70	N	A	A	C
	Gliadin α	20	12	1	1	86	2	4	0	33	70	N	A	A	C
	Phaseolin	26	12	0	2	86	7	5	0	56	73	N	N	A	N
	Lectin	24	10	4	2	63	3	3	2	33	54	C	A	A	C
7.5	Pea *rbc*	57	17	10	2	59	12	10	6	44	50	C	N	A	N
	Lemma	52	15	6	2	70	13	11	5	45	50	A	N	A	N
	Wheat *W3*	47	13	5	0	72	12	13	4	43	53	C	N	A	N
	Wheat *W54.3*	46	13	5	0	72	13	9	6	48	56	C	N	A	N
7.6	Yeast$_N$ cytochrome c_1	61	13	13	6	41	17	10	2	60	49	N	A	A	N
	Pea$_C$ cytochrome *f*	57	13	8	8	45	14	9	5	50	47	N	A	N	A
	Spinach$_C$ cytochrome *f*	59	14	12	5	45	12	12	4	43	44	N	A	N	A
	Wheat$_C$ cytochrome *f*	60	12	13	7	38	11	13	4	40	39	N	A	N	A
	Yeast *c* peroxidase	67	20	7	4	69	15	18	3	42	55	A	N	N	N
	Bovine cytochrome P-450$_{scc}$	39	16	4	3	69	9	4	3	56	63	A	A	A	N
	Yeast cytochrome *c* oxidase IV	25	9	3	4	56	3	5	1	33	48	A	A	N	N
	Bovine cytochrome *c* oxidase IV	22	9	3	3	60	2	4	1	29	49	A	C	A	N
	Yeast cytochrome *c* oxidase VI	41	11	6	4	52	4	11	5	20	36	C	N	N	C

(continued)

Table 7.11 (Continued)

Table	Gene	Number of codons	5'-Half				3'-Half				Total % Apolar	Cleavage site			
												Transit peptide		Mature sequence	
			A	N	C	%A	A	N	C	%A		Pen	Ult	Site 1	Site 2
7.7	*Schizophyllum commune* 1G2	40	19	2	1	86	7	10	1	38	60	N	N	N	C
	Fusarium solani C3	31	10	4	1	67	8	7	1	50	58	N	A	A	C
	Bombyx mori A	21	8	2	0	80	6	5	0	55	66	A	A	N	A
	Bombyx mori B	21	8	1	1	80	9	3	0	75	80	N	A	N	N
7.8	Yeast PHO3	17	5	3	1	56	7	1	0	88	70	N	A	A	N
	Yeast PHO5	17	5	3	1	56	6	2	0	75	65	N	A	A	N
	Yeast SUC2	19	10	1	0	90	6	1	1	75	84	N	A	N	A
7.9	*V.c.* toxin A1	18	7	0	1	87	6	4	0	60	72	N	A	A	C
	V.c. toxin B	21	7	2	2	64	6	4	0	60	62	N	A	A	A
	E.c. toxin LTA	18	6	2	1	67	7	2	0	78	72	N	A	N	A
	E.c. toxin LTB	21	5	4	2	47	6	4	0	60	53	N	A	N	A
	P.a. toxin A	25	9	4	0	86	9	3	0	75	72	N	A	N	C
	C.d. toxin 228	25	8	2	2	67	11	2	0	84	76	N	A	A	A
	B.p. xylanase	27	9	2	3	60	10	3	0	46	70	N	A	C	N
	E.c. *lamB*	25	12	1	2	80	8	2	0	80	80	A	A	A	C

B.s. amylase	29	10	2	3	67	15	1	0	94	86	A	A	N	A
B.l. amylase	29	9	3	4	56	11	2	0	84	68	A	A	A	N
E.a. ompA	21	7	1	2	70	9	2	0	82	76	N	A	A	A
S.t. ompA	21	7	1	2	70	9	2	0	82	76	N	A	A	A
E.c. ompF	22	9	1	2	75	8	2	0	80	78	N	A	A	C
E.c. ompC	21	8	1	2	73	9	1	0	90	80	N	A	A	C
E.c. ompT	17	8	1	0	89	4	2	2	50	70	A	C	A	N
E.c. traA	50	16	7	4	58	20	1	2	87	72	A	A	A	A
B.n. pilin	7	1	1	1	33	2	1	1	50	43	C	A	A	N
E.c. flaA	23	8	2	2	67	7	4	0	64	65	A	A	A	A
E.c. flmpA	21	9	1	2	75	6	3	0	67	72	N	A	A	A
S.a. protein A	36	12	4	5	58	11	4	0	74	64	N	A	A	N
Total 61		541	220	129		502	303	62			20A, 6C, 35N	31A, 2C, 28N	44A, 1C, 16N	23A, 13C, 25N
Mean:	30%				66%				59%	63%				

[a] Abbreviations: A, apolar (hydrophobic); C, polar, charged (hydrophilic); N, neutral (polar, uncharged); Pen, penultimate; Ult, ultimate.

[b] For additional abbreviations see the respective tables.

hydrophobicity of the 5' portion of the total of 61 sequences is 66%, whereas that of the 3' part is 59%. As a whole, codons for charged amino acids are sparse and usually characterize the 5' half, where two-thirds of their total occur; there are some cases, nevertheless, where they abound in the downstream portion, the ribulose-1,5-bisphosphate carboxylase genes of plants (Table 7.5) being particularly outstanding examples. Triplets encoding uncharged polar amino acids are more frequent in the 3' sector, being about 50% more abundant there than in the upstream part. In total hydrophobicity, the range is from the 50% level of the ribulose-1,5-bisphosphate carboxylases to 80 and 86% in two bacterial enzymes, for an overall mean rate of 63%.

In connection with the degree of hydrophobicity, one factor that has previously passed unnoticed in the literature needs to be brought out, an observation based on the number of codons for the several categories of amino acids. The apolar group of amino acids is encoded by a total of 28 codons, or 46% of the 61 that signal monomers, that is, exclusive of the three for stop combinations. Half as many, 14 (23%), code for charged polar types, and only slightly more, 19 or 31%, indicate uncharged polar varieties. Hence, a structure of 50% or less codons for apolar units is at the random-chance level and cannot validly be considered especially hydrophobic. Thus the 59% mean of 3' halves of Table 7.11 represents only a mild level of hydrophobicity; only the 65% average of the 5' halves appears to offer much conviction.

At the cleavage site, codons for hydrophobic protein monomers also are the most abundant, followed closely by those of uncharged ones, but the distribution is decidedly uneven. In the penultimate position of the presequence, those for neutral amino acids are prevalent, with a total of 35, against 20 for apolar and six for charged. But at the closing site the situation is reversed, with those encoding hydrophobic amino acids leading by a slim majority. The near absence here of triplets for charged amino acids is particularly surprising, as is their virtual nonexistence at the first location of the mature coding sequence. At the latter site codons for hydrophobic monomers are again the rule, in the ratio of 44 to 16 for neutral ones. Only at the second site of the mature sequence are codons for charged amino acids frequent, but not nearly as abundantly as those of the neutral and apolar types, which are subequal in numbers.

Importance of the Transit Peptide. Obviously the transit peptides encoded by presequences of numerous simple diplomorphic genes vary widely, as is made clear by Table 7.11. Aside from the very evident hydrophobicity that usually prevails, few common traits are in view—even the extent of hydrophobic genes is under 50% in some cases. Nevertheless, the transit peptides have been demonstrated to be requisite as a whole for the transport of proteins through various membranes (Austen *et al.*, 1984; Hannink and Donoghue, 1984; Hurt *et al.*, 1984; Horwich *et al.*, 1985a; Takahara *et al.*, 1985). But what specifically qualifies a peptide sequence to serve in this capacity remains confused at the moment. As was seen, much diversity exists among presequences, as a consequence of which experimental results are also highly varied. One series of analyses on mutations within the transit-peptide region indicated that the length of its hydrophobic sector was the major determinant of its functionality (Bankaitis *et al.*, 1984). In some cases a specific essential sequence has been identified. For instance, the β-galactosidase of *E. coli* could be targeted to the nucleus in yeast cells, provided a segment as small as 13 amino acids of the presequence was present (M. N. Hall *et al.*, 1984). An important part of this appeared to be the series lysine, isoleucine, proline, isoleucine, lysine, that is, three apolar types between two charged units. But as the tables show, few presequences possess such a series. Moreover, arginine has been stated to be essential (Horwich *et al.*, 1985b). Still another proposed requisite feature was the secondary structure, as in the appendage of the E. *coli lamB* gene (Table 7.9; Emr and Silhavy, 1983). In this investigation the results indicated that genetic shortening of the helical region between remote codons for proline and glycine induced a coiled configuration in the encoded product that prevented transit through the cell membrane. But again this arrangement of codons for amino acids that interfere with α-helix construction is far from being a universal feature.

Absence of a Presequence in Exported Gene Products. Although a transit peptide characterizes the majority of products that must pass through a membrane before being fully processed, that condition, too, lacks catholicity. Indeed, a recently identified peptide involved in the processing of the transit portion of lipoproteins in E. *coli* has proven to be devoid of this trait, in spite of its being exported (Innis *et al.*, 1984). Such E. *coli* outer membrane proteins as the products of the *ompR* and *envZ* genes and outer membrane phospholipase A, which obviously must pass through that of the cell, lack identifiable presequences (Mizuno *et al.*, 1982; Wurtzel *et al.*, 1982; de Geus *et al.*, 1984). Among the most evident members of this category are the cytochrome *c* genes, which, as seen earlier, are located in the nucleus and translated in the cytoplasm, but their products function only in the mitochondrion. Yet they universally have no presequence (Limbach and Wu, 1985a,b; Scarpulla, 1985). This absence, however, is readily understood in the present example, because cytochrome *c* does not actually penetrate the mitochondrion, remaining outside in close association with the outer membrane.

Additionally, a number of proteins that remain embedded in the cytoplasmic membrane do not possess a presequence. One such has recently come to light in rat liver, a peptide that serves as the receptor for asialoglycoprotein (Holland *et al.*, 1984). But a number of similar structures are well known, including genes for E. *coli* lactose permease and members of the *Salmonella* histidine- and E. *coli* maltose-transport systems (Ehring *et al.*, 1980; Higgins *et al.*, 1982; Froshauer and Beckwith, 1984). Several genes for viral products that become located in prokaryotic or eukaryotic membranes but lack a presequence also have been sequenced, among which are the gene III of bacteriophage f1 that infests E. *coli* and the E1 glycoprotein of a coronavirus that reproduces within the endoreticular membranes of the laboratory mouse (Boeke and Model, 1982; Armstrong *et al.*, 1984). To the contrary, the fusion protein gene of the human respiratory syncytial virus has a presequence of 26 codons, despite the fact that mature product is embedded in the cell membrane (Elango *et al.*, 1985). Consequently, it may be deduced that proteins penetrating into and becoming fixed within a membrane require factors different from those that pass through a membrane.

Multiplicity of Factors. Since the proteins that travel through membranes are so diverse in structure of the transit presequence, or, in the latter's absence, in the structure of the mature protein, it appears unrealistic to suppose that only one or a few membrane components carry out the function of recognizing and transporting proteins of all descriptions. Rather, it seems far more logical to propose that a large number of such transit-peptide-recognizing substances are present that react with one or two families of products and no other (Rapoport, 1985). Indeed, one such protein has recently been purified from endoreticulum of canine pancreas and visualized by electron microscopy. This component, which is comprised of six polypeptide subunits plus one molecule of 7 S RNA, proved to be a narrow cylinder 24 nm long and 6 nm wide (Andrews *et al.*, 1985). However, this component recognizes only the presequences of messengers that are to be processed within the endoreticulum and functions in establishing the ribosomal connection with that organelle's membrane to initiate translation. Hence, presequences and transit peptides destined for processing or functioning in other organelles would be recognized by other proteins. As a consequence of this diversity in the transit agents, no great uniformity in structure, either of the transit presequence or transported mature product, would be expected; rather, a diversity, such as that which exists, would be predicted.

The comparable multiple-factor condition that is coming to light for the peptidases that remove the transit peptide correlates well with the foregoing proposals. In *E. coli* one such peptidase that acts upon the cleavage site of precoat proteins has long been recognized (Zwizinski and Wickner, 1980; Wolfe *et al.*, 1983). As pointed out before, a second one, active only on lipoproteins, has now been characterized (Innes *et al.*, 1984). Just as the visualized multiplicity of presequence and transit-peptide-recognizing proteins may explain the structural diversity of the latter, this possible many-factored condition now being revealed affords an explanation of the inconstancy that prevails at the cleavage site.

Evolutionary Implications. If the hypothesized abundance of protein types becomes more firmly established, as seems to be the strong likelihood, then current explanations of distribution of mitochondrial and chloroplastic genes between the nuclear and organellar genomes need to be rethought. Obviously, removal of a gene from the organellar DNA for insertion into the former requires far more than is evident from the superficial data on which that type of proposal is based. The need for a preexisting presequence has already been pointed out (Chapter 6, Section 6.1.3), but now it appears that each such transplanting requires the presence in the proper location of two additional substances, a particular transit-peptide recognition protein and a specific transit peptidase. Both of the latter as well as the presequence would have to be present immediately following the translocation of the given gene, otherwise the product would be unable to reach its destination and become functional. Translocating a gene from an organelle to the nucleus may be readily performed, but making it able to carry out its usual activity in the needed site requires the simultaneous acquisition of three additional genes, if the predicted multiple-factor theory proves to be correct.

7.4. PRE- AND PROSEQUENCES OF MORE COMPLEX DIPLOMORPHS

A large number of genes, especially those for secreted products, not only encode a transit peptide in a presequence, but also have another expendable sector known as a prosequence, therefore being rather more complex than the preceding types (Figure 7.1). Like the former, this latter section is removed proteolytically, its removal bringing about the activation of the mature product. Thus, so long as the two portions are attached, it suppresses the activity of the principal protein and accordingly is here named the inhibitor peptide. In examining the structure of these two classes of appended gene sequences, the procedures parallel those of the foregoing sections, with vertebrate genes being viewed first.

7.4.1. Transit and Inhibitor Sequences of Vertebrate Genes

An unexpectedly large proportion of genes for proteins in vertebrates bear pre- and prosequences at the 5' end. Since the presence of an inhibitor peptide provides the organism with a means of controlling the activity of the substances, its occurrence on genes for such digestive enzymes as pepsinogen and chymotrypsinogen comes as no surprise. What is unexpected is that inhibitors also are found on such chemically mild substances as albumins, as is seen in the first set of examples that follows.

The Albumin Family of Genes. Included within a common protein family are four diverse major categories, albumins, α-fetoproteins (really embryonic albumins), parathyroid hormone (PTH), and lysozymes. Although the gene sequences of representatives from each of these groups have been established, the precise limits of the prosequence are not always provided. Consequently, the genes for such proteins as rat α-lactalbumin and chicken lysozyme (Jung *et al.*, 1980; Qasba and Safaya, 1984) and the α-fetoproteins of mouse and human (Gorin *et al.*, 1981; Law and Dugaiczyk, 1982; Morinaga *et al.*, 1983; Sakai *et al.*, 1985) could not be included in Table 7.12. However, the five contained therein serve well in introducing this topic, for they fall into two contrasting sets, which nevertheless share some general trends.

The structures of the three PTH genes of mammals are especially helpful because they are largely homologous. As a whole, their presequences display the same major features as did those of simpler genes, namely a high content of codons for hydrophobic amino acids and the presence of one or two triplets encoding basic monomers near the 5' end. The two codons for methionine at the extreme 5' terminus in the bovine and rat genes (Heinrich *et al.*, 1984a; Weaver *et al.*, 1984), but not in the human (Vasicek *et al.*, 1983), while not of rare occurrence, is certainly not commonplace. This same codon is highly conserved at three other places in the presequence, where it may hold functional importance. In each member of the trio, the 3' half is distinctive in having codons for acidic amino acids in the penultimate site, often preceded by one for a basic monomer two sites upstream. Their prosequences are uniformly short, consisting of only 18 nucleotide residues, and are remarkable for their basicity, each containing four triplets encoding lysine or arginine, all of which are located at the terminal portions.

The other two sequences are from human sources, the first encoding blood serum albumin (Mita *et al.*, 1984) and the second encoding that protein from liver (Morinaga *et al.*, 1983). In the 5' halves, several relationships to the PTH presequences can be perceived, although homologous sites are far from abundant. Only single codons for basic amino acids lie near the upstream termini, differently situated in each case. Distinctions between the pair of presequences exist also in the 3' halves, for whereas the liver albumin gene resembles the termini of the parathyroid hormones, the serum transit sequence does not. However, in the prosequence, that situation is reversed, with the serum protein gene more closely approaching the hormonal. In neither case is the extreme basicity of the latter approached, the sector from the serum cistron having a 50% basic ratio and that of the liver only 12½%. Thus, here it would seem, as in those of the first sections, that variability around the cleavage sites is their only constant feature.

The Trypsinogen Family of Vertebrate Genes. The present family of proteases whose active sites contain a serine residue includes trypsinogen, chymotrypsinogen,

Table 7.12

Transit and Inhibitor Peptide Genes of the Albumin Family[a]

Row 1

Presequence

Parathyroid hormone

Bovine PTH[b] ATG ATG TCT GCA AAA GAC ATG AAG GTA ATG ATT GTC ATG CTT GCC ATC TGT TTT

Rat PTH[c] ATG ATG TCT GCA AGC ATG GCT AAG GTG ATC CTC ATG CTG GCA GTT TGT CTC

Human PTH[d] ATG ATA CCT GCA AAA GAC ATG GCT AAA GTT ATC ATT GTC ATG GTC GCA ATT TGT TTT

Albumins

Human serum A[e] ATG AAG TGG GTA --- ACC --- --- TTT --- ATT --- TCC CTT --- --- CTT TTT

Human liver A[f] ATG GCT TCT CAT CGT CTG --- --- --- CTC --- CTC CTC TGC CTT --- --- GCT GGA

Row 2

Prosequence   → Mature gene

Parathyroid hormone

Bovine PTH CTT GCA AGA TCA --- GAT GGG AAG TCT --- GTT --- AA*G AAG AGA | GCT GTG

Rat PTH CTT ACC CAG GCA --- GAT GGG AAA CCC --- GTT --- AA G AAG AGA GCT GTC

Human PTH CTT ACA AAA TCG --- GAT GGG AAA TCT --- GTT --- AA*G AAG AGA TCT GTG

Albumins

Human serum A CTC TTT AGC TCG GCT TAT TCC AGG GGT --- GTG TTT --- - CGT CGA GAT GCA

Human liver A CTG GTA TTT GTG TCT GAG GCT GGC CCT ACG GGC ACC GG T GAA TCC AAG TGT

[a]Codons for hydrophobic amino acids are italicized and those for hydrophilic ones are underscored. PTH, parathyroid hormone. Asterisks indicate location of an intron; arrows mark cleavage sites.
[b]Weaver et al. (1984). [c]Heinrich et al. (1984a). [d]Vasicek et al. (1983). [e]Mita et al. (1984). [f]Morinaga et al. (1983).

elastases, and kallikreins. Typically each is represented in the genome by multiple iso-zymic species, with kallikreins being especially highly diversified. The members of that category form a distinct subfamily that processes the precursors of polypeptide hormones, each having a limited substrate specificity. One representative is a subunit of the enzyme that processes nerve growth factor, another activates epidermal growth factor, and a third removes the repressor peptide from angiotensinogen to produce angiotensin II. In the mouse the 25–30 kallikrein genes have been located on chromosome 7 (Mason *et al.*, 1983). At this early stage in the sequencing of genes, only a few members of the family have had their structures determined; however, some of these, like the trypsinogen gene and two elastase cistrons, have been sequenced, all from the rat (Craik *et al.*, 1984; Swift *et al.*, 1984), but the precise limits of the prosequences have not been established. Consequently, just three representatives of the group are included in Table 7.13, cistrons for canine chymotrypsinogen (Pinsky *et al.*, 1983) and rat and mouse kallikreins (Swift *et al.*, 1982; Mason *et al.*, 1983).

The three examples are remarkably similar in structure, including chymotryp-sinogen, despite the latter's strongly contrasting activity. Both the pre- and prosequences are rather constant in length, the most deviant of the former being that of the second species of kallikrein. Probably the greatest distinctive feature is the presence of an intron near the 5′ end of row 2 in the mouse representative, which is not indicated in the others. Undoubtedly this is the result of faulty knowledge, rather than a basic difference, since the other two sequences have been derived from mRNAs, the third alone being from the genomic DNA. Two cleavage sites for the presequences are given in the table, the more upstream one being derived from the canine and rat structures, the second from that of the mouse. Since the chemical nature of the encoded amino acids abutting the sites is similar in each case, it is not possible to determine the one more likely to prove valid. The 3′ end of the prosequence, however, is firmly established, a codon for an arginine base being uniformly at the cleavage point, with a triplet for serine preceding it in all cases. At the 5′ end of the mature coding sequence, two codons for apolar amino acids border the site.

Vertebrate Lipoproteins. The plasmolipoproteins of vertebrates are grouped into four main classes, chylomicrons, very low density, low density, and high density. At least nine species have been identified, A-I, A-II, A-IV, B, C-I to C-III, D, and E, of which the first two make up most of the high-density fraction (Shoulders *et al.*, 1983). This pair also represents the only two that are known to bear both pre- and prosequences, the remainder having only the former. Consequently, these are probably the sole representatives of the family that are potent enzymes. The A-I species is involved as a cofactor of the lecithin-cholesterol acyltransferase activity (Karathanasis *et al.*, 1983), a reaction in which A-II may also participate (Moore *et al.*, 1984).

The representatives of those two species included in Table 7.13 are from human sources, only the first of which reflects the structure of the DNA. Hence, the seeming absence of an intron in A-II results from all three of the established sequences of this species having been derived from mRNA (Knott *et al.*, 1984b; Lackner *et al.*, 1984; Moore *et al.*, 1984). Although no relationship is implied between the present proteins and the proteases also shown in the table, an unexpectedly high number of sites do display homologies in the presequence portion—even the cleavage site of that appendage shows some similarities. However, there is a codon for a basic amino acid at the immediate 5′ end that has no counterpart in the others. The prosequences are completely different from

Table 7.13
Pre- and Prosequences of the Vertebrate Trypsinogen and Liproprotein Families[a]

Row 1

Prosequence

Proteases

Canine chymotrypsinogen[b]: ATG GCT TTC --- --- CTC TGG CTC CTC TCC TGC TTC GCC CTC CTG GGC ACA GCC --- --- TTC GGC → TGC GGG

Rat kallikrein[c]: ATG CCT GTT ACC ATG TGG TTC ATC TTC TTC GCC CTG --- --- TCC --- --- CTG GGA   CGG AAT

Mouse kallikrein[d]: ATG --- --- --- --- TGG TTC CTG ATC CTG TTC CTA GCC CTG --- --- TCC --- --- CTA GGA   GGG ATT

Lipoproteins

Human A–I[e]: ATG AAA GCT GCG --- --- GTG CTG ACC TTG GCC GTG CTC TTC CTG ACG* GGG AGC CAG GCT   CGG CAT

Human A–II[f]: ATG AAG CTG --- CTC --- GCA GCA ACT GTG CTA CTC CTC ACC ATC TGC AGC CTT GAA GGA   GCT TTG

Row 2

Presequence

Proteases

Canine chymotrypsinogen: G TC CCT GCC ATC   CAG CCG GTG TTA AGT GGC CTG TCC AGG → ATC GTC

Rat kallikrein: G AT --- GCT GCA   CCT CCC GTC --- --- --- CAG TCT CGG   GTT GTT

Mouse kallikrein: G*AT --- GCT GCA   CCT CCT GTC --- --- --- CAG TCT CGA   ATA GTT

Lipoproteins

Human A–I: --- --- --- TTC   TGG --- --- --- --- --- --- CAG CAA   GAT GAA

Human A–II: --- --- --- GTT   --- --- --- --- --- --- --- CGG AGA   CAG GCA

[a]Codons for hydrophobic amino acids are italicized and those for hydrophilic ones are underscored. Asterisks indicate the location of an intron. Arrows mark cleavage sites; alt, alternative cleavage site.
[b]Pinsky et al. (1983).   [c]Swift et al. (1982).   [d]Mason et al. (1983).
[e]Shoulders et al. (1983); Law and Brewer (1984).
[f]Knott et al. (1984b); Lackner et al. (1984); Moore et al. (1984).

those of the trypsinogen family. In the first place, here they are extremely short, at most consisting of only five or six codons, whereas in the others they range from 11 to 15 triplets. Second, the chemical nature of the encoded amino acids at the downstream cleavage site differs strongly, involving the presence of at least one pair of charged monomers.

The Renin Family of Vertebrate Genes. In order to provide a more accurate picture of the vertebrate pre- and prosequences, one additional important family of their genes needs to be analyzed, that including the cistrons for renin and pepsinogen. It is at once apparent that the presequences of the pair given in Table 7.14 have no outstanding features. As in the preceding group, the second site is occupied by a codon for a charged amino acid, in one case for an acid species, in the other for a basic species (Sogawa *et al.*, 1983; Hobart *et al.*, 1984; Miyazaki *et al.*, 1984). At least five codons for leucine (CTN) are present, often in tandem fashion, and the final site at the cleavage point is occupied by a triplet for cysteine. The prosequences are strikingly longer than any that have been viewed to this point, consisting of 46–50 codons. Moreover, an unusually high percentage of these encode alkaline amino acids, ranging from 20 to 25%. These two sectors are definitely not hydrophobic, since less than half of the constituents encode amino acids of that character.

7.4.2. More Complex Diplomorphic Genes of Bacteria

Although an occasional gene for a product referred to as a preproprotein has been sequenced from seed-plant sources (e.g., Lycett *et al.*, 1983), none apparently has actually been established, nor do any appear to be known from yeast. Even the bacteria do not provide a plethora of the type sought for present purposes, although often in bacteriology the presequence is incorrectly referred to as a prosequence; all those available are from bacilli, not the usual *E. coli*. These three encode endoproteases, two of which are for the acid type known as subtilisin, the other for neutral protease. Both enzymes are largely secreted into the environment, only 5% being retained within the organism. The gene for the first of these is activated, along with those of two minor proteases, only when the bacteria are about to sporulate, but the role of the enzyme in spore formation remains unclear (Wells *et al.*, 1983; Wong *et al.*, 1984).

The Presequences. Although it is to be expected that the presequences for the same gene from two species of a given genus, such as those for subtilisin from *Bacillus subtilis* and *B. amyloliquefaciens*, should be largely homologous, what is surprising is that the structure from a different cistron, that for neutral protease, should also show a number of similarities to that pair (Table 7.15). To begin with, all have GTG as the initiation codon, translated in each case as methionine, not valine as elsewhere. Just downstream from this, all three sequences have two contiguous triplets for lysine. Then for a space no kinship between the two types is found, until beyond the middle of the first row, where a broken series of five correspondences exists, beginning with TT- triplets. The extreme 3' terminus also is constant in each case, the penultimate codon being CAG followed by GC- for alanine. The presequences are moderately short, consisting of 27–29 triplets, about two-thirds of which encode hydrophobic species of amino acids.

The Prosequences. The prosequences encoding the inhibitor peptide greatly exceed in length any that have been reported in preceding pages (Table 7.15). Of the three,

Table 7.14
Pre- and Prosequences of the Renin Family of Genes[a]

Row 1

	Presequence	Prosequence
Human renin[b]	ATG GAT GGA TGG AGA AGG ATG CCT CGC TGG GGA CTG CTG CTG CTC TGG GGC TGG TGT	ACC TTT ---
Human pepsinogen[c]	ATG AAG --- TGG --- --- CTG CTG CTG GGT CTG GCG CTC TCT --- --- --- GAG TGC	ATC ATG TAC

Row 2

Human renin	GG T --- CTC CCG ACA GAC ACC ACC ACC TTT AAA CG *G ATC TTC CTC AAG AGA ATG CCC TCA ATC CGA GAA AGC
Human pepsinogin	AA* G GTC CCC CTC ATC AGA AAG AAG TCC TTG AGG CG C ACC CTG TCC GAG CGT GGC CTG CTG --- AAG GAC TTC

Row 3

		Mature gene
Human renin	CTG AAG GAA CGA GGT GTG GAC ATG GCC --- AGG CTT GGT CCC GAG TGG AGC CAA CCC ATG AAG AGG	CTG ACA
Human pepsinogin	CTG AAG AAG CAC AAC CTC AAC CCA GCC AGA AAG TAC TTC CCC CAG TGG GAG GCT CCC --- ACC CTG	GTA GAT

[a]Codons for hydrophobic amino acids are italicized and those for hydrophilic ones are underscored. Asterisks indicate the location of an intron. Arrows mark the cleavage sites.
[b]Hobart et al. (1984); Miyazaki et al. (1984).
[c]Sogawa et al. (1983).

Table 7.15

Pre- and Prosequences from Bacterial Genes for Secreted Proteins[a]

Row 1 — Presequence

B.s. subtilisin[b]	GTG AGA --- AGC AAA AAA TTG TGG ATC AGC TTG TTT TTT GCG TTA ATC TTT ACG ATG GCA TTC AGC
B.a. subtilisin[c]	GTG AGA --- GGC AAA AAA GTA TGG ATC AGT TTG GCT TTA GCG TTA ATC TTT ACG ATG GCG TTC GGC
B.a. neutral protease[c]	GTG GGT TTA GGT AAG AAA TTG TCT GTT GTT GCC GCT TCC --- AGT TTA ACC ATC --- --- --- AGT

Row 2 — Prosequence

B.s. subtilisin	AAC ATG TCT GCG --- CAG GCT GCC GGA AAA --- --- AGC AGT ACA --- GAA AAG AAA TAC --- ATT GTC GGA
B.a. subtilisin	AGC ACA TCC TCT GCC CAG GCG GCG GGG AAA --- --- TCA AAC GGG --- GAA AAG AAA TAT --- ATT GTC GGG
B.a. neutral protease	CTG CCG GGT GTT --- CAG GCC GCT GAG AAT +351 GCG CTG GAT CAT TAT AAA GCG ATC GGC AAA TCA CCT

Row 3

B.s. subtilisin	TTT --- --- AAA CAG ACA --- ATG AGT GCC ATG AGT TCC GCC AAG AAG AAA AAG GAT GTT ATT TCT GAA AAA GGC GGA
B.a. subtilisin	TTT --- --- AAA CAG ACA --- ATG AGC ACG GCC GCC GCT AAG AAG AAG GTC GTT ATT TCT GAA AAA GGC GGG
B.a. neutral protease	CAA GCC GTT TCT AAC GGA ACC GTT GCA AAC AAA GCC GAG CTG AAA GCA GCA GCC ACA AAA GAC GGC

Row 4

B.s. subtilisin	AAG GTT CAA AAG CAA TTT AAG TAT GTT AAC GCG GCC GCA GCA ACA TTG GAT GAA AAA GCT GTA AAA AAA GAA TTG
B.a. subtilisin	AAA GTG CAA AAG CAA TTC AAA TAT GTA GAC GCA GCT TCA GCT ACA TTA AAC GAA AAA GCT GTA AAA AAA GAA TTG
B.a. neutral protease	AAA --- --- TAC CGC GCC AAA GTA ACC ATC CGC GAA ATC GAA CCG CAA CCG GAA ATC GCA AAC TGG AAA GAA GTA

Row 5

B.s. subtilisin	AAA AAA GAT AGC AGC GTT GCA TAT GTT GAA GAT ATT GCA CAT GAA TAT → GCG CAA
B.a. subtilisin	AAA AAA GAC CCG AGC GTC CCT TAC GTT GAA GAT GAT CAC GCA CAT GCG TAC GCG CAG
B.a. neutral protease	ACC GTT GAT GCG GAA ACA GGA GGA AAA ATC CTG AAA AAG GAA ATC GTG GAG CAT GCC GCC

[a] Codons for hydrophilic amino acids are underscored and those for hydrophobic ones are italicized. Arrows mark the cleavage sites.
[b] *Bacillus subtilis*; Wong et al. (1984).
[c] *B. amyloliquefaciens*; Vasantha et al. (1984).

that encoding subtilisin in *B. subtilis,* with 73 condons, is the shortest, its parallel from *B. amyloliquefaciens* being just one codon longer. In the cistron for neutral protease, the prosequence is 194 condons in length (Vasantha *et al.,* 1984), but 117 (351 nucleotides) of these are excluded from the table as meaningless in the absence of comparative material. Insofar as their general chemical natures are concerned, all are similar in having 31 codons for apolar amino acids in the regions given. The two subtilisin genes encode 15 lysines, but no arginines and either nine or ten acidic protein monomers. The neutral protease has triplets for 11 lysines and two arginines and ten for acidic components. At the 5' terminus the latter has a triplet for an acid amino acid, whereas the other two have codons for lysine. What is particularly striking, however, is that these codons very frequently lie at coinciding points in the sequence, especially beginning with the end of the third row, where three AAAs for lysine comprise a column. In row 4 this is followed almost immediately by another column of lysine codons, then by one for tyrosin (TAT), and toward the end by two sets of GAA for glutamic acid. All in all, nevertheless, this prosequence of the gene for neutral protease cannot be homologized with those of the two subtilisin cistrons.

7.5. GENES FOR CRYPTOMORPHIC PROTEINS

The several preceding sections encompassed genes for many contrasting types of proteins, which shared common distinctive structural features. All had a presequence whose product after translation was removed to release either a mature protein in simpler cases, or, in more complex instances, those that also bore an inhibitor peptide, whose removal then gave rise to the functional product. Although each pre- and prosequence was seen to be unique, they shared a number of common properties, at least functionally. In the present section, however, much more complex gene structures are considered, whose organizational traits are highly divergent from one representative subclass to another. Presequences and inhibitor sections are often present, but one or the other, or even both, may occasionally be missing. Those features thus are of minor importance in the present class of genes. What is of significance here is the posttranslational fragmentation of the product encoded by the mature coding region into one or more definitive peptides; those actual functional parts encoded by the smaller fractions of the mature region are referred to here as the ''ultimate'' genes. As a rule the sequences between their encoded peptides are called connectors. The posttranslational aspect of cleavage of the major sectors is an important consideration, for the members of one subclass grade into multigenic precursorial transcripts. In the latter, however, the several parts are cleaved while in the form of RNA and by an RNase, not after translation and by a protease. Because the sequence encoding the actual active substances is concealed within a far longer mature region, the genes of this type are said to be ''cryptomorphic,'' as pointed out in Chapter 1, Section 1.1.3. Insofar as is known, the cryptomorphic class is confined to eukaryotes, reflecting the complicated requirements of more complexly organized cells.

7.5.1. Cryptomorphic Genes Encoding Multiple Identical Proteins

Probably the simplest group of cryptomorphic genes (designated here as subclass I) is that which embraces those encoding multiple copies of a single peptide; consequently,

Table 7.16
Presequences of Simple Cryptomorphic Genes (Subclass I)[a]

Row 1

Yeast *MFα1*[b]	ATG <u>AGA</u> *TTT* *CCT* TCA *ATT* *TTT* ACT *GCA* *GTT* *TTA* *TTC* *GCA* *GCA*
Yeast *MFα2*[b]	ATG <u>AAA</u> *TTC* *ATT* TCT ACC *TTT* *CTC* ACT *TTT* *ATT* *TTA* *GCG* *GCC*
Rat enkephalin[c]	ATG *GCG* CAG *TTC* *CTG* <u>AGA</u> *CTT* TGC *ATC* TGG *CTG* *CTA* *GCG* *CTT*
Human enkephalin[d]	ATG *GCG* <u>CGG</u> *TTC* *CTG* ACA *CTT* TGC ACT TGG *CTG* *CTG* *TTG* *CTC*
Bovine enkephalin[e]	ATG *GCG* <u>CGG</u> *TTC* *CTG* *GGA* *CTC* TGC ACT TGG *CTG* *CTG* *GCG* *CTC*

Row 2

Yeast *MFα1*	TCC TCC *GCA* *TTA* *GCT* *GCT* --- --- --- CCA *GTC*	AAC ACT
Yeast *MFα2*	*GTT* TCT *GTC* ACT *GCT* --- --- --- *AGT* TCC	<u>GAT</u> <u>GAA</u>
Rat enkephalin	*GGG* *TCC* TGC *CTC* *CTG* *GCT* ACA *GTG* CAG *GCA*	<u>GAC</u> TGC
Human enkephalin	*GGC* *CCC* *GGG* *CTC* *CTG* *GCG* ACC *GTG* <u>CGG</u> *GCC*	<u>GAA</u> TGC
Bovine enkephalin	*GGC* *CCC* *GGG* *CTC* *CTG* *GCG* ACC *GTC* <u>AGG</u> *GCA*	<u>GAA</u> TGC

[a] Codons for hydrophilic amino acids are underscored and those for hydrophobic ones are italicized. Arrow indicates the cleavage site.
[b] Singh *et al.* (1983).
[c] Rosen *et al.* (1984); Yoshikawa *et al.* (1984).
[d] Legon *et al.* (1982).
[e] Gubler *et al.* (1982); Noda *et al.* (1982).

complete processing of the mature coding sector releases several active molecules of identical (or near-identical) structures. But this simplicity is only relative, even in the most primitive eukaryotes, as seen in the discussion that follows immediately.

The α-Pheromones of Yeast. Mating in yeast is coupled to the production of specific pheromones, or mating factors, there being two contrasting types, *a* and α. One population of cells produces factor *a*, and the other α, each of which is capable of arresting cells of opposite type in G_1 of the cell cycle, thereby preventing cell division and asexual multiplication (Siliciano and Tatchell, 1984). Since the gene structure of the α pheromone is by far the better documented, attention is confined entirely to that product and its gene. Recently it has been established that yeast possesses two cistrons for this substance, *MFα1* encoding four copies and *MFα2* only two (Singh *et al.*, 1983). Both are simple cryptomorphs, bearing a presequence encoding a transit peptide that enables the product to be conducted through the cell wall into the medium, where it can act upon cells of opposite type. For ease of comparison, the transit sectors are tabulated separately from the body of the cistrons, which are shown in Table 7.17.

The sequences for the two transit peptides (Table 7.16) are of nearly equal length (21 and 22 codons) and show extensive homology. Each has a codon for lysine (α1) or arginine (α2) directly after the initiation triplet. Although codons for hydrophobic amino acids predominate throughout, there is a continuous series of six near the middle of each presequence and also a run of similar length in α1 at the 3' terminus.

Table 7.17

Mature Coding Sectors of Mfα Genes of Yeast[a]

Row 1

MFα1 AAC ACT ACA ACA GAA GAT GAA ACG GCA CAA ATT CCG GCT GAA GCT GTC ATC GGT TAC TTA
MFα2 --- --- GAT GAA GAT ATC GCT CAG GTG CCA GCC ATT ATT GGA TAC TTG

Row 2

MFα1 GAT TTA GAA GGG GAT TTC GAT GTT GCT GTT TTG CCA TTT TCC AAC AGC ACA AAT AAC GGG
MFα2 GAT TTC GGA GGT GAT CAT GAC ATA GCT TTT TTT TTA CCA TTC AGT AAC GCT ACC GCC AGT GGG

Row 3

MFα1 TTA TTG TTT ATA AAT ACT ACT ATT GCC AGC ATT GCT GCT AAA GAA GGG GTA TCT TCT TTG GAT
MFα2 CTA TTG TTT ATC AAC ACT ATT GCT GAG GCG GCT GAA AAA GAG CAA CAA AAC ACT ACT TTG GCG

Row 4

          Connecting Sequence
          →
MFα1 AAA AGA GAG --- --- GCT GAA GCT   TGG CAT TGG TTC CAA CTA AAA CCT GGC CAA CCA ATG TAC   α1A
MFα2 AAA AGA GAG GCT GTT GCC GAC GCT   TGG CAC TGG TTA AAT TTG AGA CCA CAA GGC CAA CCA ATG TAC   α2A

Row 5

          Connecting Sequence
          →
MFα1 AAG AGA GCC GCA GCT CAA GCT   TGG CAT TGG CTG CAA CTA AAG CCT GGC CAA CCA ATG TAC   α1B
MFα2 AAG AGA GAG GCC AAC GCT GAT GCT   TGG CAC TGG TTG CAA CTC AAG CCA GGC CAA CCA ATG TAC   α2B

Row 6

Connecting sequence

MFα1 AAA AGA GAA GAA *GCC GAC GCT GAA GCT* TGG CAT TGG *CTG* CAA *CTA* AAG *CCT GCC* CAA *CCA ATG* TAC α1C

MFα2 TGA

Row 7

Connecting sequence

MFα1 AAA AGA GAA GAA *GCC GAC GCT GAA GCT* TGG CAT TGG *TTG* CAG *TTA* AAA *CCC GCC* CAA *CCA ATG* TAC TAA α1D

[a]Codons for hydrophilic amino acids are underscored and those for hydrophobic ones are italicized. Arrows indicate cleavage sites.
[b]Singh *et al.* (1983).

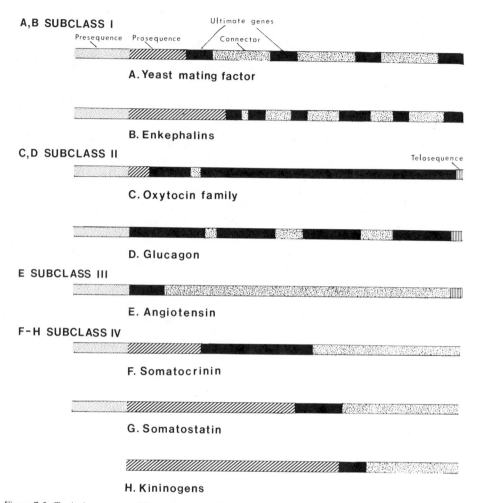

A,B SUBCLASS I

Ultimate genes

Presequence Prosequence Connector

A. Yeast mating factor

B. Enkephalins

C,D SUBCLASS II

Telosequence

C. Oxytocin family

D. Glucagon

E SUBCLASS III

E. Angiotensin

F–H SUBCLASS IV

F. Somatocrinin

G. Somatostatin

H. Kininogens

Figure 7.2. Typical structures of four subclasses of cryptomorphic genes; the fifth is shown in Figures 7.3 and 7.4.

Mature Coding Sectors. Aside from the distinctions in gene content just mentioned, the mature coding sectors of the two cistrons differ in several outstanding ways, although numerous homologies exist on a site-to-site basis (Table 7.17). The first of these is the absence of a four-codon sector at the 5′ end of α2, and later, a shorter region of two triplets that is lacking in α1 (Kurjan and Herskowitz, 1982; Singh *et al.*, 1983). But the greatest contrast between the two is in total length, α1 containing 144 codons to the 97 of α2, a condition contributing to the loss of two ultimate coding sequences from the latter. The ultimate genes that encode the actual α factors form a relatively small fraction of the whole, because each consists of only 13 codons, which is a total of just over a third of the mature coding region in α1 and about 25% in α2 (Figure 7.2A). When the mature product is activated, the processing seemingly takes place in two steps, the first proteolysis occurring 5' to the series of three charged amino acids at the left of each of rows 4–7. At present no further function is known for the lengthy peptide occupying rows 1–3 in the table that is thus removed. During the second step of processing, the six- to eight-codon-long connecting peptides are removed from the ultimate products, the genes for which are shown stippled in the table.

Distinctions between the ultimate genes of the two cistrons are few, but those that do exist present a challenging basis for speculation as to the origin of each series. The second codons, CAC in α2 and CAT in α1, provide one constant difference, but the fourth triplets, TTA and TTG in α2, TTG and CTG in α1, are problematic, a difficulty accentuated by the triplet that follows. Here one finds CAA in α1A to 1C, with CAG in D, while in α2 the signals are AAT and CAA. The next one, too, is enigmatic, but the genes thereafter are largely identical.

7.5.2. Subclass I Cryptomorphs of Vertebrates

Although only a single group of members of subclass I cryptomorphic genes has yet been sequenced from vertebrates, there can be little doubt that additions will be made to the list as knowledge of the synthesis of secreted control elements advances. Even in these representatives, the small peptides known as enkephalins, understanding of their production is limited, for their presence in various tissues has only been detected within the last decade.

The gene sequences of three enkephalins have been established, one each from the rat, cattle, and human; typically they are derived from the adrenal medullary cells, but the rat sequence is from brain neurons. Two main varieties are found in those tissues, named after the fifth (terminal) amino acid in the pentapeptides, methionine-enkephalin and leucine-enkephalin; other variants result from processing variations, one possibility of which is suggested in a later discussion. Though the actual physiological role of the group remains unsolved, they have been demonstrated to product effects similar to opiate drugs, with which they compete *in vivo* (Hughes *et al.*, 1979).

Although its final assignment to class and subclass depends upon the establishment of the complete structure, the gene encoding vitellogenin in the chicken and most likely also in *Xenopus* probably is a member of the present subclass. This is an immense cistron, embracing over 20,000 base pairs, the whole being divided into 34 exons. Recently the determination of a sector containing exons 23 and 24, which together encode phosvitin, revealed characteristics that can only be interpreted in light of the entire gene being cryptomorphic (Byrne *et al.*, 1984). The short translated sector at the 5' end of phosvitin shows no indication of representing a transit peptide, since only eight of its 20 codons are for apolar amino acids; rather, it appears more like a connector, as does also the ten-codon sector that follows the final CAG triplet (for glutamine) that marks the 3' terminus of the mature coding region. Nor are any of the usual transcriptional and translational signals present in either flanking segment. Indeed, termination of transcription of the ovalbumin gene has now been demonstrated to occur ~900 base pairs downstream from the last exon (LeMeur *et al.*, 1984).

Presequences of Mammalian Enkephalins. The three presequences of enkephalin genes (Table 7.16) are homologous to a degree expected only of those encoding regions of critical importance; consequently, despite their ephemeral existence, their peptide products must be viewed in that light. If only a predominantly hydrophobic constitution is the requirement for passage through a membrane, the question arises as to why the constancy in codon structure exhibited by this trio. The only reasonable conclusion that comes to mind is that passage either to or through the membrane requires a specific factor that reacts to and transports the particular transit peptide and the protein it bears, and no other. Thus the implication is that each cryptomorph and diplomorph is coupled to a specific factor of either the cytoplasm or membrane, as pointed out in another context earlier. In other words, for each transit-peptide-bearing species, or at least species group, of proteins, there is a membrane-transiting protein that enables it to reach its ultimate destination.

The Ultimate Enkephalin Functional Genes. Unlike the α-pheromones of yeasts, all ultimate functional enkephalin sectors of a given gene are not identical, two diverging distinctly from the rest, another having the potential for special processing, and a fourth ending in a codon for a different amino acid (Table 7.18), as already pointed out. Ultimate genes 1–3 and 5 encode the typical met-enkephalin and 6 the somewhat rarer leu-enkephalin, while the product of gene 4 is known as met-enkephalin-arg^6-gly^7-leu^8 and that of gene 7 is met-enkephalin-arg^6-phe^7. In addition, under certain, unknown conditions, ultimate gene 5 may undergo alternative processing as indicated in the table to produce met-enkephalin-arg^6-arg^7-val^8-gly^9, but the possibility remains that this route is followed only under pathological conditions.

Also, in contrast to the earlier example, the connector sequences vary in length, so that ultimate gene 2 follows almost directly after 1 (Figure 7.2B). The third one is preceded by 20 codons in addition to the cleavage sites, the fourth by a variable number exceeding 40 triplets, and the fifth to seventh by 12, 11, and 22, respectively. It is interesting that the longest connector, that between genes 3 and 4, shows a slight decrease in length with increase in phylogenetic position in the Mammalia. Perhaps this sector will prove to be successively longer in bats, marsupials, and monotremes when the gene has been sequenced from those source organisms. One feature of these genes that agrees with those of the yeast is the great length of the prosequence, running to 73 codons before the first cleavage site, although that of cattle is reduced by the loss of three codons in row 3.

Cleavage sites, each consisting of two codons for basic amino acids, are located both before and after all the ultimate genes. It should be especially noted that the paired basic amino acids following the terminus of gene 1 do dual service in also providing for cleavage at the beginning of gene 2. Consequently, at least in this instance, proteolytic cleavage must be able to occur at each end of the double basic amino acids. The pair directly preceding the several ultimate coding regions constantly consists of one triplet for lysine and a second for aginine, but those that begin the connector strands may encode any combination of two basic protein monomers.

To judge from what can be gleaned from the current literature, quite a few, mostly small hormone cistrons will be found to be representatives of this subclass. For example, the gene for thyrotropin-releasing hormone, a substance abundant in the skin of *Xenopus,* contains four or more spaced repeats of codons for the tripeptide glutamine–histidine–proline that constitutes the active hormone (Richter *et al.,* 1984). Each set is preceded and followed by triplets encoding the lysine–arginine signaling combination.

Table 7.18
Mature Coding Regions of Enkephalin Genes[a]

Row 1

```
Rat enkephalin[b]      GAC TGC AGC CAG GAC TGC GCT AAA TGC TAC AGC CGC CTG GTA CGT CCC GGC GAC ATC AAC TTC CTG* GCA
Human enkephalin[c]    GAA TGC AGC CAG GAT TGC GCG ACG TGC AGC ACG TAC CGC CCG GCC GAC ATC AAC TTC CTG GCT
Bovine enkephalin[d]   GAA TGC AGC CAG GAC TGC GCC ACG TGC AGC TAC CGC GCG CGC CCG ACT GAC CTC AAC CCG CTG GCT
```

Row 2

```
Rat enkephalin        TGC ACA CTC GAA GAA GGG CAG CTG CCT TCT TTC AAA ATC TGG GAG ACC TGC AAG GAT CTC CTG CAG
Human enkephalin      TGC GTA ATG GAA GGT AAA CTG CCT TCT CTG AAA ATT TGG GAA ACC TGC AAG GAG CTC CTG CAG
Bovine enkephalin     TGC ACT CTG GAA GTG GGG AAA CTA CCT TCT CTC AAG ACC TGG GAA ACC TGC AAG GAG CTT CTG CAG
```

Row 3

```
Rat enkephalin        GTG TCC AAG CCG GAG TTC CCT TGG GAT AAC ATC GAC ATG TAC AAA GAC AGC AAA CAG GAT GAG AGC
Human enkephalin      CTG TCC AAA CCA GAG CTT CCT CAA GAT GGC ACC AGC ACC CTC AGA GAA AAT AGC AAA CCG GAA GAA AGC
Bovine enkephalin     CTG ACC AAA CTA GAA CTT CCT CCA GAT GCC ACC AGT GCC CTC ---- ---- AGC AAA CAG GAG GAA AGC
```

Row 4

Ultimate gene #1

```
Rat enkephalin        CAC TTG CTA GCC AAG AAG  TAT GGA GGG TTC ATG
Human enkephalin      CAT TTG CTA GCC AAA AGG  TAT GGG GGC TTC ATG
Bovine enkephalin     CAC CTG CTT GCT AAG AAG  TAC GGG GGC TTC ATG
```

(continued)

Table 7.18 (Continued)

Row 5

	Connector region	→	Ultimate gene #2
Rat enkephalin	--- --- --- ---	AAA CGG	TAT *GGA GGC TTC ATG*
Human enkephalin	--- --- --- ---	AAA AGG	TAT *GGA GGC TTC ATG*
Bovine enkephalin	--- --- --- ---	AAG CGG	TAT *GGG GGC TTC ATG*

Row 6

	Connector regions	→	Ultimate gene #3
Rat enkephalin	AAG AAG (+20)	AAG AGG	TAT *GGC GGT TTC ATG*
Human enkephalin	AAG AAA (+20)	AAG CGG	TAT *GGG GGC TTC ATG*
Bovine enkephalin	AAG AAA (+20)	AAG AGA	TAT *GGG GGC TTC ATG*

Row 7

		→	Ultimate gene #4
Rat enkephalin	AAG AAG (+43)	AAG AGG	TAT *GGG TTC ATG* AGA *GGC CTC*
Human enkephalin	AAG AAG (+41)	AAG AGA	TAT *GGG GGC TTC ATG* AGA *GGC TTA*
Bovine enkephalin	AAG AAG (+40)	AAG AGA	TAC *GGG GGC TTC ATG* AGA *GGC TTA*

Row 8

		→	Ultimate gene #5 Alternative processing
Rat enkephalin	AAA AGA (+12)	AAG CGC	TAT *GGG GGC TTC ATG* AGA AGG *GTC GGG* CGC
Human enkephalin	AAG AGA (+12)	AAG CGA	TAT *GGG GGC TTC ATG* AGA AGA *GTA GGT* CGC
Bovine enkephalin	AAG AGA (+12)	AAG CGA	TAC *GGG GGT TTC ATG* AGA AGA *GTC GGT* CGT

Row 9

                                 Ultimate gene #6

Rat enkephalin   AGA AGG (+11) AAG AGA TAC *GGA GGC TTC CTG*

Human enkephalin   AGA AGA (+11) AAA CGG TAT *GGA GGT TTC CTG*

Bovine enkephalin   AGA AGA (+11) AAA AGG TAC *GGT GGC TTC CTC*

Row 10

                                 Ultimate gene #7

Rat enkephalin   AAG CGC (+22) AAA AGA TAC *GGA GGC TTT ATG CGG TTT* TGA

Human enkephalin   AAG CGC (+22) AAA AGA TAC *GGA GGA TTT ATG AGA TTT* TAA

Bovine enkephalin   AAG CGC (+22) AAA AGA TAT *GGA GGA TTT ATG AGA TTT* TAA

[a]Codons for hydrophilic amino acids are underscored and those for hydrophobic ones are italicized. Asterisk indicates the location of an intron. Arrows indicate cleavage sites. The usual ultimate genes are shaded but a section that can be added to ultimate gene 5 is indicated by lighter stippling.
[b]Rosen *et al.* (1984); Yoshikawa *et al.* (1984).
[c]Legon *et al.* (1982).
[d]Gubler *et al.* (1982); Noda *et al.* (1982).

7.5.3. Subclass II Cryptomorphic Genes

In a sense, the enkephalin family of cryptomorphs just described introduces those of subclass II, for the major distinguishing character here is the combination of two or more ultimate genes encoding different products concealed within a large primary translational product. There can be little doubt that this group is artificially composite, for, as is shown shortly, those classed here fall into two clusters, each containing closely related vertebrate genes, which lack any indication of kinship to those of the other group, although both in part encode important hormones. Here, too, belong the vasoactive intestinal polypeptide genes and their associated proteins (Hefford *et al.*, 1985; Nishizawa *et al.*, 1985).

The Oxytocin Family of Genes. Oxytocin and vasopressin, two structurally closely related nonapeptides, are synthesized by the magnocellular neurons of the supraoptic and suprachiasmatic nuclei of the hypothalamus, whence they are conducted to the neurohypophysis for release into the bloodstream. Both also are synthesized in the corpora lutea of the ovary (Ivell and Richter, 1984b). Chiefly this pair influences water balance in the body, along with cardiovascular functions and smooth muscle contractions (Majzoub *et al.*, 1984). After each presequence, described later, is a short prosequence of four codons, the first and last of which encode hydrophobic amino acids (Table 7.19). No triplet for a charged amino acid is contained here, or in the calf sequence, which is not given (Land *et al.*, 1982). Immediately following this is an ultimate gene encoding either of the hormones oxytocin or vasopressin (Figure 7.2C).

The coding properties of the four shown in the table differ at only two points, the rat vasopressin gene (Ivell and Richter, 1984a; Majzoub *et al.*, 1984) having triplets designating phenylalanine (TTC) and arginine (AGR) in place of those encoding isoleucine (ATY) and leucine (CTG), respectively, in the two oxytocin cistrons (Ivell and Richter, 1984a,b). In the short connecting section that ensues are three codons, the first specifying glycine, the second lysine, and the third arginine; the two examples from rat are identical throughout, while that of the calf oxytocin cistron has CGC in place of the AGA of the others. Beyond this sector is a long region of ~92 triplets that codes for a polypeptide called neurophycin. This protein has been stated to be involved in the transport of oxytocin or vasopressin to the posterior pituitary (Ivell and Richter, 1984a), but it is not clear whether this activity is carried out before or after translation and processing. If afterward, then that is a valid activity; if before processing, then the neurophycin is merely part of the precursor that is conducted, and its function, if any, remains unknown.

In the vasopressin gene a coding region for a glycoprotein of undetermined activity abuts against the 3' end of the neurophycin sequence, but in those for oxytocin only a single codon intervenes between the latter and the translational termination signal. This can be considered the shortest possible telosequence, a feature absent from the genes for vasopressin. In summary, it may be perceived that both the oxytocin and vasopressin genes consist of six parts, a presequence, a prosequence, an ultimate section, a connector, the neurophycin region, and either a telosequence or a glycoprotein area, in addition to the universal termination signal of all protein genes.

The connector sector, where the two peptides are cleaved proteolytically, is of particular interest in that its structure differs from those of subclass I. Here the neurophycin region is preceded by a pair of codons for lysine and arginine, as is typical; however, the hormone coding sector has only a single codon following it, one that

Table 7.19
Mature Coding Sections of Subclass II Cryptomorphic Genes[a]

Row 1

	Prosequence	Hormone	Neurophycin →

Rat oxytocin[b]
CTG ACC TCC GCC | TGC TAC ATC CAG AAC TGC CC C CTG GGC GGC AAG AGG | GCT GCG CTA GAC CTG GAT ATC

Bovine oxytocin[c]
TTG ACC TCC GCC | TGC TAC ATT CAG AAC TGC CC C CTG GGC GGC AAA CGC | GCG GTG CTG GAC CTC GAC GTG

Rat vasopressin[b]
CTC ACC TCT GCC | TGC TAC TTC CAG AAC TGC CC A AGA GGA | GCC ACA TCC GAC ATG GAG CTG

Bovine vasopressin[d]
TTC ACC TCT GCT | TGC TAC TTC CAG AAC TGC CC A AGG GGC | GCC ATG TCC GAC CTG GAG CTG

Anglerfish glucagon[e]
CGG GTT CTT ATG CTT CAG GAG GCT CCC AGC TC A AGT TTG GAG GCA GAC AGC ACA CTG AAG GAC GAG CCG

Rat glucagon[f]
CAT GCC CCT CAA GAC ACG GAG GAG AAC GCC AG*A TCA TTC CCA GCT TCC CAG ACA CCA CTT GAA GAC

Hamster glucagon[g]
CAT TCC CTT CAG GAC GAC ACG GAG GAG AAA TCC AG*A TCA TTC CCA GCT TCC CAG ACA CCA CTC GAG GAC

Human glucagon[h]
CGT TCC CTT CAA GAC GAC ACA GAG GAG AAA TCC AG*A TCA TTC TCA GCT TCC CAG CCA CTC AGT GAT

Bovine glucagon[i]
CGT TCC CTT CAG AAC ACA GAG GAG AAA TCC AG T TCA TTC CCA GCT CCG CAG ACC CCG CTC GGC GAT

← Glicentin

Row 2

Rat oxytocin
CGC AAG*TGT CTT CCC TGC GGA CCC GGC GGC AAA GGG CGC TGC TTC GGG CCG AGC ATC TGC TGC GCG GAC

Bovine oxytocin
CGC ACG TGT CTC CCC TGC GGC CCC GGG GGC AAA GGC CGC TGC TTC GGG CCC AGC ATC TGC TGC GCG GAC

Rat vasopressin
AGA CAG*TGT CTC CCC TGC GGC CCT GGC GGC AAA GGG CGC TGC TTC GGG CCG AGC ATC TGC TGC GCG GAC

Bovine vasopressin
AGA CAG CTT CTC CCC TGC GGC CCC GGG GGC AAA GGC CGC TGC TTC GGG CCC AGC ATC TGC TGC GGG GAC

Anglerfish glucagon
AGA GAG CTT TCA AAC ATG -- | AAG AGA CAC TCG GAG GGA ACT TTC TCC AAC GAC TAC AGC AAA TAC CTG

Rat glucagon
CCT GAT CAG ATA AAC GAA GAA AAA CGC CAT TCA CAG GGC ACA TTC ACC AGT GAC TAC AGC AAA TAC CTA

Hamster glucagon
CCT GAT CAA ATA AAT GAA GAC AAG CGC CAT TCA CAG GGA ACA TTC ACC AGT GAC TAC AGC AAA TAC CTG

Human glucagon
CCT GAT CAG ATG AAC GAG GAC AAG CGC CAT TCA CAG GGC ACA TTC ACC AGT GAC TAC AGC AAG TAT CTG

Bovine glucagon
CCA GAT CAG ATC AAT GAA GAT AAG CGC CAC TCG CAG GGC ACA TTC ACC AGT GAC TAC AGC AAG TAC CTG

← Glicentin   Glucagon →

(continued)

Table 7.19 (Continued)

Row 3

Rat oxytocin
GAG CTG GGC TGC TTC GTG GGC ACC GCC GAG GCG CTG TGC CGC CAG GAG GAG AAC TA C CTG CCC TCG CCC

Bovine oxytocin
GAG CTG GGC TGC TTC GTG GGC ACG GCC GAG GCG CTG TGC CGC CAA GAG GAG AAC TA C CTG CCG TCG CCC

Rat vasopressin
GAG CTG GGC TGC TTC CTG GGC ACC GCC GAG GCG CTG TGC CGC CAG GAG GAG AAC TA C CTG CCC TCG CCC

Bovine vasopressin
GAG CTG GGC TGC TTC GTG GGC ACG GCC GAG GCG CTG TGC CGC CAA GAG GAG AAC TA C CTG CCG TCG CCC

Anglerfish glucagon
GAG GAC AGG AAG GCA CAG GAG TTT GTT CTG CTG ATG AAC AAC AAG AGG AGC GG T GTG GCA GAA ---

Rat glucagon
GAC TCC CGC CGT GCT CAA GAT TTT GTG CAG TTG ATG AAC ACC AAG AGG AAC CG*G AAC AAC ATT GCC

Hamster glucagon
GAC TCC CGC CGA CCC CAA GAT TTT GTG CAG CTG ATG AAC ACC AAG AGG AAC AG G AAC AAC ATT GCC

Human glucagon
GAC TCC AGG CGT GCC CAA GAT TTT GTG CAG ATG AAT ACC AAG AGG AAC AG*G AAT AAC ATT GCC

Bovine glucagon
GAC TCC AGG CGT GCC CAG GAC TTC GTG CAG GAC TTG ATG AAT ACC AAG AGA AAC AA G AAT AAC ATT GCC

←—— Glucagon

Row 4

Rat oxytocin
TGC CAG TCT GGC CAG AAG CCT TGC GGA AGC GGC CGC TGC GCC ACC GCG GGC ATC TGC TCT AGC CCG

Bovine oxytocin
TGC CAG TCC GGC CAG AAG CCC TGC GGG AGC GGC GGC TGC GCC ACC GCG GGC ATC TGC TGC AGC CCG

Rat vasopressin
TGC CAG TCT GGC CAG AAG CCT TGC GGA AGC GGC CGC TGC GCC ACC GCG GGC ATC TGC TGT AGC GAT

Bovine vasopressin
TGC CAG TCC GGC CAG AAG CCC TGC GGG AGC GGG CGC TGC GCC ACC GCG GGC ATC TGC TGC AAC GAT

Anglerfish glucagon
AAG CGT CAC GCT GAT GGG ACC TTC ACC AGC --- --- GAT GTC AGC TCC CTC AAA GAC CAG GCA

Rat glucagon
AAA CGT CAT GAT GAA TTT GAG AGG CAT GCT GAA GGG ACC TTT ACC AGT GAT GTG AGT TCT TAC TTG GAG

Hamster glucagon
AAA CGC CAC GAT GAG TTT GAG AGG CAC GCT GAA GGG ACC TTT ACC AGC GAT GTG AGC TCT TAC TTG GAG

Human glucagon
AAA CGT CAC GAT GAA TTT GAG AGA CAT GCT GAA GGG ACC TTT ACC AGT GAT GTA AGT TCT TAT TTG GAA

Bovine glucagon
AAA CGT CAT GAT GAA TTT GAG AGA CAT GCT GAA GGG ACC TTT ACC AGT GAT GTA AGT TCT TAT TTG GAA

Glucagonlike protein I ——→

Row 5

```
                                                                                                    ——— Neurophysin
Rat oxytocin         G*AT GGC TGC CGC ACC GAC CCC GCC TGC GAC CCT GAG TCT GCC TTC TCC GAG CGC TG A
Bovine oxytocin      G AC GGC TGC CAC GAG GAC CCC GCC TGC GAC CCT GAG GCC TTC TCC CAG CAC TG A
Rat vasopressin      G*AG AGC TGC GTG GCC GAG CCC GAG TGT CGA GAG GGT TTT TTC CCC CTC ACC CGC GC T CGG --- --- GAG
Bovine vasopressin   G AG AGC TGC GTG ACC GAG CCC GAG TGC CGG GAA GGT GTC GGC TTC CCC CGC CGC GT T CGC GCC AAC GAC
Anglerfish glucagon  A TC AAA GAC TTT GTG GAC CTC AAG GCT GGA CAA GTC AGA AGA AGA GAG TAG  Glycoprotein ——→
Rat glucagon         G GC CAG GCA GCA AAG GAA TTC ATT GCT TGG CTG GTG AAA GGC CGA GGA AGG CGA GA*C TTC CCG GAA GAA
Hamster g⁻ucagon     G GC CAG GCT GCA AAG GAA TTC ATT GCT TGG CTG GTG AAA GGC AGA GGA AGG CGG GA C TTC CCA GAA GAA
Human glucagon       G GC CAA GCT GCC AAG GAA TTC ATT GCT TGG CTG GTG AAA GGC CGA GGA AGG CGA GA*T TTC CCA GAA GAG
Bovine glucagon      G GC CAA GCT GCC AAG GAA TTC ATT GCT TGG CTG GTG AAA GGC CGA GGA AGG CGA GA T TTC CCA GAA GAA
                               ——— Glucagonlike peptide I ——→
```

Row 6

```
Rat vasopressin      CAG AGC AAC GCC ACG CAG CTG GAC GGG CCA GCC CGG GAG CTG CTG CTT AGG CTG GTA CAG CTG GCT GGG
Bovine vasopressin   CGG AGC AAC GCC ACC CTG GAC GGG GCC AGC GGG CCG AGC CGG GGG GCC TTG CTG CTG CGG CTG CTG GCG GGG
Rat glucagon         GTC GCC ATA GCT GAG GAA CTT GGG CGC AGA CAT GCT GAT GGA TCC TTC TCT GAT GAG AAC ACG ATT
Hamster glucagon     GTC ACC ATT GTT GAA GAA CTC GGC CGC AGA CAT GCG GAC GGC TCC TTC TCC GAT GAG AAC ACG ATT
Human glucagon       GTC GCC ATT GTT GAA GAA CTT GGC CGC AGA CAT GCT GAT GGT TCT TTC TCT GAT GAG AAC ACC ATT
Bovine glucagon      GTC AAC ATC GTT GAA GAA CTC GGC CGC CGC AGA CAC GCC GAT GGC TCT TTC TCT GAT GAG AAC ACT GTT
                              Glucagonlike peptide II ——→
```

(continued)

Table 7.19 (Continued)

Row 7

Glycoprotein

Rat vasopressin   ACA CAA GAG TCC GTG GAT TCT GCC AAG CCC CGG GTC TAC TGA

Bovine vasopressin   GCG CCG GAG CCC GCG GAG CCC GCC CAG CCC GGC GTC TAC TGA

Rat glucagon   CTC GAT AAC CTT GCC ACC AGA GAC TTC ATC AAC TGG CTG ATT CAA ACC AAG ATC ACT GAC | AA*G AAA TAG

Hamster glucagon   CTC GAT AGT CTT GCC ACC AGG GAC TTC ATC AAC TGG CTG ATT CAA ACC AAA ATC ACT GAC AA G AAA TAA

Human glucagon   CTT GAT AAT CTT GCC GCC AGG GAC TTT ATA AAC TGG TTG ATT CAG ACC AAA ATC ACT GAC AG G --- TGA

Bovine glucagon   CTC GAT AGT CTT GCC ACC CGA GAC TTT ATA AAC TGG TTG CTT CAG ACG ACG ATT ACT GAC AG G AAG TAA

⟵——— Glucagonlike peptide II   Telopeptide

^aAsterisks indicate the location of an intron. Vertical arrows indicate cleavage sites. Codons for the basic amino acids (lysine and arginine) are underscored. Mature genes are enclosed in open boxes.
^bIvell and Richter (1984a).
^cIvell and Richter (1984b).
^dLand et al. (1982).
^eLund et al. (1983).
^fHeinrich et al. (1984b); Patzelt and Schiltz (1984).
^gBell et al. (1983a).
^hBell et al. (1983b).
ⁱLopez et al. (1983).

specifies the small hydrophobic amino acid glycine. Hence, a second enzyme may be concerned in cleavage of the connector from the sector that precedes it. As indicated by the asterisks in rows 2 and 5, the neurophycin sequences of the rat are interrupted by two introns, placed at corresponding points in the genes both for oxytocin and vasopressin, evidence in addition to the high level of homology throughout all their parts that the cistrons for the two hormones have been derived from a common ancestor in relatively recent times.

The Glucagon Family of Genes. Glucagon is a member of a family of peptide hormones that also includes secretin, vasoactive intestinal peptide, gastric inhibitory peptide, and growth hormone-releasing hormones. However, it is the only component whose gene structure has been sufficiently documented to be reported. Currently the sequence of the cistron for this substance has been established from five sources, all of which are included in Table 7.19.

Under the control of blood levels of glucose, various amino acids, and several hormones (Heinrich *et al.*, 1984b), glucagon is secreted by the A cells of the islets of the pancreas. Its chief target organ is the liver, where it plays an important role in protein and carbohydrate metabolism. Chiefly it is concerned with glucose metabolism, through actions that inhibit glycogen synthesis, accelerate glycogen breakdown, and stimulate formation of glucose. In the genes that encode this 29-amino acid peptide, a presequence but no prosequence is found, the 5'-terminal region encoding a protein called glicentin, a peptide with much the same activity as glucagon itself (Figure 7.2D). This 30-codon section is separated by a pair of codons specifying lysine and arginine from the 29-codon-long ultimate gene for the glucagon. Thus here the protease (or proteases) acts at each end of this combination of basic amino acids.

After the glucagon gene, there is a connector of ten codons, including a dual combination for lysine and arginine at each end; this is followed by a stretch of 37 triplets, the ultimate coding sector for what is referred to as "glucagonlike peptide I." Then, in mammalian cistrons, but not those from the anglerfish, a connector of 17 codons follows, provided with coding signals for two arginines at the 5' end and for two or three of the same at the other terminus. The latter serves as the apparent cleavage site from which a 33-codon-long region encoding "glucagonlike peptide II" is removed following translation. Finally there is a short telosequence of either two lysine codons or one arginine and one lysine. In the anglerfish, the glucagonlike peptide I coding region is followed by a short telosequence of three codons, beginning with two for arginine. Recently it has been proposed that the early cleavages of processing act at three points to separate glicentin, glucagon, and the intact 3' half containing the two glucagonlike peptides; however, the remaining steps in activation remain undetermined (Patzelt and Schiltz, 1984). Other genes that fall into this category are still more complexly structured than the examples cited, but current information is not sufficiently extensive to provide for a precise detailed discussion. One is of the particular note, that for a precursor known as pro-opiomelacortin, for it encodes three products, β-endorphin, melanocyte-stimulating hormone, and corticotropin (ACTH) (Oates and Herbert, 1984).

Thus the glucagon cryptomorphic gene consists of seven or eight regions, a presequence, three or four ultimate genes (one each for glicentin and glucagon, and either one or two for glucagonlike peptides), two connecting sectors (one in fish), and a telosequence. The obvious distinctions of this arrangement from that of the oxytocin provide

Table 7.20
Presequences of Cryptomorphic Genes of Subclass II[a]

Row 1

Rat oxytocin[b]	*ATG* *GCC* TGC *CCC* AGT --- *CTC* --- *GCT* TGC TGC *CTG*
Bovine vasopressin[c]	*ATG* --- *CCC* <u>GAC</u> *GCC* ACA *CTG* *CCC* *GCC* TGC *TTC* *CTC*
Rat vasopressin[b,d]	*ATG* *ATG* *CTC* AAC ACT ACG *CTC* TCT *GCT* TGC *TTC* *CTG*
Bovine glucagon[e]	*ATG* <u>AAA</u> AGC *CTT* TAC *TTT* *GTG* *GCT* *GGA* *TTG* *TTT* *GTA*
Rat glucagon[f]	*ATG* <u>AAG</u> ACC *GTT* TAC *ATC* *GTG* *GCT* *GGA* *TTG* *TTT* *GTA*
Hamster glucagon[g]	*ATG* <u>AAG</u> AAC *ATT* TAC *ATT* *GTG* *GCT* *GGA* *TTT* *TTT* TGT
Anglerfish glucagon[h]	*ATG* <u>AAA</u> <u>CGC</u> *ATC* CAC TCC *CTG* *GCT* *GGT* *ATC* *CTT* *CTG*

Row 2

Rat oxytocin	*CTT* *GGC* *CTA* --- --- --- --- *CTG* GCT ⏐ *CTG* ACC
Bovine vasopressin	--- AGC *CTG* --- --- --- --- *CTG* GCC *TTC* ACC
Rat vasopressin	--- AGC *CTG* --- --- --- --- *CTG* GCC *CTC* ACC
Bovine glucagon	*ATG* *CTG* *GTA* CAA *GGC* --- AGC TGG CAA <u>CGT</u> TCC
Rat glucagon	*ATG* *CTG* *GTA* CAA *GGC* --- AGC TGG CAG CAT *GCC*
Hamster glucagon	*GGT* *GCT* *GGT* CAA *GGC* --- AGC TGG CAG CAT TCC
Anglerfish glucagon	*GTG* *CTT* *GGT* *TTA* *ATC* CAG AGC AGC TGC <u>CGG</u> *GTT*

[a]Codons for charged amino acids are underscored and those for hydrophobic ones are italicized. The arrow marks the cleavage site.
[b]Ivell and Richter (1984a).
[c]Land *et al.* (1982).
[d]Majzoub *et al.* (1984).
[e]Lopez *et al.* (1983).
[f]Heinrich *et al.* (1984b).
[g]Bell *et al.* (1983a).
[h]Lund *et al.* (1983).

firm evidence that subclass II as here presented is not a natural grouping but merely one of convenience for immediate needs.

Subclass II Cryptomorph Presequences. The presequences of the two families examined here as subclass II of cryptomorphic genes display far more shared characteristics than do their mature coding regions (Table 7.20). In both cases these sectors are relatively short, consisting of 15–20 codons, at most one of which per sequence encodes a charged amino acid. They differ strongly, however, in the distribution of triplets for hydrophobic amino acids. Among the oxytocin family members, that type forms the larger part of the structure, 11 of the 15 of oxytocin and 10 or 11 of vasopressin falling in this category. On the other hand, the presequences of the glucagon family are less heavily equipped with such codons, but they have them grouped centrally, where ten occur without interruption, neither singly nor paired as in the rest. At the cleavage site there is no recognizable common system to signal the protease, although each family shows a high level of conservation here. The first family has two apolar codons preceding the terminus, one for a moderate-sized amino acid, the other for a small one, whereas the second group encodes the hydrophilic amino acids tryptophan and glutamine, both of rather large size. On the 3′ side of the cleavage site, the three members of the oxytocin family have codons for either leucine or phenylalanine, followed by threonine, while the glucagon cistrons vary, specifying either arginine or histidine at the first site and serine or valine at the second. Only the last two can be rated as small amino acids. Consequently, we remain without a clue as to the nature of the signal for the cleaving enzyme.

Evolutionary Notes. The availability of a glucagon gene sequence from a lower vertebrate opens an avenue permitting speculation as to the possible origin of its compound nature. Since this more primitive cistron lacks the nucleotides encoding a second glucagonlike peptide, it is obvious that the latter arose by a duplicative process in some higher form. Whether this event occurred in some amphibian or reptile or only in the lower mammals cannot be determined until its primary structure has been established from an avian source. This proposal, besides being self-evident, is along standard lines in current literature, but the rat and human genes provide data that seem to carry farther the process of development of compound structures such as the present one. In these two, the DNA, not just the mRNA, provided the basis for sequencing, so the several introns that exist have been detected. One such is found in row 5 of Table 7.19 at a point corresponding to the penultimate site in the anglerfish telosequence. Consequently, it may be that in some instances such as here an intron is associated with the point of duplication of a gene segment, although in what capacity remains unclear. That this may be the case is further suggested by the presence of an inserted region close after the 3′ end of the section encoding glucagon itself. Thus glucagonlike peptide I may be the result of a prepiscine duplication of the glucagon sequence, with a later duplication producing the coding section for glucagonlike peptide II. There is no evidence as to the origin of the insert located near the middle of the glicentin region. Establishment of the glucagon gene structure from agnathans or elasmobranchs probably will be necessary to disclose the actual steps in the phylogeny of this unusual complex gene.

7.5.4. Cryptomorphic Genes of Subclass III

Since in the glucagon family of subclass II genes just analyzed, the region encoding glicentin was actually a prosequence, it appears logical to place in the ensuing group, subclass III, other but different cistron structures that share this same feature. Here the distinctive characteristics are the existence of a single ultimate gene, which displaces the prosequence of a much larger DNA coding frame, most of whose product is without known function. In short, a relatively small ultimate functional gene is hidden in the 5′ section of a large structure, the product not becoming active until the major 3′ part is removed enzymatically. Only three types of hormones of vertebrates are currently known to represent this subclass, luteinizing hormone-releasing hormone, angiotensinogen, and ubiquitin (Lund *et al.*, 1985).

Luteinizing Hormone-Releasing Hormone. Luteinizing hormone-releasing hormone (LHRH), also called gonadotropin-releasing hormone, is an important element in the control of reproduction in vertebrates. Produced by hypothalamic neurons, it is se-

Table 7.21
Gene Structures of Subclass III Cryptomorphs[a]

Row 1

Ultimate gene

Human LHRH[b]   CAG CAC TGG TCC TAT GGA CTG CGC CCT GGA | AAG AGA GAT GCC GAA AAT TTG ATT GAT TCT TTC CAA GAG*

Rat angiotensin[c]   GAC CGC TAC ATC CAC CCC TTT CAT CTC CTC TAC TAC AGC AAG ACC ACC TGC GCC CAG CTG GAG AAC

Human angiotensin[d]   GAC CGG GTG TAC ATA CAC CCC TTC CAC CTC GTC ATC CAC AAT CAC AGT ACC TGT GAG CAG CTG GCA AAG

Row 2

Human LHRH   ATA GTC AAA GAG GTT GGT CAA CTG GCA GAA ACC CAA CGC TTC GAA TGC ACC ACG CAC CCA CGT TCT

Rat angiotensin   CCC AGT GTG GAG ACG CTC CCA GAG CCA ACC TTT GAG CCT GTG CCC ATT CAG GCC AAG ACC TCC CCC GTG

Human angiotensin   GCC AAT GCC GGG AAG CCC AAA GAC CCC ACC TTC ATA CCT CCA ATT CAG GCC AAG ACA TCC CCT GTG

Row 3

                                                                                          Telosequence
                                                                                          ↓
Human LHRH   CCC CTC CGA GAC CTG AAA GGA GCT CTG* GAA AGT CTG ATT GAA GAG GAA ACT GGG CAG | AAG AAG ATT | TAA

Rat angiotensin   GAT GAG AAG ACC CTG CGA GAT AAG CTC CTG CTG GCC ACT GAG AAG CTA GAG GAT CGG CAG CGA

Human angiotensin   GAT GAA AAG AAG GCC CTA GCC CTA CAG CAG CTG GTG CTA GTC GCT GCA AAA CTT GAC ACC GAA CAC AAG TTG AGG

Row 4

Rat angiotensin   GCT GCC CAG GTC GCG ATG ATT GCC AAC TTC ATG GGT TTC CGC ATG TAC AAG ATG CTG AGT GAG GCA AGA

Human angiotensin   GCC GCA ATG GTC GGG ATG GTG CTG GCC AAC TTC TTG GGC TTC CGT ATA TAT GGC ATG CAC AGT GAG CTA TGG

Row 5

Rat angiotensin   GGT GTA GCC ACT GGG GCC --- GTC CTC TCT CCA CCG GCC CTC TTT GGC ACC CTG TCT TTC TAC CTT

Human angiotensin   GGC GTG GTC CAT GGG GCC ACC GTC CTC TCC CCA ACA GCT GTC TTT GGC ACC CTG TCT CTC TAT CTG

Row 6

Rat angiotensin   GGA TCG TTG GAT CCC ACG GCC AGC CAG CTG CAG CTG CTG GGC GTC CCT GTG AAG GAG GGA GAC TGC

Human angiotensin GGA GCC TTG GAC CAC ACA GCT GAC AGG CTA CAG GCA ATC CTG GGT GTT CCT TGG AAG GAC AAG AAC TGC

Row 7

Rat angiotensin   ACC TCC CGG CTG GAC GGA CAT AAG GTC CTC ACT GCC CTG CAG GCT GTT CAG GGC TTG CTG GTC ACC CAG

Human angiotensin ACC TCC CGG CTG GAT GCG CAC AAG GTC CTG TCT GCC CTG CAG GGC CTG CTA GTG GCC CAG

Row 8

Rat angiotensin   GGT GGA AGC AGC CAG ACA CCC CTG CTA CAG TCC ACC GTG GTG GGC CTC TTC ACT GCC CCA GGC TTG

Human angiotensin GGC AGG GCT GAT AGC CAG CAG CCC CAG CTG CTG TCC ACG GTG GGC GTG TTC ACA GCC CCA GGC CTG

Row 9

Rat angiotensin   CGC CTA AAA CAG CCA TTT GTT GAG AGC TTG GGT CCC TTC ACC CCC GCC ATC TTC CCT CGC TCT CTG GAC

Human angiotensin CAC CTG AAG CAG CCG TTT GTG CAG GGC CTG GCT CTC TAT ACC CCT GTG CTC CCA CGC TCT CTG GAC

Row 10

Rat angiotensin   TTA TCC ACT GAC CCA GTT CTT GCT GCC CAG AAA ATC AAC AGG TTT GTG CAG GCT GTG ACA GGG TGG AAG

Human angiotensin TTC --- ACA GAA CTG GAT GTT GCT GCT GAG AAG ATT GAC AAG AGG TTC ATG CAG GCT GTG ACA GGA TGG AAG

Row 11

Rat angiotensin   ATG AAC TTG CCA CTA GAG GGG GTC AGC ACG GAC AGC ACC CTA TTT TTC AAC ACC TAC GTT CAC TTC CAA

Human angiotensin ACT GGC TGC TCC CTG ATG GGA GCC AGT GTG GAC AGC ACC CTG GCT TTC AAC ACC TAC GTC CAC TTC CAA

(continued)

Table 7.21 (Continued)

Row 12

Rat angiotensin   GGG AAG ATG AGA GGC TTC TCC CAG CTG ACT GGG CTC CAT GAG TTC TGG GTG GAC AAC AGC ACC TCA GTG

Human angiotensin   GGG AAG ATG AAG GGC TTC TCC CTG GCC GAG CCC CAG GAG TTC TGG GTG GAC AAC AGC ACC TCA GTG

Row 13

Rat angiotensin   TCT GTG CCC ATG CTC TCG GGC ACT GGC AAC TTC CAG CAC TGG AGT GAC GCC CAG AAC TTC TCC GTG

Human angiotensin   TCT GTT CCC ATG CTC TCT GGC ATG GGC ACC TTC CAG CAC TGG AGT GAC CAC CAG AAC TTC TCG GTG

Row 14

Rat angiotensin   ACA CGC GTG CCC CTG GGT GAG AGT GTC ACC CTG CTG ATC CAG CCC CAG TGC GCC TCA GAT CTC GAC

Human angiotensin   ACT CAA GTG CCC TTC ACT GAG AGC GCC TGC CTG CTG CTG ATC CAG CCT CAC TAT GCC TCT GAC CTG GAC

Row 15

Rat angiotensin   AGG GTG GAG CTC CTC TTC CAG CAC TTC CTG ACT TGG ATA AAG AAC CCG CCT CCT CGG GCC ATC

Human angiotensin   AAG GTG GAG GGT CTC ACT TTC CAG CAA AAC TCC CTC AAC TGG ATG AAG AAA CTG TCT CCC CGG ACC ATC

Row 16

Rat angiotensin   CGT CTG ACC CTG CCG CAG CTG GAA ATT CGG GGA TCC TAC AAC CTG CAG GAC CTG CTG GCT CAG GCC AAG

Human angiotensin   CAC CTG ACC ATG CCC CAA CTG GTG CTG CAA GGA TCT TAT GAC CTG CAG GAC CTG CTC GCC CAG CAG GCT GAG

Row 17

Rat angiotensin   CTG TCT ACC CTT TTG GGT GCT GAG GCA AAT CTG GGC AAG ATG GGT GAC ACC AAC CCC CGA GTG GGA GAG

Human angiotensin   CTG CCC GCC ATT CTG CAC ACC GAG CTG CAA AAA TTG AGC AAT GAC CGC ATC AGG GTG GGG GAG

Row 18

Rat angio-ensin   GTT CTC AAC AGC ATC CTC CTT GAA CTC CAA GCA GGC GAG GAG CAG CCC ACA GAG TCT GCC CAG CAG

Human angiotensin   GTG CTG AAC AGC ATT TTT GAG CTT GAA GCG --- GAT GAG AGA GAG CCC ACA GAG TCT ACC CAA CAG

Row 19

Rat angio-ensin   CCT GGC TCA CCC GAG GTG CTG GAC GTG ACC CTG AGC AGT CCG TTC CTG TTC GCC ATC TAC GAG CGG GAC

Human angiotensin   CTT AAC AAG CCT GAG GTC TTG GAG GTG ACC CTG AAC CGC CCA TTC CTG TTT GCT GTG TAT GAT CAA AGC

Row 20

Rat angiotensin   TCA GGT GCG CTG CAC TTT CTG GGC AGA GTG GAT AAC CCC CAA AAT GTG GTG TGA

Human angiotensin   GCC ACT GCC CTG CAC TTC CTG GGC CGC GTG GCC AAC CCG CTG AGC ACA GCA TGA

[a]Codons for basic amino acids are underscored. Arrows indicate cleavage sites. Asterisks indicate location of introns.
[b]Luteinizing hormone-releasing hormone; Seeburg and Adelman (1984).
[c]Ohkuba et al. (1983).
[d]Kageyama et al. (1984).

creted into capillaries to induce the release of luteinizing and follicle-stimulating hormones from the anterior pituitary. A small protein, it consists of only nine amino acids and is encoded by a region of 30 nucleotide residues at the very outset of the mature coding sector of its gene, that is, immediately following the presequence (Table 7.21). If the remainder of the 72-triplet-long gene encodes a useful product, its function remains unelucidated. That it may have some physiological activity is suggested by the presence of a possible telosequence having much the same structure as those of subclass II cryptomorphs in consisting of two codons for lysine, followed by one for an uncharged amino acid. Two introns interrupt the coding region for the unknown substance, but their significance similarly remains obscure.

The ultimate gene section is followed immediately by a combination of lysine and arginine codons typical of many vertebrate gene cleavage signals. In this case, however, it is preceded by a triplet for glycine, which amino acid is employed in amidation of the carboxyl end of the active hormone (Seeburg and Adelman, 1984), so that the peptide actually consists of only nine monomeric units, not ten as sometimes stated in the literature.

The Angiotensinogen Family of Genes. Angiotensinogen, when secreted by the liver into the bloodstream, is a molecule consisting of 453 amino acids, plus a transit peptide of 24 such residues (Figure 7.2E). Although this precursor is thus quite large, when cleaved by renin the angiotensin I that is released is only ten residues in length. This must be processed further by another protease, called dipeptidyl carboxylpeptidase, during which activity it loses two residues at the carboxyl end to produce the functional octapeptide angiotensin II (Ohkuba *et al.*, 1983). When thus mature, the hormone in the bloodstream induces arteriolar constriction and stimulates the adrenal cortex to release aldosterone. In addition, angiotensin is synthesized in the brain by like processes, where it causes thirst and is active in the control of vasopressin and corticotropin release. Briefly stated, it thereby plays an important role in control of blood pressure and water balance in the body.

As in luteinizing hormone-releasing hormone, the ultimate gene is located at the 5′ end of the mature cistron, immediately after the presequence. Also as there, this region is short, containing just 30 nucleotide residues, exactly as in the other member of this subclass. Unlike its predecessor, however, this region is not followed by codons for either lysine or arginine, nor in fact by any recognizable signaling combination. Thus, at present the processes of recognition of the cleavage site by renin are totally unknown. Whether the mature gene section is broken by introns also is unknown, since both the rat and human sequences were established from DNA complementary to the processed messengers (Ohkuba *et al.*, 1983; Kageyama *et al.*, 1984).

Whether calcitonin, a thyroid-produced enzyme associated with calcium metabolism, should be considered a member of the present or next subclass could not be established, since the precise limits of the presequence are not determined (Le Moullec *et al.*, 1984). The single ultimate gene is followed by a 23-codon-long sector that is proteolytically removed posttranslationally, but the unusually long presequence (84 codons) suggests that that part may actually be a prosequence, at least in part. In addition, the gene for gastrin-releasing hormone, a 27-amino acid peptide released by the stomach and upper intestine of mammals and the proventriculus of birds, belongs in this subclass (Spindel *et al.*, 1984).

7.5.5. Subclass IV Cryptomorphic Genes

The members of subclass IV cryptomorphic genes grade into the diplomorphic variety and could justifiably be considered extreme examples of those that bear both pre- and prosequences. As a matter of observation, one family has been referred to as pre-proproteins in the literature (Montminy *et al.*, 1984). The chief reason for classifying the components as cryptomorphs lies in the exceptionally great proportions of what otherwise would be considered a prosequence. Here the ultimate genes are short, as in subclass III, but lie at or toward the 3'-terminal region rather than at the extreme 5' end. As shown shortly, the large functionless section greatly exceeds the region encoding the ultimate product. Since the small active part is thus hidden within a huge precursorial form, the several examples currently known appear better to be treated as cryptomorphic genes.

Kininogen Genes. Despite their pharmacological and medical importance, the kinins have been relatively poorly explored at the molecular level. They are small peptides of 9–11 amino acids, in this respect resembling a number of other hormonal products of cryptomorphic genes. Of the two more abundant members of the family, bradykinin and kallidin, the sequence of the former alone has been established and that only from human and bovine liver (Kitamura *et al.*, 1983, 1985; Nawa *et al.*, 1983). In the genomes of these mammals, two cistrons appear to exist which encode quite similar kininogens, as the precursorial molecules are known. In addition, two forms of precursor occur, of high and low molecular weight, the first representing the complete translational product, the other the amino acid half, the carboxyl portion being removed proteolytically at the point represented by a gap in row 18 (Table 7.22). Since the ultimate gene region just precedes this point, in one case it is located at the middle and in the other near the 3' terminus of the mature gene. At activation, the bradykinin is released by action of the kallikreins examined earlier in this chapter (Section 7.4.1).

The gene, which is of unusually great length, is unique among cryptomorphic forms in lacking a presequence (Figure 7.2H). However, the 5' end of the mature coding region contains a high percentage of codons for hydrophobic amino acids, 11 of the first 15 triplets encoding that type of monomer; its typical presequencelike structure is further enhanced by the presence of a codon for a basic amino acid located in the second site (Table 7.22). Thus this part may serve in transit through the cell membrane during the secretory processes in lieu of a removable transit peptide. But the actual procedures employed in penetrating the cell membrane are still to be established, a statement equally applicable to all secreted products studied in this newly opened field of investigation. Toward the 3' end is a peculiarity, the significance of which is presently unknown. Beginning in row 18 there are numerous codons for histidine (CAC, CAT) that continue at a high level of frequency into row 22, where they abruptly cease. At some areas these codons occur at every two or three sites, with the CAT combination greatly preferred over the other. This region, it would seem, should offer investigators a unique opportunity for study of the particular qualities of the amino acid in a naturally occurring product. *In vivo* the activated bradykinin has functions similar to oxytocin, vasopressin, angiotensin, and the other vasoactive hormones, for it produces smooth muscle contraction, induces dilation of blood vessels (resulting in lowering of blood pressure and increased blood flow), and influences the emigration of granulocytic leukocytes.

Table 7.22
Sequences of Subclass IV Cryptomorphic Genes[a]

Row 1

```
Bovine kininogen[b]            ATG AAA TTA ATC ACC ATC CTT TT C CTT TGT TCC AGG CTG --- CTA CCA AGT TTA ACC CAA GAG TCC TCT CAA
Human somatocrinin[c]          CCA CCT CCC CCT TTG ACC CTC AG G ATG CGG CGG TAT GCA GAT GCC ATC TTC ACC AAC AGC TAC CGG AAG GTG
Human somatostatin[d]          GCT CCC TCG GAC --- CCC AGA CT C CGT CAG TTT --- CTG --- CAG AAG TCC GCT GCT GCC GCG GGG AAG
Rat somatostatin[e]            GCG CCC TCG GAC --- CCC AGA CT C CGT CAG TTT --- CTG --- CAG AAG TCT CTG GCG GCT GCC ACC GGG AAA
Anglerfish somatostatin I[f]   GGA CAG CAG AGA GAC --- TCC AAA CT C CGC --- CTG CTG --- CAC CGG --- TAC CCG CTG CAG GGC TCC AAA
Anglerfish somatostatin II[f]  GAG CAG AGC GAC AAC CAG CAG CT G GAC CTG CGT --- CAG CAC --- TCG CTG GAG AGA GCC AGA GCC CGG[j]
Anglerfish somatostatin III[g] GGA CAG CAG AGA GAC --- TCC AAA CT C CGC --- CTG CTG CTG --- CAC CGG --- TAC CCG CTG CAG GGC TCC GGG
Catfish somatostatin 22[h]     GGC --- CGG CCT --- CAT GTG GT T TTG AAT TCG --- --- --- --- --- --- --- --- GCT CTG GAG
Catfish somatostatin 14[i]     GCG CCG TCT GAT --- GCC AAA CT G CGC CAG TTC --- CTC --- CAG AGG TCG ATT CTT GCA CCG TCT GTA AAA
```

Row 2

```
Bovine kininogen            GAA ATC GAC TGC AAC GAC CAG GAT GTA TTT AAA GCT GTG GAC GCT GCT CTG ACA AA A TAC AAC AGT GAA
Human somatocrinin          CTG GGC CAG CTG TCC GCC CGC AAG CTG CTC CAG GAC ATC ATG AGC AGG CAG CAG AGC AAC CAA
Human somatostatin          CAG GAA CTG GCC AAG TAC TTC TTG GCA GAG CTG --- CTG TCT GAA CCC AAC AGC AC G GAG --- AAT GAT
Rat somatostatin            CAG GAA CTG GCC AAG TAC TTC TTG GCA GAA CTG --- CTG TCT GAG CTG CCC AAC AGC AC A GAG --- AAC GAT
Anglerfish somatostatin I   CAG GAC ATG ACT CGC TCC GCC TTG GCC CTG CTC CTG --- CTG GCT CAG ATG TCT CTG CC A GAG GCC ACG TTC
Anglerfish somatostatin II  CAG GAG TGG AGT AAA CGG GCG GCG GAG GAG CTG --- CTG GCT CTG CTG GAC CTC CTG CAG GG G GAG --- AAC GAG
Anglerfish somatostatin III CAG GAC ATG ACT CGC TCC GCC CTG CTG TCG CTG --- CTG GCT CTC CTG TCG CTG CAG GG G GAG --- AAC GAG
Catfish somatostatin 22     GAG GCT CGC AAC GTG CCG TTT GGA GAA GAG --- GTA CCA GAG CTG CTG GCC CAA CCG CAA GCC GA A AAC GAG --- --- ---
Catfish somatostatin 14     CAG GAG CTC ACC AGG TAC ACG CTC CCA GAG CTC CTG GCG CAG GAG AGA CTG GCG CAA GCC CTC CAG TGG --- --- ---
```

Row 3

Bovine kininogen   AAC AAG AGT GGC AAC CAG TTT GTA TTG TAC CGC ATA ACC GAG GTC GCC AGA ATG GAT AAT CCT GAC ACA
Human somatocrinin   GAG CGA GGA GCA AGG GCA CGG CTT GGT CGT CAG GAC GTA GAC AGC ATG TGG GCA GAA CAA AAG CAA ATG GAA
Human somatostatin   GCC CTG GAA CCT --- GAA GAT CTG TCC CAG GCT GCT GAG GAG CAG GAT GAA ATG AGG CTT GAG CTG CAG AGA
Rat somatostatin   GCC CTG GAA CCT --- GAG GAT TTG CCC CAG GCA GCT GAG GAG CAG GAC GAC GCC CAC GAG CTG GAG CTG CAG AGG
Anglerfish somatostatin I   GCT CTG GAG GAG GAG AAC TTC CCT CTG GCC GAA GGA CCC CAC GCC GCC GAC CTA GAG CGG
Anglerfish somatostatin II   CAG CGG GAG GCG GAC GCG TCC ATG GCG GCA ACA GAA GGA CGG --- --- --- ATG AAC CTA GAG CGG
Anglerfish somatostatin III   GCT CTG GAG GAG GAG AAC TTC CCT CAG GCC AGA AGG GGA ACC CGA GGA CGA CCA CGC CGA CCT AGA GCG
Catfish somatostatin 22   ATG CTC AGT GAG AAC AAC GAG CTC ACG --- CCC GTT CAG GTG GAA GAA --- --- GCC --- --- ---
Catfish somatostatin 14   GTG CTG GAC TCG GAC GAG GTG TCT CGC CGC GCC GAA AGC GGC GCG CGC CTG GAG ATG GAG CGA

Row 4

Bovine kininogen   TTT TAT TCC TTG AAG TAC CAA ATC AAG GAG GGC GAC TG T CCT TTT CAA AGT AAC AAA ACT TGG CAG GAC
Human somatocrinin   TTG GAG AGC ATC CTG GCC CTG CTG CAG AAG CAC AG* C AGG AAC TCC CAG GGA TGA
Human somatostatin   TCT GCT AAC TCA AAC CCG GCT ATG GCA CCC CGA GAA CG C AAA GCT GGC TGC AAG AAT TTC TGG AAG
Rat somatostatin   TCT GCC AAC TCG AAC CCA GCA ATG GCA CCC GAA CG C AAA GCT GGC TGC AAG AAC TTC TGG AAG
Anglerfish somatostatin I   GCC GCC AGC GGG GGG CCT CTC CTC GCC GAG AG A AAA GCC GGC TGC AAG AAC TTC TGG AAA
Anglerfish somatostatin II   TCC GTG GAC TCT ACC AAC AAC CTA CCC CCT GAG CGT CG T AAA GCT GGC TGT AAG AAC TTC TAT TGG AAG
Anglerfish somatostatin III   GCC CGC CAG CGG GGG CCT CTG CTC GCC CCC CGG GAG AG A AAG GCC GGT TGC AAG AAC TTC TGG AAA
Catfish somatostatin 22   CCT CGC AGC AGG --- --- --- CTG GAG CTG GTC AG G AGA GAC AAC k TGC ATG AAC TAC TTC TGG AAG
Catfish somatostatin 14   GCC GCC GGT --- --- CCC ATG CTG GCT CCC CCC GAG CG C AAA GCC GGC TGC AAG AAT TTC TGG AAA

(continued)

Table 7.22 (Continued)

Row 5

Bovine kininogen	TGT	GAC	TAC	AAG	GAC	TCT	GCA	CAA	GCT	GCC	ACA	GGA	GAG	TGC	ACA	GCG	ACC	GTG	GCC	AAG	AGA	GGG AAT
Human somatostatin	ACT	TTC	ACA	TCC	TGT	TAG																
Rat somatostatin	ACA	TTC	ACA	TCC	TGT	TAG																
Anglerfish somatostatin I	ACC	TTC	ACC	TCC	TGC	TGA																
Anglerfish somatostatin II	GGC	TTC	ACT	TCC	TGT	TAA																
Anglerfish somatostatin III	ACC	TTC	ACC	TCC	TGC	TGA																
Catfish somatostatin 22	TCC	AGG	ACA	GCA	TGC	TGA																
Catfish somatostatin 14	ACT	TTC	ACG	TCG	TGT	TAA																

Row 6

Bovine kininogen ATG AAG TTC TCC GTG GCT ATC CAG ACC TGC CTG ATC ACT CCA GCC GAG GGC CCC GTG GTG ACA GCC CAG

Row 7

Bovine kininogen TAT GAG TGC CTT GGC TGT GTG CAT CCC ATA TCT ACC AAG AGC CCC GAC TTG GAG CCT GTT CTG AGA TAT

Row 8

Bovine kininogen GCC ATC CAA TAT TTT AAC AAC ACC AGT CAT TCC CAC CTC TTT GAT CTG AAA GAA GTA AAA AGA GCC

Row 9

Bovine kininogen CAA AGA CAG GTG GTG TCT GGA TGG AAC TAT GAA GTT AAT TAC TCA ATT GCA CAA ACT AAT TGT TCC AAG

Row 10

Bovine kininogen   GAG CAA TTT TCA TTC TTA ACT CCA GAC *TGC* AAG TCC CTT TCA AGT GGT GAT ACT GGT GAA *TGT* ACA GAT

Row 11

Bovine kininogen   AAA CCA CAT GTA GAT GTC AAG CTA AGA ATT TCT TCC TTC TCG CAG AAA TGT GAC CTT TAT CCA GTG AAG

Row 12

Bovine kininogen   GAT TTT GTA CAA CCA CCC ACC AGG CTT *TGT* GCC GGC *TGC* CCC AAA CCT ATA CCT GTT GAC AGC CCA GAC

Row 13

Bovine kininogen   CTG GAG GAG CCT CTG AGC CAT TCC ATC GCA AAG CTT AAT GCA GAG CAT GAT GGA GCC TTC TAT TTC AAG

Row 14

Bovine kininogen   ATT GAC ACT GTG AAA AAA GCA ACA GTA CAG GTG GTA GCT GGA TTG AAG TAT TCT ATT GTG TTC ATA GCA

Row 15

Bovine kininogen   AGG GAA ACC ACA *TGT* TCT AAG GGA AGT AAT GAA GAG CTG ACC AAG AGT *TGT* GAG ATC AAT ATA CAT GGT

Row 16

Bovine kininogen   CAA ATT CTA CAC *TGT* GAT GCT AAT GTC TAT GTG GTG CCT TGG GAG GAA AAA GTT TAC CCT ACT GTC AAC

Row 17

Bovine kininogen   Bradykinin
   TGT CAA CCA CTT GGA CAG ACC TCA CTC ATG AAA AGG CCT CCG GGT TTT TCA CCT TTC CGA TCA GTT CAA

(continued)

Table 7.22 (Continued)

Row 18	
Bovine kininogen	GTG ATG AAA ACT GAA GGA AGC ACA ACT → GTA AGT CTA CCC CAC TCT GCC ATG TCA CCT GTA CAA GAT

Row 19	
Bovine kininogen	GAA GAG CGG GAT TCA GGA AAA GAA CAA GGA CCC ACT CAT GGG CAT GGC TGG GAC CAT GGA AAG CAA ATA

Row 20	
Bovine kininogen	AAA TTA CAT GGC CTT GGC CTT GGC CAT AAA CAT AAG CAT GAC CAA GGT CAT GGG CAC CAT GGA AGT CAT

Row 21	
Bovine kininogen	GGT CTT GGC CAT GGA CAT CAA AAG CAA CAT GGT CTT GGC CAT GGA CAT AAG CAT GGT CAT GGC CAC GGA

Row 22	
Bovine kininogen	AAA CAT AAA AAA GGA AAA AAC AAT GGA AAG CAT TAT GAT TGG AGG ACA CCC TAT TTG GCA AGT TCT

Row 23	
Bovine kininogen	TAT GAA GAT AGC ACT ACA TCC TCT GCA CAG ACG CAA AAG ACA GAA GAG ACA CTC TCT TCC CTA

Row 24	
Bovine kininogen	GCC CAG CCA GGT GTA GCC ATT ACC TTT CCT GAC TTT CAG GAC TCA GAT CTC ATT GCA ACT GTG ATG CCT

Row 25	
Bovine kininogen	AAT ACA CTA CCA CCT CAC ACA GAG AGT GAT GAC TGG ATC CCT GAC ATC CAG ACA GAG CCA AAT AGC

Row 26

Bovine kininogen   CTT GCA TTT AAA TTG ATT TCA GAC TTT CCA GAA ACA ACC TCC CCC AAA *TGT* CCT AGT CGC CCC TGG AAG

Row 27

Bovine kininogen   CCA GTT AAT GGA GTG AAT CCA ACT GTG GAA ATG AAA GAG TCT CAT GAT TTT GAT CTT GTT GAT GCT CTT

Row 28

Bovine kininogen   CTT TAA

aCodons for basic amino acids are underscored, those for cysteine are italicized, and those for tyrosine are stippled. Asterisk indicates the point of insertion of an intron. Mature genes are partially or completely boxed.
bKitamura et al. (1983); Nawa et al. (1983).
cGubler et al. (1983); Mayo et al. (1985).
dShen et al. (1982); Shen and Rutter (1984).
eMontminy et al. (1984); Tavianini et al. (1984).
fHobart et al. (1980).
gGoodman et al. (1980).
hMagazin et al. (1982).
iMinth et al. (1982).
jInsert AGC GCC GGA CTC CTG TCC.
kInsert ACG GTG ACG AGC AAA CCA CTC AAC.

Somatocrinin Genes. Since the somatocrinins have only recently been recognized as a distinct class of hormones, it is not surprising that but a single gene sequence has been determined, in this case, from human sources. However, it has been established twice (Gubler *et al.*, 1983; Mayo *et al.*, 1985). Structurally it is more typical than the preceding, bearing a presequence in orthodox fashion. To a degree, this example provides an intermediate step between subclasses III and IV, for the ultimate gene region is located only a short distance from the 5' terminus (Figure 7.2F). However, since the active product consists of 44 amino acid residues, it actually occupies the middle region of the precursor, with 11 sites preceding and 33 following. The ultimate coding region is preceded by the standard pair of codons for basic amino acids (in this case, both are for arginine), but the presence within the active part of two comparable pairs raises questions as to what other criteria are involved in recognition by the protease during cleavage. Immediately following activation, the somatocrinin molecule is amidized through modification of the glycine residue encoded by the GGT that follows the ultimate gene region proper. The gene is unique and is located in man on chromosome 20 (Mayo *et al.*, 1985).

The Somatostatin Gene Family. The somatostatin gene family is well on its way to becoming one of the most thoroughly explored of cryptomorphic types, for two cistrons from mammalian sources and five from fish have been determined. When additional ones from avian, reptilian, and lower vertebrates have been added to this list, a well-rounded picture of their phylogeny will emerge. Even now some details are clearly discernible, but with the report of the occurrence of several varieties of somatostatinlike products in *Bacillus subtilis* (LeRoith *et al.*, 1985), a need for investigation into possible ancient origins for this hormone is clear. However, here, before those evolutionary aspects can be understood, description of the gene structure and function is an essential prelude. Although principally in the pancreas, stomach, and small intestine, somatostatin occurs abundantly also in the nervous system, particularly the hypothalamus, and may be a neurotransmitting agent. Chiefly its effects are inhibitory, retarding the secretion of other hormones, including somatocrinin, insulin, glucagon, gastrin, and growth hormone (Shen *et al.*, 1982). Characteristically, the primary transcript of the gene has understandably been considered a "prepro"protein, with the inhibitory (pro)sequence exceptionally long (Figure 7.2G) (Shen *et al.*, 1982; Funckes *et al.*, 1983; Montminy *et al.*, 1984; Shen and Rutter, 1984; Tavianini *et al.*, 1984). At least two major forms of the ultimate product are known—somatostatin I, whose sequence is given in six of the seven provided in Table 7.22, and somatostatin II, represented by the second of the anglerfish structures (Hobart *et al.*, 1980). The latter appears to inhibit insulin secretion, but not that of glucagon (Shen *et al.*, 1982). Among its peculiarities of structure is the presence of a six-codon insertion located at the end of row 1.

These two forms of the 14-amino acid peptides just precede the 3' termini of the gene structures in rows 4 and 5 of the table. However, they encode only the more familiar representatives of this hormone. Often along with this short type, a peptide of 28 amino acid residues may be found in the bloodstream, at least of mammals. This variety is indicated in Table 7.22 as extending to the 5' end of row 4, and appears to undergo cleavage to produce the shorter type. But whether this duplex type is merely an intermediate product of processing or has activity in its own right has not been sufficiently investigated, although both species have proven to be generated by the same convertase (Gluschankof *et al.*, 1985). Nor is it established whether the double molecule occurs in anglerfish or other piscine subjects. The catfish somatostatin gene presents additional problems, for its ultimate cistron encodes a product eight residues longer than the rest, consisting of 22 amino acids instead of just 14. Although the last 12 codons of this ultimate gene may be noted to be largely homologous to the remaining examples, there are a number of differences that may affect its activity. Hence, further investigations of its functions in this fish are greatly needed.

The remainder of the mature gene of somatostatin 22 may be readily seen also to differ widely from the rest, especially in length, much of rows 1, 3, and 4 being unrepresented. With this limited material, it is not possible to attempt to establish homology on a site-to-site basis, the arrangement in Table 7.22 being entirely preliminary and suggestive only. Even its 5' end is not firmly determined, for in the original description the entire sequence, including that of the transit peptide down to the ultimate gene, was considered as a propeptide (Magazin *et al.*, 1982). Obviously extensive phylogenetic changes in structure of this gene have occurred within what is often considered to be a monophyletic class of vertebrates, in spite of evidence to the contrary (see Berg, 1940). While any deductions regarding the evolutionary history are now necessarily tentative, two trends may be noted in the material at hand. First, the entire gene, including the presequence as described shortly, appears to have undergone lengthening with phylogenetic advancement. On this basis, the gene may be expected to be still shorter in the Cyclostomata, if not the Elasmobranchia as well. Hence, among the lower vertebrates, the cistron may be diplomorphic rather than cryptomorphic. The second trend apparently is in the opposite direction, involving a reduction in size of the ultimate product, with hagfishes and relatives having a somatostatin of perhaps 30 amino acid residues.

Among the obvious additions that need to be made to this subclass when nucleotide sequences are more adequately determined is the gastrin gene, whose product stimulates the secretion of the gastric juices (Yoo *et al.*, 1982; Kato *et al.*, 1983). Additional knowledge will most likely indicate their placement here; this includes the cholecystokinins (Gubler *et al.*, 1984; Kuwano *et al.*, 1984; Deschenes *et al.*, 1985) and natriuretic factor synthesized in the cardiac atrium of mammals (Maki *et al.*, 1984; Sonnenberg and Veress, 1984; Zivin *et al.*, 1984), together with cardiodilatin (Kennedy *et al.*, 1984).

7.5.6. Presequences of Subclass IV Cryptomorphs

The presequences of subclass IV cryptomorphic genes add to the impression that has been growing increasingly firm, that other than the presence of an abundance of codons for hydrophobic amino acids, they lack general characteristics of a distinctive nature. Only half of the eight examples of Table 7.23 show a triplet encoding a basic amino acid near the 5' end. An additional sequence, along with one of the four just mentioned, displays two for arginine more centrally located, and another, that of anglerfish II, possesses a like codon at the extreme 3' terminus. Although all are, as already indicated, rich in triplets for apolar monomers, the distribution of those codons varies broadly from one sequence to another. In the human somatocrinin representative, all are situated in the 5' half, whereas in that of the rat, the opposite end is more densely supplied. More frequently, however, the midsection contains the major part, but even the three species from the anglerfish fail to be entirely consistent in this matter.

Their cleavage sites show a similar lack of constancy of structure. Five of the final

Table 7.23
Presequences of Subclass IV Cryptomorphic Genes[a]

Row 1

Human somatocrinin[b]	*ATG CCA CTC* TGG *GTG TTC TTC TTT GTG ATC CTC* ACC *CTC* --- --- ---
Human somatostatin[c]	*ATG CTG* TCC TGC <u>CGC</u> *CTC* CAG TGC *GCG CTG GCT GCG CTG* --- --- ---
Rat somatostatin[d]	*ATG CTG* TCC TGC <u>CGT</u> *CTC* CAG TGC *GCG CTG GCC GCG CTC* --- --- ---
Anglerfish somatostatin I[e]	*ATG* <u>AAG</u> *ATG GTC* TCC TCC TCG <u>CGC</u> *CTC* <u>CGC</u> TGC *CTC CTC GTG CTC CTG*
Anglerfish somatostatin II[e]	*ATG* CAG TGT *ATC* <u>CGT</u> TGT *CCC GCC ATC TTG GCT CTC CTG GCG TTG GTT*
Anglerfish somatostatin III[f]	*ATG* --- --- *GTC* TCC TCC TCG <u>CGC</u> *CTC* <u>CGC</u> TGC *CTC CTC GTG CTC CTG*
Catfish somatostatin 22[g]	*ATG* TCG TCT --- TCA *CCA CTC* <u>CGT</u> *CTC* --- *GCT CTT GCC CTC ATG* TGC
Catfish somatostatin[h]	*ATG CCC* TCC ACG <u>CGG</u> *ATC* CAG TGC *GCC CTG GCT CTC CTG GCC GTC* ---

Row 2

Human somatocrinin	--- AGC AAC AGC --- --- TCC CAC --- --- TGC TCC	*CCA CCT*
Human somatostatin	--- TCC *ATC GTC CTG GCC CTG GGC* TGT *GTC* ACC GGC	*GCT CCC*
Rat somatostatin	--- TGC *ATC GTC CTG GCT TTG GGC GGT GTC* ACC GGG	*GCG CCC*
Anglerfish somatostatin I	*CTG* TCC *CTG* ACC *GCC* TCC *ATC* AGC TGC TCC *TTC GCC*	*GGA* CAG
Anglerfish somatostatin II	*CTG* TGC *GGC CCA* AGT *GTT* TCC TCC CAG *CTC* <u>GAC</u> <u>AGA</u>	<u>GAG</u> CAG
Anglerfish somatostatin III	*CTG* TCC *CTG* ACC *GTC* TCC *ATC* AGC TGC TCC *TTC GCC*	*GGA* CAG
Catfish somatostatin 22	*CTG* --- *GTC* TCA *GCC GTC* --- *GGT GTC ATA* TCG TGC	*GGC* <u>CGG</u>
Catfish somatostatin 14	--- --- *GCG CTC* TCC *GTC* TGC AGC --- *GTC* TCA *GGC*	*GCG CCG*

[a]Codons for apolar (hydrophobic) amino acids are italicized and those for basic ones are underscored. Arrow indicates cleavage site.
[b]Gubler *et al.* (1983); Mayo *et al.* (1985).
[c]Shen *et al.* (1982).
[d]Montminy *et al.* (1984); Tavianini *et al.* (1984).
[e]Hobart *et al.* (1980).
[f]Goodman *et al.* (1980).
[g]Magazin *et al.* (1982).
[h]Minth *et al.* (1982).

codons bordering that point are for the small apolar amino acids glycine and alanine, one is for a small polar uncharged unit (serine), another encodes the large and charged arginine, and the last one specifies the large, sulfur-bearing cysteine. Moreover, the penultimate codons have equally wide ranges in the amino acids they encode. On the downstream side of the cleavage line much greater consistency is displayed, for all the triplets designate apolar types, except that of the distinctive structure of anglerfish somatostatin II (Hobart *et al.*, 1980), which codes for aspartic acid. Signals for proline are most frequently found in the second site of the mature coding sector, although the catfish somatostatin 22 gene has one for arginine, and all three of the anglerfish encode glutamine. Thus, as before, the combination of amino acids actually recognized by the enzyme that removes the transit peptide remains obscure.

7.6. STRUCTURE OF SUBCLASS V CRYPTOMORPHIC GENES

The members of the fifth and last subclass of cryptomorphic genes are structurally strongly in contrast to all the others. Instead of the mature coding region being extensive but encoding only one or more ultimate products of small size, here the final proteins occupy nearly the entire precursor. That is, a precursorial transcript, usually bearing a transit peptide, is produced, which typically is cleaved into two or three ''subunits'' that are the actual functional proteins. Thus the only unproductive regions of the mature gene are the short segments that bear the signals for the cleaving enzymes. Because the ultimate genes consequently are large molecules, whose complete sequences contribute little to a basic understanding of gene structure and nature, only the portions with a greater contribution to make toward fuller appreciation of the fundamentals of subclass V cistrons are given, along with diagrams wherever these are helpful toward this same goal.

7.6.1. Genes of Certain Types of Complement

In vertebrates and other metazoans is a large group of genes known as the major histocompatibility complex, whose products are involved in immune reactions of various sorts (Steinmetz and Hood, 1983; Kaufman et al., 1984). Of the three classes into which the members are grouped, those of class III, which encode ingredients of ''complement,'' are most frequently representatives of subclass V cryptomorphs, although several species from class II may also prove to be. In addition, a small number from classes I and II show indications of the type of complexity of organization that is treated in the next chapter, which deals with ''assembled'' genes.

The proteins of complement are secreted into the bloodstream, where they enter into an intricate cascade of interactions, which ultimately result in lysis of an invading cell (Dillon, 1983, pp. 382–384). In the principal chain of processes, called the classical pathway, nine major substances are involved, referred to as C1–C9; to this number two additional types, B and D, must be added, which are active in the second, or alternative, pathway. To judge from the manner in which the molecules are fractured during the interactions, C1–C5, C9, and B are encoded by genes that belong to the present subclass of cryptomorphs, but not all of these have been fully sequenced. Another deficiency in many relevant reports is the frequent failure to show the presequence that theoretically should be present on these secreted products. Nor are the other articles consistently clear as to the precise location of the cleavage points, timing of the processing, or functions of the resulting fractions. What has been adequately documented, nevertheless, provides sufficiently deep insight into the structure and activities of this more than usually interesting set of cistrons. Two that have been explored most thoroughly, C3 and C4, are presented first to set the stage for others also available. The gene for C5 has now had its sequence established, but it is not included, since it adds little to the total picture presented here (Lundwall et al., 1985).

The Mature Gene of Factor C3. Complement component C3 is undoubtedly the best known of the entire interacting chain, for, besides being the most abundant member, it plays critical roles in both the classic and alternative pathways. The gene consists of two primary parts, a presequence of 72 nucleotide residues and a mature coding sequence of ~2940 (Figure 7.3); immediately after translation the protein encoded by the latter is

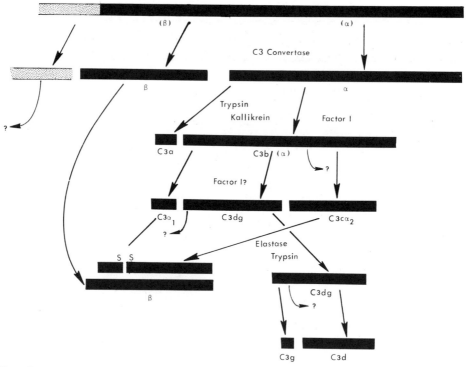

Figure 7.3. The intricately interacting behavior of complement C3. The gene structure is represented by the topmost diagram. (Based on de Bruijn and Fey, 1985.)

inert, becoming activated in the presence of foreign cells by action of either of two enzymes called convertases, one in each major pathway (de Bruijn and Fey, 1985). Following cleavage and removal of the transit peptide enzymatically into β and α chains, the latter of these ultimate products (Fey *et al.*, 1984; Wetzel *et al.*, 1984) undergoes multiple fractioning to carry out various distinctive functions, while the β remains intact (Figure 7.3). The first protease, C3 convertase, splits off a small segment, C3a, leaving the greater part as C3b, which is highly reactive. In one set of reactions, it unites with a dual combination of complement constituents C2a and C4b to form C2a·C4b·C3b. This product then may continue through the classical cascade of complement activity. Or, in the alternative pathway, C3b may unite with a different double molecule to produce pro-Bb·D·C3b to participate in the eventual lysis of the target cell, largely by acting on C5. Furthermore, it is capable of opsonizing bacteria, that is, making those organisms more susceptible to destruction by phagocytes. The small C3a sector of 78 amino acids split off by the convertase serves as an anaphylatoxin that principally binds to receptors on mast cells (macrophages) to induce the release of histamine and other factors stored in the granules of those cells. But C3b has other immune-related activities (Weigle *et al.*, 1983). The downstream portion of that chain may be acted upon by trypsin to release a particle called C3cα₁ (or kallikrein may carry out a similar action nearby), which remains attached to the B subunit. Still farther downstream, the molecule may be cleaved by factor I at two nearly adjacent points to produce both a median sector known as C3dg and a carboxyl terminal portion $C3c\alpha_2$, which is similar to $C3c\alpha_1$ in bearing a carbohydrate. The latter product is joined by a disulfide bridge to C3dg and joins it in forming a dimer with the β chain. In addition, some of these particles may be united by covalent bonds to C3b to carry out discrete immune reactions. Subsequently C3dg is further reduced by twin actions of elastase and trypsin to produce a short peptide C3g and a longer one C3d (Figure 7.3). Thus the seemingly simple apparent bipartite mature coding region is actually multifold, encoding a diversity of ultimate products.

The Mature Gene of Factor C4. The fourth component of complement, C4, is encoded in man by two separate but closely linked loci on chromosome 6 (Carroll *et al.*, 1984); polymorphism is rank, since gene *C4A* has 13 known alleles and *C4B* has 22 (Mauff *et al.*, 1983). It is synthesized in macrophages as well as in the liver as a single chain of molecular weight 200,000. Although the actual start sites of translation and transcription alike do not seem to have been established, so that the full length of the presequence is unknown, one is present nevertheless (Belt *et al.*, 1984). The mature coding region encodes three subunits, β, α, and γ, in that order $5' \rightarrow 3'$ (Schreffler *et al.*, 1984). These combine into the mature but inert protein as a simple $\alpha\beta\gamma$ trimer within the cell, the mechanism for its passage through the cytoplasmic membrane into the bloodstream remaining unexplored.

Activation is induced by a subparticle of $C\overline{1s}$, which releases the peptide C4a from the NH_2 end of the α chain, a substance apparently serving as an anaphylotoxin along with C3a. Later, factor I may further cleave this same chain, releasing the peptide C4d from the carboxyl half, an activity which results in the inactivation of the remaining C4 protein.

Both C3 and C4 show a limited degree of homology with α_2-macroglobulin, one of the plasma proteins, so it has been claimed that the three have had a common evolutionary origin (Sottrup-Jensen *et al.*, 1985). About two-thirds of C3 was shown to be similar in structure to the macroglobulin at a level between 19 and 31% homology, whereas C4 had only a comparatively short segment that displayed such a relationship. Since the overall level of identity thus was low, it may be that, rather than having common origin, the three have been exposed to similar genetic influences at the molecular level, as proposed for a certain pea nuclear protein in the preceding chapter.

7.6.2. Miscellaneous Representatives of Subclass V Cryptomorphs

Currently a small number of important proteins from a diversity of sources are known that constitute this subclass V of cryptomorphic genes, but the variety that these few established types displays strongly intimates that eventually it will prove to be a large group embracing numerous proteins of exceptional functional significance. As a whole, the structural complexities of those whose gene sequences have been determined have not been fully appreciated, for certain of the representatives undergo multifold stages of processing reminiscent of those of complement factors C3 and C4. Why they do so can scarcely be imagined at the moment. Why, for a case in point, should something as seemingly inert as a seed-storage protein need to be exposed to multistage processing? If nothing else, their structural ramifications certainly indicate that deeper investigations into their functions would doubtlessly be most profitable. But first continuity is best served if

A. Legumin

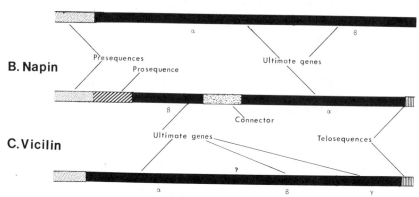

Figure 7.4. Cryptomorphic genes for seed-storage proteins.

some blood factors of vertebrates are examined, since their activities are in a cascading fashion much like that of complement.

Blood Factor X. Only a single factor from the blood-clotting cascade seems to have had its gene sequence established—and even that is incomplete at the 5' end (Leytus *et al.*, 1984). This solitary representative, for factor X, then must be taken to exemplify others, such as the blood factor IX, whose processing events are similar. This protein is encoded by a gene whose mature coding section is 1338 nucleotide residues in length. Whether that portion is preceded by a presequence has not been established, but a strong likelihood exists that it is. As shown in Figure 7.4, the mature gene provides for two subunits, referred to as light and heavy chains, that are separated by three codons in the series arginine, lysine, arginine, which signal the cleavage point. After translation and cleavage by an undetermined enzyme, the two subunits unite to form the mature inert protein, which is an αβ dimer. Activation of the factor involves cleavage by either factor IXa or VIIa, the proteolytic action removing the NH_2 end of the heavy chain, a region of about 50 amino acids (Leytus *et al.*, 1984). This reduced dimer, then known as factor Xa, converts prothrombin to thrombin in the presence of various ingredients to begin active blood-clot formation. In addition to the several cleavages by proteases, the mature product is complicated by addition of two carbohydrate moieties to the NH_2-terminal piece of the heavy chain and by conversion of 11 and 12 glutamate residues of the light and heavy chains, respectively, to γ-carboxyglutamic acid, in which process vitamin K plays an important role.

Among other blood-clotting factors whose gene structure may fall into this category is bovine protein C (Long *et al.*, 1984b), but not all vertebrate members of the subclass are of that nature. The important hormone insulin secreted by pancreatic β cells is the product of a gene whose mature coding region encodes B, C, and A chains, the C component being of unknown function. The completed hormone is in the form of the hexamer A_3B_3 (Hahn *et al.*, 1983). By coincidence, the gene encoding the insulin-receptor protein also is a member of the present subclass (Ebina *et al.*, 1985).

Seed-Storage Proteins. Unlike the simple diplomorphic examples seen at the beginning of this chapter, several types of seed-storage proteins are encoded by subclass V cryptomorphic genes and thus undergo varying degrees of posttranslational cleavages. By far the most straightforward representatives of the group are the several legumin genes from the pea, two examples of which (*legA* and *legD*) have been fully sequenced (Lycett *et al.*, 1984; Bown *et al.*, 1985). After a fairly typical presequence of 21-codon length, there is a long mature coding region, interrupted by three introns. This consists of a 996-nucleotide sequence for the α subunit at the 5′ end and one of 756 nucleotides encoding the β subunit, without any connecting elements between them, nor does there seem to be a telosequence (Figure 7.4A). Nothing has been reported as to the processing of the nascent product of translation. Homologs also are present in wheat and other grains (Robert *et al.*, 1985).

The legumin gene is unusual in having the larger subunit encoded in the 5′ portion; however, another member of this class of proteins, napin from rape (*Brassica napus*), has the more frequent arrangement. As in the foregoing, it begins with a presequence of 21 codons, but it differs in having a prosequence that codes for 17 amino acids (Crouch *et al.*, 1983). The first ultimate gene that ensues, for the β subunit, is only 36 codons in length; it is separated from the coding sequence for the α subunit by a connector of 20 codons (Figure 7.4B). At neither end of this region is there a recognizable cleavage signal, the 5′ terminus bearing triplets for proline, asparagine, and tryptophan, and the 3′ having those for proline, glutamine, and glycine, but proline may prove to be the principal element. The ultimate coding sector for the α subunit consists of 81 codons, followed by a telosequence consisting of triplets for proline, serine, and tyrosine, before the TAG translation-terminal signal. Napins constitute ~20% of the total rape seed protein and are broken down rapidly during germination. The holoenzyme, an αβ dimer of molecular weight 13,000, is highly basic, largely as the result of the high percentage (~25%) of glutamine residues it contains.

In a third member of the group, the level of complexity of the above example is equalled, but by a different means; a fourth, for ricin of castor bean, whose sequence has also been established recently (Lamb *et al.*, 1985), does not merit additional attention. Another current addition is that for glycinin, a complex product of soybean (Momma *et al.*, 1985a,b). Here in a gene for vicilin, 11 copies of which exist in the pea genome (Domoney and Casey, 1985), no inhibitory sequence intervenes between the transit peptide and mature coding sectors, the latter following the former directly (Lycett *et al.*, 1983). Moreover, the mature portion resembles that of the legumin gene in having the large subunit (α) region at the 5′ end (Figure 7.4C). This sector then leads into the β portion without any connector element, a condition repeated at its close, where the γ-coding part begins. This sector of 94 triplets is adjoined by a telosequence of unusual length, since it consists of 12 condons, including only one each for basic and acidic amino acids. However, the real complexity of the present structure lies in the apparent multifold posttranslational processing. Although details are not firmly established, it is reported that cleavage may in some cases not occur at the α–β juncture, and in others may not take place at any site, except perhaps the γ-telosequence point of contact (Lycett *et al.*, 1983). As a consequence, it is not clear whether vicilin is ever in the form of an αβγ trimeric protein, nor is it established how the precursor is processed.

7.7. COMPARISONS OF THE SEVERAL CLASSES OF COMPLEX GENES

Because of the numerous subdivisions needed to embrace the various types of diplo-morphic and cryptomorphic genes and their products, a synopsis of their principal features seems desirable. Three major types, simple, compound, and complex, are thus readily compared and their subdivisions more easily comprehended.

1. *Simple* genes encode single products that are ready for employment immediately following processing or translation.
2. *Compound* genes, such as those of many viruses (Chapter 10), encode two or more products that are useful in the cell after processing has separated them, that is, no latent period exists.
3. *Complex* genes encode two or more products, which remain intact during a latent period of varying length. Two principal classes are recognized:

   A. *Diplomorphic* genes encode either two or three different peptides, one of which is a transit (signal) sequence useful in passage through membranes. Two varieties exist:
      Simple diplomorphs, in addition to the active product, encode a transit pep-tide (presequence) which is removed from the product upon transport through a membrane.
      Complex diplomorphs also encode a transit peptide, but over and above provide for an inhibitor sequence (prosequence) which after being trans-lated remains attached to the principal gene product until the latter is needed in the cell.
   B. *Cryptomorphic* genes similarly encode a presequence, but usually lack a propeptide coding region. The ultimate gene product or products are coded within sectors from which they must be cleaved before they can be active. A highly varied assembly, they fall into at least the five following subclasses:
      Subclass I. In this category is placed those genes that encode multiple copies of the same or virtually the same protein, such as those that provide for the mating factors (α-pheromones) and glucoamylase (Yamashita *et al.*, 1985) in yeast or the enkephalins in vertebrates.
      Subclass II. This subclass embraces genes that contain ultimate coding re-gions for two or more different products which remain intact within a primary translational product for a more or less prolonged latent period.
      Subclass III. Besides the presequence, subclass III cryptomorphs contain an ultimate gene for a single product, often arranged as a prosequence, but always in the 5′ position of the original translational coding area.
      Subclass IV. The member genes resemble those of subclass III in encoding a single ultimate product, but they differ in having the coding sector for the principal protein in the 3′ portion. Often a sequence for an inhibitor peptide may be present, so that they approach the complex diplomorphs structurally.
      Subclass V. The genes that constitute this subclass each encodes a precur-sorial translational product, typically bearing a transit peptide, com-prised of two or more so-called subunits (actually functional proteins) so

that most of the coding region results in active products. Complement components of the major histocompatibility complex provide the clearest examples of this category.

It is to be anticipated that additional subclasses of cryptomorphic genes will come to light as explorations into the coding structures of the green plants, seaweeds, invertebrates, and protistans continue, for the hormones and enzymic secretions of all living things need the protective control that this device affords. In this class, as among the simpler diplomorphs, control is exercised by successive cleavages of large molecules into smaller parts. In the next chapter, the opposite convention is followed, lesser components being combined into larger ones, providing for diversity in the product rather than protection.

8

Assembled Genes

With the assembled genes that, along with possible ancestral types, provide the topic of the present chapter, the extreme pinnacle of complexity of gene organization is reached. Not even the intricacies of posttranslational cleavages that characterized the cistrons of complement factor C3, for instance, can approach those involved in the formation of the final gene structures of the immunoglobulins, the prime example of this class of genomic constituents. Despite their thus being the most elaborately constructed of all genetic determinants, those proteins nevertheless provide the best medium for introducing the present subject, for an abundance of data at the molecular level is available concerning them that cannot begin to be matched by simpler representatives of the group from whatever source.

8.1. IMMUNOGLOBULIN GENES

Since a detailed description of the structure, properties, and activities of the immunoglobulin genes has already been provided (Dillon, 1983, pp. 333–398), only the essentials need to be repeated for present purposes (Kindt and Capra, 1984). Unlike that earlier discussion, which was limited to comparisons of amino acid sequences, here the nucleotidyl structures are examined. Although there are five distinct classes of immunoglobulins, all consist of one or more tetramers having the form $\alpha_2\beta_2$, the α subunit being known as the heavy chain and the β as the light (Figure 8.1). Two varieties of the latter exist, κ and λ, either of which may be present in a particular tetramer, but not both. In a given clone of cells, moreover, only one of the two is produced, despite genes for both being present (Sun *et al.*, 1985). Because the light chains are simpler, being confined to two major types, as well as smaller, they are considered before the complexities of heavy chains are introduced.

8.1.1. The Genes of Light Chains

The genes for light chains consist of ~640 nucleotide residues for the ultimate product, plus a presequence and recognition signals, whose functions become clearer later. However, there is no single complete gene for either type, but rather each is

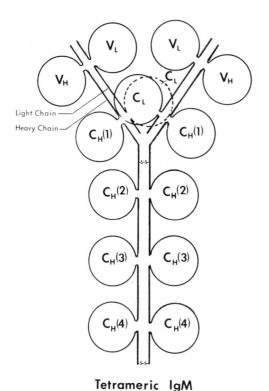

Light Chain
Heavy Chain

Tetrameric IgM

Figure 8.1. The basic structure of immunoglobulin molecules. Immunoglobulins consist
of multiple copies of the basic tetrameric subunit
depicted here.

assembled before transcription by the combining of fragmental sectors, known as genelets
(Dillon, 1983). It is for this reason that this class of cistrons is referred to here as
"assembled," for the coding structure that undergoes transcription must first be put
together by what are called "recombinational events." Since the light chains result from a
combination of only three parts, their simplicity of structure relative to that of heavy
chains, which require a larger number of genelets, becomes obvious.

In simple terms, two of the light-chain genelets encode separate domains of the
resulting protein, one for a "variable" (V) region, the other for a "constant" (C) domain.
These functional parts are held together by a third genelet, which encodes the "joining"
(J) segment. Since heavy chains have comparable parts, the several light types of genelets
of DNA or sectors of the peptides are distinguished as V_L, C_L, or J_L, if collectively, or as
V_κ or V_λ and so on when reference is made to a specific kind of light chain. The genelets
for each major variety form separate clusters, but the organization within clusters is
similar in both instances.

The Genome for κ Chains. The genomic portion encoding L_κ chains, located in
man on chromosome 2, begins with the genelets for the variable region (Figure 8.2A). In
both the laboratory mouse and man, numerous copies occur, but the precise number has
not been established. The best current estimates suggest the existence of 15–20 in the
human genome arranged in several clusters and perhaps 200 in the murine (Bentley and

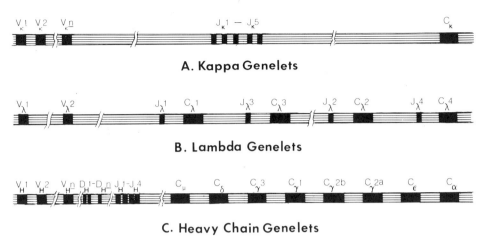

A. Kappa Genelets

B. Lambda Genelets

C. Heavy Chain Genelets

Figure 8.2. (A) Arrangement of immunoglobulin genelets in the murine genome. (Based on Hood *et al.*, 1985.) (B) The human genomic arrangement differs from that of the mouse in having more than two genelets for Cλ.

Rabbitts, 1981; Klobeck *et al.*, 1984; Pech *et al.*, 1984, 1985). However, not all of those present encode a usable polypeptide (Walfield *et al.*, 1980), some being pseudogenes. Those encoding a functional product begin with a presequence, followed by a sector consisting of ~95 codons and a spacer of undetermined length that separates these V_κ genelets from those for the joining sector. That the intergenic segment is probably 1000 or so base pairs long is suggested by the presence in the mouse of a transcribed leader 740 residues in length on the J_κ genelets. In that same mammal, and also in the rat and man, five genelets are found, designated as $J_\kappa 1$–$J_\kappa 5$, although in some mammals one copy is a pseudogene, $J_\kappa 3$ in the mouse being one example (Max *et al.*, 1981; Breiner *et al.*, 1982). These are separated in the murine genome by spacers ranging between 267 and 316 residues. This joining region is then separated from the C_κ sector by a long spacer, which in the mouse consists of 4515 nucleotides. Only a single C_κ genelet exists in the murine haploid genome; it and the rest of the κ-chain sequences are situated on chromosome 6 (Sheppard and Gutman, 1982). A like number, but not necessarily location, has been reported for the rabbit (Emorine and Max, 1983), but other laboratories describe at least four copies (Heidmann and Rugeon, 1982).

The Genes for λ Chains. In general, the arrangement of the coding sectors for λ chains resemble those for κ, consisting of V_λ, J_λ, and C_λ genelets (Figure 8.2B). Fewer copies, however, are found and the translational level is much lower, in the human blood stream being ~40% of that of the κ (Anderson *et al.*, 1984). As before, the V_λ genelets, each equipped with a presequence of ~19 codons, begin the series, in the human genome being represented by perhaps ten copies, located on chromosome 22. In the mouse still fewer exist, V_λ being duplicated only once and C_λ four times (Selsing *et al.*, 1982). Possible combinations are limited, since $V_\lambda 1$ can be expressed solely with either $C_\lambda 1$ or $C_\lambda 3$, whereas $V_\lambda 2$ is always associated with $C_\lambda 2$. The fourth constant-region genelet, $C_\lambda 4$, is a pseudogene and is nonfunctional because of a faulty splice site in the RNA. Each of the C_λ sequences is preceded by its own joining segment genelet, $J_\lambda 1$ with $C_\lambda 1$ and so

8.1.2. The Genes of Heavy Chains

As a whole, the genes for heavy chains are similar to those of the light ones, but are assembled from a total of four, or occasionally five, parts, rather than the three of the others. Variable (V_H), joining (J_H), and constant (C_H) region genelets are found here, but the first two are separated by one for an additional element, the diversity (D) region (Early *et al.*, 1980). In simple terms, the V_H, D, and J_H genelets are combined to form the variable region pregene; then an additional activity combines that resulting part to a C_H genelet and thereby creates the complete heavy-chain gene (Figure 8.3).

The Variable Region. Among mammals as a group, the V_H genelet encodes 100 amino acids, the D about eight, and J_H about 17 (Early *et al.*, 1980). As in the coding region of light chains, the V_H genelet bears a presequence, as expected of a secreted product. However, not all immunoglobulins are actually expelled from the cell; many of the molecules remain embedded in the cell membrane as receptors or identity markers. Moreover, it is peculiar that all four components of the basic $\alpha_2\beta_2$ (H_2L_2) tetramers are equipped with transit peptides, if the heterodimeric proteins are completed within the cell, as appears to be the case. In view of this, the behavior and function of the presequences of both the light and heavy chains need to be investigated more thoroughly, along with details of the processing events.

Each of the three component genelets are clustered separately from the others, those for the D region lying between those of the other two, in the series $5'-V_H1-V_Hn$, $D1-Dn$, J_H1-J_Hn-3' (Alt *et al.*, 1984). At least four clusters of V_H genelets are known to exist in the mouse. Within a cluster, the genelets are separated by spacers approaching 14,000 residues in length (Kemp *et al.*, 1981; Blankenstein *et al.*, 1984), but the total number remains undetermined. A minimum of ten D genelets are known to occur (Manser *et al.*, 1984) and just four for the J_H region, again in the murine genome (Figure 8.2C). Insofar as other vertebrate species are concerned, almost nothing firm has been established as to number and arrangement of the several types of genelets.

The Constant Region. The genelets for the heavy-chain constant region are relatively few in number, but are of special importance in that it is their product that confers the class and subclass characteristics to the completed immunoglobulin (Ig) molecule. Five major classes are recognized, IgA, IgD, IgE, IgG, and IgM, the C_H genelets for which are distinguished by corresponding lowercase Greek letters. Thus, in the mouse the genelets are arranged in the series $5'-C_\mu$, C_δ, $C_\gamma3$, $C_\gamma1$, $C_\gamma2b$, $C_\gamma2a$, C_ϵ, and $C_\alpha-3'$ (Figure 8.2C; Shimizu *et al.*, 1982).

Organization in the hamster genome is comparable to the foregoing (McGuire *et al.*, 1985), but surprisingly, the organization in the rabbit differs quite strongly. Only a single copy of C_γ is present in the latter mammal, whereas C_α is represented by four, in the sequence $5'-J_H$, C_μ, C_γ, C_ϵ, $C_\alpha1$, $C_\alpha2$, $C_\alpha3$, $C_\alpha4-3'$ (Knight *et al.*, 1985).

In the human genome, the arrangement is similar to that of the mouse, but it is distinguished by the duplication of C_α and the presence of a pseudogene for IgE: $5'-C_\mu1-$ 4, C_δ, $C_\gamma3$, $C_\gamma1$, $C_{\psi\epsilon}1$, $C_\alpha1$, $C_\gamma2$, $C_\gamma4$, $C_\alpha2-3'$ (Rabbitts *et al.*, 1981a; Flanagan and Rabbitts, 1982). The four μ genelets are separated by short intergenic spacers, and the last copy is followed downstream 1900 residues by the coding segments—actually genelets, although not recognized as such—for the sector that converts the usual secreted form of IgM to the type that is embedded in the cell membrane. The second type of genelet in the series, C_δ, similarly has dual sites of activity and accordingly has two (or perhaps even three) variant forms, one as an excreted product, another as a membrane-embedded receptor. According to present knowledge, a special genelet encodes a substitute $3'$ end for the third variant (Cheng et al., 1982).

8.1.3. Formation of Immunoglobulin Genes

Most of the steps in the assembly of the functional immunoglobulin gene from the various parts have already been suggested in the preceding sections. But to provide clarity for comparative studies, the processes are now restated in full. Additionally, a new element is introduced as the whole series of the formative and developmental steps are viewed; then, after the nucleotidyl structures have been examined, events at the molecular level are added.

Light-Chain Gene Formation. Because of the small number of genelets present in the murine genome for each given segment, the assembly of λ light chains is relatively straightforward. Further, as shown earlier, each of the C_λ genelets is closely associated with a single J_λ coding sector; consequently, $C_\lambda 1$ is consistently associated with $J_\lambda 1$, $C_\lambda 2$ with $J_\lambda 2$, and so on, as shown previously. During the early stages of development of a B lymphocyte, the J_λ and C_λ combinations become established, at least to a large extent. As the result of the limited number of V_λ genelets, only two of which exist in the mouse, the final combinational event is relatively simple, involving only the fusion of a V_λ to the $J_\lambda C_\lambda$ sector. However, even this simple process is not without some complex aspects, for, as already shown, $V_\lambda 1$ unites only with $J_\lambda 1$–$C_\lambda 1$ or $J_\lambda 3$–$C_\lambda 3$, and $V_\lambda 2$ solely with $J_\lambda 2$–$C_\lambda 2$.

As a consequence of the more numerous copies present for the variable and joining κ light-chain regions, their assembly, although fundamentally like that of λ chains, is more complicated and permits greater latitude for diversification. In the mouse any one of the four functional J_κ genelets may unite to the single one for C_κ. Since the intergenic spacer is lengthy, this recombinational event probably involves deletion of the DNA that intervenes between the two genelets. Then one of the numerous V_κ genelets is added to this J_κ–C_κ pregene to form the definitive L_κ gene, a step that similarly involves removal of intervening DNA. Within a given cell, this process is engaged in only one time, so that the same combination of V_λ–J_λ–C_λ or V_κ–J_κ–C_κ chains persists through the cell's entire existence.

Heavy-Chain Gene Formation. The total picture of heavy chain formation is com plicated, not only by the existence of more numerous segments, a greater abundance of copies of the fundamental parts, and the occurrence of membrane versus secreted variants, but also by progressive changes in the major immunoglobulin group expressed during the course of a cell life span. Furthermore, not all the steps and processes are firmly established. The first stage in the formation of a gene for a heavy chain is the union of a D genelet with one for a J_H, as shown by analysis of murine leukemia-virus-transformed fetal-liver cells, the most primitive B-lymphoid cell available for study (Alt et al., 1984). This combination is produced on both the maternal and paternal chromosomes of the diploid genome. Next a V_H genelet is attached to the DJ_H combination, accompanied by removal of intervening DNA, as is also the first step (Figure 8.3A–C). Again the second step is accomplished on both chromosomes. Once the resulting pregene $V_H DJ_H$ has become joined to the C_μ pregene on either chromosome (Figure 8.3D), however, some

A. GENOMIC ARRANGEMENT

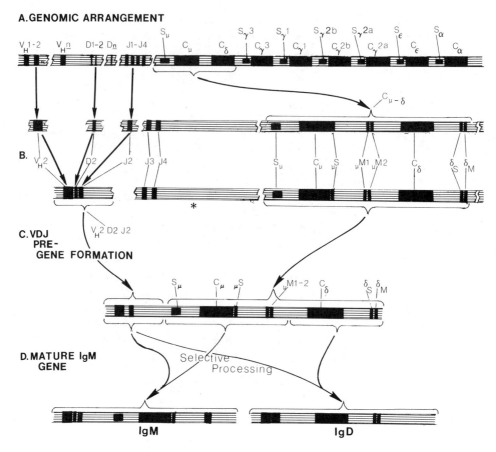

Figure 8.3. Steps in the assembly of heavy immunoglobulin chains. A complicated process, assembly of a VDJ pregene (B, C), consisting of any combination of the several V, D, and J genelets and deletion of intervening sectors, is among the first stages. That pregene then is combined to a segment containing heavy-chain genelets. A region deleted during the latter process is marked with an asterisk.

mechanism prevents further combinational events on the homolog. Thus, as stated before, a given cell only produces one particular $V_H DJ_H$ pregene, the allele being able to produce DJ_H combinations but nothing else once the full gene $V_H DJ_H C_\mu$ has been established. This suppression of one allele by a product of the other is termed ''allelic exclusion.''

Later Maturational Steps—Class Switches. As implied in the above discussion, all immunoglobulin-producing cells (the B lymphocytes and their maturational product, the plasmocyte) commence production of the substance with IgM. But upon interaction with an antigen, B cells proliferate and ultimately differentiate into plasma cells, changing the class of immunoglobulin as they do so, from the original IgM to either IgG, IgA, or IgE. In this heavy-chain class switch, as the change is called, the $V_H DJ_H$ pregene becomes joined to any one of the other constant genelets, either C_γ, C_ϵ, or C_α, accom-

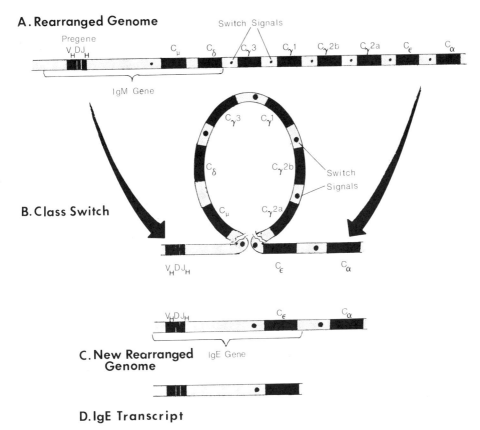

A. Rearranged Genome

B. Class Switch

C. New Rearranged Genome IgE Gene

D. IgE Transcript

Figure 8.4. Proposed stages involved in class switches. Formation of a loop brings together the two sectors to be joined, the loop then being cleaved enzymatically. Afterward the two free ends of the DNA are spliced.

panied by loss of the intervening DNA, including the coding sector of C_μ (Figure 8.4; Moore *et al.*, 1981). However, unreacted B cells produce two types of membrane immunoglobulin simultaneously, for although initially all only generate IgM, with time IgD also occurs in their cell membranes in approximately equal quantity. This dual expression is seemingly accomplished by a special mechanism in which both C_μ and C_δ are part of a single gene and thus are cotranscribed (Figure 8.3D, E). The resulting $V_H DJ_H C_\mu C_\delta$ precursorial messenger is then processed variously to produce $V_H DJ_H C_\mu$ and $V_H DJ_H C_\delta$ messengers for translation. However, in those plasma cells that synthesize IgD alone, the switch mechanism from IgM involves loss of the C_μ DNA. Consequently, IgD apparently is produced by two different mechanisms.

That the foregoing explanations are simplistic to some extent is shown by the existence of three kinds of mRNA for IgD (Maki *et al.*, 1981). One of these, of 2700 base pairs, containing six exons, may encode a membrane type, while a second, of 1800 residues, consists of three of those exons plus a fourth that is unique to it. This possibly codes for the secreted form of IgD. The function of the third kind, also of 1800 nucleotides, is unknown, but it consists of the first four exons of species 1 plus part of the fifth. The fact that the first two mRNAs have quite different 3' ends strongly suggests that some DNA rearrangement is involved, not mere processing events; thus, the several distinctive segments are not just exons; rather, they represent products of contrasting genelets, in line with the suggestion made in the discussion of genome organization.

Another Element—The Switch Region. It is the frequent suggestion that class changes in immunoglobulins are facilitated by the presence of special switch regions (Kataoka *et al.*, 1981). These "S" regions of the DNA contain multiple copies of such short sequences as GAGCT, GAGCTG, and TGGG (Kataoka *et al.*, 1983) and are located upstream of each C_H genelet except that for C_δ (Figure 8.4). When a class switch is made, the S region remains attached to the $V_H D J_H$ pregene, although the C_μ portion is deleted along with other intervening DNA. Thus its S region is free to interact with the S elements preceding any of the other C_H genelets to form a new Ig molecule (Figure 8.4B). In current theory the interaction involves the formation of an S–S combination, but specific details of the processes have not been fully clarified. Moreover, the absence of an S sector in association with the C_δ genelet presents an additional problem. Although it satisfactorily accounts for the formation of a bicistronic mRNA of C_μ and C_δ, it leaves unanswered the question of how IgD may be expressed in cells without the concomitant presence of IgM. However, it may be that in man, where IgD is synthesized alone, a vestigial S region may be present upstream of the C_δ genelet that may function in this particular class switch (Milstein *et al.*, 1984).

Another point in this connection has not received sufficient additional research to illuminate its production; this discovery pertains to the existence of what has been called "remnant DNA" (Steinmetz *et al.*, 1980; Höchtl *et al.*, 1982; Van Ness *et al.*, 1982; Selsing *et al.*, 1984). Although the generally accepted excision-deletion model is presented here, it may not prove to correspond to the actual facts when further knowledge is gained. What has been learned of remnant DNA indicates it to hold a by-product relationship to the joining of V_κ and J_κ genelets, but the relationship is complex, especially in view of its occurrence in only 50% of the myelomas (Steinmetz *et al.*, 1980; Van Ness *et al.*, 1982). The latest study (Selsing *et al.*, 1984) suggests that the recombinational events involve at least one endonuclease that cleaves the DNA precisely adjacent to the $V_H D$ and J_H recognition sequences, an exonuclease that acts upon the termini thus generated, and a ligase that then joins these ends to form the $V_H D J_H$ genelet. Among the surprising results of that investigation was the invariable association in plasmacytomas of remnant DNAs with $J_\kappa 1$.

8.2. STRUCTURES OF THE VARIABLE PREGENES

In describing the structures of the various genelets that together encode immunoglobulin molecules, it is advantageous from a number of standpoints to view them in a united fashion, rather than as components of heavy or light chains separately. The greatest gain, however, is that a superior understanding of their individual properties is provided, along with broader bases for comparative purposes. Beginning with the V genelets and then continuing in the sequence of their arrangement in the completed gene affords an opportunity to commence with a familiar part of an unfamiliar macromolecule.

8.2.1. The Genelets of the Variable Regions

As pointed out before, the genelets for the V sectors of immunoglobulins are unique in bearing a presequence, as befits the first component in the array that eventually encodes a secreted product. In the present instance, unlike many other membrane components, however, the same feature is found on those immunoglobulins that become embedded in the cell membrane. Since the structure thus displays some unique properties, it is well to examine it prior to any analysis of the genelets proper.

The Presequences of V Genelets. The presequences of eight V_H genelets and six V_L in Table 8.1 present a unique opportunity to analyze the properties of the encoded transit peptides from a goodly number of representatives of a single class of biochemicals. Here some uniformity in length is at once apparent, for only a single example, that of a murine V_κ genelet, deviates strongly in having six codons in row 1 that are consistently absent in all others. A number of other similarities among the majority of those cited are also in evidence. In the first place, all begin with the standard codon for the initiating methionine (ATG), which, in half of the representatives, is followed by a triplet (underscored) for a charged amino acid. Most of these are GACs, along with one GAT, which specify glutamic acid, although one AAG for lysine can be noted. Four, nevertheless, are for apolar amino acids (italics), mostly glycine, and two indicate uncharged polar types. At the 3′ terminus an unusual level of consistency exists, all except one of the triplets specifying polar uncharged protein monomers, with TCY for serine and TGT for cysteine being strongly predominant. Only a single codon for an apolar type is found, a statement equally applicable to the penultimate sites. At this latter position, a class distinction is observed, with all the V_κ presequences having signals for charged amino acids, whereas the large majority of the V_H representatives have either CAC for histidine or CAG for glutamine.

A second class distinction is located near the middle of row 2, where the V_H examples have either GCN, TCA, or GTC, in contrast to the TGG or TGC of those of V_L sequences. For the greater part, however, such distinctions between classes are exceptional, for homologous or related codons at corresponding sites in a large fraction of sequences is the more usual rule in these representatives. Outstanding examples of such homologies are in row 1, site 11, in which the entire column consists of codons for apolar amino acids, all of which have a central T. In row 2, site 3, CTC occurs in abundance, as does CTG in the one that follows. The site 5 column is rich in CTG, and GGN nearly fills the entire 11th. So, although the mature V sectors may differ widely, their presequences emphatically suggest their common ancestry and that not too distant in time past, a statement verified by the identical location of an intron in row 2, site 11.

The V Genelets. Those relationships identified above do not cease with the presequences, but can be noted in the mature coding genelets for the variable section in the nine examples of V_H and five of V_L provided in Table 8.2. Even that of the caiman, although a member of the reptilian line that leads to the birds, far from the mammalian line of ascent, nevertheless is largely homologous to the human and murine representatives that comprise the remainder of the table. Here the heavy- and light-chain components display more frequent class distinctions than was true of the presequences, yet there are a number of invariable points that hold for both types. At site 1, by way of illustration, CAG is found

Table 8.1
Presequences of Variable Genelets of Immunoglobulins[a]

Row 1

Human V_HG3[b]	ATG GAC TGG ACC TGG AGG GTC --- --- --- TTC --- --- --- --- --- ---
Murine V_H[c]	ATG GGA TGG AGC TTC --- ATC --- TTT CTC TTC --- --- --- --- --- ---
Murine V_H108[d]	ATG GGA TGG AGC TGG --- ATC --- TTT CTC TTC --- --- --- --- --- ---
Murine V_HBCL[e]	ATG GGT TGG AGC TGT --- ATC --- ATC TTC TTT --- --- --- --- --- ---
Murine $V_H\alpha$[f]	ATG AAG TTG --- TGG --- TTA AAC TGG GTT --- --- --- --- --- ---
Murine V_H283[g]	ATG AAT TTT GGG CTG AGC --- --- TTG ATT TTC --- --- --- --- --- ---
Murine V_H441[g]	ATG GAT TTT GGG CTG AGC ATT --- TTT --- TTT --- --- --- --- --- ---
Human V_HCESS[h]	ATG GAC ATA CTT TGT --- TCC ACG CTC CTG CTA --- --- --- --- --- ---
Human V_κ101[i]	ATG GAC ATG --- AGG GTC CTC GCT CAG CTC CTG --- --- --- --- --- ---
Human V_κ[j]	ATG GAC ATG --- AGG GTC CCC GCT CAG CTC CTG --- --- --- --- --- ---
Rabbit V_κ[k]	(Incomplete) CTG CTG --- --- --- --- --- ---
Murine V_κ[l]	ATG CAT CAC ACC --- AGC ATG GGC ATC AAA ATG GAG TCA CAG ATT CAG GTC
Murine V_κL6[m]	ATG GAC ATG AGG ACC CCT --- GCT CAG TTT CTT --- --- --- --- --- ---
Human V_λ[n]	ATG GCC TGG ACT --- CCT CTC --- --- TTT CTG --- --- --- --- --- ---

Row 2

Human V_HG3	--- --- TGC TTG CTG --- GCT GTA GCA CCA G*GT GCC CAC TCC ↓ CAG GTG
Murine V_H	--- --- CTC CTG --- --- TCA GTA ACT GCA G GT GTC CAC TCT GAG GTT
Murine V_H108	--- --- CTC CTG --- --- TCA GGA ACT GCA G*GC GTC CAC TCT GAG GTC
Murine V_HBCL	--- --- --- CTG --- GTA GCA ACA GCT ACA G GT*GTG CAC TCC CAG GTT
Murine $V_H\alpha$	TTT --- CTT TTA ACA CTT --- TTA --- CAT G*GT ATC CAG TGT GAG GTG
Murine V_H283	--- --- CTT GTC CTA ATT --- TTA --- AAA G*GT GTC CAG TGT GAA GTG
Murine V_H441	--- --- ATT GTT GCT CTT --- TTA --- AAA G GG GTC CAG TGT GAG GTG
Human V_HCESS	--- --- --- --- CTG ACT GTC CCG --- TCC*T GG GTC TTA TCC CAG GTC
Human V_κ101	GGG --- CTC CTG CTG CTC TGT TTC --- CCA*G GT GCC AGA TGT GAC ATC
Human V_κ	GGG --- CTC CTG CTA CTC TGG CTC --- CGA C*GT GTC AGA TGT GAC ATC
Rabbit V_κ	GGG --- CTC CTG CTG CTC TGG CTC --- CCA G GT GCC AGA TGT GCC CTC
Murine V_κ	TTT GTA TTC G*TG TTT CTC TGG TTG --- TCT G*GT GTT GAC GGA GAC ATT
Murine V_κL6	GGA --- ATC TTG TTG CTC TGG TTT --- CCA G*GT ATC AAA TGT GAC ATC
Human V_λ	TTC --- CTC CTC ACT TGC TGC CCA --- GGT*G GG TCC AAT TCT CAG ACT

[a]Codons for charged polar amino acids are underscored and those for hydrophobic ones are italicized. Asterisks mark the location of introns. The arrow indicates the cleavage site.
[b]Rechavi et al. (1983).   [c]Sims et al. (1982).   [d]Ohno et al. (1982).   [e]Knapp et al. (1982). [f]Kim et al. (1981).   [g]Ollo et al. (1983).   [h]Takahashi et al. (1984).   [i]Bentley and Rabbitts (1980).   [j]Klobeck et al. (1984).   [k]McCarney-Francis et al. (1984).   [l]Kelley et al. (1982). [m]Pech et al. (1981).   [n]Anderson et al. (1984).

Table 8.2
Variable Regions of Ig Genes[a]

Row 1

Heavy Chains

Caiman $V_H C3$[b]	CAG GTG CAG CTG GTG CAG TCC GGA GGA *GAT* GTG AGG AAA CCT GGA AAC TCT TTG CGC CTC TCC TGC AAA
Murine $V_H 186.2$[c]	CAG GTC CAA CTG CAG CAG CCT GGG GCT *GAG* CTT GTG AAG CCT GGG GCT TCA GTG AAG CTG TCC TGC AAG
Murine $V_H GAT$[d]	*GAG* GTT CAG CTG CAG CAG TCT GGG GCA *GAG* CTT GTG AAG CCA GGG GCC TCA GTC AAG TTG TCC TGC ACA
Murine $V_H 108A$[e]	*GAG* GTC CAG CTT CAG CAG TCA GGA CCT *GAG* CTG GTG AAA CCT GGG GCC TCA GTG AAG ATA TCC TGC AAG
Murine $V_H 107$[f]	*GAG* GTG AAG CTG GTG *GAA* TCT GGA GGA GGC TTG GTA CAG CCT GGG GGT TCT CTG AGA CTC TCC TGT GGA
Murine $V_H 283$[g]	*GAA* GTG ATG CTG GTG GAG TCT GGG GGA GGC TTA GTG AAG CCT GGA GGG TCC CTG AAA CTC TCC TGT GCA
Human $V_H CESS$[h]	CAG GTC AAC TTA AGG GAG TCT GGT CCT GCG CTG GTG AAA GCC ACA CAT ACC CTC ACA CTG ACC TGC ACC
Human $V_H G3$[i]	CAG GTC CAG CTG GTG CAG TCT GGG GCT *GAG* GTG AAG AAG CCT GGG GCC TCA GTG AAG GTT TCC TGC AAG
Human $V_H \epsilon$[j]	CAG ACG CAG TTG GTG CAG TCT GGG GCT *GAG* GTG AAG AAG CCT GGG GCA TCA GTG AGG GTC TCC TGC AAG

Light Chains

Human $V_\lambda k$[k]	CAG ACT GTG GTG ACT CAG *GAG* CCC TCA CTG ACT GTG TCC --- --- CCA GGA GGG ACA GTC ACT CTC ACC
Human $V_\kappa z$[z]	*GAA* ATT GTG TTG ACA CAG TCT CCA GCC --- ACC CTG TCT TTG TCT CCA GGG GAA AGA GCC ACC CTC TCC
Murine $V_\kappa m$[m]	*GAC* ATT GTG ATG ACC CAG *GAC* GGA GGT GTT *GAC* ATC[a] GGA GTC AGC ATC ACC
Murine $V_\kappa L6$[n]	*GAC* ATC AAG ATG ACC CAG TCC ATG TAT GCA TCT CTA GGA *GAG* AGA GTC ACT ATC ACT
Rabbit $V_\kappa o$[o]	GCC CTC GTG ATG ACC CAG ACT CCA TCC GTG TCT GCA GCT GTG GGA GGC ACA GTC ACC ATC AAG
Chicken $V_\lambda 1$[p]	GCG CTG --- --- ACT CAG CCG TCC GTG TCA GCG AAC --- --- CCG GGA *GAA* ACC GTC AAG ATC ACC

(continued)

Table 8.2 (Continued)

Row 2

Heavy Chains

Caiman V_HC3	GCC TCG GGG TTC ACC TTC GGT GGC TAC GGC ATG TTC TGG GTC CGC CAG --- GCT CCT GGG AAG GGG CTG GAC ---
Murine V_H186.2	GCT TCT GGC TAC ACC TTC ACC AGC TAC TGG ATG CAC TGG GTG AAG CAG --- AGG CCT GGA CGA GGC CTT GAG ---
Murine V_HGAT	GCT TCT GGC TTC AAC ATT AAA GAC AGC TAT ATC CAC TGG GTG AAG CAG --- AGG CCT GAA CAG GGC CTG GAG ---
Murine V_H108A	GCT TCT GGA TAC ACA TTC ACT GAC TAC AAC ATG CAC TGG GTG AAG CAG --- AGC CAT GGA AAG AGC CTT GAG ---
Murine V_H107	ACT TCT GGG TTC ACC TTC AGT GAT TTC TAC ATG GAG TGG GTC CGC CAG --- CCT CCA GGG AAG AGA CTG GAG ---
Murine V_H283	GCC TCT GGA TTC ACT TTC AGT AGC TAT ACC ATG TCT TGG GTT CGC CAG --- ACT CCG GAG AAG AGG CTG GAG ---
Human V_HCESS	TTC TCT GGG TTG TCA GTC AAC ACT CGT GGA ATG TCT GTG AGC TGG ATC --- CGT CAG CCC CCA GGG AAG GCC ---
Human V_HG3	GCA TCT GGA TAC ACC TTC AAC AGC TAC TAT ATG CAC TGG GTG CGA CAG --- GCC CCT GGA CAA GGG CTT GAG ---
Human $V_H\epsilon$	GCT TCT GGA TAC ACC TTC ATC GAC TCC TAT ATC CAC TGG GTA CGA CAG --- GCC CCT GGG CAC GGG CTT GAG ---

Light Chains

Human V_λ	TGT GCT TCC AGC ACT GGA GCA GTC ACC AGT GGT TAC TAT CCA AAC TGG --- TTC CAG CAG AAA CCT GGA CAA ---
Human V_κ	TGC AGG GCC AGT CAG AGT GTT --- AGC AGC --- TAC TTA GCC TGG --- TAC CAA CAA AAA CCT GGC CAG ---
Murine V_κ	TGC AAG GCC AGT CAG GAT GTG GGT GTT GTG AGT (Incomplete)
Murine V_κL6	TGC AAG GCG AGT CAG GAC ATT AAT AGC --- --- --- TAT TTA AGC TGG --- TTC CAG CAG AAA CCA GGG AAA ---
Rabbit V_κ	TGC CAG GCC AGT GAG AAC ATT TAC AGC --- --- TCT TTA GCC TGG --- TAT CAG CAG AAA CCA GGG CAG ---
Chicken V_λ1	TGC TCC GGG GAT AGG AGC TAC TAT GGC --- --- --- --- TGG --- TAC CAG CAG AAG GCA CCT GGC AGT

Row 3

Heavy Chains

Caiman V$_H$C3   TGG GTG GCT ACA ATT AAT ACT *GAT* GGA TCC AGC CAG TGG TAC TCC CCG GCC GTT --- CAG GGG AAA TTC ACC

Murine V$_H$186.2 TGG ATT GGA AGG ATT *GAT* CCT AAT AGT GGT GGT ACT AAG TAC AAT *GAG* AAG TTC --- AAG AGC AAG GCC ACA

Murine V$_H$GAT   TGG ATT GGA AGG ATT *GAT* CCT GCG AAT GGT AAT ACT AAA TAT *GAC* CCG AAG TTC --- CAG GGC AAG GCC ACT

Murine V$_H$108A TGG ATT GGA TAT ATT TAT CCT TAC AAT GGT GGT ACT GGC TAC AAC CAG AAG TTC --- AAG AGC AAG GCC ACA

Murine V$_H$107   TGG ATT GCT GCA AGT AGA AAC AAA GCT AAT *GAT* TAT ACA *GAG* TAC AGT GCA --- TCT GTG AAG GGT CGG

Murine V$_H$283   TGG GTC GCA ACC ATT AGT AGT GGT GGT AAC ACC TAC TAT CCA *GAC* AGT GTG --- AAG GGT CGA TTC ACC

Human V$_H$CESS   CTG *GAG* TGG CTT GCA CGC ATT *GAT* TGG *GAT* *GAT* *GAT* AAG TAC TAC GGT ACA TCT CTG *GAG* ACT AGG CTC ACC

Human V$_H$G3   TGG ATG GGA ATA ATC AAC CCT AGT GGT GGT AGC ACA AGC TAC GCA CAG AAG TTC --- CAG GGC AGA GTC ACC

Human V$_H$ε   TGG GTG GGA TGG ATC AAC CCT AAC AGT GGT GGC ACA AAC TAT GCT CCG AGA TTT --- CAG GGC AGG GTC ACC

Light Chains

Human V$_\lambda$ GCA CCC AGG GCA CTG ATT TAT AGT AGT ACA AGC AAC AAA --- CAC TCC TGG ACC CCT GCC CGG TTC TCA GGC TCC

Human V$_\kappa$ GCT CCC AGG CTC CTC ATC TAT GAT GCA TCC AAC AGG --- GCC ACT GGC ATC CCA GCC AGG TTC GGC AGT

Murine V$_\kappa$L6 TCT CCT AAG ACC CTG ATC TAT CGT GCA --- AAC AGA TTG GTA *GAT* GGG GTC CCA TCA AGG TTC AGT GGC AGT

Rabbit V$_\kappa$ CCT CCC AAG CTC CTG ATC TAT GGT GCA TCC ACT CTG GCA TCT CTG GGG GTC CCA TCG CGG AGA TTC AAA GGC AGT

Chicken V$_\lambda$1 GCC CCT GTC ACT CTG ATC TAT *GAC* AAC AAC ACC AAC AGA CCC TCG AAC --- ATC CCT TCA CGA TTC TCC GGT TCC

(continued)

Table 8.2 (Continued)

Row 4

Heavy Chains

```
Caiman V_H C3     ATC TCC AGA GGC AAC --- --- TCC CAG AAC ATG TAC CTG CAG ATG AGC AGC CTC ACA CCT GAG GAC ACA GCC
Murine V_H 186.2  CTG ACT GTA GAC AAA --- --- CCC TCC AGC ACA GCC TAC ATG CAG CTC AGC AGC CTG ACA TCT GAG GAC TCT GCG
Murine V_H GAT    ATA ACA GCA GCA ACA --- --- TCC TCC AAC ACA GCC TAC AGC CTC AGC AGC CTG ACA TCT GAG GAC ACT GCC
Murine V_H 108A   TTG ACT GTA GAC AAT --- --- TCC TCC AGC ACA GCC TAC ATG GAG CTC AGC AGC CTG ACA TCT GAG GAC TCT GCA
Murine V_H 107    TTC ATC GTC TCC AGA GAC ACT TCC CAA AGC ATC CTC TAC CTT CAG AAT GCC CTG AGA GCT GAG GAC ACT GCC
Murine V_H 283    ATC TCC AGA GAC AAT --- --- GCC AAG AAC CAG CTG TAC CTG CAA ATG AGC AGT CTG AGG TCT GAG GAC ACG GCC
Human V_H CESS    ATC TCC AAG GAC ACC --- --- TCT AAA AAC CAG GTG GTC CTT AAA GTG GAC ATG GAC CCT GCG ACA GCC
Human V_H G3      ATG ACC AGG ACG ACG --- --- TCC ACG AGC ACA GTC TAC ATG GAG CTG AGC AGC CTG AGA TCT GAG GAC ACG GCC
Human V_H ε       ATG ACC AGA GAC GCG --- --- TCC TTC AGT ACA GCC TAC ATG GAC CTG AGT CTG AGA TCT GAC GAC TCG GCC
```

Light Chains

```
Human V_λ         --- --- CTC CTT GGG GGC AAA GCT GCC CTG ACA CTG TCA GGT GTG CAG CCT GAG GAC GCT GAG TAT TAC TGC
Human V_κ         --- --- GGG TCT GGG ACA GAC TTC ACT CTC ACC ATC AGC AGC CTA GAG CCT GAA GAT TTT GCA GTT TAT TAC TGT
Murine V_κ L6     --- --- GGA TCT GGG CAA GAT TAT TCT CTC ACC ATC AGC AGC CTG GAG CTG GAA GAT GGA ATT TAT TAT TGT
Rabbit V_κ        --- --- AGA TCT GGG ACA GAG TAC ACT CTC ACC ATC AGC GGC GTG CAG CGT GAC GAT GCT GCC ACC TAC TAC TGT
Chicken V_λ 1     GGT TCC AAA TCC GGC TCC ACA GCC ACA TTA ACC ATC ACT ACT GTG CAG GCC GAC GAC GAG GCT GTC TAT TAC TGT
```

Row 5

Heavy Chains

Caiman V_HC3	ACG	TAT	TAC	TGC	GCC	AGA	
Murine V_H186.2	GTC	TAT	TAT	TGT	GCA	AGA	
Murine V_HGAT	GTC	TAT	TAC	TGT	GCT	AG–	
Murine V_H108A	GTC	TAT	TAC	TGT	GCA	AGA	
Murine V_H107	ATT	TAT	TAC	TGT	GCA	AGA	*GAT*
Murine V_H283	TTG	TAT	TAC	TGT	GCA	AGA	
Human V_HCESS	ACG	TAT	TAC	TGT	GCG	CGG	
Human V_HG3	GTG	TAT	TAC	TGT	GCG	AGA	
Human V_Hε	CTG	TTT	TAC	TGC	GCG	AAA	AGT

Light Chains

Human V_λ	CTG	CTC	---	TAC	TAT	GGT	GCT	CAG
Human V_κ	CAG	CAG	CGT	AGC	AAC	TGG	CCT	
Murine V_κL6	CTA	CAG	TAT	*GAT*	*GAG*	TTT	CCT	
Rabbit V_κ	CTA	GGC	AGT	*GAT*	AGT	AGC	*GAT*	
Chicken V_λ1	GGG	AGT	GCA	*GAC*	AGC	AGC	AGT	ACT GCT

[a]Codons for lysine and arginine are underscored, those for aspartic and glutamic acids are italicized, and those for cysteine are stippled.
[b]Litman et al. (1983). [c]Blankenstein et al. (1984). [d]Schiff et al. (1983). [e]Ohno et al. (1982). [f]Early et al. (1980). [g]Ollo et al. (1983). [h]Takahashi et al. (1984). [i]Rechavi et al. (1983). [j]Kenten et al. (1982). [k]Anderson et al. (1984). [l]Pech and Zachau (1984). [m]Kelley et al. (1982). [n]Jaenichen et al. (1984). [o]McCartney-Francis et al. (1984). [p]Reynaud et al. (1985). [q]Insert 11 codons.

in five of the V_H genelets and also in the only V_L shown, and the GAG or GAA of the remainder is reflected in the GAC and GAA of several V_κ examples. Again at site 4, the YTG of the heavy-chain set is close kin to the RTG of the light. Site 6 is another point of conservation, as is the fifth from the end of row 1, where lysine or arginine is encoded in nearly every representative. Many other similar columns of related codons occur in sufficient numbers throughout these genelets to show their common origins without necessitating the tedium of an item-by-item list. Still more would be revealed if a serious attempt at homologization were to be made. For instance, the TGY for cysteine of the penultimate column of row 1 in the heavy-chain examples doubtlessly is a homolog of the TGY in column 1 of row 2 among the light-chain genelets, for in both cases the succeeding column frequently carries codons for basic amino acids.

However, the complementarity-determining regions (CDR), formerly referred to as hypervariable sectors, do not correspond precisely between the two classes. The first one of the heavy chains spans only six codons and lies near the center of the CDR1s of the light chains, which embrace 14 triplets. In CDR2, the situation is reversed, with that of the heavy chain subtending 17 codons and that of the opposite type containing only seven. Furthermore, the members of the latter class have a third such region, located just prior to their 3' termini, whereas the corresponding sector in the former class is located in the D genelet.

Complementarity-Determining Regions. For the greater part, the CDRs fall within the loop structures of the V region (Figure 8.1). Then, when arranged in the eventual heterodimeric form, those of the heavy and light chains overlap and coil together to form structures that are viewed as the parts responsible for an immunoglobulin molecule's ability to recognize an antibody. Consequently, these regions display a greater diversity than do the remaining sectors of the V genelets, which are referred to as the "framework." The variation to be noted here is exemplified by the first site in CDR1 of the heavy chains, where seven different codons, for six amino acids (glycine, threonine, lysine, serine, asparagine, and isoleucine), are located in the nine sequences. In the second site, five distinct codons for four amino acids are found, in the third, six (for five amino acids), in the fourth, seven, specifying five monomers, and in the fifth just two (for methionine and isoleucine), but in the last position, four are present encoding four amino acids. Consequently, it is evident that within this CDR, variation is uneven, most of it occurring in columns 1 and 4, with the fifth being relatively invariant.

A similar situation occurs in CDR2 of these same chains, with columns 1, 9, and 12 being extremely variable in having eight or nine different codons each, while columns 2, 7, 11, and 17 are quite conservative in consisting of only three or four different triplets. Those of column 11 encode only the amino acids tyrosine and threonine, whereas seven distinct types are signaled by the components of columns 1, 8, and 9. Of their 150 available sites in this region of the nine sequences, well over 50% of the total are occupied by just five species of amino acid, glycine occupying 20, serine 17, asparagine 16, tyrosine 15, and threonine 13. Three, cysteine, methionine, and histidine, are not represented at all, leucine is encoded at only a single point, and aspartic acid, tryptophan, and valine at but three each. Consequently, it is obvious that variation is not completely random, but operates within distinct bounds, as pointed out elsewhere in a comparable framework, but at the level of amino acid sequences (Dillon, 1983, pp. 351–359). In the light-chain structures, variation is similarly restricted, but with the limited number of sequences that have been established, attempts at a meaningful discussion are impossible.

One final point needs to be brought out before the smaller genelets of the variable region are given attention—the lengths of the several sequences. As a whole, almost all are remarkably uniform, with the light-chain genelets usually one or two codons shorter than the heavy. Only one of the latter class is notably exceptional, that of the murine V_H107, which is three codons longer than the others. In addition, it is not unlikely that the murine V_κ sequence might prove longer than the four others in the light class, for just past the middle of row 1 there is an insert of 11 codons present in none of the others that probably increases its total length. However, the genelet has not been sequenced completely, so the contribution this added sector makes to the whole cannot be ascertained currently.

8.2.2. The Diversity and Joining Genelets

Because the D and J genelets are so closely related, both in their flanking signals and in assembly into pregenes, treating them as a unit offers many advantages. Unfortunately, the discussion suffers to some extent from the unevenness in distribution of available data among the two parts and the several classes. Of sequences for J genelets of κ chains there is nearly a plethora of material on hand, whereas for those for the same parts of λ and heavy chains data are scanty, a condition true also for the D genelets. In a few cases, as in the two human J_κ and murine $D_{SP2.1}$, one flanking signal in each is missing, as the sequences are derived from structures in which combination of J and D had already occurred. Nevertheless, those that have been established are sufficiently uniform that a firm basis for analysis is provided.

The Diversity Genelets. As indicated in the discussion of organization, light chains lack separate genelets for the diversity region, instead of which they possess a third CDR section. Each D sector proper is flanked on both sides by dual signals, only one portion of each of which is included in Table 8.3, since the remainder are more appropriate to a later discussion. It is readily noted that these sequences are well named, for the extensive variation in their lengths is immediately apparent. The shortest of those given in the table consists of only ten nucleotide residues, whereas the longest, the lone one from human cells, has 24. The nucleotide composition, too, is variable over broad limits, being much as in the CDRs of the heavy chain, to which they closely correspond functionally. In each of the first two columns, there are six different codons in the nine sequences, and five in the third. The fourth column, too, has six types of codons, although only eight sequences are represented there.

The flanking signals are of particular importance, since they are widely accepted as being involved in the assembly of the several genelets into the mature gene. That on the left flank is particularly invariant in structure, in every instance being YAC_T^AGTG. The corresponding part on the right is not nearly so stringently conserved evolutionarily. Its structure, which may be summarized in the form $CA_Y^CRR_T^AN$, displays only two sites of constancy, a point that is of basic importance in Section 8.2.3, because these signals have entered into the formulation of a theory pertaining to the assembly of parts (Early *et al.*, 1980; Alt and Baltimore, 1982). In seven of the nine structures shown in Table 8.3, the

Table 8.3
Sequence Structure of Diversity Genelets and Their Flanks

Murine $D_{NP}{}^{a}$	CACAGTG	TTG CAA CCA CAT CCT GAG AGT GT	CAGAAAACC
Murine $D_Q 52^{b,c}$	CACAGTG	CAA CTG GGA C	CACGGTG
Murine $D_{SP} 2.2^{b,d}$	TACTGTG	TCT ACT ATG ATT ACG AC	CACAGTG
Murine $D_{SP} 2.1^{b,d}$	TACTGTG	TCT ACT ATG GTA AC	-------
Murine $D_{FL} 16.1^{d}$	TACTGTG	TTT ATT ACT ACG GTA GTA GCT AC	CACAGTG
Murine $D_{SP} 2.5^{d}$	TACTGTG	TCT ACT ATG GTA AC	CACAGTG
Murine $D_{SP} 2.7^{d}$	TACTGTG	CCT ACT ATG GTA ACT 'AC	CACAGTG
Murine $D_{SP} 2.8^{d}$	TACTGTG	CCT AGT ATG GTA ACT AC	CACAGTG

[a]Blankenstein *et al.* (1984).
[b]Kurosawa *et al.* (1981).
[c]Sakano *et al.* (1981); Blackwell and Alt (1984).
[d]Kurosawa and Tonegawa (1982).

signal on the right flank forms a region of dyad symmetry with that on the left, as indicated by the arrows. The first and last examples there, however, fail to show any comparable complementarity.

The Joining Genelets in κ Chains. In each of the several mammals that have been fully investigated, five J genelets appear to be present in the κ-chain region of the genome. Those of the mouse, rat, and rabbit are given fully in Table 8.4, while only J4 and J5 of the human sources are provided, mainly for comparison with two aberrant forms of like origin. But two variations in length are to be perceived, rabbit J2, which is larger by one codon than the remainder, and rat J5r, which lacks one triplet. A single column alone is completely invariant, the penultimate, where AAA for lysine is the occupant. Several others, however, approach that condition, the central one of the ACC being especially prominent, for it is interrupted only by an ACG in the rat J3 cistron, an ACA in the murine J4 and two human J5 sequences, and an ACT in that of the rat J5r. The preceding column is completely filled by GGC, GGG, and GAG. Indeed, only the first two full columns show a high level of variation, so, as a whole, the joining segments contribute little to the variation needed in an antibody molecule.

Flanking Signals of $J_κ$. The signaling sequences that border both flanks are dual, as in the case of the diversity sectors, and also as there, only one component of each is given in the table, for the same reason. Here the left signal, typically ten residues in length, may be as short as nine or as long as 11 sites. Only a single completely constant point exists here, the T in the antepenultimate position. The seventh to ninth sites are almost always GTG, but GTA occurs twice and ATG once; however, variation elsewhere is so frequent that the summation of the signal $N_A^Y Y_{TCA}^{RGY} RTR$ is almost meaningless toward understanding how its recognition occurs. Even within a given organism, this same inconstancy exists, so that the confusion regarding the signal's structure is not merely a product of the multitudes of sequences given in the table. Nor does the member on the right flank aid in alleviating the dubious state of affairs, for it, too, is variable,

Table 8.4
Joining Sequences for Kappa Chains[a]

	Left flank		Right flank
Rat J_1^b	ACCACTGTGG	--- TGG ACG TTC GGT GGA GGC ACC AAG CTG GAA TTG AAA CGT	AAGTAGAATCC
Mouse J_1^c	ACCACTGTGG	--- TGG ACG TTC GGT GGA GGC ACC AAG CTG GAA ATC AAA CGT	AAGTAGAATCC
Rabbit J_1^d	GTCACTGTGT	--- TTG ACT TTT GGA GCT GGC ACC AAG GTA GAA ATC AAA CGT	GAGTAAAATCC
Rat J_{2b}^b	ACCAGTGTGTG	--- AAC ACG TTT GGA GCT GGG ACC AAG CTG GAA CTG AAA CGT	AAGTAGAATCC
Rat J_2^b	ACCAGTGTGTG	--- GAC ACG TTT GGA GCT GGG ACC AAG CTG GAA CTG AAA CGT	AAGTAGAATCC
Rat J_{2a}^b	ACCACTGTGTG	--- TAC ACG TTT GGA GCT GGG ACC AAG CTG GAA CTG AAA CGT	AAGTAGTCTTC
Mouse J_2^c	ACCACTGTGTG	--- TAC ACG TTC GGA GGG GGG ACC AAG CTG GAA ATA AAA CGT	AAGTAGTCTTC
Rabbit J_2^d	ACCACTGTG-	TAT AAT GCT TTC GGC GGA GGG ACC GAG GTG GTC AAA GGT	AAGTGGCCGTT
Rat J_{3r}^b	ATCACTGTAA	--- ATC ATG ATC AGT GAT GAG ACC AGA CTG GAA ATA AAA CCT	AAGTACATTTT
Rat J_3^b	ATCACTGTGA	--- TTC ACG TTC GGC TCA GGG ACG AAG TTG GAA ATA AAA CGT	AAGTAGATTTT
Mouse J_3^c	ATCACTGTAA	--- ATC ACA TTC AGT GAT GGG ACC AGA CTG GAA ATA AAA CCT	AAGTACATTTT
Rabbit J_3^d	GCCACCGTGA	--- TCC ACT CTT GGC CCA GGG ACC AAA CTG GAA ATC AAA CCT	AAGTCCCTTTC
Rat J_4^b	TTCACTGTGG	--- CTC ACG TTC GGT TCT GGG ACC AAG CTG GAG ATC AAA CGT	AAGTACATTTT
Mouse J_4^c	ATCACTGTGA	--- TTC ACG TTC GGC TCC GGG ACA AAG TTG GAA ATA AAA CGT	AAGTAGACTTT
Rabbit J_4^d	GACAGAGTGA	--- CTT ACT TTT GGC TCA GGG ACC ATG GTG GAG ATC AAA TGT	AAGTGCACTTT
Human J_4^e	CTCACTGTGG	--- CTC ACT TTC GGC GGA GGG ACC AAG GTG GAG ATC AAA CGT	AAGTGCACTTT
Human J_{4Dawdi}^f	----------	--- TTC ACT TTC GGC GGA GGG ACC AAG GTG GAC AAC AAA CGT	GAGTGCAACTT

(continued)

Table 8.4 (Continued)

Rat J_{5r} [b]	AATGCTGTGG	--- -TC ACC ATC --- CAA GAG ACT GGA CCG GAG AGT AAA CAT	AACTGGTTATT
Mouse J_5 [c]	CTCACTGTGG	--- CTC ACG TTC GGT GCT GGG ACC AAG CTG GAG CTG AAA CGT	AAGTACACTTT
Rabbit J_5 [d]	AATTCCATGA	--- ATC ACC TTT GGC GAG GAG ACC AAG CTG GAG ATC AAA CGT	AAGTACCTTTT
Human J_5 [e]	AACACTGTGG	--- ATC ACC TTC GGC CAA GGG ACA CGA CTG GAG ATT AAA CGT	AAGCATTTTTC
Human $J_{5Walker}$ [f]	----------	--- ATC ACC TTC GGC CAA GGG ACA CGA CTG GAG ATT AAA CGT	AACTAATTTTT

[a] The two components of the signals are underscored.
[b] Sheppard and Gutman (1982).
[c] Max et al. (1981).
[d] Emorine and Max (1983); Emorine et al. (1983).
[e] Hieter et al. (1982).
[f] Klobeck et al. (1984).

albeit not to the same extent as its counterpart. Indeed, the first four sites are virtually stable, the A of the first site being replaced by a G on only two occasions, and that of the second being completely invariable. Site 3 is occupied by a G in every sequence save the last, and site 4 always shows a T, with the exception of the second to last genelet, where there is a C. Thence to the end, variation is rampant. Without exception the right flank component fails to provide any region of dyad symmetry with the sequence on the left.

Joining Genelets of Other Chains. The joining genelets of λ and heavy chains have not attracted nearly so many investigations as have those of the κ just examined. However, one complete set of the four J_H sequences found in the murine genome is represented in Table 8.5, plus the pseudogene of J_H4 and the modified J3 from an immunoglobulin that reacts against phosphorylcholine (Early *et al.*, 1980). Also included there is a single example of J_λ from human cells (Anderson *et al.*, 1984).

The sequence of the latter is obviously distinct from the others. In the first place it is shorter by two codons at the 3' end, and in the second, only the first triplet at the 5' terminus is homologous with any of the rest at a corresponding position. No amount of effort at bringing out the kinships by insertions of hyphens leads to a higher level of homology than that now evident in the table. Hence, if this segment had common origins with the heavy joining segments, it must have been in the extremely remote past. A similar statement regarding relationship to the J_κ genelets is almost equally true. The dissimilarity is here weakened by the presence of TTG in site 3 in the J_λ that is clearly related to the usual TTT or TTG in the corresponding position of J_κ genelets (Table 8.4).

If any relationships between the J_H and J_κ sequences are expected, those expectations are quickly dispelled when comparisons are made. In the present case (Table 8.5), the coding sections consist of either 15 or 17 codons, compared to the 13 or 14 of the other class. Site-by-site comparisons fail to reveal any convincing level of correspondences, with the exception perhaps of the GGY in column 8 of the J_H representatives and the GGG or GGC prevalent in column 7 of the κ listing. But all the hallmarks of the latter class are noticeably absent in the heavy counterparts, including the AAA of the penultimate sites and the triple columns of glycine codons (GGN). Consequently, once again separate origins are suggested by the facts of structure.

The signaling sequences on their flanks show both kinship and lack of homology. Chiefly the former is displayed by the TGTG near the close of the left flanking series, just as in the majority of the similar run bordering on J_κ genelets. Elsewhere in this same combination, resemblances to the latter are inconsistent. At the other end of the joining segment, uniformity of structure is likewise limited, and correspondence to the κ component is completely lacking. Instead of the AA and GA of that type, here the GG combination provides the first two components of heavy-strand signals, as AG does in the single λ chain. Moreover, the third site is occupied by a T (a C in the λ example), rather than a G. No trace of regions of complementarity is found between the contrasting sets of flanking signals.

8.2.3. Steps in Generating the Variable Region Pregene

Now that the structures of the several genelets for the variable region have been analyzed, it is possible to examine the proposals that have been made regarding the mechanism of assembling them into the variable pregene. As stated in the paragraph

Table 8.5

Joining Segments of Heavy and Lambda Chains[a]

	Left Flank		Right Flank
Human J_λ[b]	GCCCTTGTTGATG	--- --- TGG AGA TTG TGT GTA TCA TAC ACA CCG AGC TCT CAA GAC --- ---	AGCCTACAT
Murine J_H1[c]	GAGACTGTGC	TAC TGG TAC TTC GAT GTC TGG GGC GCA GGG ACC ACG GTC ACC GTC TCC TCA	GGTAAGCTGGC
Murine J_H2[d,e]	GATAGTGTGAC	TAC TTT GAC TAC --- --- TGG GGC CAA GGC ACC ACT CTC ACA GTC TCC TCA	GGTGAGTCCTT
Murine J_H3[d]	CACAATGTGCC	--- --- TGG TTT GCT TAC TGG GGC CAA GGG ACT CTG GTC ACT GTC TCT GCA	GGTCAGTCCTA
Murine J_H315[f]	GATAGTGTGAC	--- --- TAC TTT GAC TAC TGG GGC CAA GGC ACC ACT CTC ACA GTC TCC TCA	GGTGAG (Incomplete)
Murine $J_H\psi$[d]	ACTCTTGTGAGAATTAG	--- --- GGG CTG ACA GTT CAT GGT GAC AAT TTC AGG GTC AGT GAC TGT CTG	GGTTTCTCTGAGG
Murine J_H4[d]	ACTATTGTGAT	TAC TAT GCT ATG GAC TAC TGG GGT CAA GGA ACC TCA GTC ACC GTC TCC TCA	GGTAAGAATGG

[a]Signals on each flank are underscored.
[b]Anderson et al. (1984).
[c]Newell et al. (1980); Sakano et al. (1981).
[d]Gough and Bernard (1981).
[e]Kurosawa and Tonegawa (1982).
[f]Early et al. (1980).

opening this chapter, these stages in building the preliminary coding section are generally referred to as rearrangements of the DNA. In this analysis, the procedure is that of devoting attention first to the uniting of D and J_H genelets, which step has already been pointed out as occurring first in the rearrangement of heavy-region DNA. Since no D genelet exists in light-chain components, the combining of the V_L and J_L genelets should be similar to the foregoing, so focus is upon that operation. Finally, the union of the V_H and DJ_H combination completes the processes of forming the pregenes for both major types.

   The Major Components Involved in Rearrangements. One major component of the rearrangement processes is provided by the double signals on the left and right flanks of the D and J genelets and on the right of the V. Since only the portions directly bordering the coding regions have been shown in preceding tables, Table 8.6 now lists both parts of possibly interacting sets—that is, the right sets of D and V_L genelets are shown together with those of the corresponding J_H or J_L coding portions. The two parts of each signal are separated by segments of more or less random nucleotides claimed as being of remarkable uniformity in length. However, one study shows that these spacers in V_λ and J_λ genelets are 23 and 12 residues in length, respectively, those of the V_κ and J_κ are 12 ± 1 and 23 ± 1, respectively, and those of the V_H and J_H are all 23 ± 1 (Kurosawa and Tonegawa, 1982). In contrast, a second report avers that all types are separated by either 11 or 22 nucleotides (Early *et al.*, 1980). Still a third investigation states all spacers are either 12 or 23 base pairs in length (Alt and Baltimore, 1982). Since there thus appears to be some variation in interpretation of the nature of their construction, the full signals and their spacers are listed in Table 8.6.

   A second major reactant in rearrangement activities is an enzymatic protein, referred to as a recombinase. One of the few suggestions that have been advanced concerning this enzyme is that one of its subunits recognizes signals that have short spacers, whereas another reacts to those with long spacers. (Kurosawa and Tonegawa, 1982). However, the possibility that several such enzymes might exist, each specialized to react with either λ, κ, or H genelet signals, has not been ruled out. One difficulty with this concept is that some of the J genelets are separated from either the D or V segment by 1000 or more nucleotides, and it scarcely seems possible that two subunits of a single molecule could subtend such a distance to react with the two signals simultaneously, as appears essential.

   Finally, the two genelets to be joined are brought together, it is believed, by formation of a loop in the DNA. Three types of loops are possible, one of which involves inversion of one of the genelets, whereas the other two do not (Figure 8.5); all probably are generated from type II, the open loop. After the genelets have been brought into proximity, the DNA is severed, the loop removed, and the two ends spliced. Consequently, the recombinase must have both DNase and splicing capabilities, together with the signal-recognizing capacity, or else several enzymes cooperate in the stated functions (Selsing *et al.*, 1984).

   The Nature of the Signals. Because the sequence structures of many J_H and D genelets have been published only after combination to one another and to V_H, the number of them affording diverse views of the nature of these signals is relatively limited. Contrastingly, numerous V_L, V_H, and J_κ flanking regions are contained in the literature, so that, together with those of restricted quantity, a substantial picture of the problem is presented. In all cases listed in Table 8.6, the sequences of both elements of each given

Table 8.6
Right- and Left-Hand Dual Signals of Interacting Genelets[a]

Right Hand (3')	Adjacent	Intervening Portion	Downstream
Murine $D_{SP}2.2^b$	CACAGTC	ATATATCCAGCA	ACAAAAACC
Murine $D_Q52^{b,c}$	CACGGTG	ACGCGTGGCTCA	ACAAAAACC
Human $D_{201}{}^d$	CACAGTG	CAAAAACCCACATCCTGAGAGTGT	CAGAAACCCT

Left Hand (5')	Upstream		Adjacent
Murine $D_Q52^{b,c}$	GGTTTTGAC	TAAGCGGAGCAC	CACAGTG
Murine $D_{SP}22^b$	GATTTTTGT	CAAGGGATCTAC	TACTGTG

Right Hand (3')	Adjacent		Downstream
Human $V_\lambda{}^e$	CACAGTG	ACAGACTCATAAGAGGAACCAAG	ACATAAACC
Human $V_\kappa101^f$	CACAGTG	TTACACACCCAA	ACATAAACC
Chicken $V_\lambda1^g$	CACGGTG	ACACAAAGCAATGGGGAAATGAT	ACAAAAACC
Human $V_\lambda26^f$	CACAGTG	AGGGAAGTCATTGTCAGCCCAG	ACACAAACC
Caiman $V_H{}^h$	CACACTG	ACTCAAACCCTATTCACGGCAAT	ACAAAATCC
Murine V_HBCL^i	---GGTG	AGTCCTTACAACCTCTCTCTTCTATT	CAGCTTAAA
Murine V_H283^j	CACAGTG	AGTGAATGTTACTGTGAGCTCAA	ACTAAAACC
Murine V_HNP^k	CACAGTG	TTGCAACCACATCCTGAGAGTGT	CAGAAAACC
Murine V_H105^l	CACAGTG	TTGTAACCACATCCTGAGTGTGT	CAGAAACCCT
Murine V_H108A^m	CAGAAAC	CCTGAGGTGCAGCAAGCTTCCTTGGGA	CTGACAAGAGT
Human V_H1^n	CACAGTG	TGAGAAACCACATCCTCAGATGT	CAGAAACCCT

Left Hand (5')	Upstream		Adjacent
Murine $J_\kappa1^o$	GGTTTTTGT	ACAGCCAGACAGTGGAGTACTAC	CACTGTGG
Murine $J_\kappa2^o$	AGTTTTTGT	ATGGGGGTTGAGTGAAGGGACAC	CAGTGTGTG
Murine $J_\kappa3^o$	GGGTTTTGT	GGAGGTAAAGTTAAAATAAAT	CACTGTAA
Murine $J_\kappa4^o$	GGTTTTTGT	AAAGGGGGGCGCAGTGATATGAAT	CACTGTGA
Murine $J_\kappa5^o$	GGTTTTTGT	AGAGAGGGGCATGTCATAGTCCT	CACTGTGG
Rat $J_\kappa1^p$	GGTTTTTGT	ACAGCCAGACAATGGAGAACTAC	CACTGTGG
Rat $J_\kappa2b^p$	AGTTTTTGT	ATGGGGGGTGAGTCAAGGGAAAC	CAGTGTGTG
Rat $J_\kappa2a^p$	AGTTTTTGT	ATGGGGGGTGAGTCAAGGGAAAC	CAGTGTGTG

(*continued*)

Table 8.6 (Continued)

Rat J_K4[p]	GGTTTTTGT	AGAGGTGCATATGTCAGAGTCTT	CACTGTGG
Chicken J_λ[q]	GGTTTTTGC	ATTGCTCCGTAT	CACTGTGT
Murine J_λ[q]	GGTTTTTGC	ATGAGTCTATAT	CACAGTGC
Murine J_H3[b,r]	ATTTATTGT	-CAGGGGTCTAATCATTGTTGTCA	CAATGTG
Murine J_H2[g]	GGTTTTTTG	TACACCCACTAAAGGGGTCTATGA	TAGTGTG
Murine J_H1[g]	AGTTTTAGT	-ATAGGAACAGAGGCAGAACAGA	GACTGTG

[a]The two active elements that comprise each signal are underscored.
[b]Kurosawa et al. (1981).
[c]Sakano et al. (1981).
[d]Takahashi et al. (1984).
[e]Anderson et al. (1984).
[f]Rabbitts et al. (1981c).
[g]Reynaud et al. (1985).
[h]Litman et al. (1983).
[i]Knapp et al. (1982).
[j]Ollo et al. (1983).
[k]Blankenstein et al. (1984).
[l]Yancopoulos and Alt (1985).
[m]Ohno et al. (1982).
[n]Rechavi et al. (1983).
[o]Max et al. (1981).
[p]Sheppard and Gutman (1982).
[q]Hollis et al. (1982).
[r]Gough and Bernard (1981).
[s]Newell et al. (1980).

dual signal are shown underscored, whereas those of the intervening segment are not underscored. If the latter are viewed first, the reason for the frequent differences in opinions regarding their precise length mentioned earlier becomes apparent—there is no absolute constancy in this feature, even within a given category of genelets. Although the presence of 12 nucleotides, both in the left- and right-hand combinations, is a uniform trait of the two D segments given there, it is doubtful that this seeming universality of occurrence will persist when representative forms are more abundant.

Even within the trio available for J_H-region signals, no two of the spacers are identical, for the nucleotide count is, in order, 23, 24, and 22. In the right-hand side of V segments the mean count is 23, but one has only 22, two have 24, one 26, and another 27. Indeed, one from man, V_K101, has 12 as in the D segments (Rabbitts et al., 1981b). Among the 11 representative J_L genelets, the range is much narrower, with 23 in most, but three show 24 and one only 21; indeed, the J_λ sequences from the mouse and chicken have but 12 (Bernard et al., 1978; Reynaud et al., 1985). Consequently, while the distance between the two elements of each of the signals may accelerate the enzymic reactions with them, it does not appear to be of limiting importance.

In connection with these spacers, it also should be noted that their nucleotidyl sequences are not completely random, as suggested by the several theories of recombination, but are so only to a restricted degree. While of no apparent functional value as far as

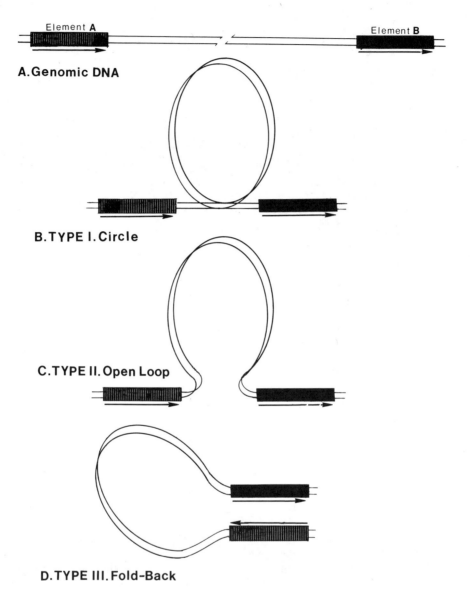

Figure 8.5. The three varieties of loops used in DNA recombinations. All are variations of type II, the open loop.

current data suggest, there is a strong tendency toward evolutionary conservation that should not be disregarded. For a striking example, comparison of the two V_H genelets (NP and 105) from the mouse genome shows that the spacers differ at only two sites. Two examples from the rat, $J_\kappa 2a$ and $J_\kappa 2b$, are likewise largely identical, and the murine $J_\kappa 5$ shares too many identical sites with the rat $J_\kappa 4$ sequence for the two not to be homologs. Furthermore, the frequent occurrence in the fifth and sixth sites of AA likewise

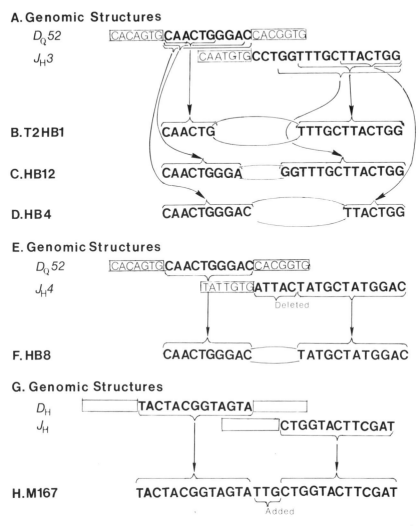

A. Genomic Structures

D_Q52 CACAGTG **CAACTGGGAC** CACGGTG

J_H3 CAATGTG **CCTGGTTTGCTTACTGG**

B. T2HB1   CAACTG   TTTGCTTACTGG

C. HB12   CAACTGGGA   GGTTTGCTTACTGG

D. HB4   CAACTGGGAC   TTACTGG

E. Genomic Structures

D_Q52 CACAGTG **CAACTGGGAC** CACGGTG

J_H4 TATTGTG **ATTACT** ATGCTATGGAC
                    Deleted

F. HB8   CAACTGGGAC   TATGCTATGGAC

G. Genomic Structures

D_H TACTACGGTAGTA

J_H CTGGTACTTCGAT

H. M167   TACTACGGTAGTA TTGCTGGTACTTCGAT
                        Added

Figure 8.6 Possible sequence changes incurred during splicing of recombined DNA in the mouse. The 3' end of D segments and 5' termini of J sectors are depicted, various portions of which may deleted during recombination, resulting in sequence diversity of immunoglobulins. In some cases bases are actually added, as in part H. (Parts A–F are based on Blackwell et al., 1984, and parts G and H on Early et al., 1980.)

intimates a degree of conservation, as do a number of other recurring combinations within given sites elsewhere in the sequences.

The Recombination Process. The actual process of recombining any two parts is, foremost, not a precise mechanical activity. In some cases, it has been demonstrated to occur in two or more ways, even when the identical genelets are involved (Kurosawa *et al.*, 1981; Blackwell and Alt, 1984). Moreover, in several instances new bases are sometimes added at the juncture of two segments (Alt and Baltimore, 1982), and in many others one to several nucleotides have been deleted (Sakano *et al.*, 1981; Lewis *et al.*, 1984). Several actual cases of deletions and one addition are illustrated in Figure 8.6.

Inversions also appear to be involved in forming some recombinants (Kurosawa *et al.*, 1981). Thus, recombination is not to be viewed as a single process, but rather as a series of related processes. Although the actual structural basis for bringing two parts into proximity may be by way of any of the three looping devices discussed earlier, here in the interest of simplicity only the open loop (type II) is considered. Additionally, only the recombining of D and J_H segments is treated, because the basic principles are the same between each interacting pair, insofar as current knowledge suggests.

In the example employed here, the J_H3 genelet of the murine genome is brought into close relationship with that of D_Q52, as in Figure 8.6A–D. This step would require two enzymes, whether two molecules of the same protein or two distinct ones is unknown, one to interact with the right-hand signals of the D genelet, the other with the left-hand ones of the J. But note that the right-hand element of the signals adjacent to the D genelet is highly homologous to that on the left, and additionally, the corresponding adjacent element of the V region is also closely related to both the foregoing. Even the downstream portions of the right-hand signal of both D and V sectors frequently are similar in nucleotidyl structure. Thus, the opening statement of lack of mechanical automaticity is doubly supported. The looping action (by an additional enzyme?) brings the two parts together to varying degrees, but typically in parallel, with the left-hand signals of J_H near those of D_Q52 at various orientations. The downstream element of the latter would then be disconnected from the coding sector and that part plus the two nucleotides lying between it and the first component of its upstream signal would, after cleavage, be joined to J_H3. Thus, the coding sequence of D_Q52 would join that of J_H3 in various combinations of fractional and entire sequences of the two components. Similar combinations have been documented between D_Q52 and J_H4 (Blackwell and Alt, 1984). In another actual case (Early *et al.*, 1980), three new nucleotides (TTG) are added to form a variant J_H (Figure 8.6G, H).

8.3. SOMATIC MUTATION IN IMMUNOGLOBULIN FORMATION

It is self-evident that the foregoing actual and visualized rearrangements often result in changes in DNA sequences of immunoglobulin genes. In these instances, it is obvious that, in part at least, the mechanisms behind those genetic alterations are enzymes involved in the cleavage and splicing of the DNA strands. But there are well-substantiated cases of changes in gene structure, known as ''somatic mutation,'' in which the specific agent has not been determined (Tonegawa, 1983). Consequently, the nature of the changes are not fully understood, but studies on the genes before and after their occurrences have demonstrated their actuality beyond all doubt (Bernard *et al.*, 1978; Rudikoff *et al.*, 1984; Darsley and Rees, 1985).

8.3.1. Somatic Mutation of Variable Genelets

Somatic mutation is not restricted to either light- or heavy-chain classes, but appears to be a widespread phenomenon; with the exception of the deletions or additions associated with the joining of two genelets as noted above, they are limited to the V sectors. Primarily they take place during the rearrangement of DNA involved in class switch after activation of the B cell by contact with an antigen. However, the mutations then generated are not directly part of the switch activities, for both the IgM and the light-chain coding sequences also become altered, although they themselves do not undergo a change in class (Selsing and Storb, 1981; Hartman and Rudikoff, 1984). Because only point mutations are generated, gene conversion or crossing-over processes cannot possibly enter into the activity.

Among the better documented cases of somatic diversification in heavy chains is the immune response of BALB/c mice to the hapten phosphorylcholine (Crews *et al.*, 1981; Kim *et al.*, 1981). In one study primary structures of 19 immunoglobulins were determined, ten of which, termed T15, proved to be identical; this group's unaltered nucleotide B sequence is given in Table 8.2 as murine $V_H 107$ (Early *et al.*, 1980). However, the other nine differed in the same region by one to eight residues, but nevertheless showed clear evidence of having been derived from T15. In this instance only IgA and IgG molecules displayed the point mutations, whereas none of IgM did. In a second example, mice treated with a different hapten, *p*-azophenylarsenate, reacted with much greater diversification, for 31 of the V_H chains showed mutations from a single germ-line sequence, that referred to as the 93G7 variable region (Sims *et al.*, 1982).

In a third illustration, also in mice, but in the strain C57B1/6, treatment was with the hapten (4-hydroxy-3-nitrophenyl)acetyl, abbreviated as NP (Bothwell *et al.*, 1981; Sablitzky and Rajewsky, 1984). On comparison of two hybridomas that secreted NP antibodies, one was found to produce an IgM during a primary response, but the other an IgG_{2a}, the result of a secondary DNA rearrangement. However, the sequences differed at only ten sites in the V_H region and had the same J segment (J2), but distinctive D sectors. Clones and sequencing of a partial library of the mouse DNA revealed the existence of seven possible candidates for the V_H regions of the NP antibodies, but none was related to the differences found between the two hybridoma products, suggesting that they result from somatic mutations. It is especially pertinent to note that data accumulated over the years have revealed that anti-NP antibodies from the C57B1/6 strain of mice all contain λ chains, although 95% of blood serum antibodies in mice in general have κ and produce a limited spectrum of isoelectric forms, although many varieties (idiotypes) exist. Thus, as indicated also in the preceding examples, the response to a given antigen by the immune system is relatively limited in the face of the immense variability inherent to its genomic structure.

8.3.2. Somatic Mutation in Light Chains

The foregoing illustrations investigated only the changes undergone by the V_H region; that concomitant researches into the corresponding sectors of the light chains might also prove profitable is shown by several studies. In the first case the V_κ coding regions of the immunoglobulins from two myelomas were examined (Selsing and Storb, 1981). The products of both tumors (MOPC167 and MOPC511) demonstrated that the same genomic V_κ genelet was rearranged in each instance, although neither was identical to the original sequence, each possessing several point mutations. As was also the situation with the preceding V_H results, these alterations were not confined to the CDR regions, but were found in the framework as well.

In a second investigation of a murine myeloma, sequencing of the V_κ genelets of two chromosomes in a differentiated B cell was employed. Comparisons of the two alleles made it clear that somatic mutation had occurred in the secondarily rearranged genelet but not in the other (Gorski *et al.*, 1983). The germ-line (inherited) DNA sequence was nearly identical to that referred to as $V_\kappa L6$ in Table 8.2, whereas that of the other differed at six single sites. Consequently, it is now well established that somatic mutation in light chains occurs under antigenic influence, often accompanying class-switch rearrangements of immunoglobulin heavy-chain genelets (Sablitzky *et al.*, 1985). They are not accidents of D–J or V–J joinings, for they are scattered throughout the length of the V genelet and do not take place in either the D, J, or C regions. Thus, it is obvious that there exists in the nongenomic portions of cells a mechanism that responds to the stimulation of antigens by producing changes in the DNA sequences of certain specific genes or genelets.

8.3.3. The Supramolecular Genetic Apparatus

Those activities described above, together with many others that control the structuring and distribution of widely diverse macromolecules in the cell, have been attributed elsewhere to what was called the "supramolecular genetic apparatus" (Dillon, 1981, 1983). The alterations in DNA sequence at or following antigenic stimulation also were proposed to be accomplished by that mechanism (Dillon, 1983, pp. 439–444). But in the preceding pages here, the account additionally shows the presence of such a controlling device when reexamined in this light.

The very act of connecting a D segment to a J clearly demands an underlying control. As was shown above, the termini of the two genelets are not merely moved together and spliced in a routine manner, as the exons of genes are by strictly mechanical processes. Too often the joining processes involve removal of chains or the addition of short sequences, such as the TTG inserted between D and J_H4 in myeloma M167 (Figure 8.6H; Early *et al.*, 1980). In fact, it has been stated in the literature that the point of splicing of D and J_H is highly variable and almost never occurs at the precise limits of both sequences [see Alt and Baltimore (1982, p. 4121) for references]. The same statement applies equally to V_κ and J_κ joinings, as well as V_H and D, one example of the latter being the GCC inserted between those two genelets of the myeloma M167 sequence just mentioned (Early *et al.*, 1980). Hence, in immune reactions changes in DNA structure take place, not only within the V regions of the immunoglobulins, but also at the points of connection of the several genelets, alterations that appear too significant for survival of the organism to be purely random-chance events.

But there is still other evidence of underlying controls in the various examples provided earlier in this section. For one, the limited number of V_H genelets that are actually employed in response to a given antigen indicates the presence of a selective force. Although there are probably 100–200 V_H genelets organized in seven or eight multigene clusters of 2–50 members each, only one or a very few more are employed in an immune response to a given antigen (Brodeur and Riblet, 1984; Hartman and Rudikoff, 1984). This is true not only in a particular clone on a one-time basis, but over years of experimentation. The murine response to the hapten NP cited in a previous paragraph is an especially good illustration of accumulated results, for the light chains were consistently of the λ variety, despite that type normally being produced at a rate just 5% that of κ. Since the antigen in this and the other instances does not enter the cell, its effects are mediated by way of an organized body of enzymes in the cytoplasm, possibly along with small RNAs, which selects particular V_H and V_L genelets for use, combines them with J and D genelets, and sets a class-switch action into motion. Fine tuning of the antibody gene is accomplished thereafter by changes in its DNA sequence through substitution of bases, that is, by removal of an existing nucleotide base and insertion of a different species. There is no means known whereby DNA can alter its nucleotidyl sequence within a given cell; enzymes alone have that capability. Thus, such point mutations are the product of an enzymatic protein or small group of such macromolecules. But the whole complex series of reactions involved in an immune response cannot be the result of individual molecules; collectively they require an entire body of highly ordered, thoroughly coordinated proteins, a system which the author treats as a second genetic mechanism, called "supramolecular" because it governs, and sometimes changes, the molecular one made of DNA.

8.4. THE CONSTANT GENELETS

In preceding sections it has been necessary at times to refer to some of the characteristics of the constant genelets, and one subdivision has been devoted to their genomic arrangement. However, their structural features have not been examined as yet; accordingly, the present discussion centers on their nucleotide sequences and flanking signals, and the associated "switch" sectors. As with the variable regions, separate analyses are necessary for the heavy and two types of light chains, with the latter being taken up first.

8.4.1. Constant Regions of κ Chains

In the laboratory mouse as well as in man, there appears to be only a single genelet for the constant region of the κ chain (Heidmann and Rugeon, 1982), which in the former mammal lies about 2500 nucleotide residues downstream of the five $J_κ$ genelets (Max *et al.*, 1981). The situation in rabbits is still confused, because of the existence of multiple alleles and by peculiarities of their expression (Emorine and Max, 1983; Emorine *et al.*, 1983, 1984). Three from this source are given in Table 8.7, along with the single one of a mouse. However, it should be noted that multiple copies of certain $C_κ$ genelets have been detected in the rabbit genome whose sequences have not yet been determined (Emorine *et al.*, 1983).

Homologies among the four examples cited in the table are remarkably few in number, considering the short phylogenetic distance between the rodents and the lagomorphs. Although there are regions of evolutionary conservation, sections of identical sites are short and intermittent. The TTC, CCA, CCA near the center of row 1 is the longest in the 5' portion, one of equal length not being encountered before the end of row 3, where the beginning of the series AGC, AGC, ACY is found, which continues into row 4. In the latter row is another sequence of three triplets, TAY, ACC, TGY, but the sole sector that exceeds this length is near the close of row 5, where the four adjacent codons AGC, TTC, AAY, AGG are to be noted. Codons for basic amino acids (underscored) are infrequent, there being but a single complete column of them, and that near the 3' terminus. In contrast, three full columns of triplets for tyrosine (stippled) are located in rows 3 and 4. Those encoding acidic amino acids (italics) are quite abundant, being represented by four complete columns and numerous singles.

Table 8.7
Constant Regions of Kappa Chains[a]

Row 1

```
Mouse C_κ^b      GGG GCT GAT GCT GCA CCA ACT GTA TCC ATC TTC CCA CCA TCC AGT GAG CAG TTA ACA TCT GGA GGT GCC
Rabbit C_κ1^c    GGT GAT CCA GTT GCA CCT ACT GTC CTC ATC TTC CCA CCA GCT GCT GAT CAG GTG GCA ACT GGA ACA GTC
Rabbit C_κ2^c    CGT GAT CCA GTT GCG CCT TCT CTC CTC TTC CCA CCA TCT AAG GAG GAG CTG ACA ACT GGA ACA GCC
Rabbit C_κb9^d   GAT CCT CCA ATT GCG CCT ACT GTC CTC GTC CTC TTC CCA CCA TCT GCT GAT CAG CTG ACA ACT GGA ACA GTC
```

Row 2

```
Mouse C_κ        TCA GTC GTG TGC TTC TTG AAC AAC TTC TAC CCC AAA GAC ATC AAT GTC AAG TGG AAG ATT GAT GGC AGT
Rabbit C_κ1      ACC ATC GTG TGT GTG GCG AAT AAA TAC TTT CCC --- GAT GTC ACC GTC ACC TGG GAG GTG GAT GGC ACC
Rabbit C_κ2      ACC ATC GTG TGC GTG GCG AAT TTC TAT CCC AGT GAC ATC ACC GTC ACC TGG AAG GTG GAT GGC ACC
Rabbit C_κb9     ACC ATC GTG TGC GTG GCA AAT TTC CGT CCC AAT GAC ATC ACC GTC ACC TGG AAG GTG GAT GAC GAA
```

Row 3

```
Mouse C_κ        GAA CGA CAA AAT GGC GTC CTG AAC AGT TGG ACT GAT CAG GAC AGC AAA GAC AGC ACC TAC AGC ATG AGC
Rabbit C_κ1      ACC CAA ACA ACT GGC ATC GAG AAC AGC CCG CAG AAT TCT GCA GAT TGT ACC TAC AAC CTC AGC
Rabbit C_κ2      ACC CAA CAG AGC GGC ATC GAG AAC ACA CCG CAG AGC CCC GAA GAC AAT ACC TAC AGC CTG AGC
Rabbit C_κb9     ATC CAA CAG AGC GGC ATC GAG AAC ACA CCG CAG AGC CCC GAA GAC TGT ACC TAC AAC CTC AGC
```

Row 4

```
Mouse C_κ        AGC ACC CTC ACG TTG ACC AAG GAC GAG TAT GAA CAT CGA AAC AGC TAT ACC TGT GAG GCC ACT CAC AAG
Rabbit C_κ1      AGC ACT CTG ACA CTG ACC AGC ACA CAG TAC AAC AGC CAC AAA GAG ACC TGC ACC GTG AAG GTG ACC CAG GGC
Rabbit C_κ2      AGC ACT CTG TCA CTG ACC AGC GCA CAG TAC AAC AGC CAC AGC GTG GAG ACC TGC ACC AGC GTG GTC CAA GGC
Rabbit C_κb9     AGC ACT CTG TCA CTG ACC AGC GCA CAG TAC AAC AGC CAC AGC GTG GAG ACC TGC ACC AGC GTG GTC CAC AAC
```

Row 5

Mouse C_K	ACA	TCA	ACT	TCA	CCC	ATT	GTC	AAG	AGC	TTC	AAC	AGG	AAT	*GAG*	TGT	TAG
Rabbit C_K1	ACG	---	ACC	TCA	---	GTC	GTC	CAG	AGC	TTC	AAT	AGG	GGT	*GAC*	TGT	TAG
Rabbit C_K2	TCA	---	GCC	TCA	CCG	ATC	GTC	CAG	AGC	TTC	AAC	AGG	GGT	*GAC*	TGC	TAG
Rabbit C_Kb9	TCG	---	GGC	TCA	GCG	ATC	GTC	CAG	AGC	TTC	AAT	AGG	GGT	*GAC*	TGT	TAG

[a] Codons for basic amino acids are underscored, those for acidic ones are italicized, and those for tyrosine are stippled.
[b] Max et al. (1981).
[c] Emorine and Max (1983).
[d] McCartney-Francis et al. (1984).

Table 8.8

Constant Regions and Flanking Signals of Lambda Chains[a]

Row 1

	Left flank																	
Mouse Cλ1[b]	CATCCTGCA	GGC CAG CCC AAG TCT TCG CCA TCA GTC ACC CTG TTT CCA CCT TCC TCT *GAA* *GAG* CTC
Mouse Cλ2[b]	ATTCGCACA	GGT CAG CCC AAG TCC ACT CCC ACT CTC ACC GTG TTT CCA CCT TCC TCT *GAG* *GAG* CTC
Mouse Cλ3[b]	ATTCACACA	GGT CAG CCC AAG TCC ACT CCC ACA CTC ACC ATG TTT CCA CCT TCC CCT *GAG* *GAG* CTC
Chicken Cλ[c]		GGC CAG CCC AAG GTG GCC CCC ACC ACC CTC TTC CCA CCG TCA AAG *GAG* *GAG* CTG
Mouse Cλ4[b]	CATCTTGCA	GGC CAA CCC AAG GCT ACA CCC TCA GTT AAT CTG TTC CCA CCT TCC TCT *GAA* *GAG* CTC

Row 2

Mouse Cλ1	*GAG* ACT --- AAC AAG GCC ACA CTG GTG TGT ACG ATC ACT *GAT* TTC **TAC** CCA GGT GTG ACA GTG *GAC* TGG
Mouse Cλ2	AAG *GAA* --- AAC AAA GCC ACA CTG GTG TGT CTG ATT TCC AAC TTT TCC CCG AGT GGT GTG ACA GTG GCC TGG
Mouse Cλ3	CAG *GAA* --- AAC AAA GCC ACA CTC GTG TGT CTG ATT TCC AAT TTT TCC CCA AGT GGT GTG ACA GTG GCC TGG
Chicken Cλ	AAC *GAA* GCC ACC AAG GCC ACC CTG GTG TGC CTG ATA AAC *GAC* TTC **TAC** CCC AGC CCA CTG ACT GTG *GAT* TGG
Mouse Cλ4	AAG ACT --- AAA AAG GCC ACA CTG GTG TGT ATG ATC ACT *GAG* TTC **TAC** CCA GCT GCT GTG AGA GTG GCC TGG

Row 3

Mouse Cλ1	AAG GTA *GAT* GGT ACC CCT GTC ACT CAG GGT ATG *GAG* ACA ACC CAG CCT TCC AAA CAG AGC AAC AAC AAG
Mouse Cλ2	AAG GCA AAT GGT ACA CCT ATC ACC CAG GGT GTG *GAC* ACT TCA AAT CCC ACC AAA *GAG* GGC AAC --- AAG
Mouse Cλ3	AAG GCA AAT GGT ACA CCT ATC ACC CAG GGT GTG *GAC* ACT TCA AAT CCC ACC AAA *GAG* *GAC* AAC --- AAG
Chicken Cλ	GTG ATC *GAT* GGC --- TCC ACC CGC TCT GGC --- *GAG* ACC ACA GCA CCA CAG CAG CGG CAG AGC AGC CAG
Mouse Cλ4	AAG GCA *GAT* GGT ACC CCT TTC ACT CAG GGT GTA *GAG* ACT ACC CAG CCT CCC AAA CAG AGG *GAC* AAC ---

Row 4

```
              Mouse C_λ 1  TAC ATG GCT AGC AGC TAC CTG ACC CTG ACA GCA AGA GCA TGG GAA AGG CAT AGC AGT TAC AGC TGC CAG
              Mouse C_λ 2  TTC ATG GCC AGC AGC TTC CTA CAT TTG ACA TCG GAC CAG TGG AGA TCT CAC AAC AGT TTT ACC TGT CAA
              Mouse C_λ 3  TAC ATG GCC AGC AGC TTC TTA CAT TTG ACA TCG GAC CAG TGG AGA TCT CAC AAC AGT TTT ACC TGC CAA
            Chicken C_λ    TAT ATG GCC AGC AGC TAT CTG TCA CTG TCT GCC AGC GAC TGG TCA AGC CAC GAG ACC TAC ACC TGC AGG
              Mouse C_λ 4  --- ATG GCT AGC AGT TAC CTG CTC TTC ACA GCA GAA GCG TGG GAA TCT CAT AGC AGT TAC AGC TGC CAT
```

Row 5
```
                                                                                                     Right flank
              Mouse C_λ 1  GTC ACT CAT GAA GGT CAC ACT GTG GAG AAG AGT TTG TCC CGT GCT GAC TGT TCC TAG    GTCATCTAA
              Mouse C_λ 2  GTT ACA CAT GAA GGG GAC ACT GTG GAG AAG AGT CTG TCT CCT GCA GAA TGT CTC TAA    GAACCCAGG
              Mouse C_λ 3  GTT ACA CAT GAA GGG GAC ACT GTG GAG AAG AGT CTG TCT CCT GCA GAA TGT CTC TAA    GAGCCCAGG
            Chicken C_λ    GTC ACA CAC GAC GGC ACC TCT ATC ACG AAG ACC CTG AAG AGG TCC GAG TGC TAA       TAGTCCCAC
              Mouse C_λ 4  GTC ACT CAT GAA GGG CAA CAT GTG GAG AAG AGT TTG TCC CGT GCT GAG TGT TCC TAG    GTCATCTAG
```

[a] Codons for tyrosine are stippled, those for the basic lysine and arginine are underscored, and those for aspartic and glutamic acids are italicized.

[b] Selsing et al. (1982).

[c] Reynaud et al. (1982).

8.4.2. Constant Genelets of λ Chains

Evolutionary kinship is more readily discernible in the C_λ genelets than in those of the previous section, along with rather distinctive chemical properties (Table 8.8). At the very 5' terminus, all five of those given, four from the mouse and one from the chicken, have the first four codons (GGY, CAG, CCC, AAG) constant, and near the close of the same row show a second region of smaller size, consisting of TTY, CCA, CC-, and still another, GAR, GAG, CT-, at the extreme end of the line. In row 2, just after its beginning, can be seen AAR, GCC, AC-, CTG, GTG, TGY, a series much longer than any found in the κ sequences. Regions of homology of any notable size are then wanting until row 4, where ATG, GCY, AGC, AGY occurs, followed in row 5 by GTY, AC-, CAY, GA-, GGN, the longest stretch of constancy in either type of C_L region.

Tyrosine does not play such an important role in the λ series as in the κ, for no conserved columns of its codons (stippled) can be perceived. In contrast, the coding signals for basic amino acids (underscored) are of greater significance; four fully conserved columns, one in each row except the fourth, are to be observed. However, fewer of them are scattered through the sequences than in the C_κ sectors. Codons for acidic elements play a somewhat reduced role, for there are only four complete columns for that type, not five as in the C_κ genelets.

8.4.3. Genelets of Constant Regions of Heavy Chains

As described in greater detail in an account of the primary structures of immunoglobulins (Dillon, 1983, pp. 347–351), the constant-region genelets of heavy chains are much longer than any of the other parts and are made up of multiple looplike domains, the number depending upon the class of these antibodies (Figure 8.1). IgM and IgA each have four loops, the rest only three. In some cases, as in IgM, a hinge region intervenes between the first and second of the domains. Each loop is formed by disulfide bonds between cysteine residues, there being a codon (heavy stippling) for one such amino acid near the beginning and end of each sequence for those sectors (Table 8.9).

Two complete genelet sequences each for C_μ and C_γ are provided in the table, two from the murine genome and one each from the human and chicken. In order to bring out any sequence similarity, the various loops are aligned under one another, the several domains being identified by numerals enclosed in parentheses. The first portion is missing from the sequence for the chicken IgM, so it necessarily begins with $C_\mu(2)$ (Dahan et al., 1983). Both it and the murine counterpart have an appended sector that is found on secreted molecules, but not when embedded in membranes; these S genelets are displayed at the bottom of the table, labeled as $C_\mu S$ (Kawakami et al., 1980). Accompanying them are two membrane genelet structures from IgM from the mouse, indicated as $C_\gamma M1$ and M2 (Wels et al., 1984).

The Looped Sectors. Comparisons between related sequences, whether the C_μ components of the mouse and chicken or the C_γ ones of the mouse and man (Ellison and Hood, 1982; Yazaki and Ohno, 1983), reveal a far greater degree of variation than might be expected of a region called constant. Part of the lack of uniformity, however, stems from the multiplicity of copies that exists in some cases, particularly in C_γ, where four of this genelet occur in the human genome, although two of the latter, $C_\gamma2$ and $C_\gamma4$, have

Table 8.9

Constant Regions of Immunoglobulin Mu Genes[a]

Row 1

Murine $C_\mu(1)^b$	GAG	AGT	CAG	---	TCC	TTC	CCA	AAT	GTC	TTC	CCC	CTC	GTC	TCC	TGC	GAG	CCC	CTG	TCT	GAT	AAG	AAT	CTG
Murine $C_\mu(2)^{b,c}$	GAG	ATG	AAC	---	CCC	AAT	GTA	AAT	GTG	TTC	GTC	---	CCA	CCA	CGG	GAT	GGC	TTC	---	TCT	GGC	CCT	GCA CCA
Chicken $C_\mu(2)^e$	CCG	AAT	GGC	---	---	ATC	CCC	CTT	TTC	GTC	ACC	ATG	CAC	CCG	TCC	CGC	GAG	GAC	TTC	GAA	GGC	CCC	TTC
Murine $C_\mu(3)^{b,c}$	AGT	CCC	TCC	ACA	GAC	ATC	CTA	ACC	TTC	---	---	ACC	ATC	CCC	CCC	TCC	TTT	GCC	GAC	ATC	TTC	CTC AGC	TCC
Chicken $C_\mu(3)^e$	TTC	GTG	CAG	GAC	GAC	ATC	GCC	CGC	ACG	GGC	TCC	TTC	---	---	GTG	GAC	ATC	TTC	ATC AGC	AAA	TCG
Murine $C_\mu(4)^{b,c}$	GTG	CAC	CAC	AAA	---	CAT	CCA	CCT	GCT	GTG	TAC	CTG	CTG	CCA	CCA	GCT	CGT	GAG	CGT	GAG	CAA CTG	AGG	GAG TCA
Chicken $C_\mu(4)^e$	AGC	AAC	GCC	---	CGC	CCC	CCA	TCC	CCA	TAC	GTC	TTC	CCC	ACG	CCC	CCC	ACG	GAA	CAA	CTG	AAC	GGC AAC	CAA CGG
Human $C_\gamma2(1)^d$	---	---	GCC	---	TCC	ACC	AAG	GGC	CCA	TCG	GTC	TTC	CCC	CTG	GCG	CCC	TGC	TCC	AGG	ACC	TCC	GAG AGC
Murine $C_\gamma3(1)^e$	---	---	GCT	---	ACA	ACA	GCC	CCA	TCT	GTC	TAT	CCC	TTG	CCT	GGC	AGT	GAC	ACA	TCT	GGA TCC
Human $C_\gamma2(2)^d$	---	GCA	CCA	---	CCT	GTG	GCA	GGA	CCG	TCA	GTC	TTC	CTC	TTC	CCC	CCA	AAA	CCC	AAG	GAC ACC	CTC ATG	ATC
Murine $C_\gamma3(2)^e$	CGT	GGT	AAC	---	ATC	TTG	GGT	GGA	CCA	TCC	GTC	TTC	ATC	TTC	CCC	CCA	AAG	CCC	AAG	GAT GCA	CTC ATC	ATC
Human $C_\gamma2(3)^d$	GGG	CAG	CCC	---	CGA	GAA	CCA	CAG	GTG	TAC	ACC	CTG	CCC	CCA	TCC	CGG	GAG	GAG	ATG	ACC	AAG	AAC CAG	GTC
Murine $C_\gamma3(3)^e$	GGA	AGA	GCC	---	CAG	ACA	CCT	CAA	GTA	TAC	ACC	ATA	CCA	CCT	CGT	GAA	CAA	ATG	TCC	AAG	AAG	AAG GTT
Murine $C_\gamma M1^e$	GAG	CTG	GAA	---	CTG	AAT	GAG	ACC	TGT	GCT	GAG	GCC	CAG	GAT	GGG	GAG	CTG	GAC	GGG	CTC	TGG	ACG ACC	ATC
Murine $C_\gamma M2^e$	GTG	AAG	TGG	---	ATC	TCC	TCA	GTG	GTG	CAG	GCC	ACG	CAG	GCC	ATC	CCT	GAC	---	---	TAC	AGG	AAC ATG

(continued)

Table 8.9 (Continued)

Chicken C$_\mu$S^a	AAA	GCA	AGT	---	GCT	GTC	AAT	GTC	TCC	TTG	GTG	TTG	GCC	GAC	TCG	GCC	GCC	GCC	TCC	TAT	TAA	
Murine C$_\mu$S^{b,c}	AAA	CCC	ACA	---	CTG	TAC	AAT	GTC	TCC	CTG	ATC	ATG	TCT	GAC	ACA	GGC	GGC	ACC	TGC	TAT	TGA	

Row 2

Murine Cμ(1)	GTG	GCC	ATG	GGC	TGC	CTG	GCC	CGG	GAC	TTC	CTG	CCC	AGC	ACC	ATT	TCC	TTC	ACC	TGG	AAC	TAC	CAG AAC
Murine C$_\mu$(2)	CGC	AAG	TCT	AAA	CTC	ATC	TGC	GAG	GCC	ACG	AAC	TTC	CCA	AAA	CCG	ATC	ACA	GTA	TCC	TGG	CTA	AAG
Chicken C$_\mu$(2)	CGC	AAC	GCC	TCC	ATC	CTC	TGC	CAG	ACC	CGC	CGC	CGC	CGT	---	CCC	ACC	ACG	TGG	TAC	AAA		
Murine C$_\mu$(3)	GCT	AAC	CTG	ACC	TGT	CTG	GTC	TCA	ACC	TAT	GAA	ACC	CTG	AAT	---	ATC	TCC	TGG	GCT	TCT		
Chicken C$_\mu$(3)	GCC	ACG	CTG	ACG	TGC	CGG	GTG	GTG	AGC	AAC	ATG	GTG	AAC	GGC	CTG	GAG	GTG	TCG	TGG	AAG	GAG	
Murine Cμ(4)	GCC	ACA	GTC	ACC	TGC	TTG	GTG	AAG	GGC	TTC	TCT	GCA	GAC	GAC	AGT	---	GTG	CAG	---	TGG	CTG	CAG
CHicken Cμ(4)	CTC	AGC	GTC	ACC	TGC	ATG	GCT	CAG	GGC	TTC	AAC	CCC	CAC	CTC	TTC	---	GTC	AGG	---	TGG	ATG	AGA
Human C$_\gamma$2(1)	---	ACA	GCC	GCC	CTG	GGC	CTG	AAG	GAC	TAC	TTC	CCC	GAA	CCG	GTG	ACG	GTG	TCG	TGG	AAC	---	
Murine C$_\gamma$3(1)	---	TCG	GTG	ACA	CTG	GGA	TGC	CTT	AAA	GGC	TTC	CCT	GAG	CCG	GTG	ACT	GTA	AAA	TGG	AAC	---	
Human C$_\gamma$2(2)	---	TCC	CGG	ACC	CCT	GAG	GTC	ACG	TGC	GTG	GTG	GTG	GAC	GTG	AGC	CAC	GAA	GAC	CCC	GAG	GTC	CAG TTC
Murine C$_\gamma$3(2)	---	TCC	CTA	ACC	AAG	CCC	AAG	GTT	ACG	TGT	GTG	GTG	GTG	GAT	GTG	AGC	GAG	GAT	GAC	CCA	GAT	GTC CAT GTC
Human C$_\gamma$2(3)	---	AGC	CTG	ACC	TGC	CTG	GTC	AAA	GGC	TTC	TAC	CCC	AGC	GAC	ATC	GCC	GTG	GAG	TGG	GAG	AGC	AAT ---
Murine C$_\gamma$3(3)	---	AGT	CTG	ACC	TGC	CTG	GTC	AAA	GGC	TTC	TTC	TCT	GAA	GCC	ATC	AGT	GTG	GTG	GAG	TGG	GAG	GAA AGG AAC ---
Murine C$_\gamma$M1	ACC	ATC	TTC	ATC	AGC	AGC	CTC	TTC	CTC	AGC	GTG	TGC	TAC	AGC	GCC	TCT	GTC	ACC	CTC	TTC	AGG*	
Murine C$_\gamma$M2	ATT	GGA	CAA	GGT	GCC	TAG																

Row 3

```
Murine  C_μ(1)    AAC ACT GAA GTC ATC CAG --- GGT ATC AGA ACC TTC CCA ACA CTG AGG ACA GGG GGC AAG TAC CTA GCC

Murine  C_μ(2)    GAT GCG AAG CTC GTC GAA TCT GGC TTC ACC ACA GAT CCG GTG ACC ATC GAG AAC AAA GGA TCC ACA CCC
Chicken C_μ(2)    AAT GCC AGC GTC CCC GTC --- --- --- GCC GCC GCC ACC ACC GCC ACC ACC GTC GGC

Murine  C_μ(3)    CAA AGT GGT GAA CCA CTG GAA ACC AAA ATT AAA ATC ATG GAA AGC CCC AAT GGC ACC TTC AGT GCT
Chicken C_μ(3)    AAG GGG GGC AAA CTG GAG ACG GCG GTG GGG AAG AGG GTC CTG CAA AGC AAC GGC CTC TAC ACG GTG GAC

Murine  C_μ(4)    AGA GGG CAA CTC TTG CCC CAA GAG AAG TAT GTG ACC AGT GCC CCG ATC CCA GAG CCT GGG GCC CCA GGC
Chicken C_μ(4)    AAC GCG GAA CCC CTC CCC CAA AGC CAA TCG GTG ACA TCG GCC CCC ATG GCG GAG GAG AAC CCC GAA AAT GAG

Human   C_γ2(1)   TCA GGC GCT CTG ACC AGC GGC GTG CAC ACC TTC CCA GCT GTC CTA CAG TCC TCA GGA CTC TAC TCC CTC
Murine  C_γ3(1)   TAT GGA GCC CTG ACC TCC AGC GGT GTG CGC ACA GTC TCT --- GGG TTC TTC TAT TCC CTC

Human   C_γ2(2)   AAC TGG TAC GTG --- GAC GGC GTG GAG CAT AAT AAT GCC AAG ACA AAG CCA CGG GAG GAG TTC AAC
Murine  C_γ3(2)   AGC TGG TTT GTG --- GAC AAC AAA GAA GTA CAC ACA GCC TGG ACA CAG CCC CGT GAA GCT CAG TAC AAG

Human   C_γ2(3)   --- GGG CAG CCG GAG AAC AAC TAC AAG ACC ACA CCT CCC ATG CTG GAC TCC GAC GGC TCC TTC TTC CTC
Murine  C_γ3(3)   --- GGA GAA CTG GAG CAG GAT TAC AAG AAC ACT CCA CCC GAT TCA GAC ATC CTG GAC GAT GGG ACC TAC TTC CTC
```

(continued)

Table 8.9 (Continued)

Row 4

Murine C$_\mu$(1)	--- ACC TCG --- --- CAG *GTC* TTG CTG TCT CCC AAG AGC ATC CTT GAA GGT TCA GAT GAA TAC CTT *GTA*
Murine C$_\mu$(2)	CAA ACC TAC AAG *GTC* ATA AGC ACA CTT ACC ATC TCT GAA ATC GAC TGG CTG AAC CTG AAT *GTG* TAC ACC
Chicken C$_\mu$(2)	CCC GAA *GTG* *GTG* GCC GAG AGC CGC ATC AGC *GTC* ACC AGC GAA GAA TGG GAC ACC GCC GCC ACC TCC AGC
Murine C$_\mu$(3)	AAG GGG *GTG* GCT AGT *GTT* *GTG* GAA GAC TGG AAT AAC AGG AAG GAA TTT *GTG* ACT *GTG* ACT CAC
Chicken C$_\mu$(3)	--- GGG *GTG* GCC ACG *GTG* TGC GCC AGC GAA TGG GAC GGG GAT GGC TAC *GTG* TCT TCT AAG *GTG* TGT AAC CAC
Murine C$_\mu$(4)	--- TTC TAC TTT ACC CAC AGC ATC CTG ACT *GTG* ACA GAG GAA TGG AAC TCC GGA GAG ACC TAT ACC
Chicken C$_\mu$(4)	--- TCC TAC *GTG* GCC TAC AGC *GTT* TTG GGG *GTG* GGG GCC GCC GAA GAG *GTC* TAC ACG
Human C$_\gamma$2(1)	AGC AGC *GTG* *GTG* ACC *GTG* CCC TCC AGC AAC TTC --- GGC ACC ACC TGC AAC *GTA* GAT CAC
Murine C$_\gamma$3(1)	AGC AGC TTG *GTG* ACT *GTA* CCC TCC AGC ACC TGG --- CCC AGC ACT *GTC* ATC TGC AAC *GTA* GCC CAC
Human C$_\gamma$2(2)	AGC ACG TTC CGT *GTG* *GTC* ACC ACC *GTT* *GTG* CAC CAG GAC TGG CTG AAC GGC AAG GAG TAC AAG
Murine C$_\gamma$3(2)	AGT ACC TTC CGA *GTC* *GTC* AGT GCC CTC CTC ATC CCC ATC CAG GAC GAC TGG ATG AGG GGC AAG GAG TTC AAA
Human C$_\gamma$2(3)	TAC AGC AAG CTC ACC *GTG* GAC AAG AGC AGG TGG --- CAG CAG GGG AAC *GTC* TTC TCA TGC TCC *GTG* ATG
Murine C$_\gamma$3(3)	TAC AGC AAG CTC ACT *GTG* GAT ACA GAC AGT *GTG* TGG --- TTG CAA GGA GAA ATT TTT ACC TGC TCC TCC *GTG* *GTG*

Row 5

Murine $C_\mu(1)$ TGC AAA ATC CAC TAC GGA GGC AAA AAC AGA GAT CTG CAT GTG CCC ATT CCA*

Murine $C_\mu(2)$ TGC CGT GTG GAT CAC AGG GGT CTC ACC TTC TTG AAG AAC GTG TCC TCC ACA TGT GCT GCC*

Chicken $C_\mu(2)$ TGC GTC GTG GAG GGG GAG GAG ATG AGG AAC AAG AGG AGG ATG GAG TGC GGA TTA GAA CCC

Murine $C_\mu(3)$ AGG GAT CTG CCT TCG CCA CAG AAA TTC ATC TCA AAA CCC AAT* GAG

Chicken $C_\mu(3)$ CCC GAT CTG CTC TTC CCC ATG GAG GAG AAG ATG AGG ACG AAA GCC

Murine $C_\mu(4)$ TGT GTT GTA GGC CAC GAG GCC CTG CCA CAC CTG GTG ACC GAG AGG ACC GTG GAC AAG TCC ACT GGT

Chicken $C_\mu(4)$ TGC CTG GTG GGC CAC GAA GCT CTG CTG GCC CTC CAG CTG GCC CAG AAG TCG GTG GAT AGG GCT TCG GGT

Human $C_\gamma 2(1)$ AAG CCC AGC AGC AAG AGC ACC AAG GTG GAC AAG ACA -- GTT* GAG CGC AAA TGT TGT GTC GAG TGC CCA CCG TGC CCA*

Murire $C_\gamma 3(1)$ CCA GCC AGC AGC AAG AGC ACT GAG TTG ATC AAG AGA ATC GGT* CCT AGA ATA CCC AAG TGC CCC AGT ACC CCC CCA GGT[f]

Human $C_\gamma 2(2)$ TGC AAG GTC TCC AAC AAA GGC CTC CCA GCC CCC ATC GAG GAG AAA ACC ATC TCC AAA ACC AAA GGT*

Murine $C_\gamma 3(2)$ TGC AAG GTC AAC AAC AAA GCC CTC CCA GCC CCC ATC GAG AGA ACC ATC TCA AAA CCC AAA GGT

Human $C_\gamma 2(3)$ CAT GAG GCT CTG CAC AAC TAC ACG CAG AAG AGC CTC TCC CTG TCT CCG GGT AAA TGA

Murine $C_\gamma 3(3)$ CaT GAG GCT CTC CAT AAC CAC CAC ACA CAG AAG AGC CTC TCT CGC TCC CCT GGT AAA TGA

---

[a] Codons for acidic amino acids are underscored and those for valine are italicized; those for phenylalanine (TTY) and proline (CCN) are lightly stippled, those for tyrosine (TAY) and tryptophan (TGG) are moderately stippled, and those for cysteine (TGY) are heavily stippled. $C_\gamma M1$ and $C_\gamma M2$ are for membrane-retained chains and $C_\mu S$ indicates sequences of the region present in secreted IgM. Asterisk indicates the location of an intron.
[b] Kawakami et al. (1980).　[c] Dahan et al. (1983).　[d] Ellison and Hood (1982).　[e] Wels et al. (1984).　[f] Insert four codons.

been demonstrated to be largely identical (Ellison and Hood, 1982). For the greater part, however, the genelet for each class is unique (Figure 8.4). The 5' ends of the several sectors, as has been noted in many gene sequences analyzed in preceding chapters, are particularly variable, but usually after the sixth codon, corresponding sites in the given loop from different sources frequently have identical occupants. Indeed, careful scrutiny of the several loops from a given genelet is highly suggestive of their having been formed evolutionarily by successive duplications. Comparison of the murine $C_\mu(1)$ and $C_\mu(2)$, for illustrative purposes, indicates the presence of four identical and three related sites in row 1, although thereafter only related triplets are found.

Within a given class of the immunoglobulins, homologous sites are found in all the multiple sectors. Surprisingly, the codons for the cysteines that close the ends of the looped regions are not consistently placed. For example, in row 2, those of $C_\mu(2)$ are located two codons downstream of the others, and in C_γ, they are not similarly placed in any two of the three different components. Interclass homologies also are apparent, but nowhere are they abundantly represented.

Chemical Features. In numerous tables of preceding chapters, codons for lysine and arginine are made prominent by use of such devices as stippling or underscoring, because their level of frequency and highly conserved locations within sequences were suggestive of those amino acids being of particular significance in the translational product. In contrast, in the present molecules, such codons are relatively infrequent and demonstrate no inclination toward being evolutionarily conserved. Those for acidic residues (GAN) are far more abundant, as are those for valine (GTN), each of which is made conspicuous in Table 8.9 by being printed underscored and italicized, respectively. Triplets specifying phenylalanine (TTY) and proline (CCN) are also frequent, and those for tyrosine (TAY) and tryptophan (TGG) are present in more than usual numbers. Accordingly, the first two are shown lightly and the latter pair moderately stippled, while those for cysteine, as already stated, are heavily stippled.

Triplets specifying proline are especially peculiar in occurring in sets of two or three, sometimes alternating with codons of a different character, but occasionally in continuous fashion, as in row 1 of the chicken $C_\mu(4)$. Sometimes they are intermixed with the TTC for phenylalanine or the TCC of serine, so that sectors rich in C nucleotides are a striking feature of the loops. Although TTT is to be noted occasionally, phenylalanine is encoded with much greater frequency by its other codon, TTC. Typically scattered, the TGG of tryptophan forms one of the few columns that is highly conserved; in the antepenultimate sites of row 2, that triplet forms a column broken only in $C_\mu(1)$, even extending through the first loops of C_γ. This constancy suggests that tryptophan must render some particular function there. The signals specifying the acidic amino acids form only incomplete columns at best; more typically, they are seen to be interspersed singly or in twos and threes throughout the sequences.

Other Regions of Constant Genelets. The other regions of the C_H genelets are too short and poorly documented to merit detailed attention. This is particularly true of the hinge regions found at the termini of the two $C_\gamma(1)$ sequences in row 5, following the intron marked by an asterisk (Table 8.9). The complete absence of homology makes evident one of the distinctive traits of these specialized sectors, their extreme variability, while their 11- or 12-codon lengths suggest a second, their brevity. All in all, they appear to be fillers that provide the necessary distance between the loops to avoid their overlapping one another.

The secretory genelets attached to the ends of the IgM components are noticeably more uniform in structure, having a number of homologous and cognate sites. One chemical property that has escaped previous notice is their general resemblance to the transit presequences examined in Chapter 7. Of course, the usual initial ATG is missing here, since these genetic elements are cotranscribed as part of a multicistronic unit, but a codon for a charged amino acid lies at the 5' end and ~50% of the sequence consists of units for hydrophobic monomers, as in typical sectors for transit peptides. Consequently, the secreted form of IgM bears four transit peptides at the 5' end of each tetramer and two equivalents at the 3' end. Since these structures end in the stop codons TAA or TGA, they obviously provide the 3' terminus to the translational products.

The two genelets for the transmembrane constituent of IgG do not live up to expectations. Both considerably exceed in length the genelet for the secretory portion, M2 consisting of 28 codons and M1 consisting of 44, but the level of hydrophobicity is relatively low. M2 encodes apolar amino acids at a level of 50%, but M1 does so to a notably lesser degree; in both instances these are concentrated to some extent within the 3' portions. Nevertheless, both do contain triplets for charged monomers at or adjacent to the 5' end. A standard translational stop signal, TAG, is at the terminus of M2, but none can be detected associated with the M1 genelet. Investigations into the mechanisms that terminate translation in the latter instance could be most productive.

8.4.4. Switch Regions and Class Switches

The last portions of the complex coding region that embraces the many parts of the ultimate immunoglobulin chain are the enhancer and several switch (S) regions involved in change from the early IgM to any other class. IgD is exceptional in lacking the latter portion; for, as has already been seen, it becomes expressed through differential processing (Figure 8.3). The S signals are not encoded by genelets, but consist of specialized regions in the intergenic spacers; they are present upstream of all classes of C_H coding sectors, except C_δ, as mentioned, and, while transcribed, they are not translated, being deleted from primary transcripts by RNA processing.

Switch Region Activity. The last word on the mechanism of action of these signals has not yet been written; much more needs to be learned about their mode of functioning. This is especially true of the enzymes that are surely involved, since none appear to have been identified. In the literature much has been made of their structural features, although there is general agreement upon few points of possible importance. Little doubt exists that much of every S region consists of multiple repeats of AGCT, often variously modified as GAGCT or GAGCTG, the whole being interrupted frequently by other short sequences, including GGGGT (Kataoka *et al.*, 1983). Important roles are played by additional combinations of nucleotides, some laboratories stressing AGGTTG (Lang *et al.*, 1982) or its modification YAGGTTG on the one hand (Stanton and Marcu, 1982; Marcu *et al.*, 1982), and others ACCAG and CGAGC (Wu *et al.*, 1984). However, the last reference has gone beyond these short structural elements in reporting the existence of four long, identical repeats, comprised of 102, 72, 98, and 109 nucleotides, respectively, each being separated from its next copy by 782 residues. Since each S region is quite lengthy, spanning ~200–500 base pairs, and is located at a considerable distance (~1500 base pairs) 5' to each C_H genelet, an extensive sector of DNA is involved in any class change.

Thus, it is evident that the recombination of a VDJ pregene from the C_μ genelet with which it was originally associated to a new C_H component involves the formation of a large loop of DNA to bring the two parts into proximity, regardless of whether it is associated with γ3,γ1, γ2b, γ2a, ε, or α (Figure 8.4). Of course, the farther downstream the particular recombinant is located, the greater the loop must be. This distance, moreover, is enhanced by the ~6500-nucleotide segment that intervenes between the pregene and the C_μ in the first union (Wu *et al.*, 1984). Formation of the loop and the combining of two remote parts consequently cannot be a random process, whether the recombination events occur within a given chromosome or between sister chromatids. Random-chance mechanisms would be just as likely to result in combinations of many diverse genes up and down the length of the genome, events that certainly do not occur. Here a particular set of coding sectors is involved, whose members must be brought into contact from remote locations and spliced in a fairly exact manner. The length of the segment that intervenes between the two parts to be united militates against the activity's being brought about by a single enzyme, although a recombinase is involved in the cleavage and subsequent splicing of the DNA. Rather, the process would seem to require action of a molecular system to span the great distance and recognize the particular components to be united. Perhaps this genetic activity needs also to be ascribed to the second, or supramolecular, genetic apparatus. This joining of once-separated parts of the DNA is not a unique process; different but related onces receive attention in the next chapter.

Enhancers. As pointed out in Chapter 1, Section 1.1.1, enhancers are sequences in the DNA that increase the level of transcription from certain promoters. The earliest representatives of these were on the leader close by that signal for the start of transcription, but more recent examples exercise their effects even when several thousand nucleotide residues away from that point (Banerji *et al.*, 1981). Now one has been found active in the synthesis of immunoglobulin heavy-chain mRNAs (Gillies *et al.*, 1983), located downstream of the J_H genelet before the S sector (Ephrussi *et al.*, 1985). The enhancing activity has been pinpointed to a fragment located in that area that is 307 base pairs in length. As a rule it seems to be effective only in B cells, especially in myelomas; thus it may serve as a signaling device for at least one tissue-specific enzyme (Mercola *et al.*, 1985). Furthermore, an enhancing element that increases the level of transcription of κ chains has been identified, located upstream of the promoter on the leader (Bergman *et al.*, 1985), and another has been documented in a similar location associated with human J_H, $C_\gamma 1$ genelets (Kudo *et al.*, 1985).

There are problems regarding these enhancers, however, for experimental results are still in a state of flux. While they may be essential for a high level of transcription initially, whether they are requisite also for maintaining Ig production has not been clearly established. The problem is illustrated by an investigation on immunoglobulin synthesis in a mouse hybridoma cell line. When the pertinent region containing the enhancer was deleted by mutation, the production of this substance was demonstrated nevertheless to be at a level comparable to unmutated tissues (Klein *et al.*, 1984).

8.5. OTHER ASSEMBLED GENES

Established instances of assembled genes are few and probably will always remain rare relative to other types, for more than likely they will prove to be largely confined to complex activities such as immune-related reactions and then primarily in advanced organisms. In bacteria, the several proteins that are known to provide immunity, like that against colicin A (Lloubes *et al.*, 1984), are encoded by simple genes, which lack presequences. However, this may not represent the complete picture, for chromosomal rearrangement has been demonstrated to occur in *Neisseria gonorrhoeae* during phasic variation in expression of the pilin gene (Segal *et al.*, 1985). In this bacterium the pilus is a major virulence factor that enables the organism to attach to host epithelial cells. During subsequent proliferation, it switches from pilus-expression (P^+) to pilus-negative (P^-) and the reverse in alternating fashion. In most P^+ to P^- changes a deletion occurs in one or both loci known to encode the protein pilin; this deletion apparently results in many instances from a recombinational event in the DNA, where directly repeated sequences are located. Additional but undetermined regulation also is associated with this phenomenon. Consequently, assembly of parts may be needed for pilin expression in this and other prokaryotes, possibly also including the flagella of *Salmonella typhimurium*.

The genes for the variable surface glycoproteins in the several species of *Trypanosoma* are also promising candidates for the assembled type, but the very unusual processes are now only beginning to be revealed. The mRNAs for these and many other proteins of *T. brucei* begin with the same sequence of 35 nucleotides, which is encoded by what are referred to as "miniexons" (Young *et al.*, 1983). These structures, actually genelets, are present in ~200 copies per genome and are densely clustered on large chromosomes. Now a gene for a variable surface glycoprotein has been localized on a small chromosome, away from all of those genelets, but still its mRNA bears the 35-nucleotide sequence (Guyaux *et al.*, 1985). Consequently, it seems evident that recombinational events may be taking place, so that assembly of genelets located on two different chromosomes is involved. However, researches on this organism are hampered by the failure of the chromatin to condense into distinct chromosomes. Whether similar events are involved in the immunogenic surface proteins of another protozoan, *Plasmodium*, remains to be determined (Godson *et al.*, 1983).

8.5.1. T-Cell Receptor Genes

The antigen receptor genes of T cells undoubtedly are the most closely related to those for the immunoglobulins, both in structure and assembly (Hood *et al.*, 1985). However, the products of the genes are somewhat simpler in being heterodimers of the $\alpha\beta$ type, not dual dimers as in the class just discussed in detail. In these leukocytes, there is an additional gene (γ) of similar constitution; since it does not enter into formation of the receptor protein, it receives attention in a separate section.

Functions of the Receptor. Among the T lymphocytes, at least three classes have been described, based on their diverse functions (Dillon, 1983, pp. 374–377, 387–397). Cytotoxic or killer cells attack and destroy any cell that carries a foreign antigen, helper lymphocytes augment immune responses, and, finally, suppressor cells retard such activities. In each case there is a correlation between the functional class and the type of molecule encoded in the major histocompatibility complex that is recognized together with the antigen, but in cytotoxic and helper cells the particular activity is not related to the structure of the receptor proteins and therefore not to those genes (Bogen *et al.*, 1985). It appears that in the suppressor class, the coding region for one of the peptide subunits (the β) may become deleted from the genome of both the homologous chromosomes that bear them, while in contrast in the cytotoxic class, another assembled gene, the one known as γ, mentioned in the foregoing paragraph, is preferentially transcribed. Overall, however, the basis for the three types of cell function still remains undeciphered.

Genomic Arrangement and Composition. Both subunits of the T-cell receptors are encoded by a series of genelets, those of the α more closely resembling the light chains of immunoglobulins and those of the β being more similar to the heavy. In neither case, however, is the comparison precise. In the mouse, the genelets for the α monomer are located on chromosome 14 and those for the β on chromosome 6 (in humans on chromosome 7) (Barker *et al.,* 1984; Caccia *et al.,* 1984; Kranz *et al.,* 1985). Since in the B cell the heavy-chain genes are on chromosome 12 and those of the major histocompatibility complex on 17, each is perceived to represent entirely distinct families.

The genomic arrangement of the α-subunit genelets has not been fully established, but on the basis of complementary DNA sequence investigations it is believed that there are V_α and J_α genelets, but no D units, that combine to form the V_α pregene (Saito *et al.,* 1984). As in immunoglobulins, that coding sequence becomes joined to a C_α genelet to produce the definitive gene for the subunit (Hood *et al.,* 1985). Thus, the illustration of the genome arrangement in Figure 8.7A is to be considered suggestive only. It is clear, nevertheless, that a number of copies of the V_α genelet are arranged in tandem fashion, with probably only a single C_α encoding segment being present. However, it is disturbing to learn that actual α-chain peptides show little resemblance to the predicted products of the two putative genes (Hannum *et al.,* 1984).

The arrangement of the β-subunit genelets has been far more firmly established. As in the immunoglobulins, there is a series of V_β genelets, whose precise number remains open to further investigation. In contrast to the heavy chains of antibodies, the D_β and J_β genelets are not arranged in clusters separately. Instead, a single copy of $D_\beta 1$ is followed downstream, first by six copies of J_β, and then by $C_\beta 1$ (Figure 8.7B; Malissen *et al.,* 1984; Siu *et al.,* 1984a,b). This is followed in turn by a $D_\beta 2$, $J_\beta 7$–12, and $C_\beta 2$; however, $J_\beta 7$ is a pseudogene (Gascoigne *et al.,* 1984). Perhaps there are additional copies of the last of these and maybe also multiple types of D_β, but these points still require additional experimental research. The C_β genelets encode a product not unlike that of corresponding Ig-producing coding regions (Malissen *et al.,* 1984).

Although thus structurally quite similar to the immunoglobulins secreted by B cells, the T-cell receptor proteins are distinct from them in recognizing antigen MHC product-bearing cells, not soluble antigens. The nucleotidyl basis for this functional difference has yet to be disclosed, but some progress toward that end has been made (Patten *et al.,* 1984). Only ten or fewer V_β genelets appear to be used frequently in the thymus, yet their sequences are more heterogeneous than are those of immunoglobulins. It was suggested that this variability possibly stemmed from the presence of additional CDRs, perhaps as many as seven in total, instead of the familiar three.

Assembly of the Genelets. No direct evidence has yet been forthcoming as to the manner in which the definitive genes are assembled, but it is likely that they will be quite similar to those of the immunoglobulins, since D_β–J_β combinations have been reported (Hood *et al.,* 1985). Furthermore, information pertaining to the production of the necessary diversity is confined strictly to the β subunit. Variability seems to be obtained in only three ways, since the somatic mutation that characterizes the immunoglobulins appears to

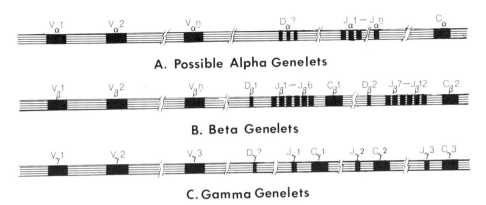

A. Possible Alpha Genelets

B. Beta Genelets

C. Gamma Genelets

Figure 8.7. Assembled genes in T cells. The genomic arrangement of the genelets for the two subunits of T-cell receptors and γ trait are shown. Knowledge is still incomplete, the presence of (A) Dα and (C) Dγ genelets being unestablished. (Based on Hood *et al.*, 1985.)

be either lacking or of rare occurrence. The first method pertains to the addition or deletion of nucleotides when two genelets become joined, such as D–J or V–D unions, with the frequent production of reading-frame mutations downstream of the juncture. The second mechanism is totally a T-cell device. Whereas in B cells the uniting of the C genelet to the V pregene is always in a single reading frame, in T cells the combination can take place in all three possible frames (Hood *et al.*, 1985), thereby instigating the production of great diversity in the translational product. The third device is the possible presence of seven, rather than three, CDRs mentioned above.

8.5.2. The T-Cell γ Gene

In the preceding section the particular association in the cytotoxic T lymphocytes and the product of the γ gene has been alluded to. Beyond its abundance there, almost nothing is known of the function of this recently discovered gene (Saito *et al.*, 1984). However, like the T-cell receptor and immunoglobulin coding regions of DNA, it must undergo assembly before transcription.

Genomic Organization. As stated earlier, the present genes are located on chromosome 13 of the mouse and are structured in a manner homologous to those of immunoglobulins. At least three V_γ genelets are present, the first two of which are located on different DNA strands, with the 5′ termini toward one another, but separated by 2500 nucleotide residues. The existence of D_γ genelets has not been established, but there are three of J_γ, one each in association with the same number of C_γ coding segments, an arrangement recalling that of the β gene (Figure 8.7C).

Relations to Immunoglobulins. Table 8.10 gives various sequences of immunoglobulins together with the corresponding parts of the γ gene that have been completely sequenced. Examination of the presequences reveals little suggestive of homology between the murine T-cell γ segment and the human Ig_γ. Those similarities found include the codons at the sixth site, which begin with TT, and those six sites later, which begin with TG; finally, three positions still farther downstream are corresponding triplets for

Table 8.10
Comparisons of Immunoglobulin and T-Cell Factor Genes[a]

Presequences

Rat OX-2[b]:
ATG GGC AGT CCG GTA TTC AGG AGA CCT TTC TGC CAT CTG TCC A CC TAC AGC[j] AGT ACA GCT CAA GTG

Human Ig_λ[c]:
ATG GCC TGG ACT CCT CTT CTG CTC CTC ACT TGC TGC C CA GGT* GGG TCC AAT TCT CAG ACT

Murine γ[d]:
ATG CTG CTC CTG AGA TGG TTC ACC TCC TGC TGC CTC TGG GTT T TT GGG CTT GGG CAG CTG GAG CAA

Murine TM86[e]:
ATG AGC TGC AGG CTT CTC CTA TAT GTT TCC CTA TGT CTT GTG G AA ACA GCA CTC ATG AAC ACT AAA

V Genelets

Row 1

Rat OX-2:
CAA GTG GAA GTG ACC CAG GAT GAA AAG CTG CAC ACA ACT GCA TCC TTA CGC TGT TCT CTA

Murine IgV_κL6[f]:
GAC ATC AAG ATG ACC CAG TCT CCA TCT --- TCC ATG TAT GCA TCT CTA GGA GAG AGA GTC ACT ATC ACT

Murine γ:
GAG CAA ACT GAA TTA TCG GTC ACC AGA GAG GAG ACA GAT GAG AAT GTG CAA ATA TCC TGT ATA GTT TAT CTT

Murine TM86:
ACT AAA ATT ACT ACT CAG ATC TTG GGA AGA GCA AAT AAG TCT TTG GAA TGT GAG ---

Row 2

Rat OX-2:
AAA ACA ACC CAG GAA GAA ACC CTT ATT GTG ACA TGG CAG AAA AAG GCC GTA GGC CCA GAA AAC ATG GTC

Murine IgV_κL6:
TGC AAG GCG AGT CAG GAC ATT AAT AAG TAT TTA AGC TGG TTC CAG CAG AAA CCA GGG AAA TCT CCT AAG

Murine γ:
CCA TAT TTC TCC AAC ACA GCT ATA CAT TGG TAC CGG CAA CAA AAC ACA CAG CAG AAG CCG CCA CAG CTC ATG ATA

Murine TM86:
CAA CAT CTG GGA CAT AAT AAT GCT ATG TAC TGG TAT AAA CAG AGC GCT GAG CAG GAG CCC CAG CTC ATG TTT

Row 3

Rat OX-2:
ACT TAC AGC AAA GCC CAT GGG GTT GTC ATT CAG CCC ACC TAC AAA GCA AGG ATA AAC ATC ACT GAG CTG

Murine IgV_κL6:
ACC CTG ATC TAT CGT GCA --- AAC AGA TTG GTA GAT GGG GTC CCA TCA AGG TTC AGT GGC AGT GGA TCT

Murine γ:
TAT GTC GCA ACA TAC AAT AAC AAT CAA CGA CCC TTA GGA GGG AAG AAA AAA ATT GAA GCA AGT AAA GAT

Murine TM86:
CTC TAC AAT CTT AAA CAG CTG ATT CGA AAT GAG ACG GTG CCC AGT CGT TTT ATA CCT GAA TGC CCA GAC

Row 4

Rat OX-2	GGA CTC TTG AAC ACA AGC ATC ACC TTC TGG AAC ACA ACC CTG --- GAT GAT GAG GGT TGC TAC ATG TGT
Murine IgV_K L6	GGG CAA GAT TAT TCT CTC ACC --- --- ATC AGC AGC CTG GAG TAT GAA GAT ATG GGA ATT TAT TAT TGT
Murine γ	TTT AAA AGT TCT ACC TCA ACC TTG GAA ATA AAT TAC TTG AAG AAA GAA GAT GAA GCC ACC ACC TAC TAC TGT
Murine TM86	--- --- AGC TCC AAG CTA CTT TTA CAT ATA TCT GCC GTG GAT CCA GAA GAC TCA GCT GTC GCT CTC TAT TTT TGT

Row 5

Rat OX-2	CTC TTC AAC ATG TTT GGA TCT GGG AAG GTC TCT GGG ACA GCT TGC CTT ACT CTC TAT GTA
Murine IgV_K L6	CTA CAG TAT GAT GAG TTT CCT
Murine γ	GCA GTC TGG ATG AG-
Murine TM86	GCC AGC AGC CAC GGA CAG GGG GTT

Joining Genelets

Human J_λ[g]	--- --- TGG AGA TTG TGT GTA TCA TAC ACA CCG AGC TCT CAA GAC --- ---
Murine $J_H 1$[h]	TAC TGG TAC TTC GAT GTC TGG GGC GCA GGG ACC ACG GTC ACC GTC TCC TCA
Murine $J_K 1$[i]	--- TGG ACG TTC GGT GGA GGC ACC AAG CTG GAA ATC AAA CGT --- ---
Murine γ	-A? AGC TCG GGC TTT CAC AAG GTA TTT GCA GAA GGA ACA AAG CTC ATA GTA ATT CCC TCC G--
Murine TM86	--- --- TCT GGA AAT ACG CTC TAT TTT GGA GAA GGA AGC CGG CTC ATT GTT GTA

[a]Codons for charged amino acids are underscored and those for apolar (hydrophobic) ones are italicized, but in the presequences only. Codons specifying tyrosine (TAC, TAT) are lightly stippled and those encoding cysteine (TGC, TGT) are heavily stippled. Asterisk indicates location of an intron.
[b]Clark et al. (1985).   [c]Knapp et al. (1982).   [d]Saito et al. (1984); Hayday et al. (1985).   [e]Hedrick et al. (1984).   [f]Jaenichen et al. (1984).   [g]Anderson et al. (1984).   [h]Newell et al. (1980); Sakano et al. (1981).   [i]Max et al. (1981).   [j]Insert ten codons.

glycine (GGN). Thus, only three of the 20 sites are homologous, that 15% rate being within the range of random chance.

A similar condition prevails in the variable region, for even the codons for cysteine, usually highly conserved, are not identically located in the 5' portions of any of the sequences given there. Just before the termination, however, where the cysteine triplets are aligned, a number of corresponding sites are brought out, the paired GAA, GAT, and TAY being especially suggestive of relationships. These same kinships are shared by the T-cell product known as TM86 (Hedrick *et al.*, 1984). However, as a whole, sequence similarity between immunoglobulins and the γ gene is slight, relationship being expressed far more convincingly by the genomic arrangement into genelets that need assembling rather than by the sequences of the several parts.

8.6. EVOLUTIONARY CONSIDERATIONS

Recently two papers discussed the evolutionary aspects of the immunoglobulins and related proteins (Clark *et al.*, 1985; Hood *et al.*, 1985), outlining logical steps in some detail. Hence, here the best approach appears to be to combine the two in a unified treatment, smoothing out some of the differences that exist and organizing the data they present into a phylogenetic scheme.

Immune-Unrelated Predecessors. As might be anticipated, the most primitive types related to immunoglobulins play no known role in antigen recognition nor other immune-associated reaction. Undoubtedly the simplest of all is a surface glycoprotein found abundantly on neurons and lymphocytes, as well as on fibroblasts and a wide assortment of other cell types. Essentially, this substance, called Thy-1, is a relatively small molecule, in structure either intermediate between the V and C regions of immunoglobulins or perhaps closer to the former (Hood *et al.*, 1985). In addition, it possesses a short transmembrane region (Figure 8.8). Another molecule that is closely kin to the foregoing is the one recently described as MRC OX-2, which is clearly more complex than Thy-1 in consisting of two regions, one approximating a V sector, the other a C. As in the preceding, the present substance is located on the surface of neurons, T lymphocytes, and many other cell varieties (Clark *et al.*, 1985). Structural relationships with immunoglobulins and others are limited, as indicated in Table 8.10. Almost no homologous sites are to be noted, except at the terminus of row 4, which must be a region of utmost functional value. Its length is remarkable, the 3'-terminal portion subtending all of what would correspond to a J segment, without showing any homologies to the latter, however.

There is a long phylogenetic gap between OX-2 and the next recognized molecule in the proposed series, polyIGR (Mostov *et al.*, 1984), since any intermediates that possibly exist have not been identified as yet. Further, as indicated in Figure 8.8, it represents a side branch away from the main line of ascent. Like the two preceding types, it is monomeric, the single chain consisting of four V sectors and one C, plus a long transmembrane portion and a lengthy cytoplasmic constituent.

Immune-Related Molecules. All the substances that react with an antigen in one way or another are dimeric. As befits the simplest known member of the group, Lyt2 (also called T-8) is a homodimer consisting of a pair of identical subunits each comprised of a V, a J, and a transmembrane sector, plus a short cytoplasmic appendage. Next in line of

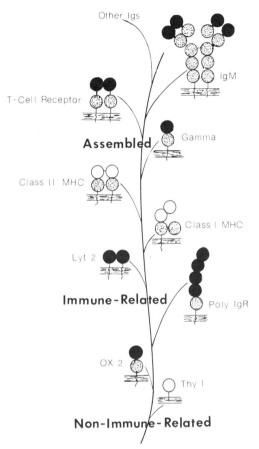

Figure 8.8. A phylogeny of immune-related molecules. (Based in part on Clark *et al.*, 1985, and Hood *et al.*, 1985.)

ascent is a heterodimeric molecule, consisting of one large subunit that bears a trans-membrane and a constant region, and two domains that do not correspond closely to any of an immunoglobulin. In this major histocompatibility complex product of class I, the β subunit is very small, being comprised entirely of a single C sector.

In remaining members, class II of that complex, the α and β subunits alike consist of a unique region, followed by a C and a transmembrane section. Thus the major distinction is in the location of the last of these areas. T-cell receptors differ but slightly, in that the unique portions are replaced by recognizable V domains, as has already been seen. These, then, together with the γ gene, are the first in this series of evolutionary events in which the coding sequence is fragmented into a series of genelets, and hence they are the pioneers among the assembled gene category. The various immunoglobulins must be placed at the tip of the branch, with the μ genelets being the earliest representative, according to present information. Researches into the immune-related molecules of the Cyclostomata and early chordates may cast light upon others of this fascinating hierarchy of unusually complicated substances and their genes.

9

Transposable Elements

Among the most exciting and informative events in the field of genetics was the discovery of elements that are moved relatively freely from one part of the genome to another. Characteristically these are highly repeated sequences in the DNA, a foretaste of which condition has already been provided in connection with the small heterogeneous RNA genes transcribed by DNA-dependent RNA polymerase III (Chapter 3, Section 3.4.1). While the specific genetic consequences of the *Alu* family discussed there are largely unknown, those transcribed by RNA polymerase II elaborated upon here have frequent and repeatable effects upon the expression of many genes. Indeed, their existence and influences in maize inheritance were reported many years prior to their final acceptance by molecular biologists (McClintock, 1949, 1951), an event that became general only after their DNA sequences made them undeniable realities.

Subsequent to their substantiation, explorations into their structures, functions, and effects on other regions of the genome have proceeded at a rapid pace, so that such repeated and mobile elements, as they are also called, have been described in great abundance. As appears inevitable in any rapidly expanding body of knowledge, the literature has been confused by misidentifications and overgeneralizations posed on insufficient data, just as in the *Alu* family. To establish discussion on as solid a basis as possible, the most thoroughly investigated types of transposable elements, certain ones of bacteria, are examined first, followed by analyses of eukaryotic types, especially those of yeast (*Saccharomyces cerevisiae*), *Drosophila,* maize, and mammals.

9.1. GENETIC ACTIVITIES OF TRANSPOSABLE ELEMENTS

While later in this chapter several concrete examples are given of the genetic results of specific elements upon the expression of genes, particularly in bacteria, plants, and *Drosophila,* preliminary statements regarding general influences on hereditary changes may add to overall clarity.

First, transposable elements are sections in DNA, sometimes rather short, but more typically of considerable length, that are moved enzymatically from one point of the genome to another on a rather frequent basis. As a whole, their position within the DNA is unstable, so the influences upon the expression of a given gene to which they have become adjacent are removed when they become excised or relocated. In some cases, outstandingly in bacteria and maize, the product (enzyme) needed for transposition is encoded within the element itself, but this is far from a universal property, as becomes evident when the individual types are examined.

Among the chief genetic effects are the following (Döring and Starlinger, 1984):

1. When transposed to a location adjacent to a gene, mobile genetic elements reduce or even totally abolish the former's expression, thus inducing a phenotype suggestive of a recessive mutation. Removal of the element where possible may restore the gene's normal function to a greater or lesser extent, depending upon the nature of the transposable sequence.
2. The above properties can be transmitted in a heritable fashion, as befits any structure in DNA.
3. The act of transpositioning characteristically leads to the formation of a short sector of nucleotides repeated at each end of the mobile structure. Actually one of the flanking "repeats," that to the left of the inserted element, is preexisting "host" (recipient) DNA, while that to the right is a copy, whose length and fidelity vary with the species of mobile unit.
4. Quite often the movable unit bears at its right terminus the typical AATAAA sequence that specifies polyadenylation, and not infrequently in the Metazoa a poly(A) or oligo(A) train is present.
5. In many cases, parts of the element are transcribed into mRNAs that encode products related or not to transposition. Such transcripts at times have been suspected of being copied by reverse transcription into a DNA segment that then becomes located at a new site in the genome (Baltimore, 1985), but this proposal has yet to be fully substantiated.
6. There are numerous families of these elements within an organism, as exemplified by the bewildering array now known from bacteria and mammalian sources.
7. In some instances the element is compound, consisting of representatives of two or more families, each of which may influence the activity of the gene in proximity to which the multiple structure becomes situated.

9.2. PRINCIPAL TRANSPOSABLE ELEMENTS OF BACTERIA

Some of the unexpected intricacies of the present topic are made clear immediately by the transposable elements (transposons) of bacteria. Two major families exist, the *Tn* (transposon) and *IS* (inserted sequence), each of which embraces numerous members that receive the greater portion of attention. In addition, many other types occur, some, like *rhs*, having only two known copies in the genome, and some being of extensive proportions, especially those sectors of DNA known as plasmids that are not part of the main continuous genomic molecules. As is becoming increasingly frequent in the literature, the terms transposon and inserted element are used synonymously for mobile (repeated) elements in general regardless of source or activity, a practice that is followed here.

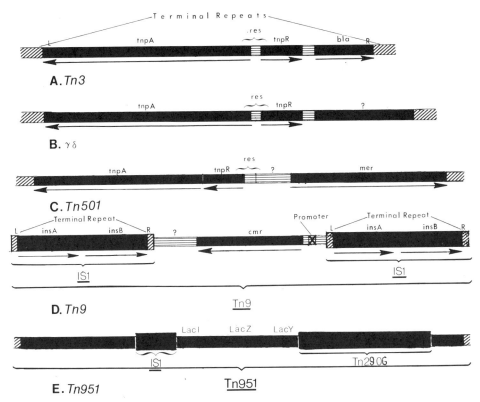

Figure 9.1. The structures of representative *Tn*-family transposons. The term transposon is here used synonymously with ''movable element.'' (Part A is based on Heffron and McCarthy, 1979; part B on Reed *et al.*, 1982; part C on Grinsted and Brown, 1984; part D on Machida *et al.*, 1983, 1984b; and part E on Michiels and Cornelius, 1984.)

9.2.1. The Transposon (Tn) Family

As a result of their convenient phenotypic markers, the *Tn* family is undoubtedly the most thoroughly explored group of mobile elements. Many members encode substances that confer antibiotic or metallic-ion resistance to the organisms or serve as toxins (Calos and Miller, 1980), properties that are readily identified in growing cultures. Although the majority of the known representative types have been given *Tn* numbers, a few, like that called γδ, have individualized names.

Transposon Tn3. Undoubtedly the best documented member of the present family is *Tn3*, whose entire 4957-nucleotide sequence has been established (Heffron *et al.*, 1979). By far the greater extent of the structure is devoted to the coding of three enzymes, the largest of which is the transposase (1015 amino acids) encoded by the gene *tnpA* (Figure 9.1A). The other two cistrons, *tnpR* and *bla,* are for a repressor-resolvase (185 amino acids) and β-lactamase (286 amino acids), respectively, which are transcribed in the opposite direction from *tnpA.* The first of these smaller polypeptides acts specifically

Table 9.1

Tn Transposable Elements of Bacteria[a]

	Left end	Right end
Tn3[b]	GGGGTCTGACGC TCAGTGGAACGA AAACTCACGTTA	TAACGTGAGTTT TCGTTCCACTGA GCGTCAGACCCC
Tn2660[c]	GGGGTCTG	CAGACCCC
Tn4[d]	GGGGGCACCTCA GAAAAACGGAAAA TAAAGCACGCTA	TAGCCGTGCTTTA TTTTCCGTTTTC TGAGACGACCCC
Tn21[e]	GGGGGCACCTCA GAAAAACGGAAAA TAAAGCACGCTA	TAGCCGTGCTTTA TTTTCCGTTTTC TGAGACGACCCC
Tn501[f]	GGGGGAACCGCA GAATTCGGAAAA AATCGTAGGCTA	TAGCCGTACGATT TTTTCCGAATTC TGCGAGCCCCCC
Tn1721[g]	GGGGGAACCGCA GAATTCGGAAAA AATCGTACGCTA	TAGCCGTACGATT TTTTCCGAATTC TGCGGGCTCCCC
Tn1771[g]	GGGGGAACCGGA GAATTCGGAAAA AATCGTACGCTA	TAGCCGTACGATT TTTTCCGAATTC TGCGGGCTCCCC
Tn5[h]	CTGACTCTTATA CACAAGTAGCGT CCTGAA	
Tn7[i]	TGTGGGCGGACA ATAAAGTCTTAA ACTGAACAAAAT	ACCCCTCCCAGT TCCCAACTATTT TGTCCGCCCACA

[a]All are from E. coli.
[b]Heffron et al. (1979).
[c]Thorpe and Clowes (1984)
[d]Hyde and Tu (1982).
[e]Zheng et al. (1981); Grinsted and Brown (1984).
[f]Brown et al. (1980).
[g]Schöffl et al. (1981).
[h]Johnson and Reznikoff (1983).
[i]Lichtenstein and Brenner (1982).

upon the genes for the other two to suppress their activity and aid in transposition, as shown later. In addition, the ends have inverted repeats, 38 base pairs long, that together are capable of forming regions of dyad symmetry in single-stranded nucleic acids. These are indicated by arrows in Table 9.1 and characterize most of the members of the *Tn* family. However, they are not always such perfect repeats as in the present case; sometimes they are interrupted, as in *Tn1721*, or quite reduced in length, as in *Tn7*. Both repeats are essential for transposition of this element (Arthur *et al.*, 1984), as is also the product of *rnpA*.

Because the transposase is thus an important factor in transposition of *Tn3*, transcription of its gene must precede or accompany the process. Two transcriptional initiation sites have been located in the transposon, one between the divergently transcribed *rnpA* and *rnpR* genes, and the second between the 3' terminus of the latter and the 5' leader of the *bla* coding sector (Wishart *et al.*, 1983). The three promoters were conjectured to be TATAATA, CATAATA, and GACAATA, in the usual 5' to 3' orientation. Although not specifically determined, the ancillary sites appeared to be AACGAAG, GTCCATT, and ATTCAAA, in the same order.

The Transpositioning Processes. The transpositioning processes, while quite different in detail, lead to an end result comparable to that observed in the immunoglobulins of the preceding chapter, namely the rearrangement of the DNA. No switch signals are to be observed here, nor are the steps so direct, for a distinctive intermediate product is the first result of recombining in prokaryotes. Since translocation is complex, involving many diverse details, this immediate discussion is designed to provide an overview, further minor points being supplied later. In stage I, the transposase is the active agent, which in *Tn3* presumedly recognizes the terminal 38-base-pair repeats and the ends of the target site, after the latter has been cleaved by an endonuclease. Unfortunately, the nature of the latter enzyme has been neglected in the literature, despite the fact that by its action on the DNA that will receive the insert it is the principal control in the genetic processes of the organisms. The activity of the transposase brings the target DNA, now freed from the genome by the cleaving enzyme (Figure 9.2), into proximity with the similarly liberated *Tn3* and associated DNA, and fuses them into what is called a cointegrate (Grindley, 1983; McCormick and Ohtsubo, 1985). Accompanying these actions is the creation of a second copy of the transposon. The process is conducted in such a fashion that the recipient and donor sections of the cointegrate are separated at each end by a copy of *Tn3*, oriented in identical directions (Figure 9.2).

In stage II of transpositioning, the repressor/resolvase plays the major role, both in recombining the two genomic fractions and suppressing the transcription of the two genes *tnpA* and *tnpR*. Although all details have not been explicitly determined, the first activity induces cleavage of the two transposons at a point referred to as the *res* site, located between the pair of genes just mentioned. The half *Tn3* molecules are then united to form a donor and a recipient circlet, the latter now modified by the presence of an intact copy of the transposon. Obviously each of these rings must then be broken and restored to its former location in the genome. In addition, it should be noted that during the act of making the copy of *Tn3*, also a five-base sector of recipient DNA on the left flank of the donor is synthesized on the latter's right end, oriented in identical fashion. These "flanking repeats" are given attention in a later discussion.

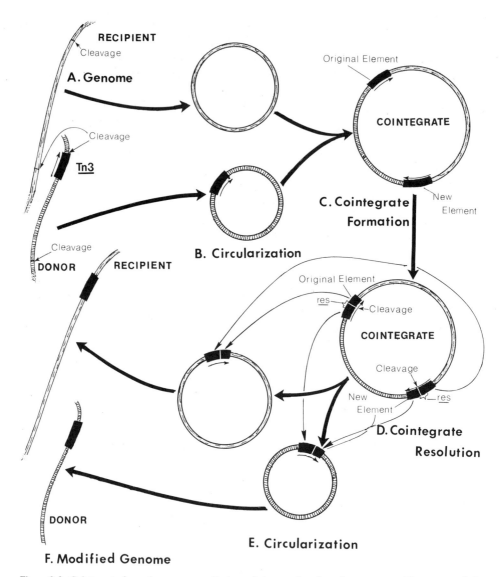

Figure 9.2. Cointegrate formation—an overall view of the translocation of transposons. The process, being complex, is shown also in Figures 9.3 and 9.4, to bring out additional details. *res*, Internal resolution site. (Based in part on Calos and Miller, 1980, and Grindley, 1983.)

A Regulated Process. Because the full picture of the events outlined above has previously escaped notice, the extent of its complexity has not been realized. But a careful examination of the multistep activity firmly demonstrates it to be highly controlled. The very first act, that of cleaving the DNA, certainly cannot be a simple mechanical process, for it involves identifying two distinct sectors of the genome, followed by cleavage of each in a precise manner at two points. Both of these sections then must be circularized.

Since their free ends have not been shown to be concatemeric, their union seemingly involves divalent bonding, possibly induced by a polymerase, together with a source for the requisite energy. The ends of the genomic sites where excision takes place have not received attention as to whether they remain loose or become spliced. If they are reconnected to form a continuous DNA molecule again, then in stage II when the completed donor and recipient circles are broken for reinsertion, they need to be properly identified and rebroken to permit the original nucleotide sequence to be restored. These steps, too, appear to require a controlling mechanism.

Certainly the transposase recognizes the inverted repeat sequences forming the termini of the mobile element, but what identifies the recipient site-to-be? Then, as pointed out by Grindley (1983), the role of that enzyme in joining the donor and recipient sectors remains undetermined, nor is there any information concerning how the ends of one are connected to those of the other. In addition, there is the problem of how the flanking repeats on mobile elements are generated. Obviously one of those repeats develops on each of the latter, when the cointegrate is produced, but formation is in opposite directions. Because resolution involves many of these same problems, but in opposite sequence, an underlying controlling mechanism, such as the supramolecular genetic apparatus, seems essential.

Figure 9.3, based in large measure on Biel *et al.* (1984), is designed to illuminate the processes involved in combining the two circlets into a single cointegrate bearing two copies of the element. In this proposal, the two rings are brought into proximity, with the intended insertion site adjacent to the transposon. Replication forks then develop, bringing about the synthesis of a twin of the mobile unit. An important aspect of the process should be especially noted, for it makes possible the union of the donor and recipient circlets. Some of the DNA strands, it suggests, are broken at the ends of the element and also at the insertion site-to-be. In the case of the donor, one on each side remains unattached until the new element is completed, while in the case of the recipient, one strand becomes attached at each end to the old element. After replication is consummated, all free ends become united to the new copy, and the cointegrate is fully formed (Figure 9.3).

Valuable as that model is in elucidating one of the important details of cointegrate formation, it fails to provide a basis for the creation of the flanking repeats. This weakness may be eliminated by the simple device of delaying reattachment of the DNA strands to the new and original elements, as in Figure 9.4. There it is hypothesized that the donor circlet remains intact during replication of the transposon, while the recipient structure becomes cleaved at the future insertion site. As the replicatory processes continue, the terminal DNA of the recipient likewise is replicated to an extent dependent on the species of transposon, in the case of *Tn3*, five base pairs. The donor DNA strands are then cleaved at one side of the element and join to the corresponding side of the new element, while those of the recipient are united appropriately to the free ends of the mobile unit (Figure 9.4). Apparently these cointegrates come into being by way of a complex knot formation, rather than simple circlets (Wasserman *et al.*, 1985). In one way or another one set of flanking repeats lies at the terminus of the original, the other at that of the new element, a condition that should exist then in cointegrates if this concept is sound. On resolution, cleavage of the elements at the *res* sites results in the transposed unit bearing the flanking

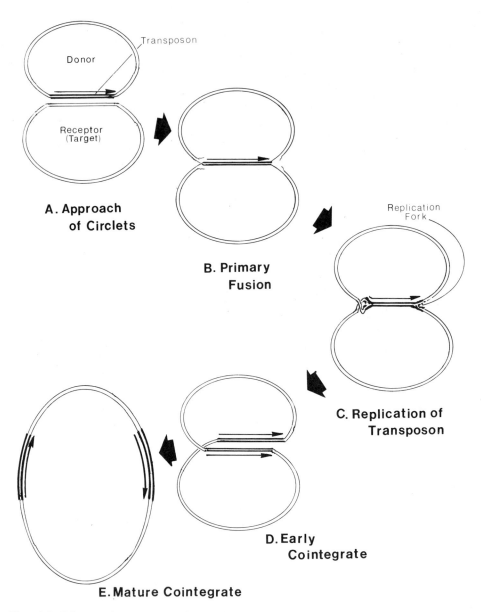

Figure 9.3. Cointegrate formation—combination of circlets to form cointegrates. In order to join into continuous circles, two ends of the DNA chain need to be relocated (C). (Based in large part on Biel *et al.*, 1984.)

repeats at each end, as seems to be universally the case. Consequently, it is amply evident that simple as the act of transpositioning appears on the surface, it has proven to be infinitely complex in the living organism and requires control at numerous points.

Relatives of Tn3. Among the closest relatives of *Tn3* is the transposon named γδ (or *Tn1000*), for its recombination system is similarly constructed and site-specific (Reed

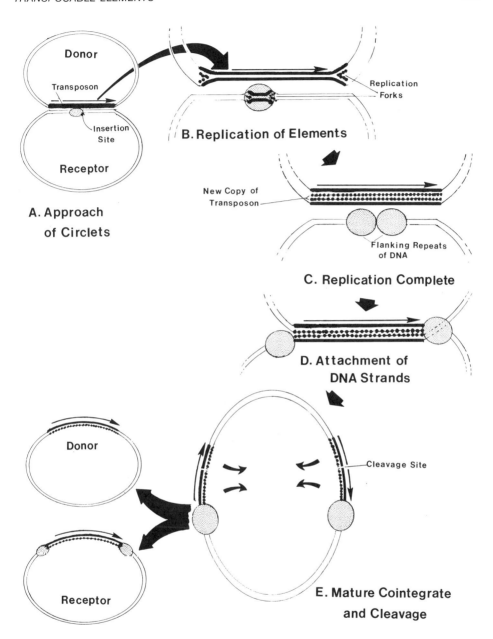

A. Approach of Circlets

B. Replication of Elements

C. Replication Complete

D. Attachment of DNA Strands

E. Mature Cointegrate and Cleavage

F. Resolved Circlets

Figure 9.4. Proposal for the formation of the flanking repeats. It is here suggested that the flanking repeats are created concurrently with the replication of the movable elements (B, C). These then become associated with the inserted element when the cointegrate is cleaved and spliced (E, F).

Table 9.2
Flanking Repeats of Bacterial Tn Elements[a]

	Left	Right
Tn2660-109[b]	GAGAC	GAGAC
Tn2660-340[b]	GAGAC	AGCAA
Tn2660-101[b]	TGATA	TGATA
Tn2660-104[b]	TTTAC	GAGAC
Tn2660-116[b]	GAGAC	TTTAC
Tn4-11[c]	GGATA	GGATA
Tn4-14[c]	TAAAG	TAAAG
Tn4-25[c]	TTATT	TTATT
Tn5-118[d]	ACCACA	
Tn5-43[d]	CACA AGCAC	
Tn5-4,8,11[d]	GCACA GGGT	
Tn5-84[d]	GCACA AGGA	
Tn5-5,79[d]	GGGTGATGG	
Tn1771[e]	TCCTT	TCCTT
Tn1771[f]	CACCA	CACCA
Tn9-3[f]	GAAGAA GGC	GAAGAA GGC
Tn9-70[f]	AGCCCG TCA	AGCCCG TCA
Tn9-8[f]	CATTAC CAG	CATTAC CAG
Tn501a[g]	TATGA	TATGA

[a]All are from *E. coli*
[b]Thorpe and Clowes (1984).
[c]Hyde and Tu (1982).
[d]Lupski *et al.* (1984).
[e]Schöffl *et al.* (1981).
[f]Galas *et al.* (1980).
[g]Brown *et al.* (1980).

et al., 1982). The two genes involved in its translocation, *tnpA* and *tnpR*, are identically situated and divergently arranged as in the other type (Figure 9.1B), with a *res* site of comparable size (170 base pairs) located between them (Newman and Grindley, 1984). Both of the references also report that the nucleotide and amino acid sequences of the resolvase of γδ are largely homologous to the corresponding parts of *Tn3*, except toward the 3' portion, where the identities of structure are at the random-chance level. One large fraction of the gene for the present mobile element remains undetermined, so that the full degree of relationship between the two transposons must await further sequence analysis. However, the third product of γδ, as yet unidentified, differs extensively from the β-lactamase product of the *Tn3 bla* gene, at least in being about 25% larger (Reed *et al.*, 1982).

Another close ally of the present mobile element is *Tn2660*, which appears to have a nucleotide structure largely homologous to that of *Tn3*, insofar as it has been determined (Thorpe and Clowes, 1984). Furthermore, a five-base-pair repeat is generated on its flanks during translocation, quite as in the more familiar species. However, in a series of experiments with mutant forms, such flanking repeats are not found to be perfect copies in all instances, some being quite aberrant (Table 9.2). At least 12 additional members of the *Tn3* family have been recognized, all with characteristics similar to the nominate form (Table 9.5).

9.2.2. Other Members of the Tn Family

The *Tn* family of mobile elements can be arranged into at least five major groups, particularly clearly on the basis of the nucleotide sequences of their termini (Table 9.1). Although in the literature only the 5′-(left)-end terminus is given, on the assumption that the opposite one is 100% complementary, the two sometimes deviate to a greater or lesser extent. Consequently, in the table only those on the right that have actually been established are cited. In the case of *Tn3* the two are fully complementary, as already seen, but those of *Tn1721* and *Tn1771* have one mismatched pair, and in *Tn7* only the first 12 sites are capable of forming Watson–Crick base pairs.

The Tn4 Subfamily. The sequence of *Tn4* is more closely related to the majority of the members cited in Table 9.1 than is the group just examined in the preceding section, for most have sites 13–24 occupied by GAA- - -GGAAAA. In the present group the three dashed sites hold AAC, whereas the next large subfamily, represented by *Tn501*, has TTC; further differences can be noted in sites 25–30 inclusive. In the case of *Tn4*, the two ends are largely complementary, but have a gap of three bases.

Tn4 ranks among the largest prokaryotic transposons known, embracing 19,000 base pairs. Included in this number are genes for three proteins conferring resistance to mercury, spectinomycin, and sulfonamide, respectively, to the organism (Hyde and Tu, 1982). Part of the size of this movable element is due to its being compound, in that it contains a copy of *Tn3* (De La Cruz and Grinsted, 1982). When transposed, flanking repeats of five nucleotide residues are created, identical in composition and rich in Ts and As, but often with one or two Gs (Table 9.2). Insertion-site characteristics are still undetermined.

A second member of this subfamily, *Tn21*, is another large transposon, containing 20,000 base pairs, part of which is attributed to a 1450-site long insert (Clerget *et al.*, 1981; Zheng *et al.*, 1981). As in the preceding member, it carries genes for resistance to mercury, sulfonamide, and streptomycin/spectromycin (De La Cruz and Grinsted, 1982), and further, its left-hand terminal repeat is identical to that of the foregoing (Table 9.1). It shows still further resemblance to that transposon in that its translocation is accompanied by the synthesis of pentanucleotide flanking repeats. As in *Tn3* and probably *Tn4*, transposition is accomplished by cointegrate intermediates (Figures 9.2–9.4). Thus it may be assumed that the element similarly contains *tnpA* and *tnpR* genes for a closely related transposase and resolvase. In fact, the amino acid sequence of the latter enzyme has been established (Diver *et al.*, 1983).

The Tn501 Subfamily. The *Tn501* subfamily of transposons is known to include *Tn1721* and *Tn1771* in addition to the nominate form (Table 9.1). The distinctive features of their left-hand terminal repeats is the presence of a G at site 10, Ts in sites 16 and 17, and AATCGT at points 25–30 inclusive. An additional common trait is that the otherwise perfect dyad symmetry between the two ends is interrupted by two or three mismatches beyond the fifth or sixth site (Brown *et al.*, 1980), not after the fourth as in the *Tn4* group. Furthermore, the terminal G on the left is unpaired. The best known member, *Tn501*, is 8500 base pairs long and carries a gene for resistance against mercuric ions (Misra *et al.*, 1985). Discoverd in a *Pseudomonas aeruginosa* strain, it has been transferred into *E. coli*, in which it has been most thoroughly investigated (Stanisich *et al.*, 1977; Bennett *et al.*, 1978). Much of the 5'-terminal region has been sequenced and the arrangement of several genes determined (Figure 9.1C; Brown *et al.*, 1983; Misra *et al.*, 1984). Its first coding sector, *merR*, encodes a regulatory product, a second, *merT*, codes for a membrane-transport function, and a third, *merA*, specifies the enzyme mercuric reductase. In addition, there are several cistrons for products whose functions remain unknown. In comparison with the corresponding parts of *Tn21*, the *mer* genes as a whole proved to be 86% homologous, whereas the *tnpR* coding sectors showed only a 72% homology level (Diver *et al.*, 1983). About 80 base pairs downstream of the 5' terminus, an 80-site-long sector corresponds almost precisely to the 5' end of *Tn21*. Considerable progress has been made in mapping the genetic structure of the entire *Tn501*, which differs in many details from others shown in Figure 9.1. However, some of the central coding structures still need to be localized, along with directions of orientation of those that have been pinpointed (Grinsted and Brown, 1984).

An additional two elements known to be members of the subfamily, *Tn1721* and *Tn1771*, resemble one another in a number of ways (Schöffl *et al.*, 1981). Both are 10,700 base pairs in length, carry genes for resistance to tetracycline, have three repeated sectors, and generate pentanucleotide flanking signals when transposed. Further, each is capable of forming multiple duplications of a 5300-base-pair section that encompasses the tetracycline-resistance (*tet*) region. The three repeated portions correspond to the usual two terminal sectors of dyad symmetry (Table 9.1), plus an extra copy following the *tet* region that is identical to the 3' terminus. This extra copy seems to provide two alternative downstream ends to the transposon, the internal one serving as the terminus during the simplification of the 5300-base-pair *tet* portion and the actual 3'-terminal during transposition.

The Tn5 Subfamily. Whether *Tn5* is unique or has still undiscovered relatives is unestablished, but at the present time it alone comprises the subfamily that bears its name. It is a compound transposon of unusual structure, for it consists of two inverted copies of an insertion element called *IS50* on each flank of a DNA sequence 2700 base pairs in length. The central portion, which is *Tn5* proper, contains several resistance genes, including one against neomycin/kanamycin, and others against streptomycin and a bleomycin-related compound (Putnoky *et al.*, 1983; Mazodier *et al.*, 1985). In addition, genes for an aminoglycoside phosphotransferase and a streptomycin-phosphotransferase have been located (Mazodier *et al.*, 1983).

The flanking insertion elements are not merely associated with this transposon, but are vital parts of it, as well as being transposable independently. That on the right (*IS50R*) codes for two enzymes essential for transposition, the first (protein 1) being a transposase, the other (protein 2) an inhibitor of transposition. In addition, some unknown factors from the cell are requisite for completion of the process (Johnson and Reznikoff, 1984). The left-hand flank (*IS50L*) differs from the other by only a single base pair, but this slight change exerts two main influences: (1) it produces an ochre mutation within the transposase, which creates a stop signal that aborts translation, both of itself and the inhibitor; and (2) it produces a promoter for the kanamycin-resistance gene of *Tn5* proper. Thus, *IS50L*, while translationally inert, is essential for transcription of the *Tn* element. Recently transposition was shown to differ from that process described in preceding pages in that cointegrates are not formed (Isberg and Syvanen, 1985). After excision, the *Tn5* molecule is circularized and then inserted in a new location, whether or not accompanied by DNA synthesis. Thus it appears that *Tn5* is actually moved from site to site, whereas in *Tn3* and others that have cointegrate intermediates, only a copy is moved, the original being returned to its former site. Since nonanucleotidyl directly repeated sectors are generated at the ends of the transposed element, some DNA synthesis seems to be requisite, despite its reported absence (Lupski *et al.*, 1984).

The terminal repeat of this transposon proper differs strongly from all others listed in Table 9.1 (Johnson and Reznikoff, 1983), opening with CTGAC instead of the multiple Gs of the rest; nor can any homology be detected in the remainder of the sequence. In *in vitro* experiments nucleotides 8–16 have been shown to be required for transposition. It has been proven to be an agent of unusual importance in studying genomic composition of numerous other bacteria, including *Caulobacter, Vibrio,* and *Rhizobium* (Belas *et al.*, 1984; Ohta *et al.*, 1984; Rostas *et al.*, 1984).

Other Transposons. A confusing array of transposons is coming to light as the plasmid structures of *E. coli* and numerous additional bacteria are explored, but the molecular aspects of most of these have received too limited attention to warrant their inclusion here. Several of considerable importance can be briefly mentioned. *Tn7* is exceptional in its attachment-site requirements, for it translocates to only a few points in the bacterial genome (Lichtenstein and Brenner, 1982; Ouartsi *et al.*, 1985). It is a large element of 14,000 bases, which includes genes for resistance to three antibiotics, trimethoprim, streptomycin, and spectinomycin. When inserted, pentanucleotide repeats are generated at the flanks, and in some variants of the element an *IS1* insert is present (Hauer and Shapiro, 1984). Although quite distinctive at their extreme ends, the left terminal repeat shows considerable homology with that of *Tn5* (Table 9.1). One very unusual cluster of transposons also is known, for instead of carrying genes for resistance factors, *Tn951* and *Tn2501* encode proteins involved in lactose fermentation (Michiels and Cornelis, 1984). Each contains a single copy of *IS1* (Figure 9.1E).

Tn9 also is associated with *IS1*, but as in *Tn5*, that element is at each flank (Figure 9.1D); however, in this case both are oriented in harmony with the central portion of the transposon (Galas *et al.*, 1980). This latter region, 2638 base pairs long, contains genetic determinant regions for resistance against chloramphenicol (Alton and Vapnek, 1979). In translocation, it shows marked preference for special parts of the genome, notably certain regions of the *lac* operon: nine-base-pair repeats are produced during translocation (Johnsrud *et al.*, 1978). *Tn6* and its close kin, *Tn2680*, are similarly organized, bearing a copy of *IS26* at each end (Mollet *et al.*, 1985); in like fashion *Tn2921*, which provides fosfomycin resistance, has a copy of *IS10* at each terminus (Navas *et al.*, 1985).

Still another species of these transposons, *Tn10*, is flanked by one of the insertion elements that provide the topic for the next section of this chapter (Kleckner, 1979). In this case, the flanking repeats are *IS10*, oriented in opposite directions (Figure 9.1; Postle *et al.*, 1984), as in *Tn2921*. Each of these is 1330 base pairs long, while *Tn10* consists of 9300 base pairs, much of its central area consisting of two genes involved with tetracycline resistance. The first of these, *tetA*, encodes the membrane-associated resistance factor, whereas the second, *tetR*, represses the synthesis of the other; the two are transcribed in opposite directions from divergent promoters located in a regulatory region. A bidirectional termination signal is also present (Schollmeier *et al.*, 1985). As in other members dependent on *IS* elements, the transposase is encoded in *IS10R* (Morisato and Kleckner, 1984). Like the foregoing, *Tn903* has an *IS* species (*IS903*) at each end, on which it is apparently dependent for transposition (Oka *et al.*, 1978). Its actual length, part of which encodes for kanamycin resistance, is 1000 base pairs, but with the 1050 base pairs of each *IS903*, its total is 3100 base pairs. Repeats, nine base pairs in length, are generated at its flanks during transpositioning, but that trait is really a feature of the *IS* species.

  Two Families of Tn Elements. From the foregoing summaries of the principal types of *Tn* elements, it is self-evident that there are actually at least two major types of these mobile units. In the first of these, the members consist of a largely independent DNA section that, as in *Tn3* and *Tn7*, carries genes for the principal enzymes involved in transposition. Since the earliest known transposons are of that type, the *Tn* designation should apply only to that group. In a new family should be placed *Tn5*, *Tn501*, *Tn7*, *Tn9*, and *Tn10*, which are dependent upon flanking *IS* units; moreover, the details of transpositioning, while incompletely known, appear to differ in the two groups. Perhaps the name of this second family could be modified to *Tnd*, the appended letter signifying their dependence on other parts for their mobility, but see Section 9.4 for a fuller discussion. This suggested change involves a minimum of confusion, yet provides for a unification of similar members that cannot fail to expedite further understanding of their functioning and relationships. The completely different nucleotide sequences of the termini in the two families also provide support for this arrangement.

9.3. THE INSERTED-SEQUENCE (IS) FAMILY

  Because of their frequent association with specific types of the transposons analyzed in the preceding section, some members of the *IS* family are already somewhat familiar. But, as in the *Tn* families, they are a highly diversified group, with many recognized species and numerous varieties of lesser importance (Iida *et al.*, 1983). As a whole, they differ from the transposons in not carrying genetic determinants for resistance factors, but they resemble those others in the nearly universal trait of short repeats being generated at each end upon translocation.

9.3.1. Insertion Element IS1

  The smallest active member of the present family, *IS1*, only 768 base pairs in length, is a natural constituent of plasmid R100 of *E. coli* and other Gram-negative bacteria (Ohtsubo and Ohtsubo, 1978). As will be recalled, it is the element found on each flank of *Tn7* and *Tn9*, and is responsible for the latters' mobility (Figure 9.1D).

Structure. Although the complete nucleotide sequence of *IS1* has long been established (Ohtsubo and Ohtsubo, 1978) and analyses of the genetic composition have been completed (Machida *et al.*, 1984a,b), the precise number of cistrons included within it has not been determined. At least two similarly oriented genes are present, *insA* and *insB*, both of which are essential for cointegration, but a total of six possible open reading frames has been reported (Machida *et al.*, 1983, 1984b). Also requisite for this event are the inverted repeats of each terminal portion of the element, which probably serve as recognition signals. Additionally, within each repeat is a promoter region, which directs transcription from the terminus toward the interior of the element (Machida *et al.*, 1984a), but the products of the transcripts remain unknown.

The Flanking Repeats. The nature of the repeated segments generated at the termini of this element during translocation has attracted much attention because of their variation in length. Although characteristically each is nine base pairs long, frequently both have been found to be shorter than that by one (Kanazawa *et al.*, 1984), a condition often considered to result from mutations of the element (Iida *et al.*, 1981). Recently, however, it was firmly established that the inconstancy in size is an intrinsic feature of *IS1*, wild-type and mutant forms alike generating either eight- or nine-base-pair products (Iida *et al.*, 1985). Frequently the flanking units are rich in A–T base pairs, but this is not consistently true (Table 9.4), for those of Q12 consist largely of C–G pairs. An additional characteristic is a trend toward termination at the 3' end in a T,G combination.

The Inverted Repeated Termini. The inverted repeats of the two terminal regions are unusual in structure in that the complementarity that exists between them is frequently interrupted by mismatched pairs. The first three pairs, for instance, are complementary (Table 9.3), followed by a single-pair interruption, then another three-site run of pairs, and so on, as indicated by the underlining arrows; the mutant form *Is1*-16 has still more limited complementarity (Kanazawa *et al.*, 1984). Within their full 70-base-pair extent, only half of which is included in the table, are contained promoter sites (Machida *et al.*, 1984a); the one in the left terminus is believed to initiate transcription of *insA* and *insB*, while that in the right component may be involved in the production of an antimessenger that negatively controls the left promoter, but this suggestion is not based on observed facts. Because of the extremely complex tertiary structure of mRNAs (Dillon, 1978, pp. 122–133), it is unlikely that any interaction could take place with a similarly folded antimessenger.

9.3.2. Other IS Element Genes

Although the *IS* elements are as diversified as the transposons (*Tn*) just discussed, they do not appear to have attracted as many investigations as that group. Consequently, the present analysis cannot be as thorough as the preceding. Yet, enough has been determined to disclose the existence of multiple subfamilies, including several with unusual characteristics.

IS1-Related Genes. Because of the frequent mutations in nucleotide structure that have been reported, relationships among major types are not easily establishable. A ready illustration of the problem is shown in Table 9.3, where numerous disparities are revealed between two variants of the supposedly same insertion element, *IS1* and *IS1*-16. The

Table 9.3
Inserted Elements (IS) of Bacteria[a]

	Left			Right		
IS1[b]	GGTGATGCTGCC	AACTTACTGATT	TAGTGTATGATG	AACATAAAACAC	TATCAATAAGTT	GGAGTCATTACC
IS1-16[c]	GGTAATGACTCC	AACTTATTGATA	GTGTTTTATGTT	CATCATACACTA	AATCAGTAAGTT	GGCAGCCATCACC
IS15[d]	GGCACTGTTGCA	AAGTTAGCGATG	AGGCAGCCTTTT	ATAAGTTTATCA	CCACCGACTATT	TGCAACAGTGCC
IS26[e]	GGCACTGTTGCA	AATAGTCGGTGG	TGATAAACTTAT	AAAAGGCTGCCT	CATCGCTAACTT	TGCAACAGTGCC
IS5[f]	GGAAGGTGCGAA	TAAGCGGGGAAA	TTCTTCTCCGCT	ACATGATCTCAT	ATCAGGGACTTG	TTCCCACCTTCC
IS2[g]	TAGACTGGCCCC	CTGAATCTCCAG	ACAACCAATATC	TGATAACAGATG	TCTGGAAATATA	GGGCAAATCCA
IS4[h]	TAATGCCGATCA	GTTAAGGATCAG	TTGACCGATCCA	AAAAGAGCCAGT	CAGTTGCTTAAC	TGACTGGCATTA
IS30[i]	TGTAGATTCAAT	TGGTCAACGCAA	CAGTTATGTGAA	TTGAAAGGGGTG	TTGCATTGACAG	ATTGAATCTACA
ISH1[j]	TGCCTTGTTTTG	CCACCGATTGAG	GGAAGTTTCAGA	CCCTACCCCGAC	GCTGTCTTGTGA	TTCAACGAGGCA
ISH1.8[k]	CTATCTTGATTC	AGCGAGAGAATC	CCGCTCTTCCCG	AGCCGCGCCGTT	CACGGCGCCGAG	GATCTCACTCCG
IS50[l]	CTGACTCTTATA	CACAAGTA (Incomplete)		(Incomplete) AAGATCTG		ATCAAGAGACAG
TL[m]	GGCTTCATGTCC	GGGAAATCTACA	TGGATCAGCAAT	CTATTCGGCGCT	AACTTTTGGTGT	GATGATGCTACT

[a]The arrows indicate regions of dyad symmetry (complementarity); TL is from Agrobacterium, ISH is from Halobacterium, and the others are from E. coli.

[b]Machida et al. (1984a). [c]Kanazawa et al. (1984). [d]Trieu-Cuot and Courvalin (1984). [e]Mollet et al. (1983). [f]Engler and van Bree (1981); Schoner and Kahn (1981). [g]Brosius and Walz (1982). [h]Mayaux et al. (1984). [i]Dalrymple et al. (1984). [j]Simsek et al. (1982). [k]Schnabel et al. (1984). [l]Sasakawa et al. (1983). [m]Gielen et al. (1984).

question such distinction poses is obviously, which of the two is the basic element in organisms to which others should be compared? Examination of those given in that table shows that the answer is both or neither! When *IS15* (Trieu-Cuot and Courvalin, 1984) is employed as the sequence whose relations are to be determined, its third base from the 5′ end is found to be a C that agrees with neither of the *IS1* structures, whereas its fourth position has an A as in IS1-16, but not *IS1* proper. Then nucleotides 9–11 are TGC as in *IS1*, differing completely from the CTC of the mutant variety. Beyond site 24, little homology is to be found with either of that pair. But *IS15* is also distinctive in having a continuous run of 14 residues that form a region of dyad symmetry with the inverted repeat at the opposite terminus. A dyad symmetric sector of identical length exists also at the termini of *IS26* (Mollet *et al.*, 1983); in fact, precisely the same nucleotide sequence is present in both species.

IS15 possesses some other features reminiscent of the first representative examined, especially in that it is part of at least one known transposon, *Tn1525* (Trieu-Cuot and Courvalin, 1984). The right-hand component, *IS15R*, is mobile in its own right; it has proven to be a duplex also, as it has in insert of a short version of itself called *IS15Δ*. Integration of this segment is accompanied by the formation of directly repeated eight-base-pair sequences flanking its ends (Table 9.4; Trieu-Cuot *et al.*, 1983). The compound structure includes genes encoding four unidentified products.

In the 820-base-pair long *IS26*, kinship to *IS1* is similarly displayed by its association with a kanamycin-resistance transposon, *Tn2680*. Although not native to *E. coli*, having been isolated originally from *Proteus vulgaris* (Mollet *et al.*, 1983), it can mediate cointegration freely in the former bacterium (Iida *et al.*, 1984), a process accompanied by formation of eight-base-pair repeats, as in the foregoing species (Table 9.4). In the present case, both *IS26L* and *IS26R* are capable of inducing formation of cointegrates carrying *IS26* duplicates at the junction of the two fused parts (Mollet *et al.*, 1983).

In spite of its 5′-terminal repeat beginning with the same sequence of nucleotides, GGC, as the last two mobile elements (Table 9.3), *TL* from *Agrobacterium tumefaciens* doubtlessly is unrelated to any of this group (Gielen *et al.*, 1984). This is located at the left end of the Ti plasmid, which is readily introduced into seed-plant cells; such insertions typically result in a transformation event, leading to the formation of crown gall, hairy root, or woolly-knot disease. As shown by the establishment of its 13,637-base-pair sequence, *TL* more closely resembles elements from advanced organisms, for it has direct repeats not inverted at its termini (Table 9.3), each 24 sites long. It encodes eight potential proteins, numbers 2, 3, 7, and 8 being oriented in the opposite direction from the others. How long the repeats are that its insertion induces at its flanks remains for future investigations to disclose, but its length and protein-encoding genes differ strongly from any definitive member of this group.

Miscellaneous Insertion Elements. The remaining better-established insertion elements are quite unrelated to the *IS1* group just described, both in structure and activity. The first of these, *IS2*, along with its kin *IS4*, serves well in exposing the differences. To begin with, the former, 1327 base pairs long, lacks any known genetic markers, but occurs in several copies in the *E. coli* genome (Hinton and Musso, 1983). Chiefly it manifests its presence through the expression of genes near which it becomes inserted, but these differ, depending on its orientation. In orientation I within operons, *IS2(I)*, as this arrangement is indicated, blocks transcriptions of genes downstream from it, whereas in

Table 9.4
Segments Flanking Bacterial Inserted Elements[a]

	Left flank	Right flank
$IS1\text{-}16^{b}$	AAA AAC GT	AAA AAC GT
$IS1\text{-}14^{b}$	AAA --C GTT G	AAA --C GTT G
$IS1\text{-}Q12^{c}$	CCT TTG CCA	CCT TTG CCA
$IS1\text{-}13^{b}$	AGA AA- GTT T	AGA AA- GTT T
$IS1\text{-}219^{c}$	CTA AAA TG	CTA AAA TG
$IS1\text{-}327^{c}$	CGA TAG TG	CGA TAG TG
$IS2^{d}$	TTA CA	TTA CA
$IS4^{e}$	CTC CCA GTG GAG	CTC CCA GTG GAG
$IS15\text{-}322^{f}$	TCT CGG GC	TCT CGG GC
$IS26\text{-}219^{g}$	GTG CAC CA	GTG CAC CA
$IS30\text{-}209^{h}$	TCA TAA GTG CC	CTC CTT TCT TA
$IS30\text{-}83^{h}$	GCG AAA -TG CCA	GTC GAA TTG CG
$ISH1^{i}$	AGT TAT TG	AGT TAT TG
$ISH1.8^{j}$	GCT GC- AAG TG	AAG TG- GCT GC
TL^{k}	GGC AGG ATA TAT ATT CAA TTG TAA AT	GGC AGG ATA TAT ACC --G TTG TAA TT

[a]The *ISH* elements are from *Halabacterium*, *TL* is from *Agrobacterium*, and the rest are from *E. coli*.
[b]Kanazawa *et al.* (1984).
[c]Iida *et al.* (1985).
[d]Brosius and Walz (1982).
[e]Mayaux *et al.* (1984).
[f]Trieu-Cuot *et al.* (1983).
[g]Mollet *et al.* (1983).
[h]Caspers *et al.* (1984).
[i]Simsek *et al.* (1982).
[j]Schnabel *et al.* (1984).
[k]Gielen *et al.* (1984).

the second arrangement, it [*IS2(II)*] augments transcription of those genes, but how it does so remains unclear. In part its insertion seems to generate promoter-bearing sequences within the operon (Hinton and Musso, 1983). By way of illustration may be cited the expression of a gene from yeast for tryptophan synthase (*TRP5*) when cloned in *E. coli* (Brosius and Walz, 1982). By itself, the cloned gene is transcribed very inefficiently, but when *IS2* is spontaneously transposed to a point 60 base pairs upstream, its expression is increased markedly. Such translocation is accompanied by the formation of five-base-pair direct repeats flanking the termini (Figure 9.4).

In the structure of its terminal inverted repeated sequences, *IS4* shows occasional homology to those of *IS2*, but unlike those of the latter, which display no region of complementarity with one another (Table 9.3), the present type shows two well-defined sectors (Klaer *et al.*, 1981). In many *E. coli* K12 strains, only a single copy exists of the 1426-base-pair element; much of this length encodes a 442-amino acid protein, whose function is unknown. The activity of *IS4* is comparable to that of its kin discussed above, but the unit has the effect of reducing transcription of the operon when placed in orientation II near the *pheS₁T* region, whose genes encode the two subunits of phenylalanyl-tRNA ligase (Mayaux *et al.*, 1984). Insertion generates a direct 12-base-pair repeat of receptor DNA at the right flank (Table 9.4).

Both *IS30* and *ISH1* terminal sequences begin with TG, but otherwise show only a low-grade homology (Table 9.3). However, the two are also comparable in length, the former, from *E. coli*, consisting of 1221 base pairs and the latter, from *Halobacterium halobium*, containing 1118 (Simsek *et al.*, 1982; Dalrymple *et al.*, 1984). The terminal repeated segments of the *E. coli* species are only partly complementary, not nearly as completely so as the second reference suggests, and still fewer complementary sites characterize the other (Table 9.3). When *IS30* is transposed, segments of only two base pairs are generated, whereas with *ISH1* eight-base-pair sectors are duplicated. The first of these two was discovered when its spontaneous insertion into the prophage gene of bacteriophage P1 caused its loss of reproductivity (Caspers *et al.*, 1984), while the second is known to inactivate the bacteriorhodopsin cistron (Simsek *et al.*, 1982). The recently sequenced *IS3* also shows kinship to *IS30* (Timmerman and Tu, 1985).

As in the preceding couple, *IS50* and *ISH1.8* share a few characteristics, such as having the first two nucleotides of the terminal repeats identical, with only limited homology existing elsewhere (Table 9.3; Sasakawa *et al.*, 1983; Schnabel *et al.*, 1984). The first base of the opening CT has been demonstrated to be essential for transposition of the element, transversion to G reducing that process to 1 or 2% of wild type (Sasakawa *et al.*, 1985). Transposition can occur in several ways with *IS50* when present in pairs as in *Tn5*, the details of which are currently under investigation (Nag *et al.*, 1985). In the first species, from *E. coli*, the repeats have been thought to be unusual in having only a few sites complementary, but actually, as has been seen, the eight matching pairs of *IS50* provide an abundance in comparison with those of *IS2*, which have only a single pair of corresponding sites, a condition also prevalent in those of *ISH1.8*. Actually the latter element leads into the topic analyzed in the next chapter, which examines viral genes, for it had its origins in the bacteriophage ΦH of *Halobacterium* (Schnabel *et al.*, 1984).

9.4. SYNOPSIS OF PROKARYOTIC TRANSPOSONS

Even though no attempt could be made to include all the transposons that have been reported from prokaryotic sources, the foregoing account provides an insight into their effects, bewildering numbers, and diversity. So that some order may emerge from that chaos, principal characteristics of the better established varieties are summarized in Table 9.5.

The Bases for the Table. Two chief aspects of the transposons provide the bases for order in the arrangement of that chart: first, the dependency* or autonomy of the element, and second, more importantly, the nucleotide sequence at the left end of each.

---

*Here the term dependency refers to the essentiality of paired auxiliary elements on the flanks for transposition.

Table 9.5
Summary of Major Bacterial Transposons[a]

Major nature	Left end	Type	Other members	Characteristic genes	Total length (base pairs)	Terminal sectors	Flank repeats (base pairs)
				CLASS D			
Autonomous	GGTRATG	IS1	--	--	768	Inverted	8
Autonomous	GGCACTG	IS15	IS26	--	--	Slightly inverted	8
Autonomous	GGCTTTG	IS903	--	--	1,050	Inverted	9
Autonomous	GGCTTCA	TL	--	Octopine synthase	13,637	Not inverted	26
Autonomous	GGAAGG	IS5	--	Three genes for unknown products[b]	1,195	Slightly inverted	4
				CLASS E			
Autonomous	TAGACTG	IS2	--	--	1,327	Not inverted	5
Autonomous	TAATGCC	IS4	--	--	1,426	Slightly inverted	11-12
				CLASS F			
Autonomous	TGY	IS30	ISH1	--	--	Slightly inverted	8-15
Autonomous	TGATCCT	IS3	--	--	1,258	--	3-4
				CLASS G			
Autonomous	CTGACT	IS50	--	--	1,531	Slightly inverted	9
Autonomous	?	ISH1.8	--	--	--	--	10
Autonomous	CTGATGA	IS10	--	--	1,329	Slightly inverted	9

				Product	Total length	Inverted repeats	Flank repeats
CLASS D							
Autonomous	GGTRATG	IS1	--	--	768	Inverted	8
Autonomous	GGCACTG	IS15	IS26	--	--	Slightly inverted	8
Autonomous	GGCTTTG	IS903	--	--	1,050	Inverted	9
Autonomous	GGCTTCA	TL	--	Octopine synthase	13,637	Not inverted	26
Autonomous	GGAAGG	IS5	--	Three genes for unknown products[b]	1,195	Slightly inverted	4
CLASS E							
Autonomous	TAGACTG	IS2	--	--	1,327	Not inverted	5
Autonomous	TAATGCC	IS4	--	--	1,426	Slightly inverted	11–12
CLASS F							
Autonomous	TGY	IS30	ISH1	--	--	Slightly inverted	8–15
Autonomous	TGATCCT	IS3	--	--	1,258	--	3–4
CLASS G							
Autonomous	CTGACT	IS50	--	--	1,531	Slightly inverted	9
Autonomous		ISH1.8	--	--	--	--	10
Autonomous	CTGATGA	IS10	--	--	1,329	Slightly inverted	9

[a] See previous tables and text for references. Total length and flank repeats are expressed in base pairs.
[b] Rak and von Reutern (1984).

When these two characters are employed together as the prime indexing keys, additional traits, such as the resistance factors and various genes carried, fall into order, as do the nature of the terminal regions and the lengths of the repeats created in the flanking DNA. Size of the molecule proves to be of little importance, for the obviously related members of the *Tn3* series range from around 5000 to 25,000 base pairs long. Consequently, classifications of these mobile units based on size, like the proposed SINE–LINE system (Singer, 1982), are artificial, although not without value for preliminary purposes.

Analysis of the Summary. When the movable structures are arranged as just outlined (Table 9.5), three well-marked groups of the *Tn* series and at least four more, probably five, in the *IS* category can be noted. Among the first of these groupings, the largest (class A) is the *Tn3* family, all the members of which commence at the left end with either four or five Gs. Whether the change in the fifth base from the T of *Tn3* and *Tn951* to the G of *Tn4* and *Tn501* is justification for subdivision into subfamilies or not depends on the outcome of further investigations. The second group (class B) of the *Tn* series has TGTG as the opening sequence, as in *Tn7* and its close kin *Tn402*, but further members possibly will be found in *Tn71* and *Tn72*, when the necessary structural investigations are completed. A similar statement applies to the dependent cluster of these elements (class C), in which the left terminal sequence (GTAA) has been determined for only a single representative, *Tn9*. In the remainder of this last group, those termini that have currently been established actually pertain to the *IS* units present on each side, not the central element proper. One of the striking points that emerges in this arrangement is the uniformity in size of the flanking repeats within each major subdivision of dependency, five in every autonomous type, and nine in each that is dependent, the latter constancy being in spite of the fact that the *IS* units are themselves variable in this trait.

As pointed out earlier, the *IS* series has been less thoroughly investigated at the molecular level, so it is difficult to perceive the extent to which the nucleotidyl structures correlate to the other fundamental characteristics. Only when the genes each one bears have been fully sequenced and the functions of the products determined will a firm basis be laid for subdivisions into families. As a preliminary attempt, however, the left terminal GGY sequence might be suggested to characterize one natural category (class D), including *IS1*, *IS15*, *IS903*, and *TL*, but it is possible that just the first two bases, GG, reflect the basic kinship. If such proves to be the case, then *IS5* is to be included in the foregoing. With a similar restriction to two residues, the next major category (class E) embraces those whose sequences begin with TA, as in *IS2* and *IS4*. In the two final groups, classes F and G, respectively, TG is the first pair of nucleotides on the one hand and CT (or even CTGA) on the other; the representatives of the former include *IS30* (*ISH1*?) and *IS3* and the latter *IS50*, *ISH1.8*, and *IS10*. Throughout the *IS* series, as just pointed out, the size of the flanking repeat is variable, appearing to be a species characteristic, not correlated to family.

9.5. MOBILE ELEMENTS IN YEAST

Transposable elements were not discovered in the common yeast (*Saccharomyces cerevisiae*) until some time after they were known in bacteria and *Drosophila*, first being reported by Cameron (1979; Cameron *et al.*, 1979). This earliest find was called *Ty1*, which subsequently became one of the most thoroughly documented members of these elements. Since then at least three other families have been described from this organism, known as τ, σ, and δ, several of which have highly distinctive behavior patterns. The processes of transpositioning yeast elements are beginning to be revealed, and what has been learned shows them to be highly distinctive from those of the bacterial types. Two basic methods are followed for relocation: recombinational, which is the more frequent, and transpositional. In the first type two Ty elements interact to recombine in a reciprocal manner, whereas in the second, an RNA intermediate is employed, so that sequence information flows from DNA to RNA and then back to DNA (Boeke *et al.*, 1985).

9.5.1. The Ty1 Movable Element

Unlike the condition among the bacterial movable elements, where the common types are simple, the best known example of yeast, *Ty1*, is compound. It is 5300 base pairs in length, but is flanked at each end by a second type of transposon 300 base pairs long (called δ) oriented in like direction (Farabaugh and Fink, 1980; Kingsman *et al.*, 1981). About 30–35 copies of *Ty1* have been reported from the haploid yeast genome and nearly 100 of δ; the latter number includes both those flanking the present transposon and independent elements. It should be noted that *Ty1* is now considered to embrace the two δ units as normal constituents, its central 5300-base-pair portion being referred to as ϵ (Winston *et al.*, 1984). Moreover, two classes of this very variable element are currently recognized (Fulton *et al.*, 1985), whose genes may encode distinctive proteins.

Behavior of Ty1. One of the more thorough explorations into the effects of *Ty1* when transposed into or adjacent to a known gene is a study of a *his4* mutation created by such an insertion (Farabaugh and Fink, 1980). This cistron encodes one of the enzymes involved in the synthesis of histidine, the mutation having the effect of suppressing its transcription. When the sequence was established, it was revealed that the *Ty1* unit preceded the *his4* promoter and in no way disrupted the latter's coding region; hence, it was conjectured that the depressed activity resulted from transcriptional competition. It was also found that insertion of the mobile unit had created the five-base-pair repeat unit TAAGA on the right flank, as usual being a copy of the receptor DNA bordering the left end of the insert. Thus the behavioral properties of the element, as well as its structural aspects, are similar to those of bacteria. When *Ty1* was excised from this location by transposition, the expression of the *his4* gene was fully restored, although a 300-base-pair portion equivalent to a δ element remained.

Repeats of similar length have been reported to be produced at other inserted sites of *Ty1* (Gafner and Philippsen, 1980), each reflecting the structure of the DNA into which the insertion was made (Table 9.6). What are more difficult to understand are the comparable duplications that exist at both ends of the ϵ portion where contact is made with the δ element, for in neither of the two examples given in the table do the repeated portions resemble adjacent DNA, either that of δ or ϵ (Farabaugh and Fink, 1980; Gafner and Philippsen, 1980).

Control of Ty1 Expression. Although the specific products of translation have not been identified, the mRNAs that are transcribed from *Ty1* have been known for some time, but the production of transcripts is complex. First of all transcription of these mobile units is strongly influenced by the mating type locus (*MAT*). In haploid cells or those

Table 9.6
Repeated Sequences at Flanks of Inserts of Yeasts and Seed Plants

	Left flank	Right flank
Yeast		
$Ty1-his4^a$	TAAGA	TAAGA
$Ty1-BIO^b$	GAAAC	GAAAC
$Ty1-D15^{b,c}$	ATTTT	ATTTT
$Ty1-\delta-his4^a$	TGGTA	TGGTA
$Ty1-\delta^b$	TACCA	TACCA
$\tau 2\delta^d$	TATTA	TATTA
Maize		
$Ac7^e$	GGTCACGC	GGTCACGC
$Ac9^f$	CATGGAGA	CATGGAGA
$Ds1^g$	GGGACTGA	GGGACTGA
$Ds1-6233^h$	CTTGTCCC	CTTGTCCC
$En1^i$	ATA	ATA
$En8^j$	GTT	GTT
$Mu1^k$	TTTTGGGGA	TTTTGGGGA
Antirrhinum		
$Tam1^l$	ATA	ATA

[a]Farabaugh and Fink (1980).
[b]Gafner and Philippsen (1980).
[c]Eibel *et al.* (1980).
[d]Genbauffe *et al.* (1984).
[e]Müller-Neumann *et al.* (1984).
[f]Pohlman *et al.* (1984).
[g]Sutton *et al.* (1984).
[h]Weck *et al.* (1984).
[i]Pereira *et al.* (1985).
[j]Schwarz-Sommer *et al.* (1984); Gierl *et al.* (1985).
[k]Barker *et al.* (1984); Taylor and Walbot (1985).
[l]Bonas *et al.* (1984).

homozygous for either the *a* or α phenotype, 5–10% of the total poly(A)-bearing mRNA present is from *Ty1,* but in heterozygotes the level is reduced by 75–95%, with similar results being recorded for other genes having *Ty1* insertions (Elder *et al.,* 1983; Williamson, 1983). Second, the product of the *SPT3* gene has been demonstrated to be involved in these processes, for mutations in that locus result in suppression of the transposon's activity (Winston *et al.*, 1984). The reduction is a consequence of a change in the transcriptional start site from the left δ unit to a new position 800 base pairs downstream. Third, the type of substance in the medium that serves as the carbon source also exercises strong influence (Taguchi *et al.*, 1984). In contrast to glycerol's typical role of alleviating depressed activity of the *Adh2* gene, which plays an important role in the production of alcohol dehydrogenase, it becomes a suppressor when *Ty1* is inserted 5' to that locus. In contrast, the normally depressor substance glucose becomes a derepressor, restoring the usual rate of activity.

In Table 9.7 the terminal sequence of an ε sector is shown to be unlike most bacterial transposons in lacking any extensive region of dyad symmetry, the two bases at the extreme ends alone being complementary. However, in view of the fact that this portion is always bordered by a δ unit, this absence is scarcely surprising, for the transposing enzyme probably recognizes those elements, not the ε part.

9.5.2. Other Transposable Elements in Yeast

Investigations of the remaining transposons of yeast are still in the early stages of development, with firm information currently relatively scanty. As a consequence, discussion is limited to the three more important units, δ, σ, and τ, and that necessarily briefly.

The δ Unit. As pointed out in the previous section, the δ unit first became known through its association with *Ty1*. Those not located on the flank of that element are referred to as solo δs, or if their homology to the first type is reduced, as divergent δs (Genbauffe *et al.*, 1984). Much has already been implied as to this unit's role in the transcription and translocation of *Ty1*. But the relationship of these two types of transposons appears to be far beyond the happenstance of their physical locations, as examination of representative terminal sequences in Table 9.7 demonstrates. The left terminal sequence of δ2 (Chisholm *et al.*, 1984), as a case in point, differs at only one site from that of the *Ty1* ε portion listed directly above it, although the right-hand region is more widely disparate, as well as slightly shorter. The two share an additional feature—the nearly complete absence of complementarity in these end pieces. Thus, a statement made in the *Ty1* discussion concerning the lack of need for enzyme-recognition structures in ε is inappropriate, for to judge from δ's location at the flanks, ε must be suspected of supplying the cognizance points, yet they, too, lack the complementary regions common to most bacterial transposons.

Further examination of the table reveals other interesting points regarding this element, the most outstanding being the broad extent of variation that occurs in the five δ sequences. Although all show extensive homology, they also are unexpectedly divergent at many points—even the very terminal bases are wanting in *GC106R* and that called *D15* (Elder *et al.*, 1983; Genbauffe *et al.*, 1984). Another sequence that has been referred to as *D15* differs completely from that given in the table (Gafner and Philippsen, 1980), but which of the two is correctly assigned this designation could not be ascertained. The bottommost structure given demonstrates that this variability is not confined to the present unit, for the incomplete *Ty1-B10* ε differs from that of *Ty1* ε at 15 of its 23 sites on the left terminus and lacks important features expected in close kin, as shown later.

Both in organization and translation, the recently discovered *Ty912* is an atypical

Table 9.7
Termini of Various Eukaryotic Transposable Elements

	Left terminal region				Right terminal region			
Yeast τ86[a]	TGTTGGAAC	GAG-AGTAA	TTAATAGTG	ACATGAGTT	TCATTAATA	CTAATTTTT	AACCTCTAA	TTATCAACA
Yeast τ106[a]	TGTTGGAAC	GAG-AGTAA	TTAATAGTG	ACATGAGTT	TCATTAATA	CTATTTTTT	TACCTCTAA	TTATCAACA
Yeast σ[b]	ATATGTGTT	TTATGAACG	TTTAGGATG	ACGTATTGT	------TAC	TGACATATC	TCATTTTGA	GATACAACA
Yeast σFD2[c]	TACGGGCTC	GAG-TAATA	CCGGAGTGT	CTTGACAAT	ATGACGTAT	TGTCATACT	GACATATCT	CATTTTGAG
Yeast σ106[a]	TGTTGCATC	TCA-AAATG	GAATACGTC	AGTATGACA	----ATACC	CACCTAAAA	TATTCATAA	AACACATAT
Yeast Ty1 ε[d]	TGTTGGAAT	AGA-AATCA	ACTATCATC	TACTAACTA	ATCCCAACA	ATTATCTCA	ACACTCACC	CATTTCTCA
Yeast δ2[b]	TGTTGGAAT	AAA-AATCA	ACTATCATC	TACTAACTA	ATCCCAACA	ATTACATCA	AAATCCA-C	ATTCTCTCA
Yeast δD15[e]	---CGGAAT	GAGGAATAA	TCGTAATAT	TAGTATCTA	TAGCCTTTA	TCAACAATG	GAATCC---	CAACAATTA
Yeast δAdh2.2[f]	TCAGAAATG	GGTCAATGT	TGAGATAAT	TGTTGGGAT	CTAGTTAGT	AGATCATAG	TTGATTTTT	ATTCCAACA
Yeast δGC106L[a]	TCAGATCTT	GGTCAATTT	TAAAATAAT	TGTTGGGAT	TAGAAGTCT	CCTCCAGAT	ATGGAATCC	ATAAAAGGA
Yeast δGC106R[a]	-GATAAATC	TAC-ATATT	GTTGTTATT	CCTTTTTCT	CTAGTTGAT	AGATGATGG	TTGATTCCT	ATTCCAAGA

Yeast *Ty1-B10* ε[g]	TGAGATATA	TGTGGGTA-	ATT (Incomplete)		(Incomplete) AG	TGGATTTTT	ATTCCAACA
Maize *Ac9*[h]	TAGGGATGA	AAACGGTCG	GTAACGGTC	GGTAAAATA	TCGGTACGG	GATTTTCCC	ATCCTACTT	TCATCCCTG
Maize *Ds1*[i]	TAGGGATGA	AAACGGTCG	GAAATCGGT	ATTTTTTCG	TTATCCTAA	CAGTTCGGA	TTACCACTT	TCATCCCTA
Maize *Ds1-6233*[j]	TAGGGATGA	AA (Incomplete)			(Incomplete) TT	TCATCCCTA
Maize *Mu1*[k]	GAGATAATT	GCCATTATG	GACGAAGAG	GGAAGGGGA	TCCGCTTCT	CTCTTCGTC	CATAATGGC	AATTATCTC
Maize *En8*[l]	CACTACAAG	AAAACGTCA	AAGGAGTGT	CAGTTAATT	TTTTGGCCG	ACACTCCTT	GCCTTTTTT	CTTGTAGTG
Antirrhinum Tam1[m]	CACTACAAC	AAAAAACCT	CTATTGGGA	CACAGAAAT	AAGAAAGTG	TCTCCAGAG	CCCAATTTT	GTTGTAGTC

[a]Genbauffe *et al.* (1984).
[b]Chisholm *et al.* (1984).
[c]Del Rey *et al.* (1982).
[d]Winston *et al.* (1984).
[e]Elder *et al.* (1983).
[f]Williamson (1983).
[g]Eibel *et al.* (1980); Gafner and Philippsen (1980).
[h]Pohlman *et al.* (1984).
[i]Müller-Neumann *et al.* (1984); Sutton *et al.* (1984).
[j]Weck *et al.* (1984).
[k]Barker *et al.* (1984); Taylor and Walbot (1985).
[l]Gierl *et al.* (1985).
[m]Bonas *et al.* (1984).

member of the family. It is compound, consisting of two units fused in such a way that their genes partly overlap (Clare and Farabaugh, 1985). The left-hand component, *Tya912*, specifies a DNA-binding protein, and the right, *Tyb912*, a reverse transcriptase. Although the latter overlaps the former by 37 sites, the product resulting from translation is compound, containing full readings of both genes, an act which is not accomplished by means of splicing, but possibly through a frameshift mechanism.

The σ Element. One of the most distinctive transposon groups of yeast was discovered almost simultaneously in two different laboratories (del Rey *et al.*, 1982; Sandmeyer and Olson, 1982). Structural features are not what make this element, known as σ, unique; rather, it is its constant association with tRNA genes. In all seven cases originally reported, this 341-base-pair unit lay 16–18 sites upstream of the 5′ end of a tRNA gene. In more recent investigations, 25 additional insertions of σ were similarly located near the tRNA coding sectors (Brodeur *et al.*, 1983). In almost all cases, orientation of the transposon was the same as that of the tRNA. As in *Ty1* and δ, transposition is accompanied by the generation of five-base-pair flanking repeated elements.

Of the three sequences in Table 9.6, one (*106*) may be perceived to be closely allied to ε and δ, both structurally and in lacking complementarity between the two terminal so-called repeats. A second one (*FD2*) is also kin to those others, but more distantly, and the third bears little semblance to any other. The lack of dyad symmetry in the two termini seems to be the only evident shared feature.

The τ Transposon. The latest addition to the major transposon list from yeast is a unit called τ (Genbauffe *et al.*, 1984). Bearing repeated sectors at each terminus, the sequence reaches a total length of 371 base pairs and is represented by ~25 copies per haploid genome. As in the other yeast transposable elements, its transposition induces the synthesis of five-base-pair repeats on the flank. Thorough comparisons of multiple variants of each of the four kinds of transposons known from this organism may help to establish any concerted differences that exist among them, but on the basis of present knowledge of their sequence structure, all four, δ, ε, σ, and τ, must be considered to be members of the same family of repeated elements (Chisholm *et al.*, 1984; Genbauffe *et al.*, 1984). This conclusion is also fully supported by the inverted repeated termini cited in Table 9.7, for upon comparison all prove to be highly homologous to one another as well as being similarly lacking in dyad symmetry.

9.6. SEED-PLANT MOVABLE ELEMENTS

Although innumerable researches on movable elements in seed plants at the genetic level have been reported since their earliest discovery (McClintock, 1949), comparable activity on a molecular basis has been lacking, so that the literature on this aspect is disappointingly sparse. As a consequence, knowledge of the genic structures and behavior lags behind that of bacteria (Döring and Starlinger, 1984).

One distinctive property of the seed-plant elements is that they frequently are associated as pairs of interacting units. Thus the *Activator* (*Ac*)/*Dissociation* (*Ds*) family has long been known to exist in maize (McClintock, 1949), in which one member, in this case *Ac*, is autonomous, whereas the other, *Ds* in this example, is dependent upon the presence of an *Ac* unit for transposition. A second well-established group is also important in plant genetic behavior, the *Enhancer (En)/Inhibitor (I)* family, the first of which is also known as the *Suppressor/Mutator (Spm)* (Gierl *et al.*, 1985). It should not be supposed, however, that all movable elements of these organisms are thus paired, for a few are known that do not display such relationships, the recently described transposon *BSI* being one example (Johns *et al.*, 1985).

A second distinction is in their mode of transposing. In the cointegrate system of bacteria and seemingly also of yeast, the end result is the insertion of a copy of the element into a new site, while the original is restored to its former location in the genome. But here in plants, the element is actually moved from its wild-type position to a new site, there being no increase in number of copies as a consequence (Saedler and Nevers, 1985). Discussion here begins with the *Ac/Ds* family because its components have been more thoroughly documented at the molecular level.

9.6.1. The Ac/Ds Family of Maize Transposons

The Activator Component. As the autonomous member of the *Ac/Ds* family of maize, *Ac* exercises the functions necessary both for its own transposition and for that of its mate, the *Ds* elements. Additionally, this component is capable of inducing chromosomal breaks adjacent to certain of the *Ds* unit, but not at its own termini. Insertion of an *Ac* element in a gene results as usual in a recessive mutation, the dominant wild-type phenotype being restored when that insertion is excised. The same statement applies equally to its associated element. Reversions resulting from excisions are often readily noticeable in the kernel phenotype when they occur during endosperm development (Müller-Neumann *et al.*, 1984), for they may lead to changes in the variegation pattern. Such alterations often supply the informational basis for inferences regarding the frequency and timing of the transpositional events.

In the 4563-base-pair long sequence of *Ac*, two open reading frames have been reported, encoding polypeptides 839 and 210 amino acid residues in length (Pohlman *et al.*, 1984). Because both proteins appear to provide activity essential to transpositioning, it has been conjectured that *Ac* is a derivative of the bacterial *Tn3* (Müller-Neumann *et al.*, 1984), but comparisons of the terminal regions given in Tables 9.1 and 9.7 fail to expose any sequence homologies that could substantiate that claim. The two references just cited are disharmonious as to the orientation of the two genes, although the sequences of the two *Ds* units (*Ds7* and *Ds9*) are virtually identical, because one shows transcription to occur divergently from a common promoter region (Pohlman *et al.*, 1984), the other as being unidirectional from isolated promoters (Müller-Neumann *et al.*, 1984). Further, the latter displays a third reading frame, for an 151-amino acid polypeptide. Although both examples given were inserted into a *waxy (Wx)* gene, they show almost no resemblances, for they were located in different regions and thus neighboring DNA sequences. Analyses of revertants reveal that the entire element is excised, but six of the eight bases of the flanking repeat remain (Pohlman *et al.*, 1984).

The Ds Element. To judge from available reports, *Ds* is not a definite element. Rather, it is an entire series of elements that have comparable behavioral patterns, for its described length varies with the locus into which it has been inserted, with a range from 405 to 2000–3000 base pairs (Courage-Tebbe *et al.*, 1983; Sutton *et al.*, 1984: Weck *et al.*, 1984). Many of the investigations are based on mutations in the *Shrunken (sh)* locus, which encodes sucrose synthase, or in the *Adh1* gene, which specifies alcohol dehydrogenase (Dennis *et al.*, 1985). The general characteristics on which identifications have been based are the ability to induce chromosomal breaks in the vicinity, an A,T-rich inserted sequence, with 11-base-pair inverted repeats at the termini, generation of direct eight-base-pair repeats on the flanks, and dependence on the presence of *Ac* for transposition and excision. Both *Ds* and Ac induce inhibition of gene activity.

In the *Adh1* mutant, the 408-base-pair insert was found 45 sites from the initiation point of that gene, accompanied by an eight-base-pair repeat on its flanks (Sutton *et al.*, 1984); the terminal inverted and flanking repeats are given in Table 9.6 and 9.7 as *Ds1*. After excision, this thorough investigation of the genomic and revertant genes disclosed that the entire right flank repeat had been retained in the latter in modified form. The original had read *AGG*GACTG, but in one revertant (*pRV·5A*) the remnant flank had been modified to *TCG*GACTG and in a second (*pRV·10A*) to *TCC*GACTG (Sutton *et al.*, 1984). In each case the modification involved the complementary base, probably by transversion of the two strands of DNA. A second study involved a double *Ds1* unit inserted into the *sh* gene (Weck *et al.*, 1984). Sequence analysis showed the transposon to be 4000 base pairs in length and to consist of two identical elements of 2000 nucleotide pairs each, one of which was inserted into the center of the other in inverted orientation. In the tables, the terminal inverted repeats and the direct flanking repeats are designated as *Ds1-6233*. Upon excision, the remnant of each flanking repeat was reduced to seven base pairs, each lacking a C, the left one on the downstream side, the right on the upstream.

9.6.2. Other Transposable Elements of Seed Plants

Results of investigations into other transposable elements of seed plants are not as clear-cut as those just cited. Researches on the second system of autonomous/dependent elements have not been as extensive at the molecular level, a deficiency compounded by the confused terminology that has resulted from its having been described under two sets of names. But what has been established about this dual system is analyzed here, along with one or two others of different organization.

The En/I Transposon System. The *Enhancer (En)/Inhibitor (I)* system was originally described and named by Peterson (1953), but the following year unknowingly was redescribed at a different locus by McClintock (1954) under the designation *Suppressor/Mutator (Spm)*. To add to the confusion, *I* is frequently referred to as ''receptor'' (Peterson, 1965). Although in current literature the synonym *Spm* still is employed along with *En*, here all are designated under the latter term to gain the necessary clarity.

En8 (Spm-I8) and *En1* (Schwarz-Sommer *et al.*, 1984; Pereira *et al.*, 1985) are considered equivalents, since their termini (but not their inner regions) have precisely the same sequence structure. They are large elements, 8400 base pairs long, that, like their *Ac/Ds* counterparts, have been found to induce mutation of the *Wx* gene. As in the latter system, their inverted terminal repeats are largely complementary. How all the plant transposon termini contrast with those of the yeast in these sectors is made obvious in Table 9.6. Insertion is accompanied by the generation of direct repeats at the flanks, each three base pairs in length (Table 9.6). *I*, the receptor component of the system, is derived from *En* by deletion of certain parts (Pereira *et al.*, 1985). Thus, it would seem that the *En* units should be considered duplex elements, consisting of a deletable and a more permanent portion. These transposons, along with all others, have been viewed as generating the DNA sequence diversity needed for evolutionary progress (Schwarz-Sommer *et al.*, 1985); however, studies demonstrating the effects of insertion and deletion of transposons in populations of organisms under a diversity of environments have yet to be conducted. In view of the frequency with which reversion occurs, it is difficult to perceive how long-term effects could be generated, especially in seed plants.

Various Seed-Plant Elements. Among the other transposable elements of importance that have been reported from maize is the *Mutator* (*Mu*) family, whose most thoroughly documented member is *Mu1* (Robertson, 1978). Originally located as a 1400-base-pair sequence inserted in an *Adh1* gene, this transposon has subsequently been found to occur as 10–60 copies per diploid genome in some, but not all, maize lines (Barker *et al.*, 1984). As in all the preceding types, its presence in a gene leads to reduced levels of production of the encoded substance. In the nucleotide sequences reported by this latter paper, four reading frames for undetermined proteins were found, two of which were oriented in the opposite direction from the others. Its reversedly repeated termini show a far higher level of complementarity than any of the preceding types, having 27 base pairs matched (Table 9.7). Its insertion is accompanied by generation of direct nine-base-pair repeats on the flanks (Table 9.6).

Although additional elements have been reported from maize and other plants, including one stated to be similar to the *copia* of *Drosophila* (Johns *et al.*, 1985), researches upon them at the molecular level are too incomplete to permit their discussion. However, one further type, in this case from the snapdragon (*Antirrhinum majus*), is of particular interest (Bonas *et al.*, 1984). In a mutation at the so-called *nivea* gene, the resulting phenotype is marked by numerous streaks of red on a white background, induced by a deficiency in an enzyme (chalcone synthase) important in the production of anthocyanin, a red pigment. This mutation has been found to result from the presence of a transposon 17,000 base pairs long. Among the especially striking features of its structure is the pronounced homology exhibited by its inverted terminal repeats to those of the maize *En8* (Table 9.7), a relationship that has received notice by other investigators (Schwarz-Sommer *et al.*, 1985). This kinship is further expressed by the degree of complementarity exhibited by these sectors, as well as by the production of trinucleotidyl direct repeats on the flanks (Table 9.6).

9.7. MOVABLE ELEMENTS OF INVERTEBRATE METAZOANS

Among invertebrate metazoans, extensive researches upon movable elements have been devoted to only two organisms, both insects, *Chironomus* and *Drosophila*. The results, however, must not be expected to overlap and confirm each other, for whereas those of *Drosophila* are comparable to those of previous sections of the text, the ones from the other dipteran have focused on Balbioni ring-associated structures. Nevertheless, these two sources, together with a sea urchin that has been investigated to a lesser degree, supply an abundance of information on this important feature of gene structure and behavior.

Table 9.8

Repeated-Element Termini from the Drosophila Genome

	Left terminus				Right terminus			
Drosophila copia[a]	TGTTGGAAT→	ATACTATTC	AACCTACAA	AAATAACGT			(Incomplete) AA	ATTACAACA
Drosophila 412[b]	TGTAGTATG	TGCCTATGC	AATATTAAG	AACAATTAA	ACGGACTTG	TGTTCTGAA	TTGGAGTTC	ATCATTACA
Drosophila B104[c]	TGTTCACAC	ATGAACACG	AATATATTT	AAAGACTTA	CTCAACGAG	TAAAGTCTT	CTTATTTGG	GATTTTACA
Drosophila 297[d]	GTGACGTAT	TTGGGTGGA	CCAAACCAG	CCACTTCCA	CATTAGTTC	AGACTCATA	AATAAAACA	ACAATTTTA
Drosophila mdg4[e]	AGTTAACAA	CTAACAATG	TATTGCTTC	GTAGCAACT	TAACATAAC	TCTGGACCT	ATTGGAACT	TATATAATT
Drosophila 17.6A[f]	AGTGACATA	TTCACATAC	AAAACCACA	TAACATAGA	CTCAAAACT	ATTTATTGC	AACCATTTA	TTTGCAATT
Drosophila opa[g]	AAGCTTGGG	AATCATCTC	GCCGACCGG	CAGCGGATTA	GATACAAAG	TTCAATGTC	CGGCTCATC	GCCGTCGAC
Drosophila F101[h]	ATCAAGGAT	TTCGATCGC	CGACGTGTG	AAGACGTTT	(Incomplete)			
Land crab EXT[i]	GACTCTGCC	TCACACCGC	CGACTGCTA	CCGAAGCCG	CTAACTCCA	AGCAAAGGG	GATAGCCTC	CATCGCACTC
Drosophila Pm25[j]	CATGATGAA	ATAACATAA	GGTGGTCCC	GTCGAAAGC	CTTGCCGAC	GGGACCACC	TTATGTTAT	TTCATCATG
Drosophila hobo[k]	CAGAGAACT	GCAGCCCCC	CACTCGCAC	TCTACGTCC	CGAGTGGTA	AAAAGTGC	CACCCTTGC	AGTTCTCTG
Drosophila FB4[l]	AGCTCAAAG	AAGCTGGGG	TCGGAAAAA	TCGAATTTT	AAAATTCGA	TTTTTCCGA	CCCAGCTT	CTTTGAGCT
Drosophila Ifm[m]	AATTAATTA	AATGTATGG	TGCAGGTCC	CTCGCCGCG	GGTGTTGCGG	ACGATCAGT	CCGTTAACT	TAGTTAACT

[a]Flavell et al. (1981); Mossie et al. (1985). [b]Shepherd and Finnegan (1984). [c]Scherer et al. (1982). [d]Ikenago and Saigo (1982). [e]Bayev et al. (1984). [f]Saigo et al. (1984). [g]Wharton et al. (1985). [h]Di Nocera et al. (1983). [i]Bonnewell et al. (1983). [j]O'Hare and Rubin (1983); Steller and Pirrotta (1985). [k]McGinnis et al. (1983). [l]Potter (1982). [m]Karlick and Fyrberg (1985).

9.7.1. Copia Transposons from Drosophila

Indeed, the many types of transposons that have been explored at the molecular level in *Drosophila* almost suffice by themselves to give a true picture of the element in general. Many belong to the *copia*-like class, mentioned in connection with the *Antirrhinum* representative in the preceding pages. Perhaps that group is the most important of all the types, but the *P* elements, *hobo*, and the *F* family all contribute to a fuller understanding of the nature of the genome.

The Copia-like Transposable Units. The *copia* family of transposable elements has been likened to the bacterial *Tn9* and yeast *Ty1* classes of transposons (Cameron *et al.*, 1979; Kleckner, 1981). As a group they range in length from 5000 to 9000 base pairs and terminate in direct repeats ranging between 200 and 500 base pairs (Mossie *et al.*, 1985), a trait quite in contrast to any in earlier portions of this chapter. These show complementarity at the ends to a limited degree and in interrupted fashion (Table 9.8). About 30 copies are scattered over the several chromosomes of *Drosophila*, which display an unusually high level of evolutionary conservation compared to other families of movable units in this insect (Rubin *et al.*, 1980). Upon integration, *copia* induces the generation of five-base-pair direct repeats at its flanks, two sets of examples of which are given in Table 9.8. In cultured cells and embryos, many copies of this structure are present in the form of circlets of DNA, unattached to the remainder of the genome (Mossie *et al.*, 1985).

Other Members of the Copia Family. Relationships among the transposable elements of *Drosophila* are not usually clear-cut, for many evolutionary changes have altered their sequence structures in these members of an ancient lineage. Fortunately, sufficient conservation has been active to permit the assignment of three additional elements to the *copia* family, the transposons known as *B104, 412,* and *297*. Of these the most closely allied is the first, the terminal direct repeats of which show distinct homologies to that of the typical form (Table 9.8; Scherer *et al.*, 1982). Moreover, the ends of these segments show only the limited amount of complementarity that characterizes *copia* itself, and, as in the latter, during insertion a five-base-pair direct repeat of host DNA is generated at one flank (Table 9.9). The 8700-base-pair *B104* is represented by 80–85 copies per haploid genome.

The transposable element *412* is also closely related to *copia* in the structure of its long terminal direct repeats (Table 9.8), which are 481 base pairs in length (Shepherd and Finnegan, 1984). During integration a short repeat of only four base pairs is generated (Table 9.9). Like *copia*, the present species has been found to occur in cell cultures in part as free circles of DNA. Among the more distantly related members of the family is that known as *297*, described from a mutant histone gene in which this element was inserted into the promoter (Ikenaga and Saigo, 1982). As in others of the family, the transposon is bounded by long terminal direct repeats, which show a limited amount of homology with the corresponding sector of *412*, but have no region of dyad symmetry at the ends (Table 9.8). In three examples, integration resulted in the same six-base direct repeat of host DNA, TATATA, suggestive that this element may have site-specific targets.

9.7.2. Miscellaneous Transposons of Drosophila

Drosophila provides a foretaste of the diversity of movable elements that may be expected among other Metazoa when their genomic parts become more thoroughly ana-

Table 9.9
Direct Repeat Flanks of Insect Movable Elements

	Left flank	Right flank
Chironomus Cla17[a]	TTATCAAA	CCCTCAAA
Chironomus Cla11[a]	TTAATTGAT	ACCACGTTT
Drosophila copia[b]	GTATT	GTATT
Drosophila copia[b]	GCCAG	GCCAG
Drosophila F101[c]	CAGTTGCCGACCA	CAGTTGCCGACCA
Drosophila F14[c]	AATAATGCGG	AATAATGCGG
Drosophila Fλ12[c]	AACTGTGCAA	AACTGTGCAA
Drosophila F19[c]	ATGTTTGAGATC	ATGTTTGAGATC
Drosophila B104[d]	GCACC	GCACC
Drosophila 412[e]	CTTG	CTTG
Drosophila 297[f]	TATATA	TATATA
Drosophila hobo[g]	GTGGGTAT	GTGGGTAT
Drosophila mdg4[h]	TATA	TATA
Drosophila mdg4[h]	TACA	TACA
Drosophila Pπ25[i]	ATACACAC	ATACACAC

[a]Schmidt (1984).
[b]Rubin et al. (1980).
[c]Di Nocera et al. (1983).
[d]Scherer et al. (1982).
[e]Shepherd and Finnegan (1984).
[f]Ikenaga and Saigo (1982).
[g]McGinnis et al. (1983).
[h]Bayev et al. (1984).
[i]O'Hare and Rubin (1983).

lyzed, for aside from the several types in the *copia* family, quite a variety of such units have already been discovered in this insect. Outstanding among these are the *F* family, *hobo*, *opa*, *Pπ25*, and the *mdg* group.

The F Family. Probably the most striking class of transposable units so far revealed in *Drosophila* is that referred to as the *F* family, the first known member of which is *F101*, whose nucleotidyl structure is unique (Di Nocera *et al.*, 1983). Subsequently, four others have been characterized, with a total representation of ~50 copies in the genome of this dipteran. Among their most outstanding traits is, first, their lack of internal and terminal repeated segments. But even more unusual is the presence at the right ends of a typical polyadenylation signal (AATAAA), followed by a run of 12–30 As. Nevertheless, when these elements are inserted, an 8- to 13-base-pair repeat of target DNA is generated (Table 9.9). The presence of this poly(A) train strongly suggests that their transpositioning is by a mechanism different from that of prokaryotes, possibly involving ordinary transcription into mRNA, followed by reverse transcription, and finally reinsertion into the genome, as appears to be the processes in yeast. Despite these highly divergent structural points, the left terminus of *F101* shows low level homology with that of *B104*, particularly beginning in the sequence at site 28 (Table 9.8). Strangely, a transposon, called *EXT*, from the Bermuda land crab (Bonnewell *et al.*, 1983), shows extensive homology in its left terminal repeat to that of *F101* (Table 9.8).

The P Element Family. Structurally the *P* family of transposons shows nothing distinctive, but genetically the members are highly unusual, being responsible for the hybrid dysgenesis mentioned in a different context in Chapter 1, Section 1.4. When males of a *P*-containing strain are crossed with M females, which lack these factors, the resulting hybrids show a series of genetic aberrations, including chromosomal rearrangements, mutations, often lethal, and a high level of gonadal sterility (Bregliano and Kidwell, 1983; Karess and Rubin, 1984). The 2900-base-pair sequence has been fully determined (O'Hare and Rubin, 1983; Steller and Pirrotta, 1985). Aside from inverted terminal sequences like those of bacterial elements (Table 9.8), the nucleotidyl structure has been found to encode a transposase needed for its own transposition and at least one other product (Karess and Rubin, 1984). Eight-base-pair direct repeats flank the termini after insertion (Table 9.9). In recent experimental studies on expression of a "glue" gene of advanced larvae, transformations induced by insertions of *P* elements proved especially valuable (Bourouis and Richards, 1985), as they had in investigations of the *white* locus (Hazelrigg *et al.*, 1984).

A Miscellany of Types. Among the lesser members of the transposon group in *Drosophila* are included the several species of the "mobile dispersed genetic" (*mdg*) elements, whose principal distinctive feature is their being scattered throughout the genome. One unusual site, where chiefly defective copies occur, is at the chromocentric region of the chromosomes (Arkhipova *et al.*, 1984; Bayev *et al.*, 1984). As in *copia* and other *Drosophila* units, the elements have long direct repeats at each end; four-base-pair direct repeats are produced on the flanks during insertion of *mdg4* (Table 9.9). The latter is approximately 7500 nucleotide pairs in length and contains a region partly complementary to a tRNA^{Lys} gene of this fly (Bayev *et al.*, 1984).

In *hobo*, the terminal repeated segments have a 12-base-pair region of dyad symmetry (Table 9.8; McGinnis *et al.*, 1983), but the height of complementarity is attained in the transposons called *foldback* (*FB*) elements (Table 9.8; Potter, 1982), where the entire terminal sequences are involved. The former type generates eight-base-pair flanking direct repeats upon insertion, whereas in the latter, the total element consists of a multiplicity of duplicated 31-site sequences (Brierley and Potter, 1985). One representative species, *FB4*, is 4089 base pairs in length; it, like others of the group, is especially active in inducing chromosomal rearrangements (Collins and Rubin, 1984). Finally, *opa* is one of a family of transcribed repeats found especially in developmentally regulated loci (Wharton *et al.*, 1985), *3S18* is associated with ribosomal RNA genes (Bell *et al.*, 1985), and *17.6A* is important as a sequence known to encode a reverse-transcriptase-like enzyme (Table 9.8: Saigo *et al.*, 1984).

9.7.3. Repetitive Sequences of Chironomus

In the midge flies of the genus *Chironomus*, an abundance of investigations have focused on the repeat units in the telomeres of the chromosomes, but it is not completely established whether any of these serve as mobile elements. Undoubtedly most are mere structural features of various genes, but until their real functional nature has been revealed, it is well to examine a broad cross section through those whose molecular structures are known. The source mentioned, telomeres, located at the tips of each arm of a chromosome, is specially constructed, permitting both the complete replication of the DNA molecules and interaction with the nuclear envelope. Although the structures are often involved in the end-to-end association of chromosomes, they possess properties that prevent permanent fusion. In addition, similar elements are derived from three Balbioni rings, known as *BR1, BR2,* and *BR6*. These occur on large puffs and bear genes for salivary proteins of unusually great proportions and yet of simple construction, in that they involve multiple repeats (Case and Byers, 1983).

Balbioni-Ring Elements. The Balbioni-ring genes just mentioned are obviously not transposons, for they encode proteins employed in constructing the larval tube. Yet the repetitive sequences of which these cistrons are comprised are reminiscent of satellite DNA, including plasmids and circular DNAs (Lendahl and Wieslander, 1984). All three are comprised of basic building blocks ranging in lengths of 150–300 nucleotide pairs. In turn, these each consist of two types of sections, one (the C region) constant in structure, the other divided into subrepeats (SR region). It is the latter, then, and its subunits that merit attention here, not only in this connection, but in throwing additional light on the nature of gene organization in general, for many other genetic elements of diverse types show remnants of building-block construction.

The subrepeats have been claimed to be unique in each of the three major species, but when they are carefully aligned in series as in Table 9.10, extensive homology is revealed. There three or four repeated segments, distinguished by numerals enclosed in parentheses, are shown from each type. When examined, it is evident that *BR2* is the most distinct from the others in having its subunits only six base pairs long, not 11 or 12 as in the others (Sümegi *et al.,* 1982). A further feature unique to that cistron is that the repetitious segments commence with AGC, which, to judge from the aligned homologous sites, have been lost from the remaining representatives. It appears warranted to say lost from them, rather than gained by *BR2,* for the subunits themselves may be seen to be double repeats of a three-base-pair block AGC, AAR, CC- (Table 9.10). On this basis, *BR2* should be viewed as the most primitive member of the family.

Although it is tempting to suggest that the remaining six codons of *BR1* are derivatives of a duplication of *BR2*'s block (Case and Byers, 1983), supportive evidence is lacking. If the TCT at the midpoint of that sequence is omitted from consideration, and the third to fifth sites of *BR2* aligned with the seventh to ninth of *BR1*, a reasonable fit is obtained, but then how are the last two codons of the sequence, CCA, GAG, explained? Partial double repeats of the two opening triplets of AA-, CC-, followed by insertion of TCT and gain of the final GAG, may be a more satisfactory possibility. Since the other three sets given appear to be derivatives of *BR1*, the latter then would be next to the most primitive of the series. The relationships of *BR2.2* and *BR240* to *BR2* proper (Case *et al.,* 1983; Höög and Wieslander, 1984) have never been fully elucidated, although all three mentioned are derived from Balbioni ring 2. The first two are quite clearly homologous to the *BR1* subrepeats and also to one another, because the two sets are nearly identical in lacking GGA in site 5 and in having that same triplet as an unusual occupant of the postmedian position. Finally, *BR6* seemingly is the most advanced member of the group,

Table 9.10
Repeating Units of Balbioni Ring Genes of Chironomus

$BR1(2)^a$	--- GAA CCT AGC AAG GGA TCT AAA CCT AGA CCA GAG
$BR1(3)$	--- AAA CCA AGT AAG GGA TCT AAA CCT AGA CCA GAG
$BR1(4)$	--- AAA CCA AGT AAG GGA TCT AAA CCT AAA CCA GAG
$BR2(1)^b$	AGC AAA CAC AGC AAA CCA
$BR2(2)$	AGC AAG CAC AGC AAG CAC
$BR2(3)$	AGC AAA CCT AGC AAG CAT
$BR2(4)$	AGC AAA CCT AGT AAA CAC
$BR2.2(1)^c$	--- AGA CCA AGC TGG --- TCA GGA ATT AGA CCA GAA
$BR2.2(2)$	--- AGA CGA AGC AGA --- TCA GGA CCA AGA CCA GAA
$BR2.2(3)$	--- GGA CCA AGC AGA --- TCA GGA TCT AGA CCA GAG
$BR240(1)^d$	--- AGA CCA AGT TGG --- TCA GGA ATT AAA CCA GAA
$BR240(2)$	--- AAA CGT AGC AAA --- TCA GGA TCA AGA CCA GAG
$BR240(3)$	--- AAA CGT AGC AAA --- TCA GGA TCT AGA CCA GAG
$BR6(2)^e$	--- AGA CCA GAA GAA CCA --- GAA --- AGA CCC GAA
$BR6(3)$	--- AGA CCA GAA AGA CCT --- GAA --- AGA CCT GAA
$BR6(4)$	--- GAA CCA GAA CGC GAA --- GAG --- --- CCC GAA

[a]Case and Byers (1983).
[b]Sümegi et al. (1982).
[c]Case et al. (1983).
[d]Höög and Wieslander (1984).
[e]Lendahl and Wieslander (1984).

because its sequence is the most strongly modified of all from the basic six subunits of *BR2* (Lendahl and Wieslander, 1984). Although a number of diverse genes have been suggested to have been developed through successive duplications and subsequent evolution, this series of closely allied cistrons provides clearer evidence of such events, because several steps possibly involved in their phylogeny are represented that are lacking in other instances.

Telomere-Associated Repeated Sectors. Brief mention needs to be made of elements that resemble the foregoing in being constructed of repeated and "constant" sequences, but are associated with telomeres, rather than chromosomal puffs or rings. One such unit, known as *Cp306* and recently sequenced (Saiga and Edström, 1985), has its 304 base pairs built of two alternating repeats, *a* and *b*, separated by spacers of variable lengths. Whereas the smaller *a* unit contains only 29 base pairs, the *b* structures are much longer, embracing 60 (Table 9.11). Upon comparison each type is found to be fairly

Table 9.11
Telomere-Associated Repeats of Chironomus[a]

Cp306 a	--AAACTGAGC	GAGCTAGAG	CAAAAAAAC	CA		
Cp306 a'	--AAACTGAGC	GAGCTAGAG	CAAAATTAC	CA		
Cp306 b	AAAAATCGTCC	TAGCTCCTC	CATTTGTCA	ACCGTTTGAGGA	GGTTTATAT	ATCGATGGAT
Cp306 b'	AAAAATCGTCA	TATCTCCTC	AATTTCTCA	ACCGTTTGAGAA	GGTTTATAT	ATCGATGGAT

[a]Saiga and Edström (1985).

constant, the second, smaller repeat, a', differing from the first at two sites, and the second copy of the longer one, b', varying at five positions relative to the b. The $a-b$ spacers are highly divergent from one to the next, whereas the $b-a$ show considerable homology among themselves.

A Movable Element of Chironomus. One movable element of *Chironomus* that has been adequately studied at the molecular level is *Cla,* named from the presence of a site recognized by the restriction enzyme ClaI (Schmidt, 1984). The members of this family are complexly organized, being arranged in clusters of 4–30 complete and partial sequences, in some subspecies being confined to the heterochromatin of centromeres, in others perhaps present at as many as 200 additional points. In some instances *Cla* elements have been found in nontranscribed spacers. As a class they are unusually small, ranging in length from 110 to 119 base pairs that are comprised of A and T residues to an extent of 80%.

Among the peculiarities of these unique transposons is the absence of repetitive termini (Table 9.12). Nor do the bases of opposite ends display any great complementarity, for only about a five-site run forms a region of dyad symmetry. Still more striking are the sequences generated at the flanks. In *Cla17* the "repeats" are preceded by three nucleotides that do not match, before a five-base-pair duplicate is found (Table 9.9), and in *Cla11* the two flanking series do not correspond at all. No reading frame for a protein appears to be present, so the enzymes involved in transpositioning this element must be located elsewhere in the genome.

9.7.4. Movable Elements of Sea Urchins

In sea urchins there is no dearth of repetitive elements, for ~500,000 in around 3000 families have been reported (Carpenter *et al.*, 1982). Although most of these are transcribed into RNA, the majority do not contain open reading frames, the transcripts largely remaining in the nucleus. However, some mRNAs from embryos have now been shown to contain short repeated elements that are absent from the unfertilized egg. One of these, recently completely sequenced, is part of an mRNA that presumably encodes embryonic ectodermal proteins (Carpenter *et al.*, 1982). This is 470 base pairs long and resembles the *Chironomus* transposon in lacking repeated terminal segments. Only five sites of these end pieces show complementarity and that is interrupted on the left side. Perhaps the flanks bear direct repeats, but none were shown to be present in the report. One peculiarity of the left terminal portion apparently has escaped previous notice, namely its being

Table 9.12
Termini of Invertebrate Movable Elements

Left end

Sea urchin *Spec1*[a]   T̲A̲A̲T̲A̲A̲ A̲A̲T ACT AGT ACT TCT ACT ACT ACT ACT ATT

Chironomus Cla[b]   G̲A̲T̲A̲T̲T̲ TTT TTT TTG AAA ATA GCC TAC TAT GAT CAA

Right end

Sea urchin *Spec1*   TTGTTTTAACAA TTGCATCTATAC ATGTGC̲T̲T̲T̲A̲T̲A̲

Chironomus Cla   ATAAATGATAAA GAATGCTAATAT AGATAT̲A̲A̲A̲A̲TC

[a]Carpenter *et al.* (1982).
[b]Schmidt (1984).

constructed largely of three-base-pair repeats. These short sectors consist of ACT, often variously modified to TCT, AGT, or ATT, as brought out in Table 9.12.

9.8. VERTEBRATE TRANSPOSABLE ELEMENTS

In the vertebrates, as might be anticipated in such highly complex organisms, a wide variety of repetitive sequences occur in the DNA, including those of such gene families as the rRNAs, tRNAs, and histones already examined in previous chapters. Then, as in the midge fly examples. the heterochromatin around the centromeres and telomeres contains 10,000 to several 100,000 copies of repeated sequences with no known function. In addition, there are multiples of others that serve as transposons. As has often been the case with previous topics, representatives derived from mammalian sources, chiefly the murine and human genomes, make up the bulk of current literature, and thus receive most of the attention here. Fortunately, however, several from *Xenopus* have been analyzed at the molecular level and permit broadening the comparative basis.

9.8.1. Movable Elements from Amphibians

Although there can be little question that the genomes of Amphibia contain numbers of movable elements comparable to those of mammals, relatively few have been established—certainly too small a number to indicate the occurrence of families of related types. The only one sequenced to indicate clearly the existence of multiple closely related species is that known as the *Rem* transposons.

The Rem Family. The three members of the *Rem* family were named originally from their containing sites reactive to the restriction enzyme *Eco*R1 (Repetitive *Eco*R1 Monomers; Hummel *et al.*, 1984). Collectively, they appear to be fairly abundant, since they constitute about 0.5% of the total DNA of *Xenopus laevis*. *Rem1* is clearly an interspersed element, whose ~25,000 copies account for 0.4% of the genome; it consists of ~490 base pairs, as does *Rem2*, which in part is similarly dispersed, but also occurs in

Table 9.13

Repetitive Sequence Termini from Vertebrate Sources

	Left terminus				Right terminus			
KpnI								
Human *F2*[a]	GAGGGCACG	CCCAAGATG	GCCGAATAG	GAACAGGTC	TGTACCCTA	GAATTTAGA	GTATAATAA	TAA-AAAAA
Human *B41*[a]	GGGCGTGGA	GCCAAGATG	ACCGAATAG	GAACAGGTC	TGTACCCTA	AAACTTGAA	GTATAATAA	TAAAAAAAA
Human *A*[b]	GGGGGAGGA	GCCAAGTTT	GCCAAATAG	GAACAGGTC	ATTGATTAA	ACACTTAAA	TGTTAAACC	TAAAACCAT
KpnN								
Human *LnA-1*[c]	TAAACTGGA	TCCCTTCCT	TATATCTTA	TACAAAAAT	TATTATTTA	CAGTGTACT	TCTTCTACT	TATTAAAAT
Miscellaneous								
Human *hinf*[d]	CCATATCGG	GCATGAATA	TCAGGAACA	CCGGCAGGT	TCGAAATCT	CACAAAGTG	TCCATAAAT	CACTCAGGG
Human *0-4*[e]	TCTATTAGT	CTGTTTTCA	CACTGCTGA	TAAAAACAT	TACAAAAAA	GGAAAAAAA	AAAAAAAAA	GAACAGAGG
Rabbit C_{BL3}^{f}	AAGCCCGCC	CCGTGGCTC	AATAGGCTA	ATCCTCCAC	CCATTGGAG	GGTGAACCG	CGGCAAAGG	AAGACCTTT
Rat *ID2249*[g]	CTGCAGGGG	GGGGGGGGG	GATTCTTAG	TATATGACT	AAAAAGAAC	CAAAAAAAA	ACCCCCCCC	CCCCTGCAG
Murine *Vr*[h]	GTCGACCAG	AAGGCTTAA	GTCCTACCC	CCCCCCCCC	GTCGGTCTT	ATCAGTTCT	CCGGGTTGT	CAGGTCGAC
Murine *B1*[i,j]	CCGGGCGTG	GTGGCGCAC	GCCTTTAAT	CCCAGCACT	AGGCAGCGC	AGGGCTACA	CAGAGAAAC	CCTGTCT--
Murine $B1_{27}$[i]	CTGGGTGTG	GTCGCACAT	GCCTTTAAT	CACAGCACT	AGGACAGCC	AGGGCTTCA	CAGAGAAAC	CCTGTCTCG
Murine *R1*[k]	GGATCCATC	TCATAATCA	GCCACTAAA	CCCAAACAC	GTGAGTGGG	TTTGGTGTC	TGTATGTGG	GATGGATCC
Murine *R103*[l]	GGATCCATC	CCATATACA	ACCACCAAA	CCCAGATAC	(Unclear)			
Murine *L1*[m]	TGAATCGAT	CCATACTTA	TCTCCTTGT	ACTAAGGTC	TCACACAAT	ATGTACTCA	CTGATAAGT	GGATACTAG
Murine *DI-2*[n]	GCACACAAG	CATTTATTC	ATTTTCTTC	GCTCGCTAG	AGATGGCAG	CCACCTTGT	AGGCAGTGC	CCAGGAAGA
Murine *Bam5*[l]	TGGTACTAC	CGGAAGATC	CAGCAATTC	CTCTCCTGG	TGGAAGGAA	GGACCAATTC	AGAGACTGC	TCCACTTGG

Murine $LTR^{i,o}$	TGAGAGAAA	GATGAAATT	TTAGGACCT	ACAATTCAT	AATCGGGAA	TGGGGGTTT	CCCCACTAG	GCTCTTTCA
Bovine LTR^{p}	TGTATGAAA	GATCATGCC	GACCTAGGA	GCCGCCACC		(Incomplete) GAC		CGGCAAACA
Chicken $CR1^{q}$	GGCCACGAG	AGGGGATGA	GGGGGCTGG	AGTACCTTC	ACCAGTCAC	TGTGTTCAG	TGTGTGCAC	AGATGGACA
Xenopus $Rem1^{r}$	TGGACTGAA	ATCCATTTC	TCAAAAGAG	CAAACAGAT	ATCTGTTTG	CTCTTTTGA	GAAATGGAT	TTCAGTGCA
Xenopus $Rem2^{r}$	TAAATGAAT	CAGATGAAA	ATTGAGCAT	AGGACTGGC	ATTCGTCCC	TTTTATAAA	ATGTATAAT	TAAGCCATA
Xenopus $SB18^{s}$	CTTCAACTT	TCCAGCATC	GTATTGTCC	ATGTATCAT	ACGGTTTTT	ATCAAATTT	CTGAAAATC	GTCACAAAG

[a] Grimaldi et al. (1984).
[b] Porter (1984).
[c] Nomiyama et al. (1984).
[d] Shimizu et al. (1983).
[e] Sun et al. (1984).
[f] Hardison and Printz (1985).
[g] Sucliffe et al. (1982).
[h] Kuehn and Arnheim (1983).
[i] Propst and Vande Woude (1984).
[j] Krayev et al. (1980).
[k] Gebhard et al. (1982).
[l] Wilson and Storb (1983).
[m] Martin et al. (1984).
[n] Feagin et al. (1983).
[o] Wirth et al. (1983).
[p] Sagata et al. (1984).
[q] Stumph et al. (1984).
[r] Hummel et al. (1984).
[s] Lam and Carroll (1983).

tandem arrays. Its 5000 repeats make up 0.08% of the DNA; *Rem3,* consisting of 463 base pairs, represents the repeated monomeric unit of satellite DNA and does not need further attention in the present context. However, all three are transcribed, apparently under developmental control. In sequence structure, only the largest element shows any resemblance to other units of this nature, its termini being occupied by long inverted repeats (Table 9.13). Neither of the others show anything of comparable construction. Among the most outstanding characters is the generation of exactly the same direct repeat, GAATTC, on the flanks of all three types (Table 9.14).

SB18 and Others of Amphibian Origin. One element, *SB18,* of controversial function, has been shown to occur in about 500 clusters per genome, each comprised of from 1 to 15 tandem repeats (Lam and Carroll, 1983). Structurally this transposon, if such it proves to be, is devoid of the usual characteristics, since the sequence shows no terminal repeated segments nor regions of dyad symmetry, except for three nucleotides at each end that are complementary (Table 9.13). By blot-hybridization no evidence for transcription of the unit was found in the total RNA from adult or embryonic tissues or oocytes. Thus, if *SB18* is transposed, it is dependent upon other sources for the necessary enzymes.

9.8.2. Movable Elements of Man

As is amply shown in Table 9.13, numerous types of movable elements have been detected and analyzed from mammals, and it would be pointless to attempt to enumerate each separately. Accordingly, one of the more thoroughly investigated families is analyzed to disclose any features that may be unique in this class of vertebrates; with that foundation others of importance are examined and any unusual characteristics disclosed. Beyond that, the tables listing the 5' and 3' termini and flanking repeats (Tables 9.13 and 9.14) suffice to show differences and relationships among the sequences.

The KpnI Family. Perhaps the best known group of movable elements in mammals is the *KpnI* family. Despite claims of its occurrence in lower primates and mice (Grimaldi and Singer, 1983; Grimaldi *et al.,* 1984; Schindler and Rush, 1985), sequence analyses have failed to reveal any from sources other than human tissues (Table 9.13). The full nucleotide structures of several complete elements have now been established (Grimaldi *et al.,* 1984), showing the family, which is represented by ~50,000 copies, to have quite well-marked characteristics. Among the sharply defined traits is the absence of repeats, either direct or inverted, at the termini, in this way being reminiscent of *SB18* from *Xenopus.* Nor are the ends of the 6000-base-pair structures capable of forming regions of dyad symmetry. At the left end, there are several evolutionarily conserved sequences, including CCAAG in sites 11–15 and RCCGAATAGGAACAGCTC from site 19 onward in each of the three examples as far as given (Table 9.13). The right end is A,T-rich and, in addition, usually bears a poly(A) tail, adding to the difficulty of determining the precise limits of the element. The character of the flanking repeats of host DNA varies with the species; in *F2* on the right flank there is a 15-base copy of a 16-base original on the left, in β*41* there is a nine-base imperfect repeat, and in *A* the repeat similarly is imperfect but is constructed of 11 base pairs (Table 9.14: Potter, 1984).

An additional element has been described recently from a human nuclear sequence that was originally assigned to this family (Nomiyama *et al.,* 1984). While it shows some resemblances to that group in lacking terminal repeated sectors, the 1800-base-pair unit

Table 9.14
Flanking Repeats at Ends of Vertebrate Inserted Segments

	Left terminus		Right terminus	
KpnI				
Human *F2*[a]	GATTAAAAT	AATATAG	GATTAA–AT	AATATAG
Human *β41*[a]	GTTATTCTA		GTTATCCTA	
Human *A*[b]	AGAAGCACC	TG	AGAAACTGC	TT
KpnN				
Human *LmA-1*[c]	AAACAAATT	GACATG	AAA–AAGCT	AAC–TG
Miscellaneous				
Human *hinF*[d]	AATT		AATT	
Human *0-4*[e]	ATGACGAAT	GA	ATGCATAAT	GA
Murine *B1*$_{2.7}$[f]	TATCTTCTT	TTG	TCTTTTTCT	G
Murine *Bam5-R103*[g]	AAGAATGCA	TGTTGA	AAGAATGAA	TGTTGA
Bovine *LTR*[h]	GACAGG		GACAGG	
Xenopus *REM1*[i]	GAATTC		GAATTC	
Xenopus *REM2*[i]	GAATTC		GAATTC	

[a]Grimaldi *et al.* (1984).
[b]Potter (1984).
[c]Nomiyama *et al.* (1984).
[d]Shimizu *et al.* (1983).
[e]Sun *et al.* (1984).
[f]Propst and Vande Woude (1984).
[g]Wilson and Storb (1983).
[h]Sagata *et al.* (1984).
[i]Hummel *et al.* (1984).

differs in being much smaller and in showing none of the conserved series near the 5' end that occur in all of the others. This unusual transposon, which generates 15-base-pair imperfect repeats on its flanks, is here placed in a new family called *KpnN*, the last letter signifying new.

Two Other Families in Man. Two additional, but much smaller families of movable elements from human sources have been sufficiently characterized to merit brief discussion. Of these the *O* family, whose ~4500 copies range in length from 370 to 440 base pairs in length, is the larger (Sun *et al.*, 1984). As in the *KpnI* and *KpnN* families, no terminal repeats are present, and the two ends display no complementarity (Table 9.13). The two representatives that have been sequenced have their 5'-terminal sectors completely homologous through site 31, so only the structure of one of them, *O-4*, is given in the table. Their 3' portions are entirely different, however, and *O-5* has a 57-base-pair insert, absent in the other. As in the *KpnI* assemblage, the length of the flanking repeat varies with the species, being 11 nucleotides in length in *O-4* but only seven in the *O-5*.

Only in the second of these additional groupings is a structure found similar to those of *Drosophila*. In this *HinF* family the termini are constructed of long direct repeats, constituting approximately two-thirds of the entire unit, which is 450 base pairs in length (Shimizu *et al.*, 1983). Here as in the remainder from man, no complementarity is displayed by the two ends (Table 9.13). Transposition generates a simple four-base-pair direct repeat on the flanks (Table 9.14).

9.8.3. Transposons from Other Mammalian Sources

As is too often the case in molecular studies, the mouse is the major source among mammals of tissue for investigations into transposon structure, as evidenced by the nine examples from this rodent in Table 9.13. But at least one sequence of a movable element has been established from each of several nonmurine species, including the rat, rabbit, and cow, all of which are to be found in the table, along with a single representative from the chicken for comparative purposes. To provide insight into the diversity that has emerged from these researches, a few of the more important families are examined in concise form.

The Murine Bam5 and R Families. Because the *Bam5* family from mice has been considered to be homologous to the *KpnI* of man (Grimaldi *et al.*, 1984), it is logical to examine it first, but also, since it is often associated with members of a separate group, the *R* family, the two need analysis together. One study that has sequenced representatives of both types is based on a compound element located near the murine immunoglobulin C_κ genelet (Wilson and Storb, 1983). As pointed out earlier, the *Bam5* unit, 535 base pairs long, shows no detectable homology with either the *KpnI* or *KpnN* corresponding regions (Table 9.13); hence, it should be considered to represent a separate and distinct type. Like those others, it lacks terminal repeated sectors, and these similarly display no evident complementarity with one another.

The *R103* movable element, with which the foregoing was associated, apparently is part of a compound structure that is transposed as a single unit, for one direct repeated flank is located at the left of *Bam5,* another at the right of *R103*. These flanking repeats are 15 base pairs long and, in differing at one site, are imperfect copies (Table 9.14). Among the features distinguishing it from *Bam5* is a smaller size (~484 base pairs) and the absence of homology in the terminal sequence (Table 9.13). Unfortunately, because of the presence of an A-rich sector at the right flank, the precise limits of *R103* could not be established. In a related example of this family, *R1*, the termini show a nine-site sector of complementarity, but this species is 1191 base pairs in length (Gebhard *et al.*, 1982). Consequently, it appears that *R103* is truncated, as probably is the case with *R2* to *R6*. All of these have a long, A,T-rich sector near the center, which seemingly provides a point for truncation; they also possess what appears to be a poly(A) tail, but that sequence is really a portion of the element proper. The type referred to as *R8* evidently is totally unrelated to the others (Gebhard and Zachau, 1983a).

The B1 Family. Perhaps the most thoroughly explored of the murine transposons are the various members of the *B1* family, which is represented by between 100,000 and 500,000 copies per haploid genome (Krayev *et al.*, 1980; Propst and Vande Woude,

1984). Recently several copies of *B1* were found associated with the oncogene (cancer-producing) *c-mos,* all of which were ~420 base pairs in length (Propst and Vande Woude, 1984) and internally had G,A regions of 50 base pairs. That described as *B2* from Ig_κ genelets is not related, but represents an entirely distinct group (Krayev *et al.,* 1982; Gebhard and Zachau, 1983b). These transcribed elements have a size of about 130–150 base pairs, and have an A-rich sector following the 3′ end (Balmain *et al.,* 1982). Four representatives have now been sequenced, two of the more divergent species being included in Table 9.13. Upon translocation, flanking repeats are generated, varying in size from species to species in the range of 11–16 base pairs, all imperfect at a minimum of two sites.

Other Transposons of Note. In addition to the *R* family members, few transposons show any considerable degree of complementarity in the terminal segments, *Vr* of the mouse and *ID224* of the rat being the most notable exceptions (Sutcliffe *et al.,* 1982; Kuehn and Arnheim, 1983). The first of these shows a sector of dyad symmetry ten base pairs long and the second 17 (Table 9.13). Among the peculiarities of the *ID224* element, which is 1110 nucleotides in length, is its confinement to brain tissue. On the other hand, the *Vr* module was discovered close to the origin of rRNA transcription; its 1750+ length consists of 13 repeated sectors, each ~135 base pairs in length, but the two terminal sections are not part of any repeat.

Only one more group of these repetitious units needs examining, that known as *LTR.* Actually that abbreviation refers to the long terminal repeats from many sources, including viruses, and consequently the term arises again in the next chapter. In the present instance, however. it refers to a specific type of transposon. Moreover, in having been found and sequenced also in the bovine genome, it provides the only case to date where the same element has been reported from more than a single source mammal (Wirth *et al.,* 1983; Sagata *et al.,* 1984). However, that from bovine sources is actually from the leukemia virus, whereas that of the mouse shows the behavior of a transposon. Nevertheless, it is clear from the homology displayed at the left-hand terminus, limited though it may be, that kinship exists between the two sequences. It will be of great interest to see how extensive such relationships are when examined in connection with the viral genome structure in the next chapter.

9.9. SYNOPSIS OF EUKARYOTIC TRANSPOSONS

Since no study has been made previously of eukaryotic transposons as a unit, the present synopsis should be viewed as experimental, a condition aggravated by the frequent gaps in current knowledge. Because of the doubtful functional nature of the *Chironomus* elements, they are not included in the table summarizing the structural features of those known to be actual transposons (Table 9.15).

The Structural Basis. For the present synoptic table of eukaryotic transposons, the same statistics are employed as for the prokaryotes, except that dependence is not indicated, since it is known to occur in only a few examples, notably certain ones from seed plants. However, the prime characteristic of the bacterial types, the sequences of the left termini, is found equally useful here. Unlike those of the previous table, in which four or five Gs began the sequence found in the largest family, here TGT or TGY is the dis-

Table 9.15
Summary of Eukaryotic Transposons[a]

Left end	Transposon family	Other members	Source	Terminal sectors Inverted	Length (base pairs)	Flanking repeats (base pairs)
			CLASS I			
TGTTGGAAC	τ26	τ106	Yeast	No	--	--
TGTTGCATC	σ106	B104	Yeast, Drosophila	No or slightly	341–8700	5
TGTTGGAAT	Ty1 ε	δ2	Yeast	Slightly	300–5900 (5300)	5
---CGGAAT	δD15	--	Yeast	No	--	--
TGTTGAAAT	copia	--	Drosophila	8 bases	5000	5
TGTAGTA	412	--	Drosophila	4 bases	--	4
TGTATTA	0-4	--	Human	No	370–440	11
TGCACTG	Rem1	--	Xenopus	Completely	490	--
			CLASS II			
TGAGA	δADH	δGC106	Yeast	No	--	--
TGAGATA	Ty1-B10 ε	--	Yeast	Slightly	--	--
TGAGAGA	LTR	--	Mouse	Slightly	--	--
TGAATCG	L1	--	Mouse	No	--	--
TGGTACT	Bam5	--	Mouse	No	535	15, imperfect

				CLASS III		
TACCCG	σFD2	--	Yeast	No	--	--
TAGGCATG	Ac9	Ds1	Maize	14 bases	405-4563	8
TAAACTG	KpnI	--	Human	No	1800	15, imperfect
TAAAIGA	Rem?	--	Xenopus	No	490	--
				CLASS IV		
ATATGTG	σ	--	Yeast	No	--	--
ATGAAGC	F10?	--	Drosophila	?	13	--
				CLASS V		
AAGCTTG	opa	--	Drosophila	No	--	--
AAGCCGG	C_BLE	--	Rabbit	Slightly	--	--
AATTAATT	Ifm	--	Drosophila	No	8800	6
AGTGACA	17.6A	--	Drosophila	No	--	--
AGTTAAC	mdg4	--	Drosophila	No	7500	4
AGCTCAA	FB4	--	Drosophila	Completely	4089	--

(continued)

Table 9.15 (Continued)

Left end	Transposon family	Other members	Source	Terminal sectors inverted	Length (base pairs)	Flanking repeats (base pairs)
			CLASS VI			
CACTACA	*En*	*Tam1*	Maize, *Antirrhinum*	13 bases	8400	3
CATGATG	*Pm25*	--	*Drosophila*	31 bases	2900	8
CAGAGAA	*hobo*	--	*Drosophila*	12 bases	--	8
CCATATC	*hinF*	--	Human	No	450	4
CCGGGCG	*B1*	*B1₂․₇*	Mouse	No	420	2, imperfect
CTGCAGG	*ID224*	--	Rat	17 bases	1110	--
CTTCAAC	*SB18*	--	*Xenopus*	Slightly	--	--
			CLASS VII			
G--GG	*KpnI*	--	Human	No	6000	9-16, imperfect
GGATCCA	*R1*	*R103*	Mouse	9 bases	484-1191	15, imperfect
GGCCACG	*CR1*	--	Chicken	--	--	--
GTCGACC	*Vr*	--	Mouse	10 bases	1750	--
GTGACGT	*297*	--	*Drosophila*	No	--	6
GCACACA	*DI-2*	--	Mouse	No	--	--
GAGATAA	*Mu1*	--	Maize	27 bases	1400	9
ACTCTG	*EXT*	--	Bermuda crab	Slightly	--	--

*See Tables 9.7–9.14 and text for references.

tinguishing series (Table 9.15). Regrettably, the application of names to individual components has been based largely on superficial resemblances, a consequence of the unsuspected complexity that is just now coming into full light.

This confusion of designations was first encountered in the yeast components, a situation that the discussion there could not treat adequately; here it becomes fully exposed when the *Ty* elements are employed as the examples. If homologous left-terminal sectors are arranged together, *Ty1* ε is found in the third row, whereas its supposed homolog, *Ty1-B10* ε, occurs seven rows later; moreover, another element, *Ty912*, claimed to be related but too incompletely known to be included in the table, differs more strongly, being functionally divided into halves. The situation with σ is even worse, for σ*FD2*, instead of being adjacent to σ*106* of row 2, falls into row 14, and plain σ is four rows later still. The same bewildering condition is extant in the seed plants, as suggested by the extreme diversity in size that exists among the transposons. Similarly in the human representatives, the need to erect a new family (*KpnN*) for a so-called *KpnI* element shows confusion to be present there, too. The *B1* and *B2* "families" are especially perplexing among mouse transposons because of the obvious lack of kinship that is often displayed, and the *Rem* units of *Xenopus* and *copia* of *Drosophila* have become convenient catchalls. In short, the molecular research on these movable units has reached a state where its further maturity depends upon a carefully composed scheme of classification, either for eukaryotic types separately, or, better, in combination with prokaryotic types. For, although none of the latter have been found present in eukaryotes, it is inconceivable that that condition will long prevail. The lack of organization is already showing its effects as elements in new locations are analyzed or sequenced. By way of illustration, a recent investigation left unnamed the *Ty* element whose entire nucleotidyl structure it had established, the transposon being referred to merely as *Ty*, with its source plasmid designation appended (Hauber *et al.*, 1985). Comparisons showed its left terminus to agree with that given in Table 9.7 as δ*2*, at least through the first 16 bases. The arrangement into classes proposed for the eukaryotic elements necessarily is not integrated with that of the prokaryotic ones, and accordingly the subdivisions are designated by roman numerals, not letters of the alphabet.

Analysis of the Summary. When the elements are arranged according to structural homologies in their left-terminal sectors, some semblance of order comes into view, but far more is necessary in order to cope with the tremendous complexity manifested by these transposons. Since the common yeasts have often been shown to be the simplest group among the Eukaryota, it is logical to begin with a common element from that source. Accordingly, the first group (class I) encountered in Table 9.15 is headed by a yeast representative and has the left terminal sequence TGTT. Alternatively, in a broader sense TGY should be viewed as the opening combination, an arrangement that includes a broad spectrum cutting through much of the entire phylogenetic tree. On this basis, *B104* of *Drosophila* is seen to be somewhat akin to σ*106* of yeast. A second group (class II) similarly begins with yeast sequences, this cluster of types being characterized by TGR. A third category (class III), with TA, and a fourth (class IV), with AT, also start with representatives from this same basic source. It is interesting that none from *Saccharomyces* are found beyond that point in the tabulation, suggesting that the remainder have been developed among higher taxa. However, it is not necessary to enlarge upon the systematic arrangement further, the sequences of events being strictly arbitrary; a more natural scheme awaits further accumulation of factual details.

10

Viral Genes—Structure and Controls

Because of their minuteness and relative simplicity, the viruses afford insights into structural arrangements and activities that might long be overlooked in higher, more complex organisms. In many cases, however, these parasites have become degenerate in part by replacement of their original genes and translational products by those of the host cells. The resulting degree of dependency varies with the species, for many, including bacteriophage T4 among numerous others, expend considerable energy synthesizing macromolecules that duplicate or augment metabolic activities already present in its cellular habitat (Schmidt, 1985). By way of illustration, the genome of that T-even phage contains genes that encode one enzyme that cleaves the bacterial tRNAs near the anticodon and another pair that together repair the resulting damage, an altogether fruitless cycle. While useless to the virus, such relict genes are important from a biological point of view in suggesting that these organisms once were more completely supplied with genes and thus were less dependent upon living cellular types for existence. Hence, degeneration and evolutionary conservation have played antagonistic roles in molding the multitudinous diversity of extant viruses from their forebears of billions of years in the past.

10.1. VIRAL GENOMES

Nothing sheds light more brightly on the extent of diversification that has occurred among the viruses than the structure of their genomes, but aside from that, knowledge of the nucleic acid content and nature is a necessary preliminary to an understanding of the viral genes and their transcriptional controls. Viral genomes fall into four great classes, double- and single-stranded DNA and the same two classes of RNA, but variation on each of these themes is extensive. Since beginning with the more familiar macromolecular organizational type offers greater clarity, that procedure is followed here. Because of their involvement in cancer production, retroviruses, including the AIDS agent among others, are largely omitted from the present chapter, being reserved for the next, which is concerned with oncogenesis.

10.1.1. Bacterial Double-Stranded-DNA Viral Genomes

As the result of their position at the tip of the viral phylogenetic tree, the double-stranded-DNA viruses should be expected to contain highly complex genomes, and such is quickly found to be the actual case. Even within the bacteria, in which viruses are referred to as bacteriophages, quite a variety of major types occur. Although far from simple, these as a whole have not attained the complexity of structure that characterizes the vaccinia (cowpox) virus of metazoans, for example, so analysis of them first is logical. Additionally, and equally importantly, these organisms have been far more thoroughly investigated than any from eukaryotic cells.

The T-Even Bacteriophages. Two series of bacteriophages have designations beginning with T, one set having odd numbers, such as T1, T3, and T7, the other having even numbers, those referred to as T2 and T4 being especially important. Since beginning discussion with the latter, better known group affords distinct advantages, this frequent practice is adhered to here also. Like the bacterial genome, that of the T-evens is circularized, the ends of the molecule in the present instance having "terminal redundancy," repetitious sequences that by their complementarity enables them to adhere in catemeric fashion. These redundant ends vary in length from several hundreds to a few thousands of base pairs (Dillon, 1978, pp. 350–362). The DNA is highly modified by the conversion of its deoxycytidines to hydroxymethylcytidines, variously substituted with α- and β-glucose and gentibiose, a disaccharide; the extent and type of glucosylation are strongly correlated to the species of bacteriophage. As a group, these organisms have relatively immense genomes, being ~166,000 base pairs long in T4 (Gerald and Karam, 1984), encoding more than 60 genes.

Among the viruses as a unit, the genomic organization is based in great measure on the timing of transcription of several sectors. At the simplest level, there are two subdivisions, "early" and "late," the former referring to the region transcribed directly after the viral genome has entered the host cell, and the latter to that part transcribed after a viral RNA polymerase has been produced or replication of the genome has begun. Sometimes the first of these stages is modified by addition of other categories, such as "immediate early," which refers to genes transcribed immediately upon entry, followed by a second stage also involving host polymerase. At times the immediate early is called "early early," and remainder, "late early." Especially in types with large genomes, such as that of T4, a phase called the "middle" may exist between the early and late portions (Pulitzer *et al.*, 1985). In such cases, the late phase is postponed until replication of the nucleic acid structure commences, but these points and others related to them become clearer when the transcriptive processes are examined later in this chapter.

The T-Odd Bacteriophage Genomes. Because the full genome of one representative (T7) has been completely sequenced, the basic features of the T-odd phages can be discussed more satisfactorily than the T-evens. In that species the genome consists of 39,936 base pairs, which embrace 53 genes (Stahl and Zinn, 1981; Dunn and Studier, 1983; Moffatt *et al.*, 1984). Unlike that of the T-evens, their double-stranded DNA molecule is linear, for the ends are not complementary, a further difference being the absence of modified bases of any sort. This structure exemplifies clearly the rule that most viruses carry genetic information very efficiently, packing a maximum number of genes into a DNA molecule restricted in size by the capsid into which it fits. Very short spacers

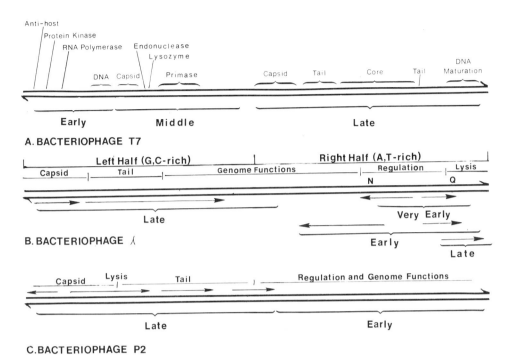

Figure 10.1. Structural arrangements of various double-stranded DNA bacteriophage genomes. (Part A is based on Studier and Dunn, 1983, and part C on Christie and Calender, 1985.)

intervene between the cistrons, as shown in detail in the discussion on transcription, so that 92% of the genome is used for coding purposes (Studier and Dunn, 1983). Although the DNA of T5 is about three times as long as that of T7, one of the smallest of the group, that of such T-evens as T4 is double that length, so the present organisms are decidedly the simpler in organization. This relative simplicity is reflected in morphological traits, for the tail of T-odd types lacks the contractility of those of the T-evens (Dillon, 1978, pp. 363–367).

The genome is divided into three major sectors (Figure 10.1A), early, middle, and late. In the first of these regions are ten genes, whose products inactivate host restriction processes, serve as viral RNA polymerase and protein kinases, or are active in replication. The middle portion is slightly the largest, carrying 22 cistrons, encoding DNA-binding protein, endo- and exonucleases, lysozyme, primase, and DNA polymerase, along with numerous unidentified products. In the late section are 21 genes, largely specifying proteins needed in construction of the capsid (referred to as the head in these viruses) and tail structures.

10.1.2. Temperate Double-Stranded-DNA Bacteriophage Genomes

Both of the preceding groups are said to be virulent, because they induce the immediate production of progeny, the accumulation of which quickly leads to the lysis of the host cell. There are numerous other types, however, that may be virulent as the foregoing ones, or they may, under certain largely unknown conditions, be "temperate," the viral genome becoming associated with that of the host and maintaining itself over long periods of time by replicating in unison with host DNA. Eventually, the phage genome, then known as a "prophage," commences to produce progeny, soon resulting in lysis as in virulent types. This delayed lysis is referred to as the "lysogenic response." It is this latter type of life history in which viral behavior resembles some of the transposons of the preceding chapter, a feature also of great significance in cancer-related forms.

Bacteriophage λ. Bacteriophage λ is one of the better known members of the temperate variety. As in the T-odds, it has a nonretractile tail, and the genome is linear and double-stranded, encoding 30–40 proteins. Among the most outstanding characteristics is that both strands are transcribed during the early stage, two subdivisions of which are recognized. Very early transcription, apparently from divergent promoters, results in the production of several enzymes, including protein N needed for the early substage. The transcription of the latter similarly proceeds bidirectionally (Figure 10.1B), generating a number of products needed for regulatory, lytic, and genomic functions, including protein Q, which is essential to late transcription. Whereas these two substages are confined to the right, A,T-rich half of the DNA, the late stage is confined to the G,C-enriched portion and is unidirectional (Figure 10.1B).

In part because of its possible relationship to comparable processes in oncogenic species, the mechanism of inserting the λ genome into that of the bacterium has been the center of much research. The processes of integrating the two genomes in the present instance, is, like many others, site-specific, taking place at definite points in each. The viral region of insertion, designated as *attP*, is 240 base pairs in length, while that of the host, called *attB*, consists of only 25 (Campbell, 1983; Griffith and Nash, 1985). In addition, two enzymes are involved in the reactions, an integrase (Int), encoded by the viral DNA, and IHF (Integration Host Factor), produced by the bacterium. Although precise data are lacking, the processes appear comparable to those of movable elements, excluding the replication of the element aspect, but including duplication of host DNA at the termini.

Other Temperate Double-Stranded DNA Phages. Although a number of additional types of temperate double-stranded DNA bacteriophages are known, most are not sufficiently investigated at the molecular level to merit space here. Among the better documented ones in this group are the several members of the P type, P1, P2, and P22 being especially important. The first of these tail-bearing forms contributes greatly to a later discussion; here the second, from *E. coli*, provides the insight into genomic organization. When illustrated in a reversed orientation as in Figure 10.1C, the DNA molecule of P2 somewhat resembles that of λ in arrangement, but fewer of the transcriptive activities take place from divergent promoters. Nor is a very early sector distinguished, there being only early and late subdivisions. The third in the series, P22, from *Salmonella*, has a genome still more similar to that of λ; it receives much attention in subsequent sections concerned with transcription.

Another type of bacteriophage is best represented by φ29, one of the better known phages infecting *B. subtilis*. Its DNA of ~18,000 base pairs is similarly linear and double-stranded, but it differs markedly from the preceding groups in having two sets of early genes, numbers 1–6 and 17 (Figure 10.2A). Moreover, all the early coding sectors are on

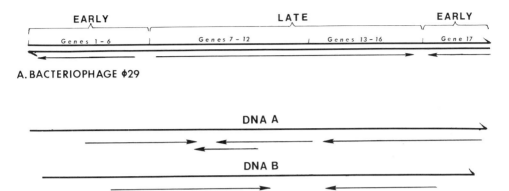

Figure 10.2. Genome organization of diverse DNA viruses. (A) Bacteriophage φ29 is one of the better known viral parasites of *Bacillus subtilis*. (After Holder and Whiteley, 1983.) (B) Like those of many plant viruses, the genome of the tomato golden mosaic virus consists of two separate parts, each single-stranded. (After Hamilton *et al.*, 1984.)

one strand and the 11 of the late section on the other, so that the transcriptive processes are strongly polarized. A close relative, M2, which also attacks this species of bacterium, receives considerable attention later.

10.2. MAMMALIAN DOUBLE-STRANDED DNA VIRUSES

Many of the viruses of mammals have genomes of double-stranded DNA, but the present discussion centers on five of the major families. These are the polyoma-, papova-, adeno-, herpes-, and poxviruses. The first three listed, being smaller and simpler, and therefore better known, receive the major portion of attention, not only here, but in subsequent sections, while the other two, despite their medical importance, sometimes are necessarily omitted altogether. In all of these groups, the DNA is in the form of a covalently closed, circular molecule twisted into a superhelix. The capsid never bears the tail universally present in the bacteriophages just considered, but is typically a skew icosahedral structure, consisting of 72 subdivisions called capsomers (Dillon, 1978, pp. 372–279).

10.2.1. Smaller DNA Viruses of Mammals

The papova- and polyomaviruses of primates rank among the most thoroughly understood forms from mammals, the simian virus 40 (SV40) representative of the second group being especially well documented. Accordingly, its genomic characteristics are given first to provide a basis for comparison.

The SV40 Genome. The genome of SV40 is relatively small, being only ~10% that of bacteriophage λ; its 5200 base pairs are sufficient to encode about ten proteins of molecular weight of 20,000 each. Apparently, the only modified nucleotide present is 5-

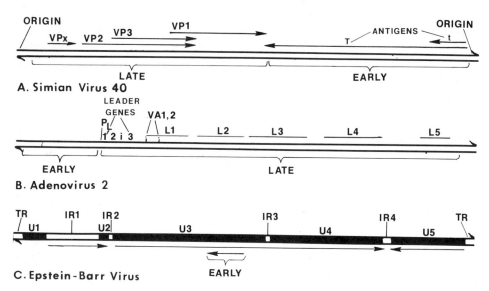

Figure 10.3. Genomes of double-stranded DNA viruses of mammals. (A) Early genes of SV40 are on the lower strand of DNA, the late ones on the upper. (Yang and Wu, 1979.) (B) Direction of transcription has not been clearly established for the genes of adenovirus 2 (P_L, major late promoter). (Based on Virtanen *et al.*, 1982.) (C) The genome of herpesvirus EBV is divided into five "unique" regions (U1–U5) and four internal repeated sequences (IR1–IR4). At the ends are terminal repeats (TR). (Based on Bodescot *et al.*, 1984; Gibson *et al.*, 1984; and Thomsen *et al.*, 1984.)

methylcytosine, and that only in small quantity. While within the virion, the DNA has a chromatinlike structure, since several species of protein are associated with that mac-romolecule. At least three of them resemble histones, but whether they are produced by the host or virus remains an unsettled issue.

Like most others from mammals, this virus is temperate, being able to insert its DNA into the host genome for an indefinite period of time. A similar nature enables a related species called the BK virus (BKV) of man to induce transformation of the inhabited cells, resulting in tumor formation. Both species have genomes of virtually identical size and are ~70% homologous, the sequence of each having been completely established (Fiers *et al.*, 1978; Reddy *et al.*, 1978; Seif *et al.*, 1979; Yang and Wu, 1979). The genes are arranged in two parts; those transcribed in the early stage are read in a counterclockwise fashion (right to left in Figure 10.3A), those of the late stage in the opposite direction. Since the two genes encoded in the early region are for antigens and the four of the late period are not known to code for any polymerases, these viruses must be dependent upon host products for transcription and DNA replication.

The Adenovirus Genome. The genome of adenoviruses, more than 50 types of which have been isolated from mammalian and avian sources, is about six times as large as that of SV40, the DNA molecule from type V (Ad5) consisting of 36,000 base pairs (Dekker and van Ormondt, 1984). Accordingly, the regular icosahedral capsid has far greater dimensions, being constructed of 252 capsomers of two types. The greater fraction of these consists of unmodified simple structures called hexons, whereas the pentons, located one at each of the 12 angles, bear long, projecting fibers having a knob at their ends.

As shown in Figure 10.3B, the genome is subdivided into early and late regions, the latter occupying by far the greater portion. The products of the early phase have not been completely characterized, but five multicistronic mRNAs are known to be generated in the late period of type II adenovirus (Ad2), one of the more thoroughly analyzed members of the group (Virtanen *et al.*, 1982). Among the products encoded here is VA1, transcribed by host RNA polymerase III (Chapter 3, Section 3.4.2; Aleström *et al.*, 1982). That substance is essential for translation of viral late mRNAs, largely through its role in transport of those molecules from the nucleus into the cytoplasm (Katze *et al.*, 1984). Probably the most distinctive feature of this group of viruses is in the modifications that occur to the messengers prior to translation. About the most peculiar is the attachment of a leader sector to the 5′ ends of mRNAs, a segment generated from three exons by elimination of three introns (Keohavong *et al.*, 1982). However, sometimes additional parts are spliced onto the leader, the i-leader being one of the more frequent additives (Chow *et al.*, 1979).

10.2.2. Larger Double-Stranded DNA Viruses of Vertebrates

Size is not the sole characteristic that distinguishes the large double-stranded DNA viruses of vertebrates from the smaller ones of the preceding section. Outstanding among the distinctions is the presence of lipids in an envelope that encases the capsid, and, second, the genomic DNA is linear rather than circular. Among the better documented of the group are the herpes- and poxviruses, the former being a complex group.

The Herpesvirus Genome. In the herpesviruses, including herpes virus 1 and 2 (HSV1 and 2) and the Epstein–Barr (EBV) virus, the genomic DNA, although linear and double-stranded, is in the form of a toroid, surrounding a central core of protein (Dillon, 1978, pp. 376–379). That of EBV, the causative agent of infectious mononucleosis, is somewhat larger than those of HSV1 and HSV2, consisting of 172,000 base pairs, compared to the 150,000 of the other two (Gibson *et al.*, 1984). In broad terms, in each case the DNA is divided into five regions containing unique sequences (U1–U5) by four internal regions of repetitious sequences (IR1–IR4) and two terminal repeats (TR) (Figure 10.3C; Bodescot *et al.*, 1984). Insofar as is established, transcription during the late stage is by way of three polycistronic messengers, one of which is oriented in opposition to the others (Figure 10.3C). Some of the internal and terminal repeats are themselves comprised of short, reiterated segments (Jones and Griffin, 1983; Costa *et al.*, 1985), and certain ones have been compared with the switch region of immunoglobulin genes (Gomez-Marquez *et al.*, 1985).

That pattern is obviously simplistic, because it has been demonstrated that two early genes occupy at least a portion of the U3 region and therefore cannot be part of the polycistronic messenger shown (Gibson *et al.*, 1984). In an additional member of the class, human cytomegalovirus, which has a genome of 240,000 base pairs, an early region is placed similarly to that shown, confirming the above statement (Thomsen *et al.*, 1984). In that virus, transcription of this early sector is oriented in reverse direction, as indicated in the EBV genome (Figure 10.3C).

The Poxvirus Genome. The vaccinia virus, which is the prototype of the poxvirus class, has a genome slightly exceeding that of EBV in size, containing ∼187,000 base pairs (Baroudy *et al.,* 1982). In part the two termini of each strand are complementary, so that the linear double-stranded DNA can fold into a single, continuous polynucleotide chain. These terminal repeated sectors, some 10,000 base pairs long, contain at least three genes that are transcribed early in infection (Blomquist *et al.,* 1984). Among the unique features of transcription is that the immediate early activities take place within the capsid, but the later ones occur within the cytoplasm of the host (Cochran *et al.,* 1985). No extensive mapping of the genome has been conducted, so the arrangement of the ∼100 genes it includes awaits clarification. Sequencing of a region to the left of center indicates that some of the early and late genes are tightly clustered, with many overlaps (Plucien-niczak *et al.,* 1985).

10.3. MISCELLANEOUS DNA VIRUSES

By far the majority of DNA viruses are members of the several double-stranded groups just described from bacteria and mammals, but there are several others that also merit attention. Two assemblies of species with this type of nucleic acid for their genomes are important plant pathogens, and another of much larger proportions has strong impact on the bacterial world.

10.3.1. Some Plant DNA-Viral Genomes

The pair of major DNA viral families infecting plants just mentioned are the cau-limoviruses and geminiviruses, the former double- and the latter typically single-stranded; however, since it is known to have double-stranded varieties (Hamilton *et al.,* 1984), the latter family appears to be a transitional form.

The Caulimoviral Genome. The caulimoviruses are best represented by the pro-totype of the group, the cauliflower mosaic virus. Its circular double-stranded genome is 8024 base pairs in length, the sequences of three strains of which have been fully established (Franck *et al.,* 1980; Gardner *et al.,* 1981; Bâlazs *et al.,* 1982). All eight of its coding areas are contained in the minus strand, mostly closely spaced, but an intergenic spacer 700 base pairs long separates genes *VI* and *VII* (Dixon and Hohn, 1984). Depend-ing upon the strain, either two or three breaks are present per genome, one in the minus, the other(s) in the plus; these represent regions of single-strand overlap and may have application in transcription. Replication appears to involve an RNA intermediate tran-script followed by reverse transcription (Pfeiffer and Hohn, 1983).

The Geminiviral Genome. Among the better known of the geminivirus family is the tomato golden mosaic virus (TGMV), whose genomic sequence has now been estab-lished, as has that of the African cassava mosaic virus (Hamilton *et al.,* 1982, 1984). In each case the genome consists of two single strands of DNA, the B component of which is just slightly shorter than the A (Figure 10.2B). In TGMV the latter consists of 2588 and the former of 2508 nucleotide residues. Four open reading frames, including one of reversed polarity, have been identified in DNA A and two convergent ones in DNA B (Figure 10.2B).

10.3.2. Single-Stranded DNA Viruses

The single-stranded DNA viruses are a relatively small group, confined almost entirely to the prokaryotes, except a few from plants, one of whose genomes was just described above. Those of bacteria fall into two major classes based on the shape of the capsid. In the icosahedral (spherical) forms, the more important representatives include ΦX174, S13, and Φ1 (Dillon, 1978, pp. 379–384), whereas in the filamentous category are placed fd, f1, M13, AE2, IKe, and Pf. Basically, the genomic characteristics of the two subdivisions are quite similar, but each possesses distinctive features of its own.

The Icosahedral Class Genome. The single-stranded DNA of the icosahedral class is in the form of a circle, the nucleotides being unmodified except for a single 5-methylcytosine per strand. In stage I immediately following infection, the genome is converted to a double-stranded circle by synthesis of a complementary (minus) strand. This "replicative form" (RF) of the molecule is present in two structural varieties, one of which is supercoiled and has both strands covalently closed, the other relaxed and with at least one single-stranded break.

ΦX174 serves admirably as a representative of genomic organization in this family, for its complete sequence has been established (Sanger *et al.*, 1978). In its single strand of DNA of 5386 residues, ten genes are found, all for proteins, many of which overlap one another, as discussed later. Preceding the first coding sector, that for protein A, is a spacer of 63 nucleotides, another of 39 sites intervenes between genes *J* and *F,* a third, 110 residues long, separates *F* and *G,* and a final one of eight sites lies between *G* and *H.* All the other cistrons either overlap or are separated by only one or two bases.

The Genome of the Filamentous Class. In these long, slender, filamentous types, whose dimensions range between 8000 and 20,000 Å in length and 50 and 60 Å in width, the genome is similar in form to those of the icosahedral types. Upon entering the bacterium by way of the male sex-pili, the single (plus) strand at once is copied into a minus strand to produce the RF molecule. Since the genomes of at least four members have been completely sequenced, those of f1, M13, fd, and IKe, their structures and gene arrangements are well known (Beck *et al.*, 1978; van Wezenbeek *et al.*, 1980: Hill and Petersen, 1982; Peeters *et al.*, 1985). These show a remarkable degree of constancy, both in size and arrangement. The shortest, that of fd, is 5781 base pairs in length, while the largest, from IKe, is 6883; Figure 10.4A–D brings out the similarities of gene arrangement. There, the intergenic region (IR) is uniformly placed to the left on an arbitrary basis, its relative position in these straight-line diagrams, whereas the DNA is circular. In each case, the order of the genes and direction of transcription are consistently the same for the four.

10.4. SINGLE-STRANDED RNA VIRUSES

The RNA viruses, attacking as they do a far greater portion of the living world than their DNA relatives, greatly outnumber the latter in species and show a much broader spectrum of diversity. Like them, however, they fall into two major categories, single- and double-stranded. Beginning with the first of these offers many advantages, because it is not only the larger class, but also much the better investigated, which reason provides the basis for viewing the bacteriophages first.

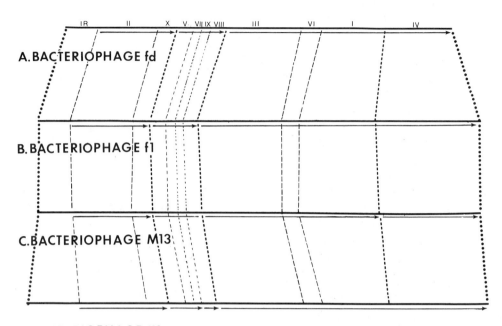

D.BACTERIOPHAGE IKe

E.YELLOW FEVER VIRUS

Figure 10.4. Genomic arrangements of four bacterial and one human double-stranded RNA viruses. (A–D) All four of these bacterial filamentous types are basically alike. (Part A is based on Beck *et al.*, 1978; Part B on Hill and Petersen, 1982; and parts C and D on Peeters *et al.*, 1985.) None of the transcriptional units have been fully established, being apparent only.

10.4.1. Genomes of Single-Stranded RNA Bacteriophages

Although four major groups have been erected for single-stranded RNA bacteriophages, only two have representatives sufficiently well characterized to justify their inclusion here. Group I contains MS2, R17, and f2 as principal members, while group III has Qβ as the sole important representative. In all cases the genome is small, consisting of between 3500 and 4500 nucleotides, and is enclosed in an icosahedral capsid constructed of 180 capsomers, plus one or more molecules of a second polypeptide, called either the A- or maturation-protein (Dillon, 1978, pp. 384–387). Before the more detailed genomic structures are discussed, it is of interest to note that comparisons of the 3'-terminal regions from 16 species representing all four groups (Inokuchi *et al.*, 1982) disclosed homology levels between members of groups I and II to be at 50–60% and of groups III and IV to be at ~50%. But only low degrees of kinships were displayed by representatives from either I or II with any from III or IV.

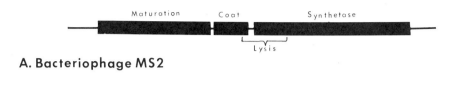

A. Bacteriophage MS2

B. Bacteriophage Qβ

Figure 10.5. Genome structure in two important single-stranded RNA bacteriophages. (Based on Atkins *et al.*, 1979.)

Group I RNA Bacteriophage Genomes. Because the genomes of members of group I are all closely allied, differing at most by less than 4% (Min Jou and Fiers, 1976), attention can be confined entirely to the best known member, MS2. The sequence of the RNA of this form was among the very first messenger species to become completely established and the intricacies of its secondary structure revealed (Fiers *et al.*, 1975; Dillon, 1978, pp. 124–131; Atkins *et al.*, 1979). Within the 3569-nucleotide molecule are sequences of four genes, one each for the A-protein, the coat protein, an overlapping one for lysis, and the β subunit of the replicase (RNA-dependent RNA polymerase), the other three subunits of which are host products. Short intergenic spacers separate the cistrons, and there are untranslated sectors at each end (Figure 10.5A). Consequently, it is evident that this virus is highly dependent on the bacterium in which it lives for most of its requirements for transcription, translation, replication, and assembly.

Group III RNA Bacteriophage Genomes. Group III single-stranded RNA bacteriophages have not been such popular models for investigations at the molecular level as those of group I, and much of what research has been conducted has centered on the nature of the replicase (Mills *et al.*, 1978; Nishihara *et al.*, 1983). Consequently, relatively little has been established concerning the structural details of the genome. That of the chief representative, Qβ, has an RNA molecule 4220 nucleotides long, which is the minus strand, not the plus as in preceding types (Figure 10.5B).

The reason for the interest displayed in the replicase is its highly selective nature. It promotes the synthesis of the genomes of Qβ and other group III RNA bacteriophages, but not that of any other viral type. Nor can it replicate any RNA strands from *E. coli* (Nishihara *et al.*, 1983). Involved in this specificity are its strict requirements for template structure; only a single internal sequence of the Qβ RNA is recognized by the replicase and then only if the genomic 3′-terminal sequence is intact (Meyer *et al.*, 1981).

10.4.2. Single-Stranded RNA Viruses of Metazoans

A great diversity of types of single-stranded RNA viruses from metazoans has been explored at a satisfactory level, far too many to enumerate in detail here. Thus in the more important groups, each of several forms whose genomes have received adequate investigation are described, the families being selected to give a view of the major variations of structure that exist.

The Picornavirus Genome. There can be little doubt that the picornavirus family deserves prominent position in this analysis, for its members include rhino- (cold-producing agents), cardio- (encephalomyocarditis and Mengo), aptho- (foot and mouth disease), and enteroviruses (polio and hepatitis A agents). In all these types, the RNA is the plus strand, serving as the polycistronic messenger for all viral structural and enzymatic proteins. The complete genome sequence of one of the first group, rhinovirus 14 (about 115 species have been identified), is 7212 nucleotides in length, of which ~6540 serve as the single reading frame (Stanway *et al.*, 1984; Callahan *et al.*, 1985). That of hepatitis A is a little longer, containing 7478 nucleotides, of which 6682 provide the single coding sector (Najarian *et al.*, 1985), and that of foot-and-mouth disease agent, with ~8500, is longer still (Carroll *et al.*, 1984). In each case three proteins are encoded, whose natures have not been determined.

The genomes of the three known serotypes of poliovirus have now had their sequences established. Their lengths of ~7500 nucleotides scarcely exceed that of rhinovirus, but their coding properties are more complex (Toyoda *et al.*, 1984). As there, much of the entire molecule serves as a single reading frame, but the resulting polyprotein is cleaved into nine polypeptides. Attached at the 5' end of the genomic RNA is a covalently bonded protein of viral origin, while at the 3' end is a poly(A) tail, which is apparently an encoded feature, not posttranscriptionally added (Nomoto *et al.*, 1982).

Despite the similarity of structure, little homology exists between the major representative types, the greatest kinship (44–65%) being displayed between the several proteins of rhino- and polioviruses (Callahan *et al.*, 1985), that is, between members of the same group.

The Flavivirus Genome. The flavivirus family includes more than 70 closely related pathogens of man and domestic animals, most of which are transmitted by blood-sucking insects or ticks. In it are included the agents for Japanese, St. Louis, and Murray Valley encephalitis, dengue, and yellow fever, the viral genome for the last of which has been completely sequenced (Rice *et al.*, 1985). Its single plus strand has proven to be considerably larger than any of the foregoing, containing 10,862 nucleotide residues and encoding nine identified genes. The first three of these are for structural proteins, the remainder for proteases and other nonstructural products (Figure 10.4E).

The Rhabdovirus Genome. The rhabdoviruses, too, have a single-stranded RNA genome, but here the nucleic acid molecule is the minus strand, needing to be transcribed to produce mRNAs. In vesicular stomatitis virus (VSV), the most familiar member of this group, the genome encodes six proteins, including a gene in the leader sequence that contains only 47 bases (Emerson, 1982). Because only a single promoter is present, located before the leader, transcription is polar and sequential, the more terminal cistrons being transcribed less frequently than those near the start site. Like all minus-stranded RNA viruses of mammals, the genome of VSV is encapsidated in ribonucleoprotein particles to form nucleocapsids, which serve as templates for both transcription and replication (Arnheiter *et al.*, 1985).

Larger RNA Genomes. In the larger RNA virus family known as the paramyxoviruses, the molecule is the minus strand as in the last group noted. It is best represented by the Sendai virus, whose 15,000-residue genome encodes seven structural proteins employed in the capsid, plus an undetermined number of nonstructural ones (Giorgi *et al.*, 1983). Two of that number, P and C, have been demonstrated by sequencing to be largely overlapping, as detailed later, and that for the complex hemagglutinin-neuraminidase has been sequenced (Miura *et al.*, 1985). Here many of the cistrons are provided with individual promoters.

In many other larger RNA viral families of metazoans, including the coronaviruses, represented by the avian infectious bronchitis virus, the genome is similar, but larger, running to 20,000 residues in the species cited (Boursnell and Brown, 1984). Still others, including the influenza A virus, have a genome comprised of multiple segments of RNA, eight in the example mentioned (Lamb *et al.*, 1985).

10.4.3. RNA Viruses of Seed Plants

A rather large variety of viruses from seed plants has been recognized, all of which have single-stranded RNA as the genomic molecule. By far the greater part of these have the genome divided into multiple strands as in the influenza virus just mentioned, but the structure of the 3' end provides a basis for recognition of three intergrading types, with a fourth group sharply distinguished from the rest. In type I, the RNA molecule (plus strand) terminates in a poly(A) tail, thus showing strong resemblance to the mRNAs of cellular organisms (Kozlov *et al.*, 1984), as well as to poliovirus. This type is represented by the como-, nepo-, poty-, and potexviruses. In type II, the RNA termini have a tRNA-like structure, as among the tymo-, tobamo-, bromo-, and cucumoviruses; in type III, the termini are devoid of both of those additions. Type IV, represented especially by the tobacco mosaic virus (TMV) family, has a unified circular single strand of RNA as the genome. Because the last of these groupings contains features of the others in a relatively simple form, discussion can expediently employ it as the model.

Type IV Viral Genomes. The type IV viral genome is adequately represented by that of the well-known TMV, whose sequence has been completely established (Goelet *et al.*, 1982; Nishiguchi *et al.*, 1985). It is a single-stranded circle of 6395 nucleotides, appearing to encode six proteins, at least four of which are known (Takamatsu *et al.*, 1983; Watanabe *et al.*, 1984). Some of the genes partly overlap others and the first cistron encodes two, the longer one resulting from readthrough of an amber stop (UAG) codon that terminates transcription of the shorter. In some strains at least, a tRNA is present at the 3' terminus (Lamy *et al.*, 1975; Dillon, 1978, pp. 346–348), as in type II members. In this case, the structure accepts histidine.

Another type IV species, turnip yellow mosaic virus (TYMV), has a genome that similarly consists of a single molecule of single-stranded RNA, the plus strand, but in addition shows a feature characteristic of members of other groups. Here, as in several additional types, the 3'-terminal region of RNA encodes the coat protein, but never gives rise to that protein directly. Instead, an RNA molecule, referred to as "subgenomic RNA," is derived from that part by unknown processes. Thus, in TYMV there is a single genomic RNA and a subgenomic variety (Morch *et al.*, 1982).

Type I Viral Genomes. Among the better known viral genomes of type I is that of the cowpea mosaic virus (Stanley and Van Kammen, 1979), a member of the comovirus family. Its single-stranded RNA is subdivided into two parts, each of which has a poly(A) sector and serves as a messenger (J. W. Davies *et al.*, 1979). Moreover, both bear proteins at the 5' termini as in the picornaviruses of metazoans and in others of type I. To determine whether the poly(A) sector is preceded by an AAUAAA polyadenylation signal, as in cellular forms, the 3' region has been sequenced, but no such signal proved to be present (J. W. Davies *et al.*, 1979). Nor should any have been suspected, for, as in polioviruses, the poly(A) here is a permanent part of the genomic structure, not a tail that is added by enzymes following transcription. Analyses and nucleotidyl structural comparisons have revealed homology to exist between comoviral RNA-dependent RNA polymerase and that of such picornaviruses as the poliovirus (Franssen *et al.*, 1984), which is likely kin to this group.

Type II Viral Genomes. As a whole the type II viruses have received relatively little attention at the molecular level; one of the few genomes from this source that has been adequately investigated is that of the barley stripe mosaic virus. In this species, the RNA is subdivided into two parts, each of which, as characteristic of this type, bears a tRNA at its 3' end (Kozlov *et al.*, 1984) that accepts tyrosine. Unexpectedly, however, this virus intergrades with type I members in that the genomic RNAs bear poly(A) sequences. The A-rich sector is in the form of an oligo(A) tract located about 235 residues from the 3' end, which is part of the tRNA^{Tyr}.

Type III Viral Genomes. Among the type III viral genomes, sequences have been established to a greater or lesser extent from at least three species, the alfalfa, brome, and cucumber mosaic viruses. In all of these the organization of the RNA is identical, there being three genomic strands, plus a subgenomic one. The sequences of all the parts have been established for the alfalfa and brome mosaic viruses and have proven to be quite similar, except in size. RNA1, the largest of the tripartite structures, is 3644 nucleotides in length in the former species, against 3234 in the second (Cornelissen *et al.*, 1983a; Ahlquist *et al.*, 1984), whereas RNA2 consists of 2593 and 2865, respectively (Cornelissen *et al.*, 1983b). The first structure has been sequenced from cucumber mosaic virus and was shown to be 3389 residues long (Rezaian *et al.*, 1985). In each case the nucleic acid encodes a single protein. The smallest, RNA3, however, is dicistronic, but the second (3') portion of 881 residues is not translated, serving only as the source for the subgenomic RNA4, as in TYMV described earlier (Brederode *et al.*, 1980). As there, the latter codes for the coat protein. In another member of the group, brome mosaic virus, which infects grasses, the genome is similarly constructed, but the three genomic RNA molecules are larger, containing in order 3300, 3000, and 2100 bases, with RNA1 the largest (Ahlquist *et al.*, 1981).

10.5. TRANSCRIPTION SIGNALS IN VIRUSES

The complex diversity of viral genomes does not cease with their overall organization, infinitely varied though it is, but extends into the details of their gene structures. Whole books would be required to cover the characteristics of the cistrons of each viral type, despite the limitations of present knowledge. Hence, the purpose now is to provide samples through the viral world to give the flavor of the subject. First, details pertaining to transcriptional initiation and termination are provided, followed by such other peculiarities as the overlapping of genes.

10.5.1. Promoters of T-Series Bacteriophage Genes

As may well be expected, the bacteriophages are outstandingly more thoroughly explored in the matter of gene structure than is any other type of virus, so here, as in so many other aspects of viral morphology, function, and evolution, they provide the firmest basis for introducing the present topic. Also not unexpectedly, the T-series of these bacterial parasites lead the others in the thoroughness of coverage by investigators, and consequently provide the basic model.

Early Promoters of T-Series Bacteriophages. Since the early region of the genomes of the T-series bacteriophages is transcribed by the host (*E. coli*) polymerase, the translation signals should be expected to resemble those structures of bacteria. And so that of T7 proves to do—but not in the usual sense! What it does resemble of the *E. coli* TA-TA- box (the −10 sequence) is in its deviation from that "consensus" (Table 10.1), as earlier chapters disclosed in the bacterial genome. One of the three more active promoters, A1, and the less responsive farther upstream, A0, are quite similar to one another in structure, but their shared nucleotidyl sequence, AT-CT-A, shows little in common with the standard bacterial sequence (Dunn and Studier, 1983). In contrast, the other two active T7 early promoters, A2 and A3, along with that of gene *63* of T4, display somewhat greater resemblance to the accepted TA-TA- structure (Pribnow, 1975; Dunn and Studier, 1983; Rand and Gait, 1984). What is of particular interest, however, is that the first two promoters of T7 share so many homologous sites between themselves (Table 10.1), but not with the last two, the reverse of which statement is likewise true. Another point of kinship, previously unnoticed, is the standard translational stop signal TAA located upstream of each promoter, the significance of which is unknown.

Mid- and Late-Stage Promoters of T7. In six of the seven midstage sequences of T7 (Table 10.2) the combination CGACTCA is consistently found at the −10 region,

Table 10.1
Early Promoters of T-Series Bacteriophages[a]

				Promoter		Start Site	
T7 Early[b]							
A0	ACCTCC	TAA	CGTCC---	ATCCTAA	AGCCAA	C	ACC
A1	AAAGTC	TAA	CCTATAGG	ATACTTA	CAGCCA	T	CGA
A2	ATGAAG	TAA	CATGCAG-	TAAGATA	CAAATC	G	CTA
A3	ATGAAG	TAA	ACACGG--	TACGATG	TACCAC	A	TGA
T4 Early[c]							
Gene *63*	TCCCTC	GTG	TTGTGT	TATAGTA	GTCTTA	C	TGA

[a]Promoters and frequent start sites of transcription are underscored.
[b]Dunn and Studier (1983).
[c]Rand and Gait (1984).

Table 10.2
Mid- and Late-Phase Promoters of T-Series
Bacteriophages[a]

				Promoter −10		Start Site	
T7 Mid[b]							
φ1.5	AGTTAA	CTG	*GTAATA*	CGACTCA	CTAAAG	G	AGG
φ1.6	TGGTCA	CGC	*TTAATA*	CGACTCA	CTAAAG	G	AGA
φ2.5	GCACCG	AAG	*TAA-TA*	CGACTCA	CTATT-	A	GGG
φ3.8	CGTGGA	TAA	*TTAATT*	GAACTCA	CTAAAG	G	GAG
φ4c	CCGACT	GAG	*ACAATC*	CGACTCA	CTAAAG	A	GAG
φ4.3	AGTCCC	ATT	*CTAATA*	CGACTCA	CTAAAG	G	AGA
φ4.7	TTCATG	AAT	*ACTATT*	CGACTCA	CTATAG	G	AGA
T4 Mid							
denV[c]	TACATC	TCC	*TGTAGG*	TATGATA	CTATAG	A	CCT
Gene 55[d]	(Incomplete)			TATGAAT	TGAGCT	A	AGA
Orf D[e]	GCTCCT	ATA	*TTGCTT*	TATAAAT	TTTT--	T	GGT
T7 Late[b]							
φ6.5	GTCCCT	AAA	*TTAATA*	CGACTCA	CTATAGG	G	AGA
φ9	GCCGGG	AAT	*TTAATA*	CGACTCA	CTATAGG	G	AGA
φ10	ACTTCG	AAA	*TTAATA*	CGACTCA	CTATAGG	G	AGA
φ13	GGCTCG	AAA	*TTAATA*	CGACTCA	CTATAGG	G	AGA
φ17	GCGTAG	GAA	*ATAATA*	CGACTCA	CTATAGG	G	AGA
φOR	CACGAT	AAA	*TTAATA*	CGACTCA	CTATAGG	G	AGA
T3 Late							
pjB10[e]	AAACAC	TGG	AAG*TAA*	TAACCCT	CACTAA	C	AGG
HpaIN[e]	TCCAAC	GTT	GTC*TAT*	TTACCCT	CACTAA	A	GGG
pjB20[e]	GAAGTG	AAA	GCC*TAA*	TTACCCT	CACTAA	A	GGG
MboI-E[f]	TCAATG	AGT	TTG*CAT*	TAACCCT	CACTAA	A	GGG
T4 Late							
Gene 67[g]	TCGTTT	CCA	AGA*CCC*	CGACCAA	GAACAA	G	AGG
Orf[h]	(Incomplete)			-AAGCTT	GCTAAG	G	AGA
P23[i]	CACTAT	TAC	TGAGAG	TATAAATA	CTCCCT	G	ATA
Gene 45[j]	TTTAAC	+15	AAATTA	GTTATAA	AATTAA	A	TCT

[a] Possible promoters and start sites are underscored; the regions adjoining the −10 sequences that may also be involved in promotion are italicized. Orf, open reading frame.
[b] Dunn and Studier (1983). [c] Valerie *et al.* (1984). [d] Gram and Rüger (1985). [e] Bailey *et al.* (1983). [f] Sarkar *et al.* (1985). [g] Völker *et al.* (1982). [h] Purohit and Mathews (1984). [i] Elliott and Geiduschek (1984). [j] Spicer *et al.* (1982).

where the promoters of prokaryotes are typically located, while that of the seventh (φ3.8) deviates only at the first two sites. Farther upstream is what could pass for the TA-TA-sequence (in italics) often considered a characteristic of the *E. coli* promoter, but it varies widely in structure from sequence to sequence. Furthermore, these runs lie more distant from the start site than in bacteria. In view of its location and conservation of structure, it is here proposed that the (CG)ACTCA just pointed out may serve as the transcriptional initiation signal in T7, either alone or in conjunction with nucleotides lying to either side. In the three sequences flanking the 5′ ends of T4 midphase genes, a structure closely akin to the bacterial standard is to be noted, properly placed around the −10 point. Downstream of the promoter in *denV* the CTATAG sequence is identical to the last of the T7 structures (φ4.7), but no uniformity of construction is to be observed in this sector, such as that marking those of T7.

Among the late genes of T7, similar constancy of construction is perceived (Table 10.2). In the six promoters from this region, precisely the same sequence is found in the −10 position as in the midsector, but this time there are no deviant forms. Indeed, the upstream hexanucleotide combination also is constant, being TTAATA in all except φ17, which deviates solely in having the initial base A. Moreover, all the downstream nucleotides are invariant to beyond the start site of transcription, the resemblance of the CTATAGG that begins this series to the CTAAAG of the midphase gene being self-evident. Again the constancy of structure and location of the CGACTCA in the −10 locality (underscored) argues for that sequence's serving as the promoter, possibly in conjunction with at least a part of the adjacent upstream and downstream nucleotides, as discussed in more detail just below.

The actual start sites of early transcription deviate widely (Table 10.1), since each of the four from T7 begins with a different nucleotide, but all are located relatively close to the translational start point, ATG. As far as can be detected, ancillary sites are absent from genes of all three phases (Elliott and Geiduschek, 1984).

A New Promoter Sequence? Because both mid- and late-phase genes are transcribed by RNA polymerase of viral origin, it is economical of space to treat them jointly in examining the foregoing proposal more extensively. First it needs to be noted that in comparison with the bacterial polymerase, the corresponding enzymes of these T-series bacteriophages are quite simple, consisting of a single polypeptide (Bailey *et al.*, 1983) that contains 884 amino acid residues (Moffatt *et al.*, 1984). Furthermore, other factors, still poorly known, also play important roles in the transcriptive processes, which are far more complex than implied by the simplicity of the polymerase (Kassavetis *et al.*, 1983; Pulitzer *et al.*, 1985). In phage T4, for instance, the products of at least four or five genes are essential.

As just seen, start and termination sites are well known in T7, and several studies have been made on promoters of T4 (Spicer *et al.*, 1982; Völker *et al.*, 1982; Rand and Gait, 1984; Valerie *et al.*, 1984; Gram and Rüger, 1985), but none seem to have been completed on other members of the two T-series. Nevertheless, the starting points of transcription of several genes of T3 have been indicated (Bailey *et al.*, 1983; Sarkar *et al.*, 1985), and through use of these, a further analysis of the promoter regions can be provided on a broader basis.

That the proposal made in connection with the initiation sectors of T7 may be justified is substantiated by those of three T3 late genes, whose promoter regions have been sequenced (Table 10.2). There the combination T-ACCCT is found, while upstream from that point only three bases show relations to the TA-TA box. Combining the two yields the decanucleotide TAAT-ACCCT, obviously related to that standard promoter of *E. coli*. The downstream series is quite as invariant as that of T7, but its CACTAA shows no kinship in structure. However, when the proposed promoter and this portion (underscored) are properly aligned as follows, with unimportant bases in lowercase, an evolutionary relationship can be detected:

(T7)	TA–cgaCTCA	CTATAGG
(T3)	<u>TAacc–CTCA</u>	CTA-A

Thus, it may be that the promoter sequence of T3 has been derived by deletion of a portion of that of T7, with parts of the neighboring upstream series becoming added to it. In T4 all constancy of structure is lacking in the 5′ leaders of four late genes, including the sequences in the −10 region and adjoining parts, so that the sequence requirements for transcriptional initiation by the polymerase cannot be fully detected at present.

10.5.2. Promoters of Genes of Other Bacteriophages

As a whole, the processes of transcription in other bacteriophages, including initiation and termination, are still in their early stages of exploration, and details are intermittently available. Accordingly, what has been established in the remaining DNA types is combined in a single table with similar information regarding RNA species (Table 10.3). Consequently, different polymerases are involved in transcription of the latter than in the former. For ease of comparison, sequences from RNA varieties have the Us for uridine replaced by Ts for thymidine, as in DNAs. For the sake of continuity and clarity, several additional DNA phages are examined before any of the RNA type.

Promoters of Bacteriophage λ. Since the entire genomic sequence of bacteriophage λ has been established (Sanger *et al.*, 1982), a number of its promoter sites are known, five that have been studied experimentally being given in Table 10.3 (Schwarz *et al.*, 1978; Hoyt *et al.*, 1982; de Haseth *et al.*, 1983; Ho *et al.*, 1983; Shih and Gussin, 1983; Hoopes and McClure, 1985). Comparisons of the −10 regions of this quintet with the bacterial consensus sequence reveal only vague resemblances and no real homology. Moreover, very little similarity is found among the five promoters themselves. In the −35 zone, considerable constancy in structure is found to exist between P_I and P_{RE} and again between P_R and P_{RH}−, but this relationship is not reflected in the promoters of the respective pairs. The DNA-dependent RNA polymerase appears to favor A and T as the start sites for its activity.

Control of certain genes is a complex process, involving the products of several cistrons; this is especially the case in coding elements concerned with the establishment of lysogeny in this temperate virus. Among the genes and products (in parentheses) concerned with this activity are the *cI* (repressor), *int* (integrase), *xis* (excisionase), and *cII* (cII protein), expression being chiefly from the promoters P_I and P_L, along with P_{aQ}, the function of which remains hypothetical (Hoopes and McClure, 1985). The first two are associated with the integrase and excisionase genes, which overlap one another, *xis* extending farther upstream, so that the promoter P_I of *int* includes its translational start

Table 10.3
Promoters of Various DNA Bacteriophage Genes[a]

	Ancillary (-35)		Promoter (-10)	Start site		
E. coli consensus	TGTT-GACANTTT		TATAATC			
Phage λ						
P_R[b,c]	GTGTTGACTATTTTA	CCTCTGGCG--	GTGATAA	TGGTTGC	A	TGT
P_I[d]	TTGC-GTGTAATTGC	GGAGACTTTGC	GATGTAC	T------	T	G--
P_{RE}[e]	TTGC-GTTTGTTTGC	(Incomplete)	GTAAGTA	T------	A	G--
P_{aQ}[f]	GCTC-GTGAACGTCA	TGGAAACGGA-	ATCATAA	AGGAAGT	T	CGA
P_{RM}[c]	GTGTTAGATATTTAT	CCCTTGCGGTG	ATAGATT	TAA-CGT	A	TG
Phage SPO1[g]						
TF1	TTTGGAGAGAAGTTT	CAAACACCC---	GATTTTT	TTATTA	C	GA
Phage Φ29[h]						
G3b (early)	GTGTTGAAAATTGT	CGAACAGGGTGA	TATAATA	AAAGAGC	T	AGA
Phage P2 (late)[i]						
F	ATAGCCTGACATCTC	CGGCGCAACT-	GAAAATA	-CCACT	C	ACC
O	ATGGCGGAGGATGCG	CATCGTCG---	GGAAACT	GATGCC	G	ACA
P	TTAGCGATCGCGGGG	CGCGACTCA--	GTAGCCT	TGCCGT	G	TAT
V	ATAGCATAACTTTTA	TATATTGT---	GCAATCT	CACATG	C	ATG
Phage Mu						
mom1 (late)[j]	TTAAGATAGTGGCGA	ATTGATGCAA-	AGGAGGTGA	GATGAA	A	TCA
mom2 (late)[k]	CACTCGACCCATGAT	GTTTTTTAAGA	TAGTGGCGA	ATTGAT	G	CAA
dam (late)[l]	GATCGAATCAATTAA	ATCGATCGG--	TAATACAG	ATCGAT	T	ATG
pC (early)[m]	GCTTTACATTAAGCT	TTTCAGTAA--	TTATCTT	TTTAGT	A	AGC
Phage ΦX174						
PG[n]	CTGTTGACAT(+11)	GTGGATTAC	TATCTGAG	TCCGAT	G	CTG
PA'[n]	AGCCTTGACCCTAAT	TTTGGTCG-	TCGCGTAC	GCAATC	G	CCG
PA1[n]	TAGCTTGCAAAA(+7)	CCTTATGGT	TACAGTATG	CCCATC	G	CAG
PA2[o]	TTGACACCCTCCCA-	ATTGTATGT	TTTCATC	CCTCC-	A	AAT
PD[o]	ACATTTTAAAAGAGC	GTGGATTAC	TATCTGA	GTCC--	G	ATG
PB1[o]	TAGCGTTGACCCTAA	TTTTGGTCG	TCGGGTA	CGCA--	A	TCG
PB2[o]	TTGCAAAATACGTGG	CCTTATGGT	TACAGTA	TGCCC-	A	TCG

(continued)

Table 10.3 (Continued)

	-35	Promoter (-10)	Start site		
Phage fd					
$P(X)^n$	TTTGATGCAATT(+6) GCTTCTGAC	TATAATA	GACAGG	G	TAA
$P(IV)^n$	ACTATTGACTCT(+7) GTCTTAATC	TAAGCTA	TCGCT-	A	TGT

[a]Experimentally confirmed sectors are underscored (promoters) or italicized (ancillary and start sites).
[b]de Haseth *et al.* (1983).
[c]Shih and Gussin (1983).
[d]R. W. Davies (1980).
[e]Ho *et al.* (1983).
[f]Hoopes and McClure (1985).
[g]Greene *et al.* (1984).
[h]Hattman and Ives (1984).
[i]Christie and Calendar (1985).
[j]Plasterk *et al.* (1984).
[k]Murray and Rabinowitz (1982).
[l]Plasterk *et al.* (1983).
[m]Krause *et al.* (1983).
[n]Otsuka and Kunisawa (1982).
[o]Sanger *et al.* (1977).

signal. However, the second promoter, P_L, is located still more strongly 5' of that point, although its identity unfortunately has not been fully established (R. W. Davies, 1980). Thus, activation of P_L results in transcription of both integrase and excisionase, whereas transcription from P_I by the product of *cII* produces only mRNAs for integrase, the aborted coding sequence of *xis* becoming incorporated into its leader sequence (Campbell, 1983). The cII protein binds DNA, selectively interacting with a repeat sequence at the −35 location on the face of the DNA molecule opposite that employed by RNA polymerase (Ho *et al.*, 1983).

Promoters of Other DNA Phages. In the DNA of SPO1 of *Bacillus subtilis*, the typical thymidine is replaced enzymatically with 5-hydroxymethyluracil, a condition that is important in the transcriptive processes of the virus (Greene *et al.*, 1984). The gene *tf1* encodes a DNA-binding protein called transcription factor 1 (TF1), which reacts preferentially with sites containing the modified base. Thus, it negatively controls transcription. Transcription in SPO1 is carried out by three different polymerases. In transcribing early genes, the host enzyme is employed, in midphase that protein is modified by phage-encoded subunit σgp28, and in the late phase, modification involves two phage subunits, σgp33 and σgp34 (Lee and Pero, 1981; Pero, 1983). The promoter, which has been experimentally determined, shows few correspondences with any of bacteriophage λ, a statement equally true for the ancillary site at the −35 location.

Another phage from this same bacillus, that known as Φ29, provides some parallels of structure between the foregoing and the early gene cluster for the gene *G3b* (Table 10.3). Here there is a closer correlation in the promoter to the *E. coli* consensus sequence than in that of SPO1, but the relationship between the two phage early regions is confined to the

−35 zone, the AATTGTCGAACA of the present species being largely homologous to the AAGTTTCAAACA of the other. An additional point of resemblance is provided by the ribosomal recognition site. It has been proposed that a possible reason for the inability of *B. subtilis* to express genes from *E. coli*, which processes the other's genes freely, is that the signal mentioned might be too short in the cistrons of the latter for the polymerase to recognize (Murray and Rabinowitz, 1982). Whereas in *E. coli* a four-nucleotide combination such as GGAC suffices, the present form requires sequences like AGAAAGTGGG. Upon examination of the SPO1 cistron and its flanking regions (Greene *et al.*, 1984), that precise combination proves to be absent, but an equivalent, AAAGGGTGG, is found at a corresponding position, suggesting the feasibility of the proposal. To the contrary, in the second set of early genes from φ29 (Holder and Whiteley, 1983), AGGAGG is advocated as serving as that signal (Escarmís and Salas, 1982).

The promoters of the four late transcriptional units of phage P2, from *E. coli*, have been the subject of a recent investigation (Christie and Calendar, 1985). In this species expression of the late genes requires the host polymerase and the product of the phage gene *ogr*, which apparently modifies the α subunit of that enzyme. Also requisite are the products of two P2 DNA replication genes, *A* and *B* (Lengyel and Calendar, 1974). Comparisons of the four late sequences demonstrate their variability among themselves (Table 10.3), in which only the occupants of the first and third sites are constant. In the first two cistrons, some resemblances to the *E. coli* consensus structure are shown, but in the remaining pair the sequences are totally different.

The processes of transcription of just one more example of double-stranded DNA phages need attention, that of bacteriophage Mu, which is one of the most active transposable elements known (Krause *et al.*, 1983). At times the viral genome may be transposed to as many as 50 different sites in the host DNA in 1 hr. One of the late genes, *mom*, which modifies certain adenine residues of its genome, is expressed only when the product of a second cistron, *dam*, is present. Also required is the protein encoded by *dad* (Hattman and Ives, 1984). Transcription of *mom* has been demonstrated to occur only after several copies of the tetranucleotide GATC situated upstream of its promoter have been methylated by *dam* (Plasterk *et al.*, 1983). The entire gene and leader have been sequenced by two laboratories independently, and, while the structures of the two are identical, different interpretations have been given as to the start site of transcription and, concomitantly, the identification of the promoters (Hattman and Ives, 1984; Plasterk *et al.*, 1984). Given in Table 10.3 as *mom1* and *mom2*, the two are seen to differ extensively between themselves and also with the corresponding sector of the third late gene, *dam*. The latter correlates most closely with the *E. coli* consensus series of nucleotides, in fact having greater resemblance than the early promoter pC cited there (Krause *et al.*, 1983).

One of the traits peculiar to Mu is a site-specific inversion of a genomic segment, called G, that carries four genes in the sequence 3′ *Sv–U–U′–S′v* 5′, the last two being in opposite orientation from the first pair (Figure 10.6A). All encode tail-fiber components involved in the infection of its host. When segment G becomes inverted, *U′* and *S′v* are transcribed, since they then lie on the same strand as the promoter and adjacent to it, while the other two become unexpressed, since no such promoter then is present in the correct orientation (Figure 10.6B; Craig, 1985). Inversions occur within 34-base-pair inverted repeats, located one at each terminus of G, and are catalyzed by the invertase product of *gin*, which gene lies just downstream of the sector; however, an unidentified host factor is

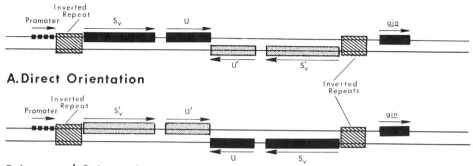

A. Direct Orientation

B. Inverted Orientation

Figure 10.6. A site-specific inversion in the double-stranded DNA bacteriophage Mu. Inversion of a DNA segment changes transcription from that of (A, solid black) the typical genes to (B, stippled) a second set. (After Craig, 1985.)

also requisite (Kahmann *et al.*, 1985). Similar inversions that control gene expression are known in the phages P1 and P7 and in the bacterium *Salmonella typhimurium*.

Promoters of Genes of Single-Stranded DNA Phages. Very little firm information is available regarding promoter structure in single-stranded DNA phages; the main exceptions pertain to two species, ΦX174 and fd (Sanger *et al.*, 1977; Otsuka and Kunisawa, 1982). In Table 10.3 seven from the first virus and two from the second are cited, whose general interrelatedness makes a combined discussion possible. One of the transcription recognition signals, P(X), proves to be virtually identical to the *E. coli* consensus, but the remainder shows at most a 50% correspondence to that sequence, with the majority having only the first nucleotide (T) like that found in the bacterium. Almost all terminate in A rather than G. Like the majority shown in the table, the polymerase can begin transcription with any of the four common nucleotides, as is especially clearly illustrated in the PA′ sequence, where initiation may be at any site in the combination ATCGC.

10.6. PROMOTERS OF EUKARYOTIC DNA VIRUSES

It is not proposed here to examine the transcriptive processes in every type of DNA virus of eukaryotes, nor later to do likewise with the RNA species. Rather, present purposes are met best by providing depth in a few types where the synthesis of mRNAs has been more fully investigated, supplemented by several samples of less well-documented representatives to create the necessary breadth. Because transcription in the relatively rare DNA viruses of seed plants has not been explored sufficiently, the DNA viral group is represented largely by forms from vertebrates. The primary exception among plant DNA viruses is the cauliflower mosaic virus, which will be recalled as a double-stranded type with two breaks in the plus and one (Δ1) in the strand that is transcribed, the minus. Six reading frames have been disclosed, all obviously in the same orientation and together covering 85% of the transcribed molecule (Guilley *et al.*, 1982). While promoters have been detected, they have not been identified, but at least one transcript originates near Δ1.

Early Promoters in Adenoviruses. The several species of adenoviruses, particularly those referred to as 2, 5, and 12, have been relatively abundantly explored as to their processes of transcription and therefore are especially useful among vertebrate viruses in introducing the subject. A degree of confusion results from the employment in the literature of two different forms of designating the genes, the letter E (early) being sometimes followed by Arabic, sometimes by Roman, numerals; the first of these alternatives is followed here. One of the pre-early genes, *E1A,* is of outstanding importance in that its product at times is involved in the transcription of early cistrons. Experiments on transcription of this coding element suggest that a promoter somewhat resembling the typical TA-TA- is present, not at its accustomed -10 position, but farther upstream at -28 (Table 10.4; Osbourne *et al.,* 1982). No region $5'$ to this sector influences transcription, nor is the promoter essential, but if it is removed by mutagenesis the level of production of mRNAs from *E1A* is reduced to 10 or 20% of the wild-type level. Other studies have yielded somewhat different results, in that the region above -231 has been reported to enhance transcription from the -28 promoter (Hen *et al.,* 1983; Sassone-Corsi *et al.,* 1983).

Expression of the early gene *E2A* is especially influenced by the 289-amino acid phosphoprotein encoded by the longer (13 S) mRNA produced from the foregoing cistron, a substance requisite for transcription from any early promoter (Gaynor and Beck, 1983). Strikingly, this stimulant of adenovirus transcription represses those processes from SV40 early promoters (Velcich and Ziff, 1985). But there may be some confusion here because of *E1A* having two promoters, for it has been demonstrated that the product of the shorter mRNA (12 S) from *E1A* represses *E2A* activity (Guilfoyle *et al.,* 1985). Thus the existence of two overlapping genes in *E1A* is clearly intimated. The postulated promoter of the early cistron *E2A* may be noted to be partly homologous to that of the pre-early gene (Table 10.4) and lies at a comparable distance from the start site. An ancillary sequence has been detected, located nearly 80 residues upstream of the start (Murthy *et al.,* 1985). It is worthy of note that two other studies of presumably this same gene of adenovirus 2 reported entirely different promoter and surrounding regions, distinguished in Table 10.4 as *E2A$_2$* and *E2A$_3$* (Langner *et al.,* 1984; Zajchowski *et al.,* 1985). In addition, the latest of these researches reported dual promoters, here distinguished as φ1 and φ2. The latter, farthest from the translational start site, bears considerable resemblance to the standard TA-TA- combination, while the first shows none at all. In addition to the promoters, a region lying between the two (italicized in *E2A$_3$* φ1) is essential for transcription from either one of them, and the ancillary site italicized in the φ2 sequence is required for expression from that promoter. This second one is also necessary for activation by the *E1A* product (Zajchowski *et al.,* 1985). In a promoter of a third gene in this category, *E3,* the sequence (Table 10.4) shows loose kinship to the others from this virus, but the ancillary site displays no similarity in any form (Lee *et al.,* 1982). Despite this variability in the signal, all transcription of the early genes results from activity of the same host-cell RNA polymerase II, possibly with the aid of various ancillary proteins.

Late Promoters in Adenovirus. Some of the problems of initiation of those adenovirus late genes that are transcribed by RNA polymerase III have already received attention in Chapter 3, Section 3.2.4, to which reference should be made for a more complete view of the subject. But those that encode an mRNA are, like the early ones, acted upon by RNA polymerase II. Very few in that category have been examined for transcriptional properties, the major late (*ML*) gene and an associated intermediate one,

Table 10.4
Promoters of DNA-Viral Genes of Mammals[a]

	Ancillary			Promoter				Start Site	
Adenovirus									
$E1A^b$	CGTTTTTATT	ATTATAGTCAGC	TGACGTGTAGTG	--TATTTATA-	CTCGGTGAG	(+10)GCC	*A*	CTC	
$E2A_1^c$	AGATGACGT	AGTTT(+23)CG	CGAAACTAGTCC	--TTAAGAGT-	CAGCGCGCA	(+8)TGA	*A*	GAG	
$E2A_2^d$		(Incomplete)	TCAGG	--TACAAATT-	TGCGAAGGT	(+9)TCC	*A*	CAG	
$E2A_3\phi1^e$	AGATGACGT	AGTTT(+21)AG	*GGCGCGAAA*CTA	--GTCCTTAA-	GAGTCAGCG	(+12)TGA	*A*	GAG	
$E2A_3\phi2^e$	CGCCG*GGTG*	*TGGCC*(+6)ACG	TAGTTTTCGCGC	--TTAAATTT-	GAGAAAGGG	(+13)CTT	*A*	AGA	
$E3^f$	GGCGCAGCT	TGCGG(+23)TC	GCCCGGGCAGGG	--TATAACTC-	ACCTGAAAA	(+9)GAG	*G*	TAT
ML (late)^g	GTGATTGGT	TTATA(+24)TC	CTGAAGGGGGGC	--TATAAAAG-	GGGGTGGGG	(+11)CTC	*A*	CTC	
$IVa2$ (mid)^g	CCCCCTAGT	GGACA(+15)CC	CACTTAGCCTCC	--TTCGTGCT-	GGCCTGGAC	(+11)GTC	*T*	CAG	
SV40									
ϕ Early^h	*CCATTCTCC*	*GCCC*CATGGCTG	ACTAATTTTTTT	--TATTTAT--	GCAGAGGCC	GA(+7)TC*G*	*G*	*CCT*
ϕ Late^i	(Incomplete) CAGCTGGTTCTT	TCCGCCTCAGAA	GGTACCTAACC	AAGTTCCTC	(+8)GTT	*A*	T		

[a]Experimentally determined promoters are underscored and initiation sites are italicized. Inc, incomplete.
[b]Osbourne *et al.* (1982).
[c]Murthy *et al.* (1985).
[d]Langner *et al.* (1984).
[e]Zajchowski *et al.* (1985).
[f]Lee *et al.* (1982).
[g]Matsui (1982).
[h]Baty *et al.* (1984).
[i]Brady *et al.* (1982).

IVa2, being among the exceptions (Matsui, 1982; Concino *et al.*, 1984). Although the *ML* leader contains a recognizable TA-TA- sequence, the midphase does not (Table 10.4), yet it is transcribed as faithfully as the other in an *in vitro* system. These two cistrons are in the form of an inverted pair, with divergently arranged promoter regions. The absence of any recognizable promoter has been reported by other laboratories, along with that same condition from *E2A* (Brady *et al.*, 1982; Natarajan and Salzman, 1985). In the later of these investigations, the results suggested the need for a product of the *E1A* along with an enhancer and promoters, but the last two elements were not specifically identified.

The Promoter of SV40 Early Genes. The early genes encoded in the SV40 genome are expressed continuously throughout lytic infection, while the late ones are suppressed until onset of viral DNA replication (Tack and Heard, 1985). After entry into the host's nucleus and after the parasite's DNA has been completely uncoated, cellular RNA polymerase II initiates transcription from the single promoter for the early region, resulting in the eventual production of mRNAs for large and small tumor (T) antigens. When those messengers have been translated, the large T antigen is returned to the nucleus, where it interacts with that same promoter and the region of origin of DNA replication (Keller and Alwine, 1984). The early promoter sequence shows no unusual traits, being identical to that of adenovirus *E1A* (Table 10.4); however, a distinctive feature is found immediately upstream of that signal in the form of an uninterrupted run of seven Ts (Benoist and Chambon, 1981; Ghosh *et al.*, 1981; Byrne *et al.*, 1983). The location of the promoter, which has been demonstrated to be essential for accurate transcription (Mathis and Chambon, 1981), is about nine base pairs closer to the start site than the adenoviral type, and hence is more similar to the corresponding element of prokaryotes.

However, the processes just described are simplistic to a degree in that they fail to include the elements involved in late-early transcription, applying only to those of the early-early stage. No specific promoter has (or, more likely, promoters have) been identified, but G,C-rich motifs contained within a trio of 21-base-pair repeated elements have proven essential (Baty *et al.*, 1984). The late-early start sites, shown underscored in Table 10.4, are located well before the TA-TA- sequence that serves in the early-early stage. Consequently, the latter is firmly excluded as a factor in this subsequent period.

An Enhancing Sequence. In addition to the promoter, a unique enhancer element has been reported to play an important role in transcription of both early and late genes. The atypical feature of this component is in its structure, which consists of a pair of identical sequences 72 base pairs long, arranged in tandem (Gruss *et al.*, 1981; Byrne *et al.*, 1983). Moreover, the repeated G,C-rich sectors extending upstream from their beginnings in the ancillary site (Table 10.4) also have proven to be of significance in the processes (Everett *et al.*, 1983). Thus, early transcription in SV40 is known to require host RNA polymerase II, the promoter and enhancer, the large T antigen, and then G,C-rich sectors located in 21-base-pair repeats. Late-early stages are known to require the same G,C-rich elements and the 72-base-pair repeated segment, but specific promoters have not been identified. Thus, as seen in earlier chapters, transcription even in these viruses is not a simple, semiautomatic process, but is highly complex, involving many interacting enzymes and structural components.

The Promoter of SV40 Late Genes. Transcription of the late genes of SV40 that encode two mRNAs (one for VP1, the other for VP2 and VP3) involves the same 72-base-pair repeats as the foregoing, but takes place in opposite orientation (Ernoult-Lange and May, 1983; Sassone-Corsi *et al.*, 1984). Also necessary to efficient mRNA synthesis is the same series of G,C-rich repeated elements used in the early phase. Only a single promoter has thus far been proposed, one that is particularly striking in length, containing 11 base pairs (Brady *et al.*, 1982, 1984). Since thus much of the early stage control mechanism also affects late stage equally, it is difficult to account for the temporal effects that have been noted. However, refinement of experimental procedures have permitted the finding that subclasses of the large T antigen exist, at least one particular type of which is active only in regulation of late transcription (Tack and Heard, 1985). Moreover, the DNA structure appears also to be modified in the late period by an unknown substance which acts on the nontranscribed strand.

10.7. TRANSCRIPTION OF SOME RNA-VIRAL GENES

In analyzing transcription initiation of RNA viruses, much the same procedure is followed as in the preceding section, attention being devoted to a few, better documented examples from different types of the parasites to provide both depth and breadth in an economical format. Here some varieties from seed plants can be included, along with those of vertebrates, thereby filling out the picture more completely.

10.7.1. Transcription of Genes of RNA Viruses

A condition made conspicuous by its absence in the foregoing statement is any mention of RNA bacteriophages. Because all the familiar members of that group have the single-stranded genome of the plus variety, transcription is unnecessary, replication of the entire nucleic acid molecule taking its place. The same statement is equally true for those RNA viruses of eukaryotes that have plus-strand genomes. On the other hand, in those with only minus strands, replication is a separate function and has received attention in the literature, albeit to a limited extent.

Transcription in Vesicular Stomatitus Virus. In all negative-strand RNA viruses. including vesicular stomatitus virus (VSV), that genome is encapsidated with ribonucleoprotein particles to form nucleocapsids, as seen in a preceding section. As there, these structures serve as templates for transcription and also genomic replication (Arnheiter *et al.*, 1985), both processes being carried out by the same RNA polymerase of viral origin, perhaps with the aid of host factors. For early use, some of the enzyme is packaged into the virion. It initiates transcription at the 3' end of the genome and continues in sequence through the five genes in their structural order 3' *N–NS–M–G–L* 5', there being only one functional promoter, located at the 3' end (Emerson, 1982). Moreover, this activity produces leader RNA from the 3' terminus, the presence of which appears to shut down host macromolecular syntheses (Wilusz *et al.*, 1983; Grinnell and Wagner, 1984). Hence, the primary transcriptive product is a long, polycistronic mRNA, which is cleaved into the several portions by unidentified processes.

Transcription of Double-Stranded RNA Viral Genes. As described earlier, three groups of eukaryotic RNA viruses are similar in having genomes consisting of multisegmented double-stranded RNA, the cytoplasmic polyhedrosis virus of the silkworm and the reo- and rotaviruses of mammals. In the first two, there are ten segments and in the last 11, but all share such other characteristics as a viral RNA polymerase stored in the capsid and the presence of caps [including $m^7GppN(m)$] at the 5' ends of their messengers. At least in the insect parasites, the production of the cap is a prerequisite to transcription (Furuichi, 1978). Each genomic segment consists of a monocistronic mRNA (plus strand) united to its complement in an end-to-end base-paired duplex, except the 5' cap of the plus strand (Imai *et al.*, 1983). Since each segment is transcribed by the viral RNA polymerase into capped mRNAs, which can either be translated or employed as templates for synthesis of minus strands, every one must have promoters and replication signals. Despite the fact that a number of genes from reoviruses and a few from rotaviruses have been fully sequenced (Cashdollar *et al.*, 1982, 1985; Richardson and Furuichi, 1983; Dyall-Smith and Holmes, 1984), numerous cap sites but no promoters have been reported.

Transcription in Influenza Viruses. Discussion of the influenza viruses, which have a minus single-strand RNA genome, might well have preceded that of the double-stranded ones, but has been reserved for this point, since their transcriptive processes may throw light on those others. Like the above, the RNA exists in multiple strands, of eight segments, however, not ten or 11 as there (Huddleston and Brownlee, 1982); an additional resemblance is that a cap of identical structure is present at the 5' end of each.

Although most segments are monocistronic, at least three, numbers 6–8, encode two proteins. The last of these carries genes for a pair of nonstructural proteins, NS_1 and NS_2 (Lamb and Choppin, 1979; Lamb and Lai, 1980; Lamb *et al.*, 1980; Porter *et al.*, 1980), whereas the first bears those for the neuriminidase (NA) and a glycoprotein (NB) (Shaw *et al.*, 1983). The third exception, segment 7, more recently has been demonstrated to encode two membrane proteins, M_1 and M_2 (Lamb *et al.*, 1985).

Transcription requires the presence of the caps, the nucleocapsid protein (NP), and three P proteins—PB_1 and PB_2 being basic, and PA being acidic (Braam *et al.*, 1983). NP, the predominant protein of the virus, comprising ~90% of the total proteins, is located along each of the eight RNA segments at about 20-nucleotide-residue intervals. Seemingly the three P proteins form a complex, which at the onset of transcription moves as a unit from the 3' ends of the viral RNA strands down the mRNAs as they are synthesized. PB_2 also interacts with the cap structure, which is then cleaved by a viral endonuclease at a purine located 10–13 residues distant, the resulting capped fragments serving as primers. Transcription proper then is initiated by the addition of a guanosine residue onto the 3' end of the primers, a reaction guided by the penultimate C residue of the viral RNA strands. Following initiation, the polymerase complex elongates the messengers in typical fashion (Braam *et al.*, 1983). Consequently, specific promoter sequences are not a feature of transcription in these and possibly others that possess capped RNA genomic strands. In one gene whose sequence has been established, that for neuraminidase (Hiti and Nayak, 1985), a TAA translational stop signal immediately precedes the ATG initiation triplet, but whether it plays any significant role is not established.

10.7.2. Transcription of Plant RNA-Viral Genes

Despite the establishment of a number of genomic sequences from RNA viruses of plants, their processes concerned with transcription and with synthesis of the characteristic subgenomic RNAs still remain completely unknown (Watanabe *et al.*, 1984). In the tobacco mosaic virus, the 5' end of the genome is capped by mGpp (Goelet *et al.*, 1982) and that structure possibly plays a role in its synthesis, no transcription of this plus-strand nucleic acid being necessary. However, the subgenomic strands are transcribed, that for the 30,000-dalton protein being initiated at a guanosine 1550 residues upstream of the terminus (Watanabe *et al.*, 1984), and that for the coat protein at 693 sites from its 3' end (Guilley *et al.*, 1979). Aside from those few data, the mechanism, enzyme(s), and other factors remain for future investigations to reveal.

10.7.3. An Evolutionary Sequence of Transcriptional Events

Although firm data regarding initiation of transcription in viruses are lacking in the desired abundance, even casual reading of the foregoing descriptions discloses differing levels of complexity in those processes. At present, the events can be arranged at best into a mere skeleton of a phylogenetic succession, but perhaps its presentation will stimulate additional studies to detail its evolutionary progress further.

The starting point in the sequence is obvious, for what can be simpler insofar as transcriptional initiation is concerned than an inheritable genome that consists of messengers ready for translation? Thus, single-stranded RNA viruses whose genome is a plus strand represent the earliest phase in the development of transcription; multiple molecules,

Table 10.5
Terminators of Bacteriophages[a]

Phage T7[b]						
TE	CGTTTATAAGGA-	GACACTTTATGT	TTAA	GAAGGTTGG	TAAATT-C	CTTGCGGCTTTG
Tφ	TGCTGA-AAGGAG	GAACTATATGCG	CTCA	TACGATATG	AACGTTGA	GACTGCCGCTGA
Phage f1						
I[c]	TAAACCGATACA	ATT*AAAGGCTCC*	TTTT	*GGAGCCTTT*	TTTTTTGG	AGATTTTCAAC-
II[d]	CCCTTTGACGTT	*GGAGTCCACGTT*	CTTT	*AATAGTGGA*	*CTCT*TGTT	CCAAACTGGAAC
Phage fd[e]						
I	TAAACCGATACA	ATT*AAAGGCTCC*	TTTT	*GGAGCCTTT*	TTTTTTGG	AGATTTTCAAC-
Phage M13						
I[f]	TAAACCGATACA	ATT*AAAGGCTCC*	TTTT	*GGAGCCTTT*	TTTTT	
II[g]		*AACCTCCCG*	CAAG	*TCGGGAGGT*	*T*CGCT	
Phage IKe[h]						
I	TTTTCAGCGTTA	TTT*AAGGGGCGC*	TATT	*GCGCCCTTT*	TTTTTACT	TTAATTCAGCTA
Phage φX174[i]						
T4	GTATGT	TTTCAT*GCCTCC*	AAAT	*CTTGGAGGC*	TTTTTTAT	GGTTCGTTCT--
T2	CAACAATTTTAA	TTGC*AGGGGC*TT	CGGC	*CCCT*TACTT	GAGGAT	
T1	ACTATA	GACCACCG*CCC*	GAAC	*GGG*ACGAAA	AATGGTTT	TTAGAGAACG
Phage P22						
T ant[j]	GATAACCAAC	GCAACG*ACCCAG*	CTTC	*GGCTGGGT*T	TTTTTATG	
T nutR[k]	CCA	ATCTGA*ACCGCC*	GACA	*ACGCGG*TAA	ACC	
R₁[k]		TCAAA*GCGCA*	*T*-CA	ACGA*ATGCG*	*C*ACAACTA	ACTAT
Phage λ[k]						
T nutR		*GCCCTG*	AAAA	*AGGGC*ATCA	AATTAAAC	CACAC
R₁		CTATGG*TGTATG*	*CA*TT	TATT*TGCAT*	*ACA*TTCAA	TCAATT
Phage Pf3[f]						
CP		TCGTT*ATAAGG*	*GGGC*	TTCGG*CTCC*	*CTTAT*TCG	TTTA

[a]Regions of dyad symmetry are italicized and noncanonical base pairs are underscored.
[b]Dunn and Studier (1983). [c]Hill and Petersen (1982). [d]La Farina and Vitale (1984). [e]Beck et al. (1978). [f]Luiten et al. (1983). [g]Smits et al. (1984). [h]Peeters et al. (1985). [i]Otsuka and Kunisawa (1982). [j]Berget et al. (1983). [k]Backhaus and Petri (1984).

each encoding a single gene, probably antedated polycistronic types, for the latter require an additional element in the form of a cleaving enzyme. The next logical step is found in those single-stranded RNA species that need to be transcribed into messengers. In the earliest of such minus-stranded forms, the polymerase of transcription is also that for genomic replication, and requires no promoter or other signal, the 5' cap alone being a prerequisite. Those that require a promoter sequence, and perhaps other factors, are at a still higher level of the processes, followed by DNA-containing species, but the steps by which transcription activities developed their greater complexity beyond that point are too dim to recount now.

10.8. TERMINATION OF TRANSCRIPTION IN VIRUSES

Although data pertaining to termination of transcription in viruses is not yet abundantly available, present knowledge is not so limited as that concerning initiation. At least it is found that many suggestions have been made as to the structure of possible terminators in a diversity of types. As expected, the terminators of bacteriophages provide the richest source of information—but not from the usual reliable forms.

10.8.1. Terminators of Bacteriophages

The T Series of Phages. As just intimated, the standard model of molecular virology, the two T series of bacteriophages, are of little value in supplying data concerning termination of transcription. Studies on the genomic sequence of T7 indicate the location of two terminators, from which the nucleotide structure of those regions given in Table 10.5 have been derived (Dunn and Studier, 1983). Obviously no series of Ts is found here, as is the frequent case in bacteria, nor is any present in the structure downstream of the initial section provided. Since regions of dyad symmetry also are lacking, how transcription is terminated here cannot even be conjectured. Nevertheless, the two examples are not without interest, for the one that lies close to the end of the genome, Tφ, is seen to be largely homologous to TE, the terminator of early transcription. The similarities of structure are striking through the first 40 residues, but they are especially impressive through the opening 24 sites. The frequent identities of occupants in corresponding positions suggest clearly that that sector may eventually prove to be involved in termination.

Terminators of Filamentous Bacteriophage Genes. The terminator sequences of the first three of the four from filamentous viruses (f1, fd, and M13) add further support to the concept that those organisms have common ancestry (Luiten *et al.*, 1983), for these structures are 100% homologous (Table 10.5). Included in the identities is the region of dyad symmetry (italicized), with a loop composed of four Ts and a run of eight Ts at the 3' end, thus greatly resembling the corresponding signal of *E. coli*. In the fourth member of that group, IKe, there is general resemblance, but greatly reduced homology, both in the dyad symmetric portion and the upstream sequence. Moreover, one of the four Ts in the hairpin loop of the others is replaced by an A, but the run of eight Ts at its downstream end is conserved. Thus IKe, although related to the other three, is more distantly so.

Apparently more than a single terminator exists per genome, for one of quite distinct structure has been determined at the end of the I region of f1 and at the 3' end of an RNA in M13, each referred to as terminator II in Table 10.5 (La Farina and Vitale, 1984; Smits *et al.*, 1984). That of f1, which has been shown to be rho-dependent, diverges from the other in having a noncanonical base pair (T,G, underscored) and an unpaired A in the stem; furthermore, the terminal run of Ts is short and interrupted. Although of the same general format as the rest, that of M13 shows almost no homology with the others and the run of Ts is absent from the 3' terminus. But what is of greater interest is that the precise sequence given elsewhere as a terminator is reported in the same paper as a promoter at the 5' end of the same RNA (Smits *et al.*, 1984), a fact that apparently escaped observation.

Still another terminator from a fifth filamentous virus, Pf3, whose host is *Pseudomonas aeruginosa*, not the *E. coli* of all the rest, shows almost nothing in common with the four just discussed insofar as sequence structure is concerned. However, it has a similar long region of dyad symmetry, with ten paired sets of bases instead of their eight, including one noncanonical T,G combination (underscored). Moreover, the downstream run of Ts is partially broken and shortened by an inserted CG combination.

Terminators of Miscellaneous Bacteriophages. In each of three viruses of a miscellany of types, the terminators of two or more genes have been investigated, presenting a further opportunity for comparisons of structures acted upon by the same enzyme. Although it is to be expected that all the terminators of a given bacteriophage should be closely similar, in these there is no meaningful level of constancy. Terminator 4 (T4) of ΦX174, for instance, parallels those of the four filamentous viruses rather closely, except in being shorter, the GCCTCC and GGAGGC of its dyad symmetric sector being recognizable modifications of the GGCTCC and TGGAGC of f1 and kin. Additionally, it is followed by a comparable run of Ts. On the other hand, its T2 sequence shows little in common in the paired sector and downstream structure, and in T1 the base pairing in the stem-and-loop element is interrupted and is followed by a series of five As (Otsuka and Kunisawa, 1982).

Similar variation in structure is seen in the phages P22 and λ, some terminators of each of which have received analysis (Backhaus and Petri, 1984). In that comparative investigation, the terminator R_1 and that of the *nutR* gene were examined from both source organisms. The two from the *nutR* gene had short loops, with only five paired sets of bases, but the two R_1 terminators differed both in length and structure. A third terminator from P22, that associated with the *ant* gene (Berget *et al.*, 1983), showed some sequence homology in the stem structure to that of the *nutR* cistron, but it is followed by a series of seven Ts absent from its mates.

10.8.2. Terminators of Genes in Viruses of Eukaryotes

To date in molecular researches on the vertebrate viruses, termination of transcription seems to have been a minor issue. In part this undoubtedly stems from the circularity of the genomes, in which transcription often is bidirectional, with broad, common initiation and termination zones for early and late gene expression. However, in a few forms, including SV40 and BKV, the typical polyadenylation signal AAUAAA has been detected, which in the first form at least may also serve for termination (Seif *et al.*, 1979; Conway and Wickens, 1985). The later study has demonstrated that the poly(A) itself is not as important in the present function as the 220-residue sequence that surrounds it.

The identical condition prevails in transcriptional termination in plant viruses. In many instances, as in the RNA2 of the tomato strain of TMV and cucumber mosaic virus (Ohno *et al.*, 1984; Rezaian *et al.*, 1984), no evidence of the standard polyadenylation signal can be traced beyond the last coding sequence, whereas in others, including TMV proper (Goelet *et al.*, 1982; Takamatsu *et al.*, 1983), one is present and correctly positioned. Tobacco streak virus RNA3 lacks that nucleotidyl combination, but three sequences in the 5' leader are repeated in the 3' region, arranged in similar fashion (Cornelissen *et al.*, 1984). Their particular role has not yet been examined, but they might prove to serve as terminators. It must also be borne in mind that a number of plant viral genomes terminate in tRNA cistrons, whose sequences themselves possibly may be sufficient to end transcription (Kozlov *et al.*, 1984).

10.9. OVERLAPPING (DUAL) GENES

A particularly characteristic feature of viral genomic structures in general is the presence of genes that overlap one another. So common is this arrangement that any attempt at covering examples even from the major types would be futile and largely an exercise in repetition. Instead, the effort here is devoted to the broader picture of this peculiarity, so that the discussion can apply to mitochondrial, chloroplastic, nuclear, and prokaryotic genomes, although the examples cited at this point are confined to the viruses.

10.9.1. Types of Overlapping Genes

Overlapping genes may be considered to fall into seven major types arranged in three great subdivisions. In those larger categories, the polarities and number of interacting sequences provide the criteria for separation. Division A contains pairs of genetic elements of opposite polarity, that is, located on different strands; division B embraces two located on the same strand and therefore having identical orientation; and division C includes sets of three or more jointly overlapping genes.

Division A. In the first of these categories are found two types. In type I only the terminal portions overlap, while in type II a smaller gene is contained within the limits of the coding region of a larger, but on the opposite strand (Figure 10.3C). The first may be considered to be divided into three subtypes, depending on whether the 5' ends of each overlap (type IA), the 3' ends do so (type IB), or the 5' end of one overlaps the 3' terminus of the other (type IC) (Figure 10.7).

Division B. Whereas the overlapping cistrons in the preceding division are necessarily confined to forms having double-stranded genomes, either RNA or DNA, those of division B can occur in species with any type of genetic apparatus, since the pair of sequences involved, being of like polarity, are situated on the same strand. Four major types appear possible, one of which has three variants. In type III the ends of the two genes overlap, three possible arrangements of which can occur. In subtype IIIA, the 5' end of one begins just before the 3' terminus of the other (Figure 10.7); subtype IIIB arrangements have the 3' end of a shorter genetic element ending somewhat beyond that of a longer component; and in subtype IIIC, the 5' end of the shorter begins prior to that of a longer one. In type IV, overlaps involve two genes that share a common initiation

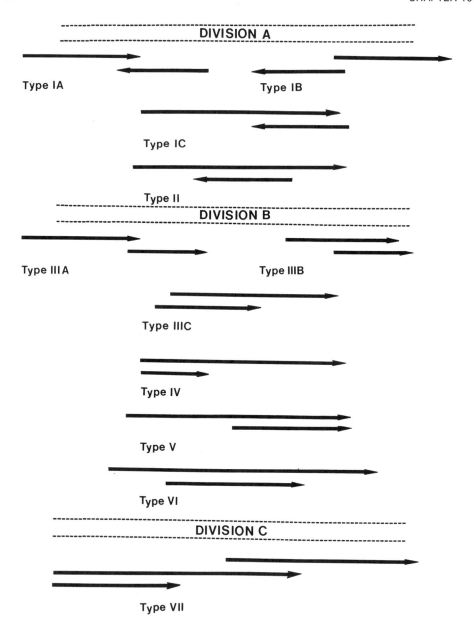

Figure 10.7. The three divisions and seven types of the major varieties of overlapping genes.

codon, but one terminates before the other. Quite in contrast, type V dual genes initiate separately, but terminate jointly, whereas in type VI a smaller genetic sequence lies totally within a larger one (Figure 10.7).

Division C. For the present it is considered expedient to place all sets of multiple overlapping genes into a single unit, type VII, until, with increasing knowledge, a greater abundance of representation than now appears evident may warrant recognition of additional groupings.

Table 10.6

Examples of Overlapping Gene Sequences[a]

T4 *frd*[b]	--- GAA TCA GTA TAT AAA <u>TGA</u>
T4 *ts*[b]	<u>A TGA</u> AAC AAT ACC AAG AT- ---
λ *xis*[c]	--- AAG AGG ATC AGA AAT GGG AAG AAG GCG AAG TCA <u>TGA</u>
λ *int*[c]	<u>AT G</u>GG AAG AAG GCG AAG TCA TGA G--
TMV *30,000*[d]	--- AGT TTG TTT ATA GAT GGC TCT AGT TGT <u>TAA</u>
TMV *28,000*[d]	<u>AT G</u>GC TCT AGT TGT TAA A-- --- ---
M13 *I*[e]	--- AAA GGT AAT TCA AAT GAA ATT GTT AAA TGT AAT <u>TAA</u>
M13 *IV*[e]	<u>AT G</u>AA ATT GTT AAA TGT AAT TAA T--
ΦX174 *C*[f]	--- ATA GGT AAG AAA TCA <u>TGA</u>
ΦX174 *D*[f]	<u>A TGA</u> GTC AAG TTA CTG AAC AAT CC
ΦX174 *A*[f]	--- ACT GCT GGC GGA AAA <u>TGA</u>
ΦX174 *K*[f]	-GG ACT GCT GGC GGA AAA TGA GAA AAT TCG ACC TAT CCT T--
ΦX174 *C*[f]	<u>A TGA</u> GAA AAT TCG ACC TAT CCT TG-

[a]Arcs indicate connections between parts of codons and primes demarcate their termini.
[b]Purohit and Mathews (1984).
[c]R. W. Davies (1980).
[d]Goelet *et al.* (1982).
[e]Van Wezenbeek and Schoenmakers (1979).
[f]Sanger *et al.* (1977).

10.9.2. Examples of Dual Genes

Unfortunately in one of the major sources of information from double-stranded DNA viruses, the genome of bacteriophage T7 (Dunn and Studier, 1983), all the genes are located on the same strand, so that it supplies no representatives of division A dual (overlapping) cistrons. However, it does yield some interesting examples of divisions B and C, as shown shortly, while phage λ supplies a few of the first.

Examples of Type III Dual Genes. By far the commonest kind of dual gene structure in viruses is type III, in which the initiation end of one of a pair of directly oriented genes overlaps the termination region of the other. One example of this condition from phage T4 is of particular interest because the downstream member, which encodes thymidylate synthase, is one of the few viral genes that contains an intron (Chu *et al.*, 1984). In this case overlap is minimal, involving only a single nucleotide, as indicated in Table 10.6, where in the downstream representative the parts of the codons are tied together by arcs, since they are out of frame with the first member (Purohit and Mathews, 1984). While thus extremely simple, this illustration nevertheless makes the problem clear: Since this is part of a polycistronic transcript, translation of the *frd* gene for dihydrofolate reductase by the usual processes concomitantly destroys the 3'-coding region *ts,* and, conversely, preservation of the latter eliminates the former. How control over expression is exercised has not been established, but it may involve posttranscriptional processing, the subject of the closing chapter. Similar single-residue overlaps are frequent, occurring between genes *IX* and *VIII* of bacteriophage f1 (Hill and Petersen, 1982) and between genes *12* and *ninA* in phage P22 (Backhaus and Petri, 1984), to cite two additional cases.

In one set of dual genes of bacteriophage λ, the overlap involving coding elements for the integrase (*int*) and the excisionase (*xis*) of this temperate form extends through eight codons (Davies, 1980). In this instance a frameshift of two nucleotides is involved, not just one, as may be noted in Table 10.6. A double structure of slightly shorter length exists in the tobacco mosaic viral genome (Goelet *et al.,* 1982), also given in that table, but for clarity the uridyls of its RNA are replaced by the thymidines of DNA in this pair, whose products are known only by the molecular weights. Other examples are given from bacteriophage M13 (van Wezenbeek and Schoenmakers, 1979) an identical set of which occurs also in phage f1 (Hill and Petersen, 1982), but is not shown. Much longer overlaps of type III are known and are not uncommon. In influenza A virus, the NS_1 and NS_2 cistrons overlap by 61 codons (Porter *et al.,* 1980), in the vaccinia genome, *F9* and *F10* share a region of 29 triplets (Plucienniczak *et al.,* 1985), and *p4* and *p3* of phage Φ29 have a 38-codon region in common (Escarmís and Salas, 1982). Here, in each case, as in the first example given, the problem is the same—the expression of one member of the dual genes results in the elimination of the other, unless a special translational device exists. Hence, it is not an efficient arrangement and involves a measure of control that bespeaks the presence of a supramolecular element. In every instance a frameshift exists between the cistrons of each pair, a device that undoubtedly serves in expression control.

Examples of Miscellaneous Types. Type VII, in which three genes form a triple direct structure with two of the components initiating separately but terminating together, has a well-established example in the BK papovavirus of man (Yang and Wu, 1979). Here *VP2* has its ATG translational initiation signal 1350 base pairs prior to that of *VP3,* but the two have a joint termination codon (TAA). Thus, the two coding sectors are in the same frame of reference. However, some 110 base pairs before that point and out of frame with it is the initiation codon for *VP1.* Consequently, three genes are seen to share an extensive coding sector, signifying that only a single member of the trio can be expressed in a given round of translation, resulting in an even lower level of synthesis efficiency than in the dual systems. A duplicate of this that occurs in SV40 is illustrated in Figure 10.3A. In phage T7 early genes (class I) a similar condition exists (Dunn and Studier, 1983), but in this set, one (gene *0.5*) of the three cistrons has a single nucleotide residue at its 3' end overlapping the initiation codon shared by the other two. The shorter of the latter, *0.6A,* terminates 58 codons before the longer, *0.6B.*

Bacteriophage T7 supplies illustrations of two other types of dual structures, each of which is located in the late (class III) region of its DNA. In the first of these, involving genes *10A* and *10B,* the initiation point is in common, but the second coding region extends 58 nucleotides beyond the first, thus exemplifying type IV duplexes. One example of a type V duality has been well-established in phage f1 (Fulford and Model, 1984). Here gene *II,* which encodes a site-specific endonuclease, has been shown to overlap gene

X, which begins at codon 300 of the former. This cistron, whose encoded product is unidentified, then terminates with gene *II*, the two sequences being completely in frame. Finally, type VI is represented by two sets of adjacent sequences in T7, genes *18.5* and *18.7* on the one hand and *19* and *19.2* on the other (Dunn and Studier, 1983). In both sets the second component listed is much shorter than the first and is enclosed entirely within the larger; each is out of phase with the sequence in which it is embedded by one position downstream. The processes by means of which the particular sequence is selected for translation should provide a fertile and challenging field for investigation by cellular and molecular biologists.

But to make the genomic structure and expression clear, it must be acknowledged that overlapping genes, although frequent, are not a predominant feature. More typically, sequential coding regions are separated by several nucleotide residues. But on abundant occasions spacing involves just one or two sites, and not rarely there may be no spacer at all between the translation stop signal of one and the initiation signal of the second. Moreover, in the numerous plus-stranded RNA species, the polycistronic messengers even lack those translational signals between components. On the other hand, in viruses with larger genomes, some gene sets may be spaced by runs of 20–100 sites, or even several hundreds.

11

The Gene in Cancer

It was through investigations of certain retroviruses that could transform cells that the first knowledge was gained of the role of specific gene sequences in the induction of tumors and malignancies. Apparently, such cistrons, which are largely homologous to particular ones of the host, act in concert with viral transcription-control elements to lead to tumor or neoplastic formation in mammals and experimental tissue cultures (R. Watson *et al.*, 1982; Feramisco *et al.*, 1985; Van Beveren *et al.*, 1985). Although for obvious reasons studies have focused mainly on vertebrates, similar processes seem to be active at times in insects and even in yeasts, fungi, and seed plants. At the present time few normal functions have been determined in the cellular counterparts of the cancer-producing, or *onc*, genes, as they are called (Vande Woude *et al.*, 1984). To distinguish between viral and cellular homologs, a *v-* or *c-* is prefixed to the gene abbreviation, as *v-onc* or *c-fas*, respectively. In some laboratories, the *c-onc* genes are referred to as proto-oncogenes, suggestive of their being nonmalignant until recombined into a viral genome (e.g., Katzen *et al.*, 1985; Sherr *et al.*, 1985). It is necessary to point out that not only are the viral genes capable of inducing cell transformation, but the *c-onc* structures themselves may do so when activated by various agents.

Viruses of many types, oncogenic or not, often acquire genes from host cells or from other viruses that may be present, and it is through this property that new viruses sometimes come into being. One specific instance of this sort that can be offered to illustrate the phenomenon is the origin of the acute transforming pathogen known as Moloney murine *sarcoma* virus. During passage of the Moloney murine *leukemia* virus in laboratory mice of the strain BALB/2, that agent recombined with a mouse chromosomal proto-oncogene referred to as the *c-mos* gene and thereby became the corresponding sarcoma virus (Oskarsson *et al.*, 1980; Coffin *et al.*, 1981). In acquiring such cellular sequences, the parasite generally loses some of its own genome, typically parts essential for replication, thus rendering them incapable of propagation. Their reproduction, and thus their survival, is dependent upon "helper" viruses (Van Beveren *et al.*, 1981). To cite a concrete example of the results of recombinatorial events, the Moloney murine sarcoma viral genome just mentioned was shown to share an ~2200-base sector at its 5' end and an ~700-base sector at its 3' terminus with its leukemia progenitor, containing a 1200-base substitution along with deletions of 1200 and 1600 base pairs.

Since an understanding of the genes involved in oncogenesis is thus intimately intertwined in many cases with those of specific viruses, what information is currently available about the forms known to be associated with malignancies is examined first. Only then can the nature of the genetic aspects be fully appreciated.

11.1. VIRAL AGENTS

Cancer and tumor induction are properties often associated with viruses of many kinds, including their occasional acquisition by forms that are usually not oncogenic. Among the latter types may be mentioned the adenoviruses, which can produce viral-encoded proteins that are active in cell transformation, as pointed out in detail in Section 11.6. On the other hand, the retroviruses that are consistently implicated in such activities provide a much broader basis for discussion, because the genomic structures and genes of a diversity of species have been well documented. Consequently, the immediate focus is upon this group and others of related structure.

11.1.1. The Retroviral Genome

Retroviruses derive their name from one of their few constant features, that of possessing a gene (*pol*) that encodes a "reverse transcriptase." This RNA-dependent DNA polymerase transcribes into double-stranded DNA the single plus-strand RNA molecule that comprises their genome, synthesis being dependent upon the presence of a tRNA molecule bound near the 5' end, and a poly(A) sector located at the opposite terminus of the RNA genome (Taylor and Illmensee, 1975; Coffin, 1979). The full details of the replicatory processes are unclear, but are thought also to involve two additional proteins, one called RNAse H and another having a stimulatory activity. Synthesis proceeds in a 5' → 3' direction along the tRNA molecule until the 5' end of the genome is attained, the resulting primer sequence then migrating to the template RNA, where it seems to attach by complementary base-pairing to a 60-base sector just upstream to the poly(A) tail. The step-by-step processes are not completely known, but in broad terms this primer is then elongated by the polymerase until a complete minus-strand copy in DNA has been made of the genomic RNA (Sutcliffe *et al.,* 1980a). How the second strand of DNA is synthesized to produce the duplex molecule remains undisclosed, as does how the terminal repeats are induced, but it is this double-stranded molecule that may be integrated into the host genome to be inherited and expressed in Mendelian fashion along with the genes of the cell.

Oncogenic retroviruses fall into two major subdivisions, the first of which are "replication-competent," having a genome that lacks cell-derived genes, but carrying all the genetic elements necessary for genomic replication. In contrast, "replication-defective" viruses contain cell-derived genes, but lack at least some of those elements whose products are necessary for synthesis of their genomes. Further differences between the two are noted in the manner of producing malignancies. In defective forms, neoplastic growth is acute and is induced in a very short time, generally within a few weeks. On the other hand, with the members of the competent group, malignant growth is chronic and follows a prolonged latent period, tumors not appearing for 3–12 months after infection. Also, unlike the first type, the competent group does not induce transformation of fibroblasts *in*

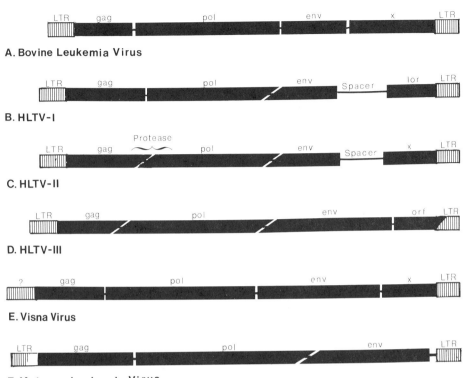

A. Bovine Leukemia Virus

B. HLTV-I

C. HLTV-II

D. HLTV-III

E. Visna Virus

F. Moloney Leukemia Virus

Figure 11.1. Genomes of replication-competent retroviruses. It is not unlikely that retroviruses capable of replication will prove to possess four basic gene sequences in their RNA genomes, or perhaps five. (A–E) Viruses from human tissues; (F) from murine cells. (Variously modified from Chess *et al.*, 1985; Derse *et al.*, 1985; Gonda *et al.*, 1985; Marx, 1985; and Shimotohno *et al.*, 1985a).

vitro. Leukemia is the most frequent result of the latter group, and sarcomas or carcinomas those of the other.

Genomes of Replication-Competent Retroviruses. It has long been assumed that the replication-competent retroviral genome contains only three genes, those called *gag, pol,* and *env,* plus a long terminal repeat (LTR) at each end (Figure 11.1), but this generalization is proving premature. More thorough explorations into certain members of the group, especially the three human T-cell leukemia viruses (HTLVs), have revealed the presence of an overlapping gene structure of type VII (Chapter 10, Section 10.9.2) involving the *gag* and *pol* sequences together with a novel cistron for a nuclease (Shimotohno *et al.*, 1985a). This has been fully established only in HTLV-II, but it may prove to be of more widespread occurrence when appropriate investigations are completed.

Furthermore, the X (or R) region that typically follows the *env* gene also is proving to bear coding elements, since an open reading frame has been discovered there, following an untranslated leader, as described later. Finally, the LTRs are becoming established as elements not simply involved structurally in combining with the host genome, but also including sections that encode essential viral products.

Compound Genes. Two of the better known viral genes, *gag* and *env*, are compound in that each encodes multiple proteins. While thus approaching the cryptomorphic types described in Chapter 7, Section 7.5, they differ in that they are processed into their respective components immediately after translation, so that the whole compound structure does not persist in the cell in a cryptic condition. The *gag* gene encodes four products of molecular weight 15,000, 12,000, 30,000, and 10,000, respectively from the 5′ to 3′ ends. All of these become located in the core of the virion, and the 30,000-dalton member (p30) is believed to be concerned with host-recognition properties (Sutcliffe *et al.*, 1980a). The second compound gene, *env*, encodes two products, which are translated as a unit to be cleaved after glycosylation, both being involved in capsid formation. The first one, a glycoprotein (gp70), is the major structural component, whereas the second, P15E, is a hydrophobic protein that is believed to anchor gp70 to viral and cellular membranes.

Overlapping Genes. Although extensive sequencing has been done to the genomes of various competent members of the retroviral family, much remains unsettled as to the precise location and number of functional coding regions. Possible overlapping genes appear to be overlooked on frequent occasions, but Sutcliffe and co-workers (1980a,b) reported the presence of an open reading frame, which they referred to as the *R* gene, arranged as a type IIIA overlap with the *env* cistron in the Moloney leukemia virus. However, no similar element seems to have been reported in others of the group. In fact, one laboratory has presented the corresponding region in HTLV-II as an untranslated spacer (Chen *et al.*, 1985), apparently unaware of the foregoing interpretation and of other reports on the gene, under the name *x*.

This 3′-terminal region of HTLV-I and -II was shown to be expressed (Wachsman *et al.*, 1984) and, later, essential for replication (Chen *et al.*, 1985). In the first virus, the encoded protein proved to have a molecular weight of 41,000, and in the second one, a molecular weight of 38,000 (Shimotohno *et al.*, 1985b). Still later the transcript of the *x* gene of both viruses was demonstrated to result from a two-step splicing, using the initiator codon of the *env* gene plus the next adjacent nucleotide and the open reading frame of X (Seiki *et al.*, 1985; Wachsman *et al.*, 1985).

The condition of the murine leukemia virus genome in the AKV strain of mice differs somewhat from the foregoing, but may be found applicable to the other members. There the *env* and *R* genes are part of a type VII overlap, the first of these being involved also in a dual structure at its 5′ end with the 3′-terminal 21 triplets of the *pol* gene (Herr *et al.*, 1982). These recent discoveries strongly suggest the need for thorough investigations of the genomes of these highly important viruses, including complete sequencings of each type.

11.1.2. The Long Terminal Repeats

The long terminal repeats (LTRs) often mentioned in connection with the genomes of oncogenic viruses are not really a property of the virion, for the full structure is created out of two partial sequences when the viral genome is integrated into the host DNA. Much importance is typically given to these repeats, nevertheless, so they must accordingly be examined critically here. The very ends of at least some of them are requisite for integration, but not for virus production (Panganiban and Temin, 1983). These LTRs vary greatly in extent, ranging from ~337 bases long in the Fujinami sarcoma virus (Shibuya

Table 11.1
Long Terminal Repeats of Oncogenic Viruses[a]

Row 1

Chicken[b]	AATGTAGTCTTA	ATCGTAGGTTAA	CATGTATATTAC	CAAATA---	AGGGAA----
Fujinami[c]	TGAAGGAAGTGT		GATGTT	GCCAGT---	TAGTCA-TCA
Simian[d]	TTT---TCAAGC	TAGCTGCAGTAA	CGCCATTTT	GCAAGG---	
Xenotropic MuLV[e]	ATTTTGCAAGGC	ATGAAAAAGTAC	CACAGCTGA	GTTTTC----	
HTLV-I[f]	TGACAATGACCA	TGAGCCCCAAAT	ATC------CCC	CGGGGGCTT	AGAGCC-TCC
HTLV-II[g]	TGACAATGGCGA	CTAGCCTCCCGA	-GCCAGCCACCC	AGGGCG---	AGT-------
Bovine[h]	TGTATGAAAGAT	CATGCCGCACCTA	GGAGCGCCACC	--GCCCCCT	AAACCA----
Abelson[i]	CCTGTA---GGT	TTG---GCA---	AGCTAGCTT	AAGTAA----	
A-Particle[j]	CGT---TCTCAC	-GC---CCGGCA	GGAAGAAGA	---CCA----	

Row 2

Chicken	TCGCCTGATGCA	CCAAATAAG	GTATTATATGAT	CCC	ATTGGTGGTCAA	GGACCGGACC	---TGAGGGCAT
Fujinami	TTGCCACTTAGT	CATCATTAC	ATAAGACAGTCT	AAA	GTCCTAAAAAGG	AAAAACAAG	--------ACAT
Simian	CACCGGAAATTAC	CCTGGTAAA	AAGCCC-----	AAA	GCATAGGGGAAG	TACAGCTAA	AGGTCAAGTCGA
Xenotropic MuLV	AAAACTTACAAA	GAAGTTTCGG	TT---------	AAA	GAATAAGGTTGA	ATAATACTG	GGACAGGGGCCA
HTLV-I	CAGTCAAAAACA	TTTCCGAGA	AACAGAAGTCTG	AAA	AGGTCAGGGCCC	AGACTAAGG	CTCTGACGTCTC
HTLV-II			CATCGA----CCC	AAA	AGGTCAGACCGT	---CTCACA	CAAAC-------
Bovine	GACAGA---GAC	GTCAGCTGC	CAG---------	AAA	AGCTGGTGACGG		------------
Abelson	CGCCAT------	TTTGCAAGG	CATGA-------	AAA	ATA-CAT-AAC-	TGAGAATAG	AGAAGTTCAGAT
A-Particle	CAGACCCAGAATC	TTC---TGC	GGC---------	AAA	ACTTTATTGCT-	TACACTTCA	GGAGCAAGAGTG

(continued)

Table 11.1 (Continued)

```
Row 3

Chicken          ATGGGCGTT   AACAGAACT     GTCGTCCTTGCG   TCATTCCTCATC   GGATCATGT    ACGCGGCAG    AGTATGATT
Fujinami         CTCGGA---   TGTCATTGG     CTGCAACCAGTA   AGGAAGTAGTGG   CGTCAGGAT    CACCCCATG    GGTGTAACC
Simian           GAAAAACAA   GGAGAACAG     GGCCAAACAGGA   TATCTGTGGTCA   TGCCTGGGC    CGGCCCAGG    GCCAAAGAC
Xenotropic MuLV  AACAGGATA   TCTGTGGTC     GAGCACCTGGGC   CCCGGCTCAGGG   CCAAGAACA    GATGTACT     CAGATAAAG
HTLV-I           CCCCCGGAG   GGCAGCTCA     GCACCGGCTCGG   GCTAGGCCCTGA   CGTGTCCCC    CTGAAGACA    AATCATAAG
HTLV-II          AATCCCAAG   TAAAGGCTC     TGACGTCTCCCC   CTTTTTTTAGGA   ACTGAAAACC   ACGGCCCTG    ACGTCCCTC
Bovine           CAGCTGGTG   GCTAGAATC     --------CCC    GTACCTCCCCAA   CTT---CCC    CT- (To row 5)
Abelson          CAAGGTCAG   GAACACAGA     AACAGCTGAATA   TGGGCCAAACAG   ---GATATG    CTGTGGTAA    GCAGTTCCT
A-Particle       TAA-----G   AAGCA-AGA     GAGAGAAAACGA   ACCCCGTCCCTT   TTT----TA    GGAGAGTT-    ATATTTC--

Row 4

Chicken          GGATAACAG   -----------   ------------   ------------   ------------  ------------  ----------
Fujinami         CAGAAGGCC   ATT---------  ------------   ------------   ------------  ------------  ----------
Simian           AGATGGTTC   CCAGAAATAGAT  GAGTCAACAGCA   GTTTCTAGGGTG   CCCCTCAACTGT  TTCAAGAACTCC[z]  ----------
Xenotropic MuLV  CGAAACTAG   CAACAGTTTCTG  GAAAGTCCCACC   TCA---AGTTCC   CCAAAAGACCGG  GAAATACCCAAG[m]  ----------
HTLV-I           CTCAGACCT   CCGGGAAGCCAC  CAAGAACCACCC   ATTTCCTCCCCA   TGTTTGTCAAGC  CGTCCTCAGGCG  ----------
HTLV-II          CCCCCTAGG   AACAGGAACAGC  TCTCCAGAAAAA   AATAGACCTCAC   CCTTACCCACTT  CCCCTAGCCGTG  ----------
Abelson          GCCCCGGCT   --CAGGGCCAAG  AACAGTTGGAA-   --CAG-------   ------------  ------------  ----------
A-Particle       GCCTAGGAC   GTGTCCACTCCCT ---GATTGGCTG   --CACCCC-ATC   GGCCGAGTTGAC  G-----------  ----------
```

Row 5

Chicken	GATGGCACC	ATTCATCGT	GGCGCATGCTGA	TTGGTGCGACTA	AGGAGTTCT	GTAACCCAC	GAA-TG----
Fujinami	GGTGGCAGC	TGATGTCGT	G--ATATCACCT	TATGGGCAAGGC	TAAAGTCGT	GCA------	TAAC------
Simian	GAGCTCACC	CCTGGCCCT	TATTTGAACTGA	CCAATTACCTTG	CTTCTCGCT	TCTGTACCC	GCGCTTTTT[72]
Xenotropic MuLV	TTTGAACTA	ACCAATCAG	CTCGCTTCTCGC	TTC-TGT-AC-C	CGCGCTTTT	TGCCCCCAG	CCCTAGCCC
HTLV-I	TTGACGACA	ACCCCTCAC	CTCAAAAAACTT	TTCATGGCACGC	ATATGGCTC	AATAAACTA	GCAGGAGTC
HTLV-II	AAAAACAAG	GCTCTGACG	ATTACCCCCTGC	CCATAAAATTTG	CCTAGTCAA	AATAAAAGA	TGCCGAGTC
Bovine	TTCCCGAAA	AATCCACAC	CCTGAG--CTGC	TG----ACCTCA	CCT--GCT-	GATAAAT--	---------
Abelson	---CTGAAT	ATGGGCCAA	ACAGGATAT--C	TG--------TG	G-TAAGCA-	GTTCCTGCC	CCGGCTCAG
A-Particle	TCA---CGG	GGAGGGCAT	---GAGCAC--A	TG--------A	AGTAGGGAA	-------CC	ACC-CTC--

← U3' R →

Row 6

Chicken	TACTTAAGC	-TTGTAG-T	TGCTAACAATAA	AGT------	GCCATTCTA	CCTCTCACCCAC	A---------
Fujinami	TATATAAGC	-CATTGTAA	CCTTCT-AATAA	AC-------	GCCATTTTA	CCATTC-ACCAC	A---------
Simian	TATAAAATG	AGCTCAGAA	ACTCCACTCGGC	---------	GCCCCAGTC	CTCCGGAGAGACT	GAGTCGCCCC
Xenotropic MuLV	TATAAAAAG	GGTAAG-AA	CTCCACACTCGC	---------	GCGCCAGTC	ATCCGATAGACT	GAGTCGCCCC
HTLV-I	TATAAAAGC	GTGGAGACA	GTTCAGGAGGGG	---------	GCTCGCATC	TCTCCTTCACGC	GCC-CGCCG
HTLV-II	TATAAAAGC	GCAAGGACA	GTTCAGGAGGTG	---------	GCTCGCTCC	CTCACCGACCCT	CTGGTCACG
Bovine	TAATAAAAT	GCCGGCCCT	GTCGAGTTAGCG	GCACCAGAA	GCGTTCTTC	TCCTGAGACCCT	CGTGCTCAG
Abelson	GGCCAAGAA	CA-GATGGT	CCCCAG--ATGC	GGTCCA---	GCCCTCAGC	AGTTTCTAGAGA	ACCATCAGA

(continued)

Table 11.1 (Continued)

```
A-Particle        GGCATATGC GCGGATTAT TTGTTTACCACT TAGAACACA GCTGTCAGC GCCATCTTGTAA CTGGCCGATG
Murine VL30-1^k   TATTAAATC TTACCTTCT ACATTTTATGTA TGGTCTCA-  GTGTCTTCT TGGGTACCGGC  TCTCCCGGG
Murine VL30-2^k   TTTAAAAGC TTGTGAAAA TTTGAGTCGTCG TCGA-----  GACTCCTCT ACCCTGTCAAA  GGTGTATGA
```

←R U5'→

Row 7

```
Chicken           ---  --TTGGTGT GCACCTGGGTTG ATCGCCCGACCG TCGATTCCCTGA  CGACTGGCA ACACCTGAATGA
Fujinami          ---  --TTGGTGT GCACCTGGGTAG ATCGACAGACCG TTGAGTCCCTAA  CGATTGGCA ACACCTGAATGA
Simian            GGG  TACCTGTCT  GTTCAATAAAAC CTCTTGCTATTT GCATCCGAAGCC  GTGGTCTCG TTGTTCCTTGGG
Xenotropic MuLV   GGG  (Incomplete)
HTLV-I            CCC  TACCTGAGG  CCGCCATCCACG CCGGTTGAGTCG ---CGTTCTGCC  GCCTCCCGC CTGTCGTCGCCTC
HTLV-II           GAG  ACTCACCTT  GGGGATCCATCC TCTCCAAGCGGC CTCGGTTGAGAC  GCCTTCCGT GGGACCGTCTCC
Bovine            CTC  TCGGTCCTG  AGCTCTCTTGCT CCCGAGCACCTTC^o CCGCCAGCTCTA TCTCCGGTC CTCTGACCGTCT
Abelson           TGT  TT-CCAGGG  TGCCCCAAGGAC CTGAAATGACCC TGTGCCTTATTT  GAACTAACC AATCAGTTCGCT
A-Particle        TGG  GCGCGGCTC  CCAACA
```

Row 8

```
Chicken     AGCTGAAGG  CTTCA
Fujinami    AGCGGAAGG  CTTCATT
Simian      AGGGTCTCT  CCTAAC-TAGTT GACTGCCCACCT CGGGGGTCT     CTTCA
HTLV-I      CTGAACTGC  GTCCGCCCTCTA GGTAAGTTTAAA GCTCAGGTC GAGACCGGG CCTTTGTCC GGCACTCCC
HTLV-II     CGGCCTCGG  CACCTCCTGAAC TGCTCCTCCCAA GGTAAGTCT CCTCTCAGG TCCAGCTCG GCTGCCCCT
Bovine      CCACGTGGA  CTCTCTCCTTTG CCTCCTGACCCC GCGCTCCAA GGGCGTCTG GCTTCGCACC CGCGTTTGT
Abelson     TCT---CGC  TTCTGTTCGCGC GCTTCTGCTCCC CGAGCTCAA TAAAAGAGC CCACAACCC CTCACTCGG
```

```
Row 9
  HTLV-I     TTGGAGCCTACC  TAGACTCAGCCG  GCTCTCCACGCT   TTGCCTGAC   CCTGCTTGCTCA   ACTCTACGTCTT
  HTLV-II    TAGGTAGTCGCT  CCCCGAGGGTCT  TTAGAGACACCC   GGGTTTCCG   CCTGCGGCTCGGC  TAGACTCTGCCT
  Bovine     TTCCTGTCTTAC  TTTCTGTTTCTC  GCGGGCCCGCGCT  CTCTCCTTC   GGCGCCCTCTAG   CGGCCAGGAGAG
  Abelson    CGGCCCAGTCCT  CCGAGTGACTGA  GTCGCCCGGGTA   CCCGTGTAT   CCAATAAACCCT   CTTGCAGTTGCA

Row 10
  HTLV-I     TGTTTCGTTTTC  TGTTCTCGCGCCG  TTACAGATC  GAAAGTTCC  ACCCCTTTC  CCTTTCATT  CACGACTGA
  HTLV-II    TAAACTTCACTT  CCGCGTTCTTGT   CTCGTTCTT  TCCTCTTCG  CCGTCACTG  AAAACGAAA  CCTCAACGC
  Bovine     ACCGGCAAACA
  Abelson    TCCGACTTGTGG  TCTCGCTGTTCC   TTGGGAGGG  TCTCCTCTG  AGTGATTGA  CTACCCGTC  AGCGGGGGT

Row 11
  HTLV-I     CTGCCGGCT  TGGCCCACG  GCCAAGTACCGG  CGACTCCGT  TGGCTCCGGAGCC  AGCGACAGC  CCATCCTAT
  HTLV-II    CGCCCTCTT  GGCAGGCGT  CCCGGGGCCAAC  ATACGCCGT  GGAGCGCAGCAA   GGGCTAGGG  CTTCCTGAA
  Abelson    CTTTCA
```

(continued)

Table 11.1 (Continued)

Row 12

HTLV-I   AGCACTCTC   AGGAGAGAAATT   TAGTACA*CA*

HTLV-II   CCTCTCCGG   GAGAGGTCTATT   GCTATAGGC   AGGCCCGCCCTA   GGAGCATTGTCT   TCCCGGGAAGACAA

Row 13

HTLV-II   *ACA*

---

[a]Terminal sectors of complementarity are italicized and possible transcriptional initiation and polyadenylation signals are underscored.
[b]Rushlow et al. (1982).
[c]Shibuya and Hanafusa (1982).
[d]Devare et al. (1983).
[e]Khan and Martin (1983).
[f]Seiki et al. (1983).
[g]Shimotohno et al. (1985a).
[h]Sagata et al. (1984, 1985).
[i]Reddy et al. (1983b).
[j]Kuff et al. (1983).
[k]Norton et al. (1984a).
[l]Insert CACATGACCG.
[m]Insert CCTTA.
[n]Insert GC.
[o]Insert 68 nucleotide residues.

and Hanafusa, 1982) to over 775 in the HTLVs (Seiki *et al.*, 1983; Shimotohno *et al.*, 1985a), as given in Table 11.1.

The Synthesis of LTRs. Since their LTRs are acquired during integration, these temperate viruses show a fundamental distinction from inserted elements, in which terminal repeated sectors are a permanent feature of many types, but are absent in others. As shown in Figure 11.2A, the termini of the virion are not repeated elements, but differ in size and molecular structure. When the RNA of the virion is transcribed into DNA, a copy of the 5' end becomes added downstream of the 3' sector and in turn a replica of the latter is added upstream of the former (Figure 11.2B). Consequently, each repeat is partly original and partly a copy, quite unlike those of movable elements viewed in Chapter 9.

In addition an R (redundant) sector is present between the 3' and 5' portions of each LTR, but its origin has never received full attention. In earlier investigations, the sector has been depicted as occurring both at the downstream end of the 5' portion and on the upstream end of the 3', but complete sequences of LTRs have revealed no redundancy in the region, which has been reported to vary extensively within some species (Christy *et al.*, 1985). However, the full details have never been completely confirmed. Integration also involves the formation of direct repeats as in transposons, that is, a copy of host DNA flanking the 5' viral terminus is added to that beyond the opposite end (Figure 11.2C); these receive attention shortly.

Transcriptional and Processing Signals. As the foregoing discussion intimates, each LTR is composed of three regions, one at the upstream end, which is derived from a sequence at the 3' terminal of the virion and referred to as unique sector, U3'. Similarly, the structure closes in a U5' portion, named after its origin, which is separated from the first part by a short R division (Figure 11.2B). A functional promoter has been identified in U3', which signal in the 5' LTR is employed in the transcription of viral RNA (Figure 11.2C; Yamamoto *et al.*, 1980; Köhrer *et al.*, 1985). Since an identical copy exists in the downstream element, it should serve equally from that point, but does not appear to do so, for unknown reasons.

Moreover, a typical signal for addition of a poly(A) train also has been identified, characteristically closely following the proposed promoter (Figure 11.2). However, its presence five codons beyond the promotion site, as in the Rous sarcoma virus (Yamamoto *et al.*, 1980) and avian sarcoma virus (Swanstrom *et al.*, 1981), or only a single nucleotide beyond it, as in the bovine leukemia virus (Sagata *et al.*, 1984), does not appear to be a reasonable location for a signal of this function. In HTLV-III, it follows at a distance of 30 codons (Starcich *et al.*, 1985), but still the location seems too close to serve a useful function. The most probable explanation of the AATAAA when present (it is often absent, as shown below) is that it serves as a triggering device in the 3' LTR for addition of the poly(A) tail, but does not do so in the 5' component. Thus, here is a still further addition to the data supporting a biomechanistic view of biotic reactions (Dillon, 1983, pp. 400–410).

Possible Enhancer Activity. Much evidence has been forthcoming supporting the existence of an enhancer element in LTRs, but specific details are only now beginning to emerge (Sodroski *et al.*, 1985). Among earlier results were those provided by a study on the induction of erythroblastosis in chickens by the avian leukemia virus (Fung *et al.*, 1983). This research demonstrated clearly that the insertion of the LTR of that virus activated a cellular oncogene known as *c-erb*. Similar results were obtained when the LTR

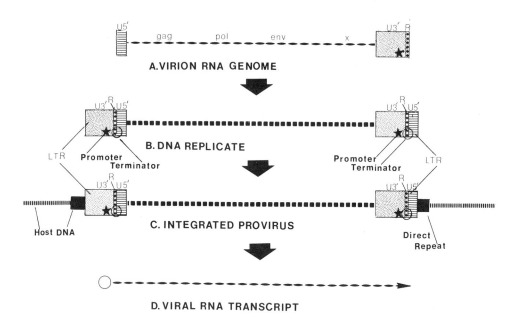

Figure 11.2. Synthesis of long terminal repeats (LTRs) of retroviruses. (A) Unlike those of transposons, the LTRs of retroviruses are synthesized during replication by combinations of copies from each end. (B) The 5′ part is added downstream of the 3′ end, and the latter, upstream of the 5′ sector. (C) In addition, when inserted into the host genome, direct repeats of the cell DNA are created flanking the viral insert. (D) Transcription involves the entire viral genome. (Based on Yamamoto *et al.*, 1980.)

of mouse mammary tumor virus was inserted, in that the ability to transform mouse fibroblasts *in vitro* was greatly increased (50- to 100-fold) by activation of the *v-ras* gene (Ostrowski *et al.*, 1984). However, in this investigation it was further shown that the presence of glucocorticoids was essential to the enhancement of transformation activity. About the same time another laboratory reported that several domains of the LTR contained enhancer properties, two of which were located within the U3′ region (Laimins *et al.*, 1984), while others of similar nature appeared to be situated in the coding region proper of the viral genome. Differences have been found to exist between the structures of various species and even among the three types of HTLVs (Sodroski *et al.*, 1984). In the latter report, the LTR of HTLV-I and -II displayed marked differences in functioning as transcriptional agents. Activation in this family of viruses, including bovine leukemia virus, involves the strand opposite that being transcribed, that is, it is a type of *trans*-activation (Sodroski *et al.*, 1985). However, at least part of the enhancing activity is now being attributed to a product of the *x* gene, often under the name of *lor* or *R* (Broome and Gilbert, 1985; Fujisawa *et al.*, 1985).

The Nature of the Signals. In by far the majority of sequence studies that include the LTR structures, no attempt has been made to determine the limits of the three subdivisions. Only those studies on the two avian and bovine leukemia viruses included in Table 11.1 are exceptional (Shibuya and Hanafusa, 1982; Sagata *et al.*, 1984, 1985), but discrepancies exist even in those. Accordingly, a new approach is taken here by beginning the alignment of the several sequences with the usually readily recognizable promoter, shown underscored in row 6 (Table 11.1). Secondarily, then, the beginning of R was arbitrarily chosen at a GC- combination characteristic of a typical cap site. No reasonable basis for the downstream terminus of R could be found, so it is almost uniformly artificial here, the CACA- of the two chicken viruses being points capable of becoming polyadenylated (Shibuya and Hanafusa, 1982), which may or may not be a valid consideration, as pointed out earlier. The remainder of the longest LTR sequences were first completed before and after those parts, and only then were the others added to homologize with the lengthier ones as far as reasonably possible. Much additional exacting research needs to be conducted before the limits of the three sectors, but especially those of R, can be thoroughly understood.

When thus aligned, an amazing degree of constancy is perceived to exist in the promoters (underscored) of nine of the 12 cited (row 6). On one occasion the second base is T instead of A (in murine VL30-2) (Norton *et al.*, 1984a), but As and Ts alone occupy that site. However, in the A-particle and Abelson viruses (Kuff *et al.*, 1983; Reddy *et al.*, 1983b), no typical structures could be found; those underscored are the closest approximations that could be located. Downstream from the promoter in the two avian viruses, the typical polyadenylation signal, AATAAA, is perceived, still within the limits of U3', but that pair is alone in showing that signal there. Indeed, the only other such sequences in the downstream region are in the simian (row 7) and Abelson viral region U5' (rows 8 and 9), but in HTLV-III, which sequence is not given, it is similarly placed (Starcich *et al.*, 1985). Upstream of the promoter, the AATAAA appears in row 5 of HTLV-I and -II (Seiki *et al.*, 1983: Shimotohno *et al.*, 1985a). In all others listed it is lacking. Consequently, the poly(A) signal, even in those few in which it is present, possibly is without functional significance, perhaps being an accident of random-chance combinations.

Other Characteristics of the Sequences. Among the distinctive features of the LTRs reported in the literature are the so-called reversed repeats at the extreme termini, more accurately described as complementary ends in these single-stranded genomes. Although not universally present, as shown by the chicken sarcoma virus sequence (Table 11.1), when they do exist, these elements vary extensively in length (italics). Almost all of those shown begin with TG- at the 5' end, with the complement, -CA, at the other. The longest set is that of the Abelson virus, which has 11 pairing bases at each terminus, whereas the two HTLV sequences have only two. In some instances, the complementary runs may be interrupted, as in the bovine leukemia structure. Since they are present in so many of the diverse forms represented in the table, they would appear to hold some functional value, but in what capacity remains undiscovered.

One additional feature is shown near the center of row 2, where all sequences except the first have a run of three to five As, the functional value of which cannot be even guessed at present. Otherwise, sequence homology is rare, occurring sporadically between such close relatives as the two HTLV and bovine leukemia structures, but elsewhere correspondences are almost exclusively at the random-chance level.

Short Direct Flanking Repeats. As in typical transposons after being inserted into the cell genome, these temperate viruses, including oncogenic forms, bear a copy at their 3' end of the host DNA located at the 5' terminus. As far as has been reported, such flanking repeats are short, typically in the range of five to seven nucleotides. Among the few concrete examples can be cited the bovine leukemia virus, in which the sequence

Table 11.2
Comparisons of Viral Gag Coding Sequences[a]

Row 1

Friend spleen v-gag[b]
ATG GGA CAG ACC GTA ACC ACC CCT CTG AGT CTG ACC CTA GAA CAC TGG GAG GAT GTC CAG CGC ACC GCG

GA-Feline sarcoma v-gag[c]
ATG GGC CAA ACT ATA ACT ACC CCC TTA AGC CTC ACC CTT GAT CAC TGG TCT GAA GTC CGG GCA CGA GCC

Murine sarcoma v-gag[d]
ATG GGC CAG ACT GTT ACC ACT CCC TTA AGT TTG ACC TTA GAT CAC TGG AAA GAT GTC GAG CGG CTC GCT

Rous sarcoma v-gag[e]
ATG GAA GCC GTC ATA AAG GTG ATT TCG TCC GCG TGT AAA ACC TAT TGC GGG AAA ACC TCT CCT TCT AAG

Fujinami sarcoma v-gag[f]
ATG GAA GCC GTC ATA AAG GTG ATT TCG TCC GCG TGT AAA ACC TAT TGC GGG AAA ACC TCT CCT TCT AAG

T-cell leukemia v-gag[g]
ATG GGC CAA ATC TTT TCC CGT AGC GCT AGC CCT ATT CCG CGA CCG CGG GGG CTG GCC CAT CAT CAC

Bovine leukemia v-gag[h]
ATG GGA AAT TCC TCC TCC TAT AAC CCC GCT AAC CCC CCC TCA GAC TGG CTC AAC CTT CTG CAA

Row 2

Friend spleen v-gag
TCC AAT CAG TCC GTA GAT GTC AAG AAG AGA CGC GTC ACC TTC TGC TCT GCC GAG TGG CCA ACT TTC

GA-Feline sarcoma v-gag
CAT AAT CAA GGT GTC GAG GTC AAA AAG AAA TGG ATT ACC TTA TGT GAG GCC GAA TGG GTG ATG ATG

Murine sarcoma v-gag
CAC AAC CAG TCG GTA GAT GTC AAG AAG AGA CGT TGG GTT ACC TTC TGC TCT GCA GAA TGG CCA ACC TTT

Rous sarcoma v-gag
AAG GAA ATA GGG GCC ATG TTG TCC CTC TTA CAA AAG GAA GGG TTG CTT ATG TCT CCC TCA GAC TTA TAT

Fujinami sarcoma v-gag
AAG GAA ATA GGG GCC ATG TTG TCC CAG TTA CAA AAG GAA GGG TTG CTT ATG TCC CTC TCA GAC TTA TAT

T-cell leukemia v-gag
TGG CTT AAC TTC CTC CAG GCG GCA TAT CGC CCC GGT CCC AGT TAC GAT TCC CTC TCA GAC TTA TAT

Bovine leukemia v-gag
AGC GCG CAA AGG CTC AAT CCG CGA CCC TCT CCT GAT TTA ACC GAT TTT ACC GAT TTA AAG TAC ATC CAT TGG

Row 3

Friend spleen v-gag
GGT GTA GGG TGG CCA CAA GAT GGT ACT TTT AAC TTG GAC ATT ATT CTA CAG GTT AAA TCT AAG GTG TTC

GA-Feline sarcoma v-gag
AAT GTG GGC TGG CCC CGA GAA GGA ACT TTT TCT CTT GAT AAC ATT TCC ACC AGT GTT CAG GAG AAG ATC TTC

Murine sarcoma v-gag
AAC GTC GGA TGG CCG CGA GAC GAC ACC TTT AAC CGA GAC CTC ATC ACC CAG GTT AAG ATC AAG GTC TTT

```
Rous sarcoma v-gag        TCC CCG GGG TCC TGG GAT CCC ATT ACC GCG GCG CTA TCC CAG CGG GCT ATG ATA CTT GGG AAA TCG GGA
Fujinami sarcoma v-gag    TCC CCG GGG TCC TGG GAT CCC ATT ACC GCG GCA CTC ACA CAG CGG GCG ATG GTA CTT GGG AAA TCG GGA
T-cell leukemia v-gag     AAA AAA TTT CTT AAA GCT ATA GCT TTA GAA ACA CCG GCT CGG ATC TGT CCC ATT AAC TAC TCC CTC CTA GCC
Bovine leukemia v-gag     TTT CAT AAG ACC CAG AAA AAA CCA TTC ACT TCT GGT GGC CCC TCA TGT CCA CCC GGG AGA

Row 4

Friend spleen v-gag       TCT GGT CCC CAC CCC GAT CAG GTC CCA TAT ATT GTC ACC TGG GAG GCT ATT GCC TAT GAA
GA-Feline sarcoma v-gag   GCC CCG GGA CCG TAT GGA CAC CCC GAC CAA GTT CCT ATT ACC ACA TGG AGA TCC TTA GCC ACA GAC
Murine sarcoma v-gag      TCA CCT GGC CCC CAT CCA GAC CAG GTC CCC ATC TAC GTG ACC TGG GAA GCC TTG GCT TTT GAC
Rous sarcoma v-gag        GAG TTA AAA ACC TGG GGA TTG GTT GGG GCA TTG AAG GCG GCT CGA GAG GAG GTT ACA TCT GAG
Fujinami sarcoma v-gag    GAG TTA AAA ACC TGG GGA TTG GTT GGG GCA TTG AAG GCG GCC CGA GAG GAG GTT ACA TCT GAG
T-cell leukemia v-gag     AGC CTA CTC CCA AAA GGA TAC CCC GGC CGG GTG AAT GAA ATT TTA CAC ATC ATC CAA ACC CAA GCC
Bovine leukemia v-gag     TTC GGC GGG GTT CCC CTC GTC TTG GCC ACC CTA AAC GAA GTA CTC TCA AAC GAA GGG GCC CCG GGT

Row 5

Friend spleen v-gag       CCC CCT CGG TGG GTA AAA CCT TTT GTC TCT CCC --- AAA CTC TCC CCC TCT CCA ACC GCT CCC ATC CTC
GA-Feline sarcoma v-gag   CCC CCT TCG TGG GTT CGT GCG TTC CTA CCC CCT CCC AAA CCT CCC ACA TCC CTC CTT CAA CCT CAC TCG
Murine sarcoma v-gag      CCC CCT CCC TGG GTC AAG CCC TTT GTA CAC CCT --- AAG CCT CCG CCT CTT CTT CCA TCC GCG CGG
Rous sarcoma v-gag        CAA GCA AAG TTT TGG TTG GGA TTA GGG GGA GGG AGG GTC TCT CCC CCA GGT CCG GAG TGC ATC GAG AAA
Fujinami sarcoma v-gag    CAA GGA AAG TTT TGG TTG GGA TTA GGG GGA GGG AGG GTC TCT CCC CCA GGT CCG GAG TGC ATC GAG AAA
T-cell leukemia v-gag     CAG ATC CCG TCC CGT CCC CGC CCA CCG CCG TCA TCC ACC CAC GAC CCC CCG GAT TCT GAT CCA
Bovine leukemia v-gag     GCA TCG GCC CCA GAA GAA CAA CCC CCC CCT TAT GAC CCC GCC ATT TTG CCA ATC ATA TCT GAA GGG
```

(continued)

Table 11.2 (Continued)

Row 6

```
Friend spleen v-gag      CCA TCC GGT CCT TCG ACT CAA CCT CCG CCC CGA TCT GCC CTT TAC CCT GCT CTT ACC CCC TCT ATA AAA
GA-Feline sarcoma v-gag  CCG CAG CCC TCC CCC CCT CTT ACC --- --- TCT TCC CTC GTT CTC CCC AAG TCA GAC CCT
Murine sarcoma v-gag     TCT CTC CCCᶻ CCT CCT CTT TCG ACC CGG CCT CAA TCC TCC CTT TAT CCA GCC CTC ACT CCT TCT TTG GGC
Rous sarcoma v-gag       CCA GCA ACG GAG CGG GGG CGA ATC GAC CCT CAA GGG GAG GAA GTG GGA GAA ACA ACT GTG CAG GAT GCG AAG
Fujinami sarcoma v-gag   CCA GCA ACG GAG CGG GGG CGA ATC GAC AAG --- --- --- GGA GAA ACA ACT GTG CAG GAT ACG AAG
T-cell leukemia v-gag    CAA ATC CCC CCT CCC TAT GTT GAG CCT GCT ACG GCC CCC CAA GTC CTT CCA GTC CAT CCA CAT GGT GCT
Bovine leukemia v-gag    AAT CGC AAC CGC CAT CGT GCT TGG GCA CTC CGA GAA TTA CAA GAT ATC AAA AAA ATT GAA AAT AAG
```

Row 7

```
Friend spleen v-gag      CCC GGA CCT TCT CCG ATA ATG GCG GAC CTC TCA CTG ACC TTC TCT CAG AAG ACC CTC CGC CGT ACG GAG
GA-feline sarcoma v-gag  CCC AAA CCG CCT GTG TTA CCG CCT GAT CCT TCT TCC CCT TTA ATT GAT CTC TTA ACA GAA GAG CCA CCT
Murine sarcoma v-gag     GCC AAA CCT AAA CCT CAA GTT CTT TCT GAC AGT GGG GGGʲ CTC ATC GAC CTA CTT ACA GAA GAC CCC CCG
Rous sarcoma v-gag       ATG GCG CCG GAG GAA ACG GCC ACA CCT AAA ACC GTT GGC ACA TCC TGC GGA ACA GCT ATT
Fujinami sarcoma v-gag   ATG GCG CCG GAG GAA ACG GCC ACA CCT AAA ACC GTT GGC ACA TCC TGC GGA ACA GCT ATT
T-cell leukemia v-gag    CCT CCT AAC CAT CGC CCA TGG CAA ATG AAG GCC ATT AAG AAA GAC CAA GTC TCC CAA GCA GCC
Bovine leukemia v-gag    GCA CCG GGT TCG CAA GTA TGG ATA CAA ACA CTA CGA CTT GCA ATC CTG CAG GCC GCC CCT ACT CCG GCT
```

Row 8

```
Friend spleen v-gag      GAC AGG GAC CGT CCT CCT CTG ACG GAG ATG GCG ACA GAG AAG AGG CCA CCT CCA CTT CTG AGA TTC CTG
GA-Feline sarcoma v-gag  CCC TAT CCA GGG GGT CAC GGG CCA CCG CCA TCA GGT CCT AGG ACC CCA --- --- --- ACC --- GCT ---
Murine sarcoma v-gag     CCT TAT AGG GAC CAC AGA CCA CCC CCT TCC GAC AGG GAC GAT AGT GGA GAA GCG ACC CCT GCG GGA
```

```
Rous sarcoma v-gag       GGC TGT AAT TGC GCC ACA GCC TCG GCT CCT CCT CCT CCT TAT GTG GGG AGT GGT TTG TAT CCT TCC CTG
Fujinami sarcoma v-gag   GGC TGT AAT TGC GCC ACA GCC TCG GCT CCT CCT CCT CCT TAT GTG GGG AGT GGT TTG TAT CCT TCC CTG
T-cell leukemia v-gag    CCT GGG AGC CCC CAG TTT ATG CAG ACC ACC ATC CGG CTT GCG GTG CAG CAG TTT GAC CCC ACT GCC AAA GAC
Bovine leukemia v-gag    GAC CTA GAA CAA CTT TGC CAA TAT ATT GCT TCC CCG GTC GAC CAA ACG GCC CAT ATG ACC AGC CTA ACG

Row 9

Friend spleen v-gag      CCC CCT CTC CCA TAG
GA-Feline sarcoma v-gag  --- --- --- --- TCC CCG ATT GCA AGC CGG CTA AGG GAA CGA GAA AAC CCT --- GCT GAA GAA
Murine sarcoma v-gag     GAG GCA CCG GAC CCC TCC CCA ATG GCA TCT CTC CGC CTG CGT GGG AGA CGG GAG CCC CTT GTG GCC GAC TCC
Rous sarcoma v-gag       GCG GGG GTG GGA GAG GAG GAG CAG CAG CAG GGT GAC ACA CCT CCG GGG GCG GCG GAA CAG CAG TCA AGG GCG GAG
Fujinami sarcoma v-gag   GCG GGG GTG GGA GAG GAG GAG CAG CAG CAG GGT GGT GAC ACA CCT CGG GGG GCG GGG GAA CAG CAG CCA AGG GCG GAG
T-cell leukemia v-gag    CTC CAA GAC CTC CTG CAG TAC CTT TGC TCC TCC TCC CTC GTG GCT TCC CTC CAT CAC CAG CAG CTA GAT
Bovine leukemia v-gag    GCA GCA ATA GCC GCT GAA GCG GCA ACA CCC TCC AGG GTT TTA ACC CCC AAA ACG GGT ACC CTA ACC

Row 10

GA-Feline sarcoma v-gag  --- --- TCT CAA GCC CTC CCC TTG AGG GAA GGC CCC AAC AAC CGA CCC CAG TAT TGG CCA TTC TCA GCT
Murine sarcoma v-gag     ACT ACC TCG CAG GCA TTC CCC CTC CGC ACA GGA GGA AAC GGA CAG CTT CAA TAC CCG TTC TCG TCT
Rous sarcoma v-gag       CCA GGG CAT GCG GGT CAG GCT CCT GGG CCG GCA GCC CTG ACT GAC TGG GCA AGG GTC AGG GAG GAG CTT GCG
Fujinami sarcoma v-gag   CCA GGG CCC ACG GGT CTG GCC CCT GGG CCG GCA GCC CTG ACT GAC TGG GCA AGG ATT AGG GAG GAG CTT GCG
T-cell leukemia v-gag    AGC CTT ATA TCA GAG GCC GAA ACC CGA GGT ATT ACA GGT --- TAT AAC CCA TTA GCC GGT CCC CTC CGT
Bovine leukemia v-gag    CAA CAA TCA GCT CAG CCC AAC GCC GGG GAT CTT AGA AGT CAA TAT CAA AAC CTC TGG CTT CAG GCC GGA
```

(continued)

Table 11.2 (Continued)

Row 11

GA-Feline sarcoma v-gag	TCA GAC CTG TAT AAC TGG AAG TCG CAT AAC CCC CCT TTC TCC CAA GAC CCA GTG GCC CTA ACT AAC CTA
Murine sarcoma v-gag	TCT GAC CTT TAC AAC TGG AAA AAT AAC CCT TCT TTT TCT GAA GAT CCA GGT AAA CTG ACA GCT CTG
Rous sarcoma v-gag	AGT ACT GGT CCG CCC GTG GCC ATG CCT GTA GTG ATT AAG ACA GAG GGA CCC TGG ACC CCT CTG
Fujinami sarcoma v-gag	AGT ACA GGT CCG CCC ATG GTG GCC ATG CCT GTA GTG ATT AAG ACA GAG GGA CCC TGG ACC CCT CTG
T-cell leukemia v-gag	GTC CAA GCC AAC AAT CCA CAA CAA GGA TTA AGG GAA TAC CAG GAA CTC --- TGG CTC GCC GCC
Bovine leukemia v-gag	AAA ATC TCC CTA CTC CTT CAG CTA CAA CCT TGG TCC ACC ATC GTC CAA GGC CCC GCC GAA AGC TCT

Row 12

GA-Feline sarcoma v-gag	ATT GAG TCC ATT TTA GTG ACG CAT CAA CCA TGG GAC GAC TGC CAG CAG CTC TTG CAG GCA CTC CTG ACA
Murine sarcoma v-gag	ATC GAG TCT GTC CTC ATC ACC CAT CAG CCC ACC TGG GAC GAC TGT CAG CAG CTG TTG GGG ACT CTG CTG ACC
Rous sarcoma v-gag	GAG CCA AAA TTG ATC ACA AGA CTG GCT GAT ACG GTC AGG ACC AAG AAC GGC TTA CGA TCC CCG --- ATT ACT ATG
Fujinami sarcoma v-gag	GAG CCA AAA TTG ATC GCA GGA CTA GCT GGT GCC GTC GGG GCC GTC GGG GGC TTG CGA TCT CCG ATC GCT GTG
T-cell leukemia v-gag	TTC GCC GCC CTG CCG GGG AGT GCC AAA GAC CCT TCC TGG GCC TCT ATC CTC CAA GGC CTG GAG GAG CCT
Bovine leukemia v-gag	GTA GAG TTT GTC AAC CGG TTA CAA ATT TCA TTA GCT GAC AAC CTT CCC GAC GGA GTC CTA AGG AAC CCA

Row 13

GA-Feline sarcoma v-gag	GGC GAA GAA AGG CAA AGG GTC CTT GAG GCC CGA AAG GTT CCA GGC GAG GAC GGA GGA CGG CCA ACC
Murine sarcoma v-gag	GGG GAA GAA AAA CAA CGG GTG CTC TTA GAG GCT AGA AAG GCG GTG CGG GGC GAT GAT GGG CGC CCC ACT
Rous sarcoma v-gag	GCA GAA GTG GAA GCG CTT ATG TCC TCC CCG CTG CTG CCG CAT GAC GTC ACG AAT CTA ATG ATG AGA GTT ATT
Fujinami sarcoma v-gag	GCA GGG GTG GAG GCG CTT ATG TCC TCC CCG CTG CCG CAT GAC GTC ACG AAT CCA ATG AGA GTT ATT

```
T-cell leukemia v-gag    TAC CAC GCC TTC GTA GAA CGC CTC AAC ATA GCT CTT GAC AAT GGG CTG CCA GAA GGC ACG GGC CCC AAA GAC
Bovine leukemia v-gag    TTA TTG ACT CCC TTA GTT ATG CAA ATG CTA ACA AGT GTC AGC AAA TTT TGC AGG GGC GAG GCC AGT

Row 14
GA-Feline sarcoma v-gag  CAA CTA CCC AAT GTC ATT GAT GAG ACT TTC CCC TTG ACC CGT CCC AAC TGG GAT TTT GCT ACG CCG GCA
Murine sarcoma v-gag     CAA CTG CCC AAT GAA GTC GAT GCC GCT TTT CCC CTC GAG CGC CCA GAC TGG GAG TAC ACC ACC CAG GCA
Rous sarcoma v-gag       TTA GGG CCT GCC CCA TAT GCC TTA TGG GAC GCT TGG ATG GTC CAA CTC CAG ACA GTT ATA GCG GCA
Fujinami sarcoma v-gag   TTA GGA CCC GCC CCA CAT GCC TTA TGG ATG GAC GCT TGG G   (fps gene follows immediately)
T-cell leukemia v-gag    CCC ATC TTA CGT TCC TTA GCC TAC TCC AAT GCA AAC GAA TGC AAA AAA TTA CTA CAG GCC CGA GGA
Bovine leukemia v-gag    GGC CGC GGT GGG GCA AAA ACT GCA GGC TTG CGC ACA ATT GGG CCC CCA AGA ATG AAA CAG CCT GCA CTT

Row 15
GA-Feline sarcoma v-gag  GGT AGG GAG CAC CTA CGC CTT TAT CGC CAG TTG CTA TTA GCG GGT CTC CGG GCT GCA AGA CGC CCC
Murine sarcoma v-gag     GGT AGG AAC CAC CTA GTC CAC TAT CGC CAG TTG CTC ATA GCG GGT CTC CAA AAC GCG GGC AGA AGC CCC
Rous sarcoma v-gag       GCC ACT CGC GAC CCC CGA CAC CCA GCG AAC GGT CAA GGG CGG GAA CGG ACT AAT TTG AAT CGC TTA
T-cell leukemia v-gag    CAC ACT AAT AGC CCT CTA GGA GAT ATG ATG CGG GCT ATG CCC GGG CGG TGT CAG ACC TGG ACC CCC AAA GAC AAA ACC ---
Bovine leukemia v-gag    CTC GTC CAC ACC GGG CCC AAG ATG CCC GGG CCC CAA CGG GCC CCC AAA AGG CCT CCC CCA GGA

Row 16
GA-Feline sarcoma v-gag  ACT AAT TTG GCA CAG GTA AAG CAA GTT GTA CAA GGG AAA GAG GAA ACG CCA GCA GCA  (fes sequence follows)
Murine sarcoma v-gag     ACC AAT TTG CCC AAG GTA AAA GGA ATA ACA CAA GGG CCC AAT GAG TCT CCC TCG GCC TTC CTA GAG AGA
Rous sarcoma v-gag       AAG GGC TTA GCT GAT GGG ATG GTG GGC AAC CCA CAG GGT GCA TTA TTA AGA CCG GGG GAA TTG
T-cell leukemia v-gag    AAA GTG TTA GTT GTC CAG CCT AAA AAA CCC CCA CAG CCG TGC TGC TTC CGG TGC GGG AAA GCA GGC
Bovine leukemia v-gag    CCA TGC TAT CGA TGC CTC AAA GAA GGC CAT TGG GCC CGG GAT TGT CCT ACC AAG GCC CCA CCT
```

(continued)

Table 11.2 (Continued)

Row 17

Murine sarcoma v-gag	CTT AAG GAA GCC TAT CGC AGG TAC ACT CCT TAT GAC CCT GAG GAC CCA GGG CAA GAA ACT AAT GTG TCT
Rous sarcoma v-gag	GTT GCT ATT ACG GCG TCG GCT CTC CAG GCG TTT AGA GAG GTT GCC CGG CTG GCG GAA CCT GCA GGT CCA
T-cell leukemia v-gag	CAC TGG AGT CGG GAC -- TGC ACT CAG CCT CGT CCC CCC GGG CCA TGC CCC CTA TGT CAA GAC CCA
Bovine leukemia v-gag	CCG GGA CCT TGC CCC ATA TGT AAA GAT CCT TCC CAT TGG AAA CGA GAC TGT CCA ACC CTC AAA TCA AAA

Row 18

Murine sarcoma v-gag	ATG TCT TTC ATT TGG CAG TCT GCC CCA GAC ATT GGG AGA AAG TTA GAG AGG TTA GAA GAT TTG AGA AAC
Rous sarcoma v-gag	TGG GCG GAC ATC ATG CAG GGA CCA TCT GAG TCC TTT GTT GAT TTT GCC AAT CGG CTT ATA AAG GCG GTT
T-cell leukemia v-gag	ACT CAC TGG AAG CGA GAC TGC CCC CGC CTA AAG CCC ACT ATC CCA GAA CCA CCA GAG GAA GAT GCC
Bovine leukemia v-gag	AAC TAA

Row 19

Murine sarcoma v-gag	AAG ACG CTT GGA GAT TTG GTT AGA GAG GCA GAA AGG ATC TTT AAT AAA CGA GAA ACC CCG GAA GAA AGA
Rous sarcoma g-gag	GAG GGG TCA GAT CTC CCG CCT TCC GCG CGG GCT GTG ATC ATT GAC TGC TTT AGG CAG AAG TCA CAG
T-cell leukemia v-gag	CTC CTA TTA GAC CTC CCC GCT GAC ATC CCA CAC CCA AAA AAC TCC ATA GGG GGG GAG GTT TAA

Row 20

Murine sarcoma v-gag	GAG GAA CGT ATC AGG AGG AGA GAG GAA AAG GAA CCC CGT AGG ACA GAG GAT GAG CAG GAG AAA AAA GAG
Rous sarcoma v-gag	CCA GAT ATT CAG CAG CTT ATA CGG ACA GCA GCA CCC TCC ACG TCC ACC CCA GGA GAG ATA ATT AAA TAT

Row 21

Murine sarcoma v-gag AAA GAA AGA GAT CGT AGG AGA CAT AGA GAG ATG AGC AGG CTA TTG GCC ACT GTC GTT AGT GGA CAG AGA

Rous sarcoma v-gag GTG CTA GAC AGG CAG AAG ACT GCC CCT CTT ACG GAT CAA GGC ATA GCC GCG GCC ATG TCG TCT GCT ATC

Row 22

Murine sarcoma v-gag CAG GAT AGA CAG GAA GAA CGA AGG AGG TCC CAA CTC GAC TGC GAC CAG TGT ACC TAC TGC GAA GAA CAA

Rous sarcoma v-gag CAG CGC TTA ATT ATG GCA GTA GTC AAT AGA GAG AGA GGG TCG GGT GGT CGT GCC CGA ---

Row 23

Murine sarcoma v-gag GGG CAC TGG GCT AAA GAT TGT CCC AAG AGA CCA CGA GGA CCT CGG GGA CCA AGA CCC CAG ACC TCC CTC

Rous sarcoma v-gag GGG CTC TGC TAC ACT TGT GGA TCC CCG GGA CAT TAT CAG GCG CAG TGC CCG AAA AAA CGG AAG TCA GGA

Row 24

Murine sarcoma v-gag CTG ACC CTA GAT GAC TAG

Rous sarcoma v-gag AAC AGC CGT GAG CGA TGT CAG TTG TGT AAC GGG ATG GGA CAC AAC GCT AAA CAG TGT AGG AAG CGG CAT

Row 25

Rous sarcoma v-gag GGC AAC CAG GGC CAA CGC CCA GGA AAA GGT CTC TCT TCG GGG CCG TGG CCC GGC CCT GAG CCA CCT GCC

Row 26

Rous sarcoma v-gag GTC TCG TTA GCG ATG ACA ATG GAA CAT AAA GAT CGC CCC TTG GTT AGG GTC ATT CTG ACT AAC ACT GGG

Row 27

Rous sarcoma v-gag AGT CAT CCG GTC AAA CAG CGT TCG GTG TAT ATC ACC GCG CTG TTG GAC TCT GGA GCG GAC ATC ACT ATT

(continued)

Table 11.2 (Continued)

Row 28

Rous sarcoma v-gag   ATT TCA GAG GAG GAT TGG CCC ACC GAT TGG CCA CTG ATG GAG GCC GCG AAC CCG CAG ATC CAT GGG ATA

Row 29

Rous sarcoma v-gag   GGA GGG GGA ATT CCC ATG CGA AAA TCT CGT GAC ATG ATA GAG TTG GGG GTT ATT AAC CGA GAC GGG TCT

Row 30

Rous sarcoma v-gag   TTG GAG CGA CCC CTG CTC CTC TTC CCC GCA GTA GCT ATG GTT AGA GGG AGT ATC CTA GGA AGA GAT TGT

Row 31

Rous sarcoma v-gag   CTG CAG GGC CTA GGG CTC CGC TTG ACA AAT TTA TAG

[a] Codons for proline (CCN) are lightly stippled, those for cysteine (TGC, TGT) are moderately stippled, and those for tryptophan are heavily stippled; triplets for basic amino acids are underscored and those for tyrosine are italicized.
[b] S. P. Clark and Mak (1983).
[c] Hampe et al. (1982).
[d] Van Beveren et al. (1981).
[e] Schwartz et al. (1983).
[f] Shibuya and Hanafusa (1982).
[g] Seiki et al. (1983).
[h] Sagata et al. (1985).
[i] Insert CTT GAA.
[j] Insert CCG.

GACAGG was described (Sagata *et al.*, 1984), and HTLV-I, where TAGTTG has been found (Seiki *et al.*, 1983). In some instances the direct repeats are tetranucleotides, as in the GATG of the murine osteosarcoma virus (Van Beveren *et al.*, 1983). Only when the proviral structure itself is examined can these repeats be identified.

11.2. TYPICAL GENES OF RETROVIRUSES

As pointed out earlier, the genomes of unmodified retroviruses contain at least four genes—*gag*, which encodes core proteins; *pol*, which specifies the RNA-directed DNA polymerase; *env*, which directs the synthesis of capsid proteins; and the recently described *x* or *R* gene in the terminal region, which provides the coding sector for a protein not yet completely identified, as discussed earlier.

11.2.1. The Gag Gene Structure

Since it has not been firmly established what capabilities each leukemia virus gene has in relation to the transformation of host cells, a comparative study of the more important might reveal common features and lay a basis for further deductions. Undoubtedly, the most thoroughly investigated is that called the *gag* gene, which encodes core proteins.

Homologous Sets of Genes. When the seven sequences of *gag* presented in Table 11.2 are examined, it is readily apparent that four sets of homologs are represented. The first three, all from mammalian sources, form one cluster, the Rous and Fujinami sarcoma viruses, both from chickens, combine into another, and each of the two leukemia viruses obviously represent separate groupings. The relationships between the pair from avian sarcomas are especially striking, for they display a high level of identical codons throughout their entire lengths. Yet strongly homologous though their sequences are, they differ pronouncedly in size as the result of recombinational events, for the Fujinami structure is truncated at its 3' end (row 14) by the insertion of a *frs* gene. In contrast, the Rous sarcoma virus sequence is the longest of all, extending through 31 rows (~700 codons).

The three mammalian sarcoma viral cistrons are not nearly as consistent in structure as are those of the fowl. Although, as just stated, they are easily perceived to be members of the same group, they vary extensively from species to species, a lack of harmony that increases toward their 3'-terminal sectors. The Friend spleen focus-forming sarcoma viral structure differs quite radically in length, for it terminates in a TAG stop signal in row 9. How the other two compare in overall size is impossible to state, since the Gardner–Arnstein feline sarcoma viral sequence is truncated (row 16) by the insertion of a *fes* cistron, while the murine sarcoma virus gene terminates (row 24) in a TAG translational stop codon.

Although the HTLV and bovine leukemia virus *gag* genes show occasional homologies intermittently throughout their sequences, such points of agreement are far too rare to consider them other than representatives of separate, distantly related groups. Among the sites of identity greater than one codon in length can be mentioned AAYGAA at the center of row 4, CC-CC-CC- before the center of row 5, YCCCCCRCC a few codons later, TCTGA near the end of the same row, GC-ATY--GCARG in row 7, and YCCTCC at the middle of row 9. The total lengths of the genes differ to some extent, the HTLV example being 42 codons longer than the bovine species.

Some Chemical Properties. Because four diverse groups are thus represented in Table 11.2, some of the most outstanding chemical properties must be viewed separately. In the three mammalian sarcoma viruses, the presence of series of identically placed codons for tryptophan (TGG; darkly stippled) is at once a noticeable feature, for each of the first five rows has at least one column of that triplet, and row 2 contains two. Then, through row 9 that codon is absent, but it occurs thereafter at a similar rate as before in the next three rows, becoming nearly absent again in the terminal sector. In the region where each of the trio of mammalian types is represented, codons for basic amino acids do not appear especially critical, for little agreement in their distribution exists from species to species. One exception is conspicuous in row 2, however, where each cistron has identically located runs of four triplets for either lysine or arginine. After the initial three rows, codons for proline (CCN; lightly stippled) are a particularly prominent aspect of these sequences, four occurring in row 4, nine or ten in row 5, five to seven in row 6, four to eight in row 7, and five or six in row 8, followed by a reduction in frequency thereafter. Since proline has a marked effect on secondary structure of a protein, its presence in such abundance in these three must hold considerable significance.

The remaining groups tend even less toward a plethora of codons for basic amino acids than the mammalian sequences. In the HTLV gene there is a total of 38, for an 8.7% level, in that of bovine leukemia, 41, or 10.4%, and in the Rous from chicken, a total of 80, for 11.3%, which compares with the 75 (14%) of the murine sequence. Moreover, triplets specifying proline are differently distributed in each group, those of HTLV being especially plentiful near the 3′ terminus, whereas those from avian sources are more frequent in the midregion. Codons for cysteine (moderate stippling) also are a marked feature of the avian types, characteristically being arranged in subadjacent pairs. Thus, it is apparent that although all the products of the various *gag* genes are core proteins, pronounced differences in their chemical properties exist from group to group.

11.2.2. The Env Gene Structure

The gene *env* is really a dicistronic compound structure that encodes two glycoproteins, one of which forms part of the capsid surface and plays a role in determining the host specificities, while the other spans the membrane. However, the exact natures and sizes of the two components have been subjected to several differing interpretations, some of which correlate to the source species. In the Kirsten murine leukemia virus and AKV, the two products are referred to as glycoprotein 70 and protein p15 (Herr *et al.*, 1982; Norton *et al.*, 1984b), whereas in the bovine leukemia virus, the designations are glycoproteins 51 and 30 (Sagata *et al.*, 1985), and in Friend spleen focus-forming virus, glycoprotein 55 (S. P. Clark and Mark, 1983). The protein p15 appears to inhibit the immune responses of various blood cells, including lymphocytes, monocytes, and macrophages (Cianciolo *et al.*, 1985).

But in many investigations no substructures are indicated, including those of the simian sarcoma, HTLV-I and -II, and Moloney murine sarcoma viruses (Seiki *et al.*, 1983; Devare *et al.*, 1983; Shimotohno *et al.*, 1985a). Moreover, substantial deletions have occurred in certain cases, so that the alignment of the seven sequences in Table 11.3 does not always bring out possible homologies to their maximum extent.

Table 11.3

Comparisons of Viral Coding Sectors for Env[a]

Row 1

HTLV-I[b]	ATG GGT AAG TTT CTC GCC ACT TTG ATT TTA TTC TTC CAG TTC TGC CCC CTC ATC TTC GGT GAT TAC AGC
HTLV-II[c]	ATG GGT AAT GTT TTC --- --- --- TTC CTA CTT TTA TTC AGT CTC ACA CAT TTT CCA CTA GCC CAG
Simian sarcoma[d]	ATG CTT CTC ACC TCA AGC CTG CAC CAC CCT CGG CAC CAG ATG AGT AGT CCT GGG AGC TGG AAA AAG CTG ATC
Murine leukemia[e]	ATG GAG AGT ACA ACG CTC TCA AAA CCC TTT AAA AAT CAG GTT AAC CCG TGG GGC CCC CTA ATT GTC CTT
Friend[f]	ATG GAA GGT CCA GCG TCC TCA AAA CCC CTT AAA GAT AAG ACT AAC CCG TGG GGC CCC CTA ATA ATC TTG
Xenotropic MuLV[g]	ATG GAA GGT TCA GCG TTC TCA AAA CCC CTT AAA GAT AAG ATT AAC CCG TGG GGC CCC CTA ATA GTT ATG
Moloney[h]	ATG GCG CGT TCA ACG CTC TCA AAA CCC CTT AAA AAT AAG GTT AAC CCG CGA GGC CGA CTA ATC CCC TTA

Row 2

HTLV-I	CCC AGC TGC TGT ACT CTC ACA ATT GGA GTC TCC TCA TAC CAC TCT AAA CCC TGC AAT CCT GCC CAG CCA
HTLV-II	CAG AGC CGA TGC ACA CTC ACG ATT GGT ATC TCC TCC TAC CAC TCC AGC CCC TGT AGC CCA ACC CAA CCC
Simian sarcoma	ATC CTC TTA AGC TGC GTA TTC GGC GGC GGA ACG AGT CTG CAA AAT AAA AAC CCC CAC CAG CCT ATG
Murine leukemia	CTG ATT CTC GGA GGG GTC AAC CCC GTT ACG TTG AGA AAC AGC CCC CAC CAG GTT TTT AAC CTC ACC TGG
Friend	CGG ATC TTA ATA AGG GCA GGA GTA TCA CAA CTT GAC AGC CCT CAT CAG GTC TCC AAT GTT ACT TGG
Xenotropic MuLV	CGG ATC TTG GTG AGG GCA GGA GCC TCG GTA CAA CGT GAC AGC CCT CAC CAG ATC TTC AAT GTT ACT TGG
Moloney	ATT CTT CTG ATG CTC AGA GGG GTC AGT GCT ACT GCT CTG TCG CCC[i] AGT CCT CAT CAA GTC TAT AAT ATC ACC TGG

(continued)

Table 11.3 (Continued)

Row 3

```
HTLV-I           GTT TGT TCG TGG ACC CTC GAC CTG GCC CTT TCA GCA GAT CAG GCC CTA CAG CCC CCC TGC CCT AAC
HTLV-II          GTC TGC ACG TGG AAC CTC GAC CTT AAT TCC CTA ACA ACG GAC CAA CTA CAC CCC TGC CCT AAC
Simian sarcoma   ACC CTC ACC TGG CAG GGG GAC CCC ATT CCT GAG GAG CTC TAT AAG ATG CTG AGT GGC CAC TCG ATT CGC
Murine leukemia  GAA GTG ACT AAT GGA GAC CGA GAA ACG GTG TGG GCA ATA ACC GGC AAT CAC CCT CTG TGG ACT TGG TGG
Friend           AGA GTT ACC AAC TTA ATG ACA GGA CAA ACA GCT AAT GCT ACC TCC CTC CTG GGG ACG ATG ACA GAG GCC
Xenotropic MuLV  AGA GTT ACC AAC CTA ATG ACA GGA CAA ACA GCT AAC GCC ACC TCC CTC CTG GGG ACG ATG ACA GAC ACC
Moloney          GAG GTA ACC AAT GGA GAT CGG CGG GAG ACG GTA TGG GCA ACT TCT GGC AAC CAC CCT CTG TGG ACC TGG TGG
```

Row 4

```
HTLV-I           CTA GTA AGT TAC TCC AGC TAC CAT GCC ACC TAT TCC CTA TAT CTA TTC CCT CAT TGG ACT AAG AAG CCA
HTLV-II          CTA ATT ACT TAC TCT GGC TTC CAT AAG ACT TAT TCC TTA TTC CCA CAT TGG ATA AAA AAG CCA
Simian sarcoma   TCC TTC GAT GAC CTC CAG CGC CTG CTG CAG GAC TCC GGA GAC AAA GAA GAT GGG GCT GAG CTG GAC CTG
Murine leukemia  CCT GAC CTC ACA CCA GAT CTC TGT ATG TTG GCC CTC CAC GGG CCG TCC TAT TGG GGC CTA GAA TAT CGG
Friend           TTT CCT AAA CTG TTT GAC TTG TGC GAT TTA ATG GGG GAC GAC TGG GAT GAG ACT GGA CTC GGG TGT
Xenotropic MuLV  TTC CCT AAA CTA TTT GAC CTG TGT GAT TTA GTA GGA GAC TAC TGG GAT GAC CCA GAA CCC GAT ---
Moloney          CCT GAC CTT ACC CCA GAT TTA TGT GCC CAC CAC CCA TCT TAT TGG GGG CTA GAA TAT CAA
```

Row 5

HTLV-I	AAC CGA AAT GGC GGA GGC *TAT* *TAT* TCA GCC TCT *TAT* TCA GAC CCT TGT TCC TTA AAG TGC CCA *TAC* CTG
HTLV-II	AAC AGA CAG GGC CTA GGG *TAC* *TAC* TCG CCT TCC *TAC* AAT GAC CCT TGC TCG CTA CAA TGC CCC *TAC* TTG
Simian sarcoma	AAC ATG ACC CGC TCC CAT TCT GGT GGC GAG CTG GAG AGC TTG GCT CGT GGG AAA AGG AGC AGC CTG GGT TCC
Murine leukemia	GCT CCT TTT TCT CCT CCG CCG CCC CCC TGC TGT TCA GGA AGC AGC GAC TCC ACG GGC CCA GGC TGT TCC
Friend	CGC ACT CCC GGG GGA AAA AGG GCA ACA TTT GAC TTC GTT TGC CCC GGG CAT ACT GTA CCA
Xenotropic MuLV	--- --- --- --- --- --- --- --- --- --- --- --- --- --- --- ATT GGG GAT GGT TGC
Moloney	TCC CCT TTT TCT CCC CCG GGG CCC CCT TGT TGC TCA GGG GGC AGC AGC --- --- CCA GGC TGT TCC

Row 6

HTLV-I	GGC TGC CAA TCA TGG ACC TGC CCC *TAT* ACA GGA GCC GTC AGC CCC *TAC* TGG AAG TTT CAA CAC GAT
HTLV-II	GGC TGC CAA GCA TGG ACA TCC GCA *TAC* ACG GGC CCC GTC TCC AGT CCA TCC TGG AAG TTT CAT TCA GAT
Simian sarcoma	CTG AGC GTT GCC GAG GCC ATG ATT GCC GAG TGC AAG ACA CGA ACC GAG GTG TTC GAG ATC TCC CGG
Murine leukemia	AGA GAT TGT GAG GAG CTG ACT TCA *TAT* ACT CCC CGG AAT ACG GCC TGG AAC AGA CTT AAG TTA
Friend	ACA GGG TGT GGA GGG CCG AGA GAG GGC *TAC* TGT GGC AAA TGG GAG ACC GGA ACC GCA GCA *TAC*
Xenotropic MuLV	CGC ACT CCC GGG GGA AGA AGA --- --- --- AGG ACA AGA CTG *TAT* GAC --- --- --- --- ---
Maloney	AGA GAC TGC GAA GAA CCT TTA ACC TCC ACC CCT CGG TGC AAC ACT GCC TGG AAC AGA CTC AAG CTA

(continued)

Table 11.3 (Continued)

Row 7

```
HTLV-I          GTC AAT TTT ACT CAA GAA GTT TCA CGC CTC AAT ATT AAT CTC CAT TTT TCA AAA TGC GGT TTT CCC TTC

HTLV-II         GTA AAT TTC ACC CAG GAA GTC AGC CAA GTG TCC CTT CGA CTA CAC TTC TCT AAG TGC GGC TCC TCC ATG

Simian sarcoma  CGC CTC ATC GAC CGC ACC AAT GCC AAC TTC CTG GTG TGG CCG CCC GTG GAG GTG CAG CGC TGC TCC

Murine leukemia TCT AAA GTG ACA CAT GCA CAC AAT GGA GGA TTC GTC TGC CCC GGG CCA CAT CGC CCC CGG TGG GCC

Friend          TGG AAG CCA TCA TCA TGG GAC CTA ATT TCC CTT AAG CGA GGA AAC ACT CCT AAG GAT CAG GGC CCC

Xenotropic MuLV --- --- --- --- --- --- --- TTC TAT GTT TGC CCC GGT --- CAT ACT GTA CCA ATA GGG

Moloney         GAC CAG ACA ACT CAT AAA TCA AAT GAG GGA TTT GTT TGC CCC GGG CCC CAC CGC CCC CGA GAA TCC
```

Row 8

```
HTLV-I          TCC CTT CTA GTC GAC GCT CCA GGA --- TAT GAC CCC ATC TGG TTC CTT AAT ACC GAA CCC AGC CAA CTG

HTLV-II         ACC CTC CTA GTA GAT GCC CCT GGA --- TAT GAT CCT TTA TGG TTC ATC ACC TCA GAA CCC ACT CAG CCT

Simian sarcoma  GGC TGT TGC AAC AAC CGC AAC GTG CAG TGC CGG CCC AGC CAA GTG CAG CTG CGG CCA GTC CAG GTG AGA

Murine leukemia CGG TCA TGT GGT GGT CCA GAA TCC TTC TAT TGT GCC TCT TGG GGC TGC GAA ACC ACA GGC CGA GCA TCC

Friend          TGT TAT GAT TCC TCG GTC TCC AGC GGC CTC GGT GCC AAA TGG GGA TGT TGG GGT CGA TGC AAC CCC CTG GTT[j]

Xenotropic MuLV --- --- TGT GGA GGG CCG GGA GAG GGC TAC TGT TGT GGC TGT GAG ACC ACT GGA CAG GCA CAG GTT

Moloney         AAG TCA TGT GGG GGT CCA CAC TCC TTC TAC TGT TGT GGC TGT GAG ACA ACC GGT AGA ACC TAC
```

Row 9

HTLV-I   CCT CCC ACC GCC CCT CTA CTC CCC CAC TCT AAC CTA GAC CAC ATC CTC GAG CCC TCT ATA CCA TGG AAA
HTLV-II   CCA CCA ACT TCT CCC CCA TTG GTC CAT GAC TCC GAC CTT GAA CAT GTC CTA ACC CCC TCC ACG TCC TGG ACC
Simian sarcoma   --- AAG ATC GAG ATT GTG CGG AAG AAG CCA ATC TTT AAG AAG ACG GTG ACG GAG GAC CAC CTG GCA
Murine leukemia   TGC AAA CCA TCC TCG TCC CAC TAC ATC ACA GTA AGC AAC AAT CTA ACC TCA --- GAC CAG GCA ACC CCA
Friend   TGG GAT GCC CCC AAA GTA TGG GGA CTG AGA CTG TAC CGA TCC ACA GGG ACC --- GAC CCG GTG ACC CGC
Xenotropic MuLV   TGG AAG CCA TCA TCA TCA TGG GAC CTA ATT TCC CTT AAG CGA GGA AAC ACT CCT AAG GAT CAG GGC CCC ---
Moloney   TGG AAG CCC TCA TCA TCA TGG GAT TTC ATC ACA GTA AAC AAC AAT CTC ACC TCT --- GAC CAG GCT GCT CAC

Row 10

HTLV-I   TCA AAA CTC CTG ACC CTT GTC CAG TTA ACC CTA CAA AGC ACT AAT TAT ACT TGC ATT GTC TGT ATC GAT
HTLV-II   ACC AAA ATA CTC AAA TTT ATC CAG CTG ACC TTA CAG AGC ACC AAT TAC TCC ATG GTT TGC ATG GTG GAT
Simian sarcoma   TGC AAG TGT GAG ATA GTG GCA GCT GCA CGG GCT GTG ACC CGA AGC CCG GGG ACT TCC CAG CAG CAG CGA
Murine leukemia   GTA TGC AAA GGT AAT GAG GAG TGG TGC AAC TCC TTA ACT ATC ATC CGG TTC ACG AGC TTT GGA AAA CAG GCC ACC
Friend   TTC TCT TTG ACC CGC CAG GTC CTC GAT ATA --- GGG CCC CGC --- --- ACA CCG --- --- --- ---
Xenotropic MuLV   --- TGT TAT GAT TCC TCG GTC TCC AGT GGC GTC CAG GGT GCC --- ACA CCG --- --- --- ---
Moloney   GTA TGC AAA GAT AAT AAG AAG TGC AAC CCC TTA GTT ATT CGG TTT ACA GAC GCC GGG AGA CGG GTT ACT

(continued)

Table 11.3 (Continued)

Row 11

HTLV-I CGT GCC AGC CTC TCC ACT TGG CAC GTC CTA TAC TCT CCC AAC GTC TCT GTT CCA TCC TCT TCT TCT ACC

HTLV-II AGA TCC AGC CTC TCA TCC TGG CAT GTA CTC TAC ACC CCC AAC ATC TCC ATT CCC CAA CAA ACC TCC TCC

Simian sarcoma GCC AAA ACG ACC CAA AGT CGG GTG ACC ATC CGG ACG GTG CGA GTC CGC CGG CCC CCC AAG GGC AAG CAC

Murine leukemia TCC TGG GTC ACA GGC CAT TGG TGG GGA TTG CGC CTA TAC GTC TCT GGA CAT GAC CCA GGG CTC ATC TTT

Friend GTT CCC ATT GGG TCT AAT CCC GTG ACT ACC GAC CAG TTA CCC CTC TCC CGA CCC GTG CAG ACC ATG CCC

Xenotropic MuLV --- --- GGG GGT CGA TGC AAC CCC CTG GTC TTA GAA TTC

Moloney TCC TGG ACC ACA GGA CAT TAC TGG GGC TTA CGT TTG TAT GTC TCC GGA CAA GAT CCA GGG CTT ACA TTT

Row 12

HTLV-I CCC CTC CTT TAC CCA TCG TTA GCG CTT CCA GCC CCC CAC CTG ACG TTA CCA TTT AAC TGG ACC CAC TGC

HTLV-II CGA ACC ATC CTC TTT CCT TCC CTT GCC CTG CCC GCT CCT CCA TCC CAA CCC TTC CCT TGG ACC CAT TGC

Simian sarcoma CGG AAA TGC AAG CAC ACG CAT GAC AAG ACG GCA CTG AAG ACG CTC GGA GCC TAA

Murine leukemia GGG ATC CGA CTT AAA ATT ACA GAC TCG GGG CCC CGG GTC GTC CCA ATA GGG CCA AAC CCC GTC TTG TCA GAC

Friend CCC AGG CCT CTT CAG CCT CCT CCA GGC GCA GCC TCT ATA GTC CCC CAG ACT GCC CCA CCT CAA

Moloney GGG ATC CGA CTC AGA TAC CAA AAT CTA GGA CCC CGC GTC CCA ATA GGG CCA AAC CCC GTT CTG GCA GAC

Row 13

HTLV-I	TTT	GAC	CCC	CAG	ATT	CAA	GCT	ATA	GTC	TCC	CCC	TGT	CAT	AAC	TCC	CTC	ATC	CTG	CCC	CCC	TTT	TCC
HTLV-II	TAC	CAA	CCT	CGC	CTA	CAG	GCG	ATA	ACA	ACA	GAT	AAC	TGC	AAC	AAC	TCC	ATT	ATC	CTC	CCC	CCT	TTT	TCC
Murine leukemia	CGA	CGA	CCA	CCT	TCC	CCT	AGA	CCC	ACC	CCG	CCT	TCA	AAC	TCC	ACC	CCA	ACC	GAG	ACA
Friend	(To row 14)
Moloney	CAA	CAG	CCA	CTC	TCC	AAG	CCC	AAA	CCT	GTT	AAG	TCG	CCT	TCA	GTC	ACC	AAA	CCA	CCC	AGT	GGG	ACT	---

Row 14

HTLV-I	TTG	TCA	CCT	GTT	CCC	ACC	CTA	GGA	TCC	CGC	TCC	CGC	CGA	GCG	GTA	CCG	GTG	GCG	GTC	GCG	GGC	GTC	TGG	CTT	GTC	TCC
HTLV-II	CTC	GCT	CCC	GTA	CCT	CCT	CCG	GCG	ACA	CGC	CGC	CGT	GCC	GTT	CCA	ATA	GCA	GTG	CTT	GTC	TCC
Murine leukemia	CCC	CTC	ACC	CTC	CCC	GAA	CCC	CCG	CCA	GCG	GGA	GTC	GAA	AAC	CGA	TTG	TTA	AAT	CTA	GTA	AAA	GGA	GCC
Friend	---	---	---	---	---	---	---	CAA	CCT	GGG	GCG	GGA	GAC	CTG	CTA	AAC	CTG	GTA	GAT	GGG	GCC
Moloney	CCT	CTC	TCC	CCT	ACC	CAA	CTT	CCA	CCG	GCG	GCG	GAA	AAT	AGG	CTG	CTA	AAC	TTA	GTA	GAC	GGA	GCG

Row 15

HTLV-I	GCC	CTG	GCC	ATG	GGA	GCC	GGA	GTG	GCT	GGC	GGG	ATT	ACC	GGC	TCC	ATG	TCC	CTC	GCC	TCA	GGA	AAG	AGC
HTLV-II	GCC	CTA	GCG	GCC	GGA	ACA	GGT	ATC	GCT	GGT	GGA	GTA	ACA	GCC	TCC	CTA	TCT	CTG	GCT	TCC	AGT	AAA	AGC
Murine leukemia	TAC	CAA	GCC	CTC	AAC	CTC	ACC	AGT	CCT	GAT	AAA	ACC	CAA	GAC	TGC	TGG	TTA	TGC	CTA	GTA	TCG	GGA	CCC
Friend	TAC	CAA	GCT	CTC	AAC	CTC	ACC	AAC	CCT	GAT	AAA	ATT	CAA	GAG	TGG	TGG	TTA	TGC	CTA	GTG	TCT	GGA	CCC
Moloney	TAC	CCA	GCC	CTC	AAC	CTC	ACC	AGT	CCT	GAC	AAA	ACC	CAA	GAG	TGC	TGG	TTG	TGT	CTA	GTA	GCG	GGA	CCC

(continued)

Table 11.3 (Continued)

Row 16

HTLV-I	CTC	CTA	CAT	GAG	GTG	GAC	AAA	GAT	ATT	TCC	CAG	TTA	ACT	CAA	GCA	ATA	GTC	AAA	AAC	CAC	AAA	AAT CTA
HTLV-II	CTT	CTC	CTC	GAG	GTT	GAC	AAA	GAC	ATC	TCC	CAC	CTT	ACC	CAG	GCC	ATA	GTG	AAA	AAT	CAT	CAA	AAC ATC
Murine leukemia	CCA	TAC	TAC	GAG	GGG	GTT	GCC	GTC	CTA	GGT	ACC	TAC	TCC	AAC	CAT	ACT	TCT	GCC	CCA	GCT	AAC	TGC TCT
Friend	CCC	TAT	TAC	GAG	GGG	GTT	GTG	GTC	CTA	GGC	ACT	AAT	TCT	AAT	CAT	ACC	TCT	GCC	(To row 25)			
Moloney	CCC	TAC	TAC	GAA	GGG	GTT	GCC	GTC	CTG	GGT	ACC	TAC	TCC	AAC	CAT	ACC	TCT	GCT	CCA	GCC	AAC	TGC TCC

Row 17

HTLV-I	CTC	AAA	ATT	GCC	CAG	TAT	GCT	GCC	CAG	AAC	AGA	CGA	GGC	CTT	GAT	CTC	CTG	TTC	TGG	GAG	CAA	GGA GGA
HTLV-II	CTC	CGG	GTT	GCA	CAG	TAT	GCA	GCC	CAA	AAT	AGA	CGA	GGA	TTA	GAC	CTC	CTA	TTC	TGG	GAA	CAA	GGG GGT
Murine leukemia	GTG	GCC	TCT	CAA	CAC	AAA	TTG	ACC	TTG	TCC	GAA	GTG	ACC	GGA	CAG	GGA	CTC	TGC	ATA	GGA	GCG	GTC CCT
Moloney	GTG	GCC	TCC	CAA	CAC	AAG	TTG	ACC	CTG	TCC	GAA	GTG	ACC	GGA	CAG	GGA	CTC	TGC	ATA	GGA	GCA	GTT CCC

Row 18

HTLV-I	TTA	TGC	AAA	GCA	TTA	CAA	GAA	CAG	TGC	CGT	TTT	CCG	AAT	ATT	ACC	AAT	TCC	CAT	GTC	CCA	ATA	CTA CAA
HTLV-II	TTG	TGC	AAG	GCC	ATA	CAG	GAG	CAA	TGT	TGC	TTC	CTC	AAC	ATC	AGT	AAC	ACT	CAT	GTA	TCC	CTC	CTC CAG
Murine leukemia	AAA	ACC	CAT	CAA	CTC	TTG	TCT	AAT	ACC	ACC	CAA	AAG	ACA	AGC	GAT	GGG	TCC	TAC	TAT	TTG	GCC	GCT CCC
Moloney	AAA	ACA	CAT	CAG	GCC	CTA	TGT	AAT	ACC	ACC	CAG	ACA	AGC	AGT	CGA	GGG	TCC	TAT	TAT	CTA	GTT	GCC CCT

Row 19

HTLV-I
GAA AGA CCC CCC CTT GAG AAT CGA GTC CTG ACT GGC GGC CTT AAC TGG GAC CTT GGC CTC TCA CAG

HTLV-II
GAA CGG CCC CCT CTT GAA AAA CGT GTC ATC ACC GGC GGG CTA AAC TGG GAT CTT GGA CTG TCC CAA

Murine leukemia
ACA GGA ACT ACC TGG GCT TGT AGT ACT GGA CTT ACT CCC ATC TCA ACC ACC ATA CTT GAC CTC ACC

Moloney
ACA GGT ACC ATG TGG GCT TGT AGT ACC GGG CTT ACT CCA TGC ATC TCC ACC ACC ATA CTG AAC CTT ACC

Row 20

HTLV-I
TCG GCT CGA GAG GCC TTA CAA ACT GGA ATC ACC CTT GTT GCG CTA CTC CTT CTT ATC CTT GCA GGA

HTLV-II
TGG GCA CGA GAA GCC CTC CAG ACA GGC ATA ACC ATT CTC GCT CTA CTC CTC GTC ATA TTG TTT GGC

Murine leukemia
ACC GAT TAC TGT GTC CTG GTC GAG CTT TGG CCA AGG GTG ACC TAC CAT TCC CCT AGT TAT GTT TAT TAC CAC

Moloney
ACT GAT TAT TGT GTT CTT GTC GAA CTC TGG CCA CTC TGG CCA GTC ACC TAT CAT TCC CCC AGC TAT GTT TAT GGC

Row 21

HTLV-I
CCA TGC ATC CTC CGT CAG CTA CGA CAC CTC CCC --- --- TCG CGC GTC AGA TAC CCC CAT TAC TCT CTT

HTLV-II
CCC TGT ATC CTC CGC CAA ATC CAG GCC CTT CCA CAG CGG TTA CAA AAC CGA CAT AAC CAG TAT TCC CTT

Murine leukemia
CAA TTT AAA AGA CGA GCC AAA TAT AAA AGA GAA CCC GTC TCA ACT CTG GCC CTA CTA TTA GGA GGA

Moloney
CTG TTT GAG AGA TCC AAC CGA CAC AAA AGA GAA CCG GTG TCG TCG GCC CTA CTA TTA TTG GGT GGA

(continued)

Table 11.3 (Continued)

Row 22

HTLV-I	ATA AAA CCT GAG TCA TCC CTG TAA
HTLV-II	ATC AAC CCA GAA ACC ATG CTA TAA

Murine leukemia	CTC ACT ATG GGC GGA ATT GCC GCT GGA GTG GGA ACA GGG ACT ACC GCC CTA GTG GCC ACT CAG CAG TTC
Moloney	CTA ACC ATG GGG GGA ATT GCC GCT GGA ATA GGA ACA GGG ACT ACT GCC CTA ATG GCC ACT CAG CAA TTC

Row 23

Murine leukemia	CAA CAA CTC CAG GCT GCC ATG CAC GAT GAC CTT AAA GAA GTT GAA AAG TCC ATC ACT AAT CTA GAA AAA
Moloney	CAG CAG CTC CAA GCC CCA GTA CAG GAT GAT CTC AGG GAG GTT GAA AAA TCA ATC TCT AAC CTA GAA AAG

Row 24

Murine leukemia	TCT TTG ACC TCC TTG TCC GAA GTA GTG TTA CAG AAT CGT AGA GGC CTA GAT CTA CTA TTC CTA AAA GAG
Moloney	TCT CTC ACT TCC CTG TCT GAA GTT GTC CTA CAG AAT CGA AGG GGC CTA GAC TTG TTA TTT CTA AAA GAA

Row 25

Murine leukemia	GGA GGT TTG TGT GCT GCC TTA AAA GAA GAA GAA TGC TGT TTC TAT GCC GAC CAC ACA GGA TTG GTA CGG GAT
Moloney	GGA GGG CTG TGT GCT GCT CTA AAA GAA GAA GAA TGT TGC TTC TAT GCG GAC CAC CAC ACA GGA CTA GTG AGA GAC
Friend spleen-focus V (from Row 16)	CTA AAA GAA AAA TGT TGT TTC TAT GCT GAC CAT ACA GGC CTA GTA AGA GAT

Row 26

Murine leukemia	AGC	ATG	GCC	AAA	CTT	AGA	GAA	AGA	TTG	AGT	CAG	AGA	CAA	AAG	CTC	TTT	AAA	TCC	CAA	CAA	GGG	TGG	TTT
Moloney	AGC	ATG	GCC	AAA	TTG	AGA	GAG	AGG	CTT	AAT	CAG	CAG	AAA	CTG	TTT	GAG	TCA	ACT	CAA	GGA	TTG	TTT
Friend spleen-focus V	AGT	ATG	GCC	AAA	TTA	AGA	AGA	AGA	CTC	ACT	CAG	CAA	AAA	CTA	TTT	GAG	TCG	AGC	CAA	GGA	TGG	TTC

Row 27

Murine leukemia	GAA	GGG	CTG	TTT	AAT	AAG	TCC	CCT	TGG	TTC	ACC	ACC	CTG	ATA	TCC	ACC	ATC	ATG	GGT	CCC	CTG	ATA	ATC
Moloney	GAG	GGA	CTG	TTT	AAC	AGA	TCC	CCT	TGG	TTT	ACC	ACC	TTG	ATA	TCT	ATG	ACC	ATT	ATG	GGA	CCC	CTC	ATT	GTA
Friend spleen-focus V	GAA	GGA	TCG	TTT	AAC	AGA	TCC	CCC	TGG	TTT	ACC	ACG	TTG	ATA	TCC	ACC	ATC	ATG	GGG	CTT	CTC	ATT	ATA

Row 28

Murine leukemia	CTC	TTG	ATT	TTA	CTC	TTT	GGG	CCT	TGT	ATT	CTC	AAT	CGC	CTC	GTC	CAG	TTT	ATC	AAA	GAC	AGG	ATT
Moloney	CTC	CTA	ATG	ATT	TTG	CTC	TTC	GGA	CCC	TGC	ATT	CTT	AAT	CGA	TTA	GTC	CAA	TTT	GTT	AAA	GAC	AGG	ATA
Friend spleen-focus V	CTC	CTA	CTC	CTA	ATT	CTG	CTT	TTG	ACC	CTG	TGG	ACC	CTG	TAT	TCT	TAA

Row 29

Murine leukemia	TCG	GTA	GTG	CAG	GCC	CTG	GTT	CTG	ACT	CAA	CAA	CTT	TAT	CAT	CAA	CTT	AAG	ACA	ATA	GAA	TGT	AAA	TCA
Moloney	TCA	GTG	GTC	CAG	GCT	CTA	GTT	TTG	ACT	CAA	CAA	CAA	TAT	CAC	CAG	CTG	AAG	CCT	ATA	GAG	TAC	GAG	CCA	TAG

Row 30

Murine leukemia	CGT GAA TAA

[a]Codons for basic amino acids are underscored and those for tyrosine are italicized; those encoding tryptophan and cysteine are stippled.
[b]Human T-cell leukemia virus-I; Seiki et al. (1983).
[c]Human T-cell leukemia virus-II; Shimotohno et al. (1985a).
[d]Devare et al. (1983). [e]Ecotropic murine leukemia virus; Herr et al. (1983).
[f]Friend spleen focus-forming virus; S. P. Clark and Mak (1983); Wolff et al. (1983).
[g]Xenotropic murine leukemia virus; Devare et al. (1983); Repaske et al. (1983).
[h]Moloney murine sarcoma virus; Shinnick et al. (1981).
[i]Insert TTA GAA TTC ACT GAC GCT GGT AGA AAG GCC AGC.
[j]Insert GGC TCC.

Sequence Structure. It is at once apparent that Table 11.3 is constructed so as to demonstrate the existence of three major homology groups, the two HTLVs representing one, the simian sarcoma virus a second, and the four remaining ones a third. However, the relationships among the members of the last of these clusters are ephemeral and extremely variable. For example, the Moloney sequence diverges temporarily from the others as early as row 2; then beginning in row 6, it becomes a distinct homolog of the ecotropic and Friend viral genes, the xenotropic becoming divergent. In row 7, the Friend cistron loses relationship with the other three, which resume the early homologies. Similar variation in one or more of the group can be noted thence throughout the remainder of their structures. Quite contrastingly, the two HTLVs display frequent regions of identity and signs of common origin consistently until their terminations. The three major groups as entities almost never show any meaningful features shared with one another.

In total length, the sequences are markedly different, the shortest by far being the xenotropic murine leukemia virus (Khan and Martin, 1983). Although this is shown to terminate in row 11, the total of 57 codons deleted from various portions makes it considerably shorter than that point suggests. Next shortest is the simian sarcoma virus, which ends in row 12 (Devare *et al.*, 1983). The Moloney and murine leukemia viral sequences, each of which consists of ~670 codons, greatly exceed all the others.

Chemical Features. Among the more prominent chemical features that first appear when the sequences (Table 11.3) are examined is the distribution of the TAC and TAT codons (italics) for tyrosine. The totals of 20 and 17 for HTLV-I and -II, respectively, are approximately equivalent to the 19 and 23 of the murine leukemia and Moloney viruses, considering the greater length of the latter two. But those of the pair from human tissues are concentrated toward the 5′ end, whereas in the other couple they are more frequent in the downstream third. Rows 4 and 5 of the first two are the richest in those triplets, while in the murine viruses rows 16 and 20 are. In contrast, the Friend virus possesses only nine, and the xenotropic murine virus seven, the simian form being totally devoid of those codons.

The TGT and TGC triplets specifying cysteine and the TGG for tryptophan also are unevenly distributed, all three being lightly stippled in Table 11.3. The Moloney and murine leukemia viruses lead in both types of codons, the former having 21 and 16, respectively, for the two amino acids, and the latter 24 and 20. These figures compare to 17 (or 18) and 13 in HTLV-I and -II. The xenotropic murine species has but four for cysteine and six for tryptophan, while the simian sarcoma virus has seven and one, respectively.

Similarly, the arrangement and number of codons for the basic amino acids (underscored) differ broadly between the homology groups. The sequence that was consistently deficient in the foregoing chemical properties here proves to lead the group in the alkalinity of its encoded product, for the simian sarcoma virus has 40 of its 272 codons for either arginine or lysine. Strikingly, the two HTLV structures are the most nearly neutral in this respect, with only 27 of the ~490 codons designating the two amino acids just mentioned, for a 5.5% level. The Friend viral product is next most basic, with a 10% rate, and the xenotropic murine form is close behind with 18 of its 192 codons, or 9.4%.

As a result of the above analysis, it is clear that the proteins encoded by these *env* cistrons not only are highly involved in host recognition, but are also capable of having widely diverse effects on the cell they parasitize. Thus, if one of these genes were to prove carcinogenic, it is most unlikely that any other one would, except those of closely allied forms, as shown with the *gag* sequences also.

Because the genes contained in the X region of these unmodified retroviruses are too poorly explored as yet for sound comparison, it is now necessary to conclude this analytical study of retroviral normal gene complements and begin an examination of the genomes of modified species and the cell components that they have acquired. They are found to support the statement made above, that, because genes from different viral sources bear like names, they should not therefore be expected to be similar in their chemical natures—each gene type may more reasonably be viewed as encoding products encompassing a wide range of diversity.

11.3. ONCOGENES FOR PROTEIN KINASES

In those viruses active in neogenesis (oncogenesis) among metazoans, the genes appear to encode three major types of proteins; the products of the first two seemingly become located in the host cell at the plasmalemma, and those of the third are conducted to the nucleus. The first group of membrane-associated products, including that of *c-ras*, bind and hydrolyze GTP and affect the activity of adenyl cyclase (McGrath *et al.*, 1984; Sweet *et al.*, 1984; Kataoka *et al.*, 1985; Toda *et al.*, 1985). The second class, of which *src* is the most prominent member, embraces tyrosine-specific protein kinases, their precise physiological functions remaining undetermined. The proteins of the third group, those that locate in the nuclear matrix, appear to be devoted to hemopoietic tissues, in which the viral gene induces tumors (Durban and Boettiger, 1981), but their normal role in the cell is still undeciphered. Outstanding among these is a product of the *c-myb* gene.

In order to facilitate the location of nearly 30 genes known to be involved in oncogenesis among metazoans, an alphabetical list is provided in Table 11.4, along with the location of their sequences, their classification, and other pertinent information. The list should not be expected to include every gene that has been found on a few occasions to result in malignancies, for that goes beyond the scope of the present analysis. However, it does point out the several new names and synonymies resulting from this study and should reduce the burdensome task of finding information concerning any given gene included here.

At least eight of the numerous oncogenes whose products are known have proven to be associated with protein kinase activities, namely, *src, yes, fgr, fps, fms, ros, abl,* and *fes* (Hampe *et al.*, 1982: Sherr *et al.*, 1985). In a majority of cases, the enzymes encoded by those genes phosphorylate the amino acid tyrosine. Whereas in the proteins of normal cells the phosphotyrosine they create is only 0.03% as abundant as such other phosphorylated amino acids as phosphoserine or phosphothreonine, in transformed cells it becomes much more abundant. When its concentration is permitted to increase greatly experimentally, as by the addition of vanadium, which inhibits cell phosphatases from dephosphorylating phosphotyrosine in proteins, the cell becomes transformed (Klarlund, 1985). Thus, *in vivo* malignancies could result from the activities of such gene products. Discussion of these kinase-encoding cistrons begins with that referred to as *src*, the most thoroughly investigated member of this class.

Table 11.4
The More Important Oncogenes of Metazoans

Gene	Class	Source	Effect	Table where sequence is given
abl	I	Mammals	B-cell leukemias	--
amv	II?	Chickens	Myeloblastosis	11.10
bas	See has	--	--	--
bcl	II?	Humans	B-cell lymphomas	--
don (new name)	II	Rats	Sarcomas	11.7
erbA, B	II?	Chickens	Erythroblastosis	--
ets	III?	Chickens	Erythroblastosis	--
fes	I	Cats; related to fps	Sarcomas	11.6
fgr	I	Cats	Fibrosarcoma	--
fms	I	Cats, especially spleen	Cell surface receptor; growth factor	11.6
fos	III	Mice	Osteosarcomas	11.9
fps	I	Chickens	Sarcomas	11.6
has	II?	Mammals	Bladder carcinomas	--
int	III?	Mammals	--	11.9
mht	II?	Avians	Carcinomas	--
mil	See raf	--	--	--
mos	II?	Mammals	Sarcomas; myelomas; plasmacytomas	11.8

myb	III	Avians	Myeloblastosis	11.9
myc	II?	Vertebrates	B- or T-lymphomas; neuroblastomas; myelocytomatoses	11.11
raf	II?	Vertebrates	Carcinomas; sarcomas	11.9
ras	II	Yeast, fungi, metazoans	Highly diverse; carcinomas, leukemias, melanomas	11.7
rel	II?	Chickens	Reticuloendotheliosis	11.9
rho	II	Snails	Unknown	11.7
ros	I	Chicken virus VR2	Sarcomas	--
sis	II	Primates	Sarcomas	11.10
src	I	Vertebrates; *Drosophila*	Sarcomas; gap junctions?	11.5
yes	I	Chicken Yamaguchi sarcoma virus	Sarcomas; epidermal growth factor	11.6
YP2	II?	Yeast	?	11.9

11.3.1. Structure of the Src Gene

How widespread among the tyrosine kinase genes the chemical features of *src* may be remains for the analyses of other such cistrons to disclose later in this section. Perhaps they shall prove unique to the present species, but, be that as it may, some are shown here to be especially remarkable. Before an intelligible discussion of those features can be presented, the relationships among the six species given in Table 11.5 must be examined. Unfortunately, those that could be presented there do not cut through a broad spectrum of the living world, three being from the chicken and the rest—all incomplete—from *Drosophila*. Those few, even with the limitations of source diversity, represent at least three, or, more likely, four, kinship groups. As a result, they show a broader view of the subject than at first appears. The gene occurs more widely than these limited resources indicate, having been reported from the Harvey and Kirsten sarcoma viruses of mice (Ellis *et al.*, 1981) as well as from *Xenopus* (Steele, 1985).

Src Gene Structures. The homology clusters are not readily interpreted, in part because all the sequences from the fruitfly are incomplete at the 5' end and two are truncated at the 3' end as well. The two sarcoma viral and chicken cytoplasmic species, however, are intact and obviously form a closely knit group of related sequences. That of the Rous sarcoma virus (RSV) is from the Schmidt–Ruppin strain (Czernilofsky *et al.*, 1980, 1983), and the avian sarcoma virus is represented by the DNA fragment rASV1441 (Takeya *et al.*, 1982). The consistency in structure they reveal suggests their origin from that of the corresponding cellular gene of the chicken (Takeya and Hanafusa, 1983), for nearly all triplets encoding amino acids considered to be of especial pertinence are situated at precisely located points. Moreover, not only are codons for the same amino acids at those positions, but they are of identical structure, even though, in the case of arginine, six different ones are available. Thus, it is obvious that the viral genes are recent derivations of the nuclear of the chicken. However, interesting variations occur. For example, in the Prague strain of the RSV, two point differences from that of the table are found in row 1, the most pronounced being the replacement of CGC in the 18th codon by CAC, and in row 2, five gene changes result in amino acid replacements at four sites (Schwartz *et al.*, 1983).

The three from *Drosophila* contrast strongly to the above. At the very beginning of the most complete sequence, that of the cellular gene (row 10 *et seq.*) (Hoffmann *et al.*, 1983), only occasional correspondences with those of the avian types from chickens suggest relationship. However, for a short distance after the commencement of the *4-* and *Dsrc* genes (Wadsworth *et al.*, 1985), all three show a fair degree of homology mutually and with the others, especially in rows 14 and 15. Beyond that point each species from *Drosophila* diverges along its own lines for a space, only to correspond to one or the other for an additional sector of variable length. Toward the latter half of row 15, for example, the cellular structure is the most divaricate, the other two being obvious homologs of the avian viral types. In the following row, only the *D* cistron shows relationship to the latter, but that kinship is completely lost by row 18. The extreme brevity of the available sequences from this insect makes further detailed comparisons meaningless. One troublesome aspect of the fruitfly's cellular gene sequence is the presence of several premature termination codons that apparently have gone unnoticed (Hoffmann *et al.*, 1983). The first of these appears in row 15 (heavy stippling), where TGA is found; the same triplet

Table 11.5

Comparison of Src Gene Coding Regions[a]

Row 1

Rous sarcoma v[b]	ATG GGG AGC AGC AAG AAG AGC CCT AAG GAC CCC AGC CAG CGC CGG CGC AGC GGC CTG GAG CCA CCC GAC AGC
Avian sarcoma v[c]	ATG GGG AGC AGC AGC AGC AGC CCT AAG GAC CCC AGC CAG CGC CGG TGC AGC CCC CTG GAG CCA CCC GAC AGC
Chicken c[d]	ATG GGG AGC AGC AAG AAG CCC AAG GAC CCC AGC CAG CGC CGG CGC AGC CCC CTG GAG CCA CCC GAC AGC

Row 2

Rous sarcoma v	ACC CAC CAC GGG GGA TTC CCA GCC TCG CAG ACC CCC GAC GAG ACA GCA GCC CCC GAC GCA CAC CGC AAC
Avian sarcoma v	ACC CAC CAC GGG GGA TTC CCA GCC TCG CAG ACC CCC AAC AAG ACA GCA GCC CCC GAC ACG CAC CGC ACC
Chicken c	ACC CAC CAC GGG GGA TTC CCA GCC TCG CAG ACC CCC AAC AAG ACA GCA GCC CCC GAC ACG CAC CGC ACC

Row 3

Rous sarcoma v	CCC AGC CGC TCC TTC GGG ACC GTG GCC ACC GAG CCC AAG CTC TTC GGG GGC TTC AAC ACT TCT GAC ACC
Avian sarcoma v	CCC AGC CGC TCC TTT GGG ACC GTG GCC ACC GAG CCC AAG CTC TTC GGG GGC TTC AAC ACT TCT GAC ACC
Chicken c	CCC AGC CGC TCC TTT GGG ACC GTG GCC ACC GAG CCC AAG CTC TTC GGG GGC TTC AAC ACT TCT GAC ACC

Row 4

Rous sarcoma v	GTT ACG TCG CCG CAG CGT GCC GGG GCA CTG GCT GGC GGC GTC ACC ACT TTC GTG GCT CTC TAC GAC TAC
Avian sarcoma v	GTC ACG TCG CCG CAG CGT GCC GGG GCA CTG GCT GGC GGC GTC ACC ACT TTC GTG GCT CTC TAC GAC TAC
Chicken c	GTT ACG TCG CCG CAG CGT GCC GGG GCA CTG GCT GGC GGC GTC ACC ACT TTC GTG GCT CTC TAC GAC TAC

(continued)

Table 11.5 (Continued)

Row 5

Rous sarcoma *v*	GAG	TCC	TGG	ATT	GAC	GAC	TTG	TCC	TTC	AAG	AAA	GGA	GAA	CGG	CTG	CAG	ATT	GTC	AAC	AAC	ACG	GAA
Avian sarcoma *v*	GAG	TCC	CGG	ACT	GAA	ACG	GAC	TTG	TCC	TTC	AAG	AAA	GGA	GAA	CGC	CTG	CAG	ATT	GTC	AAC	AAC	ACG	GAA
Chicken *c*	GAG	TCC	CGG	ACT	GAA	ACG	GAC	TTG	TCC	TTC	AAG	AAA	GGA	GAA	CGC	CTG	CAG	ATT	GTC	AAC	AAC	ACG	GAA

Row 6

Rous sarcoma *v*	GGT	AAC	TGG	TGG	CTG	GCT	CAT	TCC	CTC	ACT	ACA	CAG	ACG	GGC	TAC	ATC	CCC	AGT	AAC	TAT	GTC	GCG
Avian sarcoma *v*	GGT	GAC	TGG	TGG	CTG	GCT	CAT	TCC	CTC	ACT	ACA	GGA	CAG	ACG	GGC	TAC	ATC	CCC	AGT	AAC	TAT	GTC	GCG
Chicken *c*	GGT	GAC	TGG	TGG	CTG	GCT	CAT	TCC	CTC	ACT	ACA	GGA	CAG	ACG	GGC	TAC	ATC	CCC	AGT	AAC	TAT	GTC	GCG

Row 7

Rous sarcoma *v*	CCC	TCA	GAC	TCC	ATC	CAG	GCT	GAA	GAG	TGG	TAC	TTT	GGG	AAG	ATC	ACT	CGT	CGG	TCC	GAG	CGG	CTG
Avian sarcoma *v*	CCC	TCA	GAC	TCC	ATC	CAG	GCT	GAA	GAG	TGG	TAC	TTT	GGG	AAG	ATC	ACT	CGT	CGG	TCC	GAG	CGG	CTG
Chicken *c*	CCC	TCA	GAC	TCC	ATC	CAG	GCT	GAA	GAG	TGG	TAC	TTT	GGG	AAG	ATC	ACT	CGT	CGG	TCC	GAG	CGG	CTG

Row 8

Rous sarcoma *v*	CTG	CTC	AAC	CCC	GAA	AAC	CCC	CGG	GGA	ACC	TTC	TTG	GTC	CGG	GAG	ACG	AGC	ACA	AAA	GGT	GCC	TAT
Avian sarcoma *v*	CTG	CTC	AAC	CCC	GAA	AAC	CCC	CGG	GGA	ACC	TTC	TTG	GTC	CGG	GAG	ACG	AGC	ACA	AAA	GGT	GCC	TAT
Chicken *c*	CTG	CTC	AAC	CCC	GAA	AAC	CCC	CGG	GGA	ACC	TTC	TTG	GTC	CGG	GAG	ACG	AGC	ACA	AAA	GGT	GCC	TAT

Row 9

Rous sarcoma *v*: TGC CTC TCC GTT TCT GAC TTT GAC AAC GCC AAG GGG CTC AAT GTG AAG CAC TAC AAG ATC CGC AAG CTG

Avian sarcoma *v*: TGC CTC TCC GTT TCT GAC TTT GAC AAC GCC AAG GGG CTC AAT GTG AAG CAC TAC AAG ATC CGC AAG CTG

Chicken *c*: TGC CTC TCC GTT TCT GAC TTT GAC AAC GCC AAG GGG CTC AAT GTG AAG CAC TAC AAG ATC CGC AAG CTG

Row 10

Rous sarcoma *v*: GAC AGC GGC TTC TAC ATC ACC CGC ACA CAG TTC AGC AGC CTG CAG CAG CTG GTG GCC TAC TAC

Avian sarcoma *v*: GAC AGC GGC TTC TAC ATC ACC CGC ACA CAG TTC AGC AGC CTG CAG CAG CTG GTG GCC TAC TAC

Chicken *c*: GAC AGC GGC TTC TAC ATC ACC CGC ACA CAG TTC AGC AGC CTG CAG CAG CTG GTG GCC TAC TAC

Drosophila *c*: (Incomplete) CTT GGC CTG

Row 11

Rous sarcoma *v*: TCC AAA CAT GCT GAT GGC TTG TGC CAC CGC CTG ACC AAC GTC CCC ACG TCC AAG CCC CAG ACC CAG

Avian sarcoma *v*: TCC AAA CAC GCT GAT GGC TTG TGC CAC CGC CTG ACC AAC GTC CCC ACG TCC AAG CCC CAG ACC CAG

Chicken *c*: TCC AAA CAT GCT GAT GGC TTG TGC CAC CGC CTG ACC AAC GTC CCC ACG TCC AAG CCC CAG ACC CAG

Drosophila *c*: TGT CAC ATA TTG TCG CGT CCC CTG CAA ACC GCA GCC CCA GAT GTG GGA TTC GGG

Row 12

Rous sarcoma *v*: GGA CTC GCC AAG GAC GCG GAA ATC CCC CGG GAG TCG CTG CGG CTG GTG AAG CTG GGG CAG GGC

Avian sarcoma *v*: GGA CTC GCC AAG GAC GCG TGG GAA ATC CCC CGG GAG TCG CTG CGG CTG GTG AAG CTG GGG CAG GGC

Chicken *c*: GGA CTC GCC AAG GAC GCG TGG GAA ATC CCC CGG GAG TCG CTG CGG CTG GTG AAG CTG GGG CAG GGC

Drosophila *c*: CCG CAG CTG CGC GAT AAG TAC GAG ATT CCG CGC TCG GAG ATT CAG GTC CGC AAA GTC GGA CGC CGC

(continued)

Table 11.5 (Continued)

Row 13

Rous sarcoma *v*	TGC TTT GGA *GAG* GTC **TGG** ATG GGG ACC AAC GGC ACC ACC <u>AGA</u> GTG GCC ATA <u>AAG</u> ACT CTG <u>AAG</u> CCC
Avian sarcoma *v*	TGC TTT GGA *GAG* GTC **TGG** ATG GGG ACC AAC GGC ACC ACC <u>AGA</u> GTG GCC ATA <u>AAG</u> ACT CTG <u>AAG</u> CCC
Chicken *c*	TGC TTT GGA *GAG* GTC **TGG** ATG GGG ACC AAC GGC ACC ACC <u>AGA</u> GTG GCC ATA <u>AAG</u> ACT CTG <u>AAG</u> CCC
Drosophila 4f	(Incomplete) GCG GTC <u>AAG</u> ATG ATG <u>AAG</u> *GAA*
Drosophila c	AAC TTT GGC *GAG* GTC TTC **TAC** GGC <u>AAA</u> **TGG** CGC AAC AGC ATC CAT GTC GCG <u>AAA</u> ACG CTG <u>CGC</u> GCA
Drosophila Df	(Incomplete) GCG GTC <u>AAA</u> ACG CTG <u>CGC</u> GCA

Row 14

Rous sarcoma *v*	GGC ACC ATG TCC CCG *GAG* GCC CTG TTC CAG *GAA* GCC CAA GTG ATG <u>AAG</u> <u>AAG</u> CTC <u>CGG</u> CAT *GAG* <u>AAG</u> CTG
Avian sarcoma *v*	GGC ACC ATG TCC CCG *GAG* GCC CTG TTC CAG *GAA* GCC CAA GTG ATG <u>AAG</u> <u>AAG</u> CTC <u>CGG</u> CAT *GAG* <u>AAG</u> CTG
Chicken *c*	GGC AAC ATG TCC CCG *GAG* GCC CTG TTC CAG *GAA* GCC CAA GTG ATG <u>AAG</u> <u>AAG</u> CTC <u>CGG</u> CAT *GAG* <u>AAG</u> CTG
Drosophila 4	GGA ACC ATG TCC *GAG* GAC GAT TTC ATT *GAG* *GAG* GCC <u>AAG</u> GTG ATG ACC <u>AAG</u> CTG CAG CCA AAT CTT
Drosophila c	GGC ACC ATG TCC ACG GCT GCT TTC CTT CAG *GAG* GCG GCG <u>AAG</u> TTC ATT ATG <u>AAG</u> <u>AAG</u> <u>CGA</u> CAC AAC CGC CTG
Drosophila D	GGC ACC ATG TCC ACG GCT GCT TTC CTT CAG *GAG* GCC GCG <u>AAG</u> TTC ATT ATG <u>AAG</u> <u>AAG</u> <u>CGA</u> CAC AAC <u>CGC</u> CTG

Row 15

Rous sarcoma *v*	CTT CAA CTG **TAC** GCA GTG TCG --- *GAA* *GAG* CCC ATC **TAC** ATC GTC ATT *GAG* **TAC** ATG AGC <u>AAG</u> GGG AGC
Avian sarcoma *v*	CTA CAG CTG **TAC** GCA GTG TCG --- *GAA* *GAG* CCC ATC **TAC** ATC GTC ACT *GAG* **TAC** ATG AGC <u>AAG</u> GGG AGC
Chicken *c*	CTT CAG CTG **TAC** GCA GTG TCG --- *GAA* *GAG* CCC ATC **TAC** ATC GTC ACT *GAG* **TAC** ATG AGC <u>AAG</u> GGG AGC
Drosophila 4	CTA **TAT** GGC CTC ACC **TGC** ACC <u>AAG</u> CGG CCC ATC **TAC** ATC GTG ACC *GAG* **TAC** <u>AAG</u> CAC GGA TCC
Drosophila c	CTC **TAT** GCC CTT GCC **TGC** TCG CAG GTA --- AGT GCT ATA AAT CTA TTG TTA ATG **TGT** ATT AAC TTG
Drosophila D	CTC **TAT** GCC CTT GCC **TGC** TCG CAG *GAG* *GAG* CCC ATT TAC ATC GTG *GAG* **TAC** ATG TCC <u>AAG</u> GGC AGT

Row 16

Rous sarcoma v	---	CTC	CTG	*GAT*	TTC	CTG	AAG	GGA	*GAG*	ATG	GGC	AAG	TAC	CTG	CGG	CTG	CCA	CAG	CTC	GTT	*GAT*	ATG	GCT	GCT
Avian sarcoma v	---	CTC	CTG	*GAT*	TTC	CTG	AAG	GGA	*GAG*	ATG	GGC	AAG	TAC	CTG	CGG	CTG	CCA	CAG	CTC	GTC	*GAT*	ATG	GCT	GCT
Chicken c	---	CTC	CTG	*GAT*	TTC	CTG	AAG	GGA	*GAG*	ATG	GGC	AAG	TAC	CTG	CGG	CTG	CCA	CAG	CTC	GTC	*GAT*	ATG	GCT	GCT
Drosophila 4	TTG	TTG	AAT	TAC	TTG	CGA	CGG	CAT	*GAG*	AAG	ACC	CTG	ATT	GGT	AAT	ATG	GGT	CTA	CTC	CTT	*GAC*	ATG	TGC	ATA
Drosophila c	---	AAC	TAT	*GAA*	TCG	CTT	GCA	GGA	GCC	CAT	TTA	CAT	CGT	GCA	GGA	GTA	CAT	GTC	CAA	GGG	CAG	TCT	GCT
Drosophila D	---	CTG	CTG	AAC	TTC	TTG	CGC	GGC	GAT	*GAC*	CGT	TAC	TTG	---	---	---	CAC	TTC	*GAA*	*GAT*	CTC	ATC	TAC

Row 17

Rous sarcoma v	CAG	ATT	GCA	TCC	GGC	ATG	GCC	TAT	GTG	*GAG*	AGG	ATG	AAC	TAC	GTG	CAC	CGA	CTG	CGG	GCG	GCC	AAC
Avian sarcoma v	CAG	ATT	GCA	TCC	GGC	ATG	GCC	TAT	GTG	*GAG*	AGG	ATG	AAC	TAC	GTG	CAC	CGA	CTG	CGG	GCG	GCC	AAC
Chicken c	CAG	ATT	GCA	TCC	GGC	ATG	GCC	TAT	GTG	*GAG*	AGG	ATG	AAC	TAC	GTG	CAC	CGA	CTG	CGG	GCG	GCC	AAC
Drosophila 4	CAG	GTT	AGC	AAG	GGA	ATG	ACC	TAC	CTA	*GAG*	CGC	CAT	AAC	TAC	ATT	CAC	CGG	*GAT*	CTG	GCT	GCC	AAC^g
Drosophila c	*GAA*	CTT	CTT	GCG	CGA	GGG	CGA	AA	CCG	TTA	CTT	GCA	CTT	CGA	AGA	TCT	CAT	CTA	CAT	GCA	CAC	CCA	GGT
Drosophila D	ATG	CAC	ACC	CAG	GTG	ACC	ACC	GGT	ATG	AAG	TAT	CTA	*GAG*	TCC	AAG	CAA	GTC	ATC	CAC	CGC	*GAT*	CTG	ACG

Row 18

Rous sarcoma v	ATC	CTG	GTG	GGG	*GAG*	AAC	CTG	GTG	TGC	AAG	GTG	GCT	*GAC*	TTT	GGG	CTG	GCA	CGC	CTC	ATC	*GAG*	*GAC*	AAC
Avian sarcoma v	ATC	CTG	GTG	GGG	*GAG*	AAC	CTG	GTG	TGC	AAG	GTG	GCT	*GAC*	TTT	GGG	CTG	GCA	CGC	CTC	ATC	*GAG*	*GAC*	AAC
Chicken c	ATC	CTG	GTG	GGG	*GAG*	AAC	CTG	GTG	TGC	AAG	GTG	GCT	*GAC*	TTT	GGG	CTG	GCA	CGC	CTC	ATC	*GAG*	*GAC*	AAC

(continued)

Table 11.5 (Continued)

Drosophila 4	TAC GTT CTC GAC GAT CAA TAT ACC AGC TCG GCG GAA CCA AGT (Incomplete)
Drosophila c	GAC CAC CGG TAT GAA GTA TCT AGA GTC CAA GCA AGT CAT CCA CCG CGA TCT GAC GAC CCG TAA TGT GCT
Drosophila D	ACC CGT AAT GTG CTG ATC GGA AAT AAT GTG GCG AAG ATT TGT GAT TTT GGA CTG GCG CGT GTC ATC

Row 19

Rous sarcoma v	GAG TAC ACA GCA CGG CAA GGT GCC AAG TTC CCC ATC AAG ACA GCC CCC GAG GCA GCC CTC TAT GGC
Avian sarcoma v	GAG TAC ACA GCA CGG CAA GGT GCC AAG TTC CCC ATC AAG TGG ACA GCC CCC GAG GCA GCC CTC TAT GGC
Chicken c	GAG TAC ACA GCA CGG CAA GGT GCC AAG TTC CCC ATC AAG TGG ACA GCC CCC GAG GCA GCC CTC TAT GGC
Drosophila c	GAT CGG AGA GAA TGT GGC GAA GAT TTG TTT TGG ACT GGC GCG TGT CAT CGC GGA ███ CGA GTA
Drosophila D	GCG GAT GAC GAG TAC CGC CCC AAG CAG GGA TCC CGG (Incomplete)

Row 20

Rous sarcoma v	CGG TTC ACC ATC AAG TCG GAT GTC TGG TCC TTC GGC ATC CTG ACT GAG CTG ACC ACC AAG GGC CGG
Avian sarcoma v	CGG TTC ACC ATC AAG TCG GAT GTC TGG TCC TTC GGC ATC CTG ACT GAG CTG ACC ACC AAG GGC CGG
Chicken c	CGG TTC ACC ATC AAG TCG GAT GTC TGG TCC TTC GGC ATC CTG ACT GAG CTG ACC ACC AAG GGC CGG
Drosophila c	CCG CCC CAA GCA GGG ATC CCG GTT TCC GGT CAA GTG GAC GGC GCC CGA GGC GAT CAT CTA CGG CAA GTT

Row 21

Rous sarcoma v	GTG CCA TAC CCA GGG ATG GGC AAC GGG GAG GTG CTG GAC CGG GAG AGG GGC TAC CGC ATG CCC TGC
Avian sarcoma v	GTG CCA TAC CCA GGG ATG GGC AAC GGG GAG GTG CTG GAC CGG GAG AGG GGC TAC CGC ATG CCC TGC
Chicken c	GTG CCA TAC CCA GGG ATG GGC AAC AGG GAG GTG CTG GAC CAG GTG GAG AGG GGC TAC CGC ATG CCC TGC
Drosophila c	CTC GAT CAA GTC GGA CGT GTG CTA CAT TCT GCT GAC GGA GCT TTT CAC GTA CGG ACA AGT GCC

Row 22

Rous sarcoma v	CCG	CCC	GAG	TGC	CCC	GAG	TCG	CTG	CAT	GAC	CTT	ATG	TGC	CAG	TGC	TGG	CGG	AGG	GAC	CCT	GAG GAG CGG
Avian sarcoma v	CCG	CCC	GAG	TGC	CCC	GAG	TCG	CTG	CAT	GAC	CTT	ATG	TGC	CAG	TGC	TGG	CGG	AGG	GAC	CCT	GAG GAG CGG
Chicken c	CCG	CCC	GAG	TGC	CCC	GAG	TCG	CTG	CAT	GAC	CTC	ATG	TGC	CAG	TGC	TGG	CGG	AGG	GAC	CCT	GAG GAG CGG
Drosophila e	CTA	TCC	GGG	ACT	GCA	TCG	CGA	GGT	GAT		GAA	CAT	CGA	GCG	CGG	TTT	CCG	CAT	GCC	CAA	GCC AAC

Row 23

Rous sarcoma v	CCC	ACT	TTC	GAG	TAC	CTG	CAG	GCC	CAG	CTC	CCT	GCT	TGT	GTG	TTG	GAG	GTC	GCT	GAG	TAG
Avian sarcoma v	CCC	ACT	TTT	GAG	TAC	CTG	CAG	GCC	CAG	CTG	CTT	CCT	GCT	TGT	GTG	TTG	GAG	GTC	GCT	GAG TAG
Chicken c	CCC	ACT	TTT	GAG	TAC	CTG	CAG	GCC	TTC	CTG	GAG	GAC	TAC	TTC	ACC	TCG	ACA	GAG	CCC	CAG TAC CAG CCT
Drosophila e	GAA	TCA	CTA	CCT	CCC	GGA	CAA	CAT	TTA	TCA	GCT	GCT	CCA	GTG	CTG	GGA	TGC	TGT	GCC	CGA GAA GCG

Row 24

Chicken c	GGA	GAG	AAC	CTA	TAG																		
Drosophila e	CCC	GAC	ATT	CGA	GTT	CTT	GAA	CCA	CTA	CCT	CGA	GTC	CTT	CTC	GGT	CAC	GTC	GGA	GGT	GCC	GTA	TCG	AGA

Row 25

Drosophila e	GGT	GCA	AGA	CTA	AAC	AGA	GTC	CAG	CCT	ATT	CAC	ACA	CAT	CAC	CAA	CAC	TCT	CTC	ACA	CGC	ATA CAC ACA

Row 26

Drosophila e	CGT	CGA	TAT	ATA	TAT	ATA	TAG

aCodons for basic amino acids are underscored and those for acidic ones are italicized. Premature stop codons are heavily stippled and those for tyrosine, cysteine, and tryptophan are lightly stippled. bCzernilofsky et al. (1983). cTakeya et al. (1982). dTakeya and Hanafusa (1983). eHoffmann et al. (1983). fWadsworth et al. (1985). gInsert 18 codons.

occurs again in rows 17, 19, and 22, while TAA may be observed in rows 18 and 19, and a single TAG is situated in row 22. Whether these are typographical errors or misinterpretations of the analytical gels is an unresolvable problem at this point. A similar oversight occurs in one or two sequences in later sections of this analysis.

The Chemical Nature of Src Genes. Similarly, in a treatment of the chemical properties, the incompleteness of the *Drosophila* structures renders them useless; consequently, this analysis is based solely upon the three avian gene sequences (Table 11.5). Tyrosine (moderate stippling) is an especially important amino acid in the product of *src*, for no fewer than 21 or 22 of its codons are placed along the length of each of the three sequences, constituting ~4% of the whole. Tryptophan and cysteine codons are about half as abundant, each of which totals either nine or ten, depending upon the species. Although codons for proline are not outstandingly plentiful, with a total of only 13 (~6%), the conservation of triplet structure at nearly every point of their occurrence is remarkable. Moreover, two-thirds of the sites, where present, have the triplet CCC, the other combinations being found at only three or four positions in each sequence. This conservativeness might also be attributed to the recent origin of the viral genes from that of the avian cell.

Probably the most thought-stimulating of the chemical properties of the *src* product relates to the amino acids bearing a polar change, for, when viewed as a whole, it can be referred to as neither alkaline nor acidic, because it is both! Counts of codons specifying the two basic amino acids lysine and arginine yield totals of ~54 per gene, for a 10.2% ratio, slightly above the random-chance level of 9.4%; similarly, tallies of those encoding aspartic and glutamic acids show the presence of ~60 per sequence, or 11.3%, the random-chance level being 6.2%. Another, related condition that occurs widely in gene sequences is especially richly represented here, codons specifying a monomer having one type of charge, perhaps negative, having neighbors bearing the opposite, positive in this case. In other words, one amino acid neutralizes the effect of the other, just as the two contrasting abundant types here counteract each other. A valid question strongly in need of investigation is the value in a protein of this frequent combination of contrasts.

Possible Role in Oncogenesis. An unusual modification has been demonstrated to occur to the src protein, as well as to that encoded by *gag*. A rare, long-chain fatty acid known as myristic acid is found attached to only two known cellular proteins and these few of retroviral origin, those of the cell being a cyclic AMP-dependent protein kinase and one component of a calmoduline-binding phosphatase (Schultz *et al.*, 1985). In each case the myristylation takes place by an amide linkage to the amino-terminal glycine residue. Because of its appearance on the possible candidates for cell transformation, it has been proposed that the modification might be a preliminary essential to their being targeted to the complex mechanism of growth control.

Another possible point of involvement of the src protein in the multistep carcinogenic processes is at the gap junctions of the cell surface, where it becomes located. Such intercellular communication centers have been demonstrated to undergo alteration in transformed rat cells, reducing in size as tyrosine-protein kinase activity increased (C. C. Chang *et al.*, 1985).

11.3.2. Miscellaneous Tyrosine-Specific Kinase Genes

As a consequence of the scarcity of studies at the molecular level on the other seven known tyrosine-specific protein kinases believed to be involved in some type of neogenesis, it is advantageous to examine them in a unified fashion. Accordingly, the available sequences are presented together in Table 11.6. Unfortunately, one of the five given must have been subjected to an error during its preparation, so that through the accidental addition or omission of one residue, two adjacent termination codons were created, as indicated in row 4. Consequently, it is believed inadvisable to include the remainder of that coding structure. To these is added the 5'-terminal portion of the chicken c-src (from Table 11.5) to provide further breadth for comparison. However, any suspicions that these genes, which encode products of similar function, might have shared common origins are immediately dispelled by the apparent lack of any significant level of homology, except at the 3' ends of *fes* and *yes*, both from chickens. But perhaps analysis may disclose chemical properties among these few miscellaneous types that are shared with those of the src protein and thereby provide an insight into their oncogenetic mechanism.

The Several Genes and Their Products. The *c-fms* proto-oncogene, as cellular genes are currently designated, encodes a product of 170,000 daltons, a glycoprotein with tyrosine kinase activity that is associated with chicken macrophages from spleen and peritoneal inflammatory exudates (Roussel *et al.*, 1983). Since the receptor protein for murine blood-cell colony-stimulating factor CSF-1 is of comparable size and chemical properties and is similarly restricted to mononuclear phagocytic leukocytes, the two have been suspected as being related or even identical substances (Sherr *et al.*, 1985). Immunological tests further support the same conclusion. Like the receptor, the *fms* product is located particularly on the cell surface (Manger *et al.*, 1984), where its amino-terminal portion spans the plasmalemma, while the carboxyl part remains in the cytoplasm (Rettenmier *et al.*, 1985). In man, the gene is located on chromosome 5 (Roussel *et al.*, 1983).

The *fps* coding structure, prematurely terminated in Table 11.6, and a more recently described one referred to as *ros* also were first discovered in the chicken, the latter being the product of a newly isolated virus, UR2 (Feldman *et al.*, 1982). A third and fourth member of this group from the UR1 and Yamaguchi sarcoma viruses, whose products p150 and yes, respectively, are nearly indistinguishable from fps protein in structure and enzymatic properties, also are from the same source (Wang *et al.*, 1982). But as the alignments in Table 11.6 make manifest, a far greater homology exists between the genes *yes* and *fes*, especially in the 3'-terminal region commencing with row 18, a relationship not previously known to exist. In contrast to the published statement, no regions of homology of pertinent length can be found between *fps* and *yes*. This latter gene is continuous with the 5' end of *gag*, with which it is cotranscribed (Kitamura *et al.*, 1982). Identification of the *gag* region preceding *yes* proved it to be largely identical to the sequence of the Rous sarcoma structure (Table 11.2), deviating from the latter, however, in several sectors of rows 7–9. Consequently, the competent virus from which avian sarcoma virus Y73 was derived (in which *yes* was first discovered) should prove to be closely allied, but not identical, to the Rous forebear. In *ros*, only one tyrosine residue was found to be phosphorylated, but several serines had the same modification (Feldman *et al.*, 1982); in all cases tyrosine has been found to be the agent of transfer in the tyrosine-specific kinase activity. The gene for each of these four proteins is directly attached to the *gag* sequence, the *pol, env,* and others of the original viruses having been deleted. Subsequently, genetic elements encoding c-yes have been located on human chromosomes 18 and 6, four copies on the first and one on the latter (Semba *et al.*, 1985); that study also provided evidence that the gene products may function in the epidermal growth factor receptor-activated pathway leading to cell proliferation.

Table 11.6
Comparison of Various Tyrosine-Specific Protein Kinase Genes[a]

Row 1

```
Chicken c-src[b]   ATG GGG AGC AGC AAG AAG CCC AAG GAC CCC AGC AGC CAG CGC CGG CGC AGC CTG GAG CCA CCC GAC AGC
Avian c-fps[c]     --- CTG CGT CAG GTC AGT TGC ACC GGC CCC AAC CCC AGG AAC ACA CCA GCA CCA GCG CTG CAG CGG GGA
Feline v-fes[d]    --- CTG CGT CAG GTC AGT TGC ACC GGC CCC AGC CCC AGG AAC ACA CCA GCA CCA GCG CTG CAG CGG GGA
Feline v-fms[e]    ATG CCC AGT GGT CCC GGT CAC TAC GGA GCT AGC GCT GAG ACT CCA GGG CCT CGC CCC CTC TGC CCT
Avian v-yes[f]     CGG AGC GCC CGG CTC CCT CCT GCT CTT CCC CTC CCC CCC TTC CCT CCC
```

Row 2

```
Chicken c-src      ACC CAC CAC GGG GGA TTC CCA GCC TCG CAG ACC CCC AAC AAG AAG GCC CCC GAC ACG CAC CGC ACC
Avian c-fps        CCT GGC GGC TCA CTC AGG CTT CGG AGT CGC GGC ACA TGC CCC ACT GCT CTG CAG CCC CAT CCC ACC
Feline v-fes       GAA CAG CAG ATG CAG GCT GAG GCT CCC ACT GAG GCC ATG GCC ATG GGC ATG TGG ATG GCC CAG CGG GTC
Feline v-fms       GCA TCA TCC TGC TGC CTC CCC ACT GAG GCC ATG GGC CCA AGG GCT CTG GTC CTG CTG ATG GCC ACA
Avian v-yes        CGG GTT GCC GCG GTT CCG GGA GGA GCG GGG GCC CCG CTT CCT TCC CTC CCC TCC TTC TTC CAC
```

Row 3

```
Chicken c-src      CCC AGC CGC TCC TTT GGG ACC GTG GCC ACC GAG CCC AAG CTC TTC GGG GGC TTC AAC ACT TCT GAC ACC
Avian c-fps        AGG ATC ACA GTG CCA TGG GCT TTG GGC CGG AGC TGT GGT GCC CGA AGG GGC ACA GTG AGC TGC TGC GGC
Feline v-fes       AAG AGT GAC AGG AGG GAA GAA GCA GGG GTC CCA GTG ATA CAG CCC AGC GGC CCT GAG GAC GGC GGC ACG
Feline v-fms       GCT TGG CAT GCT CAA GGG GTC CCA GTG ATA CAG CCC AGC GGC CCT GAG CTG GTC GTG GAG GAG CCA GGC ACA
Avian v-yes        CCG CGC CGC CGA CGA CGG GCA GCA ACC GTG GGG TGC ATT AAA AGC AGG TGC AAA GAT AAA AAA AGT CCA GCC
```

Row 4

Chicken c-src   GTT ACG TCG CCG CAG CGT GCC GGG GCA CTG GCT GGC GGC GTC ACC ACT TTC GTG GCT CTC TAC TAC *GAC* **TAC**

Avian c-fps   TGC AGG ACA GCG AGT TGC GCC TCC TGG AGC TGA TGA

Feline v-fes   GGC CCC TAT AGC CCC ATC AGC CAG TCC TGG GCC ATC AGC ACG CAG GGC CTG AGC CGG TTG

Feline v-fms   ACA GTG ACC CTG CGA TGT GTG GGC AAC GGC AGC GTG *GAA* TGG *GAC* GGC CCC ATC TCC CCC CAC TGG AAC

Avian v-yes   ATG AAA TAC AGG ACT AAC ACT CCA *GAA* CCT ATT AGT TCC CAC GTC AGC CAT TAC GGG TCA *GAC* TCC

Row 5

Feline v-fes   CTG AGG CAA CAC GCG *GAG* GAT CTG AAC TCG GGG CCC CTG AGT AAG CTG GGC CTG CTG ATC CGG *GAG* CGG

Feline v-fms   CTG *GAC* CTC *GAT* CCC CCC AGC AGC ATC CTG ACC ACG AAT AAT GCC ACC TTC CAA AAC ACG GGG ACC **TAT**

Avian v-yes   AGC CAA GCA ACA CAG TCA CCG GCA ATA AAG GGA TCA GCA GTT AAT TTT AAC AGT CAT TCC ATG ACT CCT

Row 6

ST feline v-fes[d,g]   (Incomplete) CAC CCC CGG *GAC* CAG

Feline v-fes   CAG CAG CTG CGC AAG ACC TAC AGC *GAG* CAG TGG CAG CAG CTC CAG *GAG* CTC ACC AAG ACC CAC AAC

Feline v-fms   CAC TGT ACT *GAG* CCT GGA AAC CCC CGG GGG GGC AAT GCC ACC ATC CAC CTC TAT GTC *AAA* *GAC* CCT GCT

Avian v-yes   TTT GGA GGG CCC TCA GGA ATG ACA CCC TTT GGA GGA GCA TCG TCT TCA TTT TCA GCT GTG CCA AGT CCA

Row 7

ST feline v-fes   GTG CAG CTG CTG GCC AAG AAG CAG GTG TTG CAA *GAG* GCG CTG CAG GCG CTG CAG GTG GCG TTG TGC AGC

Feline v-fes   CAG *GAC* ATC *GAG* AAG AAG CTG AAG AAG AGC CAG TAC CGA GCC CTG GCA CGG *GAC* AGC GCC CAG GCC CGG CGC AAG

Feline v-fms   CGG CCT TGG AAG GTG CTG CTG GCC CAG CAG GAA GTG ACC GTG TTG ACC *GAT* GGT CAG *GAT* GCG TTG CTG CCC TGC CTG

Avian v-yes   TAT CCT AGT ACT TTA ACA GGT GGT ACT GTA TTT GTG GCC TTA *GAT* GAT TAC **TAT** *GAA* GCT AGA ACT ACA

(continued)

Table 11.6 (Continued)

Row 8

ST feline v-fe₍ CAG GCC AAG CTG CAG GCC CAG CGG GAG CTG CTG CAG (rest identical to below)

Feline v-fes TAC CAG GAG GCC AGC AGC AAA GAC CGC AAG AAG GCC AAG CTG GAG CAG CTG GGC CCC GGC CCC GAG CCC

Feline v-fms CTT ACT GAC CCA GCG TTG GAG GCA GGC GTC TCG CTG CGT GGC GTG CGG CCC GTC TTG CGC CAA

Avian v-yes GAT GAC CTT TCA TTT AAG GGG GGT GAA CGG TTC CAG ATA ATA AAC AAC ACG GAA GGC GAC GAA GGC TGG TGG GAA

Row 9

Feline v-fes CCG CCC GTC CTG CTC CTG CAG GAT GAC CGC CAC TCC ACG TCG TCC TCG GAG CAG CAG CGA GAA GGG GGA

Feline v-fms ACC AAC TAT TCC TTC TCG CCC TGG CAC GGC TTC ACC ATC CAC AAG GCC AAG TTC ATT GAG AAT CAC GTC

Avian v-yes GCA AGA TCC ATT GCT ACG GGA AAA ACA GGC TAC ATC CCA AGC AAT TAT GTA GCT CCT GCA GAC TCC ATT

Row 10

Feline v-fes AGG ACA CCC ACC TTG GAG ATC CTT AAG AGC CAC ATC TCA GGA ATC TTC CGC CCC AAG TTC TCG CTC CCT

Feline v-fms TAC CAA TGC AGT GCC CGA GTA GAC GGC AGG ACG GTG ACA TCG ATG GGC ATC TGG CTT AAA GTG CAG AAA

Avian v-yes GAA GCG GAA GAG ATG GGT GGT TTT AAA ATG GGC AGG AAG AAG GAT GCA GAA AGA CTA CTT TTA AAT CCT GGG

Row 11

Feline v-fes CCA CCC CTG CAG CTC GTA CCA GAG GTG CAG AAG CCC CTG CAC GAG GAG CTG TGG TAC CAC GGG GCC CTC

Feline v-fms GAC ATC TCC GGG CCT GCA ACC TTG ACG CTG GAG CCT GCA GGA CGG CGG ATT CAA GGA CAG GCT GCC

Avian v-yes AAC CAG CGTʰ ATT TTC TTA GTA AGA GAG AGC GAA ACC ACT AAA GGT GCT TAC TCC CTT TCC ATA CGT GAC

Row 15

Feline v-fes CAG CCC CTC ACC AAG AAG AGC GGT ATT GTC CTC AAC AGG GCT GTG CCC AAG GAC AAG TGG GTG CTA AAC

Feline v-fms   CAG ATC GTG TGC TCA GCC AGC AAC ATT GAT GTT AAC TTT GAC GTC TCC CTC CGT CAT GGA GAC ACC AAG
Avian v-yes   TGG GAT GAG GTC AGA GGT AAT AAT GTG AAG CAC AAA ATC AGA AAA CTT GAC AAT GGT GGA TAC TAT

Row 13

Feline v-fes   CAG GAA TAT GTG CTG TCG CTG GAC GGC CAG CCC CAC TTC ATC ATC CAG GCT GAC AAC
Feline v-fms   CTC ACA ATC TCT CAA CAA TCC GAC TTC CAT GAT AAC CGT TAC CAA AAA GTC CTG ACC CTC AAC CTC GAT
Avian v-yes   ATC ACA ACC AGA GCA CAA TTT GAA TCT CTC CAG AAG TTG GTG AAG CAC TCA AGA GAA CAT GCT GCT GAT GGA

Row 14

Feline v-fes   CTC TAC CGA CCG GAA GGA GAT GGC TTT GCG AGC ATC CCC TTG CTC GTC GAC CAC CTG CTG CGC TCC CAG
Feline v-fms   CAC GTG AGC TTC CAA GAT GCT GGC AAC TAC TCC TGC ACG GCC ACC AAC GCC TGG GGC AAC CAC TCC GCC
Avian v-yes   CTG TGT CAT AAG CTA ACA ACT GTA TGT CCC ACG GTG AAA CCA CAA ACA CAG GGA CTA GCA AAA GAT GCC

Row 15

Feline v-fes   CAG CCC CTC ACC AAG AAG AGC GGT ATT GTC CTC AAC AGG GCT GTG CCC AAG GAC AAG AAG TGG GTG CTA AAC
Feline v-fms   TCC ATG GTC TTC CGG GTG GTA GAG AGT GCC TCG AAC TTG ACC TCT GAG CAG AGT CTC CTC CAG GAG
Avian v-yes   TGG GAA ATT CCT AGG GAG TCT TTG AGG CTG GAA GTT AAG TTG GGC CAA GGA TGT TTT GGT GAA GTA TGG

Row 16

Feline v-fes   CAC GAG GAC CTG GTG TTG GGT GAG CAG ATC GGG CGG GGG AAC TTT GGA GAA GTG TTC AGT GGA CGC CTG
Feline v-fms   GCG ACT GTG GGT GAG AAG AAG GTT GAT GAT CTC CAA GTC AAG GTG GAG GCC TAC CCA GGC CTA GAA AGT TTT AAC
Avian v-yes   ATG GGA ACC TGG AAT GGA ACC ACA GCC ATC AAG ACA CTT GGT CTT AAA CTT GGT ACA ATG CCC GAA

(continued)

Table 11.6 (Continued)

Row 17

Feline v-fes: AGG GCC *GAC* AAC ACT CTG GTG GCC GTG AAA TCT TGT CGC *GAG* ACA CTC CCA CCT *GAC* ATC AAG GCC AAG AAG

Feline v-fms: TGG ACC TAT CTG GGA CCC TTC TCT *GAC GAC* CAG AAG CTC *GAT* TTT GTC ACC ATC AAG *GAC* ACA TAC

Avian v-yes: --- --- --- --- --- --- --- --- --- GCT ---

Row 18

Feline v-fes: TTT CTT CAG *GAA* GCA AAG ATC CTG AAG CAG TAC AGC CAC CCC AAC ATC GTG CGT CTC ATC GGC GTC TGC

Feline v-fms: AGG TAC ACC TCG ACC CTC TCC CTG CCT CGC CTG AAG CGC CGC GAG TCC GGT CGC TAC TCC TTC TTG GCC

Avian v-yes: TTC CTT CAG *GAG* GCT CAG ATC ATG AAG AAA TTA CGA CAT *GAC* AAG CTT GTT CCA CTG TAT GCC GTT ---

Row 19

Feline v-fes: ACC CAG AAG CAG CCC ATC TAC ATC GTC ATG *GAG* CTC GTG CAG GGG GGC *GAC* TTC CTG ACC TTC CTG AGG ---

Feline v-fms: CGA AAC GCT GGA GGC CAG AAT GCC CTG ACC TTT *GAG* CTC ACT CTG CGA CTC CCC CCG GTA AGG GTC ---

Avian v-yes: GTT TCT *GAG GAA* CCA ATC TAC TAC ATA GTC ACC *GAA* TTC ATG ACA AAA GGC AGC TTA CTA *GAC* TTC CTG AAG *GAA*

Row 20

Feline v-fes: ACG *GAG* GGA GCC CGC CTG CGG ATG AAG ACG CTG CTG CAG ATG GTG GGC *GAC* GCG GCC GGG ATG *GAG*

Feline v-fms: ACG ATG ACC CTC ATC AAT GGC TCT *GAC* ACC CTG CTC TGT *GAA* GCC TCC GGG TAC CCC CAG AGT GTG

Avian v-yes: GGA *GAA* GGG AAG TTC TTA AAA CTC CCA CAG CTG GTG *GAC* ATG ATG GCT GCT CAG ATT GCT *GAT* GGC ATG GCT

```
Feline v-fms   ACG TGG GTG CAG TGC AGG AGG CAC ACC GAT AGG TGT GAT GAG TCT GCA GGG CTG CTA GAG GAC TCA
Avian v-yes    TAC ATT GAA AGA ATG AAC TAC ATC CAC AGG GAT CTC CGG GCA GCC AAC ATT CTT GTA GGA GAC AAT CTT

Row 22

Feline v-fes   GCC CTG AAG ATC AGT GAC TTC GGG ATG TCC CGG GAG GCA GCC GAT GGG ATC TAC GCG GCC TCA GGG GGC
Feline v-fms   CAC TCT GAG GTC CTG AGC CAA GTG CCC TTC TAC GAG GTG ATC GTC CAT AGC CTG GCC ATC GGG ACC
Avian v-yes    GTG TGT AAA ATA GCA GAC TTC GGT CTC GCA AGG TTA ATA CAG GAC AAT GAG TAC ACT GCG AGG --- CAA

Row 23

Feline v-fes   CTC AGA CAA GTT CCG GTG AAG TGG ACG GCA CCC GAG GCT CTT AAC TAC TAC GGC CGC TAT TCC TCT GAG AGC
Feline v-fms   TTG GAA CAC AAC AGG ACA TAT GAG TCT AGA GCC TTC AAC AGC GTG GGG AAC AGC TCC CAG ACC TTC TGG
Avian v-yes    GGA GCT AAA TTT CCA ATT AAA TGG ACT GCT CCA GAA GCA GCA TTG TAT GGT CGG TTT ACA ATC AAG TCA

Row 24

Feline v-fes   CAC GTC TGG AGC TTC GGC ATC TTG CTA TGG GAG ACC TTC AGC CTG GGC GCC TCC CCC TAC CCC AAC CTC
Feline v-fms   CCC ATC TCT ATA GGA GCC CAC ACG CCG CTC CCC CAC GAG CTC CTC TTC ACG CCC GTG CTG CTC ACT TGC
Avian v-yes    GAT GTG TGG TCG TTT GGA ATT TTA CTG ACA CTG GTA ACA AAG GGG AGA GTG CCA GTG CCA TAT CCA GGA ATG

Row 25

Feline v-fes   AGC AAT CAG ACC CGC GAG TTT GTG GAA AAG GGT GGC CGC CTG CCC TGC CCC GAG CTG TGC CCC GAC
Feline v-fms   ATG TCC ATC ATG GCC TTG TTC CTG CTC CTG TTG CTT TTG TAC AAG CAG CAG AAG CCC AAG
Avian v-yes    GTG AAT CGG GAA GTT CTG GAA CAA CAA GTG GAA CGT GGA TAT AGG ATG CCT TCC CCT CAG GGC TGC CCG GAA
```

(continued)

Table 11.6 (Continued)

Row 26

Feline *v-fes*: GCT GTG TTC AGG CTC ATG *GAG* CAG TGC TGG GCC TAC *GAG* CCC GGG CAG CGG CCC AGC TTC AGC GCC ATC

Feline *v-fms*: TAC CAG GTG CGC TGG AAG ATC ATC *GAG* AGC TAC *GAG* GGC AAC AGC TAC ACC TTC ATC *GAC* CCC ACC CAG

Avian *v-yes*: TCT CTC CAC *GAG* TTA ATG AAA CTA TGT TGG AAG AAG *GAC* CCT *GAG* AGA CCA ACA TTT *GAA* TAT ATA

Row 27

Feline *v-fes*: TAC CAG *GAG* CTG CAG AGC ATC CGA AAG CGG CAT CGG TGA

Feline *v-fms*: CTG CCC TAC AAT *GAG* AAG TGG *GAG* TTC CCC CGA AAC AAC CTG CAG TTT (followed by 412 residues)

Avian *v-yes*: CAG TCT TTC CTG *GAG* *GAC* TAC ACT GCT GCA *GAA* CCG AGC GGC TAT TGA

[a] Codons for acidic amino acids are italicized and those for basic ones are underscored. Codons for tyrosine are heavily stippled, those for tryptophan are moderately stippled, and those for cysteine are lightly stippled.
[b] From Table 11.5.
[c] Shibuya and Hanafusa (1982).
[d] Hampe *et al.* (1982).
[e] Hampe *et al.* (1984).
[f] Kitamura *et al.* (1982).
[g] Snyder-Theilen feline sarcoma virus.
[h] Insert GGT.

Another component in the same category, the *abl* gene, is the transforming sequence of Abelson murine leukemia virus, the cell counterpart being situated on mouse chromosome 2, apparently in single copy (Goff *et al.*, 1982). The product of the viral structure transforms lymphocytes of the B-cell lineage. In man, the homologous gene may be involved in the induction of chronic myelogenous leukemia. Normally located on human chromosome 9, in that disease it is translocated to the Philadelphia chromosome (Collins *et al.*, 1984). But other genes, including *c-sis* described later, are also implicated in this malignancy, suggesting that a series of genetic changes are probably involved in its production. Purported homologs of *abl* have also been partially described in *Drosophila* (Hoffman-Falk *et al.*, 1982).

Chemical Properties. When the complete and partial sequences of Table 11.6 are critically examined, it is evident immediately that no sign of common origin exists, either among the miscellany of types cited fully or between these and the four rows of the chicken *c-src* sequence taken from Table 11.5. Nor is there any suggestion of common chemical properties. In the 81-codon-long sector of avian *c-fps*, an unusual richness in codons (light stippling) for cysteine exists, a total of seven being present, for an ~9% ratio. When the coding frame has been corrected and the full amino acid sequence is derived from the DNA (Huang *et al.*, 1985), it will be of great interest to discover whether that high level persists throughout its full length. Within the corresponding sectors, the *v-yes* and *v-fes* sequences have only one each and that of *v-fms* has four copies of that same triplet (Hampe *et al.*, 1982, 1984). In fact, the first has only eight, the second contains but nine, and the third 14 in the ~600-codon lengths given in the table.

No concentration of tyrosine triplets (heavy stippling) such as characterizes *src* is displayed here, for they occur only near the random-chance ratio (3.1%), the ratio being 2.7% in *v-fes*, 3.7% in *v-fms*, and 4.0% in *v-yes*. Codons for tryptophan (light stippling) are present with approximately their random-chance level of frequency (1.6%), the sequence for *v-fes* containing exactly that percentage, that of *v-yes* 1.3%, and that of *v-fms* 1.9%. In relative proportions and total content of triplets for basic (underscored) and acidic (italicized) amino acids, the genes are especially contrasting, for *v-fes* and *v-yes* have more of the former and *v-fms* a greater proportion of the latter. The first of these cistrons contains 77 basic and 72 acidic codons (12.6 and 11.7%, respectively), the second 71 and 67 (11.9 and 11.3%), and the last, 46 and 58 (7.5 and 9.5%). One particularly striking feature of the *v-yes* gene is situated in rows 1 and 2, where a nearly uninterrupted series of codons for proline, leucine, and serine combine with an occasional one for phenylalanine to create sectors consisting almost wholly of Ts and Cs (Table 11.6). That at the 5' end is especially lengthy, filling virtually the entire row 1. None of these genes, however, are enriched as a whole in that type of codon. Consequently, each of the gene products that have been fully sequenced from the tyrosine-specific protein kinase class is seen to differ strongly from the others, insofar as the chemical properties of their constituent monomers are concerned.

11.4. ADENYL CYCLASE-ACTIVATING GENES

It is not always possible to assign a given oncogene to one of the three classes outlined earlier, since all too frequently their biological activities are insufficiently ex-

Table 11.7

Comparison of Coding Regions of Ras Genes[a]

Gene	Row 1
Yeast c-ras1[b]	ATG[c] AGA GAG TAT AAG ATA GTA GTT GTC GGT GGA GGT GGC GTT GGT GGT AAA TCT GCT TTA ACA ATT CAA TTC
Yeast c-ras2[b]	ATG[d] AGA GAG TAC AAG CTA GTC GTT GGT GGT GGT GGT AAA TCT GCT TTG ACC ATA CAA TTG
Dictyostelium ras[e]	ATG ACA GAA TAT AAA TTA GTT ATT GTA GGT GGT GGT GGT AAA AGT GCA TTA ACA ATT CAA TTA
K(irsten) v-ras[f]	ATG ACT GAG TAT AAA CTT GTG GTA GTT GGA GCT AGT GGC GTA GGC AAG AGT GCC TTG ACG ATA CAG CTA
Human K1-ras[f]	ATG ACT GAA TAT AAA CTT GTG GTA GTT GGA GCT *TGT GGC GTA GGC AAG AGT GCC TTG ACG ATA CAG CTA
Mouse K-ras[g]	ATG ACT GAA TAT AAA CTT GTG GTG GTT GGA GCT GGT GGC GTA GGC AAG AGC GCC TTG ACG ATA CAG CTA
Human K2-ras[h]	ATG ACT GAA TAT AAA CTT GTG GTA GTT GGA GCT GGT GGC GTA GGC AAG AGT GCC TTG ACG ATA CAG CTA
Human N-ras[i]	ATG ACT GAG TAC AAA CTG GTG GTA GTT GGA GCA GGT GGT GGG AAA AGC GCA CTG ACA ATC CAG CTA
Drosophila ras1[j]	ATG ACG GAA TAC AAA CTG GTC GTC GGA GGC GGT GTG GGC AAG TCC GCC CTC ACC ATC CAG CTA
Harvey v-ras[k]	ATG ACA GAA TAC AAG CTT GTG GTG GGC GCT AGA GGC GTG GGC AAG AGT GCG CTG ACC ATC CAG CTG
H-T24 bladder[l]	ATG ACG GAA TAT AAG CTG GTG GTG GGC GCC GTC GGT GTG GGC AAG AGT GCG CTG ACC ATC CAG CTG
Human c-H-ras[m]	ATG ACG GAA TAT AAG CTG GTG GTG GGC GCC GGC GGT GTG GGC AAG AGT GCG CTG ACC ATC CAG CTG
Rat sarcoma v-don (v-ras)[n]	ATG GGA CAA TCG CTA ACA ACC CCC TTG AGT CTC ACT CTA GAC CAT TGG AAG GAC GTC CGA GAC CGA GCA
Drosophila ras3[o]	ATG CGT GAG TAC AAA ATC GTC CTC GGA AGC GGC GGC GGC AAA TCC GCG GTC ACG GTC CAG TTT
Aplysia rho[p]	ATG GCA GCG ATA[q] AAG CTT GTT ATA GTC GGA GAT GGT GCG GCG AAA ACA TGT CTA CTT ATT GTC TTC

Row 2

```
Yeast c-ras1        ATT CAA TCA TAC TTT GTG G AC GAA TAT GAC CCT ACT ATC ATC GAA*GAT TCT TAC AGA AAA GAA GTT GTC ATC
Yeast c-ras2        ACC CAA TCG CAC TTT GTA G AT GAA TAC GAT CCC ACA ATT GAG*GAT TCA TAC AGG AAG CAA GTG GTG ATT
Dictyostelium ras   ATT CAA*AAT CAT TTT ATT G*AT GAA TAT GAT CCA ACA ATT GAA GAT AGT TAT CGT AAA CAA GTT TCA ATT
K(irsten) v-ras     ATT CAG AAT CAC TTT GTG G AT GAA TAT GAT CCT ACG ATA CAG*GAC TCC AGG AAA CAA GTA GTA ATT
Human K1-ras        ATT CAG AAT CAT TTT GTG G AC GGA TAT GAT CCA ACA ATA GAG*GAT TCC TAC AGG AAG CAA GTA GTA ATT
Mouse K-ras         ATT CAG AAT CAC TTT GTG G AT GAG TAC GAC CCT ACG ATA GAG*(Incomplete)
Human K2-ras        ATT CAG AAT CAT TTT GTG G AC GAA TAT GAT CCA ACA ATA GAG*GAT TCC TAC AGG AAG CAA GTA GTA ATT
Human N-ras         ATC CAG AAC CAC TTT GTA G AT GAA TAT GAT CCC ACC ATA GAG*GAT TCT TAC AGA AAA CAA GTG GTT ATA
Drosophila ras1     ATC CAG AAC CAT TTC GTG G AC GAG TAC GAC CCC ACA ATC GAG GAC TCT TAC CGA AAG CAA AGG TTC ATC
Harvey v-ras1       ATC CAG AAC CAT TTT GTG G AC GAG TAT GAT CCC ACT ATA GAG GAC TCC TAC CGG AAA CAG GTA GTC ATT
H-T24 bladder       ATC CAG AAC CAT TTT GTG G AC GAA TAC GAC CCC ACT ATA GAG*GAT TCC TAC CGG AAG CAG GTG GTC ATT
Human c-H-ras       ATC CAG AAC CAT TTT GTG G AC GAA TAC GAC CCC ACT ATA GAG*GAT TCC TAC CGG AAG CAG GTG GTC ATT
Rat sarcoma v-don   CGT GAT CAG TCG GTC GAG A TC AAG AAA GGT CCT CTC CGG AGG TCG GGG ACA GTC GCG CCA GCA AGC GGT
Drosophila ras3     GTC CAG TGC ATC TTC GTG G AG AAG TAC GAT CCC ACC ATC GAG GAC AGC TAC AGG AAG CAG GTG AAG GTG
Aplysia rho         AGC AAA GAC CAG TTC CCT G AA GTT TAC TGC CCA ACA GTT TTT GAA AAT TAT GTA GCA GAC ATT GAA GTT
```

(continued)

Table 11.7 (Continued)

Row 3   59   61

Yeast c-ras1   G AT GAC AAA GTA TCC ATT TTG GAC ATT CTA GAT ACT GCT GGA CAA GAA GAG *TAT* TCT GCG ATG AGA GAA

Yeast c-ras2   G AT GAT GAA GTG TCT ATA TTG GAC ATT TTG GAT ACT GCA GGG CAG GAA GAA *TAC* TCT GCT ATG AGG GAA

Distyostelium ras   G* AT GAT GAA ACT TGT TTA TTA GAT ATT TTA GAT ACT GCA GGT CAA GAG GAA *TAT* AGT GCA ATG AGA GAT

Human K1-ras   G AT GGA GAA ACC TGT CTC TTG GAT ATT CTC GAC ACA GCA GGT CAA GAG GAG *TAC* AGT GCA ATG AGG GAC

Human K2-ras   G AT GGA GAA ACC TGT CTC TTG GAT ATT CTC GAC ACA GCA GGT CAC GAG GAG *TAC* AGT GCA ATG AGG GAC

Human N-ras   G AT GGT GAA ACC TGT TTG TTG GAC ATA CTG GAT ACA GCT GGA CAA GAA GAG *TAC* AGT GCC ATG AGA GAC

Drosophila ras1   G AT GGA GAG ACC TGC CTG ATC CTG GAC ACC GCC GGC CAA GAG GAG *TAC* TCG GCC ATG CGG GAT

Harvey v-ras1   G AT GGG GAG ACG TGT TTA CTG GAC ATC TTA GAC ACA GCA GGT CAA GAG GAG *TAT* AGT GCC ATG CGG GAC

H-T24 bladder   G AT GGG GAG ACG TGC CTG TTG GAC ATC CTG GAT ACC GCC GGC CAG GAG GAG *TAC* AGC GCC ATG CGG GAC

Human c-H-ras   G AT GGG GAG ACG TGC CTG TTG GAC ATC CTG GAT ACC GCC GGC CTG GAG GAG *TAC* AGC GCC ATG CGG GAC

Rat sarcoma v-don   G GG GCA GGA GCT CCT GGT TTG GCA GCC CCT GTA GAA GCG ATG ACA GAA *TAC* AAG CTT GTG GTG GGC

Drosophila ras3   A AC GAA CGC CAG TGC ATG CTG GAG ATC GTG AAC ACG GCG GGT ACG CAG TTC ACG GCC ATG CGG AAT

Aplysia rho   G AT GGC AAA CAG GTT GAG CTA GCT CTG TGG GAC ACA GCG GGA CAA GCG GAC *TAT* GAC AGA CTG AGG CCG

Row 4

Yeast *c-ras1*   CAG *TAC* ATG AGG ACT GGG GAA GGT TTC CTA CTG GTC *TAT* TCC GTC ACC TCT AGA AAT TCC TTT GAT GAG

Yeast *c-ras2*   CAA *TAC* ATG CGC AAC GGC GAA GGA TTC CTA TTG GTT *TAC* TCT ATA ACG TCC AAG TCG TCT CTT GAT GAG

Dictyostelium ras CAA *TAT* ATG AGA ACT GGT CAA GGA TTT TTA **TGT** GTT *TAT* TCA ATT ACA TCA AGA TCA TCA *TAT* GAT GAA

Human *K1-ras*   CAG *TAC* ATG AGG ACT GGG GAG GGC TTT CTT **TGT** GTA TTT GCC ATA AAT AAT ACT AAA TCA TTT GAA GAT

Human *K2-ras*   CAG *TAC* ATG AGG ACT GGG GAG GGC TTT CTT **TGT** GTA TTT GCC ATA AAT AAT ACT AAA TCA TTT GAA GAT

Human *N-ras*   CAA *TAC* ATG AGG ACA GGC GAA GGC TTC CTC **TGT** GTA TTT GCC ATC AAT AAT AGC AAG TCA TTT GCG GAT

Drosophila ras1   CAG *TAT* ATG CGG ACT GGC GAG GGA TTC CTG CTG GTC **TGT** GTA TTT GCC ATC AAC AGT GCG AAG TCC TTC GAG GAT

Harvey *v-ras1*   CAG *TAC* ATG CGC ACA GGG GAG GGC TTC CTC **TGT** GTA TTT GCC ATC AAC AAC ACC AAG TCC TTT GAA GAC

H-T24 bladder   CAG *TAC* ATG CGC ACC GGG GAG GGC TTC CTG **TGT** GTG TTT GCC ATC AAC AAC ACC AAG TCT TTT GAG GAC

Human *c-H-ras*   CAG *TAC* ATG CGC ACC GGG GAG GGC TTC CTG **TGT** GTG TTT GCC ATC AAC AAC ACC AAG TCT TTT GAG GAC

Rat sarcoma *v-don* GCT AGA GGC GTG GGA AAG AGT GCC CTG ACC ATC CAG ATC CAG AAC CAT TTT GTG GAC GAG *TAT* GAT

Drosophila ras3   TTG *TAC* ATG AAG AAC GGC AGC GAT TCG TGG **TGC** TCT ACT CGA TCA CGG CGC AAT CGA CGT TTA ACG ATC

Aplysia rho   CTG⁷⁷*TAC* CCT GAC ACA GAT GTC ATC CTC ATG **TGT** TTC TCT ATA GAC AGT CCA GAC AGT CTG GAG AAC ATA

(continued)

Table 11.7 (Continued)

Row 5

Yeast *c-ras1*	TTA CTG TCT *TAT TA T**CAG CAA ATT CAA AGA GTA AAA GAT TCT GAC *TAC* ATT CCT GTA GTC GTG GTA GGT
Yeast *c-ras2*	CTG ATG ACT *TAC TA T**CAA CAA ATT CCG AGA GTC AAA GAT ACC GAC *TAT* GTT CCA ATT GTG GTT GTT GGT
Dictyostelium ras	ATT GCA TCA TTT AG A GAA CAA ATT CTA AGA GTT AAA GAC AAA GAT AGA GTA CCA TTG ATT TTG GTT GGT
Human *K1-ras*	ATT CAC CAT *TAT* AG*A GAA CAA ATT AAA AGA GTT AAG GAC TCT GAA GAT GTA CCT ATG GTC CTA GTA GGA
Human *K2-ras*	ATT CAC CAT *TAT* AG A* GAA CAA ATT AAA AGA GTT AAG GAC TCT GAA GAT GTA CCT ATG GTC CTA GTA GGA
Human *N-ras*	ATT AAC CTC *TAC* AG*G GAG CAG ATT AAG CGA GTA AAA GAC TCG GAT GAT GTA CCT ATG GTG CTA GTG GGA
Drosophila ras1	ATC GGC ACC *TAC* CG T GAG CAG ATC AAG CAC GTA AAG GAT GCC GAA GAG GTG CCC ATG GTG CTG GCG GGC
Harvey *v-ras*	ATC CAT CAG *TAC* AG G GAG CAG ATC AAG CGG GTG AAA GAT TCA GAT GTC CCA ATG GTG CTG GTG GGC
H-T24 bladder	ATC CAC CAG *TAC* AG*G GAG CAG ATC AAA CGG GTG AAG GAC TCG GAT GTC CCC ATG GTG CTG GTG GGG
Human *c-H-ras*	ATC CAC CAG *TAC* AG*G GAG CAG ATC AAA CGG GTG AAG GAC TCG GAT GTC CCC ATG GTG CTG GTG GGG
Rat sarcoma *v-don*	CCC ACT ATA GAG GA C TCC CGG AAA CAG GTA GTC ATT GAT GGG GAG ACG **TGT** TTA CTG GAC ATC TTA
Drosophila ras3	**TGC** AGA ACG CGC GA G CAG ATA CTC CGG GTG AAG GAC GAC ACA GAT GTG CCC ATG GTG CTC GTG GGC AAC
Aplysia rho	CCG GAG AAG TGG AC G CCT GAG GTT CGT CAC TTT **TGT** CCA --- AAT GTT CCT ATA ATA CTT GTG GGT AAC

Row 6

Yeast *c-ras1*	AAC AAA TTG GAC CTT GAA AAT GAA AGA CAA GTC TCT *TAT* GAA GAC GGG TTA CGC TTG GCC AAG CAG TTG
Yeast *c-ras2*	AAC AAA TCT GAT TTA GAA AAC GAA CAG GTC TCT *TAC* CAG GAC GGG TTG AAC ATG GCA AAG CAA ATG
Dictyostelium ras	AAT AAA GCA GAT TTG GAT CAT GAA CGT CAA GTT AGT GTA AAT GAA GGT CAA GAA CTT GCA AAG GAT TCA
Human *K1-ras*	AAT AAA TGT GAT TTG CCT TCT AGA ACA GTA GAC ACA AAA CAG GCT CAG GAC TTA GCA AGA AGT *TAT* GGA
Human *K2-ras*	AAT AAA TGT GAT TTG CCT TCT AGA ACA GTA GAC ACA AAA CAG GCT CAG GAC TTA GCA AGA AGT *TAT* GGA
Human *N-ras*	AAC AAG TGT GAT TTG CCA ACA AGG ACA GTT GAT ACA AAA CAA GCC CAC GAA CTG GCC AAG AGT *TAC* GGG
Drosophila ras1	AAC AAG TGT GAT CTG GCC TCG TGG AAC GTT AAC AAC GAG CAG GCA AGA GAG GTG GCC AAA CAG *TAC* GGC
Harvey *v-ras1*	AAC AAG TGT GAC CTG GCC GCT CGC ACT GTT GAG TCT CGG CAG GCC CAG GAC CTT GCT CGC AGC *TAT* GGC
H-T24 bladder	AAC AAG TGT GAC CTG GCT GCA CGC ACT GTG GAA TCT CGG GAA GCT CAG GAC CTC GCC CGA AGC *TAC* GGC
Human *c-H-ras*	AAC AAG TGT GAC CTG GCT GCA CGC ACT GTG GAA TCT CGG GAA GCT CAG GAC CTC GCC CGA AGC *TAC* GGC
Rat sarcoma *v-don*	GAC ACA GCA GGT CAA GAA GAG *TAT* AGT GCC ATG CGG GAC *TAC* ATG CGC ACA GGG GAG GGC TTC CTC
Drosophila ras3	AAG TGC GAC CTC GAA GAG GAG CGC GTC GTC GGC AAG GAG CTG GGC AAG AAC TTG GCC ACC CAG TTC AAC
Aplysia rho	AAA AAG GAT CTT CGC AAC GAT GAT GAA AGT AGT ACC AAA CGT GAG CTC ATG AAA ATG AAA CAG GAA CCA GTG AGA

(*continued*)

Table 11.7 (Continued)

Row 7

Yeast c-ras1	AAT[c]	CCC	TTT	CTA	GAA	ACG	TCT	GCG	AAA	CAA	GCC	ATC*	AAC	GTA	GAC	GAG	GCC	TTT	*TAT*	AGC	CTT	ATT	CGT
Yeast c-ras2	AAC[d]	CCT	TTC	TTG	GAG	ACA	TCT	GCT	AAG	CAA	GCA	ATC*	AAC	GTG	GAA	GAG	GCG	TTT	*TAC*	ACT	CTA	GCA	CGT
Dictyostelium ras	TTG	TCC	TTT	CAT	GAG	TCA	TCT	GCT	AAA	AGT	AGA	ATT	AAT	GTT	GAA	GAG	GCA	TTT	*TAC*	TCT	TTA	GTT	CGT
Human K1-ras	ATT	CCT	TTT	ATT	GAA	ACA	TCA	GCA	AAG	ACA	AGA	CAG*	AGA	GTG	GAG	GAT	GCT	TTT	*TAT*	ACA	TTG	GTG	AGA
Human K2-ras	ATT	CCT	TTT	ATT	GAA	ACA	TCA	GCA	AAG	ACA	AGA	CAG*	AGA	GTG	GAG	GAT	GCT	TTT	*TAT*	ACA	TTG	GTG	AGA
Human N-ras	ATT	CCA	TTC	ATT	GAA	ACC	TCA	GCC	AAG	AGA	ACC	AGA	CAG*	GGT	GTT	GAA	GAT	GCT	TTT	*TAT*	ACA	CTG	GTA	AGA
Drosophila ras1	ATT	CCA	*TAC*	ATT	GAG	ACA	TCC	GCC	AAG	ACG	CGC	ATG	GGC	GTG	GAC	TTT	*TAC*	ACA	CTG	GTG	CGC		
Harvey v-ras	ATC	CCC	*TAC*	ATT	GAA	ACA	TCA	GCC	AAG	ACC	CGG	CAG	GGT	GTA	GAG	GAT	GCC	TCC	*TAC*	ACA	CTA	GTA	CGT
H-T24 bladder	ATC	CCC	*TAC*	ATC	GAG	ACC	TCG	GCC	AAG	ACC	CGG*	GGA	GTG	GAG	GAT	GCC	TTC	*TAC*	ACG	TTG	GTG	CGT	
Human c-H-ras	ATC	CCC	*TAC*	ATC	GAG	ACC	TCG	GCC	AAG	ACC	CGG*	GGA	GTG	GAG	GAT	GCC	TTC	*TAC*	ACG	TTG	GTG	CGT	
Rat sarcoma v-don	TGT	GTA	TTT	GCC	ATC	AAC	AAC	ACC	AAG	TCC	TTT	GAA	GAC	ATC	CAT	CAG	*TAC*	AGG	GAG	CAG	ATC	AAG	CGG
Drosophila ras3	TGC	GCC	TTC	ATG	GAG	GAG	ACC	TCA	GCC	AAA	GCC	AAA	GTG	AAT	GTG	AAC	GAT	ATT	TTC	*TAC*	GAC	TGG	TCC	GGC
Aplysia rho	---	CCA	GAG	GAT	GGG	CGC	GCC	ATG	GCT	GAG	AAA	ATC	AAC	AAC	GCC	*TAC*	TCT	*TAT*	CTT	GAG	**TGC**	TCT	GCT	AAA

Row 8

Yeast *c-ras1*	TTG	GTA	AGG	GAC	GAC	GGT	GGG	AAA	*TAC*	AAT	AGC	ATG	AAT	CGT	CAA	CTG	GAT	AAT	ACG	AAT	GAA	ATA	AGA	GAT TCG
Yeast *c-ras2*	TTA	GTT	AGA	GAC	GAA	GGC	GGC	AAG	*TAC*	AAC	AAG	ACT	TTG	ACG	GAA	AAT	GAC	AAC	TCC	AAG	CAA	ACT	TCT	CAA GAT
Dictyostelium ras	GAA	ATT	AGA	AAA	GAA	CTA	AAA	GGT	GAT	CAA	TCA	AGT	GGC	AAA	GCT	CAA	AAA	AAG	AAA	CAA	TGT	TTA	ATT	TTA
Human *K1-ras*	GAG	ATC	CGA	CAA	TAC	AGA	TTG	AAA	AAA	ATC	AGC	AAA	GAA	GAA	AAG	ACT	CCT	GGC	TGT	GTG	AAA	ATT	AAA	AAA TGC
Human *K2-ras*	GAG	ATC	CGA	CAA	TAC	AGA	TTG	AAA	AAA	ATC	AGC	AAA	GAA	GAA	AAG	ACT	CCT	GGC	TGT	GTG	AAA	ATT	AAA	AAA TGC
Human *N-ras*	GAA	ATA	CGC	CAG	TAC	CGA	ATG	AAA	AAA	CTC	AAC	AGC	AGT	GAT	GAT	GGG	ACT	CAG	GGT	TGT	ATG	GGA	TTG	CCA TGT
Drosophila ras1	GAA	ATT	CGC	AAG	GAC	GAC	AAG	AAG	GGG	AGG	GGC	CGC	AAA	ATG	AAC	AAG	AAG	CCG	AAT	TGT	AGA	TTT	AAA	TGT
Harvey *v-ras1*	GAG	ATT	CGG	CAG	CAT	AAA	CTG	CGG	AAA	CTG	AAC	CCG	CCT	GAT	GAG	AGT	GGC	CCT	GGC	TGC	ATG	AGC	TGC	AAG TGT
H-T24 bladder	GAG	ATC	CGG	CAG	CAC	AAG	CTG	CGG	AAG	CTG	AAC	CCT	GAT	GAG	AGT	GGC	CCC	GGC	TGC	ATG	AGC	TGC	AAG	TGT
Human *c-H-ras*	GAG	ATC	CGG	CAG	CAC	AAG	CTG	CGG	AAG	CTG	AAC	CCT	GAT	GAG	AGT	GGC	CCC	GGC	TGC	ATG	AGC	TGC	AAG	TGT
Rat sarcoma *v-don*	GTG	AAA	GAT	TCA	GAT	GAT	GTG	CCA	ATG	GTG	CTG	GTG	GGC	AAC	AAG	TGT	GAC	CTG	GCC	GCT	CAC	ACT	GTT	GAG TCT
Drosophila ras3	AGA	TCA	ACA	AGA	AGT	CGC	CCG	AGA	AGA	AAC	AGC	CGA	AAA	GTT	CCA	TGT	GTT	CTG	CTA	TAA				
Aplysia rho	ACC	AAG	GAG	GGC	GTG	AGG	GAT	GTG	TTT	GAG	ACA	GCT	ACC	ACC	AGA	GCT	GCG	CTG	CAA	GTG	AAA	AAG	AAG	AAG GGT

(continued)

Table 11.7 (Continued)

Row 9

Yeast *c-ras1*	GAG CTA ACC TCA TCT GCA ACA GCG GAT AGA GAA AAA AAG AAC AAC GGG TCT *TAT* GTA CTC GAT AAT TCT
Yeast *c-ras2*	ACA AAA GGG AGC GGT GCC AAC TCT GTG CCT AGA AAT AGC GGT GGC CTC AGG AAG ATG AGC AAT GCT GCC
Dictyostelium ras	TAA
Human *K1-ras*	ATT ATA ATG TAA
Human *K2-ras*	ATT ATA ATG TAA
Human *N-ras*	GTG GTG ATG TAA
Drosophila ras1	AAA ATG CTC TAA
Harvey *v-ras1*	GTG CTG TCC TGA
H-T24 bladder	GTG CTC TCC TGA
Human *c-H-ras*	GTG CTC TCC TGA
Rat sarcoma *v-don*	CGG CAG GCC CAG GAC CTT GCT CGC AGC *TAT* GGC ATC CCC *TAC* ATT GAA ACA TCA GCC AAG ACC CGA CCA
Aplysia rho	GGA TGT GTT GTA TTG TGA

Row 10

Yeast *c-ras1*	TTG ACC AAT GCT GGC ACT GGC TCC AGT TCA AAG TCA GCC GTT AAC CAT AAC GGT GAA ACT ACT AAA CGA
Yeast *c-ras2*	AAC GGT AAA AAT GTG AAC AGT AGC ACT GTC GTG AAT GCC AGG AAT GCA AGC ATA GAG AGT AAG ACA
Rat sarcoma *v-don*	GGT GTG GAG GAT GCC TTC *TAC* ACA CTA GTA CGT GAG ATT CGG CAG CAT AAA CTG CGG AAA CTG AAC CCG

Row 11

Yeast *c-ras1*	ACT GAT GAA AAG AAT *TAC* GTT AAT CAA AAC AAT AAC AAT GAA GGA AAT ACC AAG *TAC* TCC AGT AAC GGC
Yeast *c-ras2*	GGG TTG GCA GGC AAC CAG GCG ACA AAT GGT AAG ACA CAA ACT GAT CGC ACC AAT ATA GAC AAT TCC ACG
Rat sarcoma *v-don*	CCT GAT GAG AGT GGC CCT GGC AAG TGC ATG AGC AAG TGT GTG CTG TCC TGA

Row 12

Yeast c-ras1   AAC GGA AAT CGA AGT GAT ATT AGT CGT GGT AAT CAA AAT AAT GCC TTA AAT TCG AGA AGT AAA CAG TCT

Yeast c-ras2   GGC CAA GCT GGT CAG GCC AAC GCT CAA AGC GCT AAT ACG GTT ATT AAT CGT GTA AAT AAT AAT AGT AAG

Row 13

Yeast c-ras1   GCT GAG CCA CAA AAA AAT TCA AGC GCC AAC GCT AGA AAA GAA TCT AGT GGT GGT TGT TGT ATA ATT TGT

Yeast c-ras2   GCC GGT CAA GTT TCA AAT GCT AAA CAG GCT AGG AGC AAG CAA GCT GCA CCC GGC GGT AAC ACC AGT GAA

Row 14

Yeast c-ras1   TGA

Yeast c-ras2   GCC TCC AAG AGC GGA TCG GGT GGC TGT TGT ATT ATA AGT TAA

[a]Codons for the two basic amino acids arginine and lysine are underscored, those for cysteine are stippled, and those for tyrosine are italicized. Asterisks mark the location of introns.

[b]Dhar et al. (1984).
[c]Insert CAG GGA AAT AAA TCA ACT ATA.
[d]Insert CCT TTC AAC AAG TCG AAC ATA.
[e]Reymond et al. (1984).
[f]Shimizu et al. (1985).
[g]Tsuchida et al. (1982).
[h]Capon et al. (1983b).
[i]Brown et al. (1984); Hall and Brown (1985).
[j]Neuman-Silberberg et al. (1984).
[k]Yasuda et al. (1984).
[l]Capon et al. (1983a); Reddy (1983).
[m]Sekiya et al. (1984).
[n]Rasheed et al. (1983).
[o]Schejter and Shilo (1985).
[p]Madaule and Axel (1985).
[q]Insert CGA AAG.
[r]Insert TGT.

plored to provide the essential information. Thus, on occasion an erroneous disposition may be made here, but while this may cause some inconvenience, it is an unavoidable consequence of the present state of the art. The only other visible alternative, that of classifying the majority of the types in a miscellaneous category, seems to create equal inconvenience. However, there is no problem with classifying the basic example as a member of that group whose products are associated with the cell membrane and display a slow GTPase activity, so it, the *ras* gene, is examined before the other biologically doubtful types are discussed.

11.4.1. The Ras Gene and Its Product

That the *ras* gene family must certainly rank with the most thoroughly investigated of the oncogenes is strongly attested by the presence of 15 sequences from a wide diversity of sources (Table 11.7). Among those whose genomes have been found to contain one or more copies of this cistron are yeasts, fungi, insects, and mammals, but this oncogene still remains unidentified in the seed plants, algae, and protozoans. Although most details of its biological activity have not been determined, its binding to GTP and its hydrolysis of it to GDP and inorganic phosphate by way of adenylate cyclase are soundly established (Broek *et al.,* 1985; Fraenkel, 1985; Jurnak, 1985). Indeed, its binding properties have been specifically localized to the 12th codon in the type known as *K*(irsten)-*ras* (R. Clark *et al.,* 1985). The human genome contains at least four genes of several types (E. H. Chang *et al.,* 1982), located on four different chromosomes.

Three types of *ras* are generally recognized, the two in addition to that just mentioned being *H*(arvey)- and *N*(eural)-*ras,* but when viewed from a broad, comparative standpoint, a somewhat greater diversity is found (Table 11.7). But this point merits closer attention, as given below.

Sequence Structure of Ras Genes. A true picture of relationships among *ras* genes is dependent upon a much larger assemblage of representatives from a fuller spectrum of eukaryotic sources than is now available. Especially needed are those of unicellular species such as *Euglena, Amoeba,* and *Tetrahymena,* and the seed plants, both monocots and dicots, along with ferns, mosses, and the like. But when those of the table are surveyed and the columns of homologous codons examined, what has been called rat sarcoma *v-ras* immediately stands out as being too divergent from the rest to permit its placement in this family. Even from the second triplet, it fails to correspond on a site-to-site basis with any of the others. Actually, it is so strongly divergent that it is here named as a distinct species, *v-don,* for under its previous name its lack of conformity cannot fail to be a hindrance in investigations into kinships and physiological activities.

In contrast, the recently described *rho* from the snail, *Aplysia* (Madaule and Axel, 1985), nearly deserves to be considered as a *ras* gene, for it displays homologies at frequent points throughout its length, but especially in its 5′ sector. Nevertheless, the best disposition of the cistron is that of classifying it as a separate species within the *ras* family of genes. The two structures from yeast, although unquestionably kin to the rest, greatly exceed all others in length (Dhar *et al.,* 1982, 1984). In the first place, their 5′ ends have seven-codon inserts immediately following the ATG translational initiation signals, which are absent even from the fungal sequence. Second, their 3′ termini exceed those of most others by ∼100 codons. The rat sarcoma *v-don* gene is also elongated here, extending beyond those from mammalian sources by ~50 triplets (Rasheed *et al.*, 1983). The *c-ras* sequence (not given) from *Schizosaccharomyces pombe* (Fukui and Kaziro, 1985) differs pronouncedly from both of the yeast sequences as well as that of *Dictyostelium* (Reymond *et al.*, 1984), frequently being intermediate in structure. It approaches those of the former in having an inserted segment following the initial ATG, but in this case the added portion contains five instead of seven codons. Moreover, it is distinct from all the rest in having its single intron between the second and third codons.

The three gene types already outlined are very closely related, even the majority of introns being identically located, as marked by asterisks in Table 11.7. In fact, no basis for separating the groups is observed until the latter half of row 7. From there to the termini, group-related codons designating different amino acids are noted; in addition, at site 2 of row 7, the *H-ras* cluster has a triplet for tyrosine instead of that of phenylalanine in the others. At most of the significant points, the single *N-ras* structure (Hall and Brown, 1985) more frequently agrees with the *K-ras*, but occasionally shows alliance to the *H-ras*, as in site 10 of row 8. Apparently these three series are characteristic only of mammals, for no clear-cut basis could be found for assigning the nonmammalian species to any one of them. Perhaps when more sequences from protistans are available, proper homologies can be established.

Chemical Characteristics of Ras Genes. To avoid unnecessary belaboring of minute details, analysis of the chemical properties of *ras* genes focuses on the human *K-1* sequence, with occasional comparisons with those of other species. Fortunately the several representatives of *ras* proper deviate so slightly in sites of major codon species that the account is quite applicable to all. Unlike the tyrosine-specific kinase of the *src* gene of the preceding section, in which tyrosine codons (italics) abounded, in the present group they scarcely exceed the random-chance level (3.1%), a total of nine (4.7%) being the nearly universal frequency, a number surpassed by three in the yeast. In the corresponding area of the rat sarcoma *v-don* sequence only six occur, none of which coincides with those of the others; three more are located in its additional 50-codon 3′ sector, so that its total number agrees with that of the others, but the percentage is lower by one (3.7%). A further distinction of *ras* from the *src* types is in the complete lack of triplets specifying tryptophan; the only ones noted are in the aberrant rat sarcoma structure (row 1) and the *ras1* and *ras3* sequences of *Drosophila*, one in row 6 of the first and each of rows 4 and 7 of the second. Codons for cysteine (stippled) are relatively rare also, none occurring until row 3 in most forms, but in the rat sarcoma viral form not until row 5, and in the yeast only at the extreme 3′ end (row 13 or 14). Five are found in mammalian types.

A richer endowment is to be noted in codons (underscored) for the two basic amino acids, arginine and lysine, but at best there are only 25 per sequence, for a rate of 13%. However, in the 3′ sector of the typical mammalian representative, these species are often involved in the group differences, so the exact number and placement there vary. The totals encoding the acidic monomers, aspartic and glutamic acids, scarcely exceed the foregoing, and usually total 28, for a 15% proportion.

Carcinogenic Mutation Points in H-Ras Genes. The product of *ras* called p21 (molecular weight 21,000) is certainly the most widely occurring carcinogenic agent, being involved in colon, bladder, pancreas, and lung carcinomas, breast adenocarcinomas, myelocytic leukemias, and melanomas, to mention only the more frequent types (Capon *et al.*, 1983b; Yoakum *et al.*, 1985). Occasionally it coexists with the product of other oncogenes, notably *myc* (Murray *et al.*, 1983). Consequently, many reports have been published regarding the differences detected between the normal cellular and the transforming types. For example, the gene of the Harvey murine sarcoma virus (Harvey virus) encodes different amino acids than the human cellular homolog at only a few points, namely at codons 12 of row 1 and 59 and 61 of row 3, all the other mutations being silent. The change at site 59 from GCC (alanine) to ACA (threonine) enables the ready distinction of the two encoded proteins because of the phosphorylation properties of the second amino acid (Kasid *et al.*, 1985). When human breast cell lines had *H-v-ras* genes incorporated into their genomes, they lost their normal dependence on estradiol for growth, proliferating rapidly in the presence of the viral protein p21.

This same gene has been implicated in the growth of a melanoma in a Japanese patient (Sekiya *et al.*, 1984), but the change in the normal cell sequence at codon 61 from CAA (glutamine) to CTG (leucine) was claimed to be responsible for the tumorigenic properties. However, the CTG is the codon in the normal human cellular component, so it can scarcely have induced a malignancy. At least in yeast cells, the products of the *ras* genes are essential for survival, cells with their two sequences disrupted being inviable (DeFeo-Jones *et al.*, 1985). Mutations in the normal gene at codon 12 have been implicated both in the induction of T24 human bladder carcinoma, where a change from GTC (valine) to AGA (arginine) can be noted in the *H-v-ras* sequence (Table 11.7; Capon *et al.*, 1983a; Yasuda *et al.*, 1984), and in the *N-c-ras*, in which a spontaneous mutation of the GGT (glycine) to GAT (aspartic acid) resulted in the development of a teratocarcinoma (Tainsky *et al.*, 1984). However, mutations affecting the 12th codon have been demonstrated to occur frequently in human cancer (Feinberg *et al.*, 1983). In the bladder cancer oncogene, that mutation was accompanied by a change at site 59 from ACA (threonine) of the virus to GCC (alanine) and at site 122 from GGT to GCA (also alanine) (Capon *et al.*, 1983a).

Carcinogenic Points in K-Ras Genes. Virtually the same sites as in *H-ras* genes have been described as carcinogenic in *K-ras* structures 12, 61, and, especially, 122. Mutation in the first of these has been proposed as responsible for human lung carcinomas (Nakano *et al.*, 1984), involving GGT (glycine) → TGT (cysteine). The same mutation was described in a second study of lung cancers (Yamamoto and Perucho, 1984), but in this case, it was accompanied by a CAA (glutamine) → CAC (histidine) alteration. In still a third study on human lung cancer, a colon malignancy was also included, with more complex results (Capon *et al.*, 1983b). Site 12 in the cell oncogenes was found to be GGT and TGT as before, but in the colon growth it was GTT (valine). Precisely the same mutation (GGT → TGT) has been discovered in a human pancreas carcinoma (Hirai *et al.*, 1985), and a mutation from GGT (glycine) → AGT (serine) was shown to greatly reduce GTP hydrolysis (R. Clark *et al.*, 1985). In the mouse, γ-irradiation induced a tumor involving this oncogene; cloning and sequencing of a suitable fragment proved that only a single mutation had occurred, namely a G → A change in the second base of codon 12, altering the GGT (glycine) to GAT (aspartic acid) (Guerrero *et al.*, 1984). However, no specifics regarding genetic changes at site 122 have been published.

11.4.2. Other GTP-Associated Oncogenes

Among the oncogenes that are related to *ras* must be included the recently described *rho* of *Aplysia* (Madaule and Axel, 1985) and the very aberrant Rasheed rat sarcoma virus

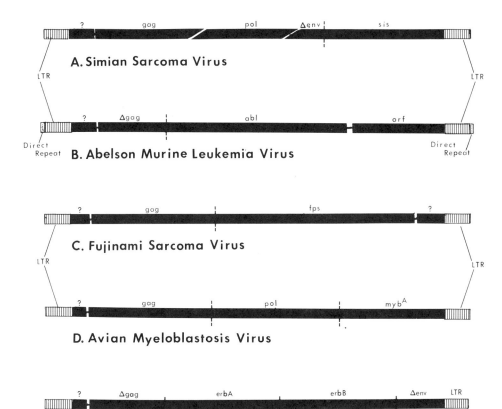

Figure 11.3. Comparisons of replication-defective retroviral genomes. (Based in part, for part A on Devare *et al.*, 1983; for part B on Reddy *et al.*, 1983a; for part C on Shibuya and Hanafusa, 1982; for part D on Nunn *et al.*, 1983; and for part E on Graf and Beug, 1983.)

renamed *don,* pointed out in the preceding section. But others that possibly belong to the present group have not been clearly delineated. Some, such as the *erbA* and *erbB* (Yamamoto *et al.,* 1983), *mil* (=*raf*) (Galibert *et al.,* 1984; Shimizu *et al.,* 1985), and *mos* (Baldwin, 1985) genes, have been reported as having structural homology with *src,* but they fail to show any tyrosine kinase activity. Since they have also been indicated to have growth-factor-related and receptor activities, they are tentatively included here.

The Erb Gene and its Product. The *erb* (erythroblastosis) gene of the avian erythroblastosis virus is dicistronic, containing the coding regions of *erbA* and *erbB* (Graf and Beug, 1983), the former being associated with a *gag* fragment and the latter with the 3′ terminal portion of *env* (Figure 11.3E). In the genome of the host cell, these two cistrons are separated by at least 15,000 base pairs and may even lie on separate chromosomes. The first of these encodes a protein of 75,000 daltons, and the second a somewhat smaller glycoprotein of 65,000 daltons. In view of the genomic distances involved, it may be difficult to perceive how the two were incorporated into the virus, a problem pointed out by Graf and Beug (1983), but the location of the genes actually is not relevant in the present context. Since *c-erbA* and *c-erbB* contain multiple introns and those of the virus do not, quite obviously the virus has gained its structures from processed messengers, possibly by reverse transcription. Apparently the products of each viral gene have transformational powers; *v-erbA* blocks the differentiation of erythroblasts while they are still immature, preparing the way for the glycoprotein of *erbB* to carry out their actual transformation (Debuire *et al.*, 1984). An additional type of *erbB*, thus far not further differentiated, induces the formation of angiosarcomas (Tracy *et al.*, 1985).

Another viral gene involved in the transformation of hematopoietic tissue in birds is the relatively little known one called *amv* (Rushlow *et al.*, 1982). Although it thus is truly oncogenic in inducing neogenesis in the form of myeloblastic leukemia, it fails to transform fibroblasts as others with such properties uniformly do. The gene sequence, which has been fully established (Table 11.9), shows few prominent features, but one is outstanding, the presence of three codons (underscored) for basic amino acids immediately following the initiator triplet. Tyrosine (italics) coding signals display an unusual distribution in that five occur in the 5'-terminal sector of 75 codons, while only three are in the remaining portions.

The Mos Oncogene and Product. Investigation into the *mos* gene and its product has been restricted entirely to mammalian sources, especially the laboratory mouse. Although known most thoroughly from the gene products inducing sarcomas, it has also been implicated in mouse myelomas and plasmacytomas (Rechavi *et al.*, 1982; Canaani *et al.*, 1983). In each of the latter, however, the presence of an inserted segment has been documented, in the first an IS-like sequence, and in the second the genome of an intracisternal A-particle. It is pertinent to note in connection with these inserted segments that the human cellular *mos* counterpart of the viral gene also has been demonstrated to exhibit transforming activity when linked to a viral LTR (Diaz *et al.*, 1985). Moreover, when the corresponding part from an intracisternal A-particle was inserted at the 5' terminus of the murine *c-mos*, a plasmacytoma was induced.

The six complete and partial sequences of Table 11.8 provide the essential features of this type of oncogene. Outstanding among the differences between them is the presence of a 31-codon-long segment at the 5' terminus of the murine cellular and sarcoma viral genes that is absent from the others. Actually, that region of the murine *c-mos* is longer than that by 16 triplets, since it has an inserted segment following the first TGG (Reddy *et al.*, 1981; Van Beveren *et al.*, 1981; Rechavi *et al.*, 1982). Beyond that point, homology among the sequences is extensive, that referred to as murine *rc-mos* being the most deviant at its 5'-terminal portion. In the first place, it is shorter there by the absence of 13 codons, and second, it diverges almost 100% from the others until the fifth codon of row 4 is reached. Thence, to the end, it is identical to the murine sarcoma *v-mos* sequence. The human *c-mos* also diverges somewhat from the five from rodents (R. Watson *et al.*, 1982), but largely in minor ways.

Prominent chemical characteristics are not in evidence, for the types of codons that have usually provided such features occur in *mos* at or below the random-chance level; for a case in point, the nine cysteine (stippled) and same number of tyrosine (italicized) codons indicate levels of 2.6%, as compared to the expected ratio of 3.1%. Similarly, the total of 33 triplets found here encoding the two basic amino acids (underscored) is very close to the 9.4% normally expected, and the 21 for glutamic and aspartic acids represent a ratio of 6.1%, right at the random-chance frequency. Only the total of nine tryptophan

Table 11.8
Comparisons of Mos Coding Sequences[a]

Row 1

Murine sarcoma v-mos[b]
ATG GCG CAT TCA ACG CCA TGC TCC CAA ACT TCC CTG GCT GTT CCT AAT CAT TTC TCC CTA GTG TCT CAT

Murine c-mos[c]
ATG TGG[g] AGC AAT CTG CCA TGC TCC CAA ACT TCC CTG GCT GTT CCT ACT CAT TTC TCC CTA GTG TCT CAT

Row 2

Murine sarcoma v-mos
GTG ACT GTC CCA TCT GAG GGT GTA ATG CCT TCG CTA AGC CTG TGT CGC CTC CCT CGT GAG CTG

Moloney murine v-mos[d]
ATG CCT TCG CCT CTA AGC CTG TGT CGC CTC CCT CGT GAG CTG

Human c-mos[d]
ATG CCC TCG CCC CTG GCC CTA CGC CCC TAC CTC CGG AGC GAG TTT

Rat c-mos[e]
ATG CCT TCG CCT CTC ATC CTG TGT CGC GTC CCT CGC GAG CTG

Murine c-mos
GTG ACT GTC CCA TCT GAG GGT GTA ATG CCT TCG CCT CTA AGC CTG TGT CGC CTC CCT CGT GAG CTG

Murine re-mos[e,f]
ATG CGC

Row 3

Murine sarcoma v-mos
TCG CCA TCG GTA GAC TCG CGG TCC TGC AGC --- ATT CCT TTG GTG GCC CCG GCC AGG AAG GCA GGG AAG CTC

Moloney murine v-mos
TCG CCA TCG GTG GAC TCG CGG TCC TGC AGC --- ATT CCT TTG GTG GCC CCG GCC AGG AAG GCA GGG AAG CTC

Human c-mos
TCC CCA TCG GTG GAC GGG GGG CCC TGC AGC --- AGT CCC TCA GAG CTA CCT --- --- --- GCG AAG CTG

Rat c-mos
TCG CCA ACG GTG GAC GTG GAC TCG AGG --- TGC AGC AGC AGC CCC TTG GTG GCC TCG CCG GGG --- GCG AAG ---

Murine re-mos
AGA TTA TTT GTT TAC CAC TTA GAA CAC AGG ATC TCA GGG CCA TCT TCT ACC GGC GAA TGT GGG CGC GGC

Row 4

Murine sarcoma v-mos
TTC CTG GGG ACC ACT CCT CCT CGG GCT CCC GGA CTG CCA CGC CGG CTG GCC TGG TTC TCC ATA GAC TGG

Moloney murine v-mos
TTC CTG GGG ACC ACT CCT CCT CGG GCT CCC GGA CTG CCA CGC CGG CTG GCC TGG TTC TCC ATA GAC TGG

(continued)

Table 11.8 (Continued)

Human c-mos   CTT CTG GGG GCC ACT CTT CCT CGG GCC CGG CGG CTG CCG CGG CTG GCC TGC TGC TCC ATT GAC TGG

Rat c-mos   TTC CTG GGG GCC ACT CGT CCT CGG GCC CGG GGC CTG TCA CGC CTG GCC TGC TGG TGC TTC ATA GAC TGG

Murine ra-mos   TCC --- CAA CAC ACT CCT CCT GCT CCC GGA CTG CCA (Rest identical to murine sarcoma v-mos)

Row 5

Murine sarcoma v-mos   GAA CAG GTA TGT CTG ATG CAT AGG CTG GGC TCT GGA GGG TTT GGC GTG TCG TAC AAA GCC ACT TAC CAC

Moloney murine v-mos   GAA CAG GTA TGT CTG ATG CAT AGG CTG GGC TCT GGA GGG TTT GGC TCG GTG TAT AAA GCC ACT TAC CAC

Human c-mos   GAG CAG GTG TGC TTG CTG CAG AGG CTG GGA GCT GGG TTT GGC TCG GTG TAC AAG GCG ACT TAC CGC

Rat c-mos   GGA CAG GTA TGC CTG CTG CAT AGG CTG GGT TCT GGA GGG TTT GGC TCG GTG TAC AAA GCC ACT TAC CAC

Row 6

Murine sarcoma v-mos   GGT GTT CCT GTG GCC ATC AAG CAA GTA AAC AAG TGC ACC GAG GAC CTA CGT GCA TCC CAG CGG AGT TTC

Moloney murine v-mos   GGT GTT CCT GTG GCC ATC AAG CAA GTA AAC AAG TGC ACC AAG GAC CTA CGT GCA TCC CAG CGG AGT TTC

Human c-mos   GGT GTT CCT GTG GCC ATA AAG AAG CAA GTG AAC AAG TGC ACC AAG AAC CGA CTA GCA TCT CGG CGG AGT TTC

Rat c-mos   GGT GTT CCT GTG GCC ATC AAG AAG CAA GTG AAC AAG TGC ACC AGA ACC CTA CGT GCA TCC CAA CGG AAT TTC

Row 7

Murine sarcoma v-mos   TGG GCT GAA CTG AAC ATT GGA GGA CTA CGC CAC GAC AAC ATA GTT CGG GTT GTG GCT GCC AGC ACG CGC

Moloney murine v-mos   TGG GCT GAA CTG AAC ATT GCA AGA CTA CGC CAC GAC AAC ATA GTT CGG GTG GCT GCC AGC AGC CGC

Human c-mos   TGG GCT GAG CTC AAC GTA GCA AGG CTG CGC CAC GAT AAC ATC GTG CGG GTG GCT GCC AGC ACG CGC

Rat c-mos   TGG GCT GAA CTG AAC ATT GCA AGG CTG CAC GAC AAC ATA ATC CGG GTT GCT GCC AGC ACG CGC

Row 8

Murine sarcoma v-mos	ACG	CCC	GAA	GAC	TCC	AAC	AGC	CTA	GGT	ACC	ATA	ATC	ATG	GAG	TTT	GGG	GGC	AAC	GTG	ACT	CTA	CAC	CAA
Moloney murine v-mos	ACG	CCC	GAA	GAC	TCC	AAC	AGC	CTA	GGT	ACC	ATA	ATC	ATG	GAG	TTT	GGG	GGC	AAC	GTG	ACT	CTA	CAC	CAA
Human c-mos	ACG	CCC	GCA	GGG	TCC	AAT	AGC	CTA	GGG	ACC	ATC	ATC	ATG	GAG	TTC	GGT	GGC	AAC	GTC	ACT	TTA	CAC	CAA
Rat c-mos	ACG	CCG	GAA	GGT	TCC	AAC	AGC	CTT	GGT	ACC	ATA	ATC	ATG	GAG	TTT	GGG	GGC	AAT	GTG	ACT	CTA	CAC	CAA

Row 9

Murine sarcoma v-mos	GTC	ATC	TAC	GAT	GCC	ACC	CGC	TCA	CCG	GAG	CCT	CTC	AGC	TGC	---	AGA	AAA	CAA	CTA	AGT	TTG	GGG	AAG
Moloney murine v-mos	GTC	ATC	TAC	GGT	GCC	ACC	CGC	TCA	CCG	GAG	CCT	CTC	AGC	TGC	---	AGA	GAA	CAA	CTG	AGT	TTG	GGG	AAG
Human c-mos	GTC	ATC	TAT	GGC	GCC	GCC	CAC	CCT	CAC	GAG^{h}	CCT	CAC	TGC	CGC	ACT	GGA	GGA	CAG	TTA	AGT	TTG	GGA	AAG
Rat c-mos	GTC	ATC	TAC	GGT	GCC	ACC	CGC	TCC	CCA	GAG	CCT	CTC	AGC	TGC	---	AGA	GAG	CAA	CTG	AGT	TTG	GGA	AAG

Row 10

Murine sarcoma v-mos	TGC	CTC	AAG	TAT	TCC	CTA	GAT	GTT	GTT	AAC	GGC	CTG	CTT	TTT	CTC	CAC	TCA	CAA	AGC	ATT	TTG	CAC	TTG
Moloney murine v-mos	TGC	CTC	AAG	TAT	TCC	CTA	GAT	GTT	GTT	AAC	GGC	CTG	CTT	TTT	CTC	CAC	TCA	CAA	AGC	ATT	TTG	CAC	TTG
Human c-mos	TGT	CTC	AAG	TAC	TCA	CTA	GAT	GTT	GTG	AAC	GGC	CTG	CTC	TTC	CTC	CAC	TCG	CAA	AGC	ATT	GTG	CAC	TTG
Rat c-mos	TGC	CTC	AAG	TAT	TCC	CTA	GAT	GTT	ATT	AAC	GGC	CTG	CTT	TTT	CTC	CAC	TCA	CAA	AGC	ATT	TTG	CAC	TTG

Row 11

Murine sarcoma v-mos	GAC	CTG	AAG	CCA	GCG	AAC	ATT	TTG	ATT	AGT	GAG	CAG	GAC	GTT	TGT	AAG	ATC	AGT	GAC	TTC	GGC	TGC	TCC

(continued)

Table 11.8 (Continued)

```
Moloney murine v-mos   GAC CTG AAG CCA AAC ATT TTG ATC AGT GAG CAA GAC GTT TGT AAG ATC AGT GAC TTC GGC TGC TCC
Human c-mos            GAC CTG AAG CCC GCG AAC ATC TTG ATC AGT GAG CAG GAT GTC TGT AAA ATT AGT GAC TTC GGT TGC TCT
Rat c-mos              GAC CTG AAG CCA GCG AAC ATT TTG ATC AGT GAG GAC GTT TGT AAG ATA AGT GAC TTC GGC TGC TCC

Row 12

Murine sarcoma v-mos   CAG AAG CTG CAG GAT CTG CGG GGC CGG CAG GCG TCC CCT CCC CAC ATA GGG GGC ACG TAC ACG CAC CAA
Moloney murine v-mos   CAG AAG CTG CAG GTT CTG CGG TGC CAG GCG TCC CCT CAC CAC ATA GGG GGC ACG ACG TAC ACG CAC CAA
Human c-mos            GAG AAG TTG GAA GAT CTG CTG TGC TTC CAG ACA CCC TCT TAC CCT CTA GGA GGC ACA TAC ACC CAC CGC
Rat c-mos              CAG AAG CTT CAG GAT CTG CGG --- CCG TCC CTT CAC CAC ATC GGG GGC ACG ACG TAC ACG CAC CAA

Row 13

Murine sarcoma v-mos   GCT CCG GAG ATC CTA AAA GGA GAG ATT GCC ACG CCC AAA GCT GAC ATC TAC TCT TTT GGA ATC ACC CTG
Moloney murine v-mos   GCT CCG GAG ATC CTG AAA GGA GAG ATT GCC ACG CCC AAA GCT GAC ATC TAC TCT TTT GGA ATC ACC CTG
Human c-mos            GCC CCG GAG CTC CTG AAA GGA GAG GAG GCC CCT ACG ACG CCT AAA GCC GAC ATT TAT TCC TTT GCC ATC ACT CTC
Rat c-mos              GCT CCG GAG CTC CTG AAA GGA GAG ATC GCC ACG CCC AAA GCC GAC ATC TAC TCT TTT GGC ATC ACC CTG

Row 14

Murine sarcoma v-mos   TGG CAG ATG ACT ACC AGA GAG GTG CCT TAC TCC GGC GAA CCT CAG TAC GTG CAG TAT GCA GTG GTA GCC
Moloney murine v-mos   TGG CAG ATG ACC ACC AGG GAG GTG CCT TAC TCC GGC GAA CCT CAG TAC GTG CAG TAT GCA GTG GTT GCC
Human c-mos            TGG CAA ATG ACT ACC AGG GAG GCG CCG TAT TCG GGG GAG CGG CAG CAC ATA CTG TAC GCG GTG GTG GCC
Rat c-mos              TGG CAG ATG ACC ACC AGG GAG GTG CCT TAC TCC GGC GAG CCT CAG TAC GTG CAG TAT GCA GTG GTA GCC
```

Row 15

Murine sarcoma v-mos	TAC AAT CTG CGT CCC TCA CTG GCA GGA GCG GTG TTC ACC GCC TCC CTG ACT GGA AAG GCA CTG CAG AAC
Moloney murine v-mos	TAC AAT CTG CGC CCC TCA CTG GCA GGA GCG GTG TTC ACC GCC TCC CTG ACT GGA AAG ACA CTG CAG AAC
Human c-mos	TAC GAC CTG CGC CCG TCC CTC TCC GCT GCC GTC TTC GAG GAC TCG CTC CCC GGG CAG CGC CTT GGG GAC
Rat c-mos	TAC AAT CTG CGC CCT CAC TGG CAG ‑‑ ‑‑ GAG GTG TTC ACC GCC TCC CTG ACT GGG AAG ACG CTG CAG AAC

Row 16

Murine sarcoma v-mos	ATC ATC CAG AGC TGC TGG GAG GCC CGC GGC CTG CAG AGG CCG ACG TGC AGA ACT GCT CCA AAG GGA CCT
Moloney murine v-mos	ATC ATC CAG AGC TGC TGG GAG GCC CGC GGC CTG CAG AGG CCG TGT GCA GAA CTG CTC CAA AGG GAC CTC
Human c-mos	GTC ATC CAG CGC TGC TGG AGA CCC AGC GCG GCG CCG AGC GCG CGG CTG CTT TTG GTG GAT CTC
Rat c-mos	AAT GTC CAG AGC TGC TGG GAG GCC CGC GGC CTG CAG AGG CCG GGT GCA GAA CTG CTC CAG AAG GAC CTG

Row 17

Murine sarcoma v-mos	CAA GGC TGT CCG AGG GAC ACT AGG CTG CCA TCG AGC CAG TGT AGA GAT AAG CTT TTG TTT CTG TTT ATT
Moloney murine v-mos	AAG GCT TTC CGA GGG GCA CTA GGC TGA
Human c-mos	ACC TCT TTG AAA GCT GAA CTC GGC TGA
Rat c-mos	AAG GCT TTC CGA GGG GCA CTG GGC TGA

Row 18

| Murine sarcoma v-mos | TTT TAT GGG ACC CCT TAT TGT ACT CCT AAT GAT TTT GCT CTT GGG ACC CTG CAT TCT TAA |

[a]Codons for basic amino acids are underscored, those for cysteine are lightly stippled, and those for tryptophan are heavily stippled; triplets for tyrosine are italicized.
[b]Reddy et al. (1981); Van Beveren et al. (1981). [c]Rechavi et al. (1982). [d]R. Watson et al. (1982). [e]Van der Hoorn and Firzlaff (1984). [f]Kuff et al. (1983). [g]Insert 16 codons. [h]Insert GGG GAC GAC GCA GGG GAG.

Table 11.9
Miscellaneous Oncogene Sequences[a]

Row 1

Chicken c-myb[b]: TGG AGG CAC ATA GAT AAG AGA ATT ATC ACT CTA CAT TCA TCT TTC TCA AAG AAT AAT CTA CTT GTG TGT

Chicken v-myb[b]: TGG AGG CAC ATA GAT AAG AGA ATT AAT CTA CTT GTG TGT

Murine c-fos[c]: ATG ATG TTC TCG GGT TTC AAC GCC GAG GCG TAC TCA TCC TCC CGC AGT AGC GCC TCC CCG GCC

Murine c-int1[d]: ATG GGG CTC TGG GCG CTG CTG CCC AGC TGG GTT TCT ACT ACG TTG CTA CTG GCA CTG ACC CTG CCC

Avian MH2 v-raf[e]: ATG --- CCA GTA GAC AGC CGG ATA ATT GAG GAT GCA ATT CGA AAC CAT GAA TCA GCT TCA CCC TCC

Yeast YP2[f]: ATG AAT AGC GAG TAC GAT TAC TTC AAA CTG CTG TTC ATC GGG AAT TCC GGT GTC GGG AAG TCC TGT

Chicken v-rel[g]: TCA GAG CCC TAC ATT GAA ATA TTT GAA CAA CCC AGG CAA AGG GGT ACG TTC AGA TAT AAA TGT GAA

Row 2

Drosophila c-myb[h]: GAT CGC CTG GAA CAG CAA GT G CAG CAG CGT TGG GCC AAA GTC CTC AAT CCG GAG

Chicken c-myb: TTT ATA TTT CAT AAT CGG ACA GAT GTT CAG TG C CAG CAC CGG TGG CAG AAA GTA TTA AAC CCA GAA

Murine c-fos: GGG GAC AGC CTT TCC TAC TAC CAT TCC CCA GCC GA C TCC TTC TCC AGC ATG GGC TCT CCT GTC AAC ACA

Murine c-int1: GCA GCC CTG GCT GCC AAC AGT AGT GGC CGA TGG TG*G GGC ATC GTG AAC ATA GCC TCC TCC ACG AAC CTG

Avian MH2 v-raf: GCT TCG TCT GGG AGT CCT AAC AAT ATG AGC CCG AC T GGC TGG TCT CAG CCC AAA ACG CCA GTC CCA GCC

Yeast YP2: TTA CTT TTG AGG TTT TCG GAC GAC ACA TAT ACC AA C GAC TAC ATC TCC ACA ATT GGA GTG GAC TTC AAG

Chicken v-rel: GGA AGA TCA GCT GGT GGT AGC ATT CCA GGA GAA CAC AG T ACT GAC AAC AAC AAG ACA TTC CCA ATA CAG

Row 3

Drosophila c-myb	CTA ATC AAG GGC CCG TGG ACG CGG GAC GAG GAC GAC ATG GTT ATT AAG CTG GTG CGC AAC TTC GGG CCC
Chicken *c-myb*	CTT ATC AAA GGT CCA ACT AAA GAG GAG GAT CAA AGG* GTA ATA GAA CTC GTG CAG AAA TAC GGT CCA
Murine *c-fos*	CAG* GAC TTT TGC GCA GAT CTG TCC GTC TCT AGT GCC AAC TTT ATC CCC ACG GTG ACA GCC ATC TCC ACC
Mouse *c-int1*	TTG ACG GAT TCC AAG AGT CTG CAG CTG GTG CTC GAG CCC AGT CTG CAG CTG AGC CGC AAG CAG CAG CGG
Murine *v-raf*	GAA AAA AAC AAA ATT AGG CCT CGT GGG CAG AGA GAC
Avian MH2 *v-raf*	CAG AGG GAG AGA GCC CCC GGA ACG AAT ACA CAG AAA AAT AAA ATT AGG CCT CGT GGA CAA AGA GAT
Yeast *YP2*	ATT AAG ACT GTA GAA CTG GAC GGC AAG AAG CTA CAG ATT TGG GAC ACT GCA GAA GAA CGT
Chicken *v-rel*	ATC CTA AAC TAT TTT GGA AAA GTC AAA ATA AGA ACT ACA TTG GTA ACA AAG AAC GAA CCC TAC AAG CCA

Row 4

Drosophila c-myb	AAG AAA TGG ACA CTA ATT GCA CGA TAT CTG AAC GGA CGA ATT GGG AAA CAG AAA TGC CGC GAG AGA TGG CAC
Chicken *c-myb*	AAG CGC TGG TCG ATT GCT AAG CAT TTG AAG GGA AGG ATT GGA AAA CAG AAA CAG AGG GAG AGG TGG CAC
Murine *c-fos*	AGC CCA GAC CTG CAG TGG CTG GTG CAG CCC ACT CTG GTC TCC CTG CAC AGC GTG AGT GGG CTC CAG AGC GCT GTG CGA GAG
Murine *c-int1*	CGA CTG ATC CGA CAG CAA CCG CGG GAT ATC CTG ACT CGG ATC GGG TCA GGT TCC TTT
Murine *v-raf*	TCG AGT TAT TAC TGG AAA ATG GAA GCC AGT GTG ATG GAG GTG CTG TCT ACT CGG ATC GGG TCA GGT TCC TTT
Avian MH2 *v-raf*	TCT AGT TAT TAC TGG GAA ATA GAA GCA AGC GAA GTC CTG CTT TCT ACC AGA ATA GGG TCA GGT TCT TTT
Yeast *YP2*	TTC CGT ACT ATC ACT TCA TCT TAC TAC CGT GGT TCG CAT GGG ATC ATT ATC GTG TAT GAT GTC ACT GAC
Chicken *v-rel*	CAC CCT CAC GAT CTA GTT GGA AAA GGC TGC AGA GAT GGC TAC TAT GAA GCA GAT TTT GGG CCC GAA CGG

(*continued*)

Table 11.9 (Continued)

Row 5

```
Drosophila c-myb    AAC CAC CTC AAT CCG AAT ATC AAG ACG GCC TGG ACC GAA AAG GAG GAG GAG ATT ATC TAC TAC CAG GCT
Chicken c-myb       AAC CAT CTG AAT CCA GAA GTG AAG AAG AAA ACC*TCC TGG ACA GAA GAG GAA GAT AGA ATT ATT TAC CAG GCA
Murine  c-fos       CCC CAT CCT TAC GGA CTC CCC ACC CAG TCT GCT GGG GCT TAC GCC AGA GCG GGA ATG GTG AAG ACC GTG
Murine  c-int1      TGC AAA TGG CAA TTC CGA AAC CGC CGC TGG AAC TGC CCC ACT GCT CCG GGG CCC CAC CTC TTC GGC AAG
Murine  v-raf       GGC ACT GTG TAC AAG GGC AAG TGG CAT GGA GAT GTT GCA GTA GTA AAG ATC CTA AAG GTG GTT GAC CCA ACT
Avian MH2 v-raf     GGA ACT GTT TAC AAA GGC AAA TGG CAT GGG GAT GTA GCA GTG AAA ATA TTA AAG GTT GTA GAT CCA ACC
Yeast YP2           CAA GAA TCC TTC AAC GGC GTG AAG GTG CTG CAA GAG ATT GAT CGG TAT GCA ACC TCA ACA GTG TTG
Chicken v-rel       CAA GTC TTG TCT TTT CAG AAT TTG GGA ATT CAA TGT GTG AAG AAA GAC CTG AAA GAA TCA ATT TCT
```

Row 6

```
Drosophila c-myb    CAC TTG GAG CTG G GC AAC CAG TGG GCA AAG ATT GCC AAA CGT CTG CCC GGA CG C ACC GAT AAC GCC ATT
Chicken c-myb       CAC AAG AGA CTG G GA AAC AGA TGG GCA GAA ATT GCA AAG TTG CTG CCT GGA CG*G ACT GAT AAC GCT ATC
Murine  c-fos       TCA GGA GGC AGA G CG CAG AGC ATC GGC AGA AGG GGC AAA GTA GAG AAA AGA TTG CTG CTG GAA GAG GAG
Murine  c-int1      ATC GTC AAC CGA G*GC TGC CGA GAA ACA GCG TTC ATC TTC GCA ATC ACC TCC GC C GGG GTC ACA CAT TCC
Murine  v-raf       CCA GAG CAA CTT C AG GCC TTC AGG AAC GAG GTG GCT GTT TTG CGC AAA ACA CG G CAT ATT AAC ATC CTG
Avian MH2 v-raf     CCA GAA CAG CAG TTT C AG GCT TTC AGA AAC GAA GTG GCT GTA TTA AGG AAG ACC CG G CAT GTT AAT ATT TTG
Yeast YP2           AAG CTA TTG GTA G GT AAC AAG TGT GAT TTA AAG GAC CGT GTC CTG GAA TA T GAC GTG GCA AAG GAA
Chicken v-rel       TTG CGA ATC TCA A AG AAA ATC AAT CCC TTT AAT GTG CCT GAG GAA CAG TTG CA T AAC ATC GAT GAG TAC
```

Row 7

```
Drosophila c-myb   AAA AAC CAT TGG AAT TCA ACA ATG CGT CGC AAA TAT GAC GTC GAA (Incomplete)
Chicken c-myb      AAG AAC CAC TGG AAT TCC ACC ATG CGC CGG AAG GTC GAG GAG GAG TCC AAA
Murine c-fos       AAA CGG AGA ATC CGA AGG GAA CGG AAT AAG ATG GCT GAG CAG CGG AGG GAG CTG
Murine c-int1      GTG GCG CGC TCC GAA GGC TCC ATC GAG TCC ACC TGC GAC TAC CGG CGG CGC GGC CCT GGG
Murine v-raf       CTG TTC ATG GGG TAC ATG ACA AAG GAC AAC CTG GCG ATT GTG ACT CAG TGG TGT GAA GGC AGT CTC
Avian MH2 v-raf    CTC TTC ATG GGC TAC ATG ACT AAA GAT AAC CTG GCC ATT GTC ACA CAG TGG TGT GAA GGC AGC AGT CTG
Yeast YP2          TTT GCG GAC GCG AAT AAG ATG CCG TTC TTA GAG ACT AGT GCT TTG GAC TCC ACC AAC GTC GAG GAT GCG
Chicken v-rel      GAT CTC AAC GTT GTC CGC CTC TGT TTC TTC CAA GCT TTC CCT GAT GAA CAT GGC AAC TAC ACA TTG GCT
```

Row 8

```
Chicken c-myb      GCC GGC CTG CCC TCG GCA ACC ACC GGC TTC CAG AAG AGC AGC CAC CTG ATG GCC TTT GCC CAC AAC CCA
Murine c-fos       ACA GAT ACA CTC CAA GCG* GAG ACA GAT CAA CTT GAA GAT GAG AAG TCT GCG TTG CAG ACT GAG ATT GCC
Murine c-int1      GGC CCC GAC TGG CAC TGG GGG GGC TGC AGT GAC AAC ATC GAT TTT GGT CGC CTC TTT GGC CGA GAG TTC
Murine v-raf       TAC AAA CAC CTG CAT GTC CAC GTT CAG GAG ACC AAA TTC CAG ATG TTC CAG ATC ATT GAC ATT GCC CGA CAG ACA
Avian MH2 v-raf    TAT AAA CAC CTG CAC CTG AGA CAA CAA ATC AAG AAG TAT GAC AAC AAC CTG AAC GAA ACC CGG CAG ACA
Yeast YP2          TTT TTA ACC ATG GCT AGA CAA CAA ATC AAG AAG CAA AGT ATG TCC CAA CAA AAC CTG AAC ACC ACT CAG AAG
Chicken v-rel      CTT CCT' TTG ATT TCC AAC CCA ATC TAT GAC AAC AAC ACG GCA GAA CTG AGG ATT TGT CGT
```

(continued)

Table 11.9 (Continued)

Row 9

```
Chicken c-myb      CCT GCA GGC CTC CCG GGG GCC GGC CAG GCC CCG CTG GGC AGT GAC TAC CCC TAC TAC CAC ATT GCT
Murine c-fos       AAT CTG CTG AAA GAG GAG GAA GAA CTG GAG TTT ATT TTG GCA GCC CAC CGA CCT GCC TGC AAG ATC CCC
Murine c-int1      GTG GAC TCC GGG GAG GAG GGG CGG GAC CTA CGC TTC CTC ATG GAC CTT CAC CTT CAC AAC GAG GCA GGG CGA
Avian MH2 v-raf    GCG CAG GGA ATG GAC TAT TTG CAT GCA AAG AAT ATC ATC CAC AGA GAC ATG AAA TCC AAT AGT ATA TTT
Yeast YP2          AAG GAA GAC AAA GGG AAC GGG AAC CTG AAG AAG GGA CAG AGT TTA ACC AAC ACC GGT GGG GGC TGC TGT TGA
Chicken v-rel      GTG AAT --- AAG AAC TGT GGA AGT GTA AAG GGA GGA GAT GAA ATT TTT CTT CTG TGT GAC AAA GTT CAA
```

Row 10

```
Chicken c-myb      GAG CCA CAA AAT* GTC CCT GGT CAG ATC CCA TAT CCA GTA GCA CTG CAT GTA AAT ATT GTC AAT GTT CCT
Murine c-fos       GAT GAC CTT GGC TTC CCA GAG GAG ATG TCT CTG GCC TCC CTG GAT TTG ACT GGA GGT CTG CCT GAG GCT
Murine c-int1      ACG*ACC GTG TTC TCT GAG ATG CGC CAA GAG TGC AAA ATA GGA GAC TTT GGT CTA GCA ACT GTA AAA TCC ACG GTG CGA
Avian MH2 v-raf    CTT CAT GAA GGC CTC ACA GTG AAA ATA GGA GAC TTT GGT CTA GCA ACT GTA AAA TCC AGG TGG AGT GGA
Chicken v-rel      AAA GAT GAC ATA GAG GTC AGA TTT GTC TTG GGC AAC TGG GAG GCA AAG GGC TCC TTC TCC CAA GCT GAT
```

Row 11

```
Chicken c-myb      CAG CCA GCT GCT GCA GCT ATT CAG* AGA CAC TAT AAT GAT GAA GAC CCT GAG AAA GAA GAA CGA ATA AAG
Murine c-fos       TCC ACC CCA GAG TCT GAG GAG GCC TTC ACC CTG CCC CTT CTC AAC GAC CCT GAG CCC AAG CCA TCC TTG
Murine c-int1      ACG TGT TGG ATG CGG CTG CCC ACG CTG CGC CGC GCT GTG GAC GTG CTG CGC CGC TTC GAC GGC GCC
Avian MH2 v-raf    TCG CAG CAG GTG GAG CAA CCC ACT GGT TCC ATT TTG TGG ATG GCA CCA GAA GTG ATA CGG ATG CAA GAC
Chicken v-rel      GTT CAT CGC CAG GTC GCA ATT TTT AGA ACA CCG TTC CTC CTC AGA GAA GAC ATC ACA GAA CCC ATC ACG
```

Row 12

Chicken c-myb	GAA TTA GAG TTG CTA CTT ATG TCG ACT GAG AAT GAA CTG <u>AAA</u> GGG CAG CAG GCA TTA CCA* ACA CAG AAC
Murine c-fos	GAG CCA GTC <u>AAG</u> AGC ATC AGC AAC GTG GAG CTG <u>AAG</u> GCA GAA CCC TTT CAT GAC TTC TTG CCG GCA
Murine c-int1	TCC <u>CGC</u> GTC CTT *TAC* GGC AAC <u>CGA</u> GGC AGC AAC <u>CGC</u> GCC TCG <u>CGG</u> GCG GAG CTG CTG <u>CGC</u> CTG GAG CCC
Avian MH2 v-raf	AGC AAT CCG TTC AGT TTT CAG TCA GAT GTC *TAC* TCC *TAT* GGA ATA GTA TTG *TAT* GAG CTA ATG ACA GGA
Chicken v-rel	GTG <u>AAG</u> ATG CAG TTA <u>CGA</u> <u>AGG</u> CCT TCA GAC GTC AGT GAA CCA GTC GAT TTC <u>AGA</u> *TAT* TTA CCA

Row 13

Chicken c-myb	CAC ACA GCA AAC *TAC* CCC GGC ▨TGG▨ CAC AGC ACC ACG GTT GCT GAC AAT ACC <u>AGG</u> ACC AGT GGT GAC AAT
Murine c-fos	TCA TCT <u>AGG</u> CCC AGT GGC TCA GAG ACC TCC <u>CGC</u> TCT GTG CCA GAT GTG GAC CTG TCC GGT TCC TTC *TAT*
Murine c-int1	GAA GAC CCC GCG CAC <u>AAG</u> CCT CCC TCC CCT CAC GAC CTC GTC *TAC* TTC GAG <u>AAA</u> TCG CCC AAC TTC ▨TGC▨
Avian MH2 v-raf	GAG CTG CCA *TAC* TCC CAC ATA AAC AAC <u>CGC</u> GAC CAG CAG ATT ATT TTC ATG GTT GGT <u>CGA</u> GGA *TAT* GCT TCT
Chicken v-rel	GAT GAA GAG GAT CCG TCT GGC AAC <u>AAA</u> GCA <u>AAA</u> <u>AGG</u> CAA <u>AGA</u> TCA ACA CTG GCT ▨TGG▨ CAA <u>AAA</u> CCC ATA

Row 14

Chicken c-myb	GCA CCT GTT TCC ▨TGT▨ TTG GGG GAA CAT CAC CAC ACT CCA TCT CCA CCA GTG GAT CAT GGT ▨TGC▨ TTA
Murine c-fos	GCA GCA GAC ▨TGG▨ GAG CCT CTG CAC AGC AAT TCC TTG GGG ATG GGG CCC ATG GTC ACA GAG CTG GAG CCC
Murine c-int1	ACG *TAC* AGT GGC <u>CGC</u> CGC CTG GGC ACA GCT GGC ACA GCT <u>CGA</u> GCT ▨TGC▨ AAC AGC TCG TCT CCC GCG CTG
Avian MH2 v-raf	CCA GAC CTC AGC <u>AAG</u> TTG *TAC* <u>AAG</u> AAC ▨TGC▨ CCC <u>AAA</u> GCA ATG <u>AAG</u> <u>AGG</u> CTC GTA GCA GAT ▨TGT▨ TTG <u>AAG</u>
Chicken v-rel	CAG GAC ▨TGC▨ GGA TCA GCT GTG ACA GAG <u>AGG</u> <u>AGG</u> CCA CAA <u>AAA</u> <u>AGG</u> GCG GCT CCT ATC CCC ACT GTC AAC CCT GAA GGA

(continued)

Table 11.9 (Continued)

Row 15

Chicken c-myb	CCT GAG GAA AGT GCG TCC CCC GCA CGG TGC ATG ATT GTT CAC CAG AGC AAC ATC CTG GAT AAT GTT AAG
Murine c-fos	CTG TGT ACT CCC GTG GTC ACC TGT ACT ACT CCG GGC TGC ACT ACT TAC ACG TCT TCC TTT GTC TTC ACC TAC
Murine c-int1	GAC GGC TGT GAG CTG CTG TGC TGT GGC GGC CAC CGC ACG CGC GTC ACG GAG CGC TGC
Avian MH2 v-raf	AAA GCT AGG GAA GAA AGA CCC TTG TTT CCG CAA ATA CTG TCT TCC ATT GAA TTG CTG CAA CAT TCT TTA
Chicken v-rel	AAG CTG AAG AAA GAA CCA AAT ATG TTT TCA CCT ACG CTG ATG CTG CCT GGG CTA GGA ACA CTG AGC TCC

Row 16

Chicken c-myb	AAT CTC TTA GAA TTT GCA GAA ACA CTC CAG TTA ATA GAC TCC*TTC TTA AAC ACA TCG TCC AAT CAC GAG
Murine c-fos	CCT GAA GCT GAC TCC TTC CCA AGC TGT GCC GCT GCC CAC CGA AAG GGC AGC AGC AAC GAG CCT TCC
Murine c-int1	AAC TGC ACC TTC CAC TGG TGC TGC CAC GTC AGC TGC CGC AAC TGC ACG CAC ACG CGC GTT CTG CAC GAG
Avian MH2 v-raf	CCC AAA ATC AAC CGG AGT GCT TCC GAA CCA TCT CTG CAC CGC GCA TCC CAT ACA GAG GAC ATA AAT TCT
Chicken v-rel	AGT CAG ATG TAC CCT GCA TGC AGC CAG ATG CCC ACC ACC CCT GCG GCG CTT GGC CCT GGG AAG CAG GAC

Row 17

Chicken c-myb	AAT CTG AAC CTG GAC AAC CCT GCA CTA ACC TCC ACG CCA GTG TGT GGC CAC AAG ATG TCT GTT ACC ACC
Murine c-fos	TCC GAC TCC CTG AGC TCA CCC ACG CTG CTG GGC CTG TGA
Murine c-int1	TGT CTA TGA
Murine v-raf	--- --- --- AC- TAC -AT CC- CCA AGG CTA CCA GTC TTC TAG
Avian MH2 v-raf	TGC ACG TTA ACA TCC -AC --- --A AGA CTG CCT GTT TTT TAG
Chicken v-rel	CTC CAT TCC TGC TGG CAG CAG CTG TAC AGC CCC CCT TCA GCC AGC AGC CTC AGC TTG CAC

Row 18

Chicken *c-myb*   CCA TTC CAC AGG GAC CAG GCT TTC[*] --- (Incomplete)

Chicken *v-rel*   CAC AGC AGC TTC ACA GCG GAA GTG CCT CAG CCT CAG GGC AGT AGC TCT CTC CCG GCC *TAT*

Row 19

Chicken *v-rel*   CCA CTG AAC TGG CCT GAT GAG AAG AAT TCC AGT TTT *TAC* AGG AAT TTT GGC AAC ACA CAT GGG ATG

Row 20

Chicken *v-rel*   GCA GCG TTG GTG TCA GCT GCA GGC ATG CAG AGT GTT TCC AGT AGC AGC ATC GTC CAG GGC ACT CAT

Row 21

Chicken *v-rel*   GCC AGT GCC ACT ACT GCA AGC ATC ATG ACC (Beginning of 3' end of *env*)

[a]Codons for basic amino acids are underscored, those for tyrosine are italicized, those for tryptophan (TGG) and cysteine are lightly stippled. Premature stop codons are darkly stippled. Asterisks indicate the locations of introns.
[b]Klempnauer et al. (1982).
[c]Van Beveren et al. (1983).
[d]Van Ooyen and Nusse (1984).
[e]Galibert et al. (1984); Sutrave et al. (1984).
[f]Gallwitz et al. (1983).
[g]Stephens et al. (1983).
[h]Katzen et al. (1985).
[i]Insert CCT.

codons (heavy stippling) exceeds the expected, indicating a frequency of 2.6%, compared to the normal 1.5%. Another distinctive feature lies in the nature of the 5' termini, where over 55% of the triplets encode hydrophobic amino acids. Moreover, a similarly high hydrophobicity is found in row 13, where 14 of the 23 are for that type of amino acid, both this and the preceding level bespeaking a condition to be expected in a membrane-embedded protein.

Miscellaneous Genes and Their Products. Consideration of the miscellany of genes that possibly are GTP-associated remaining for attention involves a matter of homology between that described from mouse cells, *raf* (Rapp *et al.*, 1983), and a more recently recognized one from the chicken, called *mil* (Sutrave *et al.*, 1984). Why a distinctive name is applied to the latter is difficult to comprehend, for their nucleotide sequences are as largely homologous as any other pair derived from mammalian and avian sources (Table 11.9). Although the species from chicken bears a 57-codon-long appendage at the 5' end, a similar condition has just been observed with the *mos* family of cistrons, in which the murine forms have a 5' appendage of considerable length (Table 11.8). Thus, *mil* is treated here as a synonym of *raf*, the latter having priority. In the mouse and human, this gene is present in one or two copies per haploid genome and lacks tyrosine-specific protein kinase activity (Rapp *et al.*, 1983).

In Table 11.9, the nucleotide sequence of *v-raf* from the chicken is given in full, along with the 5' and 3' portions of that of the mouse, since the midregions show few distinctions. In the downstream sector, the murine representative shows a deletion of one nucleotide residue, followed by an insertion of four Cs, which restores the reading frame (row 17). In keeping with its lack of tyrosine kinase activity, the cistron contains only ten codons (italicized) for that amino acid, for a 3.2%, equivalent to the expected 3.1%. Nor is it enriched in tryptophan or cysteine triplets (both stippled), since it contains only four of each. Although the 5' portion is unusually well supplied with codons for basic amino acids (underscored), the entire structure is not alkaline, containing a ratio under 10%. With the abundance it has of coding elements for hydrophobic amino acids, it appears as a whole to be constructed for encoding a typical membrane protein.

Another oncogene from yeast that has been recognized recently, called *YP2* (Gallwitz *et al.*, 1983), similarly lacks outstanding chemical properties (Table 11.9). The single copy has been mapped to chromosome VI between genes for tubulin and actin, having the same polarity as the latter. Despite the claims that the structure resembles those of *ras* and *src*, no such correspondences can be noted. It, too, appears to encode a membrane protein. Still another apparent member of class II is *bcl2*, described (but not sequenced) from human follicular lymphomas (Tsujimoto *et al.*, 1985); it was found flanking the breakpoint on chromosome 18 from a patient with acute B-lymphocytic leukemia. By use of a hybridized probe, the oncogene was detected in the cells of several rodents, so it probably occurs throughout the Mammalia.

11.4.3. The Sis Gene and Its Product

The first sarcoma-producing virus known from nonhuman primates originated from a naturally occurring virus of a pet woolly monkey (Devare *et al.*, 1982). This *sis* (simian sarcoma) gene, as it is known, is unique in another way, in that the role of the product of its normal cellular counterpart has been clearly established (Robbins *et al.*, 1982; Doolittle *et al.*, 1983; Waterfield *et al.*, 1983; Josephs *et al.*, 1984). By comparison of amino acid sequences of the *c-sis* product and platelet-derived growth factor, the former has been shown to be homologous to the larger subunit (A) of the latter (Chiu *et al.*, 1984; Betsholtz *et al.*, 1985; King *et al.*, 1985). This enzyme holds potent powers for stimulating mitosis in fibroblasts and glial and smooth muscle cells.

The gene, located in man on chromosome 22 (Favera *et al.*, 1982), probably occurs in all higher vertebrates, but so far has been detected only in the two primates. Moreover, its sequence has been established from but two sources, the cellular one of human beings and the viral from the woolly monkey, each of which has been sequenced more than once. The complete structures of both types are included in Table 11.10, along with the 5' and 3' sectors of a second one from man (*c-sisA*), the remaining portions being identical (Devare *et al.*, 1982; Chiu *et al.*, 1984; Josephs *et al.*, 1984; Ratner *et al.*, 1985). Comparisons reveal that the three differ markedly at the extreme 5' terminus, homology not commencing until the -GG of the fifth codon.

In the complete sequences, differences are quite frequent, but the majority of these are in the third nucleotide of the codons, so there is no change in the encoded amino acid. Nevertheless, some are quite marked, the change from the AAA (lysine) in row 2 of the viral sequence to the aspartic acid-specifying GAG being particularly striking. In rows 5 and 6 is found a long series of differences caused by two single point deletions that induce frameshifts. The first of these is at the extreme 5' end of row 5 and the second in codon 10 of the same row, the resulting codons being marked by arcs and primes. Finally, near the very 3' end of row 6, a third point deletion restores the original reading frame. Throughout this region then, the two nucleotide sequences are fully homologous, whereas the encoded products are completely different. The enzyme is one of the more alkaline oncogene products, containing 36 codons for lysine and arginine, for a 16% basic level. Since it has only a single triplet each for tyrosine and tryptophan (with an additional one in the simian virus) and only the random-chance number of cysteines, the protein joins others of the group in lacking distinctive features.

Included with the above in Table 11.10 is another seeming member of this large class, a gene from a virus of chickens known as *v-amv*, claimed to be the transforming cistron of avian myeloblastosis (Rushlow *et al.*, 1982). It shows no relationship to any of the *sis* family, from which it differs in its considerably greater length. Moreover, its few (seven) codons for tyrosine (italicized) are unusual in their placement, most of them being located in the 5' portion, with three clustered in row 3.

11.4.4. Myc, An Important Oncogene of Vertebrates

Although literature on the important oncogene *myc* of vertebrates is quite extensive and the protein (molecular weight 48,000) has been identified (Giallongo *et al.*, 1983), the location and function of the latter in the normal cell still remain to be established. Since the properties and structure of the gene seem to be harmonious with others of class II, it, like so many included in this group, is placed here on a tentative basis. The present oncogene is unique in that investigations are relatively evenly divided between avian and mammalian sources, with those of the latter being nearly equally devoted to murine and human malignancies. In the chicken, transformations by *myc* result variously in myelocytomatosis, bursal (B-cell) lymphomas, and carcinomas (Pachl *et al.*, 1983; Reddy *et*

Table 11.10
Comparison of Sis Gene Sequences[a]

Row 1

```
Human c-sisA[b]    --- --- --- --- --- --- GGG GAC CCC ATT CCC GAG GAG CTT TAT GAG ATG CTG AGT GAC CAC TCG ATC
Human c-sis[c]     TTG GTG TCT GCC CGG CAG GGG GAC CCC ATT CCC GAG GAG CTT TAT GAG ATG CTG AGT GAC CAC TCG ATC
Simian v-sis[d]    ATG ACC CTC ACC TGG CAG GGG GAC CCC ATT CCT GAG GAG CTC TAT AAG ATG CTG AGT GAC CAC TCG ATT
Avian v-amv[e]     ATG CGC CGG AAG GTC GAG CAG GAG GGT TAC CCG CAG TCC TCC AAA GCC GGC CCG CCC TCG CCA ACC
```

Row 2

```
Human c-sis    CGC TCC TTT GAT GAT CTC CAA CGC CTG CTC CAC GGA GAC CCC GGA G*AG GAA GAT GGG GCC GAG TTG GAC
Simian v-sis   CGC TCC TTC GAT GAC CTC CAG CGC CTG CTC CTC CAG GGA GAC TCC GGA A AA GAA GAT GGG GCT GAG CTG GAC
Avian v-amv    ACC GGC TTC CAG AAG AGC CAT CTG ATG GCC TTT GCC CAC AAC C CA CCT GCA GGC CCG CTC CCG GGG
```

Row 3

```
Human c-sis    CTG AAC ATG ACC CGC TCC CAC TCT GGA GGC GAG CTG GAG AGC TTG GCT CGT GGA AGA AGG AGC CTG G*GT
Simian v-sis   CTG AAC ATG ACC CGC TCC CAT TCT GGT GGC GAG CTG GAG AGC TTG GCT CGT GGG AAA AGG AGC CTG G GT
Avian v-amv    GCC GGC CAG GCC CCT CTG GGC AGT GAC TAC CCC TAC TAC CAC ATT GCT GAG CCA CAA AAT GTC CCT G GT
```

Row 4

```
Human c-sis    TCC CTG ACC ATT GCT GAG CCG GCC ATG ATC GCC GAG TGC AAG ACG CGC ACC GAG GTG TTC GAG ATC TCC
Simian v-sis   TCC CTG AGC GTT GCC GAG CCA GCC ATG ATT GCC GAG TGC AAG ACA CGA ACC GAG GTG TTC GAG ATC TCC
Avian v-amv    CAG ATC CCA TAT CCA GTA GCA CTG CAT ATA AAT ATT ATC AAT GTT CCT CAG CCA GCT GCA GCT ATT
```

Row 5

Human *c-sis*	CGG	CGC	CTC	ATA	GAC	CGC	ACC	AAC	GCC	AAC	TTC	CTA	GTG	TGG	CCG	CCC	TGT	GTG	GAG	GTG	CAG	CGC	TGC
Simian *v-sis*	-GG	CGC	CTC	ATC	GAC	CGC	AAT	GCC	AAC	TTC	T-G	GTG	TGG	CCG	CCC	TGC	CCC	GTC	GAG	GTG	CAG	CGC	TGC
Avian *v-onc*	CAG	AGA	CAC	*TAT*	ACT	GAT	GAA	GAC	CCT	GAG	AAA	GAA	AAA	CGA	ATA	AAG	GAA	TTA	GAG	TTG	CTA	CTT	ATG

Row 6

Human *c-sis*	TCC	GGC	TGC	TGC	AAC	AAC	CGC	AAC	GTG	CAG	TGC	CGC	CCC	ACC	CAA	GTG	CAG	CTG	CGA	CCT	GTT	CAA*	GTG
Simian *v-sis*	TCC	CGC	TGT	TGC	AAC	AAC	CGC	CAG	TGC	CGC	CCC	ACC	CAG	GTG	CGG	CCA	CGG	CCA	GT–	CCA	GTG
Avian *v-onc*	TCG	ACT	GAG	AAT	GAA	CTG	AAA	GGG	CAG	CAG	GCA	TTA	CCA	ACA	ACA	CAG	AAC	CAC	ACA	AAC	*TAC*	CCC	GGC

Row 7

Human *c-sis*	AGA	AAG	ATC	GAG	ATT	GTG	CGG	AAG	AAG	CCA	ATC	TTT	AAG	AAG	ACG	GTG	ACG	CTG	GAA	GAC	CAC	CTG
Simian *v-sis*	AGA	AAG	ATC	GAG	ATT	GTG	CGG	AAG	AAG	CCA	ATC	TTT	AAG	AAG	ACG	GTG	ACG	CTG	GAG	GAC	CAC	CTG
Avian *v-onc*	TGG	CAC	AGC	ACC	ACG	GTT	GCT	GCT	GAC	AAT	ACC	AGG	ACC	AGT	GGT	GAC	AAT	GCG	CCT	GTT	TCC	TGT	TTG	GGG

Row 8

Human *c-sis*	GCA	TGC	AAG	TGT	GAG	ACA	GTG	GCA	GCT	GCA	CGG	CCT	GTG	ACC	CGA	AGC	CCG	GGG	GGT	TCC	CAG	GAG	CAG
Simian *v-sis*	GCA	TGC	AAG	TGT	GAG	ATA	GTG	GCA	GCT	GCA	CGG	CCT	GTG	ACC	CGA	AGC	CCG	GGG	ACT	TCC	CAG	GAG	CAG
Avian *v-onc*	GAA	CAT	CAC	CAC	TGT	ACT	CCA	TCT	CCA	CCA	GTG	GAT	CAT	GGT	TGC	TTA	CCT	GAG	GAA	AGT	GCG	TCC	CCC

Row 9

Human *c-sis*	CGA	G*CC	AAA	ACG	CCC	CAA	ACT	CGG	ACC	GTG	ACG	GTG	CGA	GTC	CGC	CGG	CCC	CCC	AAG	GGC	AAG
Simian *v-sis*	CGA	G CC	AAA	ACG	ACC	CAA	AGC	CGG	ACC	GTG	ACG	GTG	CGA	GTC	CGC	CGG	CCC	CCC	AAG	GGC	AAG
Avian *v-onc*	CCA	C GG	TGC	ATG	ATT	GTT	CAC	CAC	AGC	AAC	ATC	CTG	GAT	AAT	GTT	AAG	GTT	CTC	TTA	GAA	TTT	GCA	GAA

(continued)

Table 11.10 (Continued)

Row 10

Human *c-sisA*   CAC CGG AAA TTC AAG CAC AAG ACG CAT GAC AAG ACG GCA CTG AAG GAG ACC CTT GGA GCC TAG

Human *c-sis*   CAC CGG AAA TTC AAG CAC AAG ACG CAT GAC AAG ACG GCA CTG AAG GAG ACC CTT GGA GCC TAG

Simian *v-sis*   CAC CGG AAA TGC AAG CAC AAG ACG CAT GAC AAG ACG GCA CTG AAG GAG ACC CTC GGA GCC TAA

Avian *v-amv*   ACA CTC CAG TTA ATA GAC TCC TTC TTA AAC ACA TCG TCC AAT CAC GAG AAT CTG AAC CTG GAC AAC

Row 11

Avian *v-amv*   CCT GCA CTA ACC TCC ACG CCA GTG TGT GGC CAC AAG ATG TCT GTT ACC ACC CCA TTC CAC AAG GAC CAG

Row 12

Avian *v-amv*   ACT TTC ACT GAA *TAC* AGG AAG ATG CAC GGC GGA GCA GTC TAG

[a] Codons for basic amino acids are underscored, those for tyrosine are italicized, and those for cysteine and tryptophan (TGG) are stippled. Where single-base deletions have occurred, the actual codons are united by arcs with their termini demarcated by primes.
[b] Josephs *et al.* (1984); Ratner *et al.* (1985) (only the first and last rows are included here).
[c] Chiu *et al.* (1984).
[d] Devare *et al.* (1982).
[e] Rushlow *et al.* (1982).

al., 1983a; D. K. Watson *et al.*, 1983; Hayflick *et al.*, 1985), whereas in the mouse or rat, they induce T-cell lymphomas, thymomas, and plasmacytomas (Shen-Ong *et al.*, 1982; Corcoran *et al.*, 1984; Keath *et al.*, 1984; Lemay and Jolicoeur, 1984). In the cat, the gene has been implicated in T-cell leukemias (Neil *et al.*, 1984), and in man, especially in Burkitt lymphomas and neuroblastomas (Brodeur *et al.*, 1984; Taub *et al.*, 1984; Dyson and Rabbitts, 1985).

Activation of the Myc Gene. Probably more has been uncovered regarding the activation of *myc* than that of any other oncogene. One of the most frequent methods of inducing transforming activity appears to be that of translocation. In the B-cell Burkitt lymphoma, such translocations involve movement of the *c-myc* gene from chromosome 8 to near one of the immunoglobulin loci on chromosome 2, 14, or 22 (Taub *et al.*, 1984). However, such shifts in locations are not accompanied by any significant increase in quantity of the gene transcript, the deregulation of activity being supposed to result from possible changes (somatic mutation) of the promoter region. Although no such alterations could be detected in the human malignancy (Dyson and Rabbitts, 1985), in rat thymomas induced by this cistron, novel fragments were detected from near the proviruses inserted adjacent to the *myc* sequence (Lemay and Jolicoeur, 1984). Moreover, recently it was demonstrated that translocations clustered upstream of that locus strongly influenced promoter usage (Yang *et al.*, 1985). To the contrary, in an embryonal carcinoma stem-cell line, it has been clearly demonstrated that transcription of *myc* is uniform over a period of time, with control of the product being exercised posttranscriptionally by way of an RNase (Dony *et al.*, 1985).

Alternatively, amplification of this genetic element may be associated with neoplasm production in certain types of malignancies, as suggested in a study on human neuroblastomas (Brodeur *et al.*, 1984). In this case, the *N-myc* DNA involved in transformation proved to be amplified 20- to 140-fold in ~40% of the patients, while the remainder showed no indications of such an increase in gene number.

Characteristics of the Gene Structure. The one partial and five complete sequences in Table 11.11 show a high level of homology in spite of the difference in sources, two being from mammalian tissues, the others from chickens, either cells or viruses. Nevertheless, some distinctions can be noted to correlate with the source, but these mostly have a minor impact on the encoded product, since they involve the third base of the codons. In addition, some stronger changes can be observed, particularly in the form of inserted regions. Two short ones in rows 1–3 have been acquired by the avian types, whereas in rows 4, 5, 7, 8, and 11, the mammalian representatives have gained codons. Hence, the latter exceed the avian in total length by ~37 triplets. With three or four in row 1 and one or two in row 2, the sequences are greatly enriched with codons for tyrosine (italicized) at the 5'-terminal region, but beyond that portion, those coding elements are relatively scarce. Another peculiarity is the presence in row 2 of series of four or five codons specifying acidic amino acids (heavy stippling), being uniformly GAG in the three from the fowl. An even greater, slightly interrupted series is an outstanding feature of row 12.

Unfortunately, the human *N-myc* sequence of neuroblastomas is incomplete, being represented only in the 3' region, so that those hallmarks of the rest may or may not prove to be present. In what has been established (beginning in row 12), alliance is most frequently with the avian sequences, rather than with the mammalian as might be ex-

Table 11.11

Comparison of Myc Gene Coding Regions[a]

Row 1

Murine *c-myc*[b]	ATG	CCC	CTC	AAC	GTG	AAC	TTC	ACC	AAC	AGG	AAC	*TAT*	GAC	CTC	GAC	TCC	GTA	CAG	CCC	*TAT*	TTC
Human *c-myc*[c]	ATG	CCC	CTC	AAC	GTT	AGC	TTC	ACC	AAC	AGG	AAC	*TAT*	GAC	CTC	GAC	TCG	CTG	CAG	CCC	*TAT*	TTC
Chicken *c-myc*[d]	ATG	CCG	CTC	AGC	GCC	AGC	CTC	CCC	AGC	AAG	AAC	*TAC*	GAT	*TAC*	GAC	TCG	GTG	CAG	CCC	*TAC*	TTC
MC29 *v-myc*[e]	ATG	CCG	CTC	AGC	GCC	AGC	CTC	CCC	AGC	AAG	AAC	*TAC*	GAT	*TAC*	GAC	TCG	GTG	CAG	CCC	*TAC*	TTC
MH2 *v-myc*[f]	ATG	CCG	CTC	AGC	GTC	AGC	CTC	CCC	AGC	AAG	AAC	*TAC*	GAT	*TAC*	GGC	TCG	GTG	CAG	CCC	*TAC*	TTC

Row 2

Murine *c-myc*	ATC	TGC	GAC	GAG	GAA	---	GAG	AAT	TTC	*TAT*	CAC	CAG	CAA	CAG	CAG	AGC	GAG	CTG	CAG	CCC	GCG	CCC
Human *c-myc*	*TAC*	TGC	GAC	GAG	GAG	---	GAG	AAC	TTC	*TAC*	CAG	CAG	CAG	CAG	AGC	GAG	CTG	CAG	CCC	CCC	GCG	CCC
Chicken *c-myc*	*TAC*	TTC	GAG	GAG	GAG	GAG	AAC	TTC	*TAC*	CTG	GCG	GCG	GCG	CAG	CAG	AGC	GGG	GGC	CTG	CAG	CCT	CCC
MC29 *v-myc*	*TAC*	TTC	GAG	GAG	GAG	GAG	AAC	TTC	*TAC*	CTG	GCG	GCG	GCG	CAG	CAG	AGC	GGG	GGC	CTG	CAG	CCT	CCC
MH2 *v-myc*	*TAC*	TTC	GAG	GAG	GAG	GAG	AAC	TTC	*TAC*	TCG	GCG	GCG	GCG	CAG	CAG	AGC	GGG	AGC	CTG	CAG	CCT	CCA

Row 3

Murine *c-myc*	AGT	---	---	GAG	GAT	ATC	TGG	AAG	AAA	TTC		CTG	CTT	CCC	ACC	CCG	CCC	CTG	TCC	CCG	AGC	CGC	CGC
Human *c-myc*	AGC	---	---	GAG	GAT	ATC	TGG	AAG	AAA	TTC		CTG	CCC	ACC	CCG	CCC	CTG	TCC	CCT	AGC	CGC	CGC
Chicken *c-myc*	GCC	CCG	TCC	GAG	GAC	ATC	TGG	AAG	AAG	TTT	GAG	CTC	CTG	CCC	ACG	CCG	CCC	CTC	TCG	CCC	AGC	CGC	CGC
MC29 *v-myc*	GCC	CCG	TCC	GAG	GAC	ATC	TGG	AAG	AAG	TTT	GAG	CTC	CTG	CCC	ATG	CCC	CTC	TCG	CCC	AGC	CGC	CGC
MH2 *v-myc*	GCC	CCG	TCC	GAG	GAC	ATC	TGG	AAG	AAG	TTT	GAG	CTC	CTG	CCC	CCC	CCG	CTC	TCG	CCC	AGC	TGC	CGC

```
Row 4

Murine  c-myc   TCC GGG CTC TGC TCT CCA TAT GTT GCG GTC GCT ACG TCC TTC TCC CCA AGG GAA GAC GAT GAC GGC
Human   c-myc   TCC GGG CTC TGC TGC CCC TCC TAC GTT GCG GTC --- ACA CCC TTC TCC CTT CGG GGA GAC AAC GAC GGC
Chicken c-myc   TCC AGC CTG --- --- --- GCC GCC TCC --- TGC TTC --- CCT --- --- --- --- --- --- --- ---
MC29    v-myc   TCC AGC CTG --- --- --- GCC GCC TCC --- TGC TTC --- CCT --- --- --- --- --- --- --- ---
MH2     v-myc   TCC AAC CTG --- --- --- GCC GCC TCC --- TGC TTC --- CCT --- --- --- --- --- --- --- ---

Row 5

Murine  c-myc   GGC GGT GGC AAC TTC ACC GCC GAT CAG CTG ATG ATG ACC GAG TTA CTT GGA GGA GAC ATG GTG
Human   c-myc   GGT GGC GGG AGC TTC TCC ACG GCC GAC CAG CTG ATG GTG ACC GAG CTG CTG GGA GGA GAC ATG GTG
Chicken c-myc   --- --- --- --- --- --- TCC ACC GCC GAC CAG CTG ATG GTG ACG GAG CTG CTC GGG GGG GAC ATG GTC
MC29    v-myc   --- --- --- --- --- --- TCC ACC GCC GAC CAG CTG ATG GTG ACG GAG CTG CTC GGG GGG GAC ATG GTC
MH2     v-myc   --- --- --- --- --- --- TCC ACC GCC GAC CAG CTG ATG GTG ACG GAG CTG CTC GGG GGG GAC ATG GTC

Row 6

Murine  c-myc   AAC CAG AGC TTC ATC TGC GAT CCT GAC GAG ACC TTC ATC AAG AAC ATC ATC ATC CAG GAC TGT ATG
Human   c-myc   AAC CAG AGT TTC ATC TGC GAC CCG GAC GAG ACC TTC ATC AAA AAC ATC ATC ATC CAG GAC TGT ATG
Chicken c-myc   AAC CAG AGC TTC ATC TGC GAC CCG GAC GAA TCC TTC GTC AAA TCC ATC ATC ATC CAG GAC TGC ATG
MC29    v-myc   AAC CAG AGC TTC ATC TGC GAC CCG GAC GAA TCC TTC GTC AAA TCC ATC ATC ATC CAG GAC TGC ATG
MH2     v-myc   AAC CAG AGC TCC ATC TGC GAC CCG GAC GAA TCC TTC GTC AAA TCC ATC ATC ATC CGG GAC TGC ATG
```

(continued)

Table 11.11 (Continued)

Row 7

Murine c-myc	TGG AGC GGT TCA GCC GCT GCC AAG CTG GTC TCG GAG AAG CTG GCC TCC TAC CAG GCT GCG CGC AAA
Human c-myc	TGG AGC GGC TTC TCG GCC GCC AAG CTG GTC TCA GAG AAG CTG GCC TCC TAC CAG GCT GCG CGC AAA
Chicken c-myc	TGG AGC GGC TTC TCC GCC GCC AAG CTG GTG GTG TCG GAG AAG GTG GTG TCG --- --- --- --- --- ---
MC29 v-myc	TGG AGC GGC TTC TCC GCC GCC AAG CTG GTG GTG TCG GAG AAG GTG GTG TCG --- --- --- --- --- ---
MH2 v-myc	TGG AGC GGC TTC TCC GCC GCG AAG CTG GTG GTG TCG GAG AAG GTG GTG TCG --- --- --- --- --- ---

Row 8

Murine c-myc	GAC AGC ACC AGC CCC CGC GGG CAC AGC GTC TGC TCC ACC TCC AGC CTG TAC CTG CAG GAG
Human c-myc	GAC AGC GGC AGC CCG AAC CCC GGC GCC AGC GTC TGC TCC ACC TCC AGC CTG TTG CTG CAG GAT
Chicken c-myc	--- --- GAG AAG CTC GCC ACC TAC CAA GCC TCC GAG --- --- --- --- --- --- --- GGG GGC
MC29 v-myc	--- --- GAG AAG CTC GCC ACC TAC CAA GCC TCC GAG --- --- --- --- --- --- --- GGG GGC
MH2 v-myc	--- --- GAG AAG CTC GCC ACC TAC AAA GCC TCC GAG --- --- --- --- --- --- --- GGG GGC

Row 9

Murine c-myc	CTC ACC GCC GCG GCG TCC GAG TGC ATT GAC CCC --- TCA GTG GTC TTT CCC TAC CCG CTC AAC GAC
Human c-myc	CTG AGC GCC GCC GCC TCA GAG TGC ATC GAC CCC --- TCG GTG GTC TTC CCC TAC CTC AAC GAC
Chicken c-myc	CCC --- GCC GCC GCC TCC CGA GGC CCC GGC^g GGC CCC GCC GGC CTC CAC GAC CTG
MC29 v-myc	CCC --- GCC GCC GCC TCC CGA CCC GGC^g GGC CCC GCC GGC CTC CAC GAC CTG
MH2 v-myc	CCC --- GCC GCC GCC TCC CGA CCC GGC^g GGC CCC GCC GGC CTC CAC GAC CTG

Row 10

Murine *c-myc*	AGC AGC TCG CCC AAA TCC TGT ACC TCG TCC GAT TCC ACG GCC TTC CCT TCC TCG GAC TCG CTG CTG
Human *c-myc*	AGC AGC TCG CCC AAG TCC TGC GCC TCG TCG CAA GAC TCC AGC GCC TTC TCT CCG TCC TCG GAT TCT CTG CTC
Chicken *c-myc*	CGA GCC GCG GCC GCC GCG TGC ATC GAC CCC TCG GTC GTC TTC CCC TAC CCG CTC AGC GAG CGC GCC CCG
MC29 *v-myc*	GGA GCC GCG GCC GCC GCG TGC ATC GAC CCC TCG GTC GTC TTC CCC TAC CCG CTC AGC GAG CGC GCC CCG
MH2 *v-myc*	GGA GCC GCG GCC GCC GGC TGC GGC TCC TCG GTC GTC TTC CCC TGC AGG GGC CGC GGC --- --- ---

Row 11

Murine *c-myc*	TCC TCC --- GAG TCC TCC CCA CGG GCC AGC CCT GAG CCC CTA GTG CTG CAT GAG GAG ACA CCG CCC ACC
Human *c-myc*	TCC TCG ACG GAG TCC TCC CCG CAG GCC AGC CCC GAG CCC CTG GTG CTC CAT GAG GAG ACA CCG CCC ACC
Chicken *c-myc*	CGG GCC --- --- GCC CCG --- CCC GGC GCC AAC CCC GCG GCT CTG --- CTG GGG GTC GAC ACG CCC CCC ACG
MC29 *v-myc*	CGG GCC --- --- GCC CCG --- CCC GGC GCC AAC CCC GCG GCT CTG --- CTG GGG GTC GAC ACG CCC CCC ACG
MH2 *v-myc*	--- --- --- CCG --- CCC GGC GCC GGC GCC GCG GCT CTG --- CTG GGG GTC GAC GCG CCC CCC ACG
Human *N-myc*[h]	(Incomplete) AGA GGA AGA AAT

Row 12

Murine *c-myc*	ACC AGC AGC GAG TCT C*AA GAG CAA GAG GAA GAA ATT GAT GTG GTG TCT GTG GAG AAG AGG
Human *c-myc*	ACC AGC AGC GAC TCT C*AG GAG CAA GAA GAT GAG GAA ATC GAT GTT GTT TCT GTG GAA AAG AGG
Chicken *c-myc*	ACC AGC AGC GAC TCG C*AA GAA CAA GAA GAT GAG GAA ATC GTC GTT ACA TTA GCT GAA GCG
MC29 *v-myc*	ACC AGC AGC GAC TCG G AA GAA CAA GAA GAT GAG GAA ATC GTC GTT ACA TTA GCT GAA GCG
MH2 *v-myc*	GTC GGC GGC TCG G AG GAA CAA GAA GAT GAG GAA ATC GAT GTC GTT ACA TTA GCT GAA GCG
Human *N-myc*	CGA CGT GGT CAC TGT G GA GAA GCG GCG TTC CTC CAA CAC --- CAA GGC --- --- --- --- --- TGT

(continued)

Table 11.11 (Continued)

Row 13

```
Murine c-myc   CAA ACC CCT GCC AAG AGG TCG GAG TCG GGC TCA TCT CCA TCC CGA GGC CAC CAC AGC AAA CCT CCG CAC CAC AGC
Human c-myc    CAG GCT CCT GGC AAA AGG TCA GAG TCT GGA TCA CCT TCT GCT GGA GGC CAC CAC AGC AAA CCT CCT CAC CAC AGC
Chicken c-myc  AAC GAG TCT GAA TCC AGC ACA GAG TCC AGC ACA GCA TCA GAG GAG CAC CAC TGT AAG CCC CAC CAC AGT
MC29 v-myc     AAC GAG TCT GAA TCC AGC ACA GAG TCC AGC ACA GCA TCA GAG GAG CAC CAC TGT AAG CCC CAC CAC AGT
MH2 v-myc      AAC GAG TCT GAA TCC AGC ACA GAG TCC AGC ACA GCA TCA GAG GAG CAC CAC TGT AAG CCC CAC CAC AGT
Human N-myc    CAC CAC ATT CAC CAT CAC TGT GCG TCC --- CAA GAA CGC AGC CCT GGG TCC CGG GAG GCT CAG TCC AGC
```

Row 14

```
Murine c-myc   CCA CTG CTC AAG AGG TGC CAC GTC TCC ACT CAC CAC TAC GCC GCA CCC CCC TCC ACA ---
Human c-myc    CCA CTG CTC AAG AGG TCC CAC GTC TCC ACA CAT CAG CAC CAC TAC GCA GCG CCT CCC TCC ACT ---
Chicken c-myc  CCG CTG CTC AAG CGG TGT CAC GTC AAC ATC CAC CAA CAC TAC GCT GCT CCT CCC TCC ACC ---
MC29 v-myc     CCG CTG CTC AAG CGG TGT CAC GTC AAC ATC CAC CAA CAC TAC GCT GCT CCT CCC TCC ACC ---
MH2 v-myc      CCG CTG CTC GAG CGG TGT CAC GTC AAC ATC CAC CAA CAC TAC GCT GCT CCT CCC TCC ACC ---
Human N-myc    GAG CTG ATC CTC AAA CGA TGC --- CTT CCC ATC CAC CAG CAC AAC TAT GCC GCC CCC TCT CCC TCC GTG
```

Row 15

```
Murine c-myc   --- AGG AAG GAC TAT CCA GCT GCC AAG AGG GCC AAG TTG GAC AGT GGC AGG GTC CTG --- AAG --- CAG
Human c-myc    --- CGG AAG GAC TAT CCT GCT GCC AAG AGG GTC AAG TTG GAC AGT GTC CTG --- AGA --- CAG
```

```
Chicken c-myc   ---  AAG  GTG  GAA  TAC  CCA  GCC  AAG  AGG  CTA  AAG  TTG  GAC  AGT  GGC  AGG  GTC  CTC  ---  AAA  ---  CAG
MC29 v-myc      ---  AAG  GTG  GAA  TAC  CCA  GCC  AAG  AGG  CTA  AAG  TTG  GAC  AGT  GGC  AGG  GTC  CTC  ---  AAA  ---  CAG
MH2 v-myc       ---  AAG  GTG  GAA  TAC  CCA  GCC  AAG  AGG  CTA  AAG  TTG  GAC  AGT  GGC  AGG  GTC  CTC  ---  AAA  ---  CAG
Human N-myc     GAG  AGT  GAG  GAT  GCA  CCC  CCA  AAG  AAG  ATA  AAG  AGC  GCG  GCG  TCC  CCA  CGT  CCG  CTC  AAG  AGT  GTC

Row 16

Murine c-myc    ATC  AGC  AAC  CGC  AAG  TGC  TCC  AGC  CCC  AGG  TCC  TCA  GAC  ACG  GAG  GAA  AAC  GAC  AAG  AGG  CGG  ACA
Human c-myc     ATC  AGC  AAC  AAC  CGA  AAA  TGC  ACC  AGC  CCC  AGG  TCC  TCG  GAC  ACC  GAG  GAG  AAT  GTC  AAG  AGG  CGA  ACA
Chicken c-myc   ARC  AGC  AAC  AAC  CGA  AAA  TGC  TCC  AGT  CCC  ACG  TCA  GAC  TCA  GAG  GAG  AAC  GAC  AAG  AGG  CGA  ACG
MC29 v-myc      ATC  AGC  AAC  AAC  CGA  AAA  TGC  TCC  ACT  CCC  CGC  ACG  TCA  GAC  TCA  GAG  GAG  AAC  GAC  AAG  AGG  CGA  ACG
MH2 v-myc       GTC  AGC  AAC  AAC  CGA  AAA  TGC  TCC  AGT  CCC  CGC  ACG  TCA  GAC  GAG  GTG  AAC  GAC  AAG  AGG  CGA  ACG
Human N-myc     ATC  CCC  CCA  AAG  GCT  AAG  AGC  TTG  AGC  CCC  CGA  AAC  TCT  GAC  TCG  GAC  GAC  ACT  GAC  CGT  CGC  AGA  AAC

Row 17

Murine c-myc    CAC  AAC  GTC  TTG  GAA  CGT  CAG  AGG  AGG  AAC  GAG  CTG  AAG  CGC  AGC  AGC  TTT  TTT  GCC  CTG  CGT  GAC  CAG  ATC
Human c-myc     CAC  AAC  GTC  TTG  GAG  CGC  CAG  AGG  AGG  AAC  GAG  CTA  AAA  CGG  AGC  AGC  TTT  TTT  GCC  CTG  CGT  GAC  CAG  ATC
Chicken c-myc   CAC  AAC  GTC  TTG  GAG  CGC  CAG  AGG  AGG  AAT  GAG  CTG  AAG  CTG  AGT  TTC  TTC  GCC  CTG  CGT  ACG  GAC  CAG  ATA
MC29 v-myc      CAC  AAC  GTC  TTG  GAG  CGC  CGA  AGG  AAT  GAG  CTG  AAG  CTG  CGT  TTC  TTC  GCC  CTG  CGT  GAC  CAG  ATA
MH2 v-myc       CAC  AAC  GTC  TTG  GAG  CGC  CGA  AGG  AAT  GAG  CTG  AAG  CTG  AGT  TTC  TTC  GCC  CGG  CGG  GAC  CAG  ATA
Human N-myc     CAC  AAC  ATC  CTG  GAG  CGC  CGC  ---  AAC  GAC  CTT  CGG  TCC  AGC  TTT  CTC  ACG  GAG  AGG  CAC  GTG
```

(continued)

Table 11.11 (Continued)

Row 18

Murine c-myc	CCT	GAA	TTG	GAA	AAC	AAC	AAG	GCC	CCC	AAG	GTA	GTG	ATC	CTC	AAA	AAA	GCC	ACC	GCC	*TAC*	ATC	CTG
Human c-myc	CCG	GAG	TTG	GAA	AAC	AAT	AAG	GCC	CCC	AAG	GTA	GTT	ATC	CTT	AAA	AAA	GCC	ACA	GCA	*TAC*	ATC	CTG
Chicken	CCC	GAG	GTG	GCC	AAC	AAC	AAG	GCG	CCC	AAG	GTT	GTC	ATC	CTG	AAA	AAA	GCC	ACG	ACG	GAG	*TAC*	GTT	CTG
MC29	CCC	GAG	GTG	GCC	AAC	AAC	AAG	GCC	CCC	AAG	GTT	GTC	ATC	CTG	AAA	AAA	GCC	ACG	GAG	*TAC*	GTT	CTG
MH2	CCC	GAG	GTG	GCC	AAC	AAC	AAG	GCC	CCC	AAG	GTT	GTC	ATC	CTG	AAA	AGA	GCC	ACG	GAG	*TAC*	GTT	CTG
Human	CCG	GAG	TTG	GTA	AAG	AAT	GAG	GCC	GCC	AAG	GTC	GTC	ATT	TTG	AAA	AAG	GCC	ACT	GAG	*TAT*	GTC	CAC

Row 19

Murine	TCC	ATT	CAA	GCA	GAG	GAG	CAC	AAG	CTC	ACC	TCT	GAA	AAG	GAC	TTA	TTG	AGG	AAA	CGA	CGA	GAA	CAG	TTG
Human	TCC	GTC	CAA	GCA	GCA	GAG	CAA	AAG	CTC	ATT	TCT	GAA	GAG	GAC	TTG	TTG	CGG	AAA	CGA	CGA	GAG	CAG	TTG
Chicken	TCT	ATC	CAA	TCG	GAG	GAG	CAC	AGA	CTA	ATC	GCA	AAG	AAG	GAC	CAG	TTG	AGG	CCG	AGG	AGA	GAG	CAG	TTG
MC29	TCT	CTC	CAA	TCG	GAG	GAG	CAC	AGA	CTG	ATC	GCA	GAG	AAA	GAC	CAG	TTG	AGG	CGG	AGG	AGA	GAG	CAG	TTG
MH2	TCT	ATC	CAA	TCG	GAG	GAG	CAC	AGA	CTG	ATC	GCA	GAG	AAA	GAC	CAG	TTG	AGG	CGG	AGG	AGA	CAG	CAG	TTG
Human	TCC	CTC	CAG	GCC	GAG	GAG	CAC	CAG	CTT	TTG	CTG	GAA	GAG	AAA	TTC	CAG	GCA	GCA	AAG	CAG	CAG	CAG	TTG

Row 20

Murine	AAA	CAC	AAA	CTC	GAA	CAG	CTT	CGA	AAC	TCT	GGT	GCA	TAA
Human	AAA	CAC	AAA	CTT	GAA	CAG	CTA	CGG	AAC	TCT	TGT	GCG	TAA
Chicken	AAA	CAC	AAA	CTT	GAG	CAG	CTA	AGG	AAC	TCT	CGT	GCA	TAG

MC29	AAA	CAC	AAC	CTT	GAG	CAG	CTA	AGG	AAC	TCT	CGT	GCA	TAG
MH2	AAA	CAC	AAA	CTT	GAG	CAG	CTA	AGG	AAC	TCT	CGT	GCA	TAG
Human	CTA	AAG	AAA	ATT	GAA	CAC	GCT	CGG	ACT	---	TGG	---	TAG

[a]Codons for basic amino acids are underscored and those for tyrosine are italicized; those for cysteine and tryptophan are lightly stippled and those for acidic amino acids are darkly stippled.
[b]Stantan et al. (1984).
[c]Colby et al. (1983).
[d]Bernard et al. (1983); D. K. Watson et al. (1983).
[e]Alitalo et al. (1983).
[f]Kan et al. (1984).
[g]Insert CCG CCG CCC TCG GGG CCG CCG CCT CCT CCC GCC.
[h]Michitsch and Melera (1985).

pected, but quite often it varies from all others, especially in having a number of short inserted segments. Despite the frequent divergences, the gene is clearly a member of the same family.

11.5. NUCLEAR-MATRIX-ASSOCIATED ONCOGENES

The last of the three major categories of oncogenes, that in which the products are associated with the nuclear matrix, contains only a single known member, the *myb* gene of chickens and *Drosophila*. Here another, *int1*, which may eventually prove actually to belong elsewhere, is also included on a tentative basis. In the fowl the product of the former induces myeloblastosis and in the mouse that of the latter leads to the formation of mammary tumors (van Ooyen and Nusse, 1984). Little is known of the function of the normal cellular gene *myb*, but in the fruitfly it has been established that the product is essential in early embryonic growth (Katzen *et al.*, 1985).

The Myb Gene and Product. In the avians, the picture of *myb* activity is confused somewhat by the existence of two different viruses that induce myeloblastosis, AMV (avian myeloblastosis virus) and E26, both of which have *myb* sequences in their genomes (Nunn *et al.*, 1983). The latter, however, has only a segment of that gene, but carries a complete oncogene called *ets;* this virus mainly induces erythroblastosis, along with a low level of myeloblastosis. In contrast, AMV, which contains only *myb* along with parts of *gag* and *env*, causes myeloblastosis exclusively. It is to this latter type that attention is confined here.

The viral and chicken *myb* genes are so remarkably alike (Table 11.9) that only a single line of the first of these is given; there it can be noted that the viral structure deviates only at a single site, namely the middle base of codon 17, where a T replaces the A of the cell sequence. That of the fruitfly appears at first to be truncated at the 5' end, but further examination shows the presence of a TAG translational stop signal in the structure from chickens, clearly suggesting that the preceding portion is part of another coding sequence, given as the terminus of the *pol* gene in both the cellular and viral sequences. Although the signal was pointed out in the original investigation (Klempnauer *et al.*, 1982), no final disposition of the matter was made. However, it is self-evident that the cellular structure would not contain any part of the reverse transcriptase encoding the *pol* gene of a virus, so the 5' portion of the gene has an unknown source. While the avian and insect elements are largely homologous, that of the latter is too incomplete at the 3' portion to enable meaningful comparisons to be made. No particularly prominent chemical property can be noted in the gene.

The Int1 Gene. As in the *myb* gene, codons for tyrosine and tryptophan in the *int1* cistron are at or below the frequency based on random chance, but those for cysteine (stippled) are apparently of great importance. Twenty-two of its triplets (Table 11.9) are present, for a ratio of 6%, but, more importantly, their distribution is most unusual in that ten are concentrated in the 3'-terminal portion beginning with row 15. Since the 5' end is enriched in coding elements for hydrophobic amino acids (van Ooyen and Nusse, 1984), that part of the product may be membrane-embedded *in vivo*, leaving the opposite portion free, with the cysteines capable of binding to some other protein, perhaps a subunit. The oncogenic nature of the product has been deduced from the disclosure that the gene is actively transcribed in tumor tissue, but only slightly or not at all in normal cells (Cuypers *et al.*, 1984). Moreover, in genomic studies proviruses may be found on both sides of the gene, usually in divergent orientation, suggesting that the protein of the *int1* gene is essential for tumorigenesis by the mouse mammary tumor virus (van Ooyen and Nusse, 1984).

11.6. OTHER CARCINOGENIC VIRUSES

The retroviruses, although the major type involved in carcinogenesis, are not to be viewed as the only ones known to induce malignancies, for a number of the temperate types surveyed in the preceding chapter also produce cancers, at least under certain conditions. Probably the most frequently involved are the adenoviruses, but the Epstein–Barr virus and SV40 are familiar species that also have been implicated in such diseases.

The Adenoviruses in Cancer. Three species of human adenovirus are most frequently involved in oncogenesis, Ad2, Ad5, and Ad12; cells transformed by any one of the trio carry viral DNA sequences integrated into the genome of the host cell (Stabel and Doerfler, 1982; Visser *et al.*, 1982). Although the *E1* region is essential in all cases, the degree of activity and effects vary with the viral species (Branton *et al.*, 1985; Flügel *et al.*, 1985). In Ad5, this region transforms baby rat kidney (BRK) cells with high frequency, but those cells do not produce tumors in susceptible test animals; on the other hand, in Ad12, it is less effective on the BRK cells, yet they are highly carcinogenic. As will be recalled from the preceding chapter, the *E1* region consists of two transcriptional units, *E1A* and *E1B*. It is the first of these that experiments have revealed to be concerned with frequency of transformation of BRK cells, whereas the second determines the degree of oncogenicity (Jochemsen *et al.*, 1984). In transformed cells the *E1A* transcript encodes two mRNAs, the longer of which (13 S) is chiefly involved in cancer, the protein domain encoded by the gene's first exon being particularly concerned in the activity. In some cancers produced by these viruses, amplification of the proviral DNA, in whole or part, has been noted (Nakatani *et al.*, 1985).

A Miscellany of Types. Among the miscellany of viral types that induce tumors or cancer is the well-known SV40, but its oncogenic or tumorigenic properties have not been as fully documented as those of the adenoviruses. Through investigation of a certain small nuclear RNA, called 7S-K, from normal and SV40-transformed cells, it has been determined that the host-encoded substance promoted transcription initiation of the viral genome (Sohn *et al.*, 1983). The stimulatory action of the RNA was deduced to result from its recognition of viral promoter sequences, which activity thereby facilitated the formation of the transcription-initiation complex with RNA polymerase II, perhaps by forming part of that combination. Thus this substance, which is transcribed by RNA polymerase III, is requisite for transcription by RNA polymerase II.

One final nonretroviral example that is pertinent in the present context is the human hepatitis B virus, which is widely accepted as a causative agent of primary liver carcinoma (Koch *et al.*, 1984; Rogler *et al.*, 1985). Moreover, genomes of this species have been found to be integrated in the DNA of the infected cells in both acute and chronic hepatitis and cirrhosis of the liver. Examination of the viral DNA revealed no rearrangements other than an occasional duplication, inversion, or deletion. However, there is the possibility that the integrated structure may have created an open reading frame, together with the adjacent host DNA, which may express chimeric or abnormal proteins capable of playing a role in tumorigenesis.

11.7. TRANSFORMATIONS IN NONMETAZOAN SPECIES

Metazoans are not the only organisms that have proven to be subject to genetic transformations from viral or other parasites, and by synthetic transformation the latter systems can introduce new genes into a genome, often enabling the cell to survive or to show improved growth on particular media. For example, recently a gene encoding an acetamidase taken from *Aspergillus nidulans* was integrated into the DNA of *A. niger*, a species of much industrial importance, enabling the latter thus transformed to employ acetamide as a nitrogen source (Kelly and Hynes, 1985). Similar transformation systems have been developed for various yeasts and fungi, such as *Neurospora*.

Agrobacterium Transformations. Tumor production in seed plants by bacteria or viruses is not an uncommon phenomenon, but as a whole the processes have been relatively poorly explored at the molecular level. Among the more thoroughly investigated are the diseases induced by the invasion of bacteria of the genus *Agrobacterium*. When one species of these soil bacteria, *A. tumefaciens*, infects a plant wound, a tumor is produced, whereas if *A. rhizogenes* does so, the production of adventitious roots is induced (Tepfer, 1984).

In both instances, the morphogenic effects result from the transfer of a portion of a bacterial plasmid into the plant cells, the tumor-inducing (Ti) plasmid differing from the root-inducing one (Ri). Only the latter leads to true transformation, for experimentally produced mutations in the former lead to the growth of roots or stems from the tumor (crown gall) that do not contain the Ti plasmid. On the other hand, the roots that grow under the influence of the Ri plasmid transmit that genetic structure to the progeny.

The functions of several genes of the Ti plasmid have been identified, one of which, variously named *ocs* or *nos*, encodes a substance responsible for the production of unusual amino acid derivatives known as opines (Akiyoshi *et al.*, 1985). Three others, *tmr*, *tms1*, and *tms2*, are concerned with the synthesis of various phytohormones, as is another, called *tzs*, which encodes an enzyme that produces *trans*-zeatin, a hydroxylated cytokinin. The cytokinin itself is encoded by the *tmr* cistron. In addition, three types of Ti plasmids have been recognized, two of which have been studied at the molecular level. One type, referred to as nopaline Ti plasmids, induces the formation of the substances nopaline and agrocinopine, whereas the other, the octopine variety, leads to the synthesis of octopine and agropine (Joos *et al.*, 1983). The gene that encodes the synthase that produces agrocinopine, along with at least three others that remain unidentified, has been reported to be essential to the formation of tumors. Two of those three together inhibit shoot formation, whereas the third suppresses root development and is active in the cytokinin-independent growth of transformed plant cells.

Three Plant-Tumor Genes. Sequences of three genes of *Agrobacterium tumefaciens* have been established, the *tzs* (Akiyoshi *et al.*, 1985), the *tmr* from two contrasting plasmid types, octopine and nopaline (Heidenkamp *et al.*, 1983; Goldberg *et al.*, 1984), and the nopaline synthase (Bevan *et al.*, 1983). As expected, the synthase differs entirely from the others, in both its much greater length and nucleotide structure (Table 11.12). Comparisons of the other three afford some surprises, the first being the high level of homology the *tzs* cistron displays with the two *tmr* varieties. Although more than sufficiently different to deserve recognition as a separate and distinct species, the frequent identical and cognate sites leave little room for doubt that the three have shared common ancestry not too far in the past.

The second surprise comes from an opposite observation, the unpredictable amount of distinctions in structure that is perceived between the pair of *tmr* coding regions. Although closely allied to the nopaline plasmid counterpart, that of the octopine frequently departs in nucleotide structure, on many occasions agreeing with that of *tzs* or, not unusually, with neither. Moreover, many of the point differences result in changes in coding properties, so that the octopine and nopaline proteins may have markedly distinctive properties, a point well-deserving of investigation.

Other Tumorigenic Pathogens of Seed Plants. Production of phytohormones is a common feature of tumors in seed plants, a characteristic that makes the growth of transformed tissue independent of auxin or cytokinin production by the host. This has been demonstrated to be the case in *Corynebacterium fascians,* whose infections result in those witches' brooms so characteristic of northern hackberry trees. Similarly, the virulence of infections of *Pseudomonas syringae savastanoi* is highly dependent upon the strain's ability to synthesize the auxin 3-indolacetic acid. Thus, in seed plants as in metazoans, the production of tumors and malignant growths is, at least in great measure, associated with the production of growth factors.

11.8. CHEMICAL AGENTS IN TRANSFORMATIONS

Chemical carcinogens lead to the production of malignancies in several ways, one of which involves an ability to bind covalently to DNA, RNA, or proteins, but especially the former. Many of these act upon the DNA directly, leading to the degeneracy of the genome at that site or to the production of mutations (Singer and Grunberger, 1983; Kriek *et al.,* 1984). Studies along such lines, however, while of utmost value medically, do not illuminate the role of specific genes in carcinogenesis, whereas the specific effect of chemical agents on cellular genes can do so.

In chemically induced tumor formation one of the important determinants is a genomic complex named *Ah,* to which such agents as benzo[α]pyrene and dioxin bind firmly (Nebert *et al.,* 1984). Such binding by dioxin results in the overexpression of six genes in the mouse, including those encoding cytochromes P_1-450 and P_3-450 (Chapter 6, Section 6.4.1), leading to enhanced levels of those proteins in the endoreticulum. This increase of the cytochromes gives rise to a more rapid metabolism of the chemical agent. However, the benzo[α]pyrene and dioxin themselves do not result in the malignancies, but the by-products of their breakdown are the active agents (Jaiswal *et al.,* 1985). Hence, the more cytochrome P-450s that are present, the greater the chance of the cell becoming transformed. The level of concentration of mRNA for P_1-450 in man has been shown to be directly correlated with benzo[α]pyrene metabolism in activated lymphocyte cultures.

From the foregoing discussions, the role of the gene in carcinogenesis is seen to be an exceedingly complex and variable one. Sometimes chemicals or irradiation may set a

Table 11.12
Some Genes Involved in Seed Plant Tumors[a]

Row 1

A.t. *tzs*[b]
ATG ATA CTC CAT CTC ATC *TAC* GGA CCG ACT TGC AGC GGC AAA ACG GAC ATG GCG ATC CAA ATC GCA CAA

A.t. *tmr* oct[c]
ATG GAC CTG CAT CTA ATT TTC GGT CCA ACT TGC ACA GGA AAG ACG ACG GCG ATA GCT CTT GCC CAG

A.t. *tmr* nop[d]
ATG GAT CTG CGT CTA ATT TTC GGT CCA ACT TGC ACA GGA AAG ACG TCG ACC GCG GTA GCT CTT GCC CAG

Nopaline synthase[e]
ATG GCA ATT ACC TTA TCC GCA ACT TCT TTA CCT ATT TCC GCC GCA GAT CAC CAT CCG CTT CCC TTG ACC

Row 2

A.t. *tzs*
GAA ACC GGG TGG CCG GTG GTT GCC CTT GAT CGT GTG CAA TGC TGT CCT CAA ATC GCG ACA GGT AGC GGA

A.t. *tmr* oct
CAG ACA GGG CTT CCA GTC CTT TCG CTT GAT CGG GTC CAA TGC TGT CCT CAA CTA TCA ACC GGA AGC GGA

A.t. *tmr* nop
CAG ACT GGG CTT CCA GTC CTT TCG CTC GAT CGG GTC CAA TGT TGT CCT CAG CTG TCA ACC GGA AGC GGA

Nopaline synthase
GTA GGT GTC CTC GGT TCT GGT CAC GCG GGG ACT GGA TTA GCG GCT TGG TTC GCC TCC CGG CAT GTT CCC

Row 3

A.t. *tzs*
AGA CCT TTG GAA TCG GAA TTG CAA TCA ACG AGA ATA *TAT* TTG GAT TCC CGC CCC CTC ACC GAG GGC

A.t. *tmr* oct
CGA CCA ACA GTG GAA GAA CTG AAA GGA ACG ACG CGT CTC *TAC* CTT GAT GAT CGG CCT CTG GTG GAG GGT

A.t. *tmr* nop
CGA CCA ACA GTG GAA GAA CTG AAA GGA ACG AGC CGT CTA *TAC* CTT GAT GAT CGG CCT CTG GTG GTG AAG GGT

Nopaline synthase
ACG GCG CTG TGG GCA CCA GCA GAT CAT CCA GGA TCG ATC TCA GGA ATC AAG GCC AAT GAA GGA GTT ATC

Row 4

A.t. *tzs*
ATC CTT GAC GCT GAG AGT GCC CAT CGT CGA CTC ATA TTC GAA GTG GAT TGG CGG AAG TCC GAA GAC GGT

A.t. *tmr* oct
ATC ATC GCA GCC AAG CAA GCT CAT CAT AGG CTG ATC GAG GAG GTG *TAT* AAT CAT GAG GCC AAC GGC GGG

```
A.t. tmr nop      ATC ATC GCA GCC AAG CAA GCT CAT GAA AGG CTG ATG GGG GAG GTG TAT AAT TAT GAG GCC CAC GGC GGG
Nopaline synthase ACC ACC GAG GAG GGA ATG ATT AAC GGT TCA GCC TGT GTC GAT GAC CTT GCC GCA GTT ATT CGC

Row 5
A.t. tzs          CTT ATT CTC GAG GGC GGG TCG ATT TCG CTT CTC AAT TGC ATG GCT AAA AGT CCG TTT TGG AGA TCG GGT
A.t. tmr oct      CTT ATT CTT GAG GGA GGA TCC ACC TCG TTG CTC AAC TGC ATG GCG CGA AAC AGC TAT CCG TAT TGG AGT GCA GAT
A.t. tmr nop      CTT ATT CTT GAG GGA GGA TCT ATC TCG TTG CTC AAG TGC ATG GCG CAA AGC AGT GCG AGT TAT TGG AGT GCG GAT
Nopaline synthase TCC AGC CGT GTA CTG ATT ATT GTA ACC CGT GCG GAC GTT CAC GAC AGC TTC GTC AAC GAA CTC GCC AAC

Row 6
A.t. tzs          TTT CAA TGG CAT GTC AAG CGG CTA CGT CTT GGG GAT TCG GAC GCC TTT CTC ACC CGA AAG GCC AAG CAA CGC GTT GCG
A.t. tmr oct      TTT CGT TGG CAT ATT ATT CGC CAC AAG TTA CCC GAC CAA GAG ACC TTC --- ATG AAA GGG --- GCC AAG GCC AGA
A.t. tmr nop      TTT CGT TGG CAT ATT ATT CGC CAC GAG TTA GCA GAC GAA GAG ACC TTC --- ATG AAC GTG --- GCC AAG GCC AGA
Nopaline synthase TTC AAC GGC GAA CTC GCA ACA AAG GAT ATT GTC GTC GTG TGC GGC CAT --- GGC TTC TCC --- ATC AAG TAC GAG

Row 7
A.t. tzs          GAA ATG TTT GCC ATC CGG GAA GAT --- CGC CCC TCG TTG TTG GAG GAG TTG GCG GAA CTC TGG AAC TAC CCT
A.t. tmr oct      GTT AAG CAG ATG TTG CAC CCC GCT GCT GGC CAT TCT ATT ATT CAA GAG TTG GTT CTT CTT GAT TAT TGG AAT GAA CCT
A.t. tmr nop      GTT AAG CAG ATG TTA CGC CCT GCA GGC CTT TCT ATT ATC CAA GAG TTG GTT GTT GAT CTT GAT TGG AAA GAG CCT
Nopaline synthase AGA CAG CTG CGA TTC AAG CGA ATA TTC GAG ACG AAT TCG CCC ATA ACG TCT AAG CTA TCG GAT CAA AAA
```

(continued)

Table 11.12 (Continued)

Row 8

A.t. *tzs*	GCC	GCT	CGA	CCG	ATT	TTG	GAA	GAT	ATC	GAC	GGA	*TAT*	CGC	TGC	GCA	ATT	CGT	TTT	GCG	CGC	AAA	CAC	GAT
A.t. *tmr* oct	CGG	CTG	AGG	CCC	ATT	CTG	AAA	GAG	ATC	GAT	GGA	*TAT*	CGA	*TAT*	GCC	ATG	TTG	TTT	GCT	AGC	CAG	AAC	CAG
A.t. *tmr* nop	CGG	CTG	AGG	CCC	ATA	CTG	AAA	GAG	ATC	GAT	GGA	*TAT*	CGA	*TAT*	GCC	ATG	TTG	TTT	GCT	AGC	CAG	AAC	CAG
Nopaline synthase	AAA	TGT	AAC	GTC	AAC	ATC	AAG	GAA	ATG	AAA	GCG	TCT	TTC	GGA	CTG	TCA	TGT	TTC	CCA	ATT	CAT	CGC	GAT

Row 9

A.t. *tzs*	CTC	GCA	ATC	AGC	CAG	TTG	CCA	AAT	ATT	GAT	GCA	GGG	CGG	CAC	CTA	GAG	CTC	ATA	GAG	GCC	ATA	GCT	AAT
A.t. *tmr* oct	ATC	ACG	GCA	GAT	ATG	CTA	TTG	CAG	CTT	GAC	GCA	AAT	ATG	GAA	GGT	AAG	TTG	ATT	AAT	GGG	ATC	GCT	CAG
A.t. *tmr* nop	ATC	ACA	TCC	GAT	ATG	CTA	TTG	CAG	CTT	GAC	GCA	GCA	GAT	ATG	GAG	GAT	AAG	TTG	ATT	CAT	GGG	ATC	GCT	CAG
Nopaline synthase	GAT	GCT	GGC	GTG	ATT	GAT	CTA	---	CCC	GAA	GAT	ACC	AAG	AAC	ATC	TTT	GCC	CAG	CTA	TTT	TCC	GCT	AGA

Row 10

A.t. *tzs*	GAA	*TAT*	CTT	GAA	CAT	GCG	CTC	TCG	CAG	GAG	CGC	GAT	TTT	CCT	CAG	TGG	CCA	GAA	GAT	GGC	GCA	GGA	CAG
A.t. *tmr* oct	GAG	*TAT*	TTC	ATC	CAT	GCG	CGC	CAA	CAG	CAA	CAG	AAA	TTC	CCC	CAA	GTT	AAC	GCA	GCC	GCT	TTC	GAC	GGA
A.t. *tmr* nop	GAG	*TAT*	CTC	ATC	CAT	GCA	GCA	CGC	CGA	CAA	CAG	AAA	TTC	CCT	CGA	GTT	AAC	GCA	GCC	GCT	*TAC*	GAC	GGA
Nopaline synthase	ATC	ATC	TGC	ATC	CCG	CCG	TTG	CAA	GTG	CTA	TTC	TTT	TCC	AAC	*TAT*	ATC	ACT	CAT	GCG	GTT	CCG	GCA	GTC

Row 12

Nopaline synthase CTA GAC GAG CGA ACC CCA CGA GCC GAG AAG GGC TTT TTC TTT TAT GGT GAA GGA TCC AAC ACT TAC GTT

Row 13

Nopaline synthase TGC AAC GTC CAA GAG CAA ATA GAC CAC GAA CGC CGG AAG GTT GCC GCA GCG TGT GGA TTG CGT CTC AAT

Row 14

Nopaline synthase TCT CTC TTG CAG GAA TGC AAT CAT GAA TAT GAT ACT GAC TAT GAA AGT TTG AGG GAA TAC TGC CTA GCA

Row 15

Nopaline synthase CCG TCA CCT CAT AAC GTG CAT CAT GCA TGC CCT GAC AAC ATG GAA CAT CGC TAT TTT TCT GAA GAA TTA

Row 16

Nopaline synthase TGC TCG TTG GAC GAT GTC GCG GCA ATT GCA GCT ATT GCC AAA ATC GAA ATA CCC CTC ACG CAT GCA TTC

Row 17

Nopaline synthase ATC AAT ATT ATT CAT GCG GGG AAA GGC AAG ATT AAT CCA ACT GGC AAA TCA TCC AGC GTG ATT GGT AAC

Row 18

Nopaline synthase TTC AGT TCC AGC GAC TTG ATT CGT TTT GGT GCT ACC CAC GTT TTC AAT AAG GAC GAG ATG GTG GAG TAA

[a] Codons specifying tryptophan and cysteine are stippled; triplets for tyrosine are italicized, and those encoding basic amino acids underscored.
A.t., *Agrobacterium tumefaciens*.
[b] Akiyoshi et al. (1983).
[c] oct, octopine Ti plasmid: Heidekamp et al. (1983).
[d] nop, nopaline Ti plasmid; Goldberg et al. (1984).
[e] Bevan et al. (1983).

normal gene into an aberrant behavioral pathway. In other cases, harmless movable elements may, by becoming inserted near a potentially oncogenic cistron, lead to its becoming overly expressed, ultimately resulting in a malignancy. Often viruses may be the causative agent acting through many channels. Although a single gene may be the instigating mechanism, it nevertheless is obvious that the full-blown growth involves the interplay of a number of diverse genes, among them those that encode growth-stimulating and mitogenic factors.

12

Processing the Primary Transcripts

Aside from promoters, ancillary sites, terminators, and other features involved in transcription, the mature gene structure includes signals needed for processing of the RNA copy. Indeed, there are many devices encoded into its sequence that act secondarily even after translation is completed. Among these are the clues that afford direction to cleaving enzymes, including the proteases that split transit peptides from diplomorphic gene products or release the ultimate proteins when a cryptomorphic form is to be activated. Although the amino acids involved in such recognition functions are obviously encoded in the mature coding sector and hence are a fundamental part of the gene structure, here attention is restricted to clues that are active directly in the primary transcript. These prove quite adequate to reveal that the problems associated with promoters and their associates in transcriptional functions continue into the present topic, as they doubtlessly also do with regard to the products resulting from translation—a point that becomes particularly clear in a discussion related to intron removal (Section 12.6.3).

12.1. THE U FAMILY OF NUCLEAR RNAs

The first step in a discussion of transcript processing is that of examining certain unfamiliar RNAs that have been implicated as being among the active agents, namely the small hnRNAs constituting the U family. Seven major types and numerous varieties comprise that group, a representative of each of which is provided in Table 12.1, a total of six from human tissues and one from a sea urchin. Although homologies are not altogether lacking, relationships being clearly perceptible between U1 and U2 on the one hand and U3–U5 on the other, their principal common feature is in their lengths. The longest is U3, followed in turn by U2 and U1, but none exceeds 220 nucleotide residues, while the shortest, U7, contains fewer than 60. In ensuing pages the unique features of each species are closely analyzed, but before treating each individually, the organization of their genes in the genome and other preliminary particulars need examination.

Table 12.1

Comparison of the Genes of the Several U RNAs[a]

Row 1

Human U1	ATACTTACC	TGGCAGGGGA-G	ATACCATGA	TCACGAAGG	TGGTTTTCCCAG	GGCGAGGCTTAT	CCATTGCACTCC
Human U2	ATCGCCTTCT	CGGCCTTTTGGC	-TAAGATCA	AGTGTAGTA	TCTGTTCTTATC	AGTTTAATATCT	GATACGTCCTCT
Human U3	GAAGTGACT	ATACTTTCAGGG	ATCATTTCT	ATAGTTCGT	------TACTAG	AGAAGTTTCTCT	GACTGTGTAGAG
Human U4	GAACTTTGC	AGTGGCAGT--	ATCGTAGCC	AATGAGGTT	------TATCCG	AGGCGC------	GATTATTGCTAA
Human U5	NATACTCTG	GTTTCTCTTCAG	ATCGCCATAA	ATCTTTCGC	--CTTTTACTAA	AGA--TTTC---	---CGTGGAGAG
Human U6	NGTGCCTGC	TTCGGCAGC---	ACA-TATAC	TAAAAATTGG	------AAC---	-GA--TACA---	GAGAAGATTAGC
Sea urchin U7	TCTTTCAAG	TTT-CTCTAGAA	GGGTCT---	-------CGC	---GTCCGAAGT	CGG---------	---AGGCGA---

Row 2

Human U1	GGATGTGCT-GA	CCCCTCGGATTT	CCCCAAATG	TGGGAAACTCGA	CTGCATAAT	TTGTGGTAG	TGGGGGACTGCG
Human U2	ATCCGAGGACAA	TATATTAAATGG	ATTTTTTGGA	GCAGGGAGATGG	AATAGGAGC	TTGCTCCGT	CCACTCCACGCA
Human U3	CACCCGAAACCA	CGAGGAGAGAC	GGAGCGTTC	TCTCCTGAGCGT	GAAGCCGGC	TCCTACTGT	TGCTTCCTGCCA
Human U4	TTG---AAAACT	TTTCCCAATACC	CCGCCATGA	CGACTTGAAATA	-TAGTCGGC	ATTGGCAAT	TTTTGACAGTCT
Human U5	GAA---CAACTC	TGAGTCTTAACC	CAATTTTTT	GAGCCTTGCCTT	-------GGC	AAGG-----	------------
Human U6	ATGGCCCCTGCG	CAAGGATGACAC	GCAAATTCG	TGAAGCGTTCCA	----------	------TAT	TTTT
Sea urchin U7	GTG-CCCAATT						

Row 3

Human U1	TTCGCGCTT	TCCCCTG					
Human U2	TCGACCTGG	TATTGCAGT	ACCTCCAGGAAC	GGTGCACCC			
Human U3	TTGCTATTG	GCAACTGAT	GATCGTCTTCGG	TCCTCTTGA	GGGGTGAGAGGG	AGAGAACGC	AGTCTGAGTGGT
Human U4	CTACGGAGA	----CTG					
Human U5	--------	----CTA					

[a]For sources, see Tables 12.2–12.6.

12.1.1. Occurrence and Organization

The U family of RNAs seems to be ubiquitous among eukaryotes and probably will prove to be present among such advanced bacteria as *Thermaplasma*, at least in progenitory form, as histones are appearing to be. Their occurrence throughout the Metazoa is virtually confirmed, and, as seen in later sections, at least three species have been sequenced from seed plants and several from fungi. Although these RNAs have received little or no attention in most unicellular organisms, they have been reported present in the common yeast (Wise *et al.*, 1983), the familiar ciliate *Tetrahymena* (Pedersen *et al.*, 1985), and the dinoflagellate *Crypthecodinium* (Liu *et al.*, 1984; Reddy *et al.*, 1985a), but the precise species of U RNA that occur remain unidentified. In all cases transcription is by DNA-dependent RNA polymerase II, and, as is typical of such transcripts, these small nuclear species bear capped 5′ termini (Reuter *et al.*, 1984). The cap is characteristic of the family, being 2,2,7-trimethyl guanosine (Ro-Choi *et al.*, 1975), but this is absent in U6.

Genomic Organization. As a whole, information regarding the organization of U-family RNA genes in genomes is sparse. In *Saccharomyces* the several cistrons for unidentified types have been shown to occur as unique copies (Wise *et al.*, 1983). At the other end of the scale lies the slime mold, *Dictyostelium*, in which multiples for U1 approaching 3000 have been reported (Wise and Weiner, 1981). Even greater numbers were at one time believed to be present in human DNA, for estimates of $\sim$100,000 genes for U1 had been made (Forbes *et al.*, 1983), but this figure has been gradually reduced, first to 100–150 (Manser and Gesteland, 1982), and now to $\sim$30 copies per haploid genome (Lund and Dahlberg, 1984). Moreover, an abundance of pseudogenes occurs among the functional genes, which are scattered over the chromosomes (Denison *et al.*, 1981). In avians the cistrons are similarly of category II organization (Roop *et al.*, 1981), whereas in *Xenopus* the small number of U1 coding elements are category III (tandemly arranged) (Forbes *et al.*, 1984).

Among the few other metazoans from which data are available, the genomic organization of this family of genes has been most thoroughly documented in *Drosophila*, in which multiple copies are the rule also (Saluz *et al.*, 1983). About seven genes each for U3 and U4 were found by Southern blot analyses, three for U2, and between one and three for U6. The only other of the taxon that seems to have been investigated is a sea urchin in which the low numbers of copies are organized much as in the category III of *Xenopus* (Card *et al.*, 1982).

U-Family RNA Genes in Development. A small but significant beginning has been made toward understanding the behavior of the U-family RNA genes during development, at least in *Xenopus* (Forbes *et al.*, 1983, 1984). There can be little doubt as to the importance of its members during embryogenesis, for about 8×10^8 molecules of U1 have been found stored in the transcriptionally inactive zygote and early cleavage embryo, sufficient for $\sim$6000 nuclei. Then, when transcription by RNA polymerase II commences at the 12th cleavage (4000-cell stage), five species of these RNAs, U1, U2, and U4–U6, become major products, the level of U1 increasing sevenfold in 4 hr, a rate much greater than that of mature tissue.

Developmental changes in gene expression have been revealed in the U1 genes, at least seven varieties of which are present in *Xenopus* (Forbes *et al.*, 1984), all of the same length, but differing in sequence structure. Two of them, U1a and U1b, are the predominant types in the late blastula stage and into the early gastrula, whereas an entirely different set is expressed earlier and in oogenesis. More recently a report indicated that in late oocytes U1b is transcribed more rapidly than the other variety, probably accounting for the distinction noted earlier (Krol *et al.*, 1985).

12.1.2. Control Elements of the U RNAs

The nature of promoters and terminators of transcription of the U RNAs is still in the nebulous state typical of a recently opened field of investigation, but what has been determined indicates that future results should be of great interest. In the foregoing account of their behavior during embryogenesis, it is explicit that several forms of genes for the U1s exist that are expressed differentially during oogenesis and in the cleavage to gastrula stages. This observation thus implies that different controls must exist for each variety and possibly species as well.

Examination of the leader structures of these tandemly arranged genes in human and amphibian tissue to reveal conserved and contrasting sequences that might serve as transcriptional signals have proven fruitless. In the first place no canonical TA-TA- boxes have been found in any leader, and sequence homology has been demonstrated to be very low (Skuzeski *et al.*, 1984; Krol *et al.*, 1985). Nor have any been found in *Drosophila* (Beck *et al.*, 1984). However, an element placed ~50 base pairs before the mature coding sequence was located, the TCTCCGTATG of the U1b gene being modified in the other by the replacement of the middle G by a T. Transcription was found to be dependent on gene-specific elements situated upstream of position -220, but they could not be finally identified.

Termination was suggested, but not demonstrated, to be signaled at sequences placed about 15 residues beyond the end of the mature coding sector. In U1a the combination AAAGATAGA was found there, while the closely related series AAAAGTAGA replaced it in U1b. But it is clear, even from these preliminary results, bearing in mind that these genes are transcribed by the same polymerase II that synthesizes all the mRNAs of the cell, that transcription is controlled by mechanisms far more complex than such simple signaling sequences. As pointed out in earlier chapters, real progress in learning how the cell's sensitive requirements are met cannot be made until it is fully appreciated that many interacting agents are undoubtedly involved. Among the regulating substances most probably are numerous enzymes and ancillary proteins, including distinctive ones for each major class of macromolecular product. Even the relatively simple genome of SV40 has now been shown to require a cellular factor (Sp1) for its transcription. During the early stages of RNA synthesis, it reacts with the first three of six sets of the hexanucleotide GGGCGG, whereas during the late phase it binds to the remaining three (Gidoni *et al.*, 1985). Thus it would seem that some additional factor needs to be present to control the temporal regulation of the processes.

12.2. THE SEVERAL TYPES OF U RNAs

Since the several major types of U RNAs have different roles in the processing of transcripts, a preliminary view of their structures and subtypes should be beneficial. For, as is shown in the brief discussion above of their behavior during embryogenesis, the

Table 12.2
Comparison of snRNA U1 Genes[a]

Row 1

Gene							
Pea U1[b]						(Incomplete) TGC--TTT	
Drosophila 21D[c,d]	ATACTTACCTGG	CCTAGAGGTTAA	CCGTGATCA	CGAAGGGCG	TTCCTCCGG	AGTGAGGCTTGG	CCATTGCACCTC
Drosophila 82E[c]	TCTGATTGCGTC	CTAATCGAAACC	CCTATGCCG	AGCTAAGCA	AAGCTTGTG	AGTGAGGCTTGG	TCATTGCACCTC
Sea urchin[e]	ATACTTACCTGG	CCCAGGGGTCGC	-ATTGATCA	AGAAGGATG	CACCCCCAG	GGCGAGGCTTG-	CCATTGCACTCC
Xenopus U1a[f]	ATACTTACCTGG	CAGGGGAGA-TA	CCATGATCA	CGAAGGTGG	TTCTCCCAG	GGCGAGGCTCAG	CCATTGCACTCC
Xenopus U1b[f]	ATACTTACCTGG	CAGGGGAGA-TA	CCATGATCA	CGAAGGTGG	TTCTCCCAG	GGCGAGGCTCAG	CCATTGCACTCC
Murine U1B[g]	ATACTTACCTGG	CAGGGAGA-A-TA	CCATGATCA	TGAAGGTGG	TTTTCCCAG	GGCGAGGCTCAC	CCATTGCACTCC
Rat U1A[h]	ATACTTACCTGG	CAGGGG-AGATA	CCATGATCA	CGAAGGTGG	TTTTCCCAG	GGCGAGGCTTAT	CCATTGCACTTT
Human U1[i]	ATACTTACCTGG	CAGGGGAGA-TA	CCATGATCA	CGAAGGTGG	TTTTCCCAG	GGCGAGGCTTAT	CCATTGCACTCC
Human U1.1[j]	ATATTTACTTGG	CAGGGGAGA-TA	ACATGATCA	CGAAGGTGG	TTTTCCCAG	GGCGAGGCTTAT	CCATTGCACTCC
Chicken U1.25[k]	ATACTTACCTGG	CAGGGGGAGA-CA	CCATGATCA	GGCAGGTGG	TTTTCCCAG	GGCGAGGCTCAT	CCCCTGCACTCC

Row 2

Gene							
Pea	CCTAGAG--	GTCT-ACCC	AAG---TGG---	TGG-AGCC-	TACATCATAAT-	TTGTTGCCTGAG	GGGG-CCTGCGT
Drosophila 2.1D	GGCTGAGTT	GACCTCTGC	GATTATTCCTAA	TGTGAATAA	CTCGTGCGTGTA	ATTTTTGGTAGC	CGGGAATGGCGT
Drosophila 82E	AGCTGAGTT	GACCTCTGC	GATTATTCCTAA	TGTGAATAA	CTCGTGCGTGTA	ATTTCTGGTAGC	CGGGAATGGCGT
Sea urchin	GGC-T-TGC	TGAACCTTG	CGATTCCCCAA	ACGTGGGA	ACTCGGCGCGTAC	AATTTATGGTAG	CGGAGATCTCCG
Xenopus U1A	GGC-TGTGC	TGACCCCTG	CGATTTCCCAA	ATGCGGGAA	ACTCGACTGCAT	AATTTCTCGTAG	TGGGGGACTCCG
Xenopus U1B	GGC-CGTGC	TGACCCCTG	CGATTTCCCAA	ATGCGGGAA	AGTCGACTGCAT	AATTTCTCGTAG	TGGGGGACTGCG
Murine U1B	GGGGTGTGC	TGACCCCTG	CGATTTCCCAA	ATGCGGGAA	ACTCGACTGCAT	AATTTGTCGTAG	TGGGGGACTGCG
Rat U1A	GGA-TGTGC	TGACCCCTG	CGATTTCCCAA	ATGCGGGAA	ACTCGACTGCAT	AATTTGTCGTAG	TGGGGGACTGCG

Human U1	GGA-TGTGC	TGACCCCTG	CGATTTCCCAA	ATGTGGGAA	ACTCGACTGCAT	AATTTGTCGTAG	TCGGGGACTCGCG
Human U1.1	AGA-TGTGT	TGACCCCTG	CGATTTCCCAA	ATGTGGGAA	ACTCGATTGCAT	AATTTGTCGTAG	TGGT--ACTCGG
Chicken U1.25	GGG-TGTGC	TGACCCCTG	CGATTTCCCAA	ATGCGGGAA	ACTCGACTGCAT	AATTTGTCGTAG	TCGGGGACTCGCG

Row 3

Pea	TCGGGCGGG-	CCCCCACC					
Drosophila 21D	TCGGGCCGT	CCCGA					
Drosophila 82E	TCGGGCCCGT	CCCGA					
Sea Urchin	TTCGGGCTA	TCTCCT					
Xenopus U1A	TTCGGGCTT	TCCCCTG					
Xenopus U1B	TTCGGGCTT	TCCCCTG					
Murine U1B	TTCGGGCTC	TCCCCTGAT	TTTTGTGGTGCT	AAAAGTTAG	ATGCATTCTGCT	CTTCTCATGTCT	CTTTACATGTTG
Rat U1A	TTCGGGCTC	TCCCCTG					
Human U1	TTCGGGCTT	TCCCCTG					
Human U1.1	TTCGGGCTT	TCCCCTG					
Chicken U1.25	TTCGGGCTC	TCCCCTG					

Row 4

Murine U1B	TTTGTGAGGCAT	GGCACGAATTC

a Sequences reported to be important in reactions with pre-mRNAs are shown underscored.
b Kroll et al. (1983). c Kejzlarová-Lepesant et al. (1984). d Alonso et al. (1984a). e Brown et al. (1985). f Cliberto et al. (1985). g Marzluff et al. (1983). h Watanabe-Nagasu et al. (1983). i Murphy et al. (1982). j Denison and Weiner (1982). k Earley et al. (1984).

varieties are differentially expressed, and hence may have distinctive properties—this in spite of the two U1 subtypes being distinctive at only two of their ~165-nucleotide lengths, barely over a 1% difference (Table 12.2). Since the quantity of data pertaining to the several types is inversely correlated to their numerical designations, it is logical to begin the comparative analysis with U1.

12.2.1. U1 snRNA

Although under normal conditions discussion of the U1 snRNA would include analyses of the 5' and 3' flanking sequences, the literature in this recently opened field is still too full of contradictions to make that procedure profitable. To illustrate, in some reports on the leader and its associated signals, it is affirmed that transcription commences at a site 181 nucleotides upstream of the mature coding region, as already cited in a preceding section (Lund and Dahlberg, 1984). Slightly later, in a different transcriptional environment, the same laboratory reported that the process began precisely at the 5' terminus of the mature coding portion (Htun *et al.*, 1984; Skuzeski *et al.*, 1984). Expression of the gene in the mouse appeared to require a sequence lying between 150 and 400 sites upstream from the coding sector (Moussa *et al.*, 1985). Until the point of transcriptional initiation is thoroughly established, discussions of possible promoters appear to be of little use.

Structure of U1 Mature Coding Sequences. The 11 sequences of U1 from a wide diversity of sources display a remarkable uniformity of structure (Table 12.2). Among those from metazoans, only the second example from *Drosophila*, designated 82E, differs to any extent from the rest (Alonso *et al.*, 1984a; Kejzlarová-Lepesant *et al.*, 1984), and even there, the distinctions are confined to the first 54 residues, the remaining ~110 sites displaying a high level of homology. Variation among the vertebrate forms also is slight (Denison and Weiner, 1982; Earley *et al.*, 1984); the most prominent deviations are the additions of an A at site 22 of row 1 in the rat (Watanabe-Nagasu *et al.*, 1983) and of a G in site 4 of row 2 in the murine representative (Marzluff *et al.*, 1983). However, the species partially sequenced from a seed plant (the pea) differs strongly, significant regions of homology being notable only near the end of row 2 and in row 3 (Krol *et al.*, 1983). The total lengths also are nearly invariable, although that of the pea is exceptionally short and that of the mouse greatly elongated, containing 80 more nucleotides than the others.

It is of interest to note that although the U designating this family of genes indicates a supposed richness in uridine residues (Reddy *et al.*, 1981c), counts of the composition of those in Table 12.2 show that guanidine is actually the most abundant nucleotide. In the human sequence (Branlant *et al.*, 1980; Murphy *et al.*, 1982), there are 33 As, 41 Cs, 43 Ts (Us), and 47 Gs. Nevertheless, the U designation is a convenient one for this family, despite this loss of real meaning.

Role in Processing. Although the roles of the several U RNAs receive additional attention in a later section, the comparative aspects of function are more effectively presented along with the structures. Moreover, viewing them individually provides a simple basis for what later proves to be a complex process. As for the present species, it seems definitely to be involved in the splicing of mRNAs at points where introns are to be excised during processing. Among the clearest set of experiments that have demonstrated this activity was one involving injection of SV40 pre-mRNAs into *Xenopus laevis* oocytes, either with or without antibodies to U1 (Fradin *et al.*, 1984). Actually, the antibodies were against a U1–protein complex (U1 RNP), for it is when combined with at least eight proteins that this substance is active in processing (Setyono and Pederson, 1984; Mattaj and DeRobertis, 1985). Without the antibodies, synthesis and splicing of the viral pre-mRNAs from SV40 DNA were carried out efficiently, but when the DNA and sera containing the antibodies were coinjected, transcription continued at a normal rate, but splicing of late mRNAs was dramatically inhibited. Cleavage at both ends of the introns was blocked, leading to an accumulation of late pre-mRNAs, with intact introns; these, however, bore the typical poly(A) tails at the 3′ ends. Hence, it was concluded that U1 plays an essential role in *splicing* the mRNAs *in vivo*. According to the results, this conclusion needs to be modified to roles in *cleavage* of introns and perhaps splicing, for if the latter step alone were involved, the result would have been accumulation of liberated exons, only the terminal ones of which would have borne poly(A) tails.

Further results of that same study also are illuminating. It may be noted above that only the *late* pre-mRNAs are mentioned, for the *early* ones were processed completely, the introns of the small t antigen (an early protein) being removed and the exons joined as usual. Some inhibition, nevertheless, was noted on another early product, the large T antigen. Thus, here, too, is made evident the nonmechanical nature of another detail of gene control and expression, for each class of substance encoded by genes is produced, and treated for use, by its own individual set of enzymes and ancillary substances.

When U1 is involved in processing pre-mRNAs, it has been stated to bind preferentially to a site at the 5′ end of introns (Mount *et al.*, 1983; Krämer *et al.*, 1984), but some evidence has been presented to support the idea that it could attach to the 3′ end as well (Rinke *et al.*, 1984). The nature of the interaction is comprehensible only after the structure of introns has been viewed later in this chapter, while to understand the location of this snRNA, a brief examination needs to be made of its secondary structure.

Secondary Structures of U1. Like many other classes of small RNAs, the secondary structure of U1 has been deduced to resemble a cloverleaf (Figure 12.1), the details of which vary somewhat with the source organism and subtype. Although other models have been suggested, that shown is currently the most widely accepted (Mount and Steitz, 1981). In the illustration, only in the 5′- and 3′-terminal sectors are the nucleotide base abbreviations given, since it is in those regions that the specific interactions with introns have been proposed to occur. As shown in Table 12.2, the 5′ ends of most species are particularly constant in structure, although minor variations do occur. Two of the more active portions of the free 5′ arm were shown to be UACψψAC by use of a complementary heptameric probe (AUGAAUG) and, by means of a similarly constructed pentameric sequence (AAUGG), the overlapping ψψACC (Rinke *et al.*, 1984; Tatei *et al.*, 1984). A third reactive portion was found in the free 3′ arm, where the pentameric probe CCGUC was found to bind, but no trace of the complement GGCAG appears to exist in the appropriate sector (Table 12.2). Another highly conserved sequence, ATTTNTG, is shown underscored near the 3′ terminus of row 2.

There is also an obvious problem with the second variety of *Drosophila* U1, referred to as 82E (Kejzlarová-Lepesant *et al.*, 1984); because its 5′-terminal sector includes no sequence corresponding to either the heptamer or first pentamer, it has been proposed to be a pseudogene. Since the murine gene is so strongly elongated, the entire molecule must

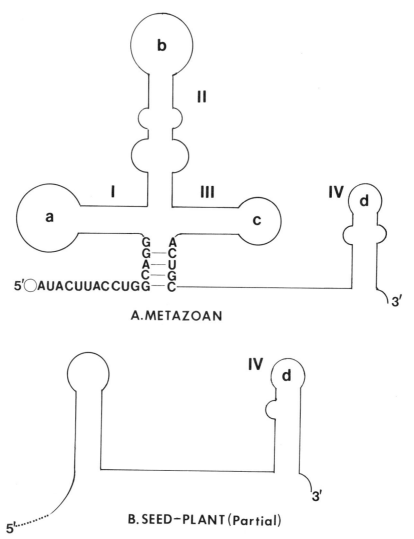

Figure 12.1. The secondary structures of U1 RNAs (A) The four-parted secondary structure is one of several models that have been proposed for the metazoan U1 molecule. (Modified from Lin and Pederson, 1984; Brown *et al.*, 1985; Krol *et al.*, 1985.) (B) The partial sequence from the pea shows considerable resemblance to the 3'-terminal region of the metazoan species. (Krol *et al.*, 1983.)

have a secondary structural configuration different from the others. The latter statement applies equally to that of the pea sequence (Figure 12.1B), but for the opposite reason, that of its being artificially abbreviated at its 5' terminus (Krol *et al.*, 1983). These deviations from the norm are found to hold considerable significance in the discussions of introns and their processing. Another complicating factor still needs resolution, however—the 5' end and much of the U1 molecule have been demonstrated to be coated with protein in the active RNP particle (Lin and Pederson, 1984), leaving little room for direct base-paring between the U1 and pre-mRNAs.

12.2.2. The snRNA U2 Gene

Although the U2 major type of snRNA doubtlessly is nearly as broadly distributed throughout the eukaryotic world as the preceding species, it has attracted far fewer investigations, as attested to by the paucity of representatives in Table 12.3. However, its discovery in the pea nucleus (Krol *et al.*, 1983) clearly substantiates that it is not confined to the metazoan world. Unlike the variation that was exhibited by the two sequences purported to represent U1 in *Drosophila,* the three varieties included here from that insect are nearly totally identical (Alonso *et al.*, 1984b). That identified as A differs at only five sites, two divergent points (italicized) being just before the 3' end of row 1, and three being in the latter half of row 2. Only the 3' half of the gene from the pea has been sequenced, but that part amply demonstrates that the seed-plant form is longer than the rest and only partly homologous with them (Krol *et al.*, 1983). In the human genome, the gene for this species is represented by ~20 copies, arranged in one or a few large clusters (Westin *et al.*, 1984b).

Chemical Constitution. Although U1 has proven to be misnamed, the present species is definitely deserving to bear the abbreviation of this family, for it is decidedly uridine-rich. When the entire molecule is considered, that base constitutes 30% of the sites, 56 residues being present in the 189-nucleotide human gene (Hammarström *et al.*, 1984; Htun *et al.*, 1985). Again in contrast to the previous member, guanosine is most poorly represented in the one at hand, with a total of only 43 residues, or 21%.

But these facts present only a part of the real picture, for the distribution of the several nucleotidyl species is quite uneven. Uridine, by way of illustration, constitutes over 40% of row 1 of that same gene, but forms minor parts of the remainder of the RNA. On the other hand, adenosine is the major component of row 2, while cytosine provides over 35% of row 3. Since this RNA most likely carries out its activities on a regional(or domain) basis rather than as a unit, the overall content of the various nucleotides appears to hold less real significance than their location.

Function of the U2 RNA. Very little has been accomplished to establish the function of U2 RNA. It is known that the biochemical does not carry out its activities until it has combined with proteins to form RNP particles. If the present species resembles the better known U1 to any extent, which is united with eight species of proteins, its particles, too, are complexly formed (Mattaj and DeRobertis, 1985). Some of the few that have been identified are common to the two species of RNA, namely "core proteins" B, D, and E (Pettersson *et al.*, 1984). Another of the shared constituents is known as Sn antigen, a protein that for binding requires a sequence including AUUUUUG, the gene counterpart of which is shown underscored near the center of row 2. In the nearly identical sequence of U1 it will be recalled as being located near the end of that same row. Another type of protein, not specifically identified, requires the two 3' loops of the U2 molecule for binding.

Since U2 is so similar in construction to U1, it is generally accepted that is also plays a part in the cleaving and splicing of mRNAs. This supposition has been confirmed to some extent by the association of the two species at major sites of gene activity in the polytene chromosomes of *Chironomus* (Sass and Pederson, 1984) where transcription was in progress.

Secondary Structure. The secondary structure of U2 as it is understood differs significantly from the cloverleaf pattern of many small RNAs in having four stem-and-

Table 12.3
Comparison of U2 Coding Regions[a]

Row 1

Drosophila A[b]	ATCGCTTCTCGG	CCTTATGGCTAA	GATCAAAGT	GTAGTATCTGTT	CTTATCAGCTTA	ACATCTGAT	AGTCCTCC
Drosophila B[b]	ATCGCTTCTCGG	CCTTATGGCTAA	GATCAAAGT	GTAGTATCTGTT	CTTATCAGCTTA	ACATCTGAT	AGTTCCTCC
Drosophila C[b]	ATCGCTTCTCGG	CCTTATGGCTAA	GATCAAAGT	GTAGTATCTGTT	CTTATCAGCTTA	ACATCTGAT	AGTTCCTCC
Human[c]	ATCGCTTCTCGG	CCTTTTGGCTAA	GATCAA-GT	GTAGTATCTGTT	CTTATCAGTTTA	ATATCTGAT	ACGTCCTCT
Rat[d]	ATCGCTTCTCGG	CCTTTTGGCTAA	GATCAA-GT	GTAGTATCTGTT	CTTATCAGTTTA	ATATCTGAT	ACGTCCTCT

Row 2

Drosophila A	ATTGGAGGACAA	CAAATGTTAAAC	TGATTTTTG-	GAATCAGAC-	GGAGTGCTA	GGAGCTTGCTCC	GCCTCTCG
Drosophila B	ATTGGAGGACAA	CAAATGTTAAAC	TGATTTTTG-	GAATCAGAC-	GGAGTGCTA	GGGGCTTGCTCC	ACCTCTGTCA
Drosophila C	ATTGGAGGACAA	CAAATGTTAAAC	TGATTTTTG-	GAATCAGAC-	GGAGTGCTA	GGGGCTTGCTCC	ACCTCTGTCA
Human	ATCCGAGGACAA	T--ATATTAAAT	GGATTTTTGG	AGCAGGGAG-	ATGGAA-TA	GGAGCTTGCTCC	GTCCACTCCA
Rat	ATCCGAGGACAA	T--ATATTAAAT	GGATTTTTG-	GAACTAGGA-	GTTGGAATA	GGAGCTTGCTCC	GTCCACTCCA
Pea[e]	(Incomplete)		ATTTCTT-	GAGGGGGAAG	AGTCACACA	GTAGCTTGCTAT	TGGGTCTCTT

Row 3

Drosophila A	CGGGTTGGCCCGGT	ATTGCAGTA	CCCCCGGGATTT	CGGCCCAACT
Drosophila B	CGGGTTGGCCCGGT	ATTGCAGTA	CGGCCGGGATTT	CGGCCCAACT
Drosophila C	CGGGTTGGCCCGGT	ATTGCAGTA	CCGCCGGGATTT	CGGCCCAACT
Human	CGCATCGACCTGGT	ATTGCAGTA	CCTCCAGGAA--	CGGTGCACCC
Rat	CGCATCGACCTGGT	ATTGCAGTA	CCTCCAGGAA--	CGGTGCACC
Pea	CGCCTCAGTCG---	CTTTTGCGT	TGCACTATAGCA	ATTGCTGGC GCACCCCAC

[a]Regions of possible functional importance are shown underscored; distinctive sites of the several types from *Drosophila* are italicized.

[b]Alonso *et al.* (1984b).

[c]Westin *et al.* (1984b).

[d]Tani *et al.* (1983).

[e]Krol *et al.* (1983).

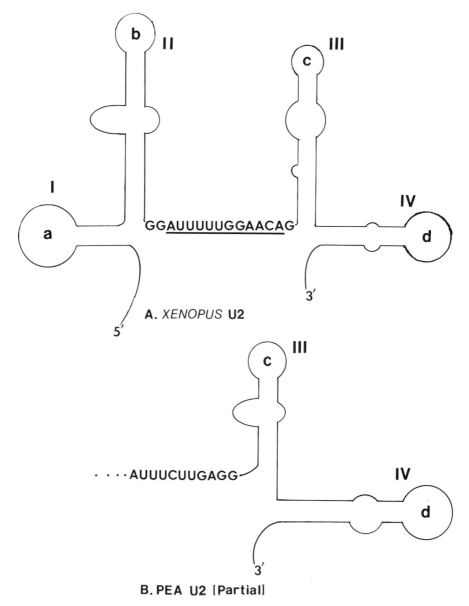

A. *XENOPUS* **U2**

B. PEA U2 (Partial)

Figure 12.2. Comparison of U2 RNA secondary structures. (A) In the *Xenopus* structure a long region between the two halves appears to hold particular functional value. (Mattaj and DeRobertis, 1985.) (B) The partial sequence from pea is strikingly similar to that of the vertebrate. (After Krol *et al.*, 1983.)

loop structures instead of just three and in lacking a petiole (Figure 12.2). Probably the most outstanding feature is the single chain in the central portion of the diagram, which bears the binding sequence for the Sn antigen (Mattaj and DeRobertis, 1985) just pointed out. Another major distinction from U1 is the absence of any suggestion of the signal UACUUAC(C) in the 5'-terminal single-stranded part supposedly reactive with introns in that species. Moreover, the 3' end is probably too short to play any role in processing.

12.2.3. The Gene of the U3 RNA

Both the genes for and the products of U3 have been subjected to fewer investigations than either of the preceding species, from which it differs in most details of structure and function. However, its transcript does agree with the others in bearing at the 5' end the typical $pm^{2,2,7}Gpp$ cap that marks the family. In function this species has been implicated in the processing of rRNAs, to the precursors of which it has been found hydrogen-bonded, and, accordingly, it is confined to the nucleolus (Reddy *et al.*, 1979, 1980).

Comparisons of U3 Gene Structures. The gene structure for this, the longest of the U family of RNAs, contains ~216 base pairs. As with the two preceding types, it is present in multiple copies in mammalian genomes, at least three variants being expressed in the human and four in the rat. Of these varieties, A and B from the first source and the fourth (D) from the latter (Reddy *et al.*, 1980; Stroke and Weiner, 1985) are given in Table 12.4. In addition, a sequence from *Dictyostelium* is also included (Wise and Weiner, 1980), in which organism the genes are dispersed in the DNA, at least five copies being detected. Numerous pseudogenes are also a marked feature of the vertebrate genome (Reddy *et al.*, 1985b).

The three varieties from mammals (Table 12.4) are perceived to be identical in composition until shortly after the 5' end of row 2, where the B gene begins to display minor variations. Nevertheless, the D variant remains nearly identical to the A form until near the opposite terminus of that same row, where it commences to deviate, remaining homologous with the other two only at scattered sites. The sequence from *Dictyostelium* has been shown to be only 40% homologous to U3 of the rat (Wise and Weiner, 1980).

Chemical Properties. From a chemical standpoint, the U3 RNA resembles U1 most closely, in that it is a guanosine-rich RNA rather than the presumed uridine-rich. Analysis of the human A gene discloses that, as in U2, the product's 5' end represented by row 1 is laden with uridine residues, 25, for a 33% ratio. The 3'-terminal portion (row 3), with U in 17 of the 66 sites, has its normal 25%, while guanosine in 25 positions approaches a 40% level. In the midregion (row 2) the latter nucleotide and cytosine are by far the predominant species, each being represented by ~30% of that sector. As might be expected from the paucity of information regarding the activity of this RNA, no suggestions exist as to what sequence may be involved in its interactions with pre-rRNAs.

Secondary Structure. Like the two preceding species, the secondary structure of U3 is basically two-looped, but in this case the two are more nearly of like length (Figure 12.3). Also as in those others, the 5' loop has a leaflike portion, but in this case with four leaflets (Bernstein *et al.*, 1983; Stroke and Weiner, 1985). The same pattern apparently prevails in all the varieties so far identified, except for slight modifications in the lower left leaflet, but reactive sites have not as yet been fully determined.

Table 12.4

Comparison of U3 Mature Genes

Row 1

Human A[a]	GAAGTGACTATA	CTTTCAGGG	ATCATTTCTATA	GTTCGTTAC	TAGAGAAGTTTC	TCTGACTGTGTA GAGCACCCG
Human B[a]	GAAGTGACTATA	CTTTCAGGG	ATCATTTCTATA	GTTCGTTAC	TAGAGAAGTTTC	TCTGACTGTGTA GAGCACCCG
Rat D[b]	AAG--ACTATA	CTTTCAGGG	ATCATTTCTATA	GTTCGTTAC	TAGAGAAGTTTC	TCTGACTGTGTA GAGCACCCG
Dictyostelium[c]	ATGACCAAACTC	TTA----GG	ATCATTTCTAGA	GTATCGTCT	ATTAAAAATTATT	CATCAATAATTT TTCCTCTTT

Row 2

Human A	AAACCACGA	GGAGGAGACGGA	GCGTTCTCT	CCTGACGGT	GAAGCCGGCTCC	TAGTGTTCCTTC CTGCCATTGCTA
Human B	AAACCACGA	GGACGAGACATA	GCGTCCCCT	CCTGACGGT	GAAGCCGGCTCT	AGGTGCTCCTTC TGCCTCTTGCCA
Rat D	AAACCACGA	GGAGGAGACGTA	GCGTTCTCT	CCTGACGGT	GAAGCCGGCTCT	TAGTGTTGCTTC CTGCAACTGCTA
Dictyostelium	----CACAG	CTAGGATGATGA	TACACA-CT	CACTATACG	AAAGGGTGAAAC	CGTTATTATCAA ATGATTCATTTA

Row 3

Human A	TTGGCAACTGAT	GATCGTCTTCGG	TCCTCTTGA	GGGGTGAGAG--GG	AGAGAACGC	AGTCGAGTGGT
Human B	TTGGCAGCTGAT	GATCGTCTTCTC	TCCTTCGGG	GGGGTAAGAG--GG	AGGGAACGC	AGTCGAGTGGA
Rat D	TTGGCCATTGAT	GATCGTTCTCGA	GCCTCTCTG	AGGTTC-GAGAGGG	AGAGAACGC	AGTCGAGTGGT
Dictyostelium	TTTGTTATTAAC	ATTGATGACCGT	CTAATTCAG	GGATGAATTG--GT	TGTATGGTG	GGATTCGTACTG GCT

[a]Reddy *et al.* (1980).
[b]Stroke and Weiner (1985).
[c]Wise and Weiner (1980).

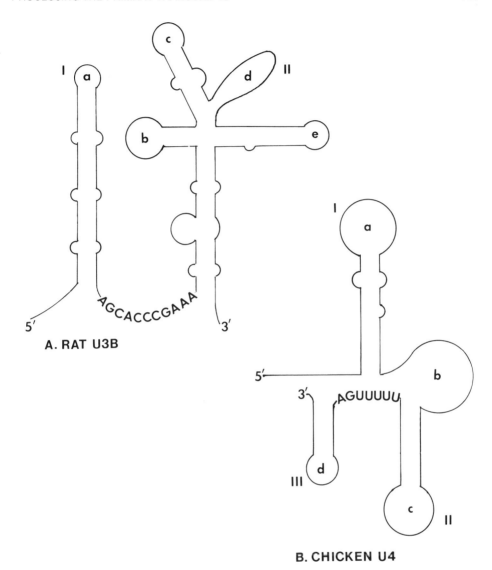

A. RAT U3B

B. CHICKEN U4

Figure 12.3. Secondary structures of U3 and U4 RNAs. (A) The several varieties of U3 that have been established deviate somewhat in details of secondary structure, but the overall pattern remains constant. As in U2, the connector holds a sequence of functional value. (Stroke and Weiner, 1985.) (B) In U4 RNA the usual bipartite arrangement is absent, and an internal region seemingly is involved in function. (Krol *et al.*, 1981.)

12.2.4. The Genes for U4 and U6

Somewhat paralleling U1 and U2, which are associated in their activity, U4 and U6 also appear to combine in the cell while carrying out their functions. But in the present instance the union is stronger, for the two species elute together in columns prepared with antibodies to the $m^{2,2,7}G$, even though U6 does not bear such a cap (Bringmann *et al.*, 1984). Thus the pair appears to be bound in the same RNA particles (Hashimoto and

Table 12.5
Comparison of U4 Genes

Row 1

Chicken[a]	-AGCTTTGCGCA	GTGGCAGTATCG	TAGCCAATG	AGTTAATC	CGAGGCGCGATT	ATTGCTAAT	TGAAAACTTTTC
Murine[b]	GAGCTTTGCGCA	GTGGCAGTATCG	TAGCCAATG	AGTTTATC	CGAGGCGCGATT	ATTGCTAAT	TGAAAACTTTTC
Human[c]	GAACTTT--GCA	GTGGCAGTATCG	TAGCCAATG	AGTTTATC	CGAGGCGCGATT	ATTGCTAAT	TGAAAACTTTTC

Row 2

Chicken	CCAATACCCC	GCCGTGACGACT	TGCAATATAGTC	GGCATTGGCAAT	TTTTGACAGTCT	CTACGGAGACTGG
Murine	CCAATACCCC	GCCATGACGACT	TGAAATATAGTC	GGCATTGGCAAT	TTTTGACAGTCT	CTACGGAGACTG
Human	CCAATACCCC	GCCGTGACGACT	TGAAATATAGTC	GGCATTGGCAAT	TTTTG----TCT	CTACGGAGACTG

[a]Krol et al. (1981).
[b]Kato and Harada (1981).
[c]Reddy et al. (1981a).

Steitz, 1984). However, U6 also exists alone in the form of perichromatin granules, possibly involved in trimming the 3' ends of histone mRNAs (Daskal, 1981), and U4 may similarly occur separately, possibly being active in adding poly(A) tails (Berget, 1984).

Gene Structure of U4. Several genes for U4 have been established, a representation of which appears in Table 12.5, one each from the chicken, mouse, and man (Kato and Harada, 1981; Krol *et al.,* 1981; Reddy *et al.,* 1981a). All display a high level of evolutionary conservation, with very few divergences, even in that of the chicken. Probably the most notable deviation in the latter is the absence of one nucleotide at the extreme 5' terminus and one addition (G) at the other, while that of the human has lost two short runs of nucleotides, one (TGC) just after its 5' beginning and ACAG preceding the downstream termination.

Since only a single nucleotide sequence of U6 has been determined (Tables 12.1 and 12.6), there is little that can be said concerning its structure; its sole striking feature is its brevity, 108 residues, compared to 139–146 in the U4.

Secondary Structure of U4. Several proposals have been made concerning the secondary structure of U4 (Kato and Harada, 1981; Krol *et al.,* 1981). Only one of the several variations of the latter reference is given in Figure 12.3. As in the others, a second stem structure is present (represented by arm IV), but it is far shorter than typical. Moreover, the middle leaflet is better described as a large loop, for it has no double-stranded stem. One portion that is of special note is one base-paired and bearing a run of four free Us in the connector between loops III and IV, a structure that may merit closer attention.

12.2.5. The Gene for U5

Greater success has been experienced with the sequencing of the U5 nuclear RNA than with the last several species, for they have been established from at least eight source organisms. Moreover, a much wider spectrum of types is represented in those given in Table 12.6, including an example from a seed plant, a euciliate, and a dinoflagellate in addition to the accustomed vertebrates. Further sequences are available from duck and pheasant (Branlant *et al.,* 1983), but they are not included, since they contribute little to the general picture.

Gene Organization. As pointed out by Branlant *et al.* (1983), several regions of this RNA are highly conserved evolutionarily, an observation equally applicable to the overall structures. The length is unusually constant, ranging from a minimum of 108 base pairs in the dinoflagellate (*Crypthecodinium*) to a maximum of 118 in the human representative (Branlant *et al.,* 1983; Liu *et al.,* 1984). Among the constant structural sequences is the dinucleotide AT at the 5' end; CT is frequently present at the opposite terminus, often with an A in the third position, especially in the dinoflagellate and *Tetrahymena.* Moreover, about 30 base pairs toward the downstream site of row 1 are also strongly conserved, but possibly the most significant region of sequence homology is displayed near the beginning of row 2, where a decanucleotide AA-TTTTTGAG (underscored) is a nearly universal feature.

Chemical Nature. Based on the pea sequence of U5 (Krol *et al.,* 1983), the chemical nature is found to be somewhat uridine-rich, in that 38 of its 122 nucleotides are of that type, for a rate of 31%. Guanidine, with 31, just slightly surpasses adenosine, with

Table 12.6
The Smaller Species of the U-Family RNAs[a]

Row 1

Pea U5[b]	GGAGCCGTGTGA	TGATGACATAGC	GAACTA-TCTTT	CGCCTTTTA	CTAAAGAAT	ACTGTGTCA	GGGTCACAATTA
Tetrahymena U5[c]	NATCACA-TAAC	TCAGCTCAATAC	GCTTTAATTTTT	CGCCTTTTA	CTAAAGATA	TCCGTGGGC	TGGGTTCTACAA
Human U5[c]	NATACTCTGGTT	TCTCTTCAGATC	GCATAAATCTTT	CGCCTTTTA	CTAAAGATT	TCCGTGGAG	AGGAACAACTCT
Rat U5[d]	NATACTCTGGTT	TCTCTTCAGATC	GTATAAATCTTT	CGCCTTTTA	CNAAAGATT	TCCGTGGAG	AGGAACAACTCT
Chicken U5[e]	GATACTCTGGTT	TCTCTTCAGATC	GTATAAATCTTT	CGCCTTTCA	CTAAAGATT	TCCGTGGAG	AGGAACAACTCT
Human U6[f]	NGTGCCTGCTTC	GGCAGCACATAT	ACTAAAATTGGA	ACGATACAG	AGAAGATTA	GCATGGCCC	CTGCGCAAGGAT
Sea urchin U7[g]	TCTTTCAAGTTT	CTCTAGAAGGGT	CTCGCGTCCGAA	GTCGGAGGC	GAGTGCCCA	ATT	

Row 2

Pea U5	GGGGCATACGCT	AG-TTTTTG	GAAGAGTTC	TCAAGTTTTGAG	GGCTCTG
Tetrahymena U5	TGTGAATTATTA	AAATTTTTG	AGGA-TT-T	GTTG-AATCCTA	
Human U5	GAGTCTTAACCC	AATTTTTTG	AG-CCTTGC	CTTGGCAAGGCTA	
Rat U5	GAGTCTTAAACC	AATTTTTTG	GAGGCCTTG	TCTTGACAAGGCT	
Chicken U5	GAGTCTTTAAACC	AATTTTTTG	AGCCTTGTT	CCGGCAAGGCTA	
Human U6	GACACGCAAATT	CGTGAAGCG	TTCCATATT	TTT	

[a]A sequence in the U5 genes that is highly conserved evolutionarily is underscored.
[b]Krol et al. (1983).
[c]Branlant et al. (1983).
[d]Okada et al. (1982).
[e]Liu et al. (1984).
[f]Epstein et al. (1980).
[g]Strub et al. (1984).

30, both being at a normal 25%. Thus, this macromolecule is as markedly cytosine-poor as it is enriched with uridine. However, the 5' half has concentrations of adenosine equal to that of uridine, each showing a 30% level, whereas in the 3' portion, guanidine is nearly as abundant as uracil, while adenosine is relatively sparse. Much of the uridine is arranged in clusters, one of which is in loop A and another in loop B, with a third constituting much of the connector between the two hairpins.

Secondary Structure. Although several acceptable models of secondary structure for U5 have been proposed (Okada *et al.*, 1982; Branlant *et al.*, 1983; Krol *et al.*, 1983; Liu *et al.*, 1984), the one that best provides for the conserved features just pointed out is the two-loop form illustrated in Figure 12.4. Although thus resembling most of the species seen earlier in this discussion, the 5' sector obviously lacks the cloverleaf configuration that characterizes most of the others. However, for ease of comparison, the small projections existing in place of leaflets are given similar designations. In possessing a highly conserved element on the link between the two major hairpins (Figure 12.4), U5 appears structurally related to U2 (Figure 12.2).

U5 as an RNP. As in all the preceding species, U5 functions in the form of an RNP particle, possibly in processing mRNAs; however, its special role remains undeciphered. Many of the proteins that are combined with the present type of RNA are identical to those of the other species, but additional varieties are present here that make the U5 RNPs unique. Further particulars of their constitution are not available, however.

12.2.6. The Gene for U7

Nowhere is the correlation between size of the U-family members and sparseness of information more clearly borne out than with the present species, for only a single sequence is available for its gene. That one, from a sea urchin, is shown in Tables 12.1 and 12.6 (Strub *et al.*, 1984). There its small size—57 nucleotide residues—is a particularly outstanding feature. While not completely confirmed, there is considerable evidence that this species may be concerned in the processing of histones.

Chemical Characteristics. In this representative of the U family of RNAs, the four nucleotides are present in nearly their normal proportions, adenosine alone being deficient, filling a total of 11 (19%) of the 57 sites. Cytidine occupies 14 (25%) of the positions, and the remaining two nucleotides are each responsible for 16 of the balance (for a 29% proportion). For the greater part, each occurs singly, the Us alone being clumped to any extent, with two trinucleotides located near the 5' end, but one cluster of three Cs is found near the opposite terminus.

12.3. PROCESSING OF mRNAs

Although details of the activities are just beginning to emerge, the processing of mRNAs is firmly established as requiring many separate steps. Among the specific changes that must be wrought before the primary transcript (pre-mRNA) of any protein-encoding gene is ready for translation are: (1) removal of parts of the leader and train, (2) formation of cap structures where and when called for on the former, (3) addition of poly(A) tails to the latter, and (4) removal of introns, accompanied by splicing of the

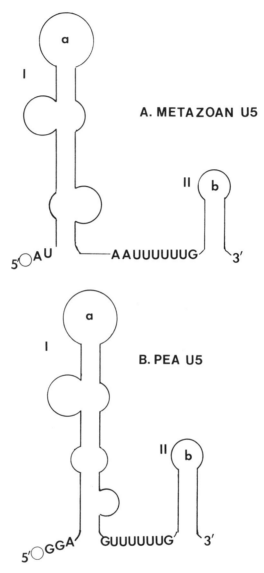

Figure 12.4. Comparison of secondary structures of U5. The fully sequenced U5 gene from (B) the pea corresponds closely in secondary structure to (A) that of the metazoan. The open circle indicates the location of the cap. (Based on Branlant *et al.*, 1983; and Krol *et al.*, 1983.)

cleaved exons. Furthermore, after translation, the product undergoes addition of various side chains, modification of the amino acids and similar specialized activities, removal of the transit peptides from diplomorphic genes, and cleavage of cryptomorphic genes into their functional parts. Hence, the processing of mRNAs is necessarily complex and accordingly difficult to explore. Moreover, thousands of types of pre-mRNAs exist in every cell, many of which possess specialized modifications that need attention from enzymes made to fit their particular requirements. As a result, the paucity of information regarding certain facets of the total process is a natural consequence of the intricacies and should cause no surprise.

12.3.1. Processing of Leaders

The processing of leaders of mRNAs is among the aspects of their posttranscriptional treatment about which little is known, but what has been established is adequate to expose some of the complexities that have evolved in cells of many types. Most of the revealed facts are found to pertain to the addition of cap structures or leader segments at the 5′ end.

Addition of 5′ Caps. The addition of a common methylated form of mRNA cap ($m^7GpppN_1mpN_2p$) illustrates the numerous variations to which a seemingly simple activity may be subjected. It will be recalled from preceding chapters that mRNAs receive this cap during the early stages of transcription in eukaryotes by polymerase II. Since the U family of snRNAs is also transcribed by this same enzyme, they should bear the same structure if the cap is an automatic product of the transcribing enzyme but, as shown in Section 12.1.1, their cap is quite different in composition. Thus, this modification of the terminal nucleotide is perceived to be a function of either an enzyme associated with the polymerase or one of the latter's subunits. Moreover, transcription by RNA polymerase II most frequently begins at a purine, in which modification of the terminal nucleotide would present no problem. To the contrary, about 25% of mRNAs have a pyrimidine at that position (Salditt-Georgieff *et al.*, 1980) and thus the cap has been suggested to be derived by endonucleolytic cleavage and subsequent addition.

In the cytoplasmic polyhedrosis virus, a double-stranded DNA species, the first step in mRNA synthesis is the formation of a preliminary cap, without which transcription of the gene is inhibited (Furuichi, 1978). This temporary additive, pppA, formed by viral polymerases, is then trimmed by a phosphohydrolase to ppA, to which the ppG is then added by a 5′–5′ linkage.

Other Leader Modifications. One type of modification that occurs to the leaders of mRNAs, already discussed briefly in Chapter 8 (Section 8.4), also relates to the foregoing topic. In connection with assembled genes, it was shown that all mRNAs of *Trypanosoma brucei* bear identical leaders 35 nucleotide residues in length. These leaders are encoded by separate sequences, really genelets, but called miniexons, present in about 200 copies per genome (DeLange *et al.*, 1984; Milhausen *et al.*, 1984; Michiels *et al.*, 1985). Analyses of the leaders of the original pre-mRNA product of RNA polymerase II transcription showed the presence of typical caps there as well as on the RNAs from the leader genelets. Since the latter replace the former by unknown processes, the final cap of the mature mRNAs is thus acquired secondarily (Laird *et al.*, 1985). It is pertinent to note that the 5 S rRNAs, and possibly others transcribed by RNA polymerase III, do not have an added leader (Lenardo *et al.*, 1985).

Knowledge is also in a preliminary state regarding the steps in formation of the leader of a familiar mitochondrial gene product, cytochrome *b* (Chapter 6, Section 6.1.2). Although confined to the organelle, the gene can be expressed only in the presence of several proteins encoded in the nucleus, including those involved in processing the transcript. Among these are a few required for the excision of introns (McGraw and Tzagoloff, 1983; Pillar *et al.*, 1983), but one other is requisite for processing the leader

(Dieckmann *et al.*, 1984a,b). In the absence of this gene, *CBP1*, the leader of cytochrome *b* pre-mRNA undergoes degradation enzymatically, so that the transcript fails to be translated. Here, then, is further substantiation of a condition pointed out especially in connection with the treatment of diplomorphic genes at the close of Chapter 7, Section 7.3, and in sections dealing with the various rRNAs—the dependence of the mitochondrion on nuclear genes is far too great even to hint at the organelle having been an independent prokaryotic invader. Although the gene for cytochrome *b* is encoded in the mitochondrial DNA, it is useless without the products of several nuclear genes—and this substance is essential for those processes for which the organelle is uniquely responsible, the respiratory activities of the cell.

12.3.2. Processing Trains of Messenger RNAs

Although the nature of transcriptional termination and polyadenylation signals received some attention in Chapter 1, Section 1.1.1, and in several chapters concerned with genes for specific classes of macromolecules, some generalities still need attention. Since nothing firm regarding terminators can be added to those discussions, aside from its remaining an open issue, the focus is solely upon the other signaling device just mentioned. Because the subject here is in the form of RNA, the Ts of DNA viewed in most preceding discussions are now replaced by Us.

The AAUAAA Sequence. Not all the results of studies on the standard polyadenylation signal, AAUAAA, are in essential agreement, because two separate processes are involved, first, cleavage, and, second, addition of the poly(A) tail. The presence of the latter on the mature mRNA is an essential feature, for all mutants deficient for polyadenylation are defective *in vivo* also for processing of the finished transcript (Moore and Sharp, 1985). But it has not been clarified whether this incapacity reflects tight coupling between the two activities or the rapid degradation of messengers lacking the poly(A) tail.

That processing of the trains is by two separate activities has been demonstrated by several procedures. If the U of the canonical sequence is mutated to a G, cleavage of the train is inhibited, whereas the adding of the tail is not affected (Montell *et al.*, 1983). On the opposite side of the coin, in an experiment in which the source of adenosine residues (ATP) in the tail was replaced by an analog, α-β-methylene-adenosine 5' triphosphate, cleavage proceeded normally, while addition of the poly(A) tail was inhibited (Moore and Sharp, 1985).

Multiple Signals. Multiple copies of the standard signal are not an uncommon occurrence on mRNAs for various types of proteins, but their usage varies with the species and the organism. One interesting example of this condition in a tropomyosin gene has been reported from *Drosophila,* in which a cluster of five poly(A) signals occurred, followed by a pair about 250 residues downstream. The first of these was a dual structure consisting of AAUAAAUAAA (Boardman *et al.,* 1985). Polyadenylation (after cleavage) seemingly took place freely between 12 and 25 nucleotides downstream of each member of the cluster, and in the embryo, beyond the two farther removed from the mature gene. Although all thus functioned equally, only two of the total, aside from the dual component, was of standard construction, the others being shortened to AUAAA or having the U replaced by a C.

In a β_1-globin gene of *Xenopus,* two AAUAAA combinations were found within an ~150-residue sector, but in this case the one located most proximal to the coding sector was responsible for polyadenylation of ~99% of the transcripts, the second one being employed in the remaining ~1% (Mason *et al.,* 1985). However, when the first was removed experimentally, the second functioned with 95% efficiency. If the upstream AAUAAA were replaced by AAUACA, ~35% of the transcripts were polyadenylated at that site. In the chicken a gene for ovomucoid similarly has been noted to possess multiple signals of this sort, all of which functioned to provide mRNAs of varying lengths (Gerlinger *et al.,* 1982).

Variations on the Theme. The several deviations in structure of the poly(A) signal just cited represent only a few of numerous examples of this kind. In the common yeast, *S. cerevisiae,* the sequence triggering the process has been found to be UAAAUAA(A or G) in the cistron for alcohol dehydrogenase, and in *S. carlsbergensis* it is UAACUAAAUAA in the cistron for the α-galactosidase (Bennetzen and Hall, 1982; Sumner-Smith *et al.,* 1985). However, the universality of these signals among the yeasts remains in doubt, for examination of at least 15 additional gene sequences for proteins selected at random from those organisms failed to show the presence of the suggested combinations in any one of them. In several the standard AAUAAA was present at suitable distances downstream of the stop codon, including the genes for carbomyl phosphate synthetase (Nyuonya and Lusty, 1984), *STA1* for extracellular glucoamylase (Yamashita *et al.,* 1985), the *GAL80* regulatory gene (Nogi and Fukusawa, 1984), and *ARG4* for argininosuccinate lyase (Beacham *et al.,* 1984). A similar condition also existed in a plasmid for the killer character of the yeast *Kluyveromyces* (Stark *et al.,* 1984). At times only variations of that sequence could be located, such as AAUAA (*HIS1;* Hinnebusch and Fink, 1983) or UAUAAAA (*CUP1;* Karin *et al.,* 1984). Consequently, until cleavage of the train and polyadenylation of a number of genes have received experimental attention, any broad generalizations are premature.

Further variants are detected in yeast mitochondrial mRNAs, in which cleavage of the train is dependent upon the dodecameric sequence AAUAAUAUUCUU in at least ten of the protein-encoding pre-mRNAs (Osinga *et al.,* 1984). No similar structure exists in the mitochondria of vertebrates from the genomes of several species of which complete sequences have been established, as seen on several preceding occasions (Anderson *et al.,* 1981, 1982; Bibb *et al.,* 1981; Roe *et al.,* 1985).

Generation of Stop Codons. Sometimes polyadenylation plays a role in generation of the translational stop codons. Since the nascent transcripts are polycistronic, they undergo cleavage prior to translation, but in many cases the close packaging characteristic of these organellar genes leaves no room for trains or even the usual termination stop codons. Hence, those pre-mRNAs whose coding sequences end in U or UA have a series of As added posttranscriptionally, thereby completing one of the standard stop triplets (Anderson *et al.,* 1981). This observation leads to two questions: First, if these organelles are direct modifications of a prokaryotic invader, as some workers visualize, how is it that they possess the enzyme necessary for polyadenylation, a strictly eukaryotic process? And second, does this simple act suggest one of the original functions of polyadenylation?

Functions of Polyadenylation. The last query in turn raises a more general one as to what roles addition of a poly(A) tract to messengers may play. Three major proposals have been made: (1) providing for stability of mRNAs, (2) assisting in transport during processing and translation, and (3) direct involvement in protein synthesis (Jacobson and Favreau, 1983). Several lines of investigation seem to substantiate the latter activity. In the first place, a given length, varying with the species being translated, is optimal, so that in the experiments reported in the last citation, addition *in vitro* of free poly(A) to poly(A)-bearing mRNAs had an inhibiting effect, whereas similar addition to poly(A)-minus messengers had a stimulating action. But an *in vivo* study of changes in the tail of the maternal mRNAs in eggs of the clam *Spisula* following fertilization has provided more direct evidence (Rosenthal *et al.,* 1983). Alterations in translational activity of five messengers were followed in the oocyte and after fertilization, four of them being inactive in the egg, the other active. After fertilization, the first four were found to be translated rapidly in the resulting zygote, and with this change, their originally short poly(A) tracts became elongated. In contrast, the fifth one, which encoded α-tubulin, had a distinct poly(A) tail while being transcribed in the oocyte, but underwent depolyadenylation when it was no longer translated after fertilization.

Ingredients of 3' Train Processing. It is self-evident that such intricate control of tail length and translational activity implicit to the last investigation above requires far more than a trigger mechanically activating an enzyme or two. Indeed, as Mason *et al.* (1985) have indicated, the canonical AAUAAA or its equivalent cannot be the sole recognition signal, nor can it work in a strictly mechanical fashion. Codon combinations that would provide that hexanucleotide must surely abound within the numerous coding regions proper for the many proteins required for eukaryotic cell function. At least one case where such an internal codon combination activates polyadenylation has been described. The Thy-1 mRNA, encoding the familiar cell-surface alloantigen, has been shown in the rat to possess no translational stop codon nor any train, but has a poly(A) tail 24 residues in length (Moriuchi *et al.,* 1983). It was suggested that the triplets for asparagine and lysine (AAT and AAA, respectively), lying six codons upstream of the 3' end, served to signal cleavage and poly(A) addition. But why they do not serve similarly in other instances where they occur is the pressing unanswered question!

Other ingredients essential to 3'-end processing are gradually being revealed. Among them is an RNP particle that contains a U7 RNA which is necessary in processing the nonpolyadenylated histone H3 mRNA. Still another RNP, incorporating a molecule of U1 RNA, has been demonstrated to be involved in the formation of the poly(A) tail in precursorial messengers (Moore and Sharp, 1984). But these two examples reflect only the beginnings of comprehending what will surely be found to be a process of considerable complexity, the control of the cleavage of trains and the length and formation of poly(A) tracts.

12.4. PROCESSING rRNAs

The macromolecules that remain untranslated in the form of various RNAs also are first transcribed into precursorial forms that need processing before they become functional. Although that statement is as applicable to transfer as to ribosomal RNAs, their nature and, hence, specific needs differ strongly. Accordingly, the latter group is examined first, while the preparation of active tRNAs is analyzed in the next major section.

Since the rRNA operons have received thorough attention in Chapter 4, Sections

4.1.1 and 4.2.1, the genomic arrangement of the genes in the sequence 5'-16 S, tRNA(s), 23 S, 5 S, tRNA(s)-3' is already familiar. Because the entire operon is often transcribed as a unit, it obviously needs treatment by cleaving enzymes at a number of points even to separate the several classes of pre-ribosomal RNAs. The complexity of the entire process necessitates the more thoroughly documented steps in the preparation of mature 5 S rRNA, particularly in bacteria, to be examined first to provide a firmer basis for comprehending those of the other, less precisely established ones.

12.4.1. Processing 5 S rRNAs

As shown in the earlier discussions of the rRNA genes (Chapter 3, Section 3.2.1; Chapter 4, Section 4.1.1), only three of the operons (*rrnC, -D,* and *-F*) of *E. coli* include cistrons for tRNAs downstream of the 5 S coding region. Despite the increased complexity resulting from these added components, most progress has been made with one of these, namely *rrnF,* so it is used as the central model. Two enzymes, RNase III and RNase E, appear to be the primary cleavage agents.

Early Steps of Processing. As will be recalled, the primary transcript of these operons usually contains all the various components; after transcription, this is cleaved into two parts, including one that bears the genes for the minor rRNA and the one or two tRNAs present on the intergenic spacer. The other portion, referred to as 25 S or 30 S RNA, depending on whether or not a tRNA occurs on the 3' end, also contains the major and supplemental rRNA cistrons (Ghora and Apirion, 1979). Here attention is confined to the smaller of those products, since preparation of the 30 S species is quite comparable. Neither RNase E nor III is involved in this initial enzymic reaction, since mutants defective in both proteins accumulate the 25 S species (Ghora and Apirion, 1978, 1979). A further cleavage by an unidentified ribonuclease leads to the formation of either 11 S or 10 S RNAs, depending upon the number of tRNAs present on the train. Because the larger train resisted further purification, additional analysis was devoted to the smaller, which included genes for the 5 S rRNA and tRNA^{Asp}.

Later Steps in Processing 5 S rRNA. The 10 S RNA depicted in Figure 12.5 shows the six stem-and-loop structures that characterize it (Ray *et al.,* 1982) generated by the action of RNase III. The steps in processing vary, depending upon the presence or absence of a tRNA in the train. In the latter case, a section called 9 S RNA, extending from the 5' end to site 210, is removed, which in addition has a 15-base termination stem from an unknown source, the enzyme involved possibly being RNase P. When processed from those with the former condition, such as *rrnF,* the section, referred to as 8 S RNA, is identical except for the terminator. Another stage reduces the size of the transcript by removing the 5' end at site 82, forming a region designated as 4 S RNA. After the precursorial tRNA is split off at site 259 (Figure 12.5), the fragment 7 S RNA remains, containing 5 S rRNA and a terminator stem (Szeberényi *et al.,* 1984, 1985). Action by RNase III then eliminates the termination stem at a site corresponding to 216 of Ray *et al.* (1982) (Figure 12.5), removal of the resulting stem in part being accomplished by RNase E. How the three residues remaining on each end are subsequently trimmed has not been determined in *E. coli,* but in *B. subtilis,* RNase M5 reportedly brings about the final maturational processing (Pace *et al.,* 1984; Stahl *et al.,* 1984).

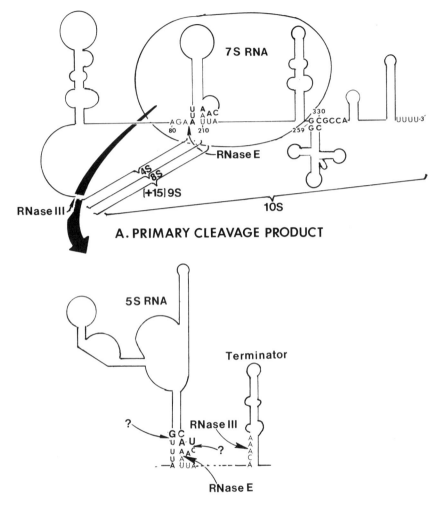

Figure 12.5. Steps in the processing of 5 S rRNA. (A) The primary cleavage product (10 S) results from the action of RNase III on the original transcript containing the sequence of the entire *rrn* operon (Ray *et al.*, 1982). (B) After removal of the sector containing the 5 S rRNA sequence and terminator, a number of enzymes, known and unestablished, complete the reductional processes. (After Szeberényi *et al.*, 1984, 1985.)

12.4.2. Processing of Other rRNAs

Investigations into the steps taken during processing of other species of rRNAs are not as advanced as those for the smaller and simpler species just discussed. As a consequence, only brief attention to the problem can be given here, and then only in regard to the major and supplemental (5.8 S) species. Because of the overall similarity of the operons for this type of macromolecule, discussion of the prokaryotic and eukaryotic activities can be in unified fashion—except for that portion pertaining to the 5.8 S variety characteristic only of eukaryotes. All seem to begin treatment of the primary transcript of the entire gene set by cleavage with RNase III or its equivalent, which breaks that large precursor between the minor and major coding sectors (Sirdeshmukh and Schlessinger, 1985).

Processing the Major Species of rRNA. In *E. coli* the primary transcript of the shorter *rrn* operons, 30 S pre-rRNA, is in the form of two loops containing the minor and major species, respectively, each closed at one side by a strongly base-paired stem, and it is these stems that are cleaved by RNase III, which acts specifically on double-stranded segments. However, this latter point is now in question, since in the absence of that enzyme the minor species is produced at normal rates, only that of the major one being totally inhibited (King and Schlessinger, 1984). *In vitro,* the action of that enzyme on the resulting 25 S product creates intermediates with two classes of 5' ends differing slightly in length. Further details of the trimming of those ends are not available, but action of the 3' terminus involves an exonuclease that proceeds concurrently with 5'-end cleavages. Moreover, these final trimmings to produce the mature 23 S rRNA apparently occur while on ribosomes that are engaged in translation (King *et al.,* 1984). Perhaps this latter observation accounts for results of studies on amino acid-starved *E. coli,* which indicated that precursorial rRNAs (25 S and the 17.5 S of the minor species) were actively employed in ribosomes when resupplementation with the amino acid took place (Mackow and Chang, 1985).

Researchers on maturation of rRNA precursors in the mitochondria of *Neurospora* have centered their focus on the removal of a long intron (~2300 base pairs) that is a consistent feature of the large (25 S) species of the fungal organelle. Apparently, the first precursor of the major rRNA is a sector sedimenting with a coefficient of 35 S, which includes a tRNA cistron on the train (Green *et al.,* 1981). In wild-type cells, the removal of the intron seems to follow cleavage of the 3' end of the 35 S RNA, for defective activity there results in aberrant folding that interferes with intron removal (Garriga *et al.,* 1984).

In eukaryotes, the processing of all rRNA gene transcripts is made complex by their first being united to ribosomal proteins. After they have been translated in the cytoplasm, those proteins are conducted into the nucleus, where they combine with the precursorial rRNA to form a large pre-rRNA molecule (Taber and Vincent, 1969). This sediments with a coefficient of 35 S in *Saccharomyces* (Mitlin and Cannon, 1984) and 45 S in rat liver (Dudov and Dabeva, 1983). In the former, modification of some of the rRNA bases has already occurred. Still in the nucleus, the large particle is split, forming 27 S and 20 S precursors, the latter being transported to the cytoplasm before processing into the 18 S minor species occurs, whereas the former is cleaved enzymatically to yield 25 S and 5.8 S mature rRNAs. In rat tissue, alternate paths exist for reduction of the primary particle, which apparently is not associated with the protein components. The most direct path has been described as commencing with cleavage in the nucleolus into a 32 S precursorial major species and a mature 18 S minor rRNA, the larger form being reduced in one step to the 28 S major species. But these stages are certainly overly simplified, for no provision is made for the 5.8 S variety, as shown below.

Processing of the 5.8 S rRNA. As in the foregoing, processing of the supplemental species of rRNA is confined to the nucleolus and involves U3 and another type of RNA, designated as 7-1 (Reddy *et al.,* 1981b). During much of its maturation, it is hydrogen-bonded to the other pre-rRNAs. Although few details have become firmly established, the first product leading to this species is a 12 S particle, followed by one sedimenting at 8 S (Reddy *et al.*, 1983a). However, some reports signify that the 3' end undergoes enzymatic treatment while in the form of a 32 S particle, with which U3 is bound to form a complex (Smith *et al.*, 1984), but the nucleotide sequence at the 5' end also appears to play an important role.

12.5. MATURATION OF tRNAs

One of the major steps in the maturation of tRNAs leads to, rather than overlaps, an important process viewed shortly in the development of mRNAs, removal of an intron. Although thus identical in many ways to those of messengers, the whole organization is so divergent from those others and so constant in the present macromolecules that discussion of the two separately leaves little room for duplication. But the relatively simple tRNAs are not as simply produced as might be expected, as the discussion shows.

12.5.1. The Introns of tRNAs

Because the structural features of the introns have already been discussed in Chapter 1, Section 1.1.2 and Chapter 2, Section 2.4.1, here only those features of possible importance in their processing need brief review. These intervening sequences are widespread among the genes of the Eukaryota, including those of their chloroplasts and certain mitochondria (Deno *et al.*, 1982), but in the Prokaryota they occur only in such advanced genera as *Sulfolobus* (Kaine *et al.*, 1983). Although variable in length, the structures are highly consistent in location, nearly always being inserted at a point one site removed downstream from the anticodon; however, in the gene for a chloroplast tRNA^{Leu}, the intron splits that latter sequence (Steinmetz *et al.*, 1982). In the secondary structure of yeast transcripts, the anticodon customarily forms part of a double-stranded region, but that of tRNA^{Pro} is exceptional in this trait (Ogden *et al.*, 1984). Nor is it consistent elsewhere, not even in *Sulfolobus* (Kaine *et al.*, 1983), and in a *Drosophila* tRNA^{Leu} the entire anticodon is situated in a large, single-stranded loop (Robinson and Davidson, 1981). Structurally the termini of the introns display no uniformity, except that the 3' nucleotide is usually A or U, although one tRNA^{Leu} of *Drosophila* has a G (Robinson and Davidson, 1981).

Basic Steps in Processing Introns. The first stage of removing the intron from monintronic tRNA transcripts is the cleavage of the tRNA coding sections from the inserted sequence, leaving the latter along with two half molecules of mature tRNAs as linear intermediates, although the intron quickly becomes circularized (Knapp *et al.*, 1979; Peebles *et al.*, 1979). The endonuclease concerned with the cleavage, still unidentified, appears highly conserved evolutionarily, for yeast tRNA genes are processed correctly in a HeLa extract *in vitro* (Standring *et al.*, 1981). As a whole the limited data indicate that tertiary structure is a prime determinant in the cleavage sites, rather than sequence identification. Mutations at the splice junctions have little effect, whereas those that alter the structure of loop II, extra arm, or the anticodon of the tRNA proper decrease efficiency or accuracy of the processes (Colby *et al.*, 1981; Willis *et al.*, 1984; M. C. Lee and Knapp, 1985). Nor does the primary sequence of the intron play a role, for extreme alterations in its length have failed to influence removal (J. D. Johnson *et al.*, 1980).

Splicing of the two half molecules seems to be more demanding in its requirements, particularly in the nature of the 3′ terminus of the 5′ (left) half and the 5′ nucleotide of the 3′ (right) part. During cleavage, the latter residue is modified by the endonuclease to a 2′,3′-cyclic phosphodiester, but prior to ligation, it is phosphorylated and activated by addition of AMP. Splicing then occurs either concurrently with or subsequent to opening of the cyclic phosphodiester to the 2′ position to produce a 3′,5′-phosphodiester,2′phosphate at the junction (Ogden *et al.*, 1984), the latter component subsequently being removed. However, the 2′ phosphate alone originates from the 2′,3′-cyclic phosphodiester, for the 3′,5′ species is derived from the γ-phosphate of ATP (Gegenheimer *et al.*, 1983).

12.5.2. Other Processing Steps of tRNAs

Actually the removal of the intron is among the last steps in the processing of tRNAs, but since the presence or absence of such sequences strongly influences the others, it has been accorded first treatment (P. F. Johnson and Abelson, 1983). All of the stages of tRNA maturation proceed in a fixed temporal order: (1) trimming of the 5′ and 3′ ends from the primary transcript, (2) addition of the universal -CCA at the 3′ terminus when not encoded in the DNA, (3) modification of the bases in the mature coding region, (4) removal of introns when present, and (5) splicing (Standring *et al.*, 1981). In fact cleavage and splicing at intron junctions has been found to be highly dependent upon the presence of the -CCA end piece, thus confirming that the triplet has to be added prior to removal of intervening segments (Guerrier-Takada *et al.*, 1984). As outlined in Chapter 2, Section 2.2.1, the primary products of tRNA transcription are quite diverse, some being monocistronic, others dimeric, and many others polycistronic. Moreover, frequently, especially in prokaryotes, the 3′-terminal -CCA is encoded in the DNA, whereas in numerous instances among eukaryotes that trimer must be added, a variation over and above the presence of introns. As a consequence, the primary transcripts are a heterogeneous lot indeed.

Basic Steps in Processing. In view of the complexity of the elementary pre-tRNAs, it is distinctly advantageous to begin with the simplest example and then add further details to that basic set of processes. Here the prokaryotes provide the model, for only in those organisms have the primary enzymes been identified. After generation of a primary transcript from a monomeric tRNA gene has been completed, one or more endonucleases reduce the size of the originally long flanking sequences, permitting the mature tRNA coding section to assume its final secondary and tertiary configuration, perhaps with the assistance of an enzyme or enzyme system. In this form, it is recognized by RNase P, which cleaves the 5′ end from the coding section proper. Afterward RNase D removes any excess nucleotides on the 3′ flanks to expose the typical -CCA (Gegenheimer and Apirion, 1981), although in many instances no extraneous bases extend downstream of that triplet.

In the case of dimeric and polycistronic transcripts, some additional steps are requisite, involving the endonucleases E and III. These two proteins act on sequences between the several gene transcripts to produce monomers and smaller multiples of the substrate ready for the activity of RNase P, and later, that of D, as before. However, it should be noted that in bacterial mutants deficient in the latter enzyme, RNase F or an equivalent can carry out the 3′ cleavages, and similarly RNase P and F can substitute for E and III to release the smaller original sectors. But nothing is now known that can carry out the activities normal to RNase P.

Special Aspects of Eukaryotic tRNA Processing. In eukaryotes, the primary transcript of tRNA genes, typically monocistronic, bears extra nucleotides at each end (Castagnoli *et al.*, 1982). Although these added residues normally play no role in the processing steps, when they are experimentally made to form base pairs, maturation is retarded. To the contrary, in the dimer containing mature coding regions of tRNA[Arg] and tRNA[Asp], base pairing is seen to exist in the extensions between the two, thereby showing that the preceding generalization is not completely valid.

Progress in studies on the whole has been slowed by the lack of data pertaining to the several enzymes involved, but a recent study, using a dimeric tRNA precursor from yeast, has added significant details (Engelke *et al.*, 1985). The major steps in processing of the dimer appear to be, first, cleavage of the extreme 5′ end of tRNA[Arg] by enconuclease P_E (eukaryotic equivalent of bacterial RNase P), to mature its 5′ sector. Second, a similar cleavage of the tRNA[Asp], possibly also by RNase P_E, releases two precursorial sequences with immature 3′ termini. However, the temporal sequence of these two events is not firm, since either one can actually occur first. Final trimming of the 3′ regions by an exonuclease completes the cleaving activities (Figure 12.6C). A third enzyme, tRNA nucleotidyltransferase, then is responsible for the addition of the 3′-terminal -CCA sequence. The name of the enzyme, which occurs in both the nucleus and cytoplasm of eukaryotic cells (Solari and Deutscher, 1982), is unfortunate, in that it implies that the trimeric combination is merely transferred from a mature tRNA undergoing degradation to a new molecule being processed. Actually, the enzyme or some system synthesizes the combination without template and adds it to the tRNA, a process that should be ascribed to the supramolecular (or second) genetic mechanism (Dillon, 1983, p. 165).

In a unique case other nucleotides are added, in this instance to the 5′ terminus (Cooley *et al.*, 1982). All tRNA[His] species so far sequenced have proven to have an additional nucleotide, which is not encoded in the gene, present at the 5′ terminus. After 5′-end trimming of the precursor by RNase P_E, a guanidine nucleotide from an unknown source is added in the presence of ATP, often followed by enzymatic modification of the newly added base. Since the 3′ terminus is an A, no complementarity is involved by base pairing, only the addition of a guanidylate residue.

Nucleotide modification in tRNAs is quite extensive, and while the steps are not without interest (Dillon, 1978, pp. 313, 316), it is beyond the scope of the present analysis. What is of particular relevance in the present context is the replacement of one residue in the tRNA sequence by another. Such replacement procedures usually center on the 5′ nucleotide of the anticodon, which reacts with the third component of the codon, and in all known cases, the exchange involves a highly modified nucleotide as the insert. For example, in the yeast $tRNA^{Asp}_{UAG}$ the first of these is replaced by a Q residue by means of the UAG enzyme called Q-insertase (Carbon *et al.*, 1983). A similar mechanism acting at the same point has been demonstrated to be responsible for the substitution of inosine (I) for adenosine in eukaryotic tRNAs (Elliott and Trewyn, 1984). In contrast, this change

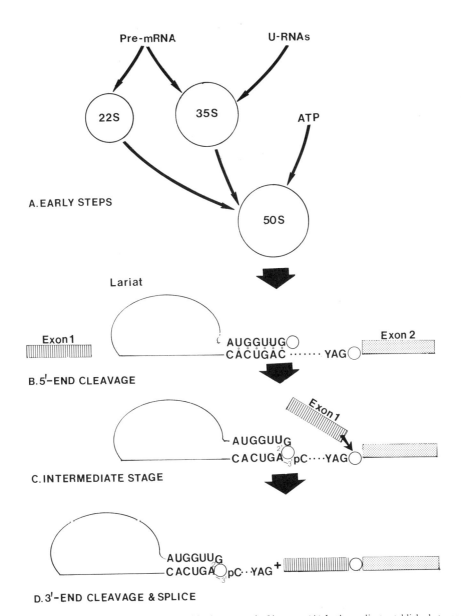

Figure 12.6. A few of the many steps involved in the removal of introns. (A) In the earliest established stages, two complexes with the pre-mRNAs are formed, one of which contains U RNAs (Frendewey and Keller, 1985); later a larger complex requiring ATP is developed. (B–D) The lariats formed from excised introns are at first continuous with downstream exons, but later are cleaved, while the upstream exon becomes spliced to the 3′ component (W. Keller, 1984). Small circles indicate phosphate bonds, as does the lowercase p.

has been ascribed to enzymic modification of the A to an I in the yeast tRNA^{Asp} and in other anticodons, an action that can occur only if the central base of the anticodon is not a pyrimidine and the third is not U (Haumont *et al.*, 1984).

12.6. INTRONS AND THEIR PROCESSING

Although it could be claimed that this section should be entitled "split genes and their processing," it seems preferable to stress the feature that distinguishes that class of genes, the presence of one or more introns. These inserted segments have elicited much interest from molecular biologists since their discovery, and, accordingly, they have provided the basis for much pertinent research. Yet, despite the accumulation of great quantities of data, much remains unlearned.

Intron Function. This is especially true with regard to the functional basis for their presence or their evolutionary *raison d'être*. Among the widely circulated hypotheses is that the intron separates the several functional or structural domains of the encoded product, and in some cases correlation between function and intron location can be perceived (Blake, 1978; Gilbert, 1978). But those relationships are occasional only and provide no firm basis for drawing general conclusions. To cite one example of contradictory evidence, the lactate dehydrogenase A cistron has six (or seven) introns dividing it into seven (or eight) exons, whereas the encoded product has only three functional domains, none of the boundaries of which coincides with those of the intervening sequences (Li *et al.*, 1985). In some diplomorphic genes, an intron separates the coding sector for the transit peptide from the ultimate coding region, as in that of the γ chain of human fibrinogen (Rixon *et al.*, 1985), but reference to the various tables of Chapter 7 reveals that when intervening segments do exist in presequence-bearing genes, they most frequently are located beyond the termination point of the latter. Although the chief reason for rejecting such a functional role is that the introns are removed well before translation produces the protein and its functional or structural subdivisions, there are others. For one, bacteria have been synthesizing proteins from intronless genes for billions of years without any apparent difficulty, and for another, intervening segments occur even in leaders on occasion (Mitra and Warner, 1984). See, for instance, the gene for ribosomal protein rp29 (Chapter 5, Section 5.5.2). Moreover, at times they split the coding sectors of transit peptides, even though those products have the single function of assisting the entire molecule in penetrating a membrane, one illustration being the human renin gene (Hobart *et al.*, 1984).

12.6.1. Structural Features of Introns

Few common structural features have been found among introns, although several have been advanced as being universal among yeast or metazoans, as discussed later. Some features must be important, at least on a taxon-to-taxon basis, because yeasts are unable to excise heterologous introns introduced into transcripts hybridized to their own (Langford *et al.*, 1983).

However, the structure of the termini of the introns has been demonstrated to be of great importance, together with that of the exons adjoining the splice sites. Thus, the intervening segment has been proposed to bear GT- at its left (5′) end and -AG at the opposite one, while the nucleotides of the exons neighboring these are, respectively, AG and G- (Lewin 1980a; Rogers and Wall, 1980). These and others in the vicinity are visualized as interacting with a sequence in U1 by loop formation and base pairing. This hypothesis has been reinforced by experimental mutation of the left-hand pair of the intron to AT in the β-globin of rabbit, with resulting failure of the signal and production of aberrant precursorial mRNAs (Wieringa et al., 1983). Similar inactivation of the splice site has been described from cloned β-thalessemia gene products (Treisman et al., 1983). In contrast, the dinucleotide GC rather than GT at the left terminus of chicken α^D-globin pre-mRNA has been reported to function at normal efficiency (Fischer et al., 1984).

Although the importance of the splice junction structure appears well established, detailed examination of a number of examples from a diversity of source organisms throws more light on that aspect later. Moreover, much remains to be understood about the splicing mechanism. By way of illustration, it is not clear how normal splice sites are distinguished from coding sequences of exons that have comparable composition, as Krainer et al. (1984) have pointed out. Nor is it understood how control over the numerous cases of alternate splicing patterns is manifested—for instance, in the *Drosophila* gene of Table 12.7, exon 4 can become attached to either exon 5 or 6, but not both (Falkenthal et al., 1985). It is toward understanding this and additional facets that other structures within the introns are receiving attention (Pikielny and Rosbash, 1985).

5′ Termini of Some Metazoan Pre-mRNA Introns. Table 12.7 presents randomly selected sequences of splice sites of 46 introns from various metazoan genes, three of which are from chicken. One of these is divided by ten introns, another by 18, and a third by five (Chang et al., 1985; Stone et al., 1985). Two others, with five and three introns, respectively, are from human tissues (Malissen et al., 1982; Fujita et al., 1983), and the last, with five, is from *Drosophila* (Falkenthan et al., 1985). In broad terms, in each case the last two codons at the 3′ end of the upstream and the pair at the 5′ end of the downstream exon are shown along with about 15 nucleotides adjoining the 5′ and 3′ termini of each intron. Thus the two cleavage/splice sites are shown for each inserted segment. At the upstream juncture the canonical GT is seen at the 5′ ends of the introns and AG at the other, with one exception. But it is also apparent at once that remarkable uniformity exists to the right of that first dinucleotide for a distance totaling three sites. Scanning through the 46 examples, it would seem that the recognition signal conceivably should be GTRRR, the purine of site 4 usually being A and that of site 5, G. No claim for absolute constancy of structure is made, for in two cases Cs are found in the third position, and in three instances Ts and in two cases Cs occupy site 4, about a 10% deviation. However, only one variation, a C, is found in the fifth position. Consequently, in metazoans a broader base for recognition of the intron's 5′ terminus is present than previously recognized.

At the codon side of the splice site, a terminal G appears to be greatly preferred; only about 20% (nine) of the representatives are exceptional, four having C, three A, and two T in place of the guanidine. Thus, as a whole, cleavage preferentially takes place between two Gs and almost always to the left of a G, but there is one other point of variation. In the third intron of the chicken glyceraldehyde phosphate dehydrogenase (*GAPDH*) gene, the intron terminates upstream in TG, so that cleavage occurs to the left of a thymidine residue. All adjoining positions included at the 5′ juncture are too variable to suggest even

Table 12.7
Standard Splice Sites of Introns of Metazoans

Intron	Exon 3'	Splice Site	Intron 5'-end	Intron 3'-end	Splice Site	Exon 5'
Chicken *GAPDH*[a]						
1	AACGG-		GTGAGTGCGGAGCCG	TTTCGTTTCTCCCAG		ATTTGGC
2	TACATG		GTGAGTTGCCTTGCA	CTCTCTCCCTGCTAG		GTTTAC
3	CAGGAG		TGAGTAGCTTTCTC	TTTGTCGTTCCTCAG		CGTGAC
4	GCTGGG		GTAAGTAGGCTACTG	TTTTTTTTTTTTCAG		GCTCAT
5	GTCAG-		GTAAAGGAGAGGAAT	TTCTTTCTTTCCCAG		CAATGCA
6	CTTATG		GTAGGTCTGGAGCTT	GCTCTCTTTCCCTAG		ACCACT
7	AATGG-		GTGAGTGTGGTCTTT	GTTCCCTTCTTGCAG		GAAGCTT
8	AAACCA		GTAAGTATTGGTGGA	TCTGCCTTTTTTCAG		GCCAAG
9	GACCAG		GTAAGTACTTTCTAT	GTCTCTCTGTTCCAG		GTTGTC
10	TCCTG-		GTATGCACAGCCTCG	CTTTCCCTTCCTCAG		GTATGAC
Chicken *c-fps*[b]						
1	ACTCAG		GTGAGCACATAT	ATCCATGCAG		GCTTCG
2	GGGGAG		GTAGGGACACT	CCCATCCTGCAG		TCGTGG
3	GCCCGG		GTAGGACCTCT	CACCTCCTCTGCAG		ACCACA
4	GCAAAG		GTGGGTGCCAGCC	CTCCCTGCAG		ACAAGG
5	TGTTCT		GTAAGAGCCC	TGCATCCCTGCAG		GAAGGA
6	CCACAG		GTACAGCGGTG	GGTCCTGCAG		GTACGA
7	GCACTC		GTGAGTCGGGCC	ATGTCCTTGCAG		CCTGAC
8	GGAGCG		GTGAGGCGAGTAG	CCATCCCCAG		GGTGCA
9	TCCACG		GTGAGTCCCCCCC	TTGCCCCCAG		GATCAG
10	TTCTCG		GTGAGGCCCTGG	TTCCACCCAG		CTGCCA
11	GCTGAC		GTGAGTGACCGTG	CTCCCTGCAG		AACCTG
12	CTCAAG		GTGAGGGTGGGC	CTGCCCGCAG		GACAAG
13	GGCCGG		GTGAGTTGGGTG	TCCTCTGCAG		GGGAAC
14	AGCCAG		GTCTGTGTGCC	TTCCCCGCAG		GATCCT
15	TGCAGG		GTGAGTCCTG	TGCGTTGCAG		GAGGGG
16	CCACAG		GTGGGCTGGGG	AGCTTCCCAG		GGACCT
17	ATTATG		GTATGGGGTGGC	GCATCTCCAG		GCCGGT
18	AGCAGG		GTGAGCGGGCAG	CCGCCCGCAG		GCGTGC

Table 12.7 (Continued)

Intron	Exon 3'	Splice Site	Intron 5'-end	Intron 3'-end	Splice Site	Exon 5'
Chicken cardiac α-actin[a]						
1	CACCAG		GTAAATTCCTGCCCG	ATTTTGTTCTTTTAG		GGTGTT
2	ACCACAG		GTAAAATAGACTTCT	ACATTTCTTTTCCAG		-GTATT
3	ACCACAG		GTCAGTCCTTCAGCT	TCTGTGTACCTGCAG		-CTGAA
4	TTTATTG		GTAAGTGCTTCAAGG	TTCTTTTTTCTCAAG		-GTATG
5	ATCAAG		GTGAGGACTTTGATT	TTCTCTTCACCCTAG		ATCATT
Human HLA[d]						
1	TGGGCGC		GTGAGTGCAGGGTCT	CCTCCTCGCTCCCAG		-GTTCC
2	GAGGGCG		GTGAGTTGACCCCGG	GCGGGGGCGGGCCAG		-GTTCT
3	CGCGCGG		GTACCAGGGCCACAG	ACTCTTTCCCGTCAG		-ACCCC
4	AGATGGG		GTAAGGAGGGAGATG	CCCCCCTTTTCCCAG		-AGCCA
5	AGCTCAG		GTAAGGAAGGGGTGA	ATTTTCTTCCCACAG		-ATAGA
Human interleukin 2[e]						
1	ATTAAT		GTAAGTATATTTCCT	ATTATACTTTCTTAG		AATTAC
2	AAGAAG		GTAAGTACAATATTT	TAATTATTATTCTAG		GCCACA
3	CTAAAG		GTAAGGCATTACTTT	ATTTTTCTTTTATAG		GGATCT
Drosophila MLC-ALK[f]						
1	AAAATG		GTGAGTGCCCAGTCG	CGATTCCGTTGACAG		GTCGAT
2	TTGAAA		GTAAGTTAAAACCCC	GTTCCTTTTGCGCAG		ATGTCG
3	CGCTTG		GTGAGTAATCAACAT	TTGTGTGCCCTATAG		GTGAGA
4	ACTCTC		GTAAGTATTCCTTCA	ATTCAATCAAAACAG		AGTTCG
5	TAAACA		GTAAGTAGGGATGGA	TCGTCTACTCCACAG		CATTCC

[a]Glyceraldehyde-3-phosphate dehydrogenase; Stone *et al.* (1985).
[b]Huang *et al.* (1985).
[c]Chang *et al.* (1985).
[d]Malissen *et al.* (1982).
[e]Fujita *et al.* (1983).
[f]Falkenthal *et al.* (1985).

distinct trends, so while a somewhat broader basis for recognition of introns at one end has been exposed by this analysis, that five-base sequence can scarcely suffice by itself for an enzyme system to recognize these extremely variable segments called introns.

The 3′-Terminal Structure of Splice Sites. A somewhat comparable, broader, but less firm, basis for enzymic recognition of the downstream splice site of introns exists than the AG that has been proposed. Although that dinucleotide is a feature of 100% of the metazoan examples of Table 12.7, it certainly cannot be adequate by itself. To this canonical combination surely can be added the pyrimidines that occur in most of the examples at the adjacent additional sites upstream, as far as the ninth, omitting the second. All constancies are broken by the rare presence of Gs or As, but always with less than a 20% deviation. The fourth from the downstream end alone is too variable to be important here. Thus, part of the recognition combination could readily be the resulting YYYYY-YAG.

The cleaving enzyme at this site is less precise in its demands than that at the upstream end, for splitting takes place with nearly equal frequency between the final G and any nucleotide that chances to follow. However, guanosine appears to be preferred, for it occurs at the initial site in the first codon of the downstream exon in about 59% of the examples, with A in over 25%. Consequently, cleavage largely centers on a GR combination.

Splice Sites of Nonmetazoan Introns. The corresponding structures from non-metazoan eukaryotes add little to the proposed combinations except a higher degree of variation. At the upstream end of the introns given in Table 12.8 from yeast, fungal, and green plant sources, the same GTR is typically present, with A being nearly the universal purine, but the Gs or As of the fourth position are too frequently displaced by pyrimidines to form part of the signaling device outside of the Metazoa. As a result, the recognition may be provided by the tetranucleotide GTA-R; however, if the examples given in the table are truly representative, a T can be added at the end to provide the pentanucleotide GTA-RT, for it is found there in 13 of the 17 introns. Although cleavage is often between two Gs, it takes place between CG or TG in three cases each, but AG does not occur.

At the downstream end, again the AG dinucleotide prevails, with two exceptions, but the Y preceding it is less consistent, being replaced on two occasions by Gs and on one by an A. Moreover, there is no notable prevalence of pyrimidines in the sites preceding the YAG combination, so that the signal can only involve that trinucleotide, aside from unknown factors needed to make it meaningful. Cleavage at this end is as variable as in the metazoans, but it is frequently between two Gs (in over 50% of the examples); any of the standard nucleotides seems acceptable as the downstream element.

Lariat Formation. Considerable progress has been made toward the understanding of some of the steps in mRNA splicing through the use of other approaches. Among the more exciting discoveries has been a precursorial step, one by-product of which is a circular RNA having an appendage that bears a short branch near its base (Figure 12.6). These branched circles or lariats, as they are called, proved to consist of the first intron attached to the second exon (Hernandez and Keller, 1983; Padgett *et al.*, 1984). First found in an adenovirus 2 system, comparable RNA configurations have subsequently been reported by studies on β-globin (Krainer *et al.*, 1984; Ruskin *et al.*, 1985); they have also been reported to occur *in vivo* in yeast pre-mRNA splicing (Domdey *et al.*, 1984). The reactions require the presence of ATP, $MgCl_2$, and monovalent cations, in a cell

Table 12.8
Splice Sites of Nonmetazoan Introns

Intron	Exon 3'	Splice Site	Intron 5'-end	Intron 3'-end	Splice Site	Exon 5'
Yeast *CYH2*[a]		↓			↓	
1	GTCTCAG		GTATGTAGTTCCATT	ATTTTTTTTGTACAG		−CCGGT
Schizosaccharomyces CDC2[b]						
1	GGGGAAG		GTAGGTGTAACTGAG	ACTGACTAATGCTAG		−GAACC
2	GTTCG−		GTAAGTTTTTACTTG	TAACCCTTTTTTAGG		ACTTTTG
3	CATGAG		GTATATTTTTTTTCT	TCGCTAACCGTTTAG		ATTGTC
4	CTCTCT		GTAAGATTTTGCATT	AACCATTTTGTCAAG		GCTATG
Aspergillus GLA[c]						
1	CCGGACT		GTATGTTTCGAGCTC	TGTCGTTTGTTGTAG		−ACTTC
2	CTGCTT		GTATGTTCTCCACCC	TAGCTGACTGGTCAG		GACAAT
3	GGATATG		GTGTGTTTGTTTATT	AACCCGCGATCGCAG		−ATCTC
4	ATTGTG		GTAAGTCTACGCTAG	TACTAACAGAAGTAG		GAAACT
Neurospora GDH[d]						
1	TACAAGG		GTACGTCTGAGAGAA	GATTTTTCACCACAG		−AGTTG
2	ACTGGCC		GTAAGTGACCGAACG	ACTCGGCGTCTCTAG		−TGAGG
Soybean *lbc*$_3$[e]						
1	ACCTC−		GTAAGTATTCTATCT	ACTAAAAATGAATAG		GATACTG
2	GGATTG		GTAAGTACTAGCCTA	TTTTTTGAATTATAG		GTACGT
3	TTTGTG		GTATGATAAATAATG	GCTGATGATTTCGAA		GTGGTT
Kidney bean Lb[e]						
1	ACCTC−		GTAAGTGTTTTGTTT	ATATAAAAATAATAG		GATACTG
2	GGATTG		GTAAGTCTCAACCAA	TTTTTGAAATTGTAG		GTGCGT
3	TTTCTG		GTAGCGTTAATGAAT	ATGTATGAATTGCAG		GTGGTA

[a]Cyclohexamide resistance; Kaufer *et al.* (1983), van der Horst and Tabak (1985).
[b]Cell division gene; Hindley and Phear (1984).
[c]Glucoamylase; Boel *et al.* (1984).
[d]Glutamate dehydrogenase; Kinnaird and Fincham (1983).
[e]Leghemoglobin; J. S. Lee and Verma (1984).

extract. Actually two types of lariats have been found, the first as described above, and a second that has lost the tail, consisting solely of the first intron. Since, along with those two types, mRNAs having exons spliced also occur (W. Keller, 1984), splicing seemingly proceeds in at least two stages, beginning at the 5' cleavage site.

Further details that have been added to the foregoing account include a demonstration that members of the U family of RNAs must be present. In time-course investigations with HeLa cells, two complexes, one that sediments at 22 S and the other at 35 S, both of which contained pre-mRNA, were the first products, the larger of which also required U RNAs. These were followed by the formation of 50 S complexes in which splicing intermediates were observed (Frendewey and Keller, 1985), the presence of ATP being essential. Similar complexes from yeasts have also been described, the largest of which sedimented at 40 S (Brody and Abelson, 1985). Analysis with mutant mRNAs has revealed that an intact 5' splice site and a polypyrimidine tract near the 3' end of the intron were absolute requirements (Frendewey and Keller, 1985); however, examination of Tables 12.6 and 12.7 shows that such tracts are frequently absent.

In yeast the 5'-terminal G of the intron has been found linked to the last A in the UACUAAC sequence that has been described as universally present in the introns of this organism (Langford and Gallwitz, 1983; Langford *et al.*, 1984). The linkage in this case is by a 2'-5'-phosphodiester bond, while the A is linked *via* a 3'-5'-phosphodiester bond to the neighboring C (Domdey *et al.*, 1984). However, some conflicting data indicate that the G-to-A linkage is as stated, but that the neighboring nucleotides vary, thus casting some doubt on the validity of the suggested heptanucleotide (Ruskin and Green, 1985; Ruskin *et al.*, 1985). This A lies at the base of the side branch and between 22 and 37 positions upstream of the 3' terminus of the intron, the distance being an important factor in location of the branch. In *Drosophila*, a corresponding sequence, YTRAY, has been described as being located at a similar point and serving in the same capacity (E. B. Keller and Noon, 1984, 1985; Reed and Maniatis, 1985), but its structure is so short and variable that like combinations can be found at numerous points in leaders, trains, and coding sectors, making its functional value questionable. In addition, the latter report showed that cleavage at the upstream end of the intron does not occur if the frequent polypyrimidine tract and AG dinucleotides at the 3' terminus are removed. It also showed that a single base substitution in the dinucleotide blocked cleavage at the downstream end, but not at the other. Studies on rabbit β-globin have indicated that either intron 1 or 2 may be removed first, but that the former more frequently receives the initial action (Zeitlin and Efstratiadis, 1984).

12.6.2. Processing mRNAs of the Chloroplast

Although the processing of split tRNA genes has been investigated extensively (Section 12.5.1), the maturational steps of other intron-bearing genes from chloroplasts have received relatively little attention. Despite the presence of at least 15 protein-encoding split genes in the chloroplast genome of *Euglena gracilis*, only three have been sequenced (M. Keller and Michel, 1985) and but two or three appear to have been established from seed plants (Karlin-Neumann *et al.*, 1985). Arranged together in Table 12.9, the intron termini display a greater diversity than those of the metazoan cytoplasmic species.

Table 12.9
Splice Site of Chloroplast Introns

Intron	Upstream exon	Splice site	5' End of intron	3' End of intron	Splice site	Downstream exon
Euglena psbA[a]		↓			↓	
1	AATATAA		GTGCGTAACTTATCT	TAGTATCTATTTTAC		-TTACT
2	GAATGG		GTGTGTCAATTAGAA	TTTACTTTAGTTTAT		TTGTAT
3	TTAAAC		GTGCGTAATTTGATC	GTTTGCTAGTGTTAT		GGTTTT
4	GTTATG		GTGCGACAAGTTTAA	AAATTTTAGTTTAAC		CACGAA
Euglena EF-Tu[b]						
1	CAAGTAG		AATAAGCTTAAAAAT	ATTCCATCAAAAACG		-ATAGT
2	GCAATAG		AAGTGTCGTTTAATA	ATAAGACATATTGGA		-AAAAA
Euglena RBPC[c]						
1			(Incomplete)	AATTTAGTTTTAT		GGTGTT
2	CAAGTG		GTGTGCGTTT	(Incomplete)		
3			(Incomplete)	TGTTTAATTTTAT		GAAGAA
4			(Incomplete)	TTATCTAATTTTAT		TGTTAC
5	GATGAA		GTGCGAATTTGTAAT	GTTTTAGTTTTAT		AATGTT
6	GAGATG		GTGTGGATTT	TACTTAATTTTAT		TACAAA
7	ATAAT-		GCATGAGAAA	(Incomplete)		
Spinach chloroplast rpl2[d]						
1	CCTTTG		ACTGCGGTTTGAACTA	GAAGAATCTACTTCA		ACCGAT
Lema chloroplast a/b[d]						
1	ATCTG		GTTAGAATCATCCTC	GGGCTTCCTGATCAG		GTACGGG

[a]M. Keller and Michel (1985).
[b]Montandon and Stutz (1983).
[c]Ribulose-1,5-biphosphate carboxylase; Koller et al. (1984).
[d]Karlin-Neumann et al. (1985).

5' Termini of Chloroplast Introns. Although the 5' termini of these chloroplast introns are quite heterogeneous, the majority of them bear the same GTG trinucleotide seen in those of cytoplasmic species. Moreover, in those of this group, the same YG combination follows downstream, so that the pentanucleotide GTGYG is a constant feature. Two others, however, have AA- as an initial dinucleotide, followed by either a T or G in the third site, but they differ so strongly in the remaining terminal portion as to indicate a probability of their being unrelated. Apparently they represent two groups, one with AATAAG as a possible signal, the second with AAGTGT. It is interesting that in both cases the adjacent regions of the preceding exon contain a single G after the last two full codons, but this could be merely a happenstance. Two additional divergent termini occur, a GCA and in the *Lemna* chlorophyll-binding protein a GTT, each of which similarly is unique in the sites that follow. Here, too, in both instances the trinucleotide is associated with an incompleted codon at the close of the adjacent exon, but this condition could, as before, be related to chance and scanty knowledge. Cleavage thus always, in the data at hand, involves a purine (usually G) on the right side, and most frequently also on the left, but on several occasions, the left component is a pyrimidine.

3' Termini of Chloroplast Introns. The 3' termini of chloroplast introns diverge 100% from those of nuclear genes. In place of the YAG found there, here the favored combination is TAY, the final pyrimidine typically being T. Upstream from that triplet are two or three Ts, preceded to the left in all cases by a TAR trinucleotide, so that a constant combination TART-TTAY exists. Moreover, these are consistently associated with the GTGYG pentanucleotide at the 5' end. In the remainder of the representative sequences, five other combinations bring the intron to a close. One of these, AAC, appears to be a variation of the TAY, since it has all the other characteristics exhibited by that type, including those of the 5' terminus. Thus, the signal just proposed probably should read TAGT-TYAY. Of the remainder, two, ACG and GGA, are respectively associated with the divergent AAT and AAG at their upstream tips, and display no shared traits. Finally, those that occupy the end in the green plant chloroplast genes, TCA and CAG, are unique in the remaining portions of the structures, as shown. Consequently, there appear to be at least five major groups of introns in genes of chloroplasts. At their downstream juncture, cleavage most frequently occurs to the right of a T, but any combination of nucleotides seemingly is acceptable. Such extreme variation in structure suggests that these organellar genes reflect a primitive condition that in the nucleus has gradually been replaced by more constant signals. It would seem that evolution slowly perfected the processing procedures through reduction of the number of enzymes needed to recognize the termini of introns through gradual loss of diversity in signal structure.

12.6.3. Processing mRNAs of the Mitochondrion

The state of knowledge concerning the processing of mRNAs from mitochondria not only is scanty, but is in a complex state of flux. In the first place, very few protein genes from this organelle contain introns; none are indicated in the mammalian types whose entire genomes have been established, including those of the mouse, man, and cattle (Anderson *et al.,* 1981, 1982; Bibb *et al.,* 1981). Moreover, a reduced number—or even absence—of introns in these organelles is suggested in other organisms by the lack of intervening segments from the cytochrome *b* gene, which coding sequence characteristically has two to five in *Aspergillus, Neurospora,* and *Saccharomyces* (Nobrega and Tzagoloff, 1980; Waring *et al.,* 1981; Burke *et al.,* 1984). Among those whose gene does not include introns are the mitochondria of maize, *Trypanosoma,* and *Drosophila* (Benne *et al.,* 1983; Clary *et al.,* 1984; Dawson *et al.,* 1984); consequently, few firm conclusions can validly be drawn in the face of the limited data.

Structures of Mitochondrial Introns. Because this section focuses principally upon the processing of protein gene transcripts, Table 12.10 is limited primarily to intron

Table 12.10
Splice Sites of Mitochondrial Introns

Intron	Upstream Exon	Splice Site	5' End of Intron	3' End of Intron	Splice Site	Downstream Exon
Cytochrome *b*						
Yeast *D273*[a] 1	GGTCAA		AATATGGCCTTATTA	TATTTATTATAAAAG		CATCCT
2	ATTGAT		ATTAAAAATATTAAT	ACATATATAAATTGT		GTACCT
Yeast *777*[b] 3	GGGTT-		TAATATAGAGGATCC	TTTAAACATTAAAAG		CTC----
Aspergillus[c] 1	GGTTT-		ATACACAGATGAACC	CTATCAACATAAATG		CTCTGTA
Neurospora[d] 1			(Incomplete)	TAATAAGTAAGTACT		TATGTT
2	GTTGATT		CAAAAATTATTACTT	TATACGTAATATATT		-TCATT
Major rRNA						
Neurospora[e] 1	AGGGAT		GTTTGTCCTTCAATA	TATGATAACAAGTTG		AACAGG
Yeast[e] 1	AGGGAT		AATTTACCCCCTTGT	ATAACAAAAAATTTG		AACAGG

[a]Nobrega and Tzagoloff (1980).
[b]Lazowska *et al.* (1981).
[c]Waring *et al.* (1981).
[d]Burke *et al.* (1984).
[e]Burke and RajBhandary (1982).

structures available from cytochrome *b* (*cob*), to which two from the major species of rRNA are added to provide a comparative basis. Brief though the representation is, it shows distinctly the variation that exists in the inserted segments from this organelle. At their 5' termini no two of the structures can be perceived to possess the same opening triplet, despite the fact that intron 3 of yeast 777 is homologous in location to that of *Aspergillus*, as indicated by the identity of the codons in the neighboring exons. As a whole, this terminal region of the protein genes consists largely of As and Ts—indeed, its counterpart at the other end and much of the interiors are of similar composition. In contrast, the upstream ends of the two rRNA introns contain relatively few As.

At the downstream end, a comparable variability prevails, since only two sets terminate in identical triplets. AAG occupies that position both in the first intron of the short form of the *cob* gene, which has only two inserted segments (strain D273-10B), and its homolog, the third intron from the long form with five (strain 777-3A). The second set involves the TTG of the two examples from the major rRNA cistrons. All the remainder are distinct, but at least in these few examples, they uniformly have the base of the final position either T or G. In the trinucleotide that precedes the termini, TAA, TAT, or AAT occurs in five sequences, with AAG or AGT in the other two. At neither cleavage site can any preferred combination be ascertained, so that excision in all examples involves any combination of two nucleotides.

Intronic Coding Sequences. An additional problem presented by the introns in a few mitochondrial genes results from their containing coding frames for proteins, but at the same time, they reveal conditions that may be applicable to nuclear split-gene processing. One such intron-encoded protein has been shown to exist on the single inserted segment of the major rRNA species of a yeast and *Neurospora* (Burke and RajBhandary, 1982; Macreadie *et al.*, 1985). In the first organism, the product of the intronic gene is essential to the transposition of the segment into corresponding genes that lack it (Jacquier and Dujon, 1985). Yet, oddly enough, it is the transcript of this rRNA that has exhibited the spectacular behavior of "self-splicing" *in vitro* in yeast and *Tetrahymena*, and in the case of the latter gene, in *E. coli* also (Zaug *et al.*, 1983; van der Horst and Tabak, 1985; Waring *et al.*, 1985). But since in the first intron of the *Neurospora cob* gene, which exhibits similar properties, a recessive mutation in a nuclear gene inhibits the splicing reaction *in vivo*, a protein seems to be involved in the processes in the intact organism (Garriga and Lambowitz, 1984).

Complex Questions. Certain other proteins that are encoded on introns in mitochondrial genes pose some complex questions that have not received attention in the literature. This is particularly so with the cytochrome *b* gene (*cob*) of yeast, which has been more thoroughly investigated than any other of the split type. Processing of its pre-mRNA involves nuclear genes as well as those of the introns, for one such (*CPB1*), given attention in an earlier context (Section 12.3.1), has been reported to be essential to the formation of stable 5′ ends on the primary transcript (Dieckmann *et al.*, 1984a,b) and another (*CPB2*) is required specifically for removal of the fifth intron (McGraw and Tzagoloff, 1983; Hill *et al.*, 1985). According to present knowledge, this pair acts solely upon the cytochrome *b* gene product, or upon it and the first intron of the cytochrome *c* oxidase (*oxi*) gene, as do five others from the nucleus, not further identified, that affect five steps in the removal of introns from this same pre-mRNA (Pillar *et al.*, 1983). But the intron-located coding sections are even more pertinent to any account of control of their processing, for the maturases that they encode are necessary for the removal of single intervening segments (Lewin, 1980b). Some of these bear typical transcription initiation and stop codons and present no further problem. Others, however, including those on *cob* introns 1–4, lack these signals, each being continuous with the exon that precedes it (Lazowska *et al.*, 1980, 1981; Bechmann *et al.*, 1981; Weiss-Brummer *et al.*, 1982; Lamb *et al.*, 1983).

Therefore it is self-evident that the coding regions in such introns must be translated before the latter are excised, thereby raising the complex questions mentioned earlier. Among the more pressing are: (1) How are the maturases removed after being translated? (2) Are the remaining portions of the introns that do not enter into maturase formation removed before or after the translation processes? (3) After removal of the maturase and nontranslated sectors, how is splicing of the cytochrome *b* exons conducted? No signals in nucleotides then exist, for amino acids provide the monomeric structures, thereby also eliminating the possibility of self-splicing and lariat formation.

It is these pressing problems and others of less broad impact that require solution before generalizations regarding intron classification and processing can validly be made. As shown in the preceding paragraphs by the yeast mitochondrial split genes *cob* and *oxi*, processing in many cases may require whole constellations of diverse gene products of both nuclear and organellar genes that act specifically on a limited number of introns. Yet the nucleotide sequences from such genes, whose requirements are unique, have often provided a major part of data in attempts at intron classification or in broad proposals for processing signals, when actually the signals would have been displaced by amino acids! Much more precise data are the real contributions needed to further understanding of the intricate problem of pre-mRNA maturation, not the obscuring effects of premature generalizations.

At one time defined merely as "untranslated intervening segments of the gene," the intron had first to be redefined as a "portion of a coding region or its flanks that is removed prior to translation." Now, with the discovery of introns that contain coding sectors continuous with those of exons, it must undergo further restatement to indicate that it is a "portion of a gene in its broadest sense that is removed before the translated product has been completely processed." This description may in turn prove faulty, for it assumes that the leader and train are not removed until after the introns have been excised, a sequence of events that has not been demonstrated as yet.

Thus one of the themes that has manifested itself repeatedly throughout this study appears one final time—cellular controls of the gene, including what at first seemed the simple act of removing an unneeded segment from a transcript, are exceedingly complex, characteristically requiring teams of proteins to carry out each step. The stages are inexplicable in terms of the purely mechanistic reactions of ordinary chemical processes, but can be interpreted only in terms of what has been called a biomechanistic point of view. Moreover, an underlying cellular means of precise control is as obvious in the processing of pre-mRNAs and other precursorial transcripts as it has been at many points in this analysis of the gene. Such a control probably involves a system of intricately coordinated proteins, possibly together with certain species of RNAs, that has been named the supramolecular, or second, genetic mechanism. But whatever name it finally receives, its existence is made undeniable by its numerous points of manifestation in gene regulation, in macromolecular behavior, somatic mutation, and throughout the cell cycle in organellar growth and function. To date only possible parts of it have been made evident; the whole is far too labyrinthal to be revealed readily. But it can only become wholly exposed if the possibility of its existence is acknowledged and a diligent search made for it. Even then, it will become known only as isolated fragments, one at a time, for there can be little doubt that it is intricately interwoven with much of the cytoplasmic content of the cell.

References

CHAPTER 1

Adams, J. M., and Cory, S. 1975. Modified nucleosides and bizarre 5′-termini in mouse myeloma mRNA. *Nature (London)* **255:**28–33.

An, G., and Friesen, J. D. 1980. Characterization of promoter-cloning plasmids: Analysis of operon structure in the *rif* region of *Escherichia coli* and isolation of an enhanced internal promoter mutant. *J. Bacteriol.* **144:**904–916.

Andrews, C., and Richardson, J. P. 1985. Transcription termination factor rho mediates simultaneous release of RNA transcripts and DNA template from complexes with *Escherichia coli* RNA polymerase. *J. Biol. Chem.* **260:**5826–5831.

Artavanis-Tsakonas, S., Schedl, P., Tschudi, C., Pirrotta, V., Steward, R., and Gehring, W. J. 1977. The 5S genes of *Drosophila melanogaster*. *Cell* **12:**1057–1067.

Astell, C. R., Thomson, M., Merchlinsky, M., and Ward, D. C. 1983. The complete DNA sequence of minute virus of mice, an autonomous parvovirus. *Nucleic Acids Res.* **11:**999–1018.

Baralle, F. E. 1983. The functional significance of leader and trailer sequences in eukaryotic mRNAs. *Int. Rev. Cytol.* **81:**71–106.

Barik, S., Bhattacharya, P., and Das, A. 1985. Autogenous regulation of transcription termination factor rho. *J. Mol. Biol.* **182:**495–508.

Barsh, G. S., Seeburg, P. H., and Gelinas, R. E. 1983. The human growth hormone gene family: Structure and evolution of the chromosomal locus. *Nucleic Acids Res.* **11:**3939–3958.

Baum, H. J., Livneh, Y., and Wensink, P. C. 1983. Homology maps of the *Drosophila* α-tubulin gene family: One of the four genes is different. *Nucleic Acids Res.* **11:**5569–5587.

Bear, D. G., Andrews, C. L., Singer, J. D., Morgan, W. D., Grant, R. A., von Hippel, P. H., and Platt, T. 1985. *Escherichia coli* transcription termination factor ρ has a two-domain structure in its activated form. *Proc. Natl. Acad. Sci. USA* **82:**1911–1915.

Bennett, P. M., Grinsted, J., Choi, C. L., and Richmond, M. H. 1978. Characterisation of Tn*501*, a transposon determining resistance to mercuric ions. *Mol. Gen. Genet.* **159:**101–106.

Benoist, C., and Chambon, P. 1981. *In vivo* sequence requirements of the SV40 early promoter region. *Nature (London)* **290:**304–310.

Berman, M. L., and Landy, A. 1979. Promoter mutations in the transfer RNA gene *tyrT* of *Escherichia coli*. *Proc. Natl. Acad. Sci. USA* **76:**4303–4307.

Beynon, J., Cannon, M., Buchanan-Wollaston, V., and Cannon, F. 1983. The *nif* promoters of *Klebsiella pneumoniae* have a characteristic primary structure. *Cell* **34:**665–671.

Birnstiel, M. L., Busslinger, M., and Strub, K. 1985. Transcription termination and 3′ processing: The end is in site! *Cell* **41:**349–359.

789

Boardman, M., Basi, G. S., and Storti, R. V. 1985. Multiple polyadenylation sites in a *Drosophila* tropomyosin gene are used to generate functional mRNAs. *Nucleic Acids Res.* **13**:1763–1776.

Bonnewell, V., Fowler, R. F., and Skinner, D. M. 1983. An inverted repeat borders a fivefold amplification in satellite DNA. *Science* **221**:862–865.

Boorstein, W. R., Vamvakopoulos, N. C., and Fiddes, J. C. 1982. Human chorionic gonadotropin β-subunit is encoded by at least eight genes arranged in tandem and inverted pairs. *Nature (London)* **300**:419–422.

Brown, D. D., Wensink, P. C., and Jordan, E. 1971. Purification and some characteristics of the 5S DNA from *Xenopus laevis. Proc. Natl. Acad. Sci. USA* **68**:3175–3179.

Buchanan-Wollaston, V., Cannon, M. C., and Cannon, F. C. 1982. The use of cloned *nif* (nitrogen fixation) DNA to investigate transcriptional regulation of *nif* expression in *Klebsiella pneumoniae. Mol. Gen. Genet.* **184**:102–106.

Calabretta, B., Robberson, D. L., Maizel, A. L., and Saunders, G. F. 1981. mRNA in human cells contains sequences complementary to the *Alu* family of repeated DNA. *Proc. Natl. Acad. Sci. USA* **78**:6003–6007.

Cassan, M., Ronceray, J., and Patte, J. C. 1983. Nucleotide sequence of the promoter region of the *E. coli lysC* gene. *Nucleic Acids Res.* **11**:6157–6166.

Childs, G., Nocente-McGrath, C., Lieber, T., Holt, C., and Knowles, J. A. 1982. Sea urchin (*Lytechinus pictus*) late-stage histone *H3* and *H4* genes: Characterization and mapping of a clustered but nontandemly linked multigene family. *Cell* **31**:383–393.

Crenet, M., Gannon, F., Hen, R., Maroteaux, F. P., and Chambon, P. 1979. Organization and sequence studies of the 17-piece chicken conalbumin gene. *Nature (London)* **282**:567–574.

Cohn, R. H., Lowry, J. C., and Kedes, L. H. 1976. Histone genes of the sea urchin (*S. purpuratus*) cloned in *E. coli:* Order, polarity, and strandedness of the five histone-coding and spacer regions. *Cell* **9**:147–161.

Courage-Tebbe, U., Döring, H. P., Federoff, N., and Starlinger, P. 1983. The controlling element *Ds* at the *Shrunken* locus in *Zea mays:* Structure of the unstable *sh-m5933* allele and several revertants. *Cell* **34**:383–393.

Davidson, E. H., and Britten, R. J. 1973. Organization, transcription, and regulation in the animal genome. *Q. Rev. Biol.* **48**:565–613.

Davidson, E. H., and Posakony, J. W. 1982. Repetitive sequence transcripts in development. *Nature (London)* **297**:633–635.

deBoer, H., Comstock, L. J., and Vasser, M. 1983. The *tac* promoter: A functional hybrid derived from the *trp* and *lac* promoters. *Proc. Natl. Acad. Sci. USA* **80**:21–25.

de la Cruz, F., and Grinsted, J. 1982. Genetic and molecular characterization of Tn*21,* a multiple resistance transposon from R100.1. *J. Bacteriol.* **151**:222–228.

Dillon, L. S. 1978. *The Genetic Mechanism and the Origin of Life,* New York, Plenum Press.

Dillon, L. S. 1983. *The Inconstant Gene,* New York, Plenum Press.

Diver, W. P., Grinsted, J., Fritzinger, D. C., Brown, N. L., Altenbucher, J., Rogowsky, P., and Schmitt, R. 1983. DNA sequences of and complementation by the *tnpR* genes of Tn 21, Tn 501, and Tn 1721. *Mol. Gen. Genet.* **191**:189–193.

Ephrussi, A., Church, G. M., Tonegawa, S., and Gilbert, W. 1985. B lineage-specific interactions of an immunoglobulin enhancer with cellular factors *in vivo. Science* **227**:134–140.

Espin, G., Alarez-Morales, A., Cannon, F., Pixon, R., and Merrick, M. 1982. Cloning of the *glnA, ntrB* and *ntrC* genes of *Klebsiella pneumoniae* and studies of their role in regulation of the nitrogen fixation (*nif*) gene cluster. *Mol. Gen. Genet.* **186**:518–524.

Etcheverry, T., Colby, D., and Guthrie, C. 1979. A precursor to a minor species of yeast tRNA[Ser] contains an intervening sequence. *Cell* **18**:11–26.

Farabaugh, P. J., and Fink, G. R. 1980. Insertion of the eukaryotic transposable element *Ty1* creates a 5-base pair duplication. *Nature (London)* **286**:352–356.

Feagin, J. E., Setzer, D. R., and Schimke, R. T. 1983. A family of repeated DNA sequences, one of which resides in the second intervening sequence of the mouse dihydrofolate reductase gene. *J. Biol. Chem.* **258**:2480–2487.

Federoff, N. V. 1979. On spacers. *Cell* **16**:697–710.

Files, J. G., Carr, S., and Hirsh, D. 1983. Actin gene family of *Caenorhabditis elegans. J. Mol. Biol.* **164**:355–375.

Finnegan, D. J., Rubin, G. M., Young, M. W., and Hogness, D. S. 1977. Repeated gene families in *Drosophila melanogaster. Cold Spring Harbor Symp. Quant. Biol.* **42**:1053–1063.

Fort, P., Marty, L., Piechaczyk, M., El Sabrouty, S., Dani, C., Jeanteur, P., and Blanchard, J. M. 1985. Various rat adult tissues express only one major mRNA species from the glyceraldehyde-3-phosphate-dehydrogenase multigenic family. *Nucleic Acids Res.* **13**:1431–1442.

Franck, A., Guilley, H., Jonard, G., Richards, K., and Hirth, L. 1980. Nucleotide sequence of cauliflower mosaic virus DNA. *Cell* **21**:285–294.

Gebhard, W., Meitinger, T., Höchtl, J., and Zachau, H. G. 1982. A new family of interspersed repetitive DNA sequences in the mouse genome. *J. Mol. Biol.* **157**:453–471.

Georgiev, O., and Birnstiel, M. L. 1985. The conserved CAAGAAAGA spacer sequence is an essential element for the formation of 3' termini of the sea urchin H3 histone mRNA by RNA processing. *EMBO J.* **4**:481–489.

Goelet, P., Lomonossoff, G. P., Butler, P. J. G., Akam, M. E., Gait, M. J., and Karn, J. 1982. Nucleotide sequence of tobacco mosaic virus RNA. *Proc. Natl. Acad. Sci. USA* **79**:5818–5822.

Goodman, H. M., Olson, M. V., and Hall, B. D. 1977. Nucleotide sequence of a mutant eukaryotic gene: The yeast tyrosine-inserting ochre suppressor *SUP4-o. Proc. Natl. Acad. Sci. USA* **74**:5453–5457.

Green, C. J., and Vold, B. S. 1983. Sequence analysis of a cluster of twenty-one tRNA genes in *Bacillus subtilis. Nucleic Acids Res.* **11**:5763–5774.

Grunstein, M., and Hogness, D. S. 1975. Colony hybridization: A method for the isolation of cloned DNAs that contain a specific gene. *Proc. Natl. Acad. Sci. USA* **72**:3961–3965.

Gruss, P., Lai, C. J., Dhar, R., and Khoury, G. 1979. Splicing as a requirement for biogenesis of functional 16S mRNA of simian virus 40. *Proc. Natl. Acad. Sci. USA* **76**:4317–4321.

Gruss, P., Efstratiadis, A., Karathanasis, S., König, M., and Khoury, G. 1981. Synthesis of stable unspliced mRNA from an intronless simian virus 40–rat preproinsulin gene recombinant. *Proc. Natl. Acad. Sci. USA* **78**:6091–6095.

Hawley, D. K., and McClure, W. R. 1983. Compilation and analysis of *Escherichia coli* promoter DNA sequences. *Nucleic Acids Res.* **11**:2237–2255.

Henikoff, S., Kelly, J. D., and Cohen, E. H. 1983. Transcription terminates in yeast distal to a control sequence. *Cell* **33**:607–614.

Hereford, L., Fahrner, K., Woolford, J., Rosbash, M., and Kaback, D. B. 1979. Isolation of yeast histone genes H2A and H2B. *Cell* **18**:1261–1271.

Hill, S., Kennedy, C., Kavanagh, E., Goldberg, R. B., and Hamm, R. 1981. Nitrogen fixation gene (*nifL*) involved in oxygen regulation of nitrogenase synthesis in *K. pneumoniae. Nature (London)* **290**:424–426.

Hinton, D. M., and Musso, R. E. 1983. Specific *in vitro* transcription of the insertion sequence IS2. *J. Mol. Biol.* **169**:53–81.

Holmes, W. M., Platt, T., and Rosenberg, M. 1983. Termination of transcription in *E. coli. Cell* **32**:1029–1032.

Hung, M. C., and Wensink, P. C. 1981. The sequence of the *Drosophila melanogaster* gene for yolk protein 1. *Nucleic Acids Res.* **9**:6407–6419.

Irani, M. H., Orosz, L., and Adhya, S. 1983. A control element within a structural gene: The *gal* operon of *Escherichia coli. Cell* **32**:783–788.

Johansen, H., Schümperli, D., and Rosenberg, M. 1984. Affecting gene expression by altering the length and sequence of the 5' leader. *Proc. Natl. Acad. Sci. USA* **81**:7698–7702.

Johnson, L. D., Henderson, A. S., and Atwood, K. C. 1974. Location of the genes for 5S RNA in the human chromosome complement. *Cytogenet. Cell Genet.* **13**:103–105.

Johnson, P. F., and Abelson, J. 1983. The yeast tRNA^tyr gene intron is essential for correct modification of its tRNA product. *Nature* **302**:681–687.

Kaine, B. P., Gupta, R., and Woese, C. R. 1983. Putative introns in tRNA genes of prokaryotes. *Proc. Natl. Acad. Sci. USA* **80**:3309–3312.

Kedes, L. H. 1979. Histone genes and histone messengers. *Annu. Rev. Biochem.* **48**:837–870.

Khoury, G., and Gruss, P. 1983. Enhancer elements. *Cell* **33**:313–314.

Kleckner, N. 1977. Translocatable elements in procaryotes. *Cell* **11**:11–23.

Kleckner, N. 1981. Transposable elements in procaryotes. *Annu. Rev. Genet.* **15**:341–404.

Klemenz, R., Stillman, D. J., and Geiduschek, E. P. 1982. Specific interactions of *Saccharomyces cerevisiae* proteins with a promoter region of eukaryotic tRNA genes. *Proc. Natl. Acad. Sci. USA* **79**:6191–6195.

Krayev, A. S., Kramerov, D. A., Skryabin, K. G., Ryskov, A. P., Bayev, A. A., and Georgiev, G. P. 1980.

The nucleotide sequence of the ubiquitous repetitive DNA sequence B1 complementary to the most abundant class of mouse fold-back RNA. *Nucleic Acids Res.* **8:**1201–1215.

Kruger, K., Grabowski, P. J., Zaug, A. J., Sands, J., Gottschling, D. E., and Cech, T. R. 1982. Self-splicing RNA: Autoexcision and autocyclization of the ribosomal RNA intervening sequence of *Tetrahymena. Cell* **31:**147–157.

Kurjan, J., and Herskowitz, I. 1982. Structure of a yeast pheromone gene (*MF*α): A putative α-factor precursor contains four tandem copies of mature α-factor. *Cell* **30:**933–943.

Lai, C. J., and Khoury, G. 1979. Deletion mutants of simian virus 40 defective in biosynthesis of late viral mRNA. *Proc. Natl. Acad. Sci. USA* **76:**71–75.

Langford, C. J., and Gallwitz, D. 1983. Evidence for an intron-contained sequence required for the splicing of yeast RNA polymerase II transcripts, *Cell* **33:**519–527.

Laski, F. A., Belagaje, R., RajBhandary, U. L., and Sharp, P. A. 1982. An amber suppressing tRNA gene derived by site-specific mutagenesis: Cloning and function in mammalian cells. *Proc. Natl. Acad. Sci. USA* **79:**5813–5817.

Lau, L. F., Roberts, J. W., and Wu, R. 1982. Transcription terminates at Λt_{R1} in three clusters. *Proc. Natl. Acad. Sci. USA* **79:**6171–6175.

Lifton, R. P., Goldberg, M. L., Karp, R. W., and Hogness, D. S. 1977. The organization of histone genes in *Drosophila melanogaster:* Functional and evolutionary implications. *Cold Spring Harbor Symp. Quant. Biol.* **42:**1047–1051.

Lopez, L. C., Frazier, N. L., Su, C. J., Kumar, A., and Saunders, G. F. 1983. Mammalian pancreatic preproglucagon contains three glucagon-related peptides. *Proc. Natl. Acad. Sci. USA* **80:**5485–5489.

Lueders, K. K., and Paterson, B. M. 1982. A short interspersed repetitive element found near some mouse structural genes. *Nucleic Acids Res.* **10:**7715–7729.

Machida, C., Machida, Y., Wang, H. C., Ishizaki, K., and Ohtsubo, E. 1983. Repression of cointegration ability of insertion element IS1 by transcriptional readthrough from flanking regions. *Cell* **34:**135–142.

Mackie, G. A., and Parsons, G. D. 1983. Tandem promoters in the gene for ribosomal protein S20. *J. Biol. Chem.* **258:**7840–7846.

Mahler, H. R. 1983. The exon: intron structure of some mitochondrial genes and its relation to mitochondrial evolution. *Int. Rev. Cytol.* **82:**1–98.

Malissen, M., Hunkapiller, T., and Hood, L. 1983. Nucleotide sequence of a light chain gene of the mouse I-A subregion: Aβ^d. *Science* **221:**750–753.

Marian, B., and Wintersberger, U. 1980. Histone synthesis during sporulation of yeast. *FEBS Lett.* **117:**63–67.

Mason, A. J., Evans, B. A., Cox, D. R., Shine, J., and Richards, R. I. 1983. Structure of mouse kallikrein gene family suggests a role in specific processing of biologically active peptides. *Nature (London)* **303:**300–307.

Mason, J. O., Williams, G. T., and Neuberger, M. S. 1985. Transcription cell type specificity is conferred by an immunoglobulin V_H gene promoter that includes a functional consensus sequence. *Cell* **41:**479–487.

Mason, P. J., Jones, M. B., Elkington, J. A., and Williams, J. G. 1985. Polyadenylation of the *Xenopus* β1-globin mRNA at a downstream minor site in the absence of the major site and utilization of an AAUACA polyadenylation signal. *EMBO J.* **4:**205–211.

McGinnis, W., Shermoen, A. W., and Beckendorf, S. K. 1983. A transposable element inserted just 5′ to a *Drosophila* glue protein gene alters gene expression and chromation structure. *Cell* **34:**75–84.

McLauchlan, J., Gaffney, D., Whitton, J. L. and Clements, J. B. 1985. The consensus sequence YGTGTTYY located downstream from the AATAAA signal is required for efficient formation of mRNA 3′ termini. *Nucleic Acids Res.* **13:**1347–1368.

McLaughlin, J. R., Chang, S.-Y., and Chang, S. 1982. Transcriptional analysis of the *Bacillus licheniformis penP* gene. *Nucleic Acids Res.* **10:**3905–3919.

McLean, P., and Dixon, R. 1981. Requirement of *nifV* gene for production of wild-type nitrogenase enzyme in *Klebsiella pneumoniae. Nature (London)* **292:**655–656.

Mercola, M., Wang, X. F., Olsen, J., and Calame, K. 1983. Transcriptional enhancer elements in the mouse immunoglobulin heavy chain locus. *Science* **221:**663–665.

Mercola, M., Goverman, J., Mirell, G., and Calame, K. 1985. Immunoglobulin heavy-chain enhancer requires one or more tissue-specific factors. *Science* **227:**266–270.

Merrick, M., Hill, S., Hennecke, H., Hahn, M., Dixon, R., and Kennedy, C. 1982. Repressor properties of the *nifL* gene product in *Klebsiella pneumoniae Mol. Gen. Genet.* **185:**75–81.

Meunier-Rotival, M., Soriano, P., Cuny, G., Strauss, F., and Bernardi, G. 1982. Sequence organization and genomic distribution of the major family of interspersed repeats of mouse DNA. *Proc. Natl. Acad. Sci. USA* **79:**355–359.

Montell, C., Fisher, E. F., Caruthers, M. H., and Berk, A. J. 1983. Inhibition of RNA cleavage but not polyadenylation by a point mutation in mRNA 3′ consensus sequence AAUAAA. *Nature (London)* **305:**600–605.

Moreno, F., Fowler, A. V., Hall, M., Silhavy, T. J., Zabin, I., and Schwartz, M. 1980. A signal sequence is not sufficient to lead β-galactosidase out of the cytoplasm. *Nature (London)* **286:**356–358.

Morgan, E. A., Ikemura, T., Lindahl, L., Fallon, A. M., and Nomura, M. 1978. Some rRNA operons in *E. coli* have tRNA genes at their distal ends. *Cell* **13:**335–344.

Morgan, W. D., Bear, D. G., and von Hippel, P. H. 1983. Rho-dependent termination of transcription. I. Identification and characterization of termination sites for transcription from the bacteriophage ΛP$_R$ promoter. *J. Biol. Chem.* **285:**9553–9564.

Nomoto, A., Omata, T., Toyoda, H., Kuge, S., Horie, H., Kataoka, Y., Genba, Y., Nakano, Y., and Imura, N. 1982. Complete nucleotide sequence of the attenuated poliovirus Sabin 1 strain genome. *Proc. Natl. Acad. Sci. USA* **79:**5793–5797.

Nudel, U., Katcoff, D., Zakut, R., Shani, M., Carmon, Y., Finer, M., Czosnek, H., Ginsburg, I., and Yaffe, D. 1982. Isolation and characterization of rat skeletal muscle and cytoplasmic actin genes. *Proc. Natl. Acad. Sci. USA* **79:**2763–2767.

Ogasawara, N., Moriya, S., and Yoshikawa, H. 1983. Structure and organization of rRNA operons in the region of the replication origin of the *Bacillus subtilis* chromosome. *Nucleic Acids Res.* **11:**6301–6318.

O'Hare, K., and Rubin, G. M. 1983. Structures of P transposable elements and their sites of insertion and excision in the *Drosophila melanogaster* genome. *Cell* **34:**25–35.

Ohkubo, H., Vogeli, G., Mudryj, M., Avvedimento, V. E., Sullivan, M., Pastan, I., and de Crombrugghe, B. 1980. Isolation and characterization of overlapping genomic clones covering the chicken α2 (type I) collagen gene. *Proc. Natl. Acad. Sci. USA* **77:**7059–7063.

Ono, Y., Onda, H., Sasada, R., Igarashi, K., Sugino, Y., and Nishioka, K. 1983. The complete nucleotide sequences of the cloned hepatitis B virus DNA: Subtypes adr and adw. *Nucleic Acids Res.* **11:**1747–1757.

Orkin, S. H., Cheng, T. C., Antonarakis, S. E., and Kazazian, H. H. 1985. Thalassemia due to a mutation in the cleavage-polyadenylation signal of the human β-globin gene. *EMBO J.* **4:**453–456.

Osheim, Y. N., and Miller, O. L. 1983. Novel amplification and transcriptional activity of chorion genes in *Drosophila melanogaster* follicle cells. *Cell* **33:**543–553.

Pearson, W. R., Mukai, T., and Morrow, J. F. 1981. Repeated DNA sequences near the 5′-end of the silk fibroin gene. *J. Biol. Chem.* **256:**4033–4041.

Peffley, D. M., and Sogin, M. L. 1981. A putative tRNA^{Trp} gene cloned from *Dictyostelium discoideum:* Its nucleotide sequence and association with repetitive DNA. *Biochemistry* **20:**4015–4021.

Pellegrini, M., Manning, J., and Davidson, N. 1977. Sequence arrangement of the rDNA of *Drosophila melanogaster*. *Cell* **10:**213–224.

Perry, R. P., and Scherrer, K. 1975. The methylated constituents of globin mRNA. *FEBS Lett.* **57:**73–78.

Piette, J., Nyunoya, H., Lusty, C. J., Cunin, R., Weyens, G., Crabeel, M., Charlier, D., Glansdorff, N., and Piérard, A. 1984. DNA sequence of the *carA* gene and the control region of *carAB:* Tandem promoters, respectively controlled by arginine and the pyrimidines, regulate the synthesis of carbamoyl-phosphate synthetase in *Escherichia coli* K-12. *Proc. Natl. Acad. Sci. USA* **81:**4134–4138.

Platt, T. 1981. Termination of transcription and its regulation in the tryptophan operon of *E. coli*. *Cell* **24:**10–23.

Plotch, S. J., Bouloy, M., Ulmanen, I., and Krug, R. M. 1981. A unique cap (m^{7}GpppXm)-dependent influenza virion endonuclease cleaves capped RNAs to generate the primer that initiates viral RNA transcription. *Cell* **23:**847–858.

Post, L. E., Arfsten, A. E., Reusser, F., and Nomura, M. 1978. DNA sequences of promoter regions for the *str* and *spc* ribosomal protein operons in *E. coli*. *Cell* **15:**215–229.

Post, L. E., Strycharz, G. D., Nomura, M., Lewis, H., and Dennis, P. P. 1979. Nucleotide sequence of the ribosomal protein gene cluster adacent to the gene for RNA polymerase subunit β in *Escherichia coli*. *Proc. Natl. Acad. Sci. USA* **76:**1697–1701.

Procunier, J. D., and Tartof, K. D. 1976. Restriction map of 5S RNA genes of *Drosophila melanogaster*. *Nature (London)* **263:**255–257.

Robinson, R. R., and Davidson, N. 1981. Analysis of a *Drosophila* tRNA gene cluster: Two tRNA[Leu] genes contain intervening sequences. *Cell* **23:**251–259.

Rosenberg, M., and Court, D. 1979. Regulatory sequences involved in the promotion and termination of RNA transcription. *Annu. Rev. Genet.* **13:**319–353.

Ruberti, I., Fragapane, P., Beccari, E., Amaldi, F., and Bozzoni, I. 1982. Characterization of histone genes isolated from *Xenopus laevis* and *Xenopus tropicalis* genomic libraries. *Nucleic Acids Res.* **10:**7543–7559.

Schibler, U., Pittet, A. C., Young, R. A., Hagenbüchle, O., Tosi, M., Gellman, S., and Wellauer, P. K. 1982. The mouse α-amylase multigene family. Sequence organization of members expressed in the pancreas, salivary gland, and liver. *J. Mol. Biol.* **155:**247–266.

Schmitt, R., Bernhard, E., and Mattes, R. 1979. Characterisation of Tn*1721*, a new transposon containing tetracycline resistance genes capable of amplification. *Mol. Gen. Genet.* **172:**53–65.

Schwartz, I., Klotsky, R. A., Elseviers, D., Gallagher, P. J., Krauskopf, M., Siddiqui, M. A. Q., Wong, J. F. H., and Roe, B. A. 1983. Molecular cloning and sequencing of *pheU* a gene for *Escherichia coli* tRNA[Phe]. *Nucleic Acids Res.* **11:**4379–4389.

Sekiya, T., Gait, M. J., Noris, K., Ramamoorthy, B., and Khorana, G. 1976. The nucleotide sequence in the promoter region of the gene for an *Escherichia coli* tyrosine transfer ribonucleic acid. *J. Biol. Chem.* **251:**4481–4489.

Sharp, P. M. 1985. Does the 'non-coding' strand code? *Nucleic Acids Res.* **13:**1389–1397.

Singer, M. F., 1982a. SINES and LINES: Highly repeated short and long interspersed sequences in mammalian genomes. *Cell* **28:**433–434.

Singer, M. F., 1982b. Highly repeated sequences in mammalian genomes. *Int. Rev. Cytol.* **76:**67–112.

Singh, R. K., and Singh, H. N. 1981. Genetic analysis of the *het* and *nif* genes in the blue-green alga *Nostoc muscorum*. *Mol. Gen. Genet.* **184:**531–535.

Smith, M. M., and Andrésson, Ó. S. 1983. DNA sequences of yeast H3 and H4 histone genes from two non-allelic gene sets encode identical H3 and H4 proteins. *J. Mol. Biol.* **169:**663–690.

Smith, M. M., and Murray, K. 1983. Yeast H3 and H4 histone messenger RNAs are transcribed from two non-allelic gene sets. *J. Mol. Biol.* **169:**641–661.

Spradling, A. C., and Mahowald, A. P. 1980. Amplification of genes for chorion proteins during oogenesis in *Drosophila melanogaster*. *Proc. Natl. Acad. Sci. USA* **77:**1096–1100.

Standring, D. N., Venegas, A., and Rutter, W. J. 1981. Yeast tRNA[Leu] gene transcribed and spliced in a HeLa cell extract. *Proc. Natl. Acad. Sci. USA* **78:**5963–5967.

Stassi, D. L., Dunn, J. J., and Lacks, S. A. 1982. Nucleotide sequence of DNA controlling expression of genes for maltosaccharide utilization in *Streptococcus pneumoniae*. *Gene* **20:**359–366.

Stephenson, E. C., Erba, H. P., and Gall, J. G. 1981. Histone gene clusters of the newt *Notophthalmus* are separated by long tracts of satellite DNA. *Cell* **24:**639–647.

Swift, G. H., Dagorn, J. C., Ashley, P. L., Cummings, S. W., and MacDonald, R. J. 1982. Rat pancreatic kallikrein mRNA: Nucleotide sequence and amino acid sequence of the encoded preproenzyme. *Proc. Natl. Acad. Sci. USA* **79:**7263–7267.

Taussig, R., and Carlson, M. 1983. Nucleotide sequence of the yeast *SUC2* gene for invertase. *Nucleic Acids Res.* **11:**1943–1954.

Treisman, R., Novak, U., Favaloro, J., and Kamen, R. 1981. Transformation of rat cells by an altered polyoma virus genome expressing only the middle-T protein. *Nature (London)* **292:**595–600.

Turner, P. C., and Woodland, H. R. 1983. Histone gene number and organization in *Xenopus; Xenopus borealis* has a homogeneous major cluster. *Nucleic Acids Res.* **11:**971–986.

Valenzuela, P., Venegas, A., Weinberg, F., Bishop, R., and Rutter, W. J. 1978. Structure of yeast phenylalanine-tRNA genes: An intervening DNA sequence within the region coding for the tRNA. *Proc. Natl. Acad. Sci. USA* **75:**190–194.

van Omman, G. J. B., Arnberg, A. C., Bass, F., Brocas, H., Sterk, A., Tegelaers, W. H. H., Vassart, G., and de Vijlder, J. J. M. 1983. The human thyroglobulin gene contains two 15–17 kb introns near its 3′-end. *Nucleic Acids Res.* **11:**2273–2285.

Venegas, A., Quiroga, M., Zaldivar, J., Rutter, W. J., and Valenzuela, P. 1979. Isolation of yeast tRNA[Leu] genes. DNA sequence of a cloned tRNA[Leu] gene. *J. Biol. Chem.* **254:**12306–12309.

Wallis, J. W., Hereford, L., and Grunstein, M. 1980. Histone H2B genes of yeast encode two different proteins. *Cell* **22:**799–805.

Walter, P., and Blobel, G. 1983. Dissassembly and reconstitution of signal recognition particle. *Cell* **34:**525–533.

Weber, F., de Villiers, J., and Schaffner, W. 1984. An SV40 "enhancer trap" incorporates exogenous enhancers or generates enhancers from its own sequences. *Cell* **36**:983–992.

Wong, H. C., Schnepf, H. E., and Whiteley, H. R. 1983. Transcriptional and translational start sites for the *Bacillus thuringiensis* crystal protein gene. *J. Biol. Chem.* **258**:1960–1967.

Woudt, L. P., Pastink, A., Kempers-Veenstra A. E., Jansen, A. E. M., Mager, W. H., and Planta, R. J. 1983. The genes coding for histones H3 and H4 in *Neurospora crassa* are unique and contain intervening sequences. *Nucleic Acids Res.* **11**:5347–5360.

Yamada, Y., Mudryj, M., Sullivan, M., and deCrombrugghe, B. 1983. Isolation and characterization of a genomic clone encoding chick α1 type III collagen. *J. Biol. Chem.* **258**:2758–2761.

Yanofski, C. 1981. Attenuation in the control of expression of bacterial operons. *Nature (London)* **289**:751–758.

Young, R. A. 1979. Transcription termination in the *Escherichia coli* ribosomal RNA operon *rrnC*. *J. Biol. Chem.* **254**:12725–12731.

CHAPTER 2

Abelson, J. 1979. RNA processing and the intervening sequence problem. *Annu. Rev. Biochem.* **48**:1035–1069.

Addison, W. R., Astell, C. R., Delaney, A. D., Gillam, I. C., Hayashi, S., Miller, R. C., Rajput, B., Smith, M., Taylor, D. M., and Tener, G. M. 1982. The structures of genes hybridizing with tRNA$_4^{Val}$ from *Drosophila melanogaster*. *J. Biol. Chem.* **257**:670–673.

Adhya, S., Gottesman, M., and deCrombrugghe, B. 1974. Release of polarity in *Escherichia coli* by gene *N* of phage lambda: Termination and antitermination of transcription. *Proc. Natl. Acad. Sci. USA* **71**:2534–2538.

Altwegg, M., and Kubli, E. 1980. The nucleotide sequence of tRNA$_\gamma$ of *Drosophila melanogaster*. *Nucleic Acids Res.* **8**: 3259–3262.

Amstutz, H., Munz, P., Heyer, W. D., Leupold, U., and Kohli, J. 1985. Concerted evolution of tRNA genes: Intergenic conversion among three unlinked serine tRNA genes in *S. pombe*. *Cell* **40**:879–886.

Anderson, S., Bankier, A. T., Barrell, B. G., deBruijn, M. H. L., Coulson, A. R., Drouin, J., Eperon, I. C., Nierlich, D. P., Roe, B. A., Sanger, F., Schreier, P. H., Smith, A. J. H., Staden, R., and Young, I. G. 1981. Sequence and organization of the human mitochondrial genome. *Nature (London)* **290**:457–465.

Anderson, S., deBruijn, M. H. L., Coulson, A. R., Eperon, I. C., Sanger, P., and Young, I. G. 1982. Complete sequence of bovine mitochondrial DNA. Conserved features of the mammalian mitochondrial genome. *J. Mol. Biol.* **156**:683–717.

Anthony, D. D., Zeszotek, E., and Goldthwait, D. A. 1966. Initiation by the DNA-dependent RNA polymerase. *Proc. Natl. Acad. Sci. USA* **56**:1026–1033.

Barreau, C., and Begueret, J. 1982. DNA-dependent RNA polymerase III from the fungus *Podospora comata*. Purification, subunit structure and comparison with the homologous enzyme of a related species. *Eur. J. Biochem.* **129**:423–428.

Barrell, D. G., and Sanger, F. 1969. The sequence of phenylalanine tRNA from *E. coli*. *FEBS Lett.* **3**:275–278.

Batts-Young, B., Maizels, N., and Lodish, H. F. 1977. Precursors of ribosomal RNA in the cellular slime mold *Dictyostelium discoideum*. *J. Biol. Chem.* **252**:3952–3960.

Bawnik, N., Beckmann, J. S., Sarid, S., and Daniel, V. 1983. Isolation of nucleotide sequence of a plant tRNA gene: Petunia asparagine tRNA. *Nucleic Acids Res.* **11**:1117–1122.

Berlani, R. E., Pentella, C., Macino, G., and Tzagoloff, A. 1980. Assembly of mitochondrial membrane systems: Isolation of mitochondrial transfer ribonucleic acid mutants and characterization of transfer ribonucleic acid genes of *Sacchromyces cerevisiae*. *J. Bacteriol.* **141**:1086–1097.

Berman, M. L., and Landy, A. 1979. Promoter mutations in the transfer RNA gene *tryT* of *Escherichia coli*. *Proc. Natl. Acad. Sci. USA* **76**:4303–4307.

Berthold, V., and Geider, K. 1976. Interaction of DNA with DNA-binding proteins. *Eur. J. Biochem.* **71**:443–449.

Bibb, M. J., Van Etten, R. A., Wright, C. T., Walberg, M. W., and Clayton, D. A. 1981. Sequence and gene organization of mouse mitochondrial DNA. *Cell* **26**:167–180.

Black, L. W., and Gold, L. M. 1971. Pre-replicative development of the bacteriophage T$_4$: RNA and protein synthesis *in vivo* and *in vitro*. *J. Mol. Biol.* **60**:365–388.

Blank, H. U., and Söll, D. 1971. The nucleotide sequence of two leucine tRNA species from *Escherichia coli.* *Biochem. Biophys. Res. Commun.* **43**:1192–1197.

Bonitz, S. G., and Tzagoloff, A. 1980. Assembly of the mitochondrial membrane system. Sequences of yeast mitochondrial tRNA genes. *J. Biol. Chem.* **255**:9075–9081.

Bonnet, J., Ebel, J. P., Dirheimer, G., Shershneva, L. P., Krutilina, A. I., Venkstern, T. V., and Bayev, A. A. 1974. The corrected nucleotide sequence of valine tRNA from baker's yeast. *Biochimie* **56**:1211–1213.

Bos, J. L., Osinga, K. A., Van der Horst, G., and Borst, P. 1979. Nucleotide sequence of the mitochondrial structural genes for cysteine-tRNA and histidine-tRNA of yeast. *Nucleic Acids Res.* **6**:3255–3266.

Buhler, J. M., Iborra, F., Sentenac, A., and Fromageot, P. 1976. Structural studies on yeast RNA polymerases. Existence of common subunits in RNA polymerases A(I) and B(II). *J. Biol. Chem.* **251**:1712–1717.

Burgess, R. R. 1976. Purification and physical properties of *E. coli* RNA polymerase. *In:* Losick, R., and Chamberlin, M., eds., *RNA Polymerase,* Cold Spring Harbor, New York, Cold Spring Harbor Laboratory, pp. 69–100.

Caillet, J., Plumbridge, J. A., and Springer, M. 1985. Evidence that *pheV,* a gene for tRNA^Phe of *E. coli,* is transcribed from tandem promoters. *Nucleic Acids Res.* **13**:3699–3709.

Camier, S., Gabrielsen, O., Baker, R., and Sentenas, A. 1985. A split binding site for transcription factor τ on the tRNA$_3^{Glu}$ gene. *EMBO J.* **4**:491–500.

Campen, R. K., Duester, G. L., Holmes, W. M., and Young, J. M. 1980. Organization of tRNA genes in the *Escherichia coli* chromosomes. *J. Bacteriol.* **144**:1083–1093.

Cantatore, P., De Benedetto, C., Gadaleta, G., Gallerani, R., Kroon, A. M., Holtrop, M., Lanave, C., Pepe, G., Quagliariello, C. and Sbisa, E. 1982. The nucleotide sequences of several tRNA genes from rat mitochondria: Common features and relatedness to homologous species. *Nucleic Acids Res.* **10**:3279–3289.

Carrara, G., Di Segni, G., Otsuka, A., and Tocchini-Valentine, G. P. 1981. Deletion of the 3' half of the yeast tRNA$_3^{Leu}$ gene does not abolish promoter function *in vitro. Cell* **27**:371–379.

Chamberlin, M. J. 1976. RNA polymerase—An overview. *In:* Losick, R., and Chamberlin, M., eds., *RNA Polymerase,* Cold Spring Harbor, New York, Cold Spring Harbor Laboratory, pp. 17–67.

Chan, J. C., Yang, J. A., Dunn, M. J., Agris, P. P., and Wong, T. W. 1982. The nucleotide sequence of a glutamate tRNA from rat liver. *Nucleic Acids Res.* **10**:4605–4608.

Ciampi, M. S., Melton, D. A., and Cortese, R. 1982. Site-directed mutagenesis of a tRNA gene: Base alterations in the coding region affect transcription. *Proc. Natl. Acad. Sci. USA* **79**:1388–1392.

Ciliberto, G., Castagnoli, L., Melton, D. A., and Cortese, R. 1982. Promoter of a eukaryotic tRNA^Pro gene is composed of three noncontiguous regions. *Proc. Natl. Acad. Sci. USA* **79**:1195–1199.

Clary, D. O., Goddard, J. M., Martin, S. C., Fauron, C. M. R., and Wolstenholme, D. R. 1982. *Drosophila* mitochondrial DNA: A novel gene order. *Nucleic Acids Res.* **10**:6619–6634.

Clary, D. O., Wahleithner, J. A., and Wolstenholme, D. R. 1983. Transfer RNA genes in *Drosophila* mitochondrial DNA: Related 5' flanking sequences and comparisons to mammalian mitochondrial tRNA genes. *Nucleic Acids Res.* **11**:2411–2425.

Colby, D., Leboy, P. S., and Guthrie, C. 1981. Yeast tRNA precursor mutated at a splice junction is correctly processed *in vivo. Proc. Natl. Acad. Sci. USA* **78**:415–419.

Cooley, L., Appel, B., and Söll, D. 1982. Post-transcriptional nucleotide addition is responsible for the formation of the 5' terminus of histidine tRNA. *Proc. Natl. Acad. Sci. USA* **79**:6475–6479.

Cory, S., and Marcker, K. A. 1970. The nucleotide sequence of methionine transfer RNA$_M$. *Eur. J. Biochem.* **12**:177–194.

Crépin, M., Cukier-Kahn, R., and Gros, F. 1975. Effect of low-molecular-weight DNA binding protein, H$_1$ factor, on the *in vitro* transcription of the lactose operon in *Escherichia coli. Proc. Natl. Acad. Sci. USA* **72**:333–337.

Cukier-Kahn, R., Jacquet, M., and Gros, F. 1972. Two heat-resistant low molecular weight proteins from *Escherichia coli* that stimulate DNA-directed RNA synthesis. *Proc. Natl. Acad. Sci. USA* **69**:3643–3647.

Dahmus, M. E. 1983. Structural relationship between the large subunits of calf thymus RNA polymerase II. *J. Biol. Chem.* **258**:3956–3960.

D'Allessio, J. M., Perna, P. J., and Paule, M. R. 1979. DNA-dependent RNA polymerases from *Acanthamoeba castellanii.* Comparative subunit structures of the homogeneous enzymes. *J. Biol. Chem.* **254**:11282–11287.

De Bruijn, M. H. L., Schreier, P. H., Eperon, I. C., Barrell, B. G., Chen, E. Y., Armstrong, P. W., Wong, J. P. H., and Roe, B. A. 1980. A mammalian mitochondrial serine transfer RNA lacking the "dihydrouridine" loop and stem. *Nucleic Acids Res.* **8**:5213–5222.

DeFranco, D., Schmidt, O., and Söll, D. 1980. Two control regions for eukaryotic tRNA gene transcription. *Proc. Natl. Acad. Sci. USA* **77**:3365–3368.

DeFranco, D., Burke, K. B., Hayashi, S., Tener, G. M., Miller, R. C., and Söll, D. 1982. Genes for tRNA$_5^{Lys}$ from *Drosophila melanogaster*. *Nucleic Acids Res.* **10**:5799–5808.

del Rey, F. J., Donahue, T. F., and Fink, G. R. 1982. *Sigma*, a repetitive element found adjacent to tRNA genes of yeast. *Proc. Natl. Acad. Sci. USA* **79**:4138–4142.

del Rey, F., Donahue, T. F., and Fink, G. R. 1983. The histidine tRNA genes of yeast. *J. Biol. Chem.* **258**:8175–8182.

Dennis, P. P. 1976. Effects of chloramphenicol on the transcriptional activities of ribosomal RNA and ribosomal protein genes in *Escherichia coli*. *J. Mol. Biol.* **108**:535–546.

Deno, H., Kato, A., Shinozaki, K., and Sugiura, M. 1982. Nucleotide sequences of tobacco chloroplast genes for elongator tRNA^{Met} and tRNA^{Val} (UAC): The tRNA^{Val} (UAC) gene contains a long intron. *Nucleic Acids Res.* **10**:7511–7520.

Dillon, L. S. 1962. Comparative cytology and the evolution of the living world. *Evolution* **16**:102–117.

Dillon, L. S. 1963. A reclassification of the major groups of organisms based upon comparative cytology. *Syst. Zool.* **12**:71–82.

Dillon, L. S. 1973. The origins of the genetic code. *Bot. Rev.* **39**:301–345.

Dillon, L. S. 1978. *The Genetic Mechanism and the Origin of Life*. New York, Plenum Press.

Dillon, L. S. 1981. *Ultrastructure, Macromolecules, and Evolution*. New York, Plenum Press.

Dillon, L. S. 1983. *The Inconstant Gene*. New York, Plenum Press.

Dingermann, T., Burke, D. J., Sharp, S., Schaack, J., and Söll, D. 1982. The 5′ flanking sequences of *Drosophila* tRNA^{Arg} genes control their *in vitro* transcription in a *Drosophila* cell extract. *J. Biol. Chem.* **257**:14738–14744.

Dube, S. K., and Marcker, K. A. 1969. The nucleotide sequence of *N*-formylmethionyl-transfer RNA. *Eur. J. Biochem.* **8**:256–262.

Dube, S. K., Marcker, K. A., and Yudelevich, A. 1970. The nucleotide sequence of a leucine transfer RNA from *E. coli*. *FEBS Lett.* **9**:168–170.

Duester, G., Campen, R. K., and Holmes, W. M. 1981. Nucleotide sequence of an *Escherichia coli* tRNA (Leu 1) operon and identification of the transcription promoter signal. *Nucleic Acids Res.* **9**:2121–2139.

Duester, G., Elford, R. M., and Holmes, W. M. 1982. Fusion of the *Escherichia coli* tRNA$_1$ promoter to the *galK* gene: Analysis of sequences necessary for growth-rate-dependent regulation. *Cell* **30**:855–864.

Dynan, W. S., and Tjian, R. 1983. Isolation of transcription factors that discriminate between different promoters recognized by RNA polymerase II. *Cell* **32**:669–680.

Eigel, A., Olah, J., and Feldmann, H. 1981. Structural comparison of two yeast tRNA$_3^{Glu}$ genes. *Nucleic Acids Res.* **9**:2961–2970.

Emmer, M., de Crombrugghe, B., Pastan, I., and Perlman, R. 1970. Cyclic AMP receptor protein of *E. coli*: Its role in the synthesis of inducible enzymes. *Proc. Natl. Acad. Sci. USA* **66**:480–487.

Engelke, D. R., Shastry, B. S., and Roeder, R. G. 1983. Multiple forms of DNA-dependent RNA polymerases in *Xenopus laevis*. Rapid purification and structural and immunological properties. *J. Biol. Chem.* **258**:1921–1931.

Eperon, I. C., Anderson, S., and Nierlich, D. P. 1980. Distinctive sequence of human mitochondrial ribosomal RNA genes. *Nature (London)* **286**:460–467.

Etcheverry, T., Colby, D., and Guthrie, C. 1979. A precursor to a minor species of yeast tRNA^{Ser} contains an intervening sequence. *Cell* **18**:11–26.

Feldmann, H., Olah, J., and Friedenreich, H. 1981. Sequence of a yeast DNA fragment containing a chromosomal replicator and a tRNA$_3^{Glu}$ gene. *Nucleic Acids Res.* **9**:2949–2959.

Fox, G. E., Stackebrandt, E., Hespell, R. B., Gibson, J., Maniloff, J., Dyer, T. A., Wolfe, R. S., Balch, W. E., Tanner, R. S., Magrun, L. J., Zablen, L. B., Blakemore, R., Gupta, R., Bonen, L., Lewis, B. J., Stahl, D. A., Luehraes, K. R., Chen, K. N., and Woese, C. R. 1980. The phylogeny of prokaryotes. *Science* **209**:457–463.

Fried, N. G., and Crothers, D. M. 1983. CAP and RNA polymerase interactions with the *lac* promoter: Binding stoichiometry and long range effects. *Nucleic Acids Res.* **11**:141–158.

Friedman, D. I., Wilgus, G. S., and Mural, R. J. 1973. Gene N regulator function of phage λ*imm* 21: Evidence that a site of N actin differs from a site of N recognition. *J. Mol. Biol.* **81**:505–516.

Fukada, K., and Abelson, J. 1980. DNA sequence of a T4 transfer RNA gene cluster. *J. Mol. Biol.* **139**:377–391.

Galli, G., Hofstetter, H., and Birnstiel, M. L. 1981. Two conserved sequence blocks within eukaryotic tRNA genes are major promoter elements. *Nature (London)* **294**:626–631.

Garber, R. L., and Gage, L. P. 1979. Transcription of a cloned *Bombyx mori* tRNA$_2^{Ala}$ gene: Nucleotide sequence of the tRNA precursor and its processing *in vitro*. *Cell* **18**:817–828.

Gauss, D. H., and Sprinzl, M. 1983a. Compilation of tRNA sequences. *Nucleic Acids Res.* **11**:r1–r53.

Gauss, D. H., and Sprinzl, M. 1983b. Compilation of sequences of tRNA genes. *Nucleic Acids Res.* **11**:r55–r103.

Ghosh, H. P., Ghosh, K., Simsek, M., and RajBhandary, U. L. 1982. Nucleotide sequence of wheat germ cytoplasmic initiator methionine transfer ribonucleic acids. *Nucleic Acids Res.* **10**:3241–3247.

Ghosh, S., and Echols, H. 1972. Purification and properties of D protein: A transcription factor of *Escherichia coli*. *Proc. Natl. Acad. Sci. USA* **69**:3660–3664.

Gilbert, W. 1976. Starting and stopping sequences for the RNA polymerase. *In:* Losick, R., and Chamberlin, M., eds., *RNA Polymerase*. Cold Spring Harbor, New York, Cold Spring Harbor Laboratory, pp. 193–205.

Gillum, A. M., Hecker, L. I., Silberklang, M., Schwartzbach, S. D., RajBhandary, U. L., and Barnett, W. E. 1977. Nucleotide sequence of *Neurosopora crassa* cytoplasmic initiator tRNA. *Nucleic Acids Res.* **4**:4109–4131.

Gilman, M. Z., Wiggs, J. L., and Chamberlin, M. J. 1981. Nucleotide sequences of two *Bacillus subtilis* promoters used by *Bacillus subtilis* sigma-28 RNA polymerase. *Nucleic Acids Res.* **9**:5991–6000.

Ginsberg, T., Rogg, H., and Staehelin, M. 1971. Nucleotide sequences of rat liver serine-tRNA. 3. The partial enzymatic digestion of serine-tRNA$_1$ and derivation of its total primary structure. *Eur. J. Biochem.* **21**:249–257.

Goddard, J. P., Squire, M., Bienz, M., and Smith, J. D. 1983. A human tRNA^{Glu} gene of high transcriptional activity. *Nucleic Acids Res.* **11**:2551–2562.

Goodman, H. M., Olson, M. V., and Hall, B. D. 1977. Nucleotide sequence of a mutant eukaryotic gene. The yeast tyrosine-inserting ochre suppressor SUP4-o. *Proc. Natl. Acad. Sci. USA* **74**:5453–5457.

Grainger, R. M., and Maizels, N. 1980. *Dictyostelium* rRNA is processed during transcription. *Cell* **20**:619–623.

Green, C. J., and Vold, B. S. 1983. Sequence analysis of a cluster of twenty-one tRNA genes in *Bacillus subtilis*. *Nucleic Acids Res.* **11**:5763–5774.

Greenblatt, J., and Li, J. 1981. The *nusA* gene protein of *Escherichia coli*. *J. Mol. Biol.* **147**:11–23.

Greenblatt, J., and Schleif, R. 1971. Arabinose C protein: Regulation of the arabinose operon *in vitro*. *Nature New Biol.* **233**:166–170.

Grosskopf, R., and Feldmann, H. 1981. tRNA genes in rat liver mitochondrial DNA. *Curr. Genet.* **4**:191–196.

Gruhl, H., and Feldmann, H. 1976. The primary structure of a non-initiating methionine-specific tRNA from brewer's yeast. *Eur. J. Biochem.* **68**:209–217.

Grummt, I. 1981. Specific transcription of mouse ribosomal DNA in a cell-free system that mimics control *in vivo*. *Proc. Natl. Acad. Sci. USA* **78**:727–731.

Guilfoyle, T. J. 1981. Purification, subunit structure, and immunological properties of chromatin-bound ribonucleic acid polymerase I from cauliflower inflorescence. *Biochemistry* **19**:5966–5972.

Hager, G., Holland, M., Valenzuela, P., Weinberg, F., and Rutter, W. J. 1976. RNA polymerases and transcriptive specificity in *Saccharomyces cerevisiae*. *In:* Losick, R., and Chamberlin, M., eds., *RNA Polymerase*, Cold Spring Harbor, New York, Cold Spring Harbor Laboratory, pp. 745–762.

Han, J. H., and Harding, J. D. 1982. Isolation and nucleotide sequence of a mouse histidine tRNA gene. *Nucleic Acids Res.* **10**:4891–4900.

Han, J. H., and Harding, J. D. 1983. Using iodinated single-stranded M13 probes to facilitate rapid DNA sequences analysis—Nucleotide sequence of lysine tRNA gene. *Nucleic Acids Res.* **11**:2053–2064.

Hanna, M. M., and Meares, C. F. 1983. Topography of transcription: Path of the leading end of nascent RNA through the *Escherichia coli* transcription complex. *Proc. Natl. Acad. Sci. USA* **80**:4238–4242.

Hansen, U. M., and McClure, W. R. 1980. Role of the σ subunit of *Escherichia coli* RNA polymerase in initiation. II. Release of σ from ternary complexes. *J. Biol. Chem.* **255**:9564–9570.

Harada, F., and Nishimura, S. 1980. tRNAs containing the G-Ψ-Ψ-C sequence. I. Two arginine tRNAs of mouse leukemia cells. *Biochem. Int.* **1**:539–564.

Harada, F., Yamaizumi, K., and Nishimura, S. 1972. Oligonucleotide sequences of RNase T₁ and pancreatic RNase digests of *E. coli* aspartic acid tRNA. *Biochem. Biophys. Res. Commun.* **49**:1605–1609.

Hawkins, A. R., and Wootton, J. C. 1981. A single DNA-binding protein from *Pseudomonas aeruginosa* homologous to proteins NS1 and NS2 (HU proteins) of *Escherichia coli* and other bacteria. *FEBS Lett.* **130**:275–278.

Hawley, D. K., and McClure, W. R. 1983. Compilation and analysis of *Escherichia coli* promoter DNA sequences. *Nucleic Acids Res.* **11**:2237–2255.

Henikoff, S., Kelly, J. D., and Cohen, E. H. 1983. Transcription terminates in yeast distal to a control sequence. *Cell* **33**:607–614.

Hershey, N. D., and Davidson, N. 1980. Two *Drosophila melanogaster* tRNA[Gly] genes are contained in a direct duplication at chromosomal locus 56F. *Nucleic Acids Res.* **8**:4899–4910.

Hill, C. W., Squires, C., and Carbon, J. 1970. Structural genes for two glycine tRNA species. *J. Mol. Biol.* **52**:557–569.

Hill, C. W., Combriato, G., Steinhart, W., Riddle, D. L., and Carbon, J. 1973. The nucleotide sequence of the GGG-specific glycine transfer ribonucleic acid of *Escherichia coli* and of *Salmonella typhimurium*. *J. Biol. Chem.* **248**:4252–4262.

Hollingsworth, M. J., and Hallick, R. B. 1982. *Euglena gracilis* choloroplast transfer RNA transcription units. Nucleotide sequence analysis of a tRNA[Tyr]-tRNA[His]-tRNA[Met]-tRNA[Glu]-tRNA[Gly] gene cluster. *J. Biol. Chem.* **257**:12795–12799.

Holmes, W. M., Platt, T., and Rosenberg, M. 1983. Termination of transcription in *E. coli*. *Cell* **32**:1029–1032.

Hosbach, H. A., Silberklang, M., and McCarthy, B. J. 1980. Evolution of a *D. melanogaster* glutamate tRNA gene cluster. *Cell* **21**:169–178.

Hovemann, B., Sharp, S. J., Yamada, H., and Söll, D. 1980a. Analysis of a *Drosophila* tRNA gene cluster. *Cell* **19**:889–895.

Hovemann, B., Schmidt, O., Yamada, H., Silverman, S., Mao, J., DeFranco, D., and Söll, D. 1980b. Arrangement and transcription of Drosophila tRNA genes. In: Söll, D., and Söll, D., Abelson, J., and Schimmel, P., eds., *tRNA; Biological Aspects*, Cold Spring Harbor, New York, Cold Spring Harbor Laboratory, pp. 325–338.

Huaifeng, M., and Hartmann, G. R. 1983. RNA polymerase: Interaction of RNA and rifampicin with the subassembly $\alpha_2\beta$. *Eur. J. Biochem.* **131**:113–118.

Hunt, C., Desai, S. M., Vaughan, J., and Weiss, S. B. 1980. Bacteriophage T5 transfer RNA. Isolation and characterization of tRNA species and refinement of the tRNA gene map. *J. Biol. Chem.* **255**:3164–3173.

Imamoto, F., and Schlessinger, D. 1975. Bearing of some recent results on the mechanisms of polarity and messenger RNA stability. *Mol. Gen. Genet.* **135**:29–38.

Indik, Z. K., and Tartof, K. D. 1982. Glutamate tRNA genes are adjacent to 5S RNA genes in *Drosophila* and reveal a conserved upstream sequence (the ACT-TA box). *Nucleic Acids Res.* **10**:4159–4172.

Ishikura, H., Yamada, Y., and Nishimura, S. 1971. The nucleotide sequence of a serine tRNA from *Escherichia coli*. *FEBS Lett.* **16**: 68–70.

Jacobs, K. A., Shen, V., and Schlessinger, D. 1978. Coupling of *lac* mRNA transcription and translation in *E. coli* cell extracts. *Proc. Natl. Acad. Sci. USA* **75**:158–161.

Jacquet, M., Cukier-Kahn, R., Pla, J., and Gros, F. 1971. A thermostable protein factor acting on *in vitro* DNA transcription. *Biochem. Biophys. Res. Commun.* **45**:1597–1607.

Jendrisak, J. J. 1980. Purification, structures and functions of the nuclear RNA polymerases from higher plants. *In:* Leaver, C. J., ed., *Genome Organization and Expression in Plants*, New York, Plenum Press, pp. 77–92.

Jendrisak, J. J., Petranyi, P. W., and Burgess, R. R. 1976. Wheat germ DNA-dependent RNA polymerase II: Purification and properties. *In:* Losick, R., and Chamberlin, R., eds., *RNA Polymerase*, Cold Spring Harbor, New York, Cold Spring Harbor Laboratory, pp. 779–791.

Johnson, W. C., Moran, C. P., and Losick, R. 1983. Two RNA polymerase sigma factors from *Bacillus subtilis* discriminate between overlapping promoters from a developmentally regulated gene. *Nature (London)* **302**:800–804.

Karabin, G. D., and Hallick, R. B. 1983. *Euglena gracilis* chloroplast transfer RNA transcription units. Nucleotide sequence analysis of a tRNA^{Thr}-tRNA^{Gly}-tRNA^{Met}-tRNA^{Ser}-tRNA^{Gln} gene cluster. *J. Biol. Chem.* **258**:5512–5518.

Kashdan, M. A., and Dudock, B. S. 1982. Structure of a spinach chloroplast threonine tRNA gene. *J. Biol. Chem.* **257**:1114–1116.

Kedinger, C., and Chambon, P. 1972. Animal DNA-dependent RNA polymerases. 3. Purification of calf-thymus BI and BII enzymes. *Eur. J. Biochem.* **28**:283–290.

Kedinger, C., Gissinger, F., and Chambon, P. 1974. Animal DNA-dependent RNA polymerases. Molecular structures and immunological properties of calf thymus AI and of calf thymus and rat liver enzymes B. *Eur. J. Biochem.* **44**:421–436.

King, G. A. M. 1977. Symbiosis and the origin of life. *Origins Life* **8**:39–43.

Kinsella, L., Hsu, C. Y. J., Schulz, W., and Dennis, D. 1982. RNA polymerase: Correlation between transcript length, abortive product synthesis, and formation of a stable ternary complex. *Biochemistry* **21**:2719–2723.

Kishi, F., Ebina, Y., Miki, T., Nakazawa, T., and Nakazawa, A. 1982. Purification and characterization of a protein from *Escherichia coli* which forms complexes with superhelical and single-stranded DNAs. *J. Biochem.* **92**:1059–1068.

Knapp, G., Ogden, R. C., Peebles, C. L., and Abelson, J. 1979. Splicing of yeast tRNA precursors: Structure of the reaction intermediates. *Cell* **18**:37–45.

Kobayashi, T., Irie, T., Yoshida, M., Takeishi, K., and Ukita, T. 1974. The primary structure of yeast glutamic acid tRNA specific to the GAA codon. *Biochim. Biophys. Acta* **366**:168–181.

Koch, W., Edwards, K., and Kossel, H. 1981. Sequencing of the 16S–23S spacer in a ribosomal RNA operon of *Zea mays* chloroplast DNA reveals two split tRNA genes. *Cell* **25**:203–213.

Köchel, H. G., Lazarus, C. M., Basak, N., and Kuntzel, H. 1981. Mitochondrial tRNA gene clusters in *Aspergillus nidulans:* Organization and nucleotide sequence. *Cell* **23**:625–633.

Kominami, K., Mishima, Y., Urano, Y., Sakai, M., and Muramatsu, M. 1982. Cloning and determination of the transcription termination site of ribosomal RNA gene of the mouse. *Nucleic Acids Res.* **10**:1963–1980.

Korn, L. J., and Yanofsky, C. 1976. Polarity suppressors defective in transcription at the attenuator of the tryptophan operon of *Escherichia coli* have altered rho factor. *J. Mol. Biol.* **106**:231–241.

Krakow, J. S., Rhodes, G., and Jovin, T. M. 1976. RNA polymerase: Catalytic mechanisms and inhibitors. *In:* Losick, R., and Chamberlin, M., eds., *RNA Polymerase*, Cold Spring Harbor, New York, Cold Spring Harbor Laboratory, pp. 127–157.

Kryukov, V. M., Ksenzenko, V. N., Kaliman, A. V., and Bayer, A. A. 1983. Cloning and DNA sequence of the genes for two bacteriophage T5 tRNAs^{Ser}. *FEBS Lett.* **158**:123–127.

Kuchino, Y., Mita, T., and Nishimura, S. 1981. Nucleotide sequence of cytoplasmic initiator tRNA from *Tetrahymena thermophila*. *Nucleic Acids Res.* **9**:4557–4562.

Kuntzel, B. Weissenbach, J., and Dirheimer, G. 1972. The sequence of nucleotides in tRNA$^{Arg}_{III}$ from brewer's yeast. *FEBS Lett.* **25**: 189–191.

Küpper, H., Sekiya, T., Rosenberg, M., Egan, J., and Landy, A. 1978. A ρ-dependent termination site in the gene coding for tyrosine tRNAsu$_3$ of *Escherichia coli*. *Nature (London)* **272**:423–428.

Lamond, A. I., and Travers, A. A. 1985. Genetically separable functional elements mediate the optimal expression and stringent regulation of a bacterial tRNA gene. *Cell* **40**:319–326.

LeGrice, S. F. J., and Sonenshein, A. L. 1982. Interaction of *Bacillus subtilis* RNA polymerase with a chromosomal promoter. *J. Mol. Biol.* **162**:551–564.

Leung, J., Addison, W. R., Delaney, A. D., MacKay, R. M., Miller, R. C., Spiegelman, G. B., Grigliatti, T. A., and Tener, G. M. 1984. *Drosophila melanogaster* tRNA$^{Val}_{3b}$ genes and their allogenes. *Gene* **34**:207–217.

Lewis, M. K., and Burgess, R. R. 1982. Eukaryotic RNA polymerases. *In:* Boyer, P.D., ed., *The Enzymes,* 3rd ed., New York, Academic Press, Vol. 15B, pp. 109–153.

Losick, R., and Chamberlin, M. eds. 1976. *RNA Polymerase,* Cold Spring Harbor, New York, Cold Spring Harbor Laboratory.

Losick, R., and Pero, J. 1976. Regulatory subunits of RNA polymerase. *In:* Losick, R., and Chamberlin, M., eds., *RNA Polymerase,* Cold Spring Harbor, New York, Cold Spring Harbor Laboratory, pp. 227–246.

Loughney, K., Lund, E., and Dahlberg, J. E. 1983. tRNA genes are found between the 16S and 23S rRNA genes in *Bacillus subtilis*. *Nucleic Acids Res.* **10**:1607–1624.

Luria, S. E., and Darnell, J. E. 1967. *General Virology,* 2nd ed., New York, Wiley.

Luzzati, D. 1970. Regulation of λ exonuclease synthesis. Role of the *N* gene product and λ repressor. *J. Mol. Biol.* **49**:515–519.

Madison, J. T., and Boguslawski, S. J. 1974. Partial digestion of a yeast lysine transfer ribonucleic acid and reconstruction of the nucleotide sequence. *Biochemistry* **13**:524–527.

Madon, J., Leser, U., and Zillig, W. 1983. DNA-dependent RNA polymerase from the extremely halophilic archaebacterium *Halococcus morrhuae*. *Eur. J. Biochem.* **135**:279–283.

Mao, J. I., Schmidt, O., and Söll, D. 1980. Dimeric transfer RNA precursors in *S. pombe. Cell* **21**:509–516.

Maréchal, L., Guillemaut, P., Grienenberger, J. M., Jeannin, G., and Weil, J. H. 1985a. Sequence and codon recognition of bean mitochondria and chloroplast tRNA^{Trp}: Evidence for a high degree of homology. *Nucleic Acids Res.* **13**:4411–4416.

Maréchal, L., Guillemaut, P., Grienenberger, J. M., Jeannin, G., and Weil, J. H. 1985b. Structure of the bean mitochondrial tRNA^{Phe} and localization of the tRNA^{Phe} gene on the mitochondrial genomes of maize and wheat. *FEBS Lett.* **184**:289–293.

Margulis, L. 1970. *Origins of Eukaryotic Cells,* New Haven, Yale University Press.

Mazzara, G. P., Seidman, J. G., McClain, W. H., Yesian, H., Abelson, J., and Guthrie, C. 1977. Nucleotide sequence of an arginine transfer ribonucleic acid from bacteriophage T4. *J. Biol. Chem.* **252**:8245–8253.

Mazzara, G. P., Plunkett, G., and McClain, W. H. 1981. DNA sequence of the transfer RNA region of bacteriophage T4: Implications for transfer RNA synthesis. *Proc. Natl. Acad. Sci. USA* **78**:889–892.

Miller, D. L., Sigurdson, C., Martin, N. C., and Donelson, J. E. 1980. Nucleotide sequence of the mitochondrial genes coding for tRNA$^{Gly}_{GGR}$ and tRNA$^{Val}_{GUR}$. *Nucleic Acids Res.* **8**:1435–1442.

Miller, K. G., and Sollner-Webb, B. 1981. Transcription of mouse rRNA genes by RNA polymerase I. *In vitro* and *in vivo* initiation and processing sites. *Cell* **27**:165–174.

Mishima, Y., Yamamoto, O., Kominami, R., and Muramatsu, M. 1981. *In vitro* transcription of a cloned mouse ribosomal RNA gene. *Nucleic Acids Res.* **9**:6773–6785.

Murao, K., Tanabe, T., Ishii, F., Namiki, M., and Nishimura, S. 1972. Primary sequence of arginine transfer RNA from *Escherichia coli. Biochem. Biophys. Res. Commun.* **47**:1332–1337.

Netzker, R., Koechel, H. G., Basak, N., and Kuentzel, H. 1982. Nucleotide sequence of *Aspergillus nidulans* mitochondrial genes coding for ATPase subunit 6, cytochrome oxidase subunit 3, seven unidentified proteins, four tRNAs and L-rRNA. *Nucleic Acids Res.* **10**:4783–4794.

Niles, E. G., Sutiphong, J., and Haque, S. 1981. Structure of the *Tetrahymena pyriformis* rRNA gene. Nucleotide sequence of the transcription initiation region. *J. Biol. Chem.* **256**:12849–12856.

Nomura, M., Morgan, E. A., and Jaskumas, S. R. 1977. Genetics of bacterial ribosomes. *Annu. Rev. Genet.* **11**:297–347.

Ogden, R. C., Beckman, J. S., Abelson, J., Kang, H. S., Söll, D., and Schmidt, O. 1979. *In vitro* transcription and processing of a yeast tRNA gene containing an intervening sequence. *Cell* **17**:399–406.

Ohme, M., Kamogashira, T., Shinozaki, K., and Sugiura, M. 1985. Structure and cotranscription of tobacco chloroplast genes for tRNA^{Glu} (UUC), tRNA^{Tyr} (GUA), and tRNA^{Asp} (GUC). *Nucleic Acids Res.* **13**:1045–1056.

Olah, J., and Feldmann, H. 1980. Structure of a yeast non-initiating methionine-tRNA gene. *Nucleic Acids Res.* **8**:1975–1986.

Olins, P. O., and Jones, D. S. 1980. Nucleotide sequence of *Scenedesmus obliquus* cytoplasmic initiator tRNA. *Nucleic Acids Res.* **8**:715–729.

Orozco, E. M., and Hallick, R. B. 1982. *Euglena gracilis* chloroplast tRNA transcription units. II. Nucleotide sequence analysis of a tRNA^{Val}–tRNA^{Asn}–tRNA^{Arg}–tRNA^{Leu} gene cluster. *J. Biol. Chem.* **257**:3265–3275.

Orozco, E. M., Rushlow, K. E., Dodd, J. R., and Hallick, R. B. 1980. *Euglena gracilis* chloroplast ribosomal RNA transcription units. II. Nucleotide sequence homology between the 16 S–23 S ribosomal RNA spacer and the 16 S ribosomal RNA leader regions. *J. Biol. Chem.* **255**:10997–11003.

Page, G. S. 1981. Characterization of the yeast tRNA$^{Ser}_2$ gene family: Genomic organization and DNA sequence. *Nucleic Acids Res.* **9**:921–934.

Panka, D., and Dennis, B. 1985. RNA polymerase. Direct evidence for two active sites involved in transcription. *J. Biol. Chem.* **260**:1427–1431.

Paule, M. R. 1981. Comparative subunit composition of the eukaryotic nuclear RNA polymerases. *Trends Biochem. Sci.* **6**:128–131.

Peffley, D. M., and Sogin, M. L. 1981. A putative tRNA^{Trp} gene cloned from *Dictyostelium discoideum:* Its nucleotide sequence and association with repetitive deoxyribonucleic acid. *Biochemistry* **20**:4015–4021.

Penswick, J. R., Martin, R., and Dirheimer, G. 1975. Evidence supporting a revised sequence for yeast alanine tRNA. *FEBS Lett.* **50**:28–31.

Pettijohn, D. E., Stonington, O. G., and Kossman, C. R. 1970. Chain termination of ribosomal RNA synthesis *in vitro*. *Nature (London)* **228**:235–239.

Piper, P. W. 1975. The nucleotide sequence of a methionine tRNA which functions in protein elongation in mouse myeloma cells. *Eur. J. Biochem.* **51**:283–293.

Piper, P. W. 1978. A correlation between a recessive lethal amber suppressor mutation in *Saccharomyces cerevisiae* and an anticodon change in a minor serine transfer RNA. *J. Mol. Biol.* **122**:217–235.

Pirtle, R., Kashdan, M., Pirtle, I., and Dudock, B. 1980. The nucleotide sequence of a major species of leucine tRNA from bovine liver. *Nucleic Acids Res.* **8**:805–815.

Prangishvilli, D., Zillig, W., Gierl, A., Biersert, L., and Holtz, I. 1982. DNA-dependent RNA polymerases of thermoacidophilic Archaebacteria. *Eur. J. Biochem.* **122**:471–477.

Pratt, K., Eden, F. C., You, K. H., O'Neill, V. A., and Hatfield, D. 1985. Conserved sequences in both coding and 5′ flanking regions of mammalian opal suppressor tRNA genes. *Nucleic Acids Res.* **13**:4765–4775.

Pribnow, D. 1975. Nucleotide sequence of an RNA polymerase binding site at an early T7 promoter. *Proc. Natl. Acad. Sci. USA* **72**:784–788.

Randerath, E., Gupta, R. C., Chia, L. L. S. Y., Chang, S. H., and Randerath, K. 1979. Yeast tRNA$_{UAG}^{Leu}$. Purification, properties, and determination of the nucleotide sequence by radioactive derivative methods. *Eur. J. Biochem.* **93**:79–94.

Raven, P. H. 1970. A multiple origin for plastids and mitochondria. *Science* **169**:641–646.

Ritossa, F. M., Atwood, K. C., and Spiegelman, S. 1966. On the redundancy of DNA complementary to amino acid transfer RNA and its absence from the nucleolar organizer region of *Drosophila melanogaster*. *Genetics* **54**:663–676.

Roberts, J. W. 1976. Transcription termination and its control in *E. coli*. *In:* Losick, R, and Chamberlin, M., eds., *RNA Polymerase,* Cold Spring Harbor, New York, Cold Spring Harbor Laboratory, pp. 247–271.

Roberts, J. W., and Carbon, J. 1975. Nucleotide sequence studies of normal and genetically altered glycine transfer ribonucleic acids from *Escherichia coli*. *J. Biol. Chem.* **25**:5530–5541.

Robinson, R. R., and Davison, N. 1981. Analysis of a *Drosophila* tRNA gene cluster: Two tRNA[Leu] genes contain intervening sequences. *Cell* **23**:251–259.

Roeder, R. G. 1974. Multiple forms of deoxyribonucleic acid-dependent ribonucleic acid polymerase in *Xenopus laevis*. *J. Biol. Chem.* **249**:241–248.

Roeder, R. G. 1976. Eukaryotic nuclear RNA polymerases. *In:* Losick, R., and Chamberlain, M., eds., *RNA Polymerase,* Cold Spring Harbor, New York, Cold Spring Harbor Laboratory, pp. 285–329.

Roeder, R. G. 1983. Multiple forms of DNA-dependent RNA polymerases in *Xenopus laevis*. Properties, purification, and subunit structure of class III RNA polymerases. *J. Biol. Chem.* **258**:1932–1941.

Rogg, H., Mueller, P, and Stoehelin, M. 1975. Nucleotide sequences of rat liver serine tRNA. Structure of serine tRNA$_3$ and partial nucleotide sequences of serine tRNA$_{2a}$. *Eur. J. Biochem.* **53**:115–127.

Rosenberg, M., and Court, D. 1979. Regulatory sequences involved in the promotion and termination of RNA transcription. *Annu. Rev. Genet.* **13**:319–353.

Rossi, J. J., and Landy, A. 1979. Structure and organization of the two tRNA[Tyr] gene clusters on the *E. coli* chromosome. *Cell* **16**:523–534.

Rouvière-Yaniv, J., and Gros, F. 1975. Characterization of a novel, low-molecular-weight DNA-binding protein from *Escherichia coli*. *Proc. Natl. Acad. Sci. USA* **72**:3428–3432.

Rouvière-Yaniv, J., Yaniv, M., and Germond, J. E. 1976. *E. coli* binding protein HU forms nucleosome-like structure with circular double-stranded DNA. *Cell* **17**:265–274.

Roy, K. L., Cooke, H., and Buckland, R. 1982. Nucleotide sequence of a segment of human DNA containing the three tRNA genes. *Nucleic Acids Res.* **10**:7313–7322.

Saccone, G., Cantatore, P., Gadaleta, G., Gallerani, R., Lanave, C., Pepe, G., and Kroon, A. M. 1981. The nucleotide sequence of the large ribosomal RNA gene and the adjacent tRNA genes from rat mitochondria. *Nucleic Acids Res.* **9**:4139–4148.

Santos, T., and Zasloff, M. 1981. Comparative analysis of human chromosomal segments bearing nonallelic dispersed tRNA$_6^{Met}$ genes. *Cell* **23**:699–709.

Schaack, J., Sharp, S., Dingermann, T., and Söll, D. 1983. Transcription of eukaryotic tRNA genes *in vitro*. II. Formation of stable complexes. *J. Biol. Chem.* **258**:2447–2453.

Schaller, H., Beck, E., and Takanami, M. 1978. Sequence and regulatory signals of the filamentous phage genome. *In:* Denhardt, D. T., Dressler, D., and Ray, D. S., eds., *Single-Stranded DNA Phages,* Cold Spring Harbor, New York, Cold Spring Harbor Laboratory, pp. 139–163.

Scherer, G., Hobom, G., and Kossel, H. 1977. DNA base sequence of the p_0 promoter region of phage lambda. *Nature (London)* **265:**117–121.

Schmidt, D., Mao, J., Ogden, R., Beckmann, J., Sakano, H., Abelson, J., and Söll, D. 1980. Dimeric tRNA precursors in yeast. *Nature (London)* **287:**750–752.

Schnabel, R., Zillig, W., and Schnabel, H., 1982. Component E of the DNA-dependent RNA polymerase of the archaebacterium *Thermoplasma acidophilum* is required for the transcription of native DNA. *Eur. J. Biochem.* **129:**473–477.

Schwartz, L. B., and Roeder, R. G. 1975. Purification and subunit structure of DNA-dependent RNA polymerase II from the mouse plasmacytoma MOPC 315. *J. Biol. Chem.* **250:**3221–3228.

Schwarz, Z., Jolly, S. O., Steinmetz, A. A., and Bogorad, L. 1981. Overlapping divergent genes in the maize chloroplast chromosome and *in vitro* transcription of the gene for tRNA[His]. *Proc. Natl. Acad. Sci. USA* **78:**3423–3427.

Seilhamer, J. J., and Cummings, D. J. 1981. Structure and sequence of the mitochondrial 20S rRNA and tRNA[Tyr] gene of *Paramecium primaurelia. Nucleic Acids Res.* **9:**6391–6406.

Sekiya, T., and Nishimura, S. 1979. Sequence of the gene for isoleucine tRNA$_1$ and the surrounding region in a ribosomal RNA operon of *Escherichia coli. Nucleic Acids Res.* **6:**575–592.

Sekiya, T., Gait, M. J., Noris, K., Ramamoorthy, B., and Khorana, H. G. 1976a. The nucleotide sequence in the promoter region of the gene for an *Escherichia coli* tyrosine transfer ribonucleic acid. *J. Biol. Chem.* **251:**4481–4489.

Sekiya, T., Takeya, T., Contreras, R., Küpper, H., Khorana, H. G., and Landy, A. 1976b. Nucleotide sequences at the two ends of the *E. coli* tyrosine tRNA genes and studies on the promoter. *In:* Losick, R., and Chamberlin, M., eds., *RNA Polymerase,* Cold Spring Harbor, New York, Cold Spring Harbor Laboratory, pp. 455–472.

Sekiya, T., Mori, M., Takahashi, N., and Nishimura, S. 1980. Sequence of the distal tRNA$_1$[Asp] gene and the transcription termination signal in the *Escherichia coli* ribosomal RNA operon *rrnF* (or *G*). *Nucleic Acids Res.* **8:**3809–3827.

Sekiya, T., Kuchino, Y., and Nishimura, S. 1981. Mammalian tRNA genes. Nucleotide sequence of rat genes for tRNA[Asp], tRNA[Gly] and tRNA[Glu]. *Nucleic Acids Res.* **9:**2239–2249.

Sekiya, T., Nishizawa, R., Matsuda, K., Taya, Y., and Nishimura, S. 1982. A rat tRNA gene cluster containing the genes for tRNA[Pro] and tRNA[Lys]. Analysis of nucleotide sequences of the genes and the surrounding regions. *Nucleic Acids Res.* **10:**6411–6419.

Sharp, S., DeFranco, D., Dingermann, T., Farrell, P., and Söll, D. 1981a. Internal control regions for transcription of eukaryotic tRNA genes. *Proc. Natl. Acad. Sci. USA* **78:**6657–6661.

Sharp, S., DeFranco, D., Silberklang, M., Hosbach, H. A., Schmidt, T., Kubli, E., Gergen, J. P., Wensink, P. C., and Söll, D. 1981b. The initiator tRNA genes of *Drosophila melanogaster:* Evidence for a tRNA pseudogene. *Nucleic Acids Res.* **9:**5867–5882.

Shibuya, K., Noguchi, S., Nishimura, S., and Sekiya, T. 1982. Characterization of a rat tRNA gene cluster containing the genes for tRNA[Asp], tRNA[Gly], and tRNA[Glu] and pseudogenes. *Nucleic Acids Res.* **10:**4441–4448.

Siebenlist, U., Simpson, R. B., and Gilbert, W. 1980. *E. coli* RNA polymerase interacts homologously with two different promoters. *Cell* **20:**269–281.

Siehnel, R. J., and Morgan, E. A. 1983. Efficient read-through of Tn9 and IS*1* by RNA polymerase molecules that initiate at rRNA promoters. *J. Bacteriol.* **153:**672–684.

Silverman, S., Schmidt, O., Söll, D., and Hovemann, B. 1979. The nucleotide sequence of a cloned *Drosophila* arginine tRNA gene and its *in vitro* transcription in *Xenopus* germinal vesicle extracts. *J. Biol. Chem.* **254:**10290–10294.

Simsek, M., and RajBhandary, U. L. 1972. The primary structure of yeast initiator transfer ribonucleic acid. *Biochem. Biophys. Res. Commun.* **49:**508–515.

Singer, C. E., and Smith, G. R. 1972. Histidine regulation in *Salmonella typhimurium.* XIII. Nucleotide sequence of histidine transfer ribonucleic acid. *J. Biol. Chem.* **247:**2989–3000.

Sprague, K. U., Larson, D., and Morton, D. 1980. 5′ flanking sequence signals are required for activity of silkworm alanine tRNA genes in homologous *in vitro* transcription systems. *Cell* **22:**171–178.

Squires, C., and Carbon, J. 1971. Normal and mutant glycine transfer RNAs. *Nature New Biol.* **233**:274–277.

Standring, D. N., Venegas, A., and Rutter, W. J. 1981. Yeast tRNA $_3^{Leu}$ gene transcribed and spliced in a HeLa cell extract. *Proc. Natl. Acad. Sci. USA* **78**:5963–5967.

Steinmetz, A. A., Gubbins, E. J., and Bogorad, L. 1982. The anticodon of the maize chloroplast gene for tRNA $_{UAA}^{Leu}$ is split by a large intron. *Nucleic Acids Res.* **10**:3027–3037.

Steinmetz, A. A., Krebbers, E. T., Schwarz, Z., Gubbins, E. J., and Bogorad, L. 1983. Nucleotide sequences of five maize chloroplast transfer RNA genes and their flanking regions. *J. Biol. Chem.* **258**:5503–5511.

Stent, G. S. 1964. The operon on its third anniversary. Modulation of transfer RNA species can provide a workable model of all operator-less operons. *Science* **144**:816–820.

Stillman, D. J., Caspers, P., and Geiduschek, E. P. 1985. Effects of temperature and single-stranded DNA on the interaction of an RNA polymerase III transcription factor with a tRNA gene. *Cell* **40**:311–317.

Stroynowski, I., Kuroda, M., and Yamofsky, C. 1983. Transcription termination *in vitro* at the tryptophan operon attenuator is controlled by secondary structures in the leader transcript. *Proc. Natl. Acad. Sci. USA* **80**:2206–2210.

Sugden, B., and Keller, W. 1973. Mammalian DNA-dependent RNA polymerases. I. Purification and properties of an α-amantin-sensitive RNA polymerase and stimulatory factors from HeLa and KB cells. *J. Biol. Chem.* **248**:3777–3788.

Swanson, M. E., and Holland, M. J. 1983. RNA polymerase I-dependent selective transcription of yeast ribosomal DNA. Identification of a new cellular ribosomal RNA precursor. *J. Biol. Chem.* **258**:3242–3250.

Takaiwa, F., and Sugiura, M. 1982. Nucleotide sequence of the 16S–23S spacer region in an rRNA gene cluster from tobacco chloroplast DNA. *Nucleic Acids Res.* **10**:2665–2676.

Tartof, K., and Perry, R. P. 1970. The 5S RNA genes of *Drosophila melanogaster*. *J. Mol. Biol.* **51**:171–183.

Thonart, P., Bechet, J., Hilger, F., and Burny, A. 1976. Thermosensitive mutations affecting ribonucleic acid polymerases in *Saccharomyces cerevisiae*. *J. Bacteriol.* **125**:25–32.

Tranquilla, T. A., Cortese, R., Melton, D., and Smith, J. D. 1982. Sequences of four tRNA genes from *Caenorhabditis elegans* and the expression of *C. elegans* tRNA^{Leu} (anticodon IAG) in *Xenopus* oocytes. *Nucleic Acids Res.* **10**:7919–7934.

Travers, A., and Cukier-Kahn, R. 1974. Effect of H1 protein on *in vitro* ribosomal RNA synthesis. *FEBS Lett.* **43**:86–88.

Uziel, M., and Weinberg, A. J. 1975. Sequence of *E. coli* tRNA$_1^{Glu}$ by automated sequential degradation. *Nucleic Acids Res.* **2**:469–476.

Valenzuela, P., Bell, G. I., and Rutter, W. J. 1976. The 24,000 dalton subunit and the activity of yeast RNA polymerases. *Biochem. Biophys. Res. Commun.* **71**:26–31.

Valenzuela, P., Venegas, A., Weinburg, F., Bishop, R., and Rutter, W. J. 1978. Structure of yeast phenylalanine-tRNA genes: An intervening DNA segment within the region coding for the tRNA. *Proc. Natl. Acad. Sci. USA* **75**:190–194.

Venegas, A., Quiroga, M., Zaldewar, J., Rutter, W. J., and Valenzuela, P. 1979. Isolation of yeast tRNA^{Leu} genes. DNA sequence of a cloned tRNA^{Leu} gene. *J. Biol. Chem.* **254**:12306–12309.

Venegas, A., Gonzalez, E., Bull, P., and Valenzuela, P. 1982. Isolation and structure of a yeast initiator tRNA^{Met} gene. *Nucleic Acids Res.* **10**:1093–1104.

Ward, D. F., and Gottesman, M. E. 1982. Suppression of transcription termination by phage lambda. *Science* **216**:946–951.

Wawrousek, E. F., and Hansen, J. N. 1983. Structure and organization of a cluster of six tRNA genes in the space between tandem ribosomal RNA gene sets in *Bacillus subtilis*. *J. Biol. Chem.* **258**:291–298.

Weber, L., and Berger, E. 1976. Base sequence complexity of the stable RNA species of *Drosophila melanogaster*. *Biochemistry* **15**:5511–5519.

Weil, P. A., Segall, J., Harris, B., Ng, S. Y., and Roeder, R. G. 1979. Faithful transcription of eukaryotic gene by polymerase III in systems reconstituted with purified DNA templates. *J. Biol. Chem.* **254**:6163–6173.

Weissenbach, J., Martin, R., and Dirheimer, G. 1975. The primary structure of tRNA$_{II}^{Arg}$ from brewer's yeast. 2. Partial digestion with ribonuclease T$_1$ and derivation of the complete sequence. *Eur. J. Biochem.* **56**:527–532.

Weissenbach, J., Kiraly, I., and Dirheimer, G. 1977. Structure primaire des tRNA$_{1a,b}$ de levure de biere. *Biochemie* **59**:381–391.

Williams, R. J., Nagel, W., Roe, B., and Dudock, B. 1974. Primary structure of *E. coli* alanine transfer RNA:

Relation to the yeast phenylalanyl-tRNA synthetase recognition site. *Biochem. Biophys. Res. Commun.* **60:**1215–1221.

Williamson, S. E., and Doolittle, W. F. 1983. Genes for tRNA[Ile] and tRNA[Ala] in the spacer between 16S and 23S rRNA genes of a blue-green alga: Strong homology to chloroplast tRNA genes and tRNA genes of *E. coli rrnD* gene cluster. *Nucleic Acids Res.* **11:**225–235.

Wilson, E. T., Larson, D., Young, L. S., and Sprague, K. U. 1985. A large region controls tRNA gene transcription. *J. Mol. Biol.* **183:**153–163.

Woese, C. R., Magrum, L. J., and Fox, G. E. 1978. Archaebacteria. *J. Mol. Evol.* **11:**245–252.

Wong, T. T., McCutchan, T., Kohli, J., and Söll, D., 1979. The nucleotide sequence of the major glutamate transfer RNA from *Schizosaccharomyces pombe*. *Nucleic Acids Res.* **6:**2057–2068.

Yaginuma, K., Kobayashi, M., Taira, M., and Koike, K. 1982. A new RNA polymerase and *in vitro* transcription of the origin of replication from rat mitochondria. *Nucleic Acids Res.* **10:**7531–7542.

Yamada, Y., and Ishikura, H. 1973. Nucleotide sequence of tRNA$_3^{Ser}$ from *Escherichia coli*. *FEBS Lett.* **29:**231–234.

Yamada, Y., and Ishikura, H. 1975. Identification of a modified nucleoside in *Escherichia coli* tRNA$_1^{Ser}$ as 2'-O-methylcytidine. *Biochim. Biophys. Acta* **402:**285–287.

Yamada, Y., Ohki, M., and Ishikura, H. 1983. The nucleotide sequence of *Bacillus subtillis* tRNA genes. *Nucleic Acids Res.* **11:**3037–3045.

Yaniv, M., and Barrell, B. G. 1969. Nucleotide sequence of *E. coli* tRNA$_1^{Val}$ *Nature (London)* **222:**278–279.

Yaniv, M., and Barrell, B. G. 1971. Sequence relationship of three valine acceptor tRNAs from *Escherichia coli*. *Nature New Biol.* **233:**113–114.

Yarus, M., and Barrell, B. G. 1971. The sequence of nucleotides in tRNA[Ile] from *E. coli* B. *Biochem. Biophys. Res. Commun.* **43:**729–734.

Yen, P. H., and Davidson, N. 1980. The gross anatomy of a tRNA gene cluster at region 42A of the *D. melanogaster* chromosome. *Cell* **22:**137–148.

Yoshida, M., 1973. The nucleotide sequence of tRNA[Gly] from yeast. *Biochem. Biophys. Res. Commun.* **50:**779–784.

Young, R. A. 1979. Transcription termination in the *Escherichia coli* ribosomal RNA operon *rrnC*. *J. Biol. Chem.* **254:**12725–12731.

Young, R. A., Macklis, R., and Steitz, J. A. 1979. Sequence of the 16S + 23S spacer region in two ribosomal RNA operons of *Escherichic coli*. *J. Biol. Chem.* **254:**3264–3271.

Zillig, W., Palm, P., and Heil, A. 1976. Functions and reassembly of subunits of DNA-dependent RNA polymerase. *In:* Losick, R., and Chamberlin, M., eds., *RNA Polymerase*, Cold Spring Harbor, New York, Cold Spring Harbor Laboratory, pp. 101–125.

Zillig, W., Stetter, K. O., and Janeković, D. 1979. DNA-dependent RNA polymerase from the Archaebacterium *Sulfolobus acidocaldarius*. *Eur. J. Biochem.* **96:**597–604.

Zillig, W., Stetter, K. O., Wunderl, S., Schulz, W., Priess, H., and Scholz, W. 1980. The *Sulfolobus–* "Caldariella" group: Taxonomy on the basis of the structure of DNA-dependent RNA polymerases. *Arch. Microbiol.* **125:**259–269.

CHAPTER 3

Akusjärvi, G., Mathews, M. B., Anderson, P., Vennström, B., and Pettersson, U. 1980. Structure of genes for virus-associated RNA$_I$ and RNA$_{II}$ of adenovirus type 2. *Proc. Natl. Acad. Sci. USA* **77:**2424–2428.

Allan, M., and Paul, J. 1984. Transcription *in vivo* of an *Alu* family member upstream from the human ε-globin gene. *Nucleic Acids Res.* **12:**1193–1200.

Allan, M., Lanyon, W. G., and Paul, J. 1983. Multiple origins of transcription in the 4.5 kb upstream of the ε-globin gene. *Cell* **35:**187–197.

Aoyama, K., Hidaka, S., Tanaka, T., and Ishikawa, K. 1982. The nucleotide sequence of 5S RNA from rat liver ribosomes. *J. Biochem.* **91:**363–367.

Balmain, A., Krumlauf, R., Vass, J. K., and Birnie, G. D. 1982. Cloning and characterisation of the abundant cytoplasmic 7S RNA from mouse cells. *Nucleic Acids Res.* **10:**4259–4277.

Baralle, F. E., Shoulders, C. C., Goodbourn, S., Jeffreys, A., and Proudfoot, N. J. 1980. The 5' flanking region of human ε-globin gene. *Nucleic Acids Res.* **8:**4393–4404.

Benhamou, J., Jourdan, R., and Jordan, B. R. 1977. Sequence of *Drosophila* 5S RNA synthesized by cultured cells and by the insect at different developmental stages. *J. Mol. Evol.* **9**:279–298.

Bhat, R. A., Metz, B., and Thimmappaya, B. 1983. Organization of the noncontiguous promoter components of adenovirus VAI RNA gene is strikingly similar to that of eukaryotic transfer RNA genes. *Mol. Cell. Biol.* **3**:1996–2005.

Bogenhagen, D. F. 1985. The intragenic control region of the *Xenopus* 5 S RNA gene contains two factor A binding domains that must be aligned properly for efficient transcription initiation. *J. Biol. Chem.* **260**:6466–6471.

Brennicke, A., Möller, S., and Blanz, P. A. 1985. The 18S and 5S ribosomal RNA genes in *Oenothera* mitochondria: Sequence rearrangements in the 18S and 5S rRNA genes of higher plants. *Mol. Gen. Genet.* **198**:404–410.

Brosius, J., Dull, T. J., Sleeter, D. D., and Noller, H. F. 1981. Gene organization and primary structure of a ribosomal RNA operon of *Escherichia coli*. *J. Mol. Biol.* **148**:107–127.

Brown, D. D., Wensink, P. C., and Jordan, E. 1971. Purification and some characteristics of 5S DNA from *Xenopus laevis*. *Proc. Natl. Acad. Sci. USA* **68**:3175–3179.

Brown, D. D., Carroll, D., and Brown, R. D. 1977. The isolation and characterization of a second oocyte 5S DNA from *Xenopus laevis*. *Cell* **12**:1045–1056.

Brownlee, G. G., Sanger, F., and Barrell, B. C. 1968. The sequence of 5S ribosomal ribonucleic acid. *J. Mol. Biol.* **34**:379–412.

Brownlee, G. G., Cartwright, E. M., and Brown, D. D. 1974. Sequence studies of the 5S DNA of *Xenopus laevis*. *J. Mol. Biol.* **89**:703–718.

Burke, D. J., Schaack, J., Sharp, S., and Söll, D. 1983. Partial purification of *Drosophila* Kc cell RNA polymerase III transcription components. Evidence for shared 5S RNA and tRNA gene factors. *J. Biol. Chem.* **258**:15224–15231.

Butler, M. H., Wall, S. M., Luehrsen, K. R., Fox, G. E., and Hecht, R. M. 1981. Molecular relationships between closely related strains and species of nematodes. *J. Mol. Evol.* **18**:18–23.

Calabretta, B., Robberson, D. L., Maizel, A. L., and Saunders, G. F. 1981. mRNA in human cells contains sequences complementary to the *Alu* family of repeated DNA. *Proc. Natl. Acad. Sci. USA* **78**:6003–6007.

Carrara, G., Di Segni, G., Otsuka, A., and Tocchini-Valentini, G. P. 1981. Deletion of the 3′ half of the yeast tRNA$_3^{Leu}$ gene does not abolish promoter function *in vitro*. *Cell* **27**:371–379.

Chao, S., Sederoff, R. R., and Levings, C. S. 1983. Partial sequence analysis of the 5S to 18S rRNA gene region of the maize mitochondrial genome. *Plant Physiol.* **71**:190–193.

Cheng, J. F., Printz, R., Callaghan, T., Shuey, D., and Hardison, R. C. 1984. The rabbit C family of short, interspersed repeats. Nucleotide sequence determination and transcriptional analysis. *J. Mol. Biol.* **176**:1–20.

Childs, G., Maxson, R., Cohn, R. H., and Kedes, L. 1981. Orphons: Dispersed genetic elements derived from tandemly repetitive genes of eucaryotes. *Cell* **23**:651–663.

Ciliberto, G., Castagnoli, L., Melton, D. A., and Cortese, R. 1982a. Promoter of a eukaryotic tRNA [Pro] gene is composed of three noncontiguous regions. *Proc. Natl. Acad. Sci. USA* **79**:1195–1199.

Ciliberto, G., Traboni, G., and Cortese, R. 1982b. Relationship between the two components of the split promoter of eukaryotic tRNA genes. *Proc. Natl. Acad. Sci. USA* **79**:1921–1925.

Ciliberto, G., Raugei, G., Constanzo, F., Dente, L., and Cortese, R. 1983. Common and interchangeable elements in the promoters of genes transcribed by RNA polymerase III. *Cell* **32**:725–733.

Coggins, L. W., Grindlay, G. J., Vass, J. K., Slater, A. A., Montague, P., Stinson, M. A., and Paul, J. 1980. Repetitive DNA sequences near three human β-type globin genes. *Nucleic Acids Res.* **8**:3319–3334.

Corry, M. J., Payne, P. I., and Dyer, T. A. 1974. The nucleotide sequence of 5S rRNA from the blue-green alga *Anacystis nidulans*. *FEBS Lett.* **46**:63–66.

Daniels, C. J., Hofman, J. D., MacWilliam, J. G., Doolittle, W. F., Woese, C. R., Luehrsen, K. R., and Fox, G. E. 1985. Sequence of 5S ribosomal RNA gene regions and their products in the archaebacterium *Halobacterium volcanii*. *Mol. Gen. Genet.* **198**:270–274.

Darlix, J. L., and Rochaix, J. D. 1981. Nucleotide sequence and structure of cytoplasmic 5S RNA and 5.8S RNA of *Chlamydomonas reinhardii*. *Nucleic Acids Res.* **9**:1291–1299.

DeFranco, D., Schmidt, O., and Söll, D. 1980. Two control regions for eukaryotic tRNA gene transcription. *Proc. Natl. Acad. Sci. USA* **77**:3365–3368.

Deininger, P. L., Jolly, D. J., Rubin, C. M., Friedmann, T., and Schmid, C. W. 1981. Base sequence studies of 300 nucleotide renatured repeated DNA clones. *J. Mol. Biol.* **151**:17–33.

Delihas, N., and Andersen, J. 1982. Generalized structures of the 5S ribosomal RNAs. *Nucleic Acids Res.* **10**:7323–7344.

Delihas, N., Andersen, J., Sprouse, H. M., Kashdan, M., and Dudock, B. 1981a. The nucleotide sequence of spinach cytoplasmic 5S ribosomal RNA. *J. Biol. Chem.* **256**:7515–7517.

Delihas, N., Andersen, J., Sprouse, H. M., and Dudock, B. 1981b. The nucleotide sequence of the chloroplast 5S ribosomal RNA from spinach. *Nucleic Acids Res.* **9**:2801–2805.

Delihas, N., Andersen, J., Andresini, W., Kaufman, S., and Lyman, H. 1981c. The 5S ribosomal RNA of *Euglena gracilis* cytoplasmic ribosomes is closely homologous to the 5S RNA of the trypanosomatid protozoa. *Nucleic Acids Res.* **9**:6627–6633.

del Rey, F. J., Donahue, T. F., and Fink, G. R. 1982. *Sigma*, a repetitive element found adjacent to tRNA genes of yeast. *Proc. Natl. Acad. Sci. USA* **79**:4138–4142.

Denis, H., and Mairy, M. 1972. Recherches biochimiques sur l'oogenèse. II. Distribution intracellulaire du RNA dans les petits oocytes du *Xenpus laevis*. *Eur. J. Biochem.* **25**:524–534.

Denis, H., and Wegnez, M. 1973. Recherches biochemiques sur l'oogenèse. 7. Synthèse et maturation du RNA 5S dans les petits oocytes de *Xenopus laevis*. *Biochimie* **55**:437–1151.

Deno, H., and Sugiura, M. 1984. Chloroplast tRNA^{Gly} gene contains a long intron in the D stem: Nucleotide sequences of tobacco chloroplast genes for tRNA^{Gly} (UCC) and tRNA^{Arg} (UCU). *Proc. Natl. Acad. Sci. USA* **81**:405–408.

Di Giovanni, L., Haynes, S. R., Misra, R., and Jelinek, W. R. 1983. *Kpn* I family of long-dispersed repeated DNA sequences of man: Evidence for entry into genomic DNA of DNA copies of poly A-terminated *Kpn* I RNAs. *Proc. Natl. Acad. Sci. USA* **80**:6533–6537.

Dillon, L. S. 1962. Comparative cytology and the evolution of life. *Evolution* **16**:102–117.

Dillon, L. S. 1963. A reclassification of the major groups of organisms based upon comparative cytology. *Syst. Zool.* **12**:71–82.

Dillon, L. S. 1981. *Ultrastructure, Macromolecules, and Evolution*, New York, Plenum Press.

Dillon, L. S. 1983. *The Inconstant Gene*, New York, Plenum Press.

Dingermann, T., Burke, D. J., Sharp, S., Schaack, J., and Söll, D. 1982. The 5′ flanking sequences of *Drosophila* tRNA^{Arg} genes control their *in vitro* transcription in a *Drosophila* cell extract. *J. Biol. Chem.* **257**:14738–14744.

Douglas, S. E., and Doolittle, W. R. 1984. Nucleotide sequence of the 5S rRNA gene and flanking regions in the cyano-bacterium, *Anacystis nidulans*. *FEBS Lett.* **166**:307–310.

Duester, G. L., and Holmes, W. M. 1980. The distal end of the ribosomal RNA operon *rrnD* of *Escherichia coli* contains a tRNA$_1^{Thr}$ gene, two 5S rRNA genes and a transcription terminator. *Nucleic Acids Res.* **8**:3793–3807.

Duncan, C. H., Biro, P. A., Choudary, P. V., Elder, J. T., Wang, R. R. C., Forget, B. G., deRiel, J. K., and Weissman, S. M. 1979. RNA polymerase III transcriptional units are interspersed among human non-α-globin genes. *Proc. Natl. Acad. Sci. USA* **76**:5095–5099.

Duncan, C. H., Jagadeesevaran, P., Wang, R. R. C., and Weissman, S. M. 1981. Structural analysis of templates and polymerase III transcripts of *Alu* family sequences interspersed among the human β-like globin genes. *Gene* **13**:185–196

Dyer, T. A., and Bowman, C. M. 1979. Nucleotide sequences of chloroplast 5S ribosomal ribonucleic acid in flowering plants. *Biochem. J.* **183**:595–604.

Elder, J. T., Pan, J., Duncan, C. H., and Weissman, S. M. 1981. Transcriptional analysis of interspersed repetitive polymerase III transcription units in human DNA. *Nucleic Acids Res.* **9**:1171–1189.

Engelke, D. R., Ng, S. Y., Shastry, B. S., and Roeder, R. G. 1980. Specific interaction of a purified transcription factor with an internal control region of 5S RNA genes. *Cell* **19**:717–728.

Erdmann, V. A., Huysmans, E., Vandenberghe, A., and De Wachter, R. 1983. Collection of published 5S and 5.8S ribosomal RNA sequences. *Nuclei Acids Res.* **11**:r107–r133.

Erdmann, V. A., Wolters, J., Huysmans, E., Vandenberghe, A., and De Wachter, R. 1984. Collection of published 5S and 5.8S ribosomal RNA sequences. *Nucleic Acids Res.* **12**(suppl.):r133–r166.

Fedoroff, N. V., and Brown, D. D. 1977. The nucleotide sequence of the repeating unit in the oocyte 5S ribosomal DNA of *Xenopus laevis*. *Cold Spring Harbor Symp. Quant. Biol.* **42**:1195–1200.

Fedoroff, N. V., and Brown, D. D. 1978. The nucleotide sequence of oocyte 5S DNA in *Xenopus laevis*. I. The AT-rich spaces. *Cell* **13**:701–716.

Fischel, J. L., and Ebel, J. P. 1975. Sequence studies on the 5S RNA of *Proteus vulgaris:* Comparison with the 5S RNA of *Escherichia coli*. *Biochimie* **57**:899–904.

Folk, W. R., Hofstetter, H., and Birnstiel, M. L. 1982. Some bacterial tRNA genes are transcribed by eukaryotic RNA polymerase III. *Nucleic Acids Res.* **10:**7153–7163.

Ford, P. J. 1971. Non-coordinated accumulation and synthesis of 5S ribonucleic acid by ovaries of *Xenopus laevis. Nature (London)* **233:**561–564.

Ford, P. J., and Brown, R. D. 1976. Sequences of 5S ribosomal RNA from *Xenopus mülleri* and the evolution of 5S gene-coding sequences. *Cell* **8:**485–493.

Forget, B. G., and Weissman, S. M. 1969. Nucleotide sequence of KB cell 5S RNA. *Science* **158:**1695–1700.

Fournier, A., Guérin, M. A., Coriet, J., and Clarkson, S. G. 1984. Structure and *in vitro* transcription of a glycine tRNA gene from *Bombyx mori. EMBO J.* **3:**1547–1552.

Fowlkes, D. M., and Shenk, T. 1980. Transcriptional control regions of the adenovirus VAI RNA gene. *Cell* **22:**405–413.

Fox, G. E., and Woese, C. R. 1975. 5S RNA secondary structure. *Nature (London)* **256:**505–507.

Fox, G. E., Luehrsen, K. R., and Woese, C. R. 1982. Archaebacterial 5S ribosomal RNA. *Zentrbl. Bakteriol. Hyg. 1 Abt. Orig.* **C3:**330–345.

Fritsch, E. F., Lawn, R. M., and Maniatis, T., 1980. Molecular cloning and characterization of the β-like globin gene cluster. *Cell* **19:**959–972.

Fuhrman, S. A., Deininger, P. L., LaPorte, P., Friedmann, T., and Geiduschek, E. P. 1981. Analysis of transcription of the human *Alu* family ubiquitous repeating element by eukaryotic polymerase III. *Nucleic Acids Res.* **9:**6439–6456.

Galli, G., Hofstetter, H., and Birnstiel, M. L. 1981. Two conserved sequence blocks within eukaryotic tRNA genes are major promoter elements. *Nature (London)* **294:**626–631.

Gamulin, V., Mao, J. I., Appel, B., Sumner-Smith, M., Yamao, F., and Söll, D. 1983. Six *Schizosaccharomyces pombe* tRNA genes including a gene for a tRNA^{Lys} with an intervening sequence which cannot base-pair with the anticodon. *Nucleic Acids Res.* **11:**8537–8546.

Garber, R. L., and Gage, L. P. 1979. Transcription of a cloned *Bombyx mori* tRNA$_2^{Ala}$ gene: Nucleotide sequence of the tRNA precursor and its processing *in vitro. Cell* **18:**817–828.

Goldsbrough, P. B., Ellis, T. H. N., and Lomonossoff, G. P. 1982. Sequence variation and methylation of the flax 5S RNA genes. *Nucleic Acids Res.* **10:**4501–4514.

Gottesfeld, J. M., Andrews, D. L., and Hoch, S. O. 1984. Association of an RNA polymerase III transcription factor with a ribonucleoprotein complex recognized by autoimmune sera. *Nucleic Acids Res.* **12:**3185–3200.

Gray, M. W., and Spencer, D. F. 1981. Is wheat mitochondrial 5S ribosomal RNA prokaryotic in nature? *Nucleic Acids Res.* **9:**3523–3529.

Gruissem, W., Kotzerke, M., and Seifart, H. K. 1981. Transcription of the cloned genes for ribosomal 5-S RNA in a system reconstituted *in vitro* from HeLa cells. *Eur. J. Biochem.* **117:**407–415.

Gruissem, W., Greenberg, B. M., Zurawski, G., Prescott, D. M., and Hallick, R. B. 1983. Biosynthesis of chloroplast transfer RNA in a spinach chloroplast transcription system. *Cell* **35:**815–828.

Guilfoyle, R., and Weinmann, R. 1981. Control region for adenovirus *VA* RNA transcription. *Proc. Natl. Acad. Sci. USA* **78:**3378–3382.

Hanas, J. S., Bogenhagen, D. F., and Wu, C. W. 1983. Cooperative model for the binding of *Xenopus* transcription factor A to the 5S RNA gene. *Proc. Natl. Acad. Sci. USA* **80:**2142–2145.

Hanas, J. S., Bogenhagen, D. F., and Wu, C. W. 1984a. DNA unwinding ability of *Xenopus* transcription factor A. *Nucleic Acids Res.* **12:**1265–1276.

Hanas, J. S., Bogenhagen, D. F., and Wu, C. W. 1984b. Binding of *Xenopus* transcription factor A to 5S RNA and to single-stranded DNA. *Nucleic Acids Res.* **12:**2745–2758.

Haynes, S. R., and Jelinek, W. R. 1981. Low molecular weight RNAs transcribed *in vitro* by RNA polymerase III from *Alu*-type dispersed repeats in Chinese hamster DNA are also found *in vivo. Proc. Natl. Acad. Sci. USA* **78:**6130–6134.

Haynes, S. R., Toomey, T. P., Leinwand, L., and Jelinek, W. R. 1981. The Chinese hamster *Alu*-equivalent sequence: A conserved, highly repetitious, interspersed deoxyribonucleic acid sequence in mammals has a structure suggesting a transposable element. *Mol. Cell. Biol.* **1:**573–583.

Hellung-Larsen, P., Kulamowica, I., and Frederiksen, S. 1980. Synthesis of low molecular weight RNA components in cells with a temperature-sensitive polymerase II. *Biochim. Biophys. Acta* **609:**201–204.

Henrick, J. P., Wolin, S. L., Rinke, J., Lerner, M. R., and Steitz, J. A. 1981. Ro small cytoplasmic ribonucleoproteins are a subclass of La ribonucleoproteins: Further characterization of the Ro and La small ribonucleoproteins from uninfected mammalian cells. *Mol. Cell. Biol.* **1:**1138–1149.

Hinnebusch, A. G., Klotz, L. C., Blanken, R. L., and Loeblish, A. R. 1981. An evaluation of the phylogenetic position of the dinoflagellate *Crypthecondinium cohnii* based on 5S rRNA characterization. *J. Mol. Evol.* **17**:334–347.

Hofstetter, H., Kressmann, A., and Birnstiel, M. L. 1981. A split promoter for a eukaryotic tRNA gene. *Cell* **24**:573–585.

Hori, H., and Osawa, S. 1979. Evolutionary change in 5S RNA secondary structure and a phylogenetic tree of 54 5S RNA species. *Proc. Natl. Acad. Sci. USA* **76**:381–385.

Hori, H., Osawa, S., Murao, K., and Ishikura, H. 1980. The nucleotide sequence of 5S ribosomal RNA from *Micrococcus lysodeikticus*. *Nucleic Acids Res.* **8**:5423–5426.

Hori, H., Lim, B. L., and Osawa, S. 1985. Evolution of green plants as deduced from 5S rRNA sequences. *Proc. Natl. Acad. Sci. USA* **82**:820–823.

Jacq, B., Jourdan, R., and Jordan, B. R. 1977. Structure and processing of precursor 5S RNA in *Drosophila melanogaster*. *J. Mol. Biol.* **117**:785–795.

Jelinek, W. R., and Schmid, C. W. 1982. Repetitive sequences in eukaryotic DNA and their expression. *Annu. Rev. Biochem.* **51**:813–844.

Jelinek, W. R., Toomey, T. P., Leinwand, L., *et al.* 1980. Ubiquitous, interspersed repeated sequences in mammalian genomes. *Proc. Natl. Acad. Sci. USA* **77**:1398–1402.

Johnson, J. D., and Raymond, G. J. 1984. Three regions of a yeast tRNA$_3^{Leu}$ gene promote RNA polymerase III transcription. *J. Biol. Chem.* **259**:5090–5094.

Jordan, B. R., Galling, G., and Jourdan, R. 1974. Sequence and conformation of 5S RNA from *Chlorella* cytoplasmic ribosomes: Comparison with other 5S RNA molecules. *J. Mol. Biol.* **87**:205–225.

Kato, N., Hoshino, H., and Harada, F. 1982. Nucleotide sequence of 4.5S RNA (C8 or hY5) from the HeLa cells. *Biochem. Biophys. Res. Commun.* **108**:363–370.

Katze, M. G., Chen, Y. T., and Krug, R. M. 1984. Nuclear–cytoplasmic transport and VAI RNA-independent translation of influenza viral messenger RNAs in late adenovirus-infected cells. *Cell* **37**:483–490.

Keus, R. J. A., Roovers, D. J., Dekker, A. F., and Groot, G. S. P. 1983. The nucleotide sequence of the 4.5S and 5S rRNA genes and flanking regions from *Spirodela oligorhiza* chloroplasts. *Nucleic Acids Res.* **11**:3405–3410.

Kingston, R. E., and Chamberlin, M. J. 1981. Pausing and attenuation of *in vitro* transcription in the *rrnB* operon of *E. coli*. *Cell* **27**:523–531.

Kjems, J., Olesen, S. O., and Garrett, R. A. 1985. Comparison of eubacterial and eukaryotic 5S RNA structures: A chemical modification study. *Biochemistry* **24**:241–250.

Kómiya, H., and Takemura, S. 1981. The nucleotide sequence of 5S ribosomal RNA from slime mold *Physarum polycephalum*. *J. Biochem.* **90**:1577–1581.

Korn, L. J., and Brown, D. D. 1978. Nucleotide sequences of *Xenopus borealis* oocyte 5S DNA: Comparison of sequences that flank several related eucaryotic genes. *Cell* **15**:1145–1156.

Koski, R. A., Allison, D. S., Worthington, M., and Hall, B. D. 1982. An *in vitro* RNA polymerase III system from *S. cerevisiae:* Effects of deletions and point mutations upon *SUP4* gene transcription. *Nucleic Acids Res.* **10**:8127–8143.

Krayev, A. S., Kramerov, D. A., Skryabin, K. G., Ryskov, A. P., Bayev, A. A., and Georgiev, G. P. 1980. The nucleotide sequence of the ubiquitous repetitive DNA sequence B1 complementary to the most abundant class of mouse fold-back RNA. *Nucleic Acids Res.* **8**:1201–1215.

Krayev, A. S., Markusheva, T. V., Kramerov, D. A., Ryskov, A. P., Skryabin, K. G., Bayev, A. A., and Georgiev, G. P. 1982. Ubiquitous transposon-like repeats B1 and B2 of the mouse genome: B2 sequencing. *Nucleic Acids Res.* **10**:7461–7475.

Kressmann, A., Hofstetter, H., Di Capua, E., Grosschedl, R., and Birnstiel, M. L. 1979. A tRNA gene of *Xenopus laevis* contains at least two sites promoting transcription. *Nucleic Acids Res.* **7**:1749–1763.

Krolewski, J. J., Schindler, C. W., and Rush, M. G. 1984. Structure of extrachromosomal circular DNAs containing both the *Alu* family of dispersed repetitive sequences and other regions of chromosomal DNA. *J. Mol. Biol.* **174**:41–54.

Kumagai, I., Digweed, M., Erdmann, V. A., Watanabe, K., and Oshima, T. 1981. The nucleotide sequence of 5S rRNA from an extreme thermophile, *Thermus thermophilus* HB8. *Nucleic Acids Res.* **9**:5159–5162.

Kumazaki, T., Hori, H., Osawa, S., Mita, T., and Higashinakagawa, T. 1982. The nucleotide sequences of 5S rRNAs from three ciliated protozoa. *Nucleic Acids Res.* **10**:4409–4412.

Lamond, A. I., and Travers, A. A. 1983. Requirement for an upstream element for optimal transcription of a bacterial tRNA gene. *Nature (London)* **305**:248–250.

Laski, F. A., Belagaje, R., RajBhandary, U. L., and Sharp, P. A. 1982. An amber suppressor tRNA gene derived by site-specific mutagenesis: Cloning and function in mammalian cells. *Proc. Natl. Acad. Sci. USA* **79**:5813–5817.

Lassar, A. B., Martin, P. L., and Roeder, R. G. 1983. Transcription of class III genes: Formation of preinitiation complexes. *Science* **222**:740–748.

Lenardo, M. J., Dorfman, D. M., Reddy, L. V., and Donelson, J. E. 1985. Characterization of the *Trypanosoma brucei* 5S RNA gene and transcript: The 5S rRNA is a spliced-leader-independent species. *Gene* **35**:131–141.

Lerner, M. R., and Steitz, J. A. 1981. Snurps and Scyrps. *Cell* **25**:298–300.

Lerner, M. R., Andrews, N. C., Miller, G., and Steitz, J. A. 1981. Two small RNAs encoded by Epstein–Barr virus and complexed with protein are precipitated by antibodies from patients with systemic lupus erythematosus. *Proc. Natl. Acad. Sci. USA* **78**:805–809.

Li, W., Reddy, R., Henning, D., Epstein, P., and Busch, H. 1982. Nucleotide sequence of 7S RNA: Homology to *Alu* DNA and 4.5S DNA. *J. Biol. Chem.* **257**:5136–5142.

Lu, A. L., Steege, D. A., and Stafford, D. W. 1980. Nucleotide sequence of a 5S ribosomal RNA gene in the sea urchin *Lytechinus variegatus. Nucleic Acids Res.* **8**:1839–1853.

Luehrsen, K. R., and Fox, G. E. 1981. Secondary structure of eukaryotic cytoplasmic 5S ribosomal RNA. *Proc. Natl. Acad. Sci. USA* **78**:2150–2154.

Luehrsen, K. R., Fox, G. E., and Woese, C. R. 1980. The sequence of *Tetrahymena thermophila* 5S ribosomal ribonucleic acid. *Curr. Microbiol.* **4**:123–126.

Luehrsen, K. R., Fox, G. E., Kilpatrick, M. W., Walker, R. T., Domdey, H., Krupp, G., and Gross, H. J. 1981. The nucleotide sequence of the 5S rRNA from the archaebacterium *Thermoplasma acidophilum. Nucleic Acids Res.* **9**:965–970.

Luoma, G. A., and Marshall, A. G. 1978a. Laser Raman evidence for a new cloverleaf secondary structure for eucaryotic 5S RNA. *J. Mol. Biol.* **125**:95–105.

Luoma, G. A., and Marshall, A. G. 1978b. Laser Raman evidence for new cloverleaf secondary structures for eukaryotic 5.8S RNA and prokaryotic 5S RNA. *Proc. Natl. Acad. Sci. USA* **75**:4901–4905.

MacKay, R. M., and Doolittle, W. F. 1981. Nucleotide sequence of *Acanthamoeba castellani* 5S and 5.8S ribosomal ribonucleic acids: phylogenetic and comparative structural analysis. *Nucleic Acids Res.* **9**:3321–3334.

MacKay, R. M., Spencer, D. F., Doolittle, W. F., and Gray, M. W. 1980. Nucleotide sequences of wheat-embryo cytosol 5-S and 5.8-S ribosomal ribonucleic acids. *Eur. J. Biochem.* **112**:561–576.

MacKay, R. M., Salgado, D., Bonen, L., Stackebrandt, E., and Doolittle, W. F. 1982. The 5S ribosomal RNAs of *Paracoccus denitrificans* and *Prochloron. Nucleic Acids Res.* **10**:2963–2970.

Maeda, N., Bliska, J. B., and Smithies, O. 1983. Recombination and balanced chromosome polymorphism suggested by DNA sequences 5′ to the human δ-globin gene. *Proc. Natl. Acad. Sci. USA* **80**:5012–5016.

Margulis, L. 1970. *Origin of Eukaryotic Cells,* New Haven, Yale University Press.

Marotta, C. A., Varricchio. F., Smith, I., Weissman, S. M., Sogin, M. L., and Pace, N. R. 1976. The primary structure of *Bacillus subtilis* and *Bacillus stearothermophilus* 5S ribonucleic acids. *J. Biol. Chem.* **251**:3122–3127.

Mattaj, I. W., Lienhard, S., Zeller, R., and DeRobertis, E. M. 1983. Nuclear exclusion of transcription factor IIIA and the 42S particle transfer-RNA-binding protein in *Xenopus* oocytes: A possible mechanism for gene control? *J. Cell Biol.* **97**:1261–1265.

Maxam, A. M., Tizard, R., Skryabin, K. G., and Gilbert, W. 1977. Promoter region for yeast 5S ribosomal RNA. *Nature (London)* **267**:643–645.

Miller, J. R. 1983. 5S ribosomal RNA genes. *In:* Maclean, N., Gregory, S. P., and Flavell, R. A., eds., *Eukaryotic Genes: Their Structure, Activity and Regulation,* London, Butterworths, pp. 225–237.

Miller, J. R., Cartwright, E. M., Brownlee, G. G., Fedoroff, N. V., and Brown, D. D. 1978. The nucleotide sequence of oocyte 5S DNA in *Xenopus laevis.* II. The GC-rich region. *Cell* **13**:717–725.

Morgens, P. H., Grabau, E. A., and Gesteland, R. F. 1984. A novel soybean mitochondrial transcript resulting from a DNA rearrangement involving the 5S rRNA gene. *Nucleic Acids Res.* **12**:5665–5684.

Morris, G. F., and Marzluff, W. F. 1983. A factor in sea urchin eggs inhibits transcription in isolated nuclei by sea urchin RNA polymerase III. *Biochemistry* **22**:645–653.

Morton, D. G., and Sprague, K. U. 1984. *In vitro* transcription of a silkworm 5S RNA gene requires an upstream signal. *Proc. Natl. Acad. Sci. USA* **81**:5519–5522.

Murphy, M. H., and Baralle, F. E. 1983. Directed semisynthetic point mutational analysis of an RNA polymerase III promoter. *Nucleic Acids Res.* **11**:7695–7716.

Newhouse, N., Nicoghosian, K., and Cedergren, R. J. 1982. The nucleotide sequence of phenylalanine tRNA and 5S RNA from *Rhodospirillum rubrum*. *Can. J. Biochem.* **59**:921–932.

Nishikawa, K., and Takemura, S. 1974. Structure and function of 5S ribosomal ribonucleic acid from *Torulopsis utilis*. II. Partial digestion from ribonucleases and derivation of the complete sequence. *J. Biochem.* **76**:935–947.

Page, G. S., Smith, S., and Goodman, H. M. 1981. DNA sequence of the rat growth hormone gene; Location of the 5′ terminus of the growth hormone mRNA and identification of an internal transposon-like element. *Nucleic Acids Res.* **9**:2087–2103.

Pan, J., Elder, J. T., Duncan, C. H., and Weissman, S. M. 1981. Structural analysis of interspersed repetitive polymerase III transcription units in human DNA. *Nucleic Acids Res.* **9**:1151–1169.

Parker, C. S., and Topol, J. 1984. A *Drosophila* RNA polymerase II transcription factor binds to the regulatory site of an hsp 70 gene. *Cell* **37**:273–283.

Pederson, D. S., Yao, M. C., Kimmel, A. R., and Gorovsky, M. A. 1984. Sequence organization and flanking clusters of 5S ribosomal RNA genes in *Tetrahymena*. *Nucleic Acids Res.* **12**:3003–3021.

Peffley, D. M., and Sogin, M. L. 1981. A putative tRNA^Trp gene cloned from *Dictyostelium discoideum*: Its nucleotide sequence and association with repetitive deoxyribonucleic acid. *Biochemistry* **20**:4015–4021.

Pelham, H. R. B., Wormington, W. M., and Brown, D. D. 1981. Related 5S RNA transcription factors in *Xenopus* oocytes and somatic cells. *Proc. Natl. Acad. Sci. USA* **78**:1760–1764.

Perez-Stable, C., Ayres, T. M., and Shen, C. K. J. 1984. Distinctive sequence organization and functional programming of an *Alu* repeat promoter. *Proc. Natl. Acad. Sci. USA* **81**:5291–5295.

Peterson, R. C., Doering, J. L., and Brown, D. D. 1980. Characterization of two *Xenopus* somatic 5S DNAs and one minor oocyte-specific 5S DNA. *Cell* **20**:131–141.

Picard, B. M., and Wegnez, M. 1979. Isolation of a 7S particle from *Xenopus laevis* oocytes: A 5S RNA–protein complex. *Proc. Natl. Acad. Sci. USA* **76**:241–245.

Picard, B. M., Maire, M., Wegnez, M., and Denis, H. 1980. Biochemical research on oogenesis. Composition of the 42S storage particles of *Xenopus laevis* oocytes. *Eur. J. Biochem.* **109**:359–368.

Piper, P. W., Lockheart, A., and Patel, N. 1984. A minor class of 5S rRNA genes in *Saccharomyces cerevisiae* X2180-1B, one member of which lies adjacent to a Ty transposable element. *Nucleic Acids Res.* **12**:4083–4096.

Poncz, M., Schwartz, E., Ballantine, M., and Surrey, S. 1983. Nucleotide sequence analysis of the δβ-globin gene region in humans. *J. Biol. Chem.* **258**:11599–11609.

Potter, S. S. 1982. DNA sequence of a foldback transposable element in *Drosophila*. *Nature (London)* **297**:201–204.

Pribula, C. D., Fox, G. E., and Woese, C. R. 1976. Nucleotide sequence of *Clostridium pasteurianum* 5S rRNA. *FEBS Lett.* **64**:350–352.

Reynolds, W. F., and Gottesfeld, J. M. 1983. 5S rRNA gene transcription factor IIIA alters the helical configuration of DNA. *Proc. Natl. Acad. Sci. USA* **80**:1862–1866.

Reynolds, W. F., Bloomer, L. S., and Gottesfeld, J. M. 1983. Control of 5S RNA transcription in *Xenopus* somatic cell chromatin: Activation with an oocyte extract. *Nucleic Acids Res.* **11**:57–75.

Robertson, H. D., and Dickson, E. 1984. Structure and distribution of *Alu* family sequences or their analogs within heterogeneous nuclear RNA of HeLa, KB, and L cells. *Mol. Cell. Biol.* **4**:310–316.

Rosa, M. D., Gottlieb, E., Lerner, M. R., and Steitz, J. A. 1981. Striking similarities are exhibited by 2 small Epstein–Barr virus-encoded RNA species and the adenovirus-associated species VAI and VAII. *Mol. Cell. Biol.* **1**:785–796.

Rosenthal, D., and Doering, J. L. 1983. The genomic organization of dispersed tRNA and 5S RNA genes in *Xenopus laevis*. *J. Biol. Chem.* **258**:7402–7410.

Roy, M. K., Singh, B., Ray, B. K., and Apirion, D. 1983. Maturation of 5-S rRNA: Ribonuclease E cleavages and their dependence on precursor sequences. *Eur. J. Biochem.* **131**:119–127.

Rubin, C. M., Houck, C. M., Deininger, P. L., Friedmann, T., and Schmid, C. W. 1980. Partial nucleotide sequence of the 300-nucleotide interspersed repeated human DNA sequences. *Nature (London)* **284**:372–374.

Ruet, A., Camier, S., Smagowicz, W., Sentenac, A., and Fromageot, P. 1984. Isolation of a class C transcription factor which forms a stable complex with tRNA genes. *EMBO J.* **3**:343–350.

Sagin, L. 1967. On the origin of mitosing cells. *J. Theor. Biol.* **14**:225–274.

Sakamoto, K., Kominami, R., Mishima, Y., and Okada, N. 1984. The 6S RNA transcribed from rodent total DNA *in vitro* is the transcript of the type 2 *Alu* family. *Mol. Gen. Genet.* **194**:1–6.

Sakonju, S. 1981. Identification of a control region that directs the initiation of transcription with a specific transcription factor. Ph.D. dissertation, Johns Hopkins University, Baltimore.

Sakonju, S., and Brown, D. D. 1982. Contact points between a positive transcription factor and the *Xenopus* 5S RNA gene. *Cell* **31**:395–405.

Sakonju, S., Brown, D. D., Engelke, D. R., Ng, S. Y., Shastry, B. S., and Roeder, R. 1981. The binding of a transcription factor to deletion mutants of a 5S ribosomal RNA gene. *Cell* **23**:665–669.

Schaack, J., Sharp, S., Dingermann, T., Burke, D. J., Cooley, L., and Söll, D. 1984. The extent of a eukaryotic tRNA gene. *J. Biol. Chem.* **259**:1461–1467.

Scherer, G., Tschudi, C., Perera, J., Delius, H., and Pirotta, V. 1982. *B104*, a new dispersed repeated gene family in *Drosophila melanogaster* and its analogies with retroviruses. *J. Mol. Biol.* **157**:435–451.

Schimenti, J. C., and Duncan, C. H. 1984. Ruminant globin gene structures suggest an evolutionary role for *Alu*-type repeats. *Nucleic Acids Res.* **12**:1641–1655.

Schmid, C. W., and Jelinek, W. R. 1982. The *Alu* family of dispersed repetitive sequences. *Science* **218**:1065–1070.

Schon, E. A., Cleary, M. L., Haynes, J. R., and Lingrel, J. B. 1981. Structure and evolution of goat γ-, β^c-, and β^A-globin genes: Three developmentally regulated genes contain inserted elements. *Cell* **27**:359–369.

Segall, J., Matsui, T., and Roeder, R. G. 1980. Multiple factors are required for the accurate transcription of purified genes by RNA polymerase III. *J. Biol. Chem.* **255**:11986–11991.

Sekiya, T., Mori, M., Takahashi, N., and Nishimura, S. 1980. Sequence of the distal tRNA$^{Asp}_1$ gene and the transcription termination signal in the *Escherichia coli* ribosomal RNA operon *rrnF*(or *G*). *Nucleic Acids Res.* **8**:3809–3827.

Setzer, D. R., and Brown, D. D. 1985. Formation and stability of the 5S RNA transcription complex. *J. Biol. Chem.* **260**:2483–2492.

Sharp, S., Dingermann, T., Schaack, J., Sharp, J. A., Burke, D. J., DeRobertis, E. M., and Söll, D. 1983. Each element of the *Drosophila* tRNA^{Arg} gene split promoter directs transcription in *Xenopus* oocytes. *Nucleic Acids Res.* **11**:8677–8690.

Shen, C. K. J., and Maniatis, T. 1982. The organization, structure, and *in vitro* transcription of *Alu* family RNA polymerase III transcription units in the human α-like globin gene cluster: Precipitation of *in vitro* transcripts by lupus anti-La antibodies. *J. Mol. Appl. Genet.* **1**:343–360.

Shi, X. P., Wingender, E., Böttrich, J., and Seifart, K. H. 1983. Faithful transcription of ribosomal 5-S RNA *in vitro* depends on the presence of several factors. *Eur. J. Biochem.* **131**:189–194.

Simoncsits, A. 1980. 3′ terminal labelling of RNA with β-^{32}P-pyrophosphate group and its application to the sequence analysis of 5S RNA from *Streptomyces griseus*. *Nucleic Acids Res.* **8**:4111–4124.

Singer, M. F. 1982. SINES and LINES: Highly repeated short and long interspersed sequences in mammalian genomes. *Cell* **28**:433–434.

Singh, B., and Apirion, D. 1982. Primary and secondary structure in a precursor of 5S rRNA. *Biochim. Biophys. Acta* **698**:252–259.

Smith, D. R., Jackson, J., and Brown, D. D. 1984. Domains of the positive transcription factor specific for the *Xenopus* 5S RNA gene. *Cell* **37**:645–652.

Spencer, D. F., Bonen, L., and Gray, M. W. 1981. Primary sequence of wheat mitochondrial 5S ribosomal ribonucleic acid: Functional and evolutionary implications. *Biochemistry* **20**:4022–4029.

Spradling, A. C., and Rubin, G. M. 1981. *Drosophila* genome organization: Conserved and dynamic aspects. *Annu. Rev. Genet.* **15**:219–264.

Sprague, K. U., Larson, D., and Morton, D., 1980. 5′ flanking sequence signals are required for activity of silkworm alanine tRNA genes in homologous *in vitro* transcription systems. *Cell* **22**:171–178.

Sprinzl, M., and Gauss, D. H. 1984. Compilation of sequences of tRNA genes. *Nucleic Acids Res.* **12**(Suppl.):r59–r131.

Stahl, D. A., Luehrsen, K. R., Woese, C. R., and Pace, N. R. 1981. An unusual 5S rRNA$_1$ from *Sulfolobus acidocaldarius*, and its implications for a general 5S rRNA structure. *Nucleic Acids Res.* **9**:6129–6137.

Stillman, D. J., and Geiduschek, E. P. 1984. Differential binding of *S. cerevisiae* RNA polymerase III transcription factors to two promoter segments of a tRNA gene. *EMBO J.* **3**:847–853.

Stillman, D. J., Sivertsen, A. L., Zentner, P. G., and Geiduschek, E. P. 1984. Correlations between transcription of a yeast tRNA gene and transcription factor–DNA interactions. *J. Biol. Chem.* **259**:7955–7962.

Stumph, W. E., Kristo, P., Tsai, M. J., and O'Malley, B. W. O. 1981. A chicken middle-repetitive DNA sequence which shares homology with mammalian ubiquitous repeats. *Nucleic Acids Res.* **9:**5383–5397.

Szeberényi, J., and Apirion, D. 1983. Initiation, processing, and termination of ribosomal RNA from a hybrid 5 S ribosomal RNA gene in a plasmid. *J. Mol. Biol.* **168:**525–561.

Tabata, S. 1980. Structure of the 5-S ribosomal RNA gene and its adjacent regions in *Torulopsis utilis*. *Eur. J. Biochem.* **110:**107–114.

Takaiwa, F., and Sugiura, M. 1980. Nucleotide sequences of the 4.5S and 5S ribosomal RNA genes from tobacco chloroplasts. *Mol. Gen. Genet.* **180:**1–4.

Takaiwa, F., and Sugiura, M. 1982. The nucleotide sequence of chloroplast 5S ribosomal RNA from a fern, *Dryopteris acuminata*. *Nucleic Acids Res.* **10:**5369–5373.

Takaiwa, F., Kusuda, M., Saga, N., and Sugiura, M. 1982. The nucleotide sequence of 5S rRNA from a red alga, *Prophyra yezoensis*. *Nucleic Acids Res.* **10:**6037–6040.

Thimmappaya, B., Weinberger, C., Schneider, R. J., and Shenk, T. 1982. Adenovirus VAI RNA is required for efficient translation of viral mRNAs at late times after infection. *Cell* **31:**543–551.

Traboni, C., Ciliberto, G., and Cortese, R. 1982. A novel method for site-directed mutagenesis: Its application to a eukaryotic tRNA[Pro] gene promoter. *EMBO J.* **1:**415–420.

Tschudi, C., and Pirrotta, V. 1980. Sequence and heterogeneity in the 5S RNA gene cluster of *Drosophila melanogaster*. *Nucleic Acids Res.* **8:**441–451.

Ullu, E., and Tschudi, C. 1984. *Alu* sequences are processed 7SL RNA genes. *Nature (London)* **312:**171–172.

Ullu, E., Murphy, S., and Melli, M. 1982. Human 7SL RNA consists of a 140 nucleotide middle-repetitive sequence inserted in an *Alu* sequence. *Cell* **29:**195–202.

Valenzuela, P., Bell, G. I., Masiarz, F. R., DeGennaro, L. J., and Rutter, W. J. 1977a. Nucleotide sequence of the yeast 5S ribosomal RNA gene and adjacent putative control regions. *Nature (London)* **267:**641–643.

Valenzuela, P., Bell, G. I., Venegas, A., Sewell, E. T., Masiarz, F. R., DeGennaro, L. J., Weinberg, F., and Rutter, W. J. 1977b. Ribosomal RNA genes of *Saccharomyces cerevisiae*. II. Physical map and nucleotide sequence of the 5S ribosomal RNA gene and adjacent intergenic regions. *J. Biol. Chem.* **252:**8126–8135.

Vandenberghe, A., Wassink, A., Raeymaekers, P., DeBaerre, R., Huysmans, E., and De Wachter, R. 1985. Nucleotide sequence, secondary structure and evolution of the 5S ribosomal RNA from five bacterial species. *Eur. J. Biochem.* **149:**537–542.

Walker, W. F., and Doolittle, W. F. 1982. Nucleotide sequences of 5S ribosomal RNA from four oomycete and chytrid water molds. *Nucleic Acids Res.* **10:**5717–5721.

Walker, R. T., Cheton, E. T. J., Kilpatrick, M. W., Rogers, M. J., and Simmons, J. 1982. The nucleotide sequence of the 5S rRNA from *Spiroplasma* species BC3 and *Mycoplasma mycoides* sp. *capri* PG3. *Nucleic Acids Res.* **10:**6363–6367.

Walter, P., and Blobel, G. 1982. Signal recognition particle contains a 7S RNA essential for protein translocation across the endoplasmic reticulum. *Nature (London)* **299:**691–698.

Watanabe, Y., Tsukada, T., Notake, M., Nakanishi, S., and Numa, S. 1982. Structural analysis of repetitive DNA sequences in the bovine corticotropin-β-lipotropin precursor gene region. *Nucleic Acids Res.* **10:**1459–1469.

Weiner, A. M. 1980. An abundant cytoplasmic 7S RNA is complementary to the dominant interspersed middle repetitive DNA sequence family in the human genome. *Cell* **22:**209–218.

Woese, C. R., and Fox, G. E. 1977. Phylogenetic structure of the prokaryotic domain: The primary kingdoms. *Proc. Natl. Acad. Sci. USA* **74:**5088–5090.

Woese, C. R., Pribula, C. D., Fox, G. E., and Zablen, L. B. 1975. The nucleotide sequence of the 5S ribosomal RNA from a photobacterium. *J. Mol. Evol.* **5:**35–46.

Woese, C. R., Luehrsen, K. R., Pribula, C. D., and Fox, G. E. 1976. Sequence characterization of 5S ribosomal RNA from eight gram positive procaryotes. *J. Mol. Evol.* **8:**143–153.

Woese, C. R. Magrum, L. J., and Fox, G. E. 1978. Archaebacteria. *J. Mol. Evol.* **11:**245–252.

Wolin, S. L., and Steitz, J. A. 1983. Genes for two small cytoplasmic Ro RNAs are adjacent and appear to be single-copy in the human genome. *Cell* **32:**735–744.

Wormington, W., Bogenhagen, D. F., Jordan, E., and Brown, D. D. 1982. A quantitative assay for *Xenopus* 5S RNA gene transcription *in vitro*. *Cell* **24:**809–818.

Yamamoto, T., Davis, C. G., Brown, M. S., Schneider, W. J., Casey, M. L., Goldstein, J. L., and Russell, D. W. 1984. The human LDL receptor: A cysteine-rich protein with multiple *Alu* sequences in its mRNA. *Cell* **39:**27–38.

Zieve, G. W. 1981. Two groups of small stable RNAs. *Cell* **25:**296–297.

CHAPTER 4

Amikam, D., Razin, S., and Glaser, G. 1982. Ribosomal RNA genes in *Mycoplasma. Nucleic Acids Res.* **10:**4215–4222.

Amikam, D., Glaser, G., and Razin, S. 1984. Mycoplasmas (*Mollicutes*) have a low number of rRNA genes. *J. Bacteriol.* **158:**376–378.

Anderson, S., Bankier, A. T., Barrell, B. G., de Bruijn, M. H. L., Coulson, A. R., Drouin, J., Eperon, I. C., Nierlick, D. P., Roe, B. A., Sanger, F., Schreier, P. H., Smith, A. J. R., Staden, R., and Young, I. G. 1981. Sequence and organization of the human mitochondrial genome. *Nature (London)* **290:**457–465.

Anderson, S., de Bruijn, M. H. L., Coulson, A. R., Eperon, I. C., Sanger, F., and Young, I. G. 1982a. Complete sequence of bovine mitochondrial DNA. Conserved features of the mammalian mitochondrial genome. *J. Mol. Biol.* **156:**683–717.

Anderson, S., Bankier, A. T., Barrell, B. G., de Bruijn, M. H. L., Coulson, A. R., Drouin, J., Eperon, I. C., Nierlich, D. P., Roe, B. A., Sanger, F., Schreier, P. H., Smith, A. J. H., Staden, R., and Young, I. G. 1982b. Comparison of the human and bovine mitochondrial genomes. *In:* Slonimski, P., Borst, P., and Attardi, G., eds., *Mitochondrial Genes,* Cold Spring Harbor, New York, Cold Spring Harbor Laboratory, pp. 5–43.

Bakken, A., Morgan, G., Sollner-Webb, B., Roan, J., Busby, S., and Reeder, R. H. 1982. Mapping of transcription initiation and termination signals on *Xenopus laevis* ribosomal DNA. *Proc. Natl. Acad. Sci. USA* **79:**56–60.

Bayev, A. A., Georgiev, O. I., Hadjiolov, A. A., Kermekchiev, M. B., Nikolaev, N., Skryabin, K. G., and Zakharyev, V. M. 1980. The structure of the yeast ribosomal RNA genes. 2. The nucleotide sequence of the initiation site for ribosomal RNA transcription. *Nucleic Acids Res.* **8:**4919–4926.

Beckingham, K. 1982. Insect rDNA. *In:* Busch, H., and Rothblum, L., eds., *The Cell Nucleus,* New York, Academic Press, Vol. X, pp. 205–269.

Bedbrook, J. R., Kolodner, R., and Bogorad, L. 1977. *Zea mays* chloroplast ribosomal RNA genes are part of a 22,000 base pair inverted repeat. *Cell* **11:**739–749.

Beilharz, M. W., Cobon, G. S., and Nagley, P. 1982. A novel species of double-stranded RNA in mitochondria of *Saccharomyces cerevisiae. Nucleic Acids Res.* **10:**1051–1070.

Bell, G. I., De Gennaro, L. J., Gelfand, D. H., Bishop, R. J., Valenzuela, P., and Rutter, W. J. 1977. Ribosomal RNA genes of *Saccharomyces cerevisiae.* I. Physical map of the repeating unit and location of the regions coding for 5S, 5.8S, 18S, and 25S ribosomal RNAs. *J. Biol. Chem.* **252:**8118–8125.

Berger, S., and Schweiger, H. G. 1982. Characterization and species differences of rDNA in algae. *In:* Busch, H., and Rothblum, L., eds., *The Cell Nucleus,* New York, Academic Press, Vol. X, pp. 31–64.

Bibb, M. J., Van Etten, R. A., Wright, C. T., Walberg, M. W., and Clayton, D. A. 1981. Sequence and gene organization of mouse mitochondrial DNA. *Cell* **26:**167–180.

Blackburn, E. H. 1982. Characterization and species differences of rDNA: Protozoans. *In:* Busch, H., and Rothblum, L., eds., *The Cell Nucleus,* New York, Academic Press, Vol. X, pp. 145–170.

Bollon, A. P. 1982. Organization of fungal ribosomal RNA genes. *In:* Busch, H., and Rothblum, L., eds., *The Cell Nucleus,* New York, Academic Press, Vol. X, pp. 67–125.

Boros, I., Csórdás-Toth, E., Kiss, A., Kiss, I., Török, I., Udvardy, A., Udvardy, K., and Venetianer, P. 1983. Identification of two new promoters probably involved in the transcription of a ribosomal RNA gene of *Escherichia coli. Biochim. Biophys. Acta* **739:**173–180.

Borst, P., and Grivell, L. A. 1978. The mitochondrial genome of yeast. *Cell* **15:**705–723.

Borst, P., Bos, J. L., Grivell, L. A., Groot, G. S. P., Heyting, C., Moorman, A. F. M., Sanders, J. P. M., Talen, J. L., Van Kreijl, C. F., and Van Ommen, G. J. B. 1977. The physical map of yeast mitochondrial DNA anno 1977. *In:* Bandlow, W., Schweyen, R. J., Wolf, K., and Kaudewitz, F., eds., *Mitochondria 1977,* Berlin, De Gruyter, pp. 213–254.

Boseley, P. G., Moss, T., Machler, M., Portmann, R., and Birnstiel, M. L. 1979. Sequence organization of the spacer DNA in a ribosomal gene unit of *Xenopus laevis. Cell* **17:**19–31.

Bowman, C. M., and Dyer, T. A. 1979. 4.5S ribonucleic acid, a novel ribosome component in the chloroplasts of flowering plants. *Biochem. J.* **183:**605–613.

Briat, J. F., Dron, M., Loiseaux, S., and Mache, R. 1982. Structure and transcription of the spinach chloroplast rDNA leader region. *Nucleic Acids Res.* **10:**6865–6878.

Briat, J. F., Dron, M., and Mache, R. 1983. Is transcription of higher plant chloroplast ribosomal operons regulated by premature termination? *FEBS Lett.* **163**:1–5.

Brosius, J., Dull, T. J., and Noller, H. F. 1980. Complete nucleotide sequence of a 23S ribosomal RNA gene from *E. coli. Proc. Natl. Acad. Sci. USA* **77**:201–204.

Brosius, J., Dull, T. J., Sleeter, D. D., and Noller, H. F. 1981. Gene organization and primary structure of a ribosomal RNA operon from *E. coli. J. Mol. Biol.* **148**:107–127.

Burke, J. M., and RajBhandary, U. L. 1982. Intron within the large rRNA gene of *N. crassa* mitochondria: A long operon reading frame and a consensus sequence important in splicing. *Cell* **31**:509–520.

Carbon, P., Ebel, J. P., and Ehresmann, C. 1981. The sequence of the ribosomal 16S RNA from *Proteus vulgaris*. Sequence comparison with *E. coli* 16S RNA and its use in secondary structure model building. *Nucleic Acids Res.* **9**:2325–2333.

Cashel, M., and Gallant, J. 1969. Two compounds implicated in the function of the *RC* gene of *Escherichia coli. Nature (London)* **221**:838–841.

Cassidy, J. R., Moore, D., Lu, B. C., and Pukkila, P. J. 1984. Unusual organization and lack of recombination in the ribosomal RNA genes of *Coprinus cinereus. Curr. Genet.* **8**:607–613.

Cech, T. R., Zaug, A. J., Grabowski, P. J., and Brehm, S. L. 1982. Transcription and splicing of the ribosomal RNA precursor of *Tetrahymena. In:* Busch, H., and Rothblum, L., eds., *The Cell Nucleus*, New York, Academic Press, Vol. X, pp. 171–204.

Cedergren, R. J., and Sankoff, D. 1976. Evolutionary origin of 5.8S ribosomal RNA. *Nature (London)* **259**:74–76.

Challoner, P. B., Ancin, A. A., Pearlman, R. E., and Blackburn, E. H. 1985. Conserved arrangements of repeated DNA sequences in nontranscribed spacers of ciliate ribosomal RNA genes: Evidence for molecular coevolution. *Nucleic Acids Res.* **13**:2661–2680.

Chan, Y. L., Olvera, J., and Wool, I. G. 1983. The structure of rat 28S ribosomal ribonucleic acid inferred from the sequence of nucleotides in a gene. *Nucleic Acids Res.* **11**:7819–7831.

Chan, Y. L., Gutell, R., Noller, H. F., and Wool, I. G. 1984. The nucleotide sequence of a rat 18S ribosomal ribonucleic acid gene and a proposal for the secondary structure of 18S ribosomal ribonucleic acid. *J. Biol. Chem.* **259**:224–230.

Chang, D. D., and Clayton, D. A. 1984. Precise identification of individual promoters for transcription of each strand of human mitochondrial DNA. *Cell* **36**:635–643.

Chao, S., Sederoff, R. R., and Levings, C. S. 1983. Partial sequence analysis of the 5S to 18S rRNA gene region of the maize mitochondrial genome. *Plant Physiol.* **71**:190–193.

Chao, S., Sederoff, R., and Levings, C. S. 1984. Nucleotide sequence and evolution of the 18S ribosomal RNA gene in maize mitochondria. *Nucleic Acids Res.* **12**:6629–6644.

Christianson, T., Edwards, J., Levens, D., Locker, J., and Rabinowitz, M. 1982. Transcriptional initiation and processing of the small ribosomal RNA of yeast mitochondria. *J. Biol. Chem.* **257**:6494–6500.

Clark, C. G., and Gerbi, S. A. 1982. Ribosomal RNA evolution by fragmentation of the 23S progenitor: Maturation pathway parallels evolutionary emergence. *J. Mol. Evol.* **18**:329–336.

Clark-Walker, G. D., and Sriprakash, K. S. 1983. Analysis of a five gene cluster and unique orientation of large genic sequences in *Torulopsis glabrata* mitochondrial DNA. *J. Mol. Evol.* **19**:342–345.

Clark-Walker, G. D., McArthur, C. R., and Sriprakash, K. S. 1983. Order and orientation of genic sequences in circular mitochondrial DNA from *Saccharomyces exiguus. J. Mol. Evol.* **19**:333–341.

Clary, D. O., and Wolstenholme, D. R. 1985. The ribosomal RNA genes of *Drosophila* mitochondrial DNA. *Nucleic Acids Res.* **13**:4029–4045.

Clary, D. O., Goddard, J. M., Martin, S. C., Fauron, C. M. R., and Wolstenholme, D. R. 1982. *Drosophila* mitochondrial DNA: A novel gene order. *Nucleic Acids Res.* **10**:6619–6637.

Coen, E. S., and Dover, G. A. 1982. Multiple Pol I initiation sequences in rDNA spacers of *Drosophila melanogaster. Nucleic Acids Res.* **10**:7017–7026.

Connaughton, J. F., Kumar, A., and Lockard, R. E. 1984a. Nucleotide sequence and structure determination of rabbit 18S ribosomal RNA. *In:* Kumar, A., ed., *Eukaryotic Gene Expression*, New York, Plenum Press, pp. 203–221.

Connaughton, J. F., Raikar, A., Lockard, R. E., and Kumar, A. 1984b. Primary structure of rabbit 18S ribosomal RNA determined by direct RNA sequence analysis. *Nucleic Acids Res.* **12**:4731–4745.

Crews, S., and Attardi, G. 1980. The sequences of the small ribosomal RNA gene and the phenylalanine tRNA gene are joined end to end in human mitochondrial DNA. *Cell* **19**:775–784.

Darlix, J. L., and Rochaix, J. D. 1981. Nucleotide sequence and structure of cytoplasmic 5S RNA and 5.8S RNA of *Chlamydomonas reinhardii*. *Nucleic Acids Res.* **9**:1291–1299.

deBoer, H. A., Gilbert, S. F., and Nomura, M. 1979. DNA sequences of promoter regions for rRNA operons *rrnE* and *rrnA* in *E. coli*. *Cell* **17**:201–209.

de la Cruz, V. F., Lake, J. A., Simpson, A. M., and Simpson, L. 1985a. A minimal ribosomal RNA: Sequence and secondary structure of the 9S kinetoplast ribosomal RNA from *Leishmania tarentolae*. *Proc. Natl. Acad. Sci. USA* **82**:1401–1405.

de la Cruz, V. F., Simpson, A. M., Lake, J. A., and Simpson, L. 1985b. Primary sequence and partial secondary structure of the 12S kinetoplast (mitochondrial) ribosomal RNA from *Leishmania tarentolae:* Conservation of peptidyl-transferase structural elements. *Nucleic Acids Res.* **13**:2337–2356.

Din, N., Engberg, J., and Gall, J. G. 1982. The nucleotide sequence at the transcription termination site of the ribosomal RNA gene in *Tetrahymena thermophila*. *Nucleic Acids Res.* **10**:1503–1512.

Doolittle, W. F., and Pace, N. R. 1971. Transcriptional organization of the ribosomal RNA cistrons in *Escherichia coli*. *Proc. Natl. Acad. Sci. USA* **68**:1786–1790.

Dorfman, D. M., Lenardo, M. J., Reddy, L. V., Van der Ploeg, L. H. T., and Donelson, J. E. 1985. The 5.8S ribosomal RNA gene of *Trypanosoma brucei:* Structural and transcriptional studies. *Nucleic Acids Res.* **13**:3533–3549.

Douglas, S. E., and Doolittle, W. F. 1984. Complete nucleotide sequence of the 23S rRNA gene of the cyanobacterium, *Anacystis nidulans*. *Nucleic Acids Res.* **12**:3373–3386.

Dron, M., Rahire, M., and Rochaix, J. D. 1982. Sequence of the chloroplast 16S rRNA gene and its surrounding regions of *Chlamydomonas reinhardii*. *Nucleic Acids Res.* **10**:7607–7620.

Dubin, D. T., Montoya, J., Timko, K. D., and Attardi, G. 1982. Sequence analysis and precise mapping of the 3' ends of HeLa cell mitochondrial ribosomal RNAs. *J. Mol. Biol.* **157**:1–19.

Duester, G. L., and Holmes, W. M. 1980. The distal end of the ribosomal RNA operon *rrnD* of *E. coli* contains a tRNA^{Thr} gene, two 5S rRNA genes, and a transcription terminator. *Nucleic Acids Res.* **8**:3793–3807.

Dujon, B. 1980. Sequence of the intron and flanking exons of the mitochondrial 21S rRNA gene of yeast strains having different alleles at the and *rib*-1 loci. *Cell* **20**:185–197.

Dyer, T. A., and Bowman, C. M. 1976. A sequence analysis of low-molecular-weight rRNA from chloroplasts of flowering plants. *In:* Bucher, T., *et al.*, eds., *Genetics and Biogenesis of Chloroplasts and Mitochondria,* Amsterdam, Elsevier/North-Holland Biomedical Press, pp. 645–651.

Edwards, K., and Kössel, H. 1981. The rRNA operon from *Zea mays* chloroplasts: Nucleotide sequence of 23S rDNA and its homology with *E. coli* 23S rDNA. *Nucleic Acids Res.* **9**:2853–2869.

Edwards, K., Bedbrook, J., Dyer, T., and Kössel, H. 1981. 4.5S rRNA from *Zea mays* chloroplasts shows structural homology with the 3' end of prokaryotic 23S rRNA. *Biochem. Int.* **2**:533–538.

Edwards, J. C., Levens, D., and Rabinowitz, M. 1982. Analysis of transcriptional initiation of yeast mitochondrial DNA in a homologous *in vitro* transcription system. *Cell* **31**:337–346.

El-Gewely, M. R., Helling, R. B., and Dibbits, J. G. T. 1984. Sequence and evolution of the regions between the *rrn* operons in the chloroplast genome of *Euglena gracilis bacillaris*. *Mol. Gen. Genet.* **94**:432–443.

Elion, E. A., and Warner, J. R. 1984. The major promoter element of rRNA transcription in yeast lies 2kb upstream. *Cell* **39**:663–673.

Engberg, J., Andersson, P., Leick, V., and Collins, J. 1976. Free ribosomal DNA molecules from *Tetrahymena pyriformis* GL are giant palindromes. *J. Mol. Biol.* **104**:455–470.

Eperon, I. C., Janssen, J. W. G., Hoeijmakers, J. H. J., and Borst, P. 1983. The major transcripts of the kinetoplast DNA of *Trypanosoma brucei* are very small ribosomal RNAs. *Nucleic Acids Res.* **11**:105–125.

Erdmann, V. A. 1975. Structure and function of 5S and 5.8S RNA. *In:* Cohn, W. E., ed., *Progress in Nucleic Acids Research and Molecular Biology,* New York, Academic Press, Vol. 18, pp. 45–90.

Erdmann, V. A., Huysmans, E., Vandenberghe, A., and DeWachter, R. 1984. Collection of published 5S and 5.8S ribosomal RNA sequences. *Nucleic Acids Res.* **12**(suppl.):r133–r166.

Feng, Y. X., Krupp, G., and Gross, H. J. 1982. The nucleotide sequence of 5.8S rRNA from the posterior silk gland of the silkworm *Philosamia cynthia ricini*. *Nucleic Acids Res.* **10**:6383–6387.

Ferris, P. J., and Vogt, V. M. 1982. Structure of the central spacer region of extrachromosomal ribosomal DNA in *Physarum polycephalum*. *J. Mol. Biol.* **159**:359–381.

Files, J. G., and Hirsh, D. 1981. Ribosomal DNA of *Caenorhabditis elegans*. *J. Mol. Biol.* **149**:223–240.

Financsek, I., Mizumoto, K., Mishima, Y., and Muramatsu, M. 1982. Human ribosomal RNA gene: Nucleotide sequence of the transcription initiation region and comparison of three mammalian genes. *Proc. Natl. Acad. Sci. USA* **79**:3092–3096.

Franz, G., Kunz, W., and Grimm, C. 1981. Determination of the region of rDNA involved in polytenization in salivary glands of *Drosophila hydei*. *Mol. Gen. Genet.* **191**:74–80.

Fujiwara, H., Kawata, Y., and Ishikawa, H. 1982. Primary and secondary structure of 5.8S rRNA from the silkgland of *Bombyx mori*. *Nucleic Acids Res.* **10**:2415–2418.

Gall, J. G., Karrer, K. M., Yao, M. C., and Grainger, R. 1977. The ribosomal RNA genes in *Tetrahymena*. *In:* Bradbury, E. M., and Javaherian, K., eds., *The Organization and Expression of the Eukaryotic Genome*, New York, Academic Press, pp. 437–444.

Gallant, J. A. 1979. Stringent control in *E. coli*. *Annu. Rev. Genet.* **13**:393–415.

Garriga, G., Bertrand, H., and Lambowitz, A. M. 1984. RNA splicing in *Neurospora* mitochondria nuclear mutants defective in both splicing and 3′ end synthesis of the large rRNA. *Cell* **36**:623–634.

Georgiev, O. I., Nikolaev, N., Hadjiolov, A. A., Skryabin, K. G., Zakharyev, V. M., and Bayev, A. A. 1981. The structure of the yeast ribosomal RNA genes. 4. Complete sequence of the 25S rRNA gene from *Saccharomyces cerevisiae*. *Nucleic Acids Res.* **9**:6953–6958.

Gerlach, W. L., and Bedbrook, J. R. 1979. Cloning and characterization of ribosomal RNA genes from wheat and barley. *Nucleic Acids Res.* **7**:1869–1885.

Gilbert, S. F., deBoer, H. A., and Nomura, M. 1979. Identification of initiation sites for the *in vitro* transcription of rRNA operons *rrnE* and *rrnA* in *E. coli*. *Cell* **17**:211–224.

Glaser, G., and Cashel, M. 1979. *In vitro* transcripts from the *rrnB* ribosomal RNA cistron originate from two tandem promoters. *Cell* **16**:111–121.

Glaser, G., Sarmientos, P., and Cashel, M. 1983. Functional interrelationship between two tandem *E. coli* ribosomal RNA promoters. *Nature (London)* **302**:74–76.

Goddard, J. M., Fauron, C. M. R., and Wolstenholme, D. R. 1982. Nucleotide sequences within the A+T-rich region and the large rRNA gene of mitochondrial DNA molecules of *Drosophila yakuba*. *In:* Slonimski, P., Borst, P., and Attardi, G., eds., *Mitochondrial Genes*, Cold Spring Harbor, New York, Cold Spring Harbor Laboratory, pp. 100–103.

Gottlieb, P., LaFauci, G., and Rudner, R. 1985. Alterations in the number of rRNA operons within the *Bacillus subtilis* genome. *Gene* **33**:259–268.

Gourse, R. L., and Nomura, M. 1984. Level of rRNA, not tRNA, synthesis controls transcription of rRNA and tRNA operons in *Escherichia coli*. *J. Bacteriol.* **160**:1022–1026.

Graf, L., Kössel, H., and Stutz, E. 1980. Sequencing of 16S–23S spacer in a ribosomal RNA operon of *Euglena gracilis* chloroplast DNA reveals two tRNA genes. *Nature (London)* **286**:908–910.

Graf, L., Roux, E., and Stutz, E. 1982. Nucleotide sequence of a *Euglena gracilis* chloroplast gene coding for the 16S rRNA: Homologies to *E. coli* and *Zea mays* chloroplast 16S rRNA. *Nucleic Acids Res.* **10**:6369–6381.

Gray, M. W., Bonen, L., Falconet, D., Huh, T. Y., Schnare, M. N., and Spencer, D. F. 1982. Mitochondrial ribosomal RNAs of *Triticum aestivum* (wheat): Sequence analysis and gene organization. *In:* Slonimski, P., Borst, P., and Attardi, G., eds., *Mitochondrial Genes*, Cold Spring Harbor, New York, Cold Spring Harbor Laboratory, pp. 483–488.

Gray, M. W., Sankoff, D., and Cedergren, R. J. 1984. On the evolutionary descent of organisms and organelles: A global phylogeny based on a highly conserved structural core in small subunit ribosomal RNA. *Nucleic Acids Res.* **12**:5837–5852.

Gray, P. W., and Hallick, R. B. 1979. Isolation of *Euglena* chloroplast 5S ribosomal RNA and mapping the 5S rRNA gene on chloroplast DNA. *Biochemistry* **18**:1820–1825.

Green, M. R., Grimm, M. F., Goewert, R. R., Collins, R. A., Cole, M. D., Lambowitz, A. M., Heckman, J. E., Yin, S., and RajBhandary, U. L. 1981. Transcripts and processing patterns for the rRNA and tRNA region of *Neurospora crassa* mitochondrial DNA. *J. Biol. Chem.* **256**:2027–2034.

Gupta, R., Lanter, J. M., and Woese, C. R. 1983. Sequence of the 16S ribosomal RNA from *Halobacterium volcanii*, an archaebacterium. *Science* **221**:656–659.

Hamming, J., Geert, A. B., and Gruber, M. 1980. *E. coli* RNA polymerase–rRNA promoter interaction and the effect of ppGpp. *Nucleic Acids Res.* **8**:3947–3963.

Harrington, C. A., and Chikaraishi, D. M. 1983. Identification and sequence of the initiaton site for rat 45S ribosomal RNA synthesis. *Nucleic Acids Res.* **11**:3317–3332.

Hassouna, N., Michot, B., and Bachellerie, J. P. 1984. The complete nucleotide sequence of mouse 28S rRNA gene. Implications for the process of size increase of the large subunit rRNA in higher eukaryotes. *Nucleic Acids Res.* **12**:3563–3583.

Hauswirth, W. W., Van de Walle, M. J., Laipis, P. J., and Olivo, P. D. 1984. Heterogeneous mitochondrial DNA D-loop sequences in bovine tissue. *Cell* **37**:1001–1007.

Heckman, J. E., and RajBhandary, U. L. 1979. Organization of tRNA and rRNA genes in *N. crassa* mitochondria: Intervening sequence in the large rRNA gene and distribution of the RNA genes. *Cell* **17**:583–595.

Henckes, G., Vannier, F., Seiki, M., Ogasawara, N., Yoshikawa, H., and Seror-Laurent, S. J. 1982. Ribosomal RNA genes in the replication origin region of *Bacillus subtilis* chromosome. *Nature (London)* **299**:268–271.

Henning, W., and Meer, B. 1971. Reduced polyteny of ribosomal RNA cistrons in giant chromosomes of *Drosophila hydei*. *Nature New Biol.* **233**:70–72.

Hill, C. W., and Harnish, B. W. 1981. Inversions between ribosomal RNA genes of *E. coli*. *Proc. Natl. Acad. Sci. USA* **78**:7069–7072.

Hindenach, B. R., and Stafford, D. W. 1984. Nucleotide sequence of the 18S–26S rRNA intergenic region of the sea urchin. *Nucleic Acids Res.* **12**:1737–1747.

Hoshikawa, Y., Iida, Y., and Iwabuchi, M. 1983. Nucleotide sequence of the transcriptional initiation region of *Dictyostelium discoideum* rRNA gene and comparison of the initiation regions of three lower eukaryotes' genes. *Nucleic Acids Res.* **11**:1725–1734.

Iams, K. P., and Sinclair, J. H. 1982. Mapping the mitochondrial DNA of *Zea mays:* Ribosomal gene localization. *Proc. Natl. Acad. Sci. USA* **79**:5926–5929.

Iida, C. T., Kownin, P., and Paule, M. R. 1985. Ribosomal RNA transcription: Proteins and DNA sequences involved in preinitiation complex formation. *Proc. Natl. Acad. Sci. USA* **82**:1668–1672.

Ingle, J., Timmis, J. N., and Sinclair, J. 1975. The relationship between satellite deoxyribonucleic acid, ribosomal ribonucleic acid gene redundancy, and genome size in plants. *Plant Physiol.* **55**:496–501.

Jacq, B. 1981. Sequence homologies between eukaryotic 5.8S rRNA and the 5' end of prokaryotic 23S rRNA: Evidences for a common evolutionary origin. *Nucleic Acids Res.* **9**:2913–2932.

Jamrich, M., and Miller, O. L. 1984. The rare transcripts of interrupted rRNA genes in *Drosophila melanogaster* are processed or degraded during synthesis. *EMBO J.* **3**:1541–1545.

Jarsch, M., Altenbuchner, J., and Böck, A. 1983. Physical organization of the genes for ribosomal RNA in *Methanococcus vannielii*. *Mol. Gen. Genet.* **189**:41–47.

Jenni, B., and Stutz, E. 1979. Analysis of *Euglena gracilis* chloroplast DNA. Mapping of a DNA sequence complementary to 16S rRNA outside of the three rRNA gene sets. *FEBS Lett.* **102**:95–99.

Jordan, B. R., Latil-Damotte, M., and Jourdan, R. 1980. Coding and spacer sequences in the 5.8 S–2 S region of *Sciara coprophila* ribosomal DNA. *Nucleic Acids Res.* **8**:3565–3573.

Jurgenson, J. E., and Bourque, D. P. 1980. Mapping of rRNA genes in an inverted repeat in *Nicotiana tabacum* chloroplast DNA. *Nucleic Acids Res.* **8**:3505–3516.

Kan, N. C., and Gall, J. G. 1982. The intervening sequence of the ribosomal RNA gene is highly conserved between two *Tetrahymena* species. *Nucleic Acids Res.* **10**:2809–2822.

Karrer, K. M., and Gall, J. G. 1976. The macronuclear ribosomal DNA of *Tetrahymena pyriformis* is a palindrome. *J. Mol. Biol.* **104**:421–453.

Keller, M., Burkard, G., Bohnert, H. J., Mubumbila, M., Gordon, K., Steinmetz, A., Heiser, D., Crouse, E. J. and Weil, J. H. 1980. Transfer RNA genes associated with the 16S and 23S rRNA genes of *Euglena* chloroplast DNA. *Biochem. Biophys. Res. Commun.* **95**:47–54.

Keus, R. J. A., Dekker, A. F., van Roon, M. A., and Groot, G. S. P. 1983. The nucleotide sequences of the regions flanking the genes coding for 23S, 16S, and 4.5S rRNA on the chloroplast DNA from *Spirodela oligorhiza*. *Nucleic Acids Res.* **11**:6465–6474.

Khan, M. S. N., and Maden, B. E. H. 1977. Nucleotide sequence relationship between vertebrate 5.8S rRNAs. *Nucleic Acids Res.* **4**:2495–2505.

Kishimoto, T., Nagamine, M., Sasaki, T., Takakusa, N., Miwa, T., Kominami, R., and Muramatsu, M. 1985. Presence of a limited number of essential nucleotides in the promoter region of mouse ribosomal RNA gene. *Nucleic Acids Res.* **13**:3515–3532.

Klemenz, R., and Geiduschek, E. P. 1980. The 5' terminus of the precursor ribosomal RNA of *Saccharomyces cerevisiae*. *Nucleic Acids Res.* **8**:2679–2689.

Klootwijk, J., Verbeet, M. P., Veldman, G. M., de Regt, V. C. H. F., van Heerikhuizen, H., Bogerd, J., and Planta, R. J. 1984. The *in vivo* and *in vitro* initiation site for transcription of the rRNA operon of *Saccharomyces carlsbergensis*. *Nucleic Acids Res.* **12**:1377–1390.

Kobayashi, H., and Osawa, S. 1982. The number of 5S rRNA genes in *Bacillus subtilis. FEBS Lett.* **141**:161–163.

Koch, W., Edwards, K., and Kössel, H. 1981. Sequencing of the 16S–23S spacer in a ribosomal RNA operon of *Zea mays* chloroplast DNA reveals two split tRNA genes. *Cell* **25**:203–213.

Kohorn, B. D., and Rae, P. M. M. 1982a. Accurate transcription of truncated ribosomal DNA templates in a *Drosophila* cell-free system. *Proc. Natl. Acad. Sci. USA* **79**:1501–1504.

Kohorn, B. D., and Rae, P. M. M. 1982b. Nontranscribed spacer sequences promote *in vitro* transcription of *Drosophila* ribosomal DNA. *Nucleic Acids Res.* **10**:6879–6886.

Kohorn, B. D., and Rae, P. M. M. 1983. A component of *Drosophila* RNA polymerase I promoter lies within the rRNA transcription unit. *Nature (London)* **304**:179–181.

Koller, B., and Delius, H. 1982. Electron microscopic analysis of the extra 16S rRNA gene and its neighbourhood in chloroplast DNA from *Euglena gracilis* strain Z. *FEBS Lett.* **139**:86–92.

Kominami, R., Mishima, Y., Urano, Y., Sakai, M., and Muramatsu, M. 1982. Cloning and determination of the transcription termination site of ribosomal RNA gene of the mouse. *Nucleic Acids Res.* **10**:1963–1980.

Kotin, R. M., and Dubin, D. T. 1984. Sequences around the 3' end of a ribosomal RNA gene of hamster mitochondria. Further support for the transcriptional attenuation "model." *Biochim. Biophys. Acta* **782**:106–108.

Kruger, K., Grabowski, P. J., Zaug, A. J., Sands, J., Gottschling, D. E., and Cech, T. R. 1982. Self-splicing RNA: Autoexcision and autocyclization of the ribosomal RNA intervening sequence of *Tetrahymena. Cell* **31**:147–157.

Kumagai, I., Pieler, T., Subramanian, A. R., and Erdmann, V. A. 1982. Nucleotide sequence and secondary structure analysis of spinach chloroplast 4.5S RNA. *J. Biol. Chem.* **257**:12924–12928.

Kumano, M., Tomioka, N., and Sugiura, M. 1983. The complete nucleotide sequence of a 23S rRNA gene from a blue-green alga, *Anacystis nidulans. Gene* **24**:219–225.

Kunz, W., Grimm, C., and Franz, G. 1982. Amplification and synthesis of rDNA: *Drosophila. In:* Busch, H., and Rothblum, L., eds., *The Cell Nucleus,* New York, Academic Press, Vol. 12, pp. 155–184.

Kusuda, J., Shinozaki, K., Takaiwa, F., and Sugiura, M. 1980. Characterization of the cloned ribosomal DNA of tobacco chloroplasts. *Mol. Gen. Genet.* **178**:1–7.

Lecanidou, R., Eickbush, T. H., and Kafatos, F. C. 1984. Ribosomal DNA genes of *Bombyx mori:* A minor fraction of the repeating units contain introns. *Nucleic Acids Res.* **12**:4703–4713.

Lee, J. C., Henry, B., and Yeh, Y. C. 1983. Binding of proteins from the large ribosomal subunits to 5.8S rRNA of *Saccharomyces cerevisiae. J. Biol. Chem.* **258**:854–858.

Leffers, H., and Garrett, R. A. 1984. The nucleotide sequence of the 16S ribosomal RNA gene of the archaebacterium *Halococcus morrhua. EMBO J.* **3**:1613–1619.

Levens, D., Morimoto, M., and Rabinowitz, M. 1981a. Mitochondrial transcription complex from *Saccharomyces cerevisiae. J. Biol. Chem.* **256**:1466–1473.

Levens, D., Ticho, B., Ackerman, E., and Rabinowitz, M. 1981b. Transcriptional initiation and 5' termini of yeast mitochondrial RNA. *J. Biol. Chem.* **256**:5226–5232.

Leweke, B., and Hemleben, V. 1982. Organization of rDNA in chromatin: Plants. *In:* Busch, H., and Rothblum, L., eds., *The Cell Nucleus,* New York, Academic Press, Vol. XI, pp. 225–253.

Liu, W., Lo, A. C., and Nazar, R. N. 1983. Structure of the ribosome-associated 5.8S ribosomal RNA. *J. Mol. Biol.* **171**:217–224.

Long, E. O., and Dawid, I. B. 1979. Expression of ribosomal DNA insertions in *Drosophila melanogaster. Cell* **18**:1185–1196.

Long, E. O., Rebbert, M. L., and Dawid, I. B. 1981. Nucleotide sequence of the initiation site for ribosomal RNA transcription in *Drosophila melanogaster:* Comparison of genes with and without insertions. *Proc. Natl. Acad. Sci. USA* **78**:1513–1517.

Loughney, K., Lund, E., and Dahlberg, J. E. 1983. Ribosomal RNA precursors of *Bacillus subtilis. Nucleic Acids Res.* **11**:6709–6721.

Machatt, M. A., Ebel, J. P., and Branlant, C. 1981. The 3'-terminal region of bacterial 23S rRNA: Structure and homology with the 3'-terminal region of eukaryotic 28S rRNA and with chloroplast 4.5S rRNA. *Nucleic Acids Res.* **9**:1533–1541.

Mankin, A. S., Teterina, N. L., Rubtsov, P. M., Baratova, L. A., and Kagramanova, V. K. 1984. Putative promoter region of rRNA operon from archaebacterium *Halobacterium halobium. Nucleic Acids Res.* **12**:6537–6546.

McCarroll, R., Olsen, G. J., Stahl, Y. D., Woese, C. R., and Sogin, M. L. 1983. Nucleotide sequence of the

Dictyostelium discoideum small-subunit ribosomal ribonucleic acid inferred from the gene sequence: Evolutionary implications. *Biochemistry* **22**:5858–5868.

Michot, B., Bachellerie, J. P., and Raynal, F. 1983. Structure of mouse rRNA precursors. Complete sequences and potential folding of the spacer regions between 18S and 28S rRNA. *Nucleic Acids Res.* **11**:3375–3391.

Miesfeld, R., and Arnheim, N. 1982. Identification of the *in vivo* and *in vitro* origin of transcription in human rDNA. *Nucleic Acids Res.* **10**:3933–3949.

Miller, J. R., Hayward, D. C., and Glover, D. M. 1983. Transcription of the "nontranscribed" spacer of *Drosophila melanogaster* rDNA. *Nucleic Acids Res.* **11**:11–19.

Miller, K. G., and Sollner-Webb, B. 1982. Transcription of mouse ribosomal RNA genes. *In:* Busch, H., and Rothblum, L., eds., *The Cell Nucleus,* New York, Academic Press, Vol. XII, pp. 69–100.

Mishima, Y., Yamamoto, O., Kominami, R., and Muramatsu, M. 1981. *In vitro* transcription of a cloned mouse ribosomal RNA gene. *Nucleic Acids Res.* **9**:6773–6785.

Montoya, J., Christianson, T., Levens, D., Rabinowitz, M., and Attardi, G. 1982. Identification of initiation sites for heavy strand and light strand transcription in human mitochondrial DNA. *Proc. Natl. Acad. Sci. USA* **79**:7195–7199.

Montoya, J., Gaines, G. L., and Attardi, G. 1983. The pattern of transcription of the human mitochondrial rRNA genes reveals two overlapping transcription units. *Cell* **34**:151–159.

Morgan, E. A. 1982. Ribosomal RNA genes in *Escherichia coli. In:* Busch, H., and Rothblum, L., eds., *The Cell Nucleus,* New York, Academic Press, Vol. X, pp. 1–29.

Morgan, G. T., Roan, J. G., Bakken, A. H., and Reeder, R. H. 1984. Variations in transcriptional activity of rDNA spacer promoters. *Nucleic Acids Res.* **12**:6043–6052.

Muto, A. 1981. Control of ribosomal RNA synthesis in *Escherichia coli.* V. Stimulation of *rrnC* gene transcription *in vitro* by a protein factor. *Mol. Gen. Genet.* **181**:69–73.

Nazar, R. N. 1980. A 5.8S rRNA-like sequence in prokaryotic 23S rRNA. *FEBS Lett.* **119**:212–214.

Nazar, R. N. 1982. The eukaryotic 5.8 and 5S ribosomal RNAs and related rDNAs. *In:* Busch, H., and Rothblum, L., eds., *The Cell Nucleus,* New York, Academic Press, Vol. XIB, pp. 1–28.

Nazar, R. N., Sitz, T. O., and Busch, H. 1975. Structural analyses of mammalian ribosomal ribonucleic acid and its precursors. Nucleotide sequence of ribosomal 5.8S RNA. *J. Biol. Chem.* **250**:8591–8597.

Nelles, L., Fang, B. L., Volckaert, G., Vandenberghe, A., and De Wachter, R. 1984. Nucleotide sequence of a crustacean 18S ribosomal RNA gene and secondary structure of eukaryotic small subunit ribosomal RNAs. *Nucleic Acids Res.* **12**:8749–8768.

Neumann, H., Gierl, A., Tu, J., Leibrock, J., Staiger, D., and Zillig, W. 1983. Organization of the genes for ribosomal RNA in archaebacteria. *Mol. Gen. Genet.* **192**:66–72.

Niles, E. G. 1978. Isolation of a high specific activity 35S ribosomal RNA precursor from *Tetrahymena pyriformis* and identification of its 5′ terminus, pppAp. *Biochemistry* **17**:4839–4844.

Niles, E. G., Sutiphong, J., and Haque, S. 1981. Structure of the *Tetrahymena pyriformis* rRNA gene. Nucleotide sequence of the transcription initiation region. *J. Biol. Chem.* **256**:12849–12856.

Nomiyama, H., Sakaki, Y., and Takagi, Y. 1981a. Nucleotide sequence of a ribosomal RNA gene intron from slime mold *Physarum polycephalum. Proc. Natl. Acad. Sci. USA* **78**:1376–1380.

Nomiyama, H., Kuhara, S., Kukita, T., Otsuka, T., and Sakaki, Y. 1981b. Nucleotide sequence of the ribosomal RNA gene of *Physarum polycephalum:* Intron 2 and its flanking regions of the 26S rRNA gene. *Nucleic Acids Res.* **9**:5507–5520.

Ohta, N., and Newton, A. 1981. Isolation and mapping of ribosomal RNA genes of *Caulobacter crescentus. J. Mol. Biol.* **153**:291–303.

Ojala, D., Merkel, C., Gelfand, R., and Attardi, G. 1980. The tRNA genes punctuate the reading of genetic information in human mitochondrial DNA. *Cell* **22**:393–403.

Ojala, D., Montoya, J., and Attardi, G. 1981. tRNA punctuation model of RNA processing in human mitochondria. *Nature (London)* **290**:470–474.

Olsen, G. J., and Sogin, M. L. 1982. Nucleotide sequence of *Dictyostelium discoideum* 5.8S ribosomal ribonucleic acid: Evolutionary and secondary structural implications. *Biochemistry* **21**:2335–2343.

Oostra, B. A., van Ooyen, A. J. J., and Gruber, M. 1977. *In vitro* transcription of three different ribosomal RNA cistrons of *E. coli:* Heterogeneity of control regions. *Mol. Gen. Genet.* **152**:1–6.

Oostra, B. A., Geert, A. B., and Gruber, M. 1980. Specific stimulation of ribosomal RNA synthesis in *E. coli* by a protein factor. *Mol. Gen. Genet.* **177**:291–295.

Orozco, E. M., Gray, P. W., and Hallick, R. B. 1980. *Euglena gracilis* chloroplast ribosomal RNA transcription units. I. The location of transfer RNA, 5S, 16S, and 23S ribosomal RNA genes. *J. Biol. Chem.* **255:**10991–10996.

Osinga, K. A., DeHaan, M., Christianson, T., and Tabak, H. F. 1982. A nonanucleotide sequence involved in promotion of ribosomal RNA synthesis and RNA priming of DNA replication in yeast mitochondria. *Nucleic Acids Res.* **10:**7993–8006.

Otsuka, T., Nomiyama, H., Sakaki, Y., and Takagi, Y. 1982. Nucleotide sequence of *Physarum polycephalum* 5.8S rRNA gene and its flanking region. *Nucleic Acids Res.* **10:**2379–2385.

Otsuka, T., Nomiyama, H., Yoshida, H., Kukita, T., Kuhara, S., and Sakaki, Y. 1983. Complete nucleotide sequence of the 26S rRNA gene of *Physarum polycephalum:* Its significance in gene evolution. *Proc. Natl. Acad. Sci. USA* **80:**3163–3167.

Ozaki, T., Hoshikawa, Y., Iida, Y., and Iwabuchi, M. 1984. Sequence analysis of the transcribed and 5' nontranscribed regions of the ribosomal RNA gene in *Dictyostelium discoideum. Nucleic Acids Res.* **12:**4171–4184.

Passananti, C., Felsani, A., Giordano, R., Metafora, S., and Spadafora, C. 1983. Cloning and characterization of the ribosomal genes of the sea-urchin *Paracentrotus lividus.* Heterogeneity of the multigene family. *Eur. J. Biochem.* **137:**233–239.

Paule, M. R., Iida, C. T., Perna, P. J., Harris, G. H., Knoll, D. A., and D'Alessio, J. M. 1984. *In vitro* evidence that eukaryotic ribosomal transcription is regulated by modification of RNA polymerase I. *Nucleic Acids Res.* **12:**8161–8180.

Pavlakis, G. N., Jordan, B. R., Wurst, R. M., and Vournakis, J. N. 1979. Sequence and secondary structure of *Drosophilia melanogaster* 5.8S and 2S rRNAs and of the processing site between them. *Nucleic Acids Res.* **7:**2213–2238.

Pene, J. J., Knight, E., and Darnell, J. E. 1968. Characterization of a new low molecular weight RNA in HeLa cell ribosomes. *J. Mol. Biol.* **33:**609–623.

Rafalski, J. A., Wiewiórowski, M., and Söll, D. 1983. Organization of ribosomal DNA in yellow lupine (*Lupinus luteus*) and sequence of the 5.8S RNA gene. *FEBS Lett.* **152:**241–246.

Raynal, F., Michot, B., and Bachellerie, J. P. 1984. Complete nucleotide sequence of mouse 18S rRNA gene: Comparison with other available homologs. *FEBS Lett.* **167:**263–268.

Reichel, R., Monstein, H., Jansen, H., Philipson, L., and Benecke, B. 1982. Small nuclear RNAs are encoded in the nontranscribed region of ribosomal spacer DNA. *Proc. Natl. Acad. Sci. USA* **79:**3106–3110.

Roberts, J. W., Grula, J. W., Posakony, J. W., Hudspeth, R., Davidson, E. H., and Britten, R. J. 1983. Comparison of sea urchin and human mtDNA: Evolutionary rearrangement. *Proc. Natl. Acad. Sci. USA* **80:**4614–4618.

Rochaix, J. D., and Malnoe, P. 1978. Anatomy of the chloroplast ribosomal DNA of *Chlamydomonas reinhardii. Cell* **15:**661–670.

Rothblum, L. I., Reddy, R., and Cassidy, B. 1982. Transcription initiation site of rat ribosomal DNA. *Nucleic Acids Res.* **10:**7345–7362.

Roux, E., Graf, L., and Stutz, E. 1983. Nucleotide sequence of a truncated rRNA operon of the *Euglena* chloroplast genome. *Nucleic Acids Res.* **11:**1957–1968.

Rubin, G. M. 1973. The nucleotide sequence of *Saccharomyces cerevisiae* 5.8S ribosomal ribonucleic acid. *J. Biol. Chem.* **248:**3860–3875.

Rubin, G. M. 1974. Three forms of the 5.8-S ribosomal RNA species in *Saccharomyces cerevisiae. Eur. J. Biochem.* **41:**197–202.

Rubtsov, P. M., Musakhanov, M. M., Zakharyev, V. M., Krayev, A. S., Skryabin, K. G., and Bayev, A. A. 1980. The structure of the yeast ribosomal RNA genes. I. The complete nucleotide sequence of the 18S ribosomal gene from *Saccharomyces cerevisiae. Nucleic Acids Res.* **8:**5779–5794.

Rungger, D., Achermann, H., and Crippa, M. 1979. Transcription of spacer sequences in genes coding for ribosomal RNA in *Xenopus* cells. *Proc. Natl. Acad. Sci. USA* **76:**3957–3961.

Saiga, H., Mizumoto, K., Matsui, T., and Higashinakagawa, T. 1982. Determination of the transcription initiation site of *Tetrahymena pyriformis* rDNA using *in vitro* capping of 35S pre-rRNA. *Nucleic Acids Res.* **10:**4223–4236.

Salim, N., and Maden, B. E. H. 1981. Nucleotide sequence of *Xenopus laevis* 18S ribosomal RNA inferred from gene sequence. *Nature (London)* **291:**205–208.

Schaak, J., Mao, J. I., and Söll, D. 1982. The 5.8S RNA gene sequence and the ribosomal repeat of *Schizosaccharomyces pombe. Nucleic Acids Res.* **10:**2851–2864.

Schwarz, Z., and Kössel, H. 1980. The primary structure of 16S rRNA from *Zea mays* chloroplast is homologous to *E. coli* 16S rRNA. *Nature (London)* **283**:739–742.

Schwarz, Z., Kössel, H., Schwarz, E., and Bogorad, L. 1981. A gene coding for tRNA^{Val} is located near 5′ terminus of 16S rRNA gene in *Zea mays* chloroplast genome. *Proc. Natl. Acad. Sci. USA* **78**:4748–4752.

Seebeck, T., and Braun, R. 1982. Organization of rDNA in chromatin: *Physarum*. *In:* Busch, H., and Rothblum, L., eds., *The Cell Nucleus,* New York, Academic Press, Vol. XI, pp. 177–191.

Seilhamer, J. J., and Cummings, D. J. 1981. Structure and sequence of the mitochondrial 20S rRNA and tRNA^{Tyr} gene of *Paramecium primaurelia*. *Nucleic Acids Res.* **9**:6391–6406.

Seilhamer, J. J., Olsen, G. J., and Cummings, D. J. 1984. *Paramecium* mitochondrial genes. I. Small subunit rRNA gene sequence and microevolution. *J. Biol. Chem.* **259**:5167–5172.

Sekiya, T., Mori, M., Takahashi, N., and Nishimura, S. 1980. Sequence of the distal tRNA$^{Asp}_1$ gene and the transcription termination signal in the *Escherichia coli* RNA operon *rrnF* (or *G*). *Nucleic Acids Res.* **8**:3809–3827.

Selker, E., and Yanofsky, C. 1979. Nucleotide sequence and conserved features of the 5.8 rRNA coding region of *Neurospora crassa*. *Nucleic Acids Res.* **6**:2561–2567.

Shen, W. F., Squires, C., and Squires, C. L. 1982. Nucleotide sequence of the *rrnG* ribosomal RNA promoter region of *Escherichia coli*. *Nucleic Acids Res.* **10**:3303–3313.

Siebenlist, U., Simpson, R. B., and Gilbert, W. 1980. *E. coli* RNA polymerase interacts homologously with two different promoters. *Cell* **20**:269–281.

Smith, S. D., Banerjee, N., and Sitz, T. O. 1984. Gene heterogeneity: A basis for alternative 5.8S rRNA processing. *Biochemistry* **23**:3648–3652.

Sollner-Webb, B., and Reeder, R. H. 1979. The nucleotide sequence of the initiation and termination sites for ribosomal RNA transcription in *X. laevis*. *Cell* **18**:485–499.

Sollner-Webb, B., Wilkinson, J. A. K., and Miller, K. G. 1982. Transcription of *Xenopus* rRNA genes. *In:* Busch, H., and Rothblum, L., eds., *The Cell Nucleus,* New York, Academic Press, Vol. XII, pp. 31–67.

Sor, F., and Fukuhara, H. 1980. Séquence nucléotidique du gêne de l'ARN ribosomique 15S mitochondrial de la levure. *C. R. Acad. Sci. (Paris)* **291**:933–936.

Sor, F., and Fukuhara, H. 1982. Nucleotide sequence of the small ribosomal RNA gene from the mitochondria of *Saccharomyces cerevisiae*. *In:* Slonimski, P., Borst, P., and Attardi, G., eds., *Mitochondrial Genes,* Cold Spring Harbor, New York, Cold Spring Harbor Laboratory, pp. 257–262.

Sor, F., and Fukuhara, H. 1983. Complete DNA sequence coding for the large ribosomal RNA of yeast mitochondria. *Nucleic Acids Res.* **11**:339–348.

Spangler, E. A., and Blackburn, E. H. 1985. The nucleotide sequence of the 17S ribosomal RNA gene of *Tetrahymena thermophila* and the identification of point mutations resulting in resistance to the antibiotics paromomycin and hygromycin. *J. Biol. Chem.* **260**:6334–6340.

Spear, B. B., and Gall, J. G. 1973. Independent control of ribosomal gene replication in polytene chromosomes of *Drosophila melanogaster*. *Proc. Natl. Acad. Sci. USA* **70**:1359–1363.

Stern, D. B., and Palmer, J. D. 1984. Recombination sequences in plant mitochondrial genomes: Diversity and homologies to known mitochondrial genes. *Nucleic Acids Res.* **12**:6141–6157.

Stewart, G. C., and Bott, K. F. 1983. DNA sequence of the tandem ribosomal RNA promoter for *B. subtilis* operon *rrnB*. *Nucleic Acids Res.* **11**:6289–6300.

Strittmatter, G., Gozdzicko-Jozefiak, A., and Kössel, H. 1985. Identification of an rRNA operon promoter from *Zea mays* chloroplasts which excludes the proximal tRNA$^{Val}_{GAC}$ from the primary transcript. *EMBO J.* **4**:599–604.

Takaiwa, F., and Sugiura, M. 1980. The nucleotide sequence for 4.5S ribosomal RNA from tobacco chloroplasts. *Nucleic Acids Res.* **8**:4125–4129.

Takaiwa, F., and Sugiura, M. 1982a. Nucleotide sequence of the 16S–23S spacer region in an rRNA gene cluster from tobacco chloroplast DNA. *Nucleic Acids Res.* **10**:2665–2676.

Takaiwa, F., and Sugiura, M. 1982b. The complete nucleotide sequence of a 23S rRNA gene from tobacco chloroplasts. *Eur. J. Biochem.* **124**:13–19.

Takaiwa, F., Kusuda, M., and Sugiura, M. 1982. The nucleotide sequence of chloroplast 4.5S rRNA from a fern, *Dryopteris acuminata*. *Nucleic Acids Res.* **10**:2257–2260.

Takaiwa, F., Oono, K., and Sugiura, M. 1984. The complete nucleotide sequence of a rice 17S rRNA gene. *Nucleic Acids Res.* **12**:5441–5448.

Tohdoh, N., and Sugiura, M. 1982. The complete nucleotide sequence of a 16S ribosomal RNA gene from tobacco chloroplasts. *Gene* **17**:213–218.

Tohdoh, N., Shinozaki, K., and Sugiura, M. 1981. Sequence of a putative promoter region for the rRNA genes of tobacco chloroplast DNA. *Nucleic Acids Res.* **9:**5399–5406.

Tomioka, N., and Sugiura, M. 1983. The complete nucleotide sequence of a 16S ribosomal RNA gene from a blue-green alga, *Anacystis nidulans. Mol. Gen. Genet.* **191:**46–50.

Torczynski, R., Bollon, A. T., and Fuke, M. 1983. The complete nucleotide sequence of the rat 18S ribosomal RNA gene and comparison with the respective yeast and frog genes. *Nucleic Acids Res.* **11:**4879–4890.

Troitskii, A. V., Bobrova, V. K., Ponomarev, A. G., and Antonov, A. S. 1984. The nucleotide sequence of chloroplast 4.5S rRNA from *Mnium rugicum* (Bryophyta): Mosses also possess this type of RNA. *FEBS Lett.* **176:**105–109.

Ulbrich, N., Kumagai, I., and Erdmann, V. A. 1984. The number of ribosomal RNA genes in *Thermus thermophilus. Nucleic Acids Res.* **12:**2055–2060.

Urano, Y., Kominami, R., Mishima, Y., and Muramatsu, M. 1980. The nucleotide sequence of the putative transcription initiation site of a cloned ribosomal RNA gene of the mouse. *Nucleic Acids Res.* **8:**6043–6058.

Valenzuela, P., Bell, G. I., Masiarz, F. R., De Gennaro, L. J., and Rutter, R. J. 1977. Nucleotide sequences of the yeast ribosomal RNA gene and adjacent putative control regions. *Nature (London)* **267:**641–643.

Van, N. T., Nazar, R. N., and Sitz, T. O. 1977. Comparative studies on the secondary structure of eukaryotic 5.8S rRNA. *Biochemistry* **16:**3754–3759.

Van Etten, R. A., Walberg, M. W., and Clayton, D. A. 1980. Precise localization and nucleotide sequence of the two mouse mitochondrial rRNA genes and three immediately adjacent novel tRNA genes. *Cell* **22:**157–170.

Van Knippenberg, P. H., Van Kimmenadi, J. M., and Heus, H. A. 1984. Phylogeny of the conserved 3′ terminal structure of the RNA of small ribosomal subunits. *Nucleic Acids Res.* **12:**2595–2604.

Vaughn, J. C., Speibeck, S. J., Ramsey, W. J., and Lawrence, C. B. 1984. A universal model for the secondary structure of 5.8S ribosomal RNA molecules, their contact sites with 28S ribosomal RNAs and their prokaryotic equivalent. *Nucleic Acids Res.* **12:**7479–7502.

Vavra, K. J., Colavito-Shepanski, M., and Gorovsky, M. A. 1982. Organization of rDNA in chromatin: *Tetrahymena. In:* Busch, H., and Rothblum, L., eds., *The Cell Nucleus,* New York, Academic Press, Vol. XI, pp. 193–223.

Verbeet, M. P., Klootwijk, J., van Heerikhuizen, H., Fontijn, R. D., Vreugdenhil, E., and Planta, R. J. 1984. A conserved element is present around the transcription initiation site for RNA polymerase A in Saccharomycetoidea. *Nucleic Acids Res.* **12:**1137–1148.

Walberg, M. W., and Clayton, D. A. 1981. Sequence and properties of the human KB cell and mouse L cell D-loop regions of mitochondrial DNA. *Nucleic Acids Res.* **9:**5411–5421.

Walker, T. A., and Pace, N. R. 1983. 5.8S ribosomal RNA. *Cell* **33:**320–322.

Ware, V. C., Tague, B. W., Clark, C. G., Gourse, R. L., Brand, R. C., and Gerbi, S. A. 1983. Sequence analysis of 28S ribosomal DNA from the amphibian *Xenopus laevis. Nucleic Acids Res.* **11:**7795–7817.

Weiner, A. M., and Emery, H. S. 1982. The rDNA of *Dictyostelium discoideum* and *Physarum polycephalum. In:* Busch, H., and Rothblum, L., eds., *The Cell Nucleus,* New York, Academic Press, Vol. X, pp. 127–143

Wellauer, P. K., and Dawid, I. B. 1974. Secondary structure maps of ribosomal RNA and DNA. I. Processing of *Xenopus laevis* ribosomal RNA and structure of single-stranded ribosomal DNA. *J. Mol. Biol.* **89:**379–395.

Wellauer, P. K., and Dawid, I. B. 1978. Ribosomal DNA in *Drosophila melanogaster.* II. Heteroduplex mapping of cloned and uncloned rDNA. *J. Mol. Biol.* **126:**769–782.

Wellauer, P. K., Dawid, I. B., and Tartof, K. D. 1978. X and Y chromosomal ribosomal DNA of *Drosophila:* Comparison of spacers and insertions. *Cell* **14:**269–278.

Whitfeld, P. R., Herrmann, R. G., and Bottomley, W. 1978a. Mapping of the ribosomal RNA genes on spinach chloroplast DNA. *Nucleic Acids Res.* **5:**1741–1751.

Whitfeld, P. R., Leaver, C. J., Bottomley, W., and Atchison, B. A. 1978b. Low-molecular-weight (4.5S) ribonucleic acid in higher plant chloroplast ribosomes. *Biochem. J.* **175:**1103–1112.

Wild, M. A., and Gall, J. G. 1979. An intervening sequence in the gene coding for 25S rRNA of *Tetrahymena pigmentosa. Cell* **16:**565–573.

Wild, M. A., and Sommer, R. 1980. Sequence of a ribosomal RNA gene intron from *Tetrahymena. Nature (London)* **283:**693–694.

Wildeman, A. G., and Nazar, R. N. 1980. Nucleotide sequence of wheat chloroplastid 4.5S ribonucleic acid sequence homologies in 4.5S RNA species. *J. Biol. Chem.* **255:**11896–11900.

Wong, S. L., and Doi, R. H. 1984. Utilization of a *Bacillus subtilis* σ^37 promoter by *Escherichia coli* RNA polymerase *in vivo*. *J. Biol. Chem.* **259:**9762–9767.

Yaginuma, K., Kobayashi, M., Taira, M., and Koike, K. 1982. A new RNA polymerase and *in vitro* transcription of the origin of replication from rat mitochondrial DNA. *Nucleic Acids Res.* **10:**7531–7542.

Yakura, K., Kato, A., and Tanifuji, S. 1983. Structural organization of ribosomal DNA in four *Trillium* species and *Paris verticillata*. *Plant Cell Physiol.* **24:**1231–1240.

Yakura, K., Kato, A., and Tanifuji, S. 1984. Length heterogeneity of the large spacer of *Vicia faba* rDNA is due to the differing number of a 325 bp repetitive sequence element. *Mol. Gen. Genet.* **193:**400–405.

Yao, M. C. 1982. Amplification of ribosomal RNA gene in *Tetrahymena. In:* Busch, H., and Rothblum, L., eds., *The Cell Nucleus,* New York, Academic Press, Vol. XII, pp. 127–153.

Yin, S., Burke, J., Chang, D. D., Browning, K. S., Heckman, J. E., Alzner-DeWeerd, B., Potter, M. J., and RajBhandary, U. L. 1982. *Neurospora crassa* mitochondrial tRNAs and rRNAs: Structure, gene organization, and DNA sequences. *In:* Slonimski, P., Borst, P., and Attardi, G., eds., *Mitochondrial Genes,* Cold Spring Harbor, New York, Cold Spring Harbor Laboratory, pp. 361–373.

Young, R. A., and Steitz, J. A. 1979. Tandem promoters direct *E. coli* ribosomal RNA synthesis. *Cell* **17:**225–234.

Young, R. A., Macklis, R., and Steitz, J. A. 1979. Sequence of the 16S–23S spacer region in two ribosomal RNA operons of *Escherichia coli*. *J. Biol. Chem.* **254:**3264–3271.

Zaug, A. J., Grabowski, P. J., and Cech, T. R. 1983. Autocatalytic cyclization of an excised intervening sequence RNA is a cleavage–ligation reaction. *Nature (London)* **301:**578–583.

CHAPTER 5

Ajiro, K., Borun, T. W., and Cohen, L. H. 1981a. Phosphorylation states of different histone 1 subtypes and their relationships to chromatin functions during HeLa S-3 cell cycle. *Biochemistry* **20:**1445–1454.

Ajiro, K., Borun, T. W., Shulman, S. D., McFadden, G. M., and Cohen, L. H. 1981b. Comparison of the structures of human histones 1A and 1B and their intramolecular phosphorylation sites during the HeLa S-3 cell cycle. *Biochemistry* **20:**1454–1464.

Albright, S. C., Wiseman, J. M., Lange, R. A., and Garrard, W. T. 1980. Subunit structures of different electrophoretic forms of nucleosomes. *J. Biol. Chem.* **255:**3673–3684.

Allan, J., Hartman, P. G., Crane-Robinson, C., and Aviles, F. X. 1980. The structure of histone H1 and its location in chromatin. *Nature (London)* **288:**675–679.

Allfrey, V. G. 1980. Molecular aspects of the regulation of eukaryotic transcription: Nucleosomal proteins and their postsynthetic modifications in the control of DNA conformation and template function. *In:* Goldstein, L., and Prescott, D. M., eds., *Cell Biology: A Comprehensive Treatise,* New York, Academic Press, Vol. 3, pp. 347–437.

Allis, C. D., Glover, C. V. C., and Gorovsky, M. A. 1979. Micronuclei of *Tetrahymena* contain two types of histone H3. *Proc. Natl. Acad. Sci. USA* **76:**4857–4861.

Allis, C. D., Glover, C. V. C., Bowen, J. K., and Gorovsky, M. A. 1980. Histone variants specific to the transcriptionally active, amitotically dividing macronucleus of the unicellular eucaryote, *Tetrahymena thermophila*. *Cell* **20:**609–617.

Alterman, R. B. M., Ganguly, S., Schulze, D. H., Marzluff, W. F., Schildkraut, C. L., and Skoultchi, A. I. 1984. Cell cycle regulation of mouse H3 histone mRNA metabolism. *Mol. Cell. Biol.* **4:**123–132.

Anderson, S., Bankier, A. T., Barrell, B. G., de Bruijn, M. H. L., Coulson, A. R., Drouin, J., Eperon, I. C., Nierlick, D. P., Roe, B. A., Sanger, F., Schreier, P. H., Smith, A. J. R., Staden, R., and Young, I. G. 1981. Sequence and organization of the human mitochondrial genome. *Nature (London)* **290:**457–465.

Anderson, S., de Bruijn, M. H. L., Coulson, A. R., Eperon, I. C., Sanger, F., and Young, I. G. 1982. Complete sequence of bovine mitochondrial DNA. Conserved features of the mammalian mitochondrial genome. *J. Mol. Biol.* **156:**683–717.

Ball, D. J., Gross, D. S., and Garrard, W. T. 1983. 5-Methylcytosine is localized in nucleosomes that contain histone H1. *Proc. Natl. Acad. Sci. USA* **80:**5490–5494.

Bannon, G. A., Bowen, J. K., Yao, M. C., and Gorovsky, M. A. 1984. *Tetrahymena* H4 genes: Structure, evolution and organization in macro- and micronuclei. *Nucleic Acids Res.* **12:**1961–1975.

Barnes, K. L., Craigie, R. A., Cattini, P. A., and Cavalier-Smith, T. 1982. Chromatin from the unicellular red alga *Porphyridium* has a nucleosome structure. *J. Cell Sci.* **57**:151–160.

Bavykin, S. G., Usachenko, S. I., Lishanskaya, A. I., Shick, V. V., Belyavsky, A. V., Undritsov, I. M., Strokov, A. A., Zalenskaya, I. A., and Mirzabekov, A. D. 1985. Primary organization of nucleosomal core particles is invariable in repressed and active nuclei from animal, plant and yeast cells. *Nucleic Acids Res.* **13**:3439–3458.

Bedwell, D., Davis, G., Gosink, M., Post, L., Nomura, M., Kestler, H., Zengel, J. M., and Lindahl, L. 1985. Nucleotide sequence of the alpha ribosomal protein operon of *Escherichia coli. Nucleic Acids Res.* **13**:3891–3903.

Berthold, V., and Geider, K. 1976. Interaction of DNA with DNA-binding proteins. The characterization of protein HD from *Escherichia coli* and its nucleic acids complexes. *Eur. J. Biochem* **71**:443–449.

Bibb, M. J., Van Etten, R. A., Wright, C. T., Walberg, M. W., and Clayton, D. A. 1981. Sequence and gene organization of mouse mitochondrial DNA. *Cell* **26**:167–180.

Bozzoni, I., Beccari, E., Luo, Z. X., Amaldi, F., Pierandrei-Amaldi, P., and Campioni, N. 1981. *Xenopus laevis* ribosomal protein genes: Isolation of recombinant and cDNA clones and study of genomic organization. *Nucleic Acids Res* **9**:1069–1086.

Brandt, W., Strickland, W., Strickland, M., Carlisle, L., Woods, D., and von Holt, C. 1979. A histone programme during the life cycle of the sea urchin. *Eur. J. Biochem.* **94**:1–10.

Branno, M., de Franciscis, V., and Tosi, L. 1983. *In vitro* methylation of histones in sea urchin nuclei during early embryogenesis. *Biochim. Biophys. Acta* **741**:136–142.

Bruschi, S., and Wells, J. R. E. 1981. Vertebrate histone gene transcription occurs from both DNA strands. *Nucleic Acids Res.* **9**:1591–1597.

Bryan, P. N., Olah, J., and Birnstiel, M. L. 1983. Major changes in the 5′ and 3′ chromatin structure of sea urchin histone genes accompany their activation and inactivation in development. *Cell* **33**:843–848.

Burlingame, R. W., Love, W. E., Wang, B. C., Hamlin, R., Xuong, N. H., and Moudrianakis, E. V. 1985. Crystallographic structure of the octameric histone core of the nucleosome at a resolution of 3.3 Å. *Science* **228**:546–553.

Busslinger, M., Portmann, R., and Birnstiel, M. L. 1979. A regulatory sequence near the 3′ end of sea urchin histone genes. *Nucleic Acids Res.* **6**:2997–3008.

Busslinger, M., Portmann, R., Irminger, J. C., and Birnstiel, M. L. 1980. Ubiquitous and gene-specific 5′ sequences in a sea urchin histone DNA clone coding for histone protein variants. *Nucleic Acids Res.* **8**:957–977.

Butler, P. J. G. 1984. A defined structure of the 30 nm chromatin fibre which accommodates different nucleosomal repeat lengths. *EMBO J.* **3**:2599–2604.

Cereghini, S., and Yaniv, M. 1984. Assembly of transfected DNA into chromatin: Structural changes in the origin-promoter-enhancer region upon replication. *EMBO J.* **3**:1243–1253.

Cerretti, D. P., Dean, D., Davis, G. R., Bedwell, D. M., and Nomura, M. 1983. The *spc* ribosomal protein operon of *Escherichia coli:* Sequence and cotranscription of the ribosomal protein genes and a protein export gene. *Nucleic Acids Res.* **11**:2599–2616.

Certa, U., Colavito-Shepanski, M., and Grunstein, M. 1984. Yeast may not contain histone H1: The only known "histone H1-like" protein in *Saccharomyces cerevisiae* is a mitochondrial protein. *Nucleic Acids Res.* **12**:7975–7985.

Chambers, S. A. M., and Shaw, B. R. 1984. Levels of histone H4 diacetylation decrease dramatically during sea urchin embryonic development and correlate with cell doubling rate. *J. Biol. Chem.* **259**:13458–13463.

Choe, J., Kolodrubetz, D., and Grunstein, M. 1982. The two yeast histone H2A genes encode similar protein subtypes. *Proc. Natl. Acad. Sci. USA* **79**:1484–1487.

Cohn, R. H., and Kedes, L. H. 1979. Nonallelic histone gene clusters of individual sea urchins (*Lytechinus pictus*): Polarity and gene organization. *Cell* **18**:843–853.

Cole, K. D., York, R. G., and Kistler, W. S. 1984. The amino acid sequence of boar H1T, a testis-specific H1 histone variant. *J. Biol. Chem.* **259**:13695–13702.

Cusick, M. E., DePamphilis, M. L., and Wassarman, P. M. 1984. Dispersive segregation of nucleosomes during replication of simian virus 40 chromosomes. *J. Mol. Biol.* **178**:249–271.

D'Andrea, R., Harvey, R., and Wells, J. R. E. 1981. Vertebrate histone genes: Nucleotide sequence of a chicken H2A gene and regulatory flanking sequences. *Nucleic Acids Res.* **9**:3119–3128.

D'Anna, J. A., Gurley, L. R., and Becker, R. R. 1981. Histone H1°a and H1°b are the same as CHO histone

H1(III) and H1(IV): New features of H1° phosphorylation during the cell cycle. *Biochemistry* **20**:4501–4505.

Dean, D., and Nomura, M. 1982. Genetics and regulation of ribosomal protein synthesis in *Escherichia coli. In:* Busch, H., and Rothblum, L., eds. *The Cell Nucleus,* New York, Academic Press, Vol. XII, pp. 185–212.

Depetrocellis, B., Parente, A., Tomei, L., and Geraci, G. 1983. An H1 histone and a protamine molecule organize the sperm chromatin of the marine worm *Chaetopterus variopedatus. Cell Diff.* **12**:129–135.

Derenzini, M., Hernandez-Verdun, D., and Bouteille, M. 1983a. Visualization of a repeating subunit organization in rat hepatocyte to chromatin fixed *in situ. J. Cell Sci.* **61**:137–149.

Derenzini, M., Pession, A., Betts-Eusebi, C. M., and Novello, F. 1983b. Relationship between the extended non-nucleosomal intranucleolar chromatin *in situ* and ribosomal RNA synthesis. *Exp. Cell Res.* **145**:127–143.

Destrée, O. H. J., Bendig, M. M., De Laaf, R. T. M., and Koster, J. G. 1984. Organization of *Xenopus* histone gene variants within clusters and their transcriptional expression. *Biochim. Biophys. Acta* **782**:132–141.

Dillon, L. S. 1962. Comparative cytology and the evolution of life. *Evolution* **16**:102–117.

Dillon, L. S. 1981. *Ultrastructure, Macromolecules, and Evolution.* New York, Plenum Press.

Dillon, L. S. 1983. *The Inconstant Gene.* New York, Plenum Press.

Djondjurov, L. P., Yancheva, N. Y., and Ivanova, E. C. 1983. Histones of terminally differentiated cells undergo continuous turnover. *Biochemistry* **22**:4095–4102.

Doenecke, D., and Tönjes, R. 1984. Conserved dyad symmetry structures at the 3' end of H5 histone genes. *J. Mol. Biol.* **178**:121–135.

Dudov, K. P., and Perry, R. P. 1984. The gene family encoding the mouse ribosomal protein L32 contains a uniquely expressed intron-containing gene and an unmutated processed gene. *Cell* **3**:457–468.

Earnshaw, W. C., Honda, B. M., Laskey, R. A., and Thomas, J. O. 1980. Assembly of nucleosomes: The reaction involving *X. laevis* nucleoplasmin. *Cell* **21**:373–383.

Egan, P. A., and Levy-Wilson, B. 1981. Structure of transcriptionally active and inactive nucleosomes from butyrate-treated and control HeLa cells. *Biochemistry* **20**:3695–3702.

Fabijanski, S., and Pellegrini, M. 1982. A *Drosophila* ribosomal protein gene is located near repeated sequences including rDNA sequences. *Nucleic Acids Res.* **10**:5979–5991.

Finch, J. T., and Klug, A. 1976. Solenoidal model for superstructure in chromatin. *Proc. Natl. Acad. Sci. USA* **73**:1897–1901.

Finch, J. T., Lutter, L. C., Rhodes, D., Brown, R. S., Rushton, B., Levitt, M., and Klug, A. 1977. Structure of nucleosome core particles of chromatin. *Nature (London)* **269**:29–36.

Franke, W. W., Scheer, U., Trendelenburg, M. F., Zentgraf, H., and Spring, H. 1978. Morphology of transcriptionally active chromatin. *Cold Spring Harbor Symp. Quant. Biol.* **42**:755–772.

Franke, W. W., Scheer, U., Spring, H., Trendelenburg, M. F., and Zentgraf, H. 1979. Organization of nucleolar chromatin. *In:* Busch, H., ed., *The Cell Nucleus,* New York, Academic Press, Vol. VII, pp. 49–95.

Franklin, S. G., and Zweidler, A. 1977. Non-allelic variants of histones 2a, 2b and 3 in mammals. *Nature (London)* **266**:273–275.

Fried, H. M., Pearson, J. J., Kim, C. H., and Warner, J. R. 1981. The genes for fifteen ribosomal proteins of *Saccharomyces cerevisiae. J. Biol. Chem.* **256**:10176–10183.

Fusauchi, Y., and Iwai, K. 1983. *Tetrahymena* histone H2A. Isolation and two variant sequences. *J. Biochem.* **93**:1487–1497.

Gallwitz, D., and Mueller, G. C. 1969. Histone synthesis *in vitro* by cytoplasmic microsomes from HeLa cells. *Science* **163**:1351–1353.

Glikin, G. C., Ruberti, I., and Worce, A. 1984. Chromatin assembly in *Xenopus* oocytes: *In vitro* studies. *Cell* **37**:33–41.

Grandy, D. K., Engel, J. D., and Dodgson, J. B. 1982. Complete nucleotide sequence of chicken H2b histone gene. *J. Biol. Chem.* **257**:8577–8580.

Green, G. R., Searcy, D. G., and DeLange, R. J. 1983. Histone-like protein in the archaebacterium *Sulfolobus acidocaldarius. Biochim. Biophys. Acta* **741**:251–257.

Green, L., Van Antwerpen, R., Stein, J., Stein, G., Tripputi, P., Emanuel, B., Selden, J., and Croce, C. 1984. A major human histone gene cluster on the long arm of chromosome 1. *Science* **226**:838–840.

Groppi, V. E., and Coffino, P. 1980. G1 and S phase mammalian cells synthesize histones at equivalent rates. *Cell* **21**:195–204.

Grosschedl, R., and Birnstiel, M. L. 1980. Spacer DNA sequences upstream of the T-A-T-A-A-A-T-A sequence are essential for promotion of H2A histone gene transcription *in vivo*. *Proc. Natl. Acad. Sci. USA* **72:**7102–7106.

Grunstein, M., Diamond, K. E., Knoppel, E., and Grunstein, J. E. 1981. Comparison of the early histone H4 gene sequence of *Strongylocentrotus purpuratus* with maternal, early, and late H4 mRNA sequences. *Biochemistry* **20:**1216–1223.

Gurley, L. R., Walton, R. A., and Tobey, R. A. 1972. The metabolism of histone fractions. IV. Synthesis of histones during the G_1-phase of the mammalian life cycle. *Arch. Biochem. Biophys.* **148:**633–641.

Harvey, R. P., Whiting, J. A., Coles, L. S., Krieg, P. A., and Wells, J. R. E. 1983. An extremely variant histone H2A sequence expressed in the chicken embryo. *Proc. Natl. Acad. Sci. USA* **80:**2819–2823.

Hatch, C. L., Bonner, W. M., and Moudrianakis, E. N. 1983. Minor histone 2A variants and ubiquinated forms in the native H2A:H2B dimer. *Science* **221:**468–470.

Heintz, N., Zernik, M., and Roeder, R. G. 1981. The structure of the human histone genes: Clustered but not tandemly repeated. *Cell* **24:**661–668.

Helms, S., Baumbach, L., Stein, G., and Stein, J. 1984. Requirement of protein synthesis for the coupling of histone mRNA levels and DNA replication. *FEBS Lett.* **168:**65–69.

Hentschel, C. C., and Birnstiel, M. L. 1981. The organization and expression of histone gene families. *Cell* **25:**301–313.

Holt, C. A., and Childs, G. 1984. A new family of tandem repetitive early histone genes in the sea urchin *Lytechinus pictus:* Evidence for concerted evolution within tandem arrays. *Nucleic Acids Res.* **12:**6455–6471.

Huang, H. C., and Cole, R. D. 1984. The distribution of H1 histone is nonuniform in chromatin and correlates with different degrees of condensation. *J. Biol. Chem.* **259:**14237–14242.

Ikemura, T., Itoh, S., Post, L. E., and Nomura, M. 1979. Isolation and characterization of stable hybrid mRNA molecules transcribed from ribosomal protein promoters in *Escherichia coli*. *Cell* **18:**895–903.

Ishimi, Y., Yasuda, H., Hirosumi, J., Hanaoka, F., and Yamada, M. A. 1983. A protein which facilitates assembly of nucleosome-like structures *in vitro* in mammalian cells. *J. Biochem.* **94:**735–744.

Jardine, N. J., and Leaver, J. L. 1978. The fractionation of histones isolated from *Euglena gracilis*. *Biochem. J.* **169:**103–111.

Jin, Y.-J., and Cole, R. D. 1985. Histone H1° is distributed unlike H1 in chromatin aggregation. *FEBS Lett.* **182:**455–458.

Jordano, J., Montero, F., and Palacián, E. 1984a. Rearrangement of nucleosomal components by modification of histone amino groups. Structural role of lysine residues. *Biochemistry* **23:**4280–4284.

Jordano, J., Montero, F., and Palacián, E. 1984b. Contribution of histones H2A and H2B to the folding of nucleosomal DNA. *Biochemistry* **23:**4285–4289.

Kedes, L. H. 1976. Histone messengers and histone genes. *Cell* **8:**321–331.

Kedes, L. H. 1979. Histone genes and histone messengers. *Annu. Rev. Biochem.* **48:**837–870.

Kelly, P. M., Schofield, P. N., and Walker, I. O. 1983. Histone gene expression in *Physarum polycephalum* 2. Coupling of histone and DNA synthesis. *FEBS Lett.* **161:**79–83.

Kinkade, J. M., and Cole, R. D. 1966. The resolution of four lysine-rich histones derived from calf thymus. *J. Biol. Chem.* **241:**5790–5797.

Klug, A., Rhodes, D., Smith, J., Finch, J., and Thomas, J. 1980. A low resolution structure for the histone core of the nucleosome. *Nature (London)* **287:**509–516.

Kornberg, R. D. 1977. Structure of chromatin. *Annu. Rev. Biochem.* **46:**931–954.

Krieg, P. A., Robins, A. J., Colman, A., and Wells, J. R. E. 1982. Chicken histone H5 mRNA: The polyadenylated RNA lacks the conserved histone 3′ terminator sequence. *Nucleic Acids Res.* **10:**6777–6785.

Krieg, P. A., Robins, A. J., D'Andrea, R., and Wells, J. R. E. 1983. The chicken H5 gene is unlinked to core and H1 histone genes. *Nucleic Acids Res.* **11:**619–627.

Krohne, G., and Franke, W. W. 1980. Immunological identification and localization of the predominant nuclear protein of the amphibian oocyte nucleus. *Proc. Natl. Acad. Sci. USA* **77:**1034–1038.

Kuo, M. T., Iyer, B., and Schwarz, R. J. 1982. Condensation of chromatin into chromosomes preserves an open configuration but alters the DNAse I hypersensitive cleavage sites of the transcribed gene. *Nucleic Acids Res.* **10:**4565–4579.

Lammi, M., Paci, M., and Gualerzi, C. O. 1984. Proteins from the prokaryotic nucleoid. The interaction between protein NS and DNA involves the oligomeric form of the protein and at least one arg residue. *FEBS Lett.* **170**:99–104.

Larkin, J. C., and Woolford, J. L. 1983. Molecular cloning and analysis of the *CRY1* gene: A yeast ribosomal protein gene. *Nucleic Acids Res.* **11**:403–420.

Lathe, R., Buc, H., Lecocq, J. P., and Bautz, E. K. F. 1980. Prokaryotic histone-like protein interacting with RNA polymerase. *Proc. Natl. Acad. Sci. USA* **77**:3548–3552.

Leer, R. J., van Raamsdonk-Duin, M. M. C., Molenaar, C. M. T., Cohen, L. H., Mager, W. H., and Planta, R. J. 1982. The structure of the gene coding for the phosphorylated ribosomal protein S10 in yeast. *Nucleic Acids Res.* **10**:5869–5878.

Leer, R. J., van Raamsdonk-Duin, M. M. C., Schoppink, P. J., Cornelissen, M. I. E., Cohen, L. H., Mager, W. H., and Planta, R. J. 1983. Yeast ribosomal protein S33 is encoded by an unsplit gene. *Nucleic Acids Res.* **11**:7759–7768.

Leer, R. J., van Raamsdonk-Duin, M. M. C., Hagendoorn, M. J. M., Mager, W. H., and Planta, R. J. 1984a. Structural comparison of yeast ribosomal protein genes. *Nucleic Acids Res.* **12**:6685–1700.

Leer, R. J., van Raamsdonk-Duin, M. M. C., Mager, W. H., and Planta, R. J. 1984b. The primary structure of the gene encoding yeast ribosomal protein L16. *FEBS Lett.* **175**:371–376.

Leer, R. J., van Raamsdonk-Duin, M. M. C., Kraakman, P., Mager, W. H., and Planta, R. J. 1985a. The genes for yeast ribosomal proteins S24 and L46 are adjacent and divergently transcribed. *Nucleic Acids Res.* **13**:701–709.

Leer, R. J., van Raamsdonk-Duin, M. M. C., Molenaar, C. M. T., Witsenboer, H. M. A., Mager, W. H., and Planta, R. J. 1985b. Yeast contains two functional genes coding for ribosomal protein S10. *Nucleic Acids Res.* **13**:5027–5039.

Lennox, R. W., and Cohen, L. H. 1983. The histone H1 complements of dividing and nondividing cells of the mouse. *J. Biol. Chem.* **258**:262–268.

Levy, S., Sures, I., and Kedes, L. 1982. The nucleotide and amino acid coding sequence of a gene for H1 histone that interacts with euchromatin. The early embryonic H1 gene of the sea urchin, *Strongylocentrotus purpuratus. J. Biol. Chem.* **257**:9438–9443.

Lifton, R. P., Goldberg, M. L., Karp, R. W., and Hogness, D. S. 1977. The organization of histone genes in *Drosophila melanogaster:* Functional and evolutionary implications. *Cold Spring Harbor Symp. Quant. Biol.* **42**:1047–1051.

Mackie, G. A. 1981. Nucleotide sequences of the gene for ribosomal protein S20 and its flanking regions. *J. Biol. Chem.* **256**:8177–8182.

Maimets, T., Remme, J., and Villems, R. 1984. Ribosomal protein L16 binds to the 3′-end of transfer RNA. *FEBS Lett.* **166**:53–56.

Marian, B., and Wintersberger, U. 1980. Histone synthesis during sporulation of yeast. *FEBS Lett.* **117**:63–67.

Martinage, A., Belaiche, D., Dupressoir, T., and Sautiere, P. 1983. Primary structure of histone H2A from gonads of the starfish *Asterias rubens. Eur. J. Biochem.* **130**:465–472.

Mathis, D., Oudet, P., and Chambon, P. 1981. Structure of transcribing chromatin. *Prog. Nucleic Acids Res. Mol. Biol.* **24**:1–55.

Matthews, D. E., Hessler, R. A., Denslow, N. D., Edwards, J. S., and O'Brien, T. W. 1982. Protein composition of the bovine mitochondrial ribosome. *J. Biol. Chem.* **257**:8788–8794.

Mauron, A., Kedes, L., Hough-Evans, B., and Davidson, E. H. 1982. Accumulation of individual histone mRNAs during embryogenesis of the sea urchin *Strongylocentrotus purpuratus. Dev. Biol.* **94**:435–440.

Maxson, R., Mohun, T., and Kedes, L. 1982. Histone genes. *In:* Maclean, N., Gregory, S. P., and Flavell, R. A., eds., *Eukaryotic Genes: Their Structure, Activity and Regulation,* London, Butterworths, pp. 277–298.

Maxson, R., Mohun, T., Gormezano, G., Childs, G., and Kedes, L. 1983. Distinct organization and patterns of expression of early and late histone gene sets in the sea urchin. *Nature (London)* **301**:120–125.

McGhee, J. D., Nickol, J. M., Felsenfeld, G., and Rau, D. C. 1983a. Higher order structure of chromatin: Orientation of nucleosomes within the 30 nm chromatin solenoid is independent of species and spacer length. *Cell* **33**:831–841.

McGhee, J. D., Nickol, J. M., Felsenfeld, G., and Rau, D. C. 1983b. Histone hyperacetylation has little effect on the higher order folding of chromatin. *Nucleic Acids Res.* **11**:4065–4075.

Miller, F. D., Dixon, G. H., Rattner, J. B., and van de Sande, J. H. 1985. Assembly and characterization of nucleosomal cores on B- vs. Z-form DNA. *Biochemistry* **24**:102–109.

Mitra, G., and Warner, J. R. 1984. A yeast ribosomal protein gene whose intron is in the 5′ leader. *J. Biol. Chem.* **259**:9218–9224.

Molenaar, C. M. T., Woudt, L. P., Jansen, A. E. M., Mager, W. H., and Planta, R. J. 1984. Structure and organization of two linked ribosomal protein genes in yeast. *Nucleic Acids Res.* **12**:7345–7358.

Monk, R. J., Meyuhas, O., and Perry, R. P. 1981. Mammals have multiple genes for individual ribosomal proteins. *Cell* **24**:301–304.

Montandon, P. E., and Stutz, E. 1984. The genes for the ribosomal proteins S12 and S7 are clustered with the gene for the EF-Tu protein on the chloroplast genome of *Euglena gracilis. Nucleic Acids Res.* **12**:2851–2859.

Moorman, A. F. M., deBoer, P. A. J., De Laaf, R. T. M., and Destrée, O. H. J. 1982. Primary structure of the H2A and H2B genes and their flanking sequences in a minor histone gene cluster of *Xenopus laevis. FEBS Lett.* **144**:235–240.

Moyne, G., Harper, F., Saragosti, S., and Yaniv, M. 1982. Absence of nucleosomes in a histone-containing nucleoprotein complex obtained by dissociation of purified SV40 virions. *Cell* **30**:123–130.

Mueller, R. D., Yasuda, H., Hatch, C. L., Bonner, W. M., and Bradbury, E. M. 1985. Identification of ubiquinated histones 2A and 2B in *Physarum polycephalum. J. Biol. Chem.* **260**:5147–5153.

Närkhammar-Meuth, M., Eliasson, R., and Magnusson, G. 1981. Discontinuous synthesis of both strands at the growing fork during polyoma DNA replication *in vitro. J. Virol.* **39**:11–20.

O'Connell, P., and Rosbash, M. 1984. Sequence, structure, and codon preferences of the *Drosophila* ribosomal protein 49 genes. *Nucleic Acids Res.* **12**:5497–5513.

Old, R. W., and Woodland, H. R. 1984. Histone genes; Not so simple after all. *Cell* **38**:624–626.

Old, R. W., Woodland, H. R., Ballantine, J. E. M., Aldridge, T. C., Newton, C. A., Bains, W. A., and Turner, P. C. 1982. Organization and expression of cloned histone gene clusters from *Xenopus laevis* and *X. borealis. Nucleic Acids Res.* **10**:7561–7580.

Olins, P. O., and Nomura, M. 1981. Regulation of the S10 ribosomal protein operon in *E. coli:* Nucleotide sequence at the start of the operon. *Cell* **26**:205–211.

Olins, A. L., and Olins, D. E. 1974. Spheroid chromatin units (bodies). *Science* **183**:330–332.

Olins, A. L., and Olins, D. E. 1979. Stereo electron microscopy of the 25nm chromatin fibers in isolated nuclei. *J. Cell Biol.* **81**:260–265.

Oudet, P., Germond, J. E., Bellard, M., Spadafora, C., and Chambon, P. 1977. Nucleosome structure. *Phil. Trans. R. Soc. Lond. B* **283**:241–258.

Overton, G. C., and Weinberg, E. S. 1978. Length and sequence heterogeneity of the histone gene repeat unit of the sea urchin, *S. purpuratus. Cell* **14**:247–257.

Palen, T. E., and Cech, T. R. 1984. Chromatin structure of the replication origins and transcription-initiations region of the ribosomal RNA genes of *Tetrahymena. Cell* **36**:933–942.

Pardon, J. F., Worcester, D. L., Wooley, J. C., Cotter, R. I., Lilley, D. M. J., and Richards, B. M. 1977. The structure of the chromatin core particle in solution. *Nucleic Acids Res.* **4**:3199–3214.

Pedersen, S., Skouv, J., Kajitani, M., and Ishihama, A. 1984. Transcriptional organization of the *rpsA* operon of *Escherichia coli. Mol. Gen. Genet.* **196**:135–140.

Pehrson, J. R., and Cole, R. D. 1981. Bovine H1° histone subfractions contain an invariant sequence which matches histone H5 rather than H1. *Biochemistry* **20**:2298–2301.

Pehrson, J. R., and Cole, R. D. 1982. Histone H1 subfractions and H1° turn over at different rates in nondividing cells. *Biochemistry* **21**:456–460.

Pierandrei-Amaldi, P., Beccaci, E., Amaldi, F., and Bozzoni, I. 1982. Ribosomal protein genes in *Xenopus laevis. In:* Busch, H., and Rothblum, L., eds. *The Cell Nucleus.* New York, Academic Press, Vol. XII, pp. 227–243.

Plumb, M., Marashi, F., Green, L., Zimmerman, A., Zimmerman, S., Stein, J., and Stein, G. 1984. Cell cycle regulation of human histone H1 mRNA. *Proc. Natl. Acad. Sci. USA* **81**:434–438.

Post, L. E., and Nomura, M. 1980. DNA sequences from the *str* operon of *Escherichia coli. J. Biol. Chem.* **255**:4660–4666.

Post, L. E., Strycharz, G. D., Nomura, M., Lewis, H., and Dennis, P. P. 1979. Nucleotide sequence of the ribosomal protein gene cluster adjacent to the gene for RNA polymerase subunit β in *Escherichia coli. Proc. Natl. Acad. Sci. USA* **76**:1697–1701.

Post, L. E., Arfsten, A. E., Davis, G. R., and Nomura, M. 1980. DNA sequence of the promoter region for the ribosomal protein operon in *Escherichia coli. J. Biol. Chem.* **255**:4653–4659.

Ramsay, N., Felsenfeld, G., Rushton, B. M., and McGhee, J. D. 1984. A 145-base pair DNA sequence that positions itself precisely and asymmetrically on the nucleosome core. *EMBO J.* **3**:2605–2611.

Reichhart, R., Jörnvall, H., Carlquist, H., Zeppezauer, M. 1985a. The primary structure of two polypeptide chains from preparations of homeostatic thymus hormone (HTH$_\alpha$ and HTH$_\beta$). *FEBS Lett.* **188**:63–67.

Reichhart, R., Zeppezauer, M., and Jörnvall, H. 1985b. Preparations of homeostatic thymus hormone consist predominantly of histones 2A and 2B and suggest additional histone functions. *Proc. Natl. Acad. Sci. USA* **82**:4871–4875.

Ring, D., and Cole, R. D. 1983. Close contacts between H1 histone molecules in nuclei. *J. Biol. Chem.* **258**:15361–15364.

Risley, M. S., and Eckhardt, R. A. 1981. H1 histone variants in *Xenopus laevis. Dev. Biol.* **84**:79–87.

Rizzo, P. J. 1981. Comparative aspects of basic chromatin proteins in dinoflagellates. *BioSystems* **14**:433–443.

Rizzo, P. J., and Morris, R. L. 1984. Some properties of the histone-like protein from *Crypthecodinium cohnii* (HCc). *BioSystems* **16**:211–216.

Rizzo, P. J., Bradley, W., and Morris, R. L. 1985. Histones of the unicellular alga *Olisthodiscus luteus. Biochemistry* **24**:1727–1734.

Robbins, E., and Borun, T. W. 1966. The cytoplasmic synthesis of histones in HeLa cells and its temporal relationship to DNA replication. *Proc. Natl. Acad. Sci. USA* **57**:409–416.

Roberts, S. B., Weisser, K. E., and Childs, G. 1984. Sequence comparisons of non-allelic late histone genes and their early counterparts. *J. Mol. Biol.* **174**:647–662.

Rodriques, J. De A., Brandt, W. F., and von Holt, C. 1979. Plant histone 2 from wheat germ, a family of histone H2A variants. *Biochim. Biophys. Acta* **578**:196–206.

Roufa, D. J., and Marchionni, M. A. 1982. Nucleosome segregation at a defined mammalian chromosomal site. *Proc. Natl. Acad. Sci. USA* **79**:1810–1814.

Rouvière-Yaniv, J., and Gros, F. 1975. Characterization of a novel, low-molecular-weight DNA binding protein from *Escherichia coli. Proc. Natl. Acad. Sci. USA* **72**:3420–3432.

Ruberti, I., Fragapane, P., Pierandrei-Amaldi, P., Beccari, E., Amaldi, F., and Bozzoni, I. 1982. Characterization of histone genes isolated from *Xenopus laevis* and *Xenopus tropicalis* genomic libraries. *Nucleic Acids Res.* **10**:7543–7559.

Ruiz-Carrillo, A., Affolter, M., and Renaud, J. 1983. Genomic organization of the genes coding for the 6 main histones of the chicken: Complete sequence of the H5 gene. *J. Mol. Biol.* **170**:843–860.

Ryoji, M., and Worcel, A. 1984. Chromatin assembly in *Xenopus* oocytes: *In vivo* studies. *Cell* **37**:21–32.

Scheer, U., Hinssen, H., Franke, W. W., and Jockusch, B. M. 1984. Microinjection of actin-binding proteins and actin antibodies demonstrates involvement of nuclear actin in transcription of lampbrush chromosomes. *Cell* **39**:111–122.

Schick, V. V., Belyaysky, A. V., Bavykin, S. G., and Mirzabekov, A. D. 1980. Primary organization of nucleosome core particles. Sequential arrangement of histones along DNA. *J. Mol. Biol.* **139**:491–518.

Schmidt, R. J., Richardson, C. B., Gillham, N. W., and Boynton, J. E. 1983. Sites of synthesis of chloroplast ribosomal proteins in *Chlamydomonas. J. Cell Biol.* **96**:1451–1463.

Schnier, J., and Isono, K. 1982. The DNA sequence of the gene *rpsA* of *E. coli* coding for ribosomal protein S1. *Nucleic Acids Res.* **10**:1857–1865.

Schnier, J., Kimura, M., Foulaki, K., Subramanian, A. R., Isono, R., and Wittmann-Liebold, B. 1982. Primary structure of *E. coli* ribosomal protein S1 and of its gene *rps A. Proc. Natl. Acad. Sci. USA* **79**:1008–1011.

Schofield, P. N., and Walker, I. O. 1982. Control of histone gene expression in *Physarum polycephalum.* I. Protein synthesis during the cell cycle. *J. Cell Sci.* **57**:139–150.

Schultz, L. D., and Friesen, J. D. 1983. Nucleotide sequence of the *tcml* gene (ribosomal protein L3) of *Saccharomyces cerevisiae. J. Bacteriol.* **155**:8–14.

Seale, R. L. 1981. *In vivo* assembly of newly synthesized histones. *Biochemistry* **20**:6432–6437.

Seyedin, S. M., and Cole, R. D. 1981. H1 histones of trout. *J. Biol. Chem.* **256**:442–444.

Shih, R. J., Smith, L. D., and Keem, K. 1980. Rates of histone synthesis during early development of *Rana pipiens. Dev. Biol.* **75**:329–342.

Sierra, F., Lichtler, A., Marashi, F., Rickles, R., Van Dyke, T., Clark, S., Wells, J., Stein, G., and Stein, J. 1982. Organization of human histone genes. *Proc. Natl. Acad. Sci. USA* **79**:1795–1799.

Sierra, F., Stein, G., and Stein, J. 1983. Structure and *in vitro* transcription of a human H4 histone gene. *Nucleic Acids Res.* **11**:7069–7086.

Simpson, R. T. 1978. Structure of the chromatosome, a chromatin particle containing 160 base pairs of DNA and all the histones. *Biochemistry* **17:**5524–5531.

Sittman, D. B., Chiu, I. M., Pan, C. J., Cohn, R. H., Kedes, L. H., and Marzluff, W. F. 1981. Isolation of two clusters of mouse histone genes. *Proc. Natl. Acad. Sci. USA* **78:**4078–4082.

Sittman, D. B., Graves, R. A., and Marzluff, W. F. 1983a. Histone mRNA concentrations are regulated at the level of transcription and mRNA degradation. *Proc. Natl. Acad. Sci. USA* **80:**1849–1853.

Sittman, D. B., Graves, R. A., and Marzluff, W. F. 1983b. Structure of a cluster of mouse histone genes. *Nucleic Acids Res.* **11:**6679–6697.

Smith, M. M., and Andrésson, Ó. S. 1983. DNA sequences of yeast H3 and H4 histone genes from two non-allelic gene sets encode identical H3 and H4 proteins. *J. Mol. Biol.* **169:**663–690.

Spiker, S. 1982. Histone variants in plants. Evidence for primary structure variants differing in molecular weight. *J. Biol. Chem.* **257:**14250–14255.

Spinelli, G., Granguzza, F., Casano, C., and Burckhardt, J. 1979. Evidence for two different sets of histone genes active during embryogenesis of the sea urchin *Paracentrotus lividus*. *Nucleic Acids Res.* **6:**545–560.

Spinelli, G., Albanese, I., Anello, L., Ciaccio, M., and Di Liegro, I. 1982. Chromatin structure of histone genes in sea urchin sperms and embryos. *Nucleic Acids Res.* **10:**7977–7991.

Stein, A., and Bina, M. 1984. A model chromatin assembly system. Factors affecting nucleosome spacing. *J. Mol. Biol.* **178:**341–363.

Stephenson, E. C., Erba, H. P., and Gall, J. G. 1981. Histone gene clusters of the newt *Notophthalmus viridescens* are separated by long tracts of satellite DNA. *Cell* **24:**639–647.

Stoeckert, C. J., Beer, M., Wiggins, J. W., and Wierman, J. C. 1984. Histone positions within the nucleosome using platinum labeling and the scanning transmission electron microscope. *J. Mol. Biol.* **177:**483–505.

Strausbaugh, L. D., and Weinberg, E. S. 1982. Polymorphism and stability in the histone gene cluster of *Drosophila melanogaster*. *Chromosoma (Berl.)* **85:**489–506.

Strickland, M., Strickland, W. N., Brandt, W. F., von Holt, C., Wittmann-Liebold, B., and Lehmann, A. 1978. The complete amino acid sequence of histone H2B$_{(3)}$ from sperm of the sea urchin *Parechinus angulosus*. *Eur. J. Biochem.* **89:**443–452.

Sugarman, B. J., Dodgson, J. B., and Engel, J. D. 1983. Genomic organization, DNA sequence, and expression of chicken embryonic histone genes. *J. Biol. Chem.* **258:**9005–9016.

Sugita, M., and Sugiura, M. 1983. A putative gene of tobacco chloroplast coding for ribosomal protein similar to *E. coli* ribosomal protein S19. *Nucleic Acids Res.* **11:**1913–1918.

Sures, I., Lowry, J., and Kedes, L. H. 1978. The DNA sequence in sea urchin (*S. purpuratus*) H2A, H2B and H3 histone coding and spacer regions. *Cell* **15:**1033–1044.

Suryanarayana, T., and Subramanian, A. R. 1984. Functions of the repeating homologous sequences in nucleic acid binding domain of ribosomal protein S1. *Biochemistry* **23:**1047–1051.

Tabata, T., Sasaki, K., and Iwabuchi, M. 1983. The structural organization and DNA sequence of a wheat histone H4 gene. *Nucleic Acids Res.* **11:**5865–5875.

Tabata, T., Fukasawa, M., and Iwabuchi, M. 1984. Nucleotide sequence and genomic organization of a wheat histone H3 gene. *Mol. Gen. Genet.* **196:**397–400.

Teem, J. L., Abovich, N., Kaufer, N. F., Schwindinger, W. F., Warner, J. R., Levy, A., Woolford, J., Leer, R. J., van Raamsdonk-Duin, M. M., Mager, W. H., Planta, R. J., Schultz, L., Friesen, J. D., Fried, H., and Rosbash, M. 1984. A comparison of yeast ribosomal protein gene DNA sequences. *Nucleic Acids Res.* **12:**8295–8312.

Thoma, F., Bergman, L. W., and Simpson, R. T. 1984. Nuclease digestion of circular TRPIARS1 chromatin reveals positioned nucleosomes separated by nuclease-sensitive regions. *J. Mol. Biol.* **177:**715–733.

Turner, P. C., and Woodland, H. R. 1982. H3 and H4 histone cDNA sequence from *Xenopus*: A sequence comparison of H4 genes. *Nucleic Acids Res.* **10:**3769–3780.

Turner, P. C., and Woodland, H. R. 1983. Histone gene number and organisation in *Xenopus: Xenopus borealis* has a homogeneous major cluster. *Nucleic Acids Res.* **11:**971–986.

Turner, P. C., Aldridge, T. C., Woodland, H. R., and Old, R. W. 1983. Nucleotide sequence of H1 histone genes from *Xenopus laevis*. A recently diverged pair of H1 genes and unusual H1 pseudogene. *Nucleic Acids Res.* **11:**4093–4107.

Van Dongen, W. M. A. M., de Laaf, L., Zaal, R., Moorman, A. F. M., and Destrée, O. H. J. 1981. The organization of the histone genes in the genome of *Xenopus laevis*. *Nucleic Acids Res.* **9:**2297–2311.

Van Dongen, W. M. A. M., Moorman, A. F. M., and Destrée, O. H. J. 1983. Histone gene expression in early development of *Xenopus laevis*. *Differentiation* **24:**226–233.

Vaslet, C. A., O'Connell, P., Izquierdo, M., and Rosbash, M. 1980. Isolation and mapping of a cloned ribosomal protein gene of *Drosophila melanogaster*. *Nature (London)* **285:**674–676.

Vayda, N. E., Rogers, A. E., and Flint, S. J. 1983. The structure of nucleoprotein cores released from adenovirions. *Nucleic Acids Res.* **11:**441–460.

Vester, B., and Garrett, R. A. 1984. Structure of a protein L23–RNA complex located at the A-site domain of the ribosomal peptidyl transferase centre. *J. Mol. Biol.* **179:**431–452.

Wallis, J. W., Hereford, L., and Grunstein, M. 1980. Histone H2B genes of yeast (*Saccharomyces cerevisiae*) encode two different proteins. *Cell* **22:**799–805.

Weisbrod, S. 1982. Active chromatin. *Nature (London)* **297:**289–295.

Weisbrod, S., and Weintraub, H. 1981. Isolation of actively transcribed nucleosomes using immobilized HMG14 and 17 and an analysis of α-globin chromatin. *Cell* **23:**391–400.

Wells, D., and Kedes, L. 1985. Structure of a human histone cDNA: Evidence that basally expressed histone genes have intervening sequences and encode polyadenylated mRNAs. *Proc. Natl. Acad. Sci. USA* **82:**2834–2838.

West, M. H. P., and Bonner, W. M. 1980. Histone 2A, a heteromorphous family of eight protein species. *Biochemistry* **19:**3238–3245.

Wilhelm, M. L., and Wilhelm, F. X. 1984. A transposon-like DNA fragment interrupts a *Physarum polycephalum* histone H4 gene. *FEBS Lett.* **168:**249–254.

Wilhelm, M. L., Toublan, B., Jalouzot, R., and Wilhelm, F. X. 1984. Histone H4 gene is transcribed in S phase but also late in G_2 phase in *Physarum polycephalum*. *EMBO J.* **3:**2659–2662.

Wittmann-Liebold, B., Ashman, K., and Dzionara, M. 1984. On the statistical significance of homologous structures among the *Escherichia coli* ribosomal proteins. *Mol. Gen. Genet.* **196:**439–448.

Woodland, H. R. 1980. Histone synthesis during the development of *Xenopus*. *FEBS Lett.* **121:**1–7.

Woolford, J., and Rosbash, M. 1981. Ribosomal protein rp39(10-78), rp39(11-40), rp51, and rpS2 are not contiguous to other ribosomal protein genes in *Saccharomyces cerevisiae* genome. *Nucleic Acids Res.* **9:**5021–5036.

Woudt, L., Pastink, A., Kempers-Veenstra, A. E., Jansen, A. E. M., Mager, W. H., and Planta, R. J. 1983. The genes coding for histone H3 and H4 in *Neurospora crassa* are unique and contain intervening sequences. *Nucleic Acids Res.* **11:**5347–5360.

Wu, R. S., and Bonner, W. M. 1981. Separation of basal histone synthesis from S-phase histone synthesis in dividing cells. *Cell* **27:**321–330.

Yaguchi, M., Roy, C., and Seligy, V. L. 1979. Complete amino acid sequence of goose erythrocyte H5 histone and the homology between H1 and H5 histones. *Biochem. Biophys. Res. Commun.* **90:**1400–1406.

Yaguchi, M., Rollin, C. F., Roy, C., and Nazar, R. N. 1984. The 5S RNA binding protein from yeast (*Saccharomyces cerevisiae*) ribosomes. An RNA binding sequence in the carboxyl-terminal region. *Eur. J. Biochem.* **139:**451–457.

Yukioka, M., Sasaki, S., Henmi, S., Matsuo, M., Hatayama, T., and Inoue, A. 1984. Transcribing chromatin is not preferentially enriched with acetylated histones. *FEBS Lett.* **158:**281–284.

Zalenskaya, I. A., Pospelov, V. A., Zalensky, A. O., and Vorob'ev, V. I. 1981. Nucleosomal structure of sea urchin and starfish sperm chromatin. Histone H2B is possibly involved in determining the length of linker DNA. *Nucleic Acids Res.* **9:**473–487.

Zassenhaus, H. P., Martin, N. C., and Butow, R. A. 1984. Origins of transcripts of the yeast mitochondrial *var1* gene. *J. Biol. Chem.* **259:**6019–6027.

Zernik, M., Heintz, N., Boime, I., and Roeder, R. G. 1980. *Xenopus laevis* histone genes: Variant H1 genes are present in different clusters. *Cell* **22:**807–816.

Zhong, R., Roeder, R. G., and Heintz, N. 1983. The primary structure and expression of four cloned human histone genes. *Nucleic Acids Res.* **11:**7409–7425.

CHAPTER 6

Alt, J., and Herrmann, R. G. 1984. Nucleotide sequence of the gene for preapocytochrome *f* in the spinach plastid chromosome. *Curr. Genet.* **8:**550–558.

Anderson, S., Bankier, A. T., Barrell, B. G., de Bruijn, M. H. L., Coulson, A. R., Drouin, J., Eperon, A. C., Nierlich, D. P., Roe, B. A., Sanger, F., Schreier, P. H., Smith, A. J. H., Staden, R., and Young, I. G.

1981. Sequence and organization of the human mitochondrial genome. *Nature (London)* **290**:457–464.

Anderson, S., de Bruijn, M. H. L., Coulson, A. R., Eperon, I. C., Sanger, F., and Young, I. G. 1982. Complete sequence of bovine mitochondrial DNA. Conserved features of the mammalian mitochondrial genome. *J. Mol. Biol.* **156**:683–717.

Araya, A., Amthauer, R., Leon, G., and Krauskopf, M. 1984. Cloning, physical mapping and genome organization of mitochondrial DNA from *Cyprinus carpio* oocytes. *Mol. Gen. Genet.* **196**:43–52.

Ausubel, F. M. 1984. Regulation of nitrogen fixation genes. *Cell* **37**:5–6.

Ausubel, F. M., and Cannon, F. C. 1981. Molecular genetic analysis of *Klebsiella pneumoniae* nitrogen fixation (*nif*) genes. *Cold Spring Harbor Symp. Quant. Biol.* **45**:487–499.

Benne, R., De Vries, B. F., Van den Burg, J., and Klaver, B. 1983. The nucleotide sequence of a segment of *Trypanosoma brucei* mitochondrial maxi-circle DNA that contains the gene for apocytochrome *b* and some unusual unassigned reading frames. *Nucleic Acids Res.* **11**:6925–6941.

Beynon, J., Cannon, M., Buchanon-Wollaston, V., and Cannon, F. 1983. The *nif* promoters of *Klebsiella pneumoniae* have a characteristic primary structure. *Cell* **34**:665–671.

Bibb, M. J., Van Etten, R. A., Wright. C. T., Walberg, M. W., and Clayton, D. A. 1981. Sequence and gene organization of mouse mitochondrial DNA. *Cell* **26**:167–180.

Bitoun, R., Berman, J., Zilberstein, A., Holland, D., Cohen, J. B., Gruol, D., and Zamir, A. 1983. Promoter mutations that allow *nifA* independent expression of the nitrogen fixation *nifHDKY* operon. *Proc. Natl. Acad. Sci. USA* **80**:5812–5816.

Boer, P. H., McIntosh, J. E., Gray, M. W., and Bonen, L. 1985. The wheat mitochondrial gene for apocytochrome *b:* Absence of a prokaryotic ribosome binding site. *Nucleic Acids Res.* **13**:2281–2292.

Bonen, L., Boer, P. H., and Gray, M. W. 1984. The wheat cytochrome oxidase subunit II gene has an intron insert and three radical amino acid changes relative to maize. *EMBO J.* **3**:2531–2536.

Bonner, W. D., and Prince, R. C. 1984. The Rieske iron–sulfur cluster of plant mitochondria. *FEBS Lett.* **177**:47–50.

Borst, P., and Grivell, L. A. 1978. The mitochondrial genome of yeast. *Cell* **15**:705–723.

Chelm, B., Hoben, P. J., and Hallick, R. B. 1977. Cellular content of chloroplast DNA and chloroplast ribosomal RNA genes in *Euglena gracilis* during chloroplast development. *Biochemistry* **16**:782–786.

Clark-Walker, G. D., and Linnane, A. W. 1967. The biogenesis of mitochondria in *Saccharomyces cerevisiae*. A comparison between cytoplasmic respiratory deficient mutant yeast and chloramphenicol-inhibited wild-type cells. *J. Cell Biol.* **34**:1–14.

Clary, D. O., and Wolstenholme, D. R. 1983. Nucleotide sequence of a segment of *Drosophila* mitochondrial DNA that contains the genes for cytochrome *c* oxidase subunits II and III and ATPase subunit 6. *Nucleic Acids Res.* **11**:4211–4227.

Clary, D. O., Wahleithner, J. A., and Wolstenholme, D. R. 1984. Sequence and arrangement of the genes for cytochrome *b*, URF1, URF4L, URF4, URF5, URF6, and 5 tRNAs in *Drosophila* mitochondrial DNA. *Nucleic Acids Res.* **12**:3747–3762.

Coruzzi, G., and Tzagoloff, A. 1979. Assembly of the mitochondrial membrane system: DNA sequence of subunit 2 of yeast cytochrome oxidase. *J. Biol. Chem.* **254**:9324–9330.

Coruzzi, G., Broglie, R., Cashmore, A., and Chua, N. H. 1983. Nucleotide sequences of two pea cDNA clones encoding the small subunit of ribulose-1,5-bisphosphate carboxylase and the major chlorophyll a/b-binding thylakoid polypeptide. *J. Biol. Chem.* **258**:1399–1402.

Curtis, S. E., and Haselkorn, R. 1983. Isolation and sequence of the gene for the large subunit of ribulose-1,5-bisphosphate carboxylase from the cyanobacterium *Anabaena* 7120. *Proc. Natl. Acad. Sci. USA* **80**:1835–1839.

Dardas, A., Gal, D., Barrelle, M., Sauret-Ignazi, G., Sterjiades, R., and Pelmont, J. 1985. The demethylation of guaiacol by a new bacterial cytochrome P-450. *Arch. Biochem. Biophys.* **236**:585–592.

Davis, K. A., Hatefi, Y., Poff, K. L., and Butler, W. L. 1973. The *b*-type chromosomes of bovine yeast mitochondria. Absorption spectra, enzymatic properties, and distribution in the electron transfer complexes. *Biochim. Biophys. Acta* **325**:341–356.

Dawson, A. J., Jones, V. P., and Leaver, C. J. 1984. The apocytochrome *b* gene in maize mitochondria does not contain introns and is preceded by a potential ribosome binding site. *EMBO J.* **3**:2107–2113.

Dewey, R. E., Schuster, A. M., Levings, C. S., and Timothy, D. H. 1985. Nucleotide sequence of F_0-ATPase proteolipid (subunit 9) gene of maize mitochondria. *Proc. Natl. Acad. Sci. USA* **82**:1015–1019.

Dieckmann, C. L., Homison, G., and Tzagoloff, A. 1984. Assembly of the mitochondrial membrane system.

Nucleotide sequence of a yeast nuclear gene (*CBP1*) involved in a 5' end processing of cytochrome *b* pre-mRNA. *J. Biol. Chem.* **259**:4732–4738.

Dillon, L. S. 1962. Comparative cytology and the evolution of life. *Evolution* **16**:102–117.

Dillon, L. S. 1981. *Ultrastructure, Macromolecules, and Evolution*, New York, Plenum Press.

Dron, M., Rahire, M., and Rochaix, J. D. 1982. Sequence of the chloroplast DNA region of *Chlamydomonas reinhardii* containing the gene of the large subunit of ribulose bisphosphate carboxylase and parts of its flanking genes. *J. Mol. Biol.* **162**:775–793.

Drummond, M., Clements, J., Merrick, M., and Dixon, R. 1983. Positive control and autogenous regulation of the *nifLA* promoter in *Klebsiella pneumoniae*. *Nature (London)* **301**:302–307.

Dunon-Bluteau, D., Volovitch, M., and Brun, G. 1985. Nucleotide sequence of a *Xenopus laevis* mitochondrial DNA fragment containing the D-loop, flanking tRNA genes and the apocytochrome *b* gene. *Gene* **36**:65–78.

Eble, K. S., and Dawson, J. H. 1984. Novel reactivity of cytochrome P-450-CAM. Methyl hydroxylation of 5,5-difluorocamphor. *J. Biol. Chem.* **259**:14389–14393.

Edwards, J. C., Osinga, K. A., Christianson, T., Hensgens, L. A. M., Janssens, P. M., Rabinowitz, M., and Tabak, H. F. 1983. Initiation of transcription of the yeast mitochondrial gene coding for ATPase subunit 9. *Nucleic Acids Res.* **11**:8269–8282.

Falk, G., Hampe, A., and Walker, J. E. 1985. Nucleotide sequence of the *Rhodospirillum rubrum atp* operon. *Biochem. J.* **228**:391–407.

Fillingame, R. H. 1981. Biochemistry and genetics of bacterial proton-translocating ATPases. *Curr. Top. Bioenerg.* **11**:35–106.

Fitch, W. M. 1976. The molecular evolution of cytochrome *c* in eukaryotes. *J. Mol. Biol.* **8**:13–48.

Fox, T. D. 1979. Five TGA "stop" codons occur within the translated sequence of the yeast mitochondrial gene for cytochrome *c* oxidase subunit II. *Proc. Natl. Acad. Sci. USA* **76**:6534–6538.

Fox, T. D., and Leaver, C. J. 1981. The *Zea mays* mitochondrial gene coding cytochrome oxidase subunit II has an intervening sequence and does not contain TGA codons. *Cell* **26**:315–323.

Fuhrmann, M., and Hennecke, H. 1984. *Rhizobium japonicum* nitrogenase Fe protein gene (*nifH*). *J. Bacteriol.* **158**:1005–1011.

Fujii-Kuriyama, Y., Mizukami, Y., Bawajiri, K., Sogawa, K., and Muramatsu, M. 1982. Primary structure of a cytochrome P-450: Coding nucleotide sequence of phenobarbital-inducible cytochrome P-450 cDNA from rat liver. *Proc. Natl. Acad. Sci. USA* **79**:2793–2797.

Gonzalez, F. J., Mackenzie, P. I., Kimura, S., and Nebert, D. W. 1984. Isolation and characterization of full-length mouse cDNA and genomic clones of 3-methylcholanthrene-inducible cytochrome P_1-450 and P_3-450. *Gene* **29**:281–292.

Gonzalez, F. J., Nebert, D. W., Hardwick, J. P., and Kasper, C. B. 1985. Complete cDNA and protein sequence of a pregnenolone 16α-carbonitrile-induced cytochrome P-450. A representative of a new gene family. *J. Biol. Chem.* **260**:7435–7441.

Hase, T., Wakabayashi, S., Nakano, T., Zumft, W. G., and Matsubara, H. 1984. Structural homologies between the amino acid sequence of *Clostridium pasteurianum* MoFe protein and the DNA sequence of *nifD* and *K* genes of phylogenetically diverse bacteria. *FEBS Lett.* **166**:39–43.

Haugland, R., and Verma, D. P. S. 1981. Interspecific plasmid and genomic DNA sequence homologies and localization of *nif* genes in effective and ineffective strains of *Rhizobium japonicum*. *J. Mol. Appl. Genet.* **1**:205–218.

Heinemeyer, W., Alt, J., and Herrmann, R. G. 1984. Nucleotide sequence of the clustered genes for apocytochrome *b6* and subunit 4 of the cytochrome *b/f* complex in the spinach plastid chromosome. *Curr. Genet.* **8**:543–549.

Hensgens, L. A. M., Brakenhoff, J., De Vries, B. F., Sloof, P., Tromp, M. C., Van Boom, J. H., and Benne, R. 1984. The sequence of the gene for cytochrome *c* oxidase subunit I, a frame-shift containing gene for cytochrome *c* oxidase subunit II and seven unassigned reading frames in *Trypanosoma brucei* mitochondrial maxi-circle DNA. *Nucleic Acids Res.* **12**:7327–7344.

Herrmann, R. G., Alt, J., Schiller, B., Widger, W. R., and Cramer, W. A. 1984. Nucleotide sequence of the gene for apocytochrome *b*-559 on the spinach plastid chromosome: Implications for the structure of the membrane protein. *FEBS Lett.* **176**:239–244.

Hesse, J. E., Wieczoick, L., Altendorf, K., Reicin, A. S., Dorus, E., and Epstein, W. 1984. Sequence homology between two membrane transport ATPases, the Kdp-ATPase of *Escherichia coli* and the Ca^{2+}-ATPase of sarcoplasmic reticulum. *Proc. Natl. Acad. Sci. USA* **81**:4746–4750.

Hiesel, R., and Brennicke, A. 1983. Cytochrome oxidase subunit II gene in mitochondria of *Oenothera* has no intron. *EMBO J.* **2**:2173–2178.

Hines, R. N., Levy, J. B., Conrad, R. D., Iversen, P. L., Shen, M.-L., Renli, A. M., and Bresnick, E. 1985. Gene structure and nucleotide sequence for rat cytochrome P-450c. *Arch. Biochem. Biophys.* **237**:465–476.

Ingelman-Sundberg, M., and Johansson, I. 1980. Cytochrome b_5 as electron donor to rabbit liver cytochrome P-450$_{LM_2}$ in reconstituted phospholipid vesicles. *Biochem. Biophys. Res. Commun.* **97**:582–589.

Jaiswal, A. K., Gonzalez, F. J., and Nebert, D. W. 1985. Human dioxin-inducible cytochrome P_1-450: Complementary DNA and amino acid sequences. *Science* **228**:80–83.

John, M. E., John, M. C., Ashley, P., MacDonald, R. J., Simpson, E. R., and Waterman, M. R. 1984. Identification and characterization of cDNA clones specific for cholesterol side-chain cleavage cytochrome P-450. *Proc. Natl. Acad. Sci. USA* **81**:5628–5632.

Kaipainen, P., Nebert, D. W., and Lang, M. A. 1984. Purification and characterization of a microsomal cytochrome P-450 with high activity of coumarin 7-hydroxylase from mouse liver. *Eur. J. Biochem.* **144**:425–431.

Kallas, T., Rebière, M. C., Rippka, R., and Tandeau de Marsac, N. 1983. The structural *nif* genes of the cyanobacteria *Gloeothece* sp. and *Calothrix* sp. share homology with those of *Anabaena* sp., but the *Gloeothece* genes have a different arrangement. *J. Bacteriol.* **155**:427–431.

Kaluza, K., and Hennecke, H. 1984. Fine structure analysis of the *nifDK* operon encoding the α and β subunits of dinitrogenase from *Rhizobium japonicum. Mol. Gen. Genet.* **196**:35–42.

Kao, T.-H., Moon, E., and Wu, R. 1984. Cytochrome oxidase subunit II gene of rice has an insertion sequence within the intron. *Nucleic Acids Res.* **12**:7305–7315.

Kawajiri, K., Gotoh, O., Sagawa, K., Tagashira, Y., Muramatsu, M., and Fujii-Kuriyama, Y. 1984. Coding nucleotide sequence of 3-methylcholanthrene-inducible cytochrome P-450d cDNA from rat liver. *Proc. Natl. Acad. Sci. USA* **81**:1649–1653.

Kimura, S., Gonzalez, F. J., and Nebert, D. W. 1984. Mouse cytochrome P_3-450: Complete cDNA and amino acid sequence. *Nucleic Acids Res.* **12**:2917–2928.

Koller, B., Gingrich, J. C., Stiegler, G. L., Farley, M. A., Delius, H., and Hallick, R. B. 1984. Nine introns with conserved boundary sequences in the *Euglena gracilis* chloroplast ribulose-1,5-bisphosphate carboxylase gene. *Cell* **36**:545–553.

Kulkarni, A. P., and Hodgson, E. 1980. Multiplicity of cytochrome P-450 in microsomal membranes from the housefly, *Musca domestica. Biochim. Biophys. Acta* **632**:573–588.

Kuwahara, S. I., Harada, N., Yoshioka, H., Miyata, T., and Omura, T. 1984. Purification and characterization of four forms of cytochrome P-450 from liver microsomes of phenobarbitol-treated and 3-methylcholanthrene-treated rats. *J. Biochem.* **95**:703–714.

Lammers, P. J., and Haselkorn, R. 1983. Sequence of the *nifD* gene coding for the α subunit of dinitrogenase from the cyanobacterium *Anabaena. Proc. Natl. Acad. Sci. USA* **80**:4723–4727.

Lang, B. F., Ohne, F., and Bonen, L. 1985. The mitochondrial genome of the fission yeast, *Schizosaccharomyces pombe.* The cytochrome *b* gene has an intron closely related to the first two introns in the *Saccharomyces cerevisiae cox1* gene. *J. Mol. Biol.* **184**:353–366.

Limbach, K. J., and Wu, R. 1983. Isolation and characterization of two alleles of the chicken cytochrome *c* gene. *Nucleic Acids Res.* **11**:8931–8950.

Lomax, M. L., Bachman, N. J., Nasoff, M. S., Caruthers, M. H., and Grossman, L. I. 1984. Isolation and characterization of a cDNA clone for bovine cytochrome *c* oxidase subunit IV. *Proc. Natl. Acad. Sci. USA* **81**:6295–6299.

Ludwig, B., Suda, K., and Cerletti, N. 1983. Cytochrome c_1, from *Paracoccus denitrificans. Eur. J. Biochem.* **137**:597–602.

Maarse, A. C., Van Loon, A. P. G. M., Riezman, H., Gregor, I., Schatz, G., and Grivell, L. A. 1984. Subunit IV of yeast cytochrome *c* oxidase: Cloning and nucleotide sequencing of the gene and partial amino acid sequencing of the mature protein. *EMBO J.* **3**:2831–2837.

Macino, G., and Tzagoloff, A. 1979. Assembly of the mitochondrial membrane system. The DNA sequence of a mitochondrial ATPase gene in *Saccharomyces cerevisiae. J. Biol. Chem.* **254**:4617–4623.

Macino, G., and Tzagoloff, A. 1980. Assembly of the mitochondrial membrane system. Sequence analysis of a yeast mitochondrial ATPase gene containing the *oli*-2 and *oli*-4 loci. *Cell* **20**:507–517.

Macreadie, I. G., Novitski, C. E., Maxwell, R. J., John, U., Ooi, B. G., McMullen, G. L., Lukins, H. B.,

Linnane, A. V., and Nagley, P. 1983. Biogenesis of mitochondria: The mitochondrial gene (*aapl*) coding for mitochondrial ATPase subunit 8 in *Saccharomyces cerevisiae*. *Nucleic Acids Res.* **11**:4435–4451.

Mazur, B. J., and Chui, C.-F. 1982. Sequence of the gene coding for the β-subunit of dinitrogenase from the blue-green alga *Anabaena*. *Proc. Natl. Acad. Sci. USA* **79**:6782–6786.

Mazur, B. J., and Chui, C. F. 1985. Sequence of a genomic DNA clone for the small subunit of ribulose bisphosphate carboxylase-oxygenase from tobacco. *Nucleic Acids Res.* **13**:2373–2386.

McIntosh, L., Poulson, C., and Bogorad, L. 1980. Chloroplast gene sequence for the large subunit of ribulose bisphosphate carboxylase of maize. *Nature (London)* **288**:556–560.

Mevarech, M., Rice, D., and Haselkorn, R. 1980. Nucleotide sequence of a cyanobacterial *nifH* gene coding for nitrogenase reductase. *Proc. Natl. Acad. Sci. USA* **77**:6476–6480.

Minami, E.-I., and Watanabe, A. 1984. Thylakoid membranes: The translational site of chloroplast DNA-regulated thylakoid polypeptides. *Arch. Biochem. Biophys.* **235**:562–570.

Miziorko, H. M., and Lorimer, G. H. 1983. Ribulose-1,5-bisphosphate carboxylase-oxygenase. *Annu. Rev. Biochem.* **52**:507–535.

Mizukami, Y., Sogawa, K., Suwa, Y., Muramatsu, M., and Fujii-Kuriyama, Y. 1983. Gene structure of a phenobarbitol-inducible cytochrome P-450 of rat liver. *Proc. Natl. Acad. Sci. USA* **80**:3958–3962.

Montgomery, J. L., Leung, D. W., Smith, M., Shalit, P., Faye, G., and Hall, B. D. 1980. Isolation and sequence of the gene for iso-2-cytochrome *c* in *Saccharomyces cerevisiae*. *Proc. Natl. Acad. Sci. USA* **77**:541–545.

Moon, E., Kao, T. H., and Wu, R. 1985. Pea cytochrome oxidase subunit II gene has no intron and generates two mRNA transcripts with different 5′-termini. *Nucleic Acids Res.* **13**:3195–3212.

Morelli, G., and Macino, G. 1984. Two intervening sequences in the ATPase subunit 6 gene of *Neurospora crassa*. *J. Mol. Biol.* **178**:491–507.

Morohashi, K., Fujii-Kuriyama, Y., Okada, Y., Sogawa, K., Hirose, T., and Inayama, S. 1984. Molecular cloning and nucleotide sequence of cDNA for mRNA of mitochondrial cytochrome P-450 (SCC) of bovine adrenal cortex. *Proc. Natl. Acad. Sci. USA* **81**:4647–4651.

Nargang, F., McIntosh, L., and Somerville, C. 1984. Nucleotide sequence of the ribulose bisphosphate carboxylase gene from *Rhodospirillum rubrum*. *Mol. Gen. Genet.* **193**:220–224.

Netzker, R., Köchel, H. G., Basak, N., and Küntzel, H. 1982. Nucleotide sequence of *Aspergillus nidulans* mitochondrial genes coding for ATPase subunit 6, cytochrome oxidase subunit 3, seven unidentified proteins for tRNAs and L-rRNA. *Nucleic Acids Res.* **10**:4783–4794.

Nierzwicki-Bauer, S. A., Curtis, S. E., and Haselkorn, R. 1984. Cotranscription of genes encoding the small and large subunits of ribulose-1,5-bisphosphate carboxylase in the cyanobacterium *Anabaena* 7120. *Proc. Natl. Acad. Sci. USA* **81**:5961–5965.

Niranjan, B. G., Wilson, N. M., Jefcoate, C. R., and Avadhani, N. G. 1984. Hepatic mitochondrial cytochrome P-450 system. *J. Biol. Chem.* **259**:12495–12501.

Nobrega, F. G., and Tzagoloff, A. 1980. Assembly of the mitochondrial membrane system. DNA sequence and organization of the cytochrome *b* gene in *Saccharomyces cerevisiae* D273-10B. *J. Biol. Chem.* **255**:9828–9837.

Okino, S. T., Quattrochi, L. C., Barnes, H. J., Osanto, S., Griffin, K. J., Johnson, E. F., and Tukey, R. H. 1985. Cloning and characterization of cDNAs encoding 2,3,7,8-tetrachlorodibenzo-*p*-dioxin-inducible rabbit mRNAs for cytochrome P-450 isozymes 4 and 6. *Proc. Natl. Acad. Sci. USA* **82**:5310–5314.

Orian, J. M., and Marzuki, S. 1981. The largest mitochondrial translation product copurifying with the mitochondrial adenosine triphosphatase of *Saccharomyces cerevisiae* is not a subunit of the enzyme complex. *J. Bacteriol.* **146**:813–815.

Orian, J. M., Murphy, M., and Marzuki, S. 1981. Mitochondrially synthesized protein subunits of the yeast mitochondrial adenosine triphosphatase. A reassessment. *Biochim. Biophys. Acta* **652**:234–239.

Ow, D. W., and Ausubel, F. M. 1983. Regulation of nitrogen metabolism genes by *nifA* gene product in *Klebsiella pneumoniae*. *Nature (London)* **301**:307–313.

Phillips, A. L., and Gray, J. C. 1984. Location and nucleotide sequence of the gene for the 15.2 kDa polypeptide of the cytochrome *b–f* complex from pea chloroplasts. *Mol. Gen. Genet.* **194**:477–484.

Reeck, G. R., and Teller, D. C. 1983. Sequence similarity between *Rhodospirillum rubrum* ribulose bisphosphate carboxylase/oxygenase and the large subunit of the plant enzyme. *FEBS Lett.* **154**:134–138.

Reichelt, B. Y., and Delaney, S. F. 1983. The nucleotide sequence for the large subunit of ribulose 1,5-bisphosphate carboxylase from a unicellular cyanobacterium, *Synechococcus* PCC6301. *DNA* **2**:121–128.

Roberts, G. P., and Brill, W. J. 1981. Genetics and regulation of nitrogen fixation. *Annu. Rev. Microbiol.* **35**:207–235.

Rossen, L., Ma, Q.-S., Mudd, E. A., Johnston, A. W. B., and Downie, J. A. 1984. Identification and DNA sequence of *fixZ*, a *nifB*-like gene from *Rhizobium leguminosarum*. *Nucleic Acids Res.* **12**:7123–7134.

Russell, P. R., and Hall, B. D. 1982. Structure of the *Schizosaccharomyces pombe* cytochrome *c* gene. *Mol. Cell. Biol.* **2**:106–116.

Sadler, I., Suda, K., Schatz, G., Kaudewitz, F., and Haid, A. 1984. Sequencing of the nuclear gene for the yeast cytochrome c_1 precursor reveals an unusually complex amino-terminal presequence. *EMBO J.* **3**:2137–2143.

Scarpulla, R. C., Agne, K. M., and Wu, R. 1981. Isolation and structure of a rat cytochrome *c* gene. *J. Biol. Chem.* **256**:6480–6486.

Scott, K. F., Rolfe, B. G., and Shine, J. 1981. Biological nitrogen fixation: Primary structure of the *Klebsiella pneuomoniae nifH* and *nifD* genes. *J. Mol. Appl. Genet.* **1**:71–81.

Scott, K. F., Rolfe, B. G., and Shine, J. 1983. Nitrogenase structural genes are unlinked in the nonlegume *Parasponia rhizobium*. *DNA* **2**:141–148.

Senior, A. E. 1973. The structure of mitochondrial ATPase. *Biochim. Biophys. Acta* **301**:249–277.

Shanmugam, K. T., O'Gara, F., Andersent, K., and Valentine, R. C. 1978. Biological nitrogen fixation. *Annu. Rev. Plant Physiol.* **29**:263–276.

Shinozaki, K., and Sugiura, M. 1982. The nucleotide sequence of the tobacco chloroplast gene for the large subunit of ribulose-1,5-bisphosphate carboxylase/oxygenase. *Gene* **20**:91–102.

Shinozaki, K., and Sugiura, M. 1983. The gene for the small subunit of ribulose-1,5-bisphosphate carboxylase/oxygenase is located close to the gene for the large subunit in the cyanobacterium *Anacystis nidulans* 6301. *Nucleic Acids Res.* **11**:6957–6964.

Shinozaki, K., Yamada, C., Takahata, N., and Sugiura, M. 1983. Molecular cloning and sequence analysis of the cyanobacterial gene for the large subunit of ribulose-1,5-bisphosphate carboxylase/oxygenase. *Proc. Natl. Acad. Sci. USA* **80**:4050–4054.

Singh, R. K., and Singh, H. N. 1981. Genetic analysis of the *het* and *nif* genes in the blue-green alga *Nostoc muscorum*. *Mol. Gen. Genet.* **184**:531–535.

Smith, M., Leung, D. W., Gillam, S., Astell, C. R., Montgomery, D. L., and Hall, B. D. 1979. Sequence of the gene for iso-1-cytochrome *c* in *Saccharomyces cerevisiae*. *Cell* **16**:753–761.

Sogawa, K., Gotoh, O., Kawajiri, K., and Fujii-Kuriyama, Y. 1984. Distinct organization of methylcholanthrene- and phenobarbital-inducible cytochrome P-450 in the rat. *Proc. Natl. Acad. Sci. USA* **81**:5066–5070.

Stern, D. B., and Palmer, J. D. 1984. Extensive and widespread homologies between mitochondrial DNA and chloroplast DNA in plants. *Proc. Natl. Acad. Sci. USA* **81**:1946–1950.

Sundaresan, V., and Ausubel, F. M. 1981. Nucleotide sequence of the gene coding for nitrogenase iron protein from *Klebsiella pneumoniae*. *J. Biol. Chem.* **256**:2808–2812.

Suwa, Y., Mizukami, Y., Sogawa, K., and Fujii-Kuriyama, Y. 1985. Gene structure of a major form of phenobarbital-inducible cytochrome P-450 in rat liver. *J. Biol. Chem.* **260**:7980–7984.

Szeto, W. W., Zimmerman, J. L., Sundaresan, V , and Ausubel, F. M. 1984. A *Rhizobium meliloti* symbiotic regulatory gene. *Cell* **36**:1035–1043.

Tabita, F. R., and McFadden, B. A. 1974. D-Ribulose-1,5-bisphosphate carboxylase from *Rhodospirillum rubrum*. Quaternary structure, composition, catalytic, and immunological properties. *J. Biol. Chem.* **249**:3459–3464.

Thalenfeld, B. E., and Tzagoloff, A. 1980. Assembly of the mitochondrial membrane system. Sequence of the *oxi2* gene of yeast mitochondrial DNA. *J. Biol. Chem.* **255**:6173–6180.

Thomas, P. E., Reik, L. M., Ryan, D. E., and Levin, W. 1981. Regulation of three forms of cytochrome P-450 and epoxide hydrolase in rat liver microsomes. *J. Biol. Chem.* **256**:1044–1052.

Thöny, B., Kaluza, K., and Hennecke, H. 1985. Structural and functional homology between the α and β subunits of the nitrogenase MoFe protein as revealed by sequencing the *Rhizobium japonicum nifK* gene. *Mol. Gen. Genet.* **198**:441–448.

Török, I., and Kondorosi, Å. 1981. Nucleotide sequence of the *R. meliloti* nitrogenase reductase (*nifH*) gene. *Nucleic Acids Res.* **9**:5711–5723.

Tybulewicz, V. L. J., Falk, G., and Walker, J. E. 1984. *Rhodopseudomonas blastica atp* operon. Nucleotide sequence and transcription. *J. Mol. Biol.* **179**:185–214.

Voordouw, G., and Brenner, S. 1985. Nucleotide sequence of the gene encoding the hydrogenase from *Desulfovibrio vulgaris* (Hildenborough). *Eur. J. Biochem.* **148**:515–520.

Wakabayashi, S., Matsubara, H.. Kim, C. H., Kawai, K., and King, T. E. 1980. The complete amino acid sequence of bovine heart cytochrome c_1. *Biochem. Biophys. Res. Commun.* **97**:1548–1554.

Wakabayashi, S., Matsubara, H., Kim, C. H., and King, T. E. 1982. Structural studies of bovine heart cytochrome c_1. *J. Biol. Chem.* **257**:9335–9344.

Walker, J. E., Saraste, M., and Gay, N. J. 1984. The *unc* operon. Nucleotide sequence, regulation and structure of ATP-synthase. *Biochim. Biophys. Acta* **768**:164–200.

Waring, R. B., Davies, R. W., Lee, S., Grisi, E., Berks, M. M., and Scazzocchio, C. 1981. The mosaic organization of the apocytochrome *b* gene of *Aspergillus nidulans* revealed by DNA sequencing. *Cell* **27**:4–11.

Weinman, J. J., Fellows, F. F., Gresshoff, P. M., Shine, J., and Scott, K. F. 1984. Structural analysis of the genes encoding the molybdenum–iron protein of nitrogenase in the *Parasponia rhizobium* strain ANU289. *Nucleic Acids Res.* **12**:8329–8344.

White, P. C., New, M. I., and Dupont, B. 1984. Cloning and expression of cDNA encoding a bovine adrenal cytochrome P-450 specific for steroid 21-hydroxylation. *Proc. Natl. Acad. Sci. USA* **81**:1986–1990.

Williams, D. E., Hale, S. E., Okita, R. T., and Masters, B. S. S. 1984. A prostaglandin ω-hydroxylase cytochrome P-450 (P-450$_{PG-\omega}$) purified from lungs of pregnant rabbits. *J. Biol. Chem.* **259**:14600–14608.

Willey, D. L., Howe, C. J., Auffret, A. D., Bowman, C. M., Dyer, T. A., and Gray, J. C. 1984a. Location and nucleotide sequence of the gene for cytochrome *f* in wheat chloroplast DNA. *Mol. Gen. Genet.* **194**:416–422.

Willey, D. L., Auffret, A. D., and Gray, J. G. 1984b. Structure and topology of cytochrome *f* in pea chloroplast membranes. *Cell* **36**:555–562.

Wright, R. M., Ko, C., Cumsky, M. G., and Poyton, R. O. 1984. Isolation and sequence of the structural gene for cytochrome *c* oxidase subunit VI from *Saccharomyces cerevisiae*. *J. Biol. Chem.* **259**:15401–15407.

Wyatt, J. T., Thelma, C. M., and Jackson, J. W. 1973. An examination of three strains of the blue-green alga genus, *Fremyella*. *Phycologia* **12**:153–161.

Yabusaki, Y., Murakami, H., Nakamura, K., Nomura, N., Shimizu, M., Oeda, K., and Ohkawa, H. 1984a. Characterization of complementary DNA clones coding for two forms of 3-methylcholanthrene-inducible rat liver cytochrome P-450. *J. Biochem.* **96**:793–804.

Yabusaki, Y., Shimizu, M., Murakami, H., Nakamura, K., Oeda, K., and Ohkawa, H. 1984b. Nucleotide sequence of a full-length cDNA coding for 3-methylcholanthrene-induced rat liver P-450MC. *Nucleic Acids Res.* **12**:2929–2938.

Zurawski, G., Perrot, B., Bottomley, W., and Whitfeld, P. R. 1981. The structure of the gene for the large subunit of ribulose 1,5-bisphosphate carboxylase from spinach chloroplast DNA. *Nucleic Acids Res.* **9**:3251–3270.

CHAPTER 7

Alt, J., and Herrmann, R. G. 1984. Nucleotide sequence of the gene for preapocytochrome *f* in the spinach plastid genome. *Curr. Genet.* **8**:550–558.

Anderson, O. D., Litts, J. C., Gautier, M. F., and Greene, F. C. 1984. Nucleic acid sequence and chromosome arrangement of a wheat storage protein gene. *Nucleic Acids Res.* **12**:8129–8144.

Andrews, D. W., Walter, P., and Ottensmeyer, F. P. 1985. Structure of the signal recognition particle by electron microscopy. *Proc. Natl. Acad. Sci. USA* **82**:785–789.

Arima, K., Oshima, T., Kubota, I., Nakamura, N., Mizunaga, T., and Tohe, A. 1983. The nucleotide sequence of the yeast *PHO5* gene: A putative precursor of repressible acid phosphatase contains a signal peptide. *Nucleic Acids Res.* **11**:1657–1672.

Armstrong, J., Niemann, H., Smeekens, S., Rottier, P., and Warren, G. 1984. Sequence and topology of a model intracellular membrane protein, E1 glycoprotein, from a coronavirus. *Nature (London)* **308**:751–752.

Austen, B. M., Hermon-Taylor, J., Kaderbhai, M. A., and Ridd, D. H. 1984. Design and synthesis of a consensus signal sequence that inhibits protein translocation into rough microsomal vesicles. *Biochem. J.* **224**:317–325.

Bajwa, W., Meyhack, B., Rudolph, H., Schweingruber, A. M., and Hinnen, A. 1984. Structural analysis of the two tandemly repeated acid phosphatase genes in yeast. *Nucleic Acids Res.* **12:**7721–7739.

Bankaitis, V. A., Rasmussen, B. A., and Bassford, P. J. 1984. Intragenic suppressor mutations that restore export of maltose binding protein with a truncated signal peptide. *Cell* **37:**243–252.

Bedouelle, H., Schmeissner, U., Hofnung, M., and Rosenberg, M. 1982. Promoters of the *malEFG* and *malK–lamB* operons in *Escherichia coli* K12. *J. Mol. Biol.* **161:**519–531.

Bell, G. I., Santerre, R. F., and Mullenbach, G. T. 1983a. Hamster preproglucagon contains the sequence of glucogon and two related peptides. *Nature (London)* **302:**716–718.

Bell, G. I., Sanchez-Pescador, R., Laybourn, P. J., and Najarian, R. C. 1983b. Exon duplication and divergence in the human preproglucagon gene. *Nature (London)* **304:**368–371.

Belt, K. T., Carroll, M. C., and Porter, R. R. 1984. The structural basis of the multiple forms of human complement component C4. *Cell* **36:**907–914.

Bensi, G., Raugei, G., Klefenz, H., and Cortese, R. 1985. Structure and expression of the human haptoglobin locus. *EMBO J.* **4:**119–126.

Berg, L. S. 1940. Classification of fishes, both recent and fossil. *Trav. Inst. Zool. Acad. Sci. USSR* **5:**1–304.

Boeke, J. D., and Model, P. 1982. A prokaryotic membrane anchor sequence: Carboxyl terminus of bacteriophage f1 gene III protein retains it in the membrane. *Proc. Natl. Acad. Sci. USA* **79:**5200–5204.

Boel, E., Schwartz, T. W., Norris, K. E., and Fiil, N. P. 1984. A cDNA encoding a small common precursor for human pancreatic polypeptide and pancreatic icosapeptide. *EMBO J.* **3:**909–912.

Boguski, M. S., Elshourbagy, N., Taylor, J. M., and Gordon, J. I. 1984. Rat apolipoprotein A-IV contains 13 tandem repetitions of a 22-amino acid segment with amphipathic helical potential. *Proc. Natl. Acad. Sci. USA* **81:**5021–5025.

Boorstein, W. R., Vamvakopoulos, N. C., and Fiddes, J. C. 1982. Human chorionic gonadotropin β-subunit is encoded by at least eight genes arranged in tandem and inverted pairs. *Nature (London)* **300:**419–422.

Bown, D., Levasseur, M., Croy, R. R. D., Boulter, D., and Gatehouse, J. A. 1985. Sequence of a pseudogene in the legumin gene family of pea (*Pisum sativum* L.). *Nucleic Acids Res.* **13:**4527–4538.

Braun, G., and Cole, S. T. 1983. Molecular characterization of the gene coding for major outer membrane protein OmpA from *Enterobacter aerogenes*. *Eur. J. Biochem.* **137:**495–500.

Broglie, R., Coruzzi, G., Lamppa, G., Keith, B., and Chua, N. H. 1983. Structural analysis of nuclear genes coding for the precursor to the small subunit of wheat ribulose-1,5-bisphosphate carboxylase. *Biotechnology* **1:**55–61.

Byrne, B. M., von het Schip, A. D., von de Klundert, J. A. M., Arnberg, A. C., Gruber, M., and AB, G. 1984. Amino acid sequence of phosvitin derived from the nucleotide sequence of part of the chicken vitellogenin gene. *Biochemistry* **23:**4275–4279.

Carroll, M. C., Belt, T., Palsdottir, A., and Porter, R. R. 1984. Structure and organization of the C4 genes. *Phil. Trans. R. Soc. Lond. B* **306:**379–388.

Chin, W. W., Kronenberg, H. M., Dee, P. C., Maloof, F., and Habener, J. F. 1981. Nucleotide sequence of the mRNA encoding the pre-α-subunit of mouse thyrotropin. *Proc. Natl. Acad. Sci. USA* **78:**5329–5333.

Chin, W. W., Godine, J. E., Klein, D. R., Chang, A. S., Tan, L. K., and Habener, J. F. 1983. Nucleotide sequence of the cDNA encoding the precursor of the β subunit of rat lutropin. *Proc. Natl. Acad. Sci. USA* **80:**4649–4653.

Cooke, N. E., Coit, D., Weiner, R. I., Baxter, J. D., and Martial, J. A. 1980. Structure of cloned DNA complementary to rat prolactin messenger RNA. *J. Biol. Chem.* **255:**6502–6510.

Cooke, N. E., Coit, D., Shine, J., Baxter, J. D., and Martial, J. A. 1981. Human prolactin. cDNA structural analysis and evolutionary comparisons. *J. Biol. Chem.* **256:**4007–4016.

Coruzzi, G., Broglie, R., Cashmore, A., and Chua, N. H. 1983. Nucleotide sequences of two pea cDNA clones encoding the small subunit of ribulose 1,5-bisphosphate carboxylase and the major chlorophyll *a*/*b*-binding thylakoid polypeptide. *J. Biol. Chem.* **258:**1399–1402.

Coruzzi, G., Broglie, R., Edwards, C., and Chua, N. H. 1984. Tissue-specific and light-regulated expression of a pea nuclear gene encoding the small subunit of ribulose-1,5-bisphosphate carboxylase. *EMBO J.* **3:**1671–1679.

Craik, C. S., Choo, Q. L., Swift, G. H., Quinto, C., MacDonald, R. J., and Rutter, W. J. 1984. Structure of two related rat pancreatic trypsin genes. *J. Biol. Chem.* **259:**14255–14264.

Crouch, M. L., Tenbarge, K. M., Simon, A. E., and Ferl, R. 1983. cDNA clones for *Brassica napus* seed storage proteins: Evidence from nucleotide sequence analysis that both subunits of napin are cleaved from a precursor polypeptide. *J. Mol. Appl. Genet.* **2:**273–283.

de Bruijn, M. H. L., and Fey, G. H. 1985. Human complement component C3: cDNA coding sequence and derived primary structure. *Proc. Natl. Acad. Sci. USA* **82:**708–712.

de Geus, P., Verheij, H. M., Riegman, N. H., Hoekstra, W. P. M., and de Haas, G. H. 1984. The pro- and mature forms of the *E. coli* K-12 outer membrane phospholipase A are identical. *EMBO J.* **3:**1799–1802.

Deschenes, R. J., Haun, R. S., Funckes, C. L., and Dixon, J. E. 1985. A gene encoding rat cholecystokinin. *J. Biol. Chem.* **260:**1280–1286.

Dillon, L. S. 1983. *The Inconstant Gene,* New York, Plenum Press.

Domoney, D., and Casey, R. 1985. Measurement of gene number for seed storage proteins in *Pisum. Nucleic Acids Res.* **13:**687–699.

Dons, J. J. M., Mulder, G. H., Rouwendal, G. J. A., Springer, J., Bremer, W., and Wessels, J. G. H. 1984. Sequence analysis of a split gene involved in fruiting from the fungus *Shizophyllum commune. EMBO J.* **3:**2101–2106.

Dugaiczyk, A., Law, S. W., and Dennison, O. E. 1982. Nucleotide sequence and the encoded amino acids of human serum albumin mRNA. *Proc. Natl. Acad. Sci. USA* **79:**71–75.

Ebina, Y., Ellis, L., Jarnagin, K., Edery, M., Graf, L., Clauser, E., Ou, J. H., Masiarz, F., Kan, Y. W., Goldfine, I. D., Roth, R. A., and Rutter, W. J. 1985. The human insulin receptor cDNA: The structural basis for hormone-activated transmembrane signalling. *Cell* **40:**747–758.

Ehring, R., Beyreuther, K., Wright, J. K., and Overath, P. 1980. *In vitro* and *in vivo* products of *E. coli* lactose permease gene are identical. *Nature (London)* **283:**537–540.

Elango, N., Satake, M., Coligan, J. E., Norrby, E., Camargo, E., and Venkatesan, S. 1985. Respiratory syncytial virus fusion glycoprotein: nucleotide sequence of mRNA, identification of cleavage activation site and amino acid sequence of N-terminus of F_1 subunit. *Nucleic Acids Res.* **13:**1559–1574.

Elleman, T. C., and Hoyne, P. A. 1984. Nucleotide sequence of the gene encoding pilin of *Bacteroides nodosus,* the causal organism of ovine footrot. *J. Bacteriol.* **160:**1184–1187.

Emr, S. D., and Silhavy, T. J. 1983. Importance of secondary structure in the signal sequence for protein secretion. *Proc. Natl. Acad. Sci. USA* **80:**4599–4603.

Erwin, C. R., Croyle, M. L., Donelson, J. E., and Maurer, R. A. 1983. Nucleotide sequence of cloned complementary deoxyribonucleic acid for the α subunit of bovine pituitary glycoprotein hormones. *Biochemistry* **22:**4856–4860.

Fey, G. H., Lundwall, A., Wetsel, R. A., Zack, B. F., de Bruijn, M. H. L., and Domdey, H. 1984. Nucleotide sequence of complementary DNA and derived amino acid sequence of murine complement protein C3. *Phil. Trans. R. Soc. Lond. B* 306:333–344.

Fiddes, J. C., and Goodman, H. M. 1979. Isolation, cloning and sequence analysis of the cDNA for the α-subunit of human chorionic gonadotropin. *Nature (London)* **281:**351–355.

Finlay, B. B., Frost, L. S., and Paranchych, W. 1984. Localization, cloning, and sequence determination of the conjugative plasmid ColB2 pilin gene. *J. Bacteriol.* **160:**402–407.

Fojo, S. S., Law, S. W., and Brewer, H. B. 1984. Human apolipoprotein C-II: complete nucleic acid sequence of preapolipoprotein C-II. *Proc. Natl. Acad. Sci. USA* **81:**6354–6357.

Forde, B. G., Kreis, M., Williamson, M. S., Fry, R. P., Pywell, J., Shewry, P. R., Bunce, N., and Miflin, B. J. 1985. Short tandem repeats shared by B- and C-hordein cDNAs suggest a common evolutionary origin for two groups of cereal storage protein genes. *EMBO J.* **4:**9–15.

Freudl, R., and Cole, S. T. 1983. Cloning and molecular characterization of the *ompA* gene from *Salmonella typhimurium. Eur. J. Biochem.* **134:**497–502.

Froshauer, S., and Beckwith, J. 1984. The nucleotide sequence of the gene for *malF* protein, an inner membrane component of the maltose transport system of *Escherichia coli. J. Biol. Chem.* **259:**10896–10903.

Frost, L. S., Paranchych, W., and Willetts, N. S. 1984. DNA sequence of the F *traALE* region that includes the gene for F pilin. *J. Bacteriol.* **160:**395–401.

Fukusaki, E., Panbangred, W., Shinonyo, A., and Okada, H. 1984. The complete nucleotide sequence of the xylanase gene (*xynA*) of *Bacillus pumilus. FEBS Lett.* **171:**197–201.

Funckes, C. L., Minth, C. D., Deschenes, R., Magazin, M., Tavianini, M. A., Sheets, M., Collier, K., Weith, H. L., Aron, D. C., Roos, B. A., and Dixon, J. E. 1983. Cloning and characterization of a mRNA-encoding rat preprosomatostatin. *J. Biol. Chem.* **258:**8781–8787.

Gennaro, M. L., and Greenaway, P. J. 1983. Nucleotide sequences within the cholera toxin operon. *Nucleic Acids Res.* **11:**3855–3861.

Gluschankof, P., Morel, A., Benoit, R., and Cohen, P. 1985. The somatostatin-28 convertase of rat brain cortex generates both somatostatin-14 and somatostatin-28 (1–12). *Biochem. Biophys. Res. Commun.* **128**:1051–1057.

Godine, F. E., Chin, W. W., and Habener, J. F. 1982. α subunit of rat pituitary glycoprotein hormones. Primary structure of the precursor determined from the nucleotide sequence of cloned cDNAs. *J. Biol. Chem.* **257**:8368–8371.

Goodall, G. J., Richardson, M., Furuichi, Y., Wodnar-Filipowicz, A., and Horecker, B. L. 1985. Sequence of a cloned 523-bp cDNA for thymosin B₄. *Arch. Biochem. Biophys.* **236**:445–447.

Goodman, R. H., Jacobs, J. W., Chin, W. W., Lund, P. K., Dee, P. C., and Habener, J. F. 1980. Nucleotide sequence of a cloned structural gene coding for a precursor of pancreatic somatostatin. *Proc. Natl. Acad. Sci. USA* **77**:5869–5873.

Gordon, G., Gayda, R. C., and Markovitz, A. 1984. Sequence of the regulatory region of *ompT*, the gene specifying major outer membrane protein α(3b) of *Escherichia coli* K-12: Implications for regulation and processing. *Mol. Gen. Genet.* **193**:414–421.

Gorin, M. B., Cooper, D. L., Eiferman, F., van de Rijn, P., and Tilghman, S. M. 1981. The evolution of α-fetoprotein and albumin. I. A comparison of the primary amino acid sequence of mammalian α-fetoprotein and albumin. *J. Biol. Chem.* **256**:1954–1959.

Gray, G. L., Smith, D. H., Baldridge, J. S., Harkins, R. N., Vasil, M. L., Chen, E. Y., and Heyneker, H. L. 1984. Cloning, nucleotide sequence and expression in *Escherichia coli* of the exotoxin A structural gene of *Pseudomonas aeruginosa*. *Proc. Natl. Acad. Sci. USA* **81**:2645–2649.

Greenberg, R., and Groves, M. L. 1984. Plasmin cleaves human β-casein. *Biochem. Biophys. Res. Commun.* **125**:463–468.

Gubbins, E. J., Maurer, R. A., Hartley, J. L., and Donelson, J. E. 1979. Construction and analysis of recombinant DNAs containing a structural gene for rat prolactin. *Nucleic Acids Res.* **6**:915–930.

Gubler, U., Seeburg, P., Hoffman, B. J., Gage, L. P., and Udenfriend, S. 1982. Molecular cloning establishes proenkephalin as precursor of enkephalin-containing peptide. *Nature (London)* **295**:206–208.

Gubler, U., Monahan, J. J., Lomedico, P. T., Bhatt, R. S., Collier, K. J., Hoffman, B. J., Böhlen, P., Esch, F., Ling, N., Zeytin, F., Brazeau, P., Pooniman, M. S., and Gage, L. P. 1983. Cloning and sequence analysis of cDNA for the precursor of human growth hormone-releasing factor, somatocrinin. *Proc. Natl. Acad. Sci. USA* **80**:4311–4314.

Gubler, U., Chua., A. O., Hoffman, B. J., Collier, K. J., and Eng, J. 1984. Cloned cDNA to cholecystokinin mRNA predicts an identical preprocholecystokinin in pig brain and gut. *Proc. Natl. Acad. Sci. USA* **81**:4307–4310.

Gurr, J. A., Catterall, J. F., and Kourides, I. A. 1983. Cloning of cDNA encoding the pre-β subunit of mouse thyrotropin. *Proc. Natl. Acad. Sci. USA* **80**:2122–2126.

Hahn, V., Winkler, J., Rapoport, T. A., Liebscher, D. H., Contelle, C., and Rosenthal, S. 1983. Carp preproinsulin cDNA sequence and evolution of insulin genes. *Nucleic Acids Res.* **11**:4541–4552.

Hall, L., Laird, J. E., and Craig, R. K. 1984. Nucleotide sequence determination of guinea-pig casein β mRNA reveals homology with bovine and rat α$_{s1}$ caseins and conservation of the non-coding regions of the mRNA. *Biochem. J.* **222**:561–570.

Hall, M. N., Hereford, L., and Herskowitz, I. 1984. Targeting of *E. coli* β-galactosidase to the nucleus in yeast. *Cell* **36**:1057–1065.

Hannink, M., and Donoghue, D. J. 1984. Requirement for a signal sequence in biological expression of the v-*sis* oncogene. *Science* **226**:1197–1199.

Hefford, M. A., Evans, R. M., Oda, G., and Kaplan, H. 1985. Unusual chemical properties of N-terminal histidine residues of glucagon and vasoactive intestinal peptide. *Biochemistry* **24**:867–874.

Heinrich, G., Kronenberg, H. M., Potts, J. T., and Habener, J. F. 1984a. Gene encoding parathyroid hormone. Nucleotide sequence of the rat gene and deduced amino acid sequence of the rat preproparathyroid hormone. *J. Biol. Chem.* **259**:3320–3329.

Heinrich, G., Gros, P., and Habener, J. F. 1984b. Glucagon gene sequence. Four of six exons encode separate functional domains of rat pre-proglucagon. *J. Biol. Chem.* **259**:14082–14087.

Heller, K., and Kadner, R. J. 1985. Nucleotide sequence of the gene for the vitamin B₁₂ receptor protein in the outer membrane of *Escherichia coli*. *J. Bacteriol.* **161**:904–908.

Higgins, C. F., Haag, P. D., Nikaido, K., Ardeshir, F., Garcia, G., and Ames, G. F. L. 1982. Complete nucleotide sequence and identification of membrane components of the histidine transport operon of *S. typhimurium*. *Nature (London)* **298**:723–727.

Hobart, P. M., Crawford, R., Shen, L. P., Pictet, R., and Rutter, W. J. 1980. Cloning and sequence analysis of cDNAs encoding two distinct somatostatin precursors found in the endocrine pancreas of anglerfish. *Nature (London)* **288:**137–141.

Hobart, P. M., Fogliano, M., O'Connor, B. A., Schaefer, I. M., and Chirgwin, J. M. 1984. Human renin gene: Structure and sequence analysis. *Proc. Natl. Acad. Sci. USA* **81:**5026–5030.

Hoffman, L. M., Ma, Y., and Barker, R. F. 1982. Molecular cloning of *Phaseolus vulgaris* lectin mRNA and use of cDNA as a probe to estimate lectin transcript levels in various tissues. *Nucleic Acids Res.* **10:**7819–7828.

Holland, E. C., Leung, J. O., and Drickamer, K. 1984. Rat liver asialoglycoprotein receptor lacks a cleavable NH$_2$-terminal signal sequence. *Proc. Natl. Acad. Sci. USA* **81:**7338–7342.

Hope, I. A., Mackay, M., Hyde, J. E., Goman, M. and Scaife, J. 1985. The gene for an exported antigen of the malaria parasite *Plasmodium falciparum* cloned and expressed in *Escherichia coli. Nucleic Acids Res.* **13:**369–379.

Horwich, A. L., Kalousek, F., Mellman, I., and Rosenberg, L. E. 1985a. A leader peptide is sufficient to direct mitochondrial import of a chimeric chain. *EMBO J.* **4:**1129–1135.

Horwich, A. L., Kalousek, F., and Rosenberg, L. E. 1985b. Arginine in the leader peptide is required for both import and proteolytic cleavage of a mitochondrial precursor. *Proc. Natl. Acad. Sci. USA* **82:**4930–4933.

Hughes, J., Smith, T. W., Kosterlitz, H. W., Fothergill, L. A., Morgan, B. A., and Morris, H. R. 1979. Identification of two related pentapeptides from the brain with potent opiate agonist activity. *Nature (London)* **258:**577–579.

Hurt, E. C., Pesold-Hurt, B., and Schatz, G. 1984. The cleavable prepiece of an imported mitochondrial protein is sufficient to direct cytosolic dihydrofolate reductase into the mitochondrial matrix. *FEBS Lett.* **178:**306–310.

Iatrou, K., Tsitilous, S. G., and Kafatos, F. C. 1984. DNA sequence transfer between two high-cysteine chorion gene families in the silkmoth *Bombyx mori. Proc. Natl. Acad. Sci. USA* **81:**4452–4456.

Innis, M. A., Tokunaga, M., Williams, M. E., Loranger, J. M., Chang, S. Y., Chang, S., and Wu, H. C. 1984. Nucleotide sequence of the *Escherichia coli* prolipoprotein signal peptidase *(lsp)* gene. *Proc. Natl. Acad. Sci. USA* **81:**3708–3712.

Inokuchi, K., Mutoh, N., Matsuyama, S., and Mizushima, S. 1982. Primary structure of the *ompF* gene that codes for a major outer membrane protein of *Escherichia coli* K-12. *Nucleic Acids Res.* **10:**6957–6968.

Ivell, R., and Richter, D. 1984a. Structure and comparison of the oxytocin and vasopressin genes from rats. *Proc. Natl. Acad. Sci. USA* **81:**2006–2010.

Ivell, R., and Richter, D. 1984b. The gene for the hypothalamic peptide hormone oxytocin is highly expressed in the bovine corpus luteum: Biosynthesis, structure and sequence analysis. *EMBO J.* **3:**2351–2354.

Jameson, L., Chin, W. W., Hollenberg, A. N., Chang, A. S., and Habener, J. F. 1984. The gene encoding the β subunit of rat luteinizing hormone. Analysis of gene structure and evolution of nucleotide sequence. *J. Biol. Chem.* **259:**15474–15480.

Jung, A., Sippel, A. E., Grez, M., and Shütz, G. 1980. Exons encode functional and structural units of chicken lysozyme. *Proc. Natl. Acad. Sci. USA* **77:**5759–5763.

Kaczorek, M., Delpeyroux, F., Chenciner, N., Streeck, R. E., Murphy, J. R. Boquet, P., and Tiollais, P. 1983. Nucleotide sequence and expression of the diptheria *tox228* gene in *Escherichia coli. Science* **221:**855–858.

Kageyama, R., Ohkubo, H., and Nakanishi, S. 1984. Primary structure of human preangiotensinogen deduced from the cloned cDNA sequence. *Biochemistry* **23:**3603–3609.

Kaput, J., Goltz, S., and Blobel, G. 1982. Nucleotide sequence of the yeast nuclear gene for cytochrome *c* peroxidase precursor. *J. Biol. Chem.* **257:**15054–15058.

Karathanasis, S. K., Zannis, V. I., and Breslow, J. L. 1983. Isolation and characterization of the human apolipoprotein A-I gene. *Proc. Natl. Acad. Sci. USA* **80:**6147–6151.

Kasarda, D. D., Okita, T. W., Bernardin, J. E., Baecker, P. A., Nimmo, C. C., Lew, E. J. L., Dietler, M. D., and Greene, F. C. 1984. Nucleic acid (cDNA) and amino acid sequences of σ-type gliadins from wheat *(Triticum aestivum). Proc. Natl. Acad. Sci. USA* **81:**4712–4716.

Kato, K., Hayashizaki, Y., Takahashi, Y., Himino, S., and Matsubara, K. 1983. Molecular cloning of the human gastrin gene. *Nucleic Acids Res.* **11:**8197–8203.

Kaufman, J. F., Auffray, C., Korman, A. J., Shackelford, D. A., and Strominger, J. 1984. The class II molecules of the human and murine major histocompatibility complex. *Cell* **36:**1–13.

Kennedy, B. P., Marsden, J. J., Flynn, T. G., de Bold, A. J., and Davies, P. L. 1984. Isolation and nucleotide sequence of a cloned cardionatrin cDNA. *Biochem. Biophys. Res. Commun.* **122:**1076–1082.

Kitamura, N., Takagaki, Y., Furuto, S., Tanaka, T., Nawa, H., and Nakanishi, S. 1983. A single gene for bovine high molecular weight and low molecular weight kininogens. *Nature (London)* **305**:545–549.

Kitamura, N., Kitagawa, H., Fukushima, D., Takagaki, Y., Miyata, T., and Nakanishi, S. 1985. Structural organization of the human kininogen gene and a model for its evolution. *J. Biol. Chem.* **260**:8610–8617.

Klemm, P. 1984. The *fimA* gene encoding the type-1 fimbrial subunit of *Escherichia coli. Eur. J. Biochem.* **143**:395–399.

Knott, T. J., Robertson, M. E., Priestley, L. M., Wallis, S., and Scott, J. 1984a. Characterisation of mRNAs encoding the precursor for human apolipoprotein CI. *Nucleic Acids Res.* **12**:3909–3915.

Knott, T. J., Priestley, L. M., Urdea, M., and Scott, J. 1984b. Isolation and characterization of a cDNA encoding the precursor for human apolipoprotein AII. *Biochem. Biophys. Res. Commun.* **120**:732–740.

Kreis, M., Rahman, S., Forde, B. G., Pywell, J., Shewry, P. R., and Miflin, B. J. 1983. Sub-families of hordein mRNA encoded at the *Hor2* locus of barley. *Mol. Gen. Genet.* **191**:194–200.

Kurjan, J., and Herskowitz, I. 1982. Structure of a yeast pheromone gene (*MFα*): A putative α-factor precursor contains four tandem copies of mature α-factor. *Cell* **30**:933–943.

Kuwano, R., Araki, K., Usui, H., Fukui, T., Othsuka, E., Ikehara, M., and Takahashi, Y. 1984. Molecular cloning and nucleotide sequence of cDNA coding for rat brain cholecystokinin precursor. *J. Biochem.* **96**:923–926.

Lackner, K. J., Law, S. W., and Brewer, H. B. 1984. Human apolipoprotein A-II: Complete nucleic acid sequence of preproapolipoprotein A-II. *FEBS Lett.* **175**:159–164.

Lamb, F. I., Roberts, L. M., and Lord, J. M. 1985. Nucleotide sequence of cloned cDNA coding for preproricin. *Eur. J. Biochem.* **148**:265–270.

Land, H., Schatz, G., Schmale, H., and Richter, D. 1982. Nucleotide sequence of cloned cDNA encoding bovine arginine vasopressin-neurophycin II precursor. *Nature (London)* **295**:299–303.

Law, S. W., and Brewer, H. B. 1984. Nucleotide sequence and the encoded amino acids of human apolipoprotein A-I mRNA. *Proc. Natl. Acad. Sci. USA* **81**:66–70.

Law, S. W., and Dugaiczyk, A. 1982. Homology between the primary structure of α-fetoprotein, deduced from a complete cDNA sequence, and serum albumin. *Nature (London)* **291**:201–205.

Legon, S., Glover, D. M., Hughes, J., Lowry, P. J., Rigby, P. W. J., and Watson, C. J. 1982. The structure and expression of the preproenkephalin gene. *Nucleic Acids Res.* **10**:7905–7918.

LeMeur, M. A., Galliot, B., and Gerlinger. P. 1984. Termination of the ovalbumin gene transcription. *EMBO J.* **3**:2779–2786.

Le Moullec, J. M., Jullienne, A., Chenais, J., Lasmoles, F., Guliana, J. M., Milhaud, G., and Moukhtar, M. S. 1984. The complete sequence of human preprocalcitonin. *FEBS Lett.* **167**:93–97.

Lentz, S. R., Birken, S., Lustbader, J., and Boime, I. 1984. Posttranslational modification of the carboxyl-terminal region of the β subunit of human chorionic gonadotropin. *Biochemistry* **23**:5330–5337.

LeRoith, D., Pickens, W., Vinik, A. I., and Shiloach, J. 1985. *Bacillus subtilis* contains multiple forms of somatostatin-like material. *Biochem. Biophys. Res. Commun.* **127**:713–719.

Leytus, S. P., Chung, D. W., Kisiel, W., Kurachi, K., and Davie, E. W. 1984. Characterization of a cDNA coding for human factor X. *Proc. Natl. Acad. Sci. USA* **81**:3699–3702.

Limbach, K. J., and Wu, R. 1985a. Characterization of a mouse somatic cytochrome *c* gene and three cytochrome *c* pseudogenes. *Nucleic Acids Res.* **13**:617–630.

Limbach, K. J., and Wu, R. 1985b. Characterization of two *Drosophila melanogaster* cytochrome *c* genes and their transcripts. *Nucleic Acids Res.* **13**:631–644.

Linzer, D. I. H., and Nathans, D. 1984. Nucleotide sequence of a growth-related mRNA encoding a member of the prolactin-growth hormone family. *Proc. Natl. Acad. Sci. USA* **81**:4255–4259.

Linzer, D. I. H., and Talamantes, F. 1985. Nucleotide sequence of mouse prolactin and growth hormone mRNAs and expression of these mRNAs during pregnancy. *J. Biol. Chem.* **260**:9574–9579.

Lockman, H. A., Galen, J. E., and Kaper, J. B. 1984. *Vibrio cholerae* enterotoxin genes: Nucleotide sequence analysis of DNA encoding ADP-ribosyltransferase. *J. Bacteriol.* **159**:1086–1089.

Lomax, M. I., Bachman, N. J., Nasoff, M. S., Caruthers, M. H., and Grossman, L. I. 1984. Isolation and characterization of a cDNA clone for bovine cytochrome *c* oxidase subunit IV. *Proc. Natl. Acad. Sci. USA* **81**:6295–6299.

Long, G. L., Chandra, T., Woo, S. L. C., Davie, E. W., and Kurachi, K. 1984a. Complete sequence of the cDNA for human a_1-antitrypsin and the gene for the S variant. *Biochemistry* **23**:4828–4837.

Long, G. L., Belagaje, R. M., and MacGillivray, R. T. A. 1984b. Cloning and sequencing of liver cDNA coding for bovine protein C. *Proc. Natl. Acad. Sci. USA* **81**:5653–5656.

Lopez, L. C., Frazier, M. L., Su, C. J., Kumar, A., and Saunders, G. F. 1983. Mammalian pancreatic preproglucagon contains three glucagon-related peptides. *Proc. Natl. Acad. Sci. USA* **80**:5485–5489.

Lund, P. K., Goodman, R. H., Montminy, M. R., Dee, P. C., and Habener, J. F. 1983. Anglerfish islet preproglucagon II. Nucleotide and corresponding amino acid sequence of the cDNA. *J. Biol. Chem.* **258**:3280–3284.

Lund, P. K., Moats-Staats, B. M., Simmons, J. G., Hoyt, E., D'Ercole, J., Martin, F., and Van Wyk, J. J. 1985. Nucleotide sequence analysis of a cDNA encoding human ubiquitin reveals that ubiquitin is synthesized as a precursor. *J. Biol. Chem.* **260**:7609–7613.

Lundwall, A. B., Wetsel, R. A., Kristensen, T., Whitehead, A. S., Woods, D. E., Ogden, R. C., Colten, H. R., and Tack, B. F. 1985. Isolation and sequence analysis of a cDNA clone encoding the fifth complement component. *J. Biol. Chem.* **260**:2108–2112.

Lycett, G. W., Delauney, A. J., Gatehouse, J. A., Gilroy, J., Croy, R. R. D., and Boulter, D. 1983. The vicilin gene family of pea (*Pisum sativum* L.): A complete cDNA coding sequence for preprovicilin. *Nucleic Acids Res.* **11**:2367–2380.

Lycett, G. W., Croy, R. R. D., Shirsat, A. H., and Boulter, D. 1984. The complete nucleotide sequence of a legumin gene from pea (*Pisum sativum* L.). *Nucleic Acids Res.* **12**:4493–4506.

Maarse, A. C., Van Loon, A. P. G. M., Riezman, H., Gregor, I., Schatz, G., and Grivell, L. A. 1984. Subunit IV of yeast cytochrome *c* oxidase: cloning and nucleotide sequencing of the gene and partial amino acid sequencing of the mature protein. *EMBO J.* **3**:2831–2837.

Magazin, M., Minth, C. D., Funckes, C. L., Deschenes, R., Tavianini, M. A., and Dixon, J. E. 1982. Sequence of a cDNA encoding pancreatic preprosomatostatin-22. *Proc. Natl. Acad. Sci. USA* **79**:5152–5156.

Majzoub, J. A., Pappey, A., Burg, R., and Habener, J. F. 1984. Vasopressin gene is expressed at low levels in the hypothalamus of the Brattleboro rat. *Proc. Natl. Acad. Sci. USA* **81**:5296–5299.

Maki, M., Parmentier, M., and Ingami, T. 1984. Cloning of genomic DNA for human atrial natriuretic factor. *Biochem. Biophys. Res. Comm.* **125**:797–802.

Martial, J. A., Hallewell, R. A., Baxter, J. D., and Goodman, H. M. 1979. Human growth hormone: Complementary DNA cloning and expression in bacteria. *Science* **205**:602–607.

Mason, A. J., Evans, B. A., Cox, D. R., Shine, J., and Richards, R. I. 1983. Structure of mouse kallikrein gene family suggests a role in specific processing of biologically active peptides. *Nature (London)* **303**:300–307.

Mauff, G., Stener, M., and Bender, K. 1983. The C4B chain: Evidence for genetically determined polymorphism. *Hum. Genet.* **64**:186–188.

Maurer, R. A., Croyle, M. L., and Donelson, J. E. 1984. The sequence of a cloned cDNA for the β subunit of bovine thyrotropin predicts a protein containing both NH_2 and COOH terminal extensions. *J. Biol. Chem.* **259**:5024–5027.

Mayo, K. E., Cerelli, G. M., Lebo, R. V., Bruce, B. D., Rosenfeld, M. G., and Evans, R. M. 1985. Gene encoding human growth hormone-releasing factor precursor: Structure, sequence, and chromosomal assignment. *Proc. Natl. Acad. Sci. USA* **82**:63–67.

Michaelis, S., and Beckwith, J. 1982. Mechanism of incorporation of cell envelope proteins in *Escherichia coli*. *Annu. Rev. Microbiol.* **36**:435–465.

Miller, W. L., Martial, J. A., and Baxter, J. D. 1980. Molecular cloning of DNA complementary to bovine growth hormone in RNA. *J. Biol. Chem.* **255**:7521–7524.

Minth, C. D., Taylor, W. L., Magazin, M., Tavianini, M., Collier, K., Weith, H. L., and Dixon, J. E. 1982. The structure of cloned DNA complementary to catfish pancreatic somatostatin-14 messenger RNA. *J. Biol. Chem.* **257**:10372–10377.

Mita, S., Maeda, S., Shimada, K., and Araki, S. 1984. Cloning and sequence analysis of cDNA for human prealbumin. *Biochem. Biophys. Res. Commun.* **124**:558–564.

Miyazaki, H., Kukamizu, A., Hirose, S., Hayashi, T., Hori, H., Ohkuba, H., Nakamishi, S., and Murakami, K. 1984. Structure of the human renin gene. *Proc. Natl. Acad. Sci. USA* **81**:5999–6003.

Mizuno, T., Wurtzel, E. T., and Inouye, M. 1982. Osmoregulation of gene expression. II. DNA sequence of the *envZ* gene of the *ompB* operon of *E. coli* and characterization of its gene product. *J. Biol. Chem.* **257**:13692–13698.

Mizuno, T., Chou, M. Y., and Inouye, M. 1983. A comparative study of the genes for three porins of the *Escherichia coli* outer membrane. DNA sequence of the osmoregulated *ompC* gene. *J. Biol. Chem.* **258**:6932–6940.

Momma, T., Negoro, T., Udaka, K., and Fukizawa, C. 1985a. A complete cDNA coding for the sequence of glycinin A_2B_{1a} subunit precursor. *FEBS Lett.* **188:**117–122.

Momma, T., Negoro, T., Hirano, H., Matsumoto, A., Udaka, K., and Fukazawa, C. 1985b. Glycinin $A_5A_4B_3$ mRNA: cDNA cloning and nucleotide sequencing of a splitting storage protein subunit of soybean. *Eur. J. Biochem.* **149:**491–496.

Montminy, M. R., Goodman, R. H., Horovitch, S. J., and Habener, J. F. 1984. Primary structure of the gene encoding rat preprosomatostatin. *Proc. Natl. Acad. Sci. USA* **81:**3337–3340.

Mooi, F. R., van Buuren, M., Koopman, G., Roosendaal, B., and de Graf, F. K. 1984. K88ab gene of *Escherichia coli* encodes a fimbria-like protein distinct from the K88ab fimbrial adhesion. *J. Bacteriol.* **159:**482–487.

Moore, M. N., Kao, F. T., Tsao, Y. K., and Chan, L. 1984. Human apolipoprotein A-II: Nucleotide sequence of a cloned cDNA, and localization of its structural gene on human chromosome 1. *Biochem. Biophys. Res. Commun.* **123:**1–7.

Morinaga, T., Sakai, M., Wegmann, T. G., and Tamaoki, T. 1983. Primary structures of human α-fetoprotein and its mRNA. *Proc. Natl. Acad. Sci. USA* **80:**4604–4608.

Morohashi, K., Fujii-Kuriyama, Y., Okada, Y., Sogawa, K., Hirose, T., Inayama, S., and Omura, T. 1984. Molecular cloning and nucleotide sequence of cDNA for mRNA of mitochondrial cytochrome P-450(SCC) of bovine adrenal cortex. *Proc. Natl. Acad. Sci. USA* **81:**4647–4651.

Nawa, H., Kitamura, N., Hirose, T., Asai, M., Inayama, S., and Nakanishi, S. 1983. Primary structures of bovine liver low molecular weight kininogen precursors and their two mRNAs. *Proc. Natl. Acad. Sci. USA* **80:**90–94.

Nishizawa, M., Hayakawa, Y., Yanaihara, N., and Okamoto, H. 1985. Nucleotide sequence divergence and functional constraint in VIP precursor mRNA evolution between human and rat. *FEBS Lett.* **183:**55–59.

Noda, M., Furutani, Y., Takahashi, H., Toyosato, M., Hirose, T., Inayama, S., Nakanishi, S., and Numa, S. 1982. Cloning and sequence analysis of cDNA for bovine adrenal preproenkephalin. *Nature (London)* **295:**202–206.

Oates, E., and Herbert, E. 1984. 5' sequence of porcine and rat proopiomelanocortin mRNA. One porcine and two rat forms. *J. Biol. Chem.* **259:**7421–7425.

Ohkuba, H., Kageyama, R., Ojihara, M., Hirose, T., Inayama, S., and Nakanishi, S. 1983. Cloning and sequence analysis of cDNA for rat angiotensinogen. *Proc. Natl. Acad. Sci. USA* **80:**2196–2200.

Ohmura, K., Nakamura, K., Yamazaki, H., Shiroza, T., Yamane, K., Jigami, Y., Tanaka, H., Yoda, K., Yamasaki, M., and Tamura, G. 1984. Length and structural effect of signal peptides derived from *Bacillus subtilis* α-amylase on secretion of *Escherichia coli* β-lactamase in *B. subtilis* cells. *Nucleic Acids Res.* **12:**5307–5319.

Okita, T. W., Cheesbrough, V., and Reeves, C. D. 1985. Evolution and heterogeneity of the α/β-type and γ-type gliadin DNA sequences. *J. Biol. Chem.* **260:**8203–8213.

Patzelt, C., and Schiltz, E. 1984. Conversion of proglucagon in pancreatic alpha cells: The major end products are glucagon and a single peptide, the major proglucagon fragment, that contains two glucagon-like sequences. *Proc. Natl. Acad. Sci. USA* **81:**5007–5011.

Pedersen, K., Devereux, J., Wilson, D. R., Sheldon, E., and Larkins, B. A. 1982. Cloning and sequence analysis reveal structural variation among related zein genes in maize. *Cell* **29:**1015–1029.

Pinsky, S. D., LaForge, K. S., Luc, V., and Scheele, G. 1983. Identification of cDNA clones encoding secretory isoenzyme forms: Sequence determination of canine pancreatic prechymotrypsinogen 2 mRNA. *Proc. Natl. Acad. Sci. USA* **80:**7486–7490.

Policastro, P., Ouitt, C. E., Hoshina, M., Fukuoka, H., Boothby, M. R., and Boime, I. 1983. The β subunit of human chorionic gonadotropin is encoded by multiple genes. *J. Biol. Chem.* **258:**11492–11499.

Prat, S., Cortadas, J., Pwigdomènech, P., and Palau, J. 1985. Nucleic acid (cDNA) and amino acid sequences of the maize endosperm protein glutelin-2. *Nucleic Acids Res.* **13:**1493–1504.

Qasba, P. K., and Safaya, S. K. 1984. Similarity of the nucleotide sequences of rat α-lactalbumin and chicken lysozyme genes. *Nature (London)* **308:**377–380.

Rafalski, J. A., Scheets, K., Motzler, M., Peterson, D. M., Hedgcoth, C., and Söll, D. G. 1984. Developmentally regulated plant genes: The nucleotide sequence of a wheat gliadin genomic clone. *EMBO J.* **3:**1409–1415.

Ramabhadran, T. V., Reitz, B. A., and Tiemeier, D. C. 1984. Synthesis and glycosylation of the common α subunit of human glycoprotein hormones in mouse cells. *Proc. Natl. Acad. Sci. USA* **81:**6701–6705.

Rapoport, T. A. 1985. Extensions of the signal hypothesis—Sequential insertion model versus amphipathic tunnel hypothesis. *FEBS Lett.* **187**:1–4.

Reeck, G. R., and Hedgcoth, C. 1985. Amino acid sequence alignment of cereal storage proteins. *FEBS Lett.* **180**:291–294.

Richter, K., Kawashima, E., Egger, R., and Kreil, G. 1984. Biosynthesis of thyrotropin releasing hormone in the skin of *Xenopus laevis:* Partial sequence of the precursor deduced from cloned cDNA. *EMBO J.* **3**:617–621.

Robert, L. S., Nozzolillo, C., and Altosaar, I. 1985. Homology between legumin-like polypeptides from cereals and pea. *Biochem. J.* **226**:847–852.

Rose, G. D., Geselowitz, A. R., Lesser, G. J., Lee, R. H., and Zehfus, M. H. 1985. Hydrophobicity of amino acid residues in globular proteins. *Science* **229**:834–838.

Rosen, H., Douglass, J., and Herbert, E. 1984. Isolation and characterization of the rat proenkephalin gene. *J. Biol. Chem.* **259**:14309–14313.

Roskam, W. G., and Rougeon, F. 1979. Molecular cloning and nucleotide sequence of the human growth hormone structural gene. *Nucleic Acids Res.* **7**:305–320.

Sadler, I., Suda, K., Schatz, G., Kaudewitz, F., and Haid, A. 1984. Sequencing of the nuclear genes for the yeast cytochrome c_1 precursor reveals an unusually complex amino-terminal presequence. *EMBO J.* **3**:2137–2143.

Sakai, M., Morinaga, T., Urano, Y., Watanabe, K., Wegmann, T. G., and Tamaoki, T. 1985. The human α-fetoprotein gene. Sequence organization and the 5' flanking region. *J. Biol. Chem.* **260**:5055–5060.

Sasavage, N. L., Nilson, J. H., Horowitz, S., and Rottman, F. M. 1981. Nucleotide sequence of bovine prolactin messenger RNA. Evidence for sequence polymorphism. *J. Biol. Chem.* **256**:678–682.

Scarpulla, R. C. 1985. Association of a truncated cytochrome *c* processed pseudogene with a similarly truncated member from a long interspersed repeat family of rat. *Nucleic Acids Res.* **13**:763–775.

Schreffler, D. C., Atkinson, J. P., Chan, A. C., Kasp, D. R., Killion, C. C., Ogata, R. T., and Rosa, P. A. 1984. The *C4* and *Slp* genes of the complement region of the murine *H-2* major histocompatibility complex. *Phil. Trans. R. Soc. Lond. B* **306**:395–403.

Schuler, M. A., Schmitt, E. S., and Beachy, R. N. 1982. Closely related families of genes code for the α and $α_1$ subunits of the soybean 7S storage protein complex. *Nucleic Acids Res.* **10**:8225–8244.

Seeburg, P. H., and Adelman, J. P. 1984. Characterization of cDNA for precursor of human luteinizing hormone releasing hormone. *Nature (London)* **311**:666–668.

Selby, M. J., Barta, A., Baxter, J. D., Bell, G. I., and Eberhardt, N. L. 1984. Analysis of a major human chorionic somatomammotropin gene. *J. Biol. Chem.* **259**:13131–13138.

Shen, L. P., and Rutter, W. J. 1984. Sequence of the human somatostatin I gene. *Science* **224**:168–170.

Shen, L. P., Pictet, R. L., and Rutter, W. J. 1982. Human somatostatin I: Sequence of the cDNA. *Proc. Natl. Acad. Sci. USA* **79**:4575–4579.

Shoulders, C. C., Kornblihtt, A. R., Munro, B. S., and Baralle, F. E. 1983. Gene structure of human apoliproprotein AI. *Nucleic Acids Res.* **11**:2827–2837.

Sibakov, M., and Palva, I. 1984. Isolation and the 5'-end nucleotide sequence of *Bacillus licheniformis* α-amylase gene. *Eur. J. Biochem.* **145**:567–572.

Siliciano, P. G., and Tatchell, K. 1984. Transcription and regulatory signals at the mating type locus in yeast. *Cell* **37**:969–978.

Singh, A., Chen, E. Y., Lugovoy, J. M., Chang, C. N., Hitzeman, R. A., and Seeburg, P. H. 1983. *Saccharomyces cerevisiae* contains two discrete genes coding for the α-factor pheromone. *Nucleic Acids Res.* **11**:4049–4063.

Slightom, J. L., Sun, S. M., and Hall, T. C. 1983. Complete nucleotide sequence of a French bean storage protein gene: Phaseolin. *Proc. Natl. Acad. Sci. USA* **80**:1897–1901.

Smith, S. M., Bedbrook, J., and Speirs, J. 1983. Characterisation of three cDNA clones encoding different mRNAs for the precursor to the small subunit of wheat ribulosebisphosphate carboxylase. *Nucleic Acids Res.* **11**:8719–8734.

Sogawa, K., Fujii-Kuriyama, Y., Mizukami, Y., Ichihara, Y., and Takahashi, K. 1983. Primary structure of human pepsinogen gene. *J. Biol. Chem.* **258**:5306–5311.

Soliday, C. L., Flurkey, W. H., Okita, T. W., and Kolattukudy, P. E. 1984. Cloning and structure determination of cDNA for cutinase, an enzyme involved in fungal penetration of plants. *Proc. Natl. Acad. Sci. USA* **81**:3939–3943.

Sonnenberg, H., and Veress, A. T. 1984. Cellular mechanism of release of atrial natriuretic factor. *Biochem. Biophys. Res. Comm.* **124**:443–449.

Sottrup-Jensen, L., Stepanik, T. M., Kristensen, T., Lønblad, P. B., Jones, C. M., Wierzbicki, D. M., Magnusson, S., Domdey, H., Wetsel, R. A., Lundwall, A., Tack, B. F., and Fey, G. H. 1985. Common evolutionary origin of α_2-macroglobulin and complement components C3 and C4. *Proc. Natl. Acad. Sci. USA* **82**:9–13.

Spena, A., Viotti, A., and Pirrotta, V. 1983. Two adjacent genomic zein sequences: Structure, organization and tissue-specific restriction pattern. *J. Mol. Biol.* **169**:799–811.

Spindel, E. R., Chin, W. W., Price, J., Rees, L. H., Besser, G. M., and Habener, J. F. 1984. Cloning and characterization of cDNAs encoding human gastrin-releasing peptide. *Proc. Natl. Acad. Sci. USA* **81**:5699–5703.

Steinmetz, M., and Hood, L. 1983. Genes of the major histocompatibility complex in mouse and man. *Science* **222**:727–733.

Stewart, A. F., Willis, I. M., and Mackinlay, A. G. 1984. Nucleotide sequences of bovine α_{s1} and κ-casein cDNAs. *Nucleic Acids Res.* **12**:3895–3907.

Stiekema, W. J., Wimpee, C. F., and Tobin, E. M. 1983. Nucleotide sequence encoding the precursor of the small subunit of ribulose-1,5-bisphosphate carboxylase from *Lemna gibba* LG3. *Nucleic Acids Res.* **11**:8051–8066.

Sumner-Smith, M., Rafalski, J. A., Sugiyama, T., Stoll, M., and Söll, D. 1985. Conservation and variability of wheat α/β-gliadin genes. *Nucleic Acids Res.* **13**:3905–3916.

Swift, G. H., Dagorn, J. C., Ashley, P. L., Cummings, S. W., and MacDonald, R. J. 1982. Rat pancreatic kallikrein mRNA: Nucleotide sequence and amino acid sequence of the encoded preproenzyme. *Proc. Natl. Acad. Sci. USA* **79**:7263–7267.

Swift, G. H., Craik, C. S., Stary, S. J., Quinto, C., Lahaie, R. G., Rutter, W. J., and MacDonald, R. J. 1984. Structure of the two related elastase genes expressed in the rat pancreas. *J. Biol. Chem.* **259**:14271–14278.

Takahara, M., Hibler, D. W., Barr, P. J., Gerlt, J. A., and Inouye, M. 1985. The *ompA* signal peptide directed secretion of staphylococcal nuclease A by *Escherichia coli*. *J. Biol. Chem.* **260**:2670–2674.

Takahashi, H., Nabeshima, Y., Nabeshima, Y.-I., Ogata, K., and Takeuchi, S. 1984. Molecular cloning and nucleotide sequence of DNA complementary to human decidual prolactin mRNA. *J. Biochem.* **95**:1491–1499.

Talbot, D. R., Adang, M. J., Slightom, J. L., and Hall, T. C. 1984. Size and organization of a multigene family encoding phaseolin, the major seed storage protein of *Phaseolus vulgaris* L. *Mol. Gen. Genet.* **198**:42–49.

Tanabe, T., Noda, M., Furutani, Y., Takai, T., Takahashi, H., Tanaka, K., Hirose, T., Inayama, S., and Numa, S. 1984. Primary sequence of β subunit precursor of calf acetylcholine receptor deduced from cDNA sequence. *Eur. J. Biochem.* **144**:11–17.

Taussig, R., and Carlson, M. 1983. Nucleotide sequence of the yeast SUC2 gene for invertase. *Nucleic Acids Res.* **11**:1943–1954.

Tavianini, M. A., Hayes, T. E., Magazin, M. D., Minth, C. D., and Dixon, J. E. 1984. Isolation, characterization, and DNA sequence of the rat somatostatin gene. *J. Biol. Chem.* **259**:11798–11803.

Truong, A. T., Duez, C., Belayev, A., Renard, A., Pictet, R., Bell, G. I., and Martial, J. A. 1984. Isolation and characterization of the human prolactin gene. *EMBO. J.* **3**:429–437.

Uhlén, M., Guss, B., Nilsson, B., Gatenbeck, S., Philipson, L., and Lindberg, M. 1984. Complete sequence of the staphylococcal gene encoding protein A. A gene evolved through multiple duplications. *J. Biol. Chem.* **259**:1695–1702.

van Die, I., and Bergmans, H. 1984. Nucleotide sequence of the gene encoding the F7$_2$ fimbrial subunit of a uropathogenic *Escherichia coli* strain. *Gene* **32**:83–90.

Van Heuverswyn, B., Streydio, C., Brocas, H., Refetoff, S., Dumont, J., and Vassart, G. 1984. Thyrotropin controls transcription of the thyroglobulin gene. *Proc. Natl. Acad. Sci. USA* **81**:5941–5945.

Vasantha, N., Thompson, L. D., Rhodes, C., Banner, C., Nagle, J., and Filpula, D. 1984. Genes for alkaline protease and neutral protease from *Bacillus amyloliquefaciens* contain a large open reading frame between the regions coding for signal sequence and mature protein. *J. Bacteriol.* **159**:811–819.

Vasicek, T. J., McDevitt, B. E., Freeman, M. W., Fennick, B. J., Hendy, G. N., Potts, J. T., Rich, A., and Kronenberg, H. M. 1983. Nucleotide sequence of the human parathyroid hormone gene. *Proc. Natl. Acad. Sci. USA* **80**:2927–2931.

Voight, J. 1985. Macromolecules released into the culture medium during the vegetative cell cycle of the unicellular green alga *Chlamydomonas reinhardii*. *Biochem. J.* **226**:259–268.

Weaver, C. A., Gordon, D. F., Kissil, M. S., Mead, D. A., and Kemper, B. 1984. Isolation and complete nucleotide sequence of the gene for bovine parathyroid hormone. *Gene* **28**:319–329.

Weigle, W. O., Goodman, M. G., Morgan, E. L., and Hugli, T. E. 1983. Regulation of immune response by components of the complement cascade and their activated fragments. *Springer Semin. Immunopathol.* **6**:173–194.

Wells, J. A., Ferrari, E., Henner, D. J., Estell, D. A., and Chen, E. Y. 1983. Cloning sequencing and secretion of *Bacillus amyloliquefaciens* subtilisin in *Bacillus subtilis*. *Nucleic Acids Res.* **11**:7911–7925.

Wetsel, R. A., Lundwall, Å, Davidson, F., Gibson, T., Tack, B. F., and Fey, G. H. 1984. Structure of a murine complement component C3.II. Nucleotide sequence of cloned complementary DNA coding for the α chain. *J. Biol. Chem.* **259**:13857–13862.

Wikström, M., and Casey, R. 1985. The oxidation of exogenous cytochrome *c* by mitochondria. Resolution of a long-standing controversy. *FEBS. Lett.* **183**:293–298.

Willey, D. L., Auffret, A. D., and Gray, J. C. 1984a. Structure and topology of cytochrome *f* in pea chloroplast membranes. *Cell* **36**:555–562.

Willey, D. L., Howe, C. J., Auffret, A. D., Bowman, C. M., Dyer, T. A., and Gray, J. C. 1984b. Location and nucleotide sequence of the gene for cytochrome *f* in wheat chloroplast DNA. *Mol. Gen. Genet.* **194**:416–422.

Wolfe, P. B., Wickner, W., and Goodman, J. M. 1983. Sequence of the leader peptidase gene of *Escherichia coli* and the orientation of leader peptidase in the bacterial envelope. *J. Biol. Chem.* **258**:12073–12080.

Wong, S. L., Price, C. W., Goldfarb, D. S., and Doi, R. H. 1984. The subtilisin *E* gene of *Bacillus subtilis* is transcribed from a σ^{37} promoter *in vivo*. *Proc. Natl. Acad. Sci. USA* **81**:1184–1188.

Wright, R. M., Ko, C., Cumsky, M. G., and Poyton, R. O. 1984. Isolation and sequence of the structural gene for cytochrome *c* oxidase subunit VI from *Saccharomyces cerevisiae*. *J. Biol. Chem.* **259**:15401–15407.

Wurtzel, E. T., Chou, M. Y., and Inouye, M. 1982. Osmoregulation of gene expression. I. DNA sequence of the *ompR* gene of the *ompB* operon of *Escherichia coli* and characterization of its gene product. *J. Biol. Chem.* **257**:13685–13691.

Yamamoto, T., Nakazawa, T., Miyata, T., Kaji, A., and Yokota, T. 1984a. Evolution and structure of two ADP-ribosylation enterotoxins, *Escherichia coli* heat-labile toxin and cholera toxin. *FEBS Lett.* **169**:241–246.

Yamamoto, T., Tamura, T., and Yokota, T. 1984b. Primary structure of heat-labile enterotoxin produced by *Escherichia coli* pathogenic for humans. *J. Biol. Chem.* **259**:5037–5044.

Yamashita, I., Suzuki, K., and Fukui, S. 1985. Nucleotide sequence of the extracellular glucoamylase gene *STA1* in the yeast, *Saccharomyces diastaticus*. *J. Bacteriol.* **161**:567–573.

Yoo, O. J., Powell, C. T., and Agarwal, K. L. 1982. Molecular cloning and nucleotide sequence of full-length cDNA coding for porcine gastrin. *Proc. Natl. Acad. Sci. USA* **79**:1049–1053.

Yoshikawa, K., Williams, C., and Sabol, S. L. 1984. Rat brain preproenkephalin mRNA, cDNA cloning, primary structure, and distribution in the central nervous system. *J. Biol. Chem.* **259**:14301–14308.

Zannis, V. I., McPherson, J., Goldberger, G., Karathanasis, S. K., and Breslow, J. L. 1984. Synthesis, intracellular processing, and signal peptide of human apolipoprotein E. *J. Biol. Chem.* **259**:5495–5499.

Zivin, R. A., Condra, J. H., Dixon, R. A. F., Seidah, N. G., Chretien, M., Nemer, M., Chamberland, M., and Dronin, J. 1984. Molecular cloning and characterization of DNA sequences encoding rat and human atrial natriuretic factors. *Proc. Natl. Acad. Sci. USA* **81**:6325–6329.

Zwizinski, C., and Wickner, W. 1980. Purification and characterization of leader (signal) peptidase from *Escherichia coli*. *J. Biol. Chem.* **255**:7973–7977.

CHAPTER 8

Alt, F. W., and Baltimore, D. 1982. Joining of immunoglobulin heavy chain gene segments: Implications from a chromosome with evidence of three D-J_H fusions. *Proc. Natl. Acad. Sci. USA* **79**:4118–4122.

Alt, F. W., Yancopoulos, G. D., Blackwell, T. K., Wood, C., Thomas, E., Boss, M., Coffman, R., Rosenberg, N., Tonegawa, S., and Baltimore, D. 1984. Ordered rearrangement of immunoglobulin heavy chain variable region segments. *EMBO J.* **3**:1209–1219.

Anderson, M. L. M., Szajnert, M. F., Kaplan, J. C., McColl, L., and Young, B. D. 1984. The isolation of a human IgV$_\lambda$ gene from a recombinant library of chromosome 22 and estimation of its copy number. *Nucleic Acids Res.* **12**:6647–6661.

Banerji, J., Rusconi, S., and Schaffner, W. 1981. Expression of a β-globin gene is enhanced by remote SV40 DNA sequences. *Cell* **27**:299–308.

Barker, P. E., Ruddle, F. H., Royer, H. D., Aaito, O., and Reinherz, E. L. 1984. Chromosomal location of human T-cell receptor gene T$_i$β. *Science* **226**:348–349.

Bentley, D. L., and Rabbitts, T. H. 1980. Human immunoglobulin variable region genes—DNA sequences of two V$_\kappa$ genes and a pseudogene. *Nature (London)* **288**:730–733.

Bentley, D. L., and Rabbitts, T. H. 1981. Human V$_\kappa$ immunoglobulin gene number: Implications for the origin of antibody diversity. *Cell* **24**:613–623.

Bergman, Y., Rice, D., Grosschedl, R., and Baltimore, D. 1985. Two regulatory elements for immunoglobulin κ light chain gene expression. *Proc. Natl. Acad. Sci. USA* **81**:7041–7045.

Bernard, O., Hozumi, N., and Tonegawa, S. 1978. Sequences of mouse immunoglobulin light chain genes before and after somatic changes. *Cell* **15**:1133–1144.

Blackwell, T. K., and Alt, F. W. 1984. Site-specific recombination between immunoglobulin D and J$_H$ segments that were introduced into the genome of a murine pre-B cell line. *Cell* **37**:105–112.

Blankenstein, T., Zoebelein, G., and Krawinkel, U. 1984. Analysis of immunoglobulin heavy chain V-region genes belonging to the V$_{NP}$-gene family. *Nucleic Acids Res.* **12**:6887–6900.

Bogen, B., Jørgensen, T., and Hannestad, K. 1985. T helper cell recognition of idiotypes on λ light chains of M315 and T952: Evidence for dependence on somatic mutations in the third hypervariable region. *Eur. J. Immunol.* **15**:278–281.

Bothwell, A. L. M., Paskind, M., Reth, M., Imanishi-kari, T., Rajewsky, K., and Baltimore, D. 1981. Heavy chain variable region contribution to the NP[b] family of antibodies: Somatic mutation evident in a γ2a variable region. *Cell* **24**:625–637.

Breiner, A. V., Brandt, C. R., Milcarek, C., Sweet, R. W., Ziv, E., Burstein, R., and Schechter, I. 1982. Somatic DNA rearrangement generates functional rat immunoglobulin κ chain genes: The J$_\kappa$ gene cluster is longer in rat than in mouse. *Gene* **18**:165–174.

Brodeur, P. H., and Riblet, R. 1984. The immunoglobulin heavy chain variable region (Igh-V) locus in the mouse. I. One hundred Igh-V genes comprise seven families of homologous genes. *Eur. J. Immunol.* **14**:922–930.

Caccia, N., Kronenberg, M., Saxe, D., Haars, R., Bruns, G. A. P., Goverman, J., Malissen, M., Willard, H., Yoshikai, Y., Simon, M., Hood, L., and Mak, T. W. 1984. The T cell receptor β chain genes are located on chromosome 6 in mice and chromosome 7 in humans. *Cell* **37**:1091–1099.

Cheng, H. L., Blattner, F. R., Fitzmaurice, L., Mushinski, J. F., and Tucker, P. W. 1982. Structure of genes for membrane and secreted murine IgD heavy chains. *Nature (London)* **296**:410–415.

Clark, M. J., Gagnon, J., Williams, A. F., and Barclay, A. N. 1985. MRC OX-2 antigen: A lymphoid/neuronal membrane glycoprotein with a structure like a single immunoglobulin light chain. *EMBO J.* **4**:113–118.

Crews, S., Griffin, J., Huang, H., Calame, K., and Hood, L. 1981. A single V$_H$ gene segment encodes response to phosphorylcholine: Somatic mutation is correlated with the class of the antibody. *Cell* **25**:59–66.

Dahan, A., Reynaud, C. A., and Weill, J. C. 1983. Nucleotide sequence of the constant region of a chicken μ heavy chain immunoglobulin mRNA. *Nucleic Acids Res.* **11**:5381–5389.

Darsley, M. J., and Rees, A. R. 1985. Nucleotide sequences of five antilysozyme monoclonal antibodies. *EMBO J.* **4**:393–398.

Dillon, L. S. 1981. *Ultrastructure, Macromolecules, and Evolution,* New York, Plenum Press.

Dillon, L. S. 1983. *The Inconstant Gene,* New York, Plenum Press.

Early, P., Huang, H., Davis, M., Calame, K., and Hood, L. 1980. An immunoglobulin heavy chain variable region gene is generated from three segments of DNA-V$_H$, D, and J$_H$. *Cell* **19**:981–992.

Ellison, J., and Hood, L. 1982. Linkage and sequence homology of two human immunoglobulin γ heavy chain constant region genes. *Proc. Natl. Acad. Sci. USA* **79**:1984–1988.

Emorine, L., and Max, E. E. 1983. Structural analysis of a rabbit immunoglobulin κ2 J-C locus reveals multiple deletions. *Nucleic Acids Res.* **11**:8877–8890.

Emorine, L., Dreher, K., Kindt, T. J., and Max, E. E. 1983. Rabbit immunoglubulin κ genes: Structure of a germline b4 allotype J-C locus and evidence for several b4-related sequences in the rabbit genome. *Proc. Natl. Acad. Sci. USA* **80:**5709–5713.

Emorine, L., Sogn, J. A., Trinh, D., Kindt, T. J., and Max, E. E. 1984. A genomic gene encoding the b5 rabbit immunoglobulin κ constant region: Implications for latent allotype phenomenon. *Proc. Natl. Acad. Sci. USA* **81:**1789–1793.

Ephrussi, A., Church, G. M., Tonegawa, S., and Gilbert, W. 1985. B lineage-specific interactions of an immunoglobulin enhancer with cellular factors *in vivo. Science* **227:**134–140.

Flanagan, J. G., and Rabbitts, T. H. 1982. Arrangement of immunoglobulin heavy-chain constant region genes implies evolutionary duplication of a segment containing γ, ε, and α genes. *Nature (London)* **300:**709–713.

Gascoigne, N. R. J., Chien, Y. H., Becker, D. M., Kavaler, J., and Davis, M. M. 1984. Genomic organization and sequence of T-cell receptor β-chain constant- and joining-region genes. *Nature (London)* **310:**387–391.

Gillies, S. D., Morrison, S. L., Oi, V. T., and Tonegawa, S. 1983. A tissue-specific transcription enhancer element is located in the major intron of a rearranged immunoglobulin heavy chain gene. *Cell* **33:**717–728.

Godson, G. N., Ellis, J., Svec, P., Schlesinger, D. H., and Nussenzweig, V. 1983. Identification and chemical synthesis of a tandemly repeated immunogenic region of *Plasmodium knowlesi* circumsporozoite protein. *Nature (London)* **305:**29–33.

Gorski, J., Rollins, P., and Mach, B. 1983. Somatic mutations of immunoglobulin variable genes are restricted to the rearranged V gene. *Science* **220:**1179–1181.

Gough, N. M., and Bernard, O. 1981. Sequences of the joining region genes for immunoglobulin heavy chains and their role in generation of antibody diversity. *Proc. Natl. Acad. Sci. USA* **78:**509–513.

Guyaux, M., Cornelissen, A. W. C. A., Pays, E., Steinert, M., and Borst, P. 1985. *Trypanosoma brucei:* A surface antigen mRNA is discontinuously transcribed from two distinct chromosomes. *EMBO J.* **4:**995–998.

Hannum, C. H., Kappler, J. W., Trowbridge, I. S., Marrack, P., and Freed, J. H. 1984. Immunoglobulin-like nature of the α-chain of a human T-cell antigen/MHC receptor. *Nature (London)* **312:**65–67.

Hartman, A. B., and Rudikoff, S. 1984. V_H genes encoding the immune response to β-(1,6)-galactan: Somatic mutation in IgM molecules. *EMBO J.* **3:**3023–3030.

Hayday, A. C., Saito, H., Gillies, S. D., Kranz, D. M., Tanigawa, G., Elsen, H. N., and Tonegawa, S. 1985. Structure, organization, and somatic rearrangement of T-cell gamma genes. *Cell* **40:**259–269.

Hedrick, S. M., Nielsen, E. A., Kavaler, J., Cohen, D. I., and Davis, M. M. 1984. Sequence relationships between putative T-cell receptor polypeptides and immunoglobulins. *Nature (London)* **308:**153–158.

Heidmann, O., and Rugeon, F. 1982. Multiple sequences related to a constant-region-kappa light chain gene in the rabbit genome. *Cell* **28:**507–513.

Hieter, P. A., Maizel, J. V., and Leder, P. 1982. Evolution of human immunoglobulin κ J region genes. *J. Biol. Chem.* **257:**1516–1522.

Höchtl, J., Müller, C. R., and Zachau, H. G. 1982. Recombined flanks of the variable and joining segments of immunoglobulin genes. *Proc. Natl. Acad. Sci. USA* **79:**1383–1387.

Hollis, G. F., Hieter, P. A., McBride, O. W., Swan, D., and Leder, P. 1982. Processed genes: a dispersed human immunoglobin gene bearing evidence of RNA-type processing. *Nature (London)* **296:**321–325.

Hood, L., Kronenberg, M., and Hunkapiller, T. 1985. T cell antigen receptors and the immunoglobulin supergene family. *Cell* **40:**225–229.

Jaenichen, H. R., Pech, M., Lindenmaier, W., Wildgruber, N., and Zachau, H. G. 1984. Composite human $V_κ$ genes and a model of their evolution. *Nucleic Acids Res.* **12:**5249–5263.

Kataoka, T., Miyata, T., and Honjo, T. 1981. Repetitive sequences in class-switch recombination regions of immunoglobulin heavy chain genes. *Cell* **23:**357–368.

Kataoka, T., Takeda, S. I., and Honjo, T. 1983. *Escherichia coli* extract-catalyzed recombination in switch regions of mouse immunoglobulin genes. *Proc. Natl. Acad. Sci. USA* **80:**2666–2670.

Kawakami, T., Takahashi, N., and Honjo, T. 1980. Complete nucleotide sequence of mouse immunoglobulin μ gene and comparison with other immunoglobulin heavy chain genes. *Nucleic Acids Res.* **8:**3933–3945.

Kelley, D. E., Coleclough, C., and Perry, R. P. 1982. Functional significance and evolutionary development of the 5'-terminal regions of immunoglobulin variable-region genes. *Cell* **29:**681–689.

Kemp, D. J., Tyler, B., Bernard, O., Gough, N., Gerondakis, S., Adams, J. M., and Cory, S. 1981.

Organization of genes and spacers within the mouse immunoglobulin V_H locus. *J. Mol. Appl. Genet.* **1**:245–261.

Kenten, J. H., Molgaard, H. V., Houghton, M., Derbyshire, R. B., Viney, J., Bell, L. O., and Gould, H. J. 1982. Cloning and sequence determination of the gene for the human immunoglobulin-ε chain expressed in a myeloma cell line. *Proc. Natl. Acad. Sci. USA* **79**:6661–6665.

Kim, S., Davis, M., Sinn, E., Patten, P., and Hood, L. 1981. Antibody diversity: Somatic hypermutation of rearranged V_H genes. *Cell* **27**:573–581.

Kindt, T. J., and Capra, J. D. 1984. *The Antibody Enigma,* New York, Plenum Press.

Klein, S., Sablitsky, F., and Radbruch, A. 1984. Deletion of the IgH enhancer does not reduce immunoglobulin heavy chain production of a hybridoma IgD class switch variant. *EMBO J.* **3**:2473–2476.

Klobeck, H. G., Combriato, G., and Zachau, H. G. 1984. Immunoglobulin genes of the κ light chain type from two human lymphoid cell lines are closely related. *Nucleic Acids Res.* **12**:6995–7006.

Knapp, M. R., Liu, C. P., Newell, N., Ward, R. B., Tucker, P. W., Strober, S., and Blattner, F. 1982. Simultaneous expression of immunoglobulin μ and δ heavy chains by a cloned B-cell lymphoma: A single copy of the V_H gene is shared by two adjacent C_H genes. *Proc. Natl. Acad. Sci. USA* **79**:2996–3000.

Knight, K. L., Burnett, R. C., and McNicholas, J. M. 1985. Organization and polymorphism of rabbit immunoglobulin heavy chain genes. *J. Immunol.* **134**:1245–1250.

Kranz, D. M., Saito, H., Disteche, C. M., Swisshelm, K., Pravtcheva, D., Ruddle, F. H., Eisen, H. N., and Tonegawa, S. 1985. Chromosomal locations of the murine T-cell receptor alpha-chain gene and the T-cell gamma gene. *Science* **227**:941–944.

Kudo, A., Ishihara, T., Nishimura, Y., and Watanabe, T. 1985. A cloned human immunoglobulin heavy chain gene with a novel direct-repeat sequence in 5' flanking region. *Gene* **33**:181–189.

Kurosawa, Y., and Tonegawa, S. 1982. Organization, structure, and assembly of immunoglobulin heavy chain diversity DNA segments. *J. Exp. Med.* **155**:201–218.

Kurosawa, Y., von Boehmer, H., Haas, W., Sakano, H., Trauneker, A., and Tonegawa, S. 1981. Identification of D segments of immunoglobulin heavy-chain genes and their rearrangement in T lymphocytes. *Nature (London)* **290**:565–570.

Lang, R. B., Stanton, L. W., and Marcu, K. B. 1982. On immunoglobulin heavy chain gene switching: Two γb genes are rearranged via switch sequences in MPC-11 cells but only one is expressed. *Nucleic Acids Res.* **10**:611–630.

Lewis, S., Gifford, A., and Baltimore, D. 1984. Joining of $V_κ$ to $J_κ$ gene segments in a retroviral vector introduced into lymphoid cells. *Nature (London)* **308**:425–428.

Litman, G. W., Berger, L., Murphy, K., Litman, R., Hinds, K., Jahn, C. L., and Erickson, B. W. 1983. Complete nucleotide sequence of an immunoglobulin V_H gene homologue from *Caiman,* a phylogenetically ancient reptile. *Nature (London)* **303**:349–352.

Lloubes, R. P., Chartier, M. J., Journet, A. M., Varenne, S. G., and Lazdunski, C. J. 1984. Nucleotide sequence of the gene for the immunity protein to colicin A. *Eur. J. Biochem.* **144**:73–78.

Maki, R., Roeder, W., Traunecker, A., Sidman, C., Wabi, M., Raschke, W., and Tonegawa, S. 1981. The role of DNA rearrangement and alternative RNA processing in the expression of immunoglobulin delta genes. *Cell* **24**:353–365.

Mallissen, M., Minard, K., Mjolsness, S., Kronenberg, M., Goverman, J., Hunkapiller, T., Prystowsky, M. B., Yoshikai, Y., Fitch, F., Mak, T. W., and Hood, L. 1984. Mouse T-cell antigen receptor: Structure and organization of constant and joining gene segments encoding the β polypeptide. *Cell* **37**:1101–1110.

Manser, T., Huang, S. Y., and Gefter, M. L. 1984. Influence of clonal selection on the expression of immunoglobulin variable region genes. *Science* **226**:1283–1288.

Marcu, K. B., Lang, R. B., Stanton, L. W., and Harris, L. J. 1982. A model for the molecular requirements of immunoglobulin heavy chain class switching. *Nature (London)* **298**:87–89.

Max, E. E., Maizel, J. V., and Leder, P. 1981. The nucleotide sequence of a 5.5-kilobase DNA segment containing the mouse κ immunoglobulin J and C region genes. *J. Biol. Chem.* **256**:5116–5120.

McCartney, Francis, N., Skurla, R. M., Mage, R. G., and Bernstein, K. E. 1984. κ-chain allotypes and isotypes in the rabbit: cDNA sequences of clones encoding *b9* suggest an evolutionary pathway and possible role of the interdomain disulfide bond in quantitative allotype expression. *Proc. Natl. Acad. Sci. USA* **81**:1794–1798.

McGuire, K. L., Duncan, W. R., and Tucker, P. W. 1985. Phylogenetic conservation of immunoglobulin heavy chains: Direct comparison of hamster and mouse $C_μ$ genes. *Nucleic Acids Res.* **13**:5611–5628.

Mercola, M., Goverman, J., Mirell, C., and Calame, K. 1985. Immunoglobulin heavy-chain enhancer requires one or more tissue-specific factors. *Science* **227**:266–270.

Milstein, C. P., Deverson, E. V., and Rabbitts, T. H. 1984. The sequence of the human immunoglobulin μ–δ intron reveals possible vestigial switch segments. *Nucleic Acids Res.* **12**:6523–6535.

Moore, K. W., Rogers, J., Hunkapiller, T., Early, P., Nottenburg, C., Weissman, I., Bazin, H., Wall, R., and Hood, L. E. 1981. Expression of IgD may use both DNA rearrangement and RNA splicing mechanisms. *Proc. Natl. Acad. Sci. USA* **78**:1800–1804.

Mostov, K. E., Friedlander, M., and Blobel, G. 1984. The receptor for transepithelial transport of IgA and IgM contains multiple immunoglobulin-like domains. *Nature (London)* **308**:37–43.

Newell, N., Richards, J. E., Tucker, P. W., and Blattner, F. R. 1980. J genes for heavy chain immunoglobulins of mouse. *Science* **209**:1128–1132.

Ohno, S., Kato, K., Hozumi, T., and Matsunaga, T. 1982. Mouse immunoglobulin coding sequences for the heavy-chain variable region arose as repeats of the two short building blocks. *Proc. Natl. Acad. Sci. USA* **79**:132–136.

Ollo, R., Sikorav, J. L., and Rougeon, F. 1983. Structural relationships among mouse and human immunoglobulin V_H genes in the subgroup III. *Nucleic Acids Res.* **11**:7887–7897.

Patten, P., Yokota, T., Rothbard, J., Chien, Y. H., Arai, K. I., and Davis, M. M. 1984. Structure, expression and divergence of T-cell receptor β-chain variable regions. *Nature (London)* **312**:40–44.

Pech, M., and Zachau, H. G. 1984. Immunoglobulin genes of different subgroups are interdigitated within the V_κ locus. *Nucleic Acids Res.* **12**:9229–9236.

Pech, M., Höchtl, J., Schnell, H., and Zachau, H. G. 1981. Differences between germ-line and rearranged immunoglobulin V_κ coding sequences suggest a localized mutation mechanism. *Nature (London)* **291**:668–670.

Pech, M., Jaenichen, H. R., Pohlenz, H. D., Neumaier, P. S., Klobeck, H. B., and Zachau, H. G. 1984. Organization and evolution of a gene cluster for human immunoglobulin variable regions of the kappa type. *J. Mol. Biol.* **176**:189–204.

Pech, M., Smola, H., Pohlenz, H. D., Straubinger, B., Gerl, R., and Zachau, H. G. 1985. A large section of the gene locus encoding human immunoglobulin variable regions of the kappa type is duplicated. *J. Mol. Biol.* **183**:291–299.

Rabbitts, T. H., Forster, A., and Milstein, C. P. 1981a. Human immunoglobulin heavy chain genes: Evolutionary comparisons of C_μ, C_δ, and C_γ genes and associated switch sequences. *Nucleic Acids Res.* **9**:4509–4524.

Rabbitts, T. H., Bentley, D. L., and Milstein, C. P. 1981b. Human antibody genes: V gene variability and C_H gene switching strategies. *Immunol. Rev.* **59**:69–91.

Rabbitts, T. H., Bentley, D. L., Dunnick, W., Forster, A., Matthyssens, G. E. A. R., and Milstein, C. 1981c. Immunoglobulin genes undergo multiple sequence rearrangements during differentiation. *Cold Spring Harbor Symp. Quant. Biol.* **45**:867–878.

Rechavi, G., Ram, D., Glazer, L., Zakut, R., and Givol, D. 1983. Evolutionary aspects of immunoglobulin heavy chain variable region (V_H) gene subgroup. *Proc. Natl. Acad. Sci. USA* **80**:855–859.

Reynaud, C. A., Dahan, A., and Weill, J. C. 1983. Complete sequence of a chicken λ light chain immunoglobulin derived from the nucleotide sequence of its mRNA. *Proc. Natl. Acad. Sci. USA* **80**:4099–4103.

Reynaud, C. A., Anquez, V., Dahan, A., and Weill, J. C. 1985. A single rearrangement event generates most of the chicken immunoglobulin light chain diversity. *Cell* **40**:283–291.

Rudikoff, S., Pawlita, M., Pumphrey, J., and Heller, M. 1984. Somatic diversification of immunoglobulins. *Proc. Natl. Acad. Sci. USA* **81**:2162–2166.

Sablitzky, F., and Rajewsky, K. 1984. Molecular basis of an isogeneic antiidiotypic response. *EMBO J.* **3**:3005–3012.

Sablitzky F., Wildner, G., and Rajewsky, K. 1985. Somatic mutation and clonal expansion of B cells in an antigen-driven immune response. *EMBO J.* **4**:345–350.

Saito, H., Kranz, D. M., Takagaki, Y., Hayday, A. C., Eisen, H. N., and Tonegawa, S. 1984. Complete primary structure of a heterodimeric T-cell receptor deduced from cDNA sequences. *Nature (London)* **309**:757–762.

Sakano, H., Kurosawa, Y., Weigert, M., and Tonegawa, S. 1981. Identification and nucleotide sequence of a diversity segment (D) of immunoglobulin heavy-chain genes. *Nature (London)* **290**:562–565.

Schiff, C., Milili, M., and Fougereau, M. 1983. Immunoglobulin diversity: Analysis of the germ line V_H gene repertoire of the murine anti-GAT response. *Nucleic Acids Res.* **11**:4007–4016.

Segal, E., Billyard, E., So, M., Storzbach, S., and Meyer, T. F. 1985. Role of chromosomal rearrangement in *N. gonorrhoeae* pilus phase variation. *Cell* **40**:293–300.

Selsing, E., and Storb, U. 1981. Somatic mutation of immunoglobulin light-chain variable-region genes. *Cell* **25**:47–58.

Selsing, E., Miller, J., Wilson, R., and Storb, U. 1982. Evolution of mouse immunoglobulin λ genes. *Proc. Natl. Acad. Sci. USA* **79**:4681–4685.

Selsing, E., Voss, J., and Storb, U. 1984. Immunoglobulin gene 'remnant' DNA—Implications for antibody gene recombination. *Nucleic Acids Res.* **12**:4229–4246.

Sheppard, H. W., and Gutman, G. A. 1982. Rat kappa-chain J-segment genes: Two recent gene duplications separate rat and mouse. *Cell* **29**:121–127.

Shimizu, A., Takahashi, N., Yaoita, Y., and Honjo, T. 1982. Organization of the constant-region gene family of the mouse immunoglobulin heavy chain. *Cell* **28**:499–506.

Sims, J., Rabbitts, T. H., Estess, P., Slaughter, C., Tucker, P. W., and Capra, J. D. 1982. Somatic mutation in genes for the variable portion of the immunoglobulin heavy chain. *Science* **216**:309–311.

Siu, G., Clark, S. P., Yoshikai, Y., Malissen, M., Yanagi, Y., Strauss, E., Mak, T. W., and Hood, L. 1984a. The human T-cell antigen receptor is encoded by variable diversity and joining gene segments that rearrange to generate a complete V gene. *Cell* **37**:393–401.

Siu, G., Kronenberg, M., Strauss, E., Haars, R., Mak, T. W., and Hood, L. 1984b. The structure, rearrangement and expression of D$_\beta$ gene segments of the murine T-cell antigen receptor. *Nature (London)* **311**:344–348.

Stanton, L. W., and Marcu, K. B. 1982. Nucleotide sequence and properties of the murine γ$_3$ immunoglobulin heavy chain gene switch region: Implications for successive C$_\gamma$ gene switching. *Nucleic Acids Res.* **10**:5993–6006.

Steinmetz, M., Altenburger, W., and Zachau, H. 1980. A rearranged DNA sequence possibly related to the translocation of immunoglobulin gene segments. *Nucleic Acids Res.* **8**:1709–1720.

Sun, L. H. K., Croce, C. M., and Showe, L. C. 1985. Cloning and sequencing of a rearranged V$_\lambda$ gene from a Burkitt's lymphoma cell line expressing kappa light chains. *Nucleic Acids Res.* **13**:4921–4934.

Takahashi, N., Noma, T., and Honjo, T. 1984. Rearranged immunoglobulin heavy chain variable region (V$_H$) pseudogene that deletes the second complementarity-determining region. *Proc. Natl. Acad. Sci. USA* **81**:5194–5198.

Tonegawa, S. 1983. Somatic generation of antibody diversity. *Nature (London)* **302**:575–581.

Van Ness, B. G., Coleclough, C., Perry, R. P., and Weigert, M. 1982. DNA between variable and joining gene segments of immunoglobulin κ light chain is frequently retained in cells that rearrange the κ locus. *Proc. Natl. Acad. Sci. USA* **79**:262–266.

Walfield, A. M., Storb, U., Selsing, E., and Zentgraf, H. 1980. Comparison of different rearranged immunoglobulin kappa genes of a myeloma by electronmicroscopy and restriction mapping of cloned DNA: Implications for "allelic exclusion." *Nucleic Acids Res.* **8**:4689–4707.

Wels, J. A., Word, C. J., Rimm, D., Der-Balan, G. P., Martinez, H. M., Tucker, P. W., and Blattner, F. R. 1984. Structural analysis of the murine IgG3 constant region gene. *EMBO J.* **3**:2041–2046.

Wu, T. T., Reid-Miller, M., Perry, H. M., and Kabat, E. A. 1984. Long identical repeats in the mouse γ2b switch region and their implications for the mechanism of class switching. *EMBO J.* **3**:2033–2040.

Yancopoulos, G. D., and Alt, F. W. 1985. Developmentally controlled and tissue-specific-expression of unrearranged V$_H$ gene segments. *Cell* **40**:271–281.

Yazaki, A., and Ohno, S. 1983. Recurrence of 49-base decamers, nonamers, and octamers within mouse C$_\mu$ gene of Ig heavy chain and its primordial building block. *Proc. Natl. Acad. Sci. USA* **80**:2337–2340.

Young, J. R., Shah, J. S., Matthyssens, G., and Williams, R. O. 1983. Relationship between multiple copies of a *T. brucei* variable surface glycoprotein gene whose expression is not controlled by duplication. *Cell* **32**:1149–1159.

CHAPTER 9

Alton, N. K., and Vapnek, D. 1979. Nucleotide sequence analysis of the chloramphenicol resistance transposon *Tn 9*. *Nature (London)* **282**:864–869.

Arkhipova, I. R., Gorelova, T. V., Alyin, Y. V., and Schuppe, N. G. 1984. Reverse transcription of *Drosophila* mobile dispersed genetic element RNAs: Detection of intermediate forms. *Nucleic Acids Res.* **12**:7533–7541.

Arthur, A., Nimmo, E., Hettle, S., and Sherratt, D. 1984. Transposition and transposition immunity of transposon Tn3 derivatives having different ends. *EMBO J.* **3**:1723–1729.

Balmain, A., Frew, L., Cole, G., Krumlauf, R., Ritchie, A., and Birnie, G. D. 1982. Transcription of repeated sequences of the mouse B1 family in Friend erythroleukaemic cells. *J. Mol. Biol.* **160**:163–179.

Baltimore, D. 1985. Retroviruses and retrotransposons. The role of reverse transcription in shaping the eukaryotic genome. *Cell* **40**:481–482.

Barker, R. F., Thompson, D. V., Talbot, D. R., Swanson, J., and Bennetzen, L. 1984. Nucleotide sequence of the maize transposable element *Mu1*. *Nucleic Acids Res.* **12**:5955–5967.

Bayev, A. A., Lyubomirskaya, N. V., Dzhumagaliev, E. B., Ananiev, E. V., Amiantova, I. G., and Ilyin, Y. V. 1984. Structural organization of transposable element *mdg4* from *Drosophila melanogaster* and a nucleotide sequence of its long terminal repeats. *Nucleic Acids Res.* **12**:3707–3723.

Belas, R., Mileham, A., Simon, M., and Silverman, M. 1984. Transposon mutagenesis of marine *Vibrio* spp. *J. Bacteriol.* **158**:890–896.

Bell, J. R., Bogardus, A. M., Schmidt, T., and Pellegrini, M. 1985. A new *copia*-like transposable element found in a *Drosophila* rDNA gene unit. *Nucleic Acids Res.* **13**:3881–3891.

Bennett, P. M., Grinsted, J., Choi, C. L., and Richmond, M. H. 1978. Characterisation of *Tn501*, a transposon determining resistance to mercuric ions. *Mol. Gen. Genet.* **159**:101–106.

Biel, S. W., Adelt, G., and Berg, D. E. 1984. Transcriptional control of *IS1* transposition in *Escherichia coli*. *J. Mol. Biol.* **174**:251–264.

Boeke, J. D., Garfinkel, D. J., Styles, C. A., and Fink, G. R. 1985. *Ty* elements transpose through an RNA intermediate. *Cell* **40**:491–500.

Bonas, U., Sommer, H., and Saedler, H. 1984. The 17-kb *Tam1* element of *Antirrhinum majus* induces a 3-bp duplication upon integration into the chalcone synthase gene. *EMBO J.* **3**:1015–1019.

Bonnewell, V., Fowler, R. F., and Skinner, D. M. 1983. An inverted repeat borders a five-fold amplification in satellite DNA. *Science* **221**:862–865.

Bourouis, M., and Richards, G. 1985. Remote regulatory sequences of the *Drosophila* glue gene *sgs3* as revealed by P-element transformation. *Cell* **40**:349–357.

Bregliano, J. C., and Kidwell, M. G. 1983. Hybrid dysgenesis determinants. *In:* Shapiro, J. A., ed.. *Mobile Genetic Elements,* New York, Academic Press, pp. 363–410.

Brierley, H. L., and Potter, S. S. 1985. Distinct characteristics of loop sequences of two *Drosophila* foldback transposable elements. *Nucleic Acids Res.* **13**:485–500.

Brodeur, G. M., Sandmeyer, S. B., and Olson, M. V. 1983. Consistent association between *sigma* elements and tRNA genes in yeast. *Proc. Natl. Acad. Sci. USA* **80**:3292–3296.

Brosius, J., and Walz, A. 1982. DNA sequences flanking an *E. coli* insertion element *IS2* in a cloned yeast *TRP5* gene. *Gene* **17**:223–228.

Brown, N. L., Choi, C. L., Grinsted, J., Richmond, M. H., and Whitehead, P. R. 1980. Nucleotide sequences at the ends of the mercury resistance transposon, *Tn501*. *Nucleic Acids Res.* **8**:1933–1945.

Brown, N. L., Ford, S. J., Pridmore, R. D., and Fritzinger, D. C. 1983. Nucleotide sequence of a gene from the *Pseudomonas* transposon *Tn501* encoding mercuric reductase. *Biochemistry* **22**:4089–4095.

Calos, M. P., and Miller, J. H. 1980. Transposable elements. *Cell* **20**:579–595.

Cameron, J. R. 1979. Evidence for transposition of dispersed repetitive DNA families in yeast. *Cell* **16**:739–751.

Cameron, J. R., Loh, E. Y., and Davis, R. W. 1979. Evidence for transposition of dispersed repetitive DNA families in yeast. *Cell* **16**:739–751.

Carpenter, C. D., Bruskin, A. M., Spain, L. M., Eldon, E. D., and Klein, W. H. 1982. The 3' untranslated regions of two related mRNAs contain an element highly repeated in the sea urchin genome. *Nucleic Acids Res.* **10**:7829–7842.

Case, S. T., and Byers, M. R. 1983. Repeated nucleotide sequence arrays in Balbiani ring 1 of *Chironomus tentans* contain internally nonrepeating and subrepeating elements. *J. Biol. Chem.* **258**:7793–7799.

Case, S. T., Summers, R. L., and Jones, A. G. 1983. A variant tandemly repeated nucleotide sequence in Balbiani ring 2 of *Chironomus tentans*. *Cell* **33**:555–562.

Caspers, P., Dalrymple, B., Iida, S., and Arber, W. 1984. IS30, a new insertion sequence of *Escherichia coli* K12. *Mol. Gen. Genet.* **196**:68–73.

Chisholm, G. E., Genbauffe, F. S., and Cooper, T. G. 1984. *tau*, a repeated DNA sequence in yeast. *Proc. Natl. Acad. Sci. USA* **81**:2965–2969.

Clare, J., and Farabaugh, P. 1985. Nucleotide sequence of a yeast Ty element: Evidence for an unusual mechanism of gene expression. *Proc. Natl. Acad. Sci. USA* **82:**2829–2833.

Clerget, M., Chandler, M., and Caro, L. 1981. The structure of R1*drd*19: A revised physical map of the plasmid. *Mol. Gen. Genet.* **181:**183–191.

Collins, M., and Rubin, G. M. 1984. Structure of chromosomal rearrangements induced by the FB transposable element in *Drosophila. Nature (London)* **308:**323–327.

Courage-Tebbe, U., Döring, H. P., Federoff, N., and Starlinger, P. 1983. The controlling element *Ds* at the *Shrunken* locus in *Zea mays:* Structure of the unstable *sh-m5933* allele and several revertants. *Cell* **34:**383–393.

Dalrymple, B., Caspers, P., and Arber, W. 1984. Nucleotide sequence of the prokaryotic genetic element IS30. *EMBO J.* **3:**2145–2149.

De La Cruz, F., and Grinsted, J. 1982. Genetic and molecular characterization of Tn*21;* A multiple resistance transposon from R100.1. *J. Bacteriol.* **151:**222–228.

del Rey, F. J., Donahue, T. F., and Fink, G. R. 1982. *Sigma,* a repetitive element found adjacent to tRNA genes of yeast. *Proc. Natl. Acad. Sci. USA* **79:**4138–4142.

Dennis, E. S., Sachs, M. M., Gerlach, W. L., Finnegan, E. J., and Peacock, W. J. 1985. Molecular analysis of the alcohol dehydrogenase 2 (*Adh2*) gene of maize. *Nucleic Acids Res.* **13:**727–743.

Dillon, L. S. 1978. *The Genetic Mechanism and the Origin of Life,* New York, Plenum Press.

Di Nocera, P. P., Digon, M. E., and Dawid, I. B. 1983. A family of oligoadenylate-terminated transposable sequences in *Drosophila melanogaster. J. Mol. Biol.* **168:**715–727.

Diver, W. P., Grinsted, J., Fritzinger, D. C., Brown, N. L., Altenbuchner, J., Rogousky, P., and Schmitt, R. 1983. DNA sequences and complementation by the *tnpR* genes of Tn*21,* Tn*501,* and Tn*1721. Mol. Gen. Genet.* **191:**189–193.

Döring, H. P., and Starlinger, P. 1984. Barbara McClintock's controlling elements: Now at the DNA level. *Cell* **39:**253–259.

Eibel, H., Gafner, J., Stotz, A., and Philippsen, P. 1980. Characterization of the yeast mobile element *Ty1. Cold Spring Harbor Symp. Quant. Biol.* **45:**609–618.

Elder, R. T., Loh, E. Y., and Davis, R. W. 1983. RNA from the yeast transposable element *Ty1* has both ends in the direct repeats, a structure similar to retrovirus. *Proc. Natl. Acad. Sci. USA* **80:**2432–2436.

Engler, J. A., and van Bree, M. P. 1981. The nucleotide sequence and protein-coding capability of the transposable element *IS5. Gene* **14:**155–163.

Farabaugh, P. J., and Fink, G. R. 1980. Insertion of the eukaryotic transposable element *Ty1* creates a 5-base pair duplication. *Nature (London)* **286:**352–356.

Feagin, J. E., Setzer, D. R., and Schimke, R. T. 1983. A family of repeated sequences, one of which resides in the second intervening sequence of the mouse dihydrofolate reductase gene. *J. Biol. Chem.* **258:**2480–2487.

Flavell, A. J., Levis, R., Simon, M. A., and Rubin, G. M. 1981. The 5' termini of RNAs encoded by the transposable element *copia. Nucleic Acids Res.* **9:**6279–6291.

Fulton, A. M., Mellor, J., Dobson, M. J., Chester, J., Warmington, J. R., Indge, K. J., Oliver, S. G., de la Paz, P., Wilson, W., Kingsman, A. J., and Kingsman, S. M. 1985. Variants within the yeast Ty sequence family encode a class of structurally conserved proteins. *Nucleic Acids Res.* **13:**4097–4112.

Gafner, J., and Philippsen, P. 1980. The yeast transposon *Ty1* generates duplications of target DNA on insertion. *Nature (London)* **286:**414–418.

Galas, D. J., Calos, M. P., and Miller, J. H. 1980. Sequence analysis of Tn*9* insertions in the *lacZ* gene. *J. Mol. Biol.* **144:**19–41.

Gebhard, W., and Zachau, H. G. 1983a. Organization of the R family and other interspersed repetitive DNA sequences in the mouse genome. *J. Mol. Biol.* **170:**255–270.

Gebhard, W., and Zachau, H. G. 1983b. Simple DNA sequences and dispersed repetitive elements in the vicinity of mouse immunoglobulin κ light chain genes. *J. Mol. Biol.* **170:**567–573.

Gebhard, W., Meitinger, T., Höchtl, J., and Zachau, H. G. 1982. A new family of interspersed repetitive DNA sequences in the mouse genome. *J. Mol. Biol.* **157:**453–471.

Genbauffe, F. S., Chisholm, G. E., and Cooper, T. G. 1984. Tau, sigma, and delta: A family of repeated elements in yeast. *J. Biol. Chem.* **259:**10518–10523.

Gielen, J., De Beuckeleer, M., Seurinck, J., Deboeck, F., De Greve, H., Lemmers, M., Van Montagu, M., and Schell, J. 1984. The complete nucleotide sequence of the TL-DNA of the *Agrobacterium tumefaciens* plasmid pTiAch5. *EMBO J.* **3**:835–846.

Gierl, A., Schwarz-Sommer, Z., and Saedler, H. 1985. Molecular interactions between the components of the En-I transposable element system of *Zea mays*. *EMBO J.* **4**:579–583.

Grimaldi, G., and Singer, M. F. 1983. Members of the *KpnI* family of long interspersed repeated sequences join and interrupt α-satellite in the monkey genome. *Nucleic Acids Res.* **11**:321–338.

Grimaldi, G., Skowronski, J., and Singer, M. F. 1984. Defining the beginning and end of *KpnI* family segments. *EMBO J.* **3**:1753–1759.

Grindley, N. D. F. 1983. Transposition of *Tn3* and related transposons. *Cell* **32**:3–5.

Grinsted, J., and Brown, N. L. 1984. A *Tn21* terminal sequence within *Tn501*: Complementation of *TnpA* gene function and transposon evolution. *Mol. Gen. Genet.* **197**:497–502.

Hardison, R. C., and Printz, R. 1985. Variability within the rabbit C repeats and sequences shared with other SINES. *Nucleic Acids Res.* **13**:1073–1088.

Hauber, J., Nelböck-Hochstetter, P., and Feldmann, H. 1985. Nucleotide sequence and characteristics of a Ty element from yeast. *Nucleic Acids Res.* **13**:2745–2758.

Hauer, B., and Shapiro, J. A. 1984. Control of Tn7 transposition. *Mol. Gen. Genet.* **194**:149–158.

Hazelrigg, T., Levis, R., and Rubin, G. M. 1984. Transformation of *white* locus DNA in *Drosophila*: Dosage compensation, zeste interaction, and position effects. *Cell* **36**:469–481.

Heffron, F., McCarthy, B. J., Ohtsubo, H., and Ohtsubo, E. 1979. DNA sequence analysis of the transposon *Tn3*: Three genes and three sites involved in transposition of *Tn3*. *Cell* **18**:1153–1163.

Hinton, D. M., and Musso, R. E. 1983. Specific *in vitro* transcription of the insertion sequence *IS2*. *J. Mol. Biol.* **169**:53–81.

Höög, C., and Wieslander, L. 1984. Different evolutionary behavior of structurally related, repetitive sequences occurring in the same Balbiani ring gene in *Chironomus tentans*. *Proc. Natl. Acad. Sci. USA* **81**:5165–5169.

Hummel, S., Meyerhof, W., Korge, E., and Knöchel, W. 1984. Characterization of highly and moderately repetitive 500 bp *Eco* R1 fragments from *Xenopus laevis* DNA. *Nucleic Acids Res.* **12**:4921–4938.

Hyde, D. R., and Tu, C. P. D. 1982. Insertion sites and the terminal nucleotide sequences of the Tn4 transposon. *Nucleic Acids Res.* **10**:3981–3993.

Iida, S., Marcoli, R., and Bickle, T. A. 1981. Variant insertion element *IS1* generates 8-base pair duplications of the target sequence. *Nature (London)* **294**:374–376.

Iida, S., Meyer, J., and Arber, W. 1983. Prokaryotic *IS* elements. *In:* Shapiro, J. A., ed., *Mobile Genetic Elements,* New York, Academic Press, pp. 159–221.

Iida, S., Mollet, B., Meyer, J., and Arber, W. 1984. Functional characterization of the prokaryotic mobile genetic element *IS26*. *Mol. Gen. Genet.* **198**:84–89.

Iida, S., Hiestand-Nauer, R., and Arber, W. 1985. Transposable element *IS1* intrinsically generates target duplications of variable length. *Proc. Natl. Acad. Sci. USA* **82**:839–843.

Ikenaga, H., and Saigo, K. 1982. Insertion of a movable element, *297*, into the T-A-T-A box for the H$_3$ histone gene in *Drosophila melanogaster*. *Proc. Natl. Acad. Sci. USA* **79**:4143–4147.

Isberg, R. R., and Syvanen, M. 1985. *Tn5* transposes independently of cointegrate resolution. Evidence for an alternative model of transposition. *J. Mol. Biol.* **182**:69–78.

Johns, M. A., Mottinger, J., and Freeling, M. 1985. A low copy number, *copia*-like transposon in maize. *EMBO J.* **4**:1093–1102.

Johnson, R. C., and Reznikoff, W. S. 1983. DNA sequences at the ends of transposon *Tn5* required for transposition. *Nature (London)* **304**:280–282.

Johnson, R. C., and Reznikoff, W. S. 1984. Role of the IS*5OR* proteins in the promotion and control of *Tn5* transposition. *J. Mol. Biol.* **177**:645–661.

Johnsrud, L., Calos, M. P., and Miller, J. H. 1978. The transposon *Tn9* generates a 9 bp repeated sequence during integration. *Cell* **15**:1209–1219.

Kanazawa, H., Kiyasu, T., Noumi, T., Futai, M., and Yamaguchi, K. 1984. Insertions of transposable elements in the promoter proximal region of the gene cluster for *Escherichia coli* H$^+$-ATPase: 8 base pair repeat generated by insertion of IS1. *Mol. Gen. Genet.* **194**:179–187.

Karess, R. E., and Rubin, G. M. 1984. Analysis of P transposable element functions in *Drosophila*. *Cell* **38**:135–146.

Karlik, C. C., and Fyrberg, E. A. 1985. An insertion within a variably spliced *Drosophila* tropomyosin gene blocks accumulation of only one encoded isoform. *Cell* **41**:57–66.

Kingsman, A. J., Gimlich, R. L., Clarke, L., Chinault, A. C., and Carbon, J. 1981. Sequence variation in dispersed repetitive sequences in *Saccharomyces cerevisiae*. *J. Mol. Biol.* **145:**619–632.

Klaer, R., Kühn, S., Tillmann, E., Fritz, H. J., and Starlinger, P. 1981. The sequence of *IS4*. *Mol. Gen. Genet.* **181:**169–175.

Kleckner, N. 1979. DNA sequence analysis of *Tn10* insertions: Origins and role of 9 bp flanking repetitions during *Tn10* translocation. *Cell* **16:**711–720.

Kleckner, N. 1981. Transposable elements in prokaryotes. *Annu. Rev. Genet.* **15:**341–404.

Krayev, A. S., Kramerov, D. A., Skryabin, K. G., Ryskov, A. P., Bayev, A. A., and Georgiev, G. P. 1980. The nucleotide sequence of the ubiquitous repetitive DNA sequence B1 complementary to the most abundant class of mouse fold-back RNA. *Nucleic Acids Res.* **8:**1201–1215.

Krayev, A. S., Markusheva, T. V., Kramerov, D. A., Ryskov, A. P., Skryabin, K. G., Bayev, A. A., and Georgiev, G. P. 1982. Ubiquitous transposon-like repeats B1 and B2 of the mouse genome: B2 sequencing. *Nucleic Acids Res.* **10:**7461–7475.

Kuehn, M., and Arnheim, N. 1983. Nucleotide sequence of the genetically labile repeated elements 5' to the origin of mouse rRNA transcription. *Nucleic Acids Res.* **11:**211–224.

Lam, B. S., and Carroll, D. 1983. Tandemly repeated sequences from *Xenopus laevis*. II. Dispersed clusters of a 388 base-pair repeating unit. *J. Mol. Biol.* **165:**587–597.

Lendahl, U., and Wieslander, L. 1984. Balbiani ring 6 gene in *Chironomus tentans:* A diverged member of the Balbiani ring gene family. *Cell* **36:**1027–1034.

Lichtenstein, C., and Brenner, S. 1982. Unique insertion site of *Tn7* in the *E. coli* chromosome. *Nature (London)* **297:**601–603.

Lupski, J. R., Gershon, P., Ozaki, L. S., and Godson, G. N. 1984. Specificity of *Tn5* insertions into a 36-bp DNA sequence repeated in tandem seven times. *Gene* **30:**99–106.

Machida, C., Machida, Y., Wang, H. C., Ishizaki, K., and Ohtsubo, E. 1983. Repression of cointegration ability of insertion element *IS1* by transcriptional readthrough from flanking regions. *Cell* **34:**135–142.

Machida, C., Machida, Y., and Ohtsubo, E. 1984a. Both inverted repeat sequences located at the ends of *IS1* provide promoter functions. *J. Mol. Biol.* **177:**247–267.

Machida, Y., Machida, C., and Ohtsubo, E. 1984b. Insertion element *IS1* encodes two structural genes required for its transposition. *J. Mol. Biol.* **177:**229–245.

Martin, S. L., Voliva, C. F., Burton, F. H., Edgell, M. H., and Hutchison, C. A. 1984. A large interspersed repeat found in mouse DNA contains a long open reading frame that evolves as if it encodes a protein. *Proc. Natl. Acad. Sci. USA* **81:**2308–2312.

Mayaux, J. F., Springer, M., Graffe, M., Fromant, M., and Fayat, G. 1984. *IS4* transposition in the attenuator region of the *Escherichia coli pheS,T* operon. *Gene* **30:**137–146.

Mazodier, P., Giraud, E., and Gasser, F. 1983. Genetic analysis of the streptomycin resistance encoded by *Tn5*. *Mol. Gen. Genet.* **192:**155–162.

Mazodier, P., Cossart, P., Giraud, E., and Gasser, F. 1985. Completion of the nucleotide sequence of the central region of *Tn5* confirms the presence of three resistance genes. *Nucleic Acids Res.* **13:**195–205.

McClintock, B. 1949. Mutable loci in maize. *Carnegie Inst. Wash. Year Book* **48:**142–154.

McClintock, B. 1951. Chromosome organization and genic expression. *Cold Spring Harbor Symp. Quant. Biol.* **16:**13–47.

McClintock, B. 1954. Mutations in maize and chromosomal aberrations in *Neurospora*. *Carnegie Inst. Wash. Year Book* **53:**254–260.

McCormick, M., and Ohtsubo, E. 1985. Cointegrates carrying two copies of a *Tn3* derivative in an inverted orientation. *Gene* **34:**197–206.

McGinnis, W., Shermoen, A. W., and Beckendorf, S. K. 1983. A transposable element inserted just 5' to a *Drosophila* glue protein gene alters gene expression and chromatin structure. *Cell* **34:**75–84.

Michiels, T., and Cornelis, G. 1984. Detection and characterization of *Tn2501*, a transposon included within the lactose transposon *Tn951*. *J. Bacteriol.* **158:**866–871.

Misra, T. K., Brown, N. L., Fritzinger, D. C., Pridmore, R. D., Barnes, W. M., Haberstroh, L., and Silver, S. 1984. Mercuric ion-resistance operons of plasmid R100 and transposon *Tn501:* The beginning of the operon including the regulatory region and the first two structural genes. *Proc. Natl. Acad. Sci. USA* **81:**5975–5979.

Misra, T. K., Brown, N. L., Haberstroh, L., Schmidt, A., Goddette, D., and Silver, S. 1985. Mercuric reductase structural genes from plasmid R100 and transposon *Tn501:* Functional domains of the enzyme. *Gene* **34:**253–262.

Mollet, B., Iida, S., Shepherd, J., and Arber, W. 1983. Nucleotide sequence of *IS26*, a new prokaryotic mobile genetic element. *Nucleic Acids Res.* **11**:6319–6330.

Mollet, B., Clerget, M., Meyer, J., and Iida, S. 1985. Organization of the *Tn6*-related kanamycin resistance transposon *Tn2680* carrying two copies of *IS26* and an *IS903* variant, *IS903B. J. Bacteriol.* **163**:55–60.

Morisato, D., and Kleckner, N. 1984. Transposase promotes double strand breaks and single strand joints at *Tn10* termini *in vivo. Cell* **39**:181–190.

Mossie, K. G., Young, M. W., and Varmus, H. E. 1985. Extrachromosomal DNA forms of *copia*-like transposable elements, F elements and middle repetitive DNA sequences in *Drosophila melanogaster. J. Mol. Biol.* **182**:31–43.

Müller-Neumann, M., Yoder, J. I., and Starlinger, P. 1984. The DNA sequence of the transposable element *Ac* of *Zea mays* L. *Mol. Gen. Genet.* **198**:19–24.

Nag, D. K., Das Gupta, U., Adelt, G., and Berg, D. E. 1985. *IS50*-mediated inverse transposition: Specificity and precision. *Gene* **34**:17–26.

Navas, J., García-Lobo, J. M., Léon, J., and Ortíz, J. M. 1985. Structural and functional analyses of the fosfomycin resistance transposon *Tn2921. J. Bacteriol.* **162**:1061–1067.

Newman, B. J., and Grindley, N. D. F. 1984. Mutants of the γ resolvase: A genetic analysis of the recombination function. *Cell* **38**:463–469.

Nomiyama, H., Tsuzuki, T., Wakasugi, S., Fukuda, M., and Shimada, K. 1984. Interruption of a human nuclear sequence homologous to mitochondrial DNA by a member of the *KpnI* 1.8 kb family. *Nucleic Acids Res.* **12**:5225–5234.

O'Hare, K., and Rubin, G. M. 1983. Structures of *P* transposable elements and their sites of insertion and excision in the *Drosophila melanogaster* genome. *Cell* **34**:25–35.

Ohta, N., Swanson, E., Ely, B., and Newton, A. 1984. Physical mapping and complementation analysis of transposon *Tn5* mutations in *Caulobacter crescentus:* Organization of transcriptional units in the hook gene cluster. *J. Bacteriol.* **158**:897–904.

Ohtsubo, H., and Ohtsubo, E. 1978. Nucleotide sequence of an insertion element, *IS1. Proc. Natl. Acad. Sci. USA* **75**:615–619.

Oka, A., Nomura, N., Sugimoto, K., Sugisaki, H., and Takanami, M. 1978. Nucleotide sequence at the insertion sites of a kanamycin transposon. *Nature (London)* **276**:845–847.

Ouartsi, A., Borowski, D., and Brevet, J. 1985. Genetic analysis of *Tn7* transposition. *Mol. Gen. Genet.* **198**:221–227.

Pereira, A., Schwarz-Sommer, Z., Gierl, A., Bertram, I., Peterson, P. A., and Saedler, H. 1985. Genetic and molecular analysis of the Enhancer (*En*) transposable element system of *Zea mays. EMBO J.* **4**:17–23.

Peterson, P. A. 1953. A mutable pale green locus in maize. *Genetics* **38**:682–683.

Peterson, P. A. 1965. A relationship between the *Spm* and *En* control systems in maize. *Am. Nat.* **99**:391–398.

Pohlman, R. F., Fedoroff, N. V., and Messing, J. 1984. The nucleotide sequence of the maize controlling element *activator. Cell* **37**:635–643.

Postle, K., Nguyen, T. T., and Bertrand, K. P. 1984. Nucleotide sequence of the repressor gene of the *Tn10* tetracycline resistance determinant. *Nucleic Acids Res.* **12**:4849–4863.

Potter, S. S. 1982. DNA sequence of a foldback transposable element in *Drosophila. Nature (London)* **297**:201–204.

Potter, S. S. 1984. Rearranged sequences of a human *KpnI* element. *Proc. Natl. Acad. Sci. USA* **81**:1012–1016.

Propst, F., and Vande Woude, G. F. 1984. A novel transposon-like repeat interrupted by an LTR element occurs in a cluster of *B1* repeats in the mouse *C-mos* locus. *Nucleic Acids Res.* **12**:8381–8392.

Putnoky, P., Kiss, G. B., Ott, I., and Kondorosi, A. 1983. *Tn5* carries a streptomycin resistance determinant downstream from the kanamycin resistance gene. *Mol. Gen. Genet.* **191**:288–294.

Rak, B., and von Reutern, M. 1984. Insertion element *IS5* contains a third gene. *EMBO J.* **3**:807–811.

Reed, R. R., Shibuya, G. I., and Steitz, J. A. 1982. Nucleotide sequence of γδ resolvase gene and demonstration that its gene product acts as a repressor of transcription. *Nature (London)* **300**:381–383.

Robertson, D. S. 1978. Characterization of a mutator system in maize. *Mutat. Res.* **51**:21–28.

Rostas, K., Sista, P. R., Stanley, J., and Verma, D. P. S. 1984. Transposon mutagenesis of *Rhizobium japonicum. Mol. Gen. Genet.* **197**:230–235.

Rubin, G. M., Brorein, W. J., Dunsmuir, P., Flavell, A. J., Levis, R., Strobel, E., Toole, J. J., and Young, E. 1980. *Copia*-like transposable elements in the *Drosophila* genome. *Cold Spring Harbor Symp. Quant. Biol.* **45**:619–628.

Saedler, H., and Nevers, P. 1985. Transposition in plants: A molecular model. *EMBO J.* **4**:585–590.

Sagata, N., Yasunaga, T., Ogawa, Y., Tsuzuku-Kawamura, J., and Ikawa, Y. 1984. Bovine leukemia virus: Unique structural features of its long terminal repeats and its evolutionary relationship to human T-cell leukemia virus. *Proc. Natl. Acad. Sci. USA* **81**:4741–4745.

Saiga, H., and Edström, J. E. 1985. Long tandem arrays of complex repeat units in *Chironomus* telomeres. *EMBO J.* **4**:799–804.

Saigo, K., Kugimiya, W., Matsuo, Y., Inouye, S., Yoshioka, K., and Yuki, S. 1984. Identification of the coding sequence for a reverse transcriptase-like enzyme in a transposable genetic element in *Drosophila melanogaster*. *Nature (London)* **312**:659–661.

Sandmeyer, S. B., and Olson, M. V. 1982. Insertion of a repetitive element at the same position in the 5' flanking regions of two dissimilar yeast tRNA genes. *Proc. Natl. Acad. Sci. USA* **79**:7674–7678.

Sasakawa, C., Carle, G. F., and Berg, D. E. 1983. Sequences essential for transposition at the termini of *IS50*. *Proc. Natl. Acad. Sci. USA* **80**:7293–7297.

Sasakawa, C., Phadnis, S. H., Carle, G. F., and Berg, D. E. 1985. Sequences essential for *IS50* transposition. The first base-pair. *J. Mol. Biol.* **182**:487–493.

Scherer, G., Tschudi, C., Perera, J., Delius, H., and Pirrotta, V. 1982. *B104*, a new dispersed repeated gene family in *Drosophila melanogaster* and its analogies with retroviruses. *J. Mol. Biol.* **157**:435–451.

Schindler, C. W., and Rush, M. G. 1985. The *KpnI* family of long interspersed nucleotide sequences is present on discrete sizes of circular DNA in monkey (BSC-1) cells. *J. Mol. Biol.* **181**:161–173.

Schmidt, E. R. 1984. Clustered and interspersed repetitive DNA sequence family of *Chironomus*. The nucleotide sequence of the Cla-elements and of the various flanking sequences. *J. Mol. Biol.* **178**:1–5.

Schnabel, H., Palm, P., Dick, K., and Grampp, B. 1984. Sequence analysis of the insertion element *ISH1.8* and of associated structural changes in the genome of phage φH of the archaebacterium *Halobacterium halobium*. *EMBO J.* **3**:1717–1722.

Schöffl, F., Arnold, W., Pühler, A., Altenbuchner, J., and Schmitt, R. 1981. The tetracycline resistance transposons *Tn1721* and *Tn1771* have three 38 base-pair repeats and generate five base-pair direct repeats. *Mol. Gen. Genet.* **181**:87–94.

Schollmeier, K., Gärtner, D., and Hillen, W. 1985. A bidirectionally active signal for termination of transcription is located between *tetA* and *orfh* on transposon *Tn10*. *Nucleic Acids Res.* **13**:4227–4237.

Schoner, B., and Kahn, M. 1981. The nucleotide sequence of *IS5* from *Escherichia coli*. *Gene* **14**:165–174.

Schwarz-Sommer, Z., Gierl, A., Klösgen, R. B., Wienaud, U., Peterson, P. A., and Saedler, H. 1984. The *Spm (En)* transposable element controls the excision of a 2-kb DNA insert at the wx^{m-8} allele of *Zea mays*. *EMBO J.* **3**:1021–1028.

Schwarz-Sommer, Z., Gierl, A., Cuypers, H., Peterson, P. A., and Saedler, H. 1985. Plant transposable elements generate the DNA sequence diversity needed in evolution. *EMBO J.* **4**:591–597.

Shepherd, B. M., and Finnegan, D. J. 1984. Structure of circular copies of the *412* transposable element present in *Drosophila melanogaster* tissue culture cells, and isolation of a free *412* long-terminal repeat. *J. Mol. Biol.* **180**:21–40.

Shimizu, Y., Yoshida, K., Ren, C. S., Fujinaga, K., Rajagopalan, S., and Chinnadurai, G. 1983. *Hinf* family: A novel repeated DNA family of the human genome. *Nature (London)* **302**:587–590.

Simsek, M., Das Sarma, S., RajBhandary, U. L., and Khorana, H. G. 1982. A transposable element from *Halobacterium halobium* which inactivates the bacteriorhodopsin gene. *Proc. Natl. Acad. Sci. USA* **79**:7268–7272.

Singer, M. F. 1982. SINEs and LINEs: Highly repeated short and long interspersed sequences in mammalian genomes. *Cell* **28**:433–434.

Stanisich, V. A., Bennett, P. M., and Richmond, M. H. 1977. Characterization of a translocation unit encoding resistance to mercuric ions that occurs on a nonconjugative plasmid in *Pseudomonas aeroginosa*. *J. Bacteriol.* **129**:1227–1233.

Steller, H., and Pirrotta, V. 1985. A transposable P vector that confers selectable G418 resistance to *Drosophila* larvae. *EMBO J.* **4**:167–171.

Stumph, W. E., Hodgson, C. P., Tsai, M. J., and O'Malley, B. W. 1984. Genomic structure and possible retroviral origin of the chicken *CR1* repetitive DNA sequence family. *Proc. Natl. Acad. Sci. USA* **81**:6667–6671.

Sümegi, J., Wieslander, L., and Daneholt, B. 1982. A hierarchic arrangement of the repetitive sequences in the Balbiani ring 2 of *Chironomus tentans*. *Cell* **30**:579–587.

Sun, L., Paulson, K. E., Schmid, C. W., Kadyk, L., and Leinwand, L. 1984. Non-*Alu* family interspersed repeats in human DNA and their transcriptional activity. *Nucleic Acids Res.* **12:**2669–2690.

Sutcliffe, J. G., Milner, R. J., Bloom, F. E., and Lerner. R. A. 1982. Common 82-nucleotide sequence unique to brain RNA. *Proc. Natl. Acad. Sci. USA* **79:**4942–4946.

Sutton, W. D., Gerlach, W. L., Schwartz, D., and Peacock, W. J. 1984. Molecular analysis of *Ds* controlling element mutations at the *Adh1* locus of maize. *Science* **223:**1265–1268.

Taguchi, A. K. W., Ciriacy, M., and Young, E. T. 1984. Carbon source dependence of transposable element-associated gene activation in *Saccharomyces cerevisiae. Mol. Cell. Biol.* **4:**61–68.

Taylor, L. P., and Walbot, V. 1985. A deletion adjacent to the maize transposable element *Mu-1* accompanies loss of *Adh1* expression. *EMBO J.* **4:**869–876.

Thorpe, P. A., and Clowes, R. C. 1984. Absence of direct repeats flanking transposons resulting from intramolecular transposition. *Gene* **28:**103–112.

Timmerman, K. P., and Tu, C. P. D. 1985. Complete sequence of *IS3. Nucleic Acids Res.* **13:**2127–2139.

Trieu-Cuot, P., and Courvalin, P. 1984. Nucleotide sequence of the transposable element *IS15. Gene* **30:**113–120.

Trieu-Cuot, P., Labigne-Roussel, A., and Courvalin, P. 1983. An *IS15* insertion generates an eight base-pair duplication of the target DNA. *Gene* **24:**125–129.

Wasserman, S. A., Dungan, J. M., and Cozzarelli, N. R. 1985. Discovery of a predicted DNA knot substantiates a model for site-specific recombination. *Science* **229:**171–174.

Weck, E., Courage, U., Döring, H. P., Federoff, N., and Starlinger, P. 1984. Analysis of *sh-m6233,* a mutation induced by the transposable element *Ds* in the sucrose synthase gene of *Zea mays. EMBO J.* **3:**1713–1716.

Wharton, K. A., Yedvobnick, B., Finnerty, V. G., and Artavanis-Tsakonas, S. 1985. *Opa:* A novel family of transcribed repeats shared by the *notch* locus and other developmentally regulated loci in *D. melanogaster. Cell* **40:**55–82.

Williamson, V. M. 1983. Transposable elements in yeast. *Int. Rev. Cytol.* **83:**1–25.

Wilson, R., and Storb, U. 1983. Association of two different repetitive DNA elements near immunoglobulin light chain genes. *Nucleic Acids Res.* **11:**1803–1817.

Winston, F., Durbin, K. J., and Fink, G. R. 1984. The *SPT3* gene is required for normal transcription of *Ty* elements in *S. cerevisiae. Cell* **39:**675–682.

Wirth, T., Glöggler, K., Baumruker, T., Schmidt, M., and Horak, I. 1983. Family of middle repetitive DNA sequences in the mouse genome with structural features of solitary retroviral long terminal repeats. *Proc. Natl. Acad. Sci. USA* **80:**3327–3330.

Wishart, W. L., Machida, C., Ohtsubo, H., and Ohtsubo, E. 1983. *Escherichia coli* RNA polymerase binding sites and transcription initiation sites in the transposon *Tn3. Gene* **24:**99–113.

Zheng, Z. X., Chandler, M., Hipskind, R., Clerget, M., and Caro, L. 1981. Dissection of the r-determinant of the plasmid R100.1: The sequence at the extremities of *Tn21. Nucleic Acids Res.* **9:**6265–6276.

CHAPTER 10

Ahlquist, P., Luckow, V., and Kaesberg, P. 1981. Complete nucleotide sequence of brome mosaic virus RNA3. *J. Mol. Biol.* **153:**23–38.

Ahlquist, P., Dasgupta, R., and Kaesberg, P. 1984. Nucleotide sequence of the brome mosaic virus genome and its implications for viral replication. *J. Mol. Biol.* **172:**369–383.

Aleström, P., Akusjärvi, G., Pettersson, M., and Pettersson, U. 1982. DNA sequence analysis of the region encoding the terminal protein and the hypothetical N-gene product of adenovirus type II. *J. Biol. Chem.* **257:**13492–13498.

Arnheiter, H., Davis, N. L., Wertz, G., Schubert, M., and Lazzarini, R. A. 1985. Role of the nucleocapsid protein in regulating vesicular stomatitis virus RNA synthesis. *Cell* **41:**259–267.

Atkins, J. F., Steitz, J. A., Anderson, C. W., and Model, P. 1979. Binding of mammalian ribosomes to MS2 phage RNA reveals an overlapping gene encoding a lysis function. *Cell* **18:**247–256.

Backhaus, H., and Petri, J. B. 1984. Sequence analysis of a region from the early right operon in phage P22, including the replication genes *18* and *12. Gene* **32:**289–303.

Bailey, J. N., Klement, J. F., and McAllister, W. T. 1983. Relationship between promoter structure and template specificities exhibited by the bacteriophage T3 and T7 RNA polymerases. *Proc. Natl. Acad. Sci. USA* **80:**2814–2818.

Balâzs, E., Guilley, H., Jonard, G., and Richards, K. 1982. Nucleotide sequence of DNA from an altered-virulence D/H of the cauliflower mosaic virus. *Gene* **19**:239–249.

Baroudy, B. M., Venkateson, S., and Moss, B. 1982. Incompletely base-paired flip-flop terminal loops link the two DNA strands of the vaccinia virus genome into one uninterrupted polynucleotide chain. *Cell* **28**:315–324.

Baty, D., Barrera-Saldana, H. A., Everett, R. D., Vigneron, M., and Chambon, P. 1984. Mutational dissection of the 21 bp repeat region of the SV40 early promoter reveals that it contains overlapping elements of the early-early and late-early promoters. *Nucleic Acids Res.* **12**:915–932.

Beck, E., Sommer, R., Auerswald, E. A., Kürz, C., Zink, B., Osterburg, G., Schaller, H., Sugimoto, K., Sugisaki, H., Okamoto, T., and Takanami, M. 1978. Nucleotide sequence of bacteriophage fd DNA. *Nucleic Acids Res.* **5**:4495–4503.

Benoist, C., and Chambon, P. 1981. *In vivo* sequence requirements of the SV40 early promoter region. *Nature (London)* **290**:304–310.

Berget, P. B., Poteete, A. R., and Sauer. R. T. 1983. Control of phage P22 tail protein expression by transcription termination. *J. Mol. Biol.* **164**:561–572.

Blomquist, M. C., Hunt, L. T., and Barker, W. C. 1984. Vaccinia virus 19-kilodalton protein: Relationship to several mammalian proteins including two growth factors. *Proc. Natl. Acad. Sci. USA* **81**:7363–7367.

Bodescot, M., Chambrand, B., Farrell, P., and Perricaudet, M. 1984. Spliced RNA from the IR1-U2 region of Epstein–Barr virus: Presence of an open reading frame for a repetitive polypeptide. *EMBO J.* **3**:1913–1917.

Boursnell, M. E. G., and Brown, T. D. K. 1984. Sequencing of coronavirus IBV genomic RNA: A 195-base open reading frame encoded by mRNA B. *Gene* **29**:87–92.

Braam, J., Ulmanen, I., and Krug, R. M. 1983. Molecular model of a eucaryotic transcription complex: Functions and movements of influenza P proteins during capped RNA-primed transcription. *Cell* **34**:609–618.

Brady, J., Radonovich, M., Vodkin, M., Natarajan, V., Thoren, M., Das, G., Janik, J., and Salzman, N. P. 1982. Site-specific base substitution and deletion mutations that enhance or suppress transcription of the SV40 major late RNA. *Cell* **31**:625–633.

Brady, J., Radonovich, M., Thoren, M., Das, G., and Salzman, N. P. 1984. Simian virus 40 major late promoter: An upstream DNA sequence required for efficient *in vitro* transcription. *Mol. Cell. Biol.* **4**:133–141.

Brederode, F. T., Koper-Zwarthoff, E. C., and Bol, J. F. 1980. Complete nucleotide sequence of alfalfa mosaic virus RNA 4. *Nucleic Acids Res.* **8**:2213–2223.

Byrne, B. J., Davis, M. S., Yamaguchi, J., Bergsma, D. J., and Subramanian, K. N. 1983. Definition of the simian virus 40 early promoter region and demonstration of a host range bias in the enhancement effect of the simian virus 40 72-base-pair repeat. *Proc. Natl. Acad. Sci. USA* **80**:721–725.

Callahan, P. L., Mizutani, S., and Colonno, R. J. 1985. Molecular cloning and complete sequence determination of RNA genome of human rhinovirus type 14. *Proc. Natl. Acad. Sci. USA* **82**:732–736.

Campbell, A. 1983. Bacteriophage λ. *In:* Shapiro, J. A., ed., *Mobile Genetic Elements,* New York, Academic Press, pp. 65–104.

Carroll, A. R., Rowlands, D. J., and Clarke, B. E. 1984. The complete nucleotide sequence of the RNA coding for the primary translation product of foot and mouth disease virus. *Nucleic Acids Res.* **12**:2461–2472.

Cashdollar, L. W., Esparza, J., Hudson, G. R., Chmelo, R., Lee, P. W. K., and Joklik, W. K. 1982. Cloning the double-stranded RNA genes for reovirus: Sequence of the cloned S2 gene. *Proc. Natl. Acad. Sci. USA* **79**:7644–7648.

Cashdollar, L. W., Chmelo, R. A., Wiener, J. R., and Joklik, W. K. 1985. Sequences of the S1 genes of the three serotypes of reovirus. *Proc. Natl. Acad. Sci. USA* **82**:24–28.

Chow, L. T., Broker, T., and Lewis, J. B. 1979. Complex splicing patterns of RNAs from the early regions of adenovirus-2. *J. Mol. Biol.* **134**:265–303.

Christie, G. E., and Calendar, R. 1985. Bacteriophage P2 late promoters. II. Comparison of the four late promoter sequences. *J. Mol. Biol.* **181**:373–382.

Chu, F. K., Maley, G. F., Maley, F., and Belfort, M. 1984. Intervening sequence in the thymidylate synthase gene of bacteriophage T4. *Proc. Natl. Acad. Sci. USA* **81**:3049–3053.

Cochran. M. A., Puckett, C., and Moss, B. 1985. *In vitro* mutagenesis of the promoter region for a vaccinia virus gene: Evidence for tandem early and late regulatory signals. *J. Virol.* **54**:30–37.

Concino, M. F., Lee, R. F., Merryweather, J. P., and Weinmann, R. 1984. The adenovirus major late promoter

TATA box and initiation site are both necessary for transcription *in vitro. Nucleic Acids Res.* **12:**7423–7433.

Conway, L., and Wickens, M. 1985. A sequence of A-A-U-A-A-A is required for formation of simian virus 40 late mRNA3 termini in frog oocytes. *Proc. Natl. Acad. Sci. USA* **82:**3949–3953.

Cornelissen, B. J. C., Brederode, F. T., Moormann, R. J. M., and Bol, J. F. 1983a. A complete nucleotide sequence of alfalfa mosaic virus RNA 1. *Nucleic Acids Res.* **11:**1253–1265.

Cornelissen, B. J. C., Brederode, F. T., Vesneman, G. H., van Boom, J. H., and Bol, J. F. 1983b. Complete nucleotide sequence of alfalfa mosaic virus RNA 2. *Nucleic Acids Res.* **11:**3019–3025.

Cornelissen, B. J. C., Janssen, H., Zuidema, D., and Bol, J. F. 1984. Complete nucleotide sequence of tobacco streak RNA 3. *Nucleic Acids Res.* **12:**2427–2437.

Costa, R. H., Draper, K. G., Kelly, T. J., and Wagner, E. K. 1985. An unusual spliced herpes simplex virus type 1 transcript with sequence homology to Epstein–Barr virus DNA. *J. Virol.* **54:**317–328.

Craig, N. L. 1985. Site-specific inversion: Enhancers, recombination proteins, and mechanism. *Cell* **41:**649–650.

Davies, J. W., Stanley, J., and Van Kammen, A. 1979. Sequence homology adjacent to the 3′ terminal poly(A) of cowpea mosaic virus RNAs. *Nucleic Acids Res.* **7:**493–500.

Davies, R. W. 1980. DNA sequence of the int–xis–P$_1$ region of the bacteriophage λ: Overlap of the *int* and *xis* genes. *Nucleic Acids Res.* **8:**1765–1782.

de Haseth, P. L., Goldman, R. A., Cech, C. L., and Caruthers, M. H. 1983. Chemical synthesis and biochemical reactivity of bacteriophage lambda P$_R$ promoter. *Nucleic Acids Res.* **11:**773–787.

Dekker, B. M. M., and van Ormondt, H. 1984. The nucleotide sequence of fragment *Hind*III-C of human adenovirus type 5 DNA (map positions 17.1–31.7). *Gene* **27:**115–120.

Dillon, L. S. 1978. *The Genetic Mechanism and the Origin of Life,* New York, Plenum Press.

Dixon, L. K., and Hohn, T. 1984. Initiation of translation of the cauliflower mosaic virus genome from a polycistronic mRNA: Evidence from deletion mutagenesis. *EMBO J.* **3:**2731–2736.

Dunn, J. J., and Studier, F. W. 1983. Complete nucleotide sequence of bacteriophage T7 DNA and the locations of T7 genetic elements. *J. Mol. Biol.* **166:**477–535.

Dyall-Smith, M. L., and Holmes, I. H. 1984. Sequence homology between human and animal rotavirus serotype-specific glycoproteins. *Nucleic Acids Res.* **12:**3973–3982.

Elliott, T., and Geiduschek, E. P. 1984. Defining a bacteriophage T4 late promoter: absence of a ''−35'' region. *Cell* **36:**211–219.

Emerson, S. U. 1982. Reconstitution studies detect a single polymerase entry site on the vesicular stomatitis virus genome. *Cell* **31:**635–642.

Ernoult-Lange, M., and May, E. 1983. Evidence of transcription from the late region of the integrated simian virus 40 genome in transformed cells: Location of the 5′ ends of late transcripts in cells abortively infected and in cells transformed by simian virus 40. *J. Virol.* **46:**756–767.

Escarmís, C., and Salas, M. 1982. Nucleotide sequence of the early genes 3 and 4 of bacteriophage φ29. *Nucleic Acids Res.* **10:**5785–5798.

Everett, R. D., Baty, D., and Chambon, P. 1983. The repeated GC-rich motifs upstream from the TATA box are important elements of the SV40 early promoter. *Nucleic Acids Res.* **11:**2447–2464.

Fiers, W., Contreras, R., Duerinck, F., Haegmean, G., Merregaert, J., Min Jou, W., Raeymakers, A., Volckaert, G., Ysebaert, M., Van de Kerckhove, J., Nolf, F., and Van Montagu, M. 1975. A-protein gene of bacteriophage MS2. *Nature (London)* **256:**273–278.

Fiers, W., Contreras, R., Haegeman, G., Rogiers, R., Van deVoorde, A., Van Heuverswyn, H., Van Herreweghe, J., Volckaert, G., and Ysebaert, M. 1978. Complete nucleotide sequence of SV40 DNA. *Nature (London)* **273:**113–119.

Franck, A., Guilley, H., Jonard, G., Richards, K., and Hirth, L. 1980. Nucleotide sequence of cauliflower mosaic virus DNA. *Cell* **21:**285–294.

Franssen, H., Leunissen, J., Goldback, R., Lomonossoff, G., and Zimmern, D. 1984. Homologous sequences in non-structural proteins from cowpea mosaic virus and picornaviruses. *EMBO J.* **3:**655–661.

Fulford, W., and Model, P. 1984. Gene X of bacteriophage f1 is required for phage DNA synthesis. Mutagenesis of in-frame overlapping genes. *J. Mol. Biol.* **178:**137–153.

Furuichi, Y. 1978. ''Pretranscriptional capping'' in the biosynthesis of cytoplasmic polyhedrosis virus mRNA. *Proc. Natl. Acad. Sci. USA* **75:**1086–1090.

Gardner, R. C., Howarth, A. J., Hahn, P., Brown-Leudi, M., Shepherd, R. J., and Messing, J. 1981. The complete nucleotide sequence of an infectious clone of cauliflower mosaic virus by M13mp7 shot gun sequencing. *Nucleic Acids Res.* **9**:2871–2887.

Gaynor, R. B., and Beck, A. J. 1983. *Cis*-acting induction of adenovirus transcription. *Cell* **33**:683–693.

Gerald, W. L., and Karam, J. D. 1984. Expression of a DNA replication gene cluster in bacteriophage T4: Genetic linkage and the control of gene product interactions. *Genetics* **107**:537–549.

Ghosh, P. K., Lebowitz, P., Frisque, R. J., and Gluzman, Y. 1981. Identification of a promoter component involved in positioning the 5′ termini of simian virus 40 early mRNAs. *Proc. Natl. Acad. Sci. USA* **78**:100–104.

Gibson, T., Stockwell, P., Ginsburg, M., and Barrell, B. 1984. Homology between two EBV early genes and HSV ribonucleotide reductase and 38K genes. *Nucleic Acids Res.* **12**:5087–5099.

Giorgi, C., Blumberg, B. M., and Kolakofsky, D. 1983. Sendai virus contains overlapping genes expressed from a single mRNA. *Cell* **35**:829–836.

Goelet, P., Lommossoff, G. P., Butler, P. J. G., Akam, M. E., Gait, M. J., and Karn, J. 1982. Nucleotide sequence of tobacco mosaic virus RNA. *Proc. Natl. Acad. Sci. USA* **79**:5818–5822.

Gomez-Marquez, J., Puga, A., and Notkins, A. L. 1985. Regions of the terminal repetitions of the herpes simplex virus type 1 genome. Relationship to immunoglobulin switch-like DNA sequences. *J. Biol. Chem.* **260**:3490–3495.

Gram, H., and Rüger, W. 1985. Genes *55, at, 47* and *46* of bacteriophage T4: The genomic organization as deduced by sequence analysis. *EMBO J.* **4**:257–264.

Greene, J. R., Brennan, S. M., Andrew, D. J., Thompson, C. C., Richards, S. H., Heinrikson, R. L., and Geiduschek, E. P. 1984. Sequence of the bacteriophage SPO1 gene coding for transcription factor 1, a viral homolog of the bacterial type II DNA-binding proteins. *Proc. Natl. Acad. Sci. USA* **81**:7031–7035.

Griffith, J. D., and Nash, H. A. 1985. Genetic rearrangement of DNA induces knots with a unique topology: Implications for the mechanism of synapsis and crossing-over. *Proc. Natl. Acad. Sci. USA* **82**:3124–3128.

Grinnell, B. W., and Wagner, R. R. 1984. Nucleotide sequence and secondary structure of VSV leader RNA and homologous DNA involved in inhibition of DNA-dependent transcription. *Cell* **36**:533–543.

Gruss, P., Dhar, R., and Khoury, G. 1981. Simian virus 40 tandem repeated sequences as an element of the early promoter. *Proc. Natl. Acad. Sci. USA* **78**:943–947.

Guilfoyle, A., Osheroff, W. P., and Rossini, M. 1985. Two functions encoded by adenovirus early region 1A are responsible for the activation and repression of the DNA-binding protein gene. *EMBO J.* **4**:707–713.

Guilley, H., Jonard, G., Kukla, B., and Richards, K. E. 1979. Sequence of 1000 nucleotides at the 3′ end of tobacco mosaic virus RNA. *Nucleic Acids Res.* **6**:1287–1308.

Guilley, H., Dudley, R. K., Jonard, G., Balázs, E., and Richards, K. E. 1982. Transcription of cauliflower mosaic virus DNA: Detection of promoter sequences, and characterization of transcripts. *Cell* **30**:763–773.

Hamilton, W. D. O., Bisaro, D. M., and Buck, K. W. 1982. Identification of novel DNA forms in tomato golden mosaic virus infected tissue. Evidence for a two component viral genome. *Nucleic Acids Res.* **10**:4901–4912.

Hamilton, W. D. O., Stein, V. E., Coutts, R. H. A., and Buck, K. W. 1984. Complete nucleotide sequence of the infectious cloned DNA components of tomato golden mosaic virus: Potential coding regions and regulatory sequences. *EMBO J.* **3**:2197–2205.

Hattman, S., and Ives, J. 1984. SI nuclease mapping of the phage Mu *mom* gene promoter: A model for the regulation of *mom* expression. *Gene* **29**:185–198.

Hen, R. Borrelli, E., Sassone-Corsi, P., and Chambon, P. 1983. An enhancer element is located 340 base pairs upstream from the adenovirus-2 E1A capsite. *Nucleic Acids Res.* **11**:8747–8760.

Hill, D. F., and Petersen, G. B. 1982. Nucleotide sequence of bacteriophage f1 DNA. *J. Virol.* **44**:32–46.

Hiti, A. L., and Nayak, D. P. 1985. Complete nucleotide sequence of the neuraminidase gene of human influenza virus A/WSN/33. *J. Virol.* **41**:730–734.

Ho, Y. S., Wulff, D. L., and Rosenberg, M. 1983. Bacteriophage λ protein cII binds promoters on the opposite face of the DNA from RNA polymerase. *Nature (London)* **304**:703–708.

Holder, R. D., and Whiteley, H. R. 1983. *In vitro* synthesis of late bacteriophage φ29 RNA. *J. Virol.* **46**:690–702.

Hoopes, B. C., and McClure, W. R. 1985. A cII-dependent promoter is located within the *Q* gene of bacteriophage λ. *Proc. Natl. Acad. Sci. USA* **82**:3134–3138.

Hoyt, M. A., Knight, D. M., Das, A., Miller, H. I., and Echols, H. 1982. Control of phage λ development by stability and synthesis of cII protein: Role of the viral cIII and host *hflA, himA,* and *himD* genes. *Cell* **31:**565–573.

Huddleston, J. A., and Brownlee, G. G. 1982. The sequence of the nucleoprotein gene of human influenza A virus, strain A/NT/60/68. *Nucleic Acids Res.* **10:**1029–1038.

Imai, M., Richardson, M. A., Ikegami, N., Shatkin, A. J., and Furuichi, Y. 1983. Molecular cloning of double-stranded RNA virus genomes. *Proc. Natl. Acad. Sci. USA* **80:**373–377.

Inokuchi, Y., Hirashima, A., and Watanabe, I. 1982. Comparison of the nucleotide sequences at the 3′-terminal region of RNAs from RNA coliphages. *J. Mol. Biol.* **158:**711–730.

Jones, M. D., and Griffin, B. E. 1983. Clustered repeat sequences in the genome of Epstein Barr virus. *Nucleic Acids Res.* **11:**3919–3937.

Kahmann, R., Rudt, F., Koch, C., and Mertens, G. 1985. G inversion in bacteriophage Mu DNA is stimulated by a site within the invertase gene and a host factor. *Cell* **41:**771–780.

Kassavetis, G. A., Elliott, T., Rabussay, D. P., and Geiduschek, E. P. 1983. Initiation of transcription at phage T4 late promoters with purified RNA polymerases. *Cell* **33:**887–897.

Katze, M. G., Chen, Y.-T., and Krug, R. M. 1984. Nuclear–cytoplasmic transport and VA1 RNA-independent translation of influenza viral messenger RNAs in late adenovirus-infected cells. *Cell* **37:**483–490.

Keller, J. M., and Alwine, J. C. 1984. Activation of the SV40 late promoter: Direct effects of T antigen in the absence of viral DNA replication. *Cell* **36:**381–389.

Keohavong, P., Gattoni, R., LeMoullec, J. M., Jacob, M., and Stévenin, J. 1982. The orderly splicing of the first three leaders of the adenovirus-2 major late transcript. *Nucleic Acids Res.* **10:**1215–1229.

Kozlov, Y. V., Rupasov, V. V., Adshey, D. M., Belgelkarskaya, S. N., Agranovsky, A. A., Mankin, A. S., Morozov, S. Y., Dolja, V. V., and Atabekov, J. G. 1984. Nucleotide sequence of the 3′-terminal tRNA-like structure in barley stripe mosaic virus genome. *Nucleic Acids Res.* **12:**4001–4009.

Krause, H. M., Rothwell, M. R., and Higgins, N. P. 1983. The early promoter of bacteriophage Mu: Definition of the site of transcript initiation. *Nucleic Acids Res.* **11:**5483–5495.

La Farina, M., and Vitale, M. 1984. Rho-dependence of the terminator active at the end of the I region of transcription of bacteriophage f1. *Mol. Gen. Genet.* **195:**5–9.

Lamb, R. A., and Choppin, P. W. 1979. Segment 8 of the influenza virus genome is unique in coding for two polypeptides. *Proc. Natl. Acad. Sci. USA* **76:**4908–4912.

Lamb, R. A., and Lai, C. J. 1980. Sequence of interrupted and uninterrupted mRNAs and cloned DNA coding for the two overlapping non-structural proteins of influenza virus. *Cell* **21:**475–485.

Lamb, R. A., Choppin, P. W., Chanock, R. M., and Lai, C. J. 1980. Mapping of the two overlapping genes for polypeptides NS_1 and NS_2 on RNA segment 8 of influenza virus genome. *Proc. Natl. Acad. Sci. USA* **77:**1857–1861.

Lamb, R. A., Zebedee, S. L., and Richardson, C. D. 1985. Influenza virus M_2 protein is an integral membrane protein expressed on the infected-cell surface. *Cell* **40:**827–833.

Lamy, D., Jonard, G., Guilley, H., and Hirth, L. 1975. Comparison between the 3′OH end RNA sequence of two strains of TMV which may be aminoacylated. *J. Biol. Chem.* **247:**4966–4974.

Langner, K. D., Vardimon, L., Renz, D., and Doerfler, W. 1984. DNA methylation of three 5′ C-C-G-G 3′ sites in the promoter and 5′ region inactivate the *E2a* gene of adenovirus type 2. *Proc. Natl. Acad. Sci. USA* **81:**2950–2954.

Lee, G., and Pero, J. 1981. Conserved nucleotide sequences in temporally controlled bacteriophage promoters. *J. Mol. Biol.* **152:**247–265.

Lee, D. C., Roeder, R. G., and Wold, W. S. M. 1982. DNA sequences affecting specific initiation of transcription *in vitro* from the EIII promoter of adenovirus 2. *Proc. Natl. Acad. Sci. USA* **79:**41–45.

Lengyel, J. A., and Calendar, R. 1974. Control of bacteriophage P2 protein and DNA synthesis. *Virology* **57:**305–313.

Luiten, R. G. M., Schoenmakers, J. G. G., and Konings, R. N. H. 1983. Major coat protein gene of the filamentous *Pseudomonas aeruginosa* phage Pf3: Absence of a N-terminal leader signal sequence. *Nucleic Acids Res.* **11:**8073–8085.

Mathis, D. J., and Chambon, P. 1981. The SV40 early region TATA box is required for accurate *in vitro* initiation of transcription. *Nature (London)* **290:**310–315.

Matsui, T. 1982. *In vitro* accurate initiation of transcription on the adenovirus type 2 IVa2 gene which does not contain a TATA box. *Nucleic Acids Res.* **10:**7089–7101.

Meyer, F., Weber, H., and Weissmann, C. 1981. Interactions of Qβ replicase with Qβ RNA. *J. Mol. Biol.* **153:**631–660.

Mills, D. R., Dobkin, C., and Kramer, F. R. 1978. Template-determined variable rate of RNA chain elongation. *Cell* **15**:541–550.

Min Jou, W., and Fiers, W. 1976. Studies on the bacteriophage MS2. XXXIII. Comparison of the nucleotide sequences in related bacteriophage RNAs. *J. Mol. Biol.* **106**:1047–1060.

Miura, N., Nakatani, Y., Ishiura, M., Uchida, T., and Okada, Y. 1985. Molecular cloning of a full-length cDNA encoding the hemaglutinin-neuraminidase glycoprotein of Sendai virus. *FEBS Lett.* **188**:112–116.

Moffatt, B. A., Dunn, J. J., and Studier, F. W. 1984. Nucleotide sequence of the gene for bacteriophage T7 RNA polymerase. *J. Mol. Biol.* **173**:265–269.

Morch, M. D., Zagórski, W., and Haenni, A. L. 1982. Proteolytic maturation of the turnip-yellow-mosaic-virus polyprotein coded *in vitro* occurs by internal catalysis. *Eur. J. Biochem.* **127**:259–265.

Murray, C. L., and Rabinowitz, J. C. 1982. Nucleotide sequences of transcription and translation initiation region in *Bacillus* phage ϕ29 early genes. *J. Biol. Chem.* **257**:1053–1062.

Murthy, S. C. S., Bhat, G. P., and Thimmoppaya, B. 1985. Adenovirus EIIA early promoter: Transcriptional control elements and induction by the viral pre-early E1A gene, which appears to be sequence independent. *Proc. Natl. Acad. Sci. USA* **82**:2230–2234.

Najarian, R., Caput, D., Gee, W., Potter, S. J., Renard, A., Merryweather, J., Van Nest, G., and Dina, D. 1985. Primary structure and gene organization of human hepatitis A virus. *Proc. Natl. Acad. Sci. USA* **82**:2627–2631.

Natarajan, V., and Salzman, N. P. 1985. *Cis* and *trans* activation of adenovirus IVa_2 gene transcription. *Nucleic Acids Res.* **13**:4067–4083.

Nishiguchi, M., Kibuchi, S., Kiho, Y., Ohno, T., Meshi, T., and Okada, Y. 1985. Molecular basis of plant viral virulence: The complete nucleotide sequence of an attenuated strain of tobacco mosaic virus. *Nucleic Acids Res.* **13**:5585–5590.

Nishihara, T., Mills, D. R., and Kramer, F. R. 1983. Localization of the Qβ replicase recognition site in MDV-1 RNA. *J. Biochem.* **93**:669–674.

Nomoto, A., Omata, T., Toyoda, H., Kuge, S., Hori, H., Kataoka, Y., Genba, Y., Nakano, Y., and Imura, N. 1982. Complete nucleotide sequence of the attenuated poliovirus Sabin 1 strain genome. *Proc. Natl. Acad. Sci. USA* **79**:5793–5797.

Ohno, T., Aoyagi, M., Yamanashi, Y., Saito, H., Ikawa, S., Meshi, T., and Okada, Y. 1984. Nucleotide sequence of the tobacco mosaic virus (tomato strain) genome and comparison with the common strain genome. *J. Biochem.* **96**:1915–1923.

Osbourne, T. F., Gaynor, R. B., and Berk, A. J. 1982. The TATA homology and the mRNA 5' untranslated sequence are not required for the expression of essential adenovirus E1A functions. *Cell* **29**:139–148.

Otsuka, J., and Kunisawa, T. 1982. Characteristic base sequence patterns of promoter and terminator sites in ϕX174 and Fd phage DNAs. *J. Theor. Biol.* **97**:415–436.

Peeters, B. P. H., Peters, R. M., Schoenmakers, J. G. G., and Konings, R. N. H. 1985. Nucleotide sequence and genetic organization of the genome of the N-specific filamentous bacteriophage IKe. *J. Mol. Biol.* **181**:27–39.

Pero, J. 1983. A prokaryotic model for the developmental control of gene expression. *In:* Subtelnig, S., and Kafatos, E. C., eds. *Gene Structure and Regulation in Development,* New York, Alan R. Liss, pp. 227–233.

Pfeiffer, P., and Hohn, T. 1983. Involvement of reverse transcription in the replication of cauliflower mosaic virus: A detailed model and test of some aspects. *Cell* **33**:781–789.

Plasterk, R. H. A., Vrieling, H., and Van de Putte, 1983. Transcription initiation of Mu *mom* depends on methylation of the promoter region and a phage-encoded transactivator. *Nature (London)* **301**:344–347.

Plasterk, R. H. A., Vollering, M., Brinkman, A., and Van de Putte, P. 1984. Analysis of the methylation-regulated Mu *mom* transcript. *Cell* **36**:189–196.

Plucienniczak, A., Schroeder, E., Zettlmeissl, G., and Streeck, R. E. 1985. Nucleotide sequence of a cluster of early and late genes in a conserved segment of the vaccinia virus genome. *Nucleic Acids Res.* **13**:985–998.

Porter, A. G., Smith, J. C., and Emtage, J. S. 1980. Nucleotide sequence of influenza virus RNA segment 8 indicates that coding regions for NS_1 and NS_2 proteins overlap. *Proc. Natl. Acad. Sci. USA* **77**:5074–5078.

Pribnow, D. 1975. Bacteriophage T7 early promoters. Nucleotide sequences of two RNA polymerase binding sites. *J. Mol. Biol.* **99**:419–443.

Pulitzer, J. F., Colombo, M., and Ciaramella, M. 1985. New control elements of bacteriophage T4 pre-replicative transcription. *J. Mol. Biol.* **182**:249–263.

Purohit, S., and Mathews, C. K. 1984. Nucleotide sequence reveals overlap between T4 phage genes encoding dihydrofolate reductase and thymidylate synthase. *J. Biol. Chem.* **259**:6261–6266.

Rand, K. N., and Gait, M. J. 1984. Sequence and cloning of bacteriophage T4 gene 63 encoding RNA ligase and tail fibre attachment activities. *EMBO J.* **3**:397–402.

Reddy, V. B., Thimmappaya, B., Dhar, R., Subramanian, K. N., Zain, B. S., Pan, J., Ghosh, P. K., Celma, M. L., and Weissman, S. M. 1978. The genome of simian virus 40. *Science* **200**:494–502.

Rezaian, M. A., Williams, R. H. V., Gordon, K. H. J., Gould, A. R., and Symons, R. H. 1984. Nucleotide sequence of cucumber-mosaic-virus RNA2 reveals a translation product significantly homologous to corresponding proteins of other viruses. *Eur. J. Biochem.* **143**:277–284.

Rezaian, M. A., Williams, R. H. V., and Symons, R. H. 1985. Nucleotide sequence of cucumber mosaic virus RNA 1. Presence of a sequence complementary to part of the viral satellite RNA and homologies with other viral RNAs. *Eur. J. Biochem.* **150**:331–339.

Rice, C. M., Lenches, E. M., Eddy, S. R., Shin, S. J., Sheets, R. L., and Strauss, J. H. 1985. Nucleotide sequence of yellow fever virus: Implications for flavivirus gene expression and evolution. *Science* **229**:726–733.

Richardson, M. A., and Furuichi, Y. 1983. Nucleotide sequence of reovirus genome segment S3, encoding nonstructural protein sigma NS. *Nucleic Acids Res.* **11**:6399–6408.

Sanger, F., Air, G. M., Barrell, B. G., Brown, N. L., Coulson, A. R., Fiddes, J. C., Hutchison, C. A., Slocombe, P. M., and Smith, H. 1977. Nucleotide sequence of bacteriophage φX174 DNA. *Nature (London)* **265**:687–695.

Sanger, F., Coulson, A. R., Friedmann, T., Air, G. M., Barrell, B. G., Brown, N. L., Fiddes, J. C., Hutchison, C. A., Slocombe, P. M., and Smith, M. 1978. The nucleotide sequence of bacteriophage φX174. *J. Mol. Biol.* **125**:225–246.

Sanger, F., Coulson, A. R., Hong, G., Hill, D., and Peterson, G. 1982. Nucleotide sequence of bacteriophage λ DNA. *J. Mol. Biol.* **162**:729–774.

Sarkar, P., Sengupta, D., Baser, S., and Maitra, U. 1985. Nucleotide sequence of a major class-III phage-T3 RNA-polymerase promoter located at 98.0% of phage-T3 genetic map. *Gene* **33**:351–355.

Sassone-Corsi, P., Hen, R., Borrelli, E., Leff, T., and Chambon, P. 1983. Far upstream sequences are required for efficient transcription from the adenovirus-2 E1A transcription unit. *Nucleic Acids Res.* **11**:8738–8745.

Sassone-Corsi, P., Dougherty, J. P., Wasylajk, B., and Chambon, P. 1984. Stimulation of *in vitro* transcription from heterologous promoters by the simian virus 40 enhancer. *Proc. Natl. Acad. Sci. USA* **81**:308–312.

Schmidt, F. J. 1985. RNA splicing in prokaryotes: Bacteriophage T4 leads the way. *Cell* **41**:339–340.

Schwarz, E., Scherer, G., Hobom, G., and Kössel, H. 1978. Nucleotide sequence of *cro*, cII and part of the *O* gene in phage λ DNA. *Nature (London)* **272**:410–414.

Seif, I., Khowry, G., and Dhar, R. 1979. The genome of human papovavirus BKV. *Cell* **18**:963–977.

Shaw, M. W., Choppin, P. W., and Lamb, R. A. 1983. A previously unrecognized influenza B virus glycoprotein from a bicistronic mRNA that also encodes the viral neuraminidase. *Proc. Natl. Acad. Sci. USA* **80**:4879–4883.

Shih, M. C., and Gussin, G. N. 1983. Mutations affecting two different steps in transcription initiation at the phage λ P_{RM} promoter. *Proc. Natl. Acad. Sci. USA* **80**:496–500.

Smits, M. A., Jansen, J., Konings, R. N. H., and Schoenmakers, J. G. G. 1984. Initiation and termination signals for transcription in bacteriophage M13. *Nucleic Acids Res.* **12**:4071–4081.

Spicer, E. K., Noble, J. A., Nossal, N. G., Konigsberg, W. H., and Williams, K. R. 1982. Bacteriophage T4 gene 48. Sequence of the structural gene and its protein product. *J. Biol. Chem.* **257**:8972–8979.

Stahl, S. J., and Zinn, K. 1981. Nucleotide sequence of the cloned gene for bacteriophage T7 RNA polymerase. *J. Mol. Biol.* **148**:481–485.

Stanley, J., and Van Kammen, A. 1979. Nucleotide sequences adjacent to the proteins covalently linked to the cowpea mosaic virus genome. *Eur. J. Biochem.* **101**:45–49.

Stanway, G., Hughes, P. J., Mountford, R. C., Minor, P. D., and Almond, J. W. 1984. The complete nucleotide sequence of a common cold virus: Human rhinovirus 14. *Nucleic Acids Res.* **12**:7859–7875.

Studier, F. W., and Dunn, J. J. 1983. Organization and expression of bacteriophage T7 DNA. *Cold Spring Harbor Symp. Quant. Biol.* **47**:999–1007.

Tack, L. C., and Heard, P. 1985. Both *trans*-acting factors and chromatin structure are involved in the regulation of transcription from the early and late promoters in simian virus 40 chromosomes. *J. Virol.* **54**:207–218.

Takamatsu, N., Ohno, T., Meshi, T., and Okada, Y. 1983. Molecular cloning and nucleotide sequence of the 30K and the coat protein cistron of TMV (tomato strain) genome. *Nucleic Acids Res.* **11**:3767–3778.

Thomsen, D. R., Stenberg, R. M., Goins, W. F., and Stinski, M. F. 1984. Promoter-regulatory region of the major immediate early gene of human cytomegalovirus. *Proc. Natl. Acad. Sci. USA* **81**:659–665.

Toyoda, H., Kohara, M., Kataoka, Y., Suganuma, T., Omata, T., Imura, N., and Nomoto, A. 1984. Complete nucleotide sequences of all three poliovirus serotype genomes. Implications for genetic relationship, gene function and antigenic determinants. *J. Mol. Biol.* **174**:561–585.

Valerie, K., Henderson, E. E., and de Riel, J. K. 1984. Identification, physical map location and sequence of the *denV* gene from bacteriophage T4. *Nucleic Acids Res.* **12**:8085–8096.

van Wezenbeek, P. M. G. F., and Schoenmakers, J. G. G. 1979. Nucleotide sequence of the genes III, VI, and I of bacteriophage M13. *Nucleic Acids Res.* **6**:2799–2818.

van Wezenbeek, P. M. G. F., Hulsebos, T. J. M., and Schoenmakers, J. G. G. 1980. Nucleotide sequence of the filamentous bacteriophage M13 DNA genome: Comparisons with phage fd. *Gene* **11**:129–148.

Velcich, A., and Ziff, E. 1985. Adenovirus E1a proteins repress transcription from the SV40 early promoter. *Cell* **40**:705–716.

Virtanen, A., Aleström, P., Persson, H., Katze, M. G., and Pettersson, U. 1982. An adenovirus agnogene. *Nucleic Acids Res.* **10**:2539–2548.

Völker, T. A., Gafner, J., Bickle, T. A., and Showe, M. K. 1982. Gene *67*, a new essential bacteriophage T4 head gene codes for a prehead core component, PIP. I. Genetic mapping and DNA sequence. *J. Mol. Biol.* **161**:479–489.

Watanabe, Y., Meshi, T., and Okada, Y. 1984. The initiation site for transcription of the TMV 30-kDa protein messenger RNA. *FEBS Lett.* **173**:247–250.

Wilusz, J., Kurilla, M. G., and Keene, J. D. 1983. A host protein (La) binds to a unique species of minus-sense leader RNA during replication of vesicular stomatitis virus. *Proc. Natl. Acad. Sci. USA* **80**:5827–5831.

Yang, R. C. A., and Wu, R. 1979. BK virus DNA: Complete nucleotide sequence of a human tumor virus. *Science* **206**:456–462.

Zajchowski, D. A., Boeuf, H., and Kédinger, C. 1985. The adenovirus-2 early EIIa transcription unit possesses two overlapping promoters with different sequence requirements for EIa-dependent stimulation. *EMBO J.* **4**:1293–1300.

CHAPTER 11

Akiyoshi, D. E., Regier, D. A., Jen, G., and Gordon, M. P. 1985. Cloning and nucleotide sequence of the *tzs* gene from *Agrobacterium tumefaciens* strain T37. *Nucleic Acids Res.* **13**:2773–2788.

Alitalo, K., Bishop, J. M., Smith, D. R., Chen, E. Y., Colby, W. W., and Levinson, A. D. 1983. Nucleotide sequence of the v-*myc* oncogene of avian retrovirus MC29. *Proc. Natl. Acad. Sci. USA* **80**:100–104.

Baldwin, G. S. 1985. Epidermal growth factor precursor is related to the translation product of the Moloney sarcoma virus oncogene *mos*. *Proc. Natl. Acad. Sci. USA* **82**:1921–1925.

Bernard, O., Cory, S., Gerondakis, S., Webb, E., and Adams, J. M. 1983. Sequence of the murine and human *myc* oncogenes and two modes of *myc* transcription resulting from chromosome translocation in B lymphoid tumours. *EMBO J.* **2**:2375–2383.

Betsholtz, C., Bywater, M., Westermark, B., Bürk, R. R., and Heldin, C. H. 1985. Expression of the c-*sis* gene and secretion of a platelet-derived growth factor-like protein by simian virus 40-transformed BHK cells. *Biochem. Biophys. Res. Commun.* **130**:753–760.

Bevan, M., Barnes, W. M., and Chilton, M. D. 1983. Structure and transcription of the nopaline synthase gene region of T-DNA. *Nucleic Acids Res.* **11**:369–385.

Branton, P. E., Bayley, S. T., and Graham, F. L. 1985. Transformation by human adenoviruses. *Biochim. Biophys. Acta* **780**:67–94.

Brodeur, G. M., Seeger, R. C., Schwab, M., Varmus, H. E., and Bishop, J. M. 1984. Amplification of N-*myc* in untreated human neuroblastomas correlates with advanced disease stage. *Science* **224**:1121–1124.

Broek, D., Samiy, N., Fasano, O., Fujiyama, A., Tamanoi, F., Northup, J., and Wigler, M. 1985. Differential activation of yeast adenylate cyclase by wild-type and mutant *RAS* proteins. *Cell* **41**:763–769.

Broome, S., and Gilbert, W. 1985. Rous sarcoma virus encodes a transcriptional activator. *Cell* **40**:537–546.

Brown, R., Marshall, C. J., Pennie, S. G., and Hall, A. 1984. Mechanism of activation of an *N-ras* gene in the human fibrosarcoma cell line HT 1080. *EMBO J.* **3**:1321–1326.

Canaani, E., Dreazen, O., Klar, A., Rechavi, G., Ram, D., Cohen, J. B., and Givol, D. 1983. Activation of the *c-mos* oncogene in a mouse plasmacytoma by insertion of an endogenous intracisternal A-particle genome. *Proc. Natl. Acad. Sci. USA* **80**:7118–7122.

Capon, D. J., Chen, E. Y., Levinson, A. D., Seeburg, P. H., and Goeddel, D. V. 1983a. Complete nucleotide sequences of the T24 human bladder carcinoma oncogene and its normal homolog. *Nature (London)* **302**:33–37.

Capon, D. J., Seeburg, P. H., McGrath, J. P., Hayflick, J. S., Edman, U., Levinson, A. D., and Goeddel, D. V. 1983b. Activation of Ki-*ras2* gene in human colon and lung carcinomas by two different point mutations. *Nature (London)* **304**:507–513.

Chang, C. C., Trosko, J. E., Kung, H. J., Bombick, D., and Matsumura, F. 1985. Potential role of the *src* gene products in inhibition of gap-junctional communication in NIH/3T3 cells. *Proc. Natl. Acad. Sci. USA* **82**:5360–5364.

Chang, E. H., Gonda, M. A., Ellis, R. W., Scolnick, E. M., and Lowy, D. R. 1982. Human genome contains four genes homologous to transforming genes of Harvey and Kirsten murine sarcoma viruses. *Proc. Natl. Acad. Sci. USA* **79**:4848–4852.

Chen, I. S. Y., Slamon, D. J., Rosenblatt, J. D., Shah, N. P., Quan, S. G., and Wachsman, W. 1985. The α gene is essential for HTLV replication. *Science* **229**:54–58.

Chiu, I. M., Reddy, E. P., Givol, D., Robbins, K. C., Tronick, S. R., and Aaronson, S. A. 1984. Nucleotide sequence analysis identifies the human c-*sis* protooncogene as a structural gene for platelet-derived growth factor. *Cell* **37**:123–129.

Christy, R. J., Brown, A. R., Gourlie, B. B., and Huang, R. C. 1985. Nucleotide sequences of murine intracisternal A-particle gene LTRs have extensive variability within the R region. *Nucleic Acids Res.* **13**:289–302.

Cianciolo, G. J., Copeland, T. D., Oroszlan, S., and Snyderman, R. 1985. Inhibition of lymphocyte proliferation by a synthetic peptide homologous to retroviral envelope proteins. *Science* **230**:453–455.

Clark, R., Wong, G., Arnheim, N., Nitecki, D., and McCormick, F. 1985. Antibodies specific for amino acid 12 of the *ras* oncogene product inhibit GTP binding. *Proc. Natl. Acad. Sci. USA* **82**:5280–5284.

Clark, S. P., and Mak, T. W. 1983. Complete nucleotide sequence of an infectious clone of Friend spleen focus-forming provirus: gp55 is an envelope fusion glycoprotein. *Proc. Natl. Acad. Sci. USA* **80**:5037–5041.

Coffin, J. M. 1979. Structure, replication, and recombination of retrovirus genomes: Some unifying hypotheses. *J. Gen. Virol.* **42**:1–26.

Coffin, J. M., Varmus, H. E., Bishop, J. M., Essex, M., Hardy, W. D., Martin, G. S., Rosenberg, N. E., Scolnick, E. M., Weinberg, R. A., and Vogt, P. K. 1981. Proposal for naming host cell-derived inserts in retrovirus genomes. *J. Virol.* **40**:953–957.

Colby, W. W., Chen, E. Y., Smith, D. H., and Levinson, A. D. 1983. Identification and nucleotide sequence of a human locus homologous to the v-*myc* oncogene of avian myelocytomatosis virus MC29. *Nature (London)* **301**:722–725.

Collins, S. J., Kubonishi, I., Miyoshi, I., and Groudine, N. T. 1984. Altered transcription of the *c-abl* oncogene in K-562 and other chronic myelogenous leukemia cells. *Science* **225**:72–74.

Corcoran, L. M., Adams, J. M., Dunn, A. R., and Cory, S. 1984. Murine T lymphomas in which the cellular *myc* oncogene has been activated by retroviral insertion. *Cell* **37**:113–122.

Cuypers, H. T., Selten, G., Quint, W., Zijlstra, M., Maandag, E. R., Boelens, W., Van Wezenbeek, P., Melief, C., and Berns, A. 1984. Murine leukemia virus-induced T-cell lymphomagenesis: Integration of proviruses in a distinct chromosomal region. *Cell* **37**:141–150.

Czernilofsky, A. P., Levinson, A. D., Varmus, H. E., Bishop, J. M., Tischer, E., and Goodman, H. M. 1980. Nucleotide sequence of an avian sarcoma virus oncogene (*src*) and proposed amino acid sequence for gene product. *Nature (London)* **287**:198–203.

Czernilofsky, A. P., Levinson, A. D., Varmus, H. E., Bishop, J. M., Tischer, E., and Goodman, H. 1983. Corrections to the nucleotide sequence of the *src* gene of Rous sarcoma virus. *Nature (London)* **301**:736–738.

Debuire, B., Henry, C., Benaissa, M., Biserte, G., Claverie, J. M., Saule, S., Martin, P., and Stehelin, D. 1984. Sequencing the *erbA* gene of avian erythroblastosis virus reveals a new type of oncogene. *Science* **224**:1456–1459.

DeFeo-Jones, D., Tatchell, K., Robinson, L. C., Sigal, I. S., Vass, W. C., Lowy, D. R., and Scolnick, E. M. 1985. Mammalian and yeast *ras* gene products: Biological function in their heterologous systems. *Science* **228**:179–184.

Derse, D., Caradonna, S. J., and Casey, J. W. 1985. Bovine leukemia virus long terminal repeat: A cell type-specific promoter. *Science* **227**:317–320.

Devare, S. G., Reddy, E. P., Robbins, K. C., Andersen, P. R., Tronick, S. R., and Aaronson, S. A. 1982. Nucleotide sequence of the transforming gene of simian sarcoma virus. *Proc. Natl. Acad. Sci. USA* **79**:3179–3182.

Devare, S. G., Reddy, E. P., Law, J. D., Robbins, K. C., and Aaronson, S. A. 1983. Nucleotide sequence of the simian sarcoma virus genome: Demonstration that its acquired cellular sequences encode the transforming gene product p28^{sis}. *Proc. Natl. Acad. Sci. USA* **80**: 731–735.

Dhar, R., Ellis, R. W., Shih, T. Y., Oroszlan, S., Shapiro, B., Maizel, J., Lowy, D., and Scolnick, E. 1982. Nucleotide sequence of the p21 transforming protein of Harvey murine sarcoma virus. *Science* **217**:934–937.

Dhar, R., Nieto, A., Koller, R., DeFeo-Jones, D., and Scolnick, E. M. 1984. Nucleotide sequence of two *ras*^{H}-related genes isolated from the yeast, *Saccharomyces cerevisiae*. *Nucleic Acids Res.* **12**:3611–3618.

Diaz, M. O., Le Beau, M. M., Rowley, J. D., Drabkin, H. A., and Patterson, D. 1985. The role of the c-*mos* gene in the 8j21 translocation acute myeloblastic leukemia. *Science* **229**:767–769.

Dillon, L. S. 1983. *The Inconstant Gene*, New York, Plenum Press.

Dony, C., Kessel, M., and Gruss, P. 1985. Post-transcriptional control of *myc* and p53 expression during differentiation of the embryonal carcinoma cell line F9. *Nature (London)* **317**:636–639.

Doolittle, R. F., Hunkapiller, M. W., Hood, L. E., Devare, S. G., Robbins, K. C., Aaronson, S. A., and Antoniades, H. N. 1983. Simian sarcoma virus *onc* gene, *v-sis*, is derived from the gene (or genes) encoding a platelet-derived growth factor. *Science* **221**:275–277.

Durban, E. M., and Boettiger, D. 1981. Differential effects of transforming avian RNA tumor viruses on avian macrophages. *Proc. Natl. Acad. Sci. USA* **78**:3600–3604.

Dyson, P. J., and Rabbitts, T. H. 1985. Chromatin structure around the *c-myc* gene in Burkitt lymphomas with upstream and downstream translocation points. *Proc. Natl. Acad. Sci. USA* **82**:1984–1988.

Ellis, R. W., DeFeo, D., Shih, T. Y., Gonda, M. A., Young, H. A., Tsuchida, N., Lowy, D. R., and Scolnick, E. M. 1981. The p21 *src* genes of Harvey and Kirsten sarcoma viruses originate from divergent members of a family of normal vertebrate genes. *Nature (London)* **292**:506–511.

Favera, R. D., Gallo, R. C., Giallongo, A., and Croce, C. M. 1982. Chromosomal localization of the human homolog (*c-sis*) of the simian sarcoma virus *onc* gene. *Science* **218**:686–688.

Feinberg, A. P., Vogelstein, B., Droller, M. J., Baylin, S. B., and Nelkin, B. D. 1983. Mutation affecting the 12th amino acid in the c-Ha-*ras* oncogene product occurs infrequently in human cancer. *Science* **220**:1175–1177.

Feldman, R. A., Wang, L. H., Hanafusa, H., and Balduzzi, R. C. 1982. Avian sarcoma virus UR2 encodes a transforming protein which is associated with a unique protein kinase activity. *J. Virol.* **42**:228–236.

Feramisco, J., Ozanne, B., and Stiles, C., eds. 1985. *Cancer Cells 3: Growth Factors and Transformation*, Cold Spring Harbor, New York, Cold Spring Harbor Laboratory.

Flügel, R. M., Bannert, H., Suhai, S., and Darai, G. 1985. The nucleotide sequence of the early region of the *Tupaia* adenovirus DNA corresponding to the oncogene region E1b of human adenovirus 7. *Gene* **34**:73–80.

Fraenkel, D. G. 1985. On *ras* gene function in yeast. *Proc. Natl. Acad. Sci. USA* **82**:4740–4744.

Fujisawa, J. I., Seiki, M., Kiyokawa, T., and Yoshida, M. 1985. Functional activation of the long terminal repeat of human T-cell leukemia virus type I by a *trans*-acting factor. *Proc. Natl. Acad. Sci. USA* **82**:2277–2281.

Fukui, Y., and Kaziro, Y. 1985. Molecular cloning and sequence analysis of a *ras* gene from *Schizosaccharomyces pombe*. *EMBO J.* **4**:687–691.

Fung, Y. K. T., Lewis, W. G., Crittenden, L. B., and Kung, H. J. 1983. Activation of the cellular oncogene c-*erbB* by *LTR* insertion: Molecular basis for induction of erythroblastosis by avian leukosis virus. *Cell* **33**:357–368.

Galibert, F., de Dinechin, S. D., Righi, M., and Stehelin, D. 1984. The second oncogene *mil* of avian retrovirus MH2 is related to the *src* gene family. *EMBO J.* **3**:1333–1338.

Gallwitz, D., Donath, C., and Sander, C. 1983. A yeast gene encoding a protein homologous to the human c-*has/bas* proto-oncogene product. *Nature (London)* **306**:704–707.

Giallongo, A., Appella, E., Ricciardi, R., Rovera, G., and Croce, C. M. 1983. Identification of the *c-myc* oncogene product in normal and malignant B cells. *Science* **222:**430–432.

Goff, S. P., D'Eustachio, P., Ruddle, F. H., and Baltimore, D. 1982. Chromosomal assignment of the endogenous proto-oncogene *C-abl*. *Science* **218:**1317–1319.

Goldberg, S. B., Flick, J. S., and Rogers, S. G. 1984. Nucleotide sequences of the *tmr* locus of *Agrobacterium tumefaciens* pTi T37 T-DNA. *Nucleic Acids Res.* **12:**4665–4677.

Gonda, M. A., Wong-Staal, F., Gallo, R. C., Clements, J. E., Narayan, O., and Gilden, R. V. 1985. Sequence homology and morphologic similarity of HTLV-III and visna virus, a pathogenic lentivirus. *Science* **227:**173–177.

Graf, T., and Beug, H. 1983. Role of the *v-erbA* and *v-erbB* oncogenes of avian erythroblastosis virus in erythroid cell transformation. *Cell* **34:**7–9.

Guerrero, I., Villasante, A., Corces, V., and Pellicer, A. 1984. Activation of a *c-K-ras* oncogene by somatic mutation in mouse lymphomas induced by gamma radiation. *Science* **225:**1159–1162.

Hall, A., and Brown, R. 1985. Human N-*ras:* cDNA cloning and gene structure. *Nucleic Acids Res.* **13:**5255–5268.

Hampe, A., Laprevotte, I., and Galibert, F. 1982. Nucleotide sequences of feline retroviral oncogene (*v-fes*) provide evidence for a family of tyrosine-specific protein kinase genes. *Cell* **30:**775–785.

Hampe, A., Gobet, M., Sherr, C. J., and Galibert, F. 1984. Nucleotide sequence of the feline retroviral oncogene *v-fms* shows unexpected homology with oncogenes encoding tyrosine-specific protein kinases. *Proc. Natl. Acad. Sci. USA* **81:**85–89.

Hayflick, J., Seeburg, P. H., Ohlsson, R., Pfeifer-Ohlsson, S., Watson, D., Papas, T., and Duesberg, P. H. 1985. Nucleotide sequence of two overlapping *myc*-related genes in avian carcinoma virus OK10 and their relation to the *myc* genes of other viruses and the cell. *Proc. Natl. Acad. Sci. USA* **82:**2718–2722.

Heidekamp, F., Dirkse, W. G., Hille, J., and van Ormond, H. 1983. Nucleotide sequence of the *Agrobacterium tumefaciens* octopine Ti plasmid-encoded *tmr* gene. *Nucleic Acids Res.* **11:**6211–6223.

Herr, W., Corbin, V., and Gilbert, W. 1982. Nucleotide sequence of the 3' half of AKV. *Nucleic Acids Res.* **10:**6931–6944.

Hirai, H., Okabe, T., Anraku, Y., Fujisawa, M., Urabe, A., and Takaku, F. 1985. Activation of the c-K-*ras* oncogene in a human pancreas carcinoma. *Biochem. Biophys. Res. Commun.* **127:**168–174.

Hoffman-Falk, H., Einal, P., Shilo, B. Z., and Hoffmann, F. M. 1982. *Drosophila melanogaster* DNA clones homologous to vertebrate oncogenes: Evidence for a common ancestor to the *src* and *abl* cellular genes. *Cell* **30:**589–597.

Hoffmann, F. M., Fresco, L. D., Hoffman-Falk, H., and Shilo, B. Z. 1983. Nucleotide sequences of the *Drosophila src* and *abl* homologs: Conservation and variability in the *src* family oncogenes. *Cell* **35:**393–401.

Huang, C. C., Hammond, C., and Bishop, J. M. 1985. Nucleotide sequence and topography of chicken c-*fps* genesis of a retroviral oncogene encoding a tyrosine-specific protein kinase. *J. Mol. Biol.* **181:**175–186.

Jaiswal, A. K., Gonzalez, F. J., and Nebert, D. W. 1985. Human P_1-450 gene sequence and correlation of mRNA with genetic differences in benzo[a]pyrene metabolism. *Nucleic Acids Res.* **13:**4503–4520.

Jochemsen, A. G., Bos, J. L., and van der Eb, A. J. 1984. The first exon of region *E1a* genes of adenoviruses 5 and 12 encodes a separate functional protein domain. *EMBO J.* **3:**2923–2927.

Joos, H., Inzé, D., Caplan, A., Sormann, M., Van Montagu, M., and Schell, J. 1983. Genetic analysis of T-DNA transcripts in nopaline crown galls. *Cell* **32:**1057–1067.

Josephs, S. F., Ratner, L., Clarke, M. F., Westin, E. H., Reitz, M. S., and Wong-Staal, F. 1984. Transforming potential of human *c-sis* nucleotide sequences encoding platelet-derived growth factor. *Science* **225:**636–639.

Jurnak, F. 1985. Structure of the GDP domain of EF-Tu and location of the amino acids homologous to *ras* oncogene proteins. *Science* **230:**32–36.

Kan, N. C., Floredellis, C. S., Mark, G. E., Duesberg, P. H., and Papas, T. S. 1984. Nucleotide sequence of avian carcinoma virus MH2: Two potential *onc* genes, one related to avian virus MC29 and the other related to sarcoma virus 3611. *Proc. Natl. Acad. Sci. USA* **81:**3000–3004.

Kasid, A., Lippman, M. E., Papageorge, A. G., Lowy, D. R., and Gelmann, E. P. 1985. Transfection of v-*ras* [H]DNA into MCF-7 human breast cancer cells bypasses dependence on estrogen for tumorigenicity. *Science* **228:**725–728.

Kataoka, T., Powers, S., Cameron, S., Fasano, O., Goldfarb, M., Broach, J., and Wigler, M. 1985. Functional homology of mammalian and yeast *RAS* genes. *Cell* **40:**19–26.

Katzen, A. L., Kornberg, T. B., and Bishop, J. M. 1985. Isolation of the protooncogene *c-myb* from *D. melanogaster. Cell* **41**:449–456.

Keath, E. J., Kelekar, A., and Cole, M. D. 1984. Transcriptional activation of the translocated *c-myc* oncogene in mouse plasmacytomas: Similar RNA levels in tumor and proliferating normal cells. *Cell* **37**:521–528.

Kelly, J. M., and Hynes, M. J. 1985. Transformation of *Aspergillus niger* by the *amdS* gene of *Aspergillus nidulans. EMBO J.* **4**:475–479.

Khan, A. S., and Martin, M. A. 1983. Endogenous murine leukemia proviral long terminal repeats contain a unique 190-base-pair insert. *Proc. Natl. Acad. Sci. USA* **80**:2699–2703.

King, C. R., Giese, N. A., Robbins, K. C., and Aaronson, S. A. 1985. *In vitro* mutagenesis of the v-*sis* transforming gene defines functional domains of its growth factor-related product. *Proc. Natl. Acad. Sci. USA* **82**:5295–5299.

Kitamura, N., Kitamura, A., Toyoshima, K., Hirayama, Y., and Yoshida, M. 1982. Avian sarcoma virus y73 genome sequence and structural similarity of its transforming gene product to that of Rous sarcoma virus. *Nature (London)* **297**:205–208.

Klarlund, J. K. 1985. Transformation of cells by an inhibitor of phosphatases acting on phosphotyrosine in proteins. *Cell* **41**:707–717.

Klempnauer, K. H., Gonda, T. J., and Bishop, J. M. 1982. Nucleotide sequence of the retroviral leukemia gene v-*myb* and its cellular progenitor *c-myb:* The architecture of a transduced oncogene. *Cell* **31**:453–463.

Koch, S., von Loringhoven, A. F., Kahmann, R., Hofschneider, P. H., and Koshy, R. 1984. The genetic organization of integrated hepatitis B virus DNA in the human hepatoma cell line PLC/PRF/5. *Nucleic Acids Res.* **12**:6871–6886.

Köhrer, K., Grummt, I., and Horak, I. 1985. Functional RNA polymerase II promoters in solitary retroviral long terminal repeats (LTR-IS elements). *Nucleic Acids Res.* **13**:2631–2645.

Kriek, E., Den Engelse, L., Scherer, E., and Westra, J. G. 1984. Formation of DNA modifications by chemical carcinogens. Identification, localization, and quantification. *Biochim. Biophys. Acta* **738**:181–201.

Kuff, E. L., Feenstra, A., Lueders, K., Rechavi, G., Givol, D., and Canaani, E. 1983. Homology between an endogenous viral LTR and sequences inserted in an activated cellular oncogene. *Nature (London)* **302**:547–548.

Laimins, L. A., Tsichlis, P., and Khoury, G. 1984. Multiple enhancer domains in the 3′ terminus of the Prague strain of Rous sarcoma virus. *Nucleic Acids Res.* **12**:6427–6442.

Lemay, G., and Jolicoeur, P. 1984. Rearrangement of a DNA sequence homologous to a cell-virus junction fragment in several Moloney murine leukemia virus-induced thymomas. *Proc. Natl. Acad. Sci. USA* **81**:38–42.

Lovinger, G. G., Mark, G., Todaro, G. J., and Schochetman, G. 1981. 5′-Terminal nucleotide noncoding sequences of retroviruses. Relatedness of two Old World primate type C viruses and avian spleen necrosis virus. *J. Virol.* **39**:238–245.

Madaule, P., and Axel, R. 1985. A novel *ras*-related gene family. *Cell* **41**:31–40.

Manger, R., Najita, L., Nichols, E. J., Hakomori, S. I., and Rohrschneider, L. 1984. Cell surface expression of the McDonough strain of feline sarcoma virus *fms* gene product (gp140[fms]). *Cell* **39**:327–337.

Marx, J. L. 1985. More about the HTLV's and how they act. *Science* **229**:37–38.

McGrath, J. P., Capon, D. J., Goeddel, D. V., and Levinson, A. D. 1984. Comparative biochemical properties of normal and activated human *ras* p21 protein. *Nature (London)* **310**:644–649.

Michitsch, R. W., and Melera, P. W. 1985. Nucleotide sequence of the 3′ exon of the human *N-myc* gene. *Nucleic Acids Res.* **13**:2545–2558.

Murray, M. J., Cunningham, J. M., Parada, L. F., Dautry, F., Lebowitz, P., and Weinberg, R. A. 1983. The HL-60 transforming sequence: A *ras* oncogene coexisting with altered *myc* genes in hematopoietic tumors. *Cell* **33**:749–757.

Nakano, H., Yamamoto, F., Neville, C., Evans, D., Mizuno, T., and Perucho, M. 1984. Isolation of transforming sequences of two lung carcinomas: Structural and functional analysis of the activated *c-K-ras* oncogenes. *Proc. Natl. Acad. Sci. USA* **81**:71–75.

Nakatani, H., Tahara, E., Sakamoto, H., Terada, M., and Sugimura, T. 1985. Amplified DNA sequences in cancer. *Biochem. Biophys. Res. Commun.* **130**:508–514.

Nebert, D. W., Eisen, H. J., and Hankinson, O. 1984. The Ah receptor: Binding specificity only for foreign chemicals? *Biochem. Pharmacol.* **33**:917–924.

Neil, J. C., Hughes, D., McFarlane, R., Wilkie, N. M., Onions, D. E., Lees, G., and Jarrett, O. 1984.

Transduction and rearrangement of the *myc* gene by feline leukemia virus in naturally occurring T-cell leukemias. *Nature (London)* **308:**814–820.

Neuman-Silberberg, F. S., Schejter, E., Hoffmann, F. M., and Shilo, B. Z. 1984. The *Drosophila ras* oncogenes: Structure and nucleotide sequence. *Cell* **37:**1027–1033.

Norton, J. D., Connor, J., and Avery, R. J. 1984a. Unusual long terminal repeat sequence of a retrovirus transmissable mouse (VL30) genetic element: Identification of functional domains. *Nucleic Acids Res.* **12:**3445–3460.

Norton, J. D., Connor, J., and Avery, R. J. 1984b. Genesis of Kirsten murine sarcoma virus: Sequence analysis reveals recombination points and potential leukaemogenic determinant on parental leukemia virus genome. *Nucleic Acids Res.* **12:**6839–6852.

Nunn, M. F., Seeburg, P. H., Moscovici, C., and Duesberg, P. H. 1983. Tripartite structure of the avian erythroblastosis virus E26 transforming gene. *Nature (London)* **306:**391–395.

Oskarsson, M., McClements, W. L., Blair, D. G., Maizel, J. V., and Vande Woude, G. F. 1980. Properties of a normal mouse cell DNA sequence (*src*) homologous to the *src* sequence of Moloney sarcoma virus. *Science* **207:**1222–1224.

Ostrowski, M. C., Huang, A. L., Kessel, M., Wolford, R. G., and Hager, G. L. 1984. Modulation of enhancer activity by the hormone responsive regulatory element from mouse mammary tumor virus. *EMBO J.* **3:**1891–1899.

Pachl, C., Schubach, W., Eisenman, R., and Linial, M. 1983. Expression of *c-myc*-RNA in bursal lymphoma cell lines: Identification of *c-myc*-encoded proteins by hybrid-selected translation. *Cell* **33:**335–344.

Panganiban, A. T., and Temin, H. M. 1983. The terminal nucleotides of retrovirus DNA are required for integration but not virus production. *Nature (London)* **306:**155–160.

Rapp, U. R., Goldsborough, M. D., Mark, G. E., Bonner, T. I., Groffen, J., Reynolds, F. H., and Stephenson, J. R. 1983. Structure and biological activity of *v-raf*, a unique oncogene transduced by a virus. *Proc. Natl. Acad. Sci. USA* **80:**4218–4222.

Rasheed, S., Norman, G. L., and Heidecker, G. 1983. Nucleotide sequence of the Rasheed rat sarcoma virus oncogene: New mutations. *Science* **221:**155–157.

Ratner, L., Josephs, S. F., Jarrett, R., Reitz, M. S., and Wong-Staal, F. 1985. Nucleotide sequence of transforming human *c-sis* cDNA clones with homology to platelet-derived growth factor. *Nucleic Acids Res.* **13:**5007–5018.

Rechavi, G., Givol, D., and Canaani, E. 1982. Activation of a cellular oncogene by DNA rearrangement: Possible involvement of an IS-like element. *Nature (London)* **300:**607–610.

Reddy, E. P. 1983. Nucleotide sequence analysis of the T24 human bladder carcinoma oncogene. *Science* **220:**1061–1063.

Reddy, E. P., Smith, M. J., and Aaronson, S. A. 1981. Complete nucleotide sequence and organization of the Moloney murine sarcoma virus genome. *Science* **214:**445–450.

Reddy, E. P., Smith, M. J., and Srinivasan, A. 1983a. Nucleotide sequence of Abelson murine leukemia virus genome: Structural similarity of its transforming gene product to other *onc* gene products with tyrosine-specific kinase activity. *Proc. Natl. Acad. Sci. USA* **80:**3623–3627.

Reddy, E. P., Reynolds, R. K., Watson, D. K., Schultz, R. A., Lautenberger, J., and Papas, T. S. 1983b. Nucleotide sequence analysis of the proviral genome of avian myelocytomatosis virus (MC29). *Proc. Natl. Acad. Sci. USA* **80:**2500–2504.

Repaske, R., O'Neill, R. R., Khan, A. S., and Martin, M. A. 1983. Nucleotide sequence of the *env*-specific segment of the NSF-Th-1 xenotropic murine leukemia virus. *J. Virol.* **46:**204–211.

Rettenmier, C. W., Roussel, M. F., Quinn, C. O., Kitchingman, G. R., Look, A. T., and Sherr, C. J. 1985. Transmembrane orientation of glycoproteins encoded by the *v-fms* oncogene. *Cell* **40:**971–981.

Reymond, C. D., Gomer, R. H., Mehdy, M. C., and Firtel, R. A. 1984. Developmental regulation of a *Dictyostelium* gene encoding a protein homologous to mammalian *ras* protein. *Cell* **39:**141–148.

Robbins, K. C., Devare, S. G., Reddy, E. P., and Aaronson, S. A. 1982. *In vivo* identification of the transforming gene product of simian sarcoma virus. *Science* **218:**1131–1133.

Rogler, C. E., Sherman, M., Su, C. Y., Shafritz, D. A., Summers, J., Shows, T. B., Henderson, A., and Kew, M. 1985. Deletion in chromosome 11p associated with a hepatitis B integration site in hepatocellular carcinoma. *Science* **230:**319–322.

Roussel, M. F., Sherr, C. J., Barker, P. E., and Ruddle, F. H. 1983. Molecular cloning of the *c-fms* locus and its assignment to human chromosome 5. *J. Virol.* **48:**770–773.

Rushlow, K. E., Lautenberger, J. A., Papas, T. S., Baluda, M. A., Perbal, B., Chirikjian, J. G., and Reddy, E. P. 1982. Nucleotide sequence of the transforming gene of avian myeloblastosis virus. *Science* **216**:1421–1423.

Sagata, N., Yasunaga, T., Ogawa, Y., Tsuzuku-Kawamura, J., and Ikaua, Y. 1984. Bovine leukemia virus: Unique structural features of its long terminal repeats and its evolutionary relationship to human T-cell leukemia virus. *Proc. Natl. Acad. Sci. USA* **81**:4741–4745.

Sagata, N., Yasunaga, T., Tsuzuku-Kawamura, J., Ohishi, K., Ogawa, Y., and Ikawa, Y. 1985. Complete nucleotide sequence of the genome of bovine leukemia virus: Its evolutionary relationship to other retroviruses. *Proc. Natl. Acad. Sci. USA* **82**:677–681.

Schejter, E. D., and Shilo, B. Z. 1985. Characterization of functional domains of p21*ras* by use of chimeric genes. *EMBO J* **4**:407–412.

Schultz, A. M., Henderson, L. E., Orozlan, S., Garber, E. A., and Hanafusa, H. 1985. Amino terminal myristylation of the protein kinase p60*src*, a retroviral transforming protein. *Science* **227**:427–429.

Schwartz, D. E., Tizard, R., and Gilbert, W. 1983. Nucleotide sequence of Rous sarcoma virus. *Cell* **32**:853–869.

Seiki, M., Hattori, S., Hirayama, Y., and Yoshida, M. 1983. Human adult T-cell leukemia virus: Complete nucleotide sequence of the provirus genome integrated in leukemia cell DNA. *Proc. Natl. Acad. Sci. USA* **80**:3618–3622.

Seiki, M., Hikikoshi, A., Tanigushi, T., and Yoshida, M. 1985. Expression of the p*X* gene of HTLV-I: General splicing mechanism in the HTLV family. *Science* **228**:1532–1534.

Sekiya, T., Fushimi, M., Hori, H., Hirohashi, S., Nishimura, S., and Sugimura, T. 1984. Molecular cloning and the total nucleotide sequence of the human c-Ha-*ras*-1 gene activated in a melanoma from a Japanese patient. *Proc. Natl. Acad. Sci. USA* **81**:4771–4775.

Semba, K., Yamanashi, Y., Nishizawa, M.. Sukegawa, J., Yoshida, M., Sasaki, M., Yamamoto, T., and Toyoshima, K. 1985. Location of the c-*yes* gene on the human chromosome and its expression in various tissues. *Science* **227**:1038–1040.

Shen-Ong, G. L. C., Keath, E. J., Piccoli, S. P., and Cole, M. D. 1982. Novel *myc* oncogene RNA from abortive immunoglobulin-gene recombination in mouse plasmacytomas. *Cell* **31**:443–452.

Sherr, C. J., Rettenmier, C. W., Sacca, R., Roussel, M. F., Look, A. T., and Stanley, E. R. 1985. c-*fms* protooncogene product is related to the receptor for the mononuclear phagocyte growth factor CSF-1. *Cell* **41**:665–676.

Shibuya, M., and Hanafusa, H. 1982. Nucleotide sequence of Fujinami sarcoma virus: Evolutionary relationship of its transforming gene with transforming genes of other sarcoma viruses. *Cell* **30**:787–795.

Shimizu, K., Nakatsu, Y., Sekiguchi, M., Hokamura, K., Tanaka, K., Terada, M., and Sugimura, L. 1985. Molecular cloning of an activated human oncogene, homologous to v-*raf* from primary stomach cancer. *Proc. Natl. Acad. Sci. USA* **82**:5641–5645.

Shimotohno, K., Takahashi, Y., Shimizu, N., Gojobori, T., Golde, D. W., Chen, I. S. Y., Miwa, M., and Sugimura, T. 1985a. Complete nucleotide sequence of an infectious clone of human T-cell leukemia virus type II: An open reading frame for the protease gene. *Proc. Natl. Acad. Sci. USA* **82**:3101–3105.

Shimotohno, K., Miwa, M., Slamon, D. J., Chen, I. S. Y., Hoshino, H. O., Takano, M., Fujino, M., and Sugimura, T. 1985b. Identification of new gene products coded from X regions of human T-cell leukemia viruses. *Proc. Natl. Acad. Sci. USA* **82**:302–306.

Singer, B., and Grunberger, D. 1983. *Molecular Biology of Mutagens and Carcinogens*, New York, Plenum Press.

Sodroski, J. G., Rosen, C. A., and Haseltine, W. A. 1984. *Trans*-acting transcriptional activation of the long terminal repeat of human T lymphotropic viruses in infected cells. *Science* **225**:381–385.

Sodroski, J., Rosen, C., Wong-Staal, F., Salahuddin, S. Z., Popovic, M., Arya, S., and Gallo, R. C. 1985. *Trans*-acting transcriptional regulation of human T-cell leukemia virus type III long terminal repeat. *Science* **227**:171–173.

Sohn, U., Styszko, J., Coombs, D., and Krause, M. 1983. 7S-K nuclear RNA from simian virus 40-transformed cells has sequence homology to the viral early promoter. *Proc. Natl. Acad. Sci. USA* **80**:7090–7094.

Stabel, S., and Doerfler, W. 1982. Nucleotide sequence at the site of junction between adenovirus type 12 DNA and repetitive hamster cell DNA in transformed cell line CLAC1. *Nucleic Acids Res.* **10**:8007–8023.

Stantan, L. W., Fahrlander, P. D., Tesser, P. M., and Marcu, K. B. 1984. Nucleotide sequence comparison of normal and translocated murine c-*myc* genes. *Nature (London)* **310**:423–425.

Starcich, B., Ratner, L., Josephs, S. F., Okamoto, T., Gallo, R. C., and Wong-Staal, F. 1985. Characterization of long terminal repeat sequence of HTLV-III. *Science* **227**:538–540.

Steele, R. E. 1985. Two divergent cellular *src* genes are expressed in *Xenopus laevis*. *Nucleic Acids Res.* **13**:1747–1761.

Stephens, R. M., Rice, N. R., Hiebach, R. R., Bose, H. R., and Gilden, R. V. 1983. Nucleotide sequence of *v-rel:* The oncogene of reticuloendotheliosis-virus. *Proc. Natl. Acad. Sci. USA* **80**:6229–6233.

Sutcliffe, J. G., Shinnick, T. M., Verma, I. M., and Lerner, R. A. 1980a. Nucleotide sequence of Moloney leukemia virus: 3′ end reveals details of replication, analogy to bacterial transposons, and an unexpected gene. *Proc. Natl. Acad. Sci. USA* **77**:3302–3306.

Sutcliffe, J. G., Shinnick, T. M., Green, N., Liu, F. T., Niman, H. L., and Lerner, R. A. 1980b. Chemical synthesis of a polypeptide predicted from nucleotide sequence allows detection of a new retroviral gene product. *Nature (London)* **287**:801–805.

Sutrave, P., Bonner, T. I., Rapp, U. R., Jansen, H. W., Patchinsky, T., and Bister, K. 1984. Nucleotide sequence of avian retroviral oncogene *v-mil:* Homologue of murine retroviral oncogene *v-raf. Nature (London)* **309**:85–88.

Swanstrom, R., DeLorbe, W. J., Bishop, J. M., and Varmus, H. E. 1981. Nucleotide sequence of cloned unintegrated avian sarcoma virus DNA: Viral DNA contains direct and inverted repeats similar to those in transposable elements. *Proc. Natl. Acad. Sci. USA* **78**:124–128.

Sweet, R. W., Yokoyama, S., Kamata, T., Feramisco, J. R., Rosenberg, M., and Gross, M. 1984. The product of *ras* is a GTPase and the T24 oncogenic mutant is deficient in this activity. *Nature (London)* **311**:273–275.

Tainsky, M. A., Cooper, C. S., Grovanella, B. C., and Vande Woude, G. F. 1984. An activated *ras*N gene: Detected in late but not early passage human PA1 teratocarcinoma cells. *Science* **225**:643–645.

Takeya, T., and Hanafusa, H. 1983. Structure and sequence of the cellular gene homologous to the RSV *src* gene and the mechanism for generating the transforming virus. *Cell* **32**:881–890.

Takeya, T., Feldman, R. A., and Hanafusa, H. 1982. DNA sequence of the viral and cellular *src* gene of chickens. I. Complete nucleotide sequence of *Eco*RI fragment of recovered avian sarcoma virus which codes for gp37 and pp60src. *J. Virol.* **44**:1–11.

Taub, R., Moulding, C., Battey, J., Murphy, W., Vasicek, T., Lenoir, G. M., and Leder, P. 1984. Activation and somatic mutation of the translocated c-*myc* gene in Burkitt lymphoma cells. *Cell* **36**:339–348.

Taylor, J. M., and Illmensee, R. 1975. Site on the RNA of avian sarcoma virus at which primer is bound. *J. Virol.* **16**:553–558.

Tepfer, D. 1984. Transformation of several species of higher plants by *Agrobacterium rhizogenes:* Sexual transmission of the transformed genotype and phenotype. *Cell* **37**:959–967.

Toda, T., Uno, I., Ishikawa, T., Powers, S., Kataoka, T., Broek, D., Cameron, S., Broach, J., Matsumoto, K., and Wigler, M. 1985. In yeast, *RAS* proteins are controlling elements of adenylate cyclase. *Cell* **40**:27–36.

Tracy, S. E., Woda, B. A., and Robinson, H. L. 1985. Induction of angiosarcoma by a c-*erbB* transducing virus. *J. Virol.* **54**:304–310.

Tsuchida, N., Ryder, T., and Ohtsuba, E. 1982. Nucleotide sequence of the oncogene encoding the p21 transforming protein of Kirsten murine sarcoma virus. *Science* **217**:937–939.

Tsujimoto, Y., Cossman, J., Jaffe, E., and Croce, C. M. 1985. Involvement of the *bil-2* gene in human follicular lymphoma. *Science* **228**:1440–1443.

Van Beveren, C., Goddard, J. G., Berns, A. J. M., and Verma, I. M. 1981. Biogenesis of a transforming gene. *Mol. Biol. Rep.* **7**:163–165.

Van Beveren, C., van Straaten, F., Curran, T., Müller, R., and Verma, I. M. 1983. Analysis of FBJ-MuSV provirus and c-*fos* (mouse) gene reveals that viral and cellular *fos* gene products have different carboxy termini. *Cell* **32**:1241–1255.

Van Beveren, C., van Straaten, F., Galleshaw, J. A., and Verma, I. M. 1985. Nucleotide sequence of the genome of a murine sarcoma virus. *Cell* **27**:97–108.

Van der Hoorn, F. A., and Firzlaff, J. 1984. Complete c-*mos* (rat) nucleotide sequence: Presence of conserved domains in c-*mos* proteins. *Nucleic Acids Res.* **12**:2147–2156.

Vande Woude, G. F., Levine, A. J., Topp, W. C., and Watson, J. D., eds. 1984. *Cancer Cells 2: Oncogenes and Viral Genes,* Cold Spring Harbor, New York. Cold Spring Harbor Laboratory.

Van Ooyen, A., and Nusse, R. 1984. Structure and nucleotide sequence of the putative mammary oncogene *int*-1: Proviral insertions leave the protein-encoding domain intact. *Cell* **39**:233–240.

Visser, L., Reemst, A. C. M. B., van Mansfeld, A. D. M., and Rozijn, T. H. 1982. Nucleotide sequence analysis of the linked left and right hand terminal regions of adenovirus type 5 DNA present in the transformed rat cell line 5RK20. *Nucleic Acids Res.* **10**:2189–2198.

Wachsman, W., Shimotohno, K., Clark, S. C., Golde, D. W., and Chen, I. S. Y. 1984. Expression of the 3' terminal region of human T-cell leukemia viruses. *Science* **226**:177–180.

Wachsman, W., Golde, D. W., Temple, P. A., Orr, E. C., Clark, S. C., and Chen, I. S. Y. 1985. HTLV *x*-gene product: Requirement for the *env* methionine initiation codon. *Science* **228**:1534–1537.

Wadsworth, S. C., Madhavan, K., and Bilodeau-Wentworth, D. 1985. Maternal inheritance of transcripts from three *Drosophila src*-related genes. *Nucleic Acids Res.* **13**:2153–2170.

Wang, L. H., Hanafusa, H., Notter, M. F. D., and Balduzzi, P. C. 1982. Genetic structure and transforming sequence of avian sarcoma virus UR2. *J. Virol.* **40**:258–267.

Waterfield, M. D., Scrace, G. T., Whittle, N., Stroobant, P., Johnson, A., Wasteson, A., Westermark, B., Heldin, C. H., Huang, J. S., and Deuel, T. F. 1983. Platelet-derived growth factor is structurally related to the putative transforming protein p28^{sis} of simian sarcoma virus. *Nature (London)* **304**:35–39.

Watson, R., Oskarsson, M., and Vande Woude, G. F. 1982. Human DNA sequence homologous to the transforming gene (*mos*) of Moloney murine sarcoma virus. *Proc. Natl. Acad. Sci. USA* **79**:4078–4082.

Watson, D. K., Reddy, E. P., Duesberg, P. H., and Papas, T. S. 1983. Nucleotide sequence analysis of the chicken c-*myc* gene reveals homologous and unique coding regions by comparison with the transforming gene of avian myelocytomatosis virus MC29. Δ*gag-myc*. *Proc. Natl. Acad. Sci. USA* **80**:2146–2150.

Wolff, L., Scolnick, E., and Ruscetti, S. 1983. Envelope gene of the Friend spleen focus-forming virus. Deletion and insertions in 3' gp^{70}/p15F-encoding region have resulted in unique features in the primary structure of its protein product. *Proc. Natl. Acad. Sci. USA* **80**:4718–4722.

Yamamoto, F., and Perucho, M. 1984. Activation of a human c-K-*ras* oncogene. *Nucleic Acids Res.* **12**:8873–8885.

Yamamoto, T., de Crombrugghe, B., and Pastan, I. 1980. Identification of a functional promoter in the long terminal repeat of Rous sarcoma virus. *Cell* **22**:787–797.

Yamamoto, T., Mishida, T., Miyajima, N., Kawai, S., Ooi, T., and Toyoshima, K. 1983. The *erbB* gene of avian erythroblastosis virus is a member of of the *src* gene family. *Cell* **35**:71–78.

Yang, J. Q., Bauer, S. R., Mushinski, J. F., and Marcu, K. B. 1985. Chromosome translocation clustered 5' of the murine c-*myc* gene qualitatively affects prime promoter usage: Implications for the site of normal c-*myc* regulation. *EMBO J.* **4**:1441–1447.

Yasuda, S., Furuichi, M., and Soeda, E. 1984. An altered DNA sequence encompassing the *ras* gene of Harvey murine sarcoma virus. *Nucleic Acids Res.* **12**:5583–5588.

Yoakum, G. H., Lechner, J. F., Gabrielson, E. W., Korba, B. E., Malan-Shibley, L., Willey, J. C., Valerio, M. G., Shamsuddin, A. M., Trump, B. F., and Harris, C. C. 1985. Transformation of human bronchial epithelial cells transfected by Harvey *ras* oncogene. *Science* **227**:1174–1180.

CHAPTER 12

Alonso, A., Jorcano, J. L., Beck, E., Hovemann, B., and Schmidt, T. 1984a. *Drosophila melanogaster* U1 snRNA genes. *J. Mol. Biol.* **180**:825–836.

Alonso, A., Beck, E., Jorcano, J. L., and Hovemann, B. 1984b. Divergence of U2 snRNA sequences in the genome of *D. melanogaster*. *Nucleic Acids Res.* **12**:9543–9550.

Anderson, S., Bankier, A. T., Barrell, B. G., de Bruijn, M. H. L., Coulson, A. R., Drouin, J., Eperon, I. C., Nierlick, D. P., Roe, B. A., Sanger, F., Schreier, P. H., Smith, A. J. H., Staden, R., and Young, I. G. 1981. Sequence and organization of the human mitochondrial genome. *Nature (London)* **290**:457–464.

Anderson, S., de Bruijn, M. H. L., Coulson, A. R., Eperon, I. C., Sange, F., and Young, I. G. 1982. Complete sequence of bovine mitochondrial DNA. Conserved features of the mammalian mitochondrial genome. *J. Mol. Biol.* **156**:683–717.

Beacham, I. R., Schweitzer, B. W., Warrick, H. M., and Carbon, J. 1984. The nucleotide sequence of the yeast *ARG4* gene. *Gene* **29**:271–279.

Bechmann, H., Haid, A., Schweyen, R. J., Mathews, S., and Kaudewitz, F. 1981. Expression of the ''split gene'' COB in yeast mtDNA. Translation of intervening sequences in mutant strains. *J. Biol. Chem.* **256:**3525–3531.

Beck, E., Jorcano, J. L., and Alonso, A. 1984. *Drosophila melanogaster* U1 and U2 small nuclear RNA genes contain common flanking sequences. *J. Mol. Biol.* **173:**539–542.

Benne, R., DeVories, B. F., Van den Burg, J., and Klaver, B. 1983. The nucleotide sequence of a segment of *Trypanosoma brucei* mitochondrial maxi-circle DNA that contains the gene for apocytochrome *b* and some unusual unassigned reading frames. *Nucleic Acids Res.* **11:**6925–6941.

Bennetzen, J. L., and Hall, B. D. 1982. Codon selection in yeast. *J. Biol. Chem.* **257:**3026–3031.

Berget, S. M. 1984. Are U4 small nuclear ribonucleoproteins involved in polyadenylation? *Nature (London)* **309:**179–181.

Bernstein, L. B., Mount, S. M., and Weiner, A. M. 1983. Pseudogenes for human small nuclear RNA U3 appear to rise by integration of self-primed reverse transcripts of the RNA into new chromosomal sites. *Cell* **32:**461–472.

Bibb, M. J., Van Etten, R. A., Wright, C. T., Walberg, M. W., and Clayton, D. A. 1981. Sequence and gene organization of mouse mitochondrial DNA. *Cell* **26:**167–180.

Blake, C. C. F. 1978. Do genes-in-pieces imply proteins-in-pieces? *Nature (London)* **273:**267.

Boardman, M., Basi, G. S., and Storti, R. V. 1985. Multiple polyadenylation sites in a *Drosophila* tropomyosin gene are used to generate functional mRNAs. *Nucleic Acids Res.* **13:**1763–1776.

Boel, E., Hansen, M. T., Hjort, I., Høegh, I., and Fiil, N. P. 1984. Two different types of intervening sequences in glucoamylase gene from *Aspergillus niger*. *EMBO J.* **3:**1581–1585.

Branlant, C., Krol, A., Ebel, J. P., Lazar, E., Gallinaro, H., Jacob, M., Sri-Widada, J., and Jeanteur, P. 1980. Nucleotide sequences of nuclear U1A RNAs from chicken, rat, and man. *Nucleic Acids Res.* **8:**4143–4154.

Branlant, C., Krol, A., Lazar, E., Haendler, B., Jacob, M., Galago-Dias, L., and Pousada, C. 1983. High evolutionary conservation of the secondary structure and of certain nucleotide sequences of U5 RNA. *Nucleic Acids Res.* **11:**8359–8368.

Bringmann, P., Appel, B., Rinke, J., Reuter, R., Theissen, H., and Lührmann, R. 1984. Evidence for the existence of snRNAs U4 and U6 in a single ribonucleoprotein complex and for their association by intermolecular base pairing. *EMBO J.* **3:**1357–1363.

Brody, E., and Abelson, J. 1985. The "spliceosome": Yeast premessenger RNA associates with a 40S complex in a splicing-dependent reaction. *Science* **228:**963–967.

Brown, D. T., Morris, G. F., Chodchoy, N., Sprecher, C., and Marzluff, W. F. 1985. Structure of the sea urchin U1 RNA repeat. *Nucleic Acids Res.* **13:**537–556.

Burke, J. M., and RajBhandary, U. L. 1982. Intron within the large rRNA gene of *N. crassa* mitochondria: A long open reading frame and a consensus sequence possibly important in splicing. *Cell* **31:**509–520.

Burke, J. M., Breitenberger, C., Heckman, J. E., Dujon, B., and RajBhandary, U. L. 1984. Cytochrome *b* gene of *Neurospora crassa* mitochondria. Partial sequence and location of introns at sites different from those in *Saccharomyces cerevisiae* and *Aspergillus nidulans*. *J. Biol. Chem.* **259:**504–511.

Carbon, P., Haumont, E., Fournier, M., de Henau, S., and Grosjean, H. 1983. Site-directed *in vitro* replacement of nucleosides in the anticodon loop of tRNA: Application to the study of structural requirements for queine insertase activity. *EMBO J.* **2:**1093–1097.

Card, C. O., Morris, G. F., Brown, D. T., and Marzluff, W. F. 1982. Sea urchin small nuclear RNA genes are organized in distinct tandemly repeating units. *Nucleic Acids Res.* **10:**7677–7688.

Castagnoli, L., Ciliberto, G., and Cortese, R. 1982. Processing of eukaryotic tRNA precursors: Secondary structure of the precursor specific sequences affects the rate but not the accuracy of the processing reactions. *Nucleic Acids Res.* **10:**4135–4148.

Chang, K. S., Rothblum, K. N., and Schwartz, R. J. 1985. The complete sequence of the chicken α-cardiac actin gene: A highly conserved vertebrate gene. *Nucleic Acids Res.* **13:**1223–1237.

Ciliberto, G., Buckland, R., Cortese, R., and Philipson, L. 1985. Transcription signals in embryonic *Xenopus laevis* U1 RNA genes. *EMBO J.* **4:**1537–1543.

Clary, D. O., Wahleithner, J. A., and Wolstenholme, D. R. 1984. Sequences and arrangement of the genes for cytochrome *b*, URF1, URF4L, URF4, URF5, URF6, and five tRNAs in *Drosophila* mitochondrial DNA. *Nucleic Acids Res.* **12:**3747–3762.

Colby, D., Leroy, P. S., and Guthrie, C. 1981. Yeast tRNA precursor mutated at a splice junction is correctly processed *in vivo*. *Proc. Natl. Acad. Sci. USA* **378:**415–419.

Cooley, L., Appel, B., and Söll, D. 1982. Post-transcriptonal nucleotide addition is responsible for the formation of the 5' terminus of histidine tRNA. *Proc. Natl. Acad. Sci. USA* **79:**6475–6479.

Daskal, Y. 1981. Perichromatin granules. *In:* Busch, H., ed., *The Cell Nucleus,* New York, Academic Press, Vol. VIII, pp. 117–138.

Dawson, A. J., Jones, V. P., and Leaver, C. J. 1984. The apocytochrome *b* gene in maize mitochondria does not contain introns and is preceded by a potential ribosome binding site. *EMBO J.* **3:**2107–2113.

DeLange, T., Berkvens, T. M., Veerman, H. J. G., Frasch, A. C. C., Barry, D. J., and Borst, P. 1984. Comparison of the genes coding for the common 5' terminal sequence of messenger RNAs in three trypanosome species. *Nucleic Acids Res.* **12:**4431–4443.

Denison, R. A., and Weiner, A. M. 1982. Human U1 RNA pseudogenes may be generated by both DNA- and RNA-mediated mechanisms. *Mol. Cell. Biol.* **2:**815–828.

Denison, R. A., Van Arsdell, S. W., Berstein, L. B., and Weiner, A. M. 1981. Abundant pseudogenes for small nuclear RNAs are dispersed in the human genome. *Proc. Natl. Acad. Sci. USA* **78:**810–814.

Deno, H., Kato, A., Shinozaki, K., and Sugiura, M. 1982. Nucleotide sequences of tobacco chloroplast genes for elongator tRNA^Met and tRNA^Val (UAC): The tRNA^Val (UAC) gene contains a long intron. *Nucleic Acids Res.* **10:**7511–7520.

Dieckmann, C. L., Koerner, T. J., and Tzagoloff, A. 1984a. Assembly of the mitochondrial membrane system. *CBP1,* a yeast nuclear gene involved in 5' end processing of cytochrome *b* pre-mRNA. *J. Biol. Chem.* **259:**4722–4731.

Dieckmann, C. L., Homison, G., and Tzagoloff, A. 1984b. Assembly of the mitochondrial membrane system. Nucleotide sequence of a yeast nuclear gene (*CBP1*) involved in 5' end processing of cytochrome *b* pre-mRNA. *J. Biol. Chem.* **259:**4732–4738.

Dillon, L. S. 1978. *The Genetic Mechanism and the Origin of Life,* New York, Plenum Press.

Dillon, L. S. 1983. *The Inconstant Gene,* New York, Plenum Press.

Domdey, H., Apostol, B., Lin, R. J., Newman, A., Brody, E., and Abelson, J. 1984. Lariat structures are *in vivo* intermediates in yeast pre-mRNA splicing. *Cell* **39:**611–621.

Dudov, K. P., and Dabeva, M. D. 1983. Post-transcriptional regulation of ribosome formation in the nucleus of regenerating rat liver. *Biochem. J.* **210:**183–192.

Earley, J. M., Roebuck, K. A., and Stumph, W. E. 1984. Three linked chicken U1 RNA genes have limited flanking DNA sequence homologies that reveal potential regulatory signals. *Nucleic Acids Res.* **12:**7411–7421.

Elliott, M. S., and Trewyn, R. W. 1984. Inosine biosynthesis in transfer RNA by an enzymatic insertion of hypoxanthine. *J. Biol. Chem.* **259:**2407–2410.

Engelke, D. R., Gegenheimer, P., and Abelson, J. 1985. Nucleolytic processing of a tRNA^Arg–tRNA^Asp dimeric precursor by a homologous component from *Saccharomyces cerevisiae. J. Biol. Chem.* **260:**1271–1279.

Epstein, P., Reddy, R., Henning, D., and Busch, H. 1980. The nucleotide sequence of U6 (4.7S) RNA. *J. Biol. Chem.* **255:**8901–8906.

Falkenthal, S., Parker, V. P., and Davidson, N. 1985. Developmental variations in the splicing pattern of transcripts from the *Drosophila* gene encoding myosin alkali light chain result in different carboxyl-terminal amino acid sequences. *Proc. Natl. Acad. Sci. USA* **82:**449–453.

Fischer, H. D., Dodgson, J. B., Hughes, S., and Engel, J. D. 1984. An unusual 5' splice sequence is efficiently utilized *in vivo. Proc. Natl. Acad. Sci. USA* **81:**2733–2737.

Forbes, D. J., Kornberg, T. B., and Kirschner, M. W. 1983. Small nuclear RNA transcription and ribonucleoprotein assembly in early *Xenopus* development. *J. Cell Biol.* **97:**62–72.

Forbes, D. J., Kirschner, M. W., Caput, D., Dahlberg, J. E., and Lund, E. 1984. Differential expression of multiple U1 small nuclear RNAs in oocytes and embryos of *Xenopus laevis. Cell* **38:**681–689.

Fradin, A., Jove, R., Hemenway, C., Keiser, H. D., Manley, J. L., and Prives, C. 1984. Splicing pathways of SV40 mRNAs in *X. laevis* oocytes differ in their requirements for snRNPs. *Cell* **37:**927–936.

Frendewey, D., and Keller, W. 1985. Stepwise assembly of a pre-mRNA splicing complex requires U-snRNPs and specific intron sequences. *Cell* **42:**355–367.

Fujita, T., Takaoka, C., Matsui, H., and Taniguchi, T. 1983. Structure of the human interleukin 2 gene. *Proc. Natl. Acad. Sci. USA* **80:**7437–7441.

Furuichi, Y. 1978. "Pretranscriptional capping" in the biosynthesis of cytoplasm polyhedrosis virus mRNA. *Proc. Natl. Acad. Sci. USA* **75:**1086–1090.

Garriga, G., and Lambowitz, A. M. 1984. RNA splicing in *Neurospora* mitochondria. Self-splicing of a mitochondrial intron *in vitro*. *Cell* **39**:631–641.

Garriga, G., Bertrand, H., and Lambowitz, A. M. 1984. RNA splicing in *Neurospora* mitochondria: Nuclear mutants defective in both splicing and 3' end synthesis of the larger rRNA. *Cell* **36**:623–634.

Gegenheimer, P., and Apirion, D. 1981. Processing of procaryotic ribonucleic acid. *Microbiol. Rev.* **45**:502–541.

Gegenheimer, P., Gabins, H. J., Peebles, C. L., and Abelson, J. 1983. An RNA ligase from wheat germ which participates in transfer RNA splicing *in vitro*. *J. Biol. Chem.* **258**:8365–8373.

Gerlinger, P., Krust, A., LeMeur, M., Perrin, F., Cochet, M., Gannon, F., Dupret, D., and Chambon, P. 1982. Multiple initiation and polyadenylation sites for the chicken ovomucoid transcription unit. *J. Mol. Biol.* **162**:345–364.

Ghora, B. K., and Apirion, D. 1978. Structural analysis and *in vitro* processing to p5 rRNA of a 9S RNA molecule isolated from an *rne* mutant of *E. coli*. *Cell* **15**:1055–1066.

Ghora, B. K., and Apirion, D. 1979. 5S rRNA is contained within a 25S rRNA that accumulates in mutants of *E. coli* defective in processing of rRNA. *J. Mol. Biol.* **127**:507–513.

Gidoni, D., Kadonaga, J., Barrera-Saldoña, H., Takahachi, K., Chambon, P., and Tijan, R. 1985. Bidirectional SV40 transcription mediated by tandem Sp1 binding interactions. *Science.* **230**:511–517.

Gilbert, W. 1978. Why genes in pieces? *Nature (London)* **271**:501.

Green, M. R., Grimm, M. F., Goewert, R. R., Collins, R. A., Cole, M. D., Lambowitz, A. M., Heckman, J. E., Yin, S., and RajBhandary, U. L. 1981. Transcripts and processing patterns for the ribosomal RNA and transfer RNA region of *Neurospora crassa* mitochondrial DNA. *J. Biol. Chem.* **256**:2027–2034.

Guerrier-Takada, C., McClain, W. H., and Altman, S. 1984. Cleavage of tRNA precursors by the RNA subunit of *E. coli* ribonuclease P(M1RNA) is influenced by 3'-proximal CCA in the substrates. *Cell* **38**:219–224.

Hammarström, K., Westin, G., Bark, C., Zabielski, J., and Petterson, U. 1984. Genes and pseudogenes for human U2 RNA. Implications for the mechanism of pseudogene formation. *J. Mol. Biol.* **179**:157–169.

Hashimoto, C., and Steitz, J. A. 1984. U4 and U6 RNAs coexist in a single small nuclear ribonucleoprotein particle. *Nucleic Acids Res.* **12**:3283–3293.

Haumont, E., Fournier, M., deHenau, S., and Grosjean, H. 1984. Enzymatic conversion of adenosine to inosine in the wobble position of yeast tRNA[Asp]: The dependence on the anticodon sequence. *Nucleic Acids Res.* **12**:2705–2715.

Hernandez, N., and Keller, W. 1983. Splicing of *in vitro* synthesized messenger RNA precursors in HeLa cell extracts. *Cell* **35**:89–99.

Hill, J., McGraw, P., and Tzagoloff, A. 1985. A mutation in yeast mitochondrial DNA results in a precise excision of the terminal intron of the cytochrome *b* gene. *J. Biol. Chem.* **260**:3235–3238.

Hindley, J., and Phear, G. A. 1984. Sequence of the cell division gene CDC2 from *Schizosaccharomyces pombe:* Patterns of splicing and homology to protein kinases. *Gene* **31**:129–134.

Hinnebusch, A. G., and Fink, G. R. 1983. Repeated DNA sequences upstream from *HIS1* also occur at several other co-regulated genes in *Saccharomyces cerevisiae*. *J. Biol. Chem.* **258**:5238–5247.

Hobart, P. M., Fogliano, M., O'Connor, B. A., Schaffer, I. M., and Chirgwin, J. M. 1984. Human renin gene: Structure and sequence analysis. *Proc. Natl. Acad. Sci. USA* **81**:5026–5030.

Htun, H., Lund, E., and Dahlberg, J. E. 1984. Human U1 RNA genes contain an unusually sensitive nuclease S1 cleavage site within the conserved 3' flanking region. *Proc. Natl. Acad. Sci. USA* **81**:7288–7292.

Htun, H., Lund, E., Westin, G., Pettersson, U., and Dahlberg, J. E. 1985. Nuclease S1-sensitive sites in multigene families: Human U2 small nuclear RNA genes. *EMBO J.* **4**:1839–1846.

Huang, C. C., Hammond, C., and Bishop. J. M. 1985. Nucleotide sequence and topography of chicken *c-fps*. Genesis of a retroviral oncogene encoding a tyrosine specific protein kinase. *J. Mol. Biol.* **181**:175–186.

Jacobson, A., and Favreau, M. 1983. Possible involvement of poly(A) in protein synthesis. *Nucleic Acids Res.* **11**:6353–6368.

Jacquier, A., and Dujon, B. 1985. An intron-encoded protein is active in a gene conversion process that spreads an intron into a mitochondrial gene. *Cell* **41**:383–394.

Johnson, J. D., Ogden, R., Johnson, P., Abelson, J., Dembeck, P., and Itakura, K. 1980. Transcription and processing of a yeast tRNA gene containing a modified intervening sequence. *Proc. Natl. Acad. Sci. USA* **77**:2564–2568.

Johnson, P. F., and Abelson, J. 1983. The yeast tRNA[Tyr] gene intron is essential for correct modification of its tRNA product. *Nature (London)* **302**:681–687.

Kaine, B. P., Gupta, R., and Woese, C. R. 1983. Putative introns in tRNA genes of prokaryotes. *Proc. Natl. Acad. Sci. USA* **80**:3309–3312.

Karin, M., Najarian, R., Haslinger, A., Valenzuela, P., Welch, J., and Fogel, S. 1984. Primary structure and transcription of an amplified genetic locus: The *CUP1* locus of yeast. *Proc. Natl. Acad. Sci. USA* **81**:337–341.

Karlin-Neumann, G. A., Kohorn, B. D., Thornber, J. P., and Tobin, E. M. 1985. A chlorophyll *a/b*-protein encoded by a gene containing an intron with characteristics of a transposable element. *J. Mol. Appl. Genet.* **3**:45–61.

Kato, N., and Harada, F. 1981. Nucleotide sequence of nuclear 5.7S RNA of mouse cells. *Biochem. Biophys. Res. Commun.* **99**:1477–1485.

Kaufer, N. F., Fried, H. M., Schwindinger, W. F., Jasin, M., and Warner, J. R. 1983. Cyclohexamide resistance in yeast: the gene and its protein. *Nucleic Acids Res.* **11**:3123–3135.

Kejzlarová-Lepesant, J., Brock, H. W., Moreau, J., Dubertret, M. L., Billault, A., and Lepesant, J. A. 1984. A complete and a truncated U1 snRNA gene of *Drosophila melanogaster* are found as inverted repeats at region 82E of the polytene chromosomes. *Nucleic Acids Res.* **12**:8835–8846.

Keller, E. B., and Noon, W. A. 1984. Intron splicing: A conserved internal signal in introns of animal pre-mRNAs. *Proc. Natl. Acad. Sci. USA* **81**:7417–7420.

Keller, E. B., and Noon, W. A. 1985. Intron splicing: A conserved internal signal in introns of *Drosophila* pre-mRNAs. *Nucleic Acids Res.* **13**:4971–4981.

Keller, M., and Michel, F. 1985. The introns of the *Euglena gracilis* chloroplast gene which codes for the 32 kDa protein of photosystem II. Evidence for structural homologies with class II introns. *FEBS Lett.* **179**:69–73.

Keller, W. 1984. The RNA lariat: A new ring to the splicing of mRNA precursors. *Cell* **39**:423–425.

King, T. C., and Schlessinger, D. 1984. S1 nuclease mapping analysis of ribosomal RNA processing in wild type and processing deficient *Escherichia coli. J. Biol. Chem.* **258**:12034–12042.

King, T. C., Sirdeshmukh, R., and Schlessinger, D. 1984. RNase III cleavage is obligate for maturation but not for function of *Escherichia coli* pre-23S rRNA. *Proc. Natl. Acad. Sci. USA* **81**:185–188.

Kinnaird, J. H., and Fincham, J. R. S. 1983. The complete nucleotide sequence of the *Neurospora crassa am* (NADP-specific glutamate dehydrogenase) gene. *Gene* **26**:253–260.

Knapp, G., Ogden, R. C., Peebles, C. L., and Abelson, J. 1979. Splicing of yeast tRNA precursors. Structure of the reaction intermediates. *Cell* **18**:37–45.

Koller, B., Gingrich, J. C., Stiegler, G. L., Farley, M. A., Delius, H., and Hallick, R. B. 1984. Nine introns with conserved boundary sequences in the *Euglena gracilis* chloroplast ribulose-1,5-bisphosphate carboxylase gene. *Cell* **36**:545–553.

Krainer, A. R., Maniatis, T., Ruskin, B., and Green, M. R. 1984. Normal and mutant human B-globin pre-mRNAs are faithfully and efficiently spliced *in vivo. Cell* **36**:993–1005.

Krämer, A., Keller, W., Appel, B., and Lührmann, R. 1984. The 5' terminus of the RNA moiety of U1 small nuclear ribonucleoprotein particles is required for the splicing of messenger RNA precursors. *Cell* **38**:299–307.

Krol, A., Branlant, C., Lazar, E., Gallinaro, H., and Jacob, M. 1981. Primary and secondary structures of chicken, rat and man nuclear U4 RNAs. Homologies with U1 and U5 RNAs. *Nucleic Acids Res.* **9**:2699–2716.

Krol, A., Ebel, J. P., Rinke, J., and Lührmann, R. 1983. U1, U2, and U5 small nuclear RNAs are found in plant cells. Complete nucleotide sequence of the U5 RNA family from pea nuclei. *Nucleic Acids Res.* **11**:8583–8594.

Krol, A., Lund, E., and Dahlberg, J. E. 1985. The two embryonic U1 RNA genes of *Xenopus laevis* have both common and gene-specific transcription signals. *EMBO J.* **4**:1529–1535.

Laird, P. W., Kooter, J. M., Loosbroeck, N., and Borst, P. 1985. Mature mRNAs of *Trypanosoma brucei* possess a 5' cap acquired by discontinuous RNA synthesis. *Nucleic Acids Res.* **13**:4253–4266.

Lamb, M. R., Anziano, P. Q., Glaus, K. R., Hanson, D. K., Klapper, H. J., Perlman, P. S., and Mahler, H. R. 1983. Functional domains in introns. RNA processing intermediates in *cis*- and *trans*-acting mutants in the penultimate intron of the mitochondrial gene for cytochrome *b. J. Biol. Chem.* **258**:1991–1999.

Langford, C. J., and Gallwitz, D. 1983. Evidence for an intron-contained sequence required for the splicing of yeast RNA polymerase II transcripts. *Cell* **33**:519–527.

Langford, C. J., Nellen, W., Niessing, J., and Gallwitz, D. 1983. Yeast is unable to excise foreign intervening sequences from hybrid gene transcripts. *Proc. Natl. Acad. Sci. USA* **80**:1496–5000.

Langford, C. J., Kling, F. J., Donath, C., and Gallwitz, D. 1984. Point mutations identify the conserved, intron-contained TACTAAC box as an essential splicing signal sequence in yeast. *Cell* **36**:645–653.

Lazowska, J., Jacq, C., and Slonimski, P. P. 1980. Sequence of introns and flanking exons in wild-type and *box* 3 mutants of cytochrome *b* reveals an interlaced splicing protein coded by an intron. *Cell* **22**:333–348.

Lazowska, J., Jacq, C., and Slonimski, P. P. 1981. Splice points of the third intron in the yeast mitochondrial cytochrome *b* gene. *Cell* **27**:12–14.

Lee, J. S., and Verma, D. P. S. 1984. Structure and chromosomal arrangement of leghemoglobin genes in kidney beans suggest divergence in soybean leghemoglobin gene loci following tetraploidization. *EMBO J.* **3**:2745–2752.

Lee, M. C., and Knapp, G. 1985. Transfer RNA splicing in *Saccharomyces cerevisiae*. Secondary and tertiary structures of the substrates. *J. Biol. Chem.* **260**:3108–3115.

Lenardo, M. J., Dorfman, D. M., Reddy, L. V., and Donelson, J. E. 1985. Characterization of the *Trypanosoma brucei* ribosomal RNA gene and transcript: The 5S rRNA is a spliced-leader-independent species. *Gene* **35**:131–141.

Lewin, B. 1980a. Alternatives for splicing: Recognizing the ends of the introns. *Cell* **22**:324–326.

Lewin, B. 1980b. Alternatives for splicing: An intron-coded protein. *Cell* **22**:645–646.

Li, S. S. L., Tiano, H. F., Fukasawa, K. M., Yagi, K., Shimizu, M., Sharief, F. S., Nakashima, Y., and Pan, Y. C. E. 1985. Protein structure and gene organization of mouse lactate dehydrogenase-A isozyme. *Eur. J. Biochem.* **149**:215–225.

Lin, W. L., and Pederson, T. 1984. Ribonucleoprotein organization of eukaryotic RNA. XXXI. Structure of the U1 small nuclear ribonucleoprotein. *J. Mol. Biol.* **180**:947–960.

Liu, M. H., Reddy, R., Henning, D., Spector, D., and Busch, H. 1984. Primary and secondary structure of dinoflagellate U5 small nuclear RNA. *Nucleic Acids Res.* **12**:1529–1542.

Lund, E., and Dahlberg, J. E. 1984. True genes for human U1 small nuclear RNA. *J. Biol. Chem.* **259**:2013–2021.

Mackow, E. R., and Chang, F. N. 1985. Processing of precursor ribosomal RNA and the presence of a modified ribosome assembly scheme in *Escherichia coli* relaxed strain. *FEBS Lett.* **182**:407–412.

Macreadie, I. G., Scott, R. M., Zinn, A. R., and Butow, R. A. 1985. Transposition of an intron in yeast mitochondria requires a protein encoded by that intron. *Cell* **41**:395–402.

Malissen, M., Malissen, B., and Jordan, B. R. 1982. Exon/intron organization and complete nucleotide sequence of an *HLA* gene. *Proc. Natl. Acad. Sci. USA* **79**:893–897.

Manser, T., and Gesteland, R. F. 1982. Human U1 loci: Genes for human U1 RNA have dramatically similar genomic environments. *Cell* **29**:257–264.

Marzluff, W. F., Brown, D. T., Lobo, S., and Wang, S. S. 1983. Isolation and characterization of two linked mouse U1b small nuclear RNA genes. *Nucleic Acids Res.* **11**:6255–6270.

Mason, P. J., Jones, M. B., Elkington, J. A., and Williams, J. G. 1985. Polyadenylation of the *Xenopus* β1 globin mRNA at a downstream minor site in the absence of the major site and utilization of an AAUACA polyadenylation signal. *EMBO J.* **4**:205–211.

Mattaj, I. W., and DeRobertis, E. M. 1985. Nuclear segregation of U2 snRNA requires binding specific snRNP proteins. *Cell* **40**:111–118.

McGraw, P., and Tzagoloff, A. 1983. Assembly of the mitochondrial membrane system. Characterization of a yeast nuclear gene involved in the processing of the cytochrome *b* pre-mRNA. *J. Biol. Chem.* **258**:9459–9468.

Michiels, F., Muyldermans, S., Hamers, R., and Matthyssens, G. 1985. Putative regulatory sequences for the transcription of mini-exons in *Trypanosoma brucei* as revealed by S1 sensitivity. *Gene* **36**:263–270.

Milhausen, M., Nelson, R. G., Sathers, S., Selkirk, M., and Agabian, N. 1984. Identification of a small RNA containing the trypanosome spliced leader: A donor of shared 5′ sequences of trypanosomatid mRNAs? *Cell* **38**:721–729.

Mitlin, J. A., and Cannon, M. 1984. Defective processing of ribosomal precursor RNA in *Saccharomyces cerevisiae*. *Biochem. J.* **220**:461–467.

Mitra, G., and Warner, J.R. 1984. A yeast ribosomal protein gene whose intron is in the 5′ leader. *J. Biol. Chem.* **259**:9218–9224.

Montandon, P. E., and Stutz, E. 1983. Nucleotide sequence of a *Euglena gracilis* chloroplast genome region for the elongation factor Tu; evidence for a spliced mRNA. *Nucleic Acids Res.* **11**:5877–5892.

Montell, C., Fisher, E. F., Caruthers, M. H., and Berk, A. J. 1983. Inhibition of RNA cleavage but not polyadenylation by a point mutation in mRNA 3′ consensus sequence AAUAAA. *Nature (London)* **305**:600–605.

Moore, C. L., and Sharp, P. A. 1984. Site-specific polyadenylation in a cell-free reaction. *Cell* **36**:581–591.

Moore, C. L., and Sharp, P. A. 1985. Accurate cleavage and polyadenylation of exongenous RNA substrate. *Cell* **41**:845–855.

Moriuchi, T., Chang, H. C., Denome, R., and Silver, J. 1983. Thy-1 cDNA sequence suggests a novel regulatory mechanism. *Nature (London)* **301**:80–82.

Mount, S. M., and Steitz, J. A. 1981. Sequence of U1 RNA from *Drosophila melanogaster:* Implications for U1 secondary structure and possible involvement in splicing. *Nucleic Acids Res.* **9**:6351–6368.

Mount, S. M., Pettersson, I., Hinterberger, M., Karmas, A., and Steitz, J. A. 1983. The U1 small nuclear RNA-protein complex selectively binds a 5′ splice site *in vitro. Cell* **33**:509–518.

Moussa, N. M., Lobo, S. M., and Marzluff, W. F. 1985. Expression of a mouse U1b gene in mouse L cells. *Gene* **36**:311–319.

Murphy, J. T., Burgess, R. R., Dahlberg, J. E., and Lund, E. 1982. Transcription of a gene for human U1 small nuclear RNA. *Cell* **29**:265–274.

Nobrega, F. G., and Tzagoloff, A. 1980. Assembly of the mitochondrial membrane system. DNA sequence and organization of the cytochrome *b* gene in *Saccharomyces cerevisiae* D293-10B. *J. Biol. Chem.* **255**:9828–9837.

Nogi, Y., and Fukasawa, T. 1984. Nucleotide sequence of the yeast regulatory gene *GAL80. Nucleic Acids Res.* **12**:9287–9298.

Nyunoya, H., and Lusty, C. J. 1984. Sequence of the small subunit of yeast carbamyl phosphate synthetase and identification of its catalytic domain. *J. Biol. Chem.* **259**:9790–9798.

Ogden, R. C., Lee, M. C., and Knapp, G. 1984. Transfer RNA splicing in *Saccharomyces cerevisiae:* Defining the substrates. *Nucleic Acids Res.* **12**:9367–9382.

Okada, N., Sakamoto, K., Itoh, Y., and Oshima, Y. 1982. Sequence determination of rat U5 RNA using a chemical modification procedure for counteracting sequence compression. *J. Biochem.* **91**:1281–1291.

Osinga, K. A., De Vries, E., Van der Horst, G., and Tabak, H. F. 1984. Processing of yeast mitochondrial messenger RNAs at a conserved dodecamer sequence. *EMBO J.* **3**:829–834.

Pace, B., Stahl, D. A., and Pace, N. R. 1984. The catalytic element of a ribosomal RNA-processing complex. *J. Biol. Chem.* **259**:11454–11458.

Padgett, R. A., Konarska, M. M., Grabowski, P. J., Hardy, S. F., and Sharp, P. A. 1984. Lariat RNA's as intermediates and products in the splicing of messenger RNA precursors. *Science* **225**:898–903.

Pedersen, N., Hellung-Larsen, P., and Engberg, J. 1985. Small nuclear RNAs in the ciliate *Tetrahymena. Nucleic Acids Res.* **13**:4203–4224.

Peebles, C. L., Ogden, R. C., Knapp, O., and Abelson, J. 1979. Splicing of yeast tRNA precursors: A two-stage reaction. *Cell* **18**:27–35.

Petterson, I., Hinterberger, M., Mimori, T., Gottlieb, E., and Steitz, J. A. 1984. The structure of mammalian small nuclear ribonucleoproteins: Identification of multiple protein components reactive with anti(U1)RNP and anti-Sm autoantibodies. *J. Biol. Chem.* **259**:5907–5914.

Pikielny, C. W., and Rosbash, M. 1985. mRNA splicing efficiency in yeast and the contribution of nonconserved sequences. *Cell* **41**:119–126.

Pillar, T., Lang, B. F., Steinberger, L. I., Vogt, B., and Kaudervitz, F. 1983. Expression of the "split gene" *cob* in yeast mtDNA. Nuclear mutations specifically block the excision of different introns from its primary transcript. *J. Biol. Chem.* **258**:7954–7959.

Ray, B. K., Singh, B., Roy, M. K., and Apirion, D. 1982. Ribonuclease E is involved in the processing of 5-S ιRNA from a number of rRNA transcription units. *Eur. J. Biochem.* **125**:283–289.

Reddy, R., Henning, D., and Busch, H. 1979. Nucleotide sequence of nucleolar U3B RNA. *J. Biol. Chem.* **254**:11097–11105.

Reddy, R., Henning, D., and Busch, H. 1980. Substitutions, insertions, and deletions in two highly conserved U3 RNA species. *J. Biol. Chem.* **255**:7029–7033.

Reddy, R., Henning, D., and Busch, H. 1981a. The primary nucleotide sequence of U4 RNA. *J. Biol. Chem.* **256**:3532–3538.

Reddy, R., Li, W. Y., Henning, D., Choi, Y. C., Nohga, K., and Busch, H. 1981b. Characterization and subcellular localization of 7–8S RNAs of Novikoff hepatoma. *J. Biol. Chem.* **256**:8452–8457.

Reddy, R., Henning, D., and Busch, H. 1981c. Pseudouridine residues in the 5'-terminus of uridine-rich nuclear RNAI (U1 RNA). *Biochem. Biophys. Res. Commun.* **98:**1076–1083.

Reddy, R., Rothblum, L. I., Subrahmanyam, C. S., Liu, M.-H., Henning, D., Cassidy, B., and Busch, H. 1983a. The nucleotide sequence of 8S RNA bound to preribosomal RNA of Novikoff hepatoma. *J. Biol. Chem.* **258:**584–589.

Reddy, R., Henning, D., Liu, M. H., Spector, D., and Busch, H. 1985a. Identification and characterization of a polyadenylated small RNA (s-polyA + RNA) in dinoflagellates. *Biochem. Biophys. Res. Commun.* **127:**552–557.

Reddy, R., Henning, D., Chirala, S., Rothblum, L., Wright, D., and Busch, H. 1985b. Isolation and characterization of three rat U3 RNA pseudogenes. *J. Biol. Chem.* **260:**5715–5719.

Reed, R., and Maniatis, T. 1985. Intron sequences involved in lariat formation during pre-mRNA splicing. *Cell* **41:**95–105.

Reuter, R., Appel, B., Bringmann, P., Rinke, J., and Lührmann, R. 1984. 5'-Terminal caps of snRNAs are reactive with antibodies specific for 2,2,7-trimethylguanosine in whole cells and nuclear matrices. *Exp. Cell Res.* **154:**548–560.

Rinke, J., Appel, B., Blöcker, H., Frank, R., and Lührmann, R. 1984. The 5'-terminal sequence of U1 RNA complementary to the consensus 5' splice site of hnRNA is single-stranded in intact U1 snRNP particles. *Nucleic Acids Res.* **12:**4111–4126.

Rixon, M. W., Chung, D. W., and Davie, E. W. 1985. Nucleotide sequence of the gene for the γ chain of human fibrinogen. *Biochemistry* **24:**2077–2086.

Robinson, R. R., and Davidson, N. 1981. Analysis of a *Drosophila* tRNA gene cluster: Two tRNA^{Leu} genes contain intervening sequences. *Cell* **22:**251–259.

Ro-Choi, T. S., Choi, Y. C., Henning, D., McCloskey, J., and Busch, H. 1975. Nucleotide sequence of U-2 ribonucleic acid. Sequence of the 5'-terminal oligonucleotide. *J. Biol. Chem.* **250:**3921–3928.

Roe, B. A., Ma, D. P., Wilson, R. K., and Wong, J. F. H. 1985. The complete nucleotide sequence of the *Xenopus laevis* mitochondrial genome. *J. Biol. Chem.* **260:**9759–9774.

Rogers, J., and Wall, R. 1980. A mechanism for RNA splicing. *Proc. Natl. Acad. Sci. USA* **77:**1877–1879.

Roop, D. R., Kristo, P., Stumph, W. E., Tsai, M. J., and O'Malley, B. W. 1981. Structure and expression of a chicken gene coding for U1 RNA. *Cell* **23:**671–680.

Rosenthal, E. T., Tansey, T. R., and Ruderman, J. V. 1983. Sequence-specific adenylations and deadenylations accompany changes in the translation of maternal messenger RNA after fertilization of *Spisula* oocytes. *J. Mol. Biol.* **166:**309–327.

Ruskin, B., and Green, M. R. 1985. An RNA processing activity that debranches RNA lariats. *Science* **229:**135–140.

Ruskin, B., Greene, J. M., and Green, M. R. 1985. Cryptic branch point activation allows accurate *in vitro* splicing of human β-globin intron mutants. *Cell* **41:**833–844.

Salditt-Georgieff, M., Harpold, M., Chen-Kiang, S., and Darnell, J. E. 1980. The addition of 5'-cap structures occurs early in hnRNA synthesis and prematurely terminated molecules are capped. *Cell* **19:**69–78.

Saluz, H. P., Schmidt, T., Dudler, R., Altwegg, M., Stumm-Zollinger, E., Kubli, E., and Chen, P. S. 1983. The genes coding for 4 snRNAs of *Drosophila melanogaster:* Localization and determination of gene numbers. *Nucleic Acids Res.* **11:**77–90.

Sass, H., and Pederson, T. 1984. Transcription-dependent localization of U1 and U2 small nuclear ribonucleoproteins at major sites of gene activity in polytene chromosomes. *J. Mol. Biol.* **180:**911–926.

Setyono, B., and Pederson, T. 1984. Ribonucleoprotein organization of eukaryotic RNA. XXX. Evidence that U1 small nuclear RNA is a ribonucleoprotein when base-paired with pre-messenger RNA *in vivo. J. Mol. Biol.* **174:**285–295.

Sirdeshmukh, R., and Schlessinger, D. 1985. Ordered processing of *Escherichia coli* 23S rRNA *in vitro. Nucleic Acids Res.* **13:**5041–5054.

Skuzeski, J. M., Lund, E., Murphy, J. T., Steinberg, T. H., Burgess, R. R., and Dahlberg, J. E. 1984. Synthesis of human U1 RNA. II. Identification of two regions of the promoter essential for transcription initiation at position + 1. *J. Biol. Chem.* **259:**8345–8352.

Smith, S. D., Banerjee, N., and Sitz, T. O. 1984. Gene heterogeneity: A basis for alternative 5.8S rRNA processing. *Biochemistry* **23:**3648–3652.

Solari, A., and Deutscher, M. 1982. Subcellular localization of the tRNA processing enzyme, tRNA nucleotidyltransferase, in *Xenopus laevis* oocytes and in somatic cells. *Nucleic Acids Res.* **10:**4397–4407.

Stahl, D. A., Pace, B., Marsh, T., and Pace, N. R. 1984. The ribonucleoprotein substrate for a ribosomal RNA-processing nuclease. *J. Biol. Chem.* **259:**11448–11453.

Standring, D. N., Venegas, A., and Rutter, W. J. 1981. Yeast tRNA$_3^{Leu}$ gene transcribed and spliced in a HeLa cell extract. *Proc. Natl. Acad. Sci. USA* **78:**5963–5967.

Stark, M. J. R., Mileham, A. J., Romanos, M. A., and Boyd, A. 1984. Nucleotide sequence and transcription analysis of a linear DNA plasmid associated with the killer character of the yeast *Kluyveromyces lactis*. *Nucleic Acids Res.* **12:**6011–6030.

Steinmetz, A., Gubbins, E. J., and Bogorad, L. 1982. The anticodon of the maize chloroplast gene for tRNA$_{UAA}^{Leu}$ is split by a large intron. *Nucleic Acids Res.* **10:**3027–3037.

Stone, E. M., Rothblum, K. N., Alevy, M. C., Kuo, T. M., and Schwartz, R. J. 1985. Complete sequence of the chicken glyceraldehyde-3-phosphate dehydrogenase gene. *Proc. Natl. Acad. Sci. USA* **82:**1628–1632.

Stroke, I. L., and Weiner, A. M. 1985. Genes and pseudogenes for rat U3A and U3B small nuclear RNA. *J. Mol. Biol.* **184:**183–193.

Strub, K., Galli, G., Busslinger, M., and Birnstiel, M. L. 1984. The cDNA sequences of the sea urchin U7 small nuclear RNA suggest specific contacts between histone mRNA precursor and U7 RNA during RNA processing. *EMBO J.* **3:**2801–2807.

Sumner-Smith, M., Bozzato, R. P., Skipper, N., Davies, R. W., and Hopper, J. E. 1985. Analysis of the inducible *MEL1* gene of *Saccharomyces carlsbergensis* and its secreted product, alpha-galactosidase (melibiase). *Gene* **36:**333–340.

Szeberényi, J., Roy, M. K., Vaidya, H. C., and Apirion, D. 1984. 7S RNA, containing 5S ribosomal RNA and the termination stem, is a specific substrate for the two RNA processing enzymes RNase cIII and RNase E. *Biochemistry* **23:**2952–2957.

Szeberényi, J., Tomcsányi, T., and Apirion, D. 1985. Maturation of the 3′ end of 5-S ribosomal RNA from *Escherichia coli*. *Eur. J. Biochem.* **149:**113–118.

Taber, R. L., and Vincent, W. S. 1969. Effects of cyclohexamide on ribosomal RNA synthesis in yeast. *Biochem. Biophys. Res. Commun.* **34:**488–494.

Tani, T., Watanabe-Nagasu, N., Okada, N., and Ohshima, Y. 1983. Molecular cloning and characterization of a gene for rat U2 small nuclear RNA. *J. Mol. Biol.* **168:**579–594.

Tatei, K., Takemura, K., Mayeda, A., Fujiwara, Y., Tanaka, H., Ishihama, A., and Oshima, Y. 1984. U1 RNA–protein complex preferentially binds to both 5′ and 3′ splice junction sequences in RNA or single-stranded DNA. *Proc. Natl. Acad. Sci. USA* **81:**6281–6285.

Treisman, R., Orkin, S. H., and Maniatis, T. 1983. Specific transcription and RNA splicing defects in five cloned β-thalessemia genes. *Nature (London)* **302:**591–596.

van der Horst, G., and Tabak, H. F. 1985. Self-splicing of yeast mitochondrial ribosomal and messenger RNA precursors. *Cell* **40:**759–766.

Waring, R. B., Davies, R. W., Lee, S., Grisi, E., Berks, M. M., and Scazzocchio, C. 1981. The mosaic organization of the apocytochrome *b* gene of *Aspergillus nidulans* revealed by DNA sequencing. *Cell* **27:**4–11.

Waring, R. B., Roy, J. A., Edwards, S. W., Scazzocchio, C., and Davies, R. W. 1985. The *Tetrahymena* rRNA intron self-splices in *E. coli*: *In vivo* evidence for the importance of key base-paired regions of RNA for RNA enzyme function. *Cell* **40:**371–380.

Watanabe-Nagasu, N., Itoh, Y., Tani, T., Okano, K., Koga, N., Okada, N., and Oshima, Y. 1983. Structural analysis of gene loci for rat U1 small nuclear RNA. *Nucleic Acids Res.* **11:**1791–1801.

Weiss-Brummer, B., Rödel, G., Schweyen, R. J., and Kaudewitz, F. 1982. Expression of the split gene *cob* in yeast: Evidence for a precursor of a "maturase" protein translated from intron 4 and preceding exons. *Cell* **29:**527–536.

Westin, G., Lund, E., Murphy, J. T., Pettersson, U., and Dahlberg, J. E. 1984a. Human U2 and U1 RNA genes use similar transcription signals. *EMBO J.* **3:**3295–3301.

Westin, G., Zabielski, J., Hammarström, K., Monstein, H. J., Bark, C., and Pettersson, U. 1984b. Clustered genes for human U2 RNA. *Proc. Natl. Acad. Sci. USA* **81:**3811–3815.

Wieringa, B., Meyer, F., Reiser, J., and Weissmann, C. 1983. Unusual splice sites revealed by mutagenic inactivation of an authentic splice site of the rabbit β-globin gene. *Nature (London)* **301:**38–43.

Willis, I., Hottinger, H., Pearson, D., Chisholm, V., Leupold, U., and Söll, D. 1984. Mutations affecting excision of the intron from a eukaryotic dimeric tRNA precursor. *EMBO J.* **3:**1573–1580.

Wise, J. A., and Weiner, A. M. 1980. *Dictyostelium* small nuclear RNA D2 is homologous to rat nucleolar

RNA U3 and is encoded by a dispersed multigene family. *Cell* **22:**109–116.

Wise, J. A., and Weiner, A. M. 1981. The small nuclear RNAs of the cellular slime mold *Dictyostelium discoideum:* Isolation and characterization. *J. Biol. Chem.* **256:**956–963.

Wise, J. A., Tollervey, D., Maloney, D., Swerdlow, H., Dunn, E. J., and Guthrie, C. 1983. Yeast contains small nuclear RNAs encoded by single copy genes. *Cell* **35:**743–751.

Yamashita, I., Suzuki, K., and Fukui, S. 1985. Nucleotide sequence of the extracellular glucoamylase gene *STA1* in the yeast *Saccharomyces cerevisiae. J. Bacteriol.* **161:**567–573.

Zaug, A. J., Grabowski, P. J., and Cech, T. R. 1983. Autocatalytic cyclization of an excised intervening sequence RNA is a cleavage–ligation reaction. *Nature (London)* **301:**578–583.

Zeitlin, S., and Efstratiadis, A. 1984. *In vivo* splicing products of the rabbit β-globin pre-mRNA. *Cell* **39:**589–602.

Appendix. The Standard Genetic Code

Arranged by Codons

AAA, AAG   Lysine
AAC, AAT   Asparagine
ACA, ACC, ACG, ACT   Threonine
AGA, AGG   Arginine
AGC, AGT   Serine
ATA, ATC, ATT   Isoleucine
ATG   Methionine
CAA, CAG   Glutamine
CAC, CAT   Histamine
CCA, CCC, CCG, CCT   Proline
CGA, CGC, CGG, CGT   Arginine
CTA, CTC, CTG, CTT   Leucine
GAA, GAG   Glutamic acid
GAC, GAT   Aspartic acid
GCA, GCC, GCG, GCT   Alanine
GGA, GGC, GGG, GGT   Glycine
GTA, GTC, GTG, GTT   Valine
TAA, TAG   Stop
TAC, TAT   Tyrosine
TCA, TCC, TCG, TCT   Serine
TGA   Stop
TGC, TGT   Cysteine
TGG   Tryptophan
TTA, TTG   Leucine
TTC, TTT   Phenylalanine

Arranged by Amino Acids

Alanine   GCA, GCC, GCG, GCT
Arginine   AGA, AGG, CGA, CGC,
    CGG, CGT
Asparagine   AAC, AAT
Aspartic acid   GAC, GAT
Cysteine   TGC, TGT
Glutamic acid   GAA, GAG
Glutamine   CAA, CAG
Glycine   GGA, GGC, GGG, GGT
Histamine   CAC, CAT
Isoleucine   ATA, ATC, ATT
Leucine   CTA, CTC, CTG, CTT, TTA,
    TTG
Lysine   AAA, AAG
Methionine   ATG
Phenylalanine   TTC, TTT
Proline   CCA, CCC, CCG, CCT
Serine   AGC, AGT, TCA, TCC, TCG,
    TCT
Threonine   ACA, ACC, ACG, ACT
Tryptophan   TGG
Tyrosine   TAC, TAT
Valine   GTA, GTC, GTG, GTT
Stops   TAA, TAG, TGA

Index

Page numbers in **bold** type refer to illustrations.